The Genetic Code

First position	Second position U	C	A	G	Third position
U	Phe	Ser	Tyr	Cys	U
	Phe	Ser	Tyr	Cys	C
	Leu	Ser	Stop	Stop	A
	Leu	Ser	Stop	Trp	G
C	Leu	Pro	His	Arg	U
	Leu	Pro	His	Arg	C
	Leu	Pro	Gln	Arg	A
	Leu	Pro	Gln	Arg	G
A	Ile	Thr	Asn	Ser	U
	Ile	Thr	Asn	Ser	C
	Ile	Thr	Lys	Arg	A
	Met	Thr	Lys	Arg	G
G	Val	Ala	Asp	Gly	U
	Val	Ala	Asp	Gly	C
	Val	Ala	Glu	Gly	A
	Val	Ala	Glu	Gly	G

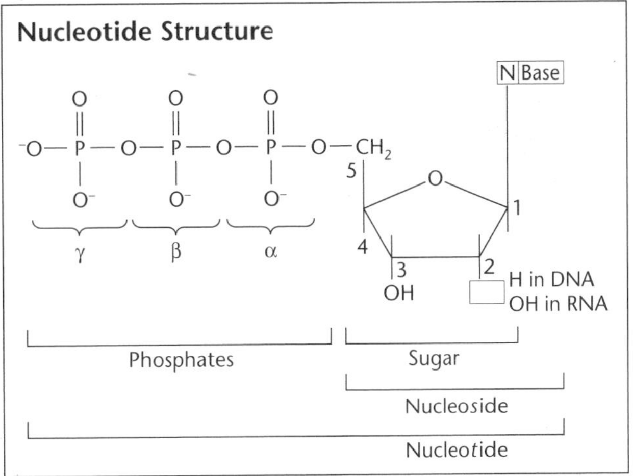

Names of Nucleic Acid Subunits

Base	Nucleoside	Nucleotide	Abbreviation RNA	Abbreviation DNA
Adenine	Adenosine	Adenosine triphosphate	ATP	dATP
Guanine	Guanosine	Guanosine triphosphate	GTP	dGTP
Cytosine	Cytidine	Cytidine triphosphate	CTP	dCTP
Thymine	Thymidine	Thymidine triphosphate		dTTP
Uracil	Uridine	Uridine triphosphate	UTP	

4TH EDITION

Molecular Genetics of Bacteria

Larry Snyder
Department of Microbiology and Molecular Genetics
Michigan State University
East Lansing, Michigan

Joseph E. Peters
Department of Microbiology
Cornell University
Ithaca, New York

Tina M. Henkin
Department of Microbiology
Ohio State University
Columbus, Ohio

Wendy Champness
Department of Microbiology and Molecular Genetics
Michigan State University
East Lansing, Michigan

Washington, DC

Library of Congress Cataloging-in-Publication Data

Molecular genetics of bacteria / Larry Snyder ... [et al.]. — 4th ed.
p. ; cm.
Rev. ed. of: Molecular genetics of bacteria / Larry Snyder and Wendy Champness. c2007.
Includes bibliographical references and index.
ISBN 978-1-55581-627-8 (hardcover : alk. paper) — ISBN 978-1-55581-716-9 (e-book)
I. Snyder, Larry. II. Snyder, Larry. Molecular genetics of bacteria.
[DNLM: 1. Bacteria—genetics. 2. Bacteriophages—genetics. 3. Chromosomes, Bacterial. 4. Genetics, Microbial—methods. 5. Molecular Biology—methods. QW 51]

572.8′293—dc23

2012027461

10 9 8 7 6 5 4 3 2
Printed in the United States of America

Address editorial correspondence to ASM Press, 1752 N St. NW, Washington, DC 20036-2904, USA
E-mail: books@asmusa.org

Send orders to ASM Press, P.O. Box 605, Herndon, VA 20172, USA
Phone: (800) 546-2416 or (703) 661-1593
Fax: (703) 661-1501
Online: estore.asm.org

doi:10.1128/9781555817169

Illustrations: Patrick Lane, ScEYEnce Studios
Cover and interior design: Susan Brown Schmidler
Cover illustration: Terese Winslow

Cover photo: Fluorescence micrograph of *Bacillus subtilis* cells showing the location of the cell membrane (red), DNA (blue), and ConE mating protein fused to green fluorescent protein (yellow-green). The ConE protein is a component of the conjugative DNA translocation channel required for horizontal transfer of the integrating conjugative element ICE*Bs1*. The ConE protein is concentrated at the cell poles, but additional protein is localized around the entire cell periphery. The lateral distribution enables cells to transfer ICEBs1 side to side, and the high concentration at the poles may contribute to the very efficient transfer of ICEBs1 observed in chains of cells from pole to pole. See chapter 5 for details. Photo courtesy of Melanie Berkmen and Alan Grossman. Modified from M. B. Berkmen et al., *J. Bacteriol.* **192:**38–45, 2010.

Contents

Preface

WITH THE ADDITION OF TWO NEW COAUTHORS, the fourth edition of the textbook *Molecular Genetics of Bacteria* has been substantially revised and some new sections have been added. We tried to do this without increasing the length of the book, which, at more than 700 pages, was already quite long. While the book retains the same number and order of chapters, many topics have been moved or integrated more completely into the text to reflect a more modern perspective. The purpose was to convey more accurately how one approaches questions in modern bacterial genetics, using the full repertoire of methods now available. Also, to make room for the new material, we made the philosophical decision to condense or eliminate descriptions of methods where they seemed unnecessarily detailed for a textbook.

Chapter 1, on DNA structure, DNA replication, and chromosome segregation, was expanded to include updates in our understanding, including how replication proceeds through obstacles typically found during normal DNA replication in bacteria, while some aspects of repair-associated replication were moved to later chapters. The chapter was also significantly expanded with new information about how numerous cell processes coordinate for the efficient processing and organizing of chromosomes after DNA replication. Scientists now more fully appreciate how sequences "hidden" in the structure guide a variety of systems that aid in repairing, segregating, packaging, and pumping the chromosome for exquisite genome stability in bacteria. In chapter 2, which covers bacterial gene expression, the translation section has been reorganized to follow the same order as the transcription section. It begins with initiation of translation and then discusses elongation followed by termination, rather than following the more historical order with the genetic code coming first. We reasoned that this order makes more sense since most students already have had some exposure to translation and the genetic code. More information on RNA degradation is now included, and the sections on gene regulation have been moved to chapter 12. The protein transport section has been moved from chapter 2 to chapter 14 (see below), where it can be better integrated with other topics of protein export. Chapter 3, on bacterial genetic analysis, also now takes a less historical approach. Rather than beginning with a review of classical genetic analysis and then contrasting it with bacterial genetic analysis as in previous editions, the chapter now begins with bacterial genetic analysis, again assuming that students have already had some general genetics. Furthermore, rather than putting more recently developed methods such as site-specific mutagenesis, recombineering, etc., into a separate section, we have integrated throughout the text all the methods available nowadays to use in a genetic analysis. The discussion of mapping by Hfr crosses has been sharply condensed, since it is likely that no one will ever again perform the laborious task of constructing the genetic map of a bacterium. The relative ease of DNA sequencing now allows the placing of mutations on the sequenced genomes of bacteria by direct comparison of sequences rather than by Hfr mapping. Transduction and transformation (including electroporation) are used extensively for genetic manipulations, so their use is still covered in some detail. Chapter 4 has been updated with more information about how plasmids are typically used in the laboratory setting in work with model organisms and beyond as well, including updates on our understanding of partitioning systems. Chapter 5 was extensively updated to illustrate the hodgepodge organization of conjugal elements and advances in our understanding of conjugation and to more fully integrate the important role of integrating conjugative elements in bacterial genomes (including a focus on one of these elements from *B. subtilis* on the front cover). Chapter 6 is updated throughout and focuses on similarities and differences between different transformation systems. The bacteriophage chapters (chapters 7 and 8) have been updated, and new material has been added. Some highlights from phage genomics are now covered, as are phage defense mechanisms, including CRISPR. The section on phage lysis is expanded, as is the text box on phage display, whose power is now demonstrated with some current uses. The interaction between lysogenic phages and genetic islands has been updated and moved into the text, as have some more recently developed techniques using lysogenic phages, for example, in detecting protein-protein interactions. Chapter 9, which covers transposable elements and site-specific recombination, has been updated to clarify the basic molecular biology of these elements, and it includes updated sections describing how they are used in the laboratory today. Chapter 10 was significantly reorganized to stress the role of homologous recombination in the repair of DNA double-strand breaks that occur at interruptions in the template DNA during replication. The role of homologous recombination in repair explains the underpinning of the evolution of the process and also clarifies how the process works in concert with DNA replication. A more comprehensive treatment for how DNA double-strand breaks are repaired across different types of bacteria, using systems found in all domains of life, is also included. Chapter 11 was updated to discuss many advances in the field of repair, including an expanded understanding of the regulation of multiple DNA polymerases found in bacteria with the SOS response. Chapters 12 and 13 have been reorganized so that chapter 12 is now focused on mechanisms of regulation of individual genes and operons and chapter 13 is mostly concerned with examples of global regulatory systems that utilize these mechanisms. There is also more emphasis on posttranscriptional regulation in both chapters, and global regulatory mechanisms in *Escherichia coli* are contrasted with those in *Bacillus subtilis*. Chapter 14 is probably the most changed chapter. It now contains our entire discussion of protein export, including the Sec and Tat systems as well as the secretion systems of gram-negative (i.e., *Proteobacteria*) and gram-positive (i.e., *Firmicutes*) bacteria. Most notably, it now contains a new section on bacterial cell biology, including cell wall synthesis and cell division and their regulation, as well as a new box on the evolution of cytoskeletal filaments, and it introduces the use of *Caulobacter crescentus* as a model system for these studies. Chapter 14 finishes with sporulation in *B. subtilis*, probably the best understood bacterial developmental system.

As in earlier editions, we do not mention the names of most investigators who have made major contributions to bacterial molecular genetics. We include only those names that have become icons in the field because they are associated with certain seminal experiments (e.g., Meselson and Stahl or Luria and Delbrück), models (e.g., Jacob and Monod), or structures (e.g., Watson and Crick). Many other names are available in the suggested reading lists, where we give some of the original references to the developments under discussion, and in the credit lines for sources of figures and tables, which are given at the end of the book.

Again we are indebted to a number of people who helped us in various ways. Some read sections of the book at our request and made valuable suggestions. Some, who have used the book for teaching, have pointed out ways to make it more useful for them and their students. Others have noticed factual errors or errors of omission and have pointed out references that helped us check our facts. In addition to those who commented on earlier editions, many of whose contributions have carried over, this list includes Dennis Arvidson, Dominique Belin, Melanie Berkmen, Helmut Bertrand, Lindsay Black, Rob Britton, Yves Brun, Rich Calendar, George Chaconas, Dhruba Chattoraj, Carton Chen, Todd Ciche, Laszlo Csonka, Gary Dunny, Marie Elliot, Laura Frost, Barbara Funnell, Peter Geiduschek,

Graham Hatfull, John Helmann, Ann Hochschild, Susan Lovett, Ken Marinas, Norman Pace, Steven Sandler, Joel Schildbach, Linda Sherwood, Chris Waters, Robert Weiss, Joanne Willey, Steve Winans, Ry Young, and Steve Zinder. Special thanks go to Lee Kroos, who agreed to update an entire section. Yet others furnished original figures that we could incorporate into the text; some of them are mentioned in the figure credits. However, in the end, any mistakes and omissions were all ours.

As with the first three editions, it was a great pleasure to work with the professionals at ASM Press. The former director of ASM Press, Jeff Holtmeier, helped us prepare for the fourth edition. We have been fortunate to continue to work with Kenneth April, the production manager, who coordinated the entire project. We have also had the good fortune to work again with two of the same professionals who did a masterful job with the first three editions: Susan Brown Schmidler, who created the book and cover design; Terese Winslow, who created the cover illustration; and Elizabeth McGillicuddy, who copyedited the manuscript. We also thank Patrick Lane of ScEYEnce Studios for bringing an attractive aestheticism to the rendering of our hand-drawn illustrations into the final figures.

Larry Snyder
Joe Peters
Tina Henkin
Wendy Champness

About the Authors

Larry (Loren R.) Snyder, PhD, is a professor emeritus of microbiology and molecular genetics at Michigan State University, where he taught microbial genetics and microbiology to undergraduate and graduate students for about 40 years. He received his BS in mathematics and zoology at the University of Minnesota in Duluth and his PhD in biophysics at the University of Chicago before doing postdoctoral work at the International Laboratory of Genetics and Biophysics in Naples, Italy, and at the Curie Institute and Faculty of Sciences at the University of Paris as a Jane Coffin Childs postdoctoral fellow. He was a visiting professor at Harvard University and the University of Tel Aviv. Most of his research was on the interaction between bacteriophage T4 and its host, *Escherichia coli*, and was supported by the National Science Foundation (NSF) and the National Institutes of Health (NIH). At Michigan State, he served as acting chair of the Department of Microbiology for one year and was a founding co-principal investigator of the NSF Center for Microbial Ecology and director of the Howard Hughes Medical Institute Undergraduate Research Program. He was awarded the College of Natural Science Alumni Association Meritorious Faculty Award in 2002.

Joseph E. Peters, PhD, is an associate professor of microbiology at Cornell University, where he has been teaching bacterial genetics and microbiology since 2002. He received his BS from Stony Brook University and his PhD from the University of Maryland at College Park. He did postdoctoral work at the Johns Hopkins University School of Medicine, in part as an NSF-Alfred P. Sloan Foundation postdoctoral research fellow in molecular evolution. His research has focused on the intersection of DNA replication, recombination, and repair and how it relates to evolution, especially in the area of transposition. Research in his laboratory is funded by the NSF and NIH. He is the chair of the advisory board for the NSF-funded *E. coli* Genetic Stock Center, the chair of the American Society for Microbiology's Division of Genetics and Molecular Biology, and the director of graduate studies for the field of microbiology at Cornell.

Tina M. Henkin, PhD, is a professor of microbiology, chair of the Department of Microbiology, and Robert W. and Estelle S. Bingham Professor of Biological Sciences at Ohio State University, where she has been teaching bacterial genetics and microbiology since 1995. She received her BA in biology at Swarthmore College and her PhD in genetics at the University of Wisconsin—Madison, and she did postdoctoral work at the Tufts University School of Medicine. Her research focuses on gene regulation and regulatory RNAs in bacteria. Research in her laboratory is funded by the NIH. She is a fellow of the American Academy of Microbiology, the American Association for the Advancement of Science, and the American Academy of Arts and Sciences; a member of the National Academy of Sciences; and a cowinner of the National Academy of Sciences Pfizer Prize in Molecular Biology for her work on riboswitch RNAs.

Wendy Champness, PhD, is a professor emerita of microbiology and molecular genetics at Michigan State University, where she taught microbial genetics and microbiology to undergraduate and graduate students for more than 25 years. She received her BS and PhD degrees at Michigan State, where she was an NSF predoctoral fellow. She did postdoctoral work at the Massachusetts Institute of Technology as a Jane Coffin Childs postdoctoral fellow and was a visiting scientist at the John Innes Research Centre in Norwich, United Kingdom, and at the University of Tel Aviv. Most of her research was on the regulation of antibiotic synthesis genes in *Streptomyces*, and research in her laboratory was supported by grants from the NSF and NIH. She was a charter member of the NSF Center for Microbial Ecology at Michigan State and was a member of the editorial board of the *Journal of Bacteriology* for 12 years as well as an associate editor of the journal *Microbiology*.

Introduction

THE GOAL OF THIS TEXTBOOK is to introduce the student to the field of bacterial molecular genetics. From the point of view of genetics and genetic manipulation, bacteria are relatively simple organisms. There also exist model bacterial organisms that are easy to grow and easy to manipulate in the laboratory. For these reasons, most methods in molecular biology and recombinant DNA technology that are essential for the study of all forms of life have been developed around bacteria. Bacteria also frequently serve as model systems for understanding cellular functions and developmental processes in more complex organisms. Much of what we know about the basic molecular mechanisms in cells, such as translation and replication, has originated with studies of bacteria. This is because such central cellular functions have remained largely unchanged throughout evolution. Ribosomes have similar structures in all organisms, and many of the translation factors are highly conserved. The DNA replication apparatuses of all organisms contain features in common, such as sliding clamps and editing functions, which were first described in bacteria and their viruses, called bacteriophages. Chaperones that help other proteins fold and topoisomerases that change the topology of DNA were first discovered in bacteria and their bacteriophages. Studies of repair of DNA damage and mutagenesis in bacteria have also led the way to an understanding of such pathways in eukaryotes. Excision repair systems, mutagenic polymerases, and mismatch repair systems are remarkably similar in all organisms, and defects in these systems are responsible for multiple types of human cancers.

In addition, as our understanding of the molecular biology of bacteria advances, we are finding a level of complexity that was not appreciated previously. Because of the small size of the vast majority of bacteria, early on, it was impossible to recognize the high level of organization that exists

doi:10.1128/9781555817169.Intro

in bacteria, leading to the misconception that bacteria were merely "bags of enzymes," where small size allowed passive diffusion to move cellular constituents around. However, it is now clear that positioning of enzymes within the bacterial cell is highly controlled. For example, despite the lack of a specialized membrane structure called the nucleus (the early defining feature of the "prokaryote" [see below]), the genome of bacteria is exquisitely organized to facilitate its repair and expression during DNA replication. In addition, advances facilitated by molecular genetics and microscopy have made it clear that many cellular processes occur in highly organized subregions within the cell. Once it was appreciated that bacteria evolved in the same basic way as all other living organisms, the relative simplicity of bacteria paved the way for some of the most important scientific advances in any field, ever. It is safe to say that a bright future awaits the fledgling bacterial geneticist, where studies of relatively simple bacteria, with their malleable genetic systems, promise to uncover basic principles of cell biology that are common to all organisms and that we can now only imagine.

However, bacteria are not just important as laboratory tools to understand other organisms; they are important and interesting in their own right. For instance, they play an essential role in the ecology of Earth. They are the only organisms that can "fix" atmospheric nitrogen, that is, convert N_2 to ammonia, which can be used to make nitrogen-containing cellular constituents, such as proteins and nucleic acids. Without bacteria, the natural nitrogen cycle would be broken. Bacteria are also central to the carbon cycle because of their ability to degrade recalcitrant natural polymers, such as cellulose and lignin. Bacteria and some types of fungi thus prevent Earth from being buried in plant debris and other carbon-containing material. Toxic compounds, including petroleum, many of the chlorinated hydrocarbons, and other products of the chemical industry, can also be degraded by bacteria. For this reason, these organisms are essential in water purification and toxic waste cleanup. Moreover, bacteria and archaea (see below) produce most of the naturally occurring so-called greenhouse gases, such as methane and carbon dioxide, which are in turn used by other types of bacteria. This cycle helps maintain climate equilibrium. Bacteria have even had a profound effect on the geology of Earth, being responsible for some of the major iron ore and other mineral deposits in Earth's crust.

Another unusual feature of bacteria and archaea is their ability to live in extremely inhospitable environments, many of which are devoid of life except for microbes. These organisms are the only ones living in the Dead Sea, where the salt concentration in the water is very high. Some types of bacteria and archaea live in hot springs at temperatures close to the boiling point of water (or above in the case of archaea), and others survive in atmospheres devoid of oxygen, such as eutrophic lakes and swamps.

Bacteria that live in inhospitable environments sometimes enable other organisms to survive in those environments through symbiotic relationships. For example, symbiotic bacteria make life possible for *Riftia* tubeworms next to hydrothermal vents on the ocean floor, where living systems must use hydrogen sulfide in place of organic carbon and energy sources. In this symbiosis, the bacteria obtain energy and fix carbon dioxide by using the reducing power of the hydrogen sulfide given off by the hydrothermal vents, thereby furnishing food in the form of high-energy carbon compounds for the worms, which lack a digestive tract. Symbiotic cyanobacteria allow fungi to live in the Arctic tundra in the form of lichens. The bacterial partners in the lichens fix atmospheric nitrogen and make carbon-containing molecules through photosynthesis to allow their fungal partners to grow on the tundra in the absence of nutrient-containing soil. Symbiotic nitrogen-fixing *Rhizobium* and *Azorhizobium* spp. in the nodules on the roots of legumes and some other types of higher plants allow the plants to grow in nitrogen-deficient soils. Other types of symbiotic bacteria digest cellulose to allow cows and other ruminant animals to live on a diet of grass. Bioluminescent bacteria even generate light for squid and other marine animals, allowing illumination and signaling in the darkness of the deep ocean.

Bacteria are also worth studying because of their role in disease. They cause many human, plant, and animal diseases, and new diseases are continuously appearing. Knowledge gained from the molecular genetics of bacteria helps in the development of new ways to treat or otherwise control old diseases that can be resistant to older forms of treatment, as well as emerging diseases.

Some bacteria that live in and on our bodies also benefit us directly. The role of our commensal bacteria in human health is only beginning to be appreciated. It has been estimated that of the 10^{14} cells in a human body, only 10% are human! Of course, bacterial cells are much smaller than our cells, but this shows how our bodies are adapted to live with an extensive bacterial flora, which helps us digest food and avoid disease, among other roles, many of which are yet to be uncovered.

Bacteria have also long been used to make many useful compounds, such as antibiotics, and chemicals, such as benzene and citric acid. Bacteria and their bacteriophages are also the source of many of the useful enzymes used in molecular biology.

In spite of substantial progress, we have only begun to understand the bacterial world around us. Bacteria are the most physiologically diverse organisms on Earth, and the importance of bacteria to life on Earth and the potential uses to which bacteria can be put can only be guessed at. Thousands of different types of bacteria are known, and new insights into their cellular mechanisms and their applications constantly emerge from research with bacteria. Moreover, it is estimated that less than 1% of the types of bacteria living in the soil and other environments have ever been isolated, including entire phyla that have been identified using culture-independent mechanisms (see Pace, Suggested Reading). Undiscovered bacteria may have all manner of interesting and useful functions. Clearly, studies of bacteria will continue to be essential to our future efforts to understand, control, and benefit from the biological world around us, and bacterial molecular genetics will be an essential tool in these efforts. However, before discussing this field, we must first briefly discuss the evolutionary relationship of the bacteria to other organisms.

The Biological Universe

The Bacteria

According to the current view, all organisms on Earth belong to three major divisions called domains: the bacteria (formerly eubacteria), the archaea (formerly archaebacteria), and the eukaryotes. Figure 1 shows the microbiologists' view of the living world, where microbes provide most of the variety and eukaryotes occupy a relatively small niche. This is not so far-fetched a concept. Sequence data show that we differ from chimpanzees by only 2% of our DNA sequence, while 25 to 50% of the genes in a typical bacterium are unique to the species. Furthermore, while mammals diverged from each other on the order of millions of years ago, the bacterial lineages diverged billions of years ago.

Bacteria can differ greatly in their physical appearance under the microscope. Although most are single celled and rod shaped or spherical, some are multicellular and undergo complicated developmental cycles. The cyanobacteria (formerly called blue-green algae) are bacteria, but they have chlorophyll and can be filamentous, which is why they were originally mistaken for algae. The antibiotic-producing actinomycetes, which include *Streptomyces* spp., are also bacteria, but they form hyphae and stalks of spores, making them resemble fungi. Another bacterial group, the *Caulobacter* spp., have both free-swimming and sessile forms that attach to surfaces through a holdfast structure. Some of the most dramatic-appearing bacteria of all belong to the genus *Myxococcus*, members of which can exist as free-living single-celled organisms but can also aggregate to form fruiting bodies, much like slime molds. As mentioned above, bacterial cells are usually much smaller than the cells of higher organisms, but a bacterium that is 1 mm long, longer than even most eukaryotic cells, has been found. Many multiply by simple division, but one very large bacterium, *Epulopiscium*, which lives in surgeonfish on the Great Barrier Reef, gives birth to multiple live progeny. Because bacteria come in so many shapes and sizes, they cannot simply be distinguished by their physical appearance, but only by biochemical criteria, such as the sequences of their ribosomal RNAs (rRNAs), whose sequences are characteristic of the three domains of life.

GRAM-NEGATIVE AND GRAM-POSITIVE BACTERIA

Bacteria have historically been divided into two major subgroups, the **gram-negative** and **gram-positive bacteria**. This division was based on the response to a test called the Gram stain. So-called "gram-negative" bacteria retain little of the dye and are pink after this staining procedure, whereas "gram-positive" bacteria retain more of the dye and turn deep blue. The difference in staining reflects the fact that gram-negative bacteria are surrounded by a thinner structure composed of both an inner and an outer membrane, while the structure surrounding gram-positive bacteria is much thicker, consisting of a single membrane surrounded by a thicker wall. However, the difference between these groups seems to be more fundamental than the possession of an outer membrane. This older form of classification is being replaced by talking about the phyla of bacteria as determined by the DNA sequence. The phyla that have historically been grouped together based on gram-positive staining can be further broken down into those with a low percentage of guanine and cytosine (low G+C) compared to adenine and thymine and those with a high percentage of guanine and cytosine (high G+C) (see chapter 1). The low-G+C gram-positive bacteria include the ***Firmicutes***, a broad group containing *Bacillus*, clostridia, lactic acid bacteria, and mycoplasmas, and the *Actinobacteria*. The high-G+C gram-positive bacteria include the actinomycetes and simpler organisms, like *Mycobacterium*. Most of the other phyla stain gram negative, with the ***Proteobacteria***, another broad group, representing the classical gram-negative bacteria, like *Escherichia coli*, *Pseudomonas*, and *Rhizobium*.

The Archaea

The **archaea** (formerly called archaebacteria) are single-celled organisms that resemble bacteria. Bacteria and archaea were previously considered one group called the **prokaryotes**, which means "before the nucleus," and are still sometimes referred to by this name. However, it has been argued that present-day bacteria and archaea have had as much time to evolve as eukaryotes since they separated and so did not "come before." Bacteria and archaea do lack a nucleus, the defined membrane structure that is found in eukaryotes that houses the vast majority of the genes of the organism. The presence or

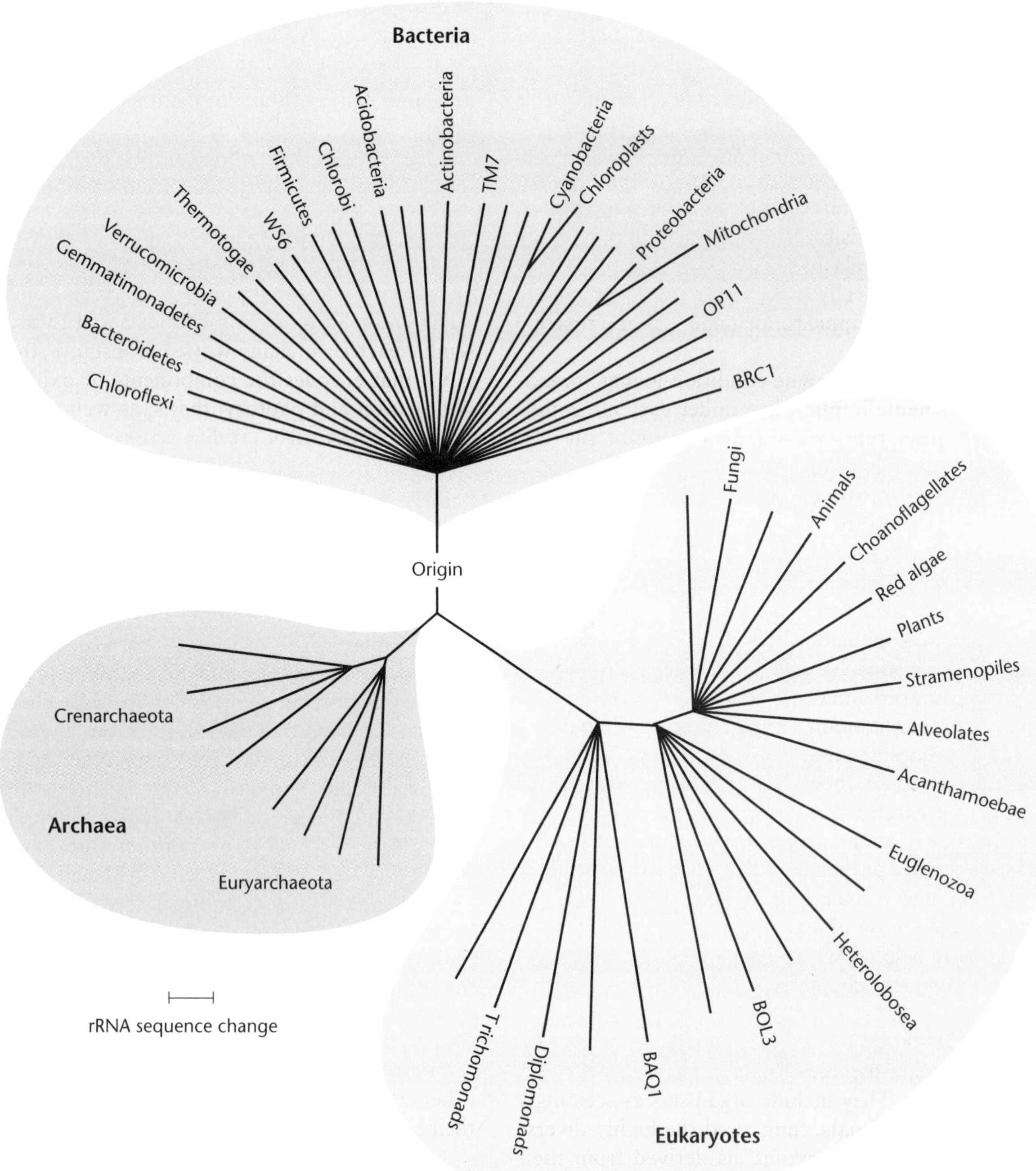

Figure 1 A molecular tree of life based on rRNA sequences emphasizing the divergence of bacteria, archaea, and eukaryotes. The organelles, mitochondria, and chloroplasts, which diverged from bacteria, are also shown. The figure represents information compiled from many rRNA sequence comparisons. Only a small subset of the actual lines of descent is shown. Names for some better-known groups are given. Representatives of some of the great many branches where there are no examples that are cultured are shown with an alphanumeric code (i.e., W56, TM7, etc.). doi:10.1128/9781555817169.Intro.f1

absence of a nuclear membrane greatly influences the mechanism used to manufacture proteins in the cell. Messenger RNA (mRNA) synthesis and translation can occur simultaneously in bacteria and archaea, since no nuclear membrane separates the ribosomes (which synthesize proteins) from the DNA. However, in most eukaryotes, mRNA made in the nucleus must be transported through the nuclear membrane before it can be translated into protein in the cytoplasm, where the ribosomes reside, and transcription and translation do not occur simultaneously.

Besides lacking a nucleus, bacterial and archaeal cells lack many other cellular constituents common to eukaryotes, including mitochondria and chloroplasts, which is not surprising, since mitochondria and chloroplasts actually originated from bacteria (see below). They also lack such visible organelles as the Golgi apparatus and the endoplasmic reticulum. The absence of most organelles generally gives bacterial and archaeal cells a much simpler appearance under the microscope than eukaryotes.

Extremophiles (or "extreme-condition-loving" organisms), as their name implies, live under extreme conditions where other types of organisms cannot survive, such as at the very high temperatures in sulfur springs, at high pressures on the ocean floor, and at very high osmolality, such as in the Dead Sea. Most extremophiles are archaea. However, it is becoming clear that archaea are important components of many environments; for example, archaea also perform unique biochemical functions, such as making methane.

The archaea themselves are a very diverse group of organisms and are sometimes divided into two phyla, the *Euryarchaeota*, containing methanogens, halophiles, and hyperthermophiles, and the *Crenarchaeota*, containing sulfur-dependent thermophiles. While substantial progress in research on archaea is being made, much less is known about the archaea than about the bacteria, although many of the components of the replication, transcription, and translation systems of archaea more closely resemble those of eukaryotes than they do those of bacteria (see below). We point out some of these similarities in subsequent chapters.

The Eukaryotes

The **eukaryotes** are members of the third domain of organisms on Earth. They include organisms as seemingly diverse as plants, animals, fungi, and the highly diverse protists. The name "eukaryotes" is derived from their nuclear membrane. They usually have a nucleus, and the word *karyon* in Greek means "nut," which is what the nucleus must have resembled to early cytologists. The eukaryotes can be unicellular, like yeasts, protozoans, and some types of algae, or they can be multicellular, like plants and animals. In spite of their widely diverse appearances, lifestyles, and relative complexity, however, all eukaryotes are remarkably similar at the biochemical level, particularly in their pathways for macromolecular synthesis.

MITOCHONDRIA AND CHLOROPLASTS

All eukaryotic cells contain something that ties them to the world of bacteria, the mitochondria. The mitochondria of eukaryotic cells are the sites of efficient adenosine triphosphate (ATP) generation through respiration. Evidence, including the sequences of many genes in their rudimentary chromosomes (see below), indicates that the mitochondria of eukaryotes are descended from free-living bacteria from the *Alphaproteobacteria* that formed a symbiosis with a primitive ancestor of eukaryotes.

Plant cells and some unicellular eukaryotic cells also contain chloroplasts, the site of photosynthesis. Like mitochondria, chloroplasts are also descended from free-living *Cyanobacteria*. Mitochondria and chloroplasts resemble bacteria in many ways. For instance, they contain DNA that encodes the components of oxidative phosphorylation and photosynthesis, as well as rRNAs and transfer RNAs (tRNAs). Even more striking, the mitochondrial and chloroplast rRNA and ribosomal proteins, as well as the membranes of the organelles, more closely resemble those of bacteria than they do those of eukaryotes. Comparisons of the sequences of highly conserved organelle genes, like the rRNAs, with those of bacteria indicates that mitochondria are descended from the *Alphaproteobacteria* and that chloroplasts are descended from the *Cyanobacteria* (Figure 1).

Mitochondria and chloroplasts may have come to be associated with early eukaryotic cells when these cells engulfed bacteria to take advantage of their superior energy-generating systems or their ability to obtain energy from light through photosynthesis. The eukaryotic cell may have contributed its ability to manage large DNAs to the symbiosis, as well as other functions. The engulfed bacteria eventually lost many of their own genes, which moved to the chromosome, from where they are expressed and their products are transported back into the organelle. The organelles had by then lost their autonomy and had become permanent symbionts of the eukaryotic cells. This process may still be going on today with intracellular bacteria in insects like *Buchnera* and others that provide amino acids and other nutrients to their hosts.

Speculations on the Origin of the Three Domains of Life

While it is clear that all known living organisms can be categorized into one of the three domains of life, bacteria, archaea, or eukaryotes, it remains unsettled how the three domains of life came to be and how they are interrelated. Bacteria and archaea are similar in that they both synthesize proteins from mRNA transcripts in a coordinated process on DNA and lack a defined nuclear membrane for their DNA. Gross similarities led to classification of bacteria and archaea together as prokaryotes until technological advances allowed analysis of the sequences of their rRNAs and the structures of their

RNA polymerases and lipids and showed that they were fundamentally different (see Pace, Suggested Reading). A convincing argument can be made that archaea are actually more closely related to eukaryotes from examining, among other functions, the molecular mechanisms of DNA replication across the three domains (see Leipe et. al., Suggested Reading) (Figure 1). The proteins responsible for recognizing where DNA replication is initiated and the proteins responsible for unwinding DNA so that it can be replicated are related to each other in archaea and eukaryotes but are obviously different from those found in bacteria. In addition, the proteins responsible for processing the DNA strands during replication are also similar or show similar chemistries in archaea and eukaryotes. One idea that has merit holds that the members of the domains found today might have originally derived their replication apparatuses from self-replicating entities that replicated more like viruses. In fact, the DNA polymerases and other replication and transcription proteins of organelles are sometimes more like those of bacterial viruses (bacteriophages) than they are like those of bacteria (see Shutt and Gray, Suggested Reading). As one theory goes, the first forms of life originated with a less stable form of genetic information through the use of RNA (see chapters 1 and 2). This ancestral use of RNA as the first polynucleotide is suggested by the central roles of RNA in cells, including the all-important role of catalyzing the synthesis of proteins from amino acids by an rRNA (see chapter 2). Viruses may have encoded the capacity to convert the founders of these three domains to a DNA-based mechanism (see Forterre, Suggested Reading). While RNA is inherently unstable, the relative stability of DNA would allow evolution to work more efficiently to catalogue beneficial changes in living organisms. All three of the domains of life may have acquired this capacity independently, greatly slowing down their evolution and allowing them to be recognizable as separate domains even today.

What Is Genetics?

Genetics can be simply defined as the manipulation of DNA to study cellular and organismal functions. Since DNA encodes all of the information needed to make the cell and the complete organism, the effects of changing this molecule can give clues to the normal functions of the cell and organism.

Before the advent of methods for manipulating DNA in the test tube, the only genetic approaches available for studying cellular and organismal functions were those of **classical genetics**. In this type of analysis, mutants (i.e., individuals that differ from the normal, or wild-type, members of the species by a certain observable attribute, or phenotype) that have alterations in the function being studied are isolated. The changes in the DNA, or mutations, responsible for the altered function are then localized in the chromosome by genetic crosses. The mutations are then grouped into genes by allelism tests to determine how many different genes are involved. The functions of the genes can then sometimes be deduced from the specific effects of the mutations on the organism. The ways in which mutations in genes involved in a biological system can alter the biological system provide clues to the normal functioning of the system.

Classical genetic analyses continue to contribute greatly to our understanding of developmental and cellular biology. A major advantage of the classical genetic approach is that mutants with a function altered can be isolated and characterized without any a priori understanding of the molecular basis of the function. Classical genetic analysis is also often the only way to determine how many gene products are involved in a function and, through suppressor analysis, to find other genes whose products may interact either physically or functionally with the products of these genes.

The development of **molecular genetic techniques** has greatly expanded the range of methods available for studying genes and their functions. These techniques include methods for isolating DNA and identifying the regions of DNA that encode particular functions, as well as methods for altering or mutating DNA in the test tube and then returning the mutated DNA to cells to determine the effect of the mutation on the organism.

The approach of first cloning a gene and then altering it in the test tube before reintroducing it into the cells to determine the effects of the alterations is sometimes called **reverse genetics** and is essentially the reverse of a classical genetic analysis. In classical genetics, a gene is known to exist only because a mutation in it has caused an observable change in the organism. With the molecular genetic approach, a gene can be isolated and mutated in the test tube without any knowledge of its function. Only after the mutated gene has been returned to the organism does its function become apparent.

Rather than one approach supplanting the other, molecular genetics and classical genetics can be used to answer different types of questions, and the two approaches often complement each other. In fact, the most remarkable insights into biological functions have often come from a combination of classical and molecular genetic approaches.

Bacterial Genetics

In bacterial genetics, genetic techniques are used to study bacteria. Applying genetic analysis to bacteria is no different in principle from applying it to other organisms. However, the methods that are available differ greatly.

Some types of bacteria are relatively easy to manipulate genetically. As a consequence, more is known about some bacteria than is known about any other type of organism. Some of the properties of bacteria that facilitate genetic experiments are described below.

Bacteria Are Haploid

One of the major advantages of bacteria for genetic studies is that they are **haploid**. This means that they have only one copy, or **allele**, of each gene. This property makes it much easier to identify cells with a particular type of mutation.

In contrast, most eukaryotic organisms are diploid, with two alleles of each gene, one on each homologous chromosome. Most mutations are recessive, which means that they do not cause a phenotype in the presence of a normal copy of the gene. Therefore, in diploid organisms, most mutations have no effect unless both copies of the gene in the two homologous chromosomes have the mutation. Backcrosses between different organisms with the mutation are usually required to produce offspring with the mutant phenotype, and even then, only some of the progeny of the backcross have the mutated gene in both homologous chromosomes. With a haploid organism such as a bacterium, however, most mutations have an immediate effect and there is no need for backcrosses.

Short Generation Times

Another advantage of some bacteria for genetic studies is that they have very short generation times. The **generation time** is the length of time the organism takes to reach maturity and produce offspring. If the generation time of an organism is too long, it can limit the number of possible experiments. Some strains of the bacterium *E. coli* can reproduce every 20 minutes under ideal conditions. With such rapid multiplication, cultures of the bacteria can be started in the morning and the progeny can be examined later in the day.

Asexual Reproduction

Another advantage of bacteria is that they multiply asexually, by cell division. Sexual reproduction, in which individuals of the same species must mate with each other to give rise to progeny, can complicate genetic experiments because the progeny are never identical to their parents. To achieve purebred lines of a sexually reproducing organism, a researcher must repeatedly cross the individuals with their relatives. However, if the organism multiplies asexually by cell division, all the progeny are genetically identical to their parent and to each other. Genetically identical organisms are called clones. Some lower eukaryotes, such as yeasts, and some types of plants, such as water hyacinths, can also multiply asexually to form clones. Identical twins, formed from the products of the division of an egg after it has been fertilized, are clones of each other. While there are a few examples where mammals have been cloned by transplanting a somatic cell into the ovary, bacteria form clones of themselves every time they divide.

Colony Growth on Agar Plates

Genetic experiments often require that numerous individuals be screened for a particular property. Therefore, it helps if large numbers of individuals of the species being studied can be propagated in a small space.

With some types of bacteria, thousands, millions, or even billions of individuals can be screened on a single agar-containing petri plate. Once on an agar plate, these bacteria divide over and over again, with all the progeny remaining together on the plate until a visible lump, or **colony**, has formed. Each colony is composed of millions of bacteria, all derived from the original bacterium and hence all clones of the original bacterium.

Colony Purification

The ability of some types of bacteria to form colonies through the multiplication of individual bacteria on plates allows colony purification of bacterial strains and mutants. If a mixture of bacteria containing different mutants or strains is placed on an agar plate, individual mutant bacteria or strains in the population each multiply to form colonies. However, these colonies may be too close together to be separable or may still contain a mixture of different strains of the bacterium. If the colonies are picked and the bacteria are diluted before replating, discrete colonies that result from the multiplication of individual bacteria may appear. No matter how crowded the bacteria were on the original plate, a pure strain of the bacterium can be isolated in one or a few steps of colony purification.

Serial Dilutions

To count the bacteria in a culture or to isolate a pure culture, it is often necessary to obtain discrete colonies of the bacteria. However, because bacteria are so small, a concentrated culture contains billions of bacteria per milliliter. If such a culture is plated directly on a petri plate, the bacteria all grow together and discrete colonies do not form. **Serial dilutions** offer a practical method for diluting solutions of bacteria before plating to obtain a measurable number of discrete colonies. The principle is that if smaller dilutions are repeated in succession, they can be multiplied to produce the total dilution. For example, if a solution is diluted in three steps by adding 1 ml of the solution to 99 ml of water, followed by adding 1 ml of this dilution to another 99 ml of water and finally by adding 1 ml of the second dilution to another

99 ml of water, the final dilution is $10^{-2} \times 10^{-2} \times 10^{-2} = 10^{-6}$, or one in a million. To achieve the same dilution in a single step, 1 ml of the original solution would have to be added to 1,000 liters (about 250 gallons) of water. Obviously, it is more convenient to handle three solutions of 100 ml each than to handle a solution of 250 gallons, which weighs about 2,000 lb!

Selections

Probably the greatest advantage of bacterial genetics is the opportunity to do **selections**, by which very rare mutants and other types of strains can be isolated. To select a rare strain, billions of the bacteria are plated under conditions where only the desired strain, not the bulk of the bacteria, can grow. In general, these conditions are called the **selective conditions**. For example, a nutrient may be required by most of the bacteria but not by the strain being selected. Agar plates lacking the nutrient then present selective conditions for the strain, since only the strain being selected multiplies to form a colony in the absence of the nutrient. In another example, the desired strain may be able to multiply at a temperature that would kill most of the bacteria. Incubating agar plates at that temperature would provide the selective condition. After the strain has been selected, a colony of the strain can be picked and colony purified away from other contaminating bacteria under the same selective conditions.

The power of selection with bacterial populations is awesome. Using a properly designed selection, a single bacterium can be selected from among billions placed on an agar plate. If we could apply such selections to humans, we could find one of the few individuals in the entire human population of Earth with a particular trait.

Storing Stocks of Bacterial Strains

Most types of organisms must be continuously propagated; otherwise, they age and die. Propagating organisms requires continuous transfers and replenishing of the food supply, which can be very time-consuming. However, many types of bacteria can be stored in a dormant state and therefore do not need to be continuously propagated. The conditions used for storage depend on the type of bacteria. Some bacteria sporulate and so can be stored as dormant spores. Others can be stored by being frozen in glycerol or being dried. Storing organisms in a dormant state is particularly convenient for genetic experiments, which often require the accumulation of large numbers of mutants and other strains. The strains remain dormant until the cells are needed, at which time they can be revived.

Genetic Exchange

Genetic experiments with an organism usually require some form of exchange of DNA or genes between members of the species. Most types of organisms on Earth are known to have some means of genetic exchange, which presumably accelerates evolution and increases the adaptability of a species.

Exchange of DNA from one bacterium to another can occur in one of three ways. In **transformation**, DNA released from one cell enters another cell of the same species. In **conjugation**, plasmids, which are small autonomously replicating DNA molecules in bacterial cells, transfer DNA from one cell to another. Finally, in **transduction**, a bacterial virus accidentally picks up DNA from a cell it has infected and injects this DNA into another cell. The ability to exchange DNA between strains of a bacterium makes possible genetic crosses and complementation tests, as well as the tests essential to genetic analysis.

Phage Genetics

Some of the most important discoveries in genetics have come from studies with viruses that infect bacteria; these viruses are called **bacteriophages**, or **phages** for short. Phages are not alive; instead, they are just genes wrapped in a protective coat of protein and/or membrane, as are all viruses. Because phages are not alive, they cannot multiply outside a bacterial cell. However, if a phage encounters a type of bacterial cell that is sensitive to that phage, the phage, or at least its DNA or RNA, enters the cell and directs it to make more phage.

Phages are usually identified by the holes, or **plaques**, they form in layers of sensitive bacteria. In fact, the name "phage" (Greek for "eat") derives from these plaques, which look like eaten-out areas. A plaque can form when a phage is mixed with large numbers of susceptible bacteria and the mixture is placed on an agar plate. As the bacteria multiply, one may be infected by the phage, which multiplies and eventually breaks open, or **lyses**, the bacterium, releasing more phage. As the surrounding bacteria are infected, the phage spread, even as the bacteria multiply to form an opaque layer called a **bacterial lawn**. Wherever the original phage infected the first bacterium, the plaque disrupts the lawn, forming a clear spot on the agar. Despite its empty appearance, this spot contains millions of the phage.

Phages offer many of the same advantages for genetics as bacteria. Thousands or even millions of phages can be put on a single plate. Also, like bacterial colonies, each plaque contains millions of genetically identical phage. By analogy to the colony purification of bacterial strains, individual phage mutants or strains can be isolated from other phages through **plaque purification**.

Phages Are Haploid

Phages are, in a sense, haploid, since they usually have only one copy of each gene. As with bacteria, this property makes isolation of phage mutants relatively easy, since all mutants immediately exhibit their phenotypes without the need for backcrosses.

Selections with Phages

Selection of rare strains of a phage is possible; as with bacteria, it requires conditions under which only the desired phage strain can multiply to form a plaque. For phages, these selective conditions may be a bacterial host in which only the desired strain can multiply or a temperature at which only the phage strain being selected can multiply. Note that the bacterial host must be able to multiply under the same selective conditions; otherwise, a plaque cannot form.

As with bacteria, selections allow the isolation of very rare strains or mutants. If selective conditions can be found for the strain, millions of phages can be mixed with the bacterial host, and only the desired strain multiplies to form a plaque. A pure strain can then be obtained by picking the phage from the plaque and plaque purifying the strain under the same selective conditions.

Crosses with Phages

Phage strains can be crossed very easily. The same cells are infected with different mutants or strains of the phage. The DNA of the two phages is then in the same cell, where the molecules can interact genetically with each other, allowing genetic manipulations, such as gene-mapping and allelism tests.

A Brief History of Bacterial Molecular Genetics

Because of the ease with which they can be handled, bacteria and their phages have long been the organisms of choice for understanding basic cellular phenomena, and their contributions to this area of study are almost countless. The following chronological list should give a feeling for the breadth of these contributions and the central position that bacteria have occupied in the development of modern molecular genetics. Some original references are given at the end of this chapter under Suggested Reading.

Inheritance in Bacteria

In the early part of the 1900s, biologists agreed that inheritance in higher organisms follows Darwinian principles. According to Charles Darwin, changes in the hereditary properties of organisms occur randomly and are passed on to the progeny. In general, the changes that happen to be beneficial to the organism are more apt to be passed on to subsequent generations.

With the discovery of the molecular basis for heredity, Darwinian evolution now has a strong theoretical foundation. The properties of organisms are determined by the sequence of their DNA, and as the organisms multiply, changes in this sequence sometimes occur randomly and without regard to the organism's environment. However, if a random change in the DNA happens to be beneficial in the situation in which the organism finds itself, the organism has an improved chance of surviving and reproducing.

As late as the 1940s, many bacteriologists thought that inheritance in bacteria was different from inheritance in other organisms. It was thought that rather than enduring random changes, bacteria could adapt to their environment by some sort of "directed" change and that the adapted organisms could then somehow pass on the change to their offspring. Such opinions were encouraged by the observations of bacteria growing under selective conditions. For example, in the presence of an antibiotic, all the bacteria in the culture soon become resistant to the antibiotic. It seemed as though the resistant bacterial mutants appeared in response to the antibiotic.

One of the first convincing demonstrations that inheritance in bacteria follows Darwinian principles was made in 1943 by Salvador Luria and Max Delbrück (see chapter 3 and Suggested Reading). Their work demonstrated that particular phenotypes, in their case resistance to a virus, occur randomly in a growing population, even in the absence of the virus. By the directed-change or adaptive-mutation hypothesis, the resistant mutants should have appeared only in the presence of the virus.

The demonstration that inheritance in bacteria follows the same principles as inheritance in eukaryotic organisms set the stage for the use of bacteria in studies of basic genetic principles common to all organisms.

Transformation

As discussed at the beginning of the Introduction, most organisms exhibit some mechanism for exchanging genes. The first demonstration of genetic exchange in bacteria was made by Fred Griffith in 1928. He was studying two variants of pneumococci, now called *Streptococcus pneumoniae*. One variant formed smooth-appearing colonies on plates and was pathogenic in mice. The other variant formed rough-appearing colonies on plates and did not kill mice. Only live, smooth-colony-forming bacteria could cause disease, since the disease requires that the bacteria multiply in the infected mice. However, when Griffith mixed dead smooth-colony formers with live rough-colony formers and injected the mixture into mice, the mice became sick and died. Moreover, he isolated live smooth-colony formers from the dead mice. Apparently, the dead smooth-colony formers were "transforming" some of the live rough-colony formers into the pathogenic, smooth-colony-forming type. The "transforming principle" given off by the dead smooth-colony formers was later shown to be DNA, since addition of purified DNA from the dead smooth-colony formers to the live rough-colony formers in a test tube transformed some members of the rough type to the smooth type (see Avery et al., Suggested Reading). This method of exchange is called transformation, and this experiment provided the first direct evidence that genes are made of DNA. Later experiments by Alfred Hershey and Martha Chase in 1952 (see Suggested Reading) showed that phage DNA alone is sufficient to direct the synthesis of more phages.

Conjugation

In 1946, Joshua Lederberg and Edward Tatum (see Suggested Reading) discovered a different type of gene exchange in bacteria. When they mixed some strains of *E. coli* with other strains, they observed the appearance of recombinant types that were unlike either parent. Unlike transformation, which requires only that DNA from one bacterium be added to the other bacterium, this means of gene exchange requires direct contact between two bacteria. It was later shown to be mediated by plasmids and is called conjugation.

Transduction

In 1953, Norton Zinder and Joshua Lederberg discovered yet a third mechanism of gene transfer between bacteria. They showed that a phage of *Salmonella enterica* serovar Typhimurium could carry DNA from one bacterium to another. This means of gene exchange is called transduction and is now known to be quite widespread.

Recombination within Genes

At the same time, experiments with bacteria and phages were also contributing to the view that genes were linear arrays of nucleotides in the DNA. By the early 1950s, recombination had been well demonstrated in higher organisms, including fruit flies. However, recombination was thought to occur only between mutations in different genes and not between mutations in the same gene. This led to the idea that genes were like "beads on a string" and that recombination is possible between the "beads," or genes, but not within a gene. In 1955, Seymour Benzer disproved this hypothesis by using the power of phage genetics to show that recombination is possible within the *r*II genes of phage T4. He mapped numerous mutations in the *r*II genes, thereby demonstrating that genes are linear arrays of mutable sites in the DNA. Later experiments with other phage and bacterial genes showed that the sequence of nucleotides in the DNA directly determines the sequence of amino acids in the protein product of the gene.

Semiconservative DNA Replication

In 1953, James Watson and Francis Crick published their structure of DNA. One of the predictions of this model is that DNA replicates by a semiconservative mechanism, in which specific pairing occurs between the bases in the old and the new DNA strands, thus essentially explaining heredity. In 1958, Matthew Meselson and Frank Stahl used bacteria to confirm that DNA replicates by this semiconservative mechanism.

mRNA

The existence of mRNA was also first indicated by experiments with bacteria and phages. In 1961, Sydney Brenner, François Jacob, and Matthew Meselson used phage-infected bacteria to show that ribosomes are the site of protein synthesis and confirmed the existence of a "messenger" RNA that carries information from the DNA to the ribosome.

The Genetic Code

Also in 1961, phages and bacteria were used by Francis Crick and his collaborators to show that the genetic code is unpunctuated, three lettered, and redundant. These researchers also showed that not all possible codons designated an amino acid and that some were nonsense. These experiments laid the groundwork for Marshall Nirenberg and his collaborators to decipher the genetic code, in which a specific three-nucleotide set encodes one of 20 amino acids. The code was later verified by the examination of specific amino acid changes due to mutations in the lysozyme gene of phage T4.

The Operon Model

François Jacob and Jacques Monod published their operon model for the regulation of the lactose utilization genes of *E. coli* in 1961, as well. They proposed that a repressor blocks RNA synthesis on the *lac* genes unless the inducer, lactose, is bound to the repressor. Their model has served to explain gene regulation in other systems, and the *lac* genes and regulatory system continue to be used in molecular genetic experiments, even in systems as far removed from bacteria as animal cells and viruses.

Enzymes for Molecular Biology

The early 1960s saw the start of the discovery of many interesting and useful bacterial and phage enzymes involved in DNA and RNA metabolism. In 1960, Arthur Kornberg demonstrated the synthesis of DNA in the test tube by an enzyme from *E. coli*. The next year, a number of groups independently demonstrated the synthesis of RNA in the test tube by RNA polymerases from bacteria. From that time on, other useful enzymes for molecular biology were isolated from bacteria and their phages, including polynucleotide kinase, DNA ligases, topoisomerases, and many phosphatases.

From these early observations, the knowledge and techniques of molecular genetics exploded. For example, in the early 1960s, techniques were developed for detecting the hybridization of RNA to DNA and DNA to DNA on nitrocellulose filters. These techniques were used to show that RNA is made on only one strand in specific regions of DNA, which later led to the discovery of promoters and other regulatory sequences. By the late 1960s, restriction endonucleases had been discovered in bacteria and shown to cut DNA in specific places (see Linn and Arber, Suggested Reading). By the early 1970s, these restriction endonucleases were being exploited to introduce foreign genes into *E. coli* (see Cohen et al., Suggested Reading), and by the late 1970s, the first human gene had been expressed in a bacterium. Also in the late 1970s, methods to sequence DNA by using enzymes from phages and bacteria were developed.

In 1988, a thermally stable DNA polymerase from a thermophilic bacterium was used to invent the technique called the **polymerase chain reaction (PCR)**. This extremely sensitive technique allows the amplification of genes and other regions of DNA, facilitating their cloning and study. Thermally stable DNA polymerases are an essential tool for genome sequencing.

More recently, advances in DNA synthesis and DNA recombination have been ushering in a new age of molecular genetics with bacteria under the name of **synthetic genomics**, where massive strands of DNA large enough to comprise entire genomes can be made from the building blocks of DNA. In 2010, a significant milestone was reached with synthetic genomics when a derivative of the entire genome of *Mycoplasma mycoides* was synthesized from scratch, assembled by recombination, and used to replace the DNA in a related species (see Gibson et al., Suggested Reading). As the techniques for synthetic genomics advance, they should allow bacterial geneticists to build completely new useful bacterial strains from the ground up. While the ability to design bacteria *de novo* will likely have to include certain safeguards and greater public understanding, these types of experiments hold great promise for industrial use, as tools in medicine, and for addressing basic scientific questions, such as what is the minimum genetic requirement for life as a free-living organism.

These examples illustrate that bacteria and their phages have been central to the development of molecular genetics and recombinant DNA technology. Contrast the timing of these developments with the timing of comparable major developments in physics (early 1900s) and chemistry (1920s and 1930s) and you can see that molecular genetics is arguably the most recent major conceptual breakthrough in the history of science.

What Is Ahead

This textbook emphasizes how molecular genetic approaches can be used to solve biological problems. As an educational experience, the methods used and the interpretation of experiments are at least as important as the conclusions drawn. Therefore, whenever possible, the experiments that led to the conclusions are presented. The first two chapters, of necessity, review the concepts of macromolecular synthesis that are essential to understanding bacterial molecular genetics. However, they also introduce more current material. Chapter 1, besides reviewing the basics of DNA replication and the techniques of molecular biology, presents some recent advances in our understanding of how replication is coordinated with other cellular processes. It also contains information on how obstacles to replication are overcome and on the structure of the bacterial chromosome. Chapter 2, in addition to reviewing the basics of protein synthesis, presents more current developments concerning protein folding and the role of RNA and protein degradation. Chapter 3, similarly, reviews basic genetic principles, but with a special emphasis on bacterial genetics. The chapter also integrates more current applications, such as marker rescue, genome sequencing, and site-specific mutagenesis with genetic analysis, as it is done now. Chapters 4 through 13 deal with more specific topics and the techniques that can be used to study them, with particular emphasis on recent evidence concerning the relatedness of seemingly disparate topics. The last chapter, chapter 14, discusses protein transport and secretion systems and contains a section on bacterial cell biology. It finishes with an important paradigm for bacterial development, sporulation in *Bacillus subtilis*. We hope that this textbook will help put modern molecular genetics in a historical perspective, bring the reader up to date on current advances in bacterial molecular genetics, and position the reader to understand future developments in this exciting and rapidly progressing field of science.

SUGGESTED READING

Ausmess, N., J. R. Kuhn, and C. Jacobs-Wagner. 2003. The bacterial cytoskeleton: an intermediate filament-like function in cell shape. *Cell* **115:**705–713.

Avery, O. T., C. M. MacLeod, and M. McCarty. 1944. Studies on the chemical nature of the substance inducing transformation of pneumococcal types. I. Induction of transformation by a desoxyribonucleic acid fraction isolated from pneumococcus type III. *J. Exp. Med.* **79**:137–158.

Brenner, S., F. Jacob, and M. Meselson. 1961. An unstable intermediate carrying information from genes to ribosomes for protein synthesis. *Nature* (London) **190**:576–581.

Cairns, J., G. S. Stent, and J. D. Watson. 1966. *Phage and the Origins of Molecular Biology*. Cold Spring Harbor Laboratory Press, Cold Spring Harbor, NY.

Cohen, S. N., A. C. Y. Chang, H. W. Boyer, and R. B. Helling. 1973. Construction of biologically functional bacterial plasmids *in vitro*. *Proc. Natl. Acad. Sci. USA* **70**:3240–3244.

Crick, F. H. C., L. Barnett, S. Brenner, and R. J. Watts-Tobin. 1961. General nature of the genetic code for proteins. *Nature* (London) **192**:1227–1232.

Forterre, P. 2006. Three RNA cells for ribosomal lineages and three DNA viruses to replicate their genomes: a hypothesis for the origin of cellular domain. *Proc. Natl. Acad. Sci. USA* **103**:3669–3674.

Gibson, D. G., J. I. Glass, C. Lartigue, V. N. Noskov, R.-Y. Chuang, M. A. Algire, G. A. Benders, M. G. Montague, L. Ma, M. M. Moodie, C. Merryman, S. Vashee, R. Krishnakumar, N. Assad-Garcia, C. Andrews-Pfannkoch, E. A. Denisova, L. Young, Z.-Q. Qi, T. H. Segall-Shapiro, C. H. Calvey, P. P. Parmar, C. A. Hutchison III, H. O. Smith, and J. C. Venter. 2010. Creation of a bacterial cell controlled by a chemically synthesized genome. *Science* **329**:52–56.

Hershey, A. D., and M. Chase. 1952. Independent functions of viral protein and nucleic acids in growth of bacteriophage. *J. Gen. Physiol.* **36**:39–56.

Jacob, F., and J. Monod. 1961. Genetic regulatory mechanisms in the synthesis of proteins. *J. Mol. Biol.* **3**:318–356.

Lederberg, J., and E. L. Tatum. 1946. Gene recombination in *Escherichia coli*. *Nature* (London) **158**:558.

Leipe, D. D., L Aravind, and E. V. Koonin. 1999. Did DNA replication evolve twice independently? *Nucleic Acids Res.* *27*:3389–3401.

Linn, S., and W. Arber. 1968. Host specificity of DNA produced by *Escherichia coli*. X. *In vitro* restriction of phage fd replicative form. *Proc. Natl. Acad. Sci. USA* **59**:1300–1306.

Luria, S., and M. Delbrück. 1943. Mutations of bacteria from virus sensitivity to virus resistance. *Genetics* **28**:491–511.

Meselson, M., and F. W. Stahl. 1958. The replication of DNA in *Escherichia coli*. *Proc. Natl. Acad. Sci. USA* **44**:671–682.

Nirenberg, M. W., and J. H. Matthei. 1961. The dependence of cell-free protein synthesis in *E. coli* upon naturally occurring or synthetic polynucleotides. *Proc. Natl. Acad. Sci. USA* **47**:1588–1602.

Olby, R. 1974. *The Path to the Double Helix*. Macmillan Press, London, United Kingdom.

Olsen, G. J., C. R. Woese, and R. Overbeek. 1994. The winds of (evolutionary) change: breathing new life into microbiology. *J. Bacteriol.* **176**:1–6.

Pace, N. R. 2009. Mapping the tree of life: progress and prospects. *Microbiol. Mol. Biol. Rev.* **73**:565–576.

Schrodinger, E. 1944. *What Is Life? The Physical Aspect of the Living Cell*. Cambridge University Press, Cambridge, United Kingdom.

Shutt, T. E., and M. W. Gray. 2006. Bacteriophage origins of mitochondrial replication and transcription protein. *Trends Genet.* *22*:90–95.

Watson, J. D. 1968. *The Double Helix*. Atheneum, New York, NY.

Woese, C. R., and G. E. Fox. 1977. Phylogenetic structure of the prokaryotic domain: the primary kingdoms. *Proc. Natl. Acad. Sci. USA* **74**:5088–5900.

Zinder, N. D., and J. Lederberg. 1952. Genetic exchange in *Salmonella*. *J. Bacteriol.* **64**:679–699.

CHAPTER 1

The Bacterial Chromosome: DNA Structure, Replication, and Segregation

DNA Structure

THE SCIENCE OF MOLECULAR GENETICS began with the determination of the structure of DNA. Experiments with bacteria and phages (i.e., viruses that infect bacteria) in the late 1940s and early 1950s, as well as the presence of DNA in chromosomes of higher organisms, had implicated this macromolecule as the hereditary material (see the introduction). In the 1930s, biochemical studies of the base composition of DNA by Erwin Chargaff established that the amount of guanine always equals the amount of cytosine and that the amount of adenine always equals the amount of thymine, independent of the total base composition of the DNA. In the early 1950s, X-ray diffraction studies by Rosalind Franklin and Maurice Wilkins showed that DNA is a double helix. Finally, in 1953, Francis Crick and James Watson put together the chemical and X-ray diffraction information in their famous model of the structure of DNA. This story is one of the most dramatic in the history of science and has been the subject of many historical treatments, some of which are listed at the end of this chapter.

Figure 1.1 illustrates the **Watson-Crick structure of DNA**, in which two strands wrap around each other to form a double helix. These strands can be extremely long, even in a simple bacterium, extending up to 1 mm—a thousand times longer than the bacterium itself. In a human cell, the strands that make up a single chromosome (which is one DNA molecule) are hundreds of millimeters, or many inches, long.

The Deoxyribonucleotides

If we think of DNA strands as chains, deoxyribonucleotides form the links. Figure 1.2 shows the basic structure of deoxyribonucleotides, called deoxynucleotides for short. Each is composed of a **base**, a **sugar**, and a **phosphate** group. The DNA bases are **adenine** (A), **cytosine** (C), **guanine**

doi:10.1128/9781555817169.ch1

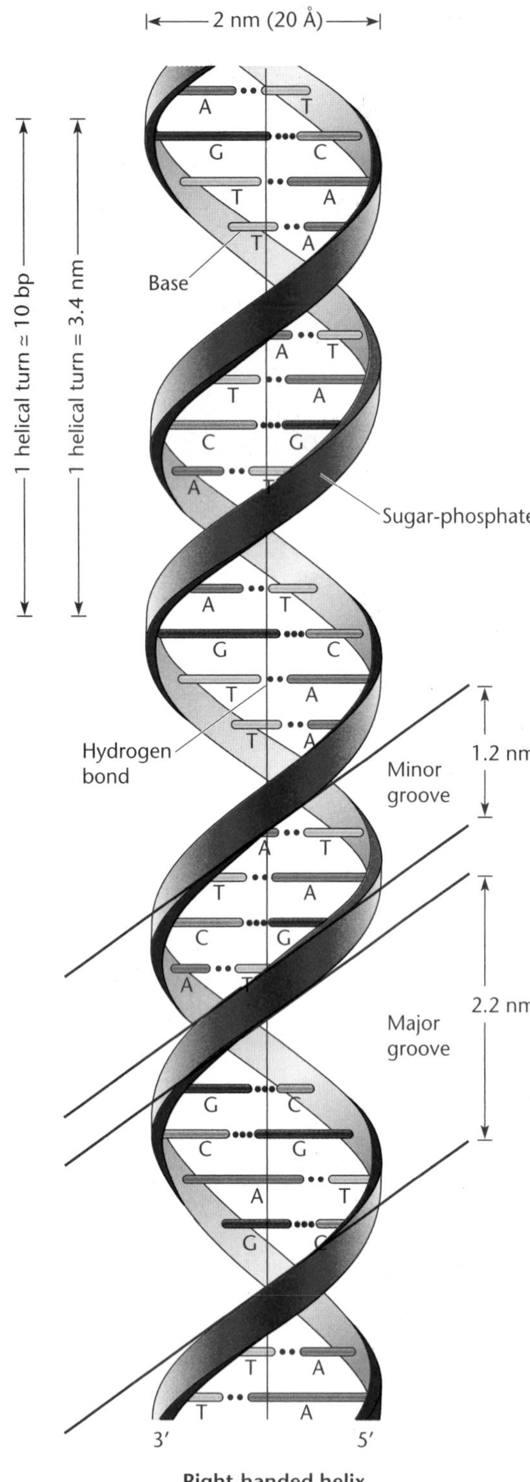

Figure 1.1 Schematic drawing of the Watson-Crick structure of DNA, showing the helical sugar-phosphate backbones of the two strands held together by hydrogen bonding between the bases. Also shown are the major and minor grooves and the dimensions of the helix.
doi:10.1128/9781555817169.ch1.f1.1

(G), and **thymine** (T), which have either one or two rings, as shown in Figure 1.2. The bases with two rings (A and G) are the purines, and those with only one ring (T and C) are pyrimidines. A third pyrimidine, uracil (U), replaces thymine in RNA. The carbons and nitrogens making up the rings of the bases are numbered sequentially, as shown in the figure. All four DNA bases are attached to the five-carbon sugar deoxyribose. This sugar is identical to ribose, which is found in RNA, except that it does not have an oxygen attached to the second carbon, hence the name deoxyribose. The carbons in the sugar of a nucleotide are also numbered 1, 2, 3, and so on, but they are labeled with "primes" to distinguish them from the carbons in the bases (Figure 1.2). The nucleotides also have one or more phosphate groups attached to a carbon of the deoxyribose sugar, as shown. The carbon to which the phosphate group is attached is indicated, although if the group is attached to the 5′ carbon (the usual situation), the carbon to which it is attached is often not stipulated.

The components of the deoxynucleotides have special names. A **deoxynucleoside** (rather than -tide) is a base attached to a sugar but lacking a phosphate. Without phosphates, the four deoxynucleosides are called **deoxyadenosine**, **deoxycytidine**, **deoxyguanosine**, and **deoxythymidine**. As shown in Figure 1.2, the deoxynucleotides have one, two, or three phosphates attached to the sugar and are known as deoxynucleoside monophosphates, diphosphates, or triphosphates, respectively. The individual deoxynucleoside monophosphates, called deoxyguanosine monophosphate, etc., are often abbreviated dGMP, dAMP, dCMP, and dTMP, where the d stands for deoxy; the G, A, C, or T stands for the base; and the MP stands for monophosphate. In turn, the diphosphates and triphosphates are abbreviated dGDP, dADP, dCDP, and dTDP and dGTP, dATP, dCTP, and dTTP, respectively. The phosphate attached to the sugar is called the α phosphate, while the next two are called the β and γ phosphates, respectively, as shown in the figure. Collectively, the four deoxynucleoside triphosphates are often referred to as dNTPs.

The DNA Chain

Phosphodiester bonds join each deoxynucleotide link in the DNA chain. As shown in Figure 1.3, the phosphate attached to the last (5′) carbon of the deoxyribose sugar of one nucleotide is attached to the third (3′) carbon of the sugar of the next nucleotide, thus forming one strand of nucleotides connected 5′ to 3′, 5′ to 3′, etc.

The 5′ and 3′ Ends

The nucleotides found at the ends of a linear piece of DNA have properties that are biochemically important and useful for orienting the DNA strand. At one end

Bases

Sugars

Nucleotides

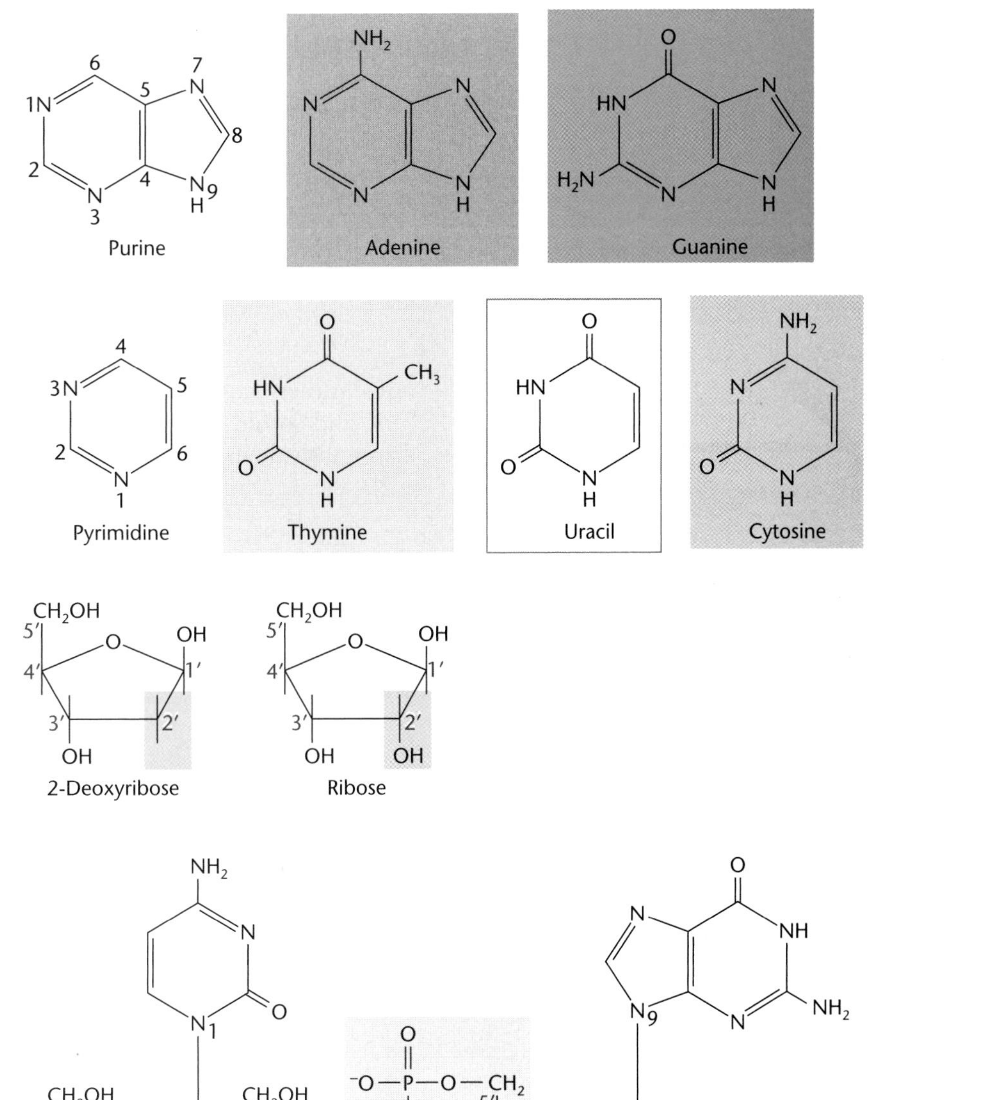

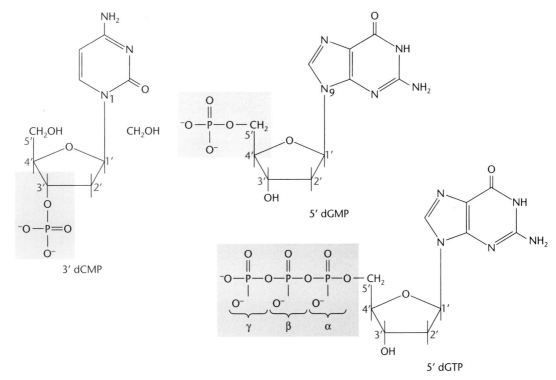

Figure 1.2 Chemical structures of deoxyribonucleotides, showing the bases and sugars and how they are assembled into a deoxyribonucleotide. doi:10.1128/9781555817169.ch1.f1.2

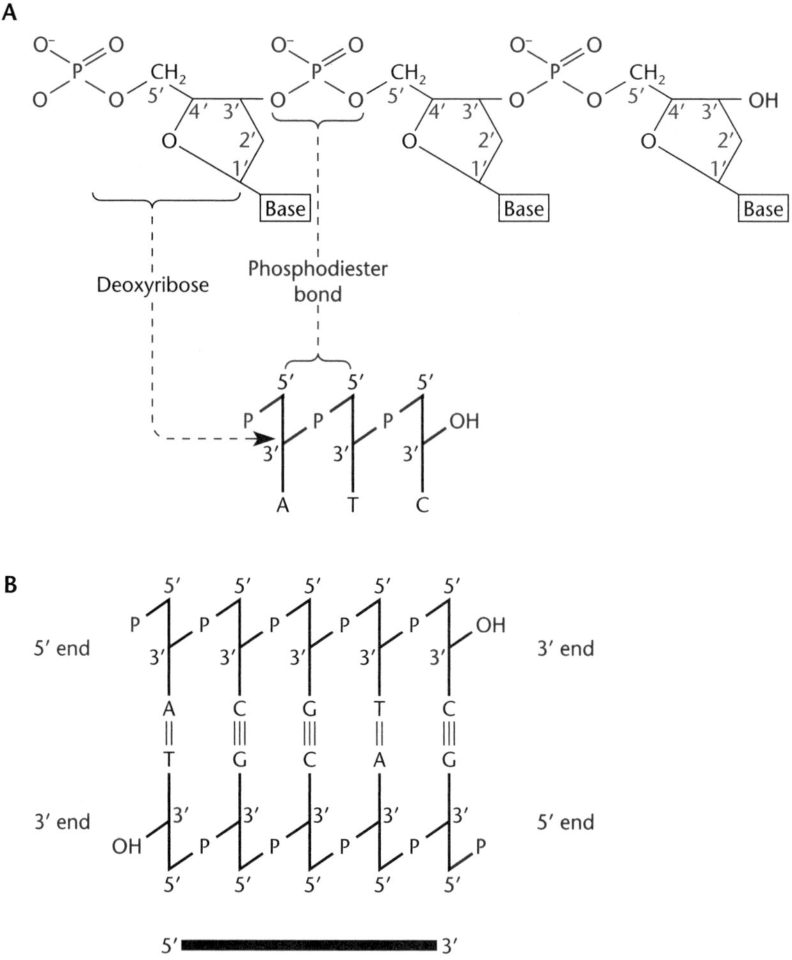

Figure 1.3 (A) Schematic drawing of a DNA chain showing the 3′-to-5′ attachment of the phosphates to the sugars, forming phosphodiester bonds. **(B)** Two strands of DNA bind at the bases in an antiparallel arrangement of the phosphate-sugar backbones.
doi:10.1128/9781555817169.ch1.f1.3

of the DNA chain, a nucleotide will have a phosphate attached to its 5′ carbon that does not connect it to another nucleotide. This end of the strand is called the 5′ end or the **5′ phosphate end** (Figure 1.3B). On the other end, the last nucleotide lacks a phosphate at its 3′ carbon. Because it has only a hydroxyl group (the OH in Figure 1.3B), this end is called the 3′ end or the **3′ hydroxyl end.**

Base Pairing

The sugar and phosphate groups of DNA form what is often called a **backbone** to support the bases, which jut out from the chain. This structure allows the bases from one single strand of DNA to form hydrogen bonds with another strand of DNA, thereby holding together two separate nucleotide chains (Figure 1.3B). The first clue that pairing between specific bases could form the basis for the structure of DNA came from Erwin Chargaff's observation about the ratios of the bases; no matter the source of the DNA, the concentration of guanine (G) always equals the concentration of cytosine (C) and the concentration of adenine (A) always equals the concentration of thymine (T). These ratios, named Chargaff's rules, gave Watson and Crick one of the essential clues to the structure of DNA. They proposed that the two strands of the DNA are held together by specific hydrogen bonding between the bases in opposite strands, as shown in Figure 1.4. Thus, the amounts of A and T and of C and G are always the same because A's pair only with T's and G's pair only with C's to hold the DNA strands together. Each A-and-T pair and each G-and-C pair in DNA is called a **complementary base pair,** and the sequences of two strands of DNA are said to be complementary if one strand always has a T where there is an A in the other strand and a G where there is a C in the other strand.

It did not escape the attention of Watson and Crick that the complementary base-pairing rules essentially explain heredity. If A pairs only with T and G pairs only with C, then each strand of DNA can replicate to make a complementary copy, so that the two replicated DNAs will be exact copies of each other. Offspring containing

Adenine Thymine

Guanine Cytosine

Figure 1.4 The two complementary base pairs found in DNA. Two hydrogen bonds form in adenine-thymine base pairs. Three hydrogen bonds form in guanine-cytosine base pairs. doi:10.1128/9781555817169.ch1.f1.4

the new DNAs would have the same sequence of nucleotides in their DNAs as their parents and thus would be exact copies of their parents.

Antiparallel Construction

As mentioned at the beginning of this section, the complete DNA molecule consists of two long chains wrapped around each other in a double helix (Figure 1.1). The double-stranded molecule can be thought of as being like a circular staircase, with the alternating phosphates and deoxyribose sugars forming the railings and the bases connected to each other forming the steps. However, the two chains run in opposite orientations, with the phosphates on one strand attached 5′ to 3′, 5′ to 3′, etc., to the sugars and those on the other strand attached 3′ to 5′, 3′ to 5′, etc. This arrangement is called **antiparallel**. In addition to phosphodiester bonds running in opposite directions, the antiparallel construction causes the 5′ phosphate end of one strand and the 3′ hydroxyl end of the other to be on the same end of the double-stranded DNA molecule (Figure 1.3B).

The Major and Minor Grooves

Because the two strands of DNA are wrapped around each other to form a double helix, the helix has two grooves between the two strands (Figure 1.1). One of these grooves is wider than the other, so it is called the **major groove**. The other, narrower groove is called the **minor groove**. Most of the modifications to DNA that are discussed in this and later chapters occur in the major groove of the helix.

The Mechanism of DNA Replication

The molecular details of DNA replication are probably similar in all organisms on Earth. The basic process of replication involves **polymerizing**, or linking, the nucleotides of DNA into long chains, or strands, using the sequence on the other strand as a guide. Because the nucleotides must be made before they can be put together into DNA, the nucleotides are an essential **precursor** of DNA synthesis.

Deoxyribonucleotide Precursor Synthesis

The precursors of DNA synthesis are the four deoxyribonucleoside triphosphates, dATP, dGTP, dCTP, and dTTP. The triphosphates are synthesized from the corresponding ribose nucleoside diphosphates by the pathway shown in Figure 1.5. In the first step, the enzyme **ribonucleotide reductase** reduces (i.e., removes an oxygen from) the ribose sugar to produce the deoxyribose sugar by changing the hydroxyl group at the 2′ position (the second carbon) of the sugar to a hydrogen. Then, an enzyme known as a **kinase** adds a phosphate to the deoxynucleoside diphosphate to make the deoxynucleoside triphosphate precursor.

The deoxynucleoside triphosphate dTTP is synthesized by a somewhat different pathway from the other three. The first step is the same. Ribonucleotide reductase synthesizes the nucleotide dUDP (deoxyuridine diphosphate) from the ribose UDP. However, from then on, the pathway differs. A phosphate is added to make dUTP, and the dUTP is converted to dUMP by a

Figure 1.5 The pathways for synthesis of deoxynucleotides from ribonucleotides. Some of the enzymes referred to in the text are identified. THF, tetrahydrofolate; DHF, dihydrofolate. doi:10.1128/9781555817169.ch1.f1.5

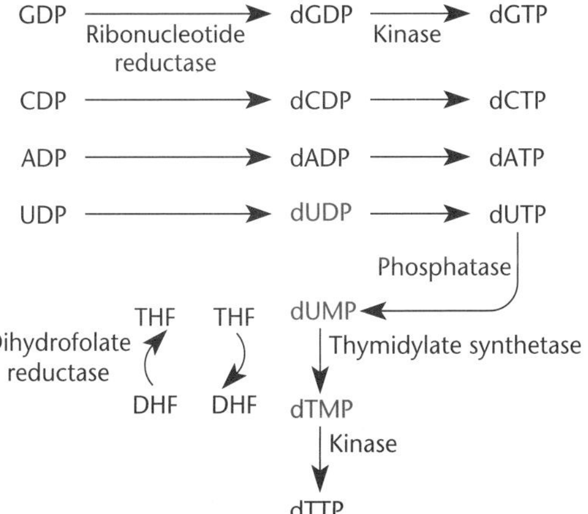

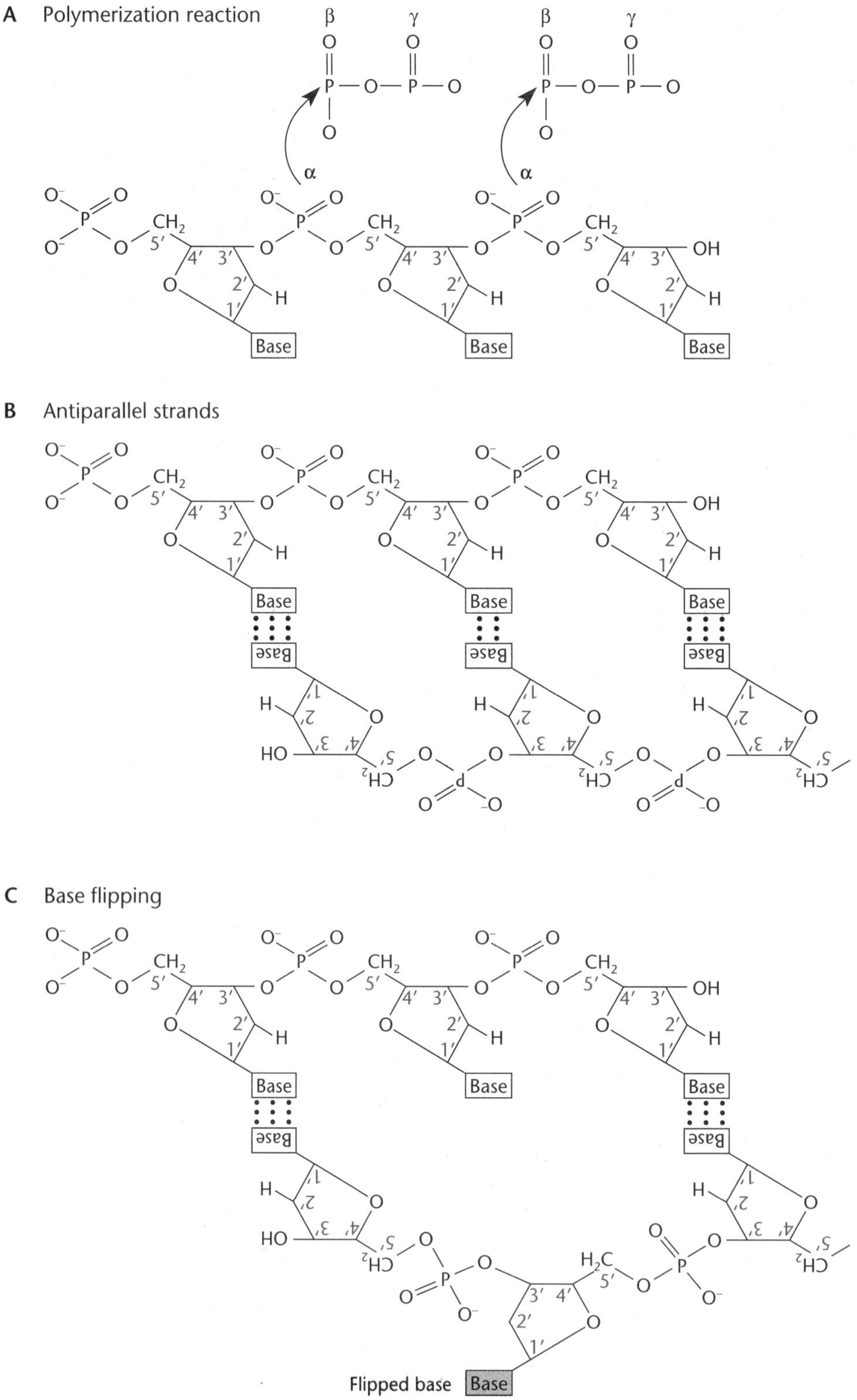

A Polymerization reaction

B Antiparallel strands

C Base flipping

Flipped base Base

Figure 1.6 Features of DNA. **(A)** Polymerization of the deoxynucleotides during DNA synthesis. The β and γ phosphates of each deoxynucleoside triphosphate are cleaved off to provide energy for the polymerization reaction. **(B)** The strands of DNA are antiparallel. **(C)** A single base can be flipped out from the double helix, which could be important in recombination and repair. doi:10.1128/9781555817169.ch1.f1.6

phosphatase that removes two of the phosphates. This molecule is then converted to dTMP by the enzyme thymidylate synthetase, using tetrahydrofolate to donate a methyl group. Kinases then add two phosphates to the dTMP to make the precursor dTTP.

Replication of the Bacterial Chromosome

Once the precursors of DNA replication are synthesized, they must be polymerized into long double-stranded DNA molecules. A very large complex of many enzymes assembles on the DNA and moves along the DNA, separating the strands and making a complementary copy of each of the strands. Thus, two strands of DNA enter the complex and four strands emerge on the other side, forming a branched structure. Each of the emerging branches contains one old "conserved" strand and one new strand, hence the name **semiconservative replication**. This branched structure where replication is occurring is called the **replication fork**. In this section, we discuss what is happening in the replication fork, including the mechanisms used to overcome obstacles that block the progression of the replication fork on the chromosome.

DNA POLYMERASES

The properties of the DNA polymerases, the enzymes that actually join the deoxynucleotides together to make the long chains, are the best guides to an understanding of the replication of DNA. These enzymes make DNA by linking one deoxynucleotide to another to generate a long chain of DNA. This process is called DNA **polymerization**, hence the name DNA polymerases.

Figure 1.6 shows the basic process of DNA polymerization by DNA polymerase. The DNA polymerase attaches the first phosphate (called α) of one deoxynucleoside triphosphate to the 3′ carbon of the sugar of the next deoxynucleoside triphosphate, in the process releasing the last two phosphates (called the β and γ phosphates) of the first deoxynucleoside triphosphate to produce energy for the reaction. Then the α phosphate of another deoxynucleoside triphosphate is attached to the 3′ carbon of this deoxynucleotide, and the process continues until a long chain is synthesized.

DNA polymerases also need a **template strand** to direct the synthesis of the new strand (Figure 1.7). As mentioned in "Base Pairing" above, complementary base pairing dictates that wherever there is a T in the template strand, an A is inserted in the strand being synthesized, and so forth according to the base-pairing rules. The DNA polymerase can move only in the 3′-to-5′ direction on the template strand, linking deoxynucleotides in the new strand in the 5′-to-3′ direction. When replication is completed, the product is a new double-stranded DNA with antiparallel strands, one of which is the old

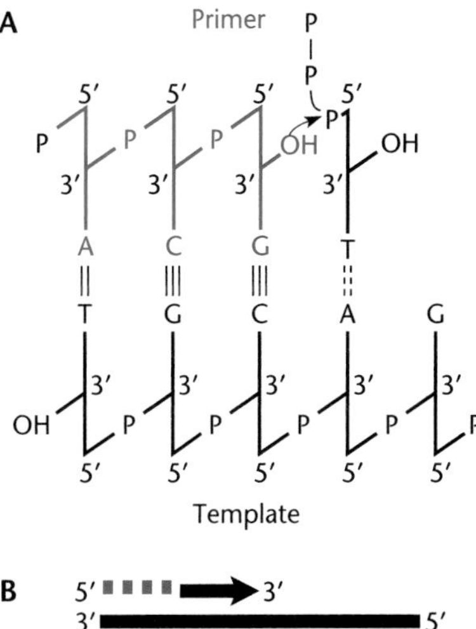

Figure 1.7 Functions of the primer and template in DNA replication. **(A)** The DNA polymerase adds deoxynucleotides to the 3′ end of the primer by using the template strand to direct the selection of each base. **(B)** Simple illustration of 5′-to-3′ DNA synthesis. The dotted line represents the primer.
doi:10.1128/9781555817169.ch1.f1.7

template strand and one of which is the newly synthesized strand.

There are two DNA polymerases that participate in normal DNA replication in *Escherichia coli*; they are called DNA polymerase III and DNA polymerase I (Table 1.1). DNA polymerase III is a large protein complex in which the enzyme that polymerizes nucleotides works with numerous accessory proteins. In *E. coli*, DNA polymerase III is responsible for the bulk of DNA replication on both DNA strands. As discussed below, DNA polymerase I has a number of features that are important because replication is continually reinitiated on one of the DNA strands. It also plays a role in DNA repair, as discussed in chapter 11. Table 1.1 lists many of the DNA replication proteins, the genes encoding them, and their functions.

PRIMASES

One type of enzyme, called a **primase**, is required during DNA replication because DNA polymerases cannot start the synthesis of a new strand of DNA; they can only attach deoxynucleotides to a preexisting 3′ OH group. The 3′ OH group to which DNA polymerase adds a deoxynucleotide is called the **primer** (Figure 1.7). The requirement for a primer for DNA polymerase creates an apparent dilemma in DNA replication. When a new strand of DNA is synthesized, there is no DNA upstream (i.e., on the 5′ side) to act as a primer. Primase makes small stretches of RNA that are complementary to the template strand, which are in turn used to initiate or prime polymerization of a new strand of DNA

Table 1.1 Proteins involved in *E. coli* DNA replication

Protein	Gene	Function
DnaA	*dnaA*	Initiator protein; primosome (priming complex) formation
DnaB	*dnaB*	DNA helicase
DnaC	*dnaC*	Delivers DnaB to replication complex
SSB	*ssb*	Binding to single-stranded DNA
Primase	*dnaG*	RNA primer synthesis
DNA ligase	*lig*	Sealing DNA nicks
DNA gyrase		Supercoiling
α	*gyrA*	Nick closing
β	*gyrB*	ATPase
DNA Pol I	*polA*	Primer removal; gap filling
DNA Pol III (holoenzyme)		
α	*dnaE*	Polymerization
ε	*dnaQ*	3′-to-5′ editing
θ	*holE*	Present in core
β	*dnaN*	Sliding clamp
τ[a]	*dnaX*	Organizes complex; joins leading and lagging DNA Pol III
γ[b]	*dnaX*	Binds clamp loaders and SSB protein
δ	*holA*	Clamp loading
δ'	*holB*	Clamp loading
χ	*holC*	Binds SSB
φ	*holD*	Holds χ to the clamp loader

[a]Full-length product of the *dnaX* gene.
[b]Shorter product of the *dnaX* gene produced by translational frameshifting (see Box 2.3).

(Figure 1.8). In some special situations, an RNA primer for DNA replication can be made by the RNA polymerase used for information processing (i.e., the RNA polymerase that makes all the other RNAs, including mRNA, tRNA, and rRNA). Unlike DNA polymerase, primase and RNA polymerases do not require a primer to initiate the synthesis of new strands. During DNA replication, special enzymes recognize and remove the RNA primer (see below).

NUCLEASES

Enzymes that degrade DNA strands by breaking the phosphodiester bonds are just as important in replication as the enzymes that polymerize DNA by forming phosphodiester bonds between the nucleotides. These bond-breaking enzymes, called **nucleases**, can be grouped into two major categories. One type can initiate breaks in the middle of a DNA strand and so are called **endonucleases**, from a Greek word meaning "within," and the other type can remove nucleotides only from the ends of DNA strands and so are called exonucleases, from a Greek word meaning "outside." **Exonucleases** can in turn be divided into two groups. Some exonucleases can degrade only from the 3′ end of a DNA strand, degrading DNA in the 3′-to-5′ direction. These are called **3′ exonucleases**; one example of their activity is their role in the editing function associated with DNA polymerases I and III, which is discussed below. Other exonucleases, called 5′ exonucleases, degrade DNA strands only from the 5′ end, an example being the 5′ exonuclease activity of DNA polymerase I, which removes RNA primers during replication (Figure 1.9).

DNA LIGASES

DNA **ligases** are enzymes that form phosphodiester bonds between the ends of separate presynthesized chains of DNA. This important function cannot be performed by any of the known DNA polymerases. During replication, ligase joins the 5′ phosphate at the end of one DNA chain to the 3′ hydroxyl at the end of another chain to make a longer, continuous chain (Figure 1.8).

ACCESSORY PROTEINS

Replication of large DNAs requires many functions that reside in separate proteins other than the subunit used for polymerizing the chain of nucleotides. They include the coordination of multiple DNA polymerases and tethering these components to the template DNA

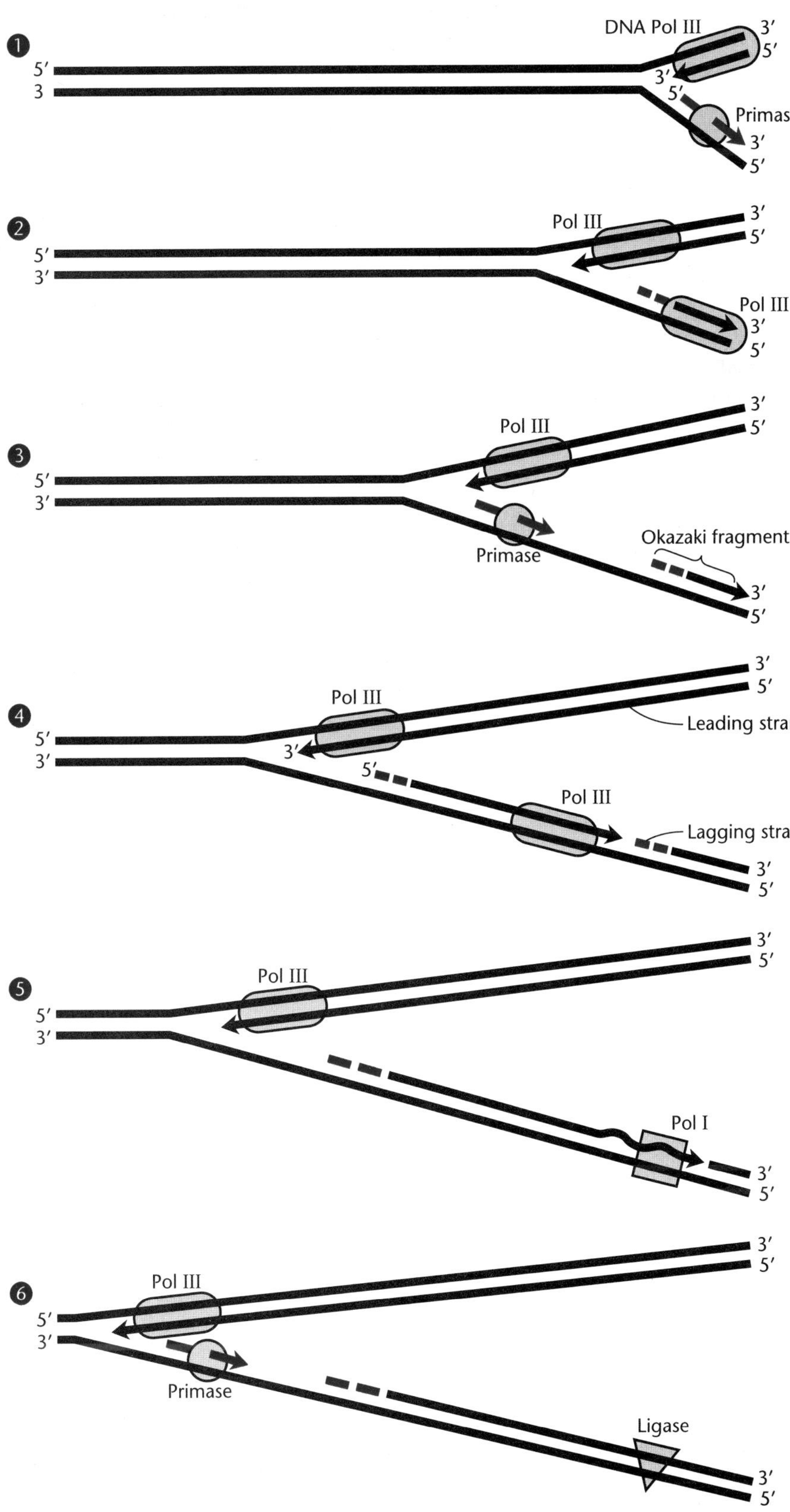

Figure 1.8 Discontinuous synthesis of one of the two strands of DNA during chromosome replication. **(1)** DNA polymerase (Pol) III replicates one strand, and the primase synthesizes RNA on the other strand in the opposite direction. **(2)** Pol III extends the RNA primer to synthesize an Okazaki fragment. **(3)** The primase synthesizes another RNA primer. **(4)** Pol III extends this primer until it reaches the previous primer. **(5)** Pol I removes the first RNA primer and replaces it with DNA. **(6)** DNA ligase seals the nick to make a continuous DNA strand, and the process continues. The strand that is synthesized continuously is the leading strand; the strand that is synthesized discontinuously is the lagging strand.
doi:10.1128/9781555817169.ch1.f1.8

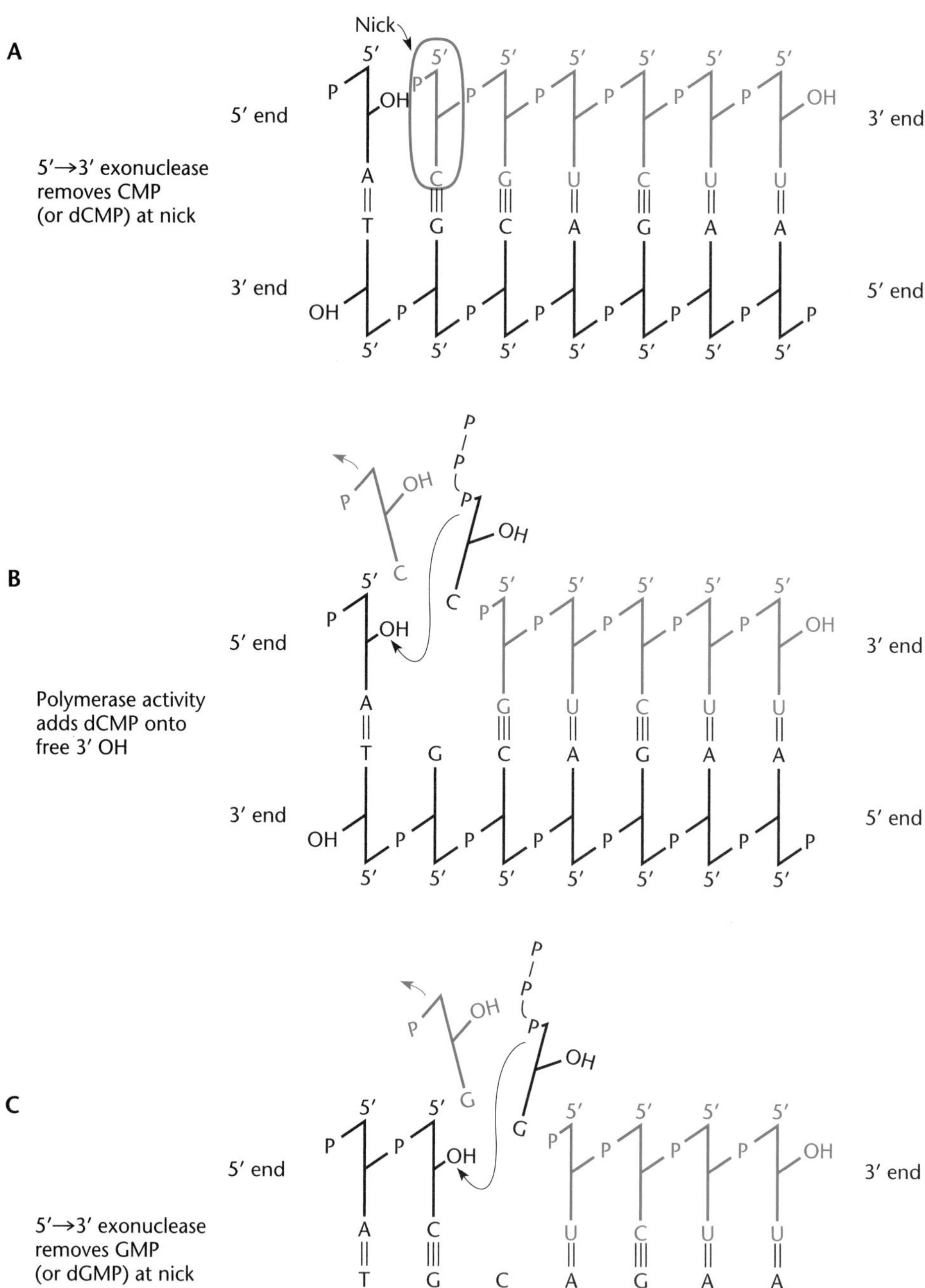

Figure 1.9 DNA polymerase I can remove the nucleotides of an RNA primer by using its "nick translation" activity. **(A)** A break, or nick, in the DNA strand occurs after the DNA polymerase III holoenzyme has incorporated the last deoxynucleotide before it encounters a previously synthesized RNA primer. **(B)** In the example, the 5′-to-3′ exonuclease activity of DNA polymerase I removes the CMP at the nick, and its DNA polymerase activity incorporates a dCMP onto the free 3′ hydroxyl. **(C)** This process continues, moving the nick in the 5′-to-3′ direction. doi:10.1128/9781555817169.ch1.f1.9

strands as a moving production platform. These processes are encoded in the accessory proteins involved in DNA replication. DNA polymerase III is the major DNA replication protein in *E. coli*, and it functions with different **DNA polymerase accessory proteins** that travel along the template strand with the molecule of DNA polymerase III responsible for actually polymerizing the new complementary DNA strands. The term **DNA polymerase III holoenzyme** can be used to describe the entire complex of proteins. The various subunits and subassemblies of the DNA polymerase III holoenzyme were originally derived from fractionation procedures and were designated by Greek letters (Table 1.1).

One of the DNA polymerase accessory proteins forms a ring around the template DNA strand and is responsible for keeping DNA polymerase from falling off. Because this ring slides freely over double-stranded DNA and will not easily come off the DNA, it is also referred to as a sliding clamp, or β clamp. The β clamp provides the foundation of the mobile platform for DNA replication, allowing it to continue for long distances without being released. In bacteria, the β clamp is a product of the *dnaN* gene, where two head-to-toe molecules form the ring around the DNA. While it was first isolated as part of the DNA polymerase III holoenzyme, the β clamp protein is important for multiple DNA transactions. A special subcomplex within DNA polymerase III is the **clamp loader**, which is responsible for loading β clamp proteins onto DNA. The clamp-loading complex is also responsible for coordinating across the DNA replication fork; the clamp loader binds the DNA polymerases on both DNA template strands and the enzyme responsible for separating the DNA strands (see below). The clamp loader is a complicated structure that consists of two τ and one γ proteins and one each of δ and δ′, which form a five-sided structure, and two additional proteins, χ and ψ. Another form of the polymerase may exist that has three τ instead of two τ and one γ proteins (see below). *E. coli* contains more δ subunits of the clamp loader than the other clamp loader constituents. While the clamp loader complex can also remove β clamps, the δ subunit is the only member of the complex that can remove β clamps by itself, without help from the others.

Replication of Double-Stranded DNA

Additional complications of DNA replication come from the fact that the DNA is double stranded and the strands are antiparallel. The replication of all bacterial chromosomes begins at one point, with the replication enzymes moving in opposite directions from this point along the chromosome. In this process, both strands of DNA are replicated at the same time with a coordinated set of proteins. Replicating the antiparallel stands is further complicated by the above-mentioned fact that DNA polymerases can replicate only in the 5′-to-3′ direction. Therefore, one DNA strand is replicated in the same direction that the replication fork is moving, and in theory, replication of this strand can continue without the need for reinitiating in a process called **leading-strand** DNA synthesis. However, replication of the other DNA strand occurs in the opposite direction from the progression of the replication machinery. Replication of this strand must continually be reinitiated in a process known as **lagging-strand** DNA synthesis. Replication of double-stranded DNA requires coordination between multiple holoenzyme subunits and DNA polymerases, as well as a host of other replication proteins.

SEPARATING THE TWO TEMPLATE DNA STRANDS

The two DNA strands must be separated to serve as templates for DNA replication, a task that DNA polymerase cannot perform on its own. This is because the bases of the DNA are inside the double helix, where they are not available to pair with the incoming deoxynucleotides to direct which nucleotide will be inserted at each step. Proteins called **DNA helicases** separate the strands of DNA (see Singleton et al., Suggested Reading). Many of these proteins form a ring around one strand of DNA and propel the strand through the ring, acting as a mechanical wedge to strip the stands apart as it moves. It takes a lot of energy to separate the strands of DNA, and helicases cleave a lot of ATP for energy, forming ADP in the process. There are about 20 different helicases in *E. coli*, and each helicase works in only one direction, either the 3′-to-5′ or the 5′-to-3′ direction. The DnaB helicase that normally separates the strands of DNA ahead of the replication fork is a large doughnut-shaped protein composed of six polypeptide products of the *dnaB* gene. It propels one strand, the template for lagging-strand DNA replication, through the center of the complex in the 5′-to-3′ direction, opening strands of DNA ahead of the replication fork (Figure 1.10). The DnaB ring cannot load onto single-stranded DNA on its own to start a DNA replication fork; it requires the loading protein DnaC. Other helicases are discussed in later chapters in connection with recombination and repair.

Once the strands of DNA have been separated, they also must be prevented from coming back together (or from annealing to themselves if they happen to be complementary over short regions). Separation of the strands is maintained by proteins called single-strand-binding (SSB) proteins or, less frequently, **helix-destabilizing proteins**. They are proteins that bind preferentially to single-stranded DNA and prevent double-stranded helical DNA from reforming prematurely.

FUNDAMENTAL DIFFERENCES IN REPLICATION OF THE TWO STRANDS

As discussed above, the antiparallel configuration of DNA requires that the two DNA polymerases travel in two different directions while still allowing the larger replication machine to travel in one direction down the chromosome (Figure 1.8). This leads to fundamental differences in the natures of leading- and lagging-strand DNA replication. While replication of the leading-strand template can occur as soon as the strands are separated by the DnaB helicase, replication of the lagging-strand

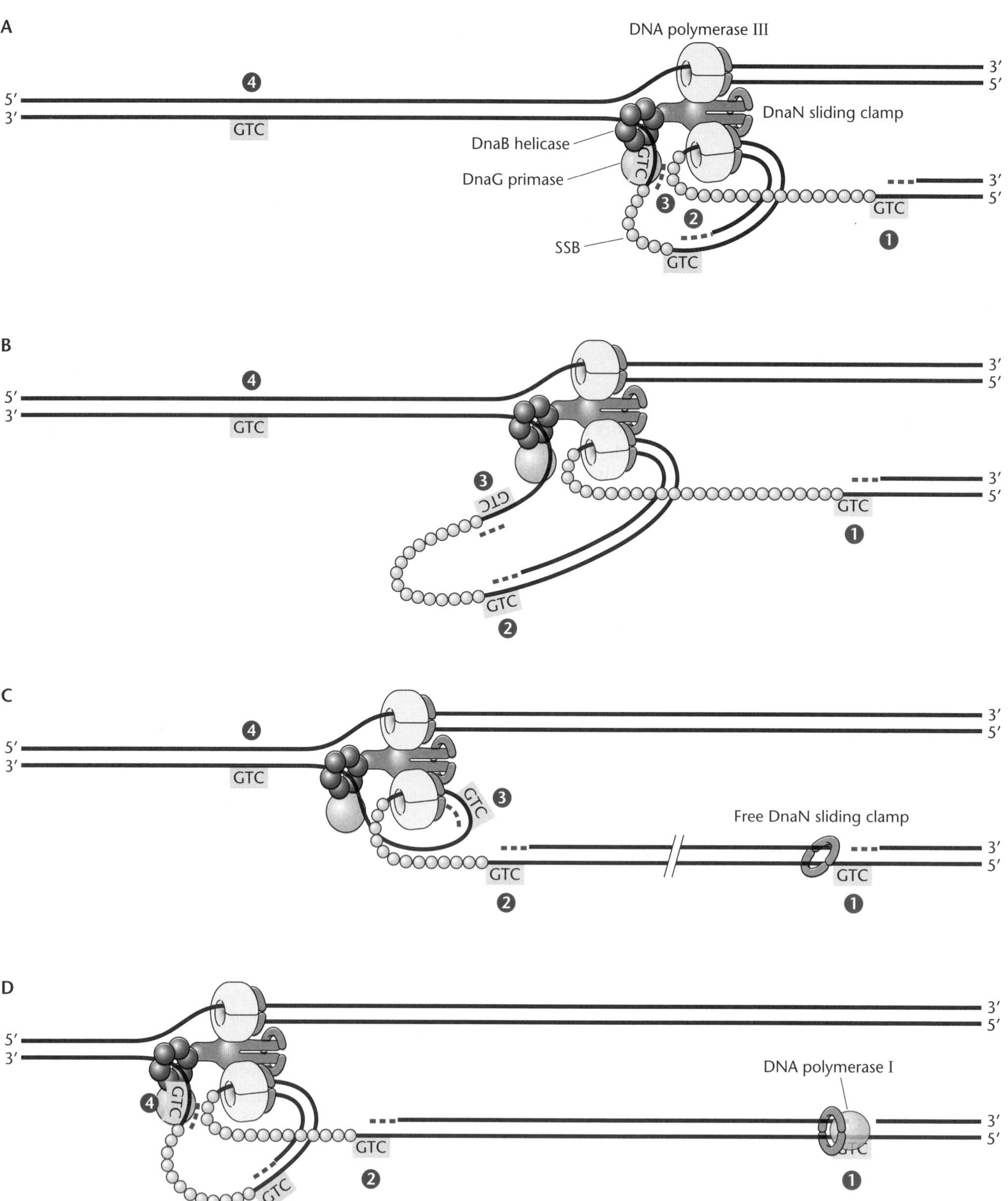

Figure 1.10 "Trombone" model for how both the leading strand and lagging strand might be simultaneously replicated at the replication fork. RNA primers are shown in blue, and their initiation sites are shown as the sequence 3′-GTC-5′ boxed in blue. **(A)** Pol III holoenzyme synthesizes lagging-strand DNA initiated from priming site 2 and runs into the primer at site 1. **(B)** The DNA strand undergoing lagging-strand replication loops out of the replication complex as the leading-strand polymerase progresses and the lagging-strand polymerase replicates toward the last Okazaki fragment. **(C)** Pol III has been released from the lagging-strand template at priming site 1 and has hopped ahead, leaving the old β clamp behind, and has reassembled with a new β clamp on the DNA at primer site 3 to synthesize an Okazaki fragment. Both the leading-strand and lagging-strand Pol III enzymes remain bound

template is not initiated until approximately 2 kilobases (kb) of single-stranded template becomes available, hence the name lagging-strand synthesis. The short pieces of DNA produced from the lagging-stand template are called **Okazaki fragments.** Synthesis of each Okazaki fragment requires a new RNA primer about 10 to 12 nucleotides in length. In *E. coli*, these primers are synthesized by DnaG primase at the template sequence 3′-GTC-5′, beginning synthesis opposite the T. These RNA primers are then used to prime DNA synthesis by DNA polymerase III, which continues until it encounters another piece that was previously synthesized further along the DNA (Figure 1.8). Before these short pieces of DNA that are annealed to the template can be joined to make a long, continuous strand of DNA, the short RNA primers must be removed. This process is carried out by DNA polymerase I using its concerted 5′ exonuclease and DNA polymerase activities (Figure 1.8). DNA polymerase I removes the RNA primer and replaces it with DNA using the upstream (i.e., 5′) Okazaki fragment as a primer. Ribonuclease (RNase) H can also contribute to this process with its ability to degrade the RNA strand of a DNA-RNA double helix. The Okazaki fragments are then joined together by DNA ligase as the replication fork moves on, as shown in Figure 1.8. By using RNA rather than DNA to prime the synthesis of new Okazaki fragments, the cell likely lowers the mistake rate of DNA replication (see below).

What actually happens at the replication fork is more complicated than is suggested by the simple picture given so far. For one thing, this picture ignores the overall topological restraints on the DNA that is replicating. The **topology** of a molecule refers to its position in space. Because the circular DNA is very long and its strands are wrapped around each other, pulling the two strands apart introduces stress into other regions of the DNA in the form of **supercoiling.** If no mechanism existed to allow the two strands of DNA to rotate around each other, supercoiling would cause the chromosome to look like a telephone cord wound up on itself, an event that has been experimentally shown to eventually halt progression of the DNA replication fork. To relieve this stress, enzymes called **topoisomerases** undo the supercoiling ahead of the replication fork. DNA supercoiling and topoisomerases are discussed below. The fork itself can also twist when the supercoiling that builds up ahead of the replication fork diffuses behind the replication fork, a process that twists the two new strands around one another and that is also sorted out by topoisomerases (see below).

COORDINATING REPLICATION OF THE TWO TEMPLATE STRANDS

The picture of the two strands of DNA replicating independently, as shown in Figure 1.8, does not take into consideration all of the coordination that must occur during DNA replication. The anatomy of the larger complex of replication factors remains unresolved; however, interactions among many of these components provide a hint as to how the larger complex functions (Figure 1.10). Rather than replicating independently, the DNA polymerases that produce the leading-strand and lagging-strand DNAs are joined to each other through the τ subunits of the holoenzyme (Table 1.1). To accommodate the fact that the two DNA polymerases must move in opposite directions and still remain tethered, the lagging-strand template loops out to pick up the slack as an Okazaki fragment is synthesized. The loop is then relaxed as the polymerase on the lagging-strand template is released from the β clamp, allowing the DNA polymerase to rapidly and efficiently "hop" ahead to the next RNA primer to begin synthesizing the next Okazaki fragment (Figure 1.10). The polymerase associates with a new β clamp assembled by the clamp loader at the site of the new RNA primer, while the old β clamp is left behind. The β clamp left on the last Okazaki fragment plays important roles in finishing synthesis and joining the fragments of lagging-strand DNA via interactions with DNA polymerase I, ligase, and repair proteins. β clamps are eventually recycled, possibly through the removal function of the δ subunit of the clamp loader. This model has been referred to as the "trombone" model of replication because the loops forming and contracting at the replication fork resemble the extension and return of the slide of the musical instrument. The situation is probably similar in all bacteria and even the other domains of life, although in some other bacteria, including *Bacillus subtilis*, and in eukaryotes, different types of DNA polymerases are used to polymerize the leading and lagging strands (see Dervyn et al., Suggested Reading).

In addition to its role in loading β clamps onto template DNA, the clamp loader also plays an important role in coordinating the various replication components. Not only does the τ subunit of the clamp loader interact with DNA polymerase on the leading-strand

to each other and the helicase through interactions with τ during the release and reassembly process. **(D)** Pol III continues synthesis of the lagging strand from priming site 3 while Pol I is removing the primer at site 1 and replacing it with DNA. Pol III holoenzyme hops to the primer at site 4 after reaching the primer at site 2. The primers and Okazaki fragments are not drawn to scale. doi:10.1128/9781555817169.ch1.f1.10

and lagging-strand templates, it also interacts with the DnaB helicase (Figure 1.10). Further coordination on the lagging-strand template is facilitated by interactions between the DnaG primase and DnaB helicase (Figure 1.10). Coordination through the clamp loader helps focus the energy from DNA polymerization and the energy that powers the helicase to allow incredibly high rates of DNA replication. The interaction between DNA polymerase III and DnaB governs the speed of unwinding so that it matches the rate of DNA polymerization to prevent undue exposure of single-stranded DNA.

THE GENES FOR REPLICATION PROTEINS

Most of the genes for replication proteins have been found by isolating mutants defective in DNA replication, but not RNA or protein synthesis. Since a mutant cell that cannot replicate its DNA will die, any mutation (for definitions of mutants and mutations, see "Replication Errors" below and chapter 3) that inactivates a gene whose product is required for DNA replication will kill the cell. Therefore, for experimental purposes, only a type of mutant called a **temperature-sensitive mutant** can be usefully isolated with mutations in DNA replication genes. These are mutants in which the product of the gene is active at one temperature but inactive at another. The mutant cells can be propagated at the temperature at which the protein is active (the permissive temperature). However, shifting to the other (nonpermissive) temperature can test the effects of inactivating the protein. The molecular basis of temperature-sensitive mutants is discussed in more detail in chapter 3.

The immediate effect of a temperature shift on a mutant with a mutation in a DNA replication gene depends on whether the product of the gene is continuously required for replication at the replication forks or is involved only in the initiation of new rounds of replication. For example, if the mutation is in a gene for DNA polymerase III or in the gene for the DnaG primase, replication ceases immediately. However, if the temperature-sensitive mutation is in a gene whose product is required only for initiation of DNA replication, for example, the gene for DnaA or DnaC (see "Initiation of Chromosome Replication" below), the replication rate for the population will slowly decline. Unless the cells have been somehow synchronized in their cell cycles, each cell is at a different stage of replication, with some cells having just finished a round of replication and other cells having just begun a new round. Cells in which rounds of chromosome replication were under way at the time of the temperature shift will complete their replication cycle but will not start a new round. Therefore, the rate of replication decreases until the rounds of replication in all the cells are completed.

Replication Errors

To maintain the stability of a species, replication of the DNA must be almost free of error. Changes in the DNA sequence that are passed on to subsequent generations are called **mutations** (see chapter 3). Depending on where these changes occur, they can severely alter the protein products of genes or other cellular functions. To avoid such instability, the cell has mechanisms that reduce the error rate.

As DNA replicates, the wrong base is sometimes inserted into the growing DNA chain. For example, Figure 1.11 shows the incorrect incorporation of a T opposite a G. Such a base pair in which the bases are paired wrongly is called a **mismatch**. Mismatches can occur when the bases take on forms called **tautomers**, which pair differently from the normal form of the base (see chapter 3). After the first replication shown in Figure 1.11, the mispaired T is usually in its normal form and pairs correctly with an A, causing a GC-to-AT change in the sequence of one of the two progeny DNAs and thus changing the base pair at that position on all subsequent copies of the mutated DNA molecule.

Editing

One way the cell reduces mistakes during replication is through editing functions. Separate proteins sometimes perform these functions for some DNA polymerases, while sometimes they are part of the DNA polymerase itself. Editing proteins are aptly named because they go back over the newly replicated DNA looking for mistakes and then recognizing and removing incorrectly inserted bases (Figure 1.12). If the last nucleotide inserted in the growing DNA chain creates a mismatch, the editing function stops replication until the offending nucleotide is removed. DNA replication then continues, inserting the correct nucleotide. Because the DNA chain grows in the 5′-to-3′ direction, the last nucleotide added is at the 3′ end. The enzyme activity found in a DNA polymerase or one of its accessory proteins that removes this nucleotide is therefore called a 3′ exonuclease. The editing proteins probably recognize a mismatch because the mispairing (between T and G in the example in Figure 1.11) causes a minor distortion in the structure of the double-stranded helix of the DNA.

DNA polymerase I is an example of a DNA polymerase in which the 3′ exonuclease editing activity is part of the DNA polymerase itself. However, in DNA polymerase III, which replicates the bacterial chromosome, the editing function resides in an accessory protein encoded by a separate gene whose product travels along the DNA with the DNA polymerase during replication. In *E. coli*, the 3′ exonuclease editing function is encoded

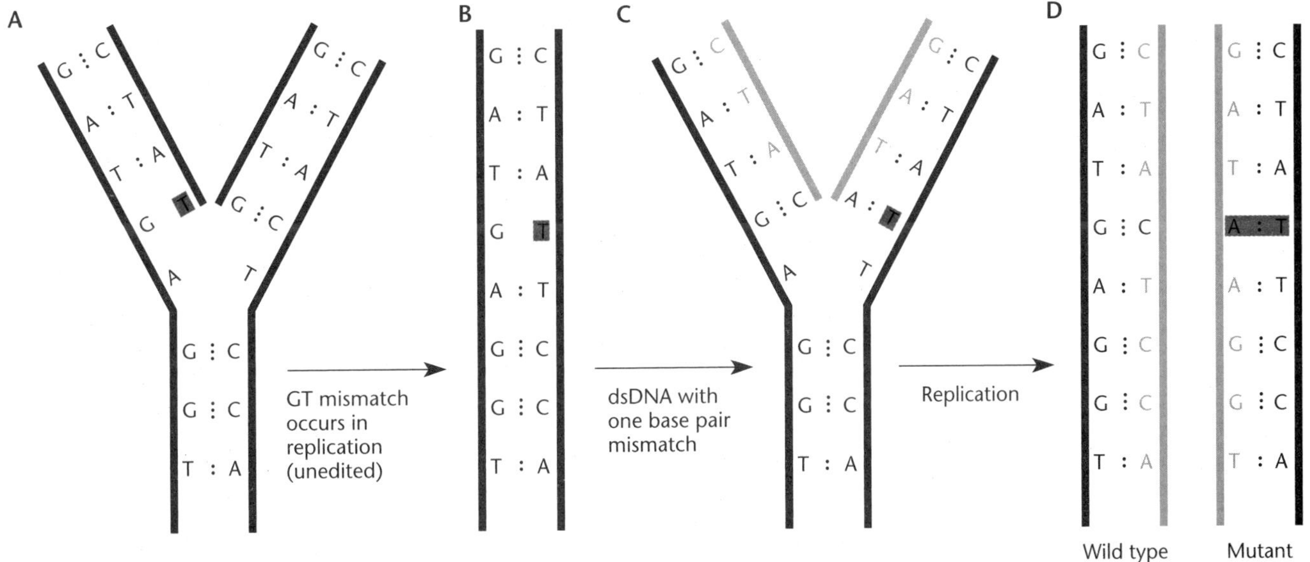

Figure 1.11 Mistakes in base pairing can lead to changes in the DNA sequence called mutations. If a T is mistakenly placed opposite a G during replication **(A)**, it can lead to an AT base pair replacing a GC base pair in the progeny DNA **(B to D)**. doi:10.1128/9781555817169.ch1.f1.11

by the *dnaQ* gene (Table 1.1), and *dnaQ* mutants, also called *mutD* mutants (i.e., cells with a mutation in this gene that inactivates the 3′ exonuclease function), show much higher rates of spontaneous mutagenesis than do cells containing the wild-type, or normally functioning, *dnaQ* gene product. Because of their high spontaneous mutation rates, *mutD* mutants of *E. coli* can be used as a tool for mutagenesis, often combined with mutations in other genes whose products normally contribute to the correction of mismatches (see chapter 11).

RNA Primers and Editing

The importance of the editing functions in lowering the number of mistakes during replication may explain why DNA replication is primed by RNA rather than by DNA. When the replication of a DNA chain has just initiated, the helix may be too short for distortions in its structure to be easily recognized by the editing proteins. The mistakes may then go uncorrected. However, if the first nucleotides inserted in a growing chain are ribonucleotides rather than deoxynucleotides, an RNA primer is synthesized rather than a DNA primer. The RNA primer can be removed and resynthesized as DNA by using preexisting upstream DNA as a primer. Under these conditions, the editing functions are active and mistakes are avoided.

Another important system that safeguards the fidelity of the replication process is responsible for fixing mismatches after the growing DNA strand leaves the polymerase. In *E. coli* and its closest relatives, this process

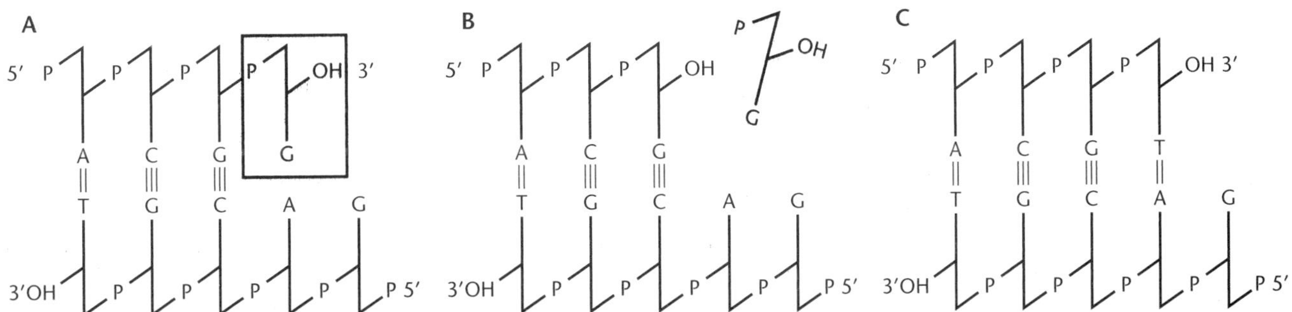

Figure 1.12 Editing function of DNA polymerase. **(A)** A G is mistakenly placed opposite an A while the DNA is replicating. **(B and C)** The DNA polymerase stops while the G is removed and replaced by a T before replication continues. doi:10.1128/9781555817169.ch1.f1.12

is guided by methylation and is termed **methyl-directed mismatch repair.** Related mismatch repair systems are used across all three domains of life, but the use of methylation signals is not widespread. The methyl-directed mismatch repair system is discussed in chapter 11.

Impediments to DNA Replication

While the picture described above and diagrammed in Figure 1.10 would suffice for pristine DNA on a template that lacked any type of physical block to the progression of the DNA replication complex, in reality, the situation in the cell is rarely this tidy. DNA polymerases frequently encounter a number of different problems. Challenges to DNA polymerases include interruptions in the DNA template, bulky adducts that cannot be replicated by DNA polymerase III, and physical blocks mediated by supercoiling and proteins bound to, or acting on, the chromosome. One extreme form of impediment to a replication fork comes from **nicks** in either the leading-strand or lagging-strand template DNAs in which the phosphate-deoxynucleotide chain is broken in one strand of the DNA. These nicks cause the destruction of the nick-containing arm of the replication fork, resulting in a broken chromosome and collapse of the DNA replication fork. Bacteria possess a highly efficient mechanism for priming repair of this broken end by using the broken DNA itself as a primer to reinitiate DNA replication. This process was likely the original driving force for the evolution of recombination and is described in chapter 10.

Damaged DNA and DNA Polymerase III

DNA polymerase III replicates DNA with incredibly high fidelity. Much of the fidelity of the enzyme comes from the structure of the catalytic pocket, where there is a **presynthetic** check for base pairing between the template strand and the incoming nucleotide. A side effect of this small binding pocket is the inability of the polymerase to tolerate **lesions** in which chemical changes have occurred in the base, the deoxyribose sugar, or even the phosphate on the DNA. There are many mechanisms for DNA replication to continue even when a cell is grown under conditions that result in highly damaged DNA. While early work suggested that the polymerization of the leading and lagging strand was so tightly coupled that a lesion on one strand would stop the entire DNA replication fork, more recent work indicates a considerable amount of flexibility. It is now clear that although the polymerases producing the leading strand and lagging strand are physically coupled, the two molecules can be momentarily uncoupled by leaving a single-strand DNA gap at the point of the lesion that blocked the DNA polymerase. Other processes can repair these gaps, and in extreme cases, where there is extensive damage in the chromosome, these gaps initiate a DNA damage response called the SOS response (see chapter 11).

Mechanisms To Deal with Impediments on Template DNA Strands

The mechanisms used for momentarily functionally uncoupling synthesis of the two strands differ depending on whether the lesion occurs on the leading-strand or lagging-strand template. The discontinuous nature of replication on the lagging-strand template affords the opportunity to circumvent lesions that halt DNA polymerase III. Typically, DNA polymerase III is recycled onto a new DNA primer when it encounters the previous Okazaki fragment (Figure 1.10). However, a stalled lagging-strand DNA polymerase III can also be recycled by premature release when it stalls at DNA damage (Figure 1.13A). The single-strand DNA gap left behind is repaired by another mechanism.

Under the generally accepted model of the function of DnaG primase, primers are placed only on the lagging-strand template. However, more recently, it has been found that in cases where the leading-strand polymerase stalls, primase can also produce an RNA primer on the leading-strand template, allowing replication to continue but leaving a gap on this strand (Figure 1.13B) (see Heller and Marians, Suggested Reading). The process of repriming DNA replication on the leading-strand template could be facilitated by a clamp loader assembly that contains three τ molecules, the holoenzyme subunit that interfaces with the polymerases. Two polymerase molecules could be engaged in DNA replication, as previously described, while the third DNA polymerase molecule would be available in the case of a stall on the leading-strand template. The γ and τ subunits of the clamp loader are encoded by the same gene, *dnaX*. Expression of the full gene results in production of the longer τ subunit, whereas a stutter in how the protein is produced from this gene, called a "frameshift," produces the shorter γ product. Interestingly, there is no requirement for γ in *E. coli*, and numerous other bacteria produce only the full-length τ subunit, which is capable of coordinating a DNA polymerase III holoenzyme containing three catalytic polymerases. The proposal that there are actually three τ subunits (and three DNA polymerases) per replication fork is supported by experiments on living cells using visually labeled proteins (see Reyes-Lamothe et. al., Suggested Reading).

While DNA polymerase III and DNA polymerase I are important for high-fidelity DNA replication, other DNA polymerases are found in *E. coli* that allow replication through damaged DNA in a process known as **translesion synthesis.** Most translesion polymerases appear to

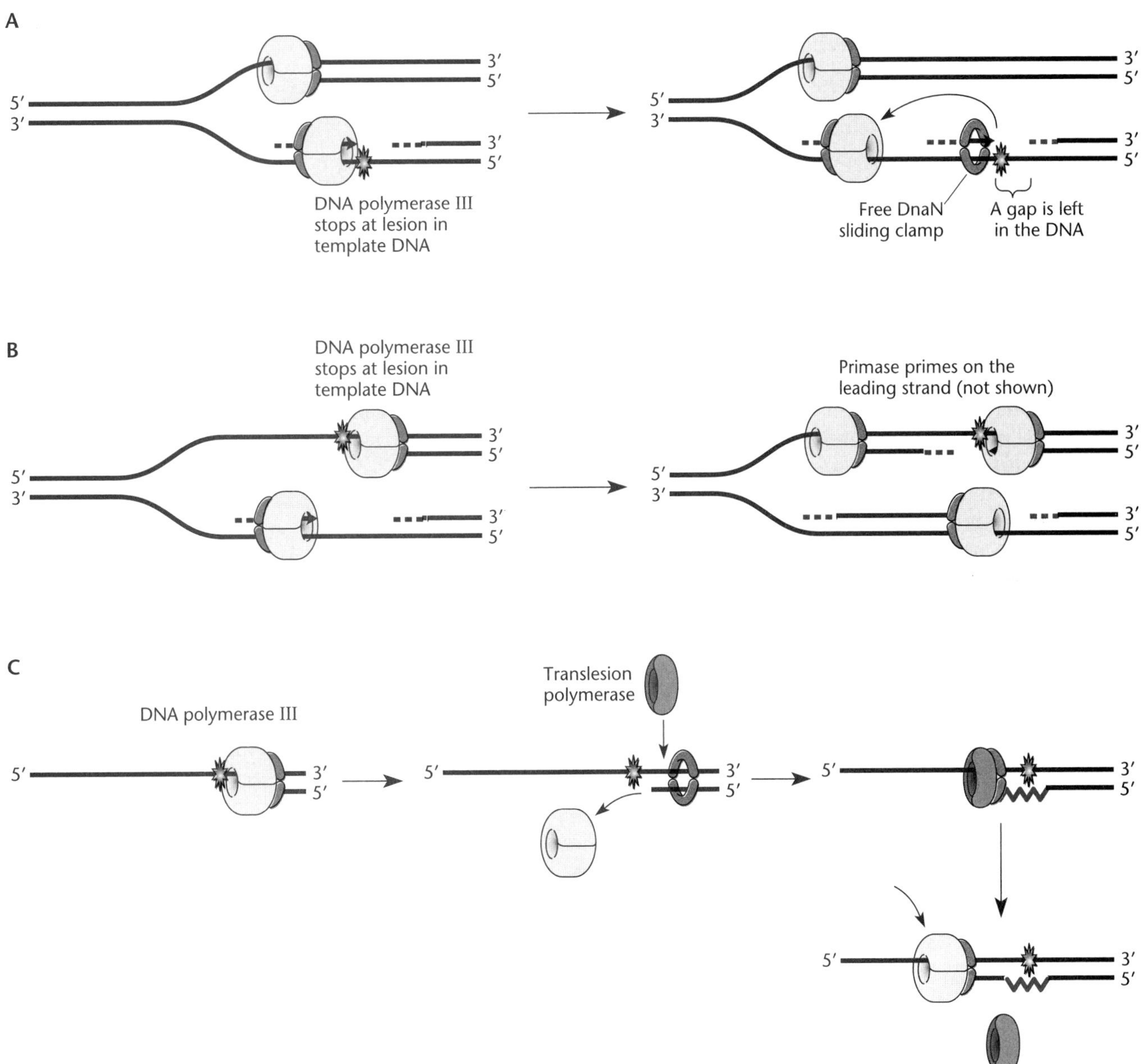

Figure 1.13 Impediments on template strands. For simplicity, only the sliding clamp and DNA polymerases are shown and not the connections between the polymerases and helicase through τ, etc. **(A)** A lesion (shown as a blue star) on the lagging-strand template that cannot be replicated by polymerase III is dealt with by restarting a new Okazaki fragment, leaving a gap that is repaired by another mechanism probably involving RecFOR (see chapter 10). **(B)** The ability of DnaG to start a new primer on the leading-strand template allows replication to continue if a lesion on the leading-strand template prevents it from being replicated by DNA polymerase III. The gap that remains can presumably be repaired by a pathway involving RecFOR (see chapter 10). **(C)** Polymerase switching allows DNA polymerase III to temporarily hand over the 3′ end of the nascent DNA strand and the β clamp to a different DNA polymerase that is able to replicate past the damage in DNA. Polymerase III reclaims its replication role within a short distance after the lesion. doi:10.1128/9781555817169.ch1.f1.13

come with a trade-off where the ability to copy damaged DNA results in very low-fidelity DNA replication. As expected, the expression of these polymerases is induced as a response to DNA damage in the cell. In addition to controlling the amount of translesion polymerase present in the cell, access to the DNA replication fork by polymerases other than DNA polymerase III is regulated by a process called **polymerase switching**, a process by which one

DNA polymerase replaces a polymerase already found at the 3′ OH end of a primed DNA template (Figure 1.13C). In *E. coli*, there are three other DNA polymerases that can be recruited to replace DNA polymerases I and III, i.e., DNA polymerases II, IV, and V (more details of this system are described in chapter 11). Each of these polymerases has different attributes, ranging from fairly accurate and highly processive (DNA polymerase II) to very inaccurate and not very processive (DNA polymerases IV and V). Processivity refers to how far a DNA polymerase moves on the template before falling off.

Having multiple DNA polymerases with different properties appears to be common in all living organisms. How accurate or processive a given DNA polymerase is may also depend on the nature of the damage found in the template DNA and/or the availability of various accessory proteins. The regulation of the use of these DNA polymerases is still incompletely understood, but there appear to be "highly evolved processes" where the system has been fine-tuned by the process of natural selection over a long time so that the most appropriate DNA-copying mechanism is used for each environmental challenge.

Physical Blocks to Replication Forks

Proteins bound to, or otherwise acting on, the chromosome can also stop the progression of DNA polymerase III. A programmed block to DNA replication occurs in some bacteria to terminate DNA replication within one region of the chromosome (see below). However, unintended blocks can occur in other situations, such as when DNA polymerase encounters RNA polymerase carrying out transcription or when RNA polymerase is stalled at sites of damage in the chromosome. Transcription appears to be the most significant impediment for DNA polymerases. In another example of the dynamic nature of the DNA polymerase III holoenzyme, when DNA polymerase III overtakes and displaces RNA polymerase on the leading-strand template, it actually uses the nascent mRNA product bound to the template strand as a new primer for DNA synthesis (Figure 1.14) (see Pomerantz and O'Donnell, Suggested Reading). Like the action of primase on the leading strand, this introduces discontinuity in the leading strand, but the process allows the replication machinery to remain intact and functional when RNA polymerase is overtaken on the leading-strand template (likely a common problem in cells). Head-on collisions between DNA polymerase and RNA polymerase are more severe and lead to replication arrest, but RNA polymerase can be removed through the action of another protein called Mfd to allow progression of DNA replication (see chapter 11).

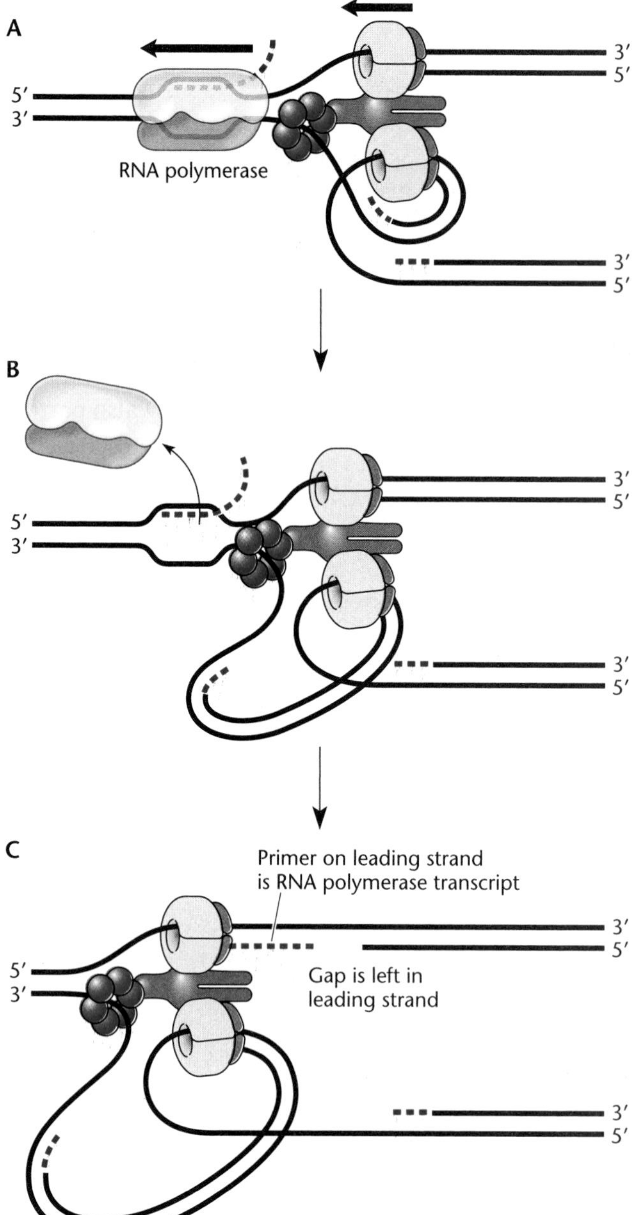

Figure 1.14 Physical blocks on template DNA. **(A)** The replication apparatus moves faster than RNA polymerase and can overtake the enzyme when replicating the leading-strand template. **(B)** DNA polymerase III can displace RNA polymerase in this situation and can reinitiate DNA polymerization using the mRNA transcript made by RNA polymerase. **(C)** The process leaves a gap on the leading strand that must be repaired by other means, probably involving RecFOR (see chapter 10). doi:10.1128/9781555817169.ch1.f1.14

In reality, protein complexes likely must always be displaced ahead of the DNA replication fork. Other helicases have been found to be required to allow DNA replication to proceed through obstacles on template DNAs. Experimentally, it was shown that two other *E. coli* helicases, Rep and UvrD, allow replication to proceed through a protein block on template DNA (see

Guy et al., Suggested Reading). These helicases travel in the 3′-to-5′ direction, and therefore, they are likely to travel on the leading-strand template while DnaB progresses forward on the lagging-strand template (i.e., the 5′-to-3′ direction). Rep interacts directly with the DnaB helicase, probably as a normal part of the **replisome**. The replisome is the collection of proteins that interact with one another (through DNA or other proteins) and that are involved in carrying out DNA replication. UvrD may be a helicase of general use for helping out when DNA replication is blocked by proteins or to remove recombination structures from the chromosome to allow DNA replication to proceed. UvrD also participates in DNA repair, as described in chapter 11.

Replication of the Bacterial Chromosome and Cell Division

So far, we have discussed the details of DNA replication, but we have not discussed how bacterial DNA as a whole is replicated, nor have we discussed how the replication process is coordinated with division of the bacterial cell. To simplify the discussion, we first consider only bacteria that grow as individual cells and divide by binary fission to form two cells of equal size, even though this is far from the only type of multiplication observed among bacteria.

The replication of the bacterial DNA occurs during the cell division cycle. The **cell division cycle** is the time during which a cell is born, grows larger, and divides into two progeny cells. Cell division is the process by which the larger cell splits into the two new cells. The **division time**, or **generation time**, is the time that elapses from the point when a cell is born until it divides. This time is usually approximately the same for all the individuals in the population under a given set of growth conditions. The original cell before cell division is called the **mother cell**, and the two progeny cells after division are called the **daughter cells**.

Structure of the Bacterial Chromosome

The DNA molecule of a bacterium that carries most of its normal genes is commonly referred to as its **chromosome**, by analogy to the chromosomes of higher organisms. This name distinguishes the molecule from plasmid DNA, which in some cases can be almost as large as chromosomal DNA but usually carries genes that are not always required for growth of the bacterium (see chapter 4).

Most bacteria have only one chromosome; in other words, there is only one unique DNA molecule per cell that carries most of the normal genes. There are exceptions, including *Vibrio cholerae*, the bacterium responsible for the disease cholera, although even in this case the second chromosome shows more characteristics of a plasmid than of a chromosome, particularly in how it initiates replication (see chapter 4). The fact that bacteria have one chromosome does not mean that there is necessarily only one copy of this chromosomal DNA in each bacterial cell. Bacterial cells whose chromosomes have replicated but that for some reason have not divided have more than one copy of this chromosomal DNA per cell. Also, as discussed below, when bacteria, such as *E. coli*, are reproducing very rapidly, new rounds of replication initiate before others are completed, increasing the DNA content of the cells. It is important to note, however, that these individual chromosomal DNAs are not unique and therefore do not represent new chromosomes, since they are directly derived from each other by replication.

The structure of bacterial DNA differs significantly from that of the chromosomes of higher organisms. One difference is that the DNA in the chromosomes of most bacteria is circular in the sense that the ends are joined to each other (for exceptions, see Box 4.1). In contrast, eukaryotic chromosomes are usually linear with free ends. As discussed in chapter 4, the circularity of bacterial chromosomal DNA allows it to replicate in its entirety without using telomeres, as eukaryotic chromosomes do, or terminally redundant ends, as some bacteriophages do. Even in cases where bacterial chromosomes are linear, they do not use the same mechanism, involving telomerases to replicate their ends, that is used by eukaryotic chromosomes. Another difference between the DNAs of bacteria and eukaryotes is that the DNA in eukaryotes is wrapped around proteins called histones to form nucleosomes. Bacteria have the proteins HU, HN-S, Fis, and IHF, around which DNA is often wrapped, and archaea do have rudimentary histones related to those of eukaryotes. However, in general, DNA is much less structured in bacteria than in eukaryotes.

Replication of the Bacterial Chromosome

The replication of the circular bacterial chromosome initiates at a unique site in the DNA called the origin of chromosomal replication, or ***oriC***, and proceeds in both directions around the circle. On the *E. coli* chromosome, *oriC* is located at 84.3 min. As mentioned above, the place in DNA at which replication occurs is known as the replication fork. The two replication forks proceed around the circle until they meet and terminate chromosomal replication. The DNA polymerases responsible for replicating the leading and lagging strands associate as a single holoenzyme. However, there is no association between the two DNA replication forks to help drive the separation of chromosomes, and therefore, other force-generating mechanisms must be at play. As discussed in "Termination of Chromosome Replication" below, some bacteria actively terminate replication at a unique site in the DNA; however, these systems are not widespread, and most bacteria may terminate replication using an unknown mechanism or simply terminate DNA replication where the two replication forks meet. Each time the two replication forks proceed around the circle and meet, a **round of replication** has been completed, and two new DNAs, called the **daughter DNAs**, are generated.

Initiation of Chromosome Replication

Much has been learned about the molecular events occurring during the initiation of replication. Some of this information has a bearing on how the initiation of chromosome replication is regulated and serves as a model for the interaction of proteins and DNA.

Two types of functions are involved in the initiation of chromosome replication. One consists of the sites or sequences on DNA at which proteins act to initiate replication. These are called ***cis*-acting sites**. The prefix *cis* means "on this side of," and these sites act only on the same DNA. The proteins involved in initiation of replication are examples of ***trans*-acting functions**. The prefix *trans* means "on the other side of," and these functions can act on any DNA in the same cell, not just the DNA from which they were made. The concepts of *cis*- and *trans*-acting properties are common in molecular genetics, and these references are used throughout this book.

ORIGIN OF CHROMOSOMAL REPLICATION

One *cis*-acting site involved in DNA replication is the *oriC* site, at which replication initiates. The sequence of *oriC* is well defined in *E. coli* and is broadly architecturally similar in most bacteria. Figure 1.15 shows the structure of the origin of replication of *E. coli*. Less than 260 base pairs (bp) of DNA is required for initiation at this site. Within *oriC* are a series of binding sites for various proteins; the most important of these binding proteins is the master initiator protein in bacteria, called DnaA (see below). The canonical DnaA-binding sequences are 9 bp in length, and these sites are termed **DnaA boxes**. While five DnaA boxes exist within *oriC*, three of these sites bind DnaA particularly strongly, i.e., are of particularly high affinity, and are always bound by DnaA (Figure 1.15). The ability of DnaA to bind these high-affinity sites at all times can be considered analogous to the origin recognition complex associated with eukaryotes. Additional sites called "I" and "τ" sites, which differ from DnaA boxes, exist in the origin region but are only occupied by DnaA that is bound to ATP and not ADP (see below) (Figure 1.15). Finally, within an AT-rich region of DNA that is opened for initiation, called the DNA-unwinding element, are three additional sites of DnaA binding that are bound only when DnaA is bound to ATP and not ADP. Binding sites for other DNA-binding proteins (IHF and Fis) are also found in this region.

INITIATION PROTEINS

Besides the *cis*-acting *oriC* site and DnaA, many *trans*-acting proteins are also required for the initiation of DNA replication, including the DnaB and DnaC proteins. DnaA is required only for initiation, allowing DnaC to load the DnaB helicase for establishing the DNA replication forks. Many proteins used in other cellular functions are also involved, such as the primase (DnaG), the normal RNA polymerase that makes most of the RNA in the cell, and the DNA-binding proteins IHF and Fis (Figure 1.15).

Figure 1.16 outlines how DnaA, DnaB, DnaC, and other proteins participate in the initiation of chromosome replication. As we will see throughout the remainder of this chapter, there are many points where the initiation of DNA replication is controlled. One important regulatory consideration concerns the nucleotide-binding state of DnaA. While DnaA always binds some of the DnaA boxes, the architecture of DnaA binding to form a larger complex that can open the DNA strands requires that DnaA be bound to ATP (DnaA-ATP). In biology, there are many examples where the nucleotide-bound state of a protein determines its activity. While these proteins have the capacity to hydrolyze nucleotides from the NTP to the NDP form, this energy is not directly used to actively carry out any particular task but instead allows the configuration of the proteins to change. In the case of DnaA, the ATP-bound form of

Figure 1.15 Structure of the *oriC* region of *E. coli*. Shown are the positions of multiple types of DnaA-binding sequences, five DnaA boxes (R1 to R5) and other DnaA-binding sites (I and τ), and an AT-rich region that is unwound to allow loading of the replication apparatus, the DNA-unwinding element (DUE). Also shown are binding sites for the IHF and Fis proteins and a large number of GATC sites (black dots) that are important in regulating initiation by acting as sites of Dam methylation. doi:10.1128/9781555817169.ch1.f1.15

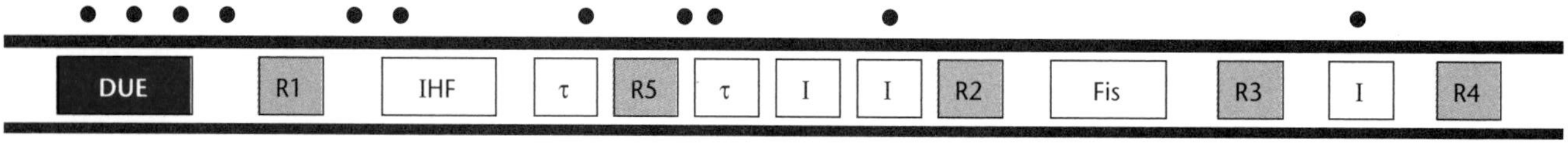

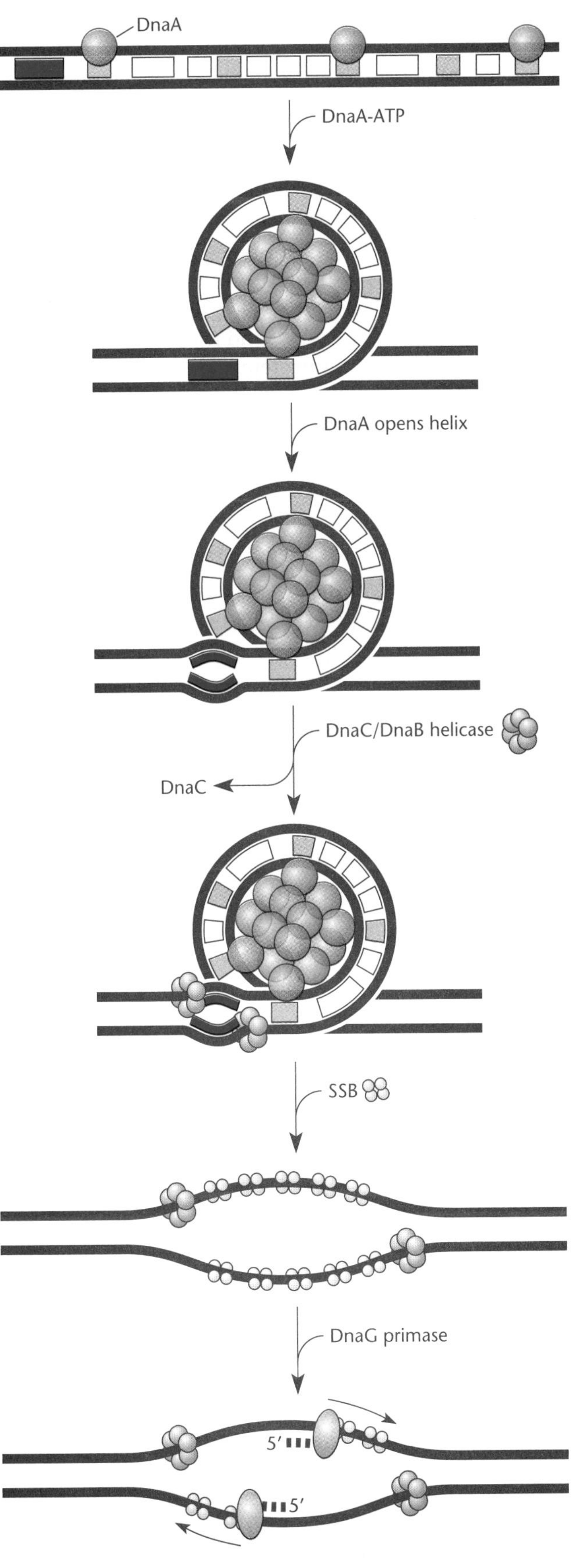

the protein allows it to form a large multimer structure composed of many molecules of DnaA protein where the DNA strands are opened through bending of the DNA by DnaA with the help of the IHF and Fis proteins. The binding and opening are also aided by supercoiling at the origin (see "Supercoiling" below) and by the SSB protein (Figure 1.16), which helps keep the helix from re-forming. DnaA binds directly to the helicase DnaB, and in a process involving DnaC, DnaB helicase is loaded into *oriC*. Action of the DnaB helicase opens the strands further for priming and replication, and DnaC leaves the complex. The complex that travels across the chromosome laying RNA primers is called the **primosome** and contains DnaB and DnaG primase. DnaG primase or another RNA polymerase may synthesize the first RNA primer to start replication.

RNA Priming of Initiation

Initiation of DNA replication requires RNA primers, but which RNA polymerase makes the primer for initiating leading-strand synthesis is not completely clear. The RNA polymerase that synthesizes most of the RNA molecules, including mRNA, in the cell (see chapter 2) is needed to initiate rounds of replication. However, its role may be to help separate the strands of DNA in the *oriC* region by transcribing through the region, because the strands may have to be separated before the DnaA protein can bind. In this case, the RNA primers themselves may be synthesized by the DnaG protein, the same RNA polymerase that regularly makes RNA primers for lagging-strand synthesis at the replication fork.

Termination of Chromosome Replication

After replication of the chromosome initiates in the *oriC* region and proceeds around the circular chromosome in both directions, the two replication forks must meet somewhere on the chromosome and the two daughter chromosomes must separate. In some bacteria, including *E. coli* and *B. subtilis*, a specific system exists to control the region where replication forks meet. What happens in other organisms is less clear. An unknown mechanism may help coordinate the termination of DNA replication; however, it remains possible that replication forks may travel at about the same rate and just meet at a point approximately equidistant from the origin in the

Figure 1.16 Initiation of replication at the *E. coli* origin (*oriC*) region. DnaA is always bound to three DnaA boxes within *oriC*, even when DnaA is in its non-ATP-bound state acting as an origin recognition complex. About a dozen DnaA-ATP proteins bind to the origin, wrapping the DNA around themselves and opening the helix. DnaC helps the DnaB helicase to bind. The DnaG primase synthesizes RNA primers, initiating replication. doi:10.1128/9781555817169.ch1.f1.16

circular chromosome. As with most cellular processes, the process of termination of chromosome replication is especially well understood in *E. coli.* In this bacterium, termination is facilitated by DNA sequences that are only 22 bp long, called *ter* sites. These sites act somewhat like the one-way gates in an automobile parking lot, allowing the replication forks to pass through in one direction but not in the other.

Figure 1.17 shows how the one-way nature of *ter* sequences can cause replication to terminate in a specific region of the chromosome. In the illustration, two *ter* sites called *terA* and *terB* bracket the termination region. Replication forks can pass the *terA* site in the clockwise direction but not in the counterclockwise direction. The opposite is true for *terB*. Thus, the clockwise-moving replication fork can pass through *terA*, but if it gets to *terB* before it meets the counterclockwise-moving fork, it stalls, because it cannot move clockwise through *terB*. Similarly, the replication fork moving in the counterclockwise direction stalls at the *terA* site and waits for the clockwise-moving replication fork. When the counterclockwise and clockwise replication forks meet, at *terA*, *terB*, or somewhere between them, the two forks terminate replication, releasing the two daughter DNAs. In *E. coli*, it is known that most DNA replication termination occurs at one *ter* site called *terC*, possibly because it is oriented to terminate replication forks traveling in the clockwise direction, which is shorter in most laboratory *E. coli* strains.

Encountering a *ter* DNA sequence, by itself, is not sufficient to stop the replication fork. A protein is also required to terminate replication at *ter* sites. The protein that works with *ter* sites, called the terminus utilization substance (Tus) in *E. coli* and the replication terminator protein (RTP) in *B. subtilis*, binds to the *ter* sites and stops the replicating helicase (DnaB in *E. coli*) that is separating the strands of DNA ahead of the replication fork. By actively terminating DNA replication forks and not letting them crash directly together, these systems may prevent overreplication through multiple mechanisms. However, in spite of these seeming advantages, most bacteria do not appear to have such systems, and deleting genes encoding the proteins responsible for terminator function has a discernible phenotype only in certain genetic backgrounds. There are probably subtle advantages to terminating chromosome replication in the *ter* region opposite the *oriC* origin of replication, but as is often the case, these advantages are not apparent in a laboratory setting.

Chromosome Segregation

While bacteria do not contain a special membrane compartment for chromosomal DNA like the nucleus of eukaryotes, even in bacteria the chromosome does not freely diffuse within the cytoplasm. In fact, as we learn more about bacterial chromosomes, we are realizing that they are maintained with an incredible amount of organization. Even with the aid of only a standard laboratory microscope and DNA stain, a mass of chromosomal DNA is easily observed in the center of the cell in a structure called the **nucleoid**, which is very compact, considering that the DNA is about a thousand times as long as the cell.

Because of its large size, the process of moving the replicating chromosomes to daughter cells, called **segregation**, is not trivial. Chromosome segregation encounters a number of obstacles. Obvious initial obstacles are viscous forces and torsional stress associated with

Figure 1.17 Termination of chromosome replication in *E. coli.* **(A)** The replication forks that start at *oriC* can traverse *terA* and *terB* in only one direction, opposite that indicated by the blue arrows. **(B)** When they meet, between or at one of the two clusters, chromosome replication terminates. f_L is the fork that initiated to the left and moved in a counterclockwise direction. f_R is the fork that initiated to the right and moved in a clockwise direction. doi:10.1128/9781555817169.ch1.f1.17

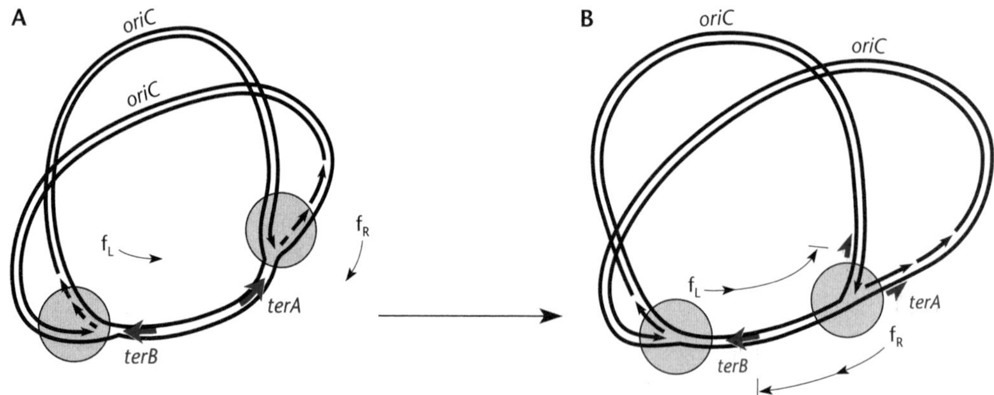

unwinding the template strands of DNA. Advances in microscopy and techniques that allow the localization of certain regions of the chromosome are revealing the choreography involved in coordinating DNA replication and chromosome segregation. Microscopy experiments using green fluorescent protein (GFP) fused to replication proteins or fused to proteins that bind to specific sites on DNA allowed localization of the origin and terminus in the cell. These experiments showed that soon after the initiation of DNA replication within the nucleoid, the origins start to move to the daughter cells. Once replication is complete, segregation would still be prevented if daughter chromosomes were joined by recombination, interlinked, or otherwise tangled during replication. Even if they were not physically joined, their separation would be very difficult if the two daughter chromosomes were randomly spread out throughout the cell. It is therefore not surprising that bacteria have a number of systems to ensure that their chromosomes segregate properly into the daughter cells during cell division. Molecular systems responsible for chromosome segregation are discussed separately below.

RESOLUTION OF DIMER CHROMOSOMES

Sometimes the two circular daughter DNAs become joined in a chromosome **dimer**, in which they are joined end to end to form a double-length circle. Dimer chromosomes result from recombination between the two replicating chromosomes and are fairly common (they are estimated to be in 15% of *E. coli* cells). Recombination involved in restarting stalled DNA replication forks from a sister chromosome probably accounts for many of these events (see chapter 10). Such dimers prevent chromosome segregation because the two daughter chromosomes are part of the same large molecule.

If dimer chromosomes can be created by recombination, they can also be resolved into the individual chromosomes by a second recombination event. The general recombination system could in theory resolve the dimers between sister chromosomes by recombination; however, the general recombination system can both create and resolve dimers depending on how many crossovers occur between the daughter DNAs. An odd number of crossovers occurring between any two sequences on the two daughter DNAs in the dimer will resolve the dimer, but an even number of crossovers will recreate a dimer.

All bacteria with a circular chromosome appear to have a system dedicated to dimer resolution. In *E. coli*, and in most bacteria, the so-called Xer recombination system is used to resolve chromosome dimers. *Streptococcus* and *Lactococcus* species have a system more closely related to bacteriophage integration systems that carry out the same function (see Le Bourgeois et al., Suggested Reading). Rather than using the general recombination system, the Xer systems involve a site-specific **recombinase** (see chapter 9) to resolve chromosome dimers. This system has evolved so that it resolves dimers into the individual chromosomes but does not create new dimers. It is also arranged so that its action is coordinated with division of the cell and other important chromosome-partitioning functions. The Xer recombination system consists of two proteins called XerC and XerD and a specific site in the chromosome called *dif*. If two copies of the *dif* site occur on the same DNA, as happens when the chromosome has formed a dimer, the Xer proteins promote recombination between the two *dif* sites, resolving the dimer into the individual chromosomes (Figure 1.18). The *dif* site is always found centrally located in the *ter* region in bacterial chromosomes. This is likely to help ensure that there is only one *dif* site in the cell until just before cell division so that it is not replicated until just before the chromosome has completed replication and just before the cell divides. As added insurance, the activity of the Xer site-specific recombination system is also made to be dependent on the formation of the **division septum** through an interaction with the FtsK protein. As diagrammed in Figure 1.18, FtsK protein is localized to the region of the cell where the division septum pinches in just prior to cell division, where it plays multiple roles, including facilitating dimer resolution. While the Xer proteins are needed for dimer resolution, they are actually active for full dimer resolution only when they interact with FtsK. As shown in Figure 1.18, the localization of FtsK at the septum limits the dimer resolution process temporally and spatially to when the daughter chromosomes are in the process of moving through the septum into daughter cells, a process that is facilitated and coordinated by FtsK itself (see below) (see Aussel et al., Suggested Reading).

Site-specific recombination at *dif* by itself would not segregate the chromosomes into the daughter cells, and they might be "guillotined" by the dividing cell if the individual chromosomes were not also moved from the center of the cell prior to division. The FtsK protein also plays this role by being a DNA translocase that can pump DNA through itself to move the two *dif* sites in a dimer chromosomal DNA to the septum in the middle of the cell before they recombine, thereby facilitating segregation into daughter cells (Figure 1.18). An obvious question is how the FtsK protein "knows" which direction to actually pump DNA to move a *dif* site to the septum. It does this by using sites on the DNA as an orientation cue. These sites are called the KOPS sites, for Fts*K*-orienting *p*olar *s*equence. Polar means that they read differently in one direction on the DNA than in the other. These sequences are oriented in the DNA so that they will only be read progressing from the origin to the *dif* site in the chromosome so that the FtsK protein, by translocating the DNA in the

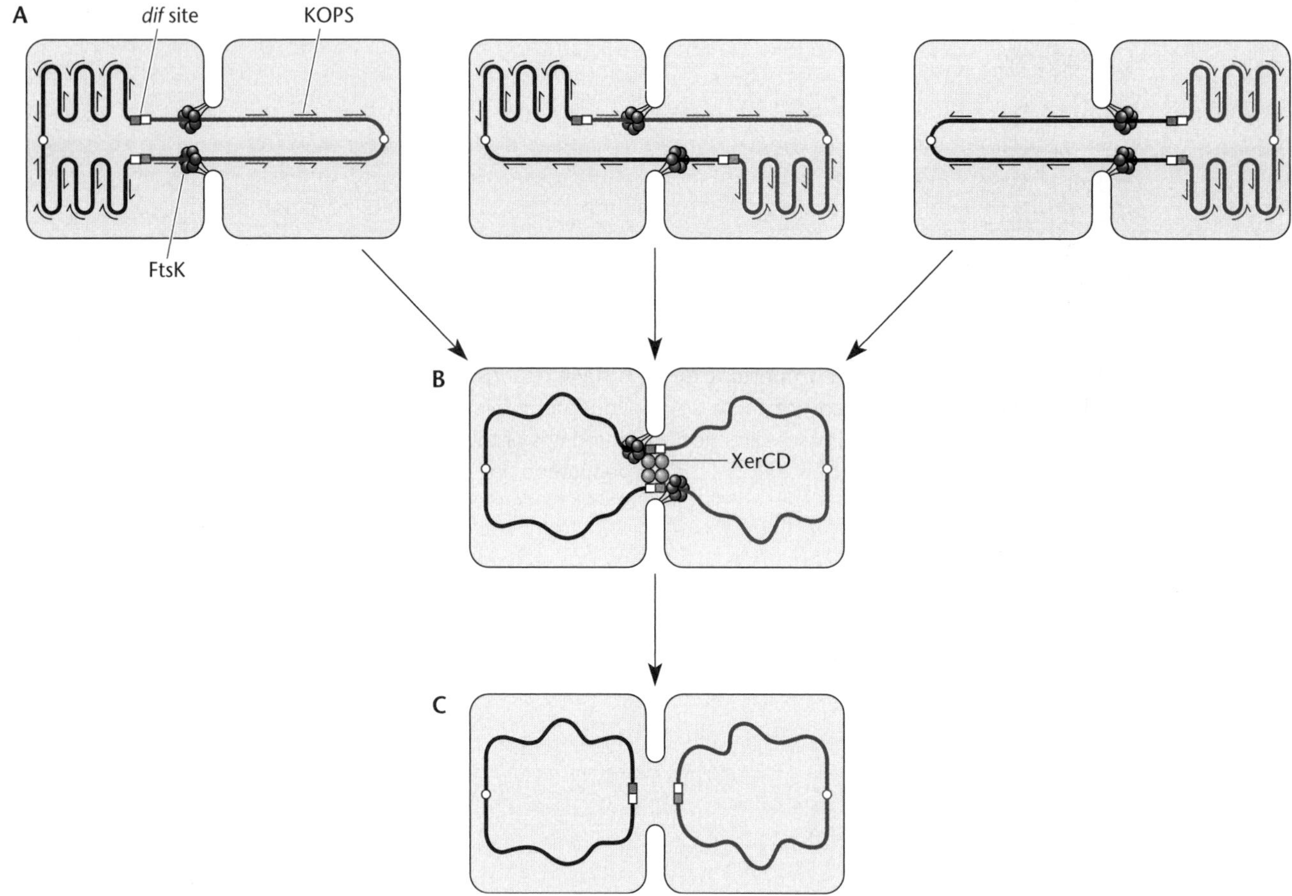

Figure 1.18 Model of the way in which chromosome translocation by FtsK coordinates chromosome segregation with dimer resolution. The two daughter chromosomes in a dimerized chromosome are shown in black and blue for emphasis. **(A)** Three possible distributions of the newly replicated dimer chromosome are shown following the start of cell septation. FtsK is a DNA translocase that can pump the chromosomes to the correct daughter cells using polar sequences in the chromosome called KOPS sites (shown as half arrows) while also moving the *dif* sites into alignment at the septum. **(B)** After aligning the *dif* sites, the FtsK protein also interacts with the XerCD enzyme, allowing it to resolve the dimer chromosomes at the *dif* sites. **(C)** The coordinated activities of the dimer resolution system and FtsK lead to monomer chromosomes that are capable of full segregation to daughter cells. doi:10.1128/9781555817169.ch1.f1.18

direction pointed to by the KOPS sites, will pump the DNA toward the *dif* sites close to the terminus of replication. Note in Figure 1.18 how the polar KOPS sites in the chromosome (shown as half-arrows) can be used as directional information by FtsK to mobilize the chromosome to the daughter cells while moving the *dif* sites toward the septum. This is just one example of how the chromosome, typically thought of as the informational storehouse of the cell, also contains structural information used by the many proteins that manage and repair the chromosome (Box 1.1). FtsK also has a domain that interacts with topoisomerase IV, a protein capable of untangling catenanes in the chromosomes (see below). Therefore, the FtsK protein coordinates a veritable clearinghouse of activities that help with chromosome management during chromosome replication and chromosome segregation.

Homologs of FtsK are widespread, and a homologous protein called SftA appears to carry out similar functions in *B. subtilis* (see Biller and Burkholder, Suggested Reading). Some bacteria have more than one FtsK-like protein, presumably for other specialized tasks. In the case of *B. subtilis*, another FtsK homolog, SpoIIIE, is responsible for translocating the final third of the chromosome into spores during spore development so the spore will get a complete copy of the chromosome (see chapter 14).

BOX 1.1

Structural Features of Bacterial Genomes

It is widely appreciated that the chromosome is the information storehouse for an organism. What is less appreciated is that the chromosome as a structure has evolved sequence motifs that allow it to be efficiently replicated, repaired, and segregated into daughter cells. The distribution and orientation of these motifs are discussed here; the molecular biology of the systems that recognize these sequences are explained in greater detail in the text. The placement of these sequence motifs in the context of the chromosome is important for their function, as is the orientation of many of these sequences. Many of the motifs are oriented in one direction, with the direction of DNA replication. DNA replication in *E. coli* and *B. subtilis* (and all bacteria studied to date) is initiated within a single *oriC* region and continues bidirectionally to a position on the chromosome equidistant from the origin (indicated by the long arrow-headed line in the figure). The *dif* site where the resolution of dimer chromosomes occurs is found near to where DNA replication normally terminates.

Certain DNA sequence motifs that guide DNA replication and DNA repair are polar in that they are not symmetrical and need to be in a specific 5′-to-3′ direction to carry out their functions. In other words, these sequences must be found in a certain orientation in the chromosome and will not work if they are flipped around. In *E. coli* and *B. subtilis*, DNA replication forks are actively terminated at specific sites called *ter* sites. The *ter* sites act as a trap for DNA replication forks, and these sites encompass a large portion of the chromosome, allowing DNA replication forks to pass when approaching from one direction but not the other. Multiple *ter* sites (10 in *E. coli* and 7 in *B. subtilis*) are found in the genome, and the redundancy of these sites may be important for catching replication events that get through the initial *ter* sites (indicated with a dashed line) or to stop replication forks that are initiated for DNA repair (see chapter 10). For unknown reasons, the central *ter* sites in *B. subtilis* are very close together while the central sites in *E. coli* are separated by hundreds of thousands of base pairs. Along the path of DNA replication are sequence motifs involved in guiding the DNA translocase proteins involved in chromosome segregation: FtsK, found in *E. coli* (which recognizes motifs called KOPS), and SpoIIIE, found in *B. subtilis* (called SRS motifs). The StpA DNA translocase from *B. subtilis* may also recognize the SRS sites. Chromosomes also include polar DNA sequences that guide the recombination machinery (see chapter 10). DNA recombination is extremely important in bacteria as a way to repair DNA double-strand breaks that occur during DNA replication. Repair of these breaks occurs when recombination reestablishes a DNA replication fork using one broken end and the sister chromosome. Reestablishment of DNA replication involves an efficient processing event that utilizes polar sites called *chi* sites. The RecBCD complex in *E. coli* or the AddAB complex in *B. subtilis* carries out this processing

doi:10.1128/9781555817169.ch1.Box1.1.f

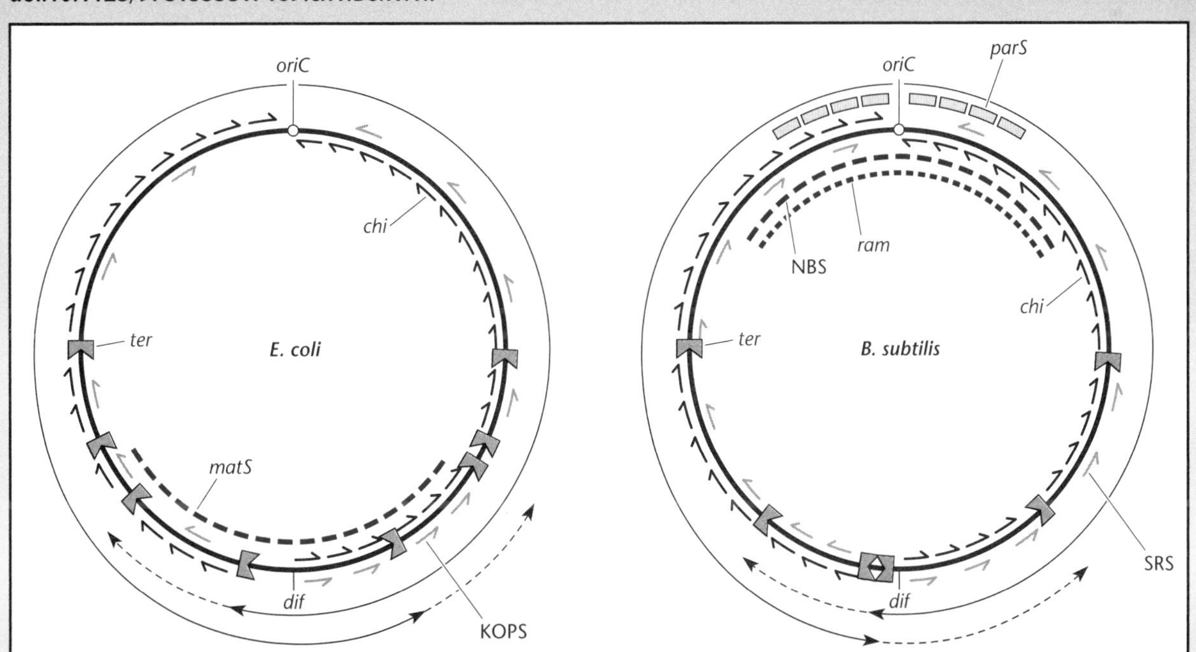

(continued)

BOX 1.1 (continued)

Structural Features of Bacterial Genomes

activity using information found in the *chi* sites. *chi* sites are species specific and are common in genomes in one orientation from the origin to the terminus region on the leading strand (found about 1 every 5 kb in the *E. coli* chromosome).

Other polar sequence biases in the chromosome include an overrepresentation of the 5′-CTG-3′ sequence that primes lagging-strand DNA synthesis (not shown). Interestingly, the most common triplet codon is the CUG (5′-CTG-3′ in DNA) that codes for leucine, comprising almost 5% of all codons in *E. coli*. The 5′-CTG-3′ sequence is found in the *chi* sequence and all of the most frequent 8-bp sequences in the chromosome. It would be difficult to argue which came first, the use of this sequence by the primase or its frequency of use as a codon. Another type of sequence bias, but one that is not polar in nature, is a general sequence bias called the G/C skew, where G and C are overrepresented in the leading strand. The trend toward A and T in the lagging strand is believed not to have an adaptive value but to be a result of the way in which repair differs on the two strands.

There are other DNA sequences that are not polar but that show biases for regions of the chromosome. Around the origin of *B. subtilis*, an area recognized by the Spo0J protein for segregation of the origin region to daughter cells, the *parS* sequence, is found 8 times (the grey rectangular boxes in the figure). Also around the origin of *B. subtilis* there is an enrichment of *ram* sites (short blue dashes), a sequence recognized by the RacA protein for maintaining segregation during sporulation, and the Noc-binding sites (NBS) (long blue dashes in the *B. subtilis* diagram) recognized by Noc, which prevents septum formation over this region of the chromosome. The GATC sites (not shown), which are important for regulating the initiation of DNA replication at *oriC* in *E. coli*, also show enrichment in the *oriC* region, with a spacing that is important for SeqA binding. The organization of the large domain comprising the terminus region of the chromosome appears to be important in *E. coli*, where the MatP protein recognizes *matS* sites (long blue dashes in the *E. coli* diagram) found across this region.

References

Blattner, F. R., G. Plunkett III, C. A. Bloch, N. T. Perna, V. Burland, M. Riley, J. Collado-Vides, J. D. Glasner, C. K. Rode, G. F. Mayhew, J. Gregor, N. W. Davis, H. A. Kirkpatrick, M. A. Goeden, D. J. Rose, B. Mau, and Y. Shao. 1997. Complete genome sequence of *Escherichia coli* K-12. *Science* **277:**1453–1462.

Touzain, F., M.-A. Petit, S. Schbath, and M. El Karoui. 2011. DNA motifs that sculpt the bacterial chromosome. *Nat. Rev. Microbiol.* **9:**15–26.

DECATENATION

DNAs also become joined to each other through the formation of **catenanes**, where the daughter DNAs become interlinked like the links on a chain. These interlinks can form as a result of DNA replication (Figure 1.19). As we discuss in "Supercoiling" below, DNA replication introduces a great deal of torsional stress ahead of the DNA replication fork (Figure 1.19A and B). This stress can be transferred across the DNA replication fork into the newly formed DNA strands in twists between the two new daughter strands called precatenanes (Figure 1.19C), because the twists will form catenanes (Figure 1.19D). Catenanes may also be caused by topoisomerases passing the strands of the two DNAs through each other (see "Topoisomerases" below). Once such interlinks are formed, the only way to unlink them is to break both strands of one of the two DNAs and pass the two strands of the other DNA through the break. The break must then be resealed. This double-strand passage, called **decatenation**, is one of the reactions performed by type II topoisomerases (see Figure 1.26). A type II topoisomerase called **topoisomerase IV** (Topo IV) plays a major role in removing most of the interlinks between the daughter DNAs in *E. coli*. The act of removing these links appears to remove one of the major cohesive forces between the daughter chromosomes prior to segregation. One of the major points of regulation appears to be spatial, where the two subunits that make Topo IV are most likely to interact through association with other proteins. In one case, this occurs following replication, when the chromosomes are being translocated across the division septum by FtsK (Figure 1.20). FtsK helps to regulate the decatenation process as it pulls the chromosome to the septum because one of the subunits of Topo IV interacts with FtsK. Interaction between Topo IV and the FtsK protein puts the enzyme in a very appropriate position for the removal of catenanes just before chromosome segregation. Topo IV is also regulated through an interaction with a protein involved in the condensation of chromosomes following DNA replication (see "Condensation" below).

CONDENSATION

By itself, just passing the two daughter DNAs through each other would not necessarily have the effect of separating the two interlinked DNAs; separating nascent chromosomes and untangled chromosomes is likely to be necessary to prevent the formation of new tangles. This

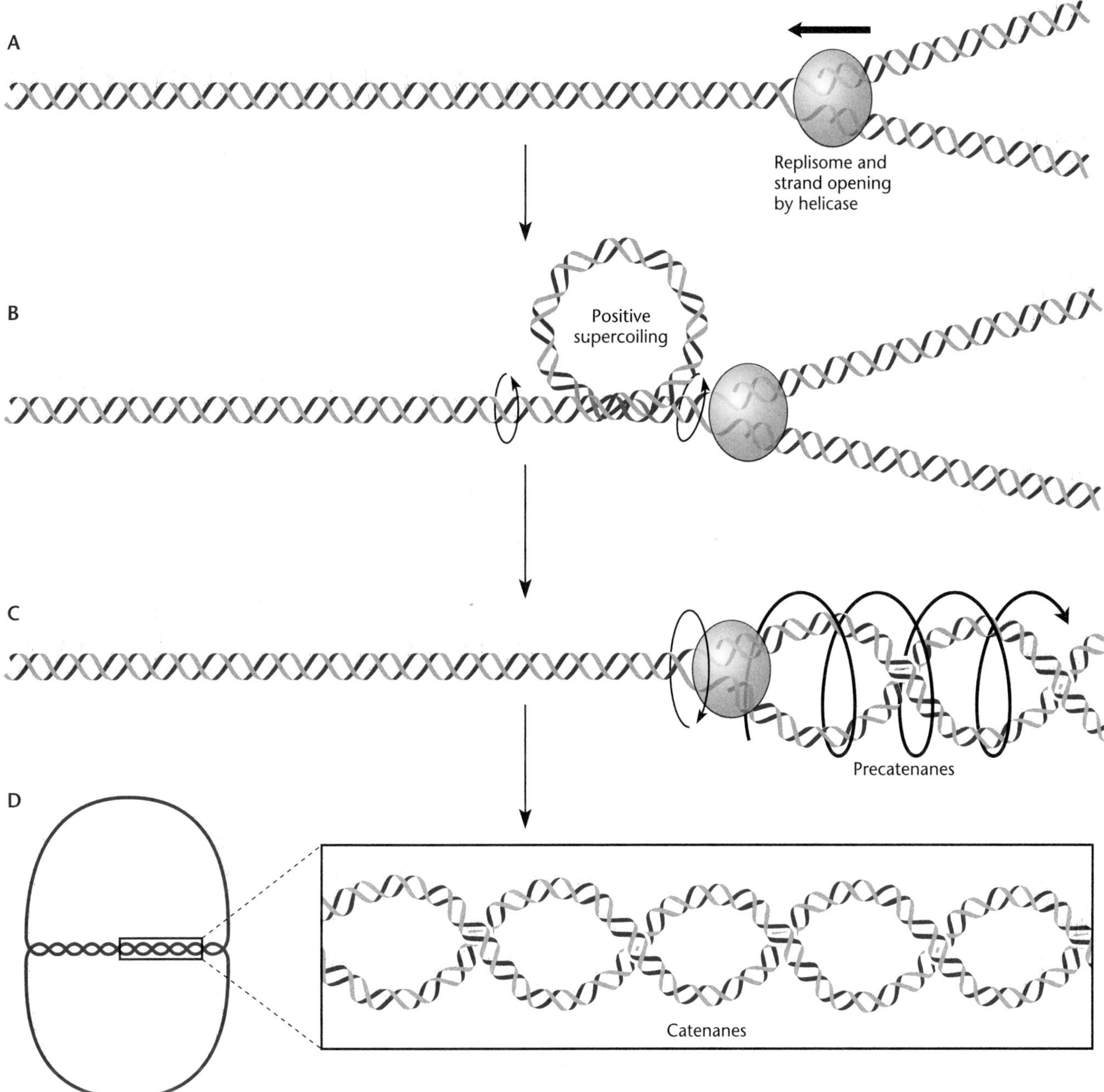

Figure 1.19 Model of the way in which unwinding of the template DNA strands can cause twists that can diffuse across the replication complex and twist the new DNA strands. **(A)** The replication machinery must open the double-stranded DNA to copy the template strands. **(B)** Unwinding the template DNA strands introduces twists (shown by black arrows) called positive supercoils ahead of the replication fork. **(C)** Some of the torsion that is generated ahead of the replication fork can be relieved by rotation of the replication complex itself (as shown by a thin black arrow). The torsional stress can spread behind the fork and intertwine the new copies of the chromosome (shown by a thick black arrow). The intertwined chromosomes are called precatenanes. **(D)** Precatenanes result in links in the daughter chromosomes called catenanes that must be unlinked for the chromosomes to separate into daughter cells.
doi:10.1128/9781555817169.ch1.f1.19

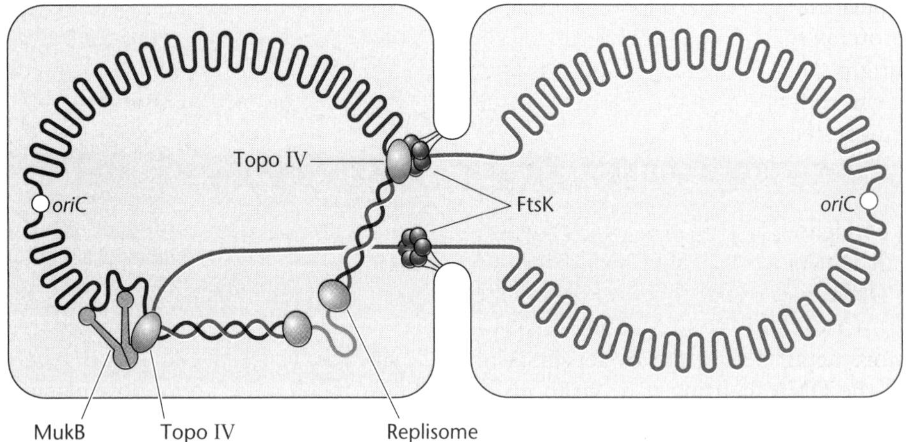

Figure 1.20 Model of the way in which chromosome decatenation by topoisomerase IV (Topo IV) is coordinated with chromosome condensation by MukB and chromosome translocation with FtsK. The daughter chromosomes are indicated by black and blue lines, and the unreplicated portion of the chromosome is shown as a grey line. Unwinding of the template DNA strands as associated with DNA replication twists the newly replicated strands of DNA, as shown in Figure 1.19, forming precatenanes that go on to become catenanes if not unlinked by the action of Topo IV. Topo IV can interact with the chromosome condensation protein MukB to remove catenanes before DNA is condensed. Topo IV also interacts with the FtsK translocase to coordinate decatenation with chromosome segregation. The replisome is shown as a circle, and double-stranded DNA is shown as a single line for simplicity.
doi:10.1128/9781555817169.ch1.f1.20

process would be facilitated by decatenating the chromosomes at the division septum immediately prior to transport of the chromosome to a daughter cell through an interaction with FtsK (see above). However, cells have another mechanism to help manage chromosomes, which is to condense them after DNA replication, a process that is also coordinated with decatenation (Figure 1.20). If the daughter chromosomes are condensed, they do not overlap as much in the cell and so are less apt to become interlinked. Condensation of chromosomes prior to mitosis has been known for a long time to occur in eukaryotic cells, where the chromosomes are only clearly visible just before mitosis. We now know that bacteria also condense their daughter DNAs to make them easier to separate prior to division, even though it is more difficult to visualize the condensation of bacterial chromosomes because of their smaller size.

Condensins

Proteins called **condensins** help to condense the DNA in the cell, making the daughter DNA molecules easier to separate. These proteins were first discovered in eukaryotes, where they help to condense DNA in chromosomes, and were named **SMC proteins** (for structural maintenance of chromosome). Condensins are long, dumbbell-shaped proteins with globular domains at the ends and a long coiled-coil region holding them together. The long coiled-coil region has a hinge so that it can fold back on itself and the two globular domains can bind together. The condensins bind to DNA through their globular domains and hold it in large loops. They can bind to DNA by themselves but also bind to other proteins called **kleisins**, which may hold them in a network-like array and condense the DNA into even smaller spaces.

Bacteria also have condensin proteins that appear to be involved in organizing the chromosome following DNA replication. The condensin of *E. coli*, called MukB, was found because mutations in its gene interfere with chromosome segregation. MukB was suspected of condensing the DNA because the protein and supercoiling of the DNA can compensate for each other in allowing proper segregation of the daughter chromosomes into daughter cells. Positive supercoils can lead to the formation of catenanes that are removed by Topo IV in a process that is actually regulated with the condensation of chromosomes by an interaction between one of the Topo IV subunits and MukB (Figure 1.20) (see Hayama and Marians, Suggested Reading). MukB interacts with DNA through association with MukF and MukE, which are structurally related to the kleisins, although they do not share amino acid sequences (see Woo et al., Suggested Reading). *B. subtilis* also has a condensin, which is more similar in amino acid sequence to the eukaryotic condensins and so was also named SMC protein (see Britton et al., Suggested Reading). It also has kleisins named ScpA and ScpB, which are also more closely related to the eukaryotic kleisins. In *B. subtilis*, the link between chromosome condensation and partitioning is becoming clearer with the finding that proteins that directly recognize the region around the origin and are involved in partitioning are able to recruit the condensin SMC in this organism (see Thanbichler, Suggested Reading).

Supercoiling

One way bacteria condense DNAs is through supercoiling. In bacteria, all DNAs are negatively supercoiled, which means that DNA is twisted in the opposite direction to the Watson-Crick helix, creating underwinds. As discussed in more detail below, the underwinds introduce stress into the DNA, causing it to wrap up on itself, much like a rope wraps up on itself if the two ends are rotated in opposite directions. This twisting occurs in loops in the DNA, causing the DNA to be condensed into a smaller space.

CHROMOSOME PARTITIONING AND SEGREGATION

Not only must the two daughter chromosomes be segregated after replication, they also must be segregated in such a way that each daughter cell gets only one of the two copies of the chromosome. Otherwise, one daughter cell would get two chromosomes and the other would be left with no chromosome and eventually would die. The apportionment of one daughter chromosome to each of the two daughter cells is called **partitioning**. Daughter cells that lack a chromosome after division are very rare, indicating that partitioning is a very efficient process in bacteria. Because of the importance of chromosome segregation, redundant mechanisms may have evolved to ensure that it occurs accurately. Indeed, many of the mechanisms that allow condensation of chromosomes can contribute to partitioning once the origin regions are located in the nascent daughter cells. While broad themes that describe partitioning across all bacteria have eluded our understanding, some important model systems are fairly well understood. In this section, we discuss what is known in the model bacteria *E. coli* and *B. subtilis*. In many ways, these processes are understood better in another model bacterium, *Caulobacter crescentus*, than they are in *E. coli* or *B. subtilis*, partly because of the relative ease of synchronizing *Caulobacter* in the cell cycle. However, we defer discussion of chromosome partitioning and segregation in *Caulobacter* until chapter 14, where we can integrate it with bacterial cell biology, the cytoskeleton, and the bacterial cell cycle.

The Par Proteins

Early work concentrated on the functions of the so-called partitioning proteins, the products of the *par* genes. The Par functions were first discovered in plasmids, which are small DNA molecules that are found in bacterial cells and that replicate independently of the chromosome (see chapter 4). Because they exist independently of the chromosome, plasmids usually must also have a system for partitioning; otherwise, they would often be lost from cells when the cells divide. The Par systems of plasmids are known to fall into two families, one represented by the Par system of plasmid R1 and the other, much larger, represented by plasmids P1, F, and many others. It is the second of these families to which the known Par functions of chromosomes belong. Because plasmids are much smaller than the chromosome, it has been easier to do experiments with them, and we know much more about how plasmid Par systems work than we do about how the corresponding chromosomal systems work. However, we can assume that the functioning of the plasmid systems gives clues to the functioning of the homologous chromosomal systems.

Plasmid Par functions usually consist of two proteins and a site on the DNA (often called *parS*) at which they act. One of these proteins, often called ParB, binds in many copies to the *parS* site on the DNA. This binding of ParB to the site often occurs only after the plasmid has replicated; it may require pairing of the two newly replicated daughter plasmid DNAs. The ParA protein polymerizes to form dynamic filaments that extend across the chromosomes and may pull the daughter plasmids across each of the segregating host chromosomes, allowing their partitioning. The ParA proteins have ATPase activity, and the cleavage of ATP affects their polymerization and depolymerization. The ParB-ParS complex affects the nucleotide-binding state of ParA, thereby affecting the extent of polymerization and depolymerization controlling partitioning (see Fig. 4.15).

Filament formation by partitioning proteins. Par proteins can form various types of filaments. The type of dynamic filament formed seems to depend on the family to which the partitioning system belongs. In the case of the R1 partitioning system, a single long filament forms that extends to the ends of the cell. Interestingly, the Par protein that forms these filaments is related to eukaryotic actin and may play similar roles in moving cellular constituents around in bacterial cells (see chapter 4). Filament formation with the other type of Par system, the one to which the known chromosomal systems are related, remains controversial. Plasmid partitioning systems are discussed in more detail in chapter 4.

Par functions in B. subtilis. As mentioned above, some types of bacteria also have Par functions that are encoded by the bacterial chromosome and that seem to perform a similar role in partitioning the chromosomes during cell division. The situation is clearer in *B. subtilis* than it is in *E. coli*. The Par proteins of *B. subtilis* are like the plasmid Par proteins in that one is an ATPase and binds to the other, but only when the other Par protein is bound to specific sites on the chromosome. In *B. subtilis*, the proteins analogous to the ParA and ParB proteins of plasmids are called Soj and Spo0J, respectively. These names come from early genetic studies of *B. subtilis* sporulation, where *spo0J* was identified as a gene required for sporulation and *soj* was identified as a suppressor of *spo0J*. There are also a number of *parS*-like sites close to the origin of chromosome replication; Spo0J has been shown to bind to these sites, as would be expected if it is analogous to ParB. The Spo0J filament is capable of recruiting the *B. subtilis* condensin protein SMC, which would presumably initiate the process of condensation in new daughter cells and facilitate chromosome segregation. Soj is likely to cooperate with Spo0J in the partitioning process and has been suggested to also coordinate partitioning with the replication initiation process by binding the condensin SMC, as mentioned above.

Surprisingly, *E. coli* does not have a recognizable Par-like system, and it seems likely that another system is responsible for the active separation of the chromosomes. *E. coli* (and some other bacteria) has an unrelated system that organizes the terminus region into a large macrodomain. This system relies on the MatP protein, which binds a *cis*-acting site, *matS*, that is distributed across the terminus region (see Mercier et. al., Suggested Reading).

Coordinating Cell Division and Chromosome Partitioning in *E. coli* and *B. subtilis*

Much has also been learned about how the bacterial division septum forms. This process is also called cytokinesis. A protein called **FtsZ**, which forms a ring around the midpoint of the cell, performs the primary step in this process. This protein is related to tubulin of eukaryotes and forms filaments that grow and shorten by adding and removing shorter filaments, called protofilaments, to its ends in the presence of GTP. Before the cell is ready to divide, the FtsZ protein may form short filaments at the center of the cell. When the cell is about to divide, these filaments may converge on the middle of the cell and form the ring at the site of the future septum. The FtsZ ring then attracts many other proteins, including the DNA translocase FtsK, discussed above. FtsZ helps form the division septum, which eventually squeezes the mother cell at its center to allow the formation of the two daughter cells. We discuss bacterial cell cytokinesis and the functions involved in more detail in chapter 14. The following major questions may be asked: why does the septum form only in the middle of the cell, and why does septum formation not occur over the nucleoid prior to chromosome segregation? The answers to these questions lie, at least in part, in two types of systems: the Min systems and the **nucleoid occlusion** systems.

The Min Proteins

In *E. coli*, three proteins called MinC, MinD, and MinE are known to be involved in selecting the site for the division septum to form. The *min* genes of *E. coli* were found because mutations in these genes can cause the division septa to form in the wrong places, sometimes pinching off smaller cells called minicells. Apparently, in the absence of the Min proteins, division septa can form in places other than the middle of the cell, for example, at the one-quarter and three-quarter positions, the sites of future division septa. When this happens, smaller minicells that lack a chromosome are pinched off, hence the name Min proteins for minicell producing. It was predicted that the Min proteins would be localized in the ends of the *E. coli* cell, where they could prevent FtsZ from forming a division septum anywhere but the middle of the cell. However, when the localization of the Min proteins was studied, using GFP fusions to the Min proteins, a very surprising result was revealed: the Min proteins oscillate from one pole of the cell to the other during the cell cycle. The MinC and MinD proteins oscillate the most, collecting at one end of the cell and then all moving to the other end. The MinE protein appears to form a ring that oscillates back and forth in the middle of the cell, apparently driving the oscillation of MinC and MinD. MinE is required for the oscillation and may drive the oscillation of MinD, which in turn drives the oscillation of MinC, which in turn inhibits the formation of the FtsZ ring by binding to FtsZ. It has been hypothesized that the purpose of the oscillation may be to ensure that the concentration of the MinC division inhibitor is highest at the poles, where it needs to inhibit FtsZ, and lowest in the middle of the cell, where it is just passing through. However, recent evidence has shown that MinCD also accumulates at the division septum just before division, perhaps to prevent additional rings from forming close to the central division septum. It is almost as though the Min proteins play the role of "division site policemen," constantly scanning the cell to make sure that FtsZ does not accumulate and form a septum in an inappropriate location.

Regulation of septum formation in *B. subtilis* differs from that found in *E. coli*. In *B. subtilis*, MinE is lacking and MinC and MinD do not oscillate. Rather, MinCD appears to tether directly to the cell poles by binding to another protein at the cell poles called DivIVA. This binding creates a gradient of concentration of MinCD in the cell and inhibits FtsZ ring formation at the poles. MinCD also concentrates briefly at the division septum in the bacterium just before cell division is completed. Therefore, these two model bacteria use somewhat different mechanisms to establish a gradient of MinCD concentration and thereby restrict FtsZ ring formation to the center of the cell.

Nucleoid Occlusion

Not only should the FtsZ ring form only in the center of the cell, but it also should not initiate the assembly of a division septum while the nucleoid is still occupying the center of the cell or it might guillotine the chromosome. In fact, it was observed in *E. coli* that FtsZ rings never formed in the center of the cell when it was still occupied by the nucleoid, which had not yet segregated. Proteins that inhibit FtsZ ring formation in the presence of the nucleoid were discovered in both *E. coli* and *B. subtilis* at about the same time and were named nucleoid occlusion proteins. Both proteins were found because they are essential only if the Min system is inactivated. The reasoning is that the nucleoid occlusion and Min systems can at least partially substitute for each other in localizing the division septum. If one or the other is missing, the FtsZ protein still forms a ring in the middle of the cell, which usually is not occupied by the nucleoid by this time. However, if both are missing, the division septum is apt to occur anywhere, even in regions that are occupied by the nucleoid. The nucleoid exclusion protein in *B. subtilis*, named Noc, was found serendipitously, because its gene, *noc*, is adjacent to the genes for the Par functions, *soj* and *spo0J*, and it was observed that mutations in this gene could not be combined with mutations in the *minD* gene without making the cells very sick. The reason *noc* mutant strains were sick was that they were forming long filaments of cells because they were not dividing properly. The nucleoid occlusion protein in *E. coli*, named SlmA, was found directly by a synthetic lethal screen (see Bernhardt and de Boer, Suggested Reading). A synthetic lethality screen is a powerful genetic tool in which mutations in genes whose products are required only if another gene product is absent are isolated. The investigators looked for mutations in genes that were required only in the absence of the products of the *min* genes. They expressed the *min* genes from an inducible promoter and looked for mutants that were sick and failed to form colonies only in the absence of inducer. Some of these mutants had mutations in a gene that was named *slmA* by the investigators. While mutants deficient in Min proteins had more Z rings, they were never over the nucleoids. However, mutants that lacked both the Min proteins and SlmA often formed Z rings over the nucleoids, as expected for a mutant deficient in nucleoid occlusion. The use of inducible promoters and other examples of synthetic phenotypes are discussed in more detail in later chapters.

The Noc and SlmA proteins seem to act by binding to DNA and then inhibiting FtsZ ring formation close to the DNA to which they are bound. In the case of Noc, there are known to be DNA sequences called Noc-binding sites that are bound by Noc and are distributed across the chromosome and concentrated in the origin region but are absent from the terminus region (Box 1.1). This allows Noc to help protect the nucleoid from the division septum until the final moments prior to the completion of DNA replication (see Wu et al., Suggested Reading).

Coordination of Cell Division with Replication of the Chromosome

It is not sufficient to know how chromosomes replicate and then segregate into the daughter cells prior to division. Something must coordinate the replication of the chromosome with division of the cells. If the cells divided before the replication of the chromosome was completed, there would not be two complete chromosomes to segregate into the daughter cells, and one cell would end up without a complete chromosome. The mechanism by which cell division is coordinated with replication of the DNA is still not completely understood, but there is a lot of relevant information.

TIMING OF REPLICATION IN THE CELL CYCLE

It is important to know when replication occurs during the cell cycle. Experiments were designed to determine the relationship between the time of chromosome replication and the cell cycle in *E. coli* (see Helmstetter and Cooper, Suggested Reading). The conclusions are still generally accepted, so it is worth going over them in some detail. The investigators recognized that if the DNA content of cells at different stages in the cell cycle could be measured, it would be possible to determine how far chromosome replication had proceeded at that time in the cell cycle. Since bacterial cells are too small to allow measurements of DNA content in a single cell by the methods they had, it was necessary to measure the DNA content in a large number of pooled cells. However, cells growing in culture are all at different stages in their cell cycles. Therefore, to know how far replication had proceeded at a certain stage in the cell cycle, it was necessary to **synchronize** the cells in the population so that all were the same age or at the same point in their life cycles at the same time.

Helmstetter and Cooper accomplished this by using what they called a bacterial "baby machine." Their idea was to first label the DNA of a growing culture of bacterial cells by adding radioactively labeled nucleosides and then fix the bacterial cells on a membrane. When the cells on the filter divided, one of the two daughter cells would no longer be attached and would be released into the medium. All of the daughter cells released at a given time would be newborns and so would be the same age. This means that cells that divided to release the daughter cells at a given time would also be the same age and would have DNA in the same replication state. The amount of radioactivity in the released cells was then a measure of how much of the chromosome had replicated in cells of that age. This experiment was done under different growth conditions to show how the timing of replication and the timing of cell division are coordinated under different growth conditions.

Figure 1.21 shows the results of these experiments. For convenience, the following letters were assigned to each of the intervals during the cell cycle. The letter I denotes the time from when the last round of chromosome replication initiated until a new round begins. C is the time it takes to replicate the entire chromosome, and D is the time from when a round of chromosome replication is completed until cell division occurs. The top of the figure shows the relationship of I, C, and D when cells are growing slowly, with a generation time of 70 min. Under these conditions, I is 70 min, C is 40 min, and D is 20 min. However, when the cells are growing in a richer medium and are dividing more rapidly, with a generation time of only 30 min, the pattern changes. The C and D intervals remain about the same, but the I interval is much shorter, only about 30 min.

Figure 1.21 Timing of DNA replication during the cell cycle, with two different generation times, 70 min **(A)** and 30 min **(B)**. The time between initiations (I) is the only time that changes. See the text for definitions of I, C, and D. doi:10.1128/9781555817169.ch1.f1.21

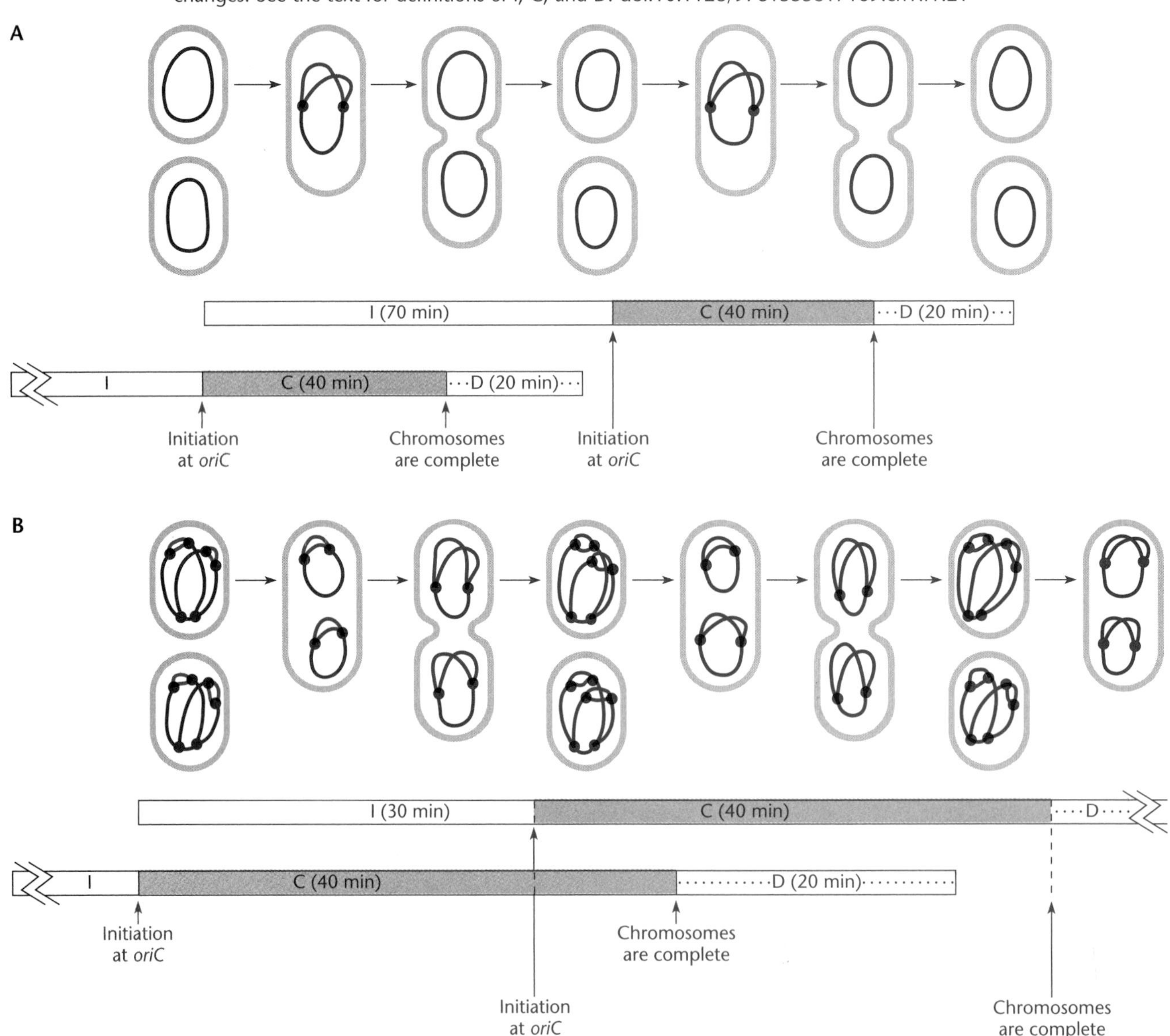

Some conclusions may be drawn from these data. One conclusion is that the C and D intervals remain about the same independent of the growth rate. At 37°C, the time it takes the chromosome to replicate is always about 40 min, and it takes about 20 min from the time a round of replication terminates until the cell divides. However, the I interval gets shorter when the cells are growing faster and have shorter generation times. In fact, the I interval is approximately equal to the generation time—the time it takes a newborn cell to grow and divide. This makes sense because, as discussed below, initiation of chromosome replication appears to occur every time the cells reach a certain size. They reach this size once every generation time, independent of how fast they are growing.

Another point apparent from the data is that in cells growing rapidly with a short generation time, the I interval can be shorter than the C interval. If I is shorter than C, a new round of chromosomal DNA replication will begin before the old one is completed. This explains the higher DNA content of fast-growing cells than of slow-growing cells. It also explains the observation that genes closer to the origin of replication are present in more copies than are genes closer to the replication terminus.

Despite providing these important results, this elegant analysis does not allow us to tell whether division is coupled to initiation or termination of chromosomal DNA replication. The fact that the I interval always equals the generation time suggests that the events leading up to division are set in motion at the time a round of chromosome replication is initiated and are completed 60 min later independent of how fast the cells are growing. However, it is also possible that the act of termination of a round of chromosome replication sets in motion a cell division 20 min later. Multiple laboratories are working to resolve these issues.

Timing of Initiation of Replication

A new round of replication must be initiated each time the cell divides, or the amount of DNA in the cell would increase until the cells were stuffed full of it or decrease until no cell had a complete copy of the chromosome. Clearly, initiation of replication is exquisitely timed. In cells growing very rapidly, in which the next rounds of replication initiate before the last ones are completed, so that the cells contain a number of origins of replication, all of the origins in a cell "fire" simultaneously, indicating tight control.

A number of attempts have been made to correlate the timing of initiation of chromosome replication with other cellular parameters during the cell cycle. Most evidence from such attempts points to initiation of replication being tied to cell mass. After cells divide, their mass, or weight, continuously increases until they divide again. The initiation of chromosome replication occurs each time the cell achieves a certain mass, the **initiation mass**. If cells are growing faster in richer medium, they are larger and achieve the initiation mass sooner than do smaller, slower-growing cells, explaining why new rounds of chromosome replication occur before the termination of previous rounds in faster-growing cells but not in slower-growing cells. However, these experiments by themselves do not explain what it is about the cell mass that triggers initiation.

ROLE OF THE DnaA PROTEIN

Initiation is regulated in multiple ways through the DnaA protein. The cellular concentration of DnaA, the ATP- versus ADP-binding state of DnaA, and access of DnaA to *oriC* are all factors involved in regulating the initiation process. The issue of DnaA concentration must also take into consideration that many molecules of DnaA protein may be needed to assemble on the origin to allow initiation (Fig. 1.16) and that during fast growth there will be many *oriC* origins per cell. Therefore, as the number of origins increases in the cell, larger amounts of the DnaA protein would be required, independent of any other factors.

The affinity of DnaA for the various binding sites within *oriC* (see "Origin of Chromosomal Replication" above) allows a mechanism for the origin to be identified but not immediately used as an origin until all conditions are in place for replication; only at higher concentrations of DnaA molecules in the cell will there be sufficient DnaA for the larger initiation complex that can provide the first step to allow strand melting and loading of the DnaB helicase (Figure 1.16). After DNA replication is initiated, the number of *oriC* sites doubles, but the concentration of DnaA protein does not increase as quickly, reducing the chance that DnaA will be available at a sufficient concentration for initiation. Other binding sites for DnaA around the chromosome further augment this dilution process for DnaA around the chromosome, as the affinity of these sites for DnaA can rival that of the origin itself, so they can compete for the amount of DNA available. Many of these 300 or so sites are distributed around the chromosome, but one region that is close to the origin, called *datA*, can bind >50 molecules of DnaA. This model involving a concentration-dependent role for DnaA is consistent with evidence that either artificially decreasing the amount of DnaA in the cell or increasing the number of copies of the DnaA-binding sites delays initiation of rounds of chromosome replication.

While the concentration of DnaA provides a method for controlling initiation, multiple other levels of control also exist. In addition to the quantity of DnaA affecting initiation, the nucleotide-binding state of the

DnaA protein itself can further control timing. As mentioned above, DnaA binds either ATP or ADP, but only the ATP-bound form can bind all of the DnaA boxes in *oriC* and initiate replication. This allows the DnaA protein to also act as a "switch" independent of its concentration because other cellular inputs can control the ATP- versus ADP-bound state of DnaA. One cellular input is the presence of the assembled replication fork immediately after initiation of replication. The presence of the β sliding-clamp protein involved in replication causes DnaA to hydrolyze ATP to ADP by interacting with another protein, a relative of DnaA called Hda (see Camara et al., Suggested Reading). This prevents immediate reinitiation while DnaA levels are still high. Two other mechanisms appear to be involved in "recycling" DnaA by encouraging it to exchange a nucleotide, putting it back into the DnaA-ATP form; one involves an exchange catalyzed by acidic phospholipids, and another involves two sites found in the chromosome called DnaA-reactivating sequences (see Fujimitsu et. al., Suggested Reading). How all these inputs are coordinated to control the amount of DnaA active for initiation is central to the regulation of the timing of initiation.

HEMIMETHYLATION AND SEQUESTRATION

In some types of bacteria, including *E. coli*, there is yet another means of delaying the initiation of new rounds of chromosome replication. As with the mechanisms described above, replication itself also plays a role in this regulatory pathway where methylation of the DNA helps delay initiation. In *E. coli* and other enteric bacteria, the A's in the symmetric sequence GATC/CTAG are methylated at the 6′ position of the larger of the two rings of the adenine base. These methyl groups are added to the bases by the enzyme **deoxyadenosine methylase** (Dam or **Dam methylase**), but this occurs only after the nucleotides have been incorporated into the DNA. Since DNA replicates by a semiconservative mechanism, the "A" in the GATC/CTAG sequence in the newly synthesized strand remains temporarily unmethylated after replication of a region containing this sequence (Figure 1.22). The DNA at this site is said to be **hemimethylated** if the bases on only one strand are methylated.

The hemimethylated state is important in the context of regulation of initiation because a *trans*-acting protein called SeqA binds only hemimethylated GATC/CTAG sequences. SeqA is an essential facilitator of hemimethylation control of replication initiation and is found only in bacteria that have Dam. The sequence GATC/CTAG is found 11 times within *oriC*, much more often than would be expected by chance alone (Figure 1.15). GATC/CTAG sequences are also associated with the low-affinity DnaA-binding sites across the origin. Furthermore, the promoter region of the *dnaA* gene, the region in which mRNA synthesis initiates for the DnaA protein, also has GATC/CTAG sequences. SeqA is able to bind all of these strategically located GATC/CTAG

Figure 1.22 Replication creates hemimethylated DNA. **(A)** The A in the sequence GATC is methylated on both strands (A_m and ^mA). **(B)** After replication, the A in GATC in the new strand is not immediately methylated by the Dam methylase. **(C)** Eventually GATC sites in the new strand are methylated, converting the DNA back to the fully methylated state.
doi:10.1128/9781555817169.ch1.f1.22

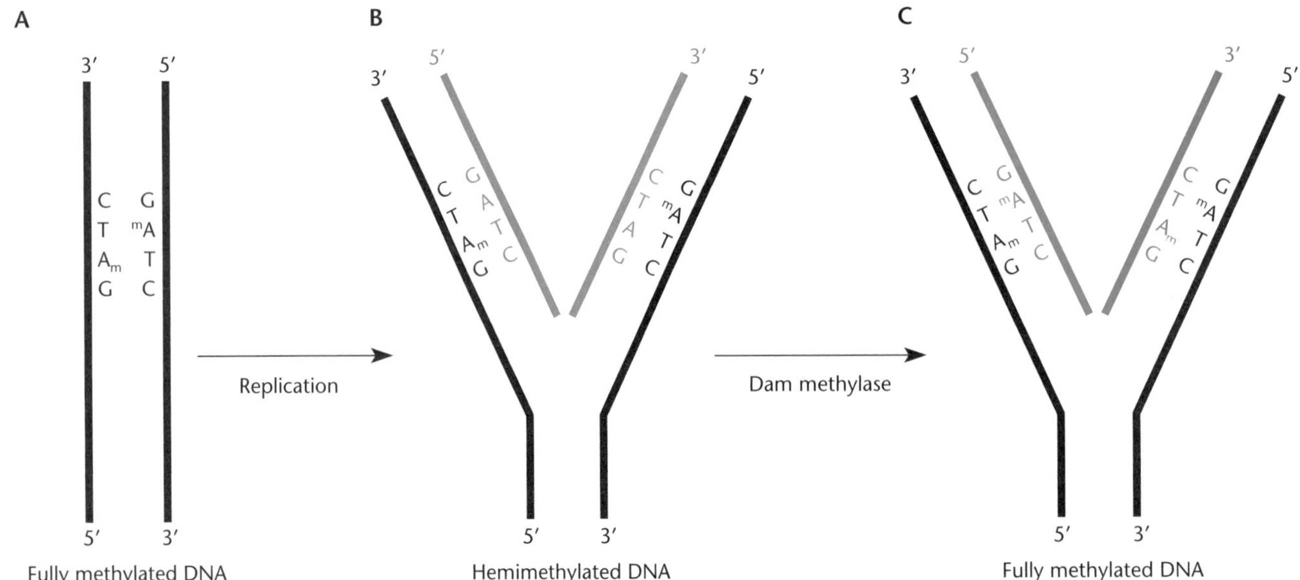

sites after they are replicated (and therefore rendered hemimethylated), which has the effect of delaying the conversion to full methylation at these sites for about one-third of a cell cycle. SeqA also blocks binding of DnaA to the low-affinity sites and inhibits expression of the *dnaA* gene. SeqA bound to hemimethylated DNA may associate with the cell membrane to sequester the *oriC* region after initiation of DNA replication (Figure 1.23). Sequestration of the hemimethylated *oriC* region is predicted to render it nonfunctional for the initiation of new rounds of replication and prevents it from being further methylated by the Dam methylase (see Slater et al., Suggested Reading).

In addition to the GATC/CTAG sequences associated with *oriC* and the *dnaA* promoter, SeqA also interacts with the GATC/CTAG sequences as the replication forks progress around the chromosome, effectively marking the location of the replisome. SeqA may play other roles in DNA replication and repair; it has been found to be essential in cells that lack the major pathways of DNA recombination involving RecA.

Figure 1.23 Model of the sequestration of the *oriC* region of the *E. coli* chromosome after initiation of chromosome replication. Before initiation, the *oriC* region is methylated in both strands. After initiation, only one of the two strands is methylated; hence, the region is hemimethylated. A protein called SeqA helps bind the hemimethylated *oriC* to the membrane, thereby sequestering it and preventing further initiation and methylation. The newly synthesized strand is shown in blue, and the methylated bases are indicated by asterisks. Only 1 of the 11 GATC/CTAG sequences in the *oriC* region is shown.
doi:10.1128/9781555817169.ch1.f1.23

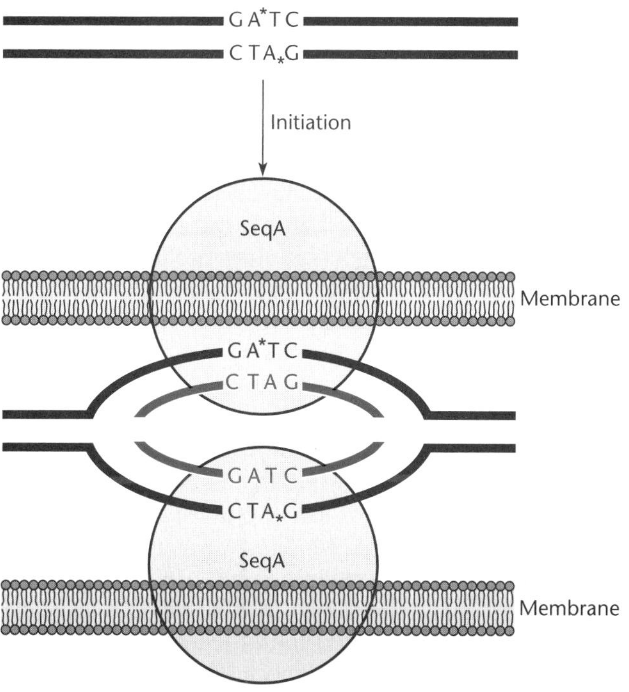

The Bacterial Nucleoid

The nucleoid was described with respect to chromosome segregation above. Indeed, experiments in many of the model systems indicate that bacteria carefully coordinate the position of the chromosome in the cell. Through techniques in which individual positions in the chromosome can be localized in whole cells, it has been shown that genes are localized in the cell in roughly the same order as one would presume by looking at the DNA sequence. A variety of techniques are providing insight into how the structure of the chromosome is maintained in the cell. The molecular mechanisms that maintain the chromosome structure remain a mystery, but specific systems are likely to exist to ensure that the chromosome is available for transcription, recombination, and other functions.

Bacterial nucleoids can be released from cells in a relatively intact state using gentle lysis procedures. DNA isolated from bacteria by such a procedure is shown in Figure 1.24. The nucleoid is composed of 30 to 50 loops of DNA emerging from a more condensed region, or core. Seeing this tangle of loops, it is difficult to imagine that the DNA in this complicated structure is actually one long, continuous circular molecule.

Supercoiling in the Nucleoid

One of the most noticeable features of the nucleoid is that most of the DNA loops are twisted up on themselves. This twisting is the result of supercoiling of the DNA, as discussed above. Figure 1.25 illustrates supercoiling. In this example, the ends of a DNA molecule have been rotated in opposite directions and the DNA has become twisted up on itself to relieve the stress. The DNA remains supercoiled as long as its ends are constrained and so cannot rotate, and a circular DNA has no free ends that can rotate. Therefore, a supercoiled circular DNA remains supercoiled but a linear DNA immediately loses its supercoiling unless regions flanking the supercoiling are somehow otherwise constrained.

Even a circular DNA loses its supercoiling if one of the strands of the DNA is cut, thereby allowing the strands to rotate around each other. The phosphodiester bond connecting the two deoxyribose sugars on the other strand serves as a swivel and rotates, resulting in relaxed (i.e., not supercoiled) DNA. A DNA with a phosphodiester bond broken in one of the two strands is said to be nicked. When nucleoids are prepared, some of the loops are usually relaxed, probably by nicks introduced during the extraction process. The fact that only some, and not all, of the loops of DNA in the nucleoid are relaxed tells us something about the structure of the nucleoid, i.e., that there are periodic barriers to rotation of the DNA. A break or nick in a circular DNA should

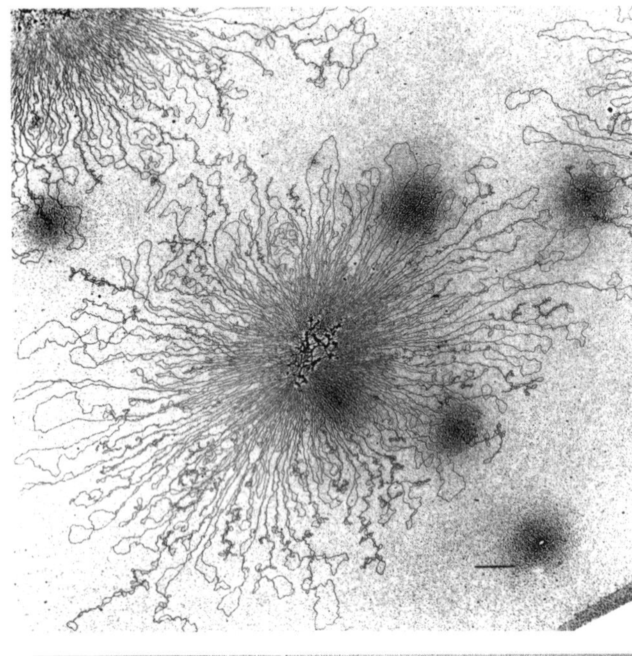

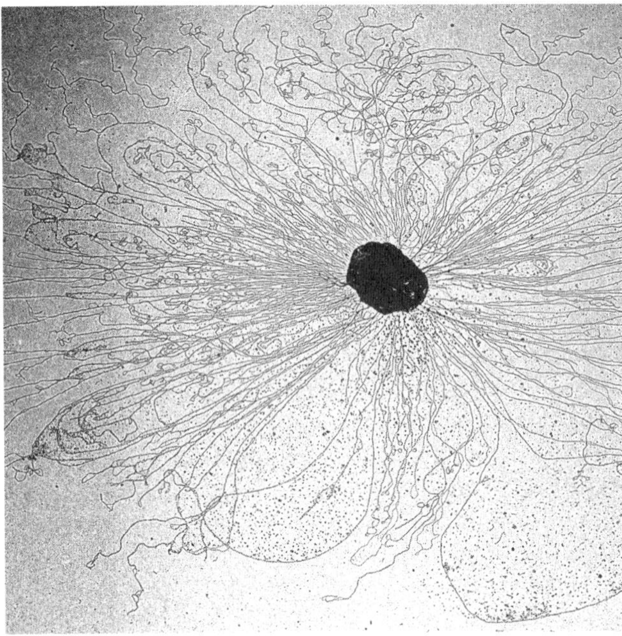

Figure 1.24 Electron micrographs of bacterial nucleoids.
doi:10.1128/9781555817169.ch1.f1.24

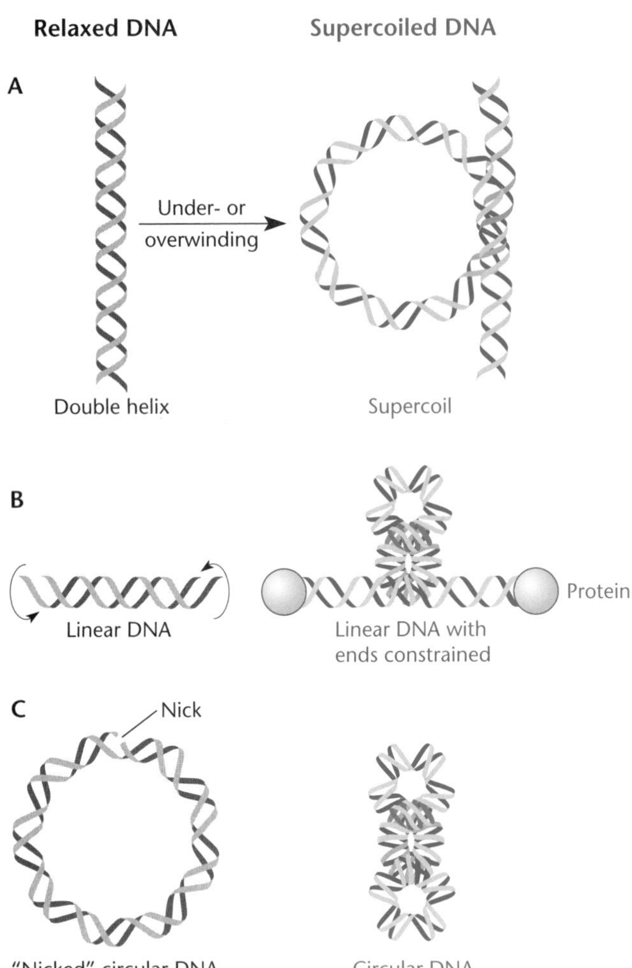

Figure 1.25 **(A)** Supercoiled DNA. **(B)** Twisting of the ends in opposite directions causes linear DNA to wrap up on itself. The supercoiling is lost if the ends of the DNA are not somehow constrained. **(C)** A break, or nick, in one of the two strands of a circular DNA relaxes the supercoils.
doi:10.1128/9781555817169.ch1.f1.25

relax the whole DNA unless portions of the molecule are periodically attached to barriers that prevent rotation of the strands. Through the examination of the expression level of over 300 genes that are sensitive to supercoiling following the introduction of breaks in the chromosome it has been estimated that the topologically isolated loops are about 10 kb in size. This is in good agreement with domain size as determined by directly measuring the lengths of loops under the microscope (see Postow et al., Suggested Reading). Although the exact mechanism or mechanisms that restrict topology are unresolved, this work indicates that the barriers to rotation are not fixed at certain places in the chromosome.

SUPERCOILING OF NATURAL DNAs

It is possible to estimate the extent of supercoiling of natural DNAs. According to the Watson-Crick structure, the two strands are wrapped around each other about once every 10.5 bp to form the double helix. Therefore, in a DNA of 2,100 bp, the two strands should be wrapped around each other about 2,100/10.5, or 200, times. In a supercoiled DNA of this size, however, the two strands are wrapped around each other either more or less than 200 times. If they are wrapped around each other more than once every 10.5 bp, the DNA is said to be **positively supercoiled**; if less than once every 10.5 bp, it is **negatively supercoiled**.

Most DNA in bacteria is negatively supercoiled, with an average of one negative supercoil for every 300 bp, although there are localized regions of higher or lower negative supercoiling. Also, in some regions, such as ahead of a transcribing RNA polymerase, the DNA may be positively supercoiled (see above).

SUPERCOILING STRESS

Some of the stress due to supercoiling of the DNA, which causes it to twist up on itself, can be relieved if the DNA is wrapped around something else, such as proteins. Sailors know about this effect: if you twist a rope in the right direction as you roll it up to store it, it does not try to unroll itself again when you are finished. Wrapping DNA around proteins in the cell is called constraining the supercoils. Unconstrained supercoils cause stress in the DNA, which can be relieved by twisting the DNA up on itself, as shown in Figure 1.25, and making the DNA more compact. The stress due to unconstrained supercoils can have other effects, as well, for example, helping to separate the strands of DNA during reactions such as replication, recombination, and initiation of RNA synthesis at promoters.

Topoisomerases

The supercoiling of DNA in the cell is modulated by topoisomerases (see Wang, Suggested Reading). Topoisomerases have been discussed above, but not the molecular details of these enzymes. All organisms have these proteins, which manage to remove the supercoils from a circular DNA without permanently breaking either of the two strands. They perform this feat by binding to DNA, breaking one or both of the strands, and passing the DNA strands through the break before resealing it. As long as the enzyme holds the cut ends of the DNA so that they do not rotate, this process, known as **strand passage**, either introduces or removes supercoils in DNA.

The topoisomerases are classified into two groups, type I and type II (Figure 1.26). These two types differ in how many strands are cut and how many strands pass through the cut. The type I topoisomerases cut one strand and pass the other strand through the break before resealing the cut. The type II topoisomerases cut both strands and pass two other strands from somewhere else in the DNA, or even another DNA, through

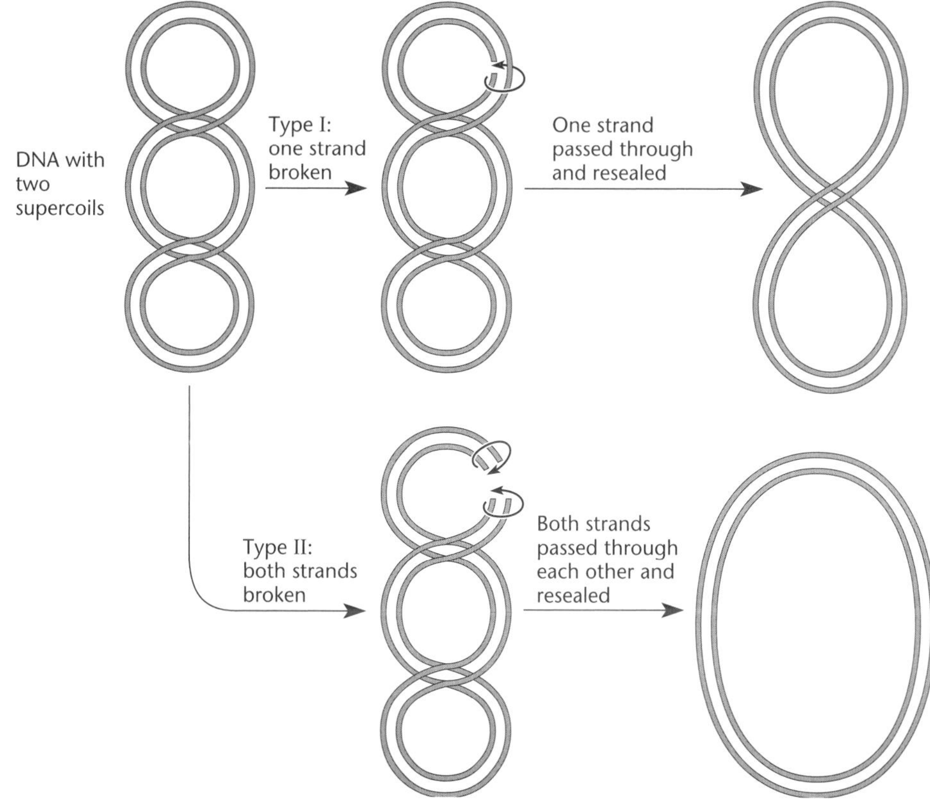

Figure 1.26 Action of the two types of topoisomerases. The type I topoisomerases break one strand of DNA and pass the other strand through the break, removing one supercoil at a time. The type II topoisomerases break both strands and pass another part of the same DNA through the breaks, introducing or removing two supercoils at a time.
doi:10.1128/9781555817169.ch1.f1.26

the break before resealing it. This basic difference changes how supercoiling is affected by these enzymes, as shown in Figure 1.26.

TYPE I TOPOISOMERASES

Bacteria have several type I topoisomerases. The major bacterial type I topoisomerase removes negative supercoils from DNA. In *E. coli* and *Salmonella enterica* serovar Typhimurium, the *topA* gene encodes this type I topoisomerase. As expected, DNA isolated from *E. coli* with a *topA* mutation is more negatively supercoiled than normal.

TYPE II TOPOISOMERASES

Bacteria also have more than one type II topoisomerase. Because type II topoisomerases can break both strands and pass two other DNA strands through the break, they can either separate two linked circular DNA molecules or link them up. Linkage sometimes happens after replication or recombination. One major type II topoisomerase in *E. coli*, Topo IV (see above), decatenates daughter chromosomes after DNA replication, releasing the major source of cohesion between the chromosomes and allowing them to be segregated into the daughter cells. Another major type II topoisomerase in bacteria is called gyrase instead of topoisomerase II because, rather than removing negative supercoils like most type II topoisomerases, this enzyme adds them. Gyrase acts by first wrapping the DNA around itself and then cutting the two strands before passing another part of the DNA through the cuts, thereby introducing two negative supercoils. Adding negative supercoils increases the stress in the DNA and so requires energy; hence, gyrase needs ATP for this reaction.

The gyrase of *E. coli* is made up of four polypeptides, two of which are encoded by the *gyrA* gene and two of which are encoded by *gyrB*. These genes were first identified by mutations that make the cell resistant to antibiotics that affect gyrase (Table 1.2). The GyrA subunits seem to be responsible for breaking the DNA and holding it as the strands pass through the cuts. The GyrB subunits have the ATP site that furnishes the energy for the supercoiling.

The Bacterial Genome

The discussion of the replication and structure of the bacterial genome ignores the complexity of the sequences and the functions they encode. Advances in DNA-sequencing technologies are drastically reducing both the time it takes to sequence DNA and the cost of DNA sequencing. We are quickly approaching a time when determining the sequences of a newly discovered bacterium might be one of the earliest experiments one does to more fully characterize the functions of a new species. In addition, there are also a number of examples where DNA from a particular environment or a consortium of organisms is chosen for DNA sequencing instead of that of a single organism derived from pure culture. Bacterial and archaeal genomes are in some ways more amenable to genome analysis than eukaryotic genomes because they are relatively small, ranging from ca. 0.5 megabases (Mb) with only about 500 genes for some obligate parasites to around 10 Mb with 10,000 genes for some free-living bacteria. They contain few introns and much less repetitive DNA than eukaryotic genomes.

Examination of genome sequences is revealing much about the nature of bacterial genomes. For example, bacterial genomes are densely packed with coding information. The average gene density in bacterial genomes is 1.1 kb; that is, on average, one gene is found in every 1.1 kb. In contrast, the mouse and human genomes have average gene densities that are about 100-fold lower, because of all their introns and other noncoding DNA. The major types of coding information evaluated at this time encode proteins or comprise rRNAs or tRNAs, but more and more regions comprising small RNAs and encoding small peptides are being discovered. Recent research on such gene products is discussed in later chapters.

Another outcome of comparing genome sequences has been the observation of the high degree of conservation in genetic linkage, called **synteny**. All bacteria share some level of synteny over stretches of chromosome sequence, but this synteny is very obvious within a genus. Major changes that break up synteny are of two types, inversions and insertions. Inversion within bacterial genomes appears to be constrained by the nature of the genomes themselves; many sequence features in the chromosome

Table 1.2 Antibiotics that block replication

Antibiotic	Source	Target
Trimethoprim	Chemically synthesized	Dihydrofolate reductase
Hydroxyurea	Chemically synthesized	Ribonucleotide reductase
5-Fluorodeoxyuridine	Chemically synthesized	Thymidylate synthetase
Nalidixic acid	Chemically synthesized	*gyrA* subunit of gyrase
Novobiocin	*Streptomyces sphaeroides*	*gyrB* subunit of gyrase
Mitomycin C	*Streptomyces caespitosus*	Cross-links DNA

must be maintained in a gradient that progresses from the origin to the terminus of replication (Box 1.1), and natural inversions seem to always occur in a way that does not cause changes in this gradient across the genome. The syntenic regions of the chromosome are also interspersed with unique insertions of DNA sequence that were originally acquired by **horizontal transfer**, i.e., by DNA transfer from other types of bacteria rather than solely by **vertical transfer** from their ancestors. These regions are composed of prophages, insertion sequence elements, and gene clusters called genetic islands. Knowledge such as this is important to our understanding of how bacteria adapt to different environments and also may suggest ways to better understand and combat bacterial diseases, particularly emergent diseases. For example, genome sequencing has revealed why *E. coli* strains can be either "intestinal friends" or "intestinal foes." Since the significance of *E. coli* as a model organism is discussed in the text, it is important to clarify what distinguishes the disease-causing *E. coli* strains, such as the O157:H7 strain that is the causative agent of deadly infections worldwide, from the harmless *E. coli* laboratory strains, such as K-12. *E. coli* K-12 and *E. coli* O157:H7 share 4.1 Mb of DNA sequence homology, but scattered throughout the genome of O157:H7 are long DNA regions that encode virulence characteristics. This additional DNA, approximately 1 Mb, is the result of horizontal transfer. We discuss the many mechanisms of horizontal transfer in other chapters, including prophages in chapter 8 and genetic islands in chapter 9. Other types of horizontally acquired genetic information can be found in the sections concerning plasmids (chapter 4) and transposons and site-specific recombination (chapter 9).

Antibiotics That Affect Replication and DNA Structure

Antibiotics are substances that block the growth of cells. Many antibiotics are naturally synthesized chemical compounds made by soil microorganisms, especially actinomycetes, that may help them compete with other soil microorganisms. There are also other ideas as to why bacteria make antibiotics, including for their use in intercellular communication, especially in highly organized structures like biofilms. Antibiotics from actinomycetes have a particularly broad spectrum of activity and target specificity, which has made them very useful in enhancing our understanding of cellular functions, as well as in treating diseases.

Many antibiotics specifically block DNA replication or change the structure of DNA. Table 1.2 lists a few representative antibiotics that affect DNA, along with their targets in the cell and their sources. Because some parts of the replication machinery have remained relatively unchanged throughout evolution, many of these antibiotics work against essentially all types of bacteria. Some even work against eukaryotic cells and so are used as antifungal agents and in antitumor chemotherapy.

Although there are many different targets for antibiotics, it appears that many antibiotics that are capable of killing cells actually do so by causing the cell to generate hydroxyl free radicals (see chapter 11) rather than merely by inhibiting their targets (see Kohanski et al., Suggested Reading). Apparently, the various functions of the cell are so closely integrated that disrupting one of them can disrupt others, including oxidation reduction reactions by flavoproteins in electron transport, leading to excessive production of reactive oxygen species (see Box 11.1).

Antibiotics That Block Precursor Synthesis

As discussed above, DNA is polymerized from the deoxynucleoside triphosphates. Any antibiotic that blocks the synthesis of these deoxynucleotide precursors will block DNA replication.

INHIBITION OF DIHYDROFOLATE REDUCTASE

Some of the most important precursor synthesis blockers are antibiotics that inhibit the enzyme dihydrofolate reductase. One such compound, trimethoprim, works very effectively in bacteria, and the antitumor drug methotrexate (amethopterin) inhibits the dihydrofolate reductase of eukaryotes. Methotrexate is used as an antitumor agent, among other uses.

Antibiotics like trimethoprim that inhibit dihydrofolate reductase kill the cell by depleting it of tetrahydrofolate, which is needed for many biosynthetic reactions. This inhibition is overcome, however, if the cell lacks the enzyme thymidylate synthetase, which synthesizes dTMP; therefore, most mutants that are resistant to trimethoprim have mutations that inactivate the *thyA* thymidylate synthetase gene. The reason is apparent from the pathway for dTMP synthesis shown in Figure 1.5. Thymidylate synthetase is solely responsible for converting tetrahydrofolate to dihydrofolate when it transfers a methyl group from tetrahydrofolate to dUMP to make dTMP. The dihydrofolate reductase is the only enzyme in the cell that can restore the tetrahydrofolate needed for other biosynthetic reactions. However, if the cell lacks thymidylate synthetase, there is no need for a dihydrofolate reductase to restore tetrahydrofolate. Therefore, inhibition of dihydrofolate reductase by trimethoprim has no effect, thus making *thyA* mutant cells resistant to the antibiotic. Of course, if the cell lacks a thymidylate synthetase, it cannot make its own dTMP from dUMP and must be provided with thymidine in the medium so that it can replicate its DNA.

There is more than one mechanism by which cells can achieve trimethoprim resistance. They can have an altered dihydrofolate reductase to which trimethoprim cannot bind, or they can have more copies of the gene so that they make more enzyme than there is trimethoprim to inhibit it. Some plasmids and transposons carry genes for resistance to trimethoprim. These genes encode dihydrofolate reductases that are much less sensitive to trimethoprim and so can act even in the presence of high concentrations of the antibiotic.

INHIBITION OF RIBONUCLEOTIDE REDUCTASE

The antibiotic hydroxyurea inhibits the enzyme ribonucleotide reductase, which is required for the synthesis of all four precursors of DNA synthesis (Figure 1.5). The ribonucleotide reductase catalyzes the synthesis of the deoxynucleoside diphosphates dCDP, dGDP, dADP, and dUDP from the ribonucleoside diphosphates, an essential step in deoxynucleoside triphosphate synthesis. Mutants resistant to hydroxyurea have an altered ribonucleotide reductase.

COMPETITION WITH DEOXYURIDINE MONOPHOSPHATE

5-Fluorodeoxyuridine and the related 5-fluorouracil have monophosphate forms resembling dUMP, the substrate for thymidylate synthetase. By competing with the natural substrate for this enzyme, they inhibit the synthesis of deoxythymidine monophosphate. Mutants resistant to these compounds have an altered thymidylate synthetase. These are useful antibiotics for the treatment of fungal, as well as bacterial, infections.

Antibiotics That Block Polymerization of Deoxynucleotides

The polymerization of deoxynucleotide precursors into DNA would also seem to be a tempting target for antibiotics. However, there seem to be surprisingly few antibiotics that directly block this process. Most antibiotics that block polymerization do so indirectly, by binding to DNA or by mimicking the deoxynucleotides and causing chain termination, rather than by inhibiting the DNA polymerase itself.

DEOXYNUCLEOTIDE PRECURSOR MIMICS

Dideoxynucleotides are similar to the normal deoxynucleotide precursors, except that they lack a hydroxyl group on the 3′ carbon of the deoxynucleotide. Consequently, they can be incorporated into DNA, but then replication stops because they cannot link up with the next deoxynucleotide. These compounds are not useful antibacterial agents, probably because they are not phosphorylated well in bacterial cells. However, this property of prematurely terminating replication has made them the basis for one of the first methods for DNA sequencing (see below).

CROSS-LINKING

Mitomycin C blocks DNA synthesis by cross-linking the guanine bases in DNA to each other. Sometimes the cross-linked bases are in opposing strands. If the two strands are attached to each other, they cannot be separated during replication. Even one cross-link in DNA that is not repaired prevents replication of the chromosome. This antibiotic is also a useful antitumor drug, probably for the same reason. DNA cross-linking also affects RNA transcription.

Antibiotics That Affect DNA Structure

ACRIDINE DYES

The acridine dyes include proflavine, ethidium, and chloroquine. These compounds insert between the bases of DNA and thereby cause frameshift mutations and inhibit DNA synthesis. Their ability to insert themselves between the bases in DNA has made acridine dyes very useful in genetics and molecular biology. Some of these applications are discussed in later chapters. In general, acridine dyes are not useful as antibiotics because of their toxicity due to their ability to block DNA synthesis in the mitochondria of eukaryotic cells. Some members of this large family of antibiotics have long been used as antimalarial drugs because of their ability to block DNA synthesis in the mitochondria (kinetoplasts) of trypanosomes. This is the basis for the antimalarial activity of the tonic water in a gin and tonic.

THYMIDINE MIMIC

5-Bromouracil (5-BU) is similar to thymine and is efficiently incorporated in its place into DNA. However, 5-BU incorporated into DNA often mispairs and increases replication errors. DNA containing 5-BU is also more sensitive to some wavelengths of ultraviolet (UV) light (which makes 5-BU useful in enrichment schemes for isolating mutants [see chapter 3]). Moreover, DNA containing 5-BU has a different density from DNA exclusively containing thymidine (another feature of 5-BU that is useful in experiments).

Antibiotics That Affect Gyrase

Many antibiotics and antitumor drugs affect topoisomerases. The type II topoisomerase, gyrase, in bacteria is a target for many different antibiotics. Because this enzyme is similar among all bacteria, these antibiotics have a broad spectrum of activity and kill many types of bacteria.

GyrA INHIBITION

Nalidixic acid specifically binds to the GyrA subunit, which is involved in cutting the DNA and in strand passage. This activity makes nalidixic acid and its many derivatives, including oxolinic acid and chloromycetin, very useful antibiotics. Another antibiotic that binds to the GyrA subunit, ciprofloxacin, is used for treating gonorrhea, anthrax, and bacterial dysentery. However, because these antibiotics can induce prophages, they may actually make some diseases worse (see chapter 8).

The mechanism of killing by these antibiotics is not completely understood. They are known to cause degradation of DNA and can cause the DNA to become covalently linked to gyrase, presumably by trapping it in an intermediate state in the process of strand passage. Bacteria resistant to nalidixic acid have an altered *gyrA* gene.

GyrB INHIBITION

Novobiocin and its more potent relative coumermycin bind to the GyrB subunit, which is involved in ATP binding. These antibiotics do not resemble ATP, but by binding to the gyrase they somehow prevent ATP cleavage, perhaps by changing the conformation of the enzyme. Mutants resistant to novobiocin have an altered *gyrB* gene.

Molecular Biology Manipulations with DNA

The meticulous studies of the mechanism of DNA replication in bacteria and the enzymes involved in DNA replication discussed above have led to many practical applications in molecular biology. These applications have had profound effects on many aspects of our everyday lives, including medicine, agriculture, and even law enforcement. We review some of these applications in this section.

Restriction Endonucleases

Among the most useful enzymes that alter DNA are the restriction endonucleases. These are enzymes that recognize specific sequences in DNA and cut the DNA in or close to the recognition sequence. They are usually accompanied by methylating activities that modify DNA by methylating the DNA in the recognition sequence, making it immune to cutting by the endonuclease activity. These enzymes are made exclusively by bacteria, and the major role for many of them may be to defend against incoming phages and other DNA elements by cleaving unmodified DNA. Restriction endonucleases have also been suggested to persist because they act as "selfish" DNAs. Given that the effect of the endonuclease lasts longer than the methylation activity, there is selection for not losing the coding information by gene deletions; loss of the protective methylating activity can leave the whole genome sensitive to the restriction capacity of the enzyme. Restriction systems can also serve as so-called toxin-antitoxin or plasmid addiction systems that help prevent the loss of plasmids by killing the cell if it is cured of the plasmid (see Box 4.2).

The restriction endonucleases are classified into three groups, types I, II, and III. The enzymes in these groups differ mostly in the relationship between their methylating and cleaving activities. The type II enzymes have proven to be most useful. Hundreds of type II enzymes are known, and many of them can be purchased from biochemical supply companies. One thing that makes them so useful is that the methylating activity can be separated from the cleaving activity. Secondly, they each recognize their own specific sequence in DNA and then cut the DNA at or close to the recognition sequence. Thirdly, the sequences recognized by many of them are palindromic and, by making staggered breaks in these sequences, they leave complementary (sticky) ends that can be used for DNA cloning (see below). The recognition sequences are often 4, 6, or 8 bp long. The sequences recognized by some restriction endonucleases are shown in Table 1.3.

USING RESTRICTION ENDONUCLEASES TO CREATE RECOMBINANT DNA

As mentioned above, one of the properties of some type II restriction endonucleases that make them so useful is that the sequences they recognize read the same in the 5′-to-3′ direction on both strands. Such a sequence is said to have twofold rotational symmetry or to be a **palindrome** because it reads the same if you rotate it through 180° and read the other strand. The word comes from a palindrome in English, where letters read the same in both directions, as in "madam, I'm Adam." Because the sequence reads the same on both strands, the restriction endonuclease binds to the identical sequence on both strands and then cuts the two strands at the same place in the sequence. For example, the restriction endonuclease HindIII (so called because it was the third restriction endonuclease found in *Haemophilus influenzae*) recognizes the 6-bp sequence 5′-AAGCTT-3′/3′-TTCGAA-5′ and cuts between the two A's on each strand (Table 1.3), which is in the same place in the sequences of the two

Table 1.3 Recognition sequences of restriction endonucleases

Enzyme	Recognition sequence[a]
Sau3A	*GATC/CTAG*
BamHI	G*GATCC/CCTAG*G
EcoRI	G*AATTC/CTTAA*G
PstI	CTGCA*G/G*ACGTC
HindIII	A*AGCTT/TTCGA*A
SmaI	CCC*GGG/GGG*CCC
NotI	GC*GGCCGC/CGCCGG*CG

[a]Asterisks indicate where the endonucleases cut.

strands read in the 5′-to-3′ direction. Such a break is called a "staggered break" because the breaks in the two strands are not exactly opposite each other in the DNA, which has the effect of leaving short single-stranded ends on both ends of the broken DNA. Because the original sequence that was cut had a twofold rotational symmetry, the two single-stranded ends are complementary to each other and so can pair with each other. More importantly, each can pair with the single-stranded end of any other DNA that was cut with the same restriction endonuclease, because they all have single-stranded ends with the same sequence. These single-stranded ends are called **sticky ends** because they can pair with ("stick to") any other single-stranded ends with the complementary sequence. Other restriction endonucleases might leave the same sticky ends even if they recognize a somewhat different sequence. Such restriction endonucleases are said to be compatible. When two sticky ends pair with each other, they leave a double-stranded DNA with staggered nicks in the two strands, which can then be sealed by DNA ligase, as illustrated in Figure 1.27. The new DNA that has been created in this way is called **recombinant DNA** because two DNAs have been recombined into new sequence combinations.

Figure 1.27 Creation of sticky complementary ends by cutting with a restriction endonuclease. The two single-stranded ends can pair with each other, and the nicks can be sealed by DNA ligase.
doi:10.1128/9781555817169.ch1.f1.27

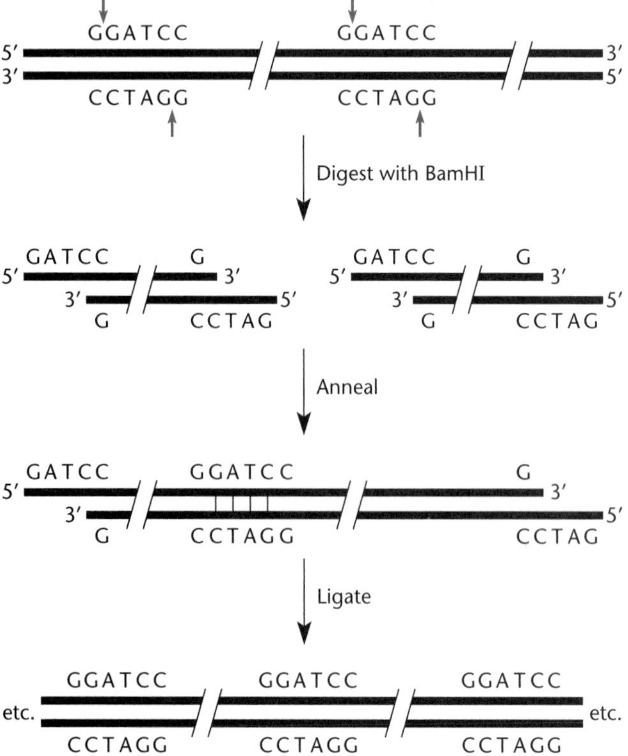

CLONING AND CLONING VECTORS

There is only one molecule of a recombinant DNA when it first forms. In order for it to be useful, many copies of the recombinant DNA molecule are needed. This is the function of cloning vectors. A **cloning vector** is a DNA that has its own origin of replication and is capable of independent replication in the cell, for example, plasmids. DNAs that have an *ori* sequence that makes them capable of independent replication in the cell are called replicons. The process of cloning a piece of DNA into a circular plasmid with its own *ori* sequence is illustrated in Figure 1.28. Once it has been joined to another piece of DNA and introduced into a cell, the cloning vector replicates itself, along with the piece of DNA to which it is joined, making many exact copies of the original DNA molecule. These exact replicas of the piece of DNA are called **DNA clones** in analogy to the genetic replicas of an organism that are made when an organism replicates itself asexually. Phages also are capable of independent replication in their bacterial hosts, and some of these have been modified to serve as convenient cloning vectors. Some examples of cloning vectors and their relative advantages are discussed in subsequent chapters.

Figure 1.28 DNA cloning. The compatible restriction endonucleases Sau3A and BamHI were used to clone a piece of DNA into a cloning vector. The DNA to be cloned was cut with Sau3A and ligated into a cloning vector cut with BamHI. The piece of DNA inserted into the cloning vector cannot replicate by itself, since it lacks an *ori* region; however, once it is inserted into the cloning vector, it replicates each time the cloning vector replicates.
doi:10.1128/9781555817169.ch1.f1.28

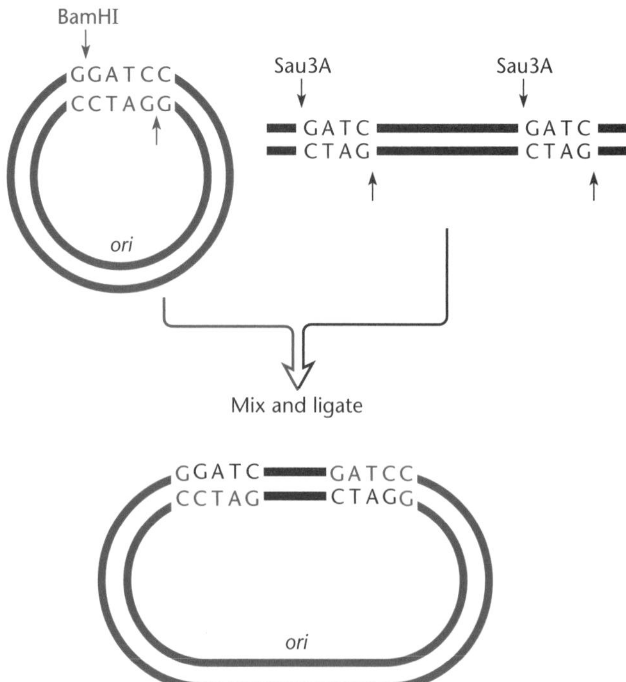

DNA LIBRARIES

A **DNA library** is a collection of DNA clones that includes all, or at least almost all, the DNA sequences of an organism. One way to make a DNA library is with restriction endonucleases. The entire DNA of an organism is cut with restriction endonucleases, and the pieces are ligated into a cloning vector cut with a compatible enzyme. The mixture is then introduced into cells by transformation; if a phage is used for cloning, the process of introducing the DNA into hosts is called transfection and the resulting genetically distinct clones are isolated as plaques (see chapter 8). If the collection is large enough, every DNA sequence of the organism is represented somewhere in the pooled clones, and the library is said to be complete. The trick then is to find the clone you want out of all the clones in the library; some methods to do this are mentioned here and in other chapters.

PHYSICAL MAPPING

Another important use of restriction endonucleases is in physical mapping. By analogy to genetic mapping, where the approximate positions of mutations are determined by genetic crosses (see the discussion of genetic analysis in chapter 3), **physical mapping** is the process of determining the exact positions of particular sequences in the nucleotide sequence of the DNA. Because restriction endonucleases cut the DNA only at specific sequences, they create unique-sized pieces for each DNA molecule. From the sizes of these pieces, it is possible to determine where the recognition sequences for the restriction endonuclease must have been on the original DNA. By comparing the sizes of the pieces left by a number of different restriction endonucleases, it is possible to order the restriction sites with respect to each other and construct a **physical map** of the DNA. Figures 1.29 and 1.30 illustrate the reasoning behind the physical mapping of restriction sites in a DNA.

RESTRICTION SITE POLYMORPHISMS

While all the members of a particular species are very similar genetically, there are minor differences, called polymorphisms, between individuals. These minor genetic

Figure 1.29 Representation of agarose gel electrophoresis of fragments of DNA. Smaller fragments migrate faster on the gel and so move farther in the same amount of time. The outside lanes are marker DNAs of known size for comparison.
doi:10.1128/9781555817169.ch1.f1.29

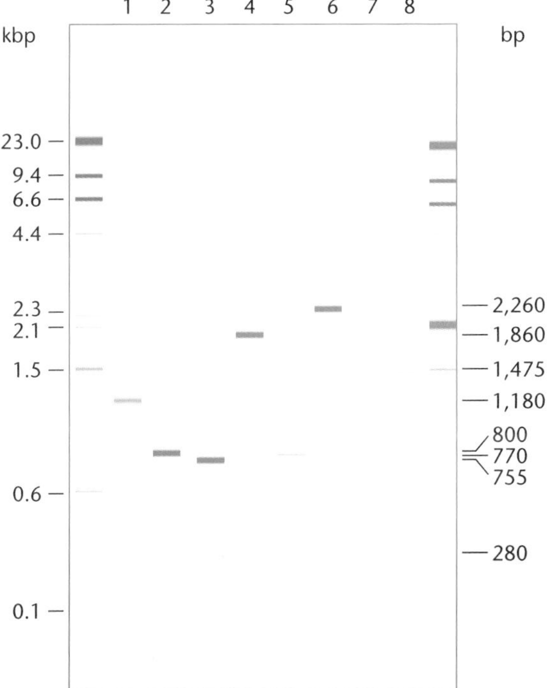

Figure 1.30 A method for mapping the restriction sites on a DNA fragment. The original fragment of 6.6 kb (6.6×10^3 bp) contains two recognition sites for the restriction endonuclease HindIII and two sites for PstI. **(A)** The fragment is digested with HindIII and PstI separately, and the isolated fragments are digested with the other restriction endonuclease (blue).
(B) From the sizes of the fragments, the order of the sites must have been as shown.
doi:10.1128/9781555817169.ch1.f1.30

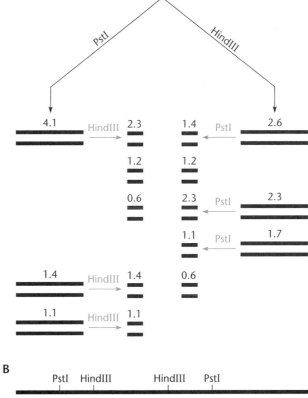

A

B

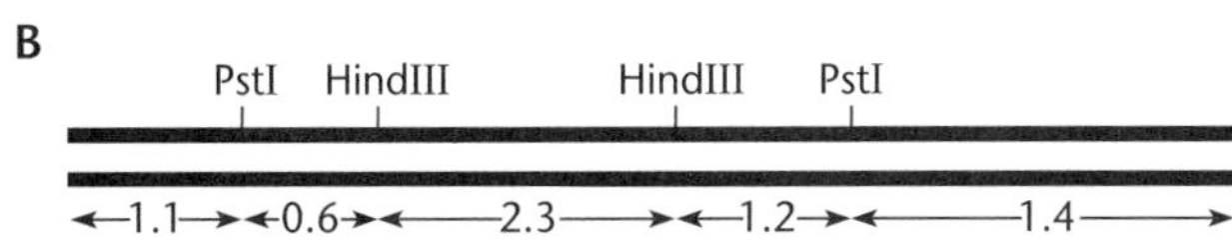

differences are reflected in differences in the sequence of DNA from each organism, in particular in differences in the locations of restriction sites. These differences are called **restriction fragment length polymorphisms** and can be due to deletions or insertions between the positions of the sites, inversions of DNA containing the sites, or mutation of the restriction site itself. Restriction fragment length polymorphisms in the DNA can be useful for determining the ancestry of a particular individual, mapping genetic diseases, and, in forensic science, identifying the person who committed a crime and left behind some blood or other material containing their DNA. Advances in DNA sequencing may lead to widespread use of direct DNA sequencing as a tool to detect single-nucleotide polymorphisms in many applications.

Hybridizations

Many applications in molecular genetics have come from our knowledge of the structure of DNA and how it is synthesized and held together. The two strands of DNA are held together in the double helix by hydrogen bonds between the complementary bases. Heating double-stranded DNA or treating it at high pH disrupts these hydrogen bonds and causes the two strands to separate. If the temperature is then lowered or the pH is returned to neutral, the complementary sequences eventually find each other and a new double helix is formed. Two strands of RNA or a strand of DNA and a strand of RNA can also be held together by such a double-stranded helix, provided that their sequences are complementary. This process is called **hybridization**. Under optimal conditions, two strands of DNA or RNA form a double-stranded helix only if their sequences are almost perfectly complementary, making hybridization very specific and sensitive and allowing the detection of RNA or DNA of a particular sequence among thousands of other sequences.

Most methods of hybridization utilize a membrane filter made of a form of nylon or some other related compound. First, a single-stranded DNA or RNA is fixed to the membrane. Then, a solution containing another DNA or RNA is added to the membrane. If the second RNA or DNA hybridizes to the RNA or DNA fixed to the membrane, it binds to that site on the membrane, as well. The hybridization can be detected provided that the second DNA or RNA has been marked or labeled somehow, for example, with radioactivity or fluorescent chemicals. Similar techniques can be used to detect the binding of specific antibodies to proteins fixed on a filter or even the binding of proteins to DNA fixed on a filter, and vice versa.

Some of the most useful techniques in molecular biology involve filter hybridization (Figure 1.31), in which the membrane filter receives a replica of the molecules on a gel. In these experiments, the filter is layered on the gel and the macromolecules (DNA, RNA, or proteins) are transferred to the same position on the filter that they were in on the gel. The transfer can be by diffusion, capillary action, or use of an electric field, depending on the application. Transfer of DNA, RNA, or proteins from a gel to a filter is called **blotting**, and the filter containing the replica is a blot. The **blot** can then be hybridized to a labeled probe to determine the locations of particular DNA or RNA sequences or proteins on the original gel.

Figure 1.31 Hybridization methods. **(A)** Method of Southern blot hybridization. In step 1, DNA is isolated and digested with a restriction endonuclease. In steps 2 and 3, after electrophoresis (step 2), the DNA is transferred and fixed to a filter (step 3). In step 4, the filter is hybridized with a probe. Only bands complementary to the probe are dark, because the signal detection procedure reveals the radioactivity or reactive chemical in the probe. **(B)** Plate hybridizations. The colonies, or plaques, on a plate are transferred to a membrane filter, and the DNA is denatured. The DNA on the filters is then hybridized to a labeled probe as in a Southern blot hybridization to identify the colonies or plaques that contain DNA complementary to the probe.
doi:10.1128/9781555817169.ch1.f1.31

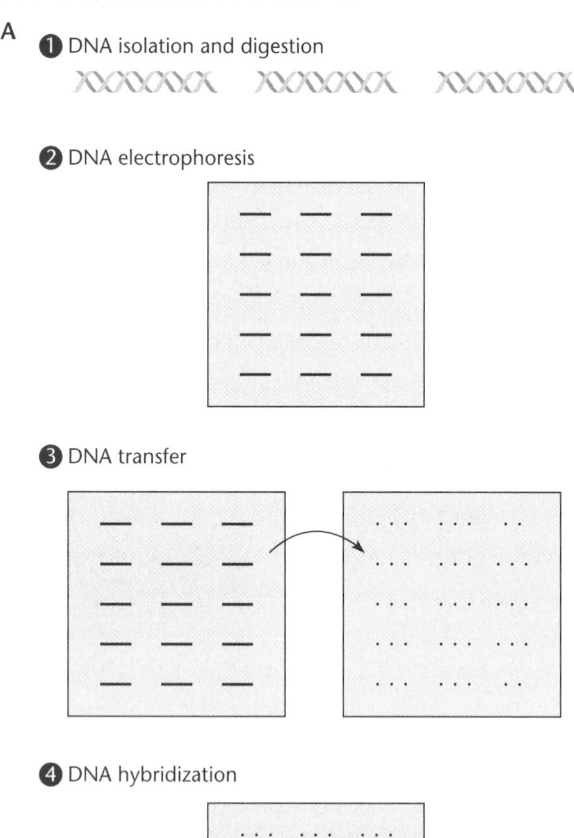

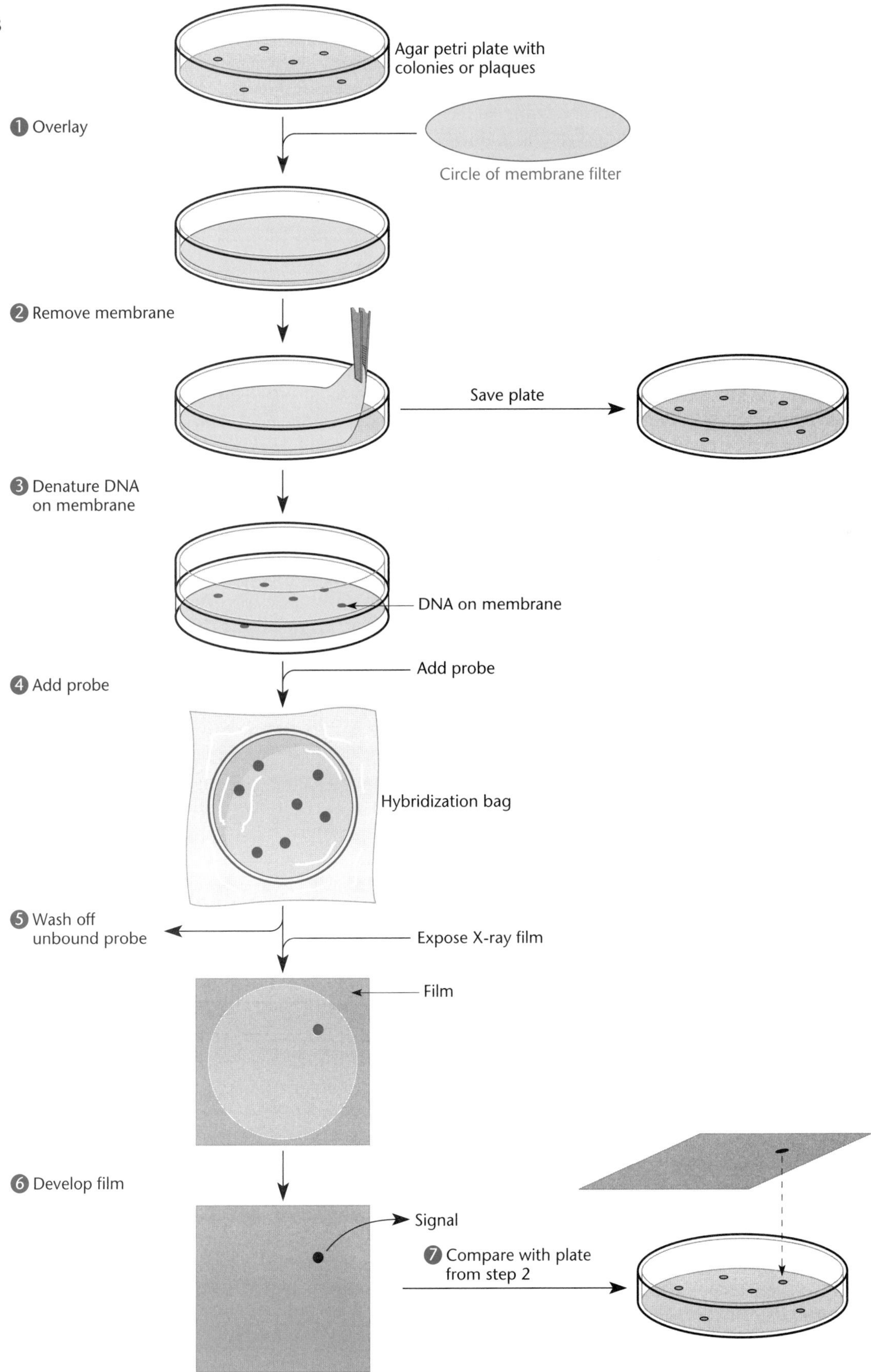
B
Agar petri plate with colonies or plaques
1 Overlay
Circle of membrane filter
2 Remove membrane
Save plate
3 Denature DNA on membrane
DNA on membrane
4 Add probe
Add probe
Hybridization bag
5 Wash off unbound probe
Expose X-ray film
Film
6 Develop film
Signal
7 Compare with plate from step 2

The first such procedure for transferring DNA bands from a gel to a filter was named **Southern blotting** because it was developed by Edwin Southern. The principle behind Southern blotting is shown in Figure 1.31. When similar procedures were developed for blotting RNA and proteins, they were whimsically given the names of other directions on the compass: **Northern blotting** for RNA blots, **Western blotting** for protein blots, and so on. For detailed protocols of this and other types of blotting, consult a cloning manual (see Sambrook and Russell, Suggested Reading).

Applications of the Enzymes Used in DNA Replication

As mentioned above, meticulous work on the properties of the enzymes involved in DNA replication has not only increased our understanding of DNA replication, but has also led directly to many important practical applications. As discussed in the introduction, many of these enzymes were first detected and purified from bacteria and phage-infected bacteria, but their applications extend to the molecular genetics of all organisms. A few of the more prominent applications are discussed here.

DNA POLYMERASES

The properties of DNA polymerases have been exploited in many applications in molecular genetics. As discussed earlier in this chapter, these enzymes all extend a primer polynucleotide chain by attaching the 5′ phosphate of an incoming deoxynucleotide triphosphate to the 3′ end of the growing primer chain. They can synthesize DNA only by extending primers that are hybridized to a template DNA, and the choice of which deoxynucleotide to add at each step is determined by complementary base pairing between the incoming deoxynucleotide and the template DNA, leading to synthesis of DNA that is a complementary copy of the template.

DNA SEQUENCING

One important application of DNA polymerases is in DNA sequencing, the process of determining the sequence of deoxynucleotides in DNA. DNA-sequencing technologies are advancing at an incredible rate. The first human genome, containing many billions of deoxynucleotides, was essentially completed in 2003, but the sequencing of 1,000 human genomes is now complete. Realistic technological pursuits aim to reduce the cost of sequencing an entire human genome to around $1,000, a cost that will in theory make DNA sequencing practical as a tool for individualized health care.

The principle behind DNA sequencing is quite simple, involving knowledge of the properties of DNA polymerases. In the original Sanger sequencing method, the chain-terminating dideoxynucleotides were used, where the absence of the 3′ OH prevents further extension of the DNA chain. Four reactions were run, each one containing a mixture of the normal deoxynucleotide and the corresponding dideoxynucleotide, with the template DNA being sequenced, as well as a primer complementary to a sequence on the DNA close to the region being sequenced. Each time the dideoxynucleotide was incorporated rather than the normal deoxynucleotide, the chain terminated. If the four reactions were then run next to each other on an acrylamide gel, a "sequencing ladder" was obtained. Which lane had the next "rung" or "step" in the ladder told you what nucleotide was next in the template DNA that was being sequenced. An advance in this strategy involved the ability to have each of the four dideoxynucleotides labeled with a different fluorescent molecule, each emitting light at a different frequency so each fragment of DNA is marked with a certain color at the terminal nucleotide. This technology allows all of the "rungs" on the ladder to be read in one lane.

DNA sequencing with the Sanger sequencing method, which takes advantage of terminating a strand of DNA using a dideoxynucleotide, continues to be widely used. However, newer high-throughput techniques are now being developed for ever-larger sequencing projects involving many thousands of much shorter templates being read at the same time, a process called massive parallel sequencing. These technologies also use the basic properties of DNA polymerases. Newer technologies use a variety of mechanisms to ascertain DNA sequence while DNA synthesis takes place, a process sometimes referred to as pyrosequencing. This general theme has been adapted in many different ways, and exciting technologies that promise to reduce the time and cost of DNA sequencing are on the horizon (Box 1.2).

Polymerase Chain Reaction

One of the most useful technical applications involving DNA polymerases is the polymerase chain reaction (PCR). This technology makes it possible to selectively amplify regions of DNA out of much longer DNAs. It is called PCR because each newly synthesized DNA serves as the template for more DNA synthesis in a sort of chain reaction until large amounts of DNA have been amplified from a single DNA molecule. The power of this method is that it can be used to detect and amplify sequences from just a few molecules of DNA from any biological specimen, for example, a drop of blood or a single hair; this has made it very useful in criminal investigations to identify the perpetrators of crimes on the basis of DNA typing. However, for our purposes here, it also has many other applications, including the physical mapping of DNA, gene cloning, mutagenesis, and DNA sequencing.

The principles behind the use of PCR to amplify a region of DNA are outlined in Figure 1.32. PCR takes

BOX 1.2

Advanced Genome-Sequencing Techniques

Initial genome-sequencing efforts involved sequencing individual fragments of the genome cloned into plasmids or phage. For larger, genome-size DNA molecules, this involved sequencing a huge collection of random clones of the genome. By collecting many times more DNA sequence than the actual size of the genome, enough overlapping DNA sequences are collected to reconstruct the virtual genome using computers. Other strategies let researchers fill in the gaps where information was missing to produce a continuous data set that represented the entire genome of the bacterium. The first genome sequence from a free-living organism was that of *H. influenzae*, published in 1995 using this strategy. Newer technologies have entirely removed the cloning step, so that DNA is sequenced directly as thousands of randomly sheared pieces of the original. While these sequencing runs are short (<200 bp), the sheer number of random sequences collected allows enough overlapping pieces of information to assemble the sequence to put entire genomes together or to address other biological questions. The strategy of sequencing thousands of short reads at one time is referred to as massive parallel sequencing and involves a variety of technologies with some commonalities.

Popular DNA-sequencing strategies involve fragmenting the DNA substrate (A and B in the figure), cleaning up the ends, and ligating short synthetic DNAs to both ends of the fragments (C). The DNAs are fixed to a solid substrate, where they are amplified in a way that yields discernible spots of clusters or wells, each with the same DNA (amplification is not shown in the figure) (D). Nucleotides flow over the spots, allowing DNA polymerase to integrate the complementary base into the substrates (E). A sensitive detection device is able to read the light or fluorescent signal from each spot that corresponds to individual bases. Again, while the read length of each individual DNA is short, many thousands of fragments can be read at the same time, allowing the collection of a massive amount of DNA sequence information. Software that can identify overlap between the sequences allows the short reads to be compiled into longer strings of sequence information.

Exciting techniques on the horizon involve more sensitive detection devices that will allow a single strand of DNA to be read. This process basically allows DNA sequencing approaching the real-time speed of the polymerase. Because the thousands of replicating polymerases are read at one time, this type of technology promises to allow much faster reads, also in a massively parallel fashion.

Technologies that reduce the cost and time needed to do DNA sequencing allow its broader application beyond genome sequencing. Massive parallel sequencing can be used in other types of experiments. For example, the gene expression profile of an organism can be determined by converting all of the RNA transcripts to DNA that can be sequenced. Massive parallel sequencing can also determine protein occupancy across a genome in a procedure called ChIP-Seq, for *ch*romatin *i*mmuno*p*recipitation and *seq*uencing. In the first step, the protein of interest is chemically cross-linked to the host chromosome in living cells. The genome is fragmented, and the protein is selectively pulled from the population of cells using an antibody that recognizes the protein. Chromosomal fragments that bound the protein can be recovered by reversing the chemical cross-link, freeing the thousands of DNA fragments, which are then sequenced using the techniques described above.

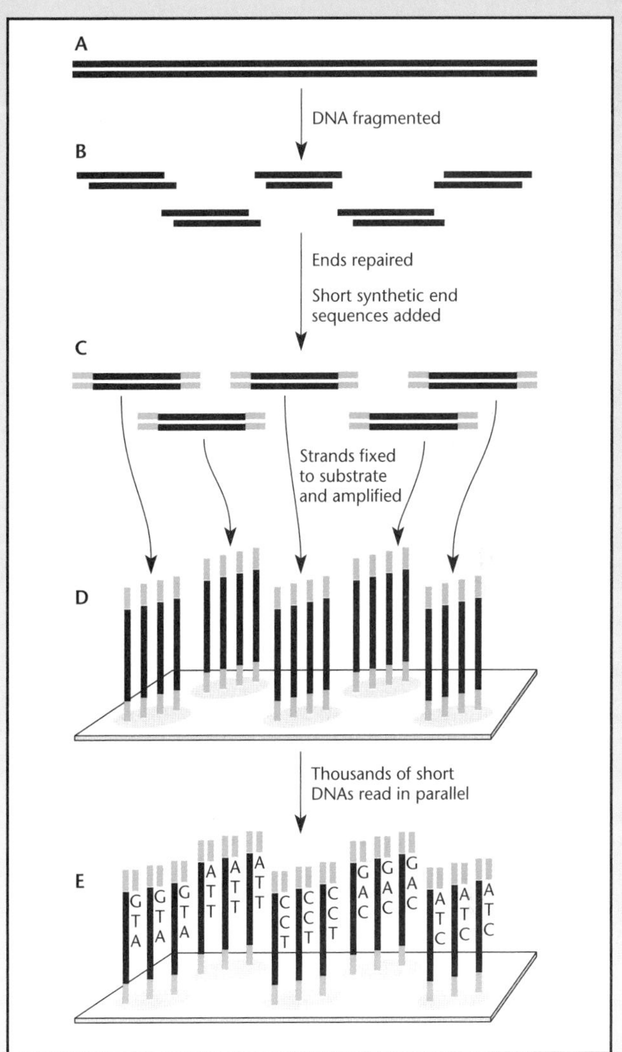

doi:10.1128/9781555817169.ch1.Box1.2.f

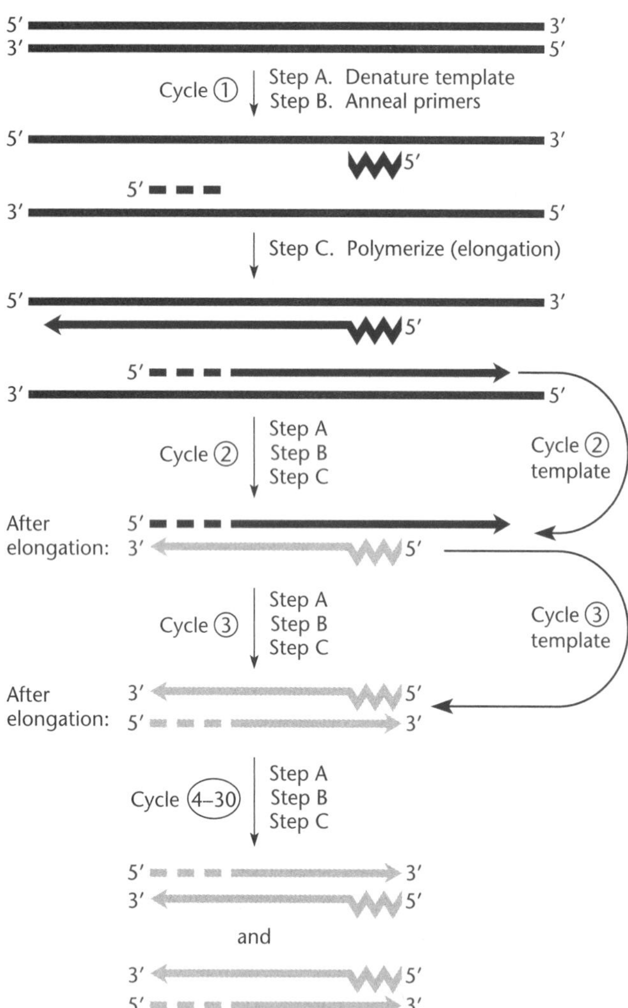

Figure 1.32 Steps in a PCR. In the first cycle, the template is denatured by heating. Primers are added; they hybridize to the separated strands for the synthesis of the complementary strand. The strands of the DNA are separated by the next heating cycle, and the process is repeated. The DNA polymerase survives the heating steps because it is from a thermophilic bacterium. The DNA sequence is amplified approximately a billionfold.
doi:10.1128/9781555817169.ch1.f1.32

advantage of the same properties of DNA polymerases that are important in other applications, i.e., their ability to make a complementary copy of a DNA template starting from the 3′ hydroxyl of a primer DNA. PCR uses two primers complementary to sequences on either side of the region to be amplified. One primer has the same sequence in the 5′-to-3′ direction as one of the strands on one side of the region to be amplified, and the other has the sequence complementary to this strand on the other side of the region to be amplified, but written in the opposite direction. Thus, the two primers prime the synthesis of DNA in opposite directions over the region to be amplified. The DNA is denatured to separate the strands and hybridized to the primers. One primer primes the synthesis of DNA over the region to be amplified and continues polymerizing past the region. If the two strands are again separated by heating and the temperature is again lowered, the other primer can then hybridize to this newly synthesized strand and prime replication back over the region. Now, however, the DNA polymerase runs off the end of the template DNA when it reaches the end of the first primer sequence, leading to the synthesis of a short piece of DNA with the primer sequences on both ends. If this DNA is then heated to separate the strands, this shorter piece can bind another primer DNA, which can then prime the synthesis of the complementary strand of the shorter DNA, and so on. This process of heating and cooling can be repeated 30 or 40 times until large numbers of copies of the particular DNA region have accumulated, beginning from one or very few longer DNA molecules containing the region.

In principle, any DNA polymerase could be used to perform PCR. However, most DNA polymerases would be inactivated by the high temperatures required to separate the strands of double-stranded DNA, making it necessary to add fresh DNA polymerase after each heating step. This is where the DNA polymerases from thermophilic bacteria, such as *Thermus aquaticus*, come to the rescue. These bacteria normally live at very high temperatures, so their DNA polymerase, called the *Taq* polymerase in the case of *T. aquaticus*, can survive the high temperatures needed to separate the strands of DNA, obviating the need to add new DNA polymerase at each step. We can just mix the primers, a tiny amount of biological material containing DNA with the region to be amplified, and the *Taq* polymerase; set a thermocycler to heat and cool over and over again; and come back a few hours later. Voilà, we should have large amounts of the region of the amplified DNA, which we can detect on a gel.

PCR MUTAGENESIS

In addition to accurate amplification of a DNA segment, PCR can be used either to make specific changes in a DNA sequence or to randomly mutagenize a region of DNA. Site-specific mutagenesis allows the investigator to make a desired change at a particular site in the sequence of DNA rather than relying on more traditional methods of mutagenesis that more or less randomly cause mutations and do not target them to a particular site (see chapter 3). Methods for site-specific mutagenesis rely on synthetic DNA primers that are mostly complementary to the sequence of the DNA being mutagenized, except for the desired mutational change. When this primer is hybridized to the DNA and used to prime the synthesis of new DNA by the

DNA polymerase, the synthesized DNA has the same sequence as the template, except for the change in the attached primer. This method of site-specific mutagenesis can be used only to make minor changes, such as single-base-pair changes, in the DNA sequence, because if the sequence of the primer is altered too much, it no longer hybridizes to the template DNA.

PCR can be used to make random changes in a sequence, because the *Taq* polymerase makes many mistakes, particularly in the presence of manganese ions, since it lacks an editing function. In fact, the mistake level during normal amplification by *Taq* polymerase is so high that new forms of thermally stable DNA polymerase are now typically used for amplifying DNA fragments that will be used for cloning. These newer polymerases are also from organisms that live at high temperature, but they possess a proofreading capacity. Regardless, clones made from PCR fragments are still normally subjected to DNA sequencing to be certain that no unwanted mutations have been introduced.

Cloning of PCR-Amplified Fragments

PCR is also useful for adding sequences, such as restriction sites for cloning, to the ends of the amplified fragment. Although the primers used for PCR amplification must be complementary to the sequence being amplified at the 3′ end, they need not be complementary at the 5′ end. Therefore, the 5′ end of the primer sequence can include, for example, the recognition sequence for a specific restriction endonuclease, making it easier to clone the PCR-amplified fragment (see above). Figure 1.33 illustrates how we can use PCR amplification to introduce restriction sites at the ends of an amplified fragment, for example, to create a fusion in an expression vector

Figure 1.33 Methods for cloning PCR-amplified fragments. **(A)** Use of PCR to add convenient restriction sites to the ends of an amplified fragment. The primers contain sequences at their 5′ ends that are not complementary to the gene to be cloned but rather contain the sequence of the cleavage site for BamHI (underlined). The amplified fragment contains a BamHI cleavage site at both ends (see the text for details). Extra random bases are added at the 5′ ends, as shown, so that the site can be cut in the amplified fragment. Some restriction endonucleases do not cut a site that begins at the end of the DNA. **(B)** The recognition site of Topo I from vaccinia virus. It breaks the phosphodiester bond 3′ of the second T in its recognition sequence (blue), transferring the phosphate bond to a tyrosine in the Topo I enzyme. This causes the dissociation of the end of the DNA, leaving a single-stranded 5′ overhang, as shown. Any other DNA with a complementary 5′ single-stranded overhang can pair with this DNA and be ligated to it by the Topo I enzyme. (C/T) means that the base can be either C or T in the recognition sequence, and N means any base. Details are given in the text. doi:10.1128/9781555817169.ch1.f1.33

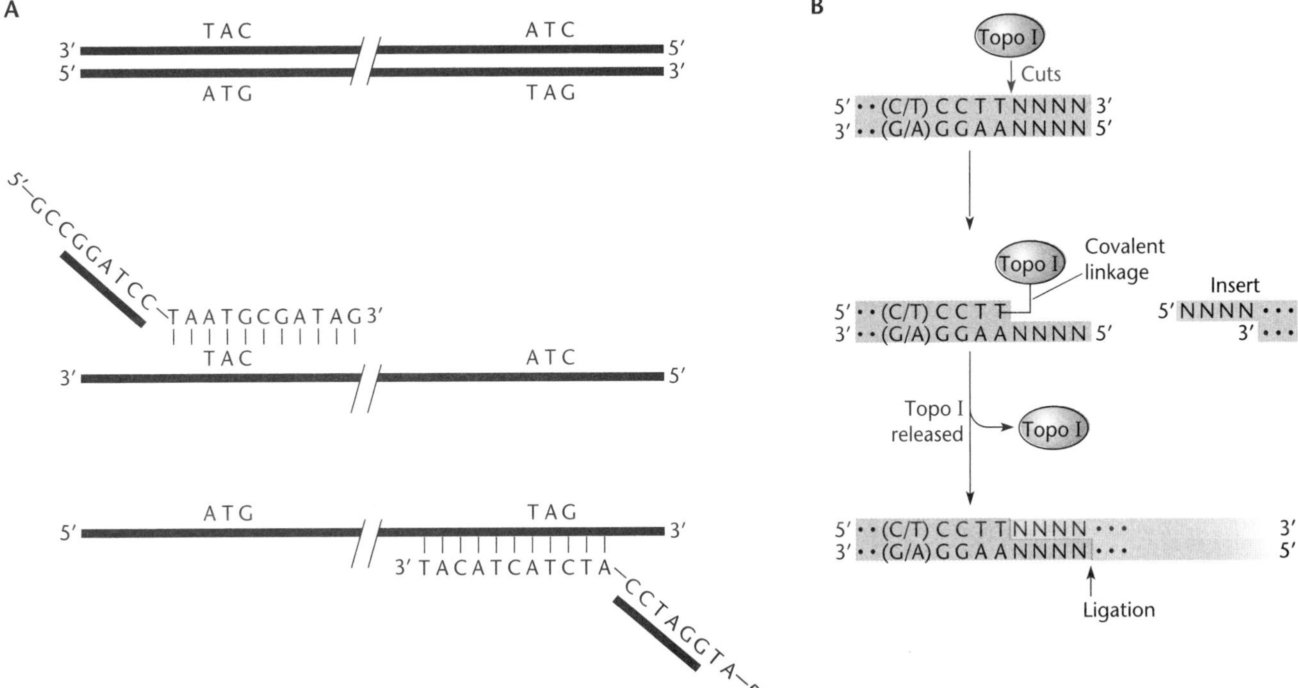

(see Figure 2.45). PCR fragments can also be cloned as blunt-ended fragments into specialized plasmid cloning vectors; this is described in chapter 4.

Topo I Cloning

One problem with the above-mentioned cloning methods is that they all involve ligation steps. Even under optimal conditions, ligation is inefficient, which limits the numbers of clones that can be obtained. This is a particular problem in some applications, such as making libraries for genomic sequencing. A more efficient system for doing cloning that does not rely on ligation is called **Topo cloning** because it relies on the type I topoisomerase (Topo I) of vaccinia virus. Like other type I topoisomerases, this enzyme makes a single-strand break in one strand of DNA and holds the broken ends while the other strand passes through the break. The break is then rejoined, thereby introducing or removing supercoils in the DNA one at a time (see above). However, unlike most topoisomerases, Topo I of vaccinia virus has strong sequence specificity and creates breaks only next to the 5-bp sequence shown in Figure 1.33B. It makes a break next to the 3′ T in the sequence and remains attached to it through the phosphate bond to one of its tyrosines to form a 3′ phosphoribosyltyrosine bond, much like Y recombinases (see Figure 9.26). Normally, the DNA would then rotate around the other strand and the topoisomerase would religate the strands of DNA to remove a supercoil. However, if the topoisomerase has cut the DNA too close to the end (within 10 bp of the end), the DNA falls apart and the topoisomerase remains joined to the T with a 5′ single-strand overhang, as shown. Note that this resembles the overhang left after a restriction endonuclease makes a staggered cut in the DNA. If another DNA with the complementary 5′ **overhang** pairs with this DNA, the topoisomerase will ligate the two ends, joining the two DNAs.

To use this technology, a plasmid cloning vector that has two recognition sites for the topoisomerase is constructed. This vector is cut on the 3′ side of the recognition sites, for example, with a restriction endonuclease, so that the desired overhang sequences remain after the topoisomerase acts. The topoisomerase is then added; it cuts the DNA, leaves the desired overhangs, and remains attached to the DNA. This activated vector can then be mixed with any DNA fragment with the same overhang sequences, and the bound topoisomerases insert the fragment into the vector with high efficiency. Once the activated vector has been prepared, it can be used to clone many fragments with high efficiency. One particularly useful application depends on the fact that the *Taq* polymerase that is commonly used in PCR amplifications naturally leaves the single base, A, as a 5′ overhang in the amplified fragment because the *Taq* polymerase also has a terminal transferase activity. If the activated vector has been constructed so that it has a single base, T, as a 5′ overhang, any PCR-amplified fragment can be effectively cloned into it.

Topo cloning has the disadvantage that preparing the activated vector is time-consuming and technical, limiting the choice of vectors that can be used. In some applications, it is necessary to clone a fragment into a number of vectors, for example, to try fusing a protein encoded by the fragment to a number of different affinity tags to see which one works best to purify that particular protein. The original vector used for Topo cloning can be engineered so that convenient restriction sites that can be used to move the cloned fragment into other vectors bracket the cloned fragment. Alternatively, the original vector used for Topo cloning can be engineered to use with other technologies, such as Gateway technology, which is based on the integrase and excisase of lysogenic phage (see chapter 8), which efficiently transfer the cloned fragment into other vectors.

SUMMARY

1. DNA consists of two strands wrapped around each other in a double helix. Each strand consists of a chain of nucleotides held together by phosphates joining their deoxyribose sugars. Because the phosphate joins the third carbon of one sugar to the fifth carbon of the next sugar, the DNA strands have directionality, or polarity, and have distinct 5′ phosphate and 3′ hydroxyl ends. The two strands of DNA are antiparallel, so that the 5′ end of one is on the same end as the 3′ end of the other.

2. DNA is synthesized from the precursor deoxynucleoside triphosphates by DNA polymerase. The first phosphate of each nucleotide is attached to the 3′ hydroxyl of the next deoxynucleotide, giving off the terminal two phosphates to provide energy for the reaction.

3. DNA polymerases require both a primer and a template strand. The pairing of the bases between the incoming deoxynucleotide and the base on the template strand dictates which deoxynucleotide will be added at each step, with A always pairing with T and G always pairing with C. The DNA polymerase synthesizes DNA in the 5′-to-3′ direction, moving in the 3′-to-5′ direction on the template.

4. DNA polymerases cannot put down the first deoxynucleotide, so RNA is usually used to prime the synthesis of

SUMMARY (continued)

a new strand. Afterward, the RNA primer is removed and replaced by DNA, using upstream DNA as a primer. The use of RNA primers helps reduce errors by allowing editing.

5. DNA polymerase does not synthesize DNA by itself but needs other proteins to help it replicate DNA. These other proteins are helicases that separate the strands of the DNA, ligases to join two DNA pieces together, primases to synthesize RNA primers, and other accessory proteins to keep the DNA polymerase on the DNA and reduce errors.

6. Both strands of double-stranded DNA are usually replicated from the same end, so that the overall direction of DNA replication is from 5′ to 3′ on one strand and from 3′ to 5′ on the other strand. Because DNA polymerase can polymerize only in the 5′-to-3′ direction, it must replicate one strand in short pieces and ligate them afterward to form a continuous strand. The short pieces are called Okazaki fragments. The two DNA polymerases replicating the leading and lagging strands remain bound to each other in a process called the trombone model of replication.

7. Lesions on DNA, proteins bound to DNA, and transcription can stall DNA replication. Primase can reinitiate replication to prevent the entire holoenzyme from stalling but leaving a gap that is repaired by other mechanisms. DNA helicases and other enzymes can be used to help DNA polymerase past bound proteins and to allow the polymerase to continue after it stalls from collisions with DNA polymerase.

8. The DNA in a bacterium that carries most of the genes is called the bacterial chromosome. The chromosome of most bacteria is a long, circular molecule that replicates in both directions from a unique origin of replication, *oriC*. Replication of the chromosomes initiates each time the cells reach a certain size. For fast-growing cells, new rounds of replication initiate before old ones are completed. This accounts for the fact that fast-growing cells have a higher DNA content than slower-growing cells.

9. Chromosome replication terminates, and the two daughter DNAs separate, when the replication forks meet. In some bacteria, multiple *ter* sites that act as "one-way gates" delay movement of the replication forks on the chromosome. Proteins that are inhibitors of the DnaB helicase stop replication at these sites.

10. To separate the daughter DNAs after replication, dimerized chromosomes, created by recombination between the daughter DNAs, are resolved by XerC and XerD, a site-specific recombination system that promotes recombination between the *dif* sites on the daughter chromosomes. The FtsK protein is a DNA translocase that promotes XerC-XerD recombination at *dif* sites to prevent dimerized chromosomal DNA from being guillotined by the forming septum. Topo IV decatenates the intertwined daughter DNAs by passing the double-stranded DNAs through each other, and its activity is also regulated temporally and spatially via an interaction with FtsK.

11. The daughter chromosomes are segregated by condensing the DNAs through supercoiling by DNA gyrase and by condensins and kleisins that hold the DNA in large supercoiled loops.

12. The FtsZ protein forms a ring at the midpoint of the cell, attracting other proteins, which form the division septum.

13. The Min proteins prevent the formation of FtsZ rings anywhere in the cell other than in the middle. Nucleoid occlusion proteins prevent the formation of FtsZ rings over nucleoids.

14. Initiation of a round of chromosome replication occurs once every time the cell divides. Initiation occurs when the ratio of DnaA protein to origins of replication reaches a critical number. After replication initiates, the DnaA protein is diluted out by having more DNA to bind to, and its ATPase is activated by interaction with the β clamp and Hda to prevent reinitiation. In some bacteria, including *E. coli* and related enteric bacteria, new initiations are prevented by hemimethylation of the newly replicated DNA at the origin and by sequestration until the replication fork has left the origin.

15. The chromosomal DNA of bacteria is usually one long, continuous circular molecule about 1,000 times as long as the cell itself. This long DNA is condensed in a small part of the cell called the nucleoid. In this structure, the DNA loops out of a central condensed core region. Some of these loops of DNA are negatively supercoiled. In *E. coli*, most DNAs have one supercoil about every 300 bases.

16. The enzymes that modulate DNA supercoiling in the cell are called topoisomerases. There are two types of topoisomerases in cells. Type I topoisomerases can remove supercoils one at a time by breaking only one strand and passing the other strand through the break. Type II topoisomerases remove or add supercoils two at a time by breaking both strands and passing another region of the DNA through the break. The enzyme responsible for adding the negative supercoils to DNA in bacteria is a type II topoisomerase called gyrase. Topo IV decatenates daughter DNAs after replication.

17. Some antibiotics block DNA replication or affect the structure of DNA. The most useful of these inhibit the synthesis of deoxynucleotides or inhibit the gyrase enzyme. Antibiotics that inhibit deoxynucleotide synthesis include inhibitors of dihydrofolate reductase (trimethoprim and methotrexate) and inhibitors of ribonucleotide reductase (hydroxyurea). Inhibitors of gyrase include novobiocin, nalidixic acid, and ciprofloxacin. Acridine dyes also affect

(continued)

SUMMARY (continued)

the structure of DNA by intercalating between the bases and are used as antimalarial drugs, as well as in molecular biology.

18. Type II restriction endonucleases recognize defined sequences in DNA and cut at a defined position in or near the recognition sites, which has made them very useful in physical mapping of DNA and in DNA cloning. Most type II restriction endonucleases cut at the same position in both strands of symmetric sequences. If the cuts are not immediately opposite each other, single-strand sticky ends will form that can hybridize to any other sticky end cut with the same or a compatible enzyme. This property has made these enzymes very useful for DNA cloning and for DNA manipulations in vitro.

19. The physical map of a DNA shows the actual location of sequences in the DNA, including the positions of recognition sequences for restriction endonucleases.

20. Cloned DNA consists of multiple copies of a DNA sequence descended from a single molecule. If a piece of DNA is inserted into a cloning vector, the DNA will replicate along with the vector, making millions of clones of the original DNA.

21. A DNA library is a collection of clones that, among themselves, contain all the DNA sequences of the organism.

22. Knowledge of the properties of the enzymes involved in DNA replication has led to many applications, including DNA sequencing, site-specific mutagenesis, and PCR.

QUESTIONS FOR THOUGHT

1. Some viruses, such as adenovirus, avoid the problem of lagging-strand synthesis by replicating the individual strands of the DNA in the leading-strand direction simultaneously from both ends so that eventually the entire molecule is replicated. Why do bacterial chromosomes not replicate in this way?

2. Why are DNA molecules so long? Would it not be easier to have many shorter pieces of DNA? What are the advantages and disadvantages of a single long DNA molecule?

3. Why do cells have DNA as their hereditary material instead of RNA, like some viruses?

4. What effect would shifting a temperature-sensitive mutant with a mutation in the *dnaA* gene for initiator protein DnaA have on the rate of DNA synthesis? Would the rate drop linearly or exponentially? Would the slope of the curve be affected by the growth rate of the cells at the time of the shift? Explain.

5. The gyrase inhibitor novobiocin inhibits the growth of almost all types of bacteria. What would you predict about the gyrase of the bacterium *Streptomyces sphaeroides*, which makes this antibiotic? How would you test your hypothesis?

6. How do you think chromosome replication and cell division are coordinated in bacteria like *E. coli*? How would you go about testing your hypothesis?

7. Why is termination of chromosome replication so sloppy that the *ter* region is nonessential for growth and there has to be more than one *ter* site in each direction to completely stop the replication fork? What are the advantages of not having a definite site on the chromosome at which replication always terminates?

PROBLEMS

1. You are synthesizing DNA on the template 5′-ACCTTACCGTAATCC-3′ from an upstream primer. You add three of the deoxynucleotides but leave out the fourth deoxynucleotide, deoxycytosine triphosphate, from the reaction mixture. What DNA would you make? Draw a picture.

2. You are synthesizing DNA from the same template and with the same upstream primer, but instead of just deoxythymidine triphosphate, you add an equal mixture of the inhibitor of replication, dideoxythymidine triphosphate, and the normal nucleotide deoxythymidine triphosphate, in addition to the other three deoxynucleoside triphosphates. What DNAs would you make? Draw a picture.

3. You are growing *E. coli* with a generation time of only 25 min. How long will the I periods, C periods, and D periods be? Draw a picture showing when the various events occur during the cell cycle.

4. You are growing *E. coli* with a generation time of 90 min. How long will the I, C, and D periods be? Draw a picture.

5. You are measuring the supercoiling of a plasmid from a *topA* mutant of *E. coli* that lacks the major type I topoisomerase and comparing it with the same plasmid from *E. coli* without the *topA* mutation. Would you expect there to be more or fewer negative supercoils in the plasmid from the mutant? Why?

6. Design a downstream PCR primer to amplify the sequences upstream of the sequence 5′-CGATCTTAAT-3′ and add an EcoRI restriction site for cloning.

SUGGESTED READING

Aussel, L., F.-X. Barre, M. Aroyo, A. Stasiak, A. J. Stasiak, and D. Sherratt. 2002. FtsK is a DNA motor protein that activates chromosome dimer resolution by switching the catalytic state of the XerC and XerD recombinases. *Cell* **108:**195–205.

Bernhardt, T. G., and P. A. J. de Boer. 2005. SlmA, a nucleoid-associated, FtsZ binding protein required for blocking septal ring assembly over chromosomes in *E. coli*. *Mol. Cell* **18:**555–564.

Biller, S. J., and W. F. Burkholder. 2009. The *Bacillus subtilis* SftA (YtpS) and SpoIIIE DNA translocases play distinct roles in growing cells to ensure faithful chromosome partitioning. *Mol. Microbiol.* **74:**790–809.

Blakely, G., G. May, R. McCulloch, L. K. Arciszewska, M. Burke, S. T. Lovett, and D. J. Sherratt. 1993. Two related recombinases are required for site-specific recombination at *dif* and *cer* in *E. coli*. *Cell* **75:**351–361.

Britton, R. A., D. C. Lin, and A. D. Grossman. 1998. Characterization of a prokaryotic SMC protein involved in chromosome partitioning. *Genes Dev.* **12:**1254–1259.

Camara, J. E., A. M. Breier, T. Brendler, S. Austin, N. R. Cozzarelli, and E. Crooke. 2005. Hda inactivation of DnaA is the predominant mechanism preventing hyperinitiation of *Escherichia coli* DNA replication. *EMBO Rep.* **6:**736–741.

Dervyn, E., C. Suski, R. Daniel, C. Bruand, J. Chapuis, J. Errington, L. Janniere, and S. D. Ehrich. 2001. Two essential DNA polymerases at the bacterial replication fork. *Science* **294:**1716–1719.

Fujimitsu, K., T. Senriuchi, and T. Katayama. 2009. Specific genomic sequences of *E. coli* promote replicational initiation by directly reactivating ADP-DnaA. *Genes Dev.* **23:**1221–1233.

Guy, C. P., J. Atkinson, M. K. Gupta, A. A. Mahdi, E. J. Gwynn, C. J. Rudolph, P. B. Moon, I. C. van Knippenberg, C. J. Cadman, M. S. Dillingham, R. G. Lloyd, and P. McGlynn. 2009. Rep provides a second motor at the replisome to promote duplication of protein-bound DNA. *Mol. Cell* **36:**654–666.

Hayama, R., and K. J. Marians. 2010. Physical and functional interaction between the condensin MukB and the decatenase topoisomerase IV in *Escherichia coli*. *Proc. Natl. Acad. Sci. USA* **107:**18826–18831.

Heller, R. C., and K. J. Marians. 2006. Replication for reactivation downstream of a blocked nascent leading strand. *Nature* **439:**557–562.

Helmstetter, C. E., and S. Cooper. 1968. DNA synthesis during the division cycle of rapidly growing *Escherichia coli* B/r. *J. Mol. Biol.* **31:**507–518.

Kohanski, M. A., D. J. Dwyer, B. Hayete, C. A. Lawrence, and J. J. Collins. 2007. A common mechanism of cellular death induced by bactericidal antibiotics. *Cell* **130:**797–810.

Le Bourgeois, P., P. Bugarel, N. Campo, M.-L. Daveran-Mingot, J. Labonté, D. Lanfranchi, T. Lautier, C. Pagès, and P. Ritzenthaler. 2007. The unconventional Xer recombination machinery of streptococci/lactococci. *PLoS Genet.* **3:**e117.

Liu, N.-J., R. J. Dutton, and K. Pogliano. 2006. Evidence that the SpoIIIE DNA translocase participates in membrane fusion during cytokinesis and engulfment. *Mol. Microbiol.* **59:**1097–1113.

Mercier, R., M. A. Petit, S. Schbath, S. Robin, M. El Karoui, F. Boccard, and O. Espéli. 2008. The MatP/MatS site-specific system organizes the terminus region of the *E. coli* chromosome into a macrodomain. *Cell* **135:**475–485.

Neidhardt, F. C., R. Curtiss III, J. L. Ingraham, E. C. C. Lin, K. B. Low, B. Magasanik, W. S. Reznikoff, M. Riley, M. Schaechter, and H. E. Umbarger (ed.). 1996. Escherichia coli *and* Salmonella: *Cellular and Molecular Biology*, 2nd ed. ASM Press, Washington, DC.

Olby, R. 1974. *The Path to the Double Helix*. Macmillan Press, London, United Kingdom.

Pomerantz, R. R., and M. O'Donnell. 2008. The replisome uses mRNA as a primer after colliding with RNA polymerase. *Nature* **456:**762–766.

Postow, L., C. D. Hardy, J. Arsuaga, and N. R. Cozzarelli. 2004. Topological domain structure of the *Escherichia coli* chromosome. *Genes Dev.* **18:**1766–1779.

Reddy. C.A., T. J. Beveridge, J. A. Breznak, G. Marzluf, T. M. Schmidt, and L. R. Snyder (ed.). 2007. *Methods for General and Molecular Microbiology*, 3rd ed. ASM Press, Washington, DC.

Reyes-Lamothe, R., D. J. Sherratt, and M. C. Leake. 2010. Stoichiometry and architecture of active DNA replication machinery in *Escherichia coli*. *Science* **328:**498–501.

Sambrook, J., and D. Russell. 2006. *The Condensed Protocols from Molecular Cloning: a Laboratory Manual*. Cold Spring Harbor Laboratory Press, Cold Spring Harbor, NY.

Shih, Y.-L., and L. Rothfield. 2006. The bacterial cytoskeleton. *Microbiol. Mol. Biol. Rev.* **70:**729–754.

Singleton, M. R., M. S. Dillingham, and D. B. Wigley. 2007. Structure and mechanism of helicase and nucleic acid translocases. *Annu. Rev. Biochem.* **76:**23–50.

Slater, S., S. Wold, M. Lu, E. Boye, K. Skarsted, and N. Kleckner. 1995. The *E. coli* SeqA protein binds *oriC* in two different methyl-modulated reactions appropriate to its roles in DNA replication initiation and origin sequestration. *Cell* **82:**927–936.

Thanbichler, M. 2009. Closing the ring: a new twist to bacterial chromosome condensation. *Cell* **137:**598–599.

Thanbichler, M., and L. Shapiro. 2006. MipZ, a spatial regulator coordinating chromosome segregation with cell division in *Caulobacter*. *Cell* **126:**147–162.

Viollier, P. H., M. Thanbichler, P. T. McGrath, L. West, M. Meewan, H. H. McAdams, and L. Shapiro. 2004. Rapid and sequential movement of individual chromosomal loci to specific subcellular locations during bacterial DNA replication. *Proc. Natl. Acad. Sci. USA* **101:**9257–9262.

Wang, J. C. 1996. DNA topoisomerases. *Annu. Rev. Biochem.* **65:**635–692.

Watson, J. D. 1968. *The Double Helix*. Atheneum, New York, NY.

Watson, J. D., and F. H. C. Crick. 1953. Molecular structure of nucleic acids. *Nature* (London) **171:**737–738.

Woo, J.-S., J.-H. Lim, H.-C. Shin, M.-K. Suh, B. Ku, K.-H. Lee, K. Joo, H. Robinson, J. Lee, S.-Y. Park, N.-C. Ha, and B.-H. Oh. 2008. Structural studies of a bacterial condensin complex reveal ATP-dependent disruption of intersubunit interactions. *Cell* **136:**85–96.

Wu, L. J., S. Ishikawa, Y. Kawai, T. Oshima, N. Ogasawara, and J. Errington. 2009. Noc protein binds to specific DNA sequences to coordinate cell division with chromosome segregation. *EMBO J.* **28:**1940–1952.

CHAPTER 2

Bacterial Gene Expression: Transcription, Translation, and Protein Folding

UNCOVERING THE MECHANISM OF PROTEIN SYNTHESIS, and therefore of gene expression, was one of the most significant accomplishments in the history of science. The process of gene expression is sometimes called the **central dogma** of molecular biology, which states that information in DNA is copied into RNA to be translated into protein. We now know of many exceptions to the central dogma. For example, information does not always flow from DNA to RNA but sometimes in the reverse direction, from RNA to DNA. The information in RNA is often changed after it has been copied from the DNA. Moreover, the information in DNA may be expressed differently depending on where it is in a gene. Despite these exceptions, however, the basic principles of the central dogma remain sound.

This chapter outlines the process of gene expression and protein synthesis. The discussion is meant to be only a broad overview, but with special emphasis on topics essential to an understanding of the chapters that follow and on subjects unique to bacteria. For more detailed treatments, consult any modern biochemistry textbook.

Overview

DNA carries the information for the synthesis of RNA and proteins in regions called **genes**. The first step in expressing a gene is to **transcribe** an RNA copy from one strand in that region. The word transcription is descriptive, because the information in RNA is copied in the same language as DNA, a language written in a sequence of nucleotides. If the gene carries information for a protein, this RNA transcript is called **messenger RNA (mRNA)**. An mRNA is a messenger because it carries the information encoded in a gene to a **ribosome**. Once on the ribosome, the information in the mRNA can be

doi:10.1128/9781555817169.ch2

translated into the protein. Translation is another descriptive word, because one language—the sequence of nucleotides in DNA and RNA—is translated into a different language—a sequence of amino acids in a protein. The mRNA is translated as it moves along the ribosome, 3 nucleotides at a time. Each 3-nucleotide sequence, called a **codon**, carries information for a specific amino acid. The assignment of each of the possible codons to amino acids is called the **genetic code.**

The actual translation from the language of nucleotide sequences to the language of amino acid sequences is performed by small RNAs called tRNAs and enzymes called **aminoacyl-tRNA synthetases (aaRS)**. The aaRS enzymes attach specific amino acids to their matching tRNAs. Then, each aminoacylated tRNA (aa-tRNA) specifically pairs with a codon in the mRNA as it moves through the ribosome, and the amino acid carried by the tRNA is inserted into the growing protein. The tRNA pairs with the codon in the mRNA through a complementary 3-nucleotide sequence called the **anticodon.** The base-pairing rules for codons and anticodons are basically the same as the base-pairing rules for DNA replication, and the pairing is antiparallel. The only major differences are that RNA has uracil (U) rather than thymine (T) and that the pairing between the last of the 3 bases in the codon and the first base in the anticodon is less stringent.

This basic outline of gene expression leaves many important questions unanswered. How does mRNA synthesis begin and end at the correct places and on the correct strand in the DNA? Similarly, how does translation start and stop at the correct places on the mRNA? What actually happens to the tRNA and ribosomes during translation? The answers to these questions and many others are important for the interpretation of genetic experiments, so we will discuss the structure of RNA and proteins and the processes by which they are synthesized in more detail.

The Structure and Function of RNA

In this section, we review the basic components of RNA and how it is synthesized. We also review how structure varies among different types of cellular RNAs and the role each type plays in cellular processes.

Figure 2.1 RNA precursors. **(A)** An rNTP (the form of NTP used as a precursor for RNA) contains a ribose sugar, a base, and three phosphates. **(B)** The four bases in RNA. **(C)** An RNA polynucleotide chain with the 5′ and 3′ ends shown in blue.
doi:10.1128/9781555817169.ch1.f2.1

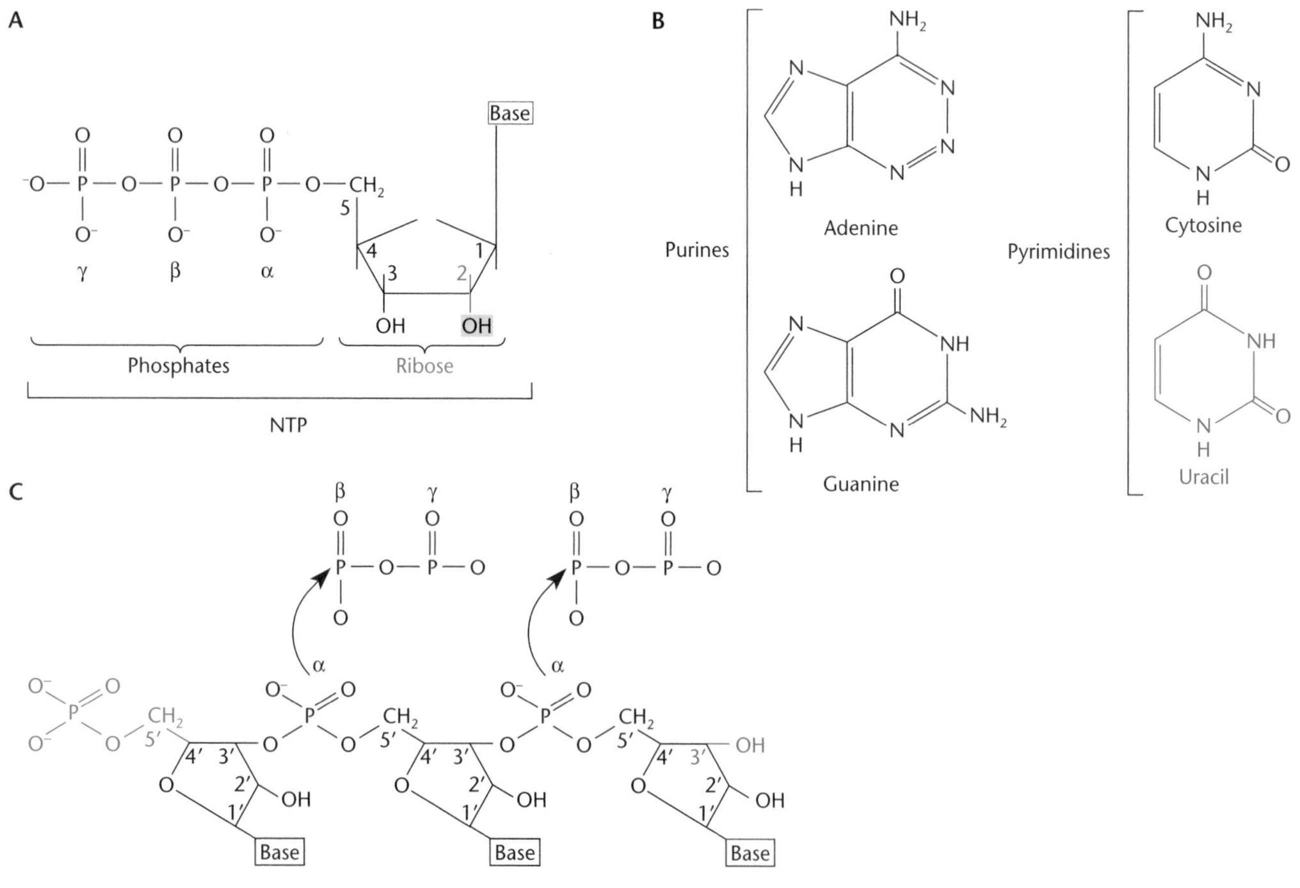

A

B

C

Types of RNA

There are several different classes of RNA in cells. Some of these, including mRNA, rRNA, and tRNA, are involved in protein synthesis. Each of these types of RNA has special properties, which are discussed below. Others are involved in regulation, replication, and protein secretion.

RNA Precursors

RNA is similar to DNA in that it is composed of a chain of nucleotides. However, RNA nucleotides contain the sugar ribose instead of deoxyribose. These five-carbon sugars differ in the second carbon, which is attached to a hydroxyl group in ribose rather than the hydrogen found in deoxyribose (see chapter 1). Figure 2.1A shows the structure of a **ribonucleoside triphosphate** (rNTP), so named because of the different sugar.

The only other difference between RNA and DNA chains when they are first synthesized is in the bases. Three of the bases—adenine, guanine, and cytosine—are the same, but RNA has uracil instead of the thymine found in DNA (Figure 2.1B). The RNA bases can also be modified after they are incorporated into an RNA chain, as discussed below.

Figure 2.1C shows the basic structure of an RNA polynucleotide chain. As in DNA, RNA nucleotides are held together by phosphates that join the 3′ carbon of one ribose sugar to the 5′ carbon of the next. This arrangement ensures that, as with DNA chains, the two ends of an RNA polynucleotide chain will be different from each other, with the 5′ end terminating in a phosphate group and the 3′ end terminating in a hydroxyl group. The 5′ end of a newly synthesized RNA chain has three phosphates attached to it because transcription initiates with an rNTP.

According to convention, the sequence of bases in RNA is given from the 5′ end to the 3′ end, which is actually the direction in which the RNA is synthesized, by addition of the 5′ phosphate of an incoming nucleoside triphosphate to the 3′ hydroxyl end of the growing RNA chain. Also by convention, regions in RNA that are closer to the 5′ end in a given sequence are referred to as **upstream** and regions that are closer to the 3′ end are referred to as **downstream**, because RNA is both made and translated in the 5′-to-3′ direction.

RNA Structure

Except for the sequence of bases and minor differences in the pitch of the helix, little distinguishes one DNA molecule from another. However, RNA chains generally have more structural properties than DNA and often are folded into complex structures. Extensive base modifications can further change the structure of the RNA molecule.

PRIMARY STRUCTURE

All RNA transcripts are synthesized in the same way, from a DNA template. Only the sequences of their nucleotides and their lengths are different. The sequence of nucleotides in RNA is the **primary structure** of the RNA. In some cases, the primary structure of an RNA is changed after it is transcribed from the DNA (see "RNA Processing and Modification" below).

SECONDARY STRUCTURE

Unlike DNA, RNA is usually single stranded. However, pairing between the bases in different regions of the molecule may cause it to fold up on itself to form double-stranded regions. Such double-stranded regions are called the **secondary structure** of the RNA. All RNAs, including mRNAs, probably have extensive secondary structure.

Figure 2.2 shows an example of RNA secondary structure in which the sequence 5′-AUCGGCA-3′ has paired with the complementary sequence 5′-UGCUGAU-3′ somewhere else in the molecule. As in double-stranded DNA, the paired strands of RNA are antiparallel, i.e., pairing occurs only when the two sequences are complementary when read in opposite directions (5′ to 3′ and 3′ to 5′) and the double-stranded RNA forms a helix, or hairpin, that is similar to a DNA-DNA helix, capped with a few unpaired residues called a loop. However, the pairing rules for double-stranded RNA are slightly different from the pairing rules for DNA (G-C/A-U; remember that T in DNA is replaced by U in RNA). In RNA, guanine can pair with uracil, as well as with cytosine. Because these G-U "wobble" pairs do not share hydrogen bonds, as indicated in the figure, they contribute less substantially to the stability of the double-stranded RNA. Additional "non-Watson-Crick" pairings are also found in

Figure 2.2 Secondary structure in an RNA. **(A)** The RNA folds back on itself to form a hairpin. The presence of a GU pair (in parentheses) does not disrupt the structure. **(B)** Different regions of the RNA can also pair with each other to form a pseudoknot. In the example, the loop of the hairpin is pairing with another region of the RNA. The blue dashes represent the phosphate-ribose backbone; the black dashes represent the hydrogen bonds.
doi:10.1128/9781555817169.ch1.f2.2

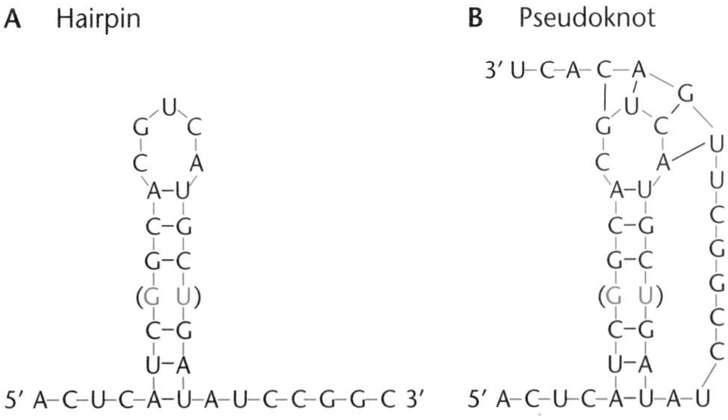

A Hairpin

5′ A-C-U-C-A-U-A-U-C-C-G-G-C 3′

B Pseudoknot

3′ U-C-A-C-A

5′ A-C-U-C-A-U-A-U

RNA; these often involve surfaces of the nucleoside other than the normal hydrogen-bonding edge. Each base pair that forms in the RNA makes the secondary structure of the RNA more stable. Consequently, the RNA generally folds so that the greatest number of continuous base pairs can form. The stability of a structure can be predicted by adding up the energy of all its hydrogen bonds that contribute to the structure. By eye, it is difficult to predict which regions of a long RNA will pair to give the most stable structure. Computer software (e.g., mfold [http://mfold.rna.albany.edu/?q=mfold]) is available that, given the sequence of bases (primary structure) of the RNA, can predict the most stable secondary structure; however, the structure of complex RNAs is very difficult to predict computationally.

TERTIARY STRUCTURE

Double-stranded regions of RNAs created by base pairing are stiffer than single-stranded regions. As a result, an RNA that has secondary structure will have a more rigid shape than one without double-stranded regions. Also, the intermingled paired regions cause the RNA to fold back on itself extensively, facilitating additional tertiary interactions. One type of tertiary interaction occurs when an unpaired region (such as the loop of a hairpin like that shown in Figure 2.2) pairs with another region of the same RNA molecule. A structure like this is called a pseudoknot, rather than a real knot, because it is held together only by hydrogen bonds, which are more easily broken than covalent chemical bonds. Together, these interactions give many RNAs a well-defined three-dimensional shape, called its **tertiary structure**. Proteins or other cellular constituents often recognize RNAs by their tertiary structures.

RNA Processing and Modification

The folding of an RNA molecule as a result of secondary and tertiary structure represents a **noncovalent change**, because only hydrogen bonds or electrostatic interactions, not chemical (covalent) bonds, are formed or broken. However, once the RNA is synthesized, covalent changes can occur during **RNA processing** and **RNA modification**.

RNA processing involves forming or breaking phosphate bonds in the RNA after it is made. For example, the terminal phosphates at the 5′ end may be removed, or the RNA may be cut into smaller pieces and even religated into new combinations, requiring the breaking and making of new phosphate bonds. In one of the most extreme cases of RNA processing, called RNA editing, nucleotides can be excised from or added to mRNA after it has been transcribed from DNA.

RNA modification, in contrast, involves altering the bases or sugars of RNA. Examples include methylation of the bases or sugars of rRNA and enzymatic alteration of the bases of tRNA. In eukaryotes, "caps" of inverted methylated nucleotides are added to the 5′ ends of some types of mRNA. In bacteria, mRNAs are not capped, and only the stable rRNAs and tRNAs are extensively modified.

Transcription

Transcription is the synthesis of RNA by copying information from a DNA template. The processes of transcription are probably fairly similar in all organisms, but it is best understood in bacteria.

Structure of Bacterial RNA Polymerase

The transcription of DNA into RNA is the work of **RNA polymerase**. In bacteria, the same RNA polymerase makes all the cellular RNAs, including rRNA, tRNA, and mRNA. There are approximately 2,000 molecules of RNA polymerase in each bacterial cell. Only the primer RNAs of Okazaki fragments are made by a different RNA polymerase (primase) during DNA replication. In contrast, eukaryotes have three nuclear RNA polymerases, as well as a mitochondrial RNA polymerase, that make their cellular RNAs.

Figure 2.3 shows a schematic structure of *Escherichia coli* RNA polymerase, which has six subunits and a molecular mass of more than 400,000 daltons (Da) (400 kDa), making it one of the largest enzymes in the *E. coli* cell. The core enzyme consists of five subunits: two identical α subunits, two very large subunits called β and β', and the ω subunit. The α, β, and β' subunits are essential parts of the RNA polymerase, and the ω subunit helps in its assembly. A sixth subunit, the σ factor, is required only for initiation and cycles off the enzyme after initiation of transcription. Without the σ factor, the RNA polymerase is called the **core enzyme**; with σ, it is called the **holoenzyme**.

Crystal structures of RNA polymerase from *Thermus aquaticus* and *E. coli* have revealed the structure shown in Figure 2.4. The overall shape resembles a crab claw. Regions of the β and β' subunits form the pincers of the crab claw (as shown in Figure 2.3). The two α subunits, αI and αII, are on the opposite end from the claw, making up the rear end when the enzyme is transcribing DNA. The carboxy-terminal regions of the α polypeptides hang out from the enzyme, where they can contact other proteins or DNA regions upstream of the promoter and stimulate transcription initiation (see below). The ω subunit is also on this end of the RNA polymerase, wrapped around the β' subunit (Figure 2.3). When the σ factor is bound to the complex to form the holoenzyme, it wraps around the front end of the core enzyme in such a way that it can contact the DNA as it enters the open claw. One of

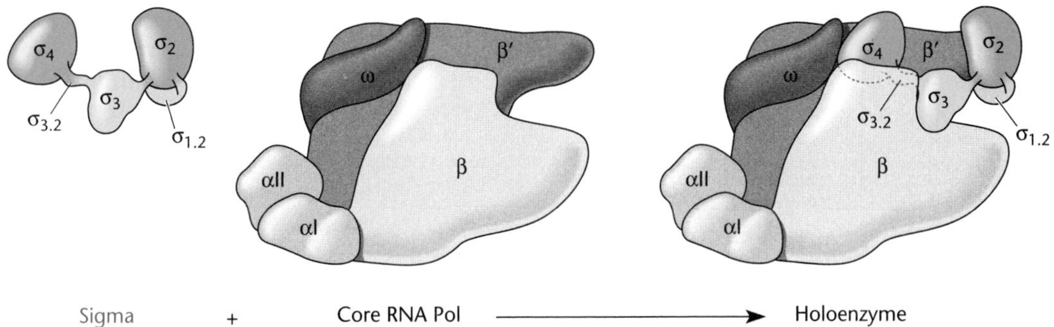

Figure 2.3 The structure of bacterial RNA polymerase. The sigma factor (σ, shown in blue), the core enzyme without σ, and the holoenzyme with σ attached are shown. The functions of some of the domains of σ, σ_1 and $\sigma_{3.2}$, are discussed in this chapter. One α subunit, αI, contacts the β subunit, and the other, αII, contacts the β′ subunit.
doi:10.1128/9781555817169.ch1.f2.3

its domains, domain σ_2, contacts the β′ pincer and is in position to bind to one part of the promoter sequence on the DNA (the –10 region), while two other domains, σ_3 and σ_4, contact the β subunit further upstream in the active-center channel in such a way that the σ_4 domain is in position to contact the –35 region of the promoter (see below). The RNA polymerases of all bacteria are probably very similar in sequence and composition. One interesting difference is that in some types of bacteria, the β and β′ subunits are attached to each other to form an even larger polypeptide. The eukaryotic and archaeal core RNA polymerases have more subunits and have very different sequences, but their basic overall structures are very similar to that of bacterial RNA polymerase.

Overview of Transcription

Much like DNA polymerase (see chapter 1), the RNA polymerase makes a complementary copy of a DNA template, building a chain of RNA by attaching the 5′ phosphate of a ribonucleotide to the 3′ hydroxyl of the one preceding it in the growing chain (Figure 2.5). However, in contrast to DNA polymerases, RNA polymerases do not need a preexisting primer to initiate the synthesis of a new chain of RNA. To begin transcription, the **RNA polymerase holoenzyme** binds to the promoter sequence and separates the strands of the DNA, exposing the bases. Unlike DNA polymerases, which require helicases to separate the strands, the RNA polymerase can complete this step by itself. Then, an rNTP complementary to the nucleotide at the transcription start site enters the complex through a channel in the RNA polymerase and pairs with the template. The second rNTP comes in, and if it is complementary to the next base in the DNA template, it is retained, but if it is not complementary, it is rejected. RNA polymerase then catalyzes the reaction in which the α phosphate of the second nucleotide joins with the 3′ hydroxyl of the first nucleotide. Then, the third nucleotide comes in and pairs with the next

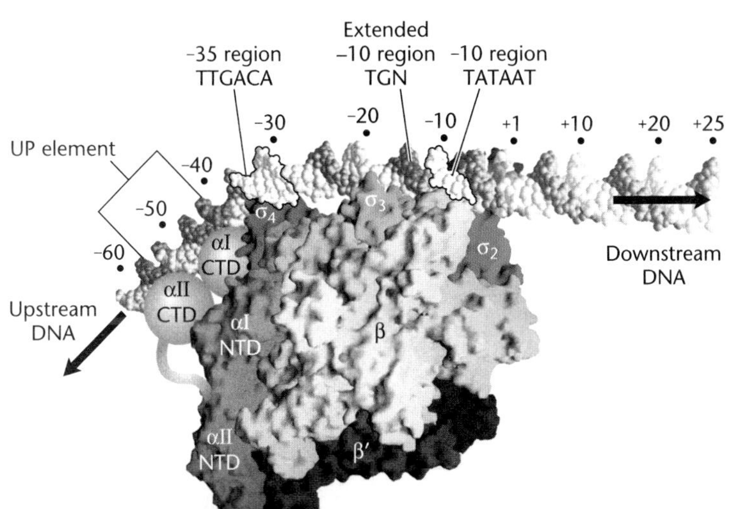

Figure 2.4 Crystal structure of bacterial RNA polymerase and σ interactions with promoter DNA. Subunits of RNA polymerase (β [light gray], β′ [dark gray], αI and αII [medium gray], and σ [blue]) are shown. The α subunits are divided into αNTD domains (which associate with the β and β′ subunits) and αCTD domains (which are connected to αNTD by a flexible linker and can bind to the UP element in the DNA); the structures of the αCTD domains are not resolved and are shown as a ball. σ is divided into four domains that extend along the promoter DNA, interacting with different regions of the promoter.
doi:10.1128/9781555817169.ch1.f2.4

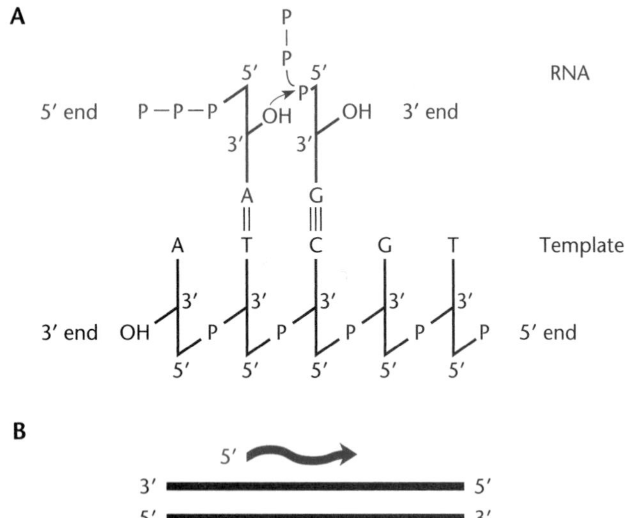

Figure 2.5 RNA transcription. **(A)** The polymerization reaction, in which incoming NTPs pair with the template strand of DNA during transcription and are joined to generate the RNA chain. The β and γ phosphates of each incoming NTP (other than the initiator NTP) are removed as pyrophosphate (PP_i). **(B)** RNA polymerase synthesizes RNA in the 5′-to-3′ direction, moving 3′ to 5′ on the template strand of the DNA. RNA is shown as a wavy blue line, and both strands of DNA are shown as straight lines. doi:10.1128/9781555817169.ch1.f2.5

A

DNA	5′ ATTACGACCTACGCAT 3′	Coding strand
	3′ TAATGCTGGATGCGTA 5′	Template strand
RNA	5′ AUUACGACCUACGCAU 3′	

B

Figure 2.6 RNA polymerase transcribes only one strand of DNA. **(A)** The coding strand (or nontemplate strand) of the DNA has the same sequence as the mRNA. The template strand is the DNA strand to which the mRNA is complementary if read in the 3′-to-5′ direction. **(B)** Schematic illustration of the DNA and RNA strands. doi:10.1128/9781555817169.ch1.f2.6

nucleotide in the template, and so forth. The RNA polymerase makes a complementary copy, i.e., **transcribes** the sequence of one strand of DNA into RNA. As shown in Figure 2.6, the strands of DNA in a region that is transcribed are named to reflect the sequence of the RNA made from that region. The **template strand** of DNA that is copied is also called the **transcribed strand.** The other strand of the DNA, which has the same sequence as the RNA copy, is called the **nontemplate strand** or **coding strand**, since it has the same sequence as the mRNA that encodes the protein (even if the RNA that is made does not actually encode a protein). The sequence of a gene is usually written as the sequence of the nontemplate, or coding, strand.

PROMOTERS

RNA transcripts are copied only from selected regions of the DNA, rather than from the whole molecule; therefore, the RNA polymerase holoenzyme can start making an RNA chain from a double-stranded DNA only at certain sites. These DNA regions are called **promoters**, and the RNA polymerase recognizes a particular nucleotide (usually a T or C) in the promoter region of the template strand as a **transcription start site,** shown as +1 in Figure 2.7. Thus, the first base in the chain is usually an A or a G laid down opposite to a T or C, respectively.

Figure 2.7 **(A)** Typical structure of a σ^{70} bacterial promoter with sequences that match the –35 and –10 consensus sequences. **(B)** The consensus sequences of such promoters. **(C)** The relationship of these sequences with respect to the start site of transcription on the two strands. RNA synthesis typically starts with an A or a G, and no primer is required. N, any nucleotide. doi:10.1128/9781555817169.ch1.f2.7

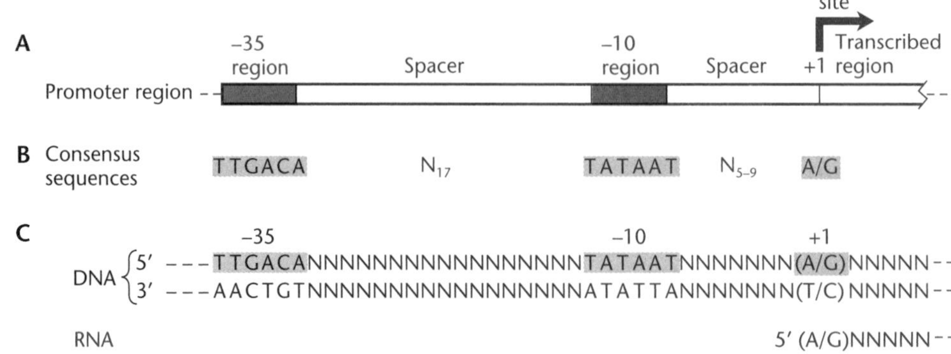

The RNA polymerase holoenzyme recognizes different types of promoters on the basis of which type of σ factor it contains. The most common promoters are those recognized by the RNA polymerase with the σ called σ^{70} in *E. coli*. The σ factors are often named for their size, and this one has a molecular mass of 70,000 Da (70 kDa). Replacement of σ^{70} with a different σ factor results in an RNA polymerase holoenzyme that recognizes a different set of promoters; this will be discussed in later chapters on gene regulation.

Promoters recognized by holoenzymes containing the same σ are not identical to each other, but they do share certain sequences, known as **consensus sequences**, by which they can be distinguished. Figure 2.6 shows the consensus sequence of promoters recognized by holoenzymes containing σ^{70} in *E. coli*, which illustrates a common pattern for promoter structure. The promoter sequence has two important regions: a short AT-rich region centered about 10 bp upstream of the transcription start site, known as the **–10 sequence**, and a second region centered about 35 bp upstream of the start site called the **–35 sequence**. The σ^{70} factor usually must bind to both sequences to start transcription (see below) but does not require a perfect match to these consensus sequences. Sequence-specific binding to the promoter determines not only the site at which transcription will initiate, but also the direction the RNA polymerase will move along the DNA (in other words, which strand of the DNA will be transcribed from a given region).

THE STEPS OF TRANSCRIPTION

Figure 2.8 shows an overview of the steps of transcription. The RNA polymerase holoenzyme recognizes a promoter and begins transcription with a nucleoside triphosphate. As the RNA chain begins to grow, the RNA polymerase holoenzyme releases its σ factor, and the five-subunit core enzyme continues moving along the template DNA strand in the 3′-to-5′ direction, synthesizing RNA in the 5′-to-3′ direction. Inside an opening in the DNA helix approximately 17 bases long, called the **transcription bubble**, the elongating RNA and the template DNA strand pair with each other to form a DNA-RNA hybrid of approximately 8 or 9 bp, which has a double-helix structure similar to that of a double-stranded DNA molecule. As RNA polymerase moves along the DNA, the upstream portion of the DNA-RNA helix separates as new ribonucleotides are incorporated into the 3′ end of the growing chain, and the 8- to 9-bp hybrid is maintained. The resulting RNA product emerges from RNA polymerase through a channel, and the DNA strands rebind to each other. The RNA polymerase moves along the DNA template until it reaches a **terminator**, which signals RNA polymerase to release both the DNA template and the RNA transcript.

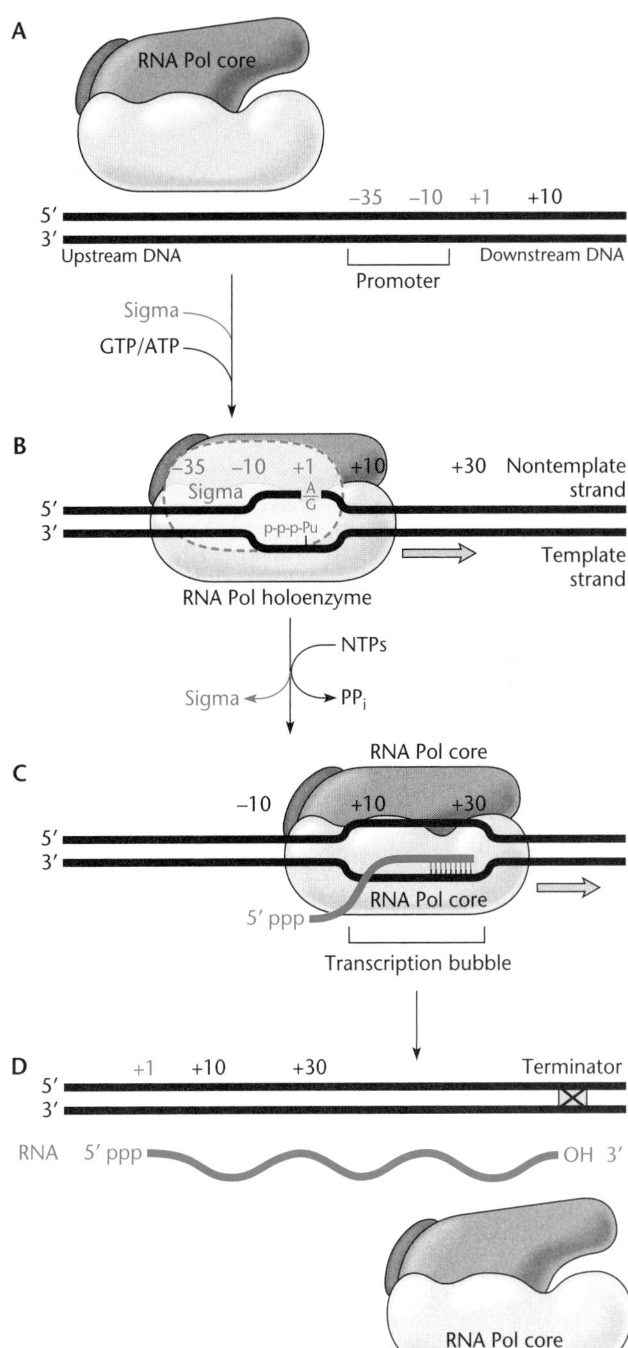

Figure 2.8 Transcription begins at a promoter and ends at a transcription terminator. **(A)** RNA polymerase core must bind σ factor to recognize a promoter. **(B)** Transcription begins when the strands of DNA are opened at the promoter, and the first rNTP, usually ATP or GTP, enters the active site. **(C)** As the RNA polymerase moves along the DNA, polymerizing ribonucleotides, it forms a transcription bubble containing an RNA-DNA double-stranded hybrid, which helps to hold the RNA polymerase on the DNA. The sigma factor is released after RNA polymerase leaves the promoter, and transcription by the RNA polymerase core continues. **(D)** The RNA polymerase encounters a transcription terminator and comes off the DNA, releasing the newly synthesized RNA.
doi:10.1128/9781555817169.ch1.f2.8

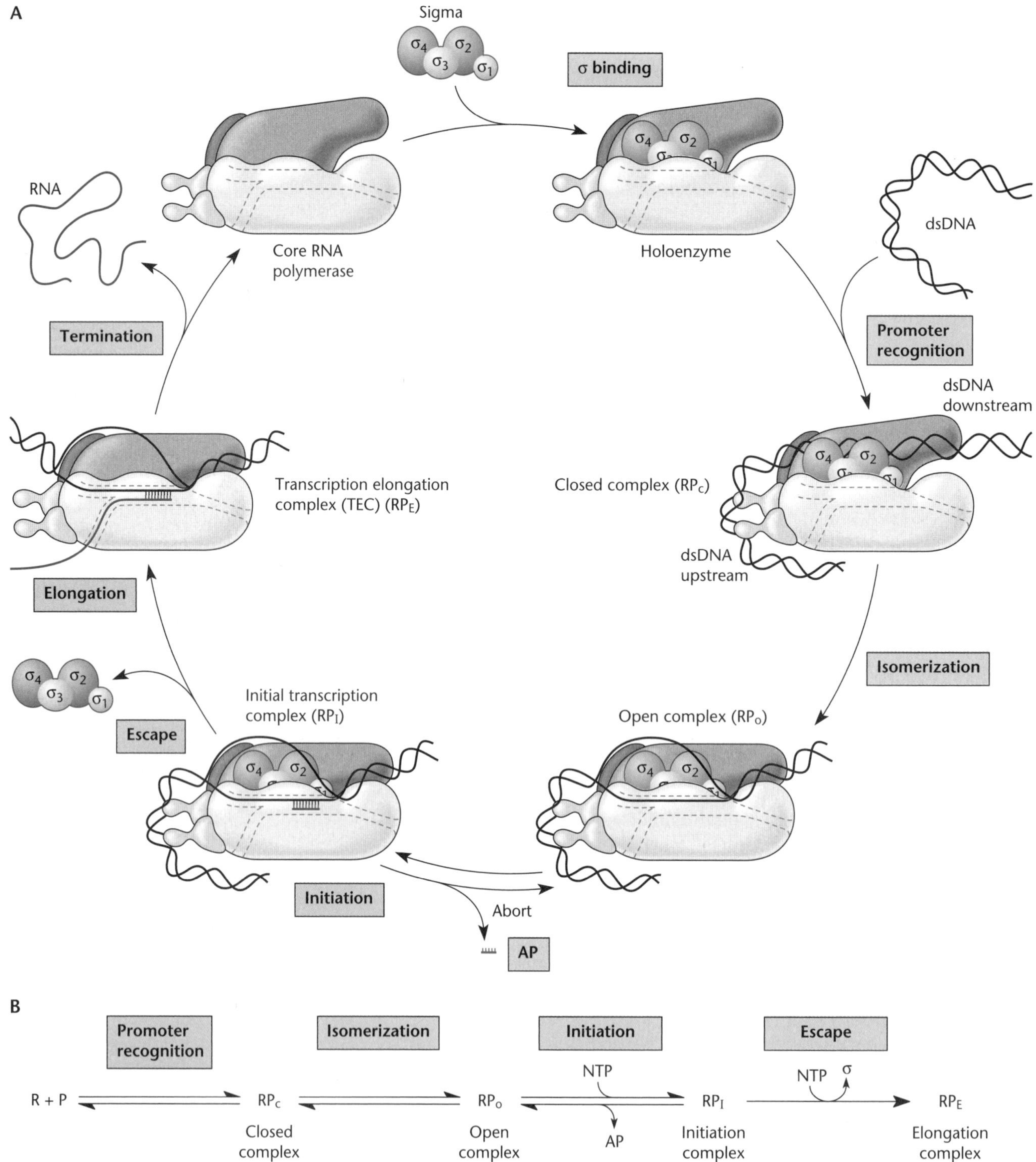

B

Promoter recognition		Isomerization		Initiation		Escape	
R + P	⇌	RP_C Closed complex	⇌	RP_O Open complex	⇌ (NTP; AP)	RP_I Initiation complex	→ (NTP; σ) RP_E Elongation complex

Figure 2.9 Overview of transcription. **(A)** The transcription cycle. Each step is discussed separately in the text. **(B)** Summary of the steps in transcription initiation. R, RNA polymerase; P, promoter DNA; AP, abortive products; dsDNA, double-stranded DNA.
doi:10.1128/9781555817169.ch1.f2.9

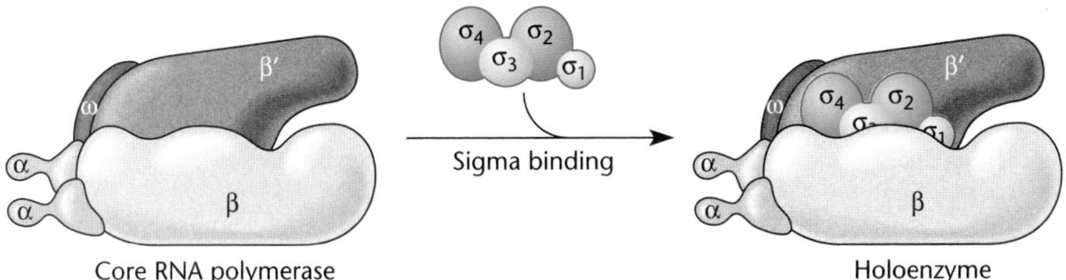

Figure 2.10 Sigma binding to RNA polymerase core. Sigma (σ) is shown in blue.
doi:10.1128/9781555817169.ch1.f2.10

Details of Transcription

It was once assumed that after initiation occurs, the RNA polymerase moves along the DNA at a uniform rate, polymerizing nucleotides into RNA. However, it is now known that the RNA polymerase often starts making RNA and then repeatedly aborts, synthesizing a number of short RNAs in a process called **abortive initiation**, before finally leaving the promoter. Even after transcription is under way, RNA polymerase often pauses and sometimes even backs up before continuing. In this section, we discuss in detail each of the steps in transcription (Figure 2.9), which have been established over many years by a large number of researchers. We discuss these steps one at a time because each of them is the basis for regulatory mechanisms that are discussed in later chapters.

PROMOTER RECOGNITION

In the first step (Figure 2.10), the **RNA polymerase core enzyme** binds to a σ factor to form the holoenzyme. The bound σ factor then directs the complex to the correct promoter in a process called **promoter recognition** or **binding** (Figure 2.11). The σ factor must be able to recognize the promoter even though the DNA in the promoter is still in a double-stranded state. Sigma factors consist of a number of domains held together by flexible linkers. Most σ factors are related to σ^{70}, and their domains play similar roles in recognizing their specific promoters. Figure 2.12 shows the conserved regions of the σ^{70} family of sigma factors and the roles played by some of the conserved domains in promoter recognition and initiation of transcription.

One domain of the bound σ, σ_4, recognizes the −35 sequence when it is still in the double-stranded state. Another σ domain, σ_2, first binds to double-stranded DNA at the AT-rich −10 sequence in what is called a closed complex (RP_c). The σ_3 domain is close to the β subunit in the active-site channel, while σ_2 is bound to the β′ subunit (Figure 2.4).

Not all σ^{70} promoters have the same features. Figure 2.13 shows some additional features of some σ^{70} promoters, and Figure 2.14 shows which regions of the RNA polymerase recognize these features. The efficiency of binding of RNA polymerase to a promoter can be enhanced by sequences upstream of the promoter, called UP (for upstream) elements, to which the carboxyl terminus of the α subunits, called αCTD (for α subunit carboxyl-terminal domain), can also bind and help stabilize the binding of RNA polymerase (Figure 2.14). The

Figure 2.11 Promoter recognition. Sigma factor, in complex with RNA polymerase core, binds to the promoter region to form an RP_c. The RP_c isomerizes to an RP_o. In the RP_o, σ_2 interacts with the nontemplate single-stranded DNA of the −10 sequence.
doi:10.1128/9781555817169.ch1.f2.11

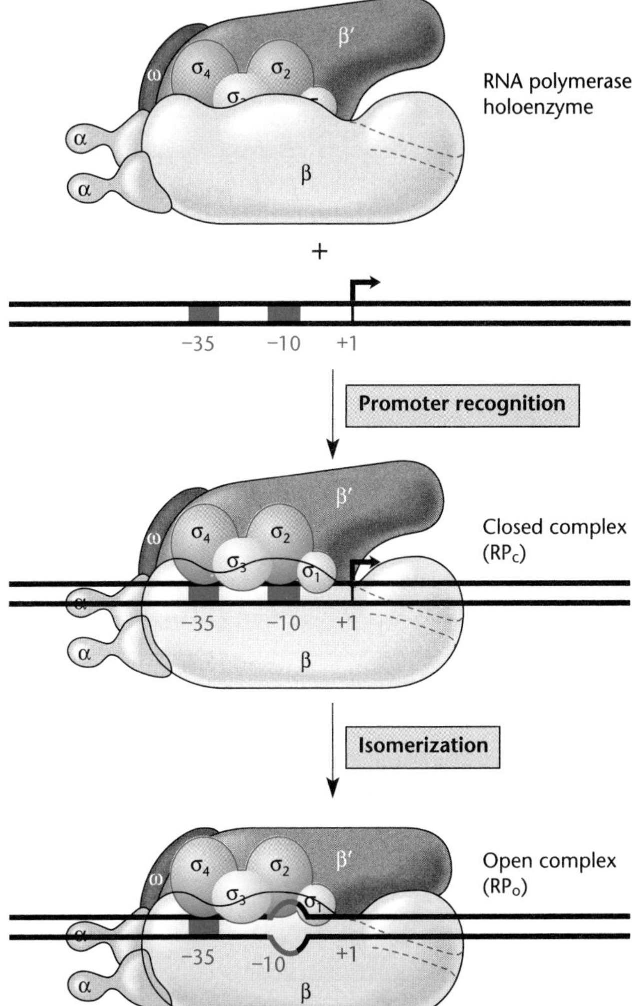

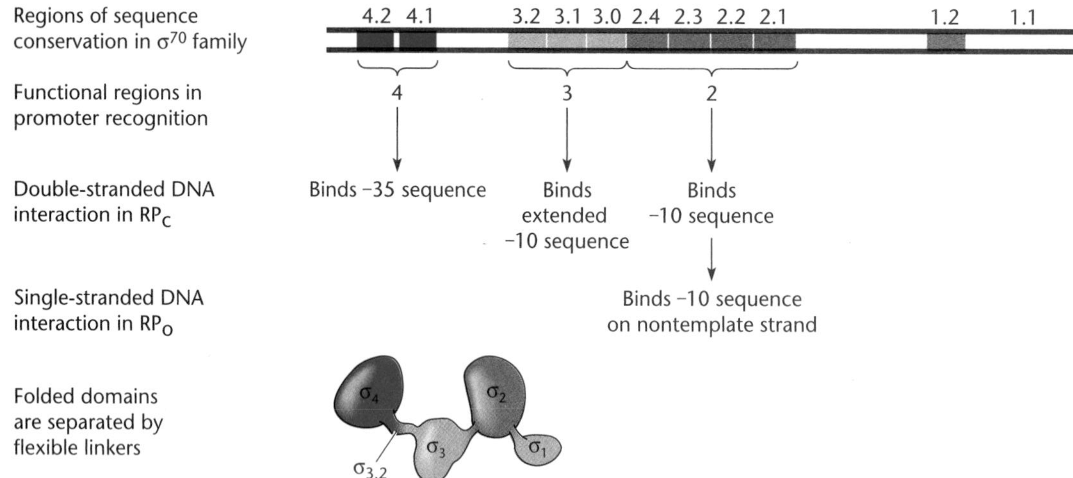

Figure 2.12 Functional regions of σ^{70}. There are four major regions of sequence conservation in the σ^{70} family; they are divided into subregions as shown. Region 1.1 is not conserved in all sigma factors. Also shown are regions functional in promoter recognition. doi:10.1128/9781555817169.ch1.f2.12

flexible linker between αCTD and the amino-terminal domain of α (αNTD) allows αCTD to reach the UP element on the DNA. Also, some promoters have what is called an extended −10 sequence (TGN, located immediately upstream of the −10 sequence to give the sequence TGNTATAAT). This sequence is recognized by the σ_3 domain and is often found in promoters that lack a −35 sequence that is efficiently recognized by σ_4. The similarity of a promoter sequence to the consensus sequences for a particular σ, in combination with other elements that interact with other domains of RNA polymerase, dictates the efficiency with which a promoter is recognized by a type of holoenzyme.

Figure 2.13 Variations on the basic σ^{70} promoter. **(A)** The consensus −10 and −35 regions; **(B)** location of the UP element; **(C)** location of the extended −10 sequence. doi:10.1128/9781555817169.ch1.f2.13

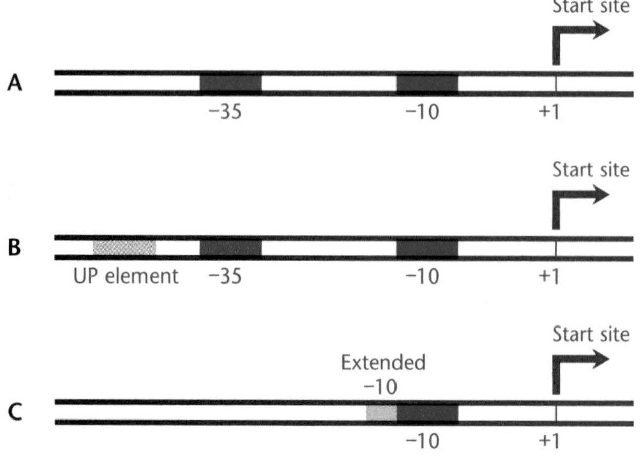

ISOMERIZATION

When RNA polymerase holoenzyme first binds to the promoter, the DNA is double stranded. The resulting complex is called the RP_c, because the DNA strands are still "closed." In the next step, the β′ pincer of the crab claw closes around the DNA to form the active-site channel around the template strand of the DNA. This allows the σ_2 region to separate the strands of DNA at the −10 region and bind to the nontemplate strand in a process called **isomerization** (Figures 2.9 and 2.11). Recall that AT base pairs are less stable than GC pairs, so the AT-rich −10 sequence is relatively easy to melt. The complex is now called the open complex (RP_o), since the strands of DNA at the −10 region of the promoter are "open." The +1 nucleotide (Figure 2.11) of the template strand is held in the active-site channel, where the polymerization reaction is about to occur.

INITIATION

In the initiation process, a single nucleoside triphosphate (usually an ATP or guanosine triphosphate [GTP]) enters through the secondary channel and pairs with nucleotide +1 (usually a T or C, respectively) in the template strand in the active site of the enzyme. Then, a second nucleoside triphosphate enters, and if that nucleoside triphosphate can base pair with the +2 nucleotide of the template strand, a phosphodiester bond forms between its α phosphate and the 3′ hydroxyl of the first nucleotide, releasing two phosphates in the form of pyrophosphate (Figure 2.5). This is called the **initiation complex** or **initial transcription complex** and is the step at which the antibiotic rifampin can block transcription (see "Antibiotic Inhibitors of Transcription" below). As shown in

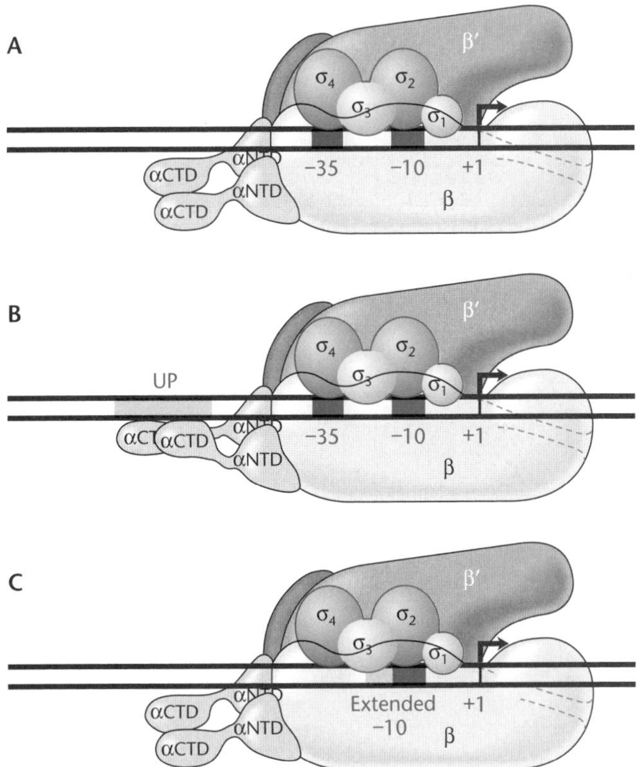

Figure 2.14 Interactions between RNA polymerase subunits and promoter elements. **(A)** Interaction of $σ^{70}$ at consensus −10 and −35 regions. **(B)** Flexible linkers between the αCTD and αNTD domains allow binding of αCTDs to UP elements in the DNA. **(C)** $σ_3$ binding to an extended −10 region. doi:10.1128/9781555817169.ch1.f2.14

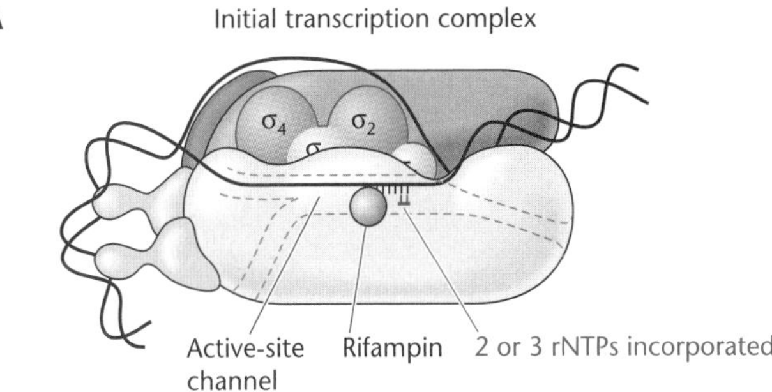

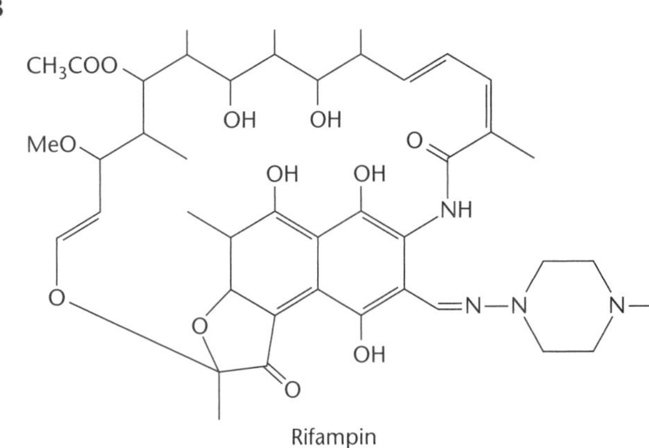

Figure 2.15 Initiation of transcription and action of the antibiotic rifampin. **(A)** Two or three rNTPs are polymerized and incorporated into the active site. Rifampin binds to the wall of the active-site channel, preventing further elongation of RNA. **(B)** Structure of the antibiotic rifampin. By convention, most carbon atoms are not shown. doi:10.1128/9781555817169.ch1.f2.15

Figure 2.15, rifampin binds to RNA polymerase in the β-subunit face of the active-site channel in such a way that the growing RNA encounters it when it reaches a length of only 2 or 3 nucleotides, preventing further growth of the RNA chain and freezing the RNA polymerase on the promoter. This explains why rifampin blocks only the initiation of transcription.

Even in the absence of the antibiotic rifampin, the RNA polymerase is not yet free to continue transcription. When the RNA chain grows to a length of about 10 nucleotides, it encounters the $σ_{3.2}$ loop, which blocks the site in RNA polymerase through which the newly synthesized RNA will emerge, a region called the exit channel (Figure 2.16). This causes transcription to stop, often releasing a short transcript about 10 nucleotides in length. This process is called abortive initiation and occurs to various degrees on many promoters for reasons that are not well understood. Eventually, a growing (or nascent) transcript pushes the $σ_{3.2}$ loop aside and enters the exit channel, causing the σ factor to be released from the core RNA polymerase. At this point, designated **promoter escape**, RNA polymerase has left the promoter site and has entered the elongation phase, during which the transcription bubble is enlarged to 17 bp, the complex is stabilized, and synthesis of RNA proceeds efficiently as the enzyme moves along the DNA template.

ELONGATION

Figure 2.17 shows the **transcription elongation complex (TEC)** in the process of elongating the RNA transcript. Most of the features are mentioned above, including the approximately 17-bp transcription bubble where the two strands of DNA are separated and the approximately 8- to 9-bp RNA-DNA hybrid that forms in the active site before the newly synthesized RNA strand separates from the DNA template strand and exits through the RNA exit channel. The RNA polymerase is capable of synthesizing RNA at a rate of 30 to 100 nucleotides per second. However, it often pauses and even slides backwards (**backtracks**). This phenomenon often occurs when hairpins form in the RNA as it exits the RNA exit channel, because the newly synthesized RNA contains inverted-repeated sequences. It is not clear why hairpins cause pausing and backtracking, but they may pull the

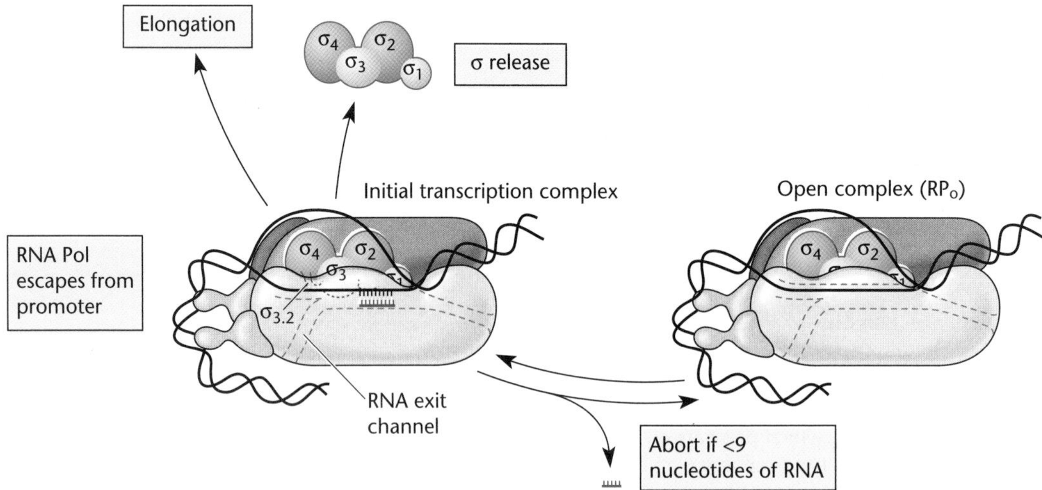

Figure 2.16 Abortive transcription and RNA polymerase escape from the promoter. RNA polymerase can escape from the promoter only if more than 10 or 11 nucleotides are polymerized. At 12 nucleotides, the RNA transcript displaces the $\sigma_{3.2}$ region, which blocks the active-site channel. Abortive transcription results when short, newly synthesized transcripts are released, and the complex returns to the RP$_o$ state. With RNA polymerase escape, σ is released, and transcription elongation can continue. doi:10.1128/9781555817169.ch1.f2.16

RNA polymerase backward or bind to it and change its conformation. Backtracking creates special problems for the TEC. When the RNA polymerase is forced backward, it pushes the 3′ end of the newly synthesized RNA forward, driving it into the secondary channel through which the nucleotides enter, as shown in Figure 2.18. It would remain this way, permanently blocked, except for the action of two proteins called GreA and GreB. These proteins insert their N termini into the secondary channel (Figure 2.18) and cleave the 3′ end of the RNA in the channel until it is in its proper place in the active center so that transcription can continue.

It is not clear why RNA polymerase pausing and backtracking are tolerated, since they reduce the rate of transcription overall and create the necessity for the Gre proteins. One possibility is that selective pausing helps the folding of the newly synthesized RNA or the protein being translated from the RNA (see below). Some genes whose RNA products must be made in large amounts, such as the rRNA genes, have special mechanisms to reduce pausing and backtracking. The rRNAs have sequences called antitermination sites that recruit protein factors that bind to the RNA polymerase and preempt the pausing effect of RNA hairpins that form

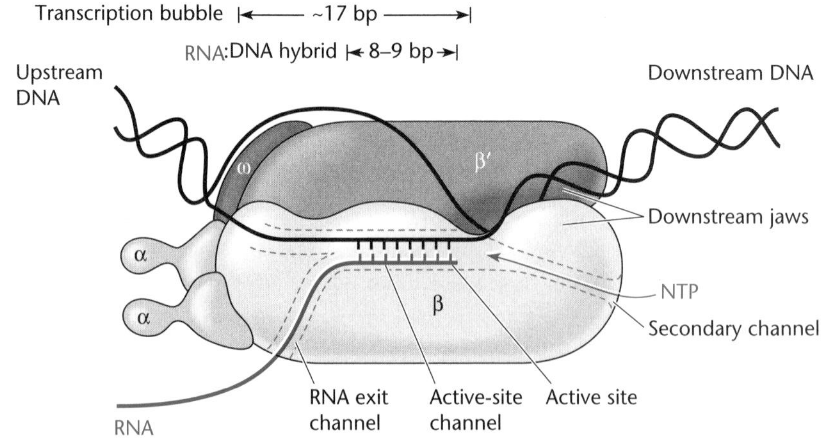

Figure 2.17 The TEC. During elongation, NTPs enter through the secondary channel and are polymerized at the active site; the nascent RNA exits through the RNA exit channel. doi:10.1128/9781555817169.ch1.f2.17

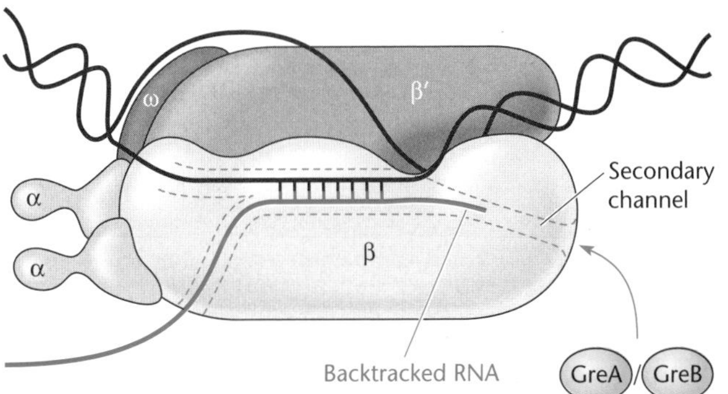

Figure 2.18 Backtracked TECs. Backward movement of RNA polymerase results in placement of the 3′ end of the nascent RNA within the secondary channel, which prevents entry of NTP substrates. GreA and GreB can enter the secondary channel to cleave the nascent RNA, which repositions the 3′ end of the transcript in the active site, allowing transcription elongation to continue.
doi:10.1128/9781555817169.ch1.f2.18

in the emerging rRNA and avoid premature termination of transcription (see below). Pausing may also play an important role in a variety of mechanisms for gene regulation (see chapter 12).

TERMINATION OF TRANSCRIPTION

Once the RNA polymerase has initiated transcription at a promoter, it continues along the DNA, polymerizing ribonucleotides, until it encounters a transcription termination signal in the DNA. These sites are not necessarily at the ends of individual genes. In bacteria, more than one gene is often transcribed into a single RNA, so a transcription termination site does not occur until the end of the cluster of genes that are transcribed together. Even if only a single gene is being transcribed, the transcription termination site may occur far downstream of the gene.

Bacterial RNA polymerase responds to two basic types of transcription termination sites, designated factor-independent and factor-dependent terminators. As their names imply, these types are distinguished by whether they work with just RNA polymerase and DNA alone or need other factors before they can terminate transcription. Both types of termination signal require participation of the newly transcribed RNA to promote termination, which means that RNA polymerase must transcribe the terminator region before termination can occur.

Factor-Independent Termination

The factor-independent (or intrinsic) transcription terminators are easy to recognize, because they have similar properties. As shown in Figure 2.19, a typical factor-independent terminator site consists of two parts. The first is an inverted repeat. When an inverted-repeat DNA sequence is transcribed into RNA, the RNA can form a hairpin because the two parts of the repeat are complementary to each other (Figure 2.19B). The inverted repeat is followed by a short string of A's in the template strand, which results in synthesis of a series of U's in the RNA. Transcription usually terminates somewhere in the string of A's in the DNA, leaving a string of U's at the 3′ end of the RNA.

Figure 2.19B shows how a factor-independent transcription terminator might work. The transcription of the U-rich RNA from the A-rich template might cause the

Figure 2.19 Transcription termination at a factor-independent termination site. **(A)** Sequence of a typical site. **(B)** The U-rich RNA causes RNA polymerase to pause, allowing a hairpin to form in the nascent RNA as it emerges from RNA polymerase, which causes RNA polymerase to dissociate from the template DNA and release the RNA product.
doi:10.1128/9781555817169.ch1.f2.19

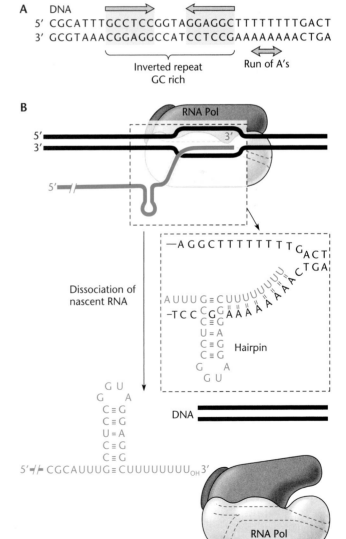

A DNA
5′ CGCATTTGCCTCCGGTAGGAGGCTTTTTTTTGACT
3′ GCGTAAACGGAGGCCATCCTCCGAAAAAAAACTGA
Inverted repeat GC rich
Run of A's

B

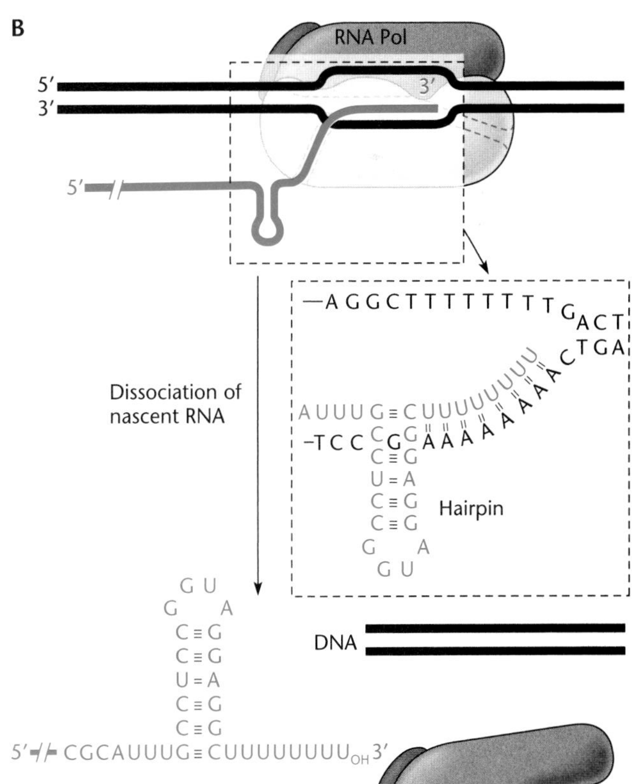

RNA polymerase to pause and allow time for the GC-rich hairpin to form. The hairpin then causes the RNA polymerase to be released by an unknown mechanism, but it may involve effects similar to those that occur during backtracking, such as pushing the 3′ end of the transcript out of the active site and hairpin-induced conformational changes. Termination is also likely to be influenced by the fact that the AU base pairs that form in the DNA-RNA hybrid are less stable. RNA polymerase and the RNA transcript fall off the DNA, terminating transcription.

Factor-Dependent Termination

While factor-independent terminators are easily recognizable, the factor-dependent transcription terminators have very little sequence in common with each other and therefore are not readily apparent. The major termination factor in *E. coli* is Rho (ρ). The ρ factor can be found in most types of bacteria, so this type of termination is probably conserved.

Any model for how the ρ factor terminates transcription at ρ-dependent termination sites has to incorporate the following facts. First, ρ usually causes the termination of RNA synthesis only if the RNA is not being translated. In bacteria, which lack a nuclear membrane, translation can begin on a nascent RNA before transcription is complete (see the introduction). Second, ρ is an RNA-dependent ATPase that cleaves ATP to get energy, but its ATPase activity is dependent on the presence of RNA. Finally, ρ is also an RNA-DNA helicase. It is similar to the DNA helicases that separate the strands of DNA during replication, but it unwinds only a double helix with RNA in one strand and DNA in the other.

Figure 2.20 illustrates a current model for how ρ terminates transcription. Recent structural evidence shows that ρ forms a hexameric (six-sided) ring made up of six subunits encoded by the *rho* gene. This ring binds to a sequence in the RNA called the *rut* (for *r*ho *ut*ilization) site. However, ρ can bind to a *rut* site in the RNA only if the RNA in this region is not occupied by ribosomes during translation of an mRNA, for example, if translation has terminated upstream at a nonsense codon, as shown in Figure 2.20. The *rut* sites are not very distinctive but are about 40 nucleotides long and have many C's and not much secondary structure. Once ρ has bound to a *rut* site through the outside of its ring, the ring moves along the RNA in the 5′-to-3′ direction, chasing the RNA polymerase. Energy for this movement is provided by the cleavage of ATP to adenosine diphosphate (ADP) by the ATPase activity of ρ. The ρ hexamer may move because the RNA binds sequentially to each of the subunits in the ring. Cleavage of ATP by the bound subunit may change the conformation of that subunit so that the RNA is transferred to the next subunit. In this way, the ρ factor ring rotates down the RNA behind the RNA

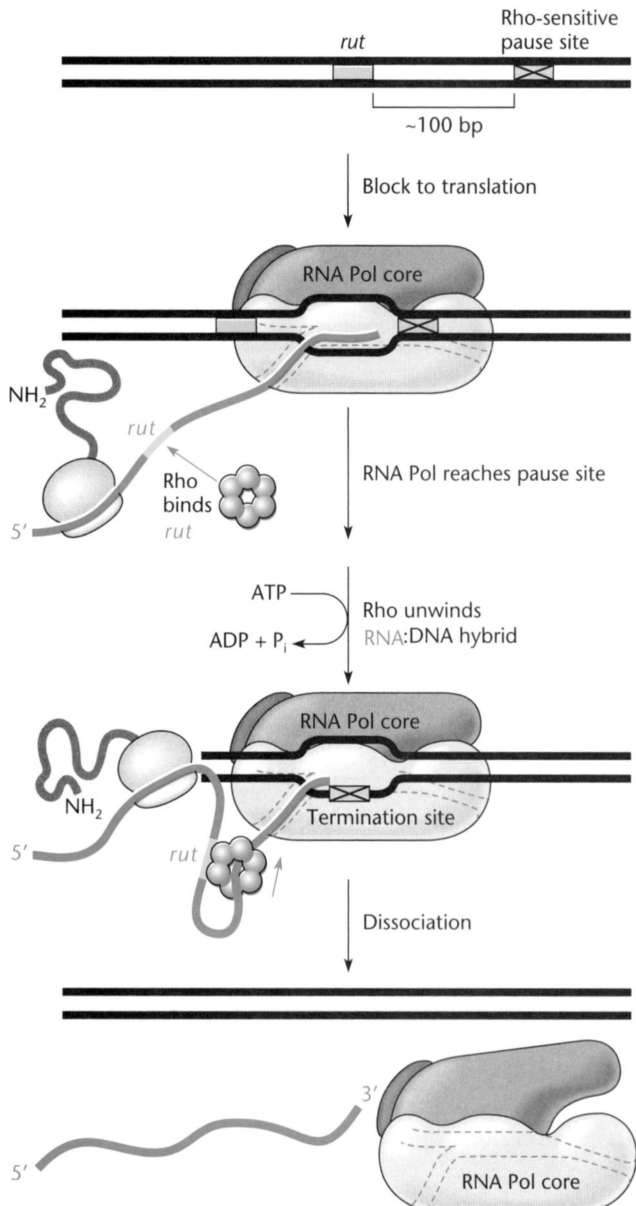

Figure 2.20 Model for factor-dependent transcription termination at a ρ-sensitive pause site. The ρ factor attaches to the RNA at a *rut* site if the RNA is not being translated (for example, if the ribosome has stopped at a nonsense codon) and forms a ring around the RNA. It then moves along the RNA with the cleavage of ATP until it catches up with paused RNA polymerase at a ρ-sensitive pause site. The helicase activity of the ρ factor then dissociates the RNA-DNA hybrid in the transcription bubble, causing the RNA polymerase and the RNA to be released.
doi:10.1128/9781555817169.ch1.f2.20

polymerase at a speed of about 60 nucleotides per second. However, the RNA polymerase is capable of transcribing at 100 nucleotides per second, so the ρ factor can catch up only if the RNA polymerase pauses at a ρ-dependent termination site. Then, the ρ factor can catch up to the RNA polymerase, and its RNA-DNA helicase activity disrupts the RNA-DNA helix in the transcription bubble, stopping transcription and releasing the RNA polymerase from the DNA template. While this model accounts for most of the known activities of ρ, it leaves unanswered the question of how ρ can access the RNA-DNA helix, which is in the inside of the RNA polymerase. One possibility is that the RNA polymerase partially opens up when it is paused at a ρ-dependent termination site. The coupling of transcription termination to translation blockage ensures that transcription of the gene will stop only when translation has terminated. However, ρ-dependent termination is not very efficient, and transcription continues through a ρ-sensitive pause site as much as 50% of the time. ρ-dependent termination not only occurs at the ends of transcribed regions, but also accounts for ρ-dependent polarity (see "Polar Effects on Gene Expression" below).

rRNAs and tRNAs

Transcription of the genes for all RNAs in the cell follows the same basic process. However, rRNAs and tRNAs play special roles in protein synthesis, so their fates after transcription differ from that of mRNAs.

The ribosomes are some of the largest structures in bacterial cells and are composed of both proteins and RNA. Bacterial ribosomes contain three types of rRNA: 16S, 23S, and 5S. The S value (from Svedberg, the name of the person who pioneered this way of measuring the sizes of molecules) is a measure of how fast a molecule sediments in an ultracentrifuge. In general, the higher the S value, the larger the RNA. The designation has persisted, even though this method of measuring molecular size is essentially no longer used.

The rRNAs are among the most highly evolutionarily conserved of all the cellular constituents, as indeed are many of the components of the translational machinery. For this reason, they have formed the basis for molecular phylogeny (Box 2.1). Comparisons of the sequences of rRNAs and other constituents of the translation apparatus from different species permit estimates to be made of how long ago these constituents separated evolutionarily.

In some strains of bacteria, the 23S rRNA is comprised of two pieces. The breaks occur because the rRNA genes contain parasitic DNAs whose sequences are cut out of the rRNA (see below). However, unlike what happens with most RNA introns, the cleavage products are not spliced back together again after the inserted segments leave the RNA. Ribosomes containing the split rRNA segments still function because the overall structure of the ribosome holds the pieces of rRNA together.

In addition to their structural role in the ribosome, the rRNAs play a direct role in translation. The 23S rRNA is the peptidyltransferase enzyme, which joins amino acids into protein on the ribosome. The 23S rRNA therefore acts as a **ribozyme**, an RNA enzyme (see below). The 16S rRNA lacks enzymatic activity but is directly involved in initiation and **termination of translation**, as well as in decoding of the sequence of the mRNA.

The rRNAs and tRNAs make up the bulk of the RNA in cells for two reasons. In a rapidly growing bacterial cell, much of the total RNA synthesis is devoted to making these RNAs. Also, the rRNAs and tRNAs are far more resistant to degradation than mRNA. With this combination of a high synthesis rate and high stability, the rRNAs and tRNAs together can amount to more than 95% of the total RNA in a bacterial cell.

Not only do the rRNAs physically associate in the ribosome, but they also are synthesized together as long precursor RNAs containing all three forms of rRNA separated by so-called spacer regions. The precursors often contain one or more tRNAs, as well (Figure 2.21). The individual rRNAs and tRNAs are released from the precursor RNAs by **ribonucleases (RNases)**. Some of these RNases participate in both rRNA and tRNA processing and RNA degradation, while others are dedicated to a single function (such as RNase P, which generates the 5′ end of tRNAs). At some point during the processing, the RNAs are also modified to make the mature rRNAs and tRNAs.

The faster a cell grows, the more protein it needs to make. Ribosomes are the site of protein synthesis; therefore, cells can increase their growth rate only if they increase the number of ribosomes. In many bacteria, the coding sequences for the rRNAs are repeated in 7 to 10 copies in the genome. Duplication of these genes leads to higher rates of rRNA synthesis in these bacteria.

Figure 2.21 Precursor of rRNA. The long molecule contains the 16S, 23S, and 5S rRNAs, as well as one or more tRNAs. RNases cut the individual RNAs out of the long precursor after it is synthesized.
doi:10.1128/9781555817169.ch1.f2.21

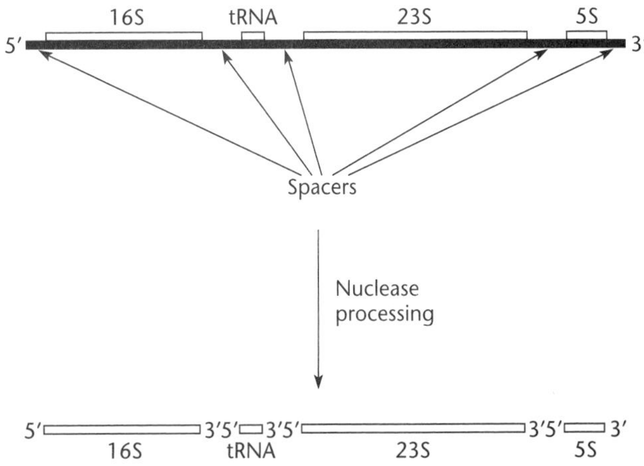

BOX 2.1

Molecular Phylogeny

The translation apparatus is the most highly conserved of all the cellular components. The structures of ribosomes, translation factors, aaRS enzymes, tRNAs, and the genetic code itself have changed remarkably little in billions of years of evolution. This is why these components have been used extensively in molecular phylogeny. By comparing the sequences of the rRNAs and other components of the translation apparatus and determining how much they have diverged, it has been possible to establish phylogenetic trees that include all organisms on Earth. The high level of conservation probably also explains why so many different antibiotics target the translation apparatus compared to other cellular components. An antibiotic designed to inhibit translation in one type of bacteria will probably inhibit translation in many other types of bacteria.

The conservation of components of the translation apparatus is so high that "rooted" evolutionary trees can be made that include eukaryotes and archaea (see the introduction). Such trees are usually not too different from what has been obtained from physiological and other comparisons, but there are sometimes surprises. For example, a 1-mm-long organism found in sea clams around thermal vents in the sea floor was shown to be a bacterium on the basis of its 16S rRNA sequence. Most bacteria are thousands of times smaller than this. Also, the sequence of the translation elongation factors led to the suggestion that the archaea are more closely related to eukaryotes than they are to other bacteria, prompting the change of their name to archaea from the original designation "archaebacteria."

Many of the initiation and elongation factors in bacteria have counterparts in archaea and eukaryotes. Nevertheless, the major differences in the translation apparatus come in the translation initiation factors. While bacteria have only three initiation factors (some of them have more than one form), archaea and eukaryotes have many more. As is the case with other cellular functions, archaea share more of their initiation factors with eukaryotes than they do with bacteria. Also, some of the initiation factors, while conserved, seem to have somewhat different functions in the three kingdoms of life. These differences may reflect differences in the initiation sites for translation.

References

Ganoza, M. C., M. C. Kiel, and H. Aoki. 2002. Evolutionary conservation of reactions in translation. *Microbiol. Mol. Biol. Rev.* **66:**460–485.

Ibawe, N., K.-I. Kuma, M. Hasegawa, S. Osawa, and T. Migata. 1989. Evolutionary relationships of archaebacteria, eubacteria and eukaryotes inferred from phylogenetic trees of duplicated genes. *Proc. Natl. Acad. Sci. USA* **86:**9355–9359.

Owens, R. J. 2004. Bacterial taxonomics: finding the wood through the phylogenetic trees. *Methods Mol. Biol.* **266:**353–383.

Pace, N. R. 2009. Mapping the tree of life: progress and prospects. *Microbiol. Mol. Biol. Rev.* **73:**565–576.

Woese, C. R. 1987. Bacterial evolution. *Microbiol. Rev.* **51:**221–271.

Although the precursor RNAs encoded by these different regions have identical rRNAs, they often contain different tRNAs and spacer regions.

MODIFICATION OF RNA

Some of the RNAs in cells are modified after they are made. For example, specific nucleotides in the rRNAs are methylated, and this sometimes confers resistance to certain antibiotics (see below). The tRNAs are probably the most highly processed and modified RNAs in cells (see Björk and Hagervall, Suggested Reading). Figure 2.22 shows a "mature" tRNA that was originally cut out of a much longer molecule that may also have included the rRNAs as well as other tRNAs. Some of the bases were modified by specific enzymes, creating altered bases, such as pseudouracil and dihydrouracil. Finally, the CCA at the 3′ end either was derived from the tRNA gene or was added posttranscriptionally by an enzyme called CCA nucleotidyltransferase. Clearly, much had to be done to this molecule after it was synthesized before it could become a functional, mature tRNA.

RNA Degradation

Different classes of RNAs have very different survival times in the cell. Highly structured RNAs, like rRNAs and tRNAs, are very stable and may persist through several rounds of cell division. The assembly of rRNAs into ribosomal particles further stabilizes them, and tRNAs are generally present either in a complex with their cognate aaRS or with **translation elongation factor Tu** (**EF-Tu**) (see below). In contrast, most bacterial mRNAs are very unstable, with an average **half-life** in *E. coli* of 1 to 3 minutes; this term refers to the time required for the amount of an RNA to decrease to 50% of its initial level. This contrasts with the situation in eukaryotes, where mRNAs are often very stable (with a half-life of hours). Efficient mRNA degradation is important for

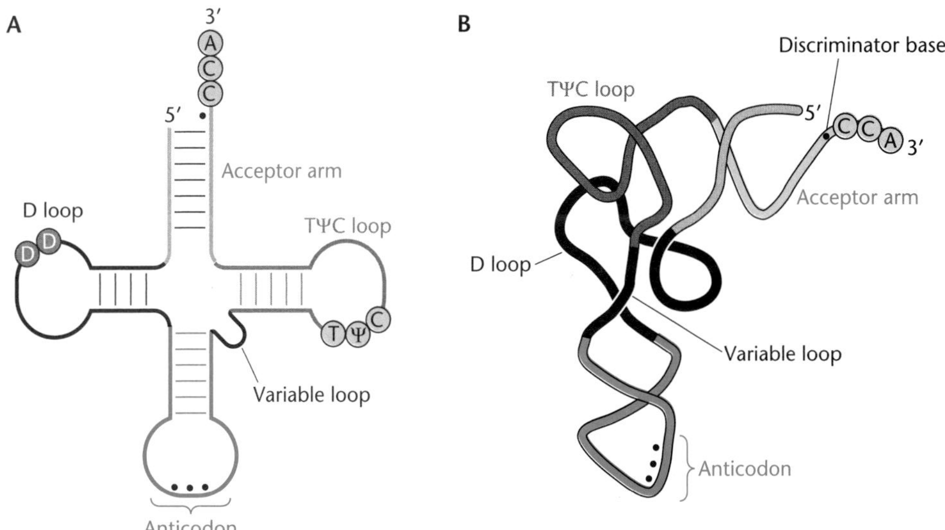

Figure 2.22 Structure of mature tRNAs. **(A)** Standard cloverleaf representation of tRNA, showing the base pairing that holds the molecule together and some of the standard modifications. D is the modified base dihydrouridine. tRNAs also contain thymine (T) and pseudouracil (ψ) among other modifications. The CCA at the 3′ end is where an amino acid attached. **(B)** Folding of tRNA into its tertiary structure. Discriminator base, the position of a base important in correct aminoacylation. doi:10.1128/9781555817169.ch1.f2.22

gene regulation and also releases nucleotides for use in new rounds of transcription. A variety of RNases participate in mRNA degradation, and the profiles of RNases vary somewhat in different groups of bacteria.

RNases

There are two major classes of RNases (Table 2.1 and Figure 2.23). **Endoribonucleases** cleave the sugar-phosphate backbone of the RNA within the RNA chain to generate two smaller RNA products, one with a 3′ hydroxyl and the other with a 5′ monophosphate. **Exoribonucleases** digest the RNA processively, removing 1 nucleotide at a time, starting at a free end. Some RNases are found in many different organisms, while others are present in only certain groups of organisms. In *E. coli*, all exoribonucleases that have been identified bind to the 3′ end of an RNA substrate and digest the RNA in a 3′-to-5′ direction. In contrast, *Bacillus subtilis* and a number of other organisms contain both 3′-5′ exoribonucleases and 5′-3′ exoribonucleases. At least one of the *B. subtilis* 5′-3′ RNases (RNase J2) has both exoribonucleolytic and endoribonucleolytic activities (Table 2.1).

MODULATION OF RNase ACTIVITY

The susceptibility of an RNA to different RNases can be affected by structural features of the RNA. RNA 3′ ends generated by **termination of transcription** at a factor-independent terminator contain an RNA hairpin, which inhibits binding of 3′-5′ exoribonucleases (Figure 2.23). Degradation of RNAs of this type is often initiated by

Table 2.1 Enzymes involved in mRNA processing and degradation

Enzyme	Substrate(s)	Description
RNase E	mRNA, rRNA, tRNA	Endonuclease, highly conserved in all gram-negative and some gram-positive bacteria (not *B. subtilis*)
RNase III	rRNA, polycistronic mRNA	Endonuclease, cleaves double-stranded RNA in some stem-loops; found in both gram-negative and gram-positive bacteria
RNase P	Polycistronic mRNA, tRNA precursors	Ribozyme, necessary to process 5′ ends of tRNAs
RNase G	5′ end of 16S rRNA, mRNA	Endonuclease, replaces RNase E in some bacteria
RNases J1 and J2	mRNA, rRNA	5′-3′ exonuclease, endonuclease; found in most gram-positive and some gram-negative bacteria (not *E. coli*)
Poly(A) polymerase	mRNA	Found in both gram-positive and gram-negative bacteria
PNPase	mRNA, poly(A) tails	3′-5′ exonuclease, found in all gram-positive and gram-negative bacteria

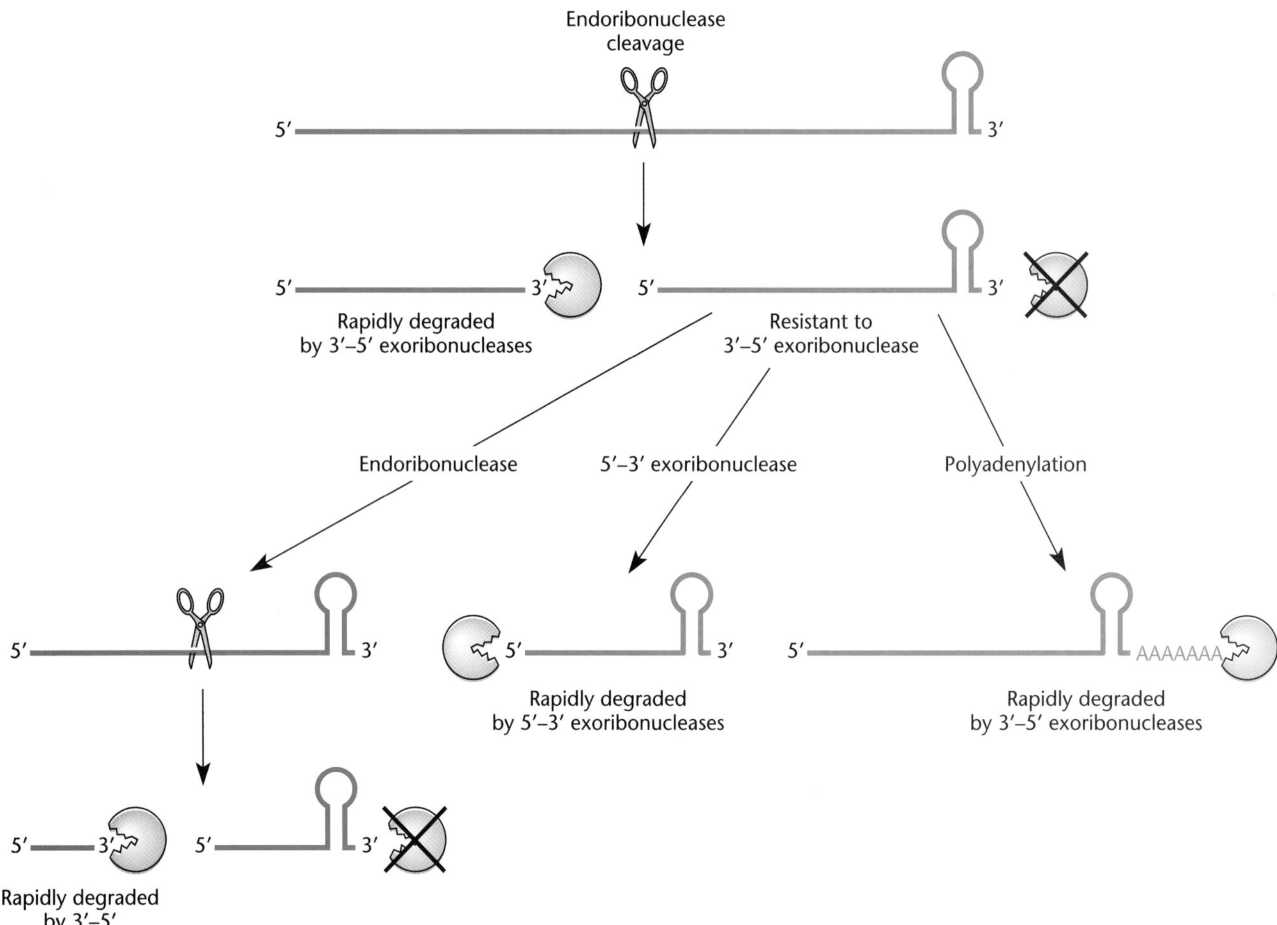

Figure 2.23 Pathways for RNA degradation. RNA transcripts that are generated by termination at a factor-independent terminator contain a hairpin at the 3′ end, which inhibits degradation by 3′-5′ exoribonucleases (exos). Degradation is often initiated by an endonucleolytic cleavage, followed by rapid exonucleolytic digestion from the new 3′ end. The stable 3′ fragment (which retains the terminator hairpin) can be cleaved again by an endoribonuclease or can be degraded by a 5′-3′ exoribonuclease in organisms that have this class of enzyme. Alternatively, a poly(A) tail can be added by poly(A) polymerase, which allows binding of a 3′-5′ exoribonuclease. doi:10.1128/9781555817169.ch1.f2.23

endonucleolytic cleavage, which removes the 3′ end of the RNA and allows the 5′ region of the molecule to be degraded. Degradation of the 3′ fragment can be initiated by polyadenylation of the 3′ end of the RNA by polyadenylate [poly(A)] polymerase, encoded by the *pcnB* gene. Addition of the poly(A) tail provides a "landing zone" for 3′-5′ exoribonucleases, which can initiate degradation of the poly(A) sequence and then continue to move through the terminator hairpin. This may be facilitated by colocalization of poly(A) polymerase and PNPase, one of the major 3′-5′ exonucleases, with other RNases into a complex called the degradosome. Note that polyadenylation of an mRNA in eukaryotes generally results in stabilization, while polyadenylation of an RNA in bacteria results in rapid degradation. Degradation of the 3′ fragment can also be directed by 5′-3′ exoribonucleases in organisms like *B. subtilis* that have this activity (see Condon, Suggested Reading).

Susceptibility to RNase degradation can be used as a mechanism to regulate gene expression, as rapid degradation of an mRNA results in reduced synthesis of its protein product. Modulation of RNA stability can occur through changes in the RNA structure that affect RNase binding, by binding of a regulatory protein to the RNA, or by binding of a regulatory RNA. Mechanisms of this type are discussed in chapter 12.

The Structure and Function of Proteins

Proteins do most of the work of the cell. While there are a few RNA enzymes (ribozymes), most of the enzymes that make and degrade energy sources and make cell constituents are proteins. Also, proteins make up much of the structure of the cell. Because of these diverse roles, there are many more types of proteins than there are types of other cell constituents. Even in a relatively simple bacterium, there are thousands of different types of proteins, and most of the DNA sequences in bacteria are dedicated to genes that encode proteins.

Protein Structure

Unlike DNA and RNA, which consist of a chain of nucleotides held together by phosphodiester bonds between the sugars and phosphates, proteins consist of chains of 20 different amino acids held together by **peptide bonds.** Figure 2.24 shows the formation of a peptide bond between two amino acids. The peptide bond is formed by joining the amino group (NH_2) of one amino acid to the carboxyl group (COOH) of the previous amino acid. These amino acids in turn are joined to other amino acids by the same type of bond, making a chain. A short chain of amino acids is called an **oligopeptide**, and a long chain is called a **polypeptide**.

Like RNA and DNA, polypeptide chains have directionality and a way to distinguish the ends of the chain from each other. In polypeptides, the direction is defined by their amino and carboxyl groups. One end of the chain, the **amino terminus** or **N terminus**, has an unattached amino group. The amino acid at this end is called the **N-terminal amino acid.** On the other end of the polypeptide, the final carboxyl group is called the **carboxyl terminus** or **C terminus**, and the amino acid is called the **C-terminal amino acid.** As we shall see, proteins are synthesized from the N terminus to the C terminus.

Protein structure terminology is the same as that for RNA structures. Proteins have primary, secondary, and tertiary structures, as well as quaternary structures. All of these are shown in Figure 2.25.

Figure 2.24 Two amino acids joined by a peptide bond. The bond connects the amino group on the second amino acid to the carboxyl group on the preceding amino acid. R is the side group of the amino acid that differs in each type of amino acid. doi:10.1128/9781555817169.ch1.f2.24

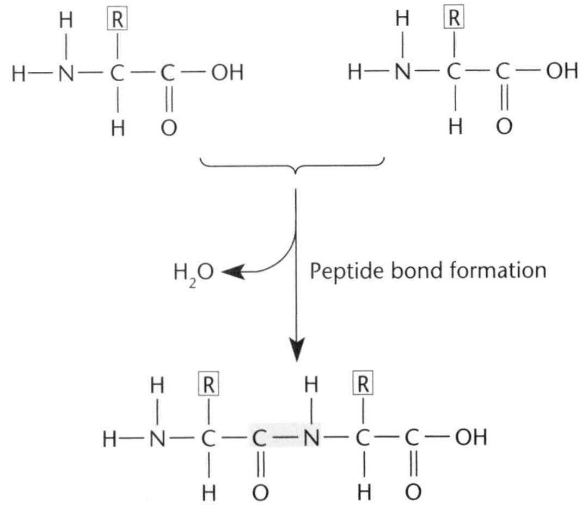

```
   H  R              H  R
   |  |              |  |
H—N—C—C—OH        H—N—C—C—OH
      |  ||              |  ||
      H  O               H  O

        H2O <— Peptide bond formation

   H  R     H  R
   |  |     |  |
H—N—C—C—N—C—C—OH
      |  ||     |  ||
      H  O      H  O
```

PRIMARY STRUCTURE

Primary structure refers to the sequence of amino acids and the length of a polypeptide. Because polypeptides are made up of 20 amino acids instead of just 4 nucleotides, as in RNA, many more primary structures are possible for polypeptides than for RNA chains. The sequence of amino acids in a polypeptide is dictated by the sequence of nucleotides in the mRNA used as the template for synthesis of that protein.

SECONDARY STRUCTURE

Also like RNA, polypeptides can have a secondary structure, in which parts of the chain are held together by hydrogen bonds. However, because many more types of interactions are possible between amino acids than between nucleotides, the secondary structure of a polypeptide is more difficult to predict. The two basic forms of secondary structures in polypeptides are α-helices, where a short region of the polypeptide chain forms a helix due to the interaction of each amino acid with the one before and the one after it, and β-sheets, in which stretches of amino acids interact with other stretches to form sheetlike structures (Figure 2.25). These types of structured regions are often joined together by more flexible regions known as linkers. Computer software is available to help predict which secondary structures of a polypeptide are possible on the basis of its primary structure. However, these programs are not entirely reliable, and techniques like X-ray crystallography and nuclear magnetic resonance spectroscopy provide much more detailed information about the secondary structure of a polypeptide.

TERTIARY STRUCTURE

Polypeptides usually also have a well-defined tertiary structure, in which they fold up on themselves with hydrophobic amino acids (such as leucine and isoleucine), which are not very soluble in water, on the inside and charged amino acids (such as glutamate and lysine), which are more water soluble, or hydrophilic, on the outside. We discuss the structure of proteins in more detail in "Protein Folding and Degradation" below.

QUATERNARY STRUCTURE

Proteins made up of more than one polypeptide chain also have **quaternary structure.** Such proteins are called multimeric proteins. When the polypeptides are the same, the protein is a **homomultimer.** When they are different, the protein is a **heteromultimer.** Other names reflect the number of polypeptides in the protein. For example, the term **homodimer** describes a protein made of two identical polypeptides, whereas **heterodimer** describes a protein

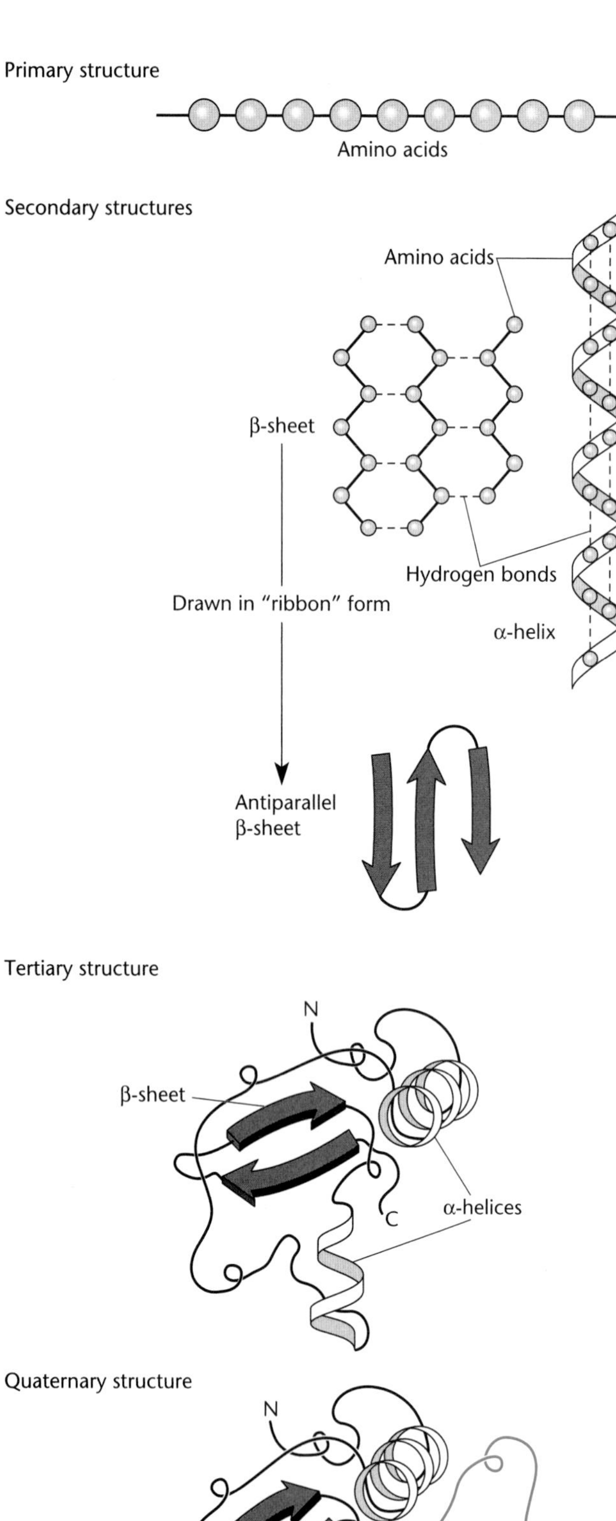

made of two different chains encoded by different genes. The names trimer, tetramer, and so on refer to increasing numbers of polypeptides. Hence, the ρ transcription termination factor is a homohexamer (see above).

The polypeptide chains in a protein are usually held together by hydrogen bonds. The only covalent chemical bonds in most proteins are the peptide bonds that link adjacent amino acids to form the polypeptide chains. As a result, if a protein is heated, it falls apart into its individual polypeptide chains. However, some proteins are unusually stable; these include extracellular enzymes, which must be able to function in the harsh environment outside the cell. Such proteins are often also held together by **disulfide bonds** between cysteine amino acids in the protein.

Translation

The translation of the sequence of nucleotides in mRNA into the sequence of amino acids in a protein occurs on the ribosome. The overall process of translation is highly conserved in bacteria, archaea, and eukaryotes, and the machinery is also highly conserved. As mentioned in "rRNAs and tRNAs" above, the ribosome is one of the largest and most complicated structures in cells, consisting of three different RNAs and over 50 different proteins in bacteria. It is also one of the major constituents of the bacterial cell, and much of the cell's biosynthetic capacity goes into making ribosomes. Each cell contains thousands of ribosomes, with the actual number depending on the growth conditions. It is also one of the most evolutionarily highly conserved structures in cells, having remained largely unchanged in shape and structure from bacteria to humans.

The ribosome is actually an enormous enzyme that performs the complicated role of polymerizing amino acids into polypeptide chains, using the information in mRNA as a guide. As such, a better name for it might have been protein polymerase, by analogy to DNA and RNA polymerases. It was given the historical name "ribosome" because it is large enough to have been visualized under the electron microscope, and so it was called a "some" (for body) and "ribosome" because it contains ribonucleotides. The recent determination of the structure of the ribosome has led to important insights into how it performs its function of polymerizing amino acids.

Structure of the Bacterial Ribosome

Figure 2.26 shows the components of a bacterial ribosome. The complete ribosome, called the 70S ribosome, consists of two subunits, the 30S subunit, which

Figure 2.25 Primary, secondary, tertiary, and quaternary structures of proteins.
doi:10.1128/9781555817169.ch1.f2.25

contains one molecule of 16S rRNA, and the 50S subunit, which contains one molecule each of 23S and 5S rRNA. Each subunit also contains **ribosomal proteins;** the 30S subunit contains approximately 21 different proteins (S1, S2, etc., where S indicates small subunit proteins), while the 50S subunit contains approximately 31 different proteins (L1, L2, etc., where L indicates large subunit proteins). Ribosomes from different bacterial species have very similar compositions, with small differences in the number of ribosomal proteins. Like the different terms for rRNA, the names of ribosomal subunits are derived from their sedimentation rates during ultracentrifugation. The 30S and 50S subunits normally exist separately in the cell and come together to form the complete 70S ribosome only when they are translating an mRNA. Note that sedimentation values are not additive; the complete ribosome is 70S, despite being composed of subunits that are 30S plus 50S in size.

The two ribosomal subunits play very different roles in translation. To initiate translation, the 30S subunit binds to the mRNA. Then, the 50S ribosome binds to the 30S subunit to make the 70S ribosome. From that point on, the 30S subunit mostly helps to select the correct aa-tRNA for each codon while the 50S subunit does most of the work of forming the peptide bonds and translocating the tRNAs from one site on the ribosome to another (see below). The 70S ribosome moves along the mRNA, allowing tRNA anticodons to pair with the mRNA codons and translate the information from the nucleic acid chain into a polypeptide. After the polypeptide chain is completed, the ribosome separates again into the 30S and 50S subunits. The role of the subunits is discussed in more detail below.

Figure 2.26 The composition of a ribosome containing one copy each of the 16S, 23S, and 5S rRNAs, as well as many proteins. The proteins of the large 50S subunit are designated L1 to L31. The proteins of the small 30S subunit are designated S1 to S21. The simple subunit shapes shown here are used to represent ribosomes in illustrations throughout the textbook. doi:10.1128/9781555817169.ch1.f2.26

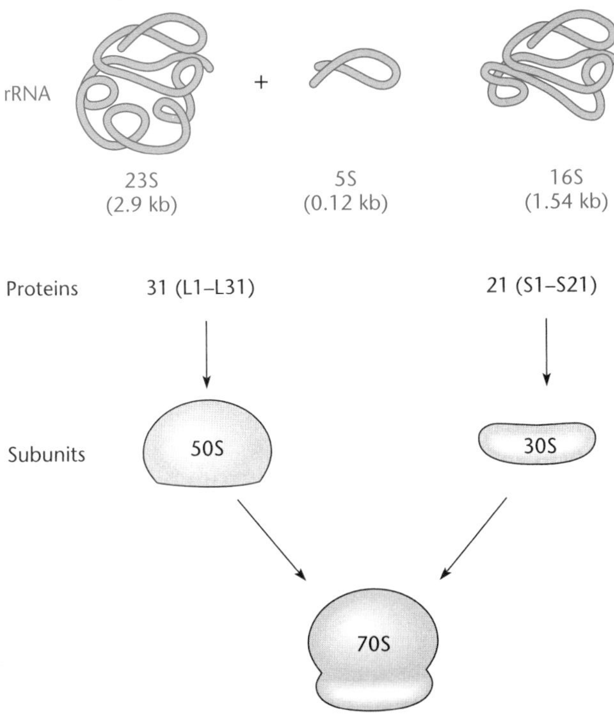

A variety of physical techniques, combined with much indirect information accumulated over the years from genetics and biochemistry, have revealed many details of the overall structure of the ribosome. The crystal structures of the individual subunits and the entire 70S ribosome have been determined and correlated with the earlier indirect information. A number of laboratories participated in this project, and this awesome achievement will go down in history as one of the major milestones in molecular biology, recognized recently by the Nobel Prize. We can review only a few of the most salient features here.

The two subunits of the ribosome are frequently represented as ovals, with a flat side that binds to the other subunit, leaving a gap between them. It is through this gap that the aa-tRNAs enter and pass through the ribosome, contributing their amino acids to the growing polypeptide chain. The polypeptide chain being synthesized passes out through a channel running through the 50S subunit. This channel is long enough to hold a chain of about 70 amino acids, so a polypeptide of this length must be synthesized before the N-terminal end of a protein first emerges from the ribosome. The 50S ribosomal subunit is rather rigid, with no moveable parts, but the 30S subunit has three domains, or regions, that can move relative to each other during translation.

The rRNAs play many of the most important roles in the ribosome, and the ribosomal proteins seem to be present mostly to give rigidity to the structure, helping to cement the rRNAs in place. This has contributed to speculation that RNAs were the primordial enzymes and that proteins came along later in the earliest stages of life on Earth. The 23S rRNA, rather than a ribosomal protein, also performs the enzymatic function that forms the peptide bonds. As mentioned above, a region of the 23S rRNA is the peptidyltransferase enzyme, which forms the peptide bonds between the carboxyl end of the growing polypeptide and the amino group of the incoming amino acid. Thus, 23S rRNA is an RNA enzyme, or ribozyme. The 23S rRNA also forms most of the channel in the 50S subunit through which the growing polypeptide passes. The 16S RNA plays crucial roles in translation initiation and in matching each incoming aa-tRNA with the mRNA. A structure of the ribosome is shown in Figure 2.27.

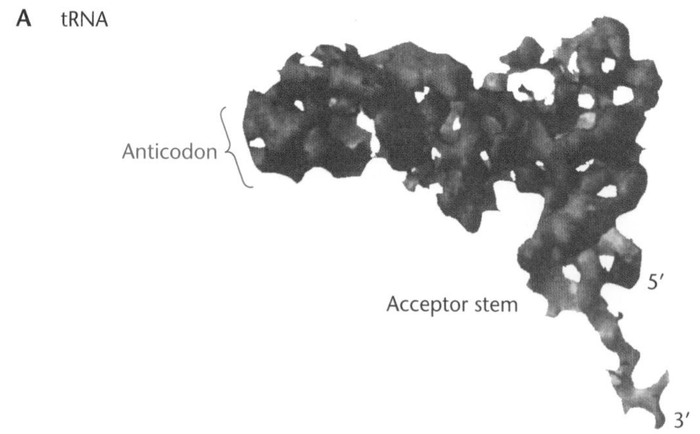

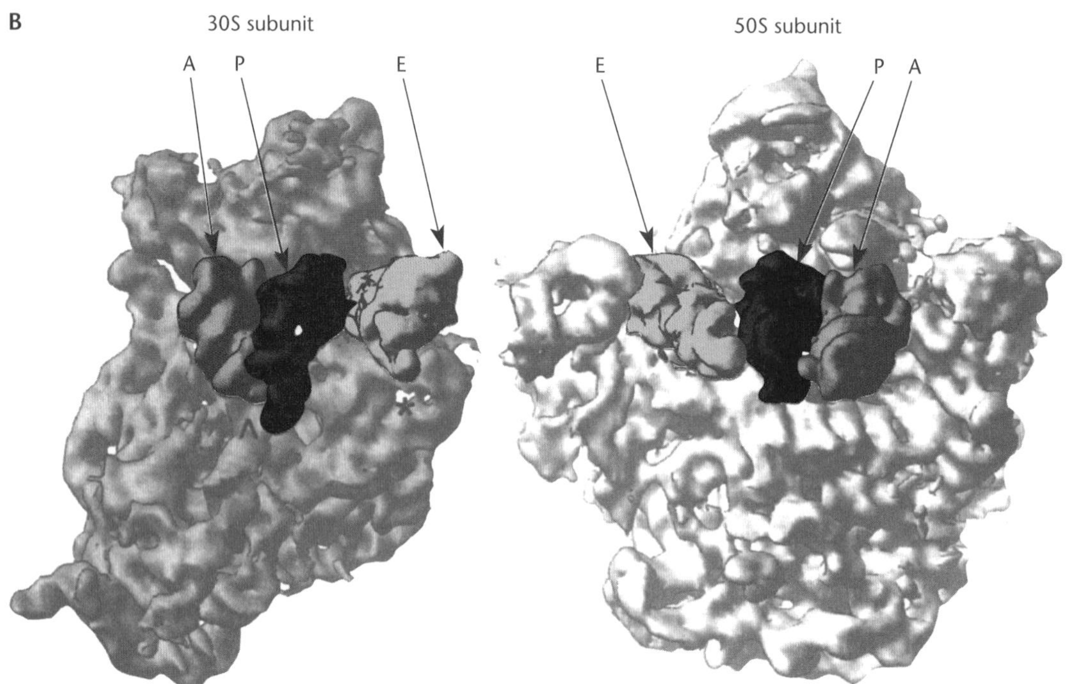

Figure 2.27 Crystal structures of a tRNA and the ribosome. **(A)** Structure of a tRNA. The anticodon loop is on the left, and the 3′ acceptor end, where the amino acid or growing polypeptide is attached, is at the bottom. **(B)** The two subunits of the ribosome separated and rotated to show the channel between them through which the tRNAs move. The 30S subunit is on the left, and the 50S subunit is on the right. The tRNAs bound at the A, P, and E sites are indicated in blue. doi:10.1128/9781555817169.ch1.f2.27

Overview of Translation

The information in mRNA is interpreted by pairing of consecutive triplets of nucleotides (codons) with the complementary anticodon sequences of the corresponding tRNAs. This pairing takes place on the ribosome, and matching of a specific tRNA (with the appropriate anticodon) to the matching amino acid is carried out by a set of enzymes called aaRS. aa-tRNAs are delivered to the ribosome by a protein factor called EF-Tu. During translation, the ribosome moves along the mRNA, and tRNAs interact with the mRNA at three distinct sites, designated the **A (aminoacyl) site**, the **P (peptidyl) site**, and the **E (exit) site** (Figure 2.28). Each aa-tRNA is brought into the A site first, where its anticodon is tested for complementarity to the mRNA codon present at that site. If the anticodon is complementary to the codon, the tRNA is retained; if the anticodon does not match the codon, the tRNA is rejected, and a new aa-tRNA is brought into the A site. A match between the anticodon and the codon in the A site triggers the

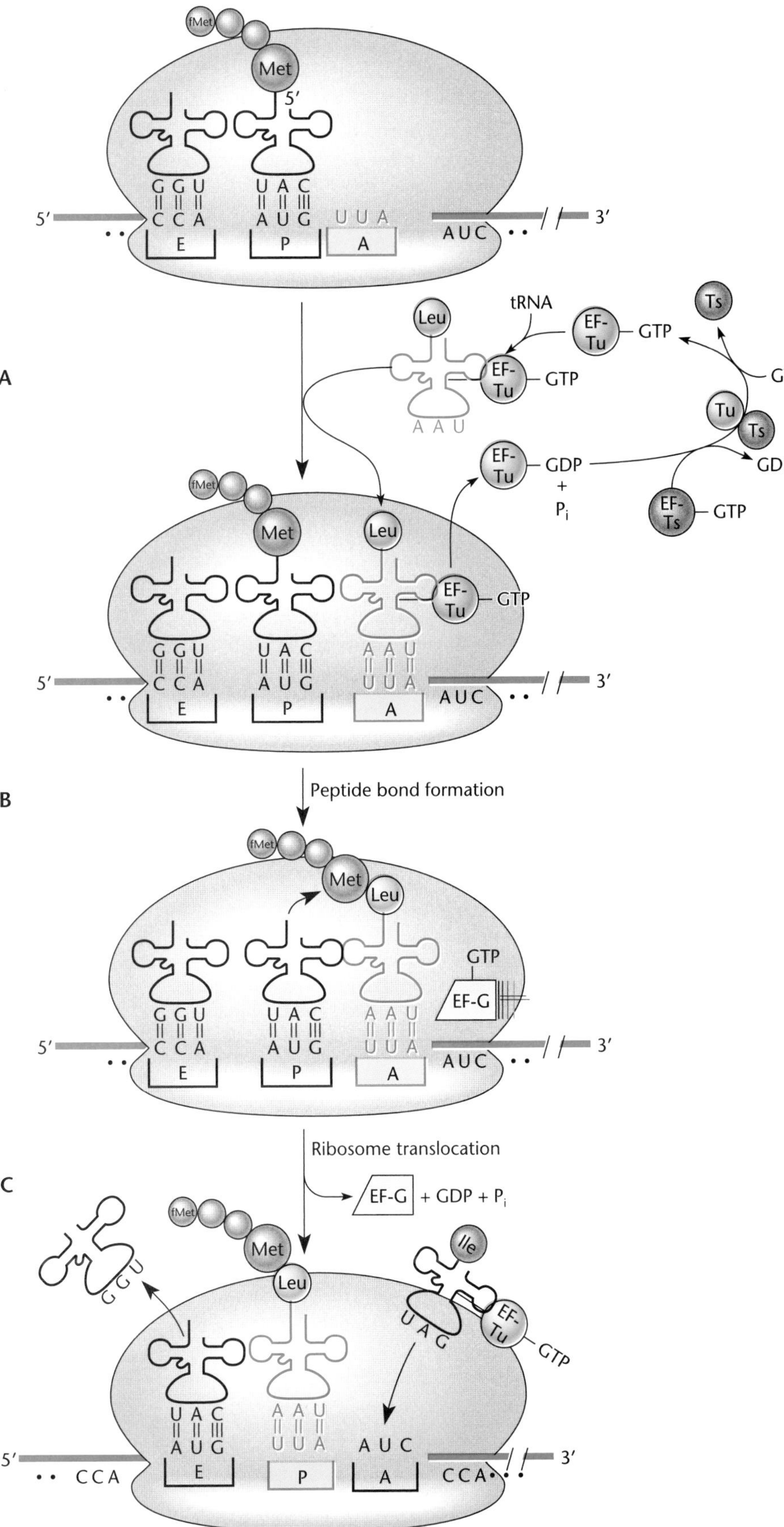

Figure 2.28 Overview of translation. **(A)** The tRNA bound to its amino acid and complexed with EF-Tu and GTP comes into the A site on the ribosome. **(B)** Peptidyltransferase (23S rRNA in the 50S ribosome) catalyzes peptide bond formation between the carboxyl end of the growing polypeptide carried by the P site tRNA to the amino end of the amino acid carried by the A site tRNA. **(C)** EF-G catalyzes translocation of the A site tRNA to the P site, making room at the A site for another aminoacyl-tRNA. Finally, the previous P site tRNA, now stripped of its polypeptide, moves to the E site before exiting the ribosome. doi:10.1128/9781555817169.ch1.f2.28

tRNA in the A site to interact with the tRNA already present in the P site. The P site tRNA carries the growing polypeptide chain, and the next step is transfer of the growing peptide chain from the P site tRNA to the A site tRNA. This results in addition of the amino acid carried on the A site tRNA to the carboxyl end of the polypeptide and attachment of the polypeptide (now 1 amino acid longer) to the A site tRNA. The now unattached P site tRNA moves to the E site of the ribosome, the A site tRNA (which now contains the polypeptide) moves to the P site of the ribosome, and the A site is now empty. Each tRNA retains contact with the mRNA through anticodon-codon pairing so that the mRNA moves through the ribosome in concert with the tRNAs. This results in placement of the next codon of the mRNA in the empty A site, which is now available for entry of another aa-tRNA. Movement of the unattached P site tRNA into the E site results in ejection of the previous unattached tRNA from the E site, which allows the next tRNA to enter the cycle. The details of this process are described below.

Details of Protein Synthesis

In this section, we discuss the process of translation in more detail. First, we discuss reading frames, which determine which nucleotides in an mRNA are recognized as codons, and tRNA aminoacylation, which is responsible for correct matching of a tRNA with its cognate amino acid. Then, we discuss translation initiation, or how the 30S subunit finds the starting point of a coding sequence in the mRNA. Next, we discuss translation elongation, or what happens as the 70S ribosome moves along the mRNA, translating the information in the mRNA in the form of nucleotides into amino acids. Finally, we discuss how translation is terminated.

READING FRAMES

Each 3-nucleotide sequence, or codon, in the mRNA encodes a specific amino acid, and the assignment of the codons is known as the genetic code. Because there are 3 nucleotides in each codon, an mRNA can be translated in three different frames in each region. Initiation of translation at a specific **initiation codon** establishes the **reading frame of translation**, so that in most cases only a single reading frame is utilized. Once translation has begun, the ribosome moves 3 nucleotides at a time through the coding part of the mRNA. If the translation is occurring in the proper frame for protein synthesis, we say the translation is in the **zero frame** for that protein. If the translation is occurring in the wrong reading frame, it can be displaced either back by 1 nucleotide in each codon (the –1 frame) or forward by 1 nucleotide (the +1 frame). In a few instances, translational frameshifts that change the reading frame even after translation has initiated can occur. Frameshift mutations (see chapter 3) cause incorrect reading of an mRNA because of the insertion or deletion of nucleotides in the DNA sequence, which is then copied into mRNA. Translational frameshifting occurs when the ribosome shifts its position on the mRNA without a change in the mRNA sequence itself.

tRNA AMINOACYLATION

Before translation can begin, a specific amino acid is attached to each tRNA by its **cognate aaRS** (Figure 2.29). Each of these enzymes specifically recognizes only 1 amino acid and one class of tRNA, hence the name cognate. How each aaRS recognizes its own tRNA varies, but the anticodon (i.e., the three tRNA nucleotides that base pair with the complementary mRNA sequence [Figure 2.30]) is not the only determinant. Other variations in the tRNA structure, such as the identity of the discriminator base immediately upstream of the CCA at the 3′ end (Figure 2.22) and the size of the variable loop, also contribute to aaRS recognition specificity. In some cases, if the anticodon in a given tRNA is mutated, the cognate aaRS still attaches the amino acid for the original tRNA, and that amino acid is inserted at a different codon in the mRNA. This is the basis of nonsense suppression, which is discussed in chapter 3. The amino acid is attached to the terminal A residue on the tRNA, and the energy for the reaction is provided by cleavage of ATP. Finally, the tRNA with its amino acid is bound to a protein called EF-Tu, which assists in delivery of the aa-tRNA to the ribosome.

Figure 2.29 Aminoacylation of a tRNA by its cognate aaRS. ATP is used as a source of energy, and the amino acid is attached to the adenosine residue at the 3′ end of the tRNA. Each amino acid utilizes a dedicated aaRS.
doi:10.1128/9781555817169.ch1.f2.29

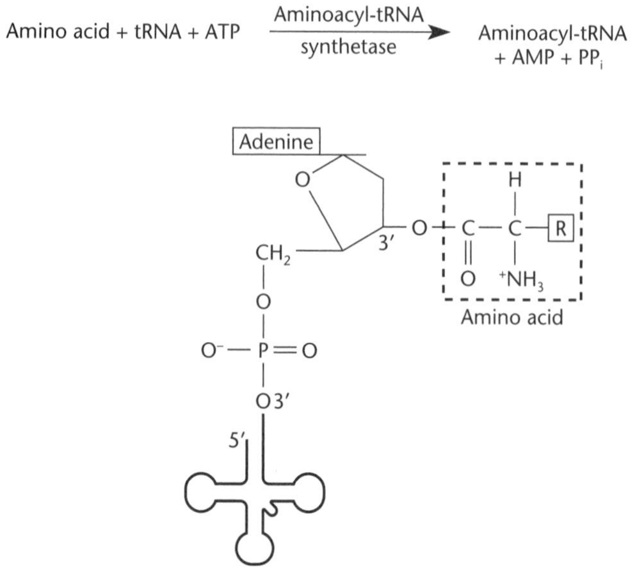

Amino acid + tRNA + ATP —(Aminoacyl-tRNA synthetase)→ Aminoacyl-tRNA + AMP + PP_i

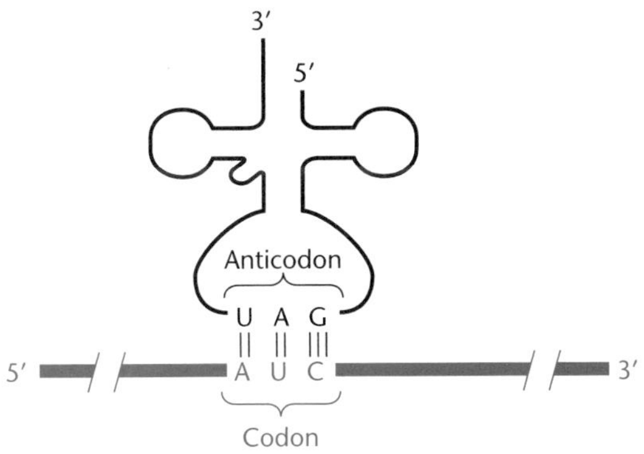

Figure 2.30 Complementary pairing between a tRNA anticodon and an mRNA codon. The codon is shown 5′-3′, and the tRNA is shown 3′-5′ to allow antiparallel pairing between the codon and the tRNA anticodon.
doi:10.1128/9781555817169.ch1.f2.30

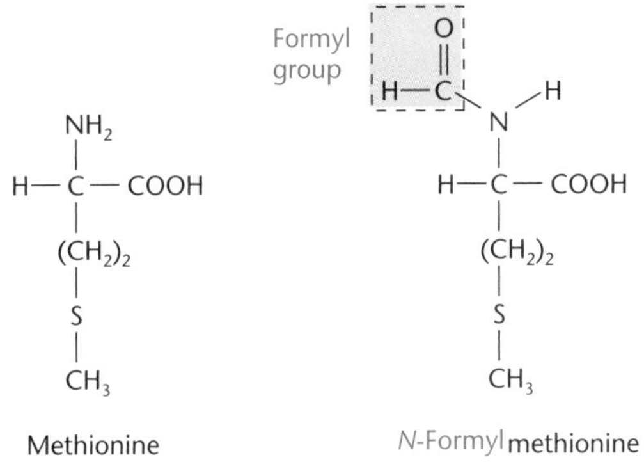

Formyl group

NH_2 — H—C—COOH — $(CH_2)_2$ — S — CH_3

Methionine

H—C(=O)—N—H — H—C—COOH — $(CH_2)_2$ — S — CH_3

N-Formylmethionine

Figure 2.31 Comparison between methionine (Met) and *N*-formylmethionine (fMet).
doi:10.1128/9781555817169.ch1.f2.31

TRANSLATION INITIATION REGIONS

In the chain of thousands of nucleotides that make up an mRNA, the ribosome must bind and initiate translation at the correct site. If the ribosome starts working at the wrong initiation codon, the protein will have the wrong N-terminal amino acids and/or the mRNA will be translated out of frame and all of the amino acids will be wrong. Hence, mRNAs have sequences called **translational initiation regions (TIRs)** or **ribosome-binding sites (RBS)** that flag the correct first codon for the ribosome. In spite of extensive research, it is still not possible to predict with 100% accuracy whether a sequence is a TIR. However, some general features of TIRs are known.

Initiation Codons

All bacterial and archaeal TIRs have an initiation codon, which is recognized by a dedicated **initiator tRNA.** This tRNA is always aminoacylated with methionine by methionyl-tRNA synthetase, and the methionine is further modified by addition of a formyl group (Figure 2.31). The three bases in initiator codons are usually AUG or GUG but in rare cases are UUG or CUG. Regardless of which amino acid these sequences call for in the genetic code (Table 2.2), if they are serving as initiation codons, they encode methionine (actually formylmethionine) as the N-terminal amino acid. After translation, this methionine is often cut off by an aminopeptidase (see below). Notice that for the initiation codons, the first position of the codon can mispair with the tRNA anticodon, which always matches AUG; this differs from "wobble" during translation elongation, which involves mismatches at the third position of the codon (see below).

The initiation codon does not have to be the first sequence in the mRNA chain. In fact, the 5′ end of the mRNA may be some distance from the TIR and the initiation codon of the first coding region in a transcript; this region is called the **5′ untranslated region** or **leader region.** In prokaryotes, many transcripts contain multiple coding regions. The translation initiation complex binds internally to the mRNA to identify the TIR for each coding region. This is different from translation in eukaryotes, where the translation initiation complex usually binds to the 5′ end of the transcript and initiates translation at the first AUG codon it encounters that is accessible (see below).

Shine-Dalgarno Sequences

Not all methionine codons serve as initiation codons. Furthermore, the fact that some of the initiation codons code for amino acids other than methionine when internal to a coding region demonstrates that the presence of one codon is clearly not enough to define a TIR. These sequences may also occur out of frame or in an mRNA sequence that is not translated at all. Obviously, other sequences in addition to these three bases must help to define them as a place to begin translation.

Many bacterial genes have a second TIR element located 5 to 10 nucleotides on the 5′ side (upstream) of the initiation codon. This sequence, named the **Shine-Dalgarno (S-D) sequence** after the two scientists who first noticed it, is complementary to a short sequence near the 3′ end of the 16S RNA. Figure 2.32 shows an example of a typical bacterial TIR with a characteristic S-D sequence. By pairing with their complementary sequences on the 16S rRNA, S-D sequences help define TIRs by properly aligning the mRNA on the ribosome. However, these sequences are not always easy to identify because they can be very short and exhibit considerable variability. Moreover, not all bacterial genes have S-D sequences. The initiation codon sometimes resides at the extreme 5′ end of the mRNA (in "leaderless mRNAs"), leaving no room for an S-D sequence. In such cases, translation initiation may occur by a somewhat different mechanism (see below).

Table 2.2 The genetic code

First position (5′ end)	Second position U	C	A	G	Third position (3′ end)
U	Phe	Ser	Tyr	Cys	U
	Phe	Ser	Tyr	Cys	C
	Leu	Ser	Stop	Stop	A
	Leu	Ser	Stop	Trp	G
C	Leu	Pro	His	Arg	U
	Leu	Pro	His	Arg	C
	Leu	Pro	Gln	Arg	A
	Leu	Pro	Gln	Arg	G
A	Ile	Thr	Asn	Ser	U
	Ile	Thr	Asn	Ser	C
	Ile	Thr	Lys	Arg	A
	Met	Thr	Lys	Arg	G
G	Val	Ala	Asp	Gly	U
	Val	Ala	Asp	Gly	C
	Val	Ala	Glu	Gly	A
	Val	Ala	Glu	Gly	G

Because of this lack of universality, often the only way to be certain that translation is initiated at a particular initiation codon is to sequence the N terminus of the polypeptide to see if the N-terminal amino acids correspond to the codons immediately adjacent to the putative initiation codon. Protein sequencing will also reveal whether the methionine encoded by the initiation codon remains on the mature protein or is removed.

INITIATOR tRNA

Translation initiation requires a dedicated aa-tRNA, the formylmethionine tRNA (**fMet-tRNAfMet**). This unique aminoacyl-tRNA has a formyl group attached to the amino group of the methionine (Figure 2.31), making it resemble a peptidyl-tRNA rather than a normal aminoacyl-tRNA. This causes it to bind to the P site rather than the A site of the ribosome, which is an important step in initiation, as discussed below, and blocks its use as an elongator tRNA. The initiator fMet-tRNA^{fMet} is synthesized somewhat differently from the other aminoacyl-tRNAs. This special tRNA uses the aminoacyl-tRNA synthetase of the normal tRNA^{Met} to attach methionine to the tRNA^{fMet}. Then, an enzyme called **transformylase** adds a formyl group to the amino group of the methionine on the tRNA^{fMet} to form fMet-tRNA^{fMet}.

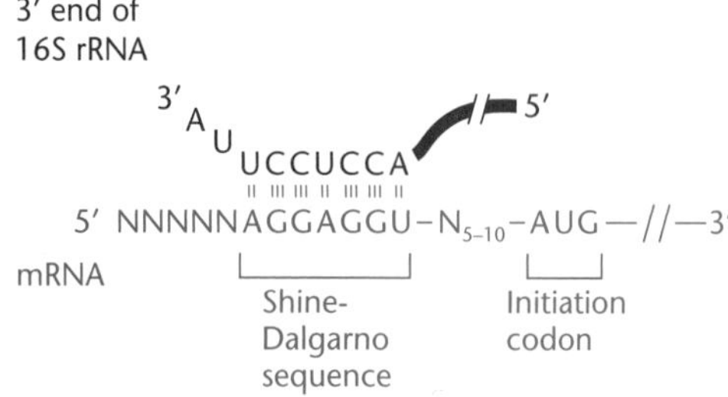

Figure 2.32 Structure of a typical bacterial TIR showing the pairing between the S-D sequence in the mRNA and a short sequence close to the 3′ end of the 16S rRNA. The initiator codon, typically AUG or GUG, is 5 to 10 bases downstream of the S-D sequence. N represents any base.
doi:10.1128/9781555817169.ch1.f2.32

STEPS IN INITIATION OF TRANSLATION

The currently accepted view of the steps in the initiation of translation at a TIR is outlined in Figure 2.33. Initiation requires three different initiation factors, IF1, IF2, and IF3, in addition to fMet-tRNA^{fMet}. These initiation factors interact mostly with the initiator tRNA and the P site of the 30S ribosomal subunit.

For initiation to occur, the 70S ribosome must first be separated or dissociated into its smaller 30S and 50S subunits. This dissociation occurs after the termination step of translation (see below) and is assisted by the IF3 initiation factor, which binds to the 30S subunit and helps to keep the subunits dissociated. Therefore, ribosomes are continuously cycling between the 70S ribosome and the 30S and 50S subunits, depending on

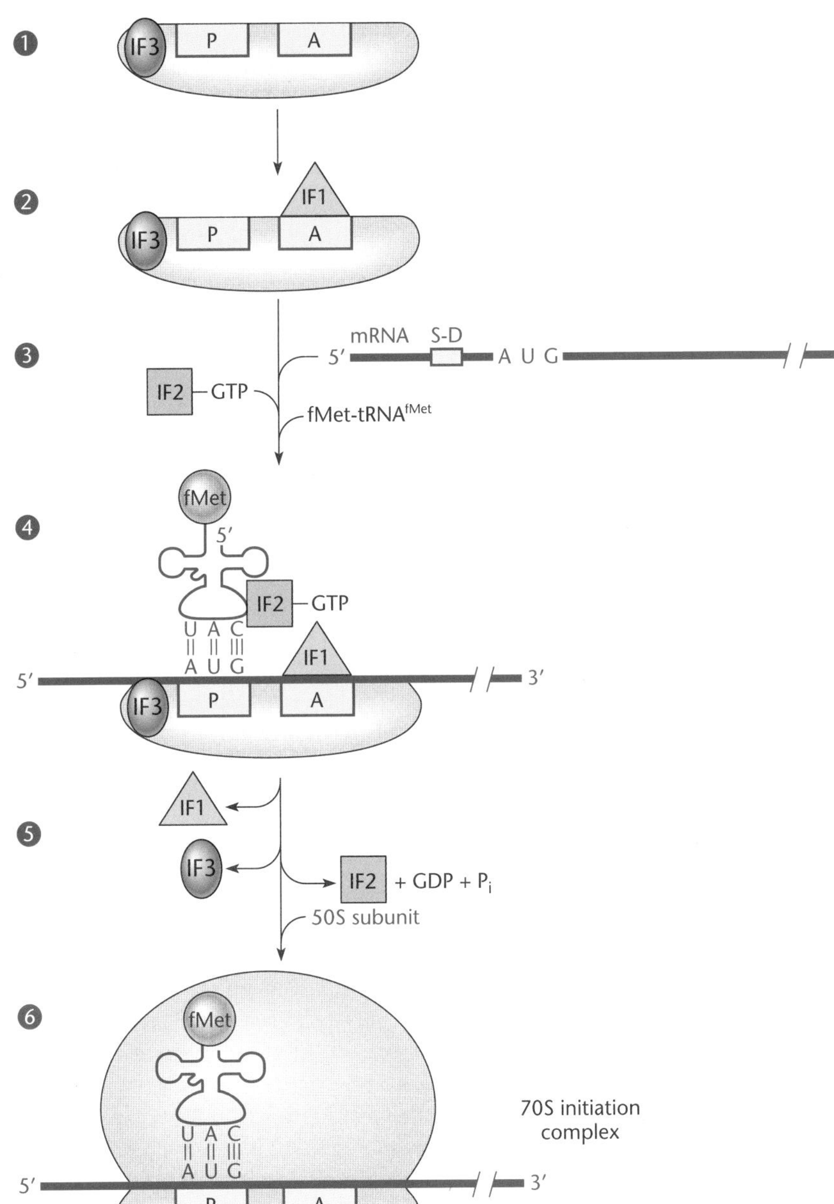

Figure 2.33 Initiation of translation. **(1)** The IF3 factor binds the 30S subunit to keep it dissociated from the 50S subunit during initiation. **(2)** IF1 binds to the A site to block the site. **(3)** A complex is formed between fMet-tRNA^{fMet}, IF2, and GTP. **(4)** The fMet-tRNA^{fMet}-IF2-GTP complex binds the P site of the 30S subunit and the mRNA TIR site. **(5)** IF1 and IF3 are released, the cleavage of GTP on IF2 correctly positions the fMet-tRNA^{fMet} on the P site initiation codon, and the 50S subunit binds. **(6)** The 70S ribosome is ready to accept another aminoacyl-tRNA at the A site.
doi:10.1128/9781555817169.ch1.f2.33

whether they have initiated translation. This is called the **ribosome cycle**.

Once the subunits are dissociated, IF1 binds to the A site on the 30S subunit to prevent the fMet-tRNA^{fMet} from inadvertently binding to this site. IF2, in conjunction with GTP, binds to fMet-tRNA^{fMet} to form a ternary (three-member) complex, which binds to the mRNA and P site of the 30S ribosomal subunit. The initial binding of fMet-tRNA^{fMet} does not depend on an initiator codon in the P site. However, IF2, with the help of IF3, adjusts the fMet-tRNA^{fMet} and the mRNA initiator codon so that the binding becomes codon-specific (see below). IF1 and IF3 are then ejected, and IF2 promotes the association of this initiation complex with the 50S subunit of the ribosome. IF2 is then released, with the cleavage of GTP to guanosine diphosphate [GDP]. The newly formed 70S ribosome is now ready for translation, and another aa-tRNA can enter the A site.

Translation Initiation from Leaderless mRNAs

As mentioned above, a few mRNAs in bacteria do not have standard TIRs with leader sequences containing S-D sequences. In these rare mRNAs, the initiator codon can be right at the 5′ end or very close to it. It is not understood how the ribosome recognizes such an initiator codon and initiates translation, but the mechanism seems to be very different from that of initiation at a normal TIR. There is some evidence that a complex first forms between fMet-tRNA^{fMet}, IF2, and the small subunit of the ribosome. This complex may then help recognize the initiation codon in the absence of upstream sequences to help distinguish the initiation codon. Other evidence suggests that the 70S ribosome itself recognizes the leaderless initiator codon. It is intriguing to think that the process of initiation of translation at initiator codons without leader sequences may resemble more closely the process used in eukaryotes and may be the remnants of a process used before these kingdoms of life separated.

TRANSLATION INITIATION IN ARCHAEA AND EUKARYOTES

Translation initiation in the archaea is similar to that in the bacteria. Like bacteria, archaea use well-defined TIRs with leader sequences and formylmethionine for initiation of translation. In contrast, eukaryotes do not seem to use special RBS elements but usually use the first AUG from the 5′ end of the mRNA as the initiation codon. This does not mean, however, that sequences around this initiator AUG are not important for its recognition. Also, secondary structure in the mRNA may mask other AUG sequences that could potentially be used as initiator codons. Although eukaryotes have a special methioninyl tRNA that responds to the first AUG codon, called Met-tRNA$_i$, the methionine attached to the eukaryotic initiator tRNA is never formylated. As in bacteria, however, the first methionine is usually removed by an aminopeptidase after the protein is synthesized. Eukaryotes and archaea also seem to use many more initiation factors and elongation factors than do bacteria. Although the exact roles of most of these initiation factors are unknown, many are obviously related to the initiation and translation factors of bacteria. It is interesting that the mechanism of translation initiation in the archaea is a sort of hybrid between that in bacteria and that in eukaryotes. The archaea use formylated methionine and S-D sequences like bacteria, but their initiation factors are more similar to those in eukaryotes.

TRANSLATION ELONGATION

During translation elongation, the ribosome moves 3 nucleotides at a time along the mRNA in the 5′-to-3′ direction, allowing tRNAs carrying amino acids (aa-tRNAs) to pair with the mRNA. Each tRNA has a specific 3-nucleotide anticodon sequence in one of its loops, and these nucleotides must be complementary to the mRNA codon for the tRNA to be bound tightly to the ribosome (Figure 2.30). As in other nucleic acids, pairing is antiparallel, so that the two RNA sequences are complementary when read in opposite directions. In other words, the 3′-to-5′ sequence of the anticodon must be complementary to the 5′-to-3′ sequence of the codon.

Entry of aa-tRNAs bound by EF-Tu into the ribosome is random. If the anticodon is complementary to the mRNA codon at the A site (Figure 2.28), codon-anticodon pairing stimulates a structural transition of the ribosome that promotes transition to the next step. A mispaired tRNA is not stabilized and is released from the ribosome. This pairing of only three bases is enough to direct the right tRNA to the A site on the ribosome; in fact, sometimes the pairing of only two bases is sufficient to direct the anticodon-codon interaction (see "Wobble" below). Accurate codon-anticodon pairing is monitored by specific residues in 16S rRNA that form the **decoding site**. The tRNA is bound between the 30S and 50S subunits of the ribosome so that the anticodon loop is in communication with the mRNA in the 30S subunit and the acceptor end of the aa-tRNA containing the bound amino acid is in communication with the 23S rRNA in the 50S subunit. The conformational change in the ribosome that is triggered by accurate codon-anticodon pairing results in hydrolysis of the GTP on EF-Tu to GDP and release of EF-Tu–GDP from the ribosome. EF-Tu–GDP is recycled into EF-Tu–GTP by the action of another protein factor, EF-Ts (Figure 2.28).

After the matching aa-tRNA is bound at the A site of the ribosome, the **peptidyltransferase** catalyzes the formation of a peptide bond between the incoming amino acid at the A site and the growing polypeptide at the adjacent P site (Figure 2.34). This reaction links the carboxyl terminus of the growing peptide chain (attached to the P site tRNA) to the amino group on the amino acid attached to the A site tRNA. Formylation of the methionine on the initiator tRNA prevents its use as an elongator tRNA, because its amino group is occupied by the formyl group. The peptidyltransferase activity is catalyzed by 23S rRNA, which indicates that this RNA acts as an RNA enzyme, or ribozyme.

After peptide bond formation, the P site tRNA no longer has anything attached to it, while the A site tRNA carries the polypeptide chain. Another enzyme, **translation elongation factor G (EF-G)** or the **translocase** (in complex with GTP), then enters the ribosome and moves (or translocates) both the polypeptide-containing tRNA and the mRNA from the A site to the P site, moving the deacylated tRNA from the P site to the E site and making room for another aa-tRNA to enter the A site. Translocation requires hydrolysis of GTP, which also results in release of EF-G from the ribosome (Figure 2.28). The deacylated tRNA in the E site later exits the ribosome (possibly in conjunction with entry of the next

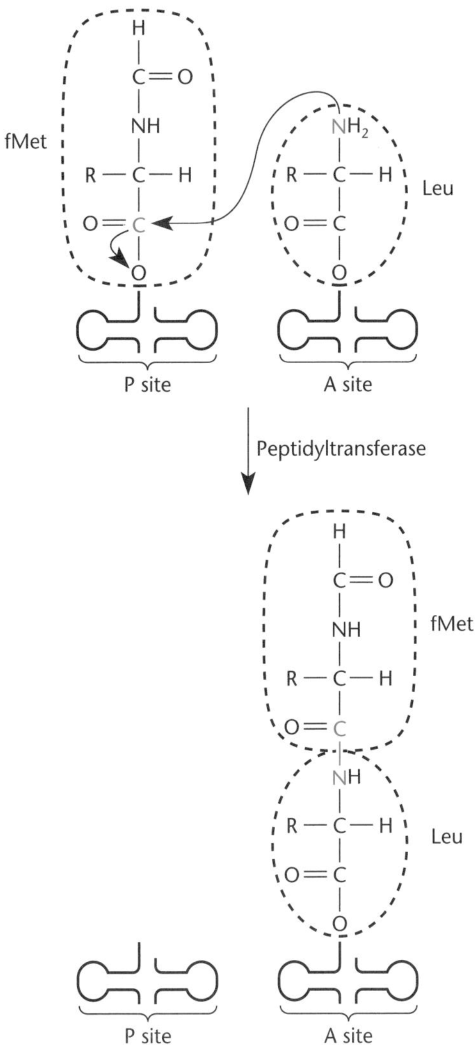

Figure 2.34 The peptidyltransferase reaction.
doi:10.1128/9781555817169.ch1.f2.34

aminoacyl-tRNA into the A site). The mRNA maintains contact with the tRNAs as they move through the ribosome and therefore also moves through the ribosome 3 nucleotides at a time.

Recent structural studies have shown that there are distinct A and P sites on both the 30S and 50S subunits of the ribosome. The anticodon end of the tRNA binds to the sites on the 30S subunit, while the acceptor CCA end, to which the amino acid or polypeptide is attached, binds to the sites on the 50S subunit. A tRNA bound to the A site on the 30S subunit and the corresponding A site on the 50S subunit is said to be bound to the A/A site, while one bound to the A site on the 30S subunit and the P site on the 50S subunit is bound to the A/P site, etc. The incoming aa-tRNA first binds to the A site on the 30S subunit through its anticodon end. The CCA end of tRNA is not properly positioned until accurate codon-anticodon pairing stimulates the structural transition, resulting in positioning of the acceptor end of the tRNA in the A site of the 50S subunit to allow interaction with the peptidyl-tRNA in the P site; the aa-tRNA is now bound to the A/A sites. The peptide bond then forms, concomitant with the movement of the CCA end of the tRNA to the P site on the 50S ribosome, so it is now bound to the A/P sites. EF-G then moves the anticodon end of the tRNA (along with the associated mRNA codon) to the P site on the 30S subunit, so the tRNA is now bound to the P/P site. There may be other dual sites, called E/E sites, to which the deacylated tRNA temporarily binds before it exits the ribosome. One attractive feature of this model is that it allows the growing polypeptide chain to stay fixed at the P site on the 50S subunit and to exit through the channel in the 50S subunit as it grows, while a progression of tRNAs move through the ribosome, making contacts with the different sites.

The translation of even a single codon in an mRNA requires a lot of energy. First, ATP must be hydrolyzed for an aaRS to attach an amino acid to a tRNA (Figure 2.29). Second, EF-Tu requires that a GTP be hydrolyzed to GDP before it can be released from the ribosome after each tRNA is bound (Figure 2.28B). Yet another GTP must be hydrolyzed to GDP for EF-G to move the tRNA with the attached polypeptide to the P site (Figure 2.28C). In all, the energy of at least three nucleoside triphosphates is required for each step of translation.

TRANSLATION TERMINATION

During elongation, translation proceeds along the mRNA, one codon at a time, until the ribosome encounters one of three special codons, UAA, UAG, or UGA, that serve as translational stop signals. These codons, designated **stop codons** or **nonsense codons** (because they do not encode an amino acid), have no corresponding tRNA (Table 2.2). When a nonsense codon enters the A site of a translating ribosome, translation stops. Similar to the positioning of translation initiation codons, the nonsense codon that terminates translation may not be at the end of the mRNA molecule. The region between the nonsense codon and the 3′ end of the mRNA (or a downstream coding sequence for another polypeptide in a transcript that encodes multiple proteins [see below]) is called the **3′ untranslated region**.

RELEASE FACTORS

In addition to a codon for which there is no tRNA, termination of translation requires **release factors**. These proteins recognize the nonsense codons in the ribosomal A site and promote the release of the polypeptide from the tRNA and the ribosome from the mRNA. In

E. coli, there are two translation release factors, called RF1 and RF2. The two release factors recognize specific nonsense codons: RF1 responds to UAA and UAG, whereas RF2 responds to UAA and UGA. Another factor, called RF3, helps to release these factors from the ribosome after termination. Eukaryotes have only one release factor, which responds to all three nonsense codons. Some types of bacteria and mitochondria also have only one release factor, but those that do generally use UGA to encode an amino acid and not as a nonsense codon (Box 2.2).

Figure 2.35 outlines the process of translation termination. After translation stops at the nonsense codon, the A site is left unoccupied because there is no tRNA to pair with the nonsense codon. The release factors bind to the A site of the ribosome instead. They then cooperate with EF-G and **ribosome release factor** (**RRF**) to cleave the polypeptide chain from the tRNA and release it and the mRNA from the ribosome. An attractive model to explain how this could happen is suggested by the observation that the release factors mimic aa-tRNA (Box 2.2). If the release factor is occupying the A site, then the peptidyltransferase might try to transfer the polypeptide chain to the release factor rather than to the amino acid on a tRNA normally occupying the A site. When EF-G then tries to translocate the release factor with the polypeptide attached to the P site, it may trigger a series of reactions that release the polypeptide. The

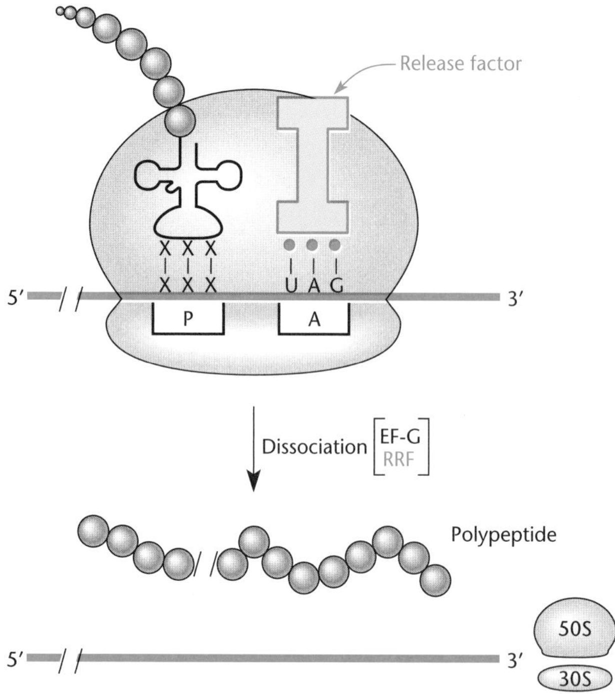

Figure 2.35 Termination of translation at a nonsense codon. A specific release factor interacts with the ribosome stalled at the nonsense codon, possibly through specific pairing between amino acids in the release factor and the nonsense codon (blue dots). Translocation by EF-G causes dissociation of the ribosome from the mRNA, with the assistance of RRF.
doi:10.1128/9781555817169.ch1.f2.35

BOX 2.2

Mimicry in Translation

The ribosome is a very busy place during translation, with numerous factors and tRNAs cycling quickly through the A and P sites. Different factors have to enter the ribosome for each of the steps and then leave when they have finished their functions. One way the complexity of the system seems to be reduced is by having the various factors and tRNAs mimic each other, which allows them to bind to the same sites on the ribosome. For example, the translation factor EF-G seems to be roughly the same shape as the translation factor EF-Tu bound to an aa-tRNA. This may allow EF-G to enter the A site, displace the tRNA (now attached to the growing polypeptide), and move it to the P site. Another example is the mimicry between the tRNAs and the release factors. The release factors resemble tRNAs in shape, but they seem to bind to specific terminator codons through interactions between amino acids in the release factors and nucleotide bases in the nonsense codon, rather than through base pairing between the codon and the anticodon on a tRNA. When the peptidyltransferase attempts to transfer the polypeptide to the release factor in the A site, it sets in motion the string of events that cause translation to be terminated and the polypeptide and mRNA to be released from the ribosome (see the text). It is an attractive idea that the release factors replaced what were once terminator tRNAs that responded to these terminator codons. Perhaps, in the earliest forms of life, everything in translation was done by RNA; now, RNA is used to make proteins, and the proteins, being more versatile, play many of the roles previously played by RNA.

References

Clark, B. F. C., S. Thirup, M. Kjeldgaard, and J. Nyborg. 1999. Structural information for explaining the molecular mechanism of protein biosynthesis. *FEBS Lett.* **452:**41–46.

Nakamura, Y., and K. Ito. 2003. Making sense of mimic in translation termination. *Trends Biochem. Sci.* **28:**99–105.

Nyborg, J., P. Nissen, M. Kjeldgaard, S. Thirup, G. Polekhina, B. F. C. Clark, and L. Reshetnikova. 1996. Structure of the ternary complex of EF-Tu: macromolecular mimicry in translation. *Trends Biochem. Sci.* **21:**81–82.

role of RRF in this process is uncertain, but it might be involved in releasing the mRNA after termination. Termination is more efficient when it occurs in the proper context, that is, when a nonsense codon is surrounded by certain sequences.

REMOVAL OF THE FORMYL GROUP AND THE N-TERMINAL METHIONINE

Normally, polypeptides do not have a formyl group attached to their N termini. In fact, they usually do not even have methionine as their N-terminal amino acid. The formyl group is removed from the completed polypeptide by a special enzyme called **peptide deformylase** (Figure 2.36). The N-terminal methionine is also often removed by an enzyme called **methionine aminopeptidase**, so that methionine is usually not the N terminal amino acid in a mature polypeptide.

trans-TRANSLATION (tmRNA)

A problem occurs when the ribosome reaches the 3′ end of an mRNA without encountering a nonsense codon. This might happen fairly often, because mRNA is constantly being degraded (often from the 3′ end) and transcription may terminate prematurely, resulting in a truncated mRNA. The release factors can function only at a nonsense codon, so the ribosome will stall on the truncated mRNA. Not only would this cause a traffic jam and sequester ribosomes in a nonfunctional state, but also, the protein that is being made will be defective because it is shorter than normal, and accumulation of defective proteins may cause problems for the cell. This is where a small RNA called **transfer-messenger RNA (tmRNA)** comes to the rescue. As the name implies, tmRNA is both a tRNA and an mRNA, as shown in Figure 2.37. It can be aminoacylated with alanine by alanyl-tRNA synthetase like a tRNA but also contains a short **open reading frame (ORF)** that terminates in a nonsense codon like an mRNA. If the ribosome reaches the end of an mRNA without encountering a nonsense codon, tmRNA (together with an accessory protein) enters the A site of the stalled ribosome, and alanine is inserted as the next amino acid of the polypeptide. Then, by a process that is not well understood, the ribosome shifts from translating the ORF on the mRNA to translating the ORF on the tmRNA, where it soon encounters the nonsense codon. The release factors then release the ribosome and the truncated polypeptide, which is now fused to a short **tag** sequence of about 10 amino acids encoded by the tmRNA. The tag sequence that has been attached to the carboxy end of the truncated polypeptide is recognized by the ClpXP protease (see below), which degrades the entire defective polypeptide. In some cases, tmRNA-mediated degradation may play a regulatory role, allowing the targeted degradation of proteins until they are needed (see Keiler, Suggested Reading).

Figure 2.36 Removal of the N-terminal formyl group by peptide deformylase **(A)** and of the N-terminal methionine by methionine aminopeptidase **(B)**.
doi:10.1128/9781555817169.ch1.f2.36

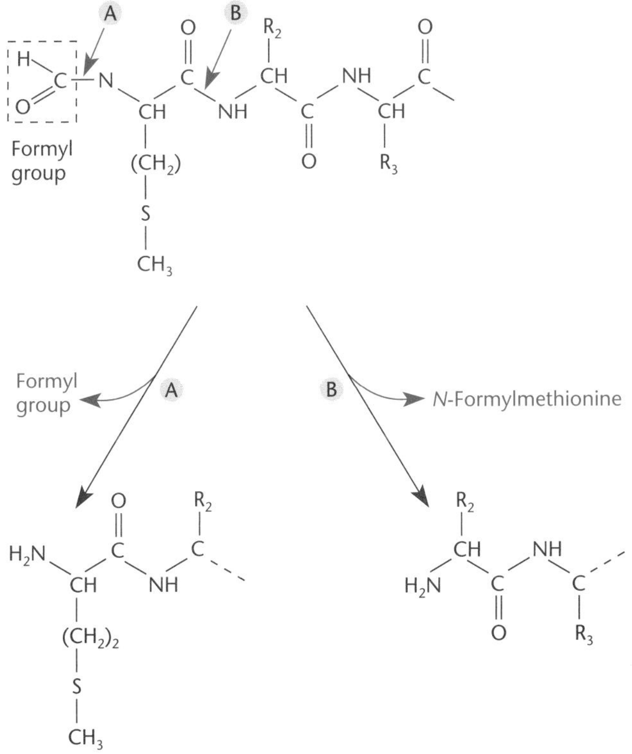

The Genetic Code

As mentioned above, the genetic code determines which amino acid will be inserted into a protein for each 3-nucleotide set, or codon, in the mRNA. More precisely, the genetic code is the assignment of each possible combination of 3 nucleotides to one of the 20 amino acids or to serve as a signal to stop translation. The code is universal, with a few minor exceptions (Box 2.3), meaning that it is the same in all organisms from bacteria to humans. The assignment of each codon to its amino acid appears in Table 2.2.

REDUNDANCY

In the genetic code, more than one codon often encodes the same amino acid. This feature of the code is called **redundancy**. There are 64 (4 × 4 × 4) possible codons that can be made of four different nucleotides taken three at a time. Thus, without redundancy, there would be far too many codons for only 20 amino acids. As shown in Table 2.2, some amino acids are encoded by a single codon (e.g., tryptophan) while others use as many as six different codons (e.g., arginine).

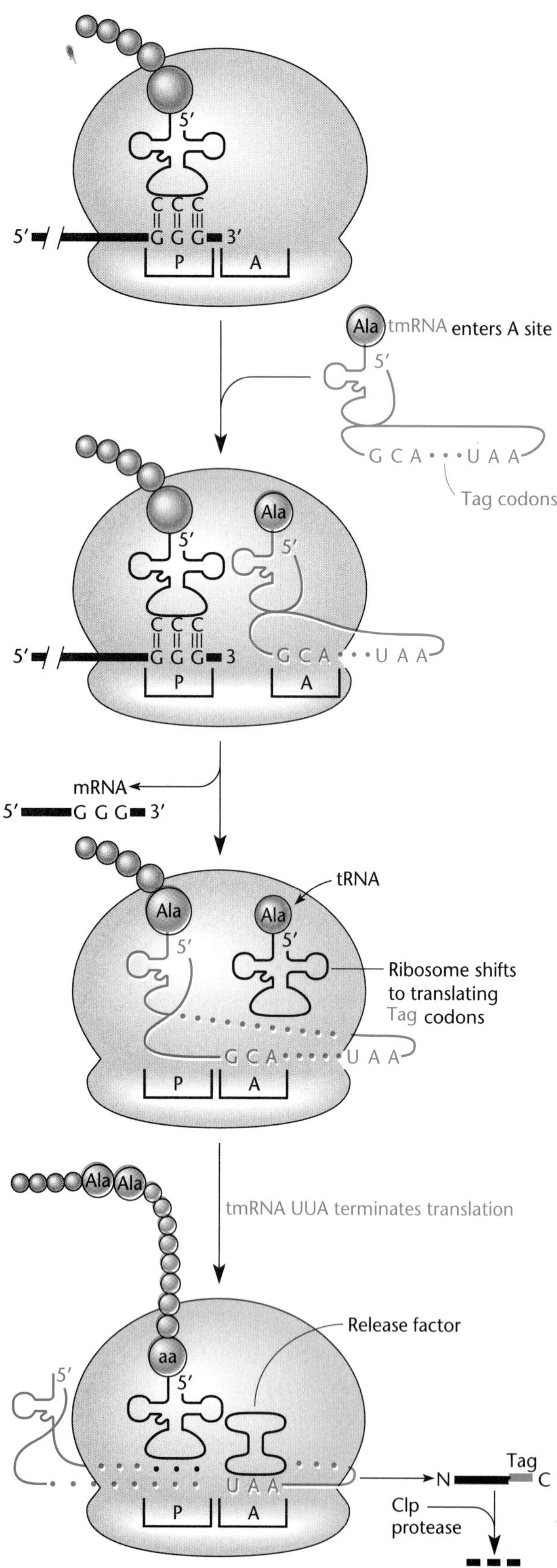

Figure 2.37 *trans*-translation by tmRNA. A ribosome translating an mRNA that lacks a nonsense codon will be stalled, because a release factor is unable to bind. tmRNA, which has features of both a tRNA and an mRNA, enters the A site, and the ribosome switches from translation of the mRNA to translation of the coding sequence in the tmRNA, which results in addition of a short polypeptide tag to the carboxy terminus of the nascent polypeptide. The ribosome and mRNA are released, and the tmRNA-encoded tag targets the polypeptide for degradation by the Clp protease system.
doi:10.1128/9781555817169.ch1.f2.37

WOBBLE

Codons that encode the same amino acid often differ only by their third base, which is why they tend to be together in the same column when the code is presented as in Table 2.2. This pattern of redundancy in the code is due to less stringent pairing, or **wobble**, between the last (3′) base in the codon on the mRNA and the first (5′) base in the anticodon on the tRNA (remember that RNA sequences are always given 5′ to 3′ and the pairing of strands of RNA, like that of DNA, is antiparallel [Figure 2.30]). As a consequence of wobble, the same tRNA can pair with more than one of the codons for a particular amino acid, so there can be fewer types of tRNA than there are codons. For example, even though there are two codons for lysine, AAA and AAG, *E. coli* has only one tRNA for lysine, which, because of wobble, can pair with both lysine codons.

The binding at the third position is not totally random, however, and certain rules apply (Figure 2.38). For example, a G in the first position of the anticodon might pair with either a C or a U in the third position of the codon but not with an A or a G, explaining why UAU and UAC, but not UAA or UAG, are codons for tyrosine. Similarly, a U in the first position of the anticodon can pair with either an A or a G in the third position of the codon (corresponding to the fact that both CAA and CAG are glutamine codons and can be recognized by a single tRNA). The rules for wobble are difficult to predict, however, because the bases in tRNA are sometimes modified, and a modified base in the first position of an anticodon can have altered pairing properties. Inosine, which is a purine base found only in tRNA, can pair with any residue, so a single tRNA with inosine at the first position of the anticodon can recognize multiple codons (UCU, UCA, and UCC in Figure 2.38, all of which encode serine).

NONSENSE CODONS

Not all codons stipulate an amino acid; of the 64 possible nucleotide combinations, only 61 actually encode an amino acid. The other three (UAA, UAG, and UGA) are nonsense codons in most organisms. The nonsense

BOX 2.3

Exceptions to the Code

One of the greatest scientific discoveries of the 20th century was that of the universal genetic code. Whether human, bacterium, or plant, for the most part, all organisms on Earth use the same three bases in nucleic acids to designate each of the amino acids. However, although the code is almost universal, there are exceptions to this general rule. In some situations, a codon can mean something else. We gave the example of initiation codons that encode different amino acids when internal to a gene than they do at the beginning of a gene, where they invariably encode methionine (see the text). Also, some organelles and primitive microorganisms use different codons for some amino acids. For example, in mammalian mitochrondria, UGA, which is normally a nonsense codon, instead designates tryptophan. Also, some protozoans use the nonsense codons UAA and UAG for glutamine. In these organisms, UGA is the only nonsense codon. Some yeasts of the genus *Candida*, the causative agent of thrush, ringworm, and vaginal yeast infections, recognize the codon CUG as serine instead of the standard leucine. In bacteria, the only known exceptions to the universal code involve the codon UGA, which encodes the amino acid glutamine in some bacteria of the genus *Mycoplasma*, which are responsible for some plant and animal diseases.

Some exceptions to the code occur only at specific sites in the mRNA. For example, UGA encodes the rare amino acid selenocysteine in some contexts. This amino acid exists at one or a very few positions in certain bacterial and eukaryotic proteins. It has its own unique aaRS, translation elongation factor (analogous to EF-Tu), and tRNA, to which the amino acid serine is added and then converted into selenocysteine. This tRNA then inserts the amino acid selenocysteine at certain UGA codons, but only at a very few unique positions in proteins and not every time a UGA appears in frame. How, then, does the tRNA distinguish between these sites and the numerous other UGAs, which usually signify the end of a polypeptide? The answer seems to be that the selenocysteine-specific EF-Tu has extra sequences that recognize the mRNA sequences around the selenocysteine-specific UGA codon, and only if the UGA codon is flanked by these particular sequences will this EF-Tu allow its tRNA to enter the ribosome. It is a mystery why the cell goes to so much trouble to insert selenocysteine in a specific site in only a very few proteins. In some instances where selenocysteine was replaced by cysteine, the mutated protein still functioned, albeit less efficiently. However, it may be required in the active center of some enzymes involved in anaerobic metabolism, and this amino acid has persisted throughout evolution, existing in organisms from bacteria to humans.

Another striking deviation from the code was discovered recently in the methanogenic archaea (archaea that make natural gas). These bacteria insert pyrrolysine for the normal nonsense codon UAG. Unlike selenocysteine, which is chemically derived from serine already on its tRNA, pyrrolysine is synthesized and then loaded onto a dedicated tRNA by a dedicated aaRS. It therefore qualifies as the 22nd amino acid. Its aa-tRNA uses the normal EF-Tu and is inserted whenever the codon UAG appears within the mRNA .

Recently, efforts have been made to purposely reprogram the genetic code to allow targeted insertion of nonnatural amino acids into specific sites of individual target proteins with the goal of generating new classes of proteins with novel activities. This work relies on engineering systems related to the selenocysteine and pyrrolysine systems to attach a new type of amino acid to a dedicated tRNA so that the tRNA will insert its amino acid only at a specific nonsense codon within the target protein. Approaches of this type will enable both specific labeling of proteins in the cell and development of new enzymatic activities.

Other exceptions violate the rule that the code is read three bases at a time until a nonsense codon is encountered. This happens with high-level frameshifting and readthrough of nonsense codons. In high-level frameshifting, the ribosome can back up one base or go forward one base before continuing translation. High-level frameshifting usually occurs where there are two cognate codons next to each other in the RNA, for example, in the sequence UUUUC, where both UUU and UUC are phenylalanine codons that are presumably recognized by the same tRNA through wobble. The ribosome with the tRNA bound can slip forward or backward by one nucleotide before it continues translating in the new reading frame, creating a frameshift. Sites at which high-level frameshifting occurs are designated "shifty sequences" and usually have common features. They often have a secondary structure, such as a pseudoknot in the RNA (see Figure 2.2) just downstream of the frameshifted region, which causes the ribosome to pause. They also may have an S-D sequence just upstream of the frameshifted site to which the ribosome then binds through its 16S rRNA, shifting the ribosome 1 nucleotide on the mRNA and causing the frameshift. Sometimes both the normal protein and the frameshifted protein, which has a different carboxyl end, can function in the cell. Examples are the *E. coli* DNA polymerase accessory proteins γ and τ, which are both products of the *dnaX* gene (see Table 1.1) but differ because of a frameshift that results in the formation of a truncated protein. Frameshifting can also allow the readthrough of nonsense codons to make "polyproteins," as occurs in

(continued)

BOX 2.3 (continued)

Exceptions to the Code

many retroviruses, such as human immunodeficiency virus (the acquired immune deficiency syndrome [AIDS] virus). Moreover, high-level frameshifting can play a regulatory role, for example, in the regulation of the RF2 gene in *E. coli*. The RF2 protein causes release of the ribosome at the nonsense codons UGA and UAA (see "Translation Termination" in the text). The gene for RF2 in *E. coli* is arranged so that its function in translation termination can be used to regulate its own synthesis through frameshifting (see chapter 12). How long the ribosome pauses at a UGA codon depends on the amount of RF2 in the cell. If there is a lot of RF2 in the cell, the pause is brief and the polypeptide is quickly released by RF2. If there is less RF2, the ribosome will pause longer, allowing time for a −1 frameshift. The RF2 protein is translated in the −1 frame, so this is the correct frame for translation of RF2, and more RF2 will be made if there is not enough for rapid termination.

In the most dramatic cases of frameshifting, the ribosome can hop over large sequences in the mRNA and then continue translating. This is known to occur in gene *60* of bacteriophage T4 and the *trpR* gene of *E. coli*. Somehow, the ribosome stops translating the mRNA at a certain codon and "hops" to the same codon further along. Presumably, the secondary and tertiary structures of the mRNA between the two codons cause the ribosome to hop. In the case of gene *60* of T4, the hopping occurs almost 100% of the time, and the protein that results is the normal product of the gene. In the *E. coli trpR* gene, the hopping is less efficient, and the physiological significance of the hopped form is unknown.

High-level readthrough of nonsense codons can also give rise to more than one protein from the same ORF. Instead of stopping at a particular nonsense codon, the ribosome sometimes continues making a longer protein, in addition to the shorter one. Examples are the synthesis of the head proteins in the RNA phage Qβ and the synthesis of Gag and Pol proteins in some retroviruses. Many plant viruses also make readthrough proteins. Again, it seems to be the sequence around the nonsense codon that promotes high-level readthrough. However, it is important to emphasize that these are all exceptions, and normally, the codons on an mRNA are translated faithfully one after the other from the TIR until a nonsense codon is encountered.

References

Baranov, P. V., O. Fayet, R. W. Hendrix, and J. F. Atkins. 2006. Recoding in bacteriophages and bacterial IS elements. *Trends Genet.* **22:**174–181.

Bock, A., K. Forchhammer, J. Heider, W. Leinfelder, G. Sawers, B. Veprek, and F. Zinoni. 1991. Selenocysteine: the 21st amino acid. *Mol. Microbiol.* **5:**515–520.

Gaston, M. A., R. Jiang, and J. A. Krzycki. 2011. Functional context, biosynthesis, and genetic encoding of pyrrolysine. *Curr. Opin. Microbiol.* **14:**342–349.

Maldonada, R., and A. J. Herr. 1998. Efficiency of T4 gene 60 translational bypassing. *J. Bacteriol.* **180:**1822–1830.

Young, T. S., I. Ahmad, J. A. Yin, and P. G. Schultz. 2010. An enhanced system for unnatural amino acid incorporation in *E. coli*. *J. Mol. Biol.* **395:**361–374.

codons are usually used to terminate translation at the end of genes (see "Translation Termination" above).

AMBIGUITY

In general, each codon specifies a single amino acid, but some can specify a different amino acid, depending on where they are in the mRNA. For example, the codons AUG and GUG encode formylmethionine if they are at the beginning of the coding region but encode methionine or valine, respectively, if they are internal to the coding region. The codons CUG, UUG, and even AUU also sometimes encode formylmethionine if they are at the beginning of a coding sequence.

The codon UGA is another exception. This codon is usually used for termination but encodes the amino acid selenocysteine in a few positions in genes (Box 2.3) and encodes tryptophan in some types of bacteria. Similarly, UAG is usually used for termination but can be used to encode the novel amino acid pyrrolysine in certain organisms (Box 2.3).

CODON USAGE

Just because more than one codon can encode an amino acid does not mean that all the codons are used equally in all organisms. The same amino acid may be preferentially encoded by different codons in different organisms. This codon preference may reflect higher concentrations of certain tRNAs or may be related to the base composition of the DNA of the organism. While mammals have an average G+C content of about 50% (so that there are about as many AT base pairs in the DNA as there are GC base pairs), some bacteria and their viruses have very high or very low G+C contents. How the G+C content can influence codon preference is illustrated by some members of the genera *Pseudomonas* and *Streptomyces*. These organisms have G+C contents of almost

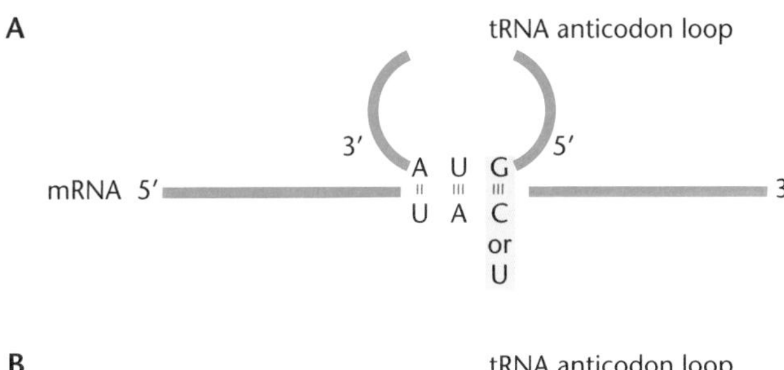

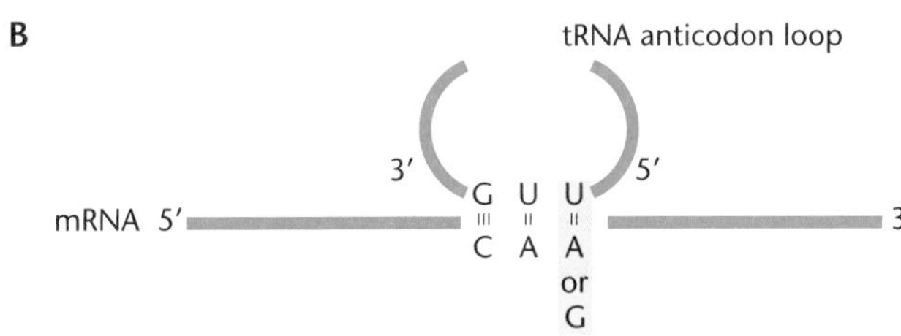

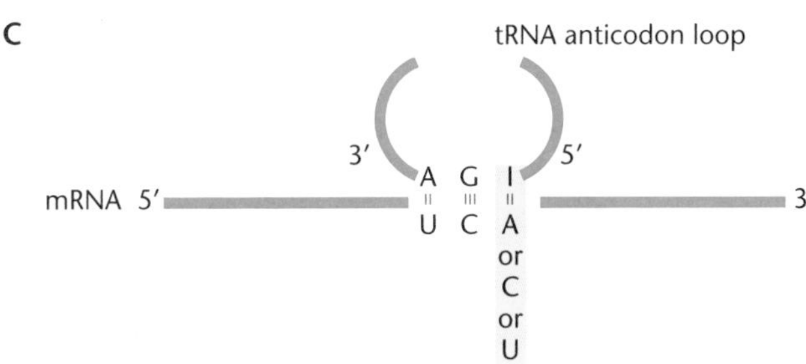

Figure 2.38 Wobble pairing between the anticodon on the tRNA and the codon in the mRNA. Many pairing interactions are possible in the third position of the codon. Alternative pairings for the anticodon base are shown: guanine **(A)**, uracil **(B)**, and inosine (a purine base found only in tRNAs) **(C)**. This allows a single tRNA to recognize multiple codons. doi:10.1128/9781555817169.ch1.f2.38

75%. To maintain such high G+C contents, the codon usage of these bacteria favors the codons that have the most G's and C's for each amino acid.

Polycistronic mRNA

In eukaryotes, each mRNA normally encodes only a single polypeptide. In contrast, in bacteria and archaea, one mRNA can encode either one polypeptide (**monocistronic mRNAs**) or more than one polypeptide (**polycistronic mRNAs**). Polycistronic mRNAs must have a separate TIR for each coding sequence to allow the simultaneous translation of each coding sequence.

The name "polycistronic" is derived from cistron, which is the genetic definition of the coding region for each polypeptide, and poly, which means many. Similarly, "monocistronic" is derived from mono, which means one. Figure 2.39 shows a typical polycistronic mRNA in which the coding sequence for one polypeptide is followed by the coding sequence for another. The space between two coding regions can be very short, and the coding sequences may even overlap. For example,

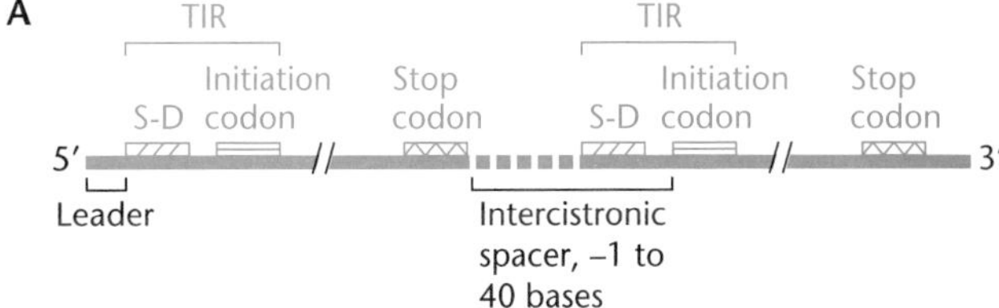

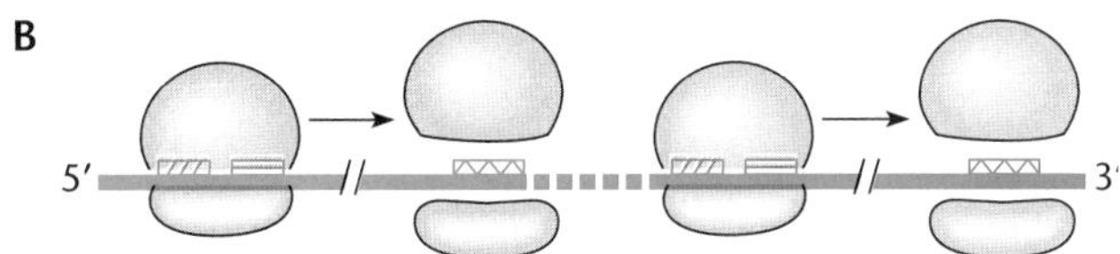

Figure 2.39 Structure of a polycistronic mRNA. **(A)** The coding sequence for each polypeptide is between the TIR and the stop codon. The region 5′ of the first initiation codon is called the leader sequence, and the untranslated region between a stop codon for one gene and the next TIR is known as the intercistronic spacer. **(B)** The association of the 30S and 50S ribosomes at a TIR and their dissociation at a stop codon. New 30S and 50S subunits associate at a downstream TIR. doi:10.1128/9781555817169.ch1.f2.39

the coding region for one polypeptide may end with the nonsense codon UAA, but the last A may be the first nucleotide of the initiator codon AUG for the next coding region. Even if the two coding regions overlap, the two coding regions on an mRNA can be translated independently by different ribosomes.

Polycistronic mRNAs do not exist in eukaryotes, in which, as described above, TIRs are much less well defined and translation usually initiates at the AUG codon closest to the 5′ end of the RNA. In eukaryotes, the synthesis of more than one polypeptide from the same mRNA usually results from differential splicing of the mRNA (Box 2.4) or from high-level frameshifting during the translation of one of the coding sequences (see "Reading Frames" above); there are also specialized events in which an RNA element called an internal ribosome entry sequence (IRES) directs binding of a ribosome to a site within the RNA. Polycistronic RNA leads to phenomena unique to bacteria, i.e., **polarity** and **translational coupling**, which are described below.

TRANSLATIONAL COUPLING

Two or more polypeptides encoded by the same polycistronic mRNA can be **translationally coupled**. Two genes are translationally coupled if translation of the upstream gene is required for translation of the gene immediately downstream.

Figure 2.40 shows an example of how two genes could be translationally coupled. The TIR including the AUG initiation codon of the second gene is sequestered in a hairpin on the mRNA, so it cannot be recognized by an initiating ribosome. However, a ribosome arriving at

BOX 2.4

Selfish DNAs: RNA Introns and Protein Inteins

The chromosomal DNA of all organisms includes a large number of parasitic DNA elements, so named because they cannot replicate themselves but can replicate only when the host DNA replicates. These parasitic DNAs often have few functions except for the ability to move from one DNA to another and thereby parasitize new hosts (see Box 10.2). When such a parasitic DNA integrates into a region of the DNA encoding a protein or RNA, it will disrupt the coding sequence, which is why these elements are sometimes called **intervening sequences**. Like all good parasites, these DNA elements do as little harm to their host as possible, which makes sense, since the parasite is dependent on the host for its own survival. Sometimes an intervening sequence inserts into the coding sequence for an essential RNA or protein. This would disrupt the coding sequence and could be lethal in a haploid organism like a bacterium, except that many of the parasitic elements minimize damage to their host by splicing their sequences out of RNAs and proteins after they are made, restoring the RNAs and proteins to functionality. Intervening sequences that splice themselves out of the RNA are called **introns**, while those that splice themselves out of the protein product of a gene are called **inteins**. The sequences upstream and downstream of the intron or intein in a gene that are rejoined following splicing are called **exons** and **exteins**, respectively.

The two types of RNA introns in bacteria are called group I and group II, based on their mechanisms of splicing (see the figure, panel A). Both groups of introns are typically self-splicing, meaning that they are capable of splicing themselves out of the RNA without the help of proteins. The RNA of an

doi:10.1128/9781555817169.ch1.Box2.4.f

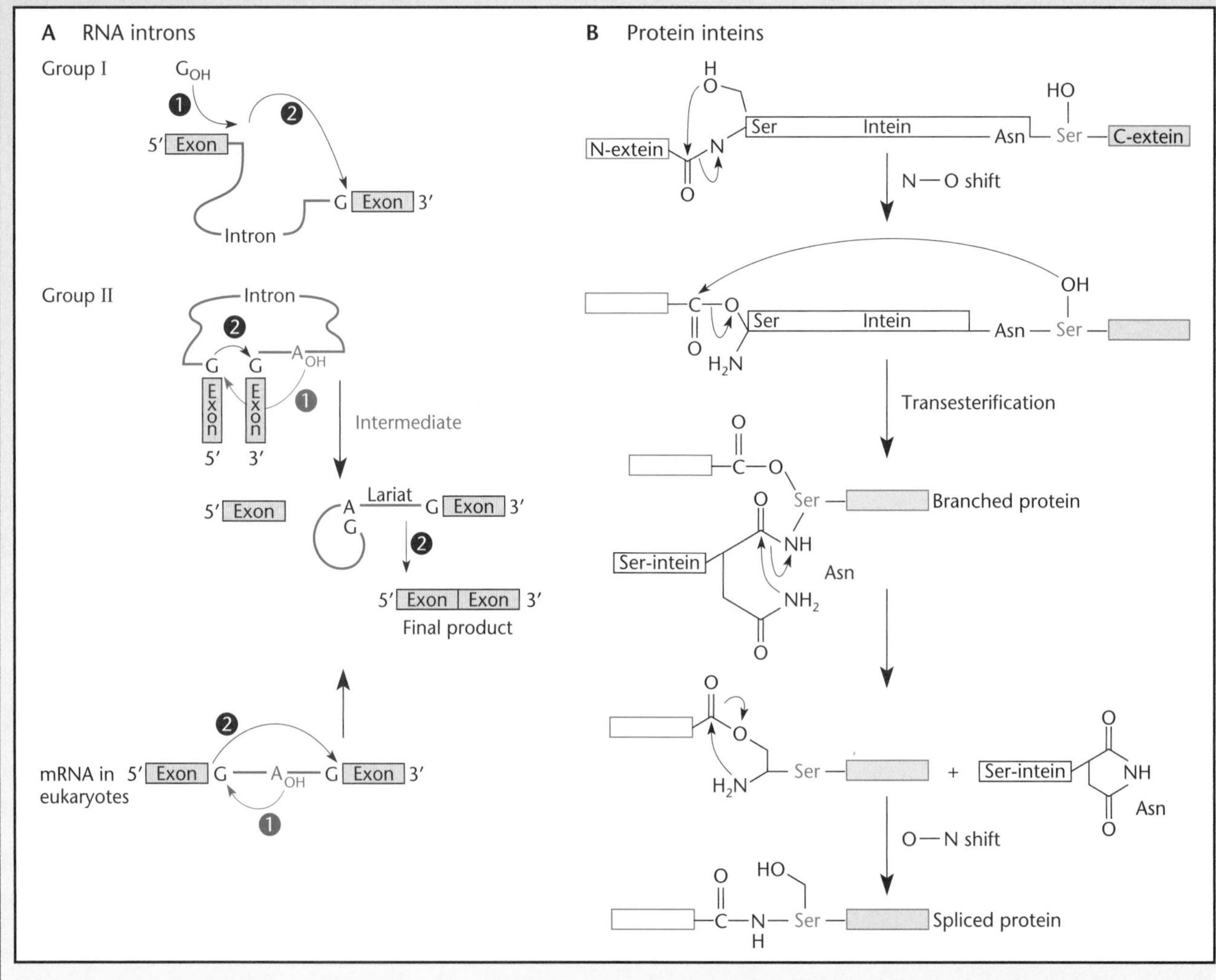

BOX 2.4 (continued)

Selfish DNAs: RNA Introns and Protein Inteins

intron is therefore an enzyme. RNA enzymes are called **ribozymes** to distinguish them from the more common protein enzymes. Other known ribozymes are some RNases and the 23S rRNA peptidyltransferase (see the text). In group I introns, a free guanosine nucleoside or nucleotide residue initiates the splicing reaction by breaking the RNA at the 5′ end of the intron (the 5′ splice site), initiating a series of phosphodiester bond transfers that complete the splicing process. The group I introns are typically found in bacteriophage protein-coding regions and in the tRNA genes of bacteria. Group I introns (and inteins [see below]) typically move to new sites by encoding a DNA endonuclease that makes a double-strand break at a specific site in a DNA that lacks them. This initiates a double-strand break repair recombination that inserts the intron in the site on the DNA. In this way, the intron can move, but it can move only into the same site on another DNA that lacks an intron at that site. It is essential that they move into the same site, because sequences flanking the intron play a role in its splicing, and they would not be able to splice themselves out of an mRNA anywhere else, where the flanking sequences would be different. This ability to move, but only into the same place in another DNA, is called **homing**, and the DNA endonuclease they encode is called a **homing endonuclease**. Homing and double-strand break repair recombination are discussed in chapter 10.

Group II intron splicing is similar to group I splicing except that the initiating nucleotide is a specific adenine base internal to the intron, creating a characteristic "lariat" structure of the intron. This type of splicing is more analogous to mRNA splicing in eukaryotes. While common in lower eukaryotes, group II introns are much rarer in bacteria and are typically found only in other movable elements, such as conjugative plasmids and transposons (see chapters 5 and 9).

Many introns encode maturase proteins that help them fold into the structure required for splicing. In the group II introns, these maturase proteins are also reverse transcriptases and, in combination with the "lariat" intron RNA, form the DNA endonuclease that cuts the target DNA. However, they move by a different process, called **retrohoming** because it goes from RNA to DNA, the reverse of the normal direction. Retrohoming is essentially the reverse of splicing in that the intron splices itself into a DNA rather than out of an RNA, as in splicing.

Protein inteins are parasitic DNAs, like self-splicing introns, except that they splice themselves out of the protein product of the gene rather than out of the mRNA. Inteins probably also exist in all organisms from bacteria to humans. Inteins self-splice themselves out of a protein by the mechanism shown in the figure (panel B). The first amino acid in the intein is always cysteine or serine. This amino acid can be rearranged so that it is attached to the amino acid upstream through its side chain, rather than by a normal peptide bond. Such a bond is called an ester bond or a thioester bond depending on whether the first amino acid in the intein is a serine or a cysteine, respectively, and is called an N-O shift because the bond to the nitrogen in the peptide bond is being shifted to the oxygen in the side chain of the serine. In the next step, this (thio)ester bond is attacked by the side chain of the first amino acid just downstream of the intein, which can be a serine, threonine, or cysteine. This replaces the side chain of the first serine or cysteine in the intein with the side chain of the first amino acid in the downstream extein and leads to the formation of a branched protein, in which one branch is the intein, as shown in the figure. This reaction is called transesterification, because the ester bond is being transferred to a different amino acid. The last amino acid in the intein now connects the intein branch to the rest of the protein, as shown. In almost all known inteins, the last amino acid is an asparagine, whose side chain can then cyclize to release the intein branch. The two exteins are now joined to each other, but they are held together by a (thio)ester bond to the side chain of the first amino acid in the downstream extein rather than a peptide bond. The peptide bond is reformed by the reversal of the original reaction (called an O-N shift) to restore the normal peptide bond, and the intein has been successfully spliced out of the protein, leaving no trace. Complicated though these reactions seem, they occur spontaneously without the help of any other proteins or energy and therefore can occur in a test tube containing just the purified protein with the intein. They also occur in any type of cell into which the gene containing the intein is inserted, independent of the original source of the intein-containing gene, which has made them very useful for some types of applications.

Not only can intein splicing be used to remove an intein from a protein that contains it, it can also be used to bring different proteins encoded by different genes together in a process called *trans*-splicing. This phenomenon was first observed with the split *dnaE* gene for the replicative DNA polymerase of a strain of the cyanobacterium *Synechocystis*, where intein splicing brings two widely separated parts of a gene product together to form an active enzyme. The *dnaE* gene for the DNA polymerase in this strain of bacteria is split into two parts separated by 745,226 bp of DNA. Apparently, an intein was once integrated into the gene for the DNA polymerase, where it could splice itself out of the protein. Some

(continued)

BOX 2.4 (continued)

Selfish DNAs: RNA Introns and Protein Inteins

time later, another large DNA was inserted into the intein. Even though the two parts of the DNA polymerase gene are now split wide apart and the DNA polymerase is made in two pieces with one or the other end of the intein attached, the two parts of the intein can still find each other and perform the splicing reaction, joining the two parts of the DNA polymerase together to make the active enzyme. The intein can even be split into three pieces and splice itself out of the protein! This technology can be exploited to assemble proteins from different clones (see Sun et al., References). RNA introns are known to perform similar *trans*-splicing reactions, and *trans*-splicing has many applications in molecular genetics.

References

Cousineau, B., S. Lawrence, D. Smith, and M. Belfort. 2000. Retrotransposition of a bacterial group II intron. *Nature* **404:**1018–1021.

Martinez-Abarca, F., and N. Toro. 2000. Group II introns in the bacterial world. *Mol. Microbiol.* **38:**917–926.

Sun, W., J. Hu, and X.-Q. Liu. 2004. Synthetic two-piece and three-piece split inteins for protein *trans*-splicing. *J. Biol. Chem.* **279:**35281–35286.

Vepritskiy, A. A., I. A. Vitol, and S. A. Nierzwicki-Bauer. 2002. Novel group I intron in the tRNA^{Leu} (UAA) gene of a proteobacterium isolated from a deep subsurface environment. *J. Bacteriol.* **184:**1481–1487.

Wu, H., Z. Hu, and X. Q. Liu. 1998. Protein *trans*-splicing by a split intein encoded in a split DnaE gene of *Synechocystis* sp. PCC6803. *Proc. Natl. Acad. Sci. USA* **95:**9226–9231.

Xu, M.-Q., and F. B. Perler. 1996. The mechanism of protein splicing and its modulation by mutation. *EMBO J.* **15:**5146–5153.

the UGA stop codon for the first gene can open up this secondary structure, allowing another ribosome to bind to the downstream TIR and initiate translation of the second gene. Thus, translation of the second gene in the mRNA depends on the translation of the first gene. Mutations that disrupt translation of the upstream coding sequence (e.g., nonsense mutations or frameshift mutations that result in premature termination because the ribosome encounters nonsense codons in the new reading frame) therefore affect not only the gene in which they are located, but also the translationally coupled downstream gene.

Figure 2.40 Model for translational coupling in a polycistronic mRNA. **(A)** The secondary structure of the RNA sequesters the TIR of the second coding sequence and blocks translation initiation. **(B)** Translation of the first coding sequence results in disruption of the secondary structure, allowing a ribosome to access the TIR to translate the second coding sequence. doi:10.1128/9781555817169.ch1.f2.40

A

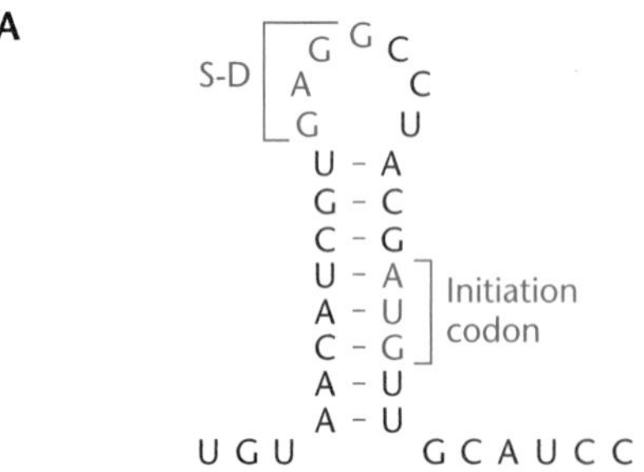

B

POLAR EFFECTS ON GENE EXPRESSION

Some mutations that affect the expression of a gene in a polycistronic mRNA can have secondary effects on the transcription of downstream genes. Such mutations are said to exert a polar effect on gene expression. Several types of mutations can result in polar effects. One type of mutation that can cause a polar effect is an insertion mutation that carries a factor-independent transcriptional terminator. For example, if a transposon "hops" into a polycistronic transcription unit, the transcriptional terminators on the transposon prevent the transcription of genes downstream of the insertion site in the same polycistronic transcription unit. Likewise, a "knockout" of a gene by insertion of an antibiotic resistance gene with a transcriptional terminator causes a polar effect on the genes downstream in the same transcription unit.

A second way in which a mutation in an upstream coding sequence can affect transcription of a downstream coding sequence is through effects on ρ-dependent termination of transcription. Recall that translation of mRNAs in bacteria normally occurs simultaneously with transcription and that the mRNA is translated in the same 5′-to-3′ direction as it is transcribed. Moreover, ribosomes often load onto a TIR as soon as it is vacated by the preceding ribosome, so the mRNA is coated with translating ribosomes. If a nonsense mutation causes dissociation of ribosomes, the abnormally naked mRNA downstream of the mutation may be targeted by the transcription termination factor ρ, which may find an exposed *rut* sequence in the mRNA and cause transcription termination, as shown in Figures 2.20 and 2.41. The nonsense mutation therefore prevents the expression of the downstream gene by preventing its transcription. Such ρ-dependent polarity effects occur only if a *rut* sequence recognizable by ρ and a ρ-dependent terminator lie between the point of the mutation and the next downstream TIR.

Superficially, translational coupling and polarity due to transcription termination have similar effects; in both cases, blocking the translation of one coding sequence affects the synthesis of another polypeptide encoded on the same mRNA. Geneticists often combine them, although the molecular bases of the two phenomena are completely different.

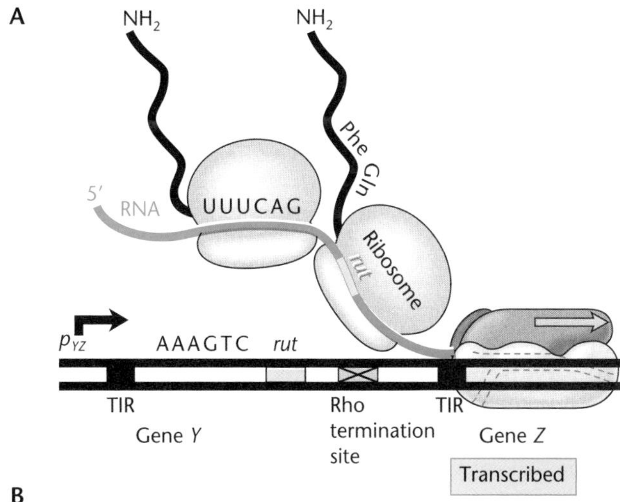

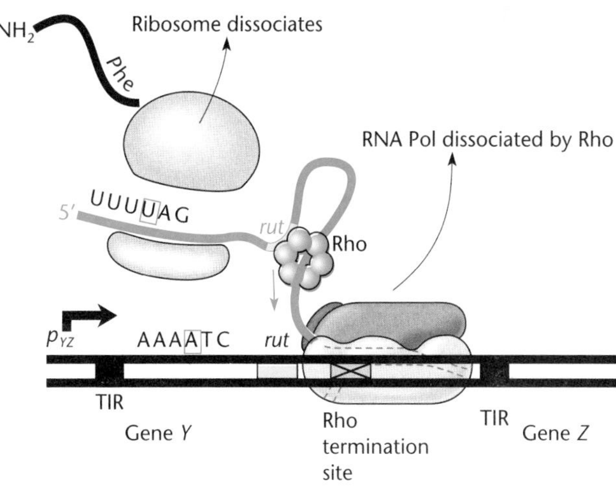

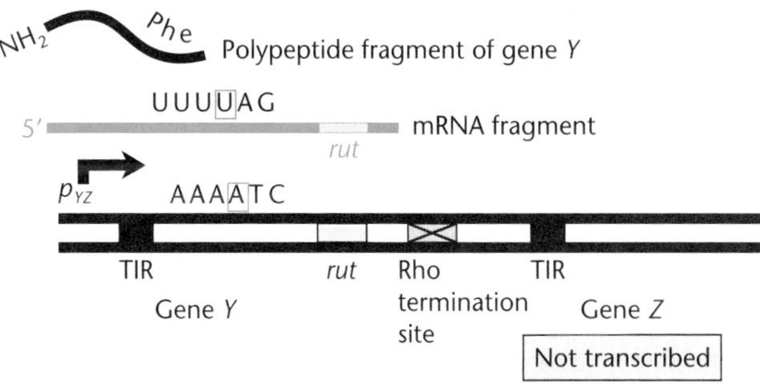

Figure 2.41 Polarity in transcription of a polycistronic mRNA transcribed from *pYZ*. **(A)** The *rut* site in gene *Y* is normally masked by ribosomes translating the gene *Y* mRNA. **(B)** If translation is blocked in gene *Y* by a mutation that changes the codon CAG to UAG (boxed in blue), the ρ factor can cause transcription termination before the RNA polymerase reaches gene *Z*. **(C)** Only fragments of the normal gene *Y* protein and mRNA are produced, and gene *Z* is not even transcribed into mRNA.
doi:10.1128/9781555817169.ch1.f2.41

Protein Folding and Degradation

Translating the information in an mRNA into a polypeptide chain is only the first step in making an active protein. To be an active protein, the polypeptide must fold into its final conformation. This is the most stable state of the protein and is determined by the primary structure of its polypeptides. Whereas some proteins efficiently fold into their active states, for other proteins, the assistance of other factors is necessary to increase the rate of folding into the active state and to prevent misfolding into an inactive state.

Protein Chaperones

Proteins called **chaperones** help other proteins fold into their final conformations. Some chaperones are dedicated to the folding of only one other protein, while others are general chaperones that help many different proteins to fold. We discuss only general chaperones here.

THE DnaK PROTEIN AND OTHER Hsp70 CHAPERONES

The **Hsp70** family of chaperones is the most prevalent and ubiquitous type of general chaperone, existing in all types of cells with the possible exception of some archaea (see Bukau and Horwich, Suggested Reading). These chaperones are also highly conserved evolutionarily. Chaperones in this family are called the Hsp70 proteins because they are about 70 kDa in size and because more of them are made (along with many other proteins) if cells are subjected to a sudden increase in temperature, or "heat shock," although other stresses that denature proteins (such as ethanol) can have the same effect. Synthesis of chaperones increases after such stresses to help refold proteins that have been denatured by the environmental stress (see chapter 13), although they also help to fold proteins under normal conditions. The Hsp70 type of chaperone was first discovered in *E. coli*, where it was given the name DnaK because it is required to assemble the DNA replication apparatus of phage λ and so is required for λ DNA replication. This name for the Hsp70 chaperone in *E. coli* is still widely used in spite of being a misnomer, as the chaperone has nothing directly to do with DNA but functions more generally in protein folding. In its role as a heat shock protein, the DnaK protein of *E. coli* also functions as a cellular thermometer, regulating the synthesis of other proteins in response to heat shock (see chapter 13).

To understand how Hsp70 chaperones, including DnaK, help fold proteins, it is necessary to understand something about the structure of most proteins. Proteins are made up of chains of amino acids that are folded up into well-defined structures, which are often rounded or globular. The amino acids that make up proteins can be charged, polar, or hydrophobic (see the inside front cover for a list). Amino acids that are charged (either acidic or basic) or polar tend to be more soluble in water and are called hydrophilic (water loving). Amino acids that are not charged or polar are hydrophobic (water hating) and tend to be in the inside of the globular protein among other hydrophobic amino acids and away from the water on the surface. If the hydrophobic amino acids are exposed, they tend to associate with hydrophobic amino acids on other proteins and cause the proteins to precipitate. This is essentially what happens when you cook an egg. High temperatures cause the proteins in the egg to unfold, exposing their hydrophobic regions, which then associate with each other, causing the proteins to precipitate into a hard white mass.

The Hsp70-type chaperones help proteins fold by binding to the hydrophobic regions in denatured proteins and nascent proteins as they emerge from the ribosome and keeping these regions from binding to each other prematurely as the protein folds. The Hsp70 proteins have an ATPase activity that, by cleaving bound ATP to ADP, helps the chaperone to sequentially bind to, and dissociate from, the hydrophobic regions of the protein they are helping to fold. The Hsp70-type chaperones are directed in their protein-folding role by smaller proteins called **cochaperones**. The major cochaperones in *E. coli* were named DnaJ and GrpE, again for historical reasons. The DnaJ cochaperone helps DnaK to recognize some proteins and to cycle on and off of the proteins by regulating its ATPase activity. It can also sometimes function as a chaperone by itself. The GrpE protein is a nucleotide exchange protein that helps regenerate the ATP-bound form of DnaK from the ADP-bound form, allowing the cycle to continue.

TRIGGER FACTOR AND OTHER CHAPERONES

Given the prevalence and central role of DnaK in the cell, it came as a surprise that *E. coli* mutants that lack DnaK still multiply, albeit slowly. In fact, the only reason they are sick at all is because they are making too many copies of the other heat shock proteins, since DnaK also regulates heat shock (see chapter 13). One reason why cells lacking DnaK are not dead is that other chaperones can substitute for it. One of these is **trigger factor**. This type of chaperone has so far been found only in bacteria, and much less is known about it. It is bound close to the exit pore of the ribosome and helps proteins fold as they emerge from the ribosome. It is also a **prolyl isomerase**. Of all the amino acids, only proline has an asymmetric carbon, which allows it to exist in two isomers. Trigger factor can convert the prolines in a protein from one isomer to the other. There are many other examples of chaperones that act as prolyl isomerases.

Another set of chaperones, including ClpA, ClpB, and ClpX, form cylinders and unfold misfolded proteins by sucking them through the cylinder. This takes energy, which is derived from cleavage of ATP. Some of them, including ClpA and ClpX, can also feed the unfolded proteins directly into an associated protease called ClpP, which degrades the unfolded protein. Association with ClpP switches the function of ClpA and ClpX from protein folding to protein degradation. ClpB, another cylindrical chaperone, does not associate with a protease but seems to cooperate with the small heat shock proteins IbpA and IbpB to help redissolve precipitated proteins so that they can be refolded by DnaK (see Mogk et al., Suggested Reading).

CHAPERONINS

Besides the relatively simple protein chaperones, cells contain much larger structures that help proteins fold. These large structures are called **chaperonins**, and they exist in all forms of life, including the archaea and eukaryotes. They are composed of two large cylinders with hollow chambers held together back to back with openings at their ends (Figure 2.42). They help fold a misfolded protein by taking it up into one of the chambers. A cap, called a **cochaperonin**, is then put on the chamber, and the protein folds within the more hospitable environment of the chamber. A more detailed model for what happens in the chamber and how this helps a protein fold is suggested by the structure (see Wang and Boisvert, Suggested Reading). When the misfolded protein is first taken up, the lining of the chamber consists of mostly hydrophobic amino acids that bind the exposed hydrophobic regions of the misfolded protein. When

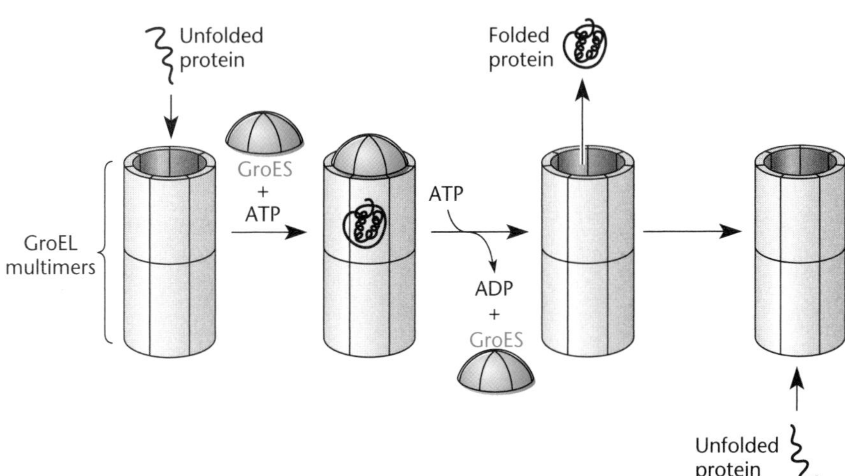

Figure 2.42 Chaperonins. The GroEL (Hsp60)-type chaperonin multimers form two connected cylinders (shown separately). A denatured protein enters the chamber in one of the cylinders, and the chamber is capped by the cochaperonin GroES (Hsp10). The denatured protein can then be helped to fold in the chamber. The other chamber plays a regulatory role but is the chamber that takes up the next unfolded protein in a sort of two-stroke engine. Details are given in the text. doi:10.1128/9781555817169.ch1.f2.42

the cochaperonin cap is put onto this chamber (the *cis*-chamber) and the bound ATP is cleaved to ADP, the lining may switch to being mostly hydrophilic amino acids, driving the more hydrophobic regions of the misfolded protein to the interior of the protein, where they usually reside in the folded protein. Binding of ATP and an unfolded protein to the other chamber (the *trans*-chamber) causes the cap to come off the *cis*-chamber, releasing the folded protein and preparing the other chamber to take up the misfolded protein and take its turn being the *cis*-chamber. This process takes a lot of energy, and a number of ATP molecules (one for each subunit of one chamber of the chaperonin, which is seven in *E. coli* [see below]) are cleaved to ADP in each cycle. This has been described as a two-stroke engine, where the folding role switches from one chamber to the other. It is a mystery why chaperonins have two chambers and why the folding has to alternate between the two chambers. Chaperonins composed of only one chamber function more poorly, although they might retain some of their activity (see Sun et al., Suggested Reading). The communication to one chamber that ATP and unfolded protein are bound to the opposite chamber is one of the most remarkable examples of **allosterism** known, especially considering that the chambers, while touching back to back, are composed of separate polypeptides.

The first chaperonin was discovered in *E. coli*, where it was named GroEL because it helps assemble the E protein of phage λ into the phage head. The GroEL chaperonin consists of 14 identical polypeptides (7 making up each cylinder) of 60 kDa. Its cochaperonin cap is called GroES, which is also made up of seven subunits, each 10 kDa in size. Unlike DnaK and the other chaperones, GroEL and GroES are required for *E. coli* growth, even at lower temperatures. The GroEL chaperonin is known to be required for folding of some essential proteins, including FtsE and the first enzyme required to make meso-A_2pm (mDAP), one of the amino acids in the peptide cross-links in the cell wall (see chapter 14), which explains why GroEL is essential. The chaperonins come in two general types called the group I and group II chaperonins. The group I chaperonins, related to GroEL, are composed of 60-kDa subunits and are found in all bacteria and in the mitochondria and chloroplasts of eukaryotes; this makes sense, since these organelles are derived from bacteria (see the introductory chapter). These chaperonins and their cochaperonins are induced by heat shock and other stresses, so they are called the **Hsp60** proteins and Hsp10 proteins for heat shock 60-kDa and 10-kDa proteins (see Bukau and Horwich, Suggested Reading). The group II chaperonins are found in the archaea and in the cytoplasm of eukaryotes. They have very little amino acid sequence in common with the group I chaperonins and are not composed of identical subunits (i.e., they are mixed multimers) and often have 8 or more polypeptide subunits per cylinder. Furthermore, if they have a cochaperonin cap, it might be attached to the opening of the chamber rather than being detachable, like GroES. They may also be more dedicated and fold only a small subset of proteins, including actin in the eukaryotic cytoplasm. Nevertheless, the two types form similar cylindrical structures and presumably use a similar two-stroke mechanism to help fold proteins. Note that this is yet another example where the archaea and eukaryotes are similar to each other and different from the bacteria (see the introductory chapter).

Protein Degradation

As noted above, protein folding can be coupled to protein degradation, through association of chaperones like ClpA and ClpX with proteases such as ClpP. Complexes such as ClpAP and ClpXP are members of the family of ATP-dependent proteases, which use cleavage of ATP to provide the energy for protein unfolding and proteolysis (see Baker and Sauer, Suggested Reading). These enzymatic machineries are important, not only for destruction of misfolded and denatured proteins, but also for regulated proteolysis to target specific substrate proteins under specific conditions. Their activity can be directed by the presence of a specific degradation tag, a short protein sequence that increases affinity for a specific protease. This tag may be present all the time within the target protein, but it becomes available for recognition only under certain conditions, which allows the target protein to be stable under some conditions and unstable under other conditions. Proteolysis can also be controlled by adaptor proteins that deliver specific target proteins to specific proteases. Regulated proteolysis is discussed as a mechanism for gene regulation in chapters 12 and 13.

Membrane Proteins and Protein Export

In order to function, proteins not only must be folded properly, but also must reach their final destinations in the cell. Often this means that they must leave the cytoplasm, where they were synthesized, and enter the membranes surrounding the cell or, in some instances, leave the cell altogether in the form of extracellular proteins. This is a conceptually simpler process in gram-positive bacteria, with only one membrane to pass through, than it is in gram-negative bacteria, with both an inner and outer membrane. Nevertheless, the processes are very similar in gram-negative and gram-positive bacteria. The primary difference is that when a protein to be secreted passes through the inner membrane of gram-negative bacteria, it must also get through the outer membrane before it is outside the cell. Also, specific cross-links, called disulfide bonds, between cysteine amino acids may form in proteins that reside in the **periplasm** or are transported outside the cell, and these bonds may make the protein more stable. Both the general mechanisms and the specialized mechanisms of protein transport in bacteria, as well as the mechanisms of disulfide bond formation, are discussed in chapter 14.

Regulation of Gene Expression

The previous sections have reviewed how a gene is expressed in the cell, from the time mRNA is transcribed from the gene until the protein product of the gene reaches its final destination in or outside the cell and has its effect. In general, genes are expressed in the cell only when the product is needed by the cell and then only as much as is required to make the amount of product needed by the cell. This saves energy and prevents the products of different genes from interfering with each other. The process by which the output of genes is changed depending on the state of the cell is called the **regulation of gene expression** and can occur at any stage in the expression of the gene.

Genes whose products regulate the expression of other genes are called **regulatory genes**. The product of a regulatory gene can either inhibit or stimulate the expression of a gene. If it inhibits expression, the regulation is negative; if it stimulates expression, the regulation is positive. Some regulatory gene products are both positive and negative regulators depending on the situation. The product of a regulatory gene can regulate the expression of only one other gene, or it can regulate the expression of many genes. The set of genes regulated by the same regulatory gene product is called a **regulon**. If a gene product regulates its own expression, it is said to be autoregulated. We discuss the molecular mechanisms of regulation of gene expression in much more detail in chapter 12, but in this chapter, we briefly review some basic concepts needed to understand the following chapters.

Transcriptional Regulation

Expression of a gene is often regulated by controlling the amount of mRNA that is made on the gene. This is called **transcriptional regulation**. It makes sense to regulate gene expression at this level, since it is wasteful to make mRNAs if the expression of the gene is going to be inhibited at a later stage. Also, bacterial genes are often arranged in a polycistronic unit, or **operon**; if the genes in this unit are involved in a related function, they can all be regulated simultaneously by regulating the synthesis of the polycistronic mRNA of that operon.

Regulation of transcription of an operon usually occurs at the start point of transcription, at the promoter. Whether or not a gene is expressed depends on whether the promoter for the gene is used to make mRNA. Transcriptional regulation at the promoter for a gene can be determined by specific recognition of the promoter by RNA polymerase holoenzyme containing an alternate sigma factor. Regulation can also use regulatory proteins that can act either negatively or positively, depending on whether the regulatory gene product is a transcriptional **repressor** or a transcriptional **activator**, respectively. The difference between regulation of transcription by repressors and activators is illustrated in Figure 2.43. A repressor binds to the DNA at an **operator** sequence close to, or even overlapping, the promoter and prevents RNA polymerase from using the promoter, often by physically obstructing access to the promoter by the RNA polymerase. An activator, in contrast, usually binds upstream of the promoter at an **activator site**, where it can help the RNA polymerase bind to the promoter or help open the promoter after the RNA polymerase binds. Sometimes,

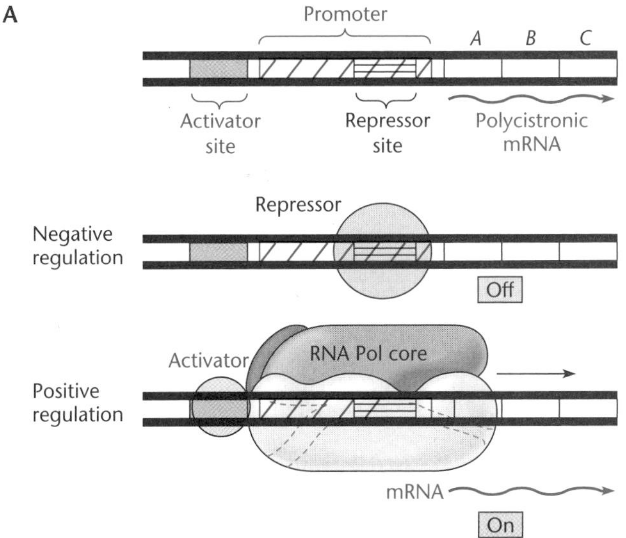

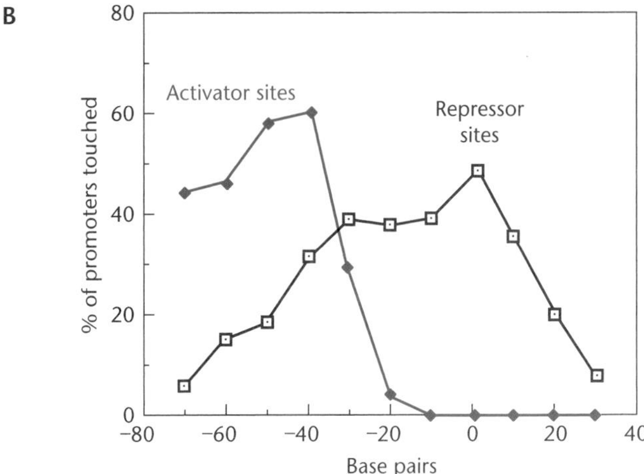

Figure 2.43 **(A)** The two general types of transcriptional regulation. In negative regulation, a repressor binds to a repressor-binding site (or operator) and turns expression of the operon off. In positive regulation, an activator protein binds upstream of the promoter and turns expression of the operon on. **(B)** Graph showing the usual locations of activator sites relative to repressor sites. Activator sites are usually farther upstream. Each datum point indicates the middle of the known region on the DNA where a regulatory protein binds. Zero on the *x* axis marks the start point of transcription. doi:10.1128/9781555817169.ch1.f2.43

a transcriptional regulator can be a repressor on some promoters and an activator on other promoters, depending on where it binds relative to the start site of transcription.

The activity of a regulatory protein can itself be modified by the binding of small molecules called **effectors**, which affects its activity. Effectors are often molecules that can be used by the cell if the regulated operon is expressed or essential metabolites that do not have to be made by the cell if their concentration is already high. If the small-molecule effector causes transcription of the operon to be turned on (for example, by binding to a repressor and changing it so that the repressor can no longer bind to the DNA), the small molecule is called an **inducer**. If binding of the effector to a repressor causes the operon to be turned off, the small molecule is called a **corepressor**. The activity of regulatory proteins can also be modulated by posttranslational modification (e.g., phosphorylation [see Box 13.4]) or by interaction with other proteins or RNAs.

Not all transcriptional regulation occurs at the promoter, however. Sometimes transcription starts and then stops prematurely, resulting in synthesis of a truncated RNA that does not include the protein-coding sequence. Such regulation is called attenuation of transcription. These and other mechanisms of transcriptional regulation are discussed in subsequent chapters.

Posttranscriptional Regulation

Expression of a gene can be regulated at later stages in gene expression (see chapter 12). For example, the mRNA may be degraded by RNases as soon as it is made, before it can be translated. In **translational regulation**, translation of the mRNA can be regulated to determine how much of the protein product is made. **Posttranslational regulation** results in regulation of the activity of a protein product of a gene by degradation by proteases or modifications, such as phosphorylation or methylation, depending on the conditions in which the cell finds itself. In addition, the product of a pathway may inhibit the activity of an enzyme in the pathway by a process called **feedback inhibition**. In general, a type of regulation of gene expression that operates after the mRNA for a gene has been made is called **posttranscriptional regulation**. Specific examples of posttranscriptional regulation are also discussed in subsequent chapters.

Genomes and Genomics

In recent years, the field of microbiology has been transformed by the availability of genome sequence data for a wide variety of microorganisms. Genome sequences provide information about the set of genes present in a particular organism, the organization of those genes into transcriptional units, possible regulatory mechanisms, and the presence of extra genetic elements, including prophages, transposons, insertion sequences, and introns and inteins (Box 2.4). This information also provides insight into relationships between different organisms and may provide evidence for horizontal gene transfer events. Utilization of this explosion of data requires careful annotation to identify these putative genes and elements.

Annotation and Comparative Genomics

For analysis of a new genome sequence, the use of bioinformatics resources can give us a profile of the similarities of DNA sequences and gene products to those of other organisms. Tools for analyzing genome sequences are rapidly increasing in number and sophistication. Examples of tools available for public use are described in Box 2.5. Newer and improved tools will be continuously developed; often a simple Web browser search can reveal useful tools and databases. However, it is important to note that, inevitably, databases contain errors; the level of inaccuracy in annotation is as high as 5 to 10%. A particularly important problem is that annotation of one genome is often based on previous annotation of related genes in other genomes; this can lead to misannotation. For example, if gene A is related to gene B, and gene B is related to gene C, gene A may be mistakenly annotated as related to gene C when in fact it is not really related to that gene. Experimental verification is necessary to sort out whether predicted relationships are in fact accurate.

What You Need To Know

We have introduced a lot of detail in this chapter, so it is worth reviewing some of the most important concepts and words. As with any field, molecular genetics has its own jargon, and in order to follow a paper or seminar that includes some molecular genetics, familiarity with this jargon is very helpful.

Figure 2.44 shows a typical gene with a promoter and transcription terminator. The mRNA is transcribed beginning at the promoter and ending at the transcription terminator. The direction on the DNA or RNA is indicated by the direction of the phosphate bonds between the carbons on the ribose or deoxyribose sugars in the backbone of the polynucleotide. These carbons are labeled with a prime to distinguish them from the carbons in the bases of the nucleotides. On one end of the RNA, the 5′ carbon of the terminal nucleotide is not joined to another nucleotide by a phosphate bond. Therefore, this is called the **5′ end.** Similarly, the other end is called the **3′ end,** because the 3′ carbon of the last nucleotide on this end is not joined to another nucleotide by a

BOX 2.5

Annotation and Comparative Genomics

Genome sequencing is one way to begin the study of a bacterium. However, this information is most useful in the context of other information about the bacterium. In this book, we show how the methods of genetic analysis and genomic analysis complement each other to permit a more complete understanding of how a bacterium functions. Figure 1 summarizes many of the types of genomic and genetic experiments available. Here, we briefly describe these experiments, many of which are more fully discussed and illustrated in subsequent chapters. In some cases, websites are included in the list; in other cases, Internet search terms are given, as websites change rapidly. For a general reference on genome annotation, see Gibson and Muse (References). For a discussion of bioinformatics, see Mount (References). Finally, for an example of a journal publication that describes a comprehensive annotation analysis of *E. coli* K-12, see Riley et al. (References).

Genome Sequence

Genome-sequencing methodology is discussed in Box 1.2.

Annotation and Comparative Genomics

For analysis of a new genome sequence, the use of bioinformatics resources, as described below, can give us a profile of the similarities of DNA sequences and gene products to those of other organisms.

Functional Annotation

Genome sequence information (e.g., Figure 2) is accumulating faster than we can understand it, but tools for analyzing genome sequences are also rapidly increasing in number and sophistication. Examples of tools available for public use are provided below.

RNA-Encoding Sequences

rRNA-encoding sequences are extremely highly conserved (Box 2.1) and so are easily recognized. For recognition of tRNA sequences, a useful bioinformatics tool is http://www.genetics.wustl.edu/eddy/tRNAscan-SE/. Methods for identifying small, noncoding, regulatory RNAs are only now being developed (see chapter 13).

Protein-Encoding Sequences

For finding genes in prokaryotes (i.e., distinguishing coding from noncoding DNA), an especially useful tool is **Glimmer** (for **Gene Locator and Interpolated Markov Modeler**) (http://www.cbcb.umd.edu/software/glimmer). A **Markov model** is a statistical tool useful in a situation in which a system, in this case a protein sequence, undergoes a series of changes in its state (i.e., amino acid substitutions) and a change from one state to the next is independent of the history of the state. A type of Markov model that is

BOX 2.5 (continued)

Annotation and Comparative Genomics

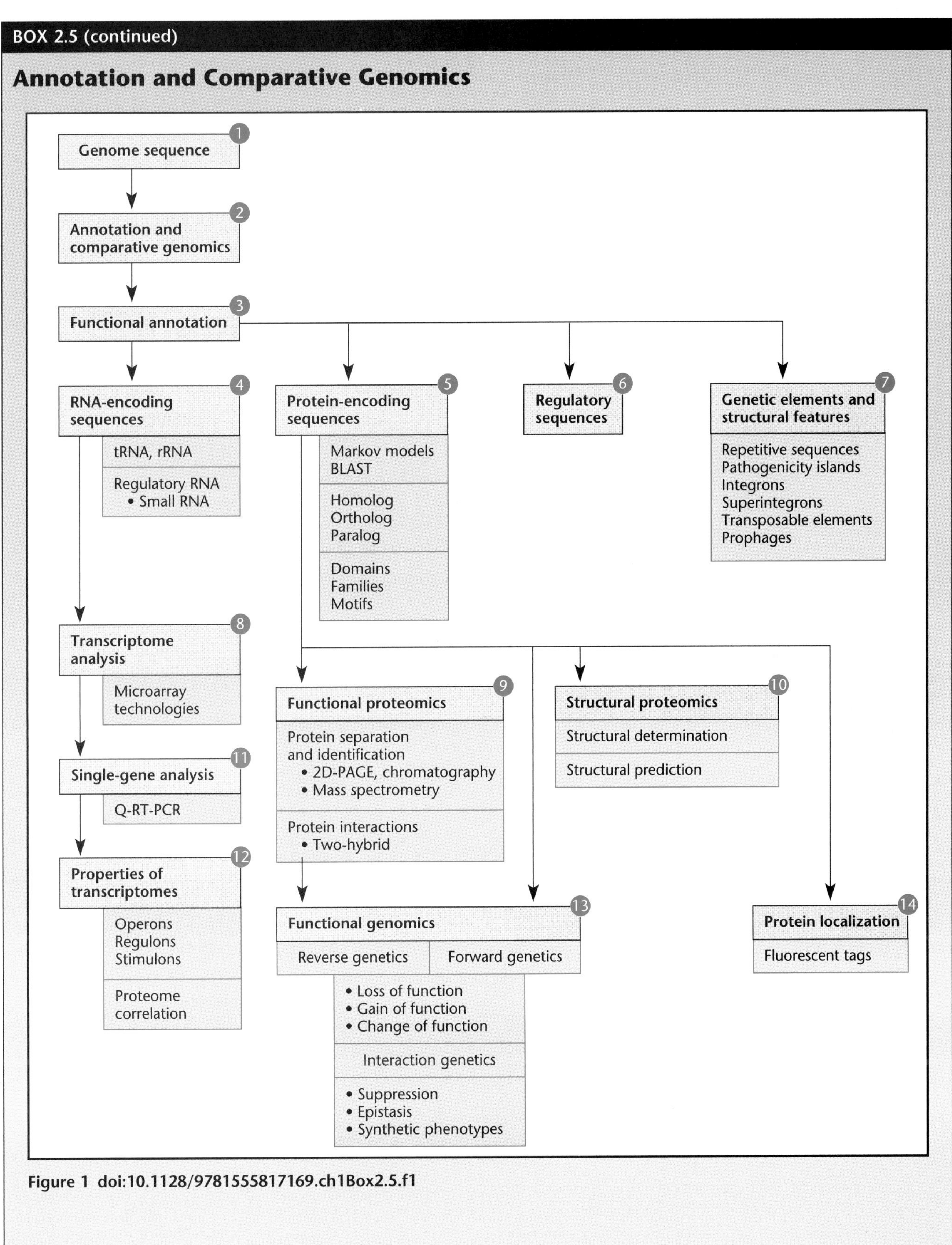

Figure 1 doi:10.1128/9781555817169.ch1Box2.5.f1

(continued)

BOX 2.5 (continued)

Annotation and Comparative Genomics

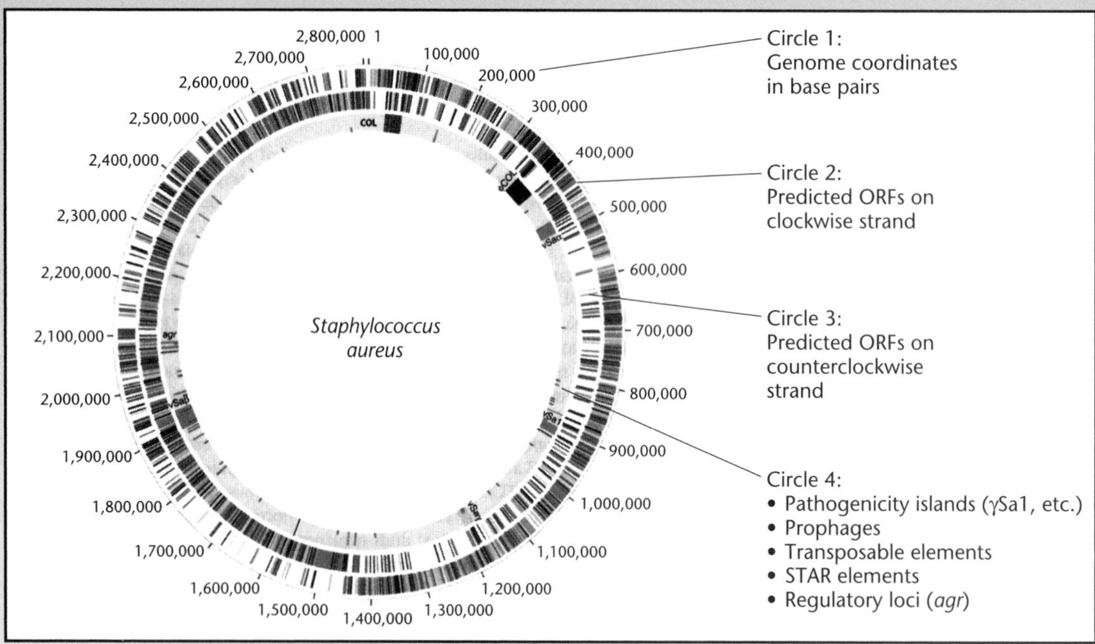

Figure 2 doi:10.1128/9781555817169.ch1.Box2.5.f2

especially useful in genome annotation is the hidden Markov model (HMM). An HMM uses previous data sets to weight an analysis; in other words, an HMM is able to "train itself" if it is "given" a set of about 50 related sequences. Once trained, the HMM places more value on states that are conserved, e.g., common amino acid substitutions. This allows HMMs to be highly sensitive. In addition, an HMM can be further trained to recognize codons or DNA sequences characteristic of a particular organism; this type of model is called an **interpolated Markov model**.

An HMM is able to consider all possible combinations of factors, such as gaps, matches, and mismatches, that could affect the alignment of a set of sequences. Thus, an HMM can pick out amino acid positions that are or are not conserved. The basis for using HMMs is the HMMER statistical tool (http://hmmer.janelia.org/).

For protein-encoding sequences, the most common sequence analysis tool for predicting function is called **BLAST (Basic Local Alignment Search Tool)**. The U.S. National Center for Biotechnology Information (NCBI) has a publicly available website with step-by-step tutorials (http://www.ncbi.nlm.nih.gov/BLAST). To use BLAST, a query of a nucleotide or amino acid sequence is submitted for comparison to the publicly available databases. Searches sometimes ask that a sequence be submitted in **FASTA** format, which means as an uninterrupted sequence, using the standard amino acid and nucleic acid codes. The website http://www.broad.mit.edu contains tutorials on FASTA, as well as many other genomics topics.

The BLAST algorithm can translate a sequence in all six possible reading frames (Figure 2.44). Moreover, the BLAST search can be performed in several ways: translated query versus protein database (blastx), protein query versus translated database (tblastn), or translated query versus translated database (tblastx). Numerous additional variations of BLAST are available at the NCBI BLAST website, including protein-protein BLAST (blastp) and position-specific iterated BLAST **(PSI-BLAST)**.

After a query is submitted, the BLAST algorithm calculates the statistical significance of any matches found. The significance of a similarity is expressed as an **E value**. The E (for expected) value is a term that indicates the significance of an alignment found between two sequences. An E value is the number of database hits of similar quality that you would expect to find by chance. One sequence would be the query sequence, and the other would be a related sequence found in a database, for example, by a BLAST search. The lower the E value, the closer the similarity found. Generally, an E value greater than 0.01 to 0.05 is considered to be insignificant.

The relatedness of gene sequences can be categorized as **homologs**, which are genes or sequences that share common ancestry. Homologs can be classified as **orthologs**,

BOX 2.5 (continued)

Annotation and Comparative Genomics

which are genes that are similar in sequence and have a common ancestor but are found in different species with (in some cases) similar functions, and **paralogs,** which are genes that arose by duplication within a given species and may have similar functions. In addition, proteins are categorized into "families," in which the individual members share certain features, as discussed below.

It is important to note that the matches that result from some BLAST searches, such as blastp, are matches to protein domains rather than to genes per se. A protein domain is generally an independently folding element of a protein; thus, proteins are mosaics of domains. Examples are the σ factor domains in Figure 2.12.

The regions of sequence conservation among proteins that are found by an HMM analysis can be used to categorize proteins into families and so can provide information about the function of a protein. The term "family" is used in many contexts and can refer to many types of categories. In a scheme that broadly defines families, proteins are divided into three types of families. One type of family contains sequences based on one or more shared protein domains. A given domain may be found in more than one functional type of protein. A second type of family is enzyme families, in which the gene products all perform the same biological function. Enzyme names are assigned by the Enzyme Commission (http://ca.expasy.org/enzyme/). A very large collection of metabolic pathways has been compiled at the website http://www.genome.ad.jp/kegg/. This site shows so-called **KEGG maps,** which provide a preliminary suggested pathway in which an enzyme might function; experimental support for the enzyme classification is important. A third family type is the "superfamily," which contains two or more proteins that are related by sequence but have not necessarily been tested for biochemical function. Thus, the importance of experimental study of protein function (see below) cannot be overstated.

Several research groups have combined the information from BLAST analyses into large databases. These include **COGs** (from the NCBI [http://www.ncbi.nlm.nih.gov/COG/]), **Pfams** (from the Sanger Centre in the United Kingdom [http://www.sanger.ac.uk/Software/Pfam/]), and **TIGRFAMs** (from the U.S. public/corporate Institute for Genomic Research [http://www.tigr.org/TIGRFAMs/]).

A COG is a **cluster of orthologous genes**. The NCBI has defined approximately 18 COGs by comparing protein-encoding regions of dozens of complete genomes of the major phylogenetic lineages; each COG is defined by proteins from at least three lineages. NCBI COGs are grouped into the following four major categories:

1. Information storage and processing, containing translation, ribosomal structure, and biogenesis functions; transcription functions; and DNA replication, recombination, and repair functions
2. Cellular processes, containing functions for cell division and partitioning; posttranslational modification, protein turnover, and chaperones; cell envelope biogenesis and the outer membrane; cell motility and secretion; inorganic ion transport and metabolism; and signal transduction mechanisms
3. Metabolism, containing functions for energy production and conversion; carbohydrate transport and metabolism; amino acid transport and metabolism; nucleotide transport and metabolism; coenzyme metabolism; lipid metabolism; and secondary-metabolite biosynthesis, transport, and catabolism
4. "Poorly characterized," containing "general function prediction only" and "function unknown"

Links to COG information can be found at the website http://www.ncbi.nlm.nih.gov/COG/.

The Pfam families are more likely to describe domains than full-length proteins, for example, indicating evidence for an ATP-binding domain in a protein.

The TIGRFAM families are a versatile resource for protein classifications and include superfamilies, which comprise all proteins that have amino acid homology but that may differ in biological function; equivalogs, which include proteins that are conserved in function; and subfamilies, which contain proteins incompletely evaluated for function.

Motifs are conserved patterns of amino acids, often the amino acids comprising the active site of a protein. Thus, a motif can indicate that a protein has biochemical activity similar to that of other proteins with a related motif.

Regulatory Sequences

Regulatory sequences are usually determined experimentally, but bioinformatics techniques are becoming more effective. For example, at the NCBI website mentioned above, **ELPH** can find motifs in DNA (or protein) sequences, **TransTerm** can find rho-independent transcriptional terminator sites, and **RBSfinder** can identify potential prokaryotic translational RBS.

Genetic Elements and Structural Features

Chapters 4 through 9 describe the features of DNA elements, such as prophages and integrons. Structural features, such as repetitive sequences, are discussed throughout the book.

(continued)

BOX 2.5 (continued)

Annotation and Comparative Genomics

Transcriptome Analysis

Parallel analysis of the expression of thousands of genes in a bacterial genome can be done by **transcriptome** analysis (**microarray** analysis), including the use of cDNA microarrays, oligonucleotide arrays, and Affymetrix GeneChip arrays (see chapter 13). A database for microarray data is GEO (for Gene Expression Omnibus). Software named MADAM (for *micro*-*array data management*) can be found at the NCBI website.

Functional Proteomics

Functional protein domains and motifs can be predicted by special algorithms and experimental methods (**proteomics**). The HMM type of algorithm described above is very useful in identifying protein domains and motifs. Some examples of useful Web search terms are Swiss-Prot, PIR (Protein Information Resource), Ensembl UniProt, ProDom, PROSITE, TRANSFAC (for transcription factors), and SENTRA (for prokaryotic signal transduction proteins).

High-throughput protein identification is now possible, using a combination of a protein separation technique, such as **two-dimensional polyacrylamide gel electrophoresis (2D-PAGE)** or **chromatography**, and fragmentation of proteins into peptides, followed by analysis of the peptides by **mass spectrometry**. By using comparisons with genome sequence data, the mass spectrometry data can be used to identify individual proteins. In another type of **proteome** analysis, interactions between proteins can be detected using **two-hybrid screens** (see Box 13.2).

Structural Proteomics

Three-dimensional protein structures are determined from the crystal structure or from nuclear magnetic resonance spectroscopy. The development of predictive algorithms has allowed some structures to be predicted from the sequence alone; predicting protein structures from amino acid sequences is a very challenging but active research area. Examples of Web search terms are "EXPASY," "Swiss-model," "Touchstone," and "COILS prediction," to mention just a few. Another set of search terms, "HMMTOP," "TMHMM," and "TmPred," are tailored for the prediction of transmembrane helices and topology.

Functional Genomics

Reverse genetics can be used to determine the function of a gene whose sequence is known. **Forward genetics** can be used to identify and sequence a gene whose function is known. Often, this requires a large repertoire of genetic techniques rather than merely knocking out the gene. **Interaction genetics** seeks to elucidate the significant interrelationships and subtle interactions of genes and gene products. Methods for all of these aspects of genetic analysis are discussed throughout the book.

Protein Localization

Gene fusion techniques using, for example, fluorescent probes can often locate the gene product within the cell. For an example, see the cover of the book.

References

Gibson, G., and S. V. Muse. 2004. *A Primer of Genome Science*, 2nd ed. Sinauer Associates, Inc., Sunderland, MA.

Gill, S. R., D. E. Fouts, G. L. Archer, E. F. Mongodin, R. T. DeBoy, J. Ravel, I. T. Paulsen, J. F. Kolonay, L. Brinkac, M. Beanan, R. J. Dodson, S. C. Daugherty, R. Madupu, S. V. Angiuoli, A. S. Durkin, D. H. Haft, J. Vamathevan, H. Khouri, T. Utterback, C. Lee, G. Dimitrov, L. Jiang, H. Qin, J. Weidman, K. Tran, K. Kang, I. R. Hance, K. E. Nelson, and C. M. Fraser. 2005. Insights on evolution of virulence and resistance from the complete genome analysis of an early methicillin-resistant *Staphylococcus aureus* strain and a biofilm-producing methicillin-resistant *Staphylococcus epidermidis* strain. *J. Bacteriol. 187*:2426–2438.

Mount, D. W. 2004. *Bioinformatics: Sequence and Genome Analysis*, 2nd ed. Cold Spring Harbor Laboratory Press, Cold Spring Harbor, NY.

Riley, M., T. Abe, M. B. Arnaud, M. K. B. Berlyn, F. R. Blattner, R. R. Chaudhuri, J. D. Glasner, T. Horiuchi, I. M. Keseler, T. Kosuge, H. Mori, N. T. Perna, G. Plunkett III, K. E. Rudd, M. H. Serres, G. H. Thomas, N. R. Thomson, D. Wishart, and B. L. Wanner. 2006. *Escherichia coli* K-12: a cooperatively developed annotation snapshot—2005. *Nucleic Acids Res.* **34**:1–9.

Typas, A., R. J. Nichols, D. A. Siegele, M. Shales, S. R. Collins, B. Lim, H. Braberg, N. Yamamoto, R. Takeuchi, B. L. Wanner, H. Mori, J. S Weissman, N J. Krogan, and C. A. Gross. 2008. High-throughput, quantitative analyses of genetic interactions in *E. coli*. *Nat. Methods* **5**:781–787.

phosphate bond. The direction on DNA or RNA from the 5′ end to the 3′ end is called the **5′-to-3′ direction**. An RNA polymerase molecule synthesizes mRNA in the 5′-to-3′ direction, moving 3′ to 5′ on the transcribed strand (or template strand) of DNA. The opposite strand of DNA from the transcribed strand has the same sequence and 5′-to-3′ polarity as the RNA, so it is called the coding strand (or nontemplate strand). Sequences of DNA in the region of a gene are usually shown as the sequence of the coding strand. A sequence that is located in the 5′ direction of another sequence on the coding strand is upstream of that sequence, while a sequence in the 3′

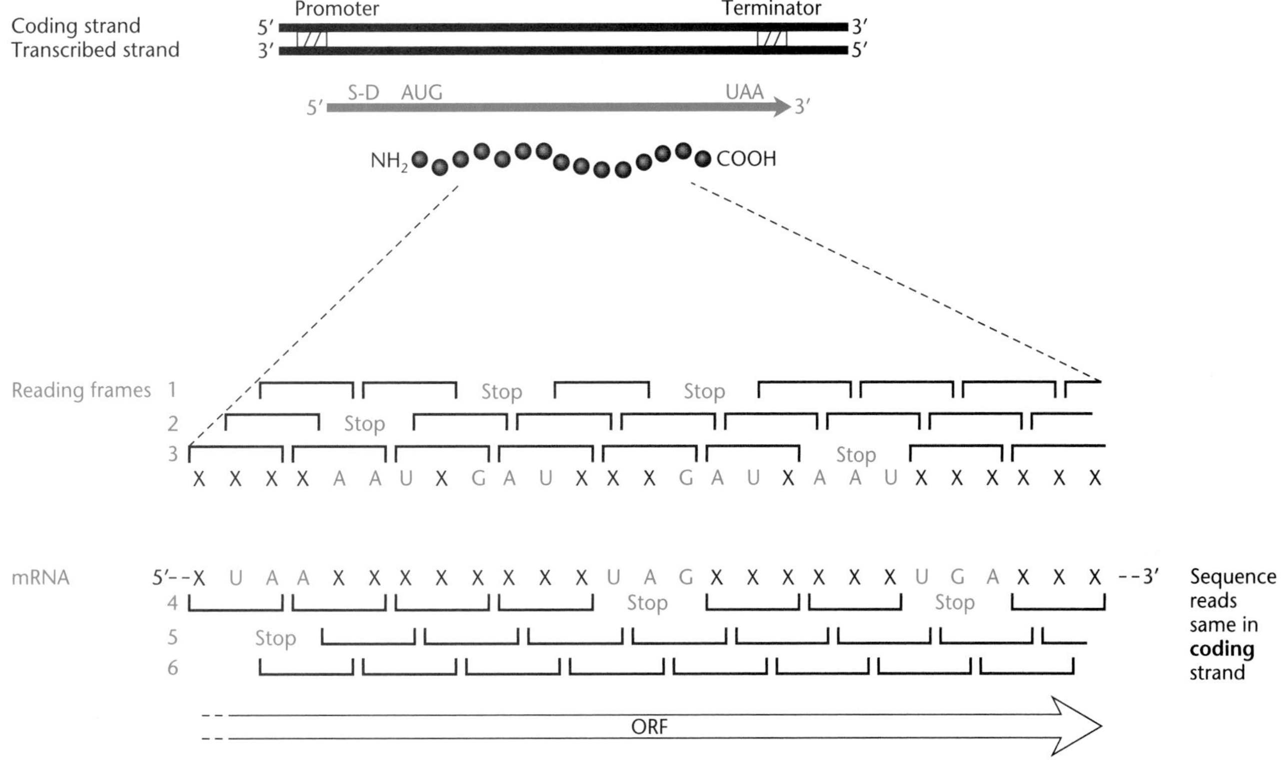

Figure 2.44 Relationship between a gene in DNA and the coding sequence in mRNA. There are a total of six different sequences in the two strands of DNA that may contain ORFs, but generally, only one ORF encodes a polypeptide in each region. doi:10.1128/9781555817169.ch1.f2.44

direction is downstream. Therefore, the promoter for a gene and the S-D sequences are both upstream of the initiation codon, while the termination codon and the transcription termination sites are both downstream.

The positions of nucleotides around a promoter are numbered as shown in Figure 2.43B. The position of the first nucleotide in the RNA is called the start point and is given the number +1; the distance in nucleotides from this point to another point is numbered negatively or positively, depending on whether the second site is upstream or downstream of the start point, respectively. We have already used this numbering system in Figure 2.7, which shows a σ^{70} promoter with consensus sequences at −10 and −35 relative to the start point of transcription. Note that these definitions can be used to describe only a region of DNA that is known to encode an RNA or protein, where we know which is the coding strand and which is the transcribed strand. Otherwise, what is upstream on one strand of DNA is downstream on the other strand.

Because mRNAs are both made and translated in the 5′-to-3′ direction, an mRNA can (and usually will) be translated while it is still being made, at least in prokaryotes, where there is no nuclear membrane separating the DNA from the cytoplasm, where the ribosomes reside. We have discussed how this can lead to phenomena unique to bacteria, such as ρ-dependent polarity, and it is used to regulate the synthesis of RNA on some genes in bacteria (see chapter 12).

It is important to distinguish promoters from TIRs and to distinguish **transcription termination** sites from translation termination sites. Figure 2.44 also illustrates this difference. Transcription begins at the promoter and defines the 5′ end of the mRNA, but the place where translation begins, the TIR, can be some distance from the 5′ end. The untranslated region on the 5′ end of an mRNA upstream of the TIR is called the 5′ untranslated region or **leader region** and can be quite long. Similarly, a nonsense codon in the reading frame for the protein is a translation terminator, not a transcription terminator. The transcription terminator, and therefore the 3′ end of the mRNA, may be some distance from the nonsense codon that terminates translation of the mRNA. The distance from the last nonsense codon to the 3′ end of the mRNA is the 3′ untranslated region. Polycistronic mRNAs encode more than one polypeptide. These mRNAs have a separate TIR and nonsense codon for each gene and can have noncoding or untranslated sequences upstream of, downstream of, and between the genes. Eukaryotes generally do not have polycistronic mRNAs, which is related to the dependence on ribosome binding to the 5′ end of the mRNA for translation initiation.

Open Reading Frames

The concept of an ORF is very important, particularly in this age of genomics. As discussed above, a reading frame in DNA is a succession of nucleotides in the DNA taken three at a time, the same way the genetic code is translated. Each DNA sequence has six reading frames, three on each strand, as illustrated in Figure 2.44. An ORF is a string of potential codons for amino acids in DNA unbroken by nonsense codons in one of the reading frames. Computer software can show where all the ORFs are in a sequence, and most DNA sequences have many ORFs on both strands, although most of them are short. The region shown in Figure 2.44 contains many ORFs, but only the longest, in frame 6, is likely to encode a polypeptide. However, the presence of even a long ORF in a DNA sequence does not necessarily indicate that the sequence encodes a protein, and fairly long ORFs often occur by chance. Furthermore, it has become evident recently that even very short ORFs can encode short peptides with important biological functions.

If an ORF does encode a polypeptide, it will begin with a TIR, but as discussed above, TIRs are sometimes difficult to identify. Clues to whether an ORF is likely to encode a protein may come from the choice of the third base in the codon for each amino acid in the ORF. Because of the redundancy of the code, an organism has many choices of codons for each amino acid, but each organism prefers to use some codons over others (see "Codon Usage" above) (Table 2.2).

A more direct way to determine if an ORF actually encodes a protein is to ask which polypeptides are made from the DNA in an in vitro transcription-translation system. These systems use extracts of cells, typically of *E. coli*, from which the DNA has been removed but the RNA polymerase, ribosomes, and other components of the translation apparatus remain. When DNA with the ORFs under investigation is added to these extracts, polypeptides can be synthesized from the added DNA. If the size of one of these polypeptides corresponds to the size of an ORF on the DNA, the ORF probably encodes a protein. Another way to determine if an ORF encodes a protein is to make a translation fusion of a reporter gene to the ORF and to determine whether the reporter gene is expressed (see below).

Transcriptional and Translational Fusions

Probably the most convenient way to determine which of the possible ORFs on the two strands of DNA in a given region are translated into proteins is to make **transcriptional** and **translational fusions** to the ORFs. These methods make use of **reporter genes**, such as *lacZ* (β-galactosidase), *gfp* (green fluorescent protein), *lux* (luciferase), or other genes whose products are easy to detect. Figure 2.45 illustrates the concepts of translational and transcription fusions.

An ORF can be translated only if it is transcribed into RNA. **Transcriptional fusions** can be used to determine whether this has occurred. To make a transcriptional fusion, a reporter gene with the sequence for a TIR but no promoter of its own is fused immediately downstream of the promoter of the gene to be tested. If the gene is transcribed into mRNA, the reporter gene will also be transcribed, and its product will be detectable in the cell. Transcriptional fusions also offer a convenient way of determining how much mRNA is made on a coding sequence. In general, the more reporter gene product that is made in the transcriptional fusion, the more mRNA was made on the upstream coding sequence. Examples of the use of transcriptional fusions in studying the regulation of operons are given in subsequent chapters.

In a translational fusion, the reporter gene lacks both a promoter and a TIR and is fused immediately downstream of the TIR of the gene under investigation; it is crucial that the two coding sequences are fused in such a way that they are translated in the same reading frame and there are no nonsense codons between them. Translation beginning at a TIR upstream of one of the coding sequences will proceed through the other coding sequence, making a **fusion protein** that contains both polypeptide sequences. The coding sequence can be fused either to the amino terminus of the reporter gene product or to its carboxyl terminus. The reporter gene product can then be assayed as before to determine how much of the fusion protein has been made. The reporter gene product must retain its activity even when fused to the potential polypeptide encoded by the ORF; otherwise, it will not be detectable. Many reporter genes have been chosen because their products remain active when fused to other polypeptides. Translational fusions are also often used to attach affinity tags to proteins to use in their purification.

Antibiotics That Block Transcription and Translation

As in chapter 1, we devote the remainder of this chapter to a discussion of antibiotics because these compounds not only are among the most useful therapeutic agents, but also allow mutants to be isolated for genetic studies. Studies of how antibiotics affect transcription and translation have greatly contributed to our understanding of these processes.

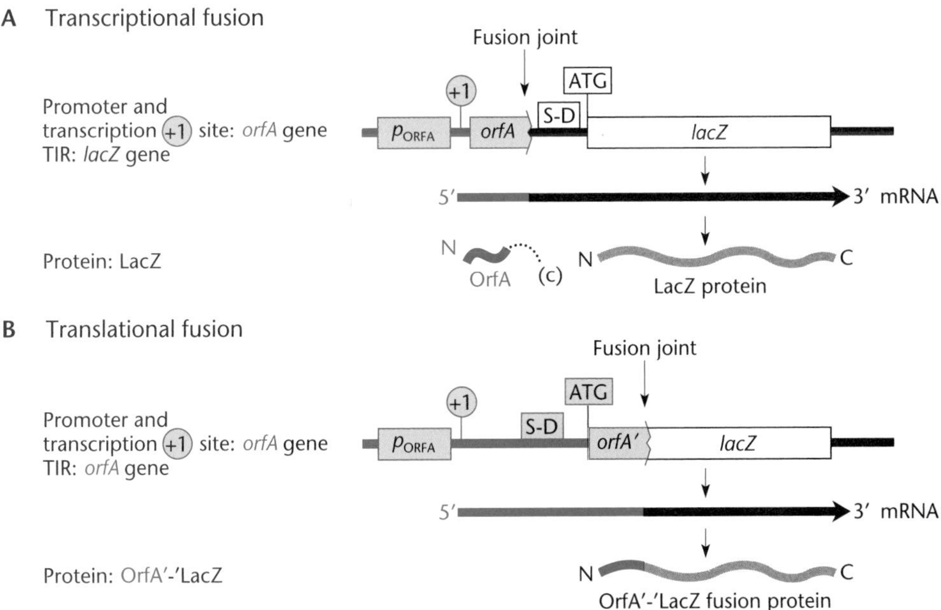

Figure 2.45 Transcriptional and translational fusions to express *lacZ* (which encodes β-galactosidase). In both types of fusion, transcription begins at the +1 site at the p_{ORFA} promoter upstream of OrfA. **(A)** In a transcriptional fusion, both the upstream OrfA coding region and the downstream *lacZ* reporter gene are translated from their own TIRs. Only the TIR for *lacZ* is shown, as S-D and ATG (boxed). The translation of the upstream OrfA continues until it encounters a nonsense codon in frame, as indicated by the dotted line. **(B)** In a translational fusion, a fusion protein is translated from the TIR upstream of OrfA to make a fusion protein containing the LacZ reporter protein fused to the remaining product of the upstream OrfA. The prime symbols indicate that part of each protein may be deleted.
doi:10.1128/9781555817169.ch1.f2.45

Antibiotic Inhibitors of Transcription

Some of the components of the transcription apparatus are the targets of antibiotics used in treatment of bacterial infections and in tumor therapy. Some of these antibiotics are made by soil bacteria and fungi, and some have been synthesized chemically.

INHIBITORS OF rNTP SYNTHESIS

Some antibiotics that inhibit transcription do so by inhibiting the synthesis of the rNTPs. An example is azaserine, which inhibits purine biosynthesis. Azaserine and other antibiotics that block the synthesis of the ribonucleotides are usually not specific to transcription, since the ribonucleotides, including ATP and GTP, have many other uses in the cell. This lack of specificity limits the usefulness of these antibiotics for studying transcription, although some of them have other uses.

INHIBITORS OF RNA SYNTHESIS INITIATION

Rifamycin and its more commonly used derivative, rifampin, block transcription by binding to the β subunit of RNA polymerase and specifically blocking the initiation of RNA synthesis. The antibiotic binds in the active-site channel of RNA polymerase and limits growth of the RNA chain to a few nucleotides (Figure 2.15). The property of blocking only transcription initiation has made these antibiotics very useful in the study of transcription. For example, they have been used to analyze the steps in initiation of RNA synthesis and to study the stability of RNA in the cell. These antibiotics are useful therapeutic agents in the treatment of tuberculosis and other difficult-to-treat bacterial infections because they inhibit the RNA polymerases of essentially all types of bacteria, but not the RNA polymerases of eukaryotes, so they are not toxic to humans and animals. Accordingly, many derivatives have been made from them.

In rifampin-resistant mutants, one or more amino acids in the β subunit of RNA polymerase lining the active-site channel have been changed so that rifampin can no longer bind but the RNA polymerase still functions. Chromosomal mutations conferring resistance to rifampin are fairly common and have limited the usefulness of these antibiotics somewhat.

INHIBITORS OF RNA ELONGATION AND TERMINATION

Streptolydigin also binds to the β subunit of the RNA polymerase of bacteria but can block RNA synthesis after it is under way. It has a weaker affinity for RNA polymerase than does rifampin, so it blocks transcription only when added at higher concentrations, which limits its usefulness. Bicyclomycin targets the transcription terminator protein ρ and prevents ρ-dependent transcription termination (but not factor-independent transcription termination). Use of bicyclomycin as an antibiotic depends on the importance of ρ-dependent transcription termination in the target organism.

INHIBITORS THAT AFFECT THE DNA TEMPLATE

Actinomycin D and bleomycin block transcription by binding to the DNA. After bleomycin binds, it also nicks the DNA. While such drugs have been useful for studying transcription in bacteria, they are not very useful in antibacterial therapy because they are not specific to bacteria and are very toxic to humans and animals. They are, however, used in antitumor therapy.

Antibiotic Inhibitors of Translation

The translation apparatus of bacteria is a particularly tempting target for antibacterial drugs because it is somewhat different from the eukaryotic translation apparatus and is highly conserved among bacteria. Antibiotics that inhibit translation are among the most useful of all the antibiotics and are listed in Table 2.3, which also lists their targets and sources. Toxicity in some cases may result from the similarity of bacterial and mitochondrial ribosomes. Some of these antibiotics are also very useful in combating fungal diseases and in cancer chemotherapy.

INHIBITORS THAT MIMIC tRNA

Puromycin mimics the 3′ end of tRNA with an amino acid attached (aa-tRNA). It enters the ribosome as does an aa-tRNA, and the peptidyltransferase attaches it to the growing polypeptide. However, it does not translocate properly from the A site to the P site, and the peptide with puromycin attached to its carboxyl terminus is released from the ribosome, terminating translation.

Studies with puromycin have contributed greatly to our understanding of translation. The model of the A and P sites in the ribosome and the concept that the 50S ribosome contains the enzyme for peptidyl bond formation, which was recently shown to be the 23S rRNA itself, came from studies with the antibiotic. Puromycin is not a very useful antibiotic for treating bacterial diseases, however, because it also inhibits translation in eukaryotes, making it toxic in humans and animals.

INHIBITORS THAT BIND TO THE 23S rRNA

Chloramphenicol

Chloramphenicol inhibits translation by binding to ribosomes and preventing the binding of aa-tRNA to the A site. It might also inhibit the peptidyltransferase reaction, preventing the formation of peptide bonds. Structural studies have shown that chloramphenicol binds to specific nucleotides in the 23S rRNA, although ribosomal proteins are also part of the binding site.

Chloramphenicol is effective at low concentrations and therefore has been one of the most useful antibiotics for studying cellular functions. For example, it has been used to determine the time in the cell cycle when proteins required for cell division and for initiation of chromosomal replication are synthesized. It is also quite useful in treating bacterial diseases, since it is not very toxic for humans and animals because it is fairly specific for the translation apparatus of bacteria. It can also cross the blood-brain barrier, making it useful for treating diseases of the central nervous system, such as bacterial meningitis. Chloramphenicol is bacteriostatic, which means that it stops the growth of bacteria without actually killing them. Such antibiotics should not be used in combination with antibiotics such as penicillin that depend on cell growth for their killing activity, since they neutralize the effect of these other antibiotics.

It takes multiple mutations in ribosomal proteins to make bacteria resistant to chloramphenicol, so resistant

Table 2.3 Antibiotics that block translation

Antibiotic	Source	Target
Puromycin	*Streptomyces alboniger*	Ribosomal A site
Kanamycin	*Streptomyces kanamyceticus*	16S rRNA
Neomycin	*Streptomyces fradiae*	16S rRNA
Streptomycin	*Streptomyces griseus*	30S ribosome
Thiostrepton	*Streptomyces azureus*	23S rRNA
Gentamicin	*Micromonospora purpurea*	16S rRNA
Tetracycline	*Streptomyces rimosus*	Ribosomal A site
Chloramphenicol	*Streptomyces venezuelae*	Peptidyltransferase
Erythromycin	*Saccharopolyspora erythraea*	23S rRNA
Fusidic acid	*Fusidium coccineum*	EF-G
Kirromycin	*Streptomyces collinus*	EF-Tu

mutants are very rare. Some bacteria have enzymes that inactivate chloramphenicol. The genes for these enzymes are often carried on plasmids and transposons, interchangeable DNA elements that are discussed in chapters 4 and 9. The best-characterized chloramphenicol resistance gene is the *cat* gene of transposon Tn9, whose product is an enzyme that specifically acetylates (adds an acetyl group to) chloramphenicol, thereby inactivating it. The *cat* gene has been used extensively as a reporter gene to study gene expression in both bacteria and eukaryotes and has been introduced into many plasmid cloning vectors.

Macrolides

Erythromycin is a member of a large group of antibiotics called the macrolide antibiotics, which have large ring structures. These antibiotics may also inhibit translation by binding to the 23S rRNA and blocking the exit channel of the growing polypeptide. This causes the polypeptide to be released prematurely at either the peptidyltransferase reaction or the translocation step, causing the peptidyl-tRNA to dissociate from the ribosome.

Erythromycin and other macrolide antibiotics have been among the most useful antibiotics. They are effective mostly against gram-positive organisms but are also useful in treating some gram-negative bacterial diseases, including *Legionella* and *Rickettsia* infections. One of the most foreboding developments in medicine is the extent to which pathogenic bacteria have become resistant to the macrolides through the misuse of these once most useful antibiotics. They achieve this resistance in a number of ways. One way is methylation of a specific adenine base in the 23S rRNA by enzymes called the Erm methylases. Methylation of this base causes a conformational change in the 23S rRNA that might prevent proper binding of the antibiotics. These enzymes are encoded by plasmids and transposons that are exchanged readily between bacteria. Others become resistant by altering preexisting efflux pumps so that they pump macrolides out of the cell or by acquiring genes encoding the pumps from resistant bacteria. Some mutational changes in the 23S rRNA can also confer resistance to these antibiotics. New derivatives of the antibiotics must be made constantly to stay ahead of the advancing bacterial resistance.

Thiostrepton

Thiostrepton and other thiopeptide antibiotics block translation by binding to 23S rRNA in the region of the ribosome involved in the peptidyltransferase reaction and preventing the binding of EF-G. Thiostrepton is specific to gram-positive bacteria; it does not enter gram-negative bacterial cells. Thiostrepton has limited usefulness because it is not very soluble. It is used mostly in veterinary medicine and agriculture.

Most thiostrepton-resistant mutants are missing the L11 ribosomal protein from the 50S ribosomal subunit. This protein seems not to be required for protein synthesis but plays a role in guanosine tetraphosphate synthesis (see chapter 13). Other mutations confer resistance by changing nucleotides 1067 and 1095 in the 23S rRNA; these nucleotides are close to where the antibiotic binds. Plasmids and transposon genes can confer thiostrepton resistance by methylating ribose sugars of the 23S rRNA in certain positions. Eukaryotes may be insensitive to the antibiotic because the analogous ribose sugars of the eukaryotic 28S rRNAs are normally extensively methylated.

INHIBITORS OF BINDING OF aa-tRNA TO THE A SITE

Tetracycline was one of the first antibiotics isolated. Recent evidence suggests that it may inhibit translation by allowing the aa-tRNA–EF-Tu complex to bind to the A site of the ribosome and allowing the GTP on EF-Tu to be cleaved to GDP but then inhibiting the next step, causing a futile cycle of binding and release of the aa-tRNA from the A site.

Tetracycline has been a very useful antibiotic for treating bacterial diseases, although it is somewhat toxic to humans because it also inhibits the eukaryotic translation apparatus. It has a very broad spectrum, acting against both gram-negative and gram-positive bacteria, as well as some protozoans, such as the one that causes amoebic dysentery. It is also used to treat acne. However, this is another case where overuse has led to the spread of resistance, and it is no longer the first choice of antibiotic against many infections.

In some types of bacteria, ribosomal mutations confer low levels of resistance to tetracycline by changing protein S10 of the ribosome. However, most clinically important resistance to tetracycline and its derivatives is acquired on plasmids and transposons. One of these genes, *tetM*, carried by the conjugative transposon Tn*916* and its relatives (see chapter 5), encodes an enzyme that confers resistance by methylating certain bases in the 16S rRNA. The *tetM* gene is ubiquitous; related genes occur in both gram-positive and gram-negative bacteria. Other tetracycline resistance genes, such as the *tet* genes carried by transposon Tn*10* and plasmid pSC101 of *E. coli*, encode membrane proteins that confer resistance by pumping tetracycline out of the cell. These tetracycline resistance genes are extensively used as reporter genes and as markers for genetic analysis in *E. coli*, and the *tetA* gene from pSC101 has been introduced into many plasmid cloning vectors (see chapter 4). However, the *tet* genes of Tn*10* and pSC101 are specific for *E. coli* and do not confer tetracycline resistance in many other types of bacteria, which limits their usefulness as genetic markers in bacteria other than *E. coli*. One of the more interesting types of resistance to tetracyclines is due to the so-called ribosome protection proteins, represented by TetO and TetQ. This is the type of resistance exhibited by the soil bacteria that make tetracycline. These proteins bind to the A site of the ribosome and release tetracycline from the A site. They may be able to bind to the A site because they mimic the translation factor EF-G.

INHIBITORS OF TRANSLOCATION

Aminoglycosides

Kanamycin and its close relatives neomycin and gentamicin are members of a larger group of antibiotics, the aminoglycoside antibiotics, which also includes streptomycin. Their mechanism of action is somewhat obscure, but they seem to affect translocation by binding to the A site of the ribosome. Aminoglycosides have a very broad spectrum of action, and some of them inhibit translation in plant and animal cells, as well as in bacteria. For example, the ability of neomycin to block translation in plants and animals has made it very useful in biotechnology, where it is used to select transgenic plants containing bacterial genes that confer resistance to these antibiotics (see below). However, their toxicity, especially during sustained use, and high rates of resistance somewhat limit their usefulness as therapeutic agents.

Bacterial mutants resistant to aminoglycosides are quite rare, and multiple mutations are often required to confer high levels of resistance. The fact that resistant mutants are rare has contributed to the usefulness of kanamycin and its relatives in biotechnology. However, most of the clinically important resistance is due to genes exchanged on transposons and plasmids. The products of some of these genes inactivate the aminoglycosides by phosphorylating, acetylating, or adenylating (adding adenosine to) them. For example, the product of the *neo* gene for kanamycin and neomycin resistance, from transposon Tn*5*, phosphorylates these antibiotics. The *neo* gene has been very important in genetics and biotechnology because it expresses kanamycin resistance in almost all gram-negative bacteria and even makes plant and animal cells resistant to a kanamycin derivative, G418, provided that the gene is transcribed and translated in the plant or animal cells.

Fusidic Acid

Fusidic acid specifically inhibits EF-G (called EF-2 in eukaryotes), probably by preventing its dissociation from the ribosome after GTP cleavage. It has been very useful in studies of the function of ribosomes. In *E. coli*, mutations that confer resistance to fusidic acid are in the *fusA* gene, which encodes EF-G. Unexpectedly, some acetyltransferases that confer resistance to chloramphenicol also bind to and inactivate fusidic acid.

SUMMARY

1. RNA is a polymer made up of a chain of ribonucleotides. The bases of the nucleotides—adenine, cytosine, uracil, and guanine—are attached to the five-carbon sugar ribose. Phosphate bonds connect the sugars to make the RNA chain, attaching the third (3′) carbon of one sugar to the fifth (5′) carbon of the next sugar. The 5′ end of the RNA is the nucleotide that has a free phosphate attached to the 5′ carbon of its sugar. The 3′ end has a free hydroxyl group at the 3′ carbon, with no phosphate attached. RNA is both made and translated from the 5′ end to the 3′ end.

2. After they are synthesized, RNAs can undergo extensive processing and modification. Processing occurs when phosphate bonds are broken or new phosphate bonds are formed. Modification occurs when the bases or the sugars of the RNA are chemically altered, for example, by methylation. In bacteria, rRNAs and tRNAs, but not mRNAs, are extensively modified.

3. The primary structure of an RNA is its sequence of nucleotides. The secondary structure is formed by hydrogen bonding between bases in the same RNA to give localized double-stranded regions. The tertiary structure is the three-dimensional shape of the RNA due to the stiffness of the double-stranded regions of secondary structure. All RNAs, including mRNA, rRNA, and tRNA, probably have secondary and tertiary structures.

4. The enzyme responsible for making RNA is called RNA polymerase. One of the largest enzymes in the cell, the bacterial RNA polymerase has five subunits plus another detachable subunit, the σ factor, which comes off after the initiation of transcription. The ω subunit helps in its assembly.

5. Transcription begins at well-defined sites on DNA called promoters. The type of promoter used depends on the type of σ factor bound to the RNA polymerase.

6. Transcription stops at sequences in the DNA called transcription terminators, which can be either factor dependent or factor independent. The factor-independent terminators have a string of A's (transcribed as U's in the RNA) that follows a sequence that forms an inverted repeat. The RNA transcribed from that region folds back on itself to form a stem-loop, or hairpin, which causes the RNA molecule to fall off the DNA template. The factor-dependent terminators do not have a well-defined sequence. The ρ protein is

SUMMARY (continued)

the best-characterized termination factor in *E. coli*. It forms a ring that binds to and encircles the RNA, moving toward the RNA polymerase. If the RNA polymerase pauses at a ρ termination site, the ρ factor causes it to dissociate, releasing the mRNA.

7. Most of the RNA in the cell falls into three classes: messenger (mRNA), ribosomal (rRNA), and transfer (tRNA). Of these, mRNA is very unstable, existing for only a few minutes before being degraded. rRNAs in bacteria are further divided into three types: 16S, 23S, and 5S. Both rRNA and tRNA are very stable and account for as much as 95% of the total RNA. Other RNAs include the primers for DNA replication and small RNAs involved in regulation or RNA processing.

8. Ribosomes, the site of protein synthesis, are made up of two subunits, the 30S subunit and the 50S subunit, as well as approximately 50 proteins. The 16S rRNA is in the 30S subunit, while the 23S and 5S rRNAs are in the 50S subunit.

9. Polypeptides are chains of the 20 amino acids that are held together by peptide bonds between the amino group of one amino acid and the carboxyl group of another. The amino terminus (N terminus) of the polypeptide has the amino acid with an unattached amino group. The carboxyl terminus (C terminus) of a polypeptide has the amino acid with a free carboxyl group.

10. Translation is the synthesis of polypeptides, using the information in mRNA to direct the sequence of amino acids. During translation, the mRNA moves in the 5′-to-3′ direction along the ribosome 3 nucleotides at a time. Three reading frames are possible, depending on how the ribosome is positioned at each triplet.

11. The genetic code is the assignment of each possible 3-nucleotide codon sequence in mRNA to 1 of 20 amino acids. The code is redundant, with more than one codon sometimes encoding the same amino acid. Because of wobble, the first position of the tRNA anticodon (written 5′ to 3′) does not have to behave by the standard base-pairing complementarity to the third position of the antiparallel codon sequence, and other pairings are possible.

12. Initiation of translation occurs at TIRs on the mRNA that consist of an initiation codon, usually AUG or GUG, and often an S-D sequence, a short sequence that is complementary to part of the 16S rRNA and precedes the initiation codon.

13. The first tRNA to enter the ribosome is a special methionyl-tRNA called fMet-$tRNA^{fMet}$, which carries the amino acid formylmethionine. After the polypeptide has been synthesized, the formyl group and often the first methionine are removed.

14. Translation termination occurs when one of the terminator or nonsense codons, UAA, UAG, or UGA, is encountered as the ribosome moves down the mRNA. Proteins called RFs are also required for release of the polypeptide.

15. The primary structure of a polypeptide is the sequence of amino acids in the polypeptide. Proteins can be made up of more than one polypeptide chain, which can be the same as or different from each other. The secondary structure results from hydrogen bonding of the amino acids to form α-helical regions and β-sheets. Tertiary structure refers to how the chains fold up on themselves, and quaternary structure refers to one or more different polypeptide chains folding up on each other.

16. Proteins that help other proteins fold are called chaperones. The most ubiquitous chaperones are the Hsp70 chaperones, called DnaK in *E. coli*, which are very similar in all types of cells from bacteria to humans. These chaperones bind to the hydrophobic regions of proteins and prevent them from associating prematurely. They are aided by their smaller cochaperones, DnaJ and GrpE, which help in binding to proteins and cycling ADP off the chaperone, respectively. Other proteins, called Hsp60 chaperonins, also help proteins fold, but by a very different mechanism. One chaperonin, called GroEL in *E. coli*, forms large cylindrical structures with internal chambers that take up unfolded proteins and help them refold properly. A cochaperonin called GroES forms a cap on the cylinder after the unfolded protein is taken up. Chaperonins like GroEL are found in bacteria and in the organelles of eukaryotes and are called group I chaperonins. Another type, group II chaperonins, is found in the cytoplasm of eukaryotes and in archaea. They have a similar structure but a very different amino acid sequence.

17. The process of passing proteins through membranes is called transport. Proteins that pass through the inner membrane into the periplasm and beyond are said to be exported. Proteins that pass out of the cell are secreted.

18. Proteins can also be held together by disulfide linkages between cysteines in the protein. Generally, only proteins that are exported into the periplasm or out of the cell have disulfide bonds. These disulfide bonds are made by oxidoreductases in the periplasm of gram-negative bacteria.

19. An ORF is a string of amino acid codons in DNA unbroken by a nonsense codon. In vitro transcription-translation systems or transcriptional and translation fusions are often required to prove that an ORF in DNA actually encodes a protein.

20. The strand of DNA from which the mRNA is made is the transcribed, or template, strand. The opposite strand, which has the same sequence as the mRNA, is the coding, or nontemplate, strand.

(continued)

SUMMARY (continued)

21. A sequence 5′ on the coding strand of DNA relative to a particular element is said to be upstream, whereas a sequence 3′ to that element is downstream.

22. The TIR sequence of a gene does not necessarily occur at the beginning of the mRNA. The 5′ end of the mRNA is called the 5′ untranslated region or leader region. Similarly, the sequence downstream of the nonsense codon is the 3′ untranslated region.

23. Because mRNA is both transcribed and translated in the 5′-to-3′ direction, translation can begin before synthesis of the mRNA is complete in bacteria, which have no nuclear membrane.

24. Bacteria often make polycistronic mRNAs with more than one polypeptide coding sequence on an mRNA. This can result in polarity of transcription and translational coupling, phenomena unique to bacteria.

25. The expression of genes is regulated, depending on the conditions in which the cell is found. This regulation can be either transcriptional or posttranscriptional. Transcriptional regulation can be either negative or positive, depending on whether the regulatory protein is a repressor or an activator, respectively. A repressor binds to an operator or operators, which are usually close to the promoter, and prevents transcription from the promoter. An activator binds to an activator sequence that is usually upstream of the promoter and increases transcription from the promoter. Transcriptional regulation can also occur after the RNA polymerase leaves the promoter, as in attenuation or antitermination of transcription. Posttranscriptional regulation can occur at the level of stability of the mRNA, translation of the mRNA, or processing, modification, or degradation of the gene product.

26. Gene fusions have many uses in modern molecular genetics. They can be either transcriptional or translational fusions. In a transcriptional fusion, the two coding regions are transcribed into the same mRNA, but each is translated from its own TIR, so expression of the downstream reporter gene is dependent on the activity of the promoter of the upstream gene. In a translational fusion, the two coding regions are fused to each other, so expression of the downstream reporter gene is dependent on the activities of both the promoter and TIR of the upstream gene.

27. Many naturally occurring antibiotics target components of the transcription and translation machinery. Some of the more useful ones are rifampin, streptomycin, tetracycline, thiostrepton, chloramphenicol, and kanamycin. In addition to their uses in treating bacterial infections, tumor chemotherapy, and biotechnology, antibiotics have also helped us understand the mechanisms of transcription and translation. In addition, the genes that confer resistance to these antibiotics have served as selectable genetic markers and reporter genes in molecular genetic studies of both bacteria and eukaryotes.

QUESTIONS FOR THOUGHT

1. Which do you think came first in the very earliest life on Earth, DNA, RNA, or protein? Why?

2. Why is the genetic code universal?

3. Why do you suppose prokaryotes have polycistronic mRNAs but eukaryotes do not?

4. Why do you suppose mitochondrial genes show differences in their genetic code from chromosomal genes in eukaryotes?

5. Why is selenocysteine inserted into proteins of almost all organisms but into only a few sites in a few proteins in these organisms?

6. Why do so many antibiotics inhibit the translation process as opposed to, say, amino acid biosynthesis?

7. Why do you think chaperonins have two linked chambers and alternate the folding of proteins between the two chambers?

8. List all the reasons you can think of why bacteria would regulate the expression of their genes.

PROBLEMS

1. What is the longest ORF in the mRNA sequence 5′-AGCUAACUGAUGUGAUGUCAACGUCCUACUCUAGCGUAGUCUAAAG-3′? Remember to look in all three frames.

2. Where do you think translation is most likely to start in the mRNA sequence 5′-UAAGUGAAAGAUGUGAAUGAAGUAGCCACCAAAGUCACUAAUGCUUCCAACA-3′? Why?

3. Which of the following is more likely to be a factor-independent transcription termination site? Note that in each case, only the transcribed strand of the DNA is shown in the 3′-to-5′ direction.

   a. 3′-AACGACTAGTACGACATACTAGTCGTTG-GCAAAAAAAATGCA-5′

   b. 3′-ACTAGCCTAAGCATCTTGCATCAGGCA-CAGAAAAAAAAATCGCA-5′

4. What are the major differences between transcription and translation? Why does translation require much more complex machinery?

5. What is the difference between an endoribonuclease and an exoribonuclease? How does the presence or absence of a 5′-3′ exonuclease affect the pathways of mRNA degradation that can occur?

6. Why is the tmRNA system important? What might happen in a cell if this system were not present?

7. What would be the effect of a mutation that inactivates the regulatory protein of an operon on the expression of the operon if:

   a. the regulation is negative?

   b. the regulation is positive?

8. Define homolog, ortholog, and paralog. Define COG and TIGRFAM equivalogs.

SUGGESTED READING

Agashe, V. R., S. Guha, H.-C. Chang, P. Genevaux, M. Hayer-Hartl, M. Stemp, C. Georgopoulos, F. Ulrich-Hartl, and J. M. Barral. 2004. Function of trigger factor and DnaK in multidomain protein folding: increase in yield at the expense of folding speed. *Cell* **117:**199–209.

Artsimovitch, I. 2005. Control of transcription termination and antitermination: the structure of bacterial RNA polymerase, p. 311–326. *In* N. P. Higgins (ed.), *The Bacterial Chromosome.* ASM Press, Washington, DC.

Baker, T. A., and R. T. Sauer. 2006. ATP-dependent proteases of bacteria: recognition logic and operating principles. *Trends Biochem. Sci.* **31:**647-653.

Ban, N., P. Nissen, J. C. Jansen, M. Capel, P. B. Moore, and T. A. Steitz. 1999. Placement of protein and RNA structures into a 5Å resolution map of the 50S ribosomal subunit. *Nature* 400:841–847.

Björk, G. R., and T. G. Hagervall. 25 July 2005, posting date. Transfer RNA modification, chapter 4.6.2. *In* A. Böck, R. Curtiss III, J. B. Kaper, F. C. Neidhardt, T. Nyström, K. E. Rudd, and C. L. Squires (ed.), EcoSal—*Escherichia coli* and *Salmonella:* cellular and molecular biology. ASM Press, Washington, D.C. http://www.ecosal.org.

Browning, D. F., and S. J. W. Busby. 2004. The regulation of bacterial transcription initiation. *Nat. Rev. Microbiol.* **2:**1–9.

Bukau, B., and A. L. Horwich. 1998. The Hsp70 and Hsp60 chaperone machines. *Cell* **92:**351–366.

Cate, J. H., M. M. Yusupov, G. Z. Yusupova, T. N. Earnest, and H. F. Noller. 1999. X-ray crystal structures of 70S ribosome functional complexes. *Science* **285:**2095–2104.

Condon, C. 2007. Maturation and degradation of RNA in bacteria. *Curr. Opin. Microbiol.* **10:**271–278.

Gabashvili, I. S., R. K. Agrawal, C. M. T. Spahn, R. A. Grassucci, D. I. Svergun, and J. Frank. 2000. Solution structure of the *E. coli* 70S ribosome at 11.5Å resolution. *Cell* **100:**537–549.

Ganoza, M. C., M. C. Kiel, and H. Aoki. 2002. Evolutionary conservation of reactions in translation. *Microbiol. Mol. Biol. Rev.* **66:**460–485.

Geszvain, K., and R. Landick. 2005. The structure of bacterial RNA polymerase, p. 286–295. *In* N. P. Higgins (ed.), *The Bacterial Chromosome.* ASM Press, Washington, D.C.

Han, M.-J., and S. Y. Lee. 2002. The *Escherichia coli* proteome: past, present, and future prospects. *Microbiol. Mol. Biol. Rev.* 66:460–485.

Keiler, K. C. 2007. Physiology of tmRNA: what gets tagged and why? *Curr. Opin. Microbiol.* **10:**169–175.

Korostelev, A., and H. F. Noller. 2007. The ribosome in focus: new structures bring new insights. *Trends Biochem. Sci. 32*:434-441.

Meinnel, T., C. Sacerdot, M. Graffe, S. Blanquet, and M. Springer. 1999. Discrimination by *Escherichia coli* initiation factor IF3 against initiation on non-canonical codons relies on complementarity rules. *J. Mol. Biol.* **290:**825–837.

Mogk, A., E. Deuerling, S. Vorderwulbecke, E. Vierling, and B. Bukau. 2003. Small heat shock proteins, ClpB and the DnaK system form a functional triad in reversing protein aggregation. *Mol. Microbiol.* **50:**585–595.

Murakami, K. S., S. Masuda, E. A. Campbell, O. Muzzin, and S. A. Darst. 2002. Structural basis of transcription initiation: an RNA polymerase holoenzyme-DNA complex. *Science* **296:**1285–1290.

Schmeing, T. M., and V. Ramakrishnan. 2009. What recent ribosome structures have revealed about the mechanism of translation. *Nature* **461:**1234–1242.

Sun, Z., D. J. Scott, and P. A. Lund. 2003. Isolation and characterisation of mutants of GroEL that are fully functional as single rings. *J. Mol. Biol.* **332:**715–728.

Wang, J., and D. C. Boisvert. 2003. Structural basis for GroEL-assisted protein folding from the crystal structure of (GroEL-KMgATP)14 at 2.0 Å resolution. *J. Mol. Biol.* **327:**843–855.

Yusupov, M. M., G. Z. Yusupova, A. Baucom, K. Lieberman, T. N. Earnest, J. H. Cate, and H. F. Noller. 2001. Crystal structure of the ribosome at 5.5 Å resolution. *Science* **292:**883–896.

Zuker, M. 2003. Mfold web server for nucleic acid folding and hybridization prediction. *Nucleic Acids Res.* **31:**3406–3415.

CHAPTER 3

Bacterial Genetic Analysis: Fundamentals and Current Approaches

As MENTIONED IN THE INTRODUCTORY CHAPTER, the relative ease with which bacteria can be handled genetically has made them very useful model systems for understanding many life processes, and much of the information on basic macromolecular synthesis discussed in the first two chapters came from genetic experiments with bacteria. In this chapter, we introduce the genetic concepts and definitions that are used in later chapters.

Definitions

In genetics, as in any field of knowledge, we need definitions. However, words do not mean much when taken out of context, so here we define only the most basic terms. We will define other important terms as we go along.

Terms Used in Genetics

The words in the next few headings are common to all types of genetic experiments, whether with prokaryotic or eukaryotic systems, with some small variations.

MUTANT

The word **mutant** refers to an organism that is the direct offspring of a normal member of the species (the **wild type**) but is genetically different, so the difference is inherited by its offspring. Organisms of the same species isolated from nature that have different properties are usually not called mutants but rather **variants** or **strains**, because even if one of the strains has recently arisen from the other in nature, we have no way of knowing which one is the mutant and which is the wild type. Organisms different from

doi:10.1128/9781555816278.ch3

the wild type because of a programmed event, such as an inversion promoted by a DNA invertase, are usually not called mutants but rather phases, since the inversion event is reversible.

PHENOTYPE

The phenotypes of an organism are all the observable properties of that organism. Usually, in genetics, the term **phenotype** means **mutant phenotype**, or the characteristics of the mutant organism that differ from those of the wild type. The corresponding normal property is sometimes referred to as the **wild-type phenotype**.

GENOTYPE

The **genotype** of an organism is the actual sequence of its DNA. If two organisms have the same genotype, they are genetically identical. Identical twins have almost the same genotype. If two organisms differ by only one mutation or a small region of their DNA, they are said to be **isogenic** except for that mutation or region.

MUTATION

A **mutation** is any heritable change in the DNA sequence. By heritable, we mean that the DNA with the changed sequence can be faithfully replicated. Practically every imaginable type of change is possible, and all changes are called mutations. However, the word "heritable" must be emphasized. Damage or changes to DNA that are not repaired or that are restored to the original sequence are not inherited and therefore are not mutations. Only a permanent change in the sequence of deoxynucleotides in DNA constitutes a mutation. For the same reason, reversible sequence changes due to programmed events, such as inversions due to DNA invertases, are usually not referred to as mutations.

ALLELE

Different forms of the same gene are called **alleles**. For example, if one form of a gene has a mutation and the other has the wild-type sequence, the two forms of the gene are different alleles of the same gene. In this case, one gene is a mutant allele and the other gene is the wild-type allele. Diploid organisms can have two different alleles of the same gene, one on each homologous chromosome. The term "allele" can also refer to genes with the same or similar sequences that appear at the same chromosomal location in closely related species. However, similar gene sequences occurring in different chromosomal locations are not alleles; rather, they are **copies** or **duplications** of the gene.

USE OF GENETIC DEFINITIONS

The following example illustrates the use of the definitions in the previous section. For an explanation of the methods, see the introductory chapter.

A culture of *Pseudomonas fluorescens* normally grows as bright-green colonies on agar plates. However, suppose that one of the colonies is colorless. It probably arose through multiplication of a mutant organism. The mutant phenotype is "colorless colony," and the corresponding wild-type phenotype is "green colony." The mutant bacterium that formed the colorless colony probably had a mutation in a gene for an enzyme required to make the green pigment. Perhaps the mutation consists of a base pair change in the gene, causing the insertion of a wrong amino acid in the polypeptide, and the resulting enzyme cannot function. Thus, the mutant and wild-type bacteria have different alleles of this gene, and we can refer to the gene in the colorless-colony-forming bacteria as the **mutant allele** and the gene in the green-colony-forming bacteria as the **wild-type allele**.

In the example above, we know that a mutation has occurred only because of the lack of color. However, recall that any heritable change in the DNA sequence is a mutation, so mutations can occur without changing the organism's phenotype. Some of them do not even change the amino acid sequence of a protein in whose gene they lie because of the redundancy of the genetic code. Many such changes, called **silent mutations**, have been found by sequencing the DNA directly.

Genetic Names

There are some commonly accepted rules for naming mutants, phenotypes, and mutations in bacteria, although different publications sometimes use different notations. We use the terms recommended by the American Society for Microbiology.

NAMING MUTANT ORGANISMS

The mutant organism can be given any name, as long as the designation does not refer specifically to the phenotype or the gene thought to have been mutated. This rule helps to avoid confusion if the gene with the mutation is introduced into another strain or if other mutations occur or are transferred into the original organism. Quite often, someone who has isolated a mutant names it after himself or herself, giving it his or her initials and a number (e.g., *Escherichia coli* AB2497). This method of naming informs others where they can obtain the mutant strain and get advice about its properties. If another mutation alters the mutant strain, the new strain is usually given a related name, such as *E. coli* AB2498. However, it is usually best not to be too descriptive in naming strains, since it will be confusing later if we were wrong about their genetic properties.

NAMING GENES

Bacterial genes are designated by three lowercase italic letters that usually refer to the function of the gene's product, when it is known. For example, the name *his* refers to a gene whose product is an enzyme required to synthesize the amino acid histidine. The name *hut* refers to a gene required to utilize the amino acid histidine as a carbon and nitrogen source. Sometimes, more than one gene encodes a product with the same function or an enzymatic pathway requires more than one type of polypeptide. In these cases, a capital letter designating each individual gene follows the three lowercase letters. For example, the *hisA* and *hisB* genes both encode polypeptides required to synthesize histidine. A mutation that inactivates either gene will make the cell unable to synthesize histidine.

NAMING MUTATIONS

Hundreds of different types of mutations can occur in a single gene, so all alleles of a particular gene are given a specific allele number. For example, *hisA4* refers to the *hisA* gene with mutation number 4, and the *hisA* gene with mutation number 4 is referred to as the *hisA4* allele. Nowadays, the actual amino acid change in a protein due to a mutation in its gene is often obtained by sequencing. Then, it is common to designate the mutation by the actual amino acid change. For example, H14Q means that the mutation has changed amino acid number 14 in the polypeptide product of the gene counting from the formylmethionine at the N terminus (see chapter 2) from the histidine it was originally to a glutamine.

If a mutation is known to completely inactivate the product of a gene, it can be called a **null mutation**. However, this designation must be used with care, because the product of a gene can sometimes retain some of its activity if the mutation is leaky or if only a part of the gene remains (see below). In some organisms, it is now possible to precisely delete all or almost all of the coding region of a gene so that little chance remains that the product has any activity. In such organisms, this has become the new standard for claiming a null mutation. Different nomenclatural rules apply if a mutation is a deletion or an insertion. We defer a discussion of these rules until we discuss these types of mutations (see below).

NAMING PHENOTYPES

Phenotypes are also denoted by three-letter names, but the letters are not italicized and the first letter is capitalized. In addition, superscripts are often used to distinguish mutant from wild-type phenotypes. For example, His$^-$ describes the phenotype of an organism with a mutated *his* gene that cannot grow without histidine in its environment. The corresponding wild-type organism grows without histidine, so it is phenotypically His$^+$. Another example, Rif^r, describes resistance to the antibiotic rifampin, which blocks RNA synthesis (see chapter 2). A mutation in the *rpoB* gene, which encodes a subunit of the RNA polymerase, makes the cell resistant to the antibiotic. The corresponding wild-type phenotype is rifampin sensitivity, or Rif^s.

Useful Phenotypes in Bacterial Genetics

What phenotypes are useful for genetic experiments depends on the organism being studied. For bacterial genetics, the properties of the colonies formed on agar plates are the most useful phenotypes, since individual cells with distinguishing characteristics are difficult to isolate, even if they can be seen under a microscope (see the introductory chapter).

The visual appearance of colonies sometimes provides useful mutant phenotypes, such as the colorless colony discussed above. Colonies formed by mutant bacteria might also be smaller than normal or smooth instead of wrinkled. The mutant bacterium may not multiply to form a colony at all under some conditions, or conversely, it may multiply when the wild type cannot.

Many mutant phenotypes have been used to study cellular processes, such as DNA recombination and repair, mutagenesis, and development. The following sections describe a few of the more commonly used phenotypes. In later chapters, we discuss many more types of mutants and demonstrate how mutations can be used to study life processes.

Auxotrophic and Catabolic Mutants

Some of the most useful bacterial mutants are **auxotrophic mutants**, or auxotrophs. These mutants have mutations in genes whose products are required to make an essential growth substance; they therefore cannot multiply unless that substance is provided as a growth supplement. The original strain from which the mutant was derived can make the substance and is sometimes called the **prototroph**. For example a His$^-$ auxotrophic mutant cannot grow unless the medium is supplemented with the amino acid histidine, while the wild type can grow without added histidine and so is a histidine prototroph. Similarly, a Bio$^-$ auxotrophic mutant cannot grow without the vitamin biotin, which is not needed by the corresponding prototroph.

Another common type of mutant is one that cannot use a particular substance for growth that can be used by the wild type. For example, the original bacterium may be able to use the sugar maltose as a sole carbon and energy source, but a Mal$^-$ mutant derived from it must be given another carbon and energy source, such as glucose, in order to grow. Other examples are mutants that can no longer use a particular amino acid as a nitrogen source or a particular phosphate-containing compound as a source of phosphate. Such mutants are sometimes called catabolic mutants because they cannot make essential catabolites or derive energy from a particular substance, although no common name is used to include all such mutants. Even though auxotrophic and catabolic mutants seem to be opposites, their molecular bases are similar. In both types, a mutation has altered a gene encoding an enzyme of a metabolic pathway, thereby inactivating the enzyme. The only difference is that in the first case, the inactivated enzyme was in a biosynthetic pathway, which is required to synthesize a substance, while in the second case, the inactivated enzyme was in a catabolic pathway, which is required to degrade a substance to use it as a carbon and energy source, a nitrogen source, or a phosphate source.

ISOLATING AUXOTROPHIC MUTANTS

Figure 3.1 shows a simple method for isolating mutants auxotrophic for histidine and biotin. In this experiment, eight bacteria were plated and allowed to form colonies on plate 1, which contains all the nutrients the bacteria need, including histidine and biotin. Bacteria in each of these colonies were picked up with a loop and transferred onto two other plates. These plates are the same as plate 1, except that plate 2 lacks biotin but has histidine and plate 3 lacks histidine but has biotin. The bacteria from most of the colonies can multiply on all three types of plates. However, the bacteria in colony 2 grow only on plate 2; they are mutants that require added histidine, that is, they are His$^-$. These mutants do not require biotin, so they are Bio$^+$. Similarly, the bacteria in colony 6 are Bio$^-$ but His$^+$, since they can grow on plate 3 but not on plate 2. Under real conditions, mutants that require histidine or biotin would not be this frequent, and hundreds of thousands of colonies would have to be tested to find one mutant that required histidine, biotin, or indeed any growth supplement not required by the wild type.

In principle, it should be possible to find auxotrophic mutants unable to synthesize any compound required for growth or unable to use any carbon and energy source. However, auxotrophic mutants must be supplied with the compound they cannot synthesize, and these compounds must enter the cell. Many bacteria cannot take in some compounds that have a high electrical charge, such as nucleotides, so some types of auxotrophs are very difficult to isolate.

Conditional-Lethal Mutants

As mentioned above, auxotrophic and catabolic mutants can be isolated because they have mutations in genes whose products are required only under certain conditions. The cells can be grown under conditions where the product of the mutated gene is not required and tested under conditions where it is required. However, many gene products of the cell are essential for growth no matter what conditions the bacteria find themselves in. The genes that encode such functions are called **essential genes**. Examples of essential genes are those for RNA polymerase, ribosomal proteins, DNA ligase, and some helicases. Cells with mutations that inactivate essential genes cannot be isolated unless the mutations inactivate the gene under only some conditions. Hence, any mutants that are isolated must have **conditional-lethal mutations**, because these DNA changes inactivate the gene product only under some conditions.

TEMPERATURE-SENSITIVE MUTANTS

The most generally useful conditional-lethal mutations in bacteria are mutations that make the mutant

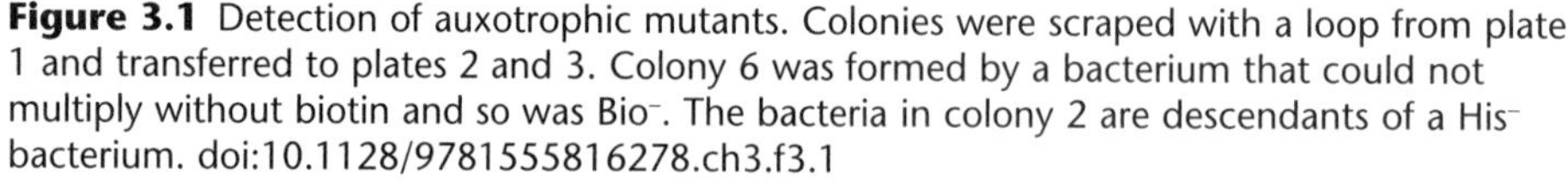

Figure 3.1 Detection of auxotrophic mutants. Colonies were scraped with a loop from plate 1 and transferred to plates 2 and 3. Colony 6 was formed by a bacterium that could not multiply without biotin and so was Bio$^-$. The bacteria in colony 2 are descendants of a His$^-$ bacterium. doi:10.1128/9781555816278.ch3.f3.1

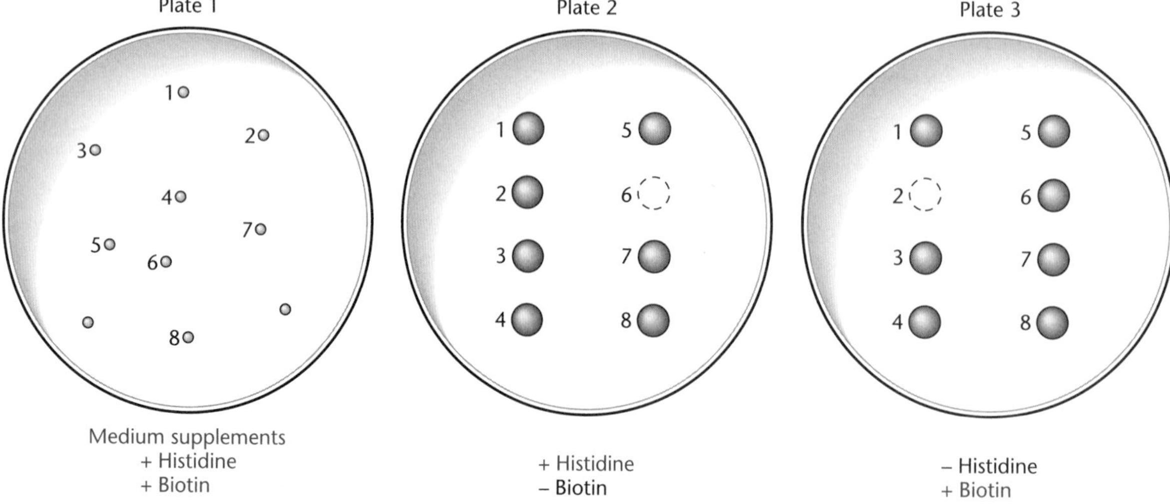

temperature sensitive for growth. Usually, such mutations change an amino acid of a protein so that the protein no longer functions at higher temperatures but still functions at lower temperatures. The higher temperatures are called the **nonpermissive temperatures** for the mutant, whereas the temperatures at which the protein still functions are the **permissive temperatures** for the mutant.

Mutations can affect the temperature stability of proteins in various ways. Often, an amino acid required for the protein's stability at the nonpermissive temperature is changed, causing the protein to unfold, or denature, partially or completely. The protein could then remain in the inactive state or be destroyed by cellular proteases that remove abnormal proteins. If the protein is not degraded, it can sometimes spontaneously renature (refold) when the temperature is lowered; growth can then resume immediately. The temperature sensitivity due to the mutation is reversible, which can be very useful for enrichments (see below) or for studying the function of the gene product. With other mutations, the protein is irreversibly denatured and must be resynthesized before growth can resume. Such mutations can be lethal if they occur in a protein required for protein synthesis, such as RNA polymerase or an aminoacyltransferase, since then the protein is required to make itself and cannot be remade.

The temperature ranges used to isolate temperature-sensitive mutations depend on the organism. Bacteria can be considered poikilothermic ("cold-blooded") organisms, since their cell temperature varies with the outside temperature. Therefore, their proteins are designed to function over a wide range of temperatures. However, different species of bacteria differ greatly in their preferred temperature ranges. For example, a "mesophilic" bacterium, such as *E. coli*, may grow well at a range of temperatures from 20 to 42°C. In contrast, a "thermophilic" bacterium, such as *Geobacillus stearothermophilus*, may grow well only between 42 and 60°C. For *E. coli*, a temperature-sensitive mutation may leave a protein functional at 33°C, the permissive temperature, but not at 42°C, the nonpermissive temperature. For *G. stearothermophilus*, the temperature-sensitive mutation may leave a protein active at the permissive temperature of 47°C but render it nonfunctional at the nonpermissive temperature of 55°C.

Temperature-sensitive mutations are extremely useful for studying the function of the product of a gene because there are few methods by which the product of a gene can be rapidly inactivated, particularly if no inhibitor of the function is known. Usually, the only other option is to stop the synthesis of the gene product, for example, by positioning the coding sequence under the control of an inducible promoter, thereby allowing expression to be turned off under specific conditions. If we grow the cells long enough, the gene product will eventually be diluted out of the cells, but any effects that are seen could be indirect long-term consequences of depriving the cell of the gene product.

Isolating Temperature-Sensitive Mutants

In principle, temperature-sensitive mutants are as easy to isolate as auxotrophic mutants. If a mutation that makes the cell temperature sensitive occurs in a gene whose protein product is required for growth, the cells stop multiplying at the nonpermissive temperature. To isolate such mutants, the bacteria are incubated on a plate at the permissive temperature until colonies appear, and then the colonies are transferred to another plate and incubated at the nonpermissive temperature. Bacteria that can form colonies at the permissive temperature but not at the nonpermissive temperature are temperature-sensitive mutants. However, temperature-sensitive mutants in a particular gene are usually much rarer than auxotrophic mutants. Many changes in a protein will inactivate it, but very few will make a protein functional at one temperature and nonfunctional at another. A worse problem is that all temperature-sensitive mutants have the same phenotype, independent of the essential gene they are in, and finding one that has a mutation in a particular gene can mean screening thousands of temperature-sensitive mutants with mutations in other essential genes. Doing some form of regional mutagenesis to limit mutations to a single gene can help, and some amino acid changes are more apt to make a protein temperature sensitive than others. Nevertheless, it has proven difficult to predict what particular amino acid changes will make a protein temperature sensitive. A better way is to randomly mutagenize the gene by a type of regional mutagenesis and then screen for mutations that make the protein temperature sensitive. The frequency of occurrence of different types of mutations and methods for doing site-specific mutagenesis are discussed later in this chapter and in subsequent chapters.

COLD-SENSITIVE MUTANTS

Cells with proteins that fail to function at low temperatures are called **cold-sensitive mutants**. Mutations that make a bacterium cold sensitive for growth are rarer than temperature-sensitive (heat-sensitive) mutations and are often in genes whose products must enter a larger complex, such as the ribosome. The increased molecular movement at the higher temperature may allow the mutated protein to wriggle into the complex despite its altered shape, but it is unable to do so at lower temperatures. Such mutations often show a phenotype only after a long delay, so they are generally less useful than mutations that make a protein unstable at higher temperatures. Both mutations are in a sense temperature sensitive, but the name is usually reserved for heat-sensitive mutations.

NONSENSE MUTATIONS

Mutations that change a codon in a gene to one of the three nonsense codons—UAA, UAG, or UGA (in most bacteria)—can also be conditional-lethal mutations. A nonsense mutation can be conditionally lethal because it causes translation to stop within the gene unless the cell has a "nonsense suppressor" tRNA, as explained later in this chapter. Because nonsense mutations are more generally useful in viral genetics than in bacterial genetics, they are discussed in more detail in chapter 7.

Resistant Mutants

Among the most useful and easily isolated types of bacterial mutants are resistant mutants. If a substance kills or inhibits the growth of a bacterium, mutants resistant to the substance can often be isolated merely by plating the bacteria in the presence of the substance. Only mutants in the population that are resistant to the substance will multiply to form a colony.

The numerous mechanisms of resistance depend on the basis for toxicity and on the options available to prevent the toxicity (examples are given in Table 3.1). For example, the mutation may destroy a cell surface receptor to which the toxic substance must bind to enter the cell. If the substance cannot enter the mutant cell, it cannot kill the cell. Alternatively, a mutation might change the "target" affected by the substance inside the cell. For example, an antibiotic might normally bind to a ribosomal protein and affect protein translation. However, if the antibiotic cannot bind to a mutant (but still functional) protein, it cannot kill the cell. An example of such a resistance mutation is a mutation to streptomycin resistance in *E. coli* (Table 3.1). The antibiotic streptomycin binds to the 16S rRNA in the 30S subunit of the ribosome and blocks translation. However, some mutations in the gene for the S12 protein, *rpsL* (for ribosomal protein small-subunit L), prevent streptomycin from binding to the ribosome but do not inactivate the S12 protein. These mutations therefore confer streptomycin resistance (Str^r) on *E. coli*. In some cases, the substance added to the cells is not toxic until one of the cell's own enzymes changes it. A mutation inactivating the enzyme that converts the nontoxic substance into the toxic one could make the cell resistant to that substance.

Inheritance in Bacteria

Salvador Luria and Max Delbrück were among the first people to attempt to study bacterial inheritance quantitatively. They published this work in their now-classic paper in the journal *Genetics* in 1943. The paper is still very much worth reading and is listed in Suggested Reading at the end of the chapter. As discussed in the introductory chapter, the experiments and reasoning of Luria and Delbrück helped debunk what was then a popular misconception among bacteriologists. At the time, it was generally thought that bacteria are different from other organisms in their inheritance. It was generally accepted that heredity in higher organisms followed "Darwinian" principles. According to Charles Darwin, random mutations occur, and if one happens to confer a desirable phenotype, organisms with the mutation are selected by the environment and become the predominant members of the population. Undesirable, as well as desirable, mutations continuously occur, but only the desirable mutations are passed on to future generations.

However, many bacteriologists thought that heredity in bacteria followed different principles. They thought that bacteria, rather than changing as the result of random mutations, somehow "adapt" to the environment by a process of directed change, after which the adapted organism would pass the adaptation on to its offspring. This process is called Lamarckian inheritance, named after the geneticist who proposed it, and acceptance of it was encouraged by the observation that all the bacteria in a culture exposed to a toxic substance seem to become resistant to that substance (Figure 3.2).

Table 3.1 Some resistance mutations

Substance	Toxicity	Resistance mutation
Bacteriophage T1	Infects and kills	Inactivates TonA or TonB outer membrane proteins; phage cannot absorb
Streptomycin	Binds to ribosomes; inhibits translation	Changes ribosomal protein S12 so that it no longer binds
Chlorate	Converted to chlorite, which is toxic	Inactivates nitrate reductase, which converts chlorate to chlorite
High concentrations of valine, no isoleucine	Feedback inhibits acetolactate synthetase; starves for isoleucine	Activates a valine-insensitive acetolactate synthetase

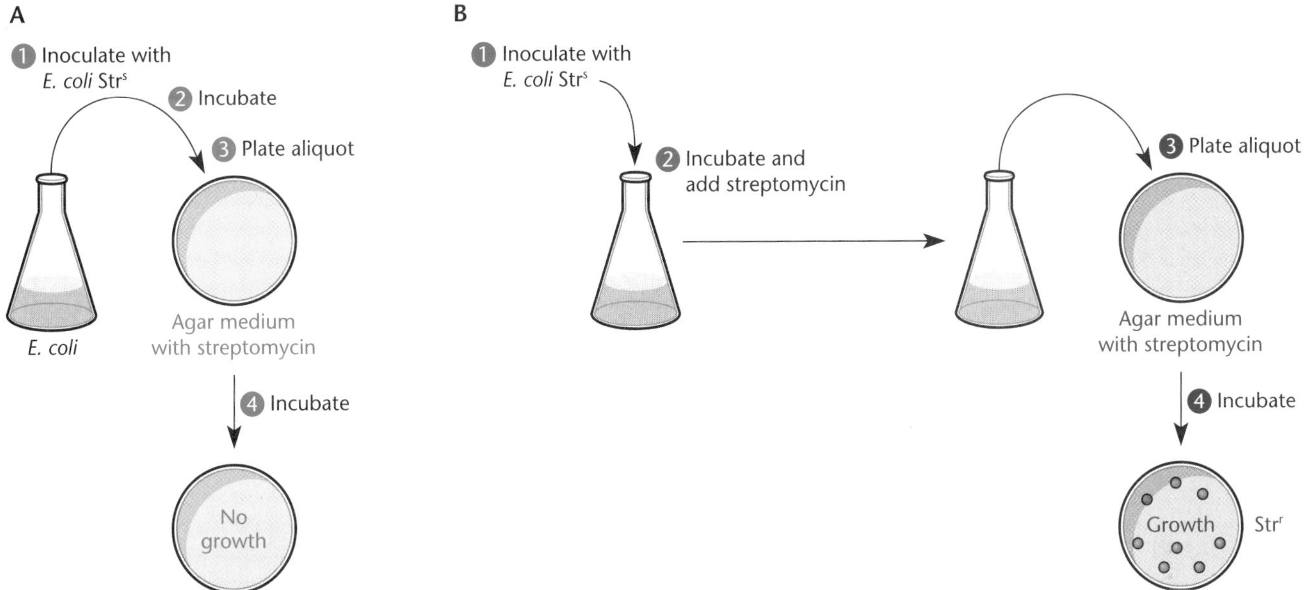

Figure 3.2 Mutants resistant to the antibiotic streptomycin and other toxic substances seem to appear in response to the presence of the substance or antibiotic. **(A)** Sensitivity of wild-type *E. coli* to the antibiotic streptomycin. Plating an *E. coli* culture on streptomycin-containing agar medium resulted in a lack of colony growth after incubation. **(B)** Emergence of streptomycin-resistant *E. coli* mutants. Streptomycin was added to a flask already containing large numbers of wild-type *E. coli*, which were further incubated in the presence of the streptomycin. When the bacteria were plated on agar medium on a plate containing streptomycin, streptomycin-resistant mutants that had accumulated in the culture formed colonies. doi:10.1128/9781555816278.ch3.f3.2

The Luria and Delbrück Experiment

The Luria and Delbrück experiment was designed to test two hypotheses for how mutants arise in bacterial cultures: the **random-mutation hypothesis** and the **directed-change hypothesis**. The random-mutation hypothesis predicts that mutants appear randomly prior to the addition of the selective agent, whereas the directed-change hypothesis predicts that mutants appear only in response to the selective agent.

One distinction between the two hypotheses is reflected in the distribution of the numbers of mutants in a series of cultures. If the random-mutation hypothesis is correct, mutations that occur early in the growth of a culture will have a disproportionate effect on the number of mutants in the culture, and the fraction of mutants should vary widely from culture to culture. Figure 3.3 illustrates this principle. In culture 1, only one mutation occurred, but this mutation gave rise to eight resistant mutants because it occurred early. In culture 2, two mutations arose, but they gave rise to only six resistant mutants because they occurred later. In their experiments, Luria and Delbrück used *E. coli* as the bacterium and bacteriophage T1 as the selective agent. As shown in Table 3.1 and Figure 3.4, phage T1 kills wild-type *E. coli*, but a mutation in either one of the genes for two outer membrane proteins called TonA and TonB can make these cells resistant to killing by the phage. If bacteria are spread on an agar plate with the phage, only those resistant to the phage multiply to form a colony. All the others are killed. The number of colonies on the plate is therefore a measure of the number of bacteria resistant to the bacteriophage in the culture.

Figure 3.5 shows the two experiments that Luria and Delbrück performed, and Table 3.2 gives some representative results. It can be seen that the two experiments give very different results, even though they seem superficially similar. In experiment 1, the authors started one culture of bacteria. After incubating it, they took out a number of small aliquots and plated them with and without phage T1 to determine the number of resistant mutants, as well as the total number of bacteria, in the culture. They then calculated the fraction of resistant mutants. In experiment 2, they started a large number of smaller cultures. After incubating these cultures, they determined the number of resistant mutants and the total number of bacteria in each culture. It can easily be seen that in experiment 1 the number of resistant mutants in each aliquot is almost the same, subject only to sampling errors and statistical fluctuations. However, in experiment 2, a very large variation in the number of resistant

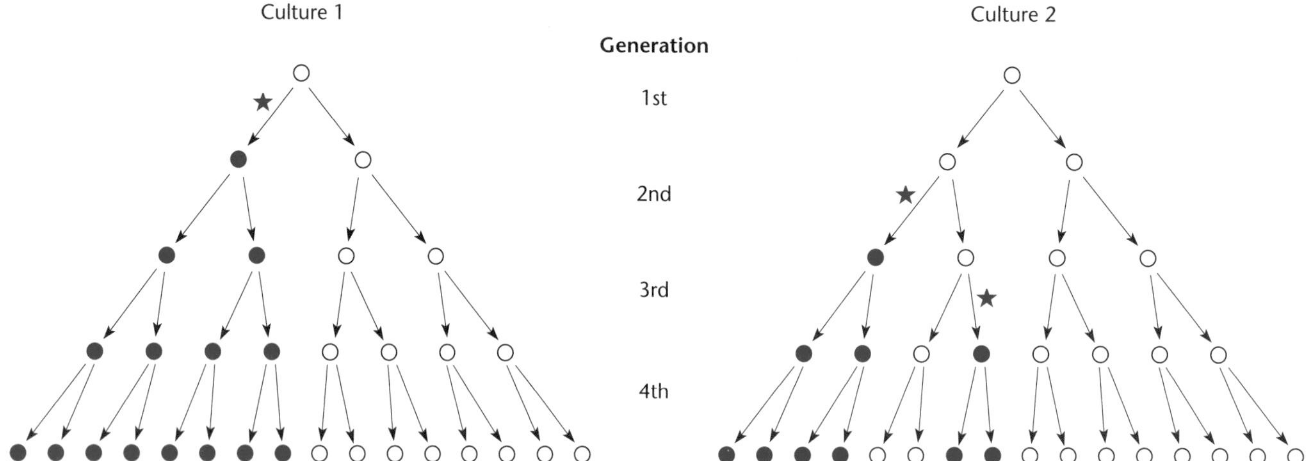

Figure 3.3 The number of mutants in a culture is not proportional to the number of mutations in the culture because earlier mutations give rise to more mutant progeny. Only one mutation occurred in culture 1, but it gave rise to eight mutant progeny because it occurred in the first generation. In culture 2, two mutations occurred, one in the second generation and one in the third. However, because these mutations occurred later, they gave rise to only six mutant progeny. The mutant cells are shaded. doi:10.1128/9781555816278.ch3.f3.3

bacteria per culture was found. Some cultures had no resistant mutants, while some had many. One culture even had 107 resistant mutants! Luria and Delbrück referred to this and the other mutant-rich cultures as "jackpot" cultures. Apparently, these are cultures in which a mutation to resistance occurred very early. Hence, these results fulfill the predictions of the random-mutation hypothesis. In contrast, the directed-change hypothesis predicts that the results of the two experiments should be the same, and certainly no jackpot cultures should appear in the second experiment. Box 3.1 presents these predictions in statistical terms.

Figure 3.4 The molecular basis for phage T1 resistance. When bacteriophage T1 infects wild-type *E. coli*, it binds to a receptor in the outer membrane containing proteins TonA and TonB (Table 3.1). After phage replication, the *E. coli* cell is lysed and new phage are released. A mutation in the *tonA* or *tonB* gene eliminates or modifies the receptor, so the cells are not infected by T1 and survive.
doi:10.1128/9781555816278.ch3.f3.4

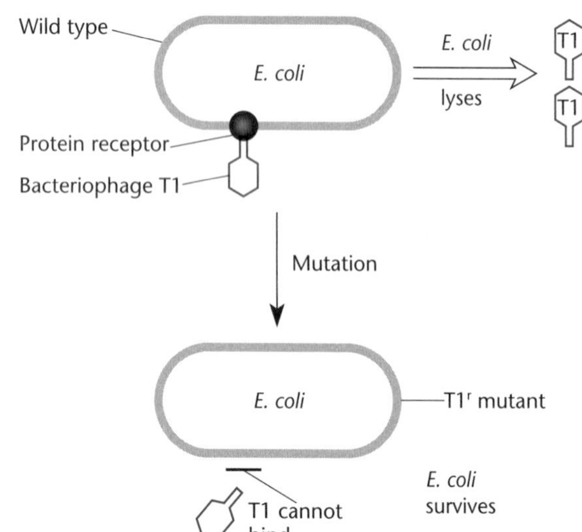

Mutants Are Clonal

The Luria and Delbrück experiments demonstrated that, even in bacteria, mutations occur randomly and are then selected upon by the environment rather than arising in response to the selective pressure. However, their analysis was fairly sophisticated mathematically and so was not generally understood. Other, simpler demonstrations showed that bacterial mutants were **clonal**, another prediction of the random-mutation hypothesis. These experiments showed that once a bacterial mutant appeared, all of the descendants of that mutant were also mutant in the same way, even in the absence of the selective pressure. One way this was demonstrated was to show that *E. coli* mutants resistant to the phage T1 grow together as a colony on a plate, even in the absence of the phage (see Newcombe, Suggested Reading). If millions of nonresistant *E. coli* bacteria were spread on a plate and colonies were allowed to develop for a short time prior to spraying the plate with the phage, a few colonies appeared, due to the multiplication of a few mutants resistant to the phage that had arisen prior to applying the phage. However, if the bacteria were again spread on the plate just prior to spraying the plate with the phage, many more resistant colonies appeared. This demonstrated that the resistant mutants were all

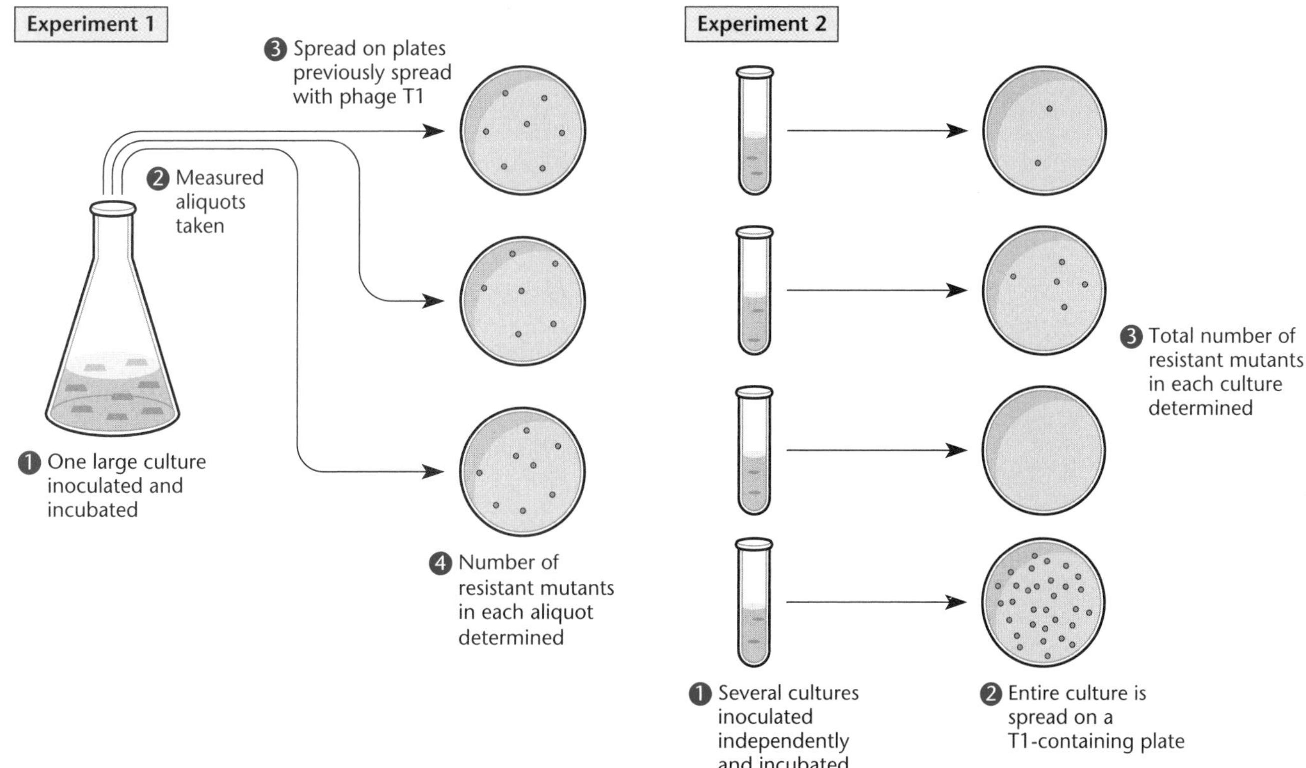

Figure 3.5 The Luria and Delbrück experiment. In experiment 1, a single flask containing standard medium was inoculated with bacteria and incubated overnight. The number of T1-resistant mutants was determined in a number of aliquots from this single culture. In experiment 2, the number of resistant mutants was determined in each of a number of smaller cultures that had been started with a few nonmutant bacteria and incubated overnight. See the text for details. doi:10.1128/9781555816278.ch3.f3.5

concentrated in a few colonies as descendents of the original mutants until they were dispersed by the spreading, i.e., that they were clonal.

The Lederbergs' Experiment

The experiments that really buried the directed-change hypothesis, at least as the sole explanation for some types of resistant bacterial mutants, were the replica-plating experiments of the Lederbergs (see Lederberg and Lederberg, Suggested Reading). These experiments showed in a simple way that bacterial mutants resistant to an antibiotic arose in the absence of the antibiotic and that the resistance was stably inherited by the descendants of the original mutant. They spread millions of bacteria on a plate without an antibiotic and allowed the bacteria to form a lawn during overnight incubation. Bacteria from this plate were then replicated onto another plate containing the antibiotic. After incubating the antibiotic-containing plate, the Lederbergs could determine where antibiotic-resistant mutants had arisen on the original plate by aligning the two plates and marking the regions on the first plate where antibiotic-resistant mutants had grown on the second. They scraped these regions off the original plate, suspended and diluted the bacteria they had scraped off this region, and plated them without antibiotic to repeat the experiment. This time, when the bacteria were replica plated onto a plate with antibiotic, there were many more resistant mutants than previously. Eventually, by repeating this process, they obtained a pure culture of bacteria, all of which were resistant to the antibiotic even though they had never been exposed to it! The only possible explanation was that the bacteria must have acquired the resistance independently of exposure to the antibiotic and passed the resistance on to their offspring. Thus, the experiments of the Lederbergs dispelled the idea that *E. coli* mutations are directed, at least as far as resistance to this antibiotic is concerned.

Mutation Rates

As defined above, a mutation is any heritable change in the DNA sequence of an organism, and we usually know that a mutation has occurred because of a phenotypic change in the organism. The **mutation rate** can be loosely defined as the chance of mutation to a particular

Table 3.2 The Luria and Delbrück experiment

Experiment 1		Experiment 2	
Aliquot no.	No. of resistant bacteria	Culture no.	No. of resistant bacteria
1	14	1	1
2	15	2	0
3	13	3	3
4	21	4	0
5	15	5	0
6	14	6	5
7	26	7	0
8	16	8	5
9	20	9	0
10	13	10	6
		11	107
		12	0
		13	0
		14	0
		15	1
		16	0
		17	0
		18	64
		19	0
		20	35

phenotype in a certain time interval. Mutation rates can differ because mutations to some phenotypes occur much more often than mutations to other phenotypes. The overall mutation rate is the sum of all the mutation rates for each mutation that can cause the phenotype. As a consequence, the mutation rate is relatively high when many different possible mutations in the DNA can give rise to a particular phenotype. However, if only a very few types of mutations can cause a particular phenotype, the mutation rate for that phenotype is relatively low. This can be illustrated by comparing the mutation rates for the phenotypes His$^-$ and streptomycin resistance (Str^r). The products of 11 genes are required for histidine biosynthesis, and each enzyme has hundreds of amino acids, many of which are required for activity. Changing any of these amino acids inactivates the enzyme, as do deletions or other disruptions of any one of the 11 genes. In contrast, streptomycin resistance results from a change in one of very few amino acids in a single ribosomal protein, S12, that changes the protein so that it no longer binds streptomycin but leaves its function intact. This makes the mutation rate for streptomycin resistance very low. Hence, mutation to Str^r occurs spontaneously in about 1 in 10^{10} to 10^{11} cells, whereas a mutation to His$^-$ occurs in about 1 in 10^6 to 10^7 cells, or 1,000 times as often. Therefore, the mutation rate for a

BOX 3.1

Statistical Analysis of the Number of Mutants per Culture

A simple statistical analysis shows that the number of mutants in experiment 2 of Luria and Delbrück does not follow a Poisson distribution. If it did, the variance would be approximately equal to the mean:

$$\text{Variance} = \sum_{i=1}^{n} \frac{(M_i - \overline{M})^2}{n-1} = \text{mean} = \overline{M} = \sum_{i=1}^{n} \frac{M_i}{n}$$

where M_i is the number of mutants in each culture and n is the number of cultures.

In experiment 1 of Luria and Delbrück, the variance was 18.23 and the mean was 16.7, so they are approximately equal. The small variation is probably due to statistical fluctuations and pipetting errors. In experiment 2, however, the variance was 752.38 and the mean was 11.35—very different values. Therefore, the number of mutants per culture does not follow a Poisson distribution, and the result is not consistent with the directed-change hypothesis; however, it is consistent with the random-mutation, or Darwinian, hypothesis.

particular phenotype gives an early clue to what kinds of mutations can cause the phenotype. An extremely high mutation rate for a particular phenotype may indicate, not a mutation, but rather the loss of a plasmid or prophage or the occurrence of some programmed recombination event, such as inversion of an invertible sequence; on the other hand, an extremely low mutation rate might suggest that two or more independent mutations are required.

Calculating Mutation Rates

To calculate anything, we must first define it precisely. The mutation rate is usually defined as the chance of a mutation each time a cell grows and divides. This is a reasonable definition because, as discussed in chapter 1, DNA replicates once each time the cell divides, and most mutations occur as mistakes are made during this process. The number of times a cell grows and divides in a culture is called the number of **cell divisions.** This is not to be confused with the division time, or **generation time,** which is the time it takes for a cell to grow and divide. Not all bacterial cells divide as they multiply, and some form long filaments, or hyphae. Nevertheless, we will refer to cell divisions as the number of times in a culture a cell has grown to maturity and doubled its mass enough to form two cells, since the DNA must have duplicated during this time. The mutation rate is then the number of mutations to a particular phenotype that have occurred in a growing culture divided by the total number of cell divisions that have occurred in the culture during the same time.

DETERMINING THE NUMBER OF CELL DIVISIONS

The total number of cell divisions that have occurred in an exponentially growing culture is easy to calculate and is simply the total number of cells in the culture minus the number of cells in the starting inoculum. To understand this, see Figure 3.6. In this illustration, a culture that was started from one cell multiplies to form eight cells in seven cell divisions. This number equals the final number of cells (8) minus the number of cells at the beginning (1). In general, the number of cell divisions that have occurred in the culture equals $N_2 - N_1$ if N_2 equals the number of cells at time 2 and N_1 equals the number of cells at time 1.

Therefore, from the definition above, the mutation rate (a) is given by the following equation:

$$a = (m_2 - m_1)/(N_2 - N_1)$$

where m_2 and m_1 are the number of mutations in the culture at time 2 and time 1, respectively.

Usually, a culture is started with only a few cells with no mutations and ends with many cells, so we can often

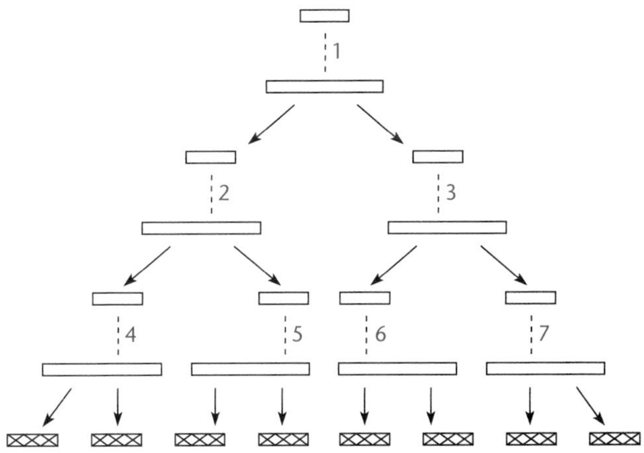

Figure 3.6 The number of cell divisions (7) equals the total number of cells in an exponentially growing culture (8) minus the number at the beginning (1).
doi:10.1128/9781555816278.ch3.f3.6

ignore the mutations and the number of cells at the beginning and just call the number of cell divisions N, where N is the total number of cells in the culture. Then, the mutation rate equation can be simplified to

$$a = m/N$$

where m is the number of mutations that have occurred in the culture and N is the number of bacteria.

DETERMINING THE NUMBER OF MUTATIONS THAT HAVE OCCURRED IN A CULTURE

From the equations above, it looks as though it might be easy to calculate the mutation rate. The total number of mutations is simply divided by the total number of cells. The problem comes in determining the number of mutational events that have occurred in a culture, because mutant cells, not mutational events, are usually what are detected. Recall from Figure 3.3 that one mutant cell resulting from a single mutational event can give rise to many mutant cells, depending on when the mutation occurred during the growth of the culture. Therefore, one cannot determine the number of mutations in a culture merely by counting the mutant cells. However, in some cases, the number of mutant cells can form the basis of a calculation of the number of mutational events and, by extension, the mutation rate. Some examples of such situations are given below.

Calculating the Mutation Rate from the Data of Luria and Delbrück

Luria and Delbrück used data like those shown in Table 3.2 to calculate the mutation rate for T1 phage resistance. They assumed that even though the number of mutants per culture does not follow a Poisson distribution, the number of mutations per culture should do so, because each cell has the same chance of acquiring a mutation to T1 phage resistance each time it grows and divides. For convenience, the **Poisson distribution** can be used to calculate the mutation rate in a case like this, where there are a lot of bacteria in the population and the probability of a mutation in each bacterium is extremely low. According to the Poisson distribution, if P_i is the probability of having i mutations in a culture, then

$$P_i = m^i e^{-m}/i!$$

where m is the average number of mutations per culture, the number they wanted to know. Therefore, if they knew how many cultures had a certain number of mutations, they could calculate the average number of mutations per culture. However, this is not as obvious as it seems. The data give the number of T1 phage-resistant mutants per culture but do not indicate how many of the cultures had one, two, three, or more mutations. For example, cultures with one mutant probably had one mutation, but others, even the one with 107 mutants, might also have had only one mutation. Only the number of cultures with zero mutations seems clear—those with zero mutants, or 11 of 20. Therefore, the probability of having zero mutations equals 11/20. Applying the formula for the Poisson distribution, the probability of having zero mutations is given by the equation

$$11/20 = m^0 e^{-m}/0! = 1 \times e^{-m}/1 = e^{-m}$$

and $m = -\ln 11/20 = 0.60$. Therefore, in this experiment, an average of 0.60 mutations occurred in each culture. From the equation for the mutation rate,

$$a = m/N = 0.60/(5.6 \times 10^8) = 1.07 \times 10^{-9}$$

Therefore, there are 1.07×10^{-9} mutations per cell division if there are a total of 5.6×10^8 total bacteria per culture. In other words, a mutation for phage T1 resistance occurs about once every 10^9, or every billion, times a cell divides.

There are a number of problems with measuring mutation rates this way, as indeed there are with any way of measuring mutation rates. One problem is **phenotypic lag,** which is a problem with T1 phage resistance in *E. coli* and with most phenotypes caused by loss of a gene product. When a mutation first inactivates either the *tonA* or *tonB* gene in the chromosome of a growing bacterium, the TonA and TonB proteins are still both in the outer membrane and the cell is still sensitive to the phage (Figure 3.4). One of them will no longer be made in the active form in the membrane as a result of the mutation, but it could take a few generations to dilute the protein out of the membrane, depending on how many copies of each protein originally exist in the membrane. This is compounded by the fact that, in an exponentially growing culture, half of all the mutations have occurred in the last generation time, so a significant proportion of all mutations could be missed. Accordingly, some of the cultures with no resistant mutants presumably had bacteria with a *tonA* or *tonB* mutation, but the resistance has not been expressed in them yet, so they are killed by the phage and not counted.

Another problem with this method for calculating the mutation rate is that it is wasteful in that it ignores most of the data and considers only the cultures with no mutants. In their classic paper, Luria and Delbrück also derived an equation to estimate the mutation rate by using the number of mutants in all of the cultures. This equation is widely used and is expressed as

$$r = aN_t \ln(N_t Ca)$$

where r is the average number of mutants per culture, C is the number of cultures, N_t is the average number of bacteria per culture, and a is the mutation rate. Note that a, the mutation rate, appears on both sides of the equation, so this equation must be solved numerically, for example, by successive approximations. In their paper, they include a table showing the solutions for different numbers of cultures that can be used to solve for a, the mutation rate. Others have subsequently proposed other methods to measure mutation rates from such data (see, e.g., Lea and Coulson and Jones et al., Suggested Reading).

Calculating the Mutation Rate from the Number of Clones of Mutants

When it is applicable, it might be possible to use the fact that the multiplication of mutants is clonal to calculate the mutation rate. As discussed above, since mutants of some types of bacteria multiplying on an undisturbed plate stay together to form a colony, the number of mutant colonies on the undisturbed plate is a measure of the number of mutations that have occurred. The bacteria on a parallel plate can then be washed off and plated to determine the total number of bacteria on each plate. Again, this way of determining the number of mutations is strongly influenced by phenotypic lag for some types of mutants.

Calculating the Mutation Rate from the Rate of Increase in the Proportion of Mutants

Another prediction of the random-mutation hypothesis is that the number of mutants should increase faster than the total population. In other words, the fraction of mutants in the population should increase as the population grows. At first, it seems surprising that the total number of mutants increases faster than the total population until one thinks about where mutants come from. If the multiplication of old mutants were the only source of mutants, the fraction of mutants would remain constant or even drop if the mutants did not multiply as rapidly as the normal type (which is often the case). However, new mutations occur constantly, and the resultant mutants are also multiplying. Therefore, new mutations are continuously adding to the total number of mutants.

This fact can also be used to measure mutation rates. The higher the mutation rate, the faster the proportion of mutants will increase as the culture multiplies. In fact, if we plot the fraction of mutants (*M*/*N*) against time (in doubling times), as in Figure 3.7, the slope of this curve is the mutation rate. The fact that other mutations are causing some mutants to become wild type again (in a process known as reversion [see below]) also affects the results shown in Figure 3.7. However, reversion of mutants becomes significant only when the number of mutants is very large and the number of mutants multiplied by the mutation rate back to the wild type (called the reversion rate) begins to approximate the rate at which new mutants are being formed, which, as we showed above, is the forward mutation rate times the number of nonmutant bacteria. At earlier stages of culture growth, the vast majority of bacteria are nonmutant, so forward mutations are occurring much more often than reversion mutations and the mutation rate is much higher than the reversion rate. Also, for reasons discussed later in this chapter, reversion rates are often much lower than forward mutation rates. Therefore, the contribution to the mutation rate of the reversion of mutants to the wild type can generally be ignored, at least in the early stages of growing a culture.

Figure 3.7 The fraction of mutants increases as a culture multiplies, and the slope is the mutation rate (*a*). *M* is the number of mutants, *N* is the total number of cells, and Time is the time in generation times, which is the total time elapsed divided by the time it takes the culture to double in mass (i.e., the doubling time [*g*]).
doi:10.1128/9781555816278.ch3.f3.7

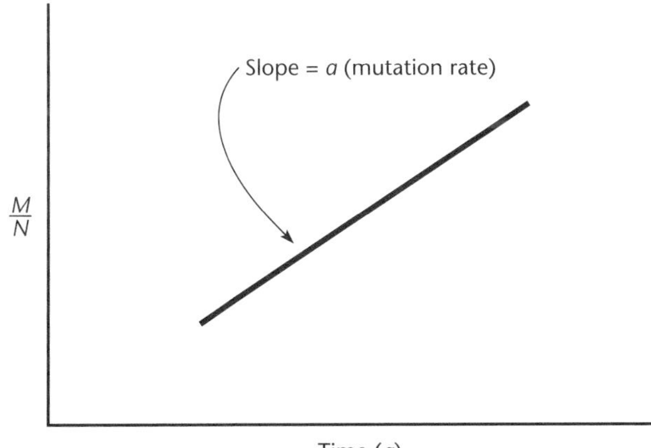

In theory, plots such as that shown in Figure 3.7 could be used in calculating mutation rates. The method does have many advantages. It is not affected by phenotypic lag, although it is affected by the relative growth rates of the mutants. In practice, however, the method is not very practical in most laboratory situations. Mutation rates are usually low, and the numbers of bacteria we can conveniently work with are relatively small, so each new mutation makes too large a contribution to the number of mutants and we do not get a straight line. To make this method practicable, we would have to work with trillions of bacteria growing in a large chemostat, in which the medium is continuously replaced and the bacteria are removed as fast as they multiply.

PRACTICAL IMPLICATIONS OF POPULATION GENETICS

The fact that the proportion of mutants of all types increases as the culture grows presents both opportunities and problems in genetics. This fact can be advantageous in the isolation of a rare mutant, such as one resistant to streptomycin. If we grow a culture from a few bacteria and plate 10^9 bacteria on agar containing streptomycin, we might not find any resistant mutants, since mutations to streptomycin resistance occur at a frequency of only about 1 in 10^{11} cell divisions. However, if we add a large number of bacteria to fresh broth, grow the broth culture to saturation, and then repeat this process a few times, the fraction of streptomycin-resistant mutants will increase. Then, when we plate 10^9 bacteria, we may find many streptomycin-resistant mutants.

Because the fraction of all types of mutants increases as the culture multiplies, if we allow a culture to go through enough generations, it will become a veritable "zoo" of different kinds of mutants—virus resistant, antibiotic resistant, auxotrophic, and so on. To deal with this problem, most researchers store cultures under nongrowth conditions (e.g., as spores or lyophilized cells or in a freezer) that still maintain cell viability. An alternative is to periodically colony purify bacteria in the culture to continuously isolate the progeny of a single cell (see the introductory chapter). The progeny of a single cell are not likely to be mutated in a way that could confound our experiments later.

Types of Mutations

As defined above, any heritable change in the sequence of nucleotides in DNA is a mutation. A single base pair may be changed, deleted, or inserted; a large number of base pairs may be deleted or inserted; or a large region of the

DNA may be duplicated or inverted. Regardless of how many base pairs are affected, a mutation is considered to be a **single mutation** if only one error in replication, recombination, or repair has altered the DNA sequence.

As discussed above, to be considered a mutation, the change in the DNA sequence must be heritable. In other words, the DNA must be able to replicate normally but with the changed sequence. Damage to DNA by itself is not a mutation, but a mutation can occur when the cell attempts to repair damage or replicate over it and a strand of DNA is synthesized that is not completely complementary to the original sequence. The wrong sequence is then replicated and thus becomes a mutation. Lethal changes in the DNA sequence (as also mentioned above) do occur but ordinarily cannot be scored as mutations, because these cells by definition do not survive, and other techniques are needed for the study of this kind of mutation. Sometimes, a situation can be established to search for mutations that are lethal only in a particular setting, for example, in the absence of another gene product. This is called a **synthetic lethality screen** and has been used to search for genes whose products can substitute for each other. We talk about these and other selections in subsequent sections of this chapter.

Properties of Mutations

The properties and causes of the different types of mutations are probably not very different in all organisms, but they are more easily studied with bacteria. A geneticist can often make an educated guess about what type of mutation is causing a mutant phenotype merely by observing some of its properties.

One property that distinguishes mutations is whether they are leaky. The term "leaky" means something very specific in genetics. It means that in spite of the mutation, the gene product still retains some activity.

Another distinguishing property of mutations is whether they **revert**. If the sequence has been changed to a different sequence, it can often be changed back to the original sequence by a subsequent mutation. The organism in which a mutation has reverted is called a **revertant**, and the **reversion rate** is the rate at which the mutated sequence in DNA returns to the original wild-type sequence.

Usually, the reversion rate is much lower than the mutation rate that gave rise to the mutant phenotype. As an illustration, consider the previously discussed example, histidine auxotrophy (His$^-$). Any mutation that inactivates any of the approximately 11 genes whose products are required to make histidine will cause a His$^-$ phenotype. Since thousands of changes can result in this phenotype, the mutation rate for His$^-$ is relatively high. However, once a *his* mutation has occurred, the mutation can revert only through a change in the mutated sequence that restores the original sequence, or at least one that inserts an acceptable amino acid at this site. Without sequencing the DNA, it is difficult to know whether the original sequence has been restored or another sequence that is acceptable at the site has been inserted, and it may be best to refer to such His$^+$ bacteria as **apparent revertants**, since we do not know yet whether they are true revertants. Everything else being equal, the reversion rate to His$^+$ revertants would be expected to be thousands of times lower than the forward mutation rate to His$^-$. If it is not, we suspect that the mutation is being suppressed rather than reverted (see below). That is why we might refer to them as apparent revertants until we know their molecular basis.

Some types of apparent revertants are very easy to detect. For example, His$^+$ apparent revertants can be obtained by plating large numbers of His$^-$ mutants on a plate with all the growth requirements except histidine. Most of the bacteria cannot multiply to form a colony. However, any His$^+$ apparent revertants in the population will multiply to form a colony (see Figure 3.20 below). The appearance of His$^+$ colonies when large numbers of a His$^-$ mutant are plated would be evidence that the *his* mutation can revert, provided the possibility of suppression has been eliminated (see below).

Base Pair Changes

A **base pair change** is when one base pair in DNA, for example, a GC pair, is changed into another base pair, for example, an AT pair. Base pair changes can be further classified as transitions or transversions (Figure 3.8). In a transition, the purine (A or G) in a base pair is replaced by the other purine and the pyrimidine (C or T) is replaced by the other pyrimidine. Thus, an AT pair would become a GC pair or a CG would become a TA. In a transversion, in contrast, the purines change into pyrimidines, and vice versa. For example, a GC could become a TA, or a CG could become an AT.

CAUSES OF BASE PAIR CHANGES

Base pair changes in DNA can occur spontaneously or be induced by external factors, such as irradiation or chemical damage. The wrong base can be inserted during replication, or a damaged base can result in mispairing during replication or repair. If mutations occur without known external factors, such as deliberate application of ionizing radiation or mutagenic chemicals, they are referred to as **spontaneous mutations**, even if they are caused by internal factors, such as reactive oxygen species generated during oxidative metabolism (see below). In this section, we discuss only spontaneous mutations; we reserve discussion of mutations due to external factors, such as irradiation and damage due to added chemicals, called **induced mutations**, for chapter 11.

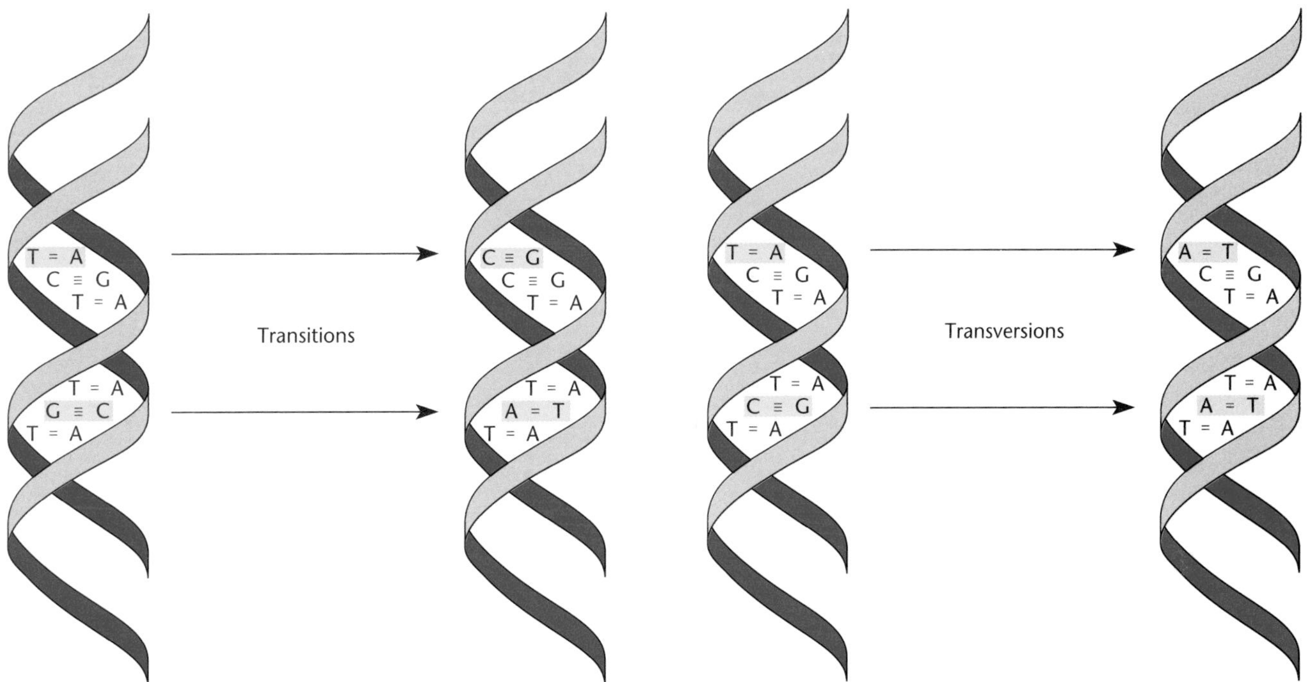

Figure 3.8 Transitions versus transversions. The mutations are shown in blue.
doi:10.1128/9781555816278.ch3.f3.8

Base Pair Changes Due to Mispairing during Replication

Base pair changes can be the result of base-pairing mistakes in replication, recombination, or repair, whenever DNA polymerase attempts to synthesize a complementary copy of DNA. Figure 3.9A shows an example of mispairing during replication. In this example, a T instead of the usual C is mistakenly placed opposite a G as the DNA replicates. In the next replication, this T usually pairs correctly with an A, causing a GC-to-AT transition in one of the two daughter DNAs. Mistakes in pairing may occur because the bases are sometimes in a different form, called the enol form, which causes them to pair differently (Figure 3.9B).

What type of base pair change occurs depends on which types of bases mispaired. Mispairing between a purine and a pyrimidine causes a transition, whereas mispairing between two purines or two pyrimidines causes a transversion. Because a pyrimidine in the enol form still pairs with a purine and a purine in the enol form still pairs with a pyrimidine, mispairing during replication usually leads to transition mutations. Furthermore, all four bases can undergo the shift to the enol form, and either the base in the DNA template or the incoming base can be in the enol form and cause mispairing. Thus, the thymine in the enol form pictured in Figure 3.9B might be in the template, in which case the transition would be AT to GC, or it could be the incoming base, resulting in a GC-to-AT transition.

Mistakes during replication leading to mutations are not random, however, and some sites are much more prone to base pair changes than are others. Mutation-prone sites are called **hot spots** and can have many causes, including the deamination of methylated bases (see below). Mispairing occurs fairly often during replication, and it is an obvious advantage for the cell to reduce the number of base pair change mutations that occur during replication. In chapter 1, we discussed some of the mechanisms used by cells to reduce these base pair changes.

Base Pair Changes Due to Spontaneous Deamination

The **deamination** of the bases in DNA, or the removal of an amino group from one of the bases in the DNA, can also cause base pair changes. This deamination can occur spontaneously, especially at higher temperatures. Cytosine is particularly susceptible to deamination, and the damage is apt to be mutagenic, since deamination of cytosine results in the formation of uracil, which pairs with adenine instead of guanine (see chapters 1 and 2 for structures). Therefore, unless the uracil due to deamination of cytosine is removed from the DNA, it will cause a CG-to-TA transition the next time the DNA replicates.

Because of the problems caused by deamination of cytosine, cells have evolved a special mechanism for removing uracil from DNA whenever it appears

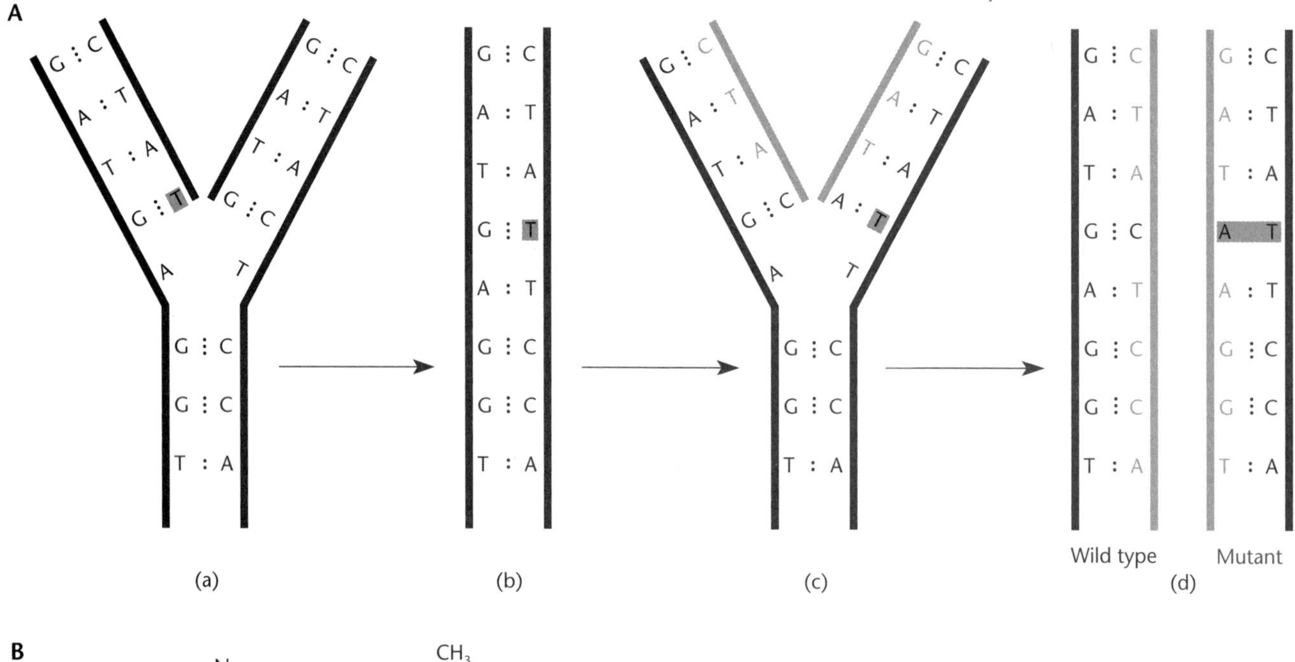

B

Deoxyribose
Deoxyribose
Guanine
Thymine (enol form)

Figure 3.9 (A) A mispairing during replication can lead to a base pair change in the DNA (shown in blue). The gray lines show the second generation of newly replicated DNA. The mispaired base is shaded in blue. **(B)** Pairing between guanine and an alternate enol form of thymine can cause G-T mismatches. doi:10.1128/9781555816278.ch3.f3.9

(Figure 3.10). An enzyme called **uracil-N-glycosylase,** the product of the *ung* gene in *E. coli*, recognizes the uracil as unusual in DNA and removes the uracil base. The DNA strand in the region where the uracil was removed is then degraded and resynthesized, and usually, the correct cytosine is inserted opposite the guanine. As expected, *ung* mutants of *E. coli* show high rates of spontaneous mutagenesis, and most of the mutations are GC-to-AT transitions. If the cytosine has a methyl group at the 5 position of the pyrimidine ring, a common modification in many organisms, the cytosine will become thymine when deaminated and will not be recognized as unusual and removed by the Ung enzyme, causing a hot spot for spontaneous mutagenesis (see above).

All organisms have the problem of cytosine deamination in their DNA, possibly partially explaining why the testicles of warm-blooded animals, including mammals, are external, where the average temperature is lower and deaminations are less frequent, since they are triggered by heat.

Base Pair Changes Due to Oxidation of Bases

Reactive forms of oxygen, such as peroxides and free radicals, are given off as by-products of oxidative metabolism by aerobic organisms, and these forms can react with and alter the bases in DNA. A common example is the altered guanine base 8-oxoG, which sometimes mistakenly pairs with adenine instead of cytosine, causing GC-to-TA or AT-to-CG transversion mutations, depending on whether the guanine has been altered in the DNA or in the incoming guanosine deoxynucleotide. We, as aerobic organisms, are also subject to this type of DNA damage, which is why consumption of antioxidants, which are oxidized in place of our DNA, may reduce the incidence of cancer and other diseases (see Box 11.1). Repair systems specific to damage such as deamination and oxidation are discussed in more detail in chapter 11.

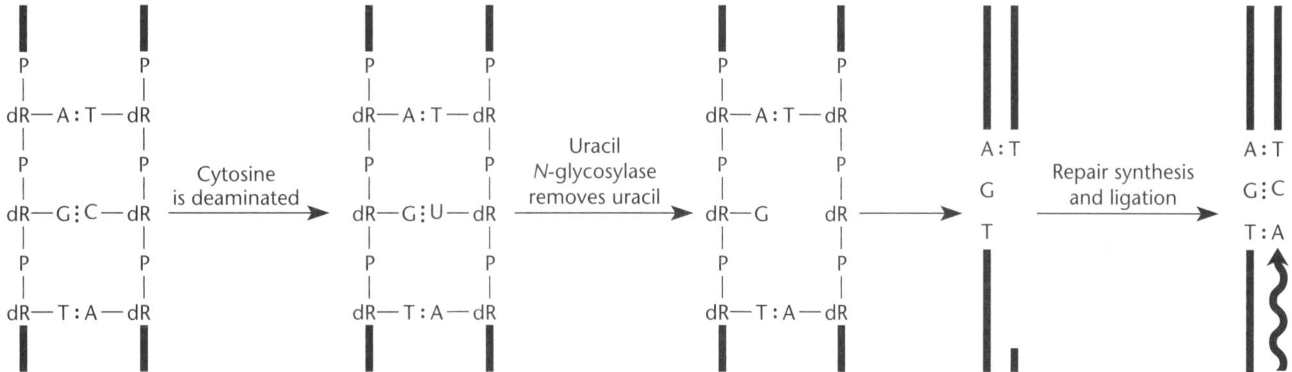

Figure 3.10 Removal of deaminated cytosine (uracil) from DNA by uracil-*N*-glycosylase. The uracil base is cleaved off, and the DNA strand is degraded and resynthesized with cytosine at that position. doi:10.1128/9781555816278.ch3.f3.10

CONSEQUENCES OF BASE PAIR CHANGES

Whether a base pair change causes a detectable phenotype depends, of course, on where the mutation occurs and what the actual change is. Even a change in an open reading frame (ORF) that encodes a polypeptide may not result in an altered protein. If the mutated base is the third position in a codon, the amino acid inserted into the protein may not be different because of the degeneracy of the code (see "The Genetic Code" in chapter 2). Mutations in the coding region of a gene that do not change the amino acid sequence of the polypeptide product are called silent mutations.

The change may also occur in a region that encodes an RNA rather than a polypeptide or in a regulatory sequence, such as an operator or promoter. Alternatively, the mutation may occur in a region that has no detectable function. We first discuss mutations that change the coding region of a polypeptide.

Missense Mutations

Because most of the DNA of bacteria is devoted to encoding proteins, most base pair changes in bacterial DNA cause one amino acid in a polypeptide to be replaced by another. These mutations are called **missense mutations** (Figure 3.11). However, even a missense mutation that changes an amino acid in a protein does not always inactivate the protein. If the original and new amino acids have similar properties, the change may have little or no effect on the activity of a protein. For example, a missense mutation that changes an acidic amino acid, such as glutamate, into another acidic amino acid, such as aspartate, may have less effect on the functioning of the protein than does a mutation that substitutes a basic amino acid, such as arginine, for an acidic one. The consequences also depend on which amino acid is changed. Certain amino acids in any given protein sequence are more essential to activity than others, and a change at one position can have a much greater effect than a change elsewhere. Investigators often use this fact to determine which amino acids are essential for activity in different proteins. Some methods to change specific amino acids in a protein, called **site-specific mutagenesis,** are discussed in chapter 1, and other methods are introduced below.

Nonsense Mutations

Instead of changing a codon into one coding for a different amino acid, base pair changes sometimes produce one of the nonsense codons, UAA, UAG, or UGA, in the ORF for the protein. These changes are called **nonsense mutations.**

Figure 3.11 Missense mutation. A mutation that changes T to C in the DNA template strand results in an A-to-G change in the mRNA. The mutant codon GUC is translated as valine instead of isoleucine.
doi:10.1128/9781555816278.ch3.f3.11

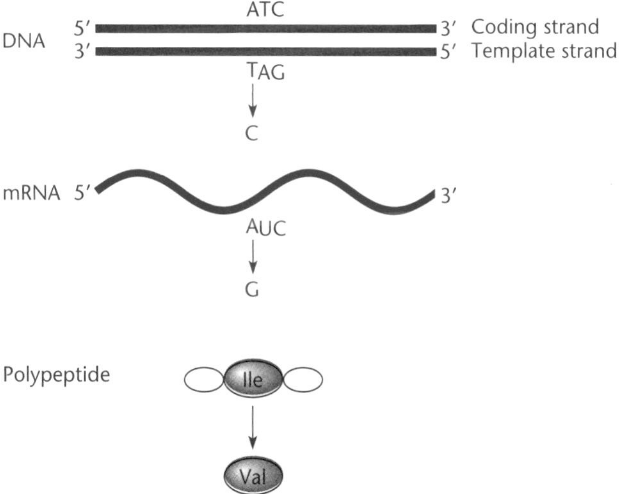

While nonsense mutations are base pair changes and have the same causes as other base pair changes, the consequences are very different. Because the nonsense codons are normally used to signify the end of a gene, these codons are normally recognized by release factors (see chapter 2), which cause release of the translating ribosome and the polypeptide chain. Therefore, if a mutation to one of the nonsense codons occurs in an ORF for a protein, the protein translation terminates prematurely at the site of the nonsense codon and the shortened or truncated polypeptide is released from the ribosome (Figure 3.12). For this reason, nonsense mutations are sometimes called "chain-terminating mutations." These mutations almost always inactivate the protein product of the gene in which they occur, especially if they occur early in the coding region for the protein. If, however, they occur in a noncoding region of the DNA or in a region that encodes an RNA rather than a protein, such as a gene for a tRNA, they are indistinguishable from other base pair changes.

The three nonsense codons—and their corresponding mutations—are sometimes referred to by color designations: **amber** for UAG, ochre for UAA, and opal for UGA. These names have nothing to do with the effects of the mutation. Rather, when nonsense mutations were first discovered at the California Institute of Technology, their molecular basis was unknown. The investigators thought that descriptive names might be confusing later on if their interpretations were wrong, so they followed the lead of physicists with their "quarks" and "barns." The first nonsense mutations to be discovered, to UAG, were called amber mutations. The legend is that they were named after a graduate student at the California Institute of Technology, Harry Bernstein (or his grandmother), who was involved in the first clear experiments describing these (his last name, Bernstein, means amber in the German language). Following suit, UAA and UGA mutations were also named after colors—ochre and opal, respectively.

Figure 3.12 Nonsense mutation. Changing the CAA codon, encoding glutamine (Gln), to UAA, a nonsense codon, causes truncation of the polypeptide gene product.
doi:10.1128/9781555816278.ch3.f3.12

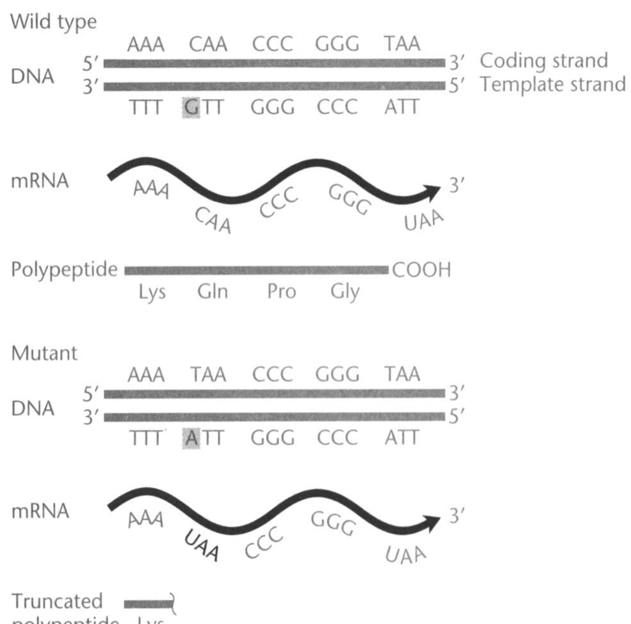

PROPERTIES OF BASE PAIR CHANGE MUTATIONS

Base pair changes are often leaky. A substituted amino acid may not work as well as the original at that position in the chain, but the protein often retains some activity. Even nonsense mutations are usually somewhat leaky, because even without a nonsense suppressor, sometimes an amino acid is inserted for a nonsense codon, albeit at a low frequency. In wild-type *E. coli*, UGA tends to be the most leaky, followed by UAG; the nonsense codon UAA tends to be the least leaky.

Base pair mutations also revert. If the base pair has been changed to a different base pair, it can also be changed back to the original base pair by a subsequent mutation (a true reversion). Being base pair changes, nonsense mutations can revert. However, as often as not, apparent revertants of nonsense mutations have suppressor mutations in tRNA genes, since these are often as common as true revertants. This property of being suppressed by nonsense suppressor mutations in tRNA genes is often used to identify nonsense mutations. Moreover, base pair changes are a type of **point mutation** because they map to a particular "point" on the DNA, as discussed below.

Frameshift Mutations

A high percentage of all spontaneous mutations are **frameshift mutations** (Figure 3.13). This type of mutation occurs when a base pair or a few base pairs are removed from or added to the DNA; if these mutations occur in an ORF that encodes a polypeptide, this causes a shift in the reading frame. Because the code is three lettered, any addition or subtraction that is not a multiple of 3 causes a frameshift in the translation of the remainder of the gene. For example, adding or subtracting 1, 2, or 4 base pairs (bp) causes a frameshift, but adding or subtracting 3 or 6 bp does not. Mutations that remove or add base pairs are often called frameshift mutations even if they do not occur in an ORF and do not actually cause a frameshift in the translation of a polypeptide. A more general name would be single (or double, etc.) base additions or deletions, since it does not imply that the mutation is in a reading frame for a protein.

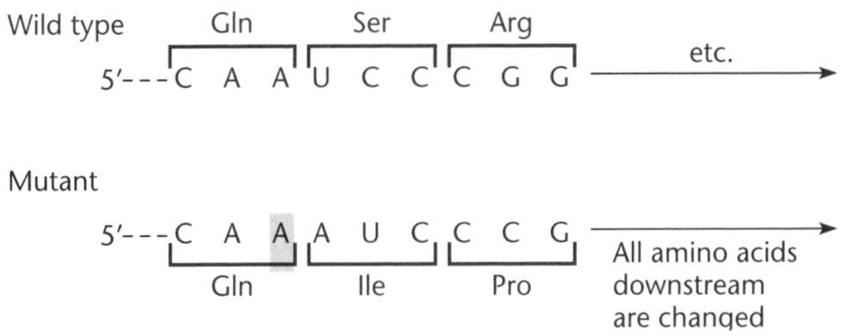

Wild type
Gln Ser Arg
5′---C A A U C C C G G etc.

Mutant
5′---C A A A U C C C G
Gln Ile Pro
All amino acids downstream are changed

Figure 3.13 Frameshift mutation. The wild-type mRNA is translated glutamine (Gln)-serine (Ser)-arginine (Arg), etc. Insertion of an A (shaded) would shift the reading frame, so that the codons would be translated glutamine (Gln)-isoleucine (Ile)-proline (Pro), etc., with all downstream amino acids being changed.
doi:10.1128/9781555816278.ch3.f3.13

CAUSES OF FRAMESHIFT MUTATIONS

Spontaneous frameshift mutations often occur where there is a short repeated sequence that can slip during DNA replication. The slippage can occur either in the template DNA or in the nucleotide being added. As an example, Figure 3.14 shows a string of AT base pairs in the DNA. Since any one of the A's in one strand can pair with any T in the other strand, the two strands could slip with respect to each other, as in the illustration. Slippage during replication could leave one T unpaired, and an AT base pair would be left out on the other strand when it replicates. Alternatively, the slippage could occur in the template before the base was added, and an extra AT base pair could appear in one strand, as shown.

PROPERTIES OF FRAMESHIFT MUTATIONS

Frameshift mutations in a coding sequence are usually not leaky and almost always inactivate the protein, because every amino acid in the protein past the point of the mutation is wrong. The protein is usually also truncated, because nonsense codons are usually encountered while the gene is being translated in the wrong frame. In general, because 3 of the 64 codons are nonsense codons, one of them should be encountered by chance about every 20 codons when the region is being translated in the wrong frame.

Another property of frameshift mutations is that they revert. If a base pair has been subtracted, one can be added to restore the correct reading frame, and vice versa. Frameshift mutations can also be suppressed (see below) by the addition or subtraction of a base pair close to the site of the original mutation that restores the original reading frame. This means of restoring the function of a protein inactivated by a frameshift mutation is discussed below and was used to determine the characteristics of the genetic code (see chapter 7). Finally, frameshift mutations are a type of point mutation in that they map to a single point on the DNA.

Some types of pathogenic bacteria apparently take advantage of the frequency and high reversion rate of frameshift mutations to avoid host immune systems. In such bacteria, genes required for the synthesis of cell surface components that are recognized by host immune systems often have repeated sequences that favor frameshift mutations. Consequently, these genes can be frequently turned off and on by frameshift mutations and subsequent reversion, allowing some members of the population to escape immune detection. For example, frameshift mutations may aid in the synthesis of virulence gene products by *Bordetella pertussis*, the causative agent of whooping cough. Frameshift mutations are also used to reversibly inactivate genes of *Haemophilus influenzae* and *Neisseria gonorrhoeae*, which cause spinal meningitis and gonorrhea, respectively.

Figure 3.14 Slippage of DNA at a repeated sequence can cause a frameshift mutation.
doi:10.1128/9781555816278.ch3.f3.14

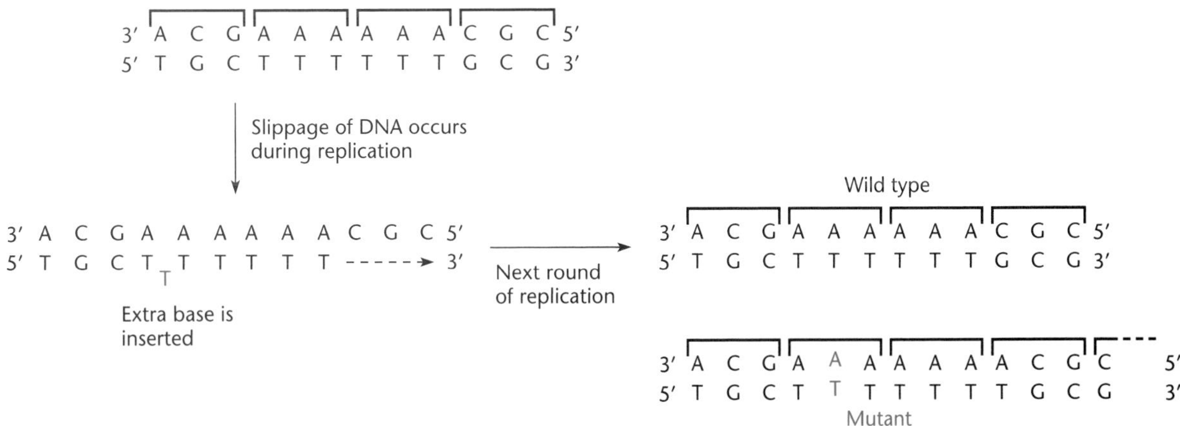

Deletion Mutations

Another common type of spontaneous mutation is the **deletion mutation**, in which entire stretches of the DNA have been cut out and removed. As much as 5% of all spontaneous mutations can be deletions in *E. coli*, and they can often be very long, removing thousands of base pairs and possibly many genes, provided none of the deleted genes encode proteins that are essential for growth.

CAUSES OF DELETIONS

Deletions are usually caused by **recombination** between directly repeated sequences in different locations of the DNA. In recombination, the strands of two DNA molecules are broken and rejoined in new combinations, hence the name recombination (see below). Recombination occurs at sites of identical sequences in the two DNAs because the strands of the two DNAs must be complementary to each other to initiate recombination. This ensures that the recombination usually occurs at the same place in the two DNAs, where their sequences are the same. However, sometimes similar or identical sequences exist in more than one location in the DNA, and recombination can sometimes mistakenly occur between these two different locations. Such recombination is called **ectopic recombination** (Greek for out of place) because it is occurring outside the correct place. Deletions result from ectopic recombination if the two similar sequences involved in ectopic recombination are **direct repeats**, that is, the two sequences are similar or identical when read in the 5′-to-3′ direction on the same strand of DNA.

As shown in Figure 3.15, ectopic recombination between directly repeated sequences in the DNA can give rise to deletions in two ways, depending on whether the directly repeated sequences are on different DNA molecules or the same DNA molecule. Part I shows how a deletion can occur when the different copies of a directly repeated sequence occur in different DNA molecules, for example, in the daughter DNAs created during replication. Different copies of the directly repeated sequences mistakenly pair with each other and recombine, deleting the sequence between them on one of the two daughter DNA molecules. Notice that after recombination the sequence between the two repeats is not deleted in both molecules but instead is repeated on the other daughter DNA molecule, creating a tandemly duplicated sequence in this DNA. We discuss tandem duplications next in this section. Recombination between different regions of two DNA molecules is sometimes called "unequal crossing over," because the two DNAs are not equally aligned during the recombination.

The outcome is different if the ectopic recombination that creates a deletion occurs between direct repeats on the same DNA, as shown in Figure 3.15, part II. Now,

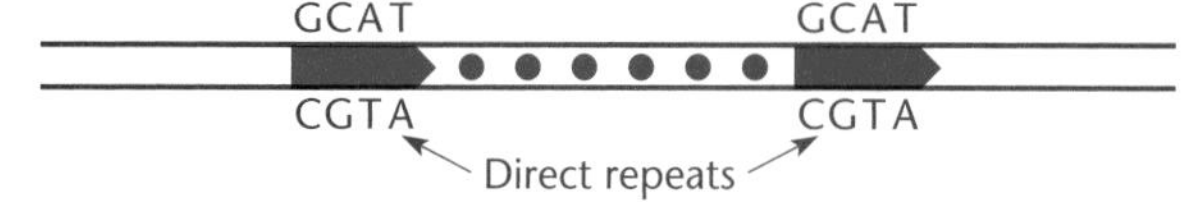

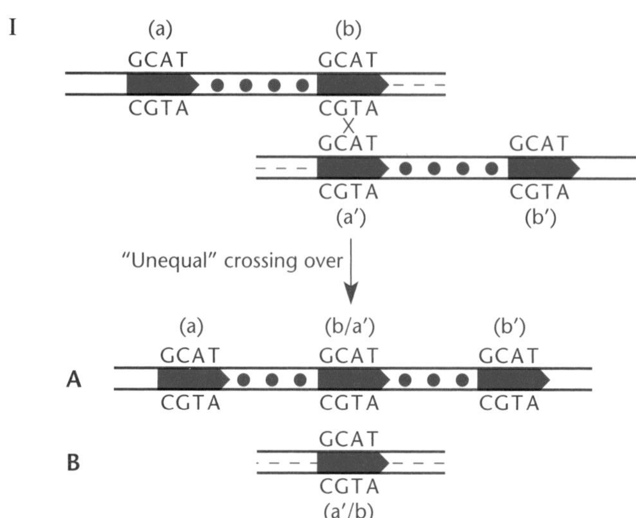

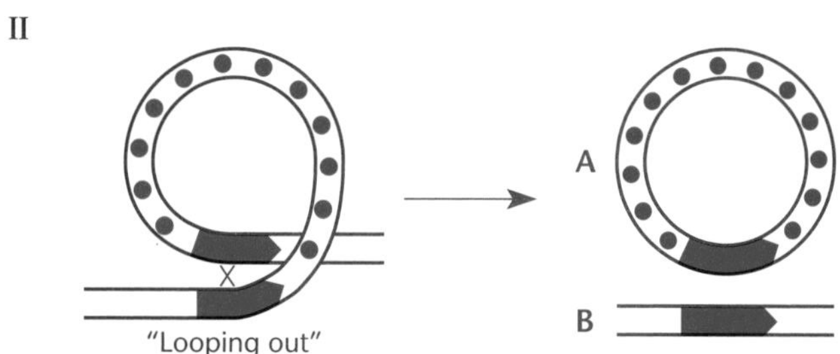

Figure 3.15 Ectopic recombination between directly repeated sequences can cause deletion and tandem-duplication mutations. **(I)** The recombination can occur between repeated sequences in different DNAs, resulting in a duplication (A) or a deletion (B). **(II)** Alternatively, it can occur between repeated sequences in the same DNA, resulting in a deletion (A) and the looped-out deleted segment (B). The direct repeats involved in recombination are often hundreds of base pairs long, but repeats of only 4 bp are shown for convenience. doi:10.1128/9781555816278.ch3.f3.15

the intervening sequence "loops out," and recombination removes the looped-out sequences between the two repeats, as shown. For purposes of illustration, the directly repeated sequences shown in the figure are much shorter than would normally be required for recombination. Usually, direct repeats that promote ectopic recombination are at least 50 bp long, at least in *E. coli*. Bacterial DNA contains several types of repeats, the longest of which include insertion sequence (IS) elements and the rRNA genes, which are thousands of base pairs long and are often repeated in many places in the DNA. IS elements are a type of transposon and are discussed in chapter 9.

PROPERTIES OF DELETION MUTATIONS

Deletions have very distinctive properties. They are usually not leaky; deleting part or all of a gene usually totally inactivates the gene product. Mutations that inactivate two or more genes simultaneously are most often deletions. Moreover, deletion mutations sometimes fuse one gene to another or put one gene under the control of different regulatory sequences. However, the most distinctive property of long deletion mutations is that they never revert. Every other type of mutation reverts at some frequency, but for a deletion to revert, the missing sequence would somehow have to be found and reinserted, which is essentially an impossible event in a culture of bacteria composed of genetically identical bacteria all harboring the deletion. Deletions also behave differently from point mutations in genetic crosses, not mapping to a single point.

Isolating Deletion Mutants

Deletion mutations have been very important historically in genetic experiments because of the ease with which they can be used to map myriad point mutations in a gene. We give a few examples later in this book, including later in this chapter, of where they were used to map point mutations in the *thyA* gene of *E. coli* and in chapter 7 of how they were used to construct mutational spectra in the *r*II genes of T4 bacteriophage. The properties of deletion mutations make them relatively easy to identify among the myriad of other types of mutations. One useful property that has been used historically is the ability of deletions to inactivate more than one gene simultaneously. Even though deletions are fairly rare, on the order of 5% of spontaneous mutations in *E. coli*, a single deletion that inactivates two nearby genes occurs much more frequently than two independent point mutations, such as base pair changes, or frameshift mutations that inactivate the same two genes. Two independent point mutations occur only as frequently as the product of the frequencies of each of the point mutations taken separately. Thus, if mutants are selected for having inactivated one gene, those that also inactivated a nearby gene are probably deletions extending into the nearby gene. This principle was used to map *E. coli lacI* mutations before DNA sequencing was available. The *lac* operon was moved close to the *tonB* gene, and mutations for T1 phage resistance due to inactivating *tonB* that also made the cells constitutive for the *lac* operon were mostly deletions extending from *tonB* into *lacI* and could be used to map *lacI* mutations isolated by various selections.

Another property of deletions that has been used historically is that they can fuse one gene to another, sometimes putting the downstream gene under the control of the regulatory region of the upstream gene. The downstream gene will then be expressed when the upstream gene is expressed and not when the downstream gene is normally expressed.

NAMING DELETION MUTATIONS

Deletion mutations are named differently from other mutations. The Greek letter Δ (delta), for *del*etion, is written in front of the gene designation and allele number, e.g., Δ*his8*. Often, deletions remove more than one gene, and so, if known, the deleted regions are shown, followed by a number to indicate the particular deletion. For example, Δ(*lac-proAB*)*195* is deletion number 195 extending through the *lac* and *proAB* genes on the *E. coli* chromosome. Often a deletion removes one or more known genes but extends into a region of unknown genes, so that the endpoints of the deletion are not known. In this case, the deletion is often named after the known gene. For example, the Δ*his8* deletion may delete the entire *his* operon but may also extend an unknown distance into neighboring genes.

With DNA sequencing, it is often possible to know the precise endpoints of a deletion. If a deletion is completely included in the coding region for one protein, the endpoints can be given in the name. For example, Δ(H4-Q16) deletes from the 4th amino acid in the protein, a histidine, to the 16th, a glutamine. If it is an in-frame deletion, i.e., has deleted a multiple of 3 bases (see above), the amino acids encoded after the deletion will be unchanged and the only change in the protein will be the precise removal of the amino acids encoded in the deleted region. This can be very useful in some types of applications, for example, finding the domains of a protein. With longer deletions including more than one gene, the endpoints of the deletions could also be precisely located, but this has not been done for many historically important deletions of *E. coli* that extend into more than one gene, even those that are still often used.

Tandem-Duplication Mutations

In a duplication mutation, a sequence in one region of the DNA is copied and one of the copies is inserted somewhere else in the DNA so that it then exists in two copies in the DNA. The most common type of duplication, a **tandem duplication**, consists of a sequence immediately followed by its duplicate. Tandem duplications occur fairly frequently and can be very long. Other types of duplications can be caused by the insertion of DNA elements containing parts of the genome, for example, transducing phages (see chapter 8). These are special cases and will be discussed in subsequent chapters.

CAUSES OF TANDEM DUPLICATIONS

Like deletions, tandem duplications can result from ectopic recombination between directly repeated sequences in different DNA molecules. In fact, as was shown in Figure 3.15, they are probably often created at the same time as a deletion. Pairing between two directly repeated sequences in different DNAs, followed by recombination, can give rise to a tandem duplication and a deletion as the two products.

PROPERTIES OF TANDEM-DUPLICATION MUTATIONS

Although tandem duplications often arise during the same ectopic recombination as deletions, their properties are very different. Tandem-duplication mutations that occur within a single gene usually inactivate the gene and are not leaky. However, if the duplicated region is long enough to include one or more genes, no genes are inactivated, even those in which the recombination occurred—the **duplication junctions**. This conclusion may seem surprising, but consider the example shown in Figure 3.16. Direct repeats in genes A and C on different DNAs pair with each other. The repeats in the two DNAs are then broken and rejoined to each other, creating a duplication in one DNA. Only part of gene A exists in the downstream duplicate copy, but an entire gene A exists in the upstream one. Conversely, only part of gene C exists in the upstream duplicate, but the entire C gene exists in the downstream duplicate. There are now two copies of gene B, both of which are unaltered. Therefore, intact genes A, B, and C still exist after the duplication, and there may be little phenotypic indication that a mutation had even occurred. Note that because duplications do not inactivate genes, bacterial cells can survive very long duplications, even those that include many essential genes.

Figure 3.16 Formation of a long tandem-duplication mutation does not inactivate any genes. **(I)** Direct repeats in genes A and C. **(II)** Ectopic recombination between the direct repeats leads to a tandem duplication. A complete gene A exists in the upstream copy and a complete gene C in the downstream copy. Gene B now exists in two copies, and one copy of gene B could be under the control of the promoter for gene C. doi:10.1128/9781555816278.ch3.f3.16

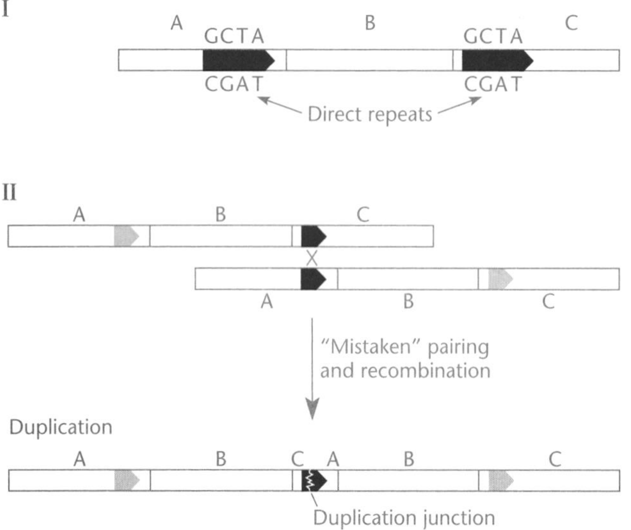

The most characteristic property of tandem duplications is that they are very unstable and revert at a high frequency. Even though the ectopic recombination events that lead to a duplication are usually rare, recombination anywhere within the duplicated segments can delete them, restoring the original sequence. The instability of tandem duplications is often the salient feature that allows their identification. If the recombination systems of the bacterium are inactivated after a tandem duplication has occurred, the tandem duplications will be stabilized, which makes it easier to study them.

Identifying Mutants with Tandem-Duplication Mutations

As mentioned above, because no genes are inactivated in longer tandem duplications, usually there is little phenotypic indication that a mutation has even occurred. However, in some cases, there may be a phenotype associated with having two copies of one or more of the genes included in the duplicated region. An excess of some gene products can affect the appearance of colonies on plates. Also, like deletions, duplications sometimes fuse two genes, putting expression of one gene under the control of the regulatory region of a different gene. In the example in Figure 3.16, part of gene A and all of gene B have been fused to the beginning of gene C, which might put genes A and B under the control of the promoter for gene C. Sometimes, by changing a gene or genes in one of the two copies, it is possible to select duplications. For example, nonsense suppressors (see below) often cannot be isolated in genes for tRNAs that exist in only one copy. If the tRNA is mutated so that it now responds to one of the nonsense codons, there will be no tRNA left to respond to the original codon for an amino acid. However, if the tRNA gene is included in a tandem duplication, then one of the two copies can be mutated and the cell will still be viable. For this reason, nonsense suppressor strains are often unstable and spontaneously lose the ability to suppress nonsense mutations at a high frequency when the duplication is removed by recombination. It also may be possible to change genes in one of the two copies of a duplication after it has formed, for example, by transduction or transformation (see below), in such a way that the cell cannot grow unless the genes in the two copies complement each other. For example, if one copy has a deletion in a gene whose product is required to make histidine and the other copy has a deletion in a different gene to make histidine, the cell cannot grow without histidine unless the duplication is conserved and the two deletions complement each other. Below, we discuss an actual case of the genetic analysis of tandem duplications in the *his* operon of *Salmonella* that uses this technique. These studies showed that tandem duplications in *Salmonella* can be very long, more than 10% of the entire genome, and they can be quite frequent, as much as one duplication every 10^4 times a cell divides, which is hundreds of times more frequent than the most frequent spontaneous mutations (see Anderson et al. Suggested Reading).

ROLE OF TANDEM-DUPLICATION MUTATIONS IN EVOLUTION

Tandem-duplication mutations may play an important role in evolution. Ordinarily, a gene cannot change without loss of its original function, and if the lost function was a necessary one, the organism will not survive. However, when a duplication has occurred, there are two copies of the genes in the duplicated region, and now one of these is free to evolve to a different function. This mechanism would allow organisms to acquire more genes and become more complex. However, how tandem duplications could persist long enough for some of the duplicated genes to evolve is not clear unless a deletion in one of the duplicates or some other genetic rearrangement makes the duplication essential.

Inversion Mutations

Sometimes a DNA sequence is not removed or duplicated but rather is flipped over, or inverted. After such an inversion, all the genes in the inverted region face in the opposite orientation relative to the other genes in the chromosome. Programmed inversions can play roles in regulation in bacteria, but because they are not true mutations but rather a type of site-specific recombination, they are considered in chapter 9.

CAUSES OF INVERSIONS

Like deletions and tandem duplications, inversions are caused by ectopic recombination between repeated sequences. However, now the repeats at which the ectopic recombination occurred are inverted rather than direct (Figure 3.17). Like direct repeats, inverted repeated sequences read the same in the 5′-to-3′ direction but on opposite strands. Also, the ectopic recombination that produces inversions must occur between two inverted-repeat regions on the same DNA.

PROPERTIES OF INVERSION MUTATIONS

Unlike deletions, inversion mutations can generally revert. Recombination between the inverted repeats that caused the mutation will "reinvert" the affected sequence, recreating the original order. However, if the ectopic recombination that originally caused the inversion

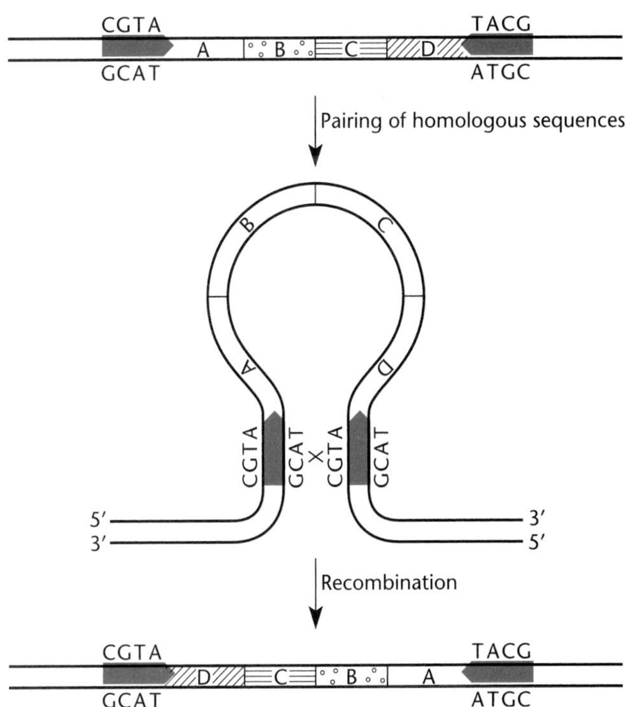

Figure 3.17 Recombination between inverted repeats can cause inversion mutations. The order of genes within the inversion is reversed after the recombination. The inverted repeats are usually ≥50 bp long, but inverted repeats of only 4 bp are shown for convenience.
doi:10.1128/9781555816278.ch3.f3.17

occurred between very short inverted repeats or repeats that are not exactly the same, then the recombination event that caused the inversion would have to occur between the exact bases involved in the first recombination to restore the correct sequence. Such a recombination could be a very rare event, and reversions of such an inversion mutation would be very rare.

Like tandem duplications, inversion mutations often cause no phenotype. If the inversion involves a longer sequence, including many genes, generally the only affected regions are those that include the **inversion junctions**, where the recombination occurred. In many cases the inverted repeats where the recombination occurred are due to repeated sequences between genes, so no genes are inactivated by the inversion. The genes in the inverted region are still intact, although they are in the opposite orientation relative to the other genes in the chromosome. While even very long inversion mutations often cause no obvious phenotypes, they probably cause other longer-term problems for the cell by inverting certain polar sequences, such as the termination and KOPS sequences discussed in chapter 1. Perhaps because of this, long inversions seem to have been selected against in nature, since the arrangements of genes in even quite distantly related bacteria are often remarkably similar (Box 3.2).

BOX 3.2

Inversions and the Genetic Map

Even a single large inversion mutation causes a dramatic change in the genetic map, or the order of genes in the DNA, of an organism. The order of all the genes is reversed between the sites of the recombination that led to the inversion. We would also expect inversions to be fairly frequent, because repeated sequences exist in many places in bacterial genomes, and many of them are inverted with respect to each other. In spite of this, inversions seem to have occurred very infrequently in evolution. As evidence, consider the genetic maps of *S. enterica* serovar Typhimurium and *E. coli*. These bacteria presumably diverged millions of years and billions of generations ago and are very different in many ways, including in the diseases they cause. Nevertheless, the maps are very similar, except for one short inverted sequence between about 25 and 27 min on the *E. coli* map. The apparent reason why large inversions have not occurred is that some sequences in the chromosome cannot be inverted without seriously disadvantaging the organism. These are sequences that are polar and do not read the same on both strands in the 5′-to-3′ direction, so if they are inverted with respect to the direction of replication, they will be nonfunctional. Some examples we know of in *E. coli* are the KOPS sequences that direct the FtsK DNA translocase toward the *dif* sequences at the terminus of replication to remove chromosome dimers after replication (see chapter 1) and chi sites that promote recombination to restart collapsed replication forks (see chapter 10).

Genetic maps of *S. enterica* serovar Typhimurium and *E. coli*, showing a high degree of conservation. The region from *hemA* to the 40-min position is inverted in *E. coli* relative to that in serovar Typhimurium. doi:10.1128/9781555816278.Box3.2.f

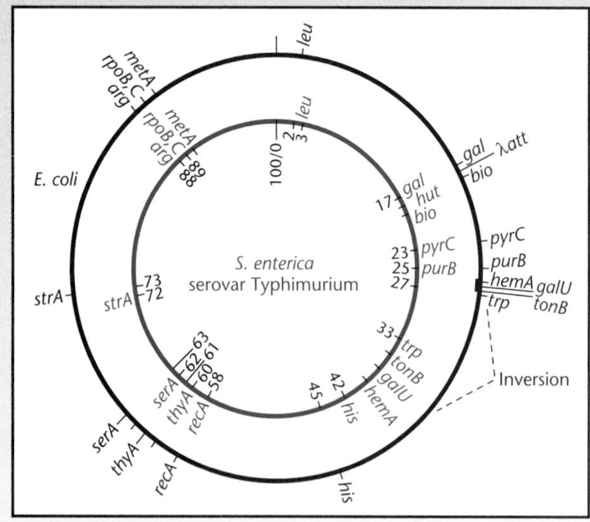

References

Mahan, M. J., and J. R. Roth. 1991. Ability of a bacterial chromosome to invert is dictated by included material rather than flanking sequences. *Genetics* **129:**1021–1032.

Miesel, L. 2011. Barriers to the formation of inversion rearrangements in *Salmonella*, p. 233–243. *In* S. Maloy, K. T. Hughes, and J. Casadesús (ed.), *The Lure of Bacterial Genetics: A Tribute to John Roth.* ASM Press, Washington, DC.

Identification of Inversion Mutations

Like deletions, inversion mutations sometimes fuse one gene to another gene. This property often provides a mechanism for detecting them. This is often why cells can use programmed inversions to regulate genes, since promoters in the inversion can transcribe a gene in one orientation but not the other or transcribe different genes depending on their orientation.

NAMING INVERSIONS

A mutation known to be an inversion is given the letters IN followed by the genes in which the inversion junctions occur, provided that they are known, followed by the number of the mutation. For example, IN(*purB-trpA*)3 is inversion number 3 in which the inverted region extends from somewhere near the gene *purB* to somewhere near the gene *trpA*, so both genes and the genes between them are included in the inversion.

Insertion Mutations

An **insertion mutation** occurs when a piece of DNA inserts into the chromosome. Smaller insertions of just a few base pairs are usually referred to as frameshift mutations, but sometimes insertions that are much too large to be caused by replication slippage occur. These mutations are usually caused by IS elements spontaneously moving, or "hopping," from another place in the DNA. IS elements are small transposable elements—usually only about 1,000 bp long—that exist in many copies in many bacterial genomes and carry genes for enzymes to promote their own movement. Because they exist in many copies, IS elements are also often the sites of ectopic recombination that creates deletion, duplication, and inversion mutations. We discuss transposons in chapter 9.

PROPERTIES OF INSERTION MUTATIONS
Insertion of DNA into a gene almost always inactivates the gene; therefore, insertion mutations are usually not leaky. Transposons, such as IS elements, also contain many transcription termination sites, so their insertion results in polarity (see chapter 2), which prevents the transcription of genes downstream of the gene into which the transposon has inserted if those genes are normally cotranscribed with the disrupted gene. They also map, like a point mutation, to a certain place in the DNA. Finally, insertion mutations do revert, but only rarely. To revert, the inserted DNA must be removed precisely, with no DNA sequences remaining. Many types of transposons copy a short sequence when they integrate into the DNA (see chapter 9), and recombination between these short repeated sequences can precisely excise the transposon. However, these repeated sequences are too short to promote recombination very frequently. The unusual properties of insertion mutations—totally inactivating the gene, mapping like a point mutation, and seldom reverting—led to their discovery (see chapter 9).

IDENTIFICATION OF INSERTION MUTATIONS
A significant percentage of all spontaneous mutations are insertion mutations, but their phenotypes are difficult to distinguish from those of other types of mutations. Often, a mutation is not known to be an insertion mutation until the DNA is isolated and/or sequenced. However, transposons can serve as useful tools in genetics experiments, because many carry a selectable gene, such as one for antibiotic resistance, making them relatively easy to map, since the insertion of such a transposon into a cell's DNA makes the mutant cell antibiotic resistant. They also significantly change the size of DNAs into which they insert, making them relatively easy to map physically. However, if the transposon has just moved from one DNA in the cell to another DNA in the same cell, the phenotype of antibiotic resistance will remain unchanged, and it is not obvious a mutation has occurred. In order to isolate transposon insertion mutations, one must introduce the transposon into the cell from outside. A common way to do this is to introduce the transposon into the cell in a DNA that cannot replicate in that cell, i.e., a **suicide vector**. Then, the cell can become resistant to the antibiotic only if the transposon hops from the original DNA into the chromosome or some other DNA that can replicate in the cell. The methods of transposon mutagenesis are central to bacterial molecular genetics and biotechnology and are discussed in some detail in chapter 9.

NAMING INSERTION MUTATIONS
An insertion mutation in a particular gene is represented by the gene name, two colons, and the name of the insertion. For example, *galK*::Tn*5* denotes the insertion of the transposon Tn*5* into the *galK* gene. Because an insertion can occur at any one of many positions within the gene (and these could result in different phenotypes), each insertion is also indicated with an allele number to distinguish them from each other (e.g., *galK35*::Tn*5*). Because of their usefulness, insertions of just a selectable gene, for example, for antibiotic resistance, are often made using recombinant DNA techniques. When insertion mutations are so constructed, they are denoted with the capital Greek letter Ω (omega) followed by the name of the insertion. For example, pBR322Ω::*kan* has a kanamycin resistance gene inserted into plasmid pBR322.

Reversion versus Suppression

Reversion mutations are often easily detected through the restoration of a mutated function. As discussed above, a true reversion actually restores the original sequence of a gene. However, sometimes the function that was lost because of the original mutation can be restored by a second mutation elsewhere in the DNA. Whenever a second mutation somewhere else in the DNA relieves the effect of a mutation, the mutation has been suppressed rather than reverted and the second mutation is called a **suppressor mutation**. The following sections present some of the mechanisms of suppression.

Intragenic Suppressors

Suppressor mutations in the same gene as the original mutation are called **intragenic suppressors**, from the Latin prefix "intra," meaning "within." These mutations can restore the activity of a mutant protein by many means. For example, the original mutation may have made an unacceptable amino acid change that inactivated the protein, but changing another amino acid somewhere else in the polypeptide could restore the protein's activity. This form of suppression is sometimes interpreted to indicate an interaction between the two amino acids in the protein, but other explanations are possible.

The suppression of one frameshift mutation by another frameshift mutation in the same gene is another example of intragenic suppression. If the original frameshift resulted from the removal of a base pair, the addition of another base pair close by could return translation to the correct frame. The second frameshift can restore the activity of the protein product, provided that ribosomes, while translating in a different frame, do not encounter any nonsense codons in that frame or insert any amino acids that significantly alter the activity of the protein.

Intergenic Suppressors

Intergenic (or **extragenic**) **suppressors** do not occur in the same gene as the original mutation. The prefix "inter" comes from the Latin for "between." There are many ways in which intergenic suppression can occur. The suppressing mutation may restore the activity of the mutated gene product or provide another gene product to take its place. Alternatively, it may alter another gene product with which the original gene product must interact in a complementary way so that the two mutated gene products can again interact properly.

One common way an intergenic suppressing mutation may restore the viability of the mutant cell is by preventing the accumulation of a toxic intermediate. If the gene for a step in a biochemical pathway is mutated, a toxic intermediate in that pathway can accumulate, causing cell death. However, a suppressing mutation in another gene of the pathway may prevent the accumulation, allowing the cell to survive even though it still has the original mutation.

The suppression of *galE* mutations by *galK* mutations provides an illustration of such an intergenic suppressor. Many types of cells, including human cells, use the pathway shown in Figure 3.18 to convert galactose into glucose in order to use the glucose as a carbon and energy source. The first step is to convert the galactose into galactose-1-PO_4 by a kinase, the product of the *galK* gene. Subsequent steps catalyzed by the products of the *galT* and *galE* genes then form uridine diphosphate (UDP)-galactose and UDP-glucose, respectively, with the release of glucose. Cells with *galE* mutations are galactose sensitive (galactosemic), and their growth is inhibited by galactose in the medium because galactose-1-PO_4 and UDP-galactose accumulate and are toxic. In humans, the hereditary human condition of galactosemia is due to a GalE deficiency.

The galactose sensitivity of *galE* mutants makes it easy to isolate *galK* mutants. Most of the apparent revertants of *galE* mutants that are no longer galactose sensitive and can grow in the presence of galactose have not actually reverted the *galE* mutation. In most of them, the *galE* mutation is still present, but it is being suppressed by a *galK* mutation, since the *galK* mutations occur much more frequently. Any mutation in the *galK* gene that inactivates the galactose kinase will prevent the accumulation of the toxic intermediates, but only very few base pair changes can revert the *galE* mutation. Furthermore, galactose-resistant mutants with *galK* suppressors can be distinguished from true *galE* revertants because the double mutants with both mutations still do not grow with galactose as the sole carbon and energy source (i.e., they are still Gal^-), and another carbon source, such as glucose, must also be provided for them to grow and form a colony.

NONSENSE SUPPRESSORS

Nonsense suppressors are another type of intergenic suppressor. A **nonsense suppressor** is usually a mutation in a tRNA gene that changes the anticodon of the tRNA product of the gene so that it recognizes a nonsense codon. In the example shown in Figure 3.19, the gene for a tRNA with the anticodon 3′-GUC-5′ (so that it normally recognizes the glutamine codon 5′-CAG-3′) mutates, causing the anticodon to become 3′-AUC-5′. This altered anticodon can pair with the nonsense codon UAG instead of CAG. However, the anticodon mutation does not significantly change the tertiary shape of the tRNA, which means that the cognate aminoacyl-tRNA synthetase still loads it with glutamine. Therefore, this mutated tRNA will insert glutamine for the nonsense codon. This can lead to synthesis of the active polypeptide and suppression of the amber mutation.

The mutated tRNA is called a **nonsense suppressor tRNA**, and nonsense suppressors themselves are referred to as amber suppressors, ochre suppressors, or opal suppressors depending on whether they suppress UAG, UAA, or UGA mutations, respectively. Table 3.3 lists

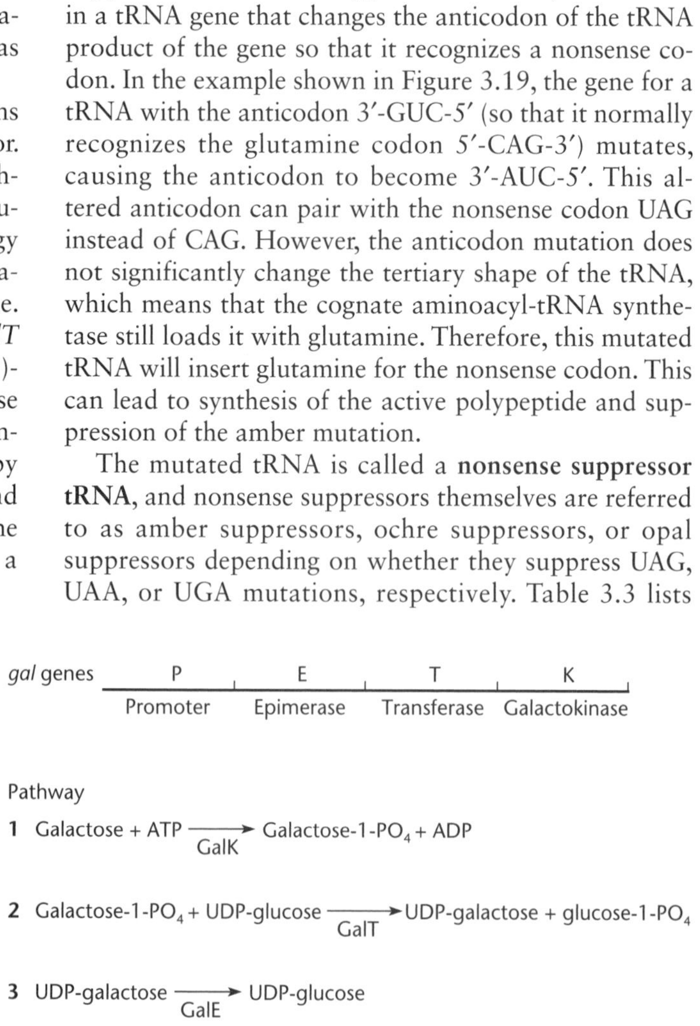

gal genes: P | E | T | K
Promoter | Epimerase | Transferase | Galactokinase

Pathway

1 Galactose + ATP $\xrightarrow[\text{GalK}]{}$ Galactose-1-PO_4 + ADP

2 Galactose-1-PO_4 + UDP-glucose $\xrightarrow[\text{GalT}]{}$ UDP-galactose + glucose-1-PO_4

3 UDP-galactose $\xrightarrow[\text{GalE}]{}$ UDP-glucose

Figure 3.18 The pathway to galactose utilization in *E. coli* and most other organisms. *galK* mutations suppress the toxicity of galactose in mutants with *galE* mutations because they prevent the accumulation of the toxic intermediates galactose 1-phosphate and UDP-galactose when cells are growing in the presence of galactose. doi:10.1128/9781555816278.ch3.f3.18

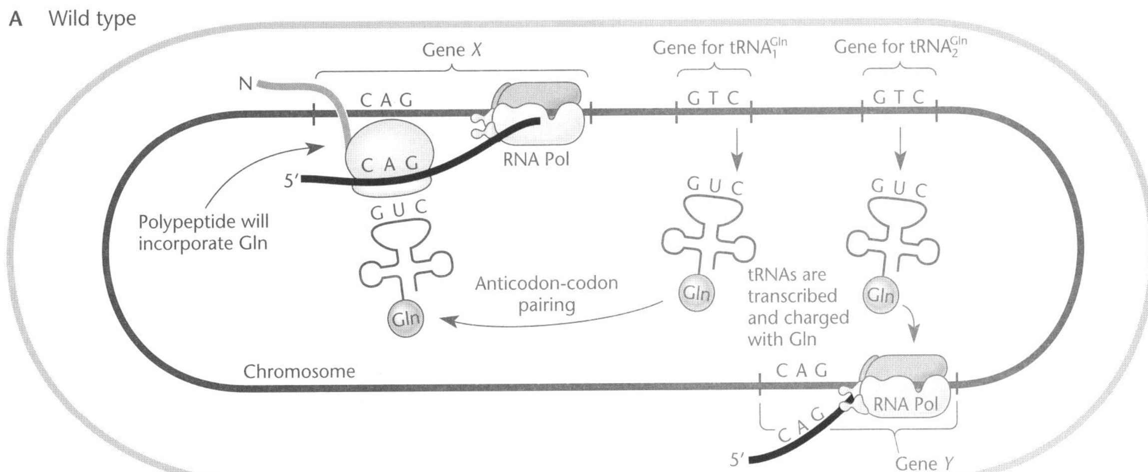

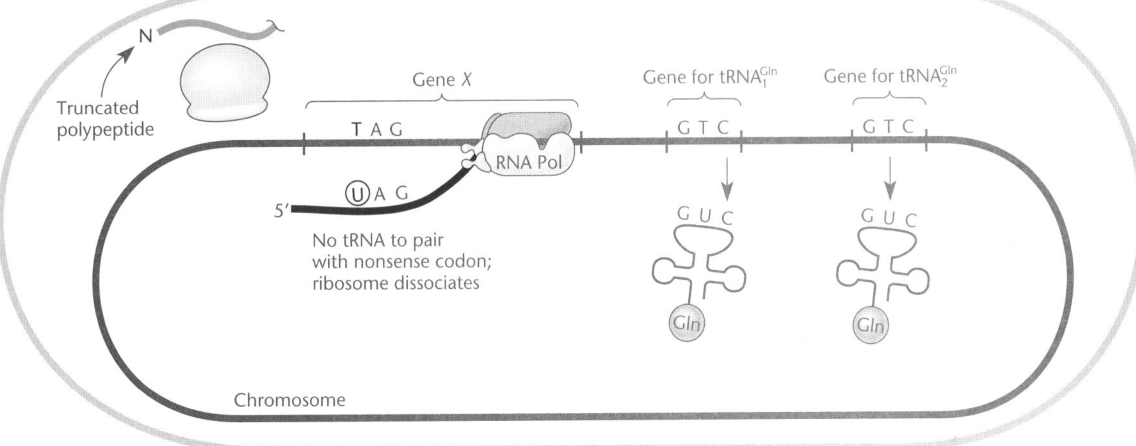

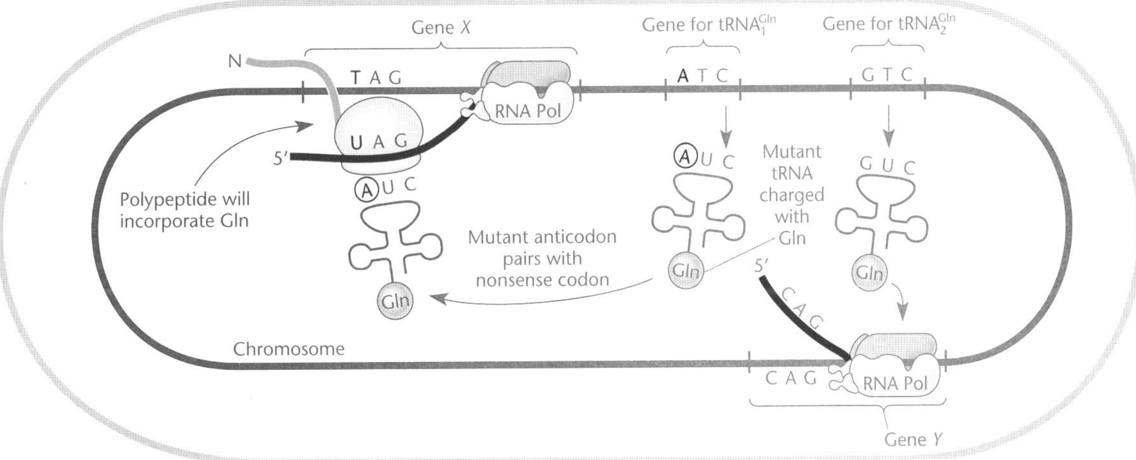

Figure 3.19 Formation of a nonsense suppressor tRNA. **(A)** Gene *X* and gene *Y* contain CAG codons, encoding glutamine. Other codons are not shown. The bacterium also has two different tRNA genes inserting glutamine for the CAG codon. Only the anticodons of the tRNAs are shown. **(B)** A mutation occurs in gene *X* (shown in blue), changing the CAG codon to UAG and causing the synthesis of a truncated polypeptide (also shown in blue). **(C)** A suppressor mutation in the gene for one of the two tRNAs changes its anticodon so that it now pairs with the nonsense codon UAG. The translational machinery now sometimes inserts glutamine for the UAG nonsense codon in gene *X*, allowing synthesis of the complete polypeptide. The anticodon of the other tRNA still pairs with the CAG codon, allowing synthesis of the gene *Y* protein and the products of other genes carrying the CAG codon.
doi:10.1128/9781555816278.ch3.f3.19

Table 3.3 All *E. coli* UAG nonsense suppressor tRNAs

Suppressor name	Gene name	tRNA	Change in anticodon	Suppressor type
supD	*serU*	$tRNA^{Ser}$	CGA→CUA	Amber
supE	*glnV*	$tRNA^{Gln}$	CUG→CUA	Amber
supF	*tyrT*	$tRNA^{Tyr}$	GUA→CUA	Amber
supB	*glnU*	$tRNA^{Gln}$	UUG→UUA	Ochre/amber
supL	*lysT*	$tRNA^{Lys}$	UUU→UUA	Ochre/amber

several *E. coli* nonsense suppressor tRNAs. Nonsense suppressors can also be classified as **allele-specific suppressors** because they suppress only one type of allele of a gene, that is, one with a particular type of nonsense mutation. In contrast, for example, the *galK*-suppressing mutations discussed above suppress any *galE* mutation and so are not allele specific.

Types of Nonsense Suppressors

A tRNA gene cannot always be mutated to form a nonsense suppressor. Generally, if there is only one type of tRNA to respond to a particular codon, the gene for that tRNA cannot be mutated to make a suppressor tRNA. The original codon to which the tRNA responded would be "orphaned," and no tRNA would respond to it wherever it appeared in an mRNA. Often, when a tRNA can be mutated to a nonsense suppressor tRNA it is because there is another tRNA that can respond to the same original codons. We have already mentioned how a tRNA gene included in a tandem duplication can be mutated to a nonsense suppressor because another copy of the tRNA gene exists in the duplication. However, such suppressors are very unstable and often revert to nonsuppressors due to recombination between the duplicated sequences. If stable nonsense suppressors can occur, it is usually because cells quite often have more than one tRNA, encoded by different genes, which can respond to the same codons. In the example in Figure 3.19, two different tRNAs encoded by different genes recognize the codon CAG, one of which continues to recognize CAG after the other has been mutated to recognize UAG.

Wobble (see Figure 2.38) offers an exception to the rule that a tRNA can be mutated to a stable nonsense suppressor only if there is another tRNA to respond to the original codon. Because of wobble, the same tRNA can sometimes respond both to its original codon and to one of the nonsense codons. For example, in a particular organism, there may be only one tRNA with the anticodon 3′-ACC-5′ that recognizes the codon for tryptophan, 5′-UGG-3′. However, from the wobble rules in Figure 2.38, if the anticodon, 3′-ACC-5′, is mutated to 3′-ACU-5′, the U in the third position of the anticodon will pair with either G or A in the third position of the codon and the tRNA might be able to recognize both the tryptophan codon UGG and the nonsense codon UGA, so that the suppressor strain could still recognize UGG and be viable. Wobble also allows the same suppressor tRNA to recognize more than one nonsense codon. In *E. coli*, all naturally occurring ochre suppressors also suppress amber mutations (Table 3.3). From the wobble rules, we know that a suppressor tRNA with the anticodon AUU should recognize both the UAG and UAA nonsense codons in mRNA. Note that in Table 3.3 anticodons are written 5′-3′ according to convention, even though they pair with the codon 3′-5′ because they are antiparallel.

Consequences of Nonsense Suppression

Strains containing a nonsense-suppressing tRNA gene are seldom completely normal for a number of reasons. Some of them occur only if the nonsense suppressor is suppressing a nonsense mutation in an essential gene, while others occur even if the cell has no nonsense mutation to be suppressed.

Missense. If the nonsense suppressor is suppressing a nonsense codon in a gene, the polypeptide synthesized as the result of a nonsense suppressor is not always fully active. Usually, the amino acid inserted at the site of a nonsense mutation is not the same amino acid that was encoded by the original gene. This changed amino acid sometimes causes the polypeptide to be less active or temperature sensitive, which can make the cell sick or temperature sensitive if the polypeptide is essential for growth.

Efficiency of suppression. Another reason cells can be sick if the suppressor is suppressing a nonsense mutation in an essential gene is that nonsense suppression is never complete. The nonsense codons are also recognized by release factors, which free the polypeptide from the ribosome (see chapter 2). Therefore, whether a nonsense codon is suppressed by the suppressor tRNA and translation continues or termination occurs and a shortened polypeptide is released depends on the outcome of a race between the release factors and the suppressor tRNA. If the tRNA can base pair with the nonsense codon before the release factors terminate translation at that point, translation will continue. Sequences around the nonsense codon (called the context [see below]) influence the outcome of this race and determine the efficiency of suppression of nonsense mutations at particular sites.

Readthrough of normal termination sites. Nonsense suppressor strains can be sick even if they are not suppressing a nonsense mutation anywhere in the genome. The major reason is that the coding regions (ORFs) for polypeptides normally end in one or more nonsense codons and the nonsense suppressor allows translation past the ends of normal ORFs. This results in longer than normal polypeptides, and the extra material at the carboxyl-terminal end of a polypeptide might interfere with its function. However, because nonsense suppressors are never 100% efficient, some of the correct polypeptides are always synthesized. Moreover, since the efficiency of suppression depends on the sequence of nucleotides around the nonsense codon, the sequences at the ends of genes presumably have a "context" that favors termination rather than suppression. As a failsafe device, more than one type of nonsense codon often lies in frame at the ends of genes. Multiple stop codons may limit the accidental readthrough of the nonsense codon or could avoid suppression by any particular tRNA suppressor. Nevertheless, tRNA suppressors can be isolated only in some single-celled organisms, including bacteria and fungi. In mammals and other eukaryotes, a still incompletely understood system called nonsense-mediated decay destroys transcripts containing in-frame nonsense codons. Presumably, this system prevents the accumulation of truncated proteins that could be toxic to the organism.

Genetic Analysis in Bacteria

One of the cornerstones of modern biological research is genetic analysis. Gregor Mendel probably performed the first definitive genetic analysis around 150 years ago, when he crossed wrinkled peas with smooth peas and counted the progeny of each type. The methods of genetic analysis have become considerably more sophisticated since then and are still central to research in cell and developmental biology. The first information about many basic cellular and developmental processes often comes from a genetic analysis of the process. The advantages of the genetic approach are that it requires few assumptions about the molecular basis for a biological process and can be applied to any type of organism, even ones about which little or nothing is known. Only recently has the biochemical basis for Mendel's smooth and wrinkled peas been elucidated.

More modern techniques have revolutionized how genetic analysis is done. Now that DNA sequencing of an entire bacterial genome takes only a few weeks and costs less than $1,000 and many hundreds of bacterial genomes have been sequenced, it is often possible to map mutations to a precise location in the genome of a bacterium about which little is known. Also, because of the large number of genes that have been sequenced and annotated, we can often tentatively identify the function of a gene product on the basis of similarities in sequence and structure to those that have already been characterized, a process called annotation (see chapter 2). Mutations such as transposon insertions can be easily located on the genome of the bacterium, and the gene in which they occur can be identified. Also, marker rescue or complementation cloning often makes it possible to locate a mutation on a sequenced genome or to prove that a mutation identified by sequencing is responsible for a phenotype. Probably, no one ever again will undergo the laborious task of constructing the genetic map of a bacterium or phage, or indeed that of any other organism, from scratch. Nevertheless, genetic analysis, including techniques of reverse genetics, is often still the only way to answer some of the most important questions in biology. For example, it is often the only way to determine how many gene products are involved in a function and to obtain a preliminary idea of the role of each gene product in the function. Suppressor analysis also offers one of the best ways to ascertain which gene products interact with each other in performing the function. Genetic analysis is covered in general genetics textbooks, and we review the basic principles here only as they apply particularly to bacteria.

Isolating Mutants

As discussed in the introductory chapter, a classical genetic analysis begins with finding mutants in which the function is altered. This process is called the **isolation of mutants** because the mutant organisms are somehow found and separated, or "isolated," from the myriad of normal or nonmutant organisms with which they are associated. As discussed in the introductory chapter, a major reason why bacteria are such excellent genetic subjects is the relative ease with which mutants can be isolated. Bacteria are generally haploid, meaning that they have only one allele of each gene. This makes the effects of even recessive mutations immediately apparent, obviating the need for backcrosses to obtain homozygous individuals that show the effects of the mutation. Bacteria also multiply asexually, not requiring crosses to make progeny. To make genetically identical bacteria, we do not need to clone them; they clone themselves when they multiply. Bacteria are also small, and for some types of bacteria, numbers equivalent to the entire human population on Earth can be placed on a single petri plate, making possible the isolation of even very rare mutants.

MUTAGENESIS

The first step in obtaining a collection of mutants for a genetic analysis is to decide whether to allow the mutations to occur spontaneously or to mutagenize the organism. Spontaneous mutations occur normally as mistakes in DNA replication, but the frequency of mutations can be greatly increased by treating the cells with certain chemicals or types of irradiation. Treatments that cause mutations are said to be mutagenic, and agents that cause mutations are **mutagens**. In general, treatments that damage DNA are mutagenic, although they differ greatly in their mutagenicity and what types of mutations they cause. The molecular basis for many types of mutagenesis is discussed in more detail in chapter 11.

Both spontaneous and induced mutations have advantages in a genetic analysis. To decide whether to mutagenize the cells and, if so, which mutagen to use, we must first ask how frequent the mutations are likely to be. Spontaneous mutations are usually rarer than induced mutations, so mutants with spontaneous mutations are more difficult to isolate. Therefore, to isolate very rare types of mutants or ones for which there is no good selection method (see below), we might have to use a mutagen. On the other hand, mutants stemming from spontaneously arising mutations are less likely to contain more than one mutation, and the presence of multiple mutations can confuse the analysis later.

One major advantage of inducing mutations by using mutagens is that a particular mutagen often induces only a particular type of mutation. Spontaneous mutations can be base pair changes, frameshifts, duplications, insertions, or deletions. However, the acridine dye mutagens, such as acriflavine, cause only frameshift mutations, and base analogs, such as 2-aminopurine, cause only base pair changes. Transposon insertion mutations (see chapter 9) have particular advantages because they are easier to map. However, they are usually inactivating, causing null mutations, which cannot be isolated in a gene that is essential. Therefore, the decision of whether to isolate spontaneous mutations or use a particular mutagen depends on how frequent the mutants are apt to be and what types of mutations are desired.

INDEPENDENT MUTATIONS

For an effective genetic analysis, mutants defective in a function should have mutations that are as representative as possible of all the mutations that can cause the phenotype. If the strains in a collection of mutants each carry a different mutation, we can get a better idea of how many genes can be mutated to give the phenotype and how many types of mutations can cause the phenotype. A general rule is that if some genes are represented by only a single mutation, then, by the Poisson distribution discussed above, there are likely to be many other genes in which mutations could give the same phenotype but which have been missed because not enough mutants have been analyzed. The attempt to identify all the genes whose products are involved in a particular phenotype is called **saturation genetics**.

There are two ways to ensure that a collection of mutations that cause a particular phenotype is as representative as possible. One way is to avoid picking **siblings**, which are organisms that are descendants of the same original mutant. If two mutants are siblings, they always have the same mutation. The best way to avoid picking siblings is to isolate only one mutant from each of a number of different cultures, all started from nonmutant bacteria. If two mutants arose in different cultures, their mutations must have arisen independently, and they could not be siblings.

A better way to avoid isolating mutants with the same mutation is to isolate mutants from cultures mutagenized with different mutagens. All mutagens have preferred hot spots and tend to mutagenize some sites more than others (see the discussion of mutational spectra in the *r*II genes of T4 phage in chapter 7). If all the mutants are spontaneous or are obtained with the same mutagen, many of them have mutations in the same hot spot, but mutants obtained with different mutagens tend to have different mutations.

Regional Mutagenesis

If the region of the chromosome to be mutagenized is known, more modern techniques can be used to introduce mutations into only that region of the DNA. Such techniques, called **regional mutagenesis**, are designed to limit mutagenesis to a selected region of the chromosome. Many of these techniques involve introducing oligonucleotides or DNA clones that have been synthesized or altered in the test tube into the cell. The oligonucleotide can either be synthesized to be completely complementary to the region of the chromosome with one desired change, if a particular mutation is wanted, or it can be synthesized to have an average of one difference per molecule, if random mutations in the region are desired. They are sometimes synthesized with trinucleotides rather than single nucleotides to avoid nonsense codons. Alternatively, double-stranded DNA fragments can be made by PCR amplification of the desired sequence, under conditions where the thermostable DNA polymerase is particularly mistake prone, to introduce random mutations into the DNA fragment. If the oligonucleotide or DNA fragment introduced into the cell is mostly complementary to a region of the chromosome, it can replace that region of the chromosome and introduce the desired change in the sequence. This process is most efficient in those types of bacteria for which **recombineering** has been developed. Recombineering systems use the recombination systems taken from bacteriophages to recombine single-stranded oligonucleotides or double-stranded DNA fragments with the chromosome at high frequency (discussed in chapter 10). However, the normal recombination systems of the cell can often be used. Such techniques can also be used to introduce a selectable marker, for example, for resistance to an antibiotic, into a gene or to delete a gene in its entirety to be certain that the gene product has no residual activity. Such methods go under the general name of **gene replacements**. We return to such techniques below.

Of course, for region-specific mutagenesis to be useful, the region to be mutagenized has to be known ahead of time. Region-specific mutagenesis cannot be used, for example, to find new genes whose products participate in a particular biological function. In that case, we must do random mutagenesis of the entire genome and identify the desired mutants among all of the mutagenized bacteria.

IDENTIFICATION OF MUTANTS

Even after mutagenesis or region-specific mutagenesis with even the most efficient recombineering system, mutants are rare and still must be found among the myriad of individuals that remain normal for the function. The process of finding mutants is called **screening**. Screening for mutants is usually the most creative part of a genetic analysis. One must anticipate the phenotypes that might be caused by mutations in the genes for a particular function in order to screen for them. This is where the geneticist earns her or his "pay," because predicting what types of mutations are possible and how to identify them often requires intuition, as well as rational thinking, but it is one of the more enjoyable aspects of genetics. For example, what do you imagine would be the phenotype of mutants defective in protein transport through the membrane? Specific examples of screening for this and other types of mutants are discussed in later chapters.

Screening for mutant bacteria usually involves finding **selective conditions** that can be used to distinguish the mutants from the original type. These are usually conditions under which either the mutant or the wild type cannot multiply to form a colony. Agar plates and media with selective conditions are called **selective plates** and **selective media**, respectively.

Selection of Mutants

Some types of mutants can be selected from among the normal or wild type. In what is sometimes called a **positive selection**, selective conditions are chosen under which the mutant but not the original wild type can multiply. This process falls under the broad category of **selectional genetics** and is one of the most powerful approaches ever devised for the study of cellular functions. It works particularly well with some types of bacteria and most phages because so many can be put on a single selective plate, so that even extremely rare mutants can be found. We have already mentioned some types of selection. Mutants with *tonA* or *tonB* mutations can be selected merely by plating *E. coli* on plates on which T1 phage have been spread. Only those cells that have a *tonA* or *tonB* mutation can multiply to form a colony. Figure 3.20 shows an example of a selection for His$^+$ apparent revertants of a *his* mutation. Cells are plated without histidine in the medium, and those that can multiply to form colonies are His$^+$ apparent revertants. These His$^+$ cells could be tested by the standard test for suppression to see if the reversion occurred at the site of the original mutation or if the *his* mutation is being suppressed by a mutation elsewhere in the genome (see Figure 3.33). Cells with a *galK* mutation are normally not selectable because *galK* mutations make cells unable to multiply with galactose as the sole carbon and energy source. However, the selection of *galK* mutants can be achieved by plating a *galE* mutant on medium containing galactose plus another carbon source, such as

Figure 3.20 Selection of a His$^+$ revertant. A His$^-$ mutant bacterium is plated on minimal medium with all the growth requirements except histidine. Any colonies that form after the plate is incubated are due to apparent His$^+$ revertants that can multiply without histidine in the medium.
doi:10.1128/9781555816278.ch3.f3.20

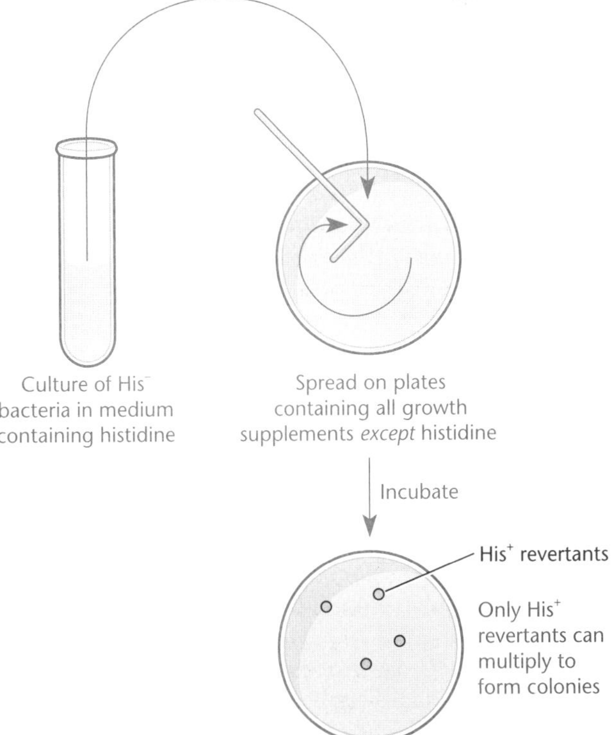
Culture of His$^-$ bacteria in medium containing histidine
Spread on plates containing all growth supplements *except* histidine
Incubate
His$^+$ revertants
Only His$^+$ revertants can multiply to form colonies

glucose (see above). Then, most of the cells that grow to form a colony are *galK* mutants. The possible applications of selectional genetics are limitless. For example, in spite of our extensive understanding of protein structure and function, it is still often impossible to predict what amino acid changes in a protein will have a particular effect on the protein, for example, make it temperature sensitive. Selectional genetics allows us to ask the cell what amino acid changes will have that effect if we can establish conditions under which the activity of the protein is toxic. Then, we can randomly mutagenize the gene for the protein, using region-specific mutagenesis, so that the protein, with almost every amino acid change possible, will be represented in some of the cells. We then plate the cells at the nonpermissive high temperature. Those cells that grow to form a colony could have a temperature-sensitive mutation in the toxic protein provided the cells do not grow at the lower permissive temperature. While selection is often not an option for many types of mutants (see below), screening for mutants is much easier with a selection, and geneticists expend much effort trying to design selections.

Screening for Mutants without a Selection

Most types of mutants cannot be selected. Most often, conditions under which the wild type but not the mutant can grow must be used. This type of mutant is the most common, because most of an organism's gene products help it to multiply, so mutations that inactivate a gene product are more likely to make the organism unable to multiply under a given set of conditions rather than able to multiply when the wild type cannot. These have been called negative selections, but they are not really selections at all; they are screens (see below), because the selective conditions are being used to screen for the mutants rather than to eliminate all other organisms that are not mutated in the same way.

To screen for mutants without a selection, the bacteria are first plated under **nonselective conditions**, that is, conditions under which both the mutant and the wild type can multiply. When the colonies have developed, some of the bacteria in each colony are transferred to selective conditions to determine which colonies contain mutant bacteria that cannot multiply to form a colony under those conditions. Once such a colony has been identified, the mutant bacterial strain can be retrieved from the corresponding colony on the original nonselective plate. Figure 3.1 shows the detection of two types of auxotrophic mutants, His^- (unable to make histidine) and Bio^- (unable to make the vitamin biotin), by such a screen.

Replica plating. Because of the general rarity of mutants, many colonies usually have to be screened to find a mutant. **Replica plating** can be used to streamline this process (illustrated in Figure 3.21). A few hundred bacteria are spread on a nonselective plate, and the plate is incubated to allow colonies to form. A replica is then made of the plate by inverting the plate and pressing it down over a piece of fuzzy cloth, such as velveteen. Then, a selective plate is inverted and pressed down over the same cloth so that the colonies are transferred from the cloth to the selective plate at the same position on the plate that they were in on the original plate. After the selective plate has been incubated, it can be held in front of the original nonselective plate to identify colonies that did not reappear on the selective plate. The missing colonies presumably contain descendants of a mutant bacterium that are unable to multiply on the selective plate.

Figure 3.21 Replica plating. **(A)** A few hundred bacteria are spread on a nonselective plate, and the plate is incubated to allow colonies to form. The plate is then inverted over velveteen cloth to transfer the colonies to the cloth. **(B)** A second plate is inverted and pressed down over the same cloth and then incubated. **(C)** Both plates after incubation. The dotted circle indicates the position of a colony missing from the selective plate. See the text for details.
doi:10.1128/9781555816278.ch3.f3.21

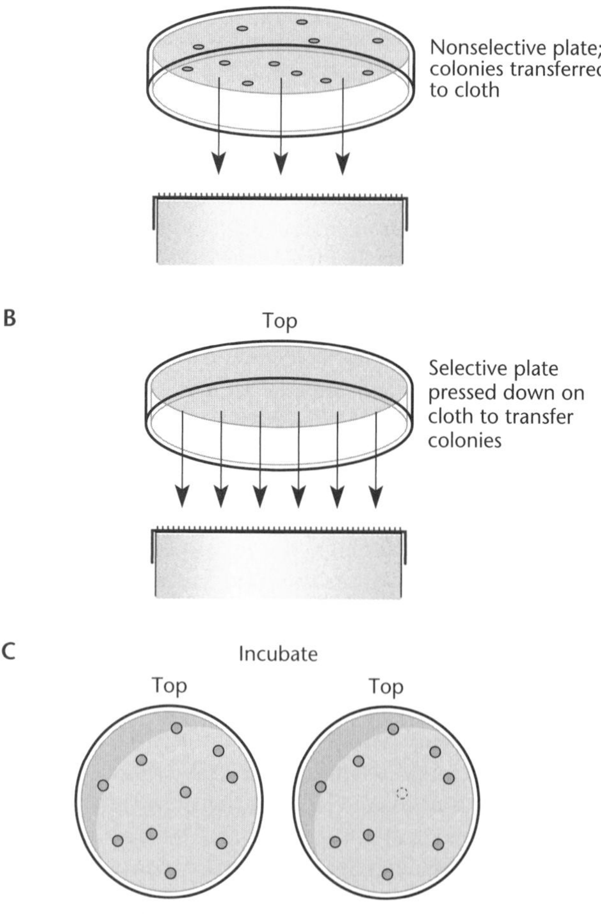

The mutant bacteria can then be taken from the colony on the original, nonselective plate. We mentioned above how replica plating was used by the Lederbergs to demonstrate that bacteria behave by the principles of Darwinian inheritance.

If a type of bacterial mutant being sought is rare, finding it by screening can be very laborious, even with replica plating. Even with the best-behaved types of bacteria, no more than about 500 bacteria can be spread on a plate and still give discrete colonies. Thus, for example, if the mutant occurs at a frequency of 1 in 10^6, more than 2,000 plates might have to be replicated to find a mutant! Anything that increases the frequency of mutants in the population will make the task easier.

Enrichment. Many fewer colonies need to be screened to find a mutant if the frequency of mutants is first increased through mutant **enrichment**. This method depends on the use of antibiotics, such as ampicillin and 5-bromouracil (5-BU), that kill growing but not nongrowing cells. Ampicillin inhibits cell wall synthesis, and when a growing bacterial cell tries to divide, it lyses (see chapter 14). A mutant cell that is not growing while the ampicillin is present does not lyse and is not killed. 5-BU also kills only growing cells, but by a very different mechanism. DNA containing 5-BU (an analog of thymine) is much more sensitive to UV light than is normal DNA containing only thymine. Cells replicate their DNA only while they are growing, so they take 5-BU into their DNA and become more UV sensitive only if they are growing while 5-BU is present in the medium.

To enrich for mutants that cannot grow under a particular set of selective conditions, the population of mutagenized cells is placed under the selective conditions in which the desired mutants stop growing. Meanwhile, the nonmutant, wild-type cells continue to multiply. The antibiotic—either ampicillin or 5-BU—is then added to kill any multiplying cells. The cells are filtered or centrifuged to remove the antibiotic and transferred to nonselective conditions. The mutant cells will survive preferentially because they were not growing in the presence of the antibiotic; therefore, they will become a higher percentage of the population. A few cells will always survive the enrichment even if they are not mutant. Cells may fail to grow for a variety of reasons. However, even if the enrichment makes the mutant only 100 times more frequent, only 1/100 as many colonies and therefore 1/100 as many plates must be replicated to find a mutant after an enrichment. In the example given above, after enrichment, we would need to replicate only 20 plates instead of 2,000 to find a mutant.

Unfortunately, enrichment cannot be applied to all types of mutants. Some mutants are killed by the selective conditions and so cannot be enriched by these procedures. To be enriched, the mutant must resume multiplying after it is removed from the selective conditions.

Genetic Characterization of Mutants

Once we have our collection of mutants with the desired phenotypes, we wish to further characterize the responsible mutations. First, we want to know where they are in the genome and what genes they are in. Locating them in the entire genome is like finding a needle in a haystack, in this case, a single base pair change or other alteration in the millions of base pairs in the entire genome of the bacterium. Until fairly recently, the only way to do this was by genetic mapping. However, genetic mapping requires that the genetic map of the organism be available. Constructing genetic maps is very laborious, and extensive genetic maps have been constructed for only a few different types of bacteria, including *E. coli*, *Bacillus subtilis*, and *Streptomyces coelicolor*. We discuss the methods of genetic mapping below. Nowadays, it is easier to have the genome of the bacterium sequenced, if it has not been sequenced already, and to identify as many genes as possible in the sequence by annotation (see Box 2.5). Mutations are then found in this annotated genome sequence. One way to locate the mutations in the genome sequence is to clone the region of the mutations by a method such as marker rescue or complementation (see below). The clones containing the region of the mutations can then be sequenced and located in the annotated sequence of the genome, or if the mutations have been obtained by transposon mutagenesis, it might be possible to sequence out from the inserted transposon into the surrounding bacterial DNA using the known sequence of the transposon to design primers. The surrounding genomic-DNA sequences can then be located on the genome sequence using available software. However, as mentioned, transposon insertions cause only null or loss-of-function mutations. If the phenotype requires an amino acid change or some other more subtle change in the sequence, mutations to that phenotype cannot be obtained by transposon mutagenesis.

Even if the entire DNA sequence of the genome of the bacterium being studied is known and it is possible to locate mutations on the genome by DNA sequencing alone, we still must do genetic analysis of the mutations. For example, it is desirable to have independent verification that the mutations identified by sequencing alone are causing the phenotypes. We also want to know that the polarity of the mutations on another gene is not responsible for the phenotypes, depending on what kind of mutations they are, or we may want to continue the analysis by seeing if there are any suppressors of the mutations to identify other proteins or functions with which the product of the mutated gene might interact, etc. All of these require genetic manipulations, so in this section we spend some time on genetic characterizations and how they are interpreted.

The two major tools of genetic characterization of mutations are recombination and complementation. While recombination and complementation can have seemingly similar outcomes, their molecular bases are completely different, and they even occur in different parts of the cell. Recombination is defined as the breakage and rejoining of two DNA molecules in new combinations, while **complementation** is the ability of the products of different mutant genes or alleles of genes to interact with or substitute for each other. While recombination can be used to locate mutations in DNA, complementation can be used to assign mutations to genes or to the parts of genes encoding individual domains of the same gene product and to give preliminary indications of the functions of gene products.

LOCATING MUTATIONS BY RECOMBINATION

We have already mentioned recombination in connection with how deletions, duplications, and inversions are formed and as a way of doing site-specific mutagenesis. Recombination can be used for mapping, region-specific mutagenesis, and gene replacements because cells have mechanisms to break and rejoin their DNA in new combinations. The site of breakage and rejoining is called a **crossover**. Recombination can be either **site-specific recombination** or general recombination. Site-specific recombination is discussed in detail in chapter 9. It uses specialized enzymes that cut and religate DNA, but only at unique sequences, so it is not generally useful for genetic characterizations and is not discussed here. Most genetic manipulations require **generalized recombination**, sometimes called **homologous recombination**, because it can occur anywhere but usually occurs only between two DNA regions that have the same or homologous sequences (*homo-logos* means "same-word" in Greek). Generalized or homologous recombination, referred to below as generalized recombination or simply recombination, probably occurs naturally in all organisms and serves the dual purpose of increasing genetic diversity within a species and restarting replication forks that have stalled at damage in the DNA (chapter 10).

Generalized recombination is quite complex and uses a number of different enzymes and pathways, depending on the situation. These details are discussed further in chapter 10. However, for now, the simplified model of recombination used in general genetics textbooks is sufficient (Figure 3.22). According to the simplified model, a strand of one or both of the two DNAs that will recombine invades the other DNA, and their opposing strands pair in a region where their nucleotide sequences are homologous (GCATA/CGTAT in Figure 3.22, although the homology usually must be much longer). This allows the two complementary opposing strands to base pair (see below). The requirement for base pairing ensures that recombination occurs in the same place in the two DNAs, since generally it is only in homologous regions that the opposing strands of the two DNAs are complementary and able to base pair. This invasion by one or both DNAs allows the paired strands of the two DNAs to cross over, bridging the two molecules, as shown. The bridged DNAs are then processed by DNA breaks, new DNA synthesis, and joining events to give rise to recombinant DNAs. The details of this process vary depending on the situation and are covered in chapter 10. Note in

Figure 3.22 A simplified diagram of recombination between two genetic markers. One strand of each or of both DNAs invades the other DNA, pairing with its complementary sequence. This joins the two DNAs. The structure is then resolved so that sequences from one DNA can become associated with the other DNA and lead to a rearrangement of genetic markers between the two DNAs. The positions of bases are shown only where the mutations have occurred and where the crossing over occurs. (–), mutation; (+), wild type. doi:10.1128/9781555816278.ch3.f3.22

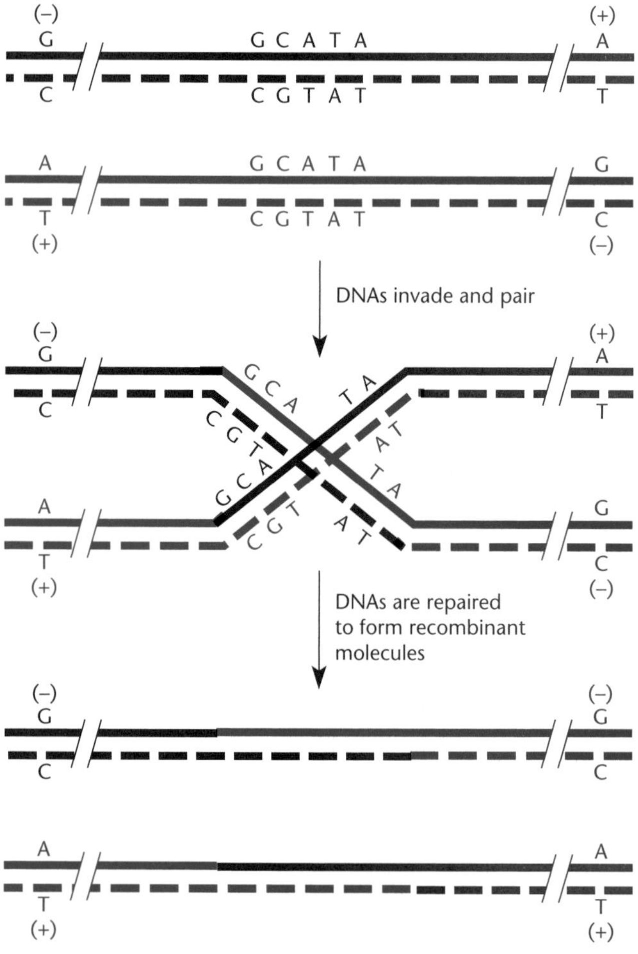

the figure that after the crossover event the distal markers are switched, although this is not a necessary outcome of such a recombination event, as we show below.

Consequences of Recombination in Bacteria

Most models of recombination shown in general genetics textbooks, such as the one in Figure 3.22, are between two complete linear DNA molecules like those in the chromosomes of eukaryotic organisms or in phages or other viruses with linear DNAs. However, in bacteria, recombination between the DNAs of different organisms usually occurs between a piece of DNA from one strain of a bacterium, called the **donor strain**, and the entire chromosome of another strain, called the **recipient strain.** If the chromosomal DNA sequences from the two strains are almost identical, which usually means they are members of the same species of bacterium, the piece of DNA from the donor is homologous to some region of the chromosome of the recipient. The two DNAs can then pair in this region, and recombination can occur in the recipient strain. The piece of DNA that comes from the donor strain can enter by any one of a number of mechanisms, including transformation, transduction, and conjugation (the molecular mechanisms of which are discussed in later chapters). The piece of chromosomal DNA from the donor strain that participates in the recombination can also be either linear, as occurs during transformation or transduction, or circular if, as is commonly the case, the fragment of the donor chromosome has been cloned into a circular plasmid or some other circular cloning vector. The consequences of the recombination are very different, depending on whether the piece of donor DNA is linear or in a circular molecule.

Figure 3.23A and B show the different consequences when the piece of chromosomal DNA from the donor strain is linear and when it is included in a circular DNA. If the DNA is linear, as in Fig. 3.23A, one crossover between the two DNAs will break the chromosome. Two crossovers will replace the region of the DNA of the recipient between the sites of the two crossovers with the DNA of the donor. In other words, it will leave a patch of DNA from the donor to replace the corresponding DNA from the recipient. In general, an odd number of crossovers (1, 3, etc.) between a linear piece of the chromosome from the donor and the entire chromosome of the recipient will break the chromosome of the recipient, and an even number (2, 4, etc.) will replace the sequence of regions of the recipient with the corresponding sequence of regions of the donor, depending on where they occur.

A very different situation prevails if the piece of donor chromosomal DNA is included in a circular DNA, such as a plasmid cloning vector, as shown in Fig 3.23B. Then, a single crossover will integrate the circular DNA into the chromosome of the recipient, and the region of chromosomal DNA from the donor that was in the circular vector is duplicated, with one copy on either side of (i.e., bracketing) the circular cloning vector. If the donor sequence differed from the recipient in one or more places, the donor sequences can be in one copy or the other, depending on where the crossover occurred. This creates a partial diploid that can be used for complementation tests (see below). However, it is an unstable situation, because the chromosomal sequence where the recombination occurred exists as a direct repeat on either side of the integrated plasmid, and a second crossover between the two direct repeats can loop out (delete) the plasmid, leaving behind a patch of DNA sequences from the donor DNA, depending on where the first and second crossovers occur. Therefore, recombination with either linear or circular DNA can be used to do gene replacements.

The example in Figure 3.24 shows a common application using recombination with a circular plasmid to do a gene replacement. In the example, a kanamycin resistance **gene cassette** including a gene for resistance to the antibiotic kanamycin (Kan^r) has been introduced into a chromosomal gene cloned in a plasmid cloning vector that carries a gene for ampicillin resistance (Amp^r). This plasmid is then introduced into the cell, and the homologous chromosomal sequences on the plasmid and in the chromosome can recombine to integrate the plasmid into the chromosome. The cell is now both Kan^r and Amp^r. A second recombination or crossover loops out the plasmid. Depending on where the second crossover occurs, it can leave the region of the gene with the Kan^r cassette, and then the cell is only Kan^r and not Amp^r. In practice, the plasmid cloning vector should not be able to replicate in the cell, i.e., it should be a suicide vector; otherwise, all of the cells become Kan^r and Amp^r even if no recombination has occurred. It is also desirable to have some way to select for cells that have lost the plasmid, in other words, to select for the second crossover; otherwise, we are faced with the laborious task of screening many cells for one in which the second crossover has occurred.

Genetic Markers

Events such as those described above can happen whenever two bacteria exchange DNA, provided the two DNAs have almost identical sequences in the region where the crossovers occur. We would have no way of detecting them if the sequences of the DNAs from the donor and recipient were identical to each other. However, if they have some differences between them in the form of mutations, whether point mutations, deletions, or insertions, we can often use these sequence differences to detect recombination and locate the sites of the

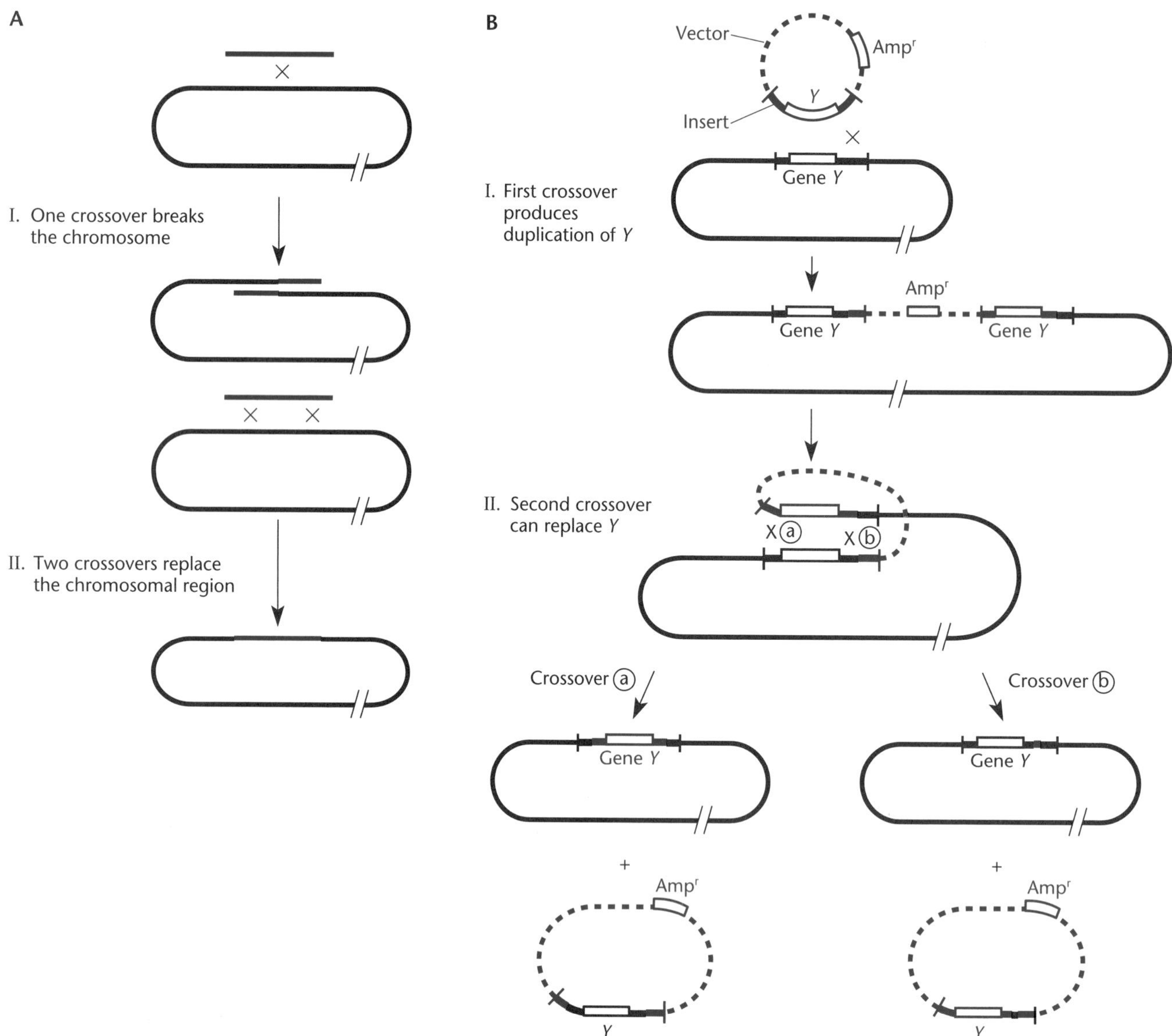

Figure 3.23 Different consequences of recombination between linear and circular DNAs in bacteria and its use for region-specific mutagenesis. **(A)** The introduced DNA from the donor is linear. (I) A single crossover breaks the chromosome and is presumably lethal. (II) Two crossovers can replace a sequence in the chromosome with the altered sequence of the donor (shown in blue). **(B)** The introduced donor DNA carrying an altered gene *Y* is cloned in a circular plasmid carrying a gene for ampicillin resistance. (I) A single crossover between the cloned DNA and the corresponding homologous region in the chromosome with a normal gene *Y* inserts the plasmid, which is now bracketed by the chromosomal region. (II) A second crossover can loop out and delete the plasmid, leaving only one copy of gene *Y*, either the mutated copy from the donor (shown in blue) or the normal gene from the recipient (shown in black). doi:10.1128/9781555816278.ch3.f3.23

mutations on the DNA. If the mutations that create the sequence differences in the DNAs cause observable phenotypes in the organism, we can use these phenotypes to determine that recombination has occurred. In so doing, we are using the sequence differences as **genetic markers**. For example, the Kan^r cassette in the donor DNA in the example shown in Figure 3.24 is a genetic marker. The sequence of the donor DNA is different as a result of the insertion of the cassette, and we use this difference as a genetic marker to determine if recombination has occurred. When the DNA that has formed as a result of a crossover segregates into progeny, these progeny are genetically different and, correspondingly, phenotypically different from the recipient. In general, progeny that are genetically different from their parents as a result of a recombination are called **recombinant types**.

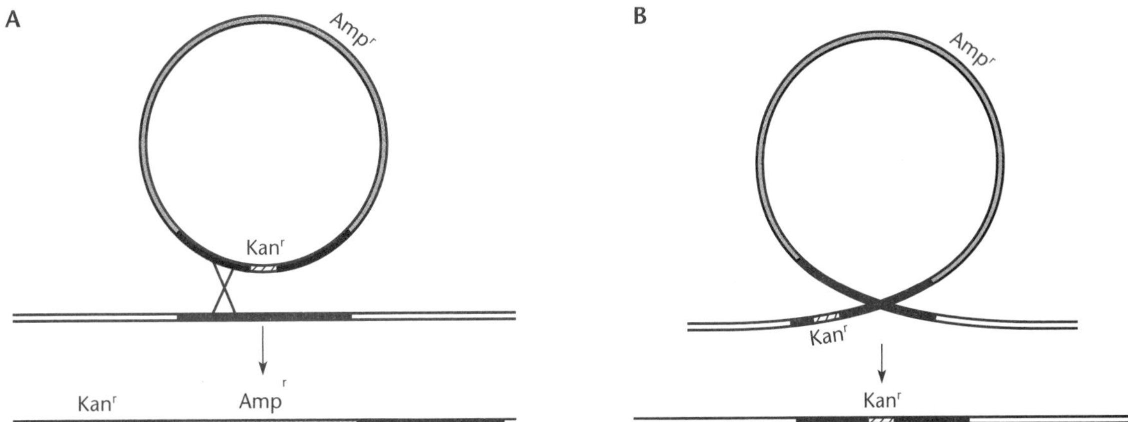

Figure 3.24 Using recombination to introduce an antibiotic resistance cassette into a chromosomal gene. **(A)** With a single crossover, the cloning vector integrates into the chromosome and the cells become ampicillin resistant (Amp^r) and kanamycin resistant (Kan^r). **(B)** With a second crossover, the cells lose the cloning vector and become Amp^s. Depending on where this second crossover occurs, the cells may be left with only the cloned sequence with the kanamycin resistance cassette and be resistant to kanamycin alone. See the text for details. Plasmid cloning vector sequences are in blue. The cloned region of the chromosome is in black. doi:10.1128/9781555816278.ch3.f3.24

Progeny that are genetically identical to one or the other of the two parents are **parental types**. Because in bacterial crosses only a part of the chromosome of the donor enters the recipient cell, recombinant types can appear only in the recipient cell. If the sequence of the recipient cell at the position of the genetic marker is replaced by the sequence of the donor cell, we know that recombination has occurred and the recipient bacterium has become recombinant for that marker. The nomenclature is much like that for the phenotypes of mutations. For example, if only the recipient has a *his* mutation that makes it His$^-$, the recombinant that gets this region from the donor will become His$^+$ and is referred to as a His$^+$ recombinant or simply a *his* recombinant. If only the recipient has a mutation in RNA polymerase that makes it rifampin resistant (Rif^r [chapter 2]) and the donor DNA comes from a strain that is rifampin sensitive (Rif^s), the recombinant for this marker will be Rif^s like the donor and the recombinants for the marker will be Rif^s recombinants, or simply *rif* recombinants. In general, when the recipient cell gets the allele of the donor, it has become recombinant for that marker. Whether recombinant types appear in the recipient cells for each of the markers that are different in the donor and recipient strains tells us something about where the mutations that caused the genetic markers are on the DNA of that species. Later in the chapter, we discuss how such genetic evidence is interpreted for the different types of crosses in bacteria.

Marker rescue. One of the most important current uses of recombination is in **marker rescue**, in which a genetic marker in the chromosome is "rescued," or becomes recombinant, by recombination with a piece of chromosomal DNA that has entered the cell that does not have that marker. How the piece of DNA enters the cell depends on the type of bacterium being studied and could include transformation or transduction. Marker rescue is useful for mapping and cloning, because if some of the cells become recombinant for a genetic marker, it tells us that the piece of chromosomal DNA from the donor must have come from the same region of the chromosome that has the mutation. In order to do marker rescue, we must have a way of selecting recombinants for the marker, because they are usually rare. For example, in the case of auxotrophic mutations, this would mean growth media lacking the required supplement. This method can be used to map mutations on the entire chromosome of a bacterium when the chromosome sequence is known or within individual genes if we have pieces of DNA deleted for regions of the gene. Figure 3.25 shows how **nested deletions** of the *thyA* gene of *E. coli* were used to map mutations within the *thyA* gene by marker rescue. This method is particularly useful to map mutations in a gene that affect different functions of the protein product and therefore define different domains of the protein. It can also be used to identify clones that come from a particular region of the chromosome, e.g., to clone a gene or region. Marker rescue has advantages and disadvantages over complementation for cloning chromosomal genes in cloning vectors. We contrast **cloning by complementation** and by marker rescue below.

Complementation Tests

The other general method for characterizing mutants is complementation. Rather than depending on breaking and joining DNA in new combinations, complementation

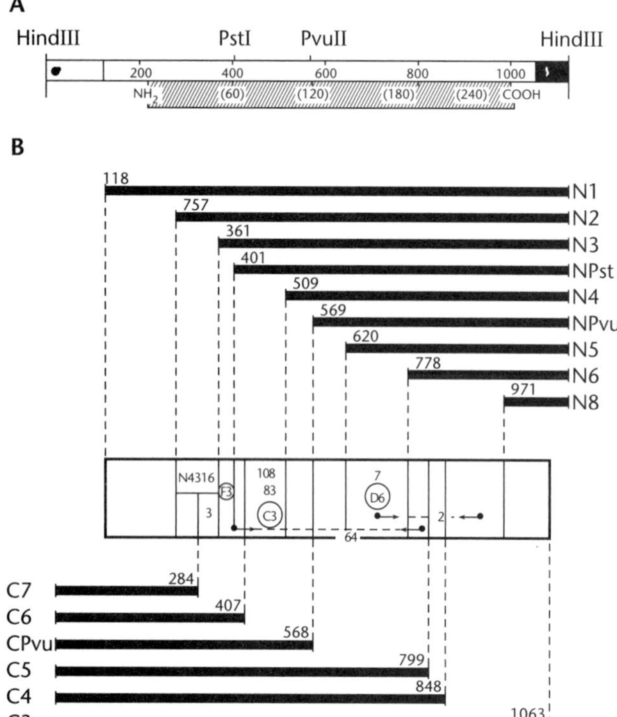

Figure 3.25 Using marker rescue to locate mutations in the physical map of the *thyA* gene of *E. coli*. **(A)** Physical map of the gene showing the positions of some restriction sites and the numbers of base pairs and amino acids from the 5′ and amino-terminal ends, respectively. **(B)** Mutations in the *thyA* gene in the chromosome were mapped using nested deletions extending various distances into the cloned *thyA* gene of *E. coli* from the N-terminal and C-terminal coding sequences. The clones were introduced into the mutant cells, and the positions of the point mutations were determined by whether Thy$^+$ recombinants could be selected. The solid bars show the regions deleted in each of the constructs. N is the amino terminus of the gene, and C is the carboxyl terminus.
doi:10.1128/9781555816278.ch3.f3.25

depends on the functional interaction of gene products made from different DNAs. Complementation allows us to determine how many gene products are represented by a collection of mutations and to obtain preliminary information about the functions affected by the mutations. Basically, we are asking whether a copy of a region of the chromosome can provide the function inactivated by a mutation, i.e., complement the mutation. If two mutations in different DNAs inactivate different functions, each can provide the function the other cannot, and the two mutations complement each other.

To perform a complementation test, we must have two copies of the regions of DNA containing one or more mutations in the same cell and see what effect this has on the phenotypes of the mutations. With a **diploid** organism, like us and most other multicellular organisms that contain two homologous chromosomes of each type, this is no problem, since diploids normally have two copies of each gene. With phages and other viruses, it is also no problem, because we can infect cells with two different mutant viruses simultaneously. However, with bacteria, which are naturally **haploid**, with only one copy of each gene, complementation tests are more difficult. Rather than being made complete diploids, bacteria can be made **partial diploids** by introducing a small region of the chromosome of one strain into another strain, using plasmids or prophages that can stably coexist with the chromosome. A common way of doing this is to introduce a plasmid cloning vector containing a clone of the region of interest into the cell. If the plasmid can replicate in the cell and if the gene or genes on the plasmid clone are expressed, they can be tested for the ability to complement mutations in the chromosome.

Plasmid clones can sometimes be used for complementation tests even if the plasmid vector cannot replicate in the cell. As we showed in Figure 3.23, if the plasmid integrates into the chromosome by a single crossover between the cloned sequence and the corresponding chromosomal sequence, the region of the chromosome will be duplicated. This will create a partial diploid of the cloned region and might give apparent complementation, but care must be taken that the mutations have not been lost by the recombination that integrated the plasmid. A safer way is to clone the chromosomal region of interest into a lysogenic phage or a transposon that then can be integrated into a different place in the chromosome. We can also sometimes see complementation between mutations on tandem duplications, provided that a way can be found to introduce different mutations into the copies of the duplication (see "Isolation of Tandem Duplications of the *his* Operon in *Salmonella*" below). Also, in some cases, it might be possible to observe complementation by an introduced DNA that exists only temporarily in the cell after some types of matings; the resulting strains are called **transient diploids**. Transient diploids are, by definition, not stable, but they might last long enough for some types of complementation tests.

RECESSIVE OR DOMINANT

Complementation can be used to tell if a mutation is recessive or dominant to the wild-type allele. A recessive mutation is subordinate to the wild type in the sense that an allele with the mutation does not exert its phenotype if the wild-type allele of the gene is present in the same cell. In contrast, a dominant mutation exerts its phenotype even if the wild-type allele is present. Recessive mutations generally inactivate the gene product, i.e., they are loss-of-function mutations, while dominant mutations are often gain-of-function mutations if they subtly change the gene product so that it can perform a function that the wild type cannot perform. If the mutant allele can function in a situation where the wild type cannot function, it is referred to as a gain-of-function allele. In eukaryotic genetics, alleles with altered function are typically called hypomorphic or hypermorphic alleles if they decrease or increase the activity of the allele, respectively, but this nomenclature has not caught on in bacterial genetics. Recessive mutations are much more common than dominant mutations because many more types of changes in the DNA inactivate the gene product than change it in some subtle way. Whether a particular type of mutation in a gene that gives a particular phenotype is dominant or recessive can tell us something about the normal functioning of the gene product.

To determine whether a mutation is recessive or dominant, we need to make a partially diploid cell that has both the wild-type allele and the mutant allele and determine whether the wild-type phenotype or the mutant phenotype prevails. To illustrate the difference between recessive and dominant mutations, we can return to the example of the *his* pathway. Most mutations that make the cell His$^-$ have inactivated one of the enzymes required to make histidine. These mutations are all recessive to the wild type because, in the presence of the wild-type allele for each of the genes, all the enzymes required to make histidine are made, and the cell will be His$^+$, the phenotype of the wild-type alleles. However, assume that there is an inhibitor of the pathway, such as an analog of histidine that binds to the first enzyme of the pathway and inactivates it. In the presence of this inhibitor, the cell is also unable to make histidine and will be His$^-$, so the phenotype caused by the wild-type allele in the presence of the inhibitor is His$^-$. However, a mutation in the gene for the enzyme can make the enzyme insensitive to the inhibitor, so that the mutant cell can make histidine even in the presence of the inhibitor. The phenotype of the cell containing this mutation is His$^+$ even in the presence of the inhibitor. Furthermore, the mutant enzyme might continue to function and make histidine in the presence of the inhibitor even if the sensitive wild-type enzyme is present. If so, in a diploid containing both the mutant and wild-type alleles, the phenotype would be His$^+$ in the presence of the inhibitor, the phenotype of the mutant allele, and the resistance mutation is dominant.

CIS-TRANS TESTS

One of the most important uses of complementation is to determine whether a mutation is *trans* acting or *cis* acting. These prefixes come from Latin and mean "on the other side" and "on this side," respectively. A ***trans*-acting mutation** usually affects a diffusible gene product, either a protein or an RNA. If the mutation affects a protein or RNA product, it can be complemented, and it does not matter which DNA has the mutation in a complementation test because the gene product is free to diffuse around the cell (i.e., the mutation acts in *trans*). A ***cis*-acting mutation** usually changes a site on the DNA, such as a promoter or an origin of replication. If the mutation affects a site on the DNA, it affects only that DNA and cannot be complemented (i.e., it acts in *cis*). In our example of the histidine synthesis genes, mutations, either recessive or dominant, that affect the enzymes that make histidine are *trans* acting and can be complemented, while a promoter mutation that prevents transcription of the genes for histidine synthesis is *cis* acting and cannot be complemented. In subsequent chapters, we discuss how *cis-trans* tests have been used to analyze the regulation of gene expression and other cellular functions.

ALLELISM TESTS

A common use of complementation is to determine how many different genes (or regions encoding a particular gene product) can be mutated to give a particular phenotype. Another name for this is an **allelism test**, because we are asking whether any two mutations are allelic, i.e., whether they affect the same gene (see above). Allelism tests can be performed only with recessive mutations and not with dominant mutations or *cis*-acting mutations, because the last two types of mutations cannot be complemented.

Returning to our example of histidine biosynthesis, assume that we have isolated a collection of mutants, all of which exhibit the His$^-$ phenotype, and want to know how many genes they represent. This should tell us how many enzymes, or more accurately, separate polypeptides, are required to make the enzymes that make the amino acid histidine in the cell and therefore allow the cell to multiply in the absence of histidine in the medium. Some proteins are composed of more than one different polypeptide, each encoded by a different gene. A different gene or complementation group, composed of all the mutations that do not complement each other, encodes each polypeptide, and if our collection of mutations is large and varied enough, each of these genes should be inactivated by at least one of our mutations. The allelism test is performed on the mutations two at a time, as illustrated in Figure 3.26A and B. If the two mutations are in different genes, as shown in Figure 3.26A, each DNA can furnish the polypeptide that cannot be furnished by the other, so that all the polypeptides are present and the diploid cell is phenotypically His$^+$. If, however, the two mutations are allelic in a region coding for the same gene product, as shown in Figure 3.26B, neither DNA can make that gene product, the two mutations cannot complement each other,

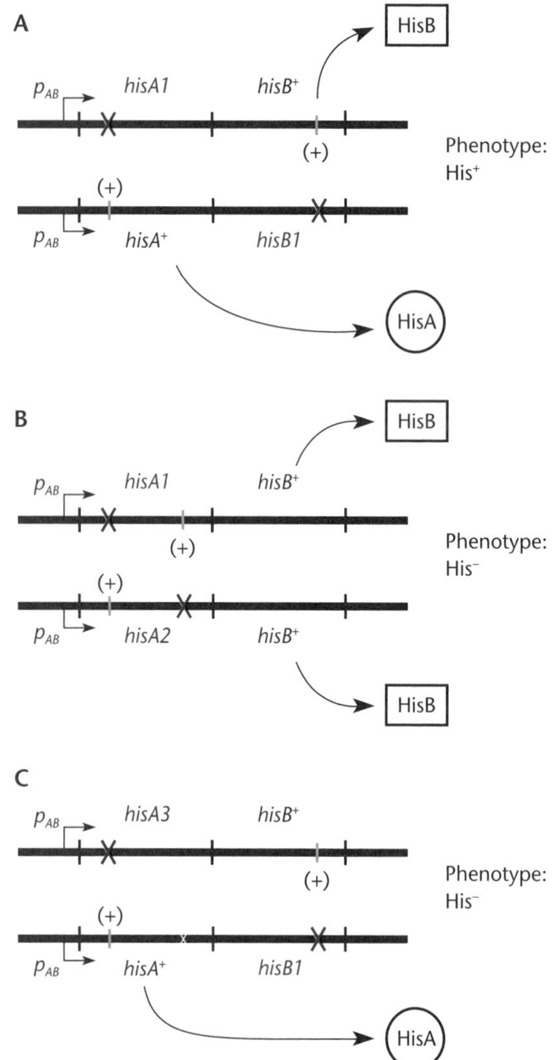

Figure 3.26 Complementation tests for allelism. Three mutations, *hisA1*, *hisA2*, and *hisB3*, that make the cell require histidine are being tested to determine whether they are allelic or are in different genes. The mutations are in different DNAs in the same cell in different combinations. **(A)** The *hisA1* and *hisB3* mutations are in different genes whose products are required for synthesis of histidine. The DNA with the *hisA1* mutation can make HisB, and the DNA with the *hisB3* mutation can make HisA; hence, the two mutations complement each other, and the cell is His$^+$. **(B)** The *hisA1* and *hisA2* mutations are in the same gene, i.e., are allelic. Neither DNA can make HisA; hence, the cell is His$^-$. **(C)** Polarity can complicate the allelism test. If mutation *hisA3* is a nonsense mutation in gene *A* that is polar on gene *B*, neither DNA will make HisB, and the two mutations will not complement each other. Nevertheless, the two mutations are not allelic.
doi:10.1128/9781555816278.ch3.f3.26

and the cells remain phenotypically His$^-$. We can then extend this analysis to include the other mutations in our collection, two at a time, to place them in complementation groups and determine how many total genes or complementation groups are represented in the collection of mutations.

Usually, the interpretation of complementation tests is straightforward and the number of complementation groups equals the number of gene products required for a biological function. However, sometimes, the interpretation of complementation tests for allelism is not so simple, and mutations that complement each other can be in the same gene and those that do not complement each other might be in different genes. Table 3.4 outlines the interpretation of complementation experiments and their possible complications.

Intragenic Complementation

When complementation occurs between two mutations that are in the same gene, it is called **intragenic complementation.** This can usually occur only with some types of mutations and if the protein product of the gene is a homodimer or higher homomultimer that contains more than one polypeptide product of the same gene (see chapter 2 for definitions of homodimer and homomultimer). Also, the polypeptide must have more than one functional domain, with different activities of the polypeptide residing in the different domains.

The classic example of intragenic complementation is in the *c1* gene encoding the repressor of λ phage (chapter 8), which is a homodimer composed of two identical polypeptides. Each polypeptide has two domains, one that binds to the operator sequences in DNA (called the DNA-binding domain) and another that forms the dimer by binding to the other subunit (called the dimerization domain). Sometimes, a mutation in the DNA-binding domain will complement a mutation in the dimerization domain, even though they are both in the *c1* gene. If both alleles of the *c1* gene are present in the same cell, hybrid repressor molecules sometimes assemble that have two different mutated subunits, one with the mutation in the DNA-binding domain and the other with the mutation in the dimerization domain. Some combinations of the hybrid repressor molecules are both able to

Table 3.4 Interpretation of complementation tests

Test result	Possible explanations
x and *y* complement	Mutations are in different genes
	Intragenic complementation has occurred[a]
x and *y* do not complement	Mutations are in the same gene
	One of the mutations is dominant
	One of the mutations affects a regulatory site or is polar

[a]See the text for an explanation of intragenic complementation. This is a less likely explanation than the mutations being in different genes.

bind to DNA and to form dimers, so the repressor is at least partially active. However, intragenic complementation is rare and occurs only between certain mutations in the gene.

Polarity

Another potential complication of complementation tests is **polarity** (see chapter 2 for a description of polarity). Polarity can have an effect opposite that of intragenic complementation in that it can prevent complementation between two mutations, even though they are in different genes and thus are not allelic. As discussed in chapter 2, a mutation in one gene that terminates translation or transcription can prevent the expression of (be polar on) another gene if the two genes are in the same operon so that they are transcribed onto the same mRNA.

We can use our hypothetical *his* operon to illustrate how polarity affects complementation tests (Fig. 3.26C). Assume that mutation *hisA3* in gene A is a nonsense mutation that is polar on gene B, because the two genes are cotranscribed into the same mRNA starting at a promoter upstream of gene A. If the *hisA3* mutation is a nonsense mutation that is polar on the *hisB* gene, it will not complement the *hisB1* mutation on a different DNA because neither DNA can make the HisB polypeptide. Whether the polarity is due to transcriptional termination in gene A downstream of the *hisA3* mutation or to translational coupling between *hisA* and *hisB*, the effect will be the same, i.e., no HisB polypeptide is made. This complicates our interpretation of the experiment in a number of ways. For one thing, from this experiment alone, we cannot even conclude that the HisA polypeptide is required for histidine biosynthesis. The *hisA3* mutation makes the cell His$^-$, but because of polarity, it could be the polypeptide encoded by the *hisB* gene that is required and the *hisA3* mutation is only preventing its synthesis by being polar on it. Alternatively, both HisA and HisB could be required to synthesize histidine, since neither is being made. Further complementation tests are required to answer these questions. If we introduce a plasmid containing a clone that expresses only the HisA polypeptide and the cell is still His$^-$, the HisB polypeptide must be required, because the *hisA3* mutation is preventing the synthesis of both the HisA and the HisB polypeptides from the chromosome but the clone provides the HisA polypeptide by complementation. On the other hand, if the complementing clone expresses only HisB and the cell is still His$^-$, the HisA polypeptide must also be required. Such complementation tests are the standard way of establishing that a mutation in a gene is responsible for a particular phenotype and not the absence of the product of another gene downstream of it in the operon due to polarity. To avoid the complications of polarity, genes are often inactivated by in-frame deletions or insertions or other mutations that cannot be polar on other genes. Transposons have been engineered so that transposon insertion mutations can be converted into short in-frame insertions (scars) lacking transcription termination sites by using site-specific recombination systems to delete most of the transposon sequences after the transposon has inserted.

Note that the effect of polarity on the expression of downstream genes is *cis* acting. In our example, the *hisA3* mutation can prevent the synthesis of the HisB polypeptide from the same DNA, the chromosome, but not from the complementing clone expressing HisB from the plasmid.

CLONING BY MARKER RESCUE AND COMPLEMENTATION

There are a number of reasons why we may want to clone a gene in which mutations give a certain phenotype. We have mentioned how the clone could direct us to the region in the genome in which the mutations lie, i.e., map the mutations. We could have the entire mutant genome sequenced to locate the mutation. However, cloning the mutated gene would give independent confirmation that we have located the responsible mutation. Once we have located the gene, comparisons to other gene sequences might give us clues to which biochemical activities it has. However, many genes in a bacterial chromosome have no known homologs, and even if homologs can be found by searches such as BLAST, such annotation can carry us only so far. To determine directly the biochemical activity and structure of a protein, we usually have to clone the gene to express and purify its product. The easiest way to go about the cloning depends on the situation. If we already know the sequence of the gene, we can use PCR to amplify the gene directly from the chromosome and clone the amplified fragment using methods outlined in chapter 1. However, in many cases, we need to identify clones from a library of the entire chromosome.

Either marker rescue or complementation can be used to identify clones of a gene for which we have mutations affecting a particular phenotype. Both approaches have their advantages and disadvantages, depending on the situation. Figures 3.27 and 3.28 contrast the use of complementation and marker rescue to identify clones of the gene for thymidylate synthetase (*thyA*) in a bacterium for which we have a *thyA* mutant. This enzyme is needed to synthesize dTMP from dUMP, so a *thyA* mutant will not be able to replicate its DNA and multiply to form a colony unless thymine is provided in the medium (see chapter 1). As shown in Figure 3.27, we obtained or made a library of the chromosome of the wild-type (*thyA*$^+$) bacterium and wanted to use complementation to identify

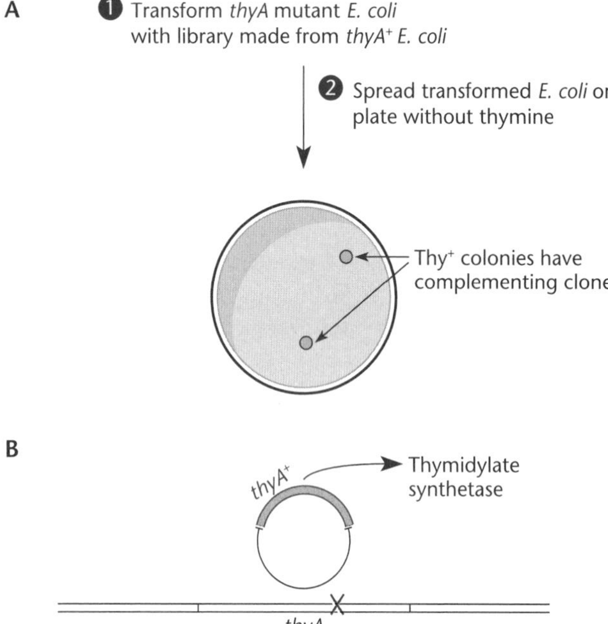

Figure 3.27 Identification of clones of the *thyA* gene of *E. coli* by complementation. **(A)** A *thyA* mutant of *E. coli* is transformed by a library of DNA from wild-type Thy$^+$ *E. coli*, and the transformants are selected directly on plates lacking thymidine but containing the antibiotic to which the cloning vector confers resistance. **(B)** Thy$^+$ transformants contain a clone of the *thyA* gene synthesizing the thymidylate synthetase from the plasmid, thus complementing the *thyA* mutation in the chromosome.
doi:10.1128/9781555816278.ch3.f3.27

which of the clones contains the *thyA* gene. We introduced this library into a mutant strain and plated the bacteria on medium lacking thymine. Any bacteria that multiplied to form a colony might contain a plasmid clone that was expressing the *thyA* gene, which complemented the *thyA* mutation in the chromosome.

Identifying clones by marker rescue recombination usually requires an addition step. If we plate the bacteria containing the library directly on plates lacking thymine, the clone will have been rearranged by the recombination that gave rise to the Thy$^+$ recombinants. One way to avoid this problem is to first plate the bacteria on medium with thymine so the bacteria will grow to form colonies whether or not the mutation is being marker rescued or complemented. These plates could then be replicated or patched onto selective plates containing all the necessary growth supplements but without thymine. Any replicated patches that contain many more Thy$^+$ colonies than average may be due to bacteria that received a clone containing at least part of the *thyA* gene from the donor, as shown in Figure 3.28. In the case of marker rescue, only some of the cells that took up the clone became Thy$^+$ by recombination with the clone, but if the clone complemented the *thyA* mutation in the chromosome, all of the cells that took up the clone would have become Thy$^+$. In the example, only a few of the cells in the colony were able to grow on the selective plate, so marker rescue was suspected.

Both approaches to cloning have their advantages and disadvantages. One advantage of marker rescue is that the piece of DNA need not be in a cloning vector that replicates in that cell and the gene need not be expressed in the cell. The DNA fragment in the clone also

Figure 3.28 Use of marker rescue to identify a clone containing at least part of the *thyA* gene of *E. coli*. **(A)** A plasmid library of clones from wild-type Thy$^+$ *E. coli* is transformed into a *thyA* mutant of *E. coli*, and the transformants are plated on nonselective plates with medium containing thymidine. After the colonies develop, the plate is replicated onto a selective plate without thymidine. Colonies that show some Thy$^+$ recombinants on the selective plate can be picked from the original nonselective plate (boxed). **(B)** The part of the *thyA* gene on the cloning vector has recombined with the *thyA* mutant gene on the chromosome to produce Thy$^+$ recombinants.
doi:10.1128/9781555816278.ch3.f3.28

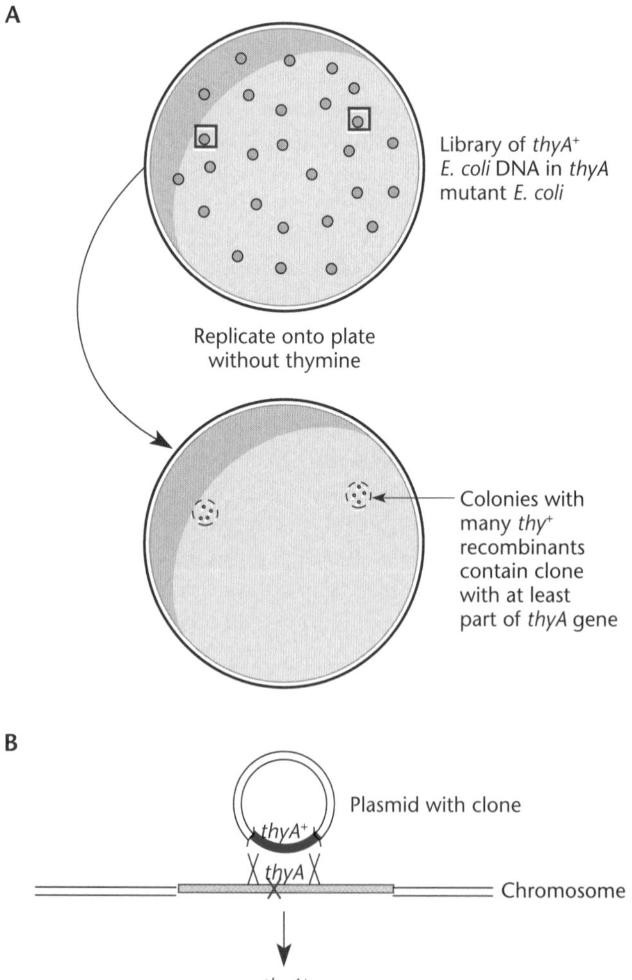

need not include the entire gene, only the part of it containing the region of the mutation. The disadvantage is that the fragment of the chromosome that participated in the marker rescue no longer exists in the cell, having been irreversibly altered by the recombination that rescued the mutation. If the piece of DNA was cloned in a multicopy cloning vector that can replicate in that cell, we can get the clone from a different plasmid in that cell that did not participate in the recombination. Otherwise, we need to return to the original nonselective plate to get the clone that demonstrated marker rescue from the original transformant before it was altered by the recombination. Also, if the clone does not contain the entire *thyA* gene, we may have to construct it from neighboring genome sequences in order to express the gene product. The major advantage of cloning by complementation is that we can introduce our library directly into the cells by transformation and select cells in which the mutation is complemented. But the major advantage of cloning by complementation is also its major disadvantage. To identify clones by complementation, the DNA fragments must be cloned in a vector that can replicate in those cells. Most plasmid cloning vectors have a fairly narrow host range, but some broad-host-range vectors that will replicate in many types of bacteria have been developed. We talk about the host range of plasmids in chapter 4. Also, the entire gene must be included in the clone, and the gene must be expressed in those cells. This means that the gene must be transcribed from a functional promoter and translated from a functional translation initiation site in the cells. However, the characteristics of a functional promoter and translation initiation site can differ between species of bacteria, and genes from distantly related organisms often cannot be expressed in a particular host, so this method is generally useful only for identifying clones in the "host of origin," for example, a bacterial gene in the bacterium it came from.

Another potential complication of complementation cloning is copy number effects. If we express the gene from a multicopy plasmid or other vector, we might make many more copies of the gene product than normally exist in the cell. This can affect the ability of the cloned gene to complement and may even change the phenotypes in unpredictable ways. Sometimes, a gene product other than the one that was mutated can give apparent complementation when it is expressed from a cloning vector. In our example, there might be another gene for thymidylate synthetase in the chromosome that is normally not expressed, perhaps because it lacks a functional promoter. A promoter in the cloning vector might direct transcription of the gene, leading to apparent complementation of the *thyA* mutation even though the clone is of a different gene. Sometimes the overproduction of a different gene product, such as occurs from a multicopy cloning vector, can give apparent complementation even if its product seems unrelated. If a mutation is complemented by a clone of a different gene whose product is being overproduced, this is called **multicopy suppression**. Identification of other proteins that can substitute for a mutated protein when expressed in greater than normal amounts is interesting and can give clues to the function of the mutated protein. However, this does not result in cloning the gene of interest, our original goal. For reasons such as these, it is often preferable to express cloned genes for complementation from a cloning vector that integrates in single copy somewhere else in the chromosome, for example, a lysogenic phage. In any case, care must be taken that any apparent clone isolated by complementation actually contains a wild-type copy of the mutated gene.

Genetic Crosses in Bacteria

We have discussed the molecular basis for genetic mapping and complementation tests without going into any detail about how DNA can be introduced into bacterial cells and how data can be analyzed using the different means of genetic exchange. As discussed in the introductory chapter, the three means of genetic exchange in bacteria are **transformation**, **transduction**, and **conjugation**. We could add **electroporation** to this list, but for genetic purposes, it is just another means of doing transformation in bacteria that are not naturally transformable. We talk about the molecular mechanisms of each of these forms of genetic transfer in chapters 5, 6, and 7. Here, we discuss their uses in genetic experiments and how data obtained by using them are analyzed.

SELECTED AND UNSELECTED MARKERS

What all the means of gene exchange in bacteria have in common is that they are not 100% efficient. Unlike us and most other known multicellular eukaryotes, bacteria do not have an obligate sexual cycle and do not need to mate to reproduce. They just divide. If one of the means of genetic exchange is available, a few of them might exchange DNA with each other and become recombinants. As in searches for mutants, we have to find the few bacteria that have recombined among the many that have not.

Fortunately, bacteria that have participated in a genetic transfer event can often be detected, even if they are very rare. The process is much like selecting mutants. Conditions are established under which only bacteria recombinant for one of the markers can multiply, independent of whether they are recombinant for any of the other markers. The recombinants for this marker, called the **selected marker**, are then tested to see if they are also recombinant for the other markers, called the **unselected markers**. Which marker is chosen to be the selected marker is a matter of convenience. The example in Figure 3.29 provides an illustration. In this example, a piece of DNA from one strain that has one of the mutations has been transferred into another strain that has another mutation. Both mutations inactivate a gene whose product is required for growth under some conditions. Therefore, the recipient strain is not able to grow under the selective conditions, because it has the m_y mutation. However, if crossovers occur between the incoming DNA and the homologous region in the recipient DNA, recombinant-type progeny can arise in which the sequence of the donor DNA, which does not have the m_y mutation (indicated by "+" in Figure 3.29), has replaced the sequence of the recipient strain, which did have the mutation, allowing the recipient strain to grow under the selective conditions. We decided to use the m_y marker as the selected marker to select recipient bacteria in which crossovers have occurred that make the strain recombinant for the selected marker, because they can be selected. Bacteria selected for being recombinant for this marker must have participated in a mating event, no matter how rare these mating events may be.

Once we have selected for recipient bacteria that have participated in a mating event because they have become recombinant for the m_y marker, we can test them to see if they are also recombinant for the m_z marker, the unselected marker. The recipient will have become recombinant for the other marker if it now has the m_z mutation and is unable to grow without the growth supplement that m_z mutants require, since that is the sequence of the donor DNA, which had the mutation. In general, a recipient has become recombinant for a marker if it has the marker sequence of the donor. If any of the recipients that were selected for being recombinant for the selected marker have also become recombinant for the unselected marker, the donor DNA from both regions must have entered the same recipient cell and both regions must have recombined with the recipient chromosome. If the mating events are rare enough, it is highly unlikely that two DNAs transferred in separately, and the two regions must have come in on the same DNA molecule. Therefore, the frequency with which bacteria selected for one marker have become recombinant for other markers gives us information about where the mutations that gave rise to the phenotypic differences were on the chromosome relative to each other. Note that once we have selected for one marker, we do not need to screen nearly as many bacteria to get meaningful data about the frequency of recombinants for the unselected markers and we can test them directly for the other markers without a selection. Specific examples of this process, using the various methods of gene exchange, which differ in how data are interpreted, are given below.

Mapping of Bacterial Markers by Transduction and Transformation

The transfer of DNA from one bacterial cell to another by transformation (electroporation) or transduction can be used to map genetic markers. Recipient bacteria that have received DNA from the donor by transformation or transduction are called **transformants** or **transductants**, respectively. Even though transformation and transduction occur by very different mechanisms, the processes by which genetic data are processed with them are quite similar, so they are treated together here. The major difference is that single-stranded DNA enters the cell during natural transformation, whereas double-stranded DNA enters during transduction. Even this difference may not matter, because the double-stranded DNA introduced by transduction is probably quickly converted into single-stranded DNA by an enzyme called the RecBCD nuclease in *E. coli*, as discussed in chapter 10. More details of how transformation and transduction occur are given in chapters 6 and 7, respectively.

TRANSFORMATION

In bacterial transformation, only the DNA from one bacterium enters another bacterium. This happens normally in some types of bacteria, which therefore are said to be **naturally competent** or **naturally transformable**. This means that they have the natural ability to pick up DNA under some conditions. Some of them pick up only DNA of the same species, while others can take up any DNA. While the purpose of natural transformation is the subject of intense debate, it may have a number of roles, including DNA repair, increasing genetic or antigenic variability, or even serving as a food source during development. In any case, it is very useful in the genetics of the bacteria that have this capability, including *Bacillus subtilis*, one of the major model systems.

Even if a type of bacterium is not naturally competent, it can sometimes be made competent by chemical treatments or electroporation. Developing a transformation system is a high priority for the study of any type of bacterium at the molecular level. The advantages of transformation are that DNA that has been manipulated in the test tube, or even synthesized chemically, can be introduced into cells, which makes it essential for many types of molecular genetic applications.

TRANSDUCTION

In transduction, DNA from the donor bacterium is packaged, or encapsulated, in a bacteriophage head, so that when the bacteriophage infects the recipient bacterium, the donor bacterial DNA is injected. There are two types of transduction, restricted and generalized. In restricted transduction, part of the phage DNA is replaced with

A Genotypes of strains
Strain 1
Strain 2
m_y
m_z
m_y will be the selected marker
m_z will be the unselected marker
B Transfer from donor to recipient
Recipient
Donor
1 DNA fragment from donor transferred to recipient
2 Recombination can occur in recipient
3 Rare bacteria that are recombinant for the m_y marker can be obtained from selective media
Recombinant type I
Recombinant type II
Result of crossovers at ① and ②
Result of crossovers at ① and ③
C Test for unselected marker
Recombinant type I
Recombinant type II
1 Are the recombinants for marker m_y also recombinant for marker m_z (the unselected marker)?
OR
Type I: Not recombinant for unselected marker
Type II: Recombinant for unselected marker

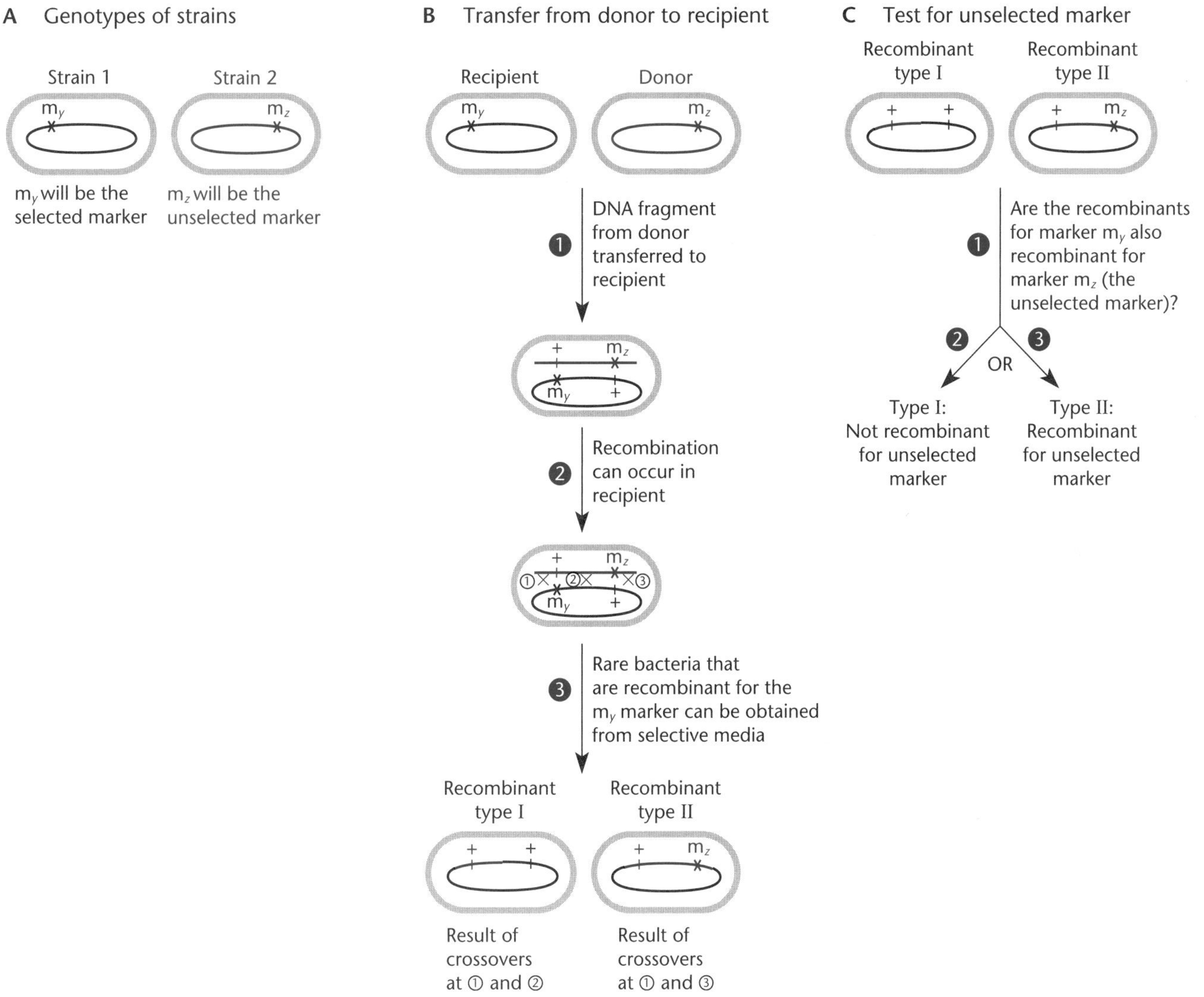

Figure 3.29 Selected versus unselected markers in a bacterial cross. Replacement by recombination of the sequence of the recipient DNA by the sequence of the donor DNA creates a recombinant type. Recipient bacteria, recombinant for one marker, the selected marker, are tested to see how many are also recombinant for a second marker. In the example, m_y is the selected marker and m_z is the unselected marker, chosen because it is easier to select recombinants for the m_y marker. **(A)** Genotypes of the strains. **(B)** A fragment of the donor DNA is transferred into the recipient cell. Recombination occurs between the incoming donor DNA and the recipient DNA, replacing regions of the recipient DNA with donor DNA. The recipients in which the m_y region of the donor has replaced that of the recipient are selected and purified on selective plates. **(C)** The recipients that have been selected for being recombinant for the m_y marker are tested to see if they are also recombinant for the m_z marker. For details, see the text. doi:10.1128/9781555816278.ch3.f3.29.

chromosomal DNA, so the transducing particles that transduce chromosomal genes also carry phage DNA and only certain regions of the chromosome are carried. In generalized transduction, the transducing particles have only chromosomal DNA, which can come from any region of the chromosome. Restricted transduction is generally less useful for mapping most markers, since only some regions of the chromosome are packaged in a phage. It is discussed in chapter 8. We discuss only mapping by generalized transduction here and refer to it simply as transduction.

An illustration of transduction creating a recombinant is shown in Figure 3.30. Very few of the phages pick up chromosomal DNA, and even fewer have picked

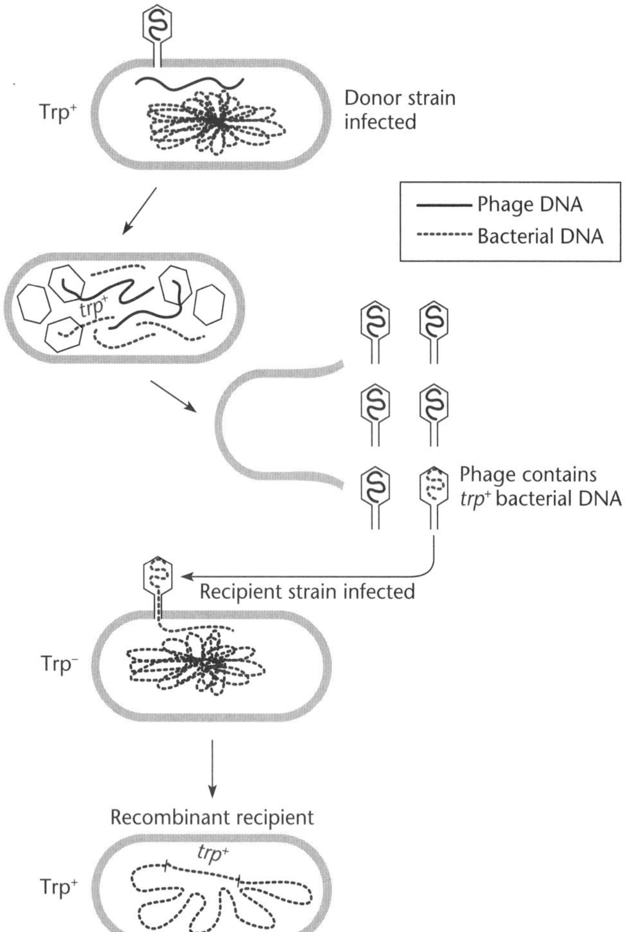

Figure 3.30 Example of generalized transduction. A phage infects a Trp+ bacterium, and in the course of packaging DNA into heads, the phage mistakenly packages some bacterial DNA containing the *trp* region instead of its own DNA into a head. In the next infection, this transducing phage injects the Trp+ bacterial DNA instead of phage DNA into the Trp− bacterium. If the incoming DNA recombines with the chromosome, a Trp+ recombinant transductant may arise. Only one strand of the DNA is shown.
doi:10.1128/9781555816278.ch3.f3.30

up the region of the genetic marker. Once the DNA is injected into the recipient cell, it still has to recombine with the corresponding region of the chromosome to make a recombinant for that marker, all of which makes recombinants for a particular marker very rare. However, even though they are very rare, recombinants can be selected with a powerful selection. Not all types of phages work well for generalized transduction, although sometimes, a nontransducer can be converted to a transducing phage by genetic manipulation. The properties of a good general transducing phage are discussed in chapter 7. Because of their continued usefulness, much effort is expended to find transducing phages for types of bacteria for which they are not available.

ANALYZING DATA OBTAINED BY TRANSFORMATION AND TRANSDUCTION

Mapping by transformation or transduction is based on whether the regions of markers can be carried in on the same piece of DNA, i.e., can be cotransformed or cotransduced. If a strain that has become recombinant for the selected marker sometimes also becomes recombinant for another unselected marker, the regions of the two markers are **cotransformable** or **cotransducible**, respectively. Both methods transfer only a relatively small piece of DNA and are inefficient enough that the donor regions of markers that are cotransformable or cotransducible have usually come in together on the same piece of DNA and therefore are very close to each other on the DNA relative to the length of the bacterial genome.

Not only does the appearance of cotransformants or cotransductants that are recombinant for two markers signify that the markers are close to each other in the DNA, but also, the higher the percentage of recombinants for the selected marker that are also recombinant for the unselected marker, the closer together the two markers are likely to be. The percentage of the total transformants or transductants selected for one marker that are also recombinant for the other marker is called the **cotransformation frequency** or **cotransduction frequency** of the two markers, respectively. In principle, this frequency between two markers should be a constant for any two markers and should be independent of which of the two markers is the selected marker and which is the unselected marker. A cross with the selected and unselected markers reversed is called a **reciprocal cross**. In the next section, we illustrate how mapping data from transductional crosses is interpreted by using an actual example. Similar reasoning would apply to transformational crosses.

MAPPING BY COTRANSDUCTION FREQUENCIES

In our example, phage P1 is used in *E. coli* for transduction to map a mutation to rifampin resistance in the gene for the β subunit of RNA polymerase (chapter 2) relative to a mutation in the *argH* gene, whose product is required to make the amino acid arginine, and a mutation in the *metA* gene required to make the amino acid methionine. The first step is to select for one of the markers and to determine if any of the other markers are cotransducible with the selected marker. For practical reasons, it is easier to use either the *argH* or the *metA* marker and not the *rif* marker as the selected marker. Rifampin resistance is difficult to select, because it takes many generations to be expressed. Any transductants to rifampin resistance will remain sensitive for many generations, because the rifampin-sensitive RNA polymerase molecules bind to promoters and block the resistant ones until the sensitive ones are diluted out by many cell divisions. For our example, we shall use a donor that is Arg^+ and Met^- and a recipient that is Rif^r and select the *argH* marker by plating on medium without arginine or rifampin but with methionine, so that recombinants can grow, whether or not they are Met^- or Met^+ or rifampin sensitive or rifampin resistant. We then test the Arg^+ transductants to see if any are also recombinant for the other markers.

Figure 3.31 illustrates the cotransduction of the *argH* and *rif* markers, and Table 3.5 gives some representative data. In this example, 33 (22 + 11) of 96, or 34%, of the Arg^+ transductants are also Rif^s and thus have the allele of the donor. Thus, the two markers are cotransducible, and the cotransduction frequency is about 34%. We can estimate how close together on the *E. coli* DNA the *argH* and *rif* markers would have to be in order to be cotransducible by the P1 phage, a phage that is commonly used with *E. coli*. The chromosome is 100 minutes (min) long, and the P1 phage head accounts for only about 2% of that length (see Table 7.3); therefore, the two markers must be less than 2 min apart. Translating this distance into base pairs of DNA, the *E. coli* chromosome is about 4.5×10^6 bp long, and the P1 phage head holds only $0.02 \times 4.5 \times 10^6$, or 90,000, bp. Thus, to be cotransducible, two markers in the DNA must be less than 90,000 bp, or 2 min, apart on the 100-min map shown below.

Figure 3.31 Cotransduction of two bacterial genetic markers. The regions of the *argH* and *rif* mutations are in close enough proximity that both regions can be carried on a piece of DNA fitting into a phage head (not shown to scale). After transduction, some of the Arg^+ transductants are also rifampin sensitive. The thick lines indicate phage genomes in progeny phage; the thin lines indicate bacterial DNA in transducing particles. See the text for details.
doi:10.1128/9781555816278.ch3.f3.31

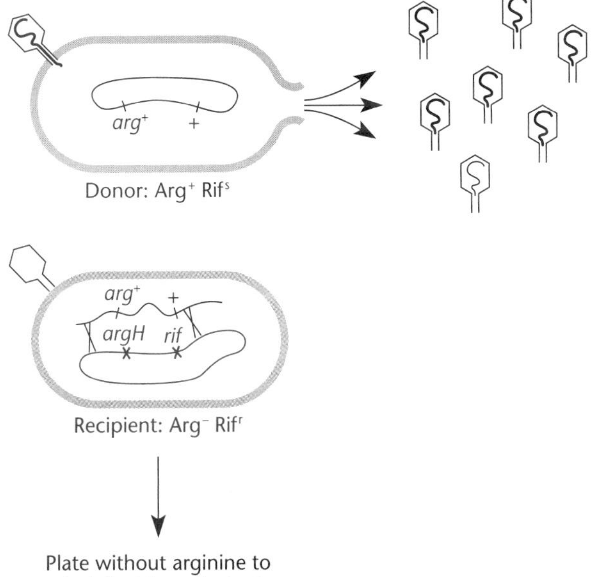

Table 3.5 Typical transductional data from a three-factor cross

Recombinant phenotype	No. of recombinants
Arg^+ Met^+ Rif^r	61
Arg^+ Met^+ Rif^s	22
Arg^+ Met^- Rif^r	2
Arg^+ Met^- Rif^s	11

From the data in Table 3.5, we can use the relative cotransduction frequencies to begin to order the *argH* and *rif* markers with respect to the *metA* marker. As mentioned above, the closer two markers are to each other, the more apt they are to be carried in the same phage head and the less likely there is to be a crossover between them. Therefore, the closer to each other they are, the higher the cotransduction frequency. In the example, 13 (2 plus 11) of the Arg^+ transductants are Met^-, so the cotransduction frequency of the *argH* and *metA* markers is ~14%, which is less than the 34% cotransduction frequency of the *arg* and *rif* markers. Therefore, the *arg* marker is closer to the *rif* marker than it is to the *met* marker, and the order seems to be *argH-rif-metA*. It also seems possible that the order is *rif-argH-metA*, with *argH* in the middle. However, both the *rif* marker and the *metA* marker seem to be on the same side of the *argH* marker, because most of the Met^- transductants are also Rif^s (11 of 13), as though the ones that received the *metA* region were more apt to also receive the *rif* region, rather than less apt, as they would be if they were on opposite sides.

Ordering Mutations by Three-Factor Crosses

A careful determination of cotransduction frequencies can reveal the order of markers in the DNA. However, **three-factor crosses** offer a less ambiguous way to determine marker order. This analysis is based on how many crossovers are required to make a certain recombinant type with a given order of markers. We have already pointed out that a single crossover between a short linear piece of the chromosome and the entire chromosome will break the chromosome and be lethal. Therefore, in a transductional cross with a linear DNA from the donor, a minimum of two crossovers are required to replace the chromosome sequence with the sequence on the incoming donor DNA and form a recombinant type. In general, in such a cross, any viable recombinant types must have originated from an even number of crossovers.

We can also use the data in Table 3.5 to illustrate the ordering of bacterial markers by a three-factor-cross analysis. There are four recombinant types possible in a cross of this type, listed in Table 3.5. With any particular order of the three markers, three of the four recombinant types listed require only two crossovers and the fourth requires four crossovers. The recombinant type that requires four crossovers should be rarer than the others. In Table 3.5, the rarest recombinant type is clearly Arg$^+$ Met$^-$ Rif^r, since only 2 of the 96 Arg$^+$ transductants were of this type. Figure 3.32 shows how many crossovers should be required to make this recombinant type if the order is *argH-metA-rif* or if the order is, as we suspect, *argH-rif-metA*. Only two crossovers are required to make this recombinant type if the order is *argH-metA-rif* (with the *metA* marker in the middle), while four crossovers are required if the order is *argH-rif-metA*. Thus, it seems clear that the order is *argH-rif-metA*, which is also consistent with the order we obtained based on cotransduction frequencies alone. Since resistance to rifampin is due to a mutation in the gene for the β subunit of RNA polymerase (see chapter 2), these data indicate that the gene for the β subunit of RNA polymerase lies between the *argH* and *metA* genes in the chromosome of *E. coli*.

Other Uses of Transformation and Transduction

There are many uses for transformation and transduction in bacterial genetics besides genetic mapping. They are all indispensible in molecular genetic studies in bacteria and explain the desirability of developing a transformation system and/or a transducing phage for the type of bacterium being studied.

STRAIN CONSTRUCTION

One of the major uses of transformation and transduction is in constructing isogenic bacterial strains, which differ only in a small region of the chromosome. Even different strains of the same species can differ at a number of genetic loci. To compare two strains in an experiment, it is necessary to compare isogenic strains; otherwise, there is no way to be certain that a phenotypic difference between two strains is due to a particular genetic difference. Therefore, meaningful experiments often require constructing isogenic strains. Transformation and transduction introduce only a small region of the chromosome, so any differences between the original recipient strain and a transductant must have been carried in on the same piece of DNA. If other genetic differences are contributing to phenotypic differences between the two strains, they must be very closely linked to the mutation being introduced.

Sometimes, we can use transformation or transduction to move mutations into a strain, even if the mutation has no easily selectable phenotype. For example, we might use a closely linked transposon carrying an antibiotic resistance gene to move such a mutation. For this purpose, collections of *E. coli* strains have been assembled that have transposon insertions around the genome that are all cotransducible with at least one other transposon insertion in the collection. Therefore, the site of any mutation is cotransducible with at least one of the transposon insertions. Using this collection to move a mutation into new genetic backgrounds is a three-step process. First, we use the collection as a donor for transduction, selecting the antibiotic resistance gene on the transposon and testing a number of the transductants for the mutation. A recipient strain that has lost the mutation due to cotransduction will probably have the transposon integrated close to the former site of the mutation. We can then use that strain as a donor to transduce the mutant strain carrying the mutation again. This time, if the transposon insertion is cotransducible with the mutation, some of the antibiotic-resistant transductants will have lost the mutation and some will still have it. We choose one of those that still have the mutation as a donor to easily transduce the mutation into as many other strains as we wish, even if the mutation has no easily selectable phenotype, merely by selecting for the antibiotic resistance on the transposon.

Figure 3.32 The number of crossovers required to make the rarest recombinant type, Arg$^+$ Rif^r Met$^-$, with two different orders of the three markers. Since order II, *argH-rif-metA*, requires four crossovers to make this recombinant type, this is probably the order of the three markers. See the text for details.
doi:10.1128/9781555816278.ch3.f3.32

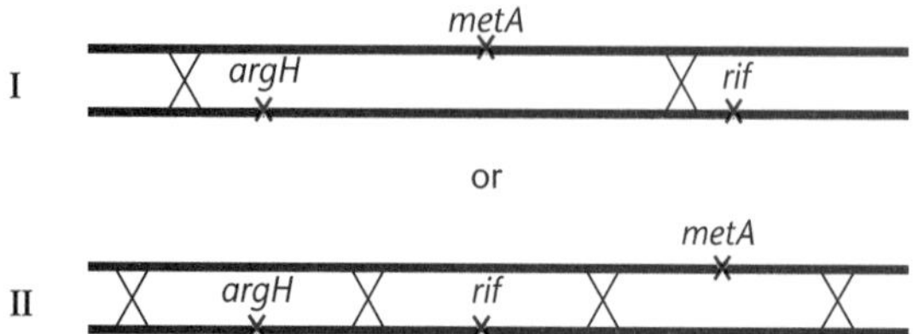

REVERSION VERSUS SUPPRESSION

Another important use of transformation or transduction is to distinguish reversion mutations from suppressor mutations. As mentioned above, suppressor mutations can be very revealing of the role of a gene product and can help identify other proteins and functions with which it might interact, but reversion and suppression can be difficult to distinguish based on their phenotypes alone. Both reversion mutations and suppressor mutations can allow a mutant strain to multiply under conditions where it could not multiply previously, but they can be distinguished by a simple genetic test. Until their molecular basis is known, it is best to refer to them as apparent revertants, since on the surface they appear to be revertants.

Figure 3.33 shows the generic recombination test for suppression, applied to apparent revertants of a *his*

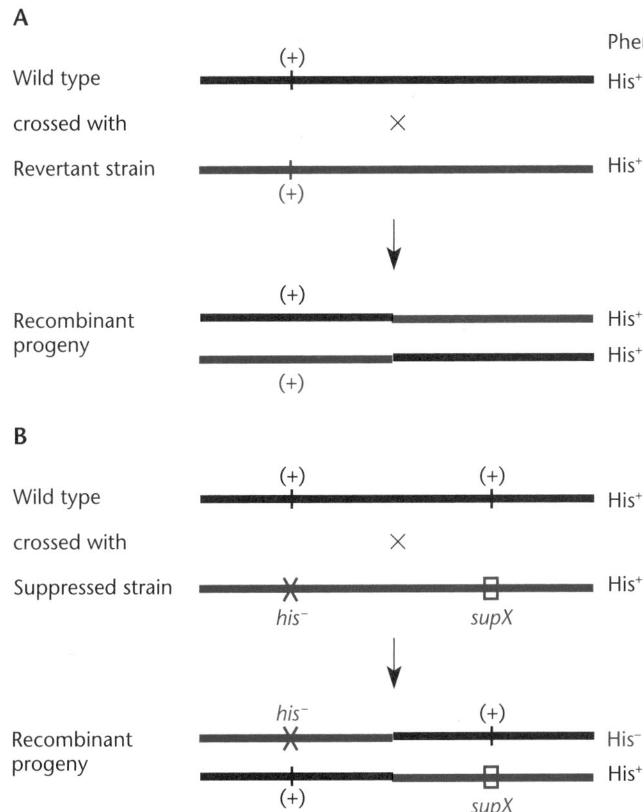

Figure 3.33 Generic test for reversion versus suppression. **(A)** The mutation had reverted, giving the His$^+$ phenotype. The blue plus sign shows the site of the reversion mutation. When the revertant strain is crossed with the wild type (in black), no His$^-$ recombinants appear in the progeny. **(B)** A suppressor mutation, *supX*, has suppressed the mutation, giving the His$^+$ phenotype. When the suppressed strain is crossed with the wild type (in black), some His$^-$ recombinants arise. The sites of the *his* mutation and suppressor mutation in the wild type are shown as pluses, and the site of the suppressor mutation is shown by the blue box.
doi:10.1128/9781555816278.ch3.f3.33

mutation obtained as shown in Figure 3.20. The apparent revertant is crossed with another strain that has never had the mutation and presumably does not have the possible suppressor mutation. Sometimes recombination occurs between the site of the original mutation and the site of the possible suppressor mutation. If the His$^+$ apparent revertants are true revertants, the original *his* mutation no longer exists, and all the recombinants are still His$^+$. However, if the original mutation has been suppressed, the suppressor mutation will sometimes be separated from the original mutation, which still exists, and some of the recombinants will be phenotypically His$^-$.

Any of the means of genetic exchange in bacteria can be used to distinguish reversion from suppression. However, transduction and transformation are particularly useful in this regard. Basically, we use the apparent revertant as a donor, selecting for a nearby cotransducible marker. If any of the transductants have the phenotype of the original mutation, the original mutation has been suppressed rather than having reverted. In the example shown in Figure 3.34, the *metA* mutation in the donor strain has apparently been reverted by being plated on medium without methionine, and the *argH* marker is being selected to test for suppression. If any of the Arg$^+$ transductants are Met$^-$ and cannot grow without methionine, the *metA* mutation has been suppressed rather than having reverted. For example, the *metA* mutation might have been a nonsense mutation and a tRNA mutation elsewhere in the chromosome is making the cell Met$^+$ by creating a nonsense suppressor.

Genetic Mapping by Hfr Crosses

The other way bacteria can exchange chromosomal DNA between strains is by conjugation. We present what is known about the molecular mechanism of conjugation in chapter 5, but here, we review how it has been used for genetic mapping. Conjugation is the transfer of DNA from one bacterium to another by a **plasmid**. Plasmids that can transfer themselves from one cell to another are called **self-transmissible plasmids**. Bacteria that have obtained DNA by conjugation are called **transconjugants**. Usually, conjugation only transfers the self-transmissible plasmid from one cell to another, but transmission of chromosomal DNA can occur if the plasmid has integrated into the chromosome or if chromosomal DNA has somehow been incorporated into the plasmid. A bacterial strain with a self-transmissible plasmid integrated into its chromosome is called an **Hfr strain**, for high-frequency recombination. As mentioned in the introductory chapter, this phenomenon was first detected in 1947 by Joshua Lederberg and Edward Tatum when they observed recombinant types after mixing some

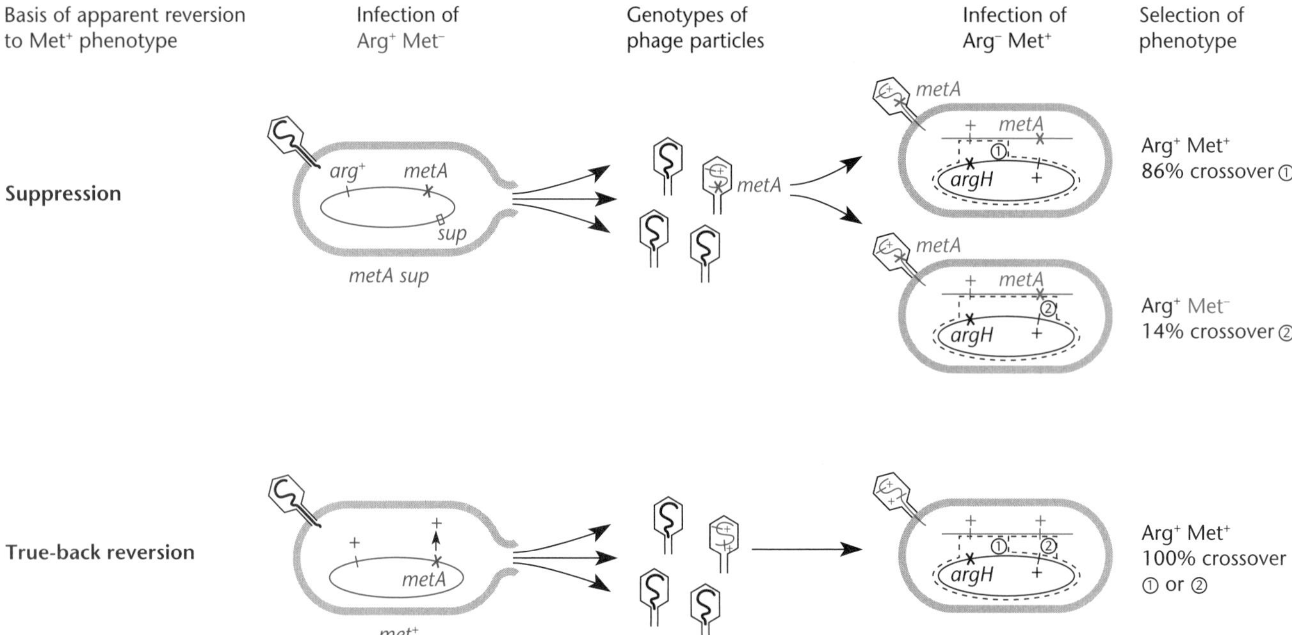

Figure 3.34 Using transduction to distinguish reversion from suppression. If the *metA* mutation has been suppressed, about 14% of the Arg^+ transductants will be Met^-. If the mutation has reverted, the *metA* mutation no longer exists, and none of the Arg^+ transductants will be Met^-. doi:10.1128/9781555816278.ch3.f3.34

strains of *E. coli* with other strains. We now know that one of the strains contained a self-transmissible plasmid called the F plasmid. In a few bacteria in the population, the F plasmid had integrated into the chromosome, and these bacteria were transferring chromosomal DNA into the other strain, leading to a high frequency of recombinant types. In retrospect, it was doubly fortuitous that some of the strains used by Lederberg and Tatum contained the F plasmid. Their experiment would not have succeeded if none of the strains they used had contained a self-transmissible plasmid. Also, any plasmid other than the F plasmid would not have worked as well, since, as discussed in chapter 5, the F plasmid they used is a mutant that is always ready to transfer.

Figure 3.35 illustrates the process by which chromosomal DNA is transferred in a mating between a donor Hfr strain containing an integrated F plasmid and a recipient strain that does not contain the F plasmid (F^-). On contact with a recipient cell, the DNA in the donor is nicked at a site in the integrated plasmid, and one strand is displaced into the recipient cell. Normally only plasmid DNA would be transferred, but because the plasmid is integrated into the chromosome, the chromosomal DNA is also transferred, beginning with chromosomal sequences at one side of the integrated plasmid and continuing around the chromosome in one direction, with the direction depending on the orientation of the integrated plasmid. In theory, if the transfer continued long enough (in *E. coli*, approximately 100 min at 37°C) it would eventually come full circle and transfer the entire 1-mm-long chromosome, finally reaching the chromosomal sequences on the other side of the plasmid and continuing into the plasmid sequences, ending at the *oriT* site in the plasmid where it started. However, transfer of the entire chromosome is nearly impossible, probably because the union between the cells is frequently broken and/or because the DNA is often broken during conjugation at nicks in the transferred strand. As a consequence, markers further from the origin of transfer are transferred less frequently. The ordered transfer of DNA markers and the decay in their frequency of transfer based on where they are in the genome allows us to use conjugation to map markers, as we discuss below.

Mapping by Hfr crosses requires a genetic map of the bacterium being mapped, and genetic maps have been constructed for only a few model bacteria, including *E. coli*, *Salmonella*, and *S. coelicolor.* Figure 3.36 shows a partial genetic map of *E. coli* and the location and direction of transfer of some of the integrated F plasmids in the Hfr strains that were used to construct the map. By convention, the direction of transfer for each Hfr strain is indicated by the point of the arrow, and the donor chromosome enters the recipient cell like an arrow starting at this point. Hfr crosses are seldom used anymore to map mutations on the chromosome, even if a genetic map is available for the bacterium being studied, since

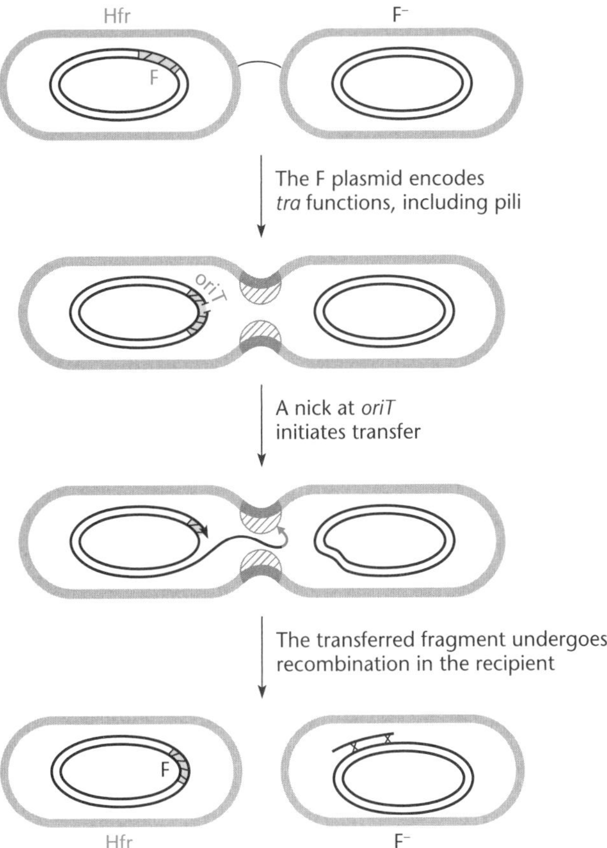

Figure 3.35 Transfer of chromosomal DNA by an integrated plasmid. Formation of mating pairs, nicking of the F *oriT* sequence, and transfer of the 5′ end of a single strand of DNA proceed as in transfer of the plasmid. Transfer of the covalently linked chromosomal DNA also occurs as long as the mating pair is stable. Because complete chromosome transfer rarely occurs, most recipient cells remain F⁻, even after long matings. Replication in the donor usually accompanies DNA transfer. Some replication of the transferred single strand may also occur. Once in the recipient cell, the transferred DNA may recombine with homologous sequences in the recipient chromosome.
doi:10.1128/9781555816278.ch3.f3.35

it is easier to use the DNA sequence of the entire bacterial genome and map mutations on the genome sequence by techniques such as marker rescue with clones or with PCR-generated fragments. Nevertheless, it is worth summarizing the procedures that have been used to construct genetic maps using Hfr crosses.

MAPPING BY GRADIENT OF TRANSFER

Mapping by Hfr crosses is like mapping by transformation and transduction in that only part of the donor strain is transferred into the recipient and recombination can occur only in the recipient strain. Also, one marker is selected, and recombinants for that marker are tested to see if they are also recombinant for the other markers. However, we must select a marker that is transferred before the other markers, and the data are interpreted differently. Rather than depending on the fact that only a small piece of DNA is transferred, genetic mapping by Hfr crosses depends on the fact that the DNA is transferred from the *oriT* on the integrated plasmid and the transfer is periodically disrupted. Therefore, the further a marker is from the integrated plasmid in the direction of transfer, the less frequently it is transferred, creating a **gradient of transfer**. Furthermore, there is an equal probability that each surviving mating pair will be disrupted in the next time interval. This creates an exponential "decay" in the transfer of markers, based on how far they are from the origin of transfer, so if we plot the frequencies of recombination of known markers versus their distances from the site of a selected marker in the direction of transfer on semilog paper, we get a more or less straight standard curve or line. If we then put the frequencies of recombinant types for an unmapped marker on this standard curve and read down to the distance from the selected marker, we will have determined the distance of the unknown marker from the selected point on the genome.

To show how this works in practice, Table 3.6 lists some data for a typical Hfr cross, and Figures 3.37 and 3.38 illustrate how these data are analyzed. In this example, we used the PK191 Hfr strain shown in the genetic map in Figure 3.36, which transfers counterclockwise from about 42 minutes to map the *rif* genetic marker, which, as mentioned above, confers resistance to the antibiotic rifampin (Rif^r). As our known markers, we use the *hisG*, *argH*, and *trpA* mutations, all of which are in the recipient, as is the unknown *rif* mutation. In an Hfr cross, the Hfr donor strain must be counterselected by plating the cross on selective medium on which the donor cannot grow; otherwise, the donor would grow on the plates and obscure any recombinants. We use the fact that the donor is Pro⁻ and requires proline for growth, due to the *lac-pro* deletion shown in Figure 3.37, to counterselect it. The partial *E. coli* genetic map in Figure 3.36 shows the positions on the chromosome of all of the markers we are using except the *rif* marker, which we are trying to map. We first select the *his* marker, since that is transferred first by this Hfr strain, followed by the *argH* marker

Table 3.6 Typical results of an Hfr cross

Selected marker	% Recombinant for unselected marker:			
	hisG	*trpA*	*argH*	*rif*
hisG	100	1	7	6
trpA	33	100	29	31
argH	28	12	100	89

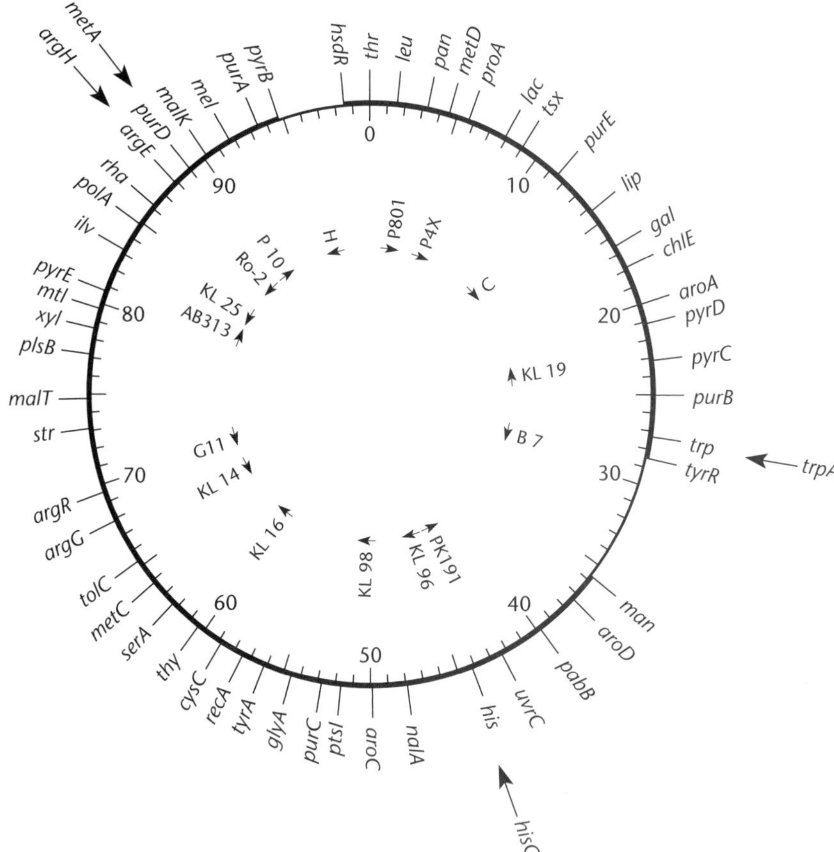

Figure 3.36 Partial genetic linkage map of *E. coli* showing the positions (large arrows) of the known markers used for the Hfr gradient of transfer in Figure 3.38. The small arrows indicate the positions of integration and directions of transfer of the F plasmid in some Hfr strains, including PK191 (located near the position of *hisG* at 44 min and transferring counterclockwise). By convention, the chromosomal DNA is shown transferred like an arrow being shot into the recipient cell, beginning from the tip of the arrow.
doi:10.1128/9781555816278.ch3.f3.36

and then the *trp* marker, as shown in Figure 3.37. We then test recombinants for the *hisG* marker to see what percentage are also recombinant for the *argH* marker (Arg$^+$) and what percentage are also recombinant for the *trpA* marker (Trp$^+$). If we take the frequency of recombinants for the *his* marker to be 100% and plot the frequency with which these recombinants have also become recombinant for each of the other markers versus their distances on the genetic map from the *his* marker, we get a more or less straight line, as illustrated in Figure 3.38. The frequency of recombinants for the *rif* marker (i.e., the frequency of His$^+$ recombinants that are also Rif^s) is then plotted on the standard curve, reading down to its position on the genetic map. As expected, it maps close to the *argH* marker, since we had determined earlier that it is cotransducible with *argH*.

We can also see from the data in Table 3.6 what would have happened if we selected one of the other markers, for example, the *trpA* marker that came in after the others in this particular Hfr cross. Each bacterium that became recombinant for the *trpA* marker must have also received the regions of the other markers, since they were transferred before the *trpA* region (Figure 3.37). Now, approximately 30% of the other markers are recombinant, independent of where they are on the genome, which gives us little specific information about their map positions. Apparently, 30% is the frequency with which a piece of DNA will recombine with the chromosome once it enters the cell. If we selected the *argH* marker, a very high percentage (89%) are recombinant for the *rif* marker, because they are close to each other or closely genetically **linked.** The distances on the DNA between most of these markers is so great that many crossovers occur between them. Only if they are very close, as is the case with the *argH* and *rif* markers, are crossovers between them infrequent enough that we can see any genetic linkage.

Isolation of Tandem Duplications of the *his* Operon in *Salmonella*

We finish this section with an example of genetic analysis in bacteria, the isolation and analysis of the frequency and length of tandem-duplication mutations in *Salmonella*. This analysis illustrates both the properties of the different types of mutations and how genetic data

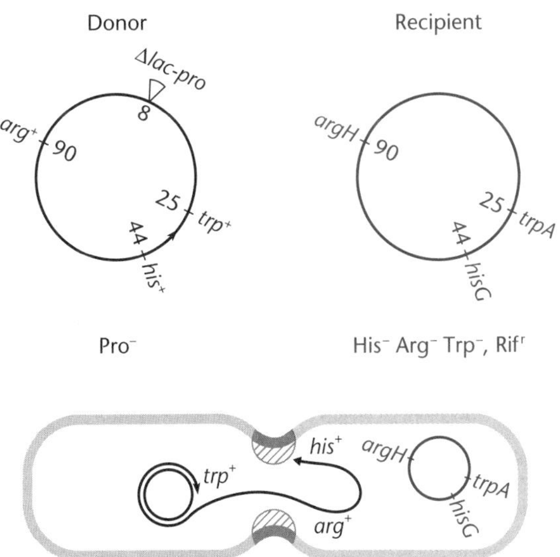

Figure 3.37 Mapping by Hfr crosses. The phenotypes and positions of the markers used for mapping in the genetic maps of the donor and recipient bacteria are shown. The chromosome is transferred from the donor to the recipient, starting at the position of the integrated self-transmissible plasmid in PK191 (arrowhead). The supplements to the plating media used to select and test for recombinants of each of the markers are also shown.
doi:10.1128/9781555816278.ch3.f3.37

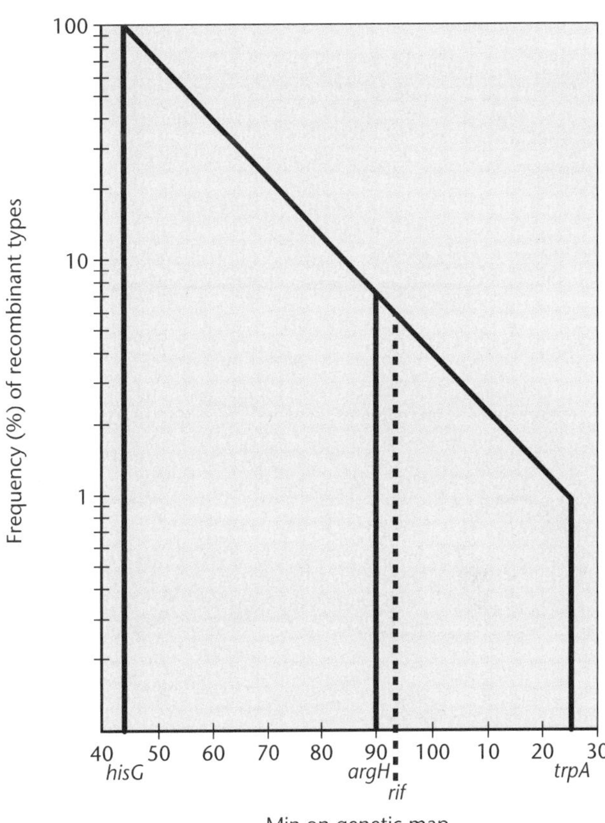

Figure 3.38 Mapping by gradient of transfer during an Hfr cross. The ordinate shows the frequency of each unselected marker, with *his* as the selected marker. The abscissa is the distance in minutes of each unselected marker from the *his* marker from the map in Figure 3.36. The dashed line shows an estimate of the position of the *rif* marker based on plotting the percentage of Rif^s recombinants from the data in Table 3.6 and reading down to the abscissa.
doi:10.1128/9781555816278.ch3.f3.38

are interpreted, and it is hard to imagine how it could be done any other way than by classical genetic techniques, even today.

As discussed above, tandem-duplication mutations can occur by recombination between directly repeated sequences, causing the duplication of the DNA between the repeated sequences. Such ectopic recombination is fairly rare, because repeated sequences are not very common in bacteria. However, once they form, tandem-duplication mutations are usually very unstable, because recombination anywhere within the duplicated segments can destroy the duplication. Also, as mentioned, most long tandem-duplication mutations do not cause easily detectable phenotypes, because they do not inactivate any genes. Even the genes in which the mistaken recombination occurred to create the duplication exist in a functional copy at the other end of the duplication. Therefore, special methods must be used to select bacteria with duplication mutations.

In our chosen example, transduction was used to select tandem duplications of the *his* region of *Salmonella enterica* serovar Typhimurium (see Anderson et al., Suggested Reading). Their selection depends on the properties of two deletion mutations in the *his* region, Δ*his2236* and Δ*his2527* (Figure 3.39A). These deletion mutations complement each other, because one ends in *hisC* and the other ends in *hisB*, so they inactivate different genes. However, because the endpoints of the two deletion mutations are very close to each other, crossovers between them occur very infrequently.

Transduction was used to detect the cells that had a preexisting *his* duplication. The transducing phage P22 was propagated on a strain with one of the deletion mutations, in the example, Δ*his2236*, and used to transduce a strain with the other deletion mutation, Δ*his2527*. The His$^+$ transductants were then selected by plating on minimal plates without histidine. A few His$^+$ transductants arose, even though there should be very little recombination between the deletions. Moreover, many of the His$^+$ transductants that arose were very

A

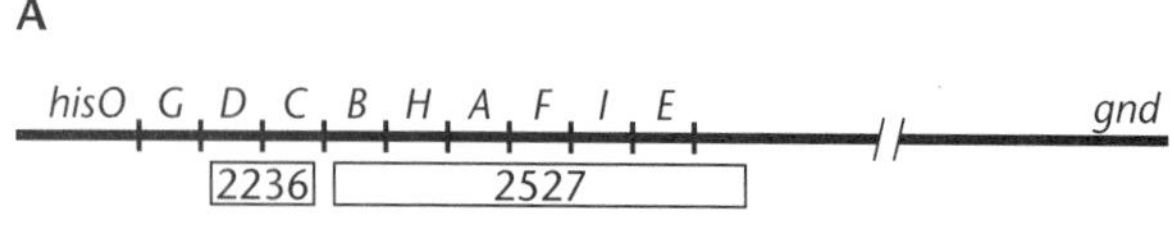

B

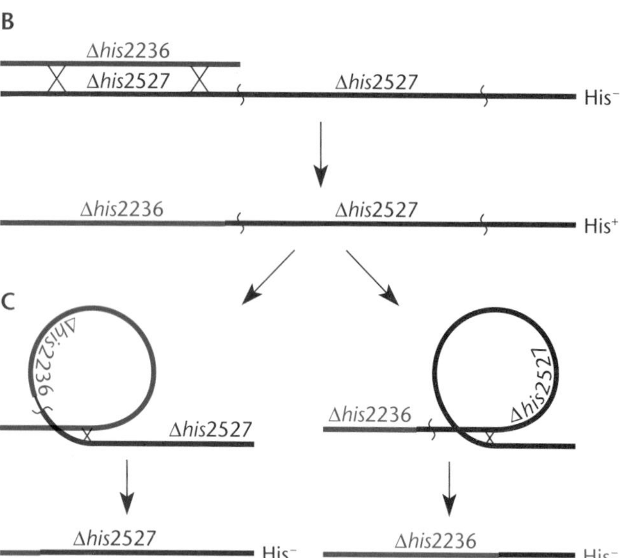

Figure 3.39 Mechanism for generation and destruction of *his* duplications. **(A)** Map of the *his* operon showing the positions of the two deletion mutations used for the analysis (boxed). The two deletion endpoints are very close together, so little recombination can occur between them, but they can complement each other because one ends in *hisB* and the other ends in *hisC*. **(B)** A duplication has occurred in some of the recipient cells, duplicating the *his* region with the Δ*his2527* deletion. One of the two copies of the duplication is transduced with the *his* region from the donor containing the Δ*his2236* deletion, making the cell His$^+$, since the two deletions complement each other. **(C)** The duplication can be destroyed by recombination anywhere within the duplicated segments, looping out one of the two duplicated segments and giving rise to His$^-$ haploid segregants with one or the other of the two deletions.
doi:10.1128/9781555816278.ch3.f3.39

unstable, spontaneously giving off His$^-$ segregants at a high frequency detected when they were propagated in the presence of histidine and without selective pressure. Also, some of the His$^+$ transductants were slimy (mucoid) in appearance, but the His$^-$ segregants had lost this mucoidy. On the basis of these observations, the investigators concluded that the unstable His$^+$ transductants had tandem duplications of the *his* region (Figure 3.39B) in which one copy has the Δ*his2236* deletion whereas the other has the Δ*his2527* deletion. Some of the recipient bacteria must have had tandem duplications in which the *his* region was duplicated and one of the two copies was transduced, so it now had the other deletion of the donor, as shown in the figure. Each of the copies now had a different deletion, and the two deletion mutations were complementing each other, making the transductant His$^+$. This type of His$^+$ transductant, although rare, is more frequent than normal His$^+$ transductants produced by recombination in the small region between the two deletions (Figure 3.39A).

The investigators could confirm that the instability was due to recombination by crossing in a mutation in the *recA* gene whose product is required for recombination and showing that this made the His$^+$ transductants stable. They could also get an idea of the actual structure of some of the duplications by determining the frequency of segregation of each of the haploid types with one or the other deletion. Which of the two deletions predominates in the haploid segregants depends on the relative distances upstream and downstream of the *his* region where the initial recombination occurred to create the duplication and which of the two copies had been transduced, the upstream or the downstream copy. The mucoidy (see above) of some of the His$^+$ transductants could also be explained if the duplicated region in some of the His$^+$ transductants contained another gene that was now also duplicated and so made twice as much of its gene product, which made the colonies appear mucoid. However, in the haploid segregants, there is only one copy of this "mucoidy" gene, and the colonies appear normal.

Lengths of Tandem Duplications

The investigators also used genetic experiments to determine the length of the duplicated segment in some of the strains. In particular, they wanted to know if the duplicated regions in some of the strains ever extended as far as the *metG* gene, about 2 min (~100,000 bp) away from the *his* region in the *E. coli* chromosome (Figure 3.40). To test this, they selected *his* duplications in a recipient strain that also had a *metG* mutation. The investigators then transduced the strains a second time with phage propagated on a *metG*$^+$ donor, selecting for Met$^+$ transductants. The Met$^+$ transductants were then allowed to segregate into haploids by growing them with histidine and methionine in the medium. Isolated His$^-$ segregants were then tested to determine if any were also Met$^-$. The reasoning was that if the *metG* gene is included in the duplicated region in a particular duplication, there should be two copies of the *metG* gene, and only one of them would have been transduced to *metG*$^+$, as shown in the figure. When the duplication segregates into the haploid segregants, some of them lose the copy with the *metG*$^+$ allele, and they become Met$^-$. The results of this work revealed that many duplications of the *his* operon did include the *metG* gene. In fact, some duplications extended much farther, even as far as the *aroD* gene, which is 10 min (or 10% of the entire genome) away from *his*. Some tandem-duplication mutations are very long and include many hundreds of genes.

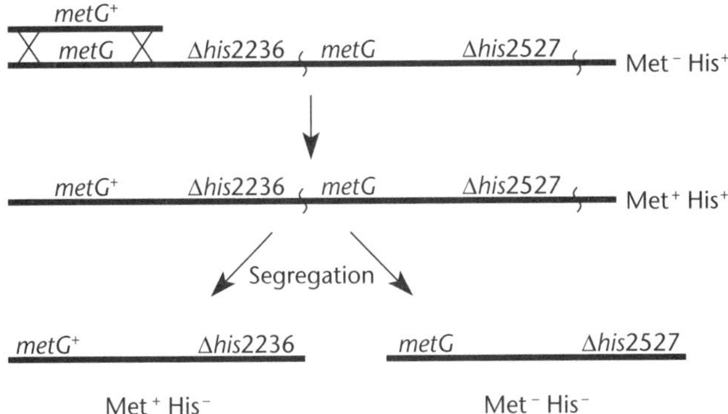

Figure 3.40 Determining if the duplicated segment in a *his* duplication can extend as far as the *metG* gene 10 min away. A duplication is made as in Figure 3.39 but in a recipient strain that also has a *metG* mutation. After selecting for His+ transductants on plates containing methionine, a strain with a *his* duplication is transduced with phage grown on Met+ bacteria, selecting for Met+ transductants. If the *metG* gene is also duplicated, only one of the two copies of the *metG* gene will be transduced, and when the strain spontaneously gives rise to His− haploid segregants, some of them will be also Met−. doi:10.1128/9781555816278.ch3.f3.40

Frequency of Spontaneous Duplications

Lastly, the investigators attempted to estimate the frequency of spontaneous tandem-duplication mutations in a growing population of bacteria. They did this by comparing transduction frequencies when the donor and recipient bacteria had the different deletion mutations mentioned above with the normal transduction frequencies when only the recipient had one of the deletions. In the first case, most of the His+ transductants must have had a tandem duplication of the *his* region, whereas in the second case, most of the His+ transductants had only one copy of the *his* region. From this comparison, the investigators estimated that duplications occur as frequently as once every 10^4 times a cell divides, which is hundreds of times greater than normal mutation frequencies. Apparently, spontaneous duplications in bacteria occur quite often during cell multiplication and can occur over very large regions of the chromosome. However, because they cause few phenotypic changes, most of these large duplications have little effect on the organism. Nevertheless, they are probably important in evolution, as mentioned above.

SUMMARY

1. A mutation is any heritable change in the sequence of the DNA of an organism. A **leaky mutation** in a coding part of the DNA leaves some residual activity of the gene product; a null mutation totally inactivates the product of the gene. A gain-of-function mutation gives the gene product an activity it did not have previously or leads to the expression of a silent gene. The organism with a mutation is called a mutant, and that organism's mutant phenotype includes all of the characteristics of the mutant organism that are different from the wild-type, or normal, organism.

2. The mutation rate is the chance of occurrence of a mutation to a particular phenotype each time a cell divides. The mutation rate offers clues to the molecular basis of the phenotype, e.g., whether it can be caused by a null mutation or only by a gain-of-function mutation. Quantitative determination of mutation rates is often difficult, because it is necessary to determine how many mutations have occurred in a growing culture, which is usually not equal to the number of mutants in the culture. One mutation can give rise to a large number of mutants, depending on when it occurred in the growth of the culture. Also, most ways of determining mutation rates are influenced by phenotypic lag and/or the relative growth rates of mutants.

3. One important conclusion from population genetics is that the fraction of mutants of all types increases as the population grows. This causes practical problems in genetics, which can be partially overcome by storing organisms without growth or by periodically colony purifying one or very few organisms before mutations have had time to accumulate.

4. The type of mutation causing a phenotype can often be ascertained from the properties of the mutation. Base pair changes revert and are often leaky. Frameshift mutations, due to the addition or deletion of one or very few base pairs, also revert but are seldom leaky. Longer deletion mutations do not revert and are seldom leaky. They also often inactivate more than one gene simultaneously and can fuse one gene to another. Tandem-duplication mutations revert at a high frequency and often have no observable phenotypes, except that they can fuse one gene to another or cause phenotypes due to the overproduction of the product of a gene included in the duplication. Inversion mutations also often have no observable phenotype and often revert. Large insertion mutations seldom revert, are usually not leaky, and are usually due to the transposition of an insertion element or other type of transposon.

5. Deletions, inversions, and tandem-duplication mutations can be caused by recombination between repeated sequences in the DNA called ectopic recombination. Deletion and tandem-duplication mutations are caused by recombination between directly repeated sequences in the DNA, and inversions are caused by recombination between inverted repeats. Deletions and tandem duplications arise as reciprocal recombination products between directly repeated sequences in different DNA molecules. Sometimes, deletions and inversions are caused by site-specific

(continued)

SUMMARY (continued)

recombinases. If so, they are not referred to as mutations, but as programmed deletion or inversion events.

6. If a secondary mutation restores the original phenotype by returning the DNA to its original sequence, the original mutation has reverted. If a secondary mutation somewhere else in the DNA restores function, the original mutation has been suppressed. Suppressors can be either intragenic or intergenic, depending on whether they occur in the same gene or in a different gene from the original mutation, respectively. An example of an intragenic suppressor is a frameshift mutation within the same gene that restores the reading frame shifted by another frameshift mutation. An example of an intergenic suppressor is a mutation in a tRNA gene that changes the tRNA so that it recognizes one or more of the nonsense codons, allowing translation of nonsense mutations in other genes. Suppressors can also be allele specific if they suppress only a particular type of mutation, for example, an amber suppressor, which suppresses only UAG mutations, independent of where they occur.

7. To isolate a mutant means to separate the mutant strain from the many other members of the population that are normal for the phenotype. Bacteria have advantages in genetic analysis, because of the ease of isolating mutants. They multiply asexually, they are usually haploid, and large numbers can multiply on a single petri plate.

8. Mutations can be either spontaneous or induced. Spontaneous mutations often occur as mistakes while the DNA is replicating, while induced mutations are deliberately caused by using mutagenic chemicals, irradiation, or transposons. Induction of mutations with mutagens has the advantage that mutations are more frequent and that a specific mutagen often causes only a specific type of mutation. Region-specific mutagenesis can be used to mutate a particular region of the chromosome if the sequence is known. Recombination can be used to replace chromosomal sequences with sequences manipulated in the test tube. In the types of bacteria for which they have been developed, recombineering systems, which use the recombination enzymes of phages to increase recombination frequencies, can be used for region-specific mutagenesis.

9. In saturation genetics, attempts are made to assemble a collection of mutants with mutations in all the genes whose products are involved in a function. To ensure that mutations are as representative as possible, different mutagens should be used, and the isolation of siblings should be avoided by mutagenizing separate cultures and using only one mutant from each culture.

10. Screening for a mutant means devising a way to distinguish the mutant from the normal, or wild, type. Selecting a mutant means devising conditions under which either only the mutant or the wild type can multiply. Most types of mutants cannot be selected. However, enrichment can sometimes be used to increase the frequency of the mutant in a population by killing the wild-type multiplying cells under the nonpermissive conditions, provided the arrest of growth of the mutant is reversible.

11. The three means of genetic exchange in bacteria are transformation (electroporation), transduction, and conjugation. In transformation and electroporation, only DNA is taken up by cells. In transduction, bacterial DNA is encapsulated in a phage head during multiplication of the phage and then injected into another bacterium upon reinfection. In conjugation, a self-transmissible plasmid transfers DNA from one bacterium to another. Sometimes, bacterial DNA can also be transferred by conjugation if the plasmid has taken up the bacterial DNA or if the plasmid has integrated into the chromosome (an Hfr strain). Gene exchange makes it possible to do genetic mapping and other types of genetic analysis with bacteria. Because only a small piece of DNA is transferred by transformation and transduction, these methods are useful for constructing isogenic strains that differ at only one locus and distinguishing suppressors from revertants. Hfr mapping has been used to locate a mutation on the entire chromosome where a genetic map is available but has largely been replaced by other methods.

12. In recombination, two DNAs are broken and rejoined in new combinations. Generalized, or homologous, recombination occurs only between two DNAs with the same sequence. Progeny organisms that are different genetically from their parents as a result of recombination are called recombinant types, while progeny that are the same as one of the parents are parental types. Marker rescue recombination can be used to tell if a mutation lies within a particular region of the chromosome or whether a clone includes the region of a mutation. Recombination frequencies can also be used to map genetic markers on the chromosome using the various means of genetic exchange. Recombinants for one of the markers (the selected marker) are tested for recombination of the other markers (the unselected markers). After mating, a marker in the recipient that is replaced by the corresponding sequence of the donor is said to be recombinant for that marker. The frequency of recombination for the unselected markers can give us information about where the markers are on the chromosome.

13. In complementation, a gene product encoded by one copy of a gene in the cell can substitute for the product of another copy of the same gene that has been inactivated by a mutation. Complementation tests reveal whether different mutations inactivate different gene products and how many separate gene products contribute to a phenotype. It can also be used to clone genes, to identify regions of genes encoding separate domains of a protein, and to determine if a mutation is dominant or recessive or is *cis* acting or *trans* acting.

QUESTIONS FOR THOUGHT

1. A single-inversion mutation greatly alters a genetic map or the order of genes in the DNA, yet the genetic maps of *S. enterica* serovar Typhimurium and *E. coli* are similar. We now know that some sequences, such as KOPS sequences and chi sequences, are polar and so do not function if they are inverted. Can you think of any other reasons why genetic maps might be preserved in evolution?

2. Do you think that tandem-duplication mutations play an important role in evolution? If so, why are they not always destroyed by recombination as quickly as they form? Give some possible mechanisms that might preserve a duplication mutation.

3. Nonsense suppressors are possible in lower organisms, but not in higher organisms. One possible explanation is that readthrough of nonsense codons at the end of coding sequences might be too much of a burden for higher organisms. Can you think of any other possible explanations?

4. Luria and Delbrück showed that mutations to T1 phage resistance in *E. coli* were not directed and occurred randomly while the bacteria were multiplying. However, this does not mean that all types of mutations are random. Can you propose a mechanism by which "directed" mutations might occur?

PROBLEMS

1. In a collection of exponentially growing cultures, all started from a few wild-type bacteria, which cultures are most likely to have had the earliest mutation to a particular mutant phenotype: those with a few mutant bacteria or those with many mutant bacteria?

2. Mutations to which phenotype would you expect to have the higher mutation rate, rifampin resistance or Arg^- (arginine auxotrophy)?

3. Luria and Delbrück grew 100 cultures of 1 ml each to 2×10^9 bacteria per ml. They then determined the number of bacteria resistant to T1 phage in each culture: 20 cultures had no resistant bacteria, 35 had one resistant mutant, 20 had two resistant mutants, and 25 had three or more resistant mutants. Calculate the mutation rate to T1 resistance by using the Poisson distribution.

4. Newcombe spread an equal number of bacteria on each of four plates. After 4 h of incubation, he sprayed plate 1 with T1 and put it back in the incubator. At the same time, he washed the bacteria off plate 2, diluted them 10^7-fold, and replated them to determine the total number of bacteria. After a further 2 h of incubation, he sprayed plate 3, washed the bacteria off plate 4, and diluted them 10^8-fold before replating them. The next morning, he counted the colonies on each plate. He found that plate 1 had 10 colonies, plate 2 had 20 colonies, plate 3 had 120 colonies, and plate 4 had 22 colonies. Calculate the rate of mutation to T1 resistance.

5. You have isolated Arg^- auxotrophs of *Klebsiella pneumoniae*.

   a. If you plate a few cells with a mutation, *arg-1*, on plates without arginine, they multiply to make tiny colonies. If you plate 10^8 cells, you get some large, rapidly growing colonies. What kind of mutation is *arg-1* likely to be?

   b. If you plate another mutant with a different mutation, *arg-2*, you get no growth on plates without arginine. Even if you plate large numbers of mutant bacteria ($>10^8$), you get no colonies. What type of mutation is *arg-2* likely to be?

   c. What are some other possible explanations for the results in paragraphs a and b?

6. Why is it necessary to isolate mutants from different cultures to be certain of getting independent mutations, i.e., ones that are not siblings?

7. Design a positive selection for each of the following types of mutants. Discuss what kind of selective plates and/or conditions you would use.

   a. Mutants resistant to the antibiotic coumermycin.

   b. Revertants of a *trp* mutation that no longer require tryptophan for growth.

   c. Double mutants with a suppressor mutation that relieves the temperature sensitivity due to a mutation in *dnaA*.

   d. Mutants with a suppressor in *araA* that relieves the sensitivity to arabinose due to a mutation in *araD*.

   e. Mutants with a mutation in *suA* (the gene for the transcription terminator protein Rho) that relieves the polarity of a *hisC* mutation on the *hisB* gene. Hint: make a partial diploid to perform complementation tests.

8. Design an enrichment procedure for reversible, temperature-sensitive mutants with mutations in genes whose products are required for cell growth.

9. Are nonsense suppressor mutations dominant or recessive? Why?

10. You have a strain of *Pseudomonas* that requires arginine, histidine, and serine as a result of a mutation in each of the genes encoding enzymes to make these amino acids. When you plate large amounts of this strain on medium lacking all three of these amino acids, a few colonies arise that are Arg^+ His^+ Ser^+. These mutations are almost as frequent as mutations that revert each of the mutations

separately. What kind of mutation do you think caused the original auxotrophy in each of the genes? What kind of mutation caused the apparent reversion of all three of these mutations? How would you test your hypothesis?

11. You wish to use transduction to determine if the order of three *E. coli* markers is *metB1–argH5–rif-8* or *argH5–metB1–rif-8*, so you do a three-factor cross. The donor for the transduction has the *rif-8* mutation, and the recipient has the *metB1* and *argH5* mutations. You select the *argH5* marker by plating on minimal plates plus methionine, purify 100 of the Arg$^+$ transductants, and test for the other markers. You find that 17 are Arg$^+$ Met$^+$ Rif^r, 20 are Arg$^+$ Met$^+$ Rif^s, 60 are Arg$^+$ Met$^($ Rif^s, and 3 are Arg$^+$ Met$^-$ Rif^r. What is the cotransduction frequency of the *argH* and *metB* markers? What is the cotransduction frequency of the *argH* and *rif-8* markers? What is the order of the three markers deduced from the three-factor cross data? Are the results consistent?

12. In the test for suppression versus reversion in Figure 3.34, what would you expect in the two cases if you used the Met$^+$ apparent revertant as a donor and a strain with the *metA* and *argH* mutations as a recipient, selecting the *metA* marker? Would you expect any Met$^+$ transductants if the mutation had been suppressed? If you did get Met$^+$ transductants, what percentage of them would you expect to be Arg$^-$? What percentage would be Arg$^+$?

13. You have isolated a *hemA* mutant of *E. coli* that requires δ-aminolevulinic acid for growth, so you can only use negative selection to identify strains that have the mutation. You wish to move this mutation into other genetic backgrounds. You obtain from the Yale Stock Collection a strain of *E. coli* that has a Tn*10* transposon conferring resistance to tetracycline inserted only 0.5 min away from the *hemA* gene. Explain the steps involved in using transduction and positive selection to move your *hemA* mutation into other *E. coli* strains.

14. You wish to use your *hemA* mutant to clone the *hemA* gene. You make a library of *E. coli* DNA and transform the clones of the library separately into a strain with the *hemA* mutation. One of the transformant colonies you test contains a few bacteria that no longer require δ-aminolevulinic acid, but most of the bacteria in the colony still require it. Is the HemA$^+$ phenotype in these few transformants due to marker rescue recombination or complementation?

15. Outline how you would replace the *argH* gene, responsible for making an enzyme for arginine synthesis, with the corresponding gene into which you have inserted a gene for chloramphenicol resistance (Cm^r) in *E. coli*. What would you expect the phenotypes of your mutation to be?

16. An *E. coli* Hfr strain that is His$^+$ Trp$^+$ but has an *argH* mutation is crossed with a recipient that is Arg$^+$ but has *hisG* and *trpA* mutations, and the cross is plated on minimal plates containing histidine and arginine but no tryptophan. Which is the selected marker, and which are the unselected markers?

17. In the Hfr cross described above, we want to map the site of insertion on a Tn*10* transposon carrying tetracycline resistance (Tet^r) in the recipient strain. If we select the *hisG* marker and test 100 transconjugants, we find that 7 are Arg$^-$, 1 is Trp$^+$, and 4 are Tet^s. Approximately where is the Tn*10* transposon inserted in the *E. coli* chromosome?

SUGGESTED READING

Anderson, R. P., C. G. Miller, and J. R. Roth. 1976. Tandem duplications of the histidine operon observed following generalized transduction in *Salmonella typhimurium. J. Mol. Biol.* **105:**201–218.

Hill, C. W., J. Foulds, L. Soll, and P. Berg. 1969. Instability of a nonsense suppressor resulting from a duplication of genes. *J. Mol. Biol.* **39:**563–581.

Jones, M. E., S. M. Thomas, and K. Clarke. 1999. The application of linear algebra to the analysis of mutation rates. *J. Theor. Biol.* **199:**11–23.

Lea, D. E., and C. A. Coulson. 1949. The distribution of numbers of mutants in bacterial populations. *J. Genet.* **49:**264–285.

Lederberg, J., and E. M. Lederberg. 1952. Replica plating and the indirect selection of bacterial mutants. *J. Bacteriol.* **63:**399.

Lederberg, J., and E. L. Tatum. 1946. Gene recombination in *E. coli. Nature* (London) **158:**558.

Low, K. B. 1996. Genetic mapping, p. 2511–2517. *In* F. C. Neidhardt, R. Curtiss III, J. L. Ingraham, E. C. C. Lin, K. B. Low, B. Magasanik, W. S. Reznikoff, M. Riley, M. Schaechter, and H. E. Umbarger (ed.), Escherichia coli *and* Salmonella: *Cellular and Molecular Biology*, 2nd ed. ASM Press, Washington, DC.

Luria, S., and M. Delbrück. 1943. Mutations of bacteria from virus sensitivity to virus resistance. *Genetics* **28:**491–511.

Maloy, S., K. T. Hughes, and J. Casadesus (ed.). 2010. *The Lure of Bacterial Genetics: A Tribute to John Roth.* ASM Press, Washington, DC.

Masters, M. 1996. Generalized transduction, p. 2421–2441. *In* F. C. Neidhardt, R. Curtiss III, J. L. Ingraham, E. C. C. Lin, K. B. Low, B. Magasanik, W. S. Reznikoff, M. Riley, M. Schaechter, and H. E. Umbarger (ed.), Escherichia coli *and* Salmonella: *Cellular and Molecular Biology*, 2nd ed. ASM Press, Washington, DC.

Newcombe, H. 1949. Origin of bacterial variants. *Nature* (London) **164:**150–151.

Yanofsky, C., B. C. Carlton, J. R. Guest, D. R. Helinski, and U. Henning. 1964. On the colinearity of gene structure and protein structure. *Proc. Natl. Acad. Sci. USA* **51:**266–272.

Zinder, N. D., and J. Lederberg. 1952. Genetic exchange in *Salmonella. J. Bacteriol.* **64:**679–699.

CHAPTER 4

Plasmids

What Is a Plasmid?

IN ADDITION TO A CHROMOSOME, BACTERIAL cells often contain plasmids. These DNA molecules are found in essentially all types of bacteria and, as discussed below, play a significant role in bacterial adaptation and evolution. They also serve as important tools in studies of molecular biology. We address such uses later in this chapter.

Plasmids, which vary widely in size from a few thousand to hundreds of thousands of base pairs (bp) (a size comparable to that of the bacterial chromosome), are most often circular molecules of double-stranded DNA. However, some bacteria have linear plasmids, and some plasmids, most often those from gram-positive bacteria, can accumulate single-stranded DNA owing to aberrant rolling-circle replication (discussed below). The number of copies also varies among plasmids, and bacterial cells can harbor more than one type. Thus, a cell can harbor two or more different types of plasmids, with hundreds of copies of some plasmid types and only one or a few copies of other types.

Like chromosomes, plasmids encode proteins and RNA molecules and replicate as the cell grows, and the replicated copies are usually distributed into each daughter cell when the cell divides. Plasmids even share some of the same types of functions for accurate partitioning (Par functions) and site-specific recombinases with the host chromosome (see below). By one definition, any independently replicating element in the cell that does not contain genes essential for bacterial growth (the so-called housekeeping genes) is called a plasmid. Plasmids probably persist because they very often provide gene products that can benefit the bacterium under certain circumstances. Consequently, isolates of bacteria taken from the environment often will lose some or all plasmids over time when cultured in the laboratory. There are a number of examples where a plasmid has taken on many

doi:10.1128/9781555817169.ch4

of the attributes of a chromosome, such as larger size and encoding multiple housekeeping genes. For example, the pSymB plasmid of some *Sinorhizobium* species is about half as big as the chromosome and carries essential genes, including a gene for an arginine tRNA and the *minCDE* genes involved in division site selection. Also, *Vibrio cholerae* has two large DNA molecules, both of which carry essential genes. *Agrobacterium tumefaciens* has two large DNA molecules, one circular and the other linear, both of which carry essential genes. In cases where a plasmid is almost as big as a chromosome and carries essential genes, which one is the chromosome and which is a plasmid? Probably a better criterion for whether a DNA molecule is a plasmid or the chromosome is the nature of its origin of replication. In all known cases, one of the large DNA molecules has a typical bacterial origin of replication, with an *oriC* site and closely linked *dnaA*, *dnaN*, and *gyrA* genes, among others, while the other DNA molecule has a typical plasmid origin, with *repABC*-like genes more characteristic of plasmids.

Naming Plasmids

Before methods for physical detection of plasmids became available, plasmids made their presence known by conferring phenotypes on the cells harboring them. Consequently, many plasmids were named after the genes they carry. For example, R-factor plasmids contain genes for resistance to several antibiotics (hence the name R, for resistance). These were the first plasmids discovered, when *Shigella* and *Escherichia coli* strains resistant to a number of antibiotics were isolated from the fecal flora of patients in Japan in the late 1950s. The ColE1 plasmid, from which many of the cloning vectors were derived, carries a gene for the protein colicin E1, a bacteriocin that kills bacteria that do not carry the plasmid. The Tol plasmid contains genes for the degradation of toluene, and the Ti plasmid of *A. tumefaciens* carries genes for tumor induction in plants (see Box 5.1). This system of nomenclature has led to some confusion, because plasmids carry various genes besides the ones for which they were originally named and because similar plasmids can contain very different sets of genes. Many plasmids have also been altered beyond recognition in the laboratory to make plasmid cloning vectors (see below) and for other purposes.

To avoid further confusion, the naming of plasmids is now standardized. Plasmids are given number-and-letter names much like bacterial strains. A small "p," for plasmid, precedes capital letters that describe the plasmid or sometimes give the initials of the person or persons who isolated or constructed it. These letters are often followed by numbers to identify the particular construct. When the plasmid is further altered, a different number is assigned to indicate the change. For example, plasmid pBR322 was constructed by Bolivar and Rodriguez from the ColE1 plasmid and is derivative number 322 of the plasmids they constructed. pBR325 is pBR322 with a chloramphenicol resistance gene inserted. The new number, 325, distinguishes the plasmid from pBR322.

Functions Encoded by Plasmids

Plasmids can encode a few or hundreds of different proteins, resulting in vast differences in their sizes. However, as mentioned above, plasmids rarely encode gene products that are always essential for growth, such as RNA polymerase, ribosomal subunits, or enzymes of the tricarboxylic acid cycle. Instead, plasmid genes usually give bacteria a selective advantage under only some conditions.

Table 4.1 lists a few naturally occurring plasmids and some traits they encode, as well as the host in which they were originally found. As mentioned above, gene products encoded by plasmids include enzymes for the utilization of unusual carbon sources, such as toluene; resistance to substances such as heavy metals and antibiotics; synthesis of antibiotics; and synthesis of toxins and proteins that allow the successful infection of other organisms. Plasmids, combined with their hosts, have also been an invaluable tool to investigate other organisms. Table 4.2 lists the major classes of plasmids used in *E. coli* and some of their relevant features for molecular genetics.

Table 4.1 Some naturally occurring plasmids and the traits they carry

Plasmid	Trait	Original source
ColE1	Bacteriocin which kills *E. coli*	*E. coli*
Tol	Degradation of toluene and benzoic acid	*Pseudomonas putida*
Ti	Tumor initiation in plants	*A. tumefaciens*
pJP4	2,4-D (2,4-dichlorophenoxyacetic acid) degradation	*Alcaligenes eutrophus*
pSym	Nodulation on roots of legume plants	*Sinorhizobium meliloti*
SCP1	Antibiotic methylenomycin biosynthesis	*Streptomyces coelicolor*
RK2	Resistance to ampicillin, tetracycline, and kanamycin	*Klebsiella aerogenes*

Table 4.2 General classes of plasmids commonly used in *E. coli*

Founding replicon	Common examples	Host range	Comments
pMB1/ColEI	pBR322, pUC vectors, pGEM vectors, pBluescript vectors	Narrow	pBR322 is a low-copy-number vector (~20 copies/cell) that has been adapted as very-high-copy-number vectors (>300 copies/cell).
p15A	pACYC177, pACYC184	Narrow	The pACYC vectors are low-copy-number vectors (~15/cell). p15A is similar to but compatible with the pMB1/ColEI replicon.
pSC101	pSC101	Narrow	Low-copy-number vector (~5 copies/cell) good for toxic genes; temperature-sensitive derivatives exist.
F plasmid	pBeloBAC11	Narrow	The original fertility (F) plasmid; the replication origin is utilized in BACs.
RK2 (RP4)	pSP329, pCM62, pCM66	Broad	IncP group
RSF1010	pJRD215, pSUP104, pSUP204	Broad	IncQ group
pSa	pUCD2	Broad	IncW group
R6K	R6K	Broad	IncX group
pBBR1MCS	pBBR1MCS-2, pBBR1MCS-3, etc.	Broad	Undefined Inc group

It is interesting to speculate about why certain types of genes are found on plasmids and others are found on the chromosome. It is easy to understand why genes that directly favor the plasmid would be encoded on the element, and this chapter goes into the details of some of these systems (see Thomas, Suggested Reading). However, there are many genes that favor the host and plasmid equally, and it is curious that certain broad classes of genes are normally encoded on the chromosome while others are on plasmids. One idea holds that plasmids tend to harbor genes that are locally advantageous (see Eberhard, Suggested Reading). For example, genes that encode such things as heavy metal resistance or antibiotic resistance may be advantageous only in certain transient situations in certain specific places. This selection might be sufficient to allow them to be maintained in some hosts in the environment even though selection is not pervasive enough to allow them to become associated with a lineage of bacteria. It would be impossible for every bacterium to maintain genes that could be advantageous for every environment. However, there is a strong selective advantage for plasmids to accumulate in hosts in these environments. Across nature, plasmids do allow bacteria to occupy a larger variety of ecological niches. This also explains why a large number of plasmids exist with the capacity to direct their own transfer between bacteria in a process called conjugation (see chapter 5) and why many plasmids are capable of replicating in different types of bacteria. As we will see below, some plasmids are also likely to be maintained, not because of the benefits they provide the host bacterium, but because the plasmid has a system to actively harm hosts that lose the plasmid (see "Toxin-Antitoxin Systems and Plasmid Maintenance" below).

Plasmid Structure

Most plasmids are circular, although linear plasmids also exist. In a circular plasmid, all of the nucleotides in each strand are joined to another nucleotide on each side by covalent bonds to form continuous strands that are wrapped around each other. Such DNAs are said to be **covalently closed circular** and, because there are no ends to rotate, the plasmid can maintain supercoiled stress. As discussed in chapter 1, nonsupercoiled DNA strands exhibit a helical periodicity of about 10.5 bp, as predicted from the Watson-Crick double-helical structure of DNA. In contrast, in a DNA that is supercoiled, the two strands are wrapped around each other either more or less often than once in about 10.5 bp. If they are wrapped around each other more often than once every 10.5 bp, the DNA is positively supercoiled; if they are wrapped around each other less often, the DNA is negatively supercoiled. Like the chromosome, covalently closed circular plasmid DNAs are usually negatively supercoiled (see chapter 1). Because DNA is stiff, the negative supercoiling introduces stress, and this stress is partially relieved by the plasmid wrapping up on itself, as illustrated in Figure 4.1A. This makes the plasmid more compact, so that it migrates more quickly in an agarose gel (Figure 4.1B). In the cell, the DNA wraps around proteins, which relieves some of the stress. The remaining stress facilitates some reactions involving the plasmid, such as separation of the two DNA strands for replication or transcription.

PURIFYING PLASMIDS

The structure of plasmids can be used to purify them away from chromosomal DNA in the cell. Cloning manuals give detailed protocols for these methods (see chapter 1, Suggested Reading), but we review them briefly

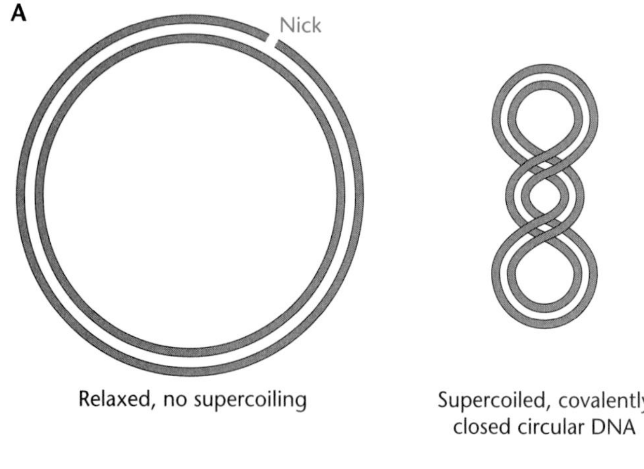

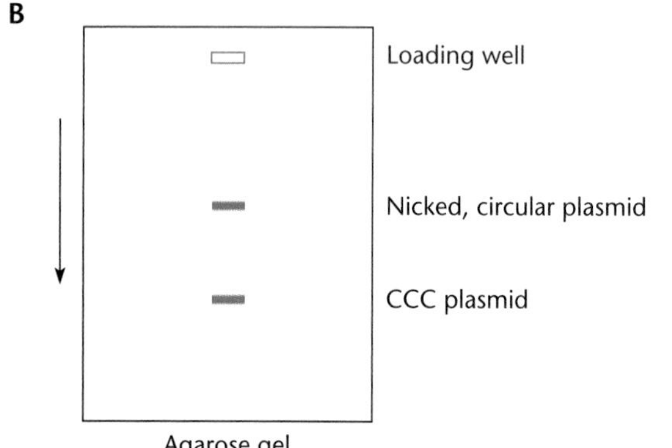

Figure 4.1 Supercoiling of a covalently closed circular (CCC) plasmid. **(A)** A break in one strand relaxes the DNA, eliminating the supercoiling and making the DNA less compact. **(B)** Schematic diagram of an agarose gel showing that the covalently closed supercoiled circles run faster on a gel than the nicked relaxed circles. Depending on the conditions, linear DNA and covalently closed circular DNA run in approximately the same position as nicked relaxed circles of the same length. The arrow shows the direction of migration.
doi:10.1128/9781555817169.ch4.f4.1

here. Many purification procedures are based on the relatively small size of most plasmids. Bacterial cells are usually lysed with a combination of a strong base (sodium hydroxide), the detergent sodium dodecyl sulfate (SDS) to solubilize the membranes, and a component to denature proteins. Following this treatment, a high-salt (potassium acetate) solution is added. Under these conditions, the chromosome precipitates with the cell debris and SDS upon centrifugation, while the small supercoiled plasmids remain in solution. Following these steps, a variety of procedures are used to further purify the plasmid DNA from the proteins and to concentrate the plasmid preparation. Historically, a solution of phenol and chloroform was added to the solution to separate the DNA from the proteins. After mixing of the aqueous lysate with phenol and chloroform, the components quickly separate into phases, similar in concept to oil and water. Plasmid DNA is separated from the proteins because the polar DNA molecules remain in the aqueous phase while the proteins remain at the interface of the solutions. Following extraction, plasmid DNA is concentrated using precipitation with a salt and alcohol solution and finally resuspended in water or a buffered solution. For convenience today, commercially available spin columns are usually used instead of separation using the phenol-chloroform solution. The plasmid solution is added to a resin matrix in a column, which binds DNA. An alcohol solution is used to wash away the residual proteins and salts from the column, and the plasmid DNA is finally eluted from the column using water or a buffered solution. The isolated plasmid DNA can then be visualized on a gel, digested with a restriction endonuclease, or sequenced, among other applications (see chapter 1)

The methods discussed above work well with plasmids that have many copies per cell and are not too large. However, large, low-copy-number plasmids are more difficult to isolate or even to detect. In some cases, plasmids are discovered by whole-genome sequencing of a bacterium when a series of DNA sequence reads assemble into a unique contiguous DNA structure separate from the host genome. Visual methods for detecting large plasmids involve separating them from the chromosome directly by electrophoresis on agarose gels (see chapter 1). The cells are often lysed directly on the agarose gel to avoid breaking the large plasmid DNA. The plasmid, because of its unique size, makes a sharp band on the gel, distinct from that due to chromosomal DNA, which is usually fragmented and so gives a more diffuse band. Also, methods such as pulsed-field gel electrophoresis have been devised to allow the separation of long pieces of DNA based on size. These methods depend on periodic changes in the direction of the electric field. The molecules attempt to reorient themselves each time the field shifts, and the longer molecules take longer to reorient than the shorter ones and thus move more slowly through the gel. Such methods have allowed the separation of DNA molecules hundreds of thousands of base pairs long and the detection of very large plasmids.

Properties of Plasmids

Replication

To exist free of the chromosome, plasmids must have the ability to replicate independently. DNA molecules that can replicate autonomously in the cell are called **replicons**. Plasmids, phage DNA, and the chromosomes are all replicons, at least in some types of cells.

To be a replicon in a particular type of cell, a DNA molecule must have at least one origin of replication, or *ori* site, where replication begins (see chapter 1). In addition, the cell must contain the proteins that enable replication to initiate at this site. Plasmids encode only a few of the proteins required for their own replication. In fact, many encode only one of the proteins needed for initiation at the *ori* site. All of the other required proteins—DNA polymerases, ligases, primases, helicases, and so on—are borrowed from the host.

Each type of plasmid replicates by one of two general mechanisms, which is determined along with other properties by its *ori* region (see "Functions of the *ori* Region" below). The plasmid replication origin is often named *oriV* for *ori* vegetative, to distinguish it from *oriT*, which is the site at which DNA transfer initiates in plasmid conjugation (see chapter 5). Most of the evidence for the mechanisms described below came from electron microscope observations of replicating plasmid DNA.

THETA REPLICATION

Some plasmids begin replication by opening the two strands of DNA at the *ori* region, creating a structure that looks like the Greek letter θ—hence the name **theta replication** (Figure 4.2A and B). In this process, an RNA primer begins replication, which can proceed in one or both directions around the plasmid. In the first case, a single replication fork moves around the molecule until it returns to the origin, and then the two daughter DNAs separate. In the other case (bidirectional replication), two replication forks move out from the *ori* region, one in either direction, and replication is complete (and the two daughter DNAs separate) when the two forks meet somewhere on the other side of the molecule.

The theta mechanism is the most common form of DNA replication, especially in gram-negative bacteria like the proteobacteria. Commonly used plasmids, including ColE1, RK2, and F, as well as the bacteriophage P1, use this type of replication. While not called the theta mechanism in chapter 1, the form of replication initiated in the chromosome at *oriC* also occurs by this mechanism (see chapter 1).

ROLLING-CIRCLE REPLICATION

Other types of plasmids replicate by very different mechanisms. One type of replication is called **rolling-circle (RC) replication** because it was first discovered in a type of phage where the template circle seems to "roll" while a copy of the plasmid is made and processed to unit length. Plasmids that replicate by this mechanism are sometimes called **RC plasmids**. This type of plasmid is widespread and is found in the largest groups of bacteria, as well as archaea.

In an RC plasmid, replication occurs in two stages. In the first stage, the double-stranded circular plasmid DNA replicates to form another double-stranded circular DNA and a single-stranded circular DNA. This stage is analogous to the replication of the DNA of some single-stranded DNA phages (see chapter 7) and to DNA transfer during plasmid conjugation (see chapter 5). In the second stage, the complementary strand is synthesized on the single-stranded DNA to make another double-stranded DNA.

The details of the RC mechanism of plasmid replication are shown in Figure 4.2C. First, the Rep protein recognizes and binds to a palindromic sequence that contains the double-strand origin (DSO) on the DNA. Binding of the Rep protein to this sequence might allow the formation of a cruciform structure by base pairing between the inverted-repeated sequences in the cruciform. Once the cruciform forms, the Rep protein can make a nick in the sequence. It is important for the models that the Rep protein is also known to function as a dimer, at least in some plasmids. After the Rep protein has made a break in the DSO sequence, it remains covalently attached to the phosphate at the 5′ end of the DNA at the nick through a tyrosine in one copy of the Rep protein in the dimer, as shown. DNA polymerase III (the replicative polymerase [see chapter 1]) uses the free 3′ hydroxyl end at the break as a primer to replicate around the circle, displacing one of the strands. It may use a host helicase to help separate the strands, or the Rep protein itself may have the helicase activity, depending on the plasmid. Once the circle is complete, the 5′ phosphate is transferred from the tyrosine on the Rep protein to the 3′ hydroxyl on the other end of the displaced strand, producing a single-stranded circular DNA. This process is called a phosphotransferase reaction and requires little energy. The same reaction is used to re-form a circular plasmid after conjugational transfer (see chapter 5).

It is less certain what happens to the newly formed double-stranded DNA when the DNA polymerase III has made its way all around the circle and gets back to the site of the DSO. Why does it not just keep going, making a longer molecule with individual genomes linked head to tail in a structure called a concatemer? Such structures are created when some phage DNAs replicate by an RC mechanism (see chapters 7 and 8), and it may also occur in some plasmids. One idea is that the DNA polymerase III does keep going past the DSO for a short distance, creating another double-stranded DSO. The other copy of the Rep protein in the dimer may then nick the newly created DSO, transferring the 5′ end to itself as described above. This might inactivate the Rep protein, releasing it with a short oligonucleotide attached. Other reactions, probably involving host DNA ligase, then cause the nick to be resealed, resulting in a circular double-stranded DNA molecule.

The displaced circular single-stranded DNA now replicates by a completely different mechanism using only host-encoded proteins. The RNA polymerase first makes

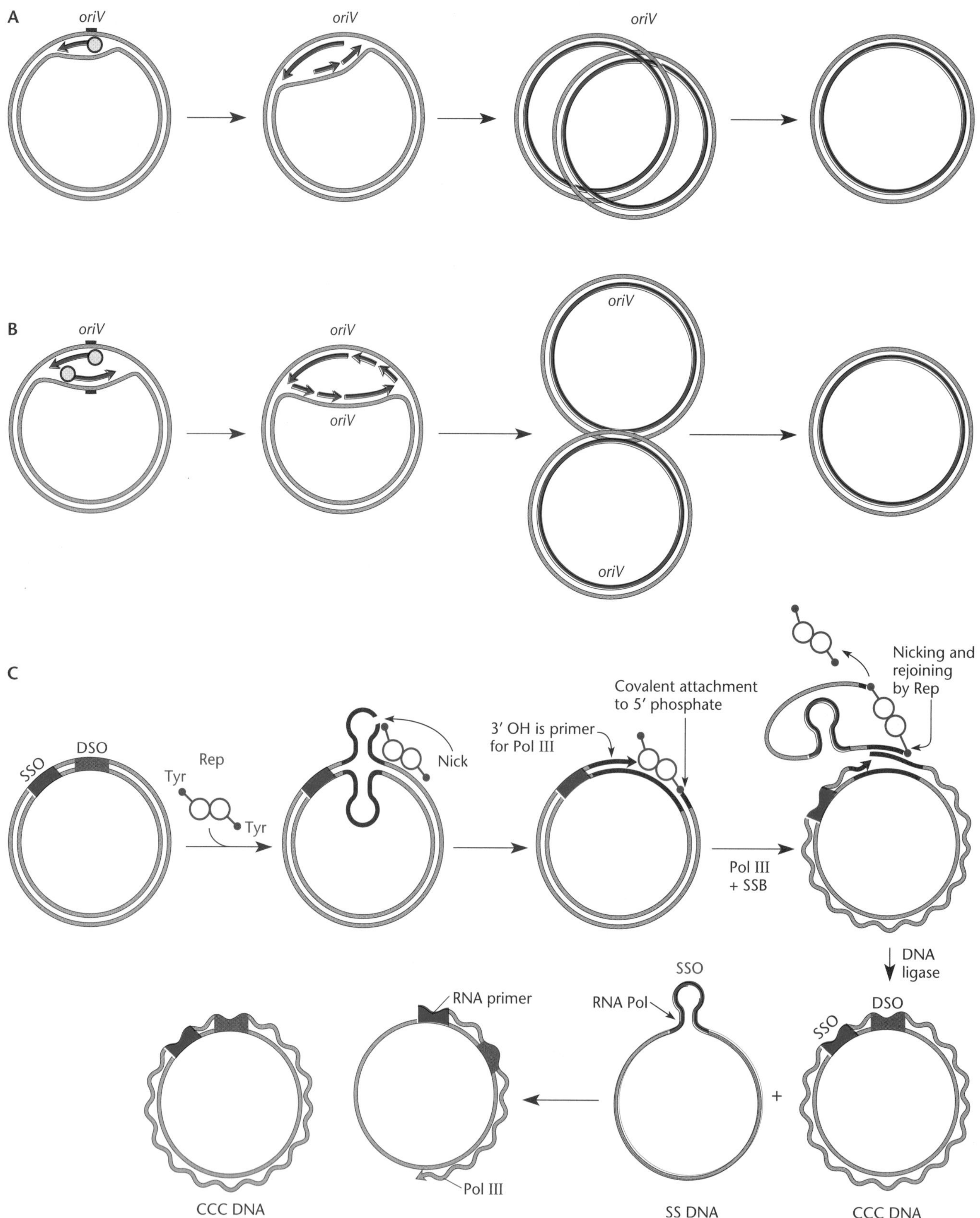

Figure 4.2 Some common schemes of plasmid replication. **(A)** Unidirectional replication. The origin region is designated *oriV.* Replication terminates when the replication fork gets back to the origin. **(B)** Bidirectional replication. Replication terminates when the replication forks meet somewhere on the DNA molecule opposite the origin. **(C)** RC replication. A nick is made at the DSO by the plasmid-encoded Rep protein, which remains bound to the 5′ phosphate end at the nick. The free 3′ OH end then serves as a primer for the DNA polymerase III (Pol III) that replicates around the circle, displacing one of the old strands as a single-stranded DNA. The Rep protein then makes another nick, releasing the single-stranded circle, and also joins the

a primer at a different origin, the single-strand origin (SSO), and this RNA then primes replication around the circle by DNA polymerase III. However, the RNA polymerase does not make this primer until the single-stranded DNA is completely displaced during the first stage of replication. This delay is accomplished by locating the SSO immediately counterclockwise from the DSO (Figure 4.2C), so that the SSO does not appear in the displaced DNA until the displacement of the single-stranded DNA is almost complete. After the entire complementary strand has been synthesized, the DNA polymerase I removes the RNA primer with its 5′ exonuclease activity while simultaneously replacing it with DNA, and host DNA ligase joins the ends to make another double-stranded plasmid. The net result is two new double-stranded plasmids synthesized from the original double-stranded plasmid.

In order for the complementary strand of the displaced single-stranded DNA to be synthesized, the RNA polymerase of the host cell must recognize the SSO on the DNA. In some hosts, the SSO is not well recognized, and single-stranded DNA accumulates. For this reason, some RC plasmids were originally called single-stranded DNA plasmids, although we now know that this is not their normal state. Broad-host-range RC plasmids presumably have an SSO that is recognized by the RNA polymerases of a wide variety of hosts, which allows them to make the complementary strand of the displaced single-stranded DNA in a variety of hosts.

The Rep protein is used only once for every round of plasmid DNA replication and is destroyed after the round is completed. This allows the replication of the plasmid to be controlled by the amount of Rep protein in the cell and keeps the total number of plasmid molecules in the cell within narrow limits. A little later in this chapter, we discuss how the copy numbers of other types of plasmids are controlled.

REPLICATION OF LINEAR PLASMIDS

As mentioned above, some plasmids and bacterial chromosomes are linear rather than circular (Box 4.1). In general, linear DNAs face two problems in all organisms. One issue with linear DNAs is that the cell must have a way to distinguish the "normal" DNA ends at the ends of the linear fragments from ends formed when DNA double-strand breaks occur, which would otherwise be lethal to the cell and must quickly be repaired. A second problem with linear plasmids and chromosomes has to do with replicating the lagging-strand template, the strand that ends with a 5′ phosphate, all the way to the end of the DNA. This has been called the "primer problem" because DNA polymerases cannot initiate the synthesis of a new strand of DNA. They can only add nucleotides to a preexisting primer, and in a linear DNA, there is no upstream primer on this strand from which to grow. Different linear DNAs solve the primer problem in different ways. Some linear plasmids have hairpin ends, with the 5′ and 3′ ends joined to each other. These plasmids replicate from an internal origin of replication to form dimeric circles composed of two plasmids joined head to tail to form a circle, as shown in the figure in Box 4.1. These dimeric circles are then resolved into individual linear plasmid DNAs with closed hairpins at the ends. The hairpin ends are presumably not recognized as DNA double-strand breaks, because they are not targets for exonucleases in the cell. A completely different mechanism is also used to maintain linear plasmids in some systems. With this mechanism, a special enzyme called a terminal protein attaches to the 5′ ends of the plasmid DNA (Box 4.1). It is interesting that bacteria with linear plasmids also often have linear chromosomes, and the two DNAs replicate by similar mechanisms.

Functions of the *ori* Region

In most plasmids, the genes for proteins required for replication are located very close to the *ori* sequences at which their products act. Thus, only a very small region surrounding the plasmid *ori* site is required for replication. As a consequence, the plasmid still replicates if most of its DNA is removed, provided that the *ori* region remains and the plasmid DNA is still circular. Smaller plasmids are easier to use as cloning vectors, as discussed below, so often the only part of the original plasmid that remains in a cloning vector is the *ori* region.

In addition, the genes in the *ori* region often determine many other properties of the plasmid. Therefore, any DNA molecule with the *ori* region of a particular plasmid will have most of the characteristics of that plasmid. The following sections describe the major plasmid properties determined by the *ori* region.

---

ends to form a circle by a phosphotransferase reaction (see the text). The DNA ligase then joins the ends of the new DNA to form a double-stranded circle. The host RNA polymerase makes a primer on the single-stranded DNA origin (SSO), and Pol III replicates the single-stranded (SS) DNA to make another double-stranded circle. DNA Pol I removes the primer, replacing it with DNA, and ligase joins the ends to make another double-stranded circular DNA. CCC, covalently closed circular; SSB, single-strand-DNA-binding protein.
doi:10.1128/9781555817169.ch4.f4.2

BOX 4.1

Linear Chromosomes and Plasmids in Bacteria

Not all bacteria have circular chromosomes and plasmids. Some, including *Borrelia burgdorferi* (the causative agent of Lyme disease), *Streptomyces* spp., *A. tumefaciens*, and *Rhodococcus fasciens*, have linear chromosomes and often multiple linear plasmids. As mentioned in the text, the replication of the ends of linear DNAs presents special problems because DNA polymerases cannot prime their own replication. This means that they cannot replicate all the way to a 3′ end in a linear DNA. If RNA at the end of a linear DNA primes the last Okazaki fragment and the RNA primer is then removed, there is no upstream primer DNA to prime its replacement with DNA as there is in a circular DNA. Eukaryotic chromosomes, which are linear, solve this problem by having special DNA regions called telomeres at their ends. Most telomeres do not need complementary sequences to be synthesized from the template as during normal DNA replication. Most use an enzyme called telomerase. The Nobel Prizes in physiology and medicine in 2010 went to people who determined how these enzymes work. Telomerase contains an RNA that is complementary to the repeated sequences at the ends of the DNA. This enzyme makes reiterated copies of the repeated telomeric sequences at the ends. When the linear chromosome replicates, some of these repeated sequences at the 3′ end are lost, but this is not a problem, because they will be resynthesized by the telomerase before the DNA replicates again.

Telomeres solve the problem of replicating linear DNAs without losing sequence information. However, bacteria with linear chromosomes are not known to have telomeres made by telomerases and must use different strategies. Two different strategies for dealing with linear chromosomes in bacteria are exemplified in *Streptomyces* and *Borrelia*. In both examples, the chromosome replicates from an *ori* sequence located toward the middle of the chromosome, from where replication occurs bidirectionally using the DnaA initiator protein in a system that is likely similar to those of other bacteria. However, the ways in which the linear ends are maintained in these organisms are very different.

The very large linear chromosome of *Streptomyces* has inverted-repeat sequences at its ends and a protein, terminal protein (TP), attached to the 5′ ends. Replication to the 3′ ends of the chromosome is thought to involve both these inverted repeats and TP in a process called "patching." After the linear DNA replicates, the 3′ end of each DNA remains single stranded, which allows the inverted-repeat sequences to form hairpins. Replication of these hairpins, combined with some sort of slippage, then allows complete replication of the ends by a process that is not completely understood.

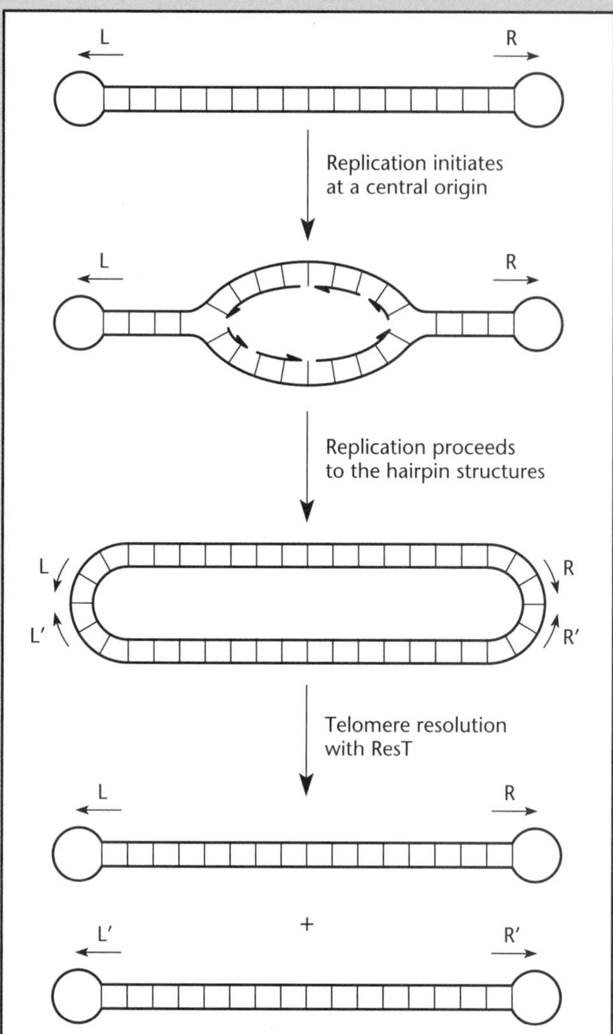

doi:10.1128/9781555817169.ch4.Box4.1.f

The mechanism used by *Borrelia* to replicate its linear chromosome is very different from that found in *Streptomyces* spp. and is illustrated in the figure. In *Borrelia*, the 5′ phosphate and 3′ OH at each end of its linear chromosome are joined to each other to form hairpins, as shown. When replication initiated in the center of the chromosome gets to the ends, the linkage between the 5′ phosphate and 3′ OH hairpins forms a dimerized chromosome, with two copies of the chromosome forming a circle containing two copies of the chromosome linked end to end, as shown. An enzyme called telomere resolvase protein (ResT) (in analogy to the telomerase of eukaryotes, even though it does not work in the same way) then recreates the original hairpin ends from

BOX 4.1 (continued)

Linear Chromosomes and Plasmids in Bacteria

these double circles by making a staggered break in the two strands where the original ends were and then rejoining the 3′ end of one strand to the 5′ end of the other strand to form a hairpin. The ResT enzyme works somewhat like some topoisomerases and tyrosine recombinases (see chapter 9) in that the breaking and rejoining process goes through a 3′ phosphoryltyrosine intermediate, where the 3′ phosphate end is covalently joined to a tyrosine (Y [see inside front cover]) before it is joined to the 5′ hydroxyl end.

Bacteria that have linear chromosomes may contain linear or circular plasmids, and the same is true for bacteria with circular chromosomes. This suggests that minimal cellular adaptation may be needed to maintain a chromosome in either the linear or circular form. Experiments in *E. coli* have provided a clever and clear way to help address this question. A lysogenic *Siphoviridae* (lambda-like) bacteriophage called N15 that was found in an isolate of *E. coli* uses the same mechanism for maintaining its linear genome as that found in *Borrelia*, with a *cis*-acting telomere region called *tos* and a *trans*-acting resolvase protein, TelN. Amazingly, it was found that the normally circular *E. coli* chromosome could be converted into a linear chromosome simply by moving the *tos* region into the terminus region of *E. coli* and expressing the TelN protein; no other adaptations were needed. The *E. coli* strain replicated normally and showed essentially no changes in gene expression. As would be predicted, the *E. coli* strain with a linear genome no longer required the Xer dimer resolution system that is normally needed to resolve circular dimer chromosomes. Supporting the finding that supercoiling is constrained into multiple domains in the chromosome, the *E. coli* strain with the linear chromosome still required the topoisomerases Topo IV and gyrase for its replication.

Linear chromosome ends also need to be differentiated from lethal double-strand break damage that must be recognized and repaired in the chromosome. Having a hairpin structure and/or a terminal protein at the end of the chromosome likely allows the cellular machinery to distinguish normal linear chromosome ends from random double-stranded breaks in *Borrelia* and *Streptomyces*.

References

Cui, T., N. Moro-Oka, K. Ohsumi, K. Kodama, T. Ohshima, N. Ogasawara, H. Mori, B. Wanner, H. Niki, and T. Horiuchi. 2007. *Escherichia coli* with a linear genome. *EMBO Rep.* **8:**181–187.

Kobryn, K., and G. Chaconas. 2002. ResT, a telomere resolvase encoded by the Lyme disease spirochete. *Mol. Cell* **9:**195–201.

Yang, C. C., C. H. Huang, C. Y. Li, Y. G. Tsay, S. C. Lee, and C. W. Chen. 2002. The terminal proteins of linear *Streptomyces* chromosome and plasmids: a novel class of replication priming proteins. *Mol. Microbiol.* **43:**297–305.

HOST RANGE

The **host range** of a plasmid includes the types of bacteria in which the plasmid can replicate; it is usually determined by the *ori* region. Some plasmids, such as those with *ori* regions of the ColE1 plasmid type, including pBR322, pET, and pUC, have a **narrow host range** (Table 4.2). These plasmids replicate only in *E. coli* and some other closely related bacteria, such as *Salmonella* and *Klebsiella* species. Work with plasmids has historically been biased by work in *E. coli*, but presumably there are also types of plasmids that will replicate only in other closely related groups of bacteria that would also technically qualify as possessing a narrow host range. In contrast, plasmids with a **broad host range** include the RK2 and RSF1010 plasmids, as well as the RC plasmids, like pBBR1MCS (Table 4.2). The host ranges of these plasmids are truly remarkable. Plasmids with the *ori* region of RK2 can replicate in most types of gram-negative proteobacteria, and RSF1010-derived plasmids even replicate in some types of gram-positive bacteria, like the *Firmicutes*. There are also plasmids that are used for work in the low-G+C gram-positive *Firmicutes*. Some plasmids used for expressing genes in various examples from the *Firmicutes* are shown in Table 4.3.

It is perhaps surprising that the same plasmid can replicate in bacteria that are so distantly related to each other. Broad-host-range plasmids must encode all of their own proteins required for initiation of replication, and therefore, they do not have to depend on the host cell for any of these functions. They also must be able to express these genes in many types of bacteria. Apparently, the promoters and ribosome initiation sites for the replication genes of broad-host-range plasmids have evolved so that they can be recognized in a wide variety of bacteria.

Determining the Host Range

The actual host ranges of most plasmids are unknown because it is sometimes difficult to determine if a plasmid can replicate in other hosts. First, we must have a way of introducing the plasmid into other bacteria. Transformation systems (see chapter 6) have been developed for

Table 4.3 Plasmids used in *B. subtilis*

Plasmid	Use	*B. subtilis ori*	*E. coli ori*	*B. subtilis* drug resistance	*E. coli* drug resistance
pUB110	Cloning vector	pUB110		Neo^r	
pMK3	Cloning vector	pMK3		Cam^r	
pDG148	Shuttle vector Inducible expression	pUB110	pBR322	Neo^r	Amp^r
pMUTIN	Integration vector Gene disruption Inducible expression *lacZ* fusion		pBR322	Erm^r	Amp^r

some, but not all, types of bacteria, and if one is available, it can be used to introduce plasmids into the bacterium. Electroporation can often be used to introduce DNA into cells. Plasmids that are self-transmissible or mobilizable (see chapter 5) can sometimes be introduced into other types of bacteria by conjugation, a process in which DNA is transferred from one cell to another using plasmid-encoded transfer functions.

Even if we can introduce the plasmid into other types of bacteria, we still must be able to select cells that have received the plasmid. Most plasmids, as isolated from nature, are not known to carry a convenient selectable gene, such as one for resistance to an antibiotic, and even if they do, the selectable gene may not be expressed in the other bacterium, since most genes are not expressed in bacteria distantly related to those in which they were originally found. Sometimes we can introduce a selectable gene, chosen because it is expressed in many hosts, into the plasmid. For example, the kanamycin resistance gene, first found in the Tn*5* transposon, is expressed in most gram-negative bacteria, making them resistant to the antibiotic kanamycin. We can either clone a marker gene into the plasmid or introduce a transposon carrying a selectable marker into the plasmid by methods discussed in chapter 9.

If all goes well and we have a way to introduce the plasmid into other bacteria, and the plasmid carries a marker that is likely to be expressed in other bacteria, we can see if the plasmid can replicate in bacteria other than its original host. Care must also be taken to ensure that the plasmid has not recombined into the host chromosome. Since the mechanisms for introducing DNA into different types of bacteria differ and because there are many barriers to plasmid transfer between species, determining the host range of a new plasmid is a very laborious process. Therefore, the host ranges of plasmids are often extrapolated from only a few examples.

REGULATION OF COPY NUMBER

Another characteristic of plasmids that is determined mostly by their *ori* region is the **copy number**, or the average number of that particular plasmid per cell. More precisely, we define the copy number as the number of copies of the plasmid in a newborn cell immediately after cell division. Copy number control must have been an important early step in the evolution of plasmids. All plasmids must regulate their replication; otherwise, they would fill up the cell and become too great a burden for the host, or their replication would not keep up with the cell replication and they would be progressively lost during cell division. Some plasmids, such as the F plasmid of *E. coli*, replicate only about once during the cell cycle. Naturally, all plasmids have a somewhat low copy number, but plasmids have also been engineered to allow a much higher copy number per cell to facilitate biotechnology. Copy number information for some plasmids is shown in Table 4.2.

The regulation mechanisms used by plasmids with higher copy numbers often differ greatly from those used by plasmids with lower copy numbers. Plasmids that have high copy numbers, such as the modified derivatives of the ColE1 plasmid origin, need only have a mechanism that inhibits the initiation of plasmid replication when the number of plasmids in the cell reaches a certain level. Consequently, these molecules are called **relaxed plasmids**. In contrast, low-copy-number plasmids, such as F, must replicate only once or very few times during each cell cycle and so must have a tighter mechanism for regulating their replication. Hence, they are called **stringent plasmids**. Much more is understood about the regulation of replication of relaxed plasmids than about the regulation of replication of stringent plasmids.

The regulation of relaxed plasmids falls into three general categories. Some plasmids are regulated by an antisense RNA, sometimes called a countertranscribed RNA (ctRNA) because it is transcribed from the same region of the plasmid DNA as an RNA essential for plasmid replication but from the opposite strand. The ctRNA is therefore complementary to the essential RNA and is able to hybridize to the essential RNA and inhibit its function. In many cases, the ctRNA inhibits the translation of a protein essential for replication. The ctRNA of these plasmids is often assisted in its inhibitory role by a protein. Other plasmids are regulated by a ctRNA alone. Still others are regulated by a protein alone, which binds to repeated sequences in the plasmid DNA called iterons, thereby inhibiting plasmid replication. Examples of these three types of regulation are discussed below.

INCOMPATIBILITY

Another function of plasmids that is controlled by the *ori* region is **incompatibility**. Incompatibility refers to the inability of two plasmids to coexist stably in the same cell. Many bacteria, as they are isolated from nature, contain more than one type of plasmid. These plasmid types coexist stably in the bacterial cell and remain there even after many cell generations. In fact, bacterial cells containing multiple types of plasmids are not cured of each plasmid any more frequently than if the other plasmids were not there.

However, sometimes two plasmids of different types cannot coexist stably in the same cell. In this case, one or the other plasmid is lost as the cells multiply; this loss is more frequent than would occur if the plasmids were not occupying the cells together with the other plasmid. If two plasmids cannot coexist stably, they are said to be members of the same **incompatibility (Inc) group**. If two plasmids can coexist stably, they belong to different Inc groups. There are a number of ways in which plasmids can be incompatible. One way is if they can each regulate the other's replication. Another way is if they share the same partitioning (*par*) functions, which are often closely associated with replication control in the *ori* region. There may be hundreds of different Inc groups, and plasmids are usually classified by the Inc group to which they belong. For example, RP4 (also called RK2) is an IncP (incompatibility group P) plasmid. In contrast, RSF1010 is an IncQ plasmid; it can therefore be stably maintained with RP4 because it belongs to a different Inc group but cannot be stably maintained with another IncQ plasmid.

Determining the Inc Group

To classify a plasmid by its Inc group, we must determine if it can coexist with other plasmids of known Inc groups. In other words, we must measure how frequently cells are cured of the plasmid when it is introduced into cells carrying another plasmid of a known Inc group. However, we can know that cells have been cured of a plasmid only when it encodes an easily testable trait, such as resistance to an antibiotic. Then, the cells become sensitive to the antibiotic if the plasmid is lost.

The experiment shown in Figure 4.3 is designed to measure the curing rate of a plasmid that contains the Cam^r gene, which makes cells resistant to the antibiotic chloramphenicol. To measure the frequency of plasmid curing, we grow the plasmid-containing cells in medium with all the growth supplements and no chloramphenicol. At different times, we take a sample of the cells, dilute it, and plate the dilutions on agar containing the same growth supplements but, again, no chloramphenicol. After incubation of the plates, we replicate the plate onto another plate containing chloramphenicol (see chapter 3). If we do not observe any growth of a colony "copied" from the master plate, the bacteria in that colony must all have been sensitive to the antibiotic, and hence, the original bacterium that had multiplied to form the colony must have been cured of the plasmid.

Figure 4.3 Measurement of the curing of a plasmid carrying resistance to chloramphenicol. See the text for details. doi:10.1128/9781555817169.ch4.f4.3

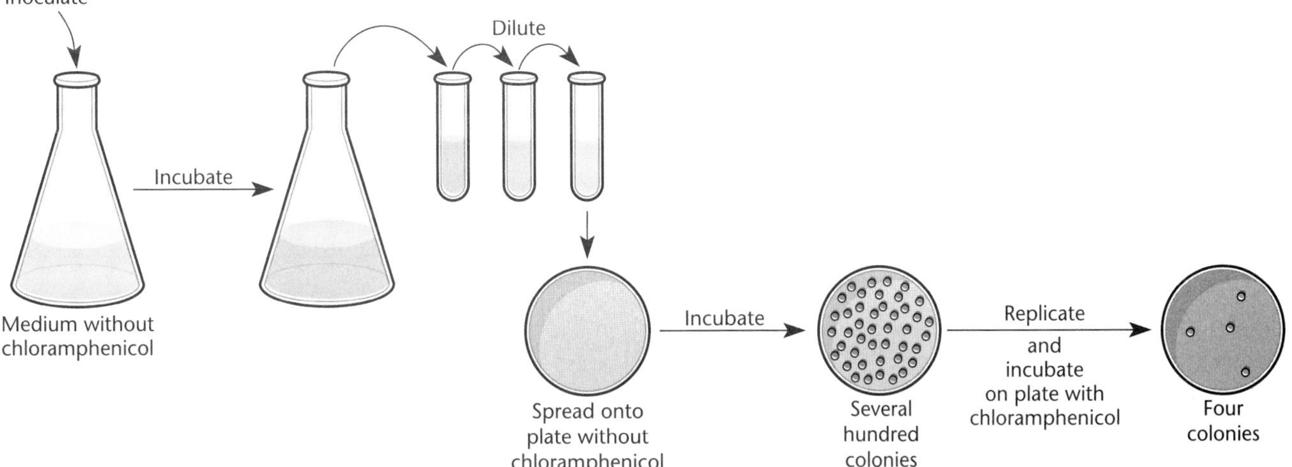

The percentage of colonies that contain no resistant bacteria is the percentage of bacteria that were cured of the plasmid at the time of plating.

To apply this test to determine if two plasmids are members of the same Inc group, the two plasmids must contain different selectable genes, for example, genes encoding resistance to different antibiotics. Then, one plasmid is introduced into cells containing the other plasmid. Resistance to both antibiotics is selected for. Then, cells containing both plasmids are incubated without either antibiotic and finally tested on antibiotic-containing plates, as described above in the example with chloramphenicol. The only difference is that the colonies are transferred onto two plates, each containing one or the other antibiotic. If the percentage of cells cured of one or the other plasmid is no higher than the percentage cured of either plasmid when it was alone, the plasmids are members of different Inc groups. We continue to apply this test until we find a known plasmid, if any, that is a member of the same Inc group as our unknown plasmid.

Technically, even two plasmids in the same Inc group could be maintained if they were maintained with a high copy number and had distinct antibiotic resistances that could be selected at the same time. However, this is not advisable, because it can select for plasmid fusion events via recombination between the vectors.

Incompatibility Due to Shared Replication Control

One way in which two plasmids can be incompatible is if they share the same mechanism of replication control. The replication control system does not recognize the two as different, so either plasmid may be randomly selected for replication. At the time of cell division, the total copy number of the two plasmids will be the same, but one may be represented much less than the other. Figure 4.4 illustrates this by contrasting the distributions at cell division of plasmids of the same Inc group with plasmids of different Inc groups. Figure 4.4A shows a cell containing two types of plasmids that belong to different Inc groups and use different replication control systems. In the illustration, the two plasmids exist in equal numbers before cell division, but after division, the two daughter cells are not likely to get the same number of each plasmid. However, in the new cells, each plasmid replicates to reach its copy number, so that at the time of the next division, both cells again have the same numbers of the plasmids. This process is repeated each generation, so very few cells will be cured of either plasmid.

Now, consider the situation illustrated in Figure 4.4B, in which the cell has two plasmids that belong to the same Inc group and therefore share the same replication control system. As in the first example, both plasmids originally exist in equal numbers, but when the cell divides, it is unlikely that the two daughter cells will receive the same number of the two plasmids. Note that in the original cell, the copy number of each plasmid is only half its normal number; both plasmids contribute to the total copy number, since they both have the same *ori* region and inhibit each other's replication. After cell division, the two plasmids replicate until the total number of plasmids in each cell equals the copy number. The underrepresented plasmid (recall that the daughters may not receive the same number of plasmids if the plasmid is high copy number) does not necessarily replicate more than the other plasmid, so that the imbalance of plasmid numbers might remain or become even worse. At the next cell division, the underrepresented plasmid has less chance of being distributed to both daughter cells, since there are fewer copies of it. Consequently, in subsequent cell divisions, the daughter cells are much more likely to be cured of one or the other of the two plasmid types by chance alone.

Incompatibility due to copy number control is probably more detrimental to low-copy-number plasmids than to high-copy-number plasmids. If the copy number is only 1, then only one of the two plasmids can replicate; each time the cell divides, a daughter is cured of one of the two types of plasmids.

Incompatibility Due to Partitioning

Two plasmids can also be incompatible if they share the same Par (partitioning) system. Par systems help segregate plasmids or chromosomes into daughter cells upon cell division (see below). Normally this helps ensure that both daughter cells get at least one copy of the plasmid and neither daughter cell is cured of the plasmid. If coexisting plasmids share the same Par system, one or the other is always distributed into the daughter cells during division. However, sometimes one daughter cell receives one plasmid type and the other cell gets the other plasmid type, producing cells cured of one or the other plasmid. We discuss what is understood about the mechanisms of partitioning below.

Plasmid Replication Control Mechanisms

The mechanisms used by some plasmids to regulate their copy number have been studied in detail. Some of the better-understood mechanisms are reviewed in this section.

ColE1-DERIVED PLASMIDS: REGULATION OF PROCESSING OF PRIMER BY COMPLEMENTARY RNA

The mechanism of copy number regulation of the plasmid ColE1 was one of the first to be studied. Figure 4.5 shows a partial genetic map of the original ColE1 plasmid. This plasmid has been put to use in numerous molecular biology studies, and many vectors have been

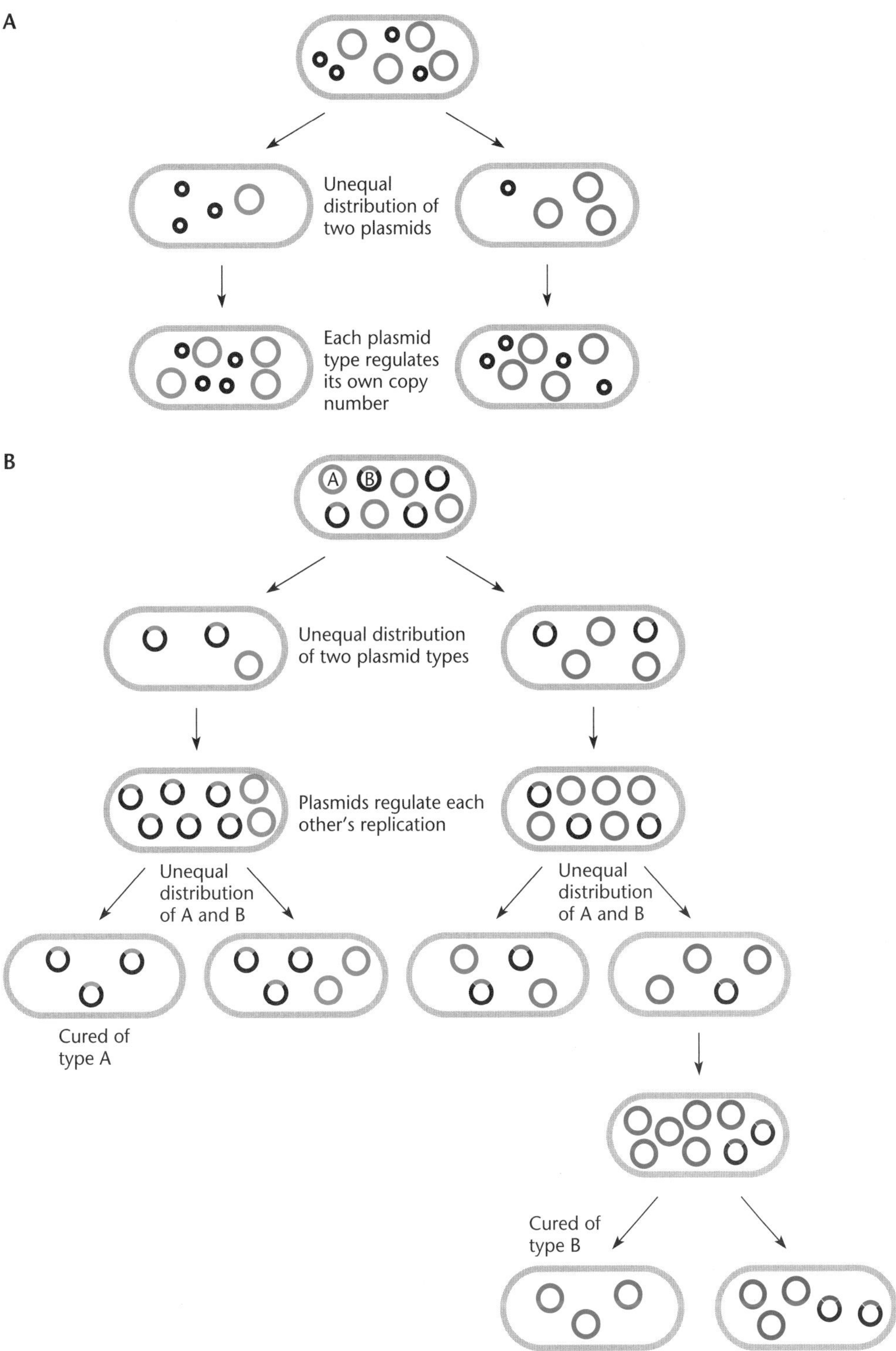

Figure 4.4 Coexistence of two plasmids from different Inc groups. **(A)** After division, both plasmids replicate to reach their copy numbers. **(B)** Curing of cells of one of two plasmids when they are members of the same Inc group. The sum of the two plasmids is equal to the copy number, but one may be underrepresented and lost in subsequent divisions. Eventually, most of the cells contain only one or the other plasmid. The light-blue region in the smaller plasmid indicates that it shares the *ori* region with the larger plasmid. doi:10.1128/9781555817169.ch4.f4.4

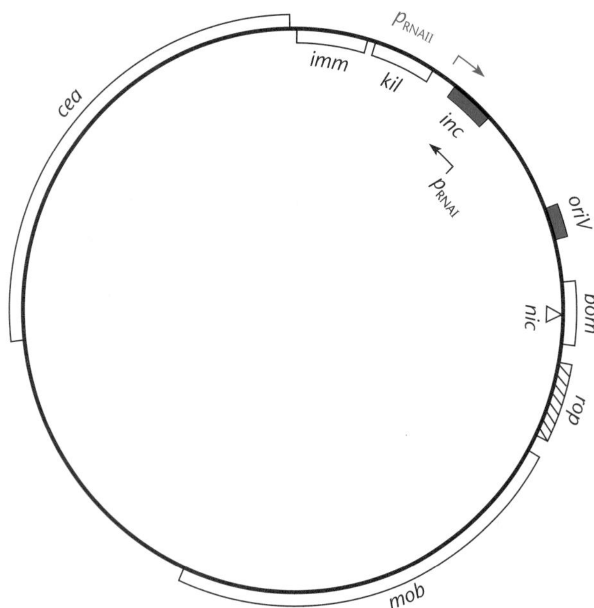

Figure 4.5 Genetic map of plasmid ColE1. The plasmid is 6,646 bp long. On the map, *oriV* is the origin of replication; p_{RNAII} is the promoter for the primer RNA II, *inc* encodes RNA I, *rop* encodes a protein that helps regulate the copy number, *bom* is a site that is nicked at *nic*, *cea* encodes colicin ColE1, and *mob* encodes functions required for mobilization (discussed in chapter 5).
doi:10.1128/9781555817169.ch4.f4.5

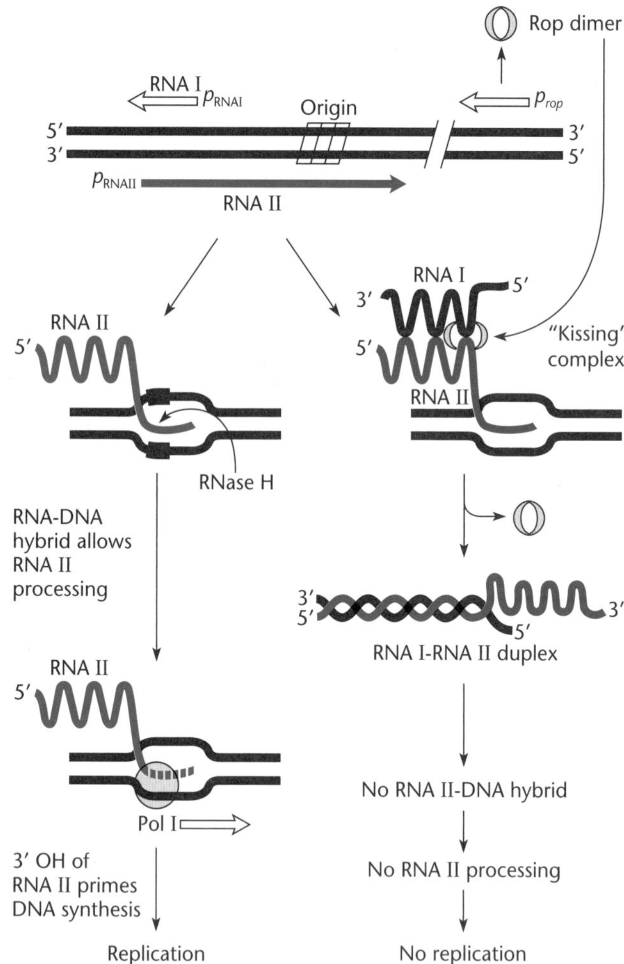

Figure 4.6 Regulation of the replication of ColE1-derived plasmids. RNA II must be processed by RNase H before it can prime replication. "Origin" indicates the transition point between the RNA primer and DNA. RNA I binds to RNA II and inhibits the processing, thereby regulating the copy number. p_{RNAI} and p_{RNAII} are the promoters for RNA I and RNA II transcription, respectively. RNA II is shown in blue. The Rop protein dimer enhances the initial pairing of RNA I and RNA II.
doi:10.1128/9781555817169.ch4.f4.6

derived from it or its close relative, pMB1 (Table 4.2). These vectors include the commonly used pBR322 plasmid and plasmids with modified forms of the pMB1/ColE1 origin, like the pUC plasmids, the pBAD plasmids, and the pET series of plasmids discussed below and in chapter 7. Expression control using the p_{BAD} promoter that is found in the pBAD plasmids is described in chapter 12. Although the genetic maps of these cloning vectors have been changed beyond recognition for the pMB1/ColE1 vectors, they all retain the basic properties of the original ColE1 *ori* region, and hence, they share many of its properties, including the mechanism of replication regulation. However, these derivatives often have modifications to vastly increase the copy number of the vector to allow greater amounts of plasmid DNA to be easily isolated (Table 4.2).

The mechanism of regulation of ColE1-derived plasmids is shown in Figure 4.6. Replication is regulated mostly through the effects of a small plasmid-encoded RNA called RNA I. This small RNA inhibits plasmid replication by interfering with the processing of another RNA called RNA II, which forms the primer for plasmid DNA replication. In the absence of RNA I, RNA II forms an RNA-DNA hybrid at the replication origin. RNA II is then cleaved by the RNA endonuclease RNase H, releasing a 3′ hydroxyl group that serves as the primer for replication first catalyzed by DNA polymerase I. Unless RNA II is processed properly, it does not function as a primer, and replication does not ensue.

RNA I inhibits DNA replication through interference with RNA II primer formation by forming a double-stranded RNA with it, as illustrated in Figure 4.6. It can do this because the two RNAs are transcribed from opposite strands in the same region of DNA. Figure 4.7 illustrates how any two RNAs transcribed from the same region of DNA but from opposite strands are complementary. The regulatory capacity of small complementary RNAs was first shown in this system, but small RNAs are now known to be very important as mechanisms for controlling gene expression in many groups of

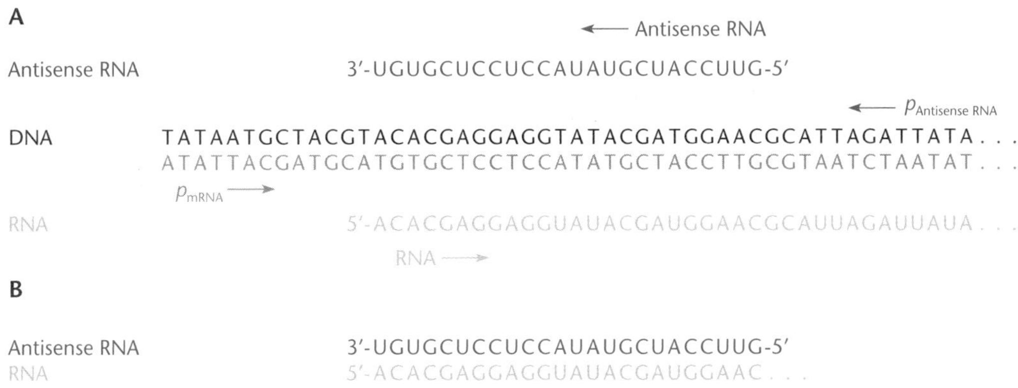

Figure 4.7 Pairing between an RNA and its antisense RNA. **(A)** An antisense RNA is made from the opposite strand of DNA in the same region. **(B)** The two RNAs are complementary and can base pair with each other to make a double-stranded RNA.
doi:10.1128/9781555817169.ch4.f4.7

bacteria (see chapter 12). Initially, the pairing between RNA I and RNA II occurs through short exposed regions on the two RNAs that are not occluded by being part of secondary structures. This initial pairing is very weak and therefore has been called a "kissing complex." The protein named Rop (Fig. 4.5) helps stabilize the kissing complex, although it is not essential. The kissing complex can then extend into a "hug," with the formation of the double-stranded RNA, as shown. Formation of the double-stranded RNA prevents RNA II from forming the secondary structure required for it to hybridize to the DNA before being processed by RNase H to form the mature primer.

Even though Rop (sometimes called Rom) is known to help RNA I to pair with RNA II and therefore help inhibit plasmid replication, it is not clear how Rop works, nor is the protein essential to maintain the copy number. Mutations that inactivate Rop cause only a moderate increase in the plasmid copy number.

This mechanism provides an explanation for how the copy number of ColE1 plasmids is maintained. Since RNA I is synthesized from the plasmid, more RNA I is made when the concentration of the plasmid is high. A high concentration of RNA I interferes with the processing of most of the RNA II, and replication is inhibited. The inhibition of replication is almost complete when the concentration of the plasmid reaches about 16 copies per cell, the copy number of the original ColE1 plasmid.

We can predict from the model what the effect of mutations in RNA I should be. Formation of the kissing complex involves pairing between very small regions of RNA I and RNA II. However, these regions must be completely complementary for this pairing to occur and for plasmid replication to be inhibited. Changing even a single base pair in this short sequence makes the mutated RNA I no longer complementary to the RNA II of the original nonmutant ColE1 plasmid, so it is no longer able to "kiss" it and regulate its replication. However, a mutation in the region of the plasmid DNA encoding RNA I also changes the sequence of RNA II made by the same plasmid in a complementary way, since they are encoded in the same region of the DNA, but from the opposite strands. Therefore, the mutated RNA I should still form a complex with the mutated RNA II made from the same mutated plasmid and prevent its processing; it just cannot interfere with the processing of RNA II from the original nonmutant plasmid. Therefore, a single-base-pair mutation in the RNA I coding region of the plasmid should effectively change the Inc group of the plasmid to form a new Inc group, of which the mutated plasmid is conceivably the sole member! In fact, the naturally occurring plasmids ColE1 and its close relative p15A, from which other cloning vectors, such as pACYC177 and pACYC184 have been derived, are members of different Inc groups, even though they differ by only 1 base in the kissing regions of their RNA I and RNA II.

R1 AND ColIb-P9 PLASMIDS: REGULATION OF TRANSLATION OF Rep PROTEIN BY COMPLEMENTARY RNA

The ColE1-derived plasmids are unusual in that they do not require a plasmid-encoded protein to initiate DNA replication at the *oriV* region, only an RNA primer synthesized from the plasmid. Most plasmids require a plasmid-encoded protein, often called Rep, to initiate replication. The Rep protein is required to separate the strands of DNA at the *oriV* region, often with the help of host proteins, including DnaA (see chapter 1). Opening the strands is a necessary first step that allows the replication apparatus to assemble at the origin. The Rep proteins are very specific in that they bind only to the *oriV* of the same type of plasmid because they bind to certain specific DNA sequences within *oriV*. The amount of Rep protein is usually limiting for replication, meaning that there is never more than is needed to initiate replication. Therefore, the copy number of the plasmid can be controlled, at least partially, by controlling the synthesis of the Rep protein.

The R1 Plasmid

One type of plasmid that regulates its copy number by regulating the amount of a Rep protein is the R1 plasmid, a member of the IncFII family of plasmids. Like ColE1 plasmids, this plasmid uses a small complementary RNA to regulate its copy number, and this small RNA forms a kissing complex with its target RNA (see Kolb et al., Suggested Reading). Also like ColE1 plasmids, the more copies of the plasmid in the cell, the more of this antisense RNA is made and the more plasmid replication is inhibited. However, rather than inhibiting primer processing, the R1 plasmid uses its complementary RNA to inhibit the translation of the mRNA that encodes the Rep protein and thereby to inhibit the replication of the plasmid DNA.

Figure 4.8 illustrates the regulation of R1 plasmid replication in more detail. The plasmid-encoded protein RepA is the only plasmid-encoded protein that is required for the initiation of replication. The *repA* gene can be transcribed from two promoters. One of these promoters, called p_{copB}, transcribes both the *repA* and *copB* genes, making an mRNA that can be translated into the proteins RepA and CopB. The second promoter, p_{repA}, is in the *copB* gene and so makes an RNA that can encode only the RepA protein. Because the p_{repA} promoter is repressed by the CopB protein, it is turned on only immediately after the plasmid enters a cell and before any CopB protein is made. The short burst of synthesis of RepA from p_{repA} after the plasmid enters a cell causes the plasmid to replicate until it reaches its copy number. Then, the p_{repA} promoter is repressed by CopB protein, and the *repA* gene can be transcribed only from the p_{copB} promoter.

Once the plasmid has attained its copy number, the regulation of synthesis of RepA, and therefore the replication of the plasmid, is regulated by the small CopA RNA. The *copA* gene is transcribed from its own promoter, and the RNA product affects the stability of the mRNA made from the p_{copB} promoter. Because the CopA RNA is made from the same region encoding the translation initiation region (TIR) for the *repA* gene, but from the other strand of the DNA, the two RNAs are complementary and can pair to make double-stranded RNA. Then, an RNase called RNase III, a chromosomally encoded enzyme that cleaves some double-stranded RNAs (see chapter 2), cleaves the CopA-RepA duplex RNA.

A Plasmid genetic organization

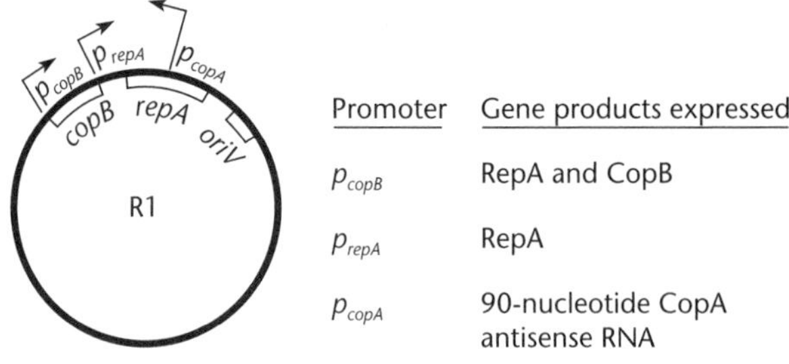

B Replication occurs after plasmid enters cells

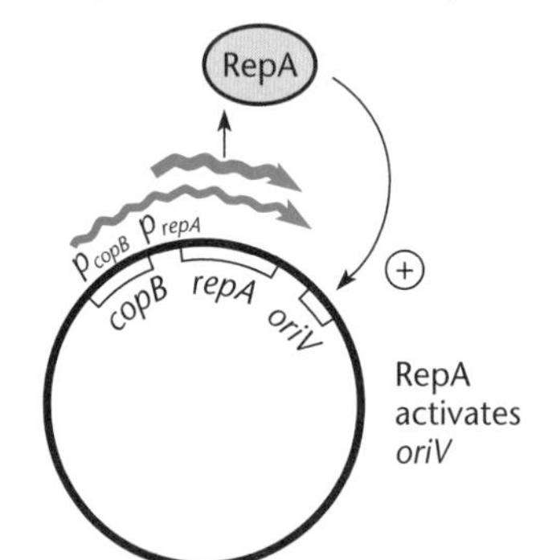

C Replication shutdown

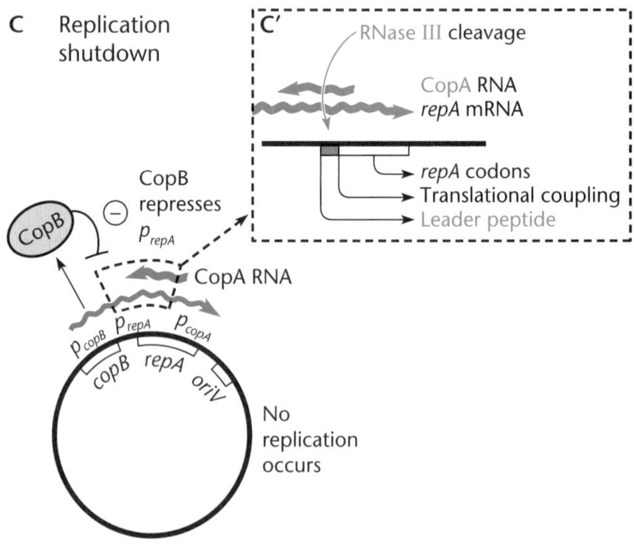

Figure 4.8 Regulation of replication of the IncFII plasmid R1. **(A)** Locations of promoters, genes, and gene products involved in the regulation. **(B)** Immediately after the plasmid enters the cell, most of the *repA* mRNA is made from promoter p_{repA} until the plasmid reaches its copy number. **(C)** Once the plasmid reaches its copy number, the CopB protein represses transcription from p_{repA}. Now, *repA* is transcribed only from p_{copB}. **(C′)** The antisense RNA CopA hybridizes to the leader peptide coding sequence in the *repA* mRNA, and the double-stranded RNA is cleaved by RNase III. This prevents translation of RepA, which is translationally coupled to the translation of the leader peptide.
doi:10.1128/9781555817169.ch4.f4.8

The reasons why cleavage of this RNA prevents the synthesis of RepA are a little complicated. The 5′ leader region of the mRNA, upstream of where the RepA protein is encoded, encodes a short leader polypeptide that has no function of its own but simply exists to be translated. The translation of RepA is coupled to the translation of this leader polypeptide (see chapter 2 for an explanation of translational coupling). Cleavage of the mRNA by RNase III in the leader region interferes with the translation of this leader polypeptide and, by blocking its translation, also blocks translation of the downstream RepA. Therefore, by having the CopA RNA activate cleavage of the mRNA for the RepA protein upstream of the RepA coding sequence, the plasmid copy number is controlled by the amount of CopA RNA in the cell, which in turn depends on the concentration of the plasmid. The higher the concentration of the plasmid, the more CopA RNA is made and the less RepA protein is synthesized, maintaining the concentration of the plasmid around the plasmid copy number.

The ColIb-P9 Plasmid

Yet another level of complexity of the regulation of the copy number by a complementary RNA is provided by the ColIb-P9 plasmid (Figure 4.9) (see Azano and Mizobuchi, Suggested Reading). As in the R1 plasmid, the Rep protein-encoding gene (called *repZ* in this case) is translated downstream of a leader peptide open reading frame, called *repY*, and the two are also translationally coupled. The translation of *repY* opens an RNA secondary structure that normally occludes the Shine-Dalgarno (S-D) sequence of the TIR of *repZ*. A sequence in the secondary structure then can pair with the loop of a hairpin upstream of *repY*, forming a pseudoknot (see Figure 2.2 for an example of a pseudoknot), thus permanently disrupting the secondary structure and leaving the S-D sequence for *repZ* exposed. A ribosome can then bind to the TIR for *repZ* and translate the initiator protein. The small complementary Inc RNA pairs with the loop of the upstream hairpin and prevents hairpin formation, leaving the S-D sequence of the *repZ* coding sequence blocked and preventing translation of *repY*.

THE pT181 PLASMID: REGULATION OF TRANSCRIPTION OF THE *rep* GENE BY A SMALL COMPLEMENTARY RNA

Not all plasmids that have an antisense RNA to regulate their copy numbers use it to inhibit translation or primer processing. Some plasmids of gram-positive bacteria, including the *Staphylococcus* plasmid pT181, use antisense RNAs to regulate transcription of the *rep* gene, in this case called *repC*, through a process called attenuation (Figure 4.10) (see Novick et al., Suggested Reading). The pT181 plasmid replicates by an RC mechanism, and the RepC protein is required to initiate replication of the leading strand at *oriV*. Also, the RepC protein is inactivated each time the DNA replicates (see above). This makes the RepC protein rate limiting for replication, i.e., the more RepC protein there is, the more plasmids are made. The antisense RNA binds to the mRNA for the RepC protein as the mRNA is being made and prevents formation of a secondary structure. This secondary structure would normally prevent the formation of a hairpin that is part of a factor-independent transcriptional terminator (see chapter 2). Therefore, if the secondary structure does not form, the hairpin forms and transcription terminates (i.e., is attenuated). Transcriptional regulation by attenuation is discussed in more detail in chapter 12.

This regulation works well only because the antisense RNA is so unstable that its concentration drops quickly if the copy number of the plasmid decreases, allowing fine-tuning of the replication of the plasmid with the copy number. In other plasmids of gram-positive bacteria, the antisense RNA is much more stable. These plasmids also use a transcriptional repressor to regulate transcription of the *rep* gene.

THE ITERON PLASMIDS: REGULATION BY COUPLING

Many commonly studied plasmids use a very different mechanism to regulate their replication. These plasmids are called iteron plasmids because their *oriV* regions contain several repeats of a certain set of DNA bases called an **iteron sequence**. The iteron plasmids include pSC101, F, R6K, P1, and the RK2-related plasmids (Table 4.2). The iteron sequences of these plasmids are typically 17 to 22 bp long and exist in about three to seven copies in the *ori* region. In addition, there are usually additional copies of these repeated sequences a short distance away that contribute to lowering the plasmid copy number.

One of the simplest of the iteron plasmids is pSC101. For our purposes, the essential features of the *ori* region of this plasmid (Figure 4.11) are the gene *repA*, which encodes the RepA protein required for initiation of replication, and three repeated iteron sequences, R1, R2, and R3, where RepA binds to regulate the copy number. The RepA protein is the only plasmid-encoded protein required for the replication of the pSC101 plasmid and many other iteron plasmids. It serves as a positive activator of replication, much like the RepA protein of the R1 plasmid. The host chromosome encodes the other proteins that either bind to this region or otherwise act to allow initiation of replication; they include DnaA, DnaB, DnaC, and DnaG (see chapter 1).

Iteron plasmid replication is regulated by two superimposed mechanisms. The first is control of RepA synthesis.

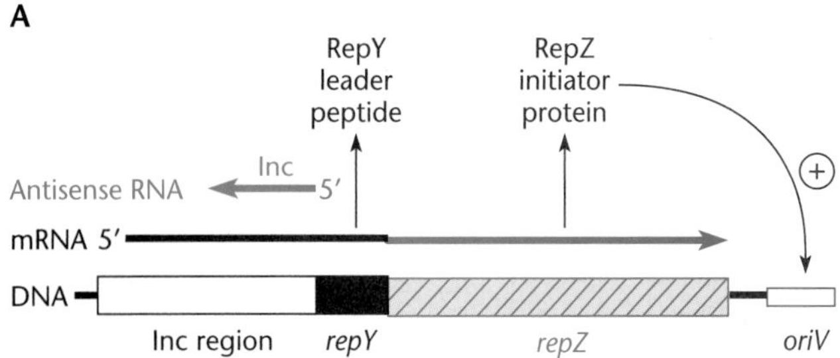

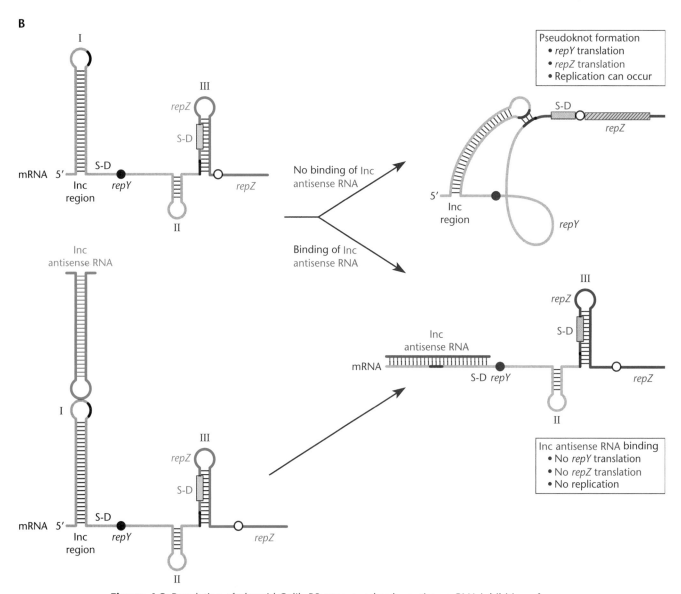

Figure 4.9 Regulation of plasmid ColIb-P9 copy number by antisense RNA inhibition of pseudoknot formation. **(A)** The minimal replicon with the *repY* (leader peptide) (black box) and *repZ* genes is shown. The Inc region encodes both the 5′ end of the *repYZ* mRNA and the antisense RNA. **(B)** The *repY* and *repZ* genes are translationally coupled. On the mRNA, the *repY* S-D sequence is exposed, whereas structure III sequesters the *repZ* S-D sequence (blue rectangle) and thereby prevents *repZ* translation. Also shown by thick black bars are regions in structures I and III that are complementary and so can pair, resulting in pseudoknot formation. The solid circle indicates the *repY* start codon; the open circle indicates the *repY* stop codon. Unfolding of structure II by the ribosome stalling at the *repY* stop codon results in the formation of a pseudoknot by base pairing between the complementary sequences and allows the ribosome to access the *repZ* S-D sequence. Binding of Inc antisense RNA to the loop of structure I directly inhibits formation of the pseudoknot, and the subsequent Inc RNA-mRNA duplex inhibits RepY translation, and consequently RepZ translation, since the two are translationally coupled. doi:10.1128/9781555817169.ch4.f4.9

Most commonly, the RepA protein represses its own synthesis by binding to its own promoter region and blocking transcription of its own gene. Therefore, the higher the concentration of plasmid, the more RepA protein is made and the more it represses its own synthesis. Thus, the concentration of RepA protein is maintained within narrow limits and the initiation of replication is strictly regulated. This type of regulation, known as **transcriptional autoregulation**, is discussed in chapter 12.

However, this mechanism of regulation by itself is not sufficient to regulate the copy number of the plasmid, especially that of low-copy-number stringent plasmids,

Figure 4.10 Regulation of plasmid pT181 copy number by antisense RNA regulation of transcription of the *repC* gene. **(A)** The genetic structure of the minimal replicon of pT181. Shown are the mRNA that encodes the RepC protein that initiates leading-strand replication at *oriV*; the Cop region that encodes the antisense RNA (RNA I) that regulates the copy number; and regions in the mRNA and antisense RNA, indicated by arrows labeled I, II, III, and IV, that can pair to form alternative secondary structures. **(B)** Formation of an antisense RNA-mRNA duplex regulates RepC expression by a transcriptional attenuation mechanism. The antisense RNA I can form a duplex with the 5′ end of the mRNA that encodes RepC and disrupt a secondary structure, allowing instead the formation of a terminator loop that causes transcription termination. doi:10.1128/9781555817169.ch4.f4.10

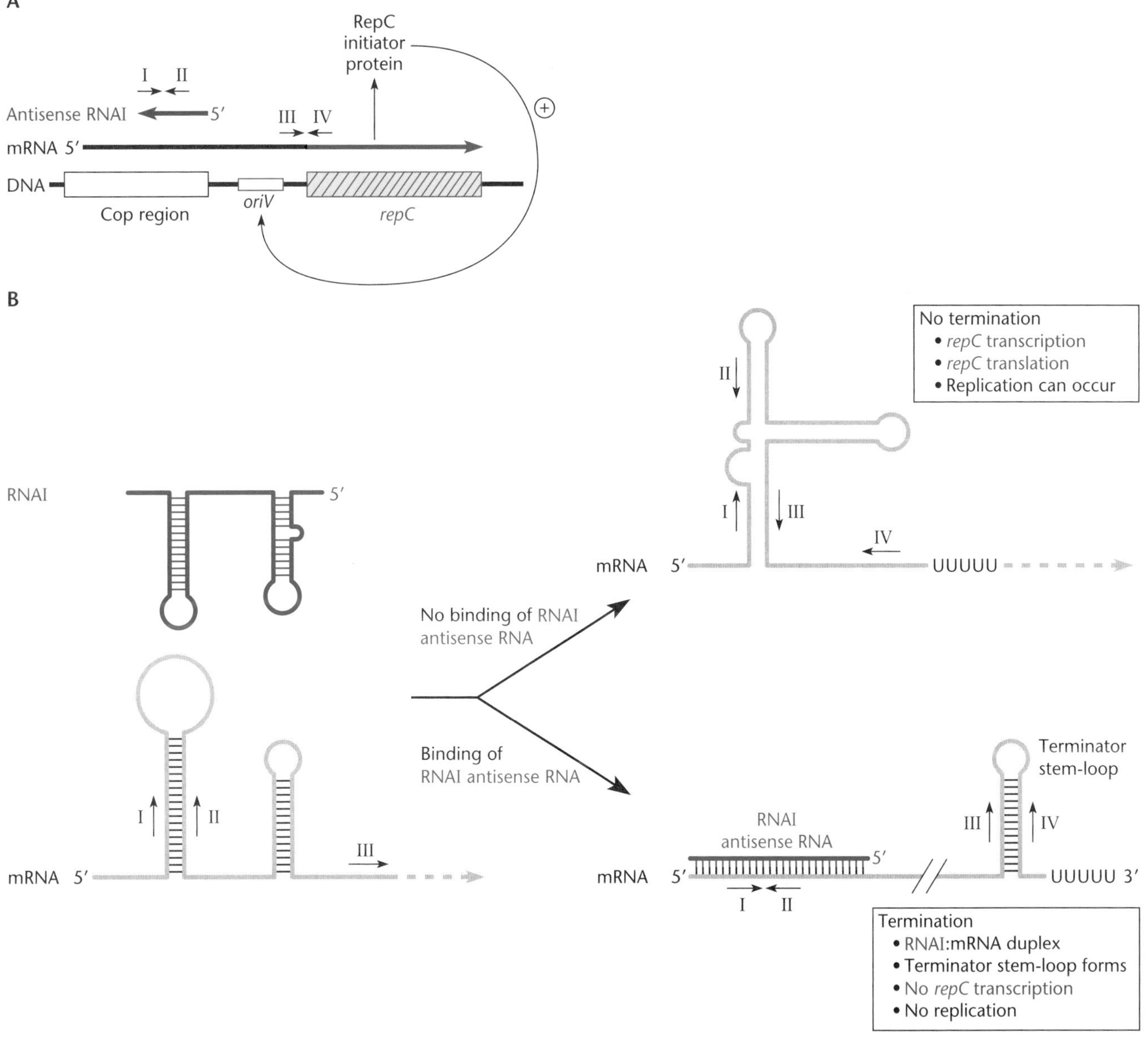

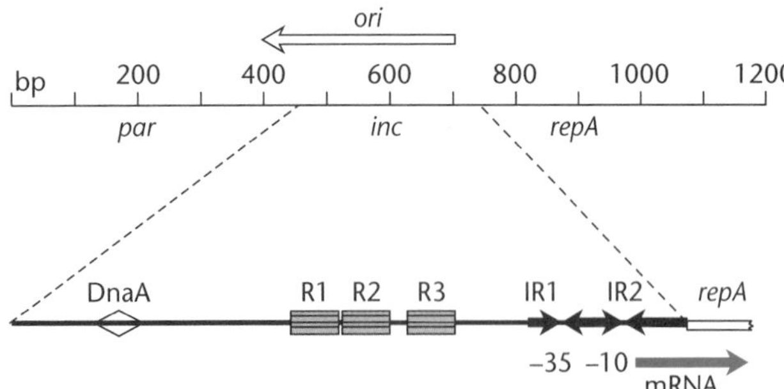

Figure 4.11 The *ori* region of pSC101. R1, R2, and R3 are the three iteron sequences (CAAAGGTCTAGCAGCAGAATTTACAGA for R3) to which RepA binds to handcuff two plasmids. RepA autoregulates its own synthesis by binding to the inverted repeats IR1 and IR2. The location of the partitioning site, *par* (see "Partitioning"), and the binding sites for the host protein DnaA are also shown.
doi:10.1128/9781555817169.ch4.f4.11

such as F and P1. Iteron plasmids must have another mechanism to regulate their copy numbers within narrow limits. This other form of regulation has been hypothesized to be the coupling of plasmids through the Rep protein and their iteron sequences (see McEachern et al., Suggested Reading). The **coupling hypothesis** for regulation of plasmid replication is illustrated in Figure 4.12. When the concentration of plasmids is high enough, they couple with each other via bound RepA proteins. This inhibits the replication of both coupled plasmids. The coupling mechanism allows plasmid replication to be controlled, not only by how much RepA protein is present in the cell, but also by the concentration of the plasmid itself or, more precisely, the concentration of the iteron sequences on the plasmid. Direct support for the coupling model in the replication control of iteron plasmids has come from electron micrographs of purified iteron plasmids mixed with the purified RepA protein for that plasmid. In these pictures, two plasmid molecules can often be seen coupled by RepA protein. In vitro and in vivo work also supports coupling as an important mechanism to prevent plasmid overreplication (see Das et al., Suggested Reading).

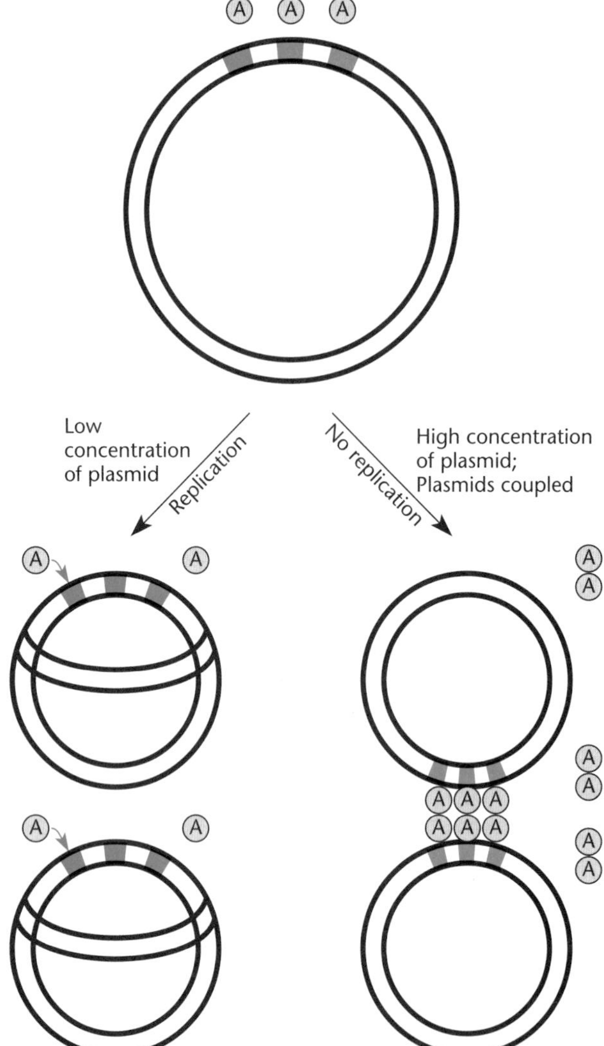

Figure 4.12 The "handcuffing" or "coupling" model for regulation of iteron plasmids. At low concentrations of plasmids, the RepA protein binds to only one plasmid at a time, initiating replication. At high plasmid and RepA concentrations, the RepA protein may dimerize and bind to two plasmids simultaneously, handcuffing them and inhibiting replication.
doi:10.1128/9781555817169.ch4.f4.12

HOST FUNCTIONS INVOLVED IN REGULATING PLASMID REPLICATION

As mentioned above, in addition to Rep, many plasmids require host proteins to initiate replication. For example, some plasmids require the DnaA protein, which is normally involved in initiating replication of the chromosome, and have *dnaA* boxes in their *oriV* regions to which DnaA binds (Figure 4.11). The DnaA protein may also directly interact with the Rep proteins of some plasmids. This may explain why some broad-host-range plasmids, such as RP4, make two Rep proteins. The different forms of the Rep protein might better interact with the DnaA proteins of different species of bacteria (see Caspi et al., Suggested Reading). The DnaA protein is involved in coordinating replication of the chromosome with cell division (see chapter 1); making their replication dependent on DnaA may allow plasmids to better coordinate their own replication with cell division. Like the chromosome origin (*oriC*), some *E. coli* plasmids also have Dam methylation sites close to their *oriV*. These methylation sites presumably help to further coordinate their replication with cell division. As with the chromosomal origin of replication, both strands of DNA at these sites must be fully methylated for initiation to occur. Immediately after initiation, only one strand of these sites is methylated (hemimethylation), delaying new initiations at the sites (see the discussion of sequestration of chromosome origins in chapter 1). Despite substantial progress, however, the method by which the replication of very stringent plasmids, such as P1 and F (with a copy number of only 1), is controlled to within such narrow limits is still something of a mystery and the object of current research.

Mechanisms To Prevent Curing of Plasmids

Cells that have lost a plasmid during cell division are said to be **cured** of the plasmid. Several mechanisms prevent curing, including toxin-antitoxin systems (Box 4.2), site-specific recombinases that resolve multimers, and partitioning systems. The last two are reviewed below.

RESOLUTION OF MULTIMERIC PLASMIDS

The possibility that a cell will lose a plasmid during cell division is increased if the plasmids form dimers or higher multimers during replication. A dimer consists of two individual copies of the plasmid molecules linked head to tail to form a larger circle, and a multimer links more than two such monomers. Such dimers and multimers probably occur as a result of recombination between monomers. Recombination between two monomers forms a dimer, and subsequent recombination can form higher and higher multimers. Also, RC replication of RC plasmids can form multimers if termination after each round of replication is not efficient. Multimers may replicate more efficiently than monomers, perhaps because they have more than one origin of replication, so they tend to accumulate if the plasmid has no effective way to remove them. The formation of multimers creates a particular problem when the plasmid attempts to segregate into the daughter cells on cell division. One reason is that multimers lower the effective copy number. Each multimer segregates into the daughter cells as a single plasmid, and if all of the plasmid is taken up in one large multimer, it can segregate into only one daughter cell. Also, the presence of more than one *par* site on the multimer may cause it to be pulled to both ends of the cell at once, much like a dicentric chromosome can lead to nondisjunction in higher organisms. Therefore, multimers greatly increase the chance of a plasmid being lost during cell division.

To avoid this problem, many plasmids have site-specific recombination systems that resolve multimers as soon as they form. These systems can be either chromosomally encoded or encoded by the plasmid itself (see chapter 9). A site-specific recombination system promotes recombination between specific sites on the plasmid if the same site occurs more than once in the molecule, as it would in a dimer or multimer. This recombination has the effect of resolving multimers into separate monomeric plasmid molecules.

A well-studied example of a plasmid-encoded site-specific recombination system is the Cre-*loxP* system encoded by P1 phage. This phage is capable of lysogeny, and its prophage form is a plasmid, subject to all the problems faced by other plasmids, including multimerization due to recombination. The Cre protein, a tyrosine (Y) recombinase, promotes recombination between two *loxP* sites on a dimerized plasmid, resolving the dimer into two monomers. This system is very efficient and relatively simple and has been useful in a number of studies, including demonstrations of the interspecies transfer of proteins (see chapter 5). It has also been used as a model for Y recombinases, since the recombinase has been crystallized with its *loxP* DNA substrate. Y recombinases and their mechanism of action are discussed in chapter 9.

The best-known examples of host-encoded site-specific recombination systems used to resolve plasmid dimers are the *cer*-XerCD and the *psi*-XerCD site-specific recombination systems used by the ColE1 plasmid and the pSC101 plasmid, respectively. The XerCD system is mentioned in chapter 1 in connection with segregation of the chromosome of *E. coli*. The XerC and XerD proteins are part of a site-specific recombinase that acts on a site, *dif*, close to the terminus of replication of the chromosome to resolve chromosome dimers created during chromosome replication. The plasmids have commandeered this site-specific system of the host to resolve their own dimers by having sites at which the recombinase can act. The site on the ColE1 plasmid is called *cer*, and the site on pSC101 is called *psi*. As described in chapter 1, resolution of dimer chromosomes at *dif* sites requires an interaction between XerCD and the FtsK protein. In contrast, sites on plasmids do not share this requirement because of differences in the sequence where recombination occurs; however, these plasmid base sites have a different requirement. The *cer* site on the ColE1 plasmid is not recognized as such but is recognized only if two other host proteins, called PepA and ArgR, are bound close by in the DNA, as shown in Figure 4.13. A similar situation occurs in the pSC101 plasmid, but there the auxiliary host proteins are PepA and ArcA~P (phosphorylated ArcA), which binds close to where ArgR binds in *cer*. Apparently, these other proteins bind to XerCD recombinase at the plasmid *cer* or *psi* sites and help orient it for the recombination process. However, it is not clear how these particular host proteins came to play this role. The only thing these accessory proteins have in common is that they all normally bind to DNA because they are involved in regulating transcription in the host. This is yet another example of a case where plasmids commandeer host functions for their own purposes, in this case, a site-specific recombination system normally used for resolving chromosome dimers, as well as transcriptional

BOX 4.2

Toxin-Antitoxin Systems and Plasmid Maintenance

As extrachromosomal elements, plasmids can easily be lost during division. As described in the text, numerous adaptations beyond encoding beneficial functions have evolved in plasmids to guard against loss, including copy number control, dimer resolution systems, and partitioning systems. However, plasmids have an additional trick for being maintained on plasmids, called toxin-antitoxin systems. In what seems like revenge, some plasmids encode proteins that kill a cell if it is cured of the plasmid. Such functions are relatively common on mobile plasmids, including the F plasmid, the R1 plasmid, and the P1 prophage, which replicates as a plasmid (see chapter 8). These systems have been called plasmid addiction systems because they cause the cell to undergo severe withdrawal symptoms and die if it is cured of the plasmid to which it is addicted.

Plasmid addiction systems all use basically the same strategy. They consist of two components, which can be either proteins or RNA. One component functions as a toxin, and the other functions as an antitoxin or antidote. While the cell contains the plasmid, both the toxin and the antitoxin are made, and the antitoxin somehow inactivates the effect of the toxin, either by binding to the toxin and inactivating it directly or by somehow indirectly alleviating its effect. The PhD-Doc system is an example of the former type (see the figure). The PhD protein is the antitoxin that binds to the Doc toxin and inactivates it. Restriction-modification systems are an example of the latter type. The restriction endonuclease component of the restriction system is the toxin that cuts the chromosome and kills the cell, but not if the modification component, the antitoxin, has methylated a base in its recognition sequence (see chapter 1). Once the cell is cured of the plasmid, neither the toxin nor the antitoxin continues to be made. The toxin, however, is more stable than the antitoxin, so eventually, the antitoxin is degraded and the cured cells contain only the toxin. Without the antitoxin to counteract it, the toxin interrupts one of the essential processes in the cell.

The toxic proteins can affect any one of many essential processes in bacteria, depending on the system. For example, the toxic protein of the F plasmid, Ccd, alters DNA gyrase so that it causes double-strand breaks in the DNA. The Hok protein of plasmid R1 destroys the cellular membrane potential, causing loss of cellular energy. The Doc toxin of P1 inhibits translation elongation through and interaction with the 30S ribosomal subunit. Revenge aside, it is hard to understand the evolutionary advantage of killing cells after plasmid loss. There may be two interrelated answers to why mobile plasmids may contain toxin-antitoxin systems. One is that plasmid loss would likely lead to better growth of the plasmidless cell. Therefore, the newly cured cells could otherwise overgrow the plasmid-containing cells in the same population, and the toxin-antitoxin systems take away this advantage. A second benefit to having the toxin-antitoxin system on a mobile element is that the cells affected by the toxin could be saved from the toxic effect of plasmid loss if they again obtained the mobile plasmid from an adjacent cell. Therefore, part of the advantage may not be killing by the toxin, but instead, a type of paralysis until the mobile plasmid can return to the cell that lost the plasmid.

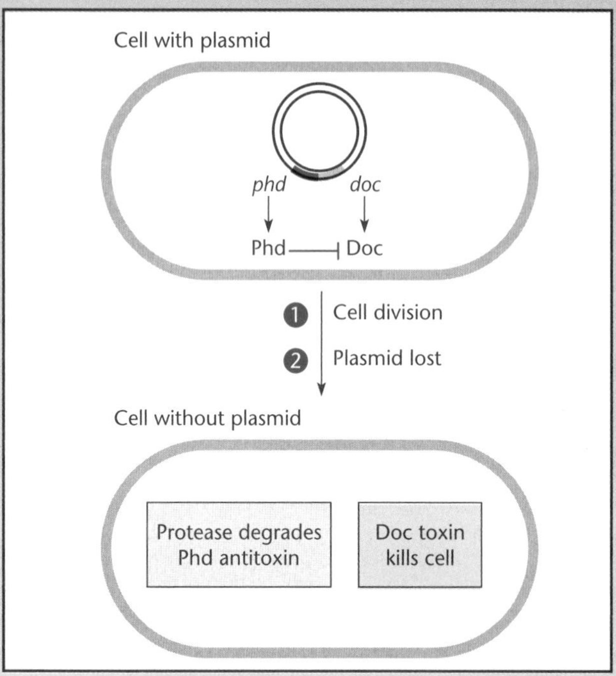

The P1 phage-encoded plasmid addiction system Phd-Doc. Cells containing the plasmid contain both Phd and Doc; Phd is the antidote to Doc, binding to it and inactivating it. If the cell is cured of the plasmid, neither Phd nor Doc is made, but Doc is more stable than Phd and outlives it. Once Phd is degraded by a protease, Doc kills the cell (actually, Doc only inhibits translation, and MazE kills the cell in response).
doi:10.1128/9781555817169.ch4.Box4.2.f

There is a good rationale for plasmid addiction systems in that they prevent cells cured of the plasmid from accumulating and thus help ensure survival of the plasmid. Therefore, it was a surprise to discover that similar toxin-antitoxin systems also occur in the chromosome. Some of these are on exchangeable DNA elements, such as genetic islands and superintegrons (see chapter 9), and may play a role similar to that of the addiction systems of plasmids, preventing loss of the DNA element. They could be considered selfish genes

BOX 4.2 (continued)

Toxin-Antitoxin Systems and Plasmid Maintenance

that prevent themselves, and therefore the DNA element in which they reside, from being lost from the cell. However, other toxin-antitoxin modules seem to be encoded by normal genes in the chromosome. Two examples of these are the MazEF and RelBE systems, both found in *E. coli* K-12. These two systems work in remarkably similar ways. MazF and RelE are the toxins and are RNases that cleave mRNA in the ribosome and block translation, killing the cell. MazF and RelE are the antitoxins, which bind to the toxins MazF and RelE, respectively, and inactivate them. The toxins and antitoxins are not made if translation is inhibited, but the toxin is more stable and longer lived than the antitoxin. These systems could therefore be considered suicide modules in that they cause the cell to kill itself if translation is inhibited, for example, by antibiotics or a plasmid addiction module, such as PhD-Doc, if the cell is cured of the P1 plasmid. A number of hypotheses have been proposed to explain the existence of these suicide modules. One is that they help prevent the spread of phages that inhibit host translation. Another is that they help shut down cellular metabolism in response to starvation and help ensure the long-term survival of some of the cells. It might be relevant that bacteria with a free-living lifestyle in the environment tend to have many more such suicide modules than do obligate parasitic bacteria, which can live only in the more stable, nurturing environment of a eukaryotic host.

Another apparent suicide system in *B. subtilis* lends credence to the idea that the purpose of suicide systems might be to kill some bacteria so that others may live (see Ellermeier et al., References). This system is much more complex than the others we have mentioned and consists of many genes in two operons, *skf* and *sdp*. To summarize, when a population of *B. subtilis* cells is starved for nutrients, some of them begin to sporulate. These cells produce a toxin that kills other cells that were slow to start sporulating. The killed cells are then devoured by the cells producing the toxin, which can then reverse their sporulation process. This buys the sporulating cells more time, in case the situation changes and they do not really need to sporulate, allowing them to avoid the need to employ a drastic measure to ensure survival.

References

Ellermeier, C. D., E. C. Hobbs, J. E. Gonzales-Pastor, and R. Losick. 2006. A three-protein signaling pathway governing immunity to a bacterial cannibalism toxin. *Cell* **124**:549–559.

Engelberg-Kulka, H., R. Hazan, and S. Amitai. 2005. *mazEF*: a chromosomal toxin-antitoxin module that triggers programmed cell death in bacteria. *J. Cell Sci.* **118**:4327–4332.

Greenfield, T. J., E. Ehli, T. Kirshenmann, T. Franch, K. Gerdes, and K. E. Weaver. 2000. The antisense RNA of the *par* locus of pAD1 regulates the expression of a 33-amino-acid toxic peptide by an unusual mechanism. *Mol. Microbiol.* **37**:652–660.

Hazan, R., B. Sat, M. Reches, and H. Engelberg-Kulka. 2001. Postsegregational killing by the P1 phage addiction module *phd-doc* requires the *Escherichia coli* programmed cell death system *mazEF*. *J. Bacteriol.* **183**:2046–2050.

regulatory proteins that are used for host cell gene regulation. XerCD is also a Y recombinase, and its mechanism of action is discussed in more detail in the section on tyrosine recombinases in chapter 9.

PARTITIONING

A very effective mechanism that plasmids have to avoid being lost from dividing cells is **partitioning** systems. These systems ensure that at least one copy of the plasmid is present in each daughter cell after cell division. The functions involved in these systems are called **Par functions**, and in many ways they are analogous to the Par functions involved in chromosome segregation. In fact, the discovery of Par systems in plasmids preceded their discovery in chromosome segregation.

The Par Systems of Plasmids

The Par systems have been studied extensively in plasmids. The Par systems of low-copy-number plasmids fall into at least two groups whose members are related to each other by sequence and function. At least one of these groups of plasmid Par functions is also related to the putative chromosomal Par systems of some bacteria, as described in chapter 1. One group is represented by the Par system of the R1 plasmid of *E. coli*, and the other, much larger group is represented by the Par systems of the F, P1, and broad-host-range RK2 plasmids, among others. The latter is also the group to which the chromosomal partitioning systems from the bacteria *Bacillus subtilis* and *Caulobacter crescentus* belong. The two groups of partitioning systems differ in the details of how they achieve the feat of plasmid partitioning, but evidence is accumulating that they do this by forming dynamic filaments in the cell. The R1 plasmid partitioning system is addressed first, since it seems to be the better understood of the two systems. It is also the best example to date of a dynamic filament-forming structure in bacteria that can move cellular constituents around, analogously to the actin filaments of eukaryotes (see "Bacterial Cell Biology and the Cell Cycle" in chapter 14).

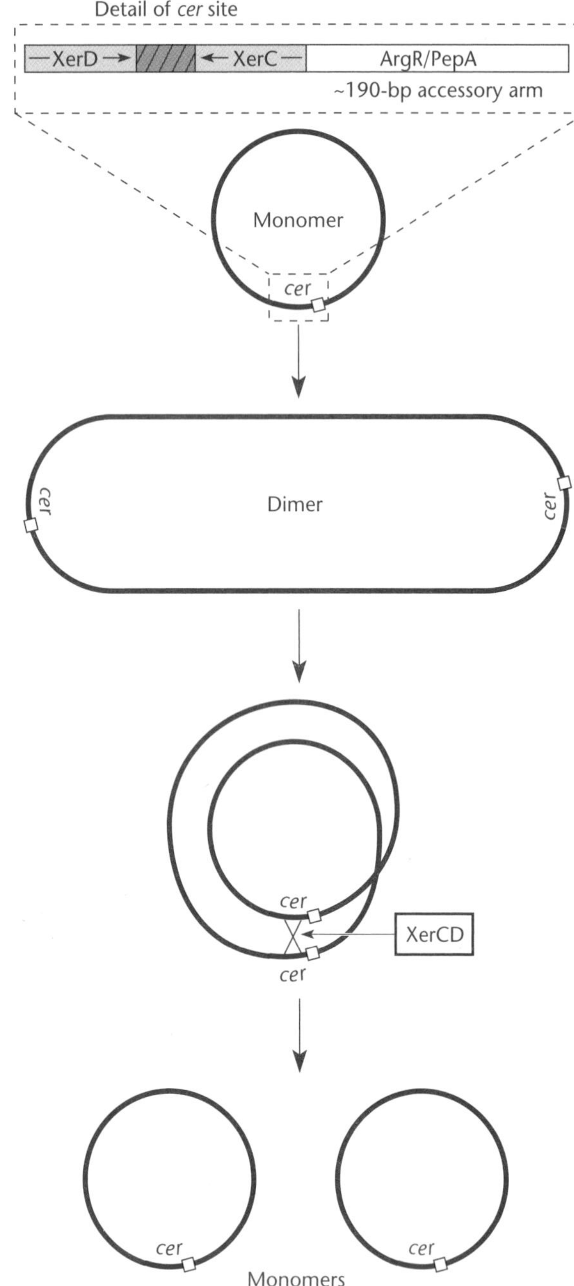

Figure 4.13 The Xer functions of *E. coli* catalyze site-specific recombination at the ColE1 plasmid *cer* site to resolve plasmid dimers. The sites of binding of the host proteins ArgR and PepA are shown.
doi:10.1128/9781555817169.ch4.f4.13

THE R1 PLASMID Par SYSTEM

The mechanism of partitioning by the R1 plasmid is illustrated in Figure 4.14. The partitioning system of the R1 plasmid consists of two protein-coding genes, *parM* and *parR*, as well as a centromere-like *cis*-acting site, *parC*. The actin-like ParM protein forms a filament that pushes the plasmids to the cell poles (see Campbell and Mullins, Suggested Reading). The polymerization process occurs quickly, rapidly pushing the plasmids to the cell poles (Figure 4.14). This is followed by depolymerization, leaving the plasmids at the poles. The polymerization and depolymerization process is regulated by the ATP-bound state of ParM; in the test tube, ParM proteins form a stable filament in the presence of nonhydrolyzable ATP, indicating that the shift from the ATP- to the ADP-bound state is important in the depolymerization process and the dynamic nature of the filaments. The prevailing model that supports the available data holds that polymerization of ParM-ATP occurs under the cap-like ParR-*parC* complex, propelling the complex toward the cell pole. Depolymerization would initiate from the base of the filament as ParM-bound ATP is hydrolyzed to ADP. Unlike other ATP-binding proteins that are also found in ADP-bound form, it appears that ParM can also be found in a third form, bound to ADP plus inorganic phosphate (P_i). The transition state of ParM-ADP plus P_i is thought to help produce a fragile intermediate state that completely collapses once ParM-ADP forms at the base of the filament.

THE P1 AND F PLASMID Par SYSTEMS

The larger group of plasmid-partitioning systems is related to the systems of the P1 plasmid prophage and the F plasmid. Also, some bacteria, including *B. subtilis* and *C. crescentus*, seem to use a similar Par system to aid in partitioning and organizing their chromosomes (see chapter 1). The composition of these Par systems is grossly similar to that of the Par system of the R1 plasmid in that they usually consist of two proteins and an adjacent *cis*-acting site on the DNA to which one of the proteins, often called ParB, binds. Also, the other protein, often called ParA, has ATP-binding motifs. However, the ATP-binding protein has no homology to actin or any other cytoskeletal protein and belongs to a special family of proteins called the P-loop ATPases. It is distantly related to MinD, a protein that helps select the division site in bacteria and, in *E. coli* at least, oscillates from one end of the cell to the other during the cell cycle (see chapter 1). In the F plasmid, the corresponding proteins are called SopA and SopB, and the site is called *sopC* (*sop*, for stability of plasmid). For simplicity, we refer to them as ParA, ParB, and *parS*. In some plasmids, the Par systems are encoded in the same region of the plasmid as the replication proteins; in rare instances, the system may lack an autonomous ParB protein, and perhaps a larger ParA-related protein may play both roles.

The mechanism of action of these Par systems appears to be fundamentally different from that of the system described above. Recent microscopic analyses of plasmid partitioning with these systems indicated that it involves directed movement across the nucleoid

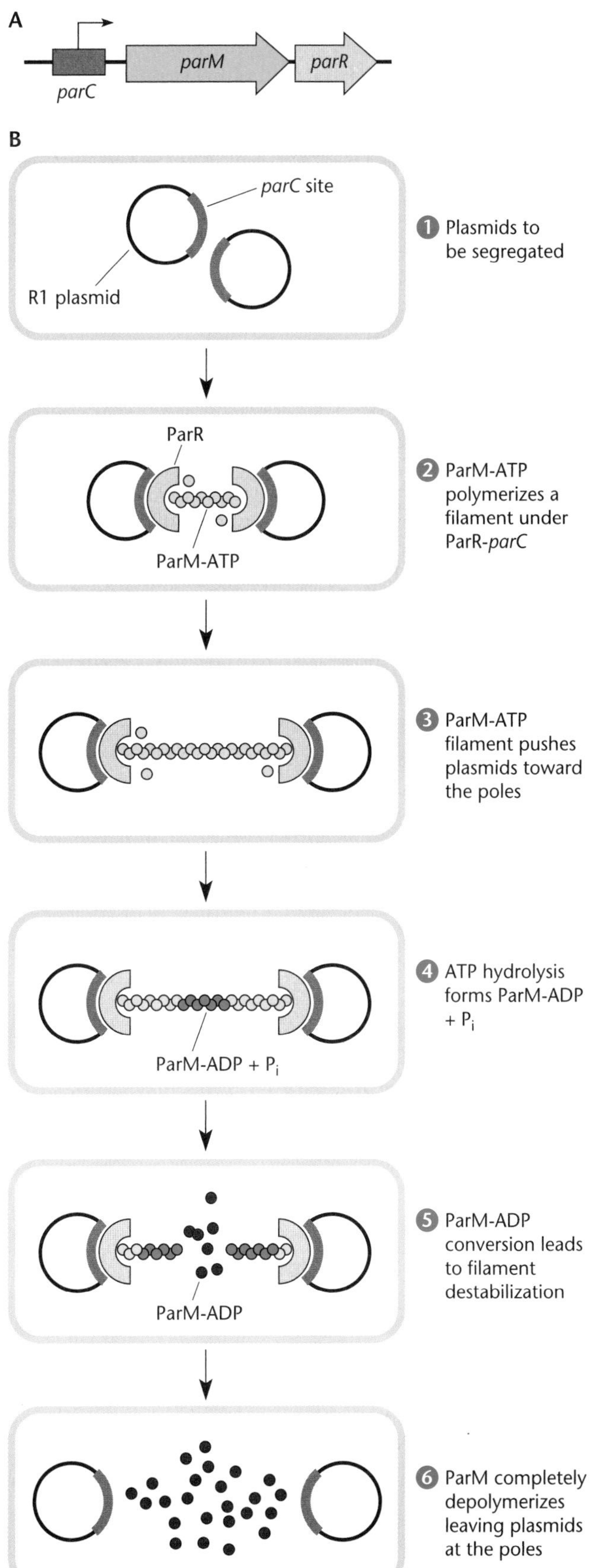
A
parC
parM
parR
B
parC site
R1 plasmid
1 Plasmids to be segregated
ParR
ParM-ATP
2 ParM-ATP polymerizes a filament under ParR-parC
3 ParM-ATP filament pushes plasmids toward the poles
ParM-ADP + P
4 ATP hydrolysis forms ParM-ADP + P
ParM-ADP
5 ParM-ADP conversion leads to filament destabilization
6 ParM completely depolymerizes leaving plasmids at the poles

for plasmid segregation. A model that accounts for the available observations holds that instead of being pushed, as in the R1 system, the ParB-*parS* complex is actually pulled by an attraction to ParA that polymerizes at random places on the nucleoid. In one of the models, this would involve the ParB-*parS* complex being pulled as it depolymerizes ParA filaments (see Ringgaard et al., Suggested Reading) (Figure 4.15). Repeated cycles of ParA filament assembly and disassembly would tend to maximize the distribution of plasmid copies to opposite cell halves, so that upon cell division, the plasmids are appropriately segregated. In this model, the partition system can be separated into two interacting subsystems. In the first subsystem, ATP-bound ParA dimers bind chromosomal DNA at random positions nonspecifically but cooperatively leading to ParA filament formation. In the second part of the system, dimers of ParB bind specifically to the *parS* site on the plasmid and nucleate the binding of new ParB dimers around this site on the plasmid. The ParB-*parS* complex on the plasmid then becomes competent for associating with a ParA filament on the chromosome via one of the ends of the filament. The interaction with ParB signals ParA dimers to hydrolyze ATP to ADP and release of ParA from DNA. As the first dimer of ParA is released, the ParB-*parS* complex then binds to the next ParA dimer on the filament. This behavior would allow the ParB-*parS* complex to be pulled along the nucleoid as it goes through cycles of ParA binding on the receding ParA filament. When the end of the filament is reached, the plasmid dissociates and is free to move to a new filament at another location. The ParA dimers released from disassociating filaments can also move to another location on the chromosome and make new filaments, but the exchange of ADP with ATP is a slow process that favors ParA diffusion to a place different from where it was released. Such a system will naturally distribute *parS*-containing plasmids away from

Figure 4.14 Model for partitioning of the R1 plasmid. **(A)** Structure of the *par* locus of R1, showing the positions of the *parM* and *parR* genes, as well as the *cis*-acting *parC* site. The transcription start site is in the *parC* site. **(B)** ParR binds to *parC*, making a site of ParM-ATP nucleation. Filaments grow by adding successive ParM-ATP subunits under the ParR-*parC* "cap," where growth of the filament pushes the plasmid containing *parC* to the cell poles. The filament is destabilized as ParM hydrolysis converts it to ParM-ADP, possibly through a ParM-ADP plus P_i intermediate, where rapid loss of the filament occurs as a wave of ParM-ADP-mediated instability passes through the filament, destabilizing it. ParM can then be recharged with ATP before the plasmids are partitioned again prior to the next cell division. The *parC* site on the plasmid is shown in grey.
doi:10.1128/9781555817169.ch4.f4.14

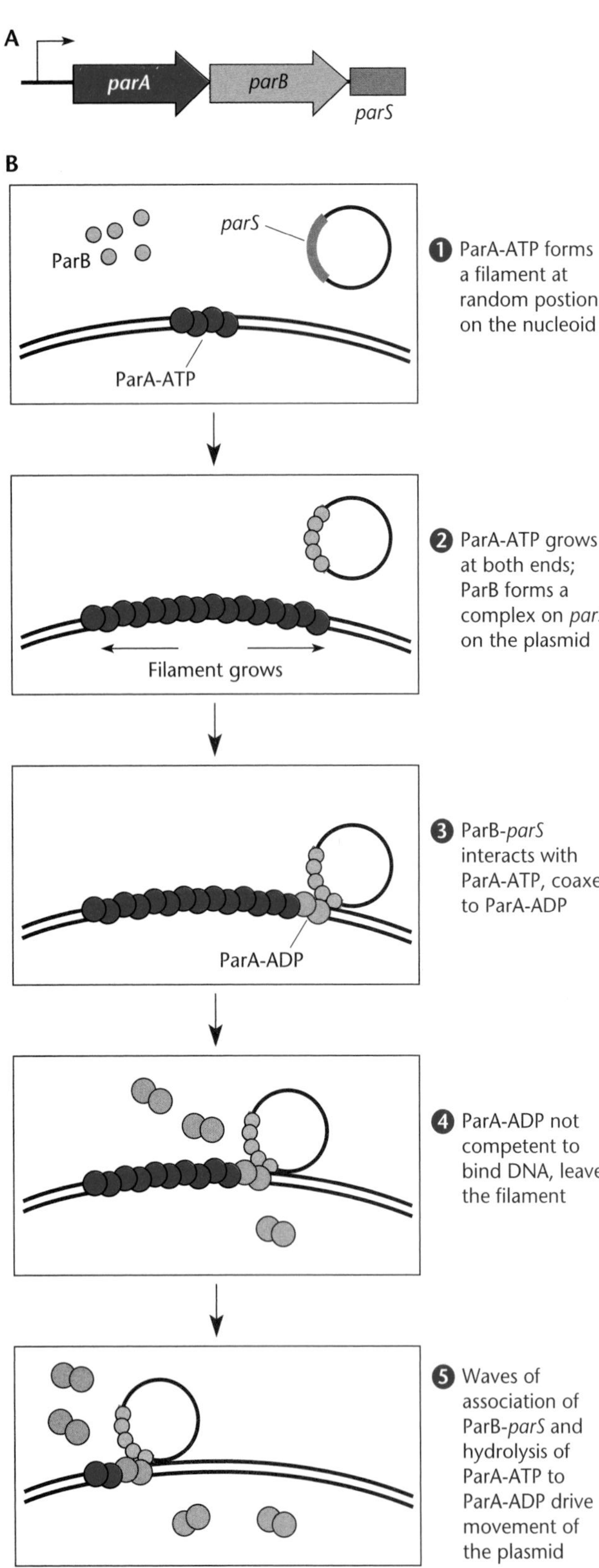
A
parA
parB
parS
B
ParB
parS
ParA-ATP
1 ParA-ATP forms a filament at random postions on the nucleoid
Filament grows
2 ParA-ATP grows at both ends; ParB forms a complex on parS on the plasmid
ParA-ADP
3 ParB-parS interacts with ParA-ATP, coaxes to ParA-ADP
4 ParA-ADP not competent to bind DNA, leaves the filament
5 Waves of association of ParB-parS and hydrolysis of ParA-ATP to ParA-ADP drive movement of the plasmid

one another in the cell. ParA filament formation is difficult to demonstrate *in vitro*, and a variation on this model has been suggested in which the ParA protein accumulates under some other form of polymerization on the nucleoid (see Vecchiarelli et al., Suggested Reading). The ParB-*parS* complex would be attracted to the region where ParA accumulated and, in turn, act to mobilize ParA off the nucleoid by encouraging ATP hydrolysis. In this way, the Par proteins would go through a type of chasing behavior reminiscent of the activity of the MinE and MinD proteins but on a local level across the nucleoid instead of an oscillation across the entire length of the cell.

INCOMPATIBILITY DUE TO PLASMID PARTITIONING

If two plasmids share the same partitioning system, they will be incompatible, even if their replication control systems are different. Incompatibility due to shared partitioning systems makes sense, considering the models presented above. If two plasmids that are otherwise different share the same Par system, one plasmid of each type can be directed to opposite ends of the cell before the cell divides. In this way, one daughter cell can get a plasmid of one type while the other daughter cell gets the plasmid of the other type, and cells are cured of one or the other plasmid. However, even though shared partitioning systems can cause incompatibility, this is usually not the sole cause of their incompatibility. Usually, cells with the same partitioning system also share the same replication control system, since the two are often closely associated on the plasmid; therefore, the incompatibility is due to both systems. In fact, in some cases, the replication control genes and the partitioning genes are intermingled around the origin of plasmid replication.

Figure 4.15 Model for partitioning by *par* systems on P1, F, and RK2. **(A)** Structure of the *par* locus on P1, F, and RK2 showing the positions of the *parA* and *parB* genes and the *cis*-acting *parS* site. Transcription for the promoter (P*parAB*) is controlled by the ParA protein. **(B)** Plasmids containing the *parS* site bound by ParB are pulled across the chromosome as they mediate the depolymerization of ParA filaments across the nucleoid. ParA-ATP binds randomly as a dimer to the nucleoid and polymerizes in both directions, while ParB specifically associates with *parS* sites on the plasmid. Interaction between the ParB-*parS* complex and the ParA filament stimulates ParA to hydrolyze ATP, converting it to ParA-ADP, a form that dissociates from the end of the filament. The ParB-*parS* complex is then free to associate with the next dimer of ParA-ATP in the filament, causing net movement of the ParB-*parS* complex-containing plasmid. ParA-ADP can then convert to its ParA-ATP form and associate at the other end of the filament or elsewhere across the nucleoid.
doi:10.1128/9781555817169.ch4.f4.15

Plasmid Cloning Vectors

As discussed in chapter 1, a cloning vector is an autonomously replicating DNA (replicon) into which other DNAs can be inserted. Any DNA inserted into the cloning vector replicates passively with the vector, so that many copies (clones) of the original piece of DNA can be obtained. Plasmids offer many advantages as cloning vectors, and many types of plasmids have been engineered to serve as plasmid cloning vectors.

As described above, the physical properties of plasmids allow them to be purified easily for manipulation and reintroduction into bacteria. Because very few functions need to reside on the plasmid itself, they can be small, which, in addition to allowing them to be manipulated more easily, also reduces the burden on the cell. Given the universality of the genetic code, they offer a way of expressing proteins from other bacteria or those from other types of organisms. In fact, in one of the first cloning experiments, a frog gene was cloned into plasmid pSC101 (see Cohen et al., Suggested Reading). Cloning vectors have a wide variety of uses that make them important for the study of bacteria or as tools for a better understanding of a broad range of other organisms. For example, sometimes they are used as expression constructs to express a protein for study directly in the host, while at other times they may be used to overexpress a protein to facilitate isolating large amounts of the protein for work outside the cell, such as for structural studies, biochemical studies, or commercial applications (see chapters 2 and 7). Plasmid constructs are very commonly constructed using *E. coli* as a host; these constructs may later be moved into less tractable organisms, where they can also replicate, in what are called shuttle vectors (see below).

ANATOMY OF A PLASMID CLONING VECTOR

Most plasmids, as they are isolated from nature, are too large to be convenient as cloning vectors and/or do not contain easily selectable genes that can be used to move them from one host to another. Commonly used vectors have a number of attributes that make them more convenient to work with. Some of these features are explained below.

ORIGINS OF REPLICATION

The most basic feature of a plasmid cloning vector is an origin of replication that allows it to replicate independently of the chromosome. In early genetic experiments, the minimal features required for autonomous replication could be isolated with the help of a selectable gene product. For example, the origin of DNA replication region that allows autonomous replication can be cloned when it resides on the same DNA fragment as a selectable marker, such as a gene allowing antibiotic resistance. As shown in Figure 4.16, the plasmid is cut into several pieces with a restriction endonuclease (indicated by the arrows), and the pieces are ligated (joined) to another piece of DNA that has a selectable marker, such as resistance to ampicillin (Amp^r). For the experiment to work, the second piece of DNA cannot have a functional origin of replication. The ligated mixture is then introduced into bacterial cells by transformation, and the antibiotic-resistant transformants are selected by plating the mixture on agar plates containing growth medium and the antibiotic. The only DNA molecules able to replicate and also to confer antibiotic resistance on the cells are hybrids with both the *ori* region of the plasmid and the piece of DNA with the antibiotic resistance gene. Therefore, only cells harboring these hybrid molecules can grow on the antibiotic-containing medium. Using similar techniques, segregation systems could also be isolated as DNA regions that when cloned into these plasmid constructs would stabilize their maintenance after many generations of growth. Given

Figure 4.16 Finding the origin of replication (*ori*) in a plasmid. Random pieces of the plasmid are ligated to a piece of DNA containing a selectable gene but no origin of replication and introduced into cells. Cells that can form a colony on the selective plates contain the selectable gene ligated to the piece of DNA containing the origin.
doi:10.1128/9781555817169.ch4.f4.16

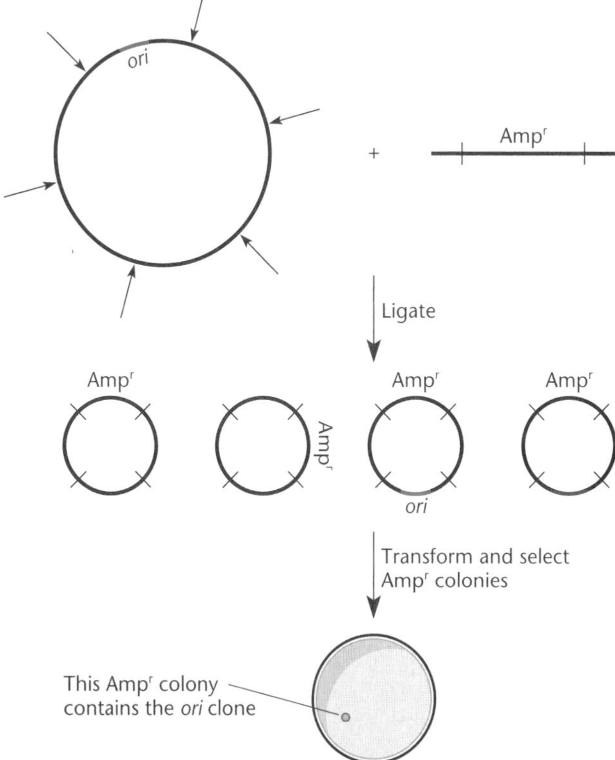

the broad availability of inexpensive DNA sequencing, DNA amplification using PCR has streamlined the manipulation of DNA, but these techniques are likely to continue to be important as more tools are needed for newly identified types of bacteria that lack a system that has been previously characterized.

SELECTABLE GENES

All plasmids exert some kind of load on the host and require a selectable gene to allow maintenance of the vector. In addition, a selectable marker is needed to select for cells that have received the plasmid, because of the low efficiency of the procedures used to introduce DNA. In some cases, it is useful to have plasmids with different selectable genetic markers, because more than one plasmid may need to be maintained at one time. Antibiotic resistance genes have historically been taken from transposons and other plasmids. Some antibiotic resistance genes that have been introduced into cloning vectors are the chloramphenicol resistance (Cam^r) gene of transposon Tn*9*; the tetracycline resistance (Tet^r) gene of plasmid pSC101; the Amp^r gene of transposon Tn*3*; and the kanamycin resistance (Kan^r) gene of transposon Tn*5*. The antibiotic resistance gene that is chosen depends on the uses to which the cloning vector will be put. Some antibiotic resistance genes, such as the Tet^r gene from pSC101, are expressed only in some types of bacteria closely related to *E. coli*, while others, such as the Kan^r gene from Tn*5*, are expressed in most gram-negative bacteria.

INTRODUCING UNIQUE RESTRICTION SITES

Since many applications of plasmid cloning vectors require that clones be introduced into restriction endonuclease cleavage sites, it is necessary that a cloning vector have some restriction sites that are unique. If a site is unique, the cognate restriction endonuclease cuts the vector at only that one site when it is used to cut the plasmid. One can then clone pieces of foreign DNA into the unique site, and the cloning vector will remain intact. Restriction sites for typical restriction endonucleases recognize a 6-bp sequence that on average will occur about once every 1,000 bp (see chapter 1), so that a cloning vector of 3,000 to 4,000 bp is apt to have more than one site for the restriction enzyme. Historically, many tricks have been used to eliminate the extra sites in the plasmid.

In many plasmid cloning vectors, the unique sites are located in a selectable gene, so that insertion of a foreign piece of DNA in the site inactivates the selectable gene. This is called **insertional inactivation** and is discussed below. Normally, during a cloning operation, only a small percentage of the cloning vector molecules pick up a foreign DNA insert. If those that have picked up an insert no longer confer the selectable trait, for example, resistance to an antibiotic, they can be more easily identified. Many cloning vectors also have the unique restriction sites for many different restriction endonucleases all grouped into one small region on the plasmid called a polyclonal or multiple restriction site. This offers the convenience of choosing among a variety of restriction endonucleases for cloning, and the cloned DNA is always inserted at the same general place in the vector, independent of the restriction endonuclease used. **Polyclonal sites** can also be used for **directional cloning**. If the cloning vector is cut by two different restriction endonucleases with unique sites within the polyclonal site, the resulting overhangs cannot pair to recyclize the plasmid. The plasmid can recyclize only if it picks up a piece of foreign DNA. If the piece of DNA to be cloned has overhangs for the two different sites at its ends, it is usually cloned in only one orientation into the polyclonal site.

The unique restriction sites can also be placed so that genes cloned into them will be expressed from promoters and TIRs on the plasmid. Plasmids with these features are called expression vectors and can be used to express foreign genes in *E. coli* and other convenient hosts. Such vectors can also be used to attach affinity tags to proteins to aid in their purification. Expression vectors and affinity tags are discussed in connection with translational and transcriptional fusions in chapter 2. Expression vectors have also been adapted from bacteriophage regulation systems, as discussed in chapter 7.

Examples of Plasmid Cloning Vectors

A number of plasmid cloning vectors have been engineered for special purposes. Almost all of these plasmids have at least some of the features mentioned above for a desirable cloning vector.

1. They are small, so that the plasmid can be easily isolated and introduced into various bacteria.
2. They have relatively high copy numbers, so that the plasmid DNA can be easily purified in sufficient quantities.
3. They carry easily selectable traits, such as a gene conferring resistance to an antibiotic, which can be used to select cells that contain the plasmid.
4. They have one or a few sites for specific restriction endonucleases, which cut DNA and allow the insertion of foreign DNA segments. Also, these sites usually occur in genes that can be easily screened for to facilitate the detection of plasmids with foreign DNA inserts by insertional inactivation.

Many plasmid cloning vectors have other special properties that aid in particular experiments. For example, some contain the sequences recognized by phage-packaging systems (*pac* or *cos* sites), so that they can be packaged into phage heads (see chapters 7 and 8). Expression vectors can be used to produce foreign proteins in bacteria. Mobilizable plasmids have mobilization (*mob*) sites, so they can be transferred by conjugation to other cells (see chapter 5). Some broad-host-range vectors have *ori* regions that allow them to replicate in many types of bacteria or even in organisms from different domains. Shuttle vectors contain more than one type of replication origin, so they can replicate in multiple unrelated organisms. These and some other types of specialty plasmid cloning vectors are discussed in more detail below and in later chapters.

pUC PLASMIDS

Some of the most commonly used plasmid cloning vectors are the pUC vectors and vectors derived from them. One pUC vector, pUC18, is shown in Figure 4.17. The pUC plasmids are very small (only ~2,700 bp of DNA) and, as explained above, have been modified to have copy numbers in the hundreds, making them relatively easy to purify. They also have the easily selectable Amp^r gene.

Figure 4.17 pUC expression vector. A gene cloned into one of the restriction sites in the multiple-cloning site almost invariably disrupts the coding sequence for the *lacZ* α-peptide. If it is inserted in the correct orientation, the gene is transcribed from the *lac* promoter (p_{lac}). If the open reading frame for the gene is in the same reading frame as that for the *lacZ* coding sequence, the gene is also translated from the *lacZ* TIR, and the N-terminal amino acids of *lacZ* become fused to the polypeptide product of the gene. doi:10.1128/9781555817169.ch4.f4.17

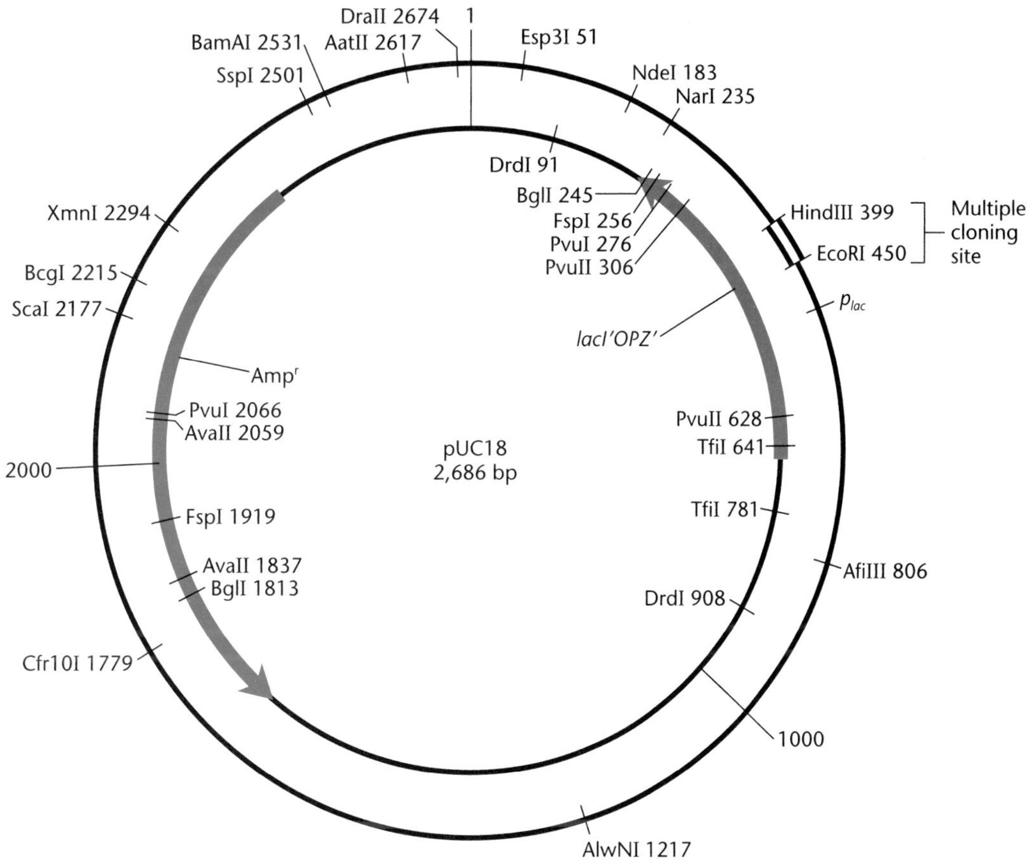

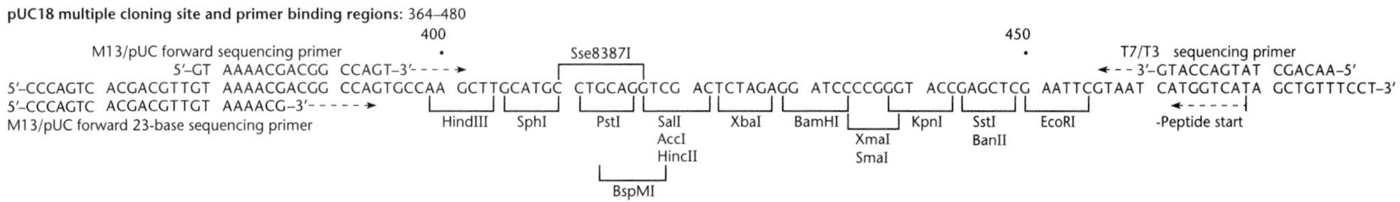

One of the most useful features of these plasmids is the ease with which they can be used for insertional inactivation. They encode the N-terminal region of the *lacZ* gene product, called the α-peptide, which is not active in the cell by itself but complements the C-terminal portion of the protein, called the *lacZ* ω-polypeptide, to make active β-galactosidase by a process termed α-complementation. This enzyme cleaves the indicator dye 5-bromo-4-chloro-3-indolyl-β-D-galactopyranoside (X-Gal) to produce a blue color. Some host strains, such as *E. coli* JM109, have been engineered to produce the ω-polypeptide of *lacZ* but not the α-peptide. As a consequence, *E. coli* JM109 lacking a plasmid forms white colonies on plates containing X-Gal, but cells containing a pUC plasmid form blue colonies on X-Gal plates because the two segments, the α-peptide encoded in the plasmid and the ω-polypeptide encoded in the chromosome, are capable of interacting to form a functional product capable of cleaving the X-Gal substrate. The pUC plasmids have a multicloning site containing the recognition sequences for many different restriction endonucleases within the coding region for the α-peptide (Figure 4.17). If a foreign DNA fragment is cloned into any one of these sites, the bacterium does not make the α-peptide and the colonies are white on X-Gal plates; bacteria containing plasmids with inserts are therefore easy to identify. These plasmids are also transcription vectors, because the promoter, p_{lac} (Figure 4.17), that drives expression of the α-peptide can also be used to express what is encoded in a piece of DNA directionally cloned into the multicloning site on the plasmid. The *lac* promoter is also inducible and only expressed if an inducer, such as isopropyl-β-D-thiogalactopyranoside (IPTG) or lactose, is added (see chapter 12). Thus, the cells can be propagated before the synthesis of the gene product is induced, a feature that is particularly desirable if the gene product is toxic to the cell. Genes cloned into one of the multicloning sites in the *lacZ* gene can also be translated from the *lacZ* TIR on the plasmid, provided that there are no intervening nonsense codons and the gene is cloned in the same reading frame as the upstream *lacZ* sequences.

CONDITIONAL VECTORS

In some cases, it is useful to have a vector that can replicate only under certain conditions. Low-copy-number vectors with temperature-sensitive replication are often useful because they facilitate plasmid curing. The vector with temperature-sensitive replication is introduced at the permissive temperature of 30°C in *E. coli*. When the cell population is shifted to the nonpermissive temperature of 40°C, replication stops and the plasmid is lost by dilution. Temperature-sensitive vectors can be useful as tools for integrating DNA sequences into the chromosome for gene fusions, as described in chapter 3. In some cases, it is useful to have a plasmid system in which replication occurs only in a certain host background. For example, as described in chapter 9, an in vitro transposition system in which a mobile DNA transposon containing an antibiotic resistance gene moves from a donor DNA plasmid into a different target plasmid as a way to subject it to insertional mutagenesis has been developed. In such a procedure, it is convenient to transform all of the DNA products under conditions where only the donor plasmid can replicate. One popular system utilizes the plasmid R6K origin of replication, called γ (see Metcalf et al., Suggested Reading). The γ origin requires the *trans*-acting protein π for replication. Normally the π protein is encoded by a gene called *pir* on the plasmid. However, in this system, only the *cis*-acting origin remains on the plasmid. Because the π protein is not normally found in *E. coli*, these vectors replicate only in a conditional host where the *pir* product is expressed from the host chromosome. The *pir* gene is either inserted into a neutral site or is introduced through the use of phage λ. Under this system, the plasmid is maintained at a copy number of about 15 with the wild-type *pir* gene. However, a strain can also be used with a mutant *pir* gene, *pir-116*, which maintains the plasmid vector with a copy number of about 250, making it more amenable to plasmid purification.

BACTERIAL ARTIFICIAL CHROMOSOME VECTORS

One problem with using high-copy-number cloning vectors such as pUC vectors is that the clones are very unstable, particularly if they contain large DNA inserts. If the clone exists in many copies, ectopic recombination between repeated sequences in the copies can rearrange the sequences in the clone (see chapter 3). This is a particular problem in some applications, for example, the sequencing of large genomes, such as the human genome, where it is necessary to obtain plasmid libraries containing very large inserts. The DNA of higher eukaryotes, including humans, contains many repeated sequences. For this reason, **bacterial artificial chromosome (BAC) cloning vectors** have been designed (Figure 4.18). These plasmid vectors are based on the F plasmid origin of replication and have a copy number of only 1 in *E. coli*. They can also accommodate very large inserts, on the order of 300,000 bp of DNA. This was expected, since it was known that F′ factors can be very large and are quite stable, especially in a $RecA^-$ host. An F′ factor is a naturally occurring plasmid in which the F plasmid has incorporated a large region of the *E. coli* chromosome (see chapter 5).

The original pBAC vector, shown in Figure 4.18 (see Shizuya et al., Suggested Reading), contains the F plasmid origin of replication and partitioning functions and a selectable chloramphenicol resistance gene.

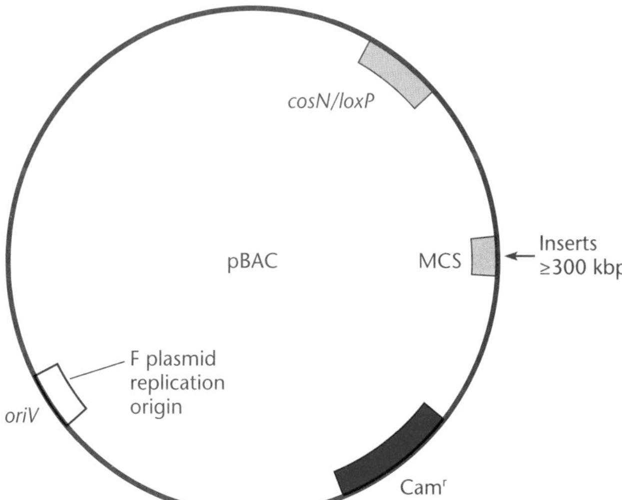

Figure 4.18 pBAC cloning vector for cloning large pieces of DNA for genome sequencing. The multiple cloning site (MCS) where clones are inserted is shown. Also shown are the *loxP* and *cosN* sites, where the plasmid can be cut by the Cre recombinase or λ terminase, respectively, for restriction mapping of the insert. These recognition sites are long enough that they almost never occur by chance in the insert.
doi:10.1128/9781555817169.ch4.f4.18

It also contains unique HindIII and BamHI restriction endonuclease cleavage sites into which large DNA fragments can be introduced, as well as a number of other features that are helpful in the cloning and sequencing of large fragments. Surrounding the cloning sites are the sites for other restriction endonucleases, chosen to be very GC rich so that they are not likely to exist in human DNA, which is relatively low in GC base pairs. This allows the DNA inserts to be excised from the cloning vector without (usually) cutting the DNA insert, as well. More contemporary vectors use sites recognized by an unusual class of enzymes called homing endonucleases, which have recognition sequences of around 30 bp. Even though these sites are not as stringently recognized as those recognized by restriction endonucleases, they still virtually eliminate the possibility that the site will be found in the DNA that is intended to be cloned. Two sites, *loxP* and *cosN*, allow the plasmid to be cut at unique sites for restriction site mapping of the clone.

Broad-Host-Range Cloning Vectors

As explained above, many of the common *E. coli* cloning vectors, such as pBR322, the pUC plasmids, and the pET plasmids, have modified *ori* regions derived from ColE1. These modified ColE1 derivatives maintain the very narrow host range found with the original plasmid (Table 4.2). They replicate only in *E. coli* and a few of its close relatives. However, some cloning applications require a plasmid cloning vector that replicates in other gram-negative bacteria, so cloning vectors have been derived from the broad-host-range plasmids RSF1010 and RK2, which replicate in most gram-negative bacteria. In addition to the broad-host-range *ori* region, these cloning vectors sometimes contain a *mob* site, which can allow them to be mobilized into other bacteria (see chapter 5). This trait is very useful, because ways of introducing DNA other than conjugation have not been developed for many types of bacteria, although electroporation works for many (see chapter 6).

SHUTTLE VECTORS

Sometimes, an experiment requires that a plasmid cloning vector be transferred from one organism into another. If the two organisms are not related, the same plasmid *ori* region is not likely to function in both organisms. Such applications require the use of **shuttle vectors**, so named because they can be used to "shuttle" genes between the two organisms. A shuttle vector has two origins of replication, one that functions in each organism. Shuttle vectors also must contain selectable genes that can be expressed in each organism.

In most cases, one of the organisms in which the shuttle vector can replicate is *E. coli*. The genetic tests can be performed with the other organism, but the plasmid can be purified and otherwise manipulated by the refined methods developed for *E. coli*.

Most plasmid replication functions and antibiotic resistance genes derived from *E. coli* are nonfunctional in gram-positive bacteria, such as *B. subtilis*, which has led to the development of a series of shuttle vectors (Table 4.3). Other shuttle vectors have been developed for other gram-positive bacteria and *E. coli*, whereas still others can be used in a wide variety of eukaryotes. For example, plasmid YEp13 (Figure 4.19) has the replication origin

Figure 4.19 Shuttle plasmid YEp13. The plasmid contains origins of replication that function in the yeast *S. cerevisiae* and the bacterium *E. coli*. It also contains genes that can be selected in *S. cerevisiae* and *E. coli*.
doi:10.1128/9781555817169.ch4.f4.19

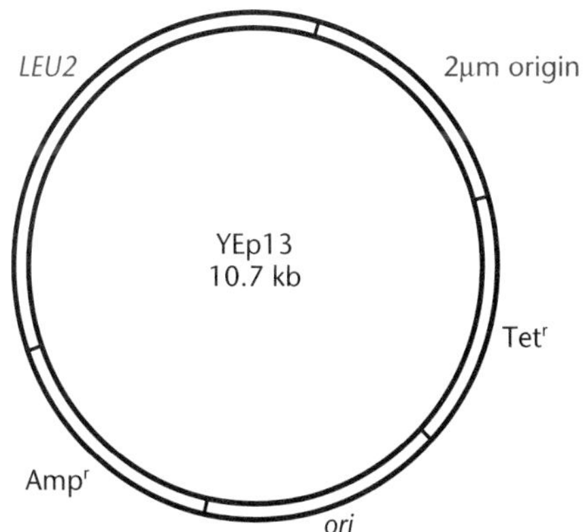

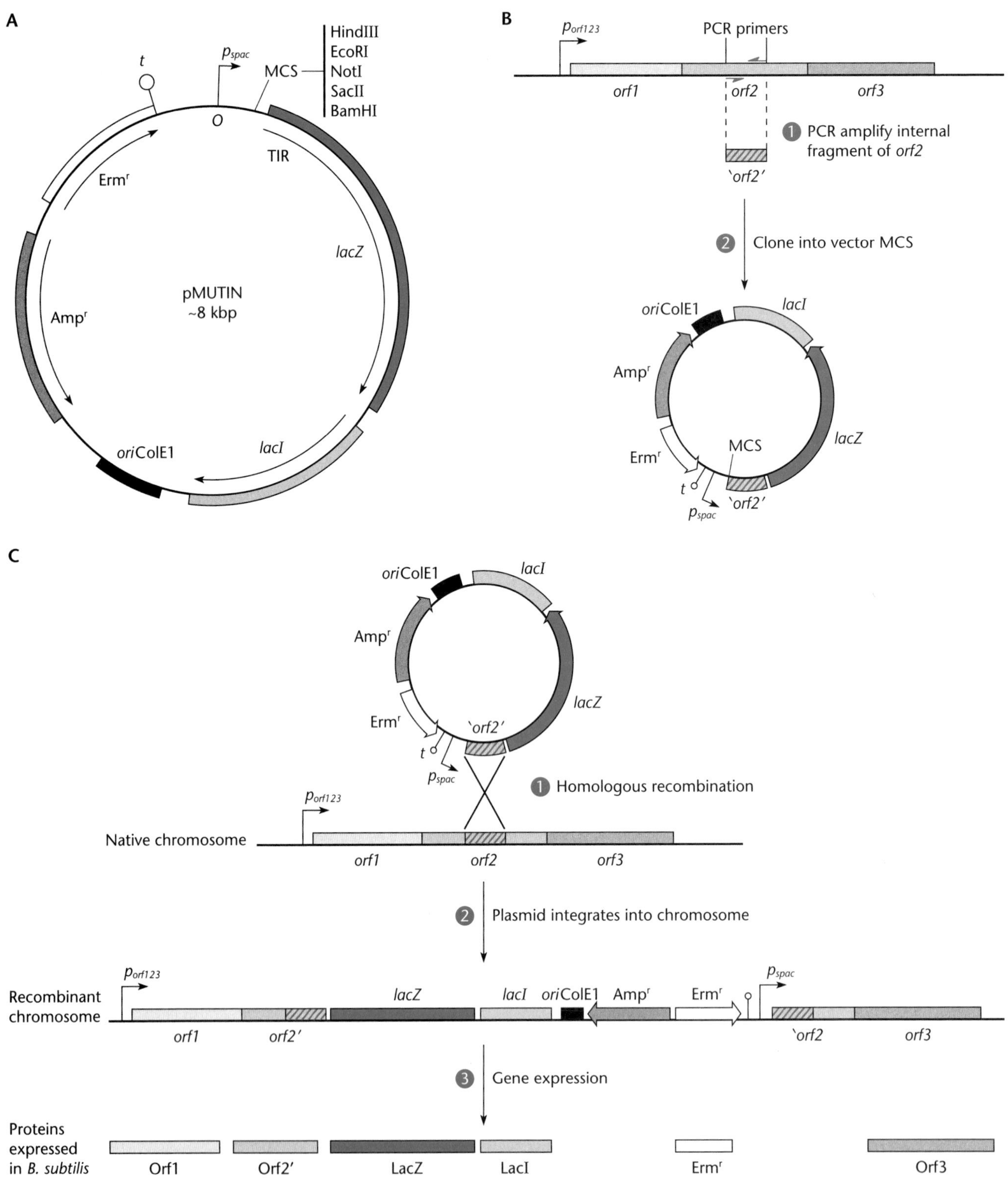

Figure 4.20 A plasmid-based method for genome-wide gene disruption in *B. subtilis.* **(A)** Map of the pMUTIN vector showing the Amp^r gene for selection in *E. coli* and the Erm^r gene for selection in *B. subtilis.* Also shown is the ColE1 origin of replication, which allows replication in *E. coli*, but not *B. subtilis.* The *lacZ* reporter gene includes a translation TIR of a *B. subtilis* gene and the multiple cloning site (MCS) into which PCR fragments can be cloned. The LacI repressor is made from the *lacI* gene of *E. coli*, modified so that it can be expressed in *B. subtilis.* p_{spac} is an inducible hybrid promoter that contains sequences of a promoter from the

of the 2μm circle, a plasmid found in the yeast *Saccharomyces cerevisiae*, so it can replicate in *S. cerevisiae*. It also has the pBR322 *ori* region and thus can replicate in *E. coli*. In addition, the plasmid contains the yeast gene *LEU2*, which can be selected in yeast, as well as an Amp^r gene, which confers ampicillin resistance on *E. coli*. Similar shuttle vectors that can replicate in mammalian or insect cells and *E. coli* have been constructed. Some of these plasmids have the replication origin of the animal virus simian virus 40 and the ColE1 origin of replication.

A VECTOR FOR GENOME-WIDE GENE DISRUPTION IN *B. SUBTILIS*

The advent of whole-genome sequencing has spurred the development of methods to determine systematically the function of each of the thousands of open reading frames of an organism. This requires methods to inactivate each of the open reading frames to begin to determine the function of the product of each of them. A plasmid vector, pMUTIN, that was used for such a systematic analysis has been developed for *B. subtilis* (see Vagner et al., Suggested Reading). Besides allowing gene-by-gene disruption, this vector allows measurement of the expression of the gene, as well as allowing the expression of downstream genes, thereby preventing polarity effects on the downstream genes (see "Polycistronic mRNA" in chapter 2). This analysis was facilitated by the highly efficient natural transformation system of *B. subtilis* (see chapter 6).

A diagram of pMUTIN is shown in Figure 4.20A. The plasmid can be grown in *E. coli*, with selection for the Amp^r gene on the plasmid. However, if the plasmid is transformed into *B. subtilis*, the plasmid is unable to replicate because the ColE1 origin of replication is not functional in the organism. The only way it can be maintained in *B. subtilis* is if it recombines with the chromosome. Recombination into the chromosome requires that the plasmid contain a region of DNA that is homologous to a region on the chromosome which in the experiment comes from the gene of interest. Selecting for the presense of the Erm^r gene by growth on medium containing erythromycin, which is expressed in *B. subtilis*, selects for the rare event where the plasmid has been introduced into the bacterium and recombined into the chromosome. Once integrated, a *lacZ* reporter gene on the plasmid, which has a TIR from *B. subtilis*, is transcriptionally fused to the promoter for the gene into which it has integrated and therefore makes β-galactosidase only under conditions where the target gene is normally transcribed (see chapter 2). An inducible p_{spac} promoter on the integrated plasmid also allows transcription of downstream genes in the operon. This promoter, which contains the *lac* operator, is active only in the presence of the inducer IPTG, because the *E. coli lacI* gene for the Lac repressor (modified for expression in *B. subtilis*) was also introduced into the plasmid (see chapter 12).

Figure 4.20B illustrates the cloning steps needed to use the pMUTIN vector to disrupt the middle gene, *orf2*, of a three-gene operon. A fragment internal to *orf2* is PCR amplified and cloned into the multiple cloning site just downstream of the p_{spac} promoter. Figure 4.20C illustrates what happens when the plasmid containing this clone is introduced into *B. subtilis* and Erm^r transformants are selected. Recombination between the cloned sequences on the plasmid and *orf2* promotes integration of the plasmid by a single crossover, making the cells Erm^r. Because the fragment was internal to *orf2*, both the upstream and downstream copies of *orf2* are incomplete and presumably inactive. If β-galactosidase is expressed, the promoter for the operon, called p_{orf123} (Figure 4.20C), must be active under the growth conditions used. If IPTG is added, *orf3* is also expressed, preventing polarity. Therefore, the only gene that is disrupted in the presence of IPTG is *orf2*, so that any phenotypes observed in the presence of IPTG are due to the disruption of *orf2*, allowing the function of the *orf2* gene to be deduced. An important caveat to this approach is that *orf2* is disrupted but not totally deleted, and the N terminus could still retain some activity. A second caveat is that the downstream *orf3* is expressed from a different promoter, so that its product could be made in larger or smaller amounts than normal; this also has the potential to cause phenotypes.

This vector has been used to disrupt more than 4,100 annotated open reading frames in *B. subtilis*. The effort involved a consortium of laboratories worldwide,

B. subtilis phage SP01 and three *lac* operators (*o*) to which the LacI repressor binds to make it inducible by IPTG. *t* is a strong hybrid transcriptional terminator from λ phage and an rRNA operon. **(B)** Cloning into pMUTIN. **(1)** A fragment internal to the target gene is PCR amplified with primers that add restriction sites compatible with those in the MCS for directional cloning (see chapter 1). **(2)** The fragment is cloned into the MCS on the plasmid. **(C)** Integration of a recombinant vector into the *B. subtilis* chromosome. **(1)** Homologous recombination into the native chromosomal locus. **(2)** Structure of the recombinant chromosome after plasmid integration. **(3)** Products of gene expression. The prime before or after an *orf* or protein indicates that only part of the *orf* or protein remains.
doi:10.1128/9781555817169.ch4.f4.20

especially in Europe and Japan. One outcome was to define a set of approximately 270 essential genes (see Kobayashi et al., Suggested Reading). Approximately 70% of the essential genes were found to have homologs in eukaryotes and archaea. This analysis would have missed essential genes that are redundant or expressed in media or under growth conditions different from those that were used, but more recent analyses have specifically studied duplicated genes (see Thomaides et al., Suggested Reading).

SUMMARY

1. Plasmids are DNA molecules that exist free of the chromosome in the cell. Most plasmids are circular, but some are linear. The sizes of plasmids range from a few thousand base pairs to almost the length of the chromosome itself. Probably the best distinguishing characteristic of a plasmid is that it has a typical plasmid origin of replication with an adjacent gene for a Rep protein rather than a chromosome origin with an *oriC* gene, along with a *dnaA* gene and other genes typical of the chromosomal origin of replication.

2. Plasmids usually carry genes for proteins that are necessary or beneficial to the host under some situations but are not essential under all conditions. Evolution probably selected for plasmids carrying nonessential or locally beneficial genes because it allows the chromosome to remain smaller but still allows bacterial populations to respond quickly to changes in the environment.

3. Plasmids replicate from a unique origin of replication, or *oriV* region. Many of the characteristics of a given plasmid derive from this *ori* region. They include the mechanism of replication, copy number control, partitioning, and incompatibility. If other genes are added to or deleted from the plasmid, it will retain most of its original characteristics, provided that the *ori* region remains.

4. Many plasmids replicate by a theta mechanism, with replication forks moving from a unique origin with leading and lagging template strands, much like circular bacterial chromosomes. Others use an RC mechanism, similar to that used to replicate some phage DNAs and during bacterial conjugation. In RC replication, the plasmid is cut at a unique site, and the Rep protein remains attached to the 5′ end at the cut through one of its tyrosines. The free 3′ end is used as a primer to replicate around the circle, displacing one of the strands. When the circle is complete, the 5′ phosphate is transferred from the Rep protein to the 3′ hydroxyl to form a single-stranded circle. A strand complementary to the single-stranded circle is made, using a different origin, and then the host ligase joins the ends to form two double-stranded circular DNAs. Linear plasmids replicate by more than one mechanism. Some have hairpin ends and replicate from an internal origin around the ends to form dimeric circles that are then processed to form two linear plasmids. Others have a terminal protein at both 5′ ends and extensive inverted-repeat sequences at their ends. They may replicate the ends by some sort of slippage mechanism.

5. The copy number of a plasmid is the number of copies of the plasmid per cell immediately after cell division.

6. Different types of plasmids use different mechanisms to regulate their initiation of replication and therefore their copy numbers. Some plasmids use small complementary RNAs (ctRNAs) transcribed from the other strand in the same region (countertranscribed) to regulate their copy numbers. In ColE1-derived plasmids, the ctRNA, called RNA I, interferes with the processing of the primer for leading-strand replication, called RNA II. In other cases, including the R1, ColIB-P9, and pT181 plasmids, the ctRNA interferes with the expression of the Rep protein required to initiate plasmid DNA replication.

7. Iteron plasmids regulate their copy numbers by two interacting mechanisms. They control the amount of the Rep protein required to initiate plasmid replication, and the Rep protein also couples plasmids through their iteron sequences.

8. Some plasmids have a special partitioning mechanism to ensure that each daughter cell gets one copy of the plasmid as the cells divide. These partitioning systems usually consist of two genes for proteins and a *cis*-acting centromere-like site.

9. If two plasmids cannot stably coexist in the cells of a culture, they are said to be incompatible or to be members of the same Inc group. They can be incompatible if they have the same copy number control system or the same partitioning functions.

10. The host range for replication of a plasmid is defined as all the different organisms in which the plasmid can replicate. Some plasmids have very broad host ranges and can replicate in a wide variety of bacteria. Others have very narrow host ranges and can replicate only in very closely related bacteria.

11. Many plasmids have been engineered for use as cloning vectors. They make particularly desirable cloning vectors for some applications because they do not kill the host, can be small, and are easy to isolate. Some plasmids can carry large amounts of DNA and are used to make BACs for eukaryotic-genome sequencing. Plasmids have been adapted to be used to express cloned genes in bacteria, to do gene replacements in the chromosome, and to perform systematic gene inactivation for functional genomics.

QUESTIONS FOR THOUGHT

1. Why are genes whose products are required for normal growth not carried on plasmids? List some genes that you would not expect to find on a plasmid and some genes you might expect to find on a plasmid.

2. Why do you suppose some plasmids have broad host ranges for replication? Why is it that not all plasmids have broad host ranges?

3. How do you imagine a partitioning system for a single-copy plasmid, such as F, could work? How might a copy number control mechanism work?

4. How would you find the genes required for replication of the plasmid if they are not all closely linked to the *ori* site?

5. How would you determine which of the replication genes of the host *E. coli* (e.g., *dnaA* and *dnaC*) are required for replication of a plasmid you have discovered?

6. The R1 plasmid has a leader polypeptide translated upstream of the gene for RepA, and cleavage of the mRNA by RNase III occurs in the coding sequence for this leader polypeptide. This blocks the translation of the leader polypeptide and also the translation of the downstream *repA* gene to which it is translationally coupled. Do you think it would have been easier just to have the cleavage occur in the coding sequence for the RepA protein itself? Why or why not?

7. Try to design a mechanism that uses inverted-repeat sequences at the ends to replicate to the ends of a linear plasmid without the DNA getting shorter each time it replicates.

PROBLEMS

1. The IncQ plasmid RSF1010 carries resistance to the antibiotics streptomycin and sulfonamide. Suppose you have isolated a plasmid that carries resistance to kanamycin. Outline how you would determine whether your new plasmid is an IncQ plasmid.

2. A plasmid has a copy number of 6. What fraction of the cells are cured of the plasmid each time the cells divide if the plasmid has no partitioning mechanism?

3. The ampicillin resistance gene of plasmid RK2 is unregulated. The more copies of the gene a bacterium has, the more gene product is made. In this case, the resistance of the cell to ampicillin is higher the more of these genes it has. Use this fact to devise a method to isolate mutants of RK2 that have a higher than normal copy number (copy-up mutations). Determine whether your mutants have mutations in the Rep-encoding gene.

4. Outline how you would determine whether a plasmid has a partitioning system.

5. What would be the effect of mutating one of the two complementary sequences in structures I and III of the ColIb-P9 plasmid origin region on the copy number of the plasmid? What would it be in the presence and absence of the Inc antisense RNA?

SUGGESTED READING

Azano, K., and K. Mizobuchi. 2000. Structural analysis of late intermediate complex formed between plasmid ColIb-P9 Inc RNA and its target RNA. How does a single antisense RNA repress translation of two genes at different rates? *J. Biol. Chem.* 275:1269–1274.

Bagdasarian, M., R. Lurz, B. Ruckert, F. C. H. Franklin, M. M. Bagdasarian, J. Frey, and K. N. Timmis. 1981. Specific purpose plasmid cloning vectors. II. Broad host, high copy number, RSF1010-derived vectors, and a host vector system for cloning in *Pseudomonas. Gene* 16:237–247.

Bao, K., and S. N. Cohen. 2001. Terminal proteins essential for the replication of linear plasmids and chromosomes in *Streptomyces. Genes Dev.* 15:1518–1527.

Campbell, C. S., and R. D. Mullins. 2007. In vitro visualization of type II plasmid segregation: bacterial actin filaments pushing plasmids. *J. Cell. Biol.* 179:1059–1066.

Caspi, R., M. Pacek, G. Consiglieri, D. R. Helinski, A. Toukdarian, and I. Konieczny. 2001. The broad host range replicon with different requirements for replication initiation in three bacterial species. *EMBO J.* 20:3262–3271.

Cohen, S. N., A. C. Y. Chang, H. W. Boyer, and R. B. Helling. 1973. Construction of biologically functional bacterial plasmids *in vitro. Proc. Natl. Acad. Sci. USA* 70:3240–3244.

Das, N., M. Valjavec-Gratian, A. N. Basuray, R. A. Fekete, P. P. Papp, J. Paulsson, and D. K. Chattoraj. 2005. Multiple homeostatic mechanisms in the control of P1 plasmid replication. *Proc. Natl. Acad. Sci. USA* 22:2856–2861.

Eberhard, W. G. 1989. Why do bacterial plasmids carry some genes and not others? Plasmid **21:**167–174.

Funnell, B. E., and G. J. Phillips (ed.). 2004. *Plasmid Biology.* ASM Press, Washington, DC.

Hamilton, C. M., M. Aldea, B. K. Washburn, P. Babitzke, and S. R. Kushner. 1989. A new method for generating deletions and gene replacements in *Escherichia coli. J. Bacteriol.* **171:** 4617–4622.

Khan, S. A. 2000. Plasmid rolling circle replication: recent developments. *Mol. Microbiol.* 37:477–484.

Kim, G. E., A. I. Derman, and J. Pogliano. 2005. Bacterial DNA segregation by dynamic SopA polymers. *Proc. Natl. Acad. Sci. USA* 102:17658–17663.

Kobayashi, K., S. D. Ehrlich, A. Albertini, G. Amati, K. K. Andersen, M. Arnaud, et al. 2003. Essential *Bacillus subtilis* genes. *Proc. Natl. Acad. Sci. USA* 100:4678–4683.

Kolb, F. A., C. Malmgren, E. Westof, C. Ehresmann, B. Ehresmann, E. G. H. Wagner, and P. Romby. 2000. An unusual structure formed by anti-sense target binding involves an extended kissing complex and a four-way junction and side-by-side helical alignment. *RNA* **6**:311–324.

McEachern, M. J., M. A. Bott, P. A. Tooker, and D. R. Helinski. 1989. Negative control of plasmid R6K replication: possible role of intermolecular coupling of replication origins. *Proc. Natl. Acad. Sci. USA* **86**:7942–7946.

Meacock, P. A., and S. N. Cohen. 1980. Partitioning of bacterial plasmids during cell division: a *cis*-acting locus that accomplishes stable plasmid maintenance. *Cell* **20**:529–542.

Metcalf, W. W., W. Jiang, L. L. Daniels, S. K. Kim, A. Haldimann, and B. L. Wanner. 1996. Conditionally replicated and conjugative plasmids carrying *lacZ* alpha for cloning, mutagenesis, and allele replacement in bacteria. *Plasmid* **35**:1–13

Novick, R. P., and F. C. Hoppensteadt. 1978. On plasmid incompatibility. *Plasmid* **1**:421–434.

Novick, R. P., S. Iordanescu, S. J. Projan, J. Kornblum, and I. Edelman. 1989. pT181 plasmid regulation is regulated by a countertranscript-driven transcriptional attenuator. *Cell* **59**:395–404.

Peters, J. E. 2007. Gene transfer in gram-negative bacteria, p. 735–755. *In* C. A. Reddy, T. J. Beveridge, J. A. Breznak, G. A. Marzluf, T. M. Schmidt, and L. R. Snyder (ed.), *Methods for General and Molecular Microbiology*, 3rd ed. ASM Press, Washington, DC.

Radloff, R., W. Bauer, and J. Vinograd. 1967. A dye-buoyant-density method for the detection and isolation of closed circular duplex DNA: the closed circular DNA in HeLa cells. *Proc. Natl. Acad. Sci. USA* **57**:1514–1521.

Ringgaard, S., J. von Zon, M. Howard, and K. Gerdes. 2009. Movement and equipositioning of plasmids by ParA filament disassembly. *Proc. Natl. Acad. Sci. USA* **106**:19369–19374.

Shizuya, H., B. Birren, U.-J. Kim, V. Mancino, T. Slepak, Y. Tachiiri, and M. Simon. 1992. Cloning and stable maintenance of 300-kilobase-pair fragments of human DNA in *Escherichia coli* using a F-factor-based vector. *Proc. Natl. Acad. Sci. USA* **89**:8794–8797.

Thomaides, H. B., E. J. Davison, L. Burston, H. Johnson, D. R. Brown, A. C. Hunt, J. Errington, and L. Czaplewski. 2007. Essential bacterial functions encoded by gene pairs. *J. Bacteriol. 189*:591–602.

Thomas, C. M. 2000. Paradigms of plasmid organization. *Mol. Microbiol.* **37**:485–491.

Vagner, V., E. Dervyn, and S. D. Ehrlich. 1998. A vector for systematic gene inactivation in *Bacillus subtilis*. *Microbiology* **144**:3097–3104.

Vecchiarelli, A. G., Y. W. Han, X. Tan, M. Mizuuchi, R. Ghirlando, C. Biertümpfel, B. E. Funnell, and K. Mizuuchi. 2010. ATP control of dynamic P1 ParA-DNA interactions: a key role for the nucleoid in plasmid partition. *Mol. Microbiol.* **78**:78–91.

Yao, S., D. R. Helinski, and A. Toukdarian. 2007. Localization of the naturally occurring plasmid ColE1 at the cell pole. *J. Bacteriol. 189*:1946–1953.

CHAPTER 5

Conjugation

A REMARKABLE FEATURE OF MANY PLASMIDS IS the ability to transfer themselves and other DNA elements from one cell to another in a process called **conjugation**. Joshua Lederberg and Edward Tatum first observed this process in 1947, when they found that mixing certain strains of *Escherichia coli* with others resulted in strains that were genetically unlike either of the originals. As discussed later in this chapter, Lederberg and Tatum suspected that bacteria of the two strains exchanged DNA—that is, two parental strains mated to produce progeny unlike themselves but with characteristics of both parents. At that time, however, plasmids were unknown, and it was not until later that the basis for the mating was understood.

The plasmid that Lederberg and Tatum were unknowingly working with, called the fertility, or F, plasmid, has been the focus of most of the research about the process of bacterial conjugation. The central role conjugation systems played in the development of bacterial genetics is contrasted with the more notorious role these plasmids play as the almost exclusive vectors for the spread of antibiotic resistance between bacteria. While historically studied with plasmids, conjugation is now also known to be used in the transfer of elements that are not maintained separate from the chromosome but instead are integrated into the chromosome and are therefore called integrating conjugative elements (ICE). Their integration into the host chromosome uses the same type of recombination that is used for the integration of many bacteriophages.

Overview

During plasmid conjugation, the two strands of an element separate in a process resembling rolling-circle replication (see "Mechanism of DNA Transfer during Conjugation in Gram-Negative Bacteria" below), and one

doi:10.1128/9781555817169.ch5

strand moves from the bacterium originally containing the plasmid—the donor—into a **recipient** bacterium. The two single strands serve as templates for DNA replication concurrent with the process of DNA transfer to yield double-stranded DNA molecules in both the donor cell and the recipient cell. A recipient cell that has received DNA as a result of conjugation is called a **transconjugant.** A simplified view of conjugation is shown in Figure 5.1.

Many naturally occurring plasmids can transfer themselves. If so, they are said to be **self-transmissible.** The prevalence of conjugation systems suggests that plasmid conjugation is advantageous for plasmids without placing an undue burden on the bacterial host. Self-transmissible plasmids encode all the functions they need to move among cells, and sometimes they also aid in the transfer of mobilizable plasmids. Mobilizable plasmids encode some, but not all, of the proteins required for transfer and consequently need the help of self-transmissible plasmids in the same cell to move.

Any bacterium harboring a self-transmissible plasmid is a potential donor, because it can transfer DNA to other bacteria. In gram-negative bacteria, such cells produce a structure, called a **sex pilus**, that facilitates conjugation (discussed below). Bacteria that lack the self-transmissible plasmid are potential recipients, and conjugating bacteria are known as **parents.** Potential donor strains were historically referred to as **male strains.**

Self-transmissible plasmids appear to exist in all types of bacteria (Guglielmini et al., Suggested Reading), but those that have been studied most extensively are from the gram-negative genera *Escherichia* and *Pseudomonas* and the gram-positive genera *Enterococcus*, *Streptococcus*, *Bacillus*, *Staphylococcus*, and *Streptomyces*. The best-known transfer systems are those of plasmids isolated from *Escherichia* and *Pseudomonas* species, so we focus our attention first on these gram-negative systems and the original conjugal plasmid, the F plasmid. ICE and conjugal plasmids found in gram-positive bacteria will be covered later in this chapter.

Figure 5.1 A simplified view of conjugation by a self-transmissible plasmid, the F plasmid. A replica of the plasmid is transferred from the donor to the recipient cell so that both the donor and recipient cells have the plasmid. A cell that has received the plasmid by conjugation is a transconjugant.
doi:10.1128/9781555817169.ch5.f5.1

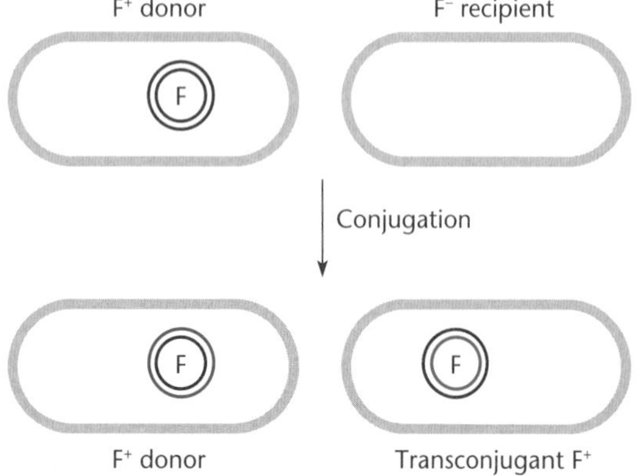

Classification of Self-Transmissible Plasmids

Bacterial plasmids have many different types of transfer systems, which are encoded by the plasmid *tra* genes [see "Transfer (*tra*) Genes" below]. These groups are based on mechanistic similarities and the use of homologous protein sets. These groupings were originally aligned with the aforementioned plasmid incompatibility groups, which are based on replication and segregation systems discussed in chapter 4. The F plasmid defines the F group.

The Fertility Plasmid

Our understanding of conjugation is largely based on work with the F plasmid, which was discovered first, and systems related to F. As mentioned above, the F plasmid has also played an essential role in bacterial genetics. The entire sequence of the ~100-kilobase (kb) F plasmid is known, and examining this sequence provides a window into the complexity of the process of conjugation (Figure 5.2). The genes required for transfer are called the *tra* genes, which make up a significant portion of the self-transmissible F plasmid genome. In addition to the *tra* gene products, there are many gene products that play supporting roles in the conjugation process or are otherwise advantageous to the host (Table 5.1). The functions of many of the gene products encoded by genes on the F plasmid remain unknown.

The hodgepodge organization of the F plasmid underscores the evolutionary path that horizontally transferred elements take. Examination of the F plasmid indicates that the sequences originated from many different sources, but beneath the patchwork is a complex and highly evolved system for transfer and maintenance of the plasmid (Figure 5.2). There is evidence for three replication origins on the F plasmid, RepFIA, RepFIB, and RepFIC. The origin responsible for replication of the plasmid by theta replication is the RepFIA origin, which is also known as *oriV*. The RepFIB/*oriS* origin is not sufficient for reliably maintaining a plasmid, and the RepFIC origin is inactivated by a Tn*1000* insertion sequence (IS) element (Figure 5.2). There are many systems encoded by genes on the plasmid that help stabilize the F plasmid. A partitioning system similar to those found on some bacterial chromosomes helps ensure that daughter cells each receive a copy of the plasmid, while multiple "postsegregational killing systems" sicken cells that lose the plasmid, which often die unless they receive

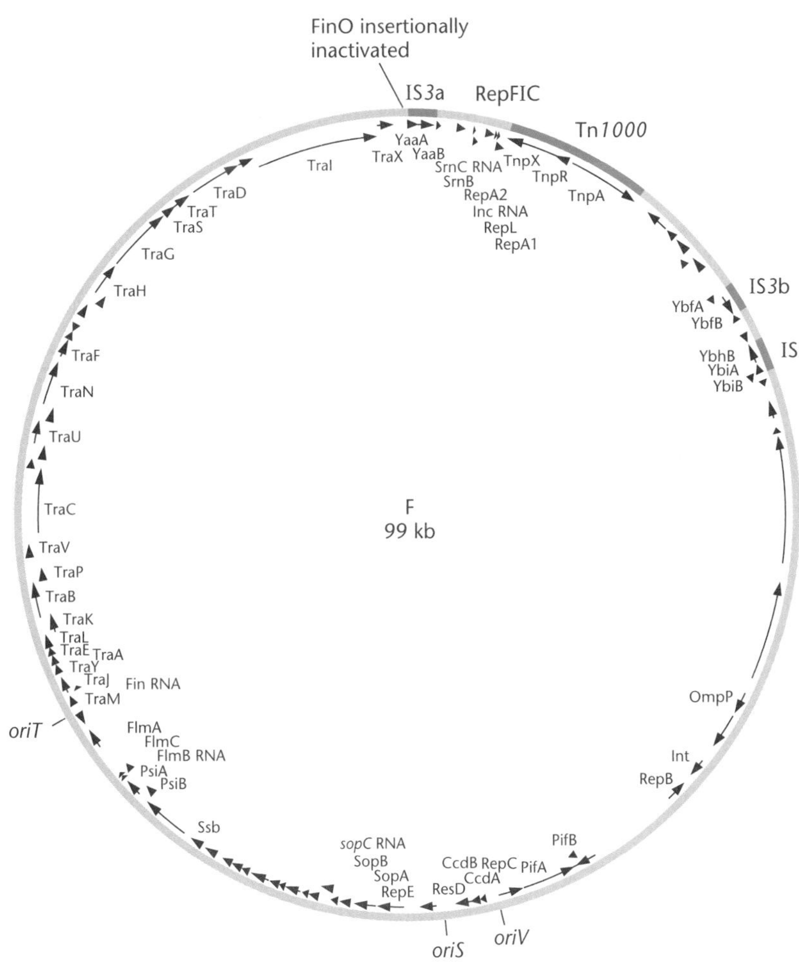

Figure 5.2 Partial genetic map of the ~100-kilobase pair (kbp) self-transmissible plasmid F. The regions of the insertion elements IS*3* (IS*3*a and IS*3*b) and IS*2* and transposon Tn*1000* are shown. *oriV* is the origin of replication; *oriT* is the origin of conjugal transfer. Multiple genes whose products' functions are known are shown and are categorized in Table 5.1. Details are given in the text.
doi:10.1128/9781555817169.ch5.f5.2

a copy of the same plasmid (see Box 4.2). Other systems that likely encourage maintenance of the plasmid are a system that blocks the development of bacteriophage T7 and a system that helps transfer by preventing induction of the host SOS DNA damage response during conjugation.

Mechanism of DNA Transfer during Conjugation in Gram-Negative Bacteria

Much has been learned over the years about the detailed mechanism of plasmid conjugation, but many fundamental questions still remain.

Transfer (*tra*) Genes

About a third of the F plasmid is comprised of *tra* genes (Figure 5.2). With so many genes required for conjugation and other transfer-related processes, it is understandable that self-transmissible plasmids are so large and that selection drives the evolution of mobilizable elements that can parasitize this machinery in other vectors for their own transfer.

The *tra* genes of a self-transmissible plasmid required for plasmid transfer can be divided into two components (Table 5.1). Some of the *tra* genes encode proteins involved in the processing of the plasmid DNA to prepare it for transfer. This is called the **Dtr component** (for *D*NA *t*ransfer and conjugal *r*eplication). The bulk of the *tra* genes encode proteins of the **Mpf component** (for *m*ating *p*air *f*ormation). This large membrane-associated structure includes the pilus that is responsible for holding the mating cells together and the channel that forms between the mating cells through which the plasmid is transferred. Below, we outline what is known of these two components.

THE Mpf COMPONENT

The function of the Mpf component is to hold a donor cell and a recipient cell together during the mating process and to form a channel through which proteins and DNA are transferred during mating. It also includes the protein that communicates "news" of mating pair formation to the Dtr system, beginning the transfer of plasmid DNA. A representation of the entire Mpf system of

Table 5.1 Some F plasmid genes and sites

Function	Protein, site, or antisense RNA
Vegetative replication	Origin RepFIA/*oriV* site (RepC, RepE)
	Origin RepFIB/*oriS* site (RepB)
	Origin RepFIC/inactivated origin [RepA2, RepL, Inc (RNA), RepA1]
Regulation of conjugation	FinO, FinP (RNA), TraJ
Mating pair formation (Mpf component)	TraA, TraB, TraC, TraD, TraE, TraF, TraG, TraH, TraK, TraL, TraN, TraP, TraQ, TraV, TraW, TraX
DNA transfer and replication (Dtr component)	*oriT* site, TraI, TraM, TraU, TraY
Plasmid SOS inhibition	PsiA, PsiB
Surface exclusion	TraS, TraT
Plasmid partitioning	SopA, SopB, *sopC* site
Postsegregational killing	SrnC (RNA) and SrnB
	CcdA and CcdB
	FlmA, FlmC, FlmB (RNA)
Exclusion of T7	PifA, PifB
Transposon functions	IS*3*a (YaaA, YaaB)
	Tn*1000* (TnpX, TnpR, TnpA)
	IS*3*b (YbfA, YbfB)
	IS*2* (YbhB, YbiA, TbiB)
Known function, but unknown relation to conjugation	OmpP, Int, ResD, SsB

a self-transmissible plasmid, the F plasmid, is shown in Figure 5.3.

The Pilus

The most dramatic feature of the Mpf structure is the **pilus**, a tube-like structure that sticks out of the cell surface (Figure 5.3). These pili are 10 nm or more in diameter with a central channel. Each pilus is constructed of many copies of a single protein called the **pilin** protein. The assembly of a pilus is shown in Figure 5.4. The pilin protein is synthesized with a long signal sequence that is removed as it passes through the membrane to assemble on the cell surface. The pilin protein is also cyclized, with its head attached to its tail, which is unusual among proteins (see Eisenbrandt et al., Suggested Reading).

The structure of the pilus itself differs markedly among plasmid transfer systems. Differences in the length, rigidity, and diameter of the pilus are likely evolutionary adaptations to the specific environments where conjugation naturally takes place, but in the laboratory, these characteristics determine the conditions needed to allow mating pair formation and plasmid transfer to occur (i.e., in liquid cultures or on solid media). Plasmids like CoIIb-P9 may be adapted to multiple environments; they have the capacity to make two types of pili, a long thin one and a short rigid one, with the former increasing the frequency of mating in liquid environments and the latter increasing the frequency of mating on a solid surface.

In the F plasmid, and probably other conjugal elements, the pilus is important for drawing cells together for mating pair formation. Once retracted, this will become the portal used for DNA transfer. Remarkably, it is still under debate whether DNA naturally travels through the extended pilus. The idea that transfer, at least under some circumstances, can occur without direct contact between cells has been supported by work with a novel system allowing transfer to be indirectly visualized under the microscope in real time (see Babic et. al., 2008, Suggested Reading).

Coupling Proteins

The Mpf component of a donor cell is the first to make contact with a recipient cell. Then, the information that it has contacted another cell is communicated to the Dtr component before DNA transfer occurs. The communication between the Dtr and Mpf systems is provided by a **coupling protein**, which is part of the Mpf system. Coupling proteins provide the specificity for the transport process, so that only some plasmids are transferred. Information that the Mpf system has encountered a recipient cell is somehow communicated through the coupling protein, which in turn signals the Dtr component to initiate the transport process. Coupling proteins (TraD in the case of the F plasmid) are distantly related to the FtsK/SpoIIIE translocases discussed in chapter 1. The coupling protein is bound to the membrane channel and acts as a DNA translocator that pumps DNA into recipient cells (Figure 5.5). The coupling proteins appear to be an adaptation that allows the pilus, which usually functions as a protein transfer machine (see below), to function as a highly efficient DNA transfer device.

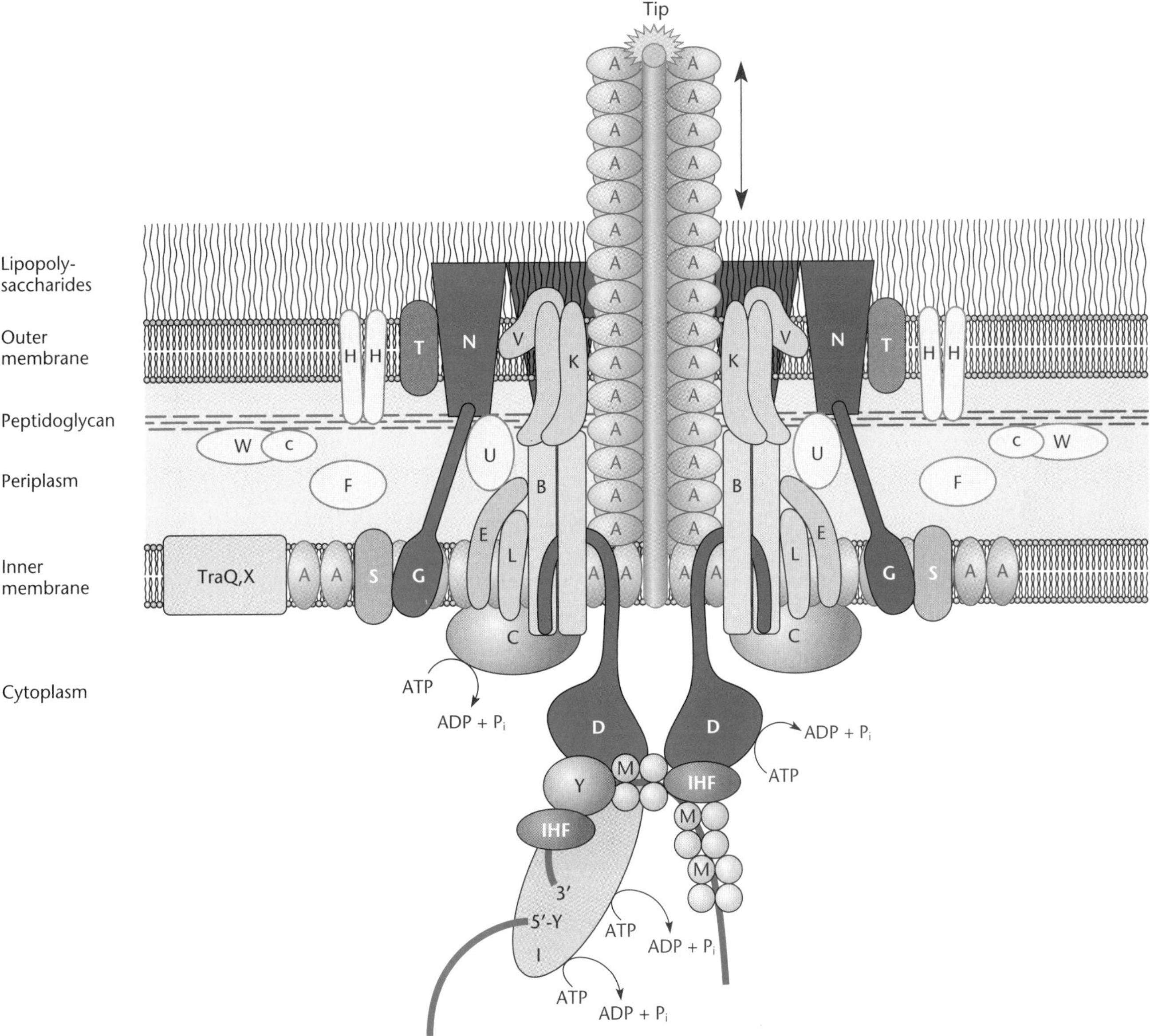

Figure 5.3 Representation of the F transfer apparatus drawn from available information. The pilus is assembled with five TraA (pilin) subunits per turn that are inserted into the inner membrane via TraQ and acetylated by TraX. The pilus is shown extending through a pore constructed of TraB and TraK, a secretin-like protein anchored to the outer membrane by the lipoprotein TraV. TraB is an inner membrane protein that extends into the periplasm and contacts TraK. TraL seeds the site of pilus assembly and attracts TraC to the pilus base, where it acts to drive assembly in an energy-dependent manner. A channel formed by the lumen is indicated, as is a specialized structure at the pilus tip that remains uncharacterized. The two-headed arrow indicates the opposing processes of pilus assembly and retraction. The Mpf proteins include TraG and TraN, which aid in mating pair stabilization (Mps), and TraS and TraT, which disrupt mating pair formation through entry and surface exclusion, respectively. TraF, TraH, TraU, TraW, and TrbC, which together with TraN are specific to F-like systems, appear to play a role in pilus retraction, pore formation, and mating pair stabilization. The relaxosome, consisting of TraY, TraM, TraI, and host-encoded IHF bound to the nicked DNA in *oriT*, is shown interacting with the coupling protein, TraD, which in turn interacts with TraB. The 5′ end of the nicked strand is shown bound to a tyrosine in TraI, and the 3′ end is shown as being associated with TraI in an unspecified way. The retained, unnicked strand is not shown. TraC, TraD, and TraI have ATP utilization motifs, represented by curved arrows, with ATP being split into ADP and inorganic phosphate (P_i). Two curved arrows are shown with TraI, which utilizes ATP separately in two different regions of the protein for relaxase and helicase activities. doi:10.1128/9781555817169.ch5.f5.3

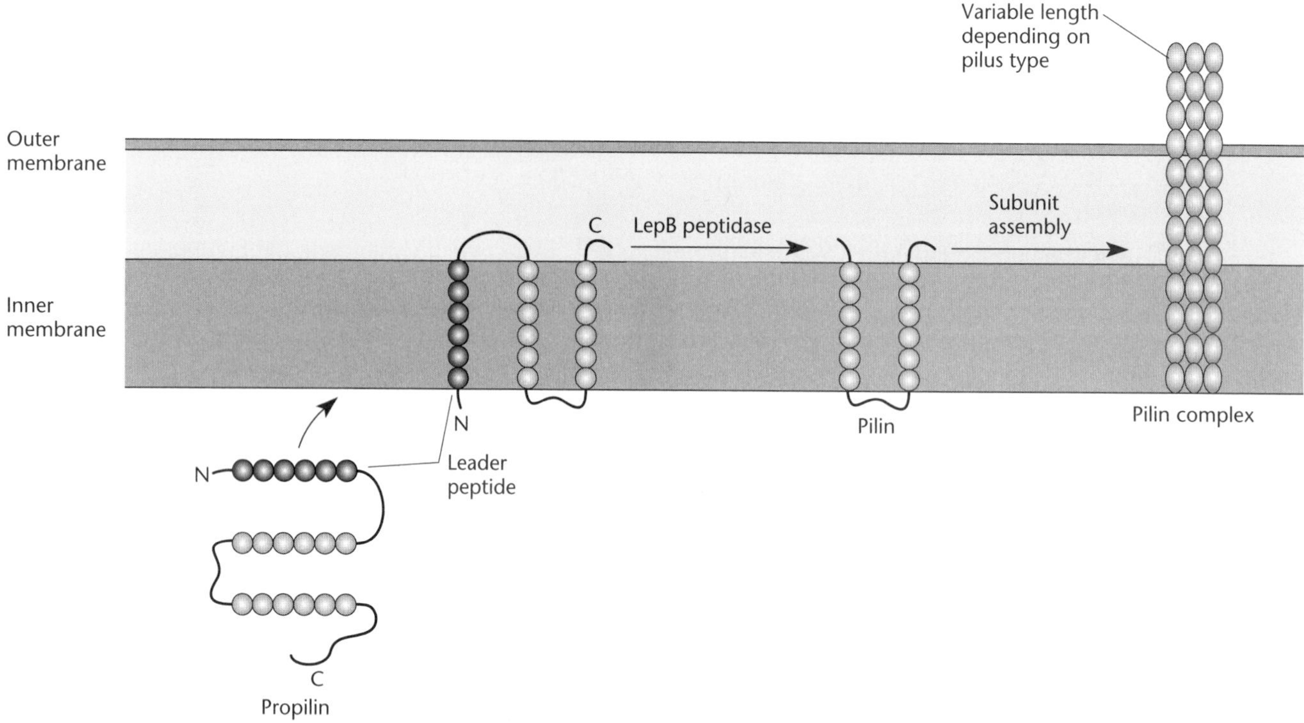

Figure 5.4 Assembly of the pilus on the cell surface. The propilin is processed by LepB peptidase as it passes through the membrane, and then it assembles between the inner and outer membranes (details are given in the text). doi:10.1128/9781555817169.ch5.f5.4

THE Dtr COMPONENT

The Dtr (or DNA-processing) component of a self-transmissible plasmid is involved in preparing the plasmid DNA for transfer. A number of proteins make up this component, and the functions of many of them are known.

Relaxase

A central part of the Dtr component of plasmids is the **relaxase**, which is TraI in the F plasmid. It is a site-specific DNA endonuclease that makes a single-strand break, or "nick," at a specific site within the *oriT* sequence (see below). Following a signal from the Mpf component, the relaxase initiates the transfer process. The way in which the relaxase works is illustrated in Figure 5.6. The relaxase breaks a phosphodiester bond at the *nic* site by transferring the bond from a deoxynucleotide to one of its own tyrosines. While covalently bound to the 5′ end of the DNA, the relaxase protein can pilot the strand into the recipient cell. The nicking reaction is called a transesterification reaction and requires very little energy because there is no net breakage or formation of new chemical bonds. In the recipient cell, the TraI relaxase protein recircularizes the plasmid by doing essentially the reverse of what it had done in the donor cell. It binds to the two halves of the cleaved *oriT* sequence, holding them together while it transfers the phosphate bond from its tyrosine back to the 3′ hydroxyl deoxynucleotide in the DNA (Figure 5.6). This transesterification reaction reseals the nick in the DNA and releases the relaxase, which has done its job and is degraded. Transfer of a relaxase protein to recipient cells appears to be maintained in all conjugation systems studied to date (see below).

The Relaxosome

The relaxase protein in the donor cell is part of a larger structure called the **relaxosome**, which is made up of a number of proteins that are normally bound to the *oriT* sequence of the plasmid. The relaxase of the F plasmid uses accessory proteins, like TraU, TraM, and TraY, for recognizing and coordinating activity with the Mpf proteins, although the exact mechanism of this signaling is still unclear. TraY and the host protein IHF are essential accessory proteins for nicking at *oriT* in the test tube, while TraM and TraD likely help coordinate activity with the transfer apparatus in living cells. In the F plasmid, the TraI protein that nicks at *oriT* also has a helicase activity that is required for conjugation. This helicase activity appears to play multiple roles during conjugation. In the donor cell, the helicase activity separates the strands for presentation to the Mpf component but also likely aids DNA polymerase III so the polymerase can replicate the second strand of the plasmid.

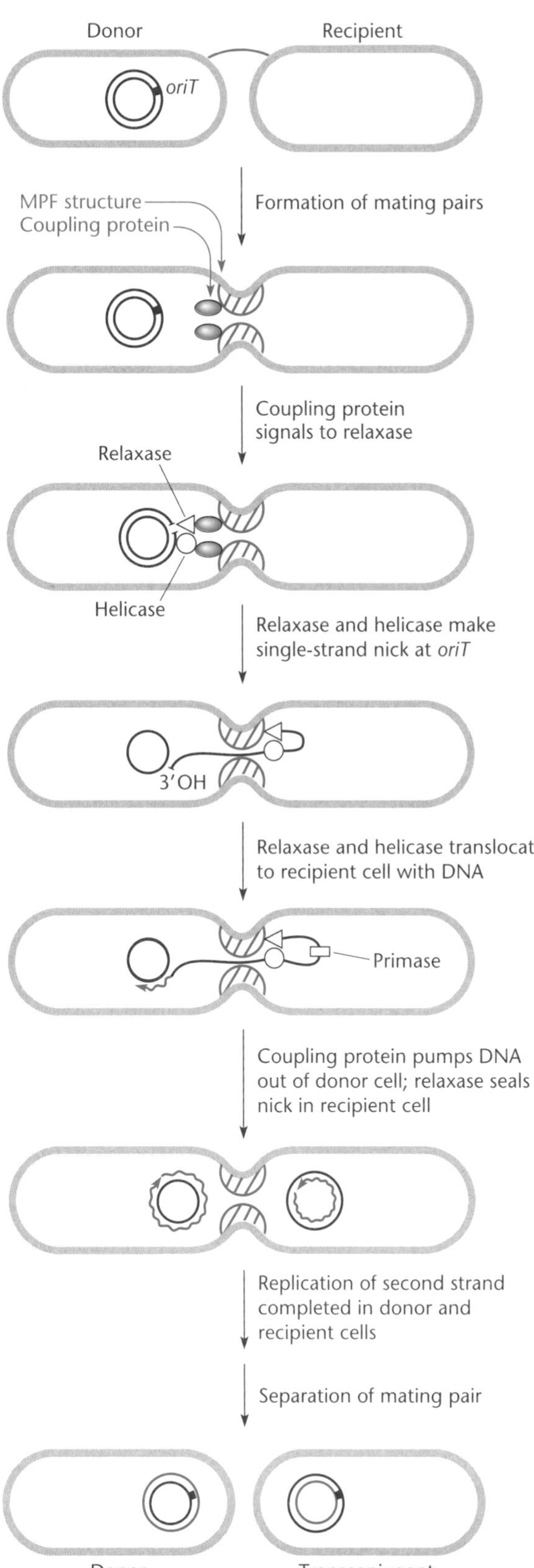

The helicase may also play additional roles in the recipient cell.

The *oriT* Sequence

The *oriT* site is not only the site at which plasmid transfer initiates, but also the site at which the DNA ends rejoin to recircularize the plasmid. Plasmid transfer initiates specifically at the *oriT* site, because the specific relaxase encoded by one of the *tra* genes nicks DNA only at this sequence. Also, presumably, the plasmid-encoded helicase enters DNA only at this sequence to separate the strands. Regulation of *oriT* activity is coordinated through binding sites for IHF, TraY, and TraM. Cloning the minimal *oriT* sequence into a plasmid allows mobilization of the plasmid if the self-transmissible F plasmid also resides in the cell, as discussed below.

PRIMASE

In many conjugation systems, a component of the Dtr system made in the donor is the **primase**. Primases are needed for chromosomal DNA replication to make an RNA primer to initiate the synthesis of the lagging strand of DNA replication (see chapter 1) and to prime plasmid replication (see chapter 4). However, at first, the role that a primase would play in the donor was not clear. A primase may not be necessary to prime replication in the donor cell, since the free 3′ hydroxyl end of DNA created at the nick in *oriT* could provide the primer for replication during transfer, similar to the priming of the first stage of replication of rolling-circle plasmids (see

Figure 5.5 Mechanism of DNA transfer during conjugation, showing the Mpf functions in blue. The donor cell produces a pilus, which forms on the cell surface and which may contact a potential recipient cell and bring it into close contact or may help hold the cells in close proximity after contact has been made, depending on the type of pilus. A pore then forms in the adjoining cell membranes. On receiving a signal from the coupling protein that contact with a recipient has been made, the relaxase protein initiates transfer starting at the *oriT* site in the plasmid. A plasmid-encoded helicase then separates the strands of the plasmid DNA. The relaxase protein, which has remained attached to the 5′ end of the single-stranded DNA, is then transported out of the donor cell through the channel directly into the recipient cell, dragging the attached single-stranded DNA along with it. The coupling protein pumps DNA out of the donor cell. Once in the recipient, the relaxase protein helps recyclize the single-stranded DNA. A primase, encoded either by the host or by the plasmid and injected with the DNA, then primes replication of the complementary strand to make the double-stranded circular plasmid DNA in the recipient. The 3′ end at the nick made by the relaxase in the donor can also serve as a primer, making a complementary copy of the single-stranded plasmid DNA remaining in the donor. Therefore, after transfer, both the donor and the recipient bacteria have a double-stranded circular copy of the plasmid. Details are given in the text. doi:10.1128/9781555817169.ch5.f5.5

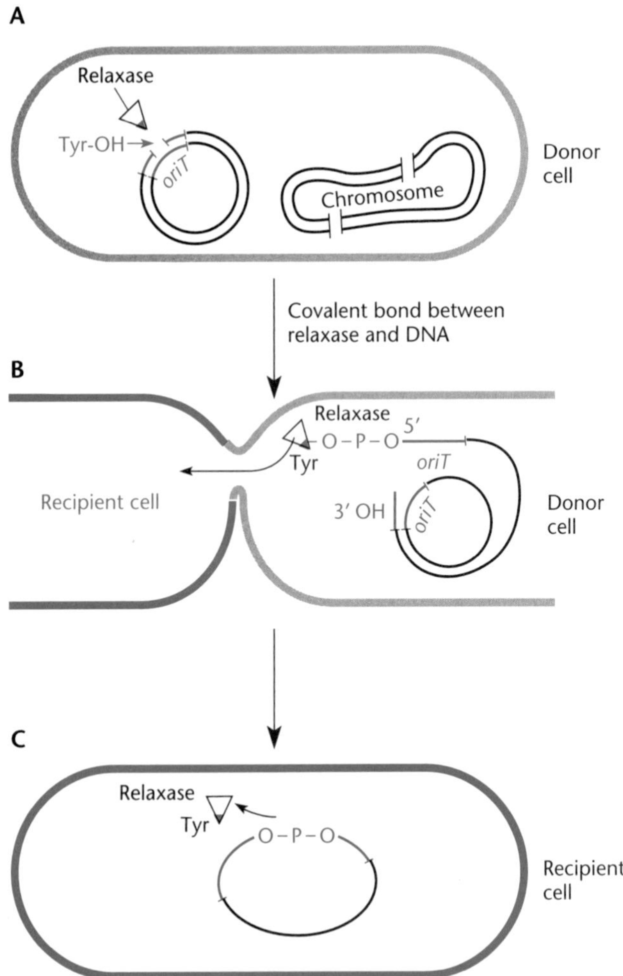

Figure 5.6 Reactions performed by the relaxase. **(A)** The relaxase nicks the DNA at a specific site in *oriT*, and the 5′ phosphate is transferred to one of its tyrosines in a transesterification reaction. **(B)** The relaxase is transferred to the recipient cell, dragging the DNA along with it. **(C)** In a reversal of the original transesterification reaction, the phosphate is transferred back to the 3′ hydroxyl of the other end of the DNA, recyclizing the DNA and releasing the relaxase.
doi:10.1128/9781555817169.ch5.f5.6

chapter 4). The second strand of the F plasmid is only made in the recipient, as it is spooled into a recipient cell in a process that is similar to lagging-strand DNA synthesis. Interestingly, priming of the F plasmid in recipient cells can occur without a primase protein expressed within the donor. Apparently the primase can be made from the single-stranded DNA as it is transferred to the recipient. While this at first appears to violate what we learned about gene expression in chapter 2, the process is allowed by one or more promoters that are capable of functioning when DNA is in the single-strand form. In other plasmids, like RP4 and ColIb-P9, a protein primase produced in the donor cell can be transported into the recipient cell, where it presumably primes lagging-strand DNA replication of the plasmid genome.

Why would a plasmid bother to make its own RNA primer if it can use the host cell primase instead to initiate DNA replication? The answer may be that it does this to make it more promiscuous and able to transfer into a wider variety of bacterial species. Sometimes a promiscuous plasmid may find it has transferred itself into a type of bacterium that is so distantly related to its original host that the primase in the bacterium does not recognize the sequences on the plasmid DNA necessary to prime the replication of the complementary strand.

Male-Specific Phages

Some types of phages can infect only cells that express a certain type of conjugation-associated pilus on their surfaces. All phages adsorb to specific sites on the cell surface to initiate infection (see chapter 7), and some phages use the pilus of a self-transmissible plasmid as their adsorption site. Phages that adsorb to the sex pilus of a self-transmissible plasmid are called **male-specific phages** because they infect only donor, or "male" cells capable of DNA transfer. Examples of male-specific phages are M13 and R17, which infect only cells carrying the F plasmid, and Pf3 and PRR1, which infect cells containing plasmid RP4 and related plasmids.

Because male-specific phages infect only cells expressing a pilus, mutations in any *tra* gene required for pilus assembly prevent infection by the phage. This offers a convenient way to determine which of the *tra* genes of a plasmid encode proteins required to express a pilus on the cell surface and which *tra* genes encode other functions required for DNA transfer. To apply this test to a particular *tra* gene, the phages are used to infect cells containing the plasmid with a mutation in the *tra* gene. If the phage can multiply in the host cell, the *tra* gene that has been mutated must not be one of those that encode a protein required for pilus expression and assembly.

Incidentally, the susceptibility of pilus-expressing cells to some phages may in part explain why the *tra* genes of plasmids are usually tightly regulated. Most self-transmissible plasmids express a pilus only immediately after entering a cell and only intermittently thereafter (see "An Example: Regulation of *tra* Genes in F Plasmids" below). If cells containing the plasmid always expressed the pilus, a male-specific phage could spread quickly through the population, destroying many of the cells and with them the plasmids they contain. By only intermittently expressing a pilus, cells containing a self-transmissible plasmid limit their susceptibility to phages that use their pilus as an adsorption site. For bacteria that reside in animal hosts, pili might also be expected to be highly immunogenic; therefore, regulating the expression of proteins on the cell surface should also limit detection and clearance of the plasmid's bacterial host in the host animal. Expression of the *tra* genes could also be a significant burden on cellular resources and reduce host competitiveness if not regulated.

Efficiency of Transfer

One of the striking features of many transfer systems is their efficiency. Under optimal conditions, some plasmids can transfer themselves into other cells in almost 100% of cell contacts. This high efficiency has been exploited in the development of methods for transferring cloned genes between bacteria and in transposon mutagenesis, both of which require highly efficient transfer of DNA. Such methods are discussed in subsequent chapters.

REGULATION OF THE *tra* GENES

As mentioned above, many naturally occurring plasmids transfer with high efficiency for only a short time after they are introduced into cells and transfer only sporadically thereafter. Most of the time, the *tra* genes are repressed, and without the synthesis of pilin and other *tra* gene functions, the pilus is lost. For unknown reasons, the repression is relieved occasionally in some of the cells, allowing this small percentage of cells to transfer their plasmids at a given time.

This property of only periodically expressing their *tra* genes probably does not prevent the plasmids from spreading quickly through a population of bacteria that does not contain them. When a plasmid-containing population of cells encounters a population that does not contain the plasmid, the plasmid *tra* genes in one of the plasmid-containing cells are eventually expressed, and the plasmid transfers to another cell. Then, when the plasmid first enters a new cell, efficient expression of the *tra* genes leads to a cascade of plasmid transfer from one cell to another (a process sometimes referred to as zygotic induction). As a result, the plasmid soon occupies most of the cells in the population. This is the rationale behind triparental matings (discussed below).

AN EXAMPLE: REGULATION OF *tra* GENES IN F PLASMIDS

Regulation of the *tra* genes in F plasmids has been studied more extensively than that of other types. This regulation is illustrated in Figure 5.7. Transfer of these plasmids depends on TraJ, a transcriptional activator. A **transcriptional activator** is a protein required for initiation of RNA synthesis at a particular promoter (see chapter 2). If TraJ were always made, the other *tra* gene products would always be made, and the cell would always have a pilus. However, the translation of TraJ is normally blocked by the concerted action of the products of two plasmid genes, *finP* and *finO*, which encode an RNA and a protein, respectively. The FinP RNA is an antisense RNA that is transcribed constitutively from a promoter within and in opposite orientation to the *traJ* gene. Complementary pairing of the FinP RNA and the *traJ* transcript negatively regulates the operon by two mechanisms: the complementary pairing prevents translation of TraJ by blocking the ribosome binding site, but the RNA also makes the transcript a productive target for degradation by RNase E (see chapter 2). The FinO protein stabilizes the FinP antisense RNA. When the plasmid first enters a cell, neither FinP RNA nor FinO protein is present, so TraJ and the other *tra* gene products are made. Consequently, a pilus appears on the cell, and the plasmid can be transferred. Initially, the transferred plasmid is in a single-stranded state. However, priming in the recipient cell can synthesize the complementary strand to make the double-stranded form. After the plasmid has become established in the double-stranded state, the FinP RNA and FinO protein can be synthesized, the *tra* genes are repressed, and the plasmid can no longer transfer. Later, the *tra* genes are expressed only intermittently. The F plasmid is highly adapted to its host in that many host factors appear to help regulate conjugation. Regulation has been linked to supercoiling of the plasmid and many other systems. Much of the F plasmid is also sensitive to silencing by HN-S, and HN-S may be an antagonist of TraJ.

The original discovery of the F plasmid may have resulted from a happy coincidence involving its *finO* gene. Because of an insertion mutation in the gene (IS*3*a in Figure 5.2), the F plasmid is itself a mutant that always expresses the *tra* genes. Consequently, a sex pilus almost always extends from the surfaces of cells harboring this F plasmid, and the F plasmid can always transfer, provided that recipient cells are available, increasing the efficiency of transfer and facilitating the discovery of conjugation. Mutations that increase the efficiency of plasmid transfer, thereby increasing their usefulness in gene cloning and other applications, have been isolated in other commonly used transfer systems.

Interspecies Transfer of Plasmids

Many plasmids have transfer systems that enable them to transfer DNA between unrelated species. They are known as **promiscuous plasmids** and include R388, RP4, and pKM101. The RP4 plasmid can transfer itself or mobilize other plasmids from *E. coli* into essentially any gram-negative bacterium. Amazingly, plasmids of this group transfer at a low frequency into cyanobacteria; gram-positive bacteria, such as *Streptomyces* species; and even plant cells. The F plasmid, which was not known to be particularly promiscuous, can transfer itself from *E. coli* into eukaryotic yeast cells.

Transfer of DNA by promiscuous plasmids probably plays an important role in evolution. Such transfer could

A Genetic organization of *tra* region

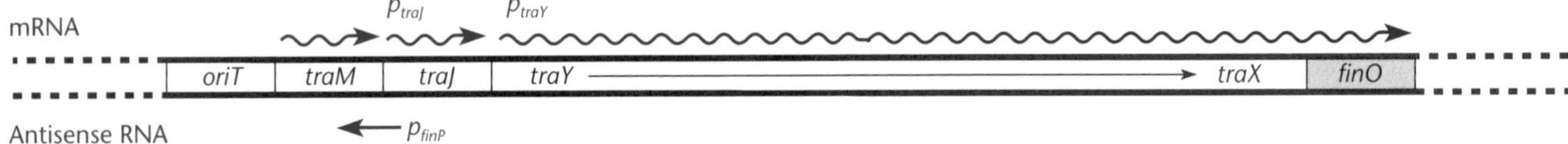

B Immediately after entry into cell

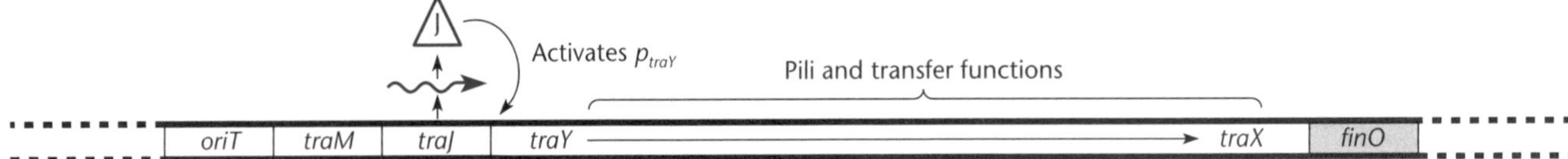

C After plasmid establishment

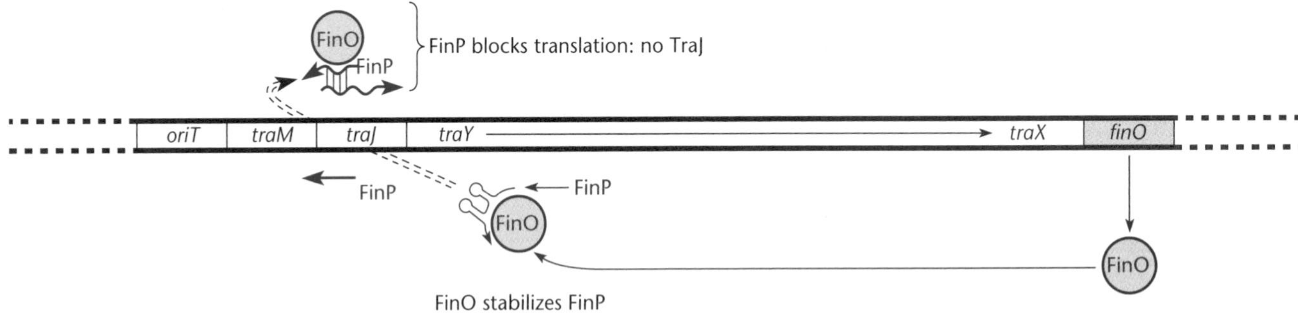

Figure 5.7 Fertility inhibition of the F plasmid. Only the relevant *tra* genes discussed in the text are shown. **(A)** Genetic organization of the *tra* region. **(B)** The *traJ* gene product is a transcriptional activator that is required for transcription of the other *tra* genes, *traY* to *traX*, and *finO* from the promoter p_{traY}. **(C)** Translation of the *traJ* mRNA is blocked by hybridization of an antisense RNA, FinP, which is transcribed in the same region from the complementary strand. A protein, FinO, stabilizes the FinP RNA. Details are given in the text. doi:10.1128/9781555817169.ch5.f5.7.

explain why genes with related functions are often very similar to each other regardless of the organisms that harbor them. These genes could have been transferred by promiscuous plasmids fairly recently in evolution, which would account for their similarity relative to the other genes of the organisms.

The interspecies transfer of plasmids also has important consequences for the use of antibiotics in treating human and animal diseases. Many of the most promiscuous plasmids, including those of the IncP replication incompatibility group, such as RP4, and the IncW replication incompatibility group plasmid R388, were isolated in hospital settings. These large plasmids (commonly called R plasmids because they carry genes for antibiotic resistance) presumably have become prevalent in response to the indiscriminate use of antibiotics in medicine and agriculture. Antibiotic resistance genes are remarkably common in the environment, and a plausible source of resistance genes could be the bacteria that make antibiotics or other bacteria that live in close proximity to these organisms (see D'Costa et al., Suggested Reading). In chapter 9, we discuss transposons and other DNA elements that are usually responsible for bringing antibiotic resistance genes found on plasmids.

Whatever their source, the emergence of R plasmids illustrates why antibiotics should be used only when they are absolutely necessary. In humans or animals treated indiscriminately with antibiotics, bacteria that carry R plasmids are selected from the normal flora. R plasmids can be quickly transferred into an invading pathogenic bacterium, making it antibiotic resistant. Consequently, the infection will be difficult to treat.

Conjugation and Type IV Protein Secretion

A surprising interrelationship between conjugation and type IV protein secretion has been revealed by experimentation and bioinformatics. In the case of conjugation, it appears that the coupling protein acts like a DNA pump to adapt a protein translocation machine for DNA transfer. One early method of determining if proteins expressed in donor cells can be transferred into recipient cells involved labeling the donor proteins with radioactivity prior to the mating. Recipient cells could be specifically examined for labeled donor proteins using two methods: by killing the donor cells with a male-specific bacteriophage or by using "minicells," which are devoid of nuclear material, as recipients (see Rees and Wilkins, Suggested Reading). Minicells are derived from a bacterial strain with a partitioning defect due to a mutation in the *min* genes, where cells without chromosomes occur at high frequency. In the case of the RP4 conjugal plasmid, the large (118-kilodalton [kDa]) Pri protein can be clearly detected in the cytoplasm of recipient cells. In other systems, other, more sensitive genetic techniques were needed to show that the relaxase is targeted for translocation to recipient cells through the pilus apparatus via specific recognition sequences found in the protein (see Parker and Meyer, Suggested Reading).

Conjugation is one form of type IV secretion system. Type IV secretion systems are also used for DNA export and import systems that transport DNA between the cytoplasm and the environment. In the case of bacterial pathogens, type IV secretion systems are often responsible for translocating proteins or other effector molecules to eukaryotic cells. In some cases, the other cell can be a eukaryotic plant cell, as in the case of the Ti plasmid, the causative agent of crown gall disease (Box 5.1). When type IV secretion is used in virulent bacteria, the pili or another adhesin is used to hold the cells together during the transfer, which also involves special membrane structures through which the macromolecules must pass. As with DNA conjugation, the processes are very specific, and only some types of proteins or plasmid DNAs can be transferred. Nevertheless, it came as a surprise how closely related these two types of systems can be. The relatedness of type IV secretion to conjugation is dramatically illustrated by comparison of the T-DNA transfer system of *Agrobacterium tumefaciens*, called VirB, to some type IV secretion systems (Figure 5.8). As discussed in Box 5.1, the T-DNA transfer system transfers the T-DNA part of a plasmid from bacterial to plant cells. It also transfers a protein that forms a channel in the plant membrane and the relaxase protein, which doubles as a protein that can target the T-DNA to the plant nucleus, where it can enter the plant DNA. This transfer system shares features with other plasmid conjugation systems in that it encodes a pilus, relaxase, coupling proteins, and chaperones, as well as many other proteins involved in the transfer process. In fact, most of the proteins of the Tra systems of the F and R388 plasmids can be assigned

Figure 5.8 Gene arrangements of type IV secretion loci. Genes encoding VirB system homologs are similarly shaded, and those encoding proteins unrelated to those in the VirB system are not shaded. **(Top)** *A. tumefaciens* is the first species shown to carry three distinct type IV secretion systems whose substrates are defined; e.g., the VirB system transfers T-DNA and protein effectors to plants, whereas the Avh system transfers pATC58 and the Trb system transfers the Ti plasmid, respectively, to other bacteria. **(Middle)** Representative type IV secretion systems of other species that direct conjugal DNA transfer. **(Bottom)** Representative type IV secretion systems that direct protein transfer during the course of infection.
doi:10.1128/9781555817169.ch5.f5.8

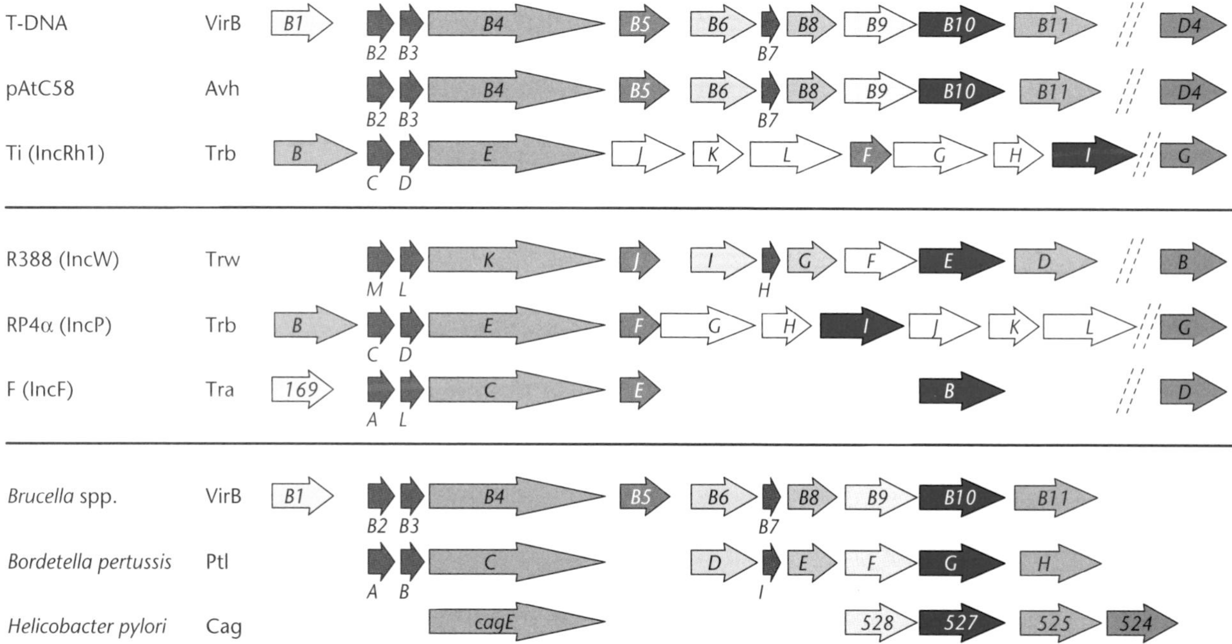

BOX 5.1

Gene Exchange between Domains

Not only can some plasmids transfer themselves into other types of bacteria, they can also sometimes transfer themselves into eukaryotes, that is, into organisms of a different domain.

A. tumefaciens and Crown Gall Tumors in Plants

The first discovery of transfer of bacterial plasmids into eukaryotes occurred in the plant disease crown gall. Crown gall disease is caused by *A. tumefaciens*; it is identified by a tumor that appears on the plant, usually where the roots join the stem (the crown). Virulent strains of *A. tumefaciens* contain a plasmid called the tumor-inducing (or Ti) plasmid. The Ti plasmids of *A. tumefaciens* are in most respects normal bacterial self-transmissible plasmids. A typical Ti plasmid is shown in Figure 1. Like other self-transmissible plasmids, Ti plasmids (*tra* and *trb* genes) encode conjugative-transfer functions that enable them to transfer themselves between bacteria. What makes these plasmids remarkable is that they can also transfer part of themselves, called the T-DNA region, into plants.

The functions required for transfer of the T-DNA into plants are encoded by a region called the *vir* region (Figure 1). This region is distinct from the *tra* and *trb* regions, which are required for the transfer of the plasmid into other bacteria, but its functions are remarkably similar to those of both of the other **Tra functions** and of other type IV secretion systems. DNA sequence analysis clearly demonstrates close similarity between many Vir proteins and Tra proteins. Figure 2 shows the structure of the type IV secretion system encoded by the *vir* region of the Ti plasmid. Like the *tra* region, the *vir* region encodes both an Mpf system, which elaborates a pilus, and a Dtr system, which processes the DNA for transfer. The pilus is composed of the pilin protein, which is likely the product of the *virB2* gene and is like the pilins of the pili of other type IV secretion systems. The Mpf system also includes a coupling protein, the product of the *virD4* gene, which communicates with the relaxosome, which includes the specific relaxase. The relaxase, the product of the *virD2* gene, cleaves the sequences bordering the T-DNA in the plasmid and remains covalently attached to

Figure 1 The structure of a Ti plasmid, showing the following regions (clockwise from the top): the T-DNA, bracketed by the border sequences, which resemble *oriT* sequences of conjugation systems (borders are not transferred in their entirety); the T-DNA contains the genes that are expressed in the plant to make opines and plant hormones (not shown); the *noc* genes, encoding enzymes for the catabolism of the opine nopaline in the bacterium; some *tra* and *trb* genes, for transfer into other bacteria; the *oriV* region, for replication of the plasmid in the bacterium; *oriT* and *tra* function genes, for transfer into other bacteria; *acc* genes, for catabolism of another opine; and *vir* genes, for transfer into plants.
doi:10.1128/9781555817169.ch5.Box5.1.f1

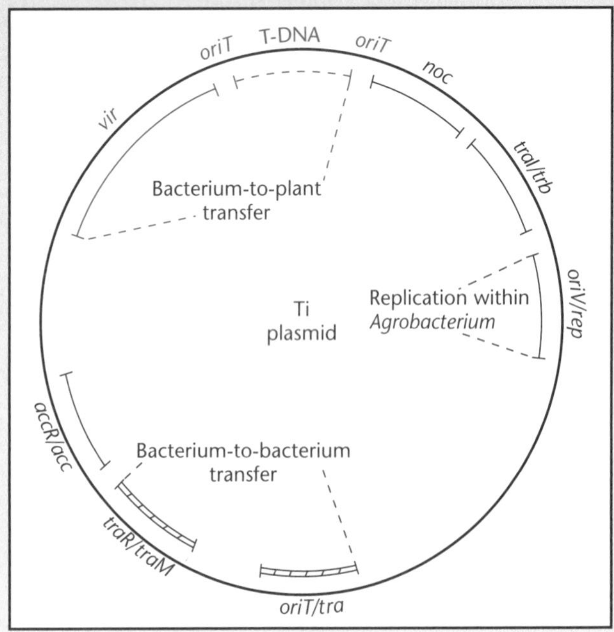

Figure 2 Structure of the type IV secretion system for transfer of T-DNA into plants (for details, see the text).
doi:10.1128/9781555817169.ch5.Box5.1.f2

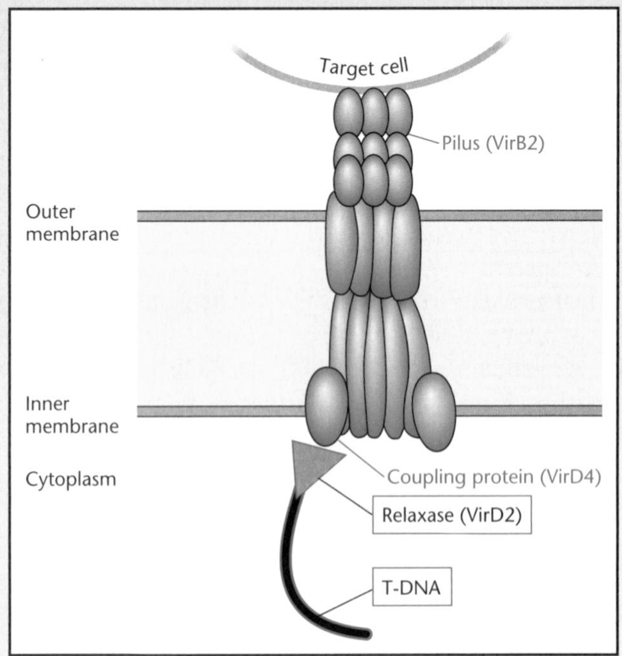

BOX 5.1 (continued)

Gene Exchange between Domains

the 5′ ends of the single-stranded T-DNAs during transfer into the plant. The sequences at which the relaxase cuts these border sequences are similar to the *oriT* sequences of IncP plasmids, and the relaxase makes a cut in exactly the same place in the sequences where the relaxases of IncP plasmids cut their *oriT*.

This is where the T-DNA transfer system begins to differ from ordinary conjugative-transfer systems. In addition to its role as a relaxase, the VirD2 protein contains amino acid sequences that target it to the plant cell nucleus once it is in the plant. These sequences, called nuclear localization signals, are essentially "passwords" that tell the plant that a particular plant nuclear protein should be transported into the nucleus after it has been translated in the cytoplasm. By imitating the password, the VirD2 protein tricks the plant into transporting it into the nucleus, dragging the attached T-DNA with it. Another plasmid-encoded protein, VirE2, is also transported into the plant cytoplasm (separately from T-DNA), where it coats the T-DNA and assists in delivery to the nucleus. Once in the nucleus, the T-DNA can enter the plant DNA by recombination. Once integrated into the plant DNA, the T-DNA of the plasmid encodes the synthesis of plant hormones that induce the plant cells to multiply and form tumors (galls) on the plant, hence the name "crown gall tumors." The T-DNA also encodes enzymes that synthesize unusual small molecules composed of an amino acid, such as arginine, joined to a carbohydrate, such as pyruvate. These compounds, called opines, are excreted from the tumor. The plant is able to express the genes on the T-DNA and make these compounds because the genes on the T-DNA have plant promoters and plant translational initiation regions, so they can be expressed once they are in the plant. Meanwhile, back in the bacterium, the Ti plasmid carries genes that allow it to use the particular opine made by that strain as a carbon and nitrogen source. Very few types of bacteria can degrade opines, which gives the *Agrobacterium* species containing this particular Ti plasmid an advantage. In this way, the bacterium creates its own special "ecological niche" at the expense of the plant.

The discovery of T-DNA, made in the 1970s, allowed the construction of transgenic plants because any foreign genes cloned into one of the T-DNA regions of the Ti plasmid will be transferred into the plant along with the T-DNA and integrated into the plant DNA. The integrated foreign genes can alter the plant, provided that they are transcribed and translated in the plant. Figure 3 shows the general procedure for making a transgenic plant. A "disarmed" strain of *A. tumefaciens* in which the native T-DNA has been deleted is used.

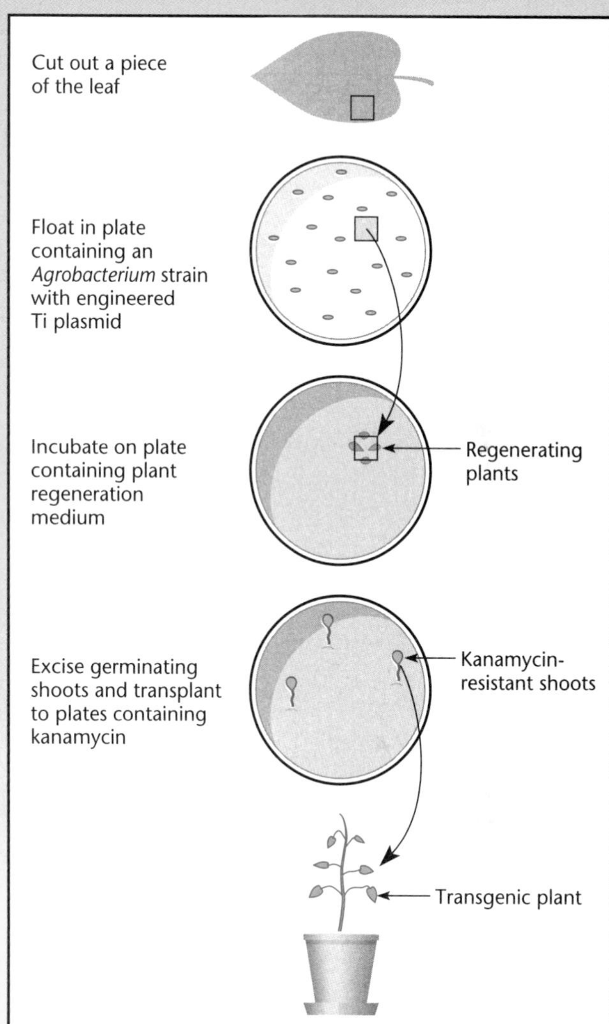

Figure 3 A procedure for making a transgenic plant (for details, see the text).
doi:10.1128/9781555817169.ch5.Box5.1.f3

A small plasmid containing the gene to be transferred, a gene for kanamycin resistance, and a DNA sequence similar to the *oriT* site of conjugal plasmids is constructed. Single plant cells are inoculated, and after allowing time for T-DNA transfer, all bacteria and all nontransformed plant cells are killed using antibiotics. Transgenic plants with the T-DNA inserted in their chromosomes can be regenerated from the kanamycin-resistant cells. An entire industry has developed around this technology, and agrobacteria have been used to genetically engineer plants to make their own insecticides, to be more nutritious, and to survive more severe growth conditions.

(continued)

BOX 5.1 (continued)

Gene Exchange between Domains

Transfer of Broad-Host-Range Plasmids into Eukaryotes

The Ti plasmid evolved to transfer part of itself into plant cells. The surprising result is that other bacterial plasmids can also transfer themselves or mobilize other plasmids into eukaryotic cells. One striking example is the mobilization of other plasmids into plant cells by the Ti plasmid. As mentioned, the sequences bracketing the T-DNA in the Ti plasmid can be thought of as *oriT* sites, and the Vir functions of the Ti plasmid can be thought of as mobilizing the T-DNA into plant cells. Plasmid RSF1010 and plasmids derived from it can also be mobilized into plant cells by the Ti plasmid, provided that they contain the correct *mob* sequence.

Not only do plasmids transfer into plants, they can also sometimes transfer into lower fungi. This observation is very surprising because the cell surfaces of plants and fungi are very different. There is no apparent reason why bacterial plasmids should have evolved to transfer into other domains. Whatever the reason, the transfer of genes between eukaryotes and bacteria may play an important role in evolution.

References

Bates, S., A. M. Cashmore, and B. M. Wilkins. 1998. IncP plasmids are unusually effective in mediating conjugation of *Escherichia coli* and *Saccharomyces cerevisiae*: involvement of the Tra2 mating system. *J. Bacteriol.* **180:**6538–6543.

Brencic, A., and S. C. Winans. 2005. Detection of and response to signals involved in host-microbe interactions by plant-associated bacteria. *Microbiol. Mol. Biol. Rev.* **69:**155–194.

Buchanan-Wollasten, U., J. E. Passiatore, and F. Cannon. 1987. The *mob* and *oriT* mobilization functions of a bacterial plasmid promote its transfer to plants. *Nature* (London) **328:**172–175.

homologs in the T-DNA transfer system (Figure 5.8). Moreover, some pathogenic bacteria that transfer proteins into eukaryotic cells as part of the disease-causing process also have analogous functions. One of the most striking similarities is to the CagA toxin-secreting system of *Helicobacter pylori*, implicated in some types of gastric ulcers. This type IV toxin-secreting system has at least five protein homologs to the T-DNA system of the Ti plasmid of *Agrobacterium*. The system delivers a toxin directly through the bacterial membranes and into the eukaryotic cell, where the toxin is phosphorylated on one of its tyrosines. In the phosphorylated state, the toxin causes many changes in the cell, including alterations in its actin cytoskeleton. Another pathogenic bacterium, *Bordetella pertussis*, the causative agent of whooping cough, also has a type IV secretion system that has nine proteins homologous to proteins in the T-DNA system. This system secretes the pertussis toxin through the outer membrane of the bacterial cell. Once outside the cell, the pertussis toxin assembles into a form that can enter the eukaryotic cell, where it can ADP-ribosylate G proteins, thereby interfering with signaling pathways and causing disease symptoms.

However, the most striking evidence that conjugation is related to type IV secretion has come from the virulence system of *Legionella pneumophila*, which causes Legionnaires' disease. Like many pathogenic bacteria, this bacterium can multiply in macrophages, specialized white blood cells that are designed to kill them (see Vogel et al., Suggested Reading). The bacteria are taken up by the macrophage but then secrete proteins that disarm the phagosomes that have engulfed them. Remarkably, the components of this type IV secretion system are analogous to the Tra functions of some self-transmissible plasmids, and this type IV system can even mobilize the plasmid RSF1010 at a low frequency (Figure 5.8).

Mobilizable Plasmids

Some plasmids are not self-transmissible but can be transferred by another, self-transmissible plasmid residing in the same cell. Plasmids that cannot transfer themselves but can be transferred by other plasmids are said to be mobilizable, and the process by which they are transferred is called **mobilization**. The simplest mobilizable plasmids merely contain the *oriT* sequence of a self-transmissible plasmid, since any plasmid that contains the *oriT* sequence of a self-transmissible plasmid can be mobilized by that plasmid. Expressed in genetic terminology, the Mpf and Dtr systems of the self-transmissible plasmid can act in *trans* on the *cis*-acting *oriT* site of the plasmid and mobilize it.

The ability to allow mobilization can be used to locate the *oriT* sequence in a plasmid (Figure 5.9). Random clones of the DNA of the self-transmissible plasmid are introduced into a nonmobilizable cloning vector, and the mixture is introduced into a cell containing the self-transmissible plasmid. Any vector plasmids that are mobilized into recipient cells probably contain a DNA insert including the *oriT* sequence. Transposons and plasmid cloning vectors containing the *oriT* site of a self-transmissible plasmid have many applications in molecular genetics because they can be mobilized into other cells.

While we can construct such a plasmid, and they are mobilizable, minimal mobilizable plasmids containing

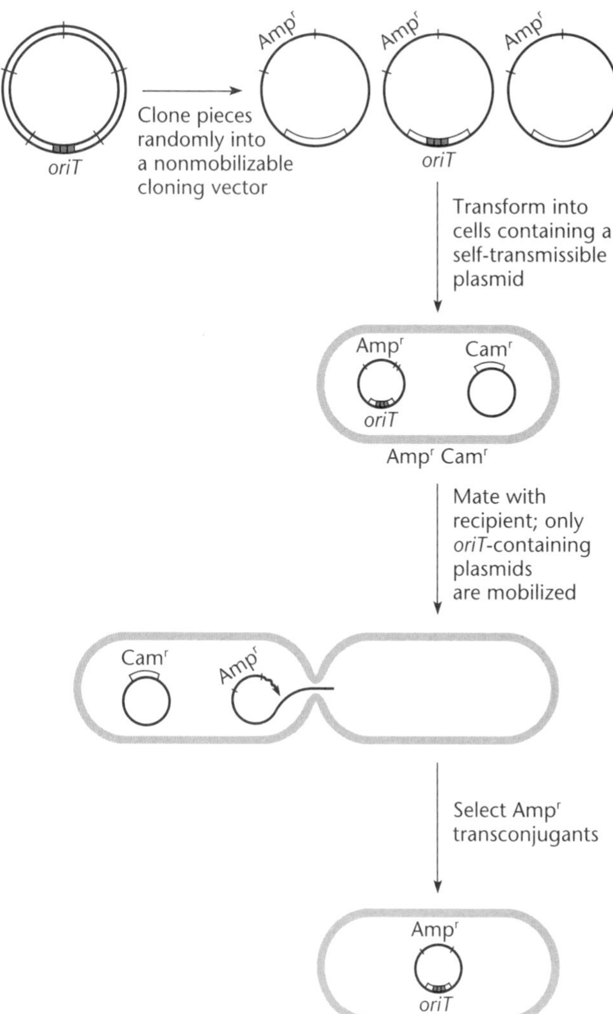

Figure 5.9 Identifying the *oriT* site on a plasmid. Pieces of the plasmid are cloned randomly into a nonmobilizable cloning vector. The mixture is transformed into cells containing the self-transmissible plasmid and mixed with a proper recipient. Pieces of DNA that allow the cloning vector to be mobilized contain the *oriT* site of the plasmid.
doi:10.1128/9781555817169.ch5.f5.9

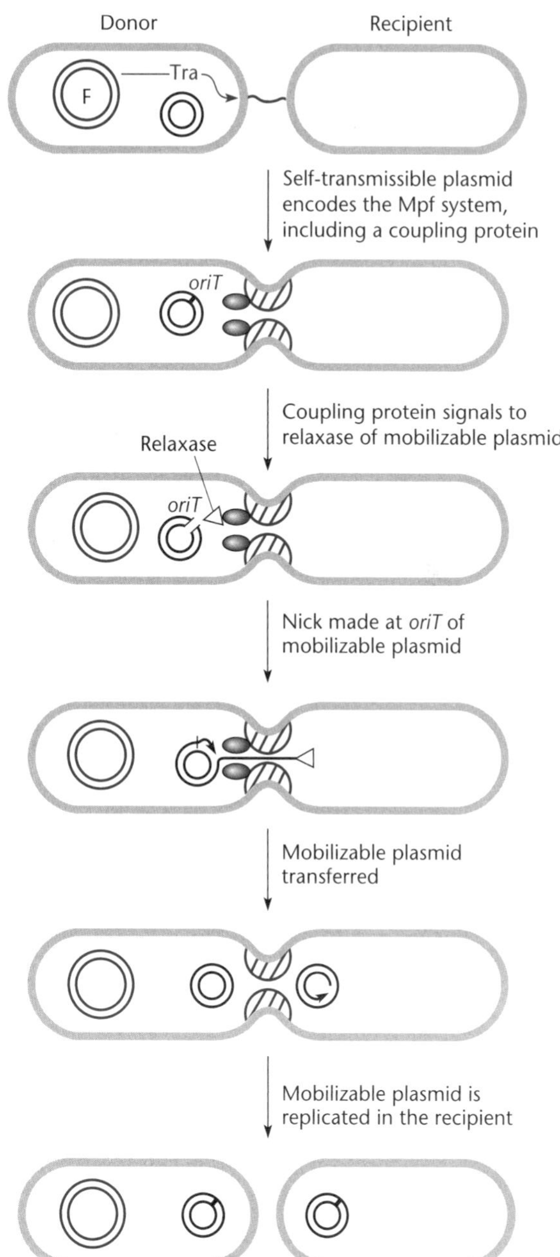

Figure 5.10 Mechanism of plasmid mobilization. The donor cell carries two plasmids, a self-transmissible plasmid, F, that encodes the *tra* functions that promote cell contact and plasmid transfer, and a mobilizable plasmid (blue). The *mob* functions encoded by the mobilizable plasmid make a single-stranded nick at *oriT* in the *mob* region. Transfer and replication of the mobilizable plasmid then occur. The self-transmissible plasmid may also transfer. Details are given in the text.
doi:10.1128/9781555817169.ch5.f5.10

only the *oriT* site of a self-transmissible plasmid do not seem to occur naturally. All mobilizable plasmids isolated so far encode their own Dtr systems, including their own relaxase and helicase. For historical reasons, the *tra* genes of the Dtr system of mobilizable plasmids are called the ***mob* genes**, and the region required for mobilization is called the ***mob* region**. The function of the *mob* gene products of mobilizable plasmids seems to be to expand the range of self-transmissible plasmids by which they can be mobilized (see below). A plasmid containing only the *oriT* sequence of a self-transmissible plasmid can be mobilized only by the *tra* system of that self-transmissible plasmid and not by those of other self-transmissible plasmids that do not share the same *oriT* site, while naturally occurring mobilizable plasmids can often be mobilized by a number of *tra* systems.

The process of mobilization of a plasmid by a self-transmissible plasmid is illustrated in Figure 5.10. The process is identical to the transfer of a self-transmissible plasmid, except that the Mpf system of the self-transmissible plasmid acts not only on its own Dtr system, but also on the Dtr system of the mobilizable plasmid. The self-transmissible plasmid forms a mating bridge with a recipient cell and communicates this information via its coupling protein, not only to its own relaxase, but also to the relaxase of a mobilizable plasmid that happens to be in the same cell. The relaxase of the mobilizable plasmid is responsible for processing at the *oriT* site and presentation to the Mfp system for transport to the recipient cell, as described above. The self-transmissible plasmid is also often transferred at the same time that it mobilizes other plasmids. However, generally, either one plasmid or the other is transferred into a particular recipient cell, due to competition between the two plasmids for coupling protein.

The secret of being mobilized by another plasmid is to be recognized by the coupling protein of the other plasmid. Any plasmid encoding its own Dtr system can be mobilized by a coresident self-transmissible plasmid, provided that its relaxase can communicate with the coupling protein of the Mpf system of the coresident plasmid. Accordingly, the relaxases of mobilizable plasmids likely evolved to communicate with a broader range of coupling proteins of self-transmissible plasmids so that they can take advantage of a number of different Mpf systems, unlike the relaxases of self-transmissible plasmids, which seem to be more specific.

PLASMID MOBILIZATION IN BIOTECHNOLOGY

Mobilization plays an important role in biotechnology because it can be used to efficiently introduce foreign DNA into bacteria. A *mob* site is often introduced into cloning vectors so that they can be efficiently transferred into cells (see Bagdasarian et al., Suggested Reading). Smaller is better in cloning vectors (see chapter 4), and a mobilizable plasmid can be much smaller than a self-transmissible plasmid because it does not need the 15 or so Mpf genes required to assemble the mating bridge, only the 4 or so Dtr genes required for DNA processing. Once foreign DNA has been cloned into such a cloning vector, it can be introduced into even distantly related bacteria by the Mpf system of a larger, promiscuous, self-transmissible plasmid. In addition, some self-transmissible plasmids have been crippled so that they cannot transfer themselves but can transfer only mobilizable plasmids. Then, the recipient cell receives only the mobilized cloning vector in such a transfer and not the self-transmissible plasmid that mobilized it.

A common application of plasmid mobilization technology is in transposon mutagenesis. These methods are most highly developed for gram-negative bacteria. Some plasmids, such as RP4, are so promiscuous that they can transfer themselves into essentially any gram-negative bacterium. If such a plasmid is used to introduce a small mobilizable plasmid that has a narrow-host-range origin of replication, like the ColE1 origin (see chapter 4), the smaller plasmid is mobilized into the bacterium but probably cannot replicate there and is eventually lost (i.e., is a suicide vector). If the smaller plasmid also contains a transposon, such as Tn*5*, containing the selectable kanamycin resistance gene and with a broad host range for transposition, the only way the recipient cell can become resistant to kanamycin is if the transposon hops into the chromosome of the recipient strain, causing random insertion mutations. The transposon insertion mutants facilitate cloning and can even be used for Hfr mapping if an *oriT* sequence has been introduced on the transposon. We return to such methods of transposon mutagenesis in chapter 9. Mobilizable plasmids can also be used to detect the transferability of naturally occurring plasmids for which no selection is available. A resident plasmid may have transfer functions and be self-transmissible if it can mobilize another plasmid carrying an easily selectable marker.

TRIPARENTAL MATINGS

Mobilization of a plasmid into a recipient cell is often used for cloning, transposon mutagenesis, or other procedures. As mentioned above, mobilizable plasmids have an advantage over self-transmissible plasmids in being smaller. Nevertheless, difficulties can be encountered before these plasmids can be mobilized. For example, the self-transmissible plasmid and the plasmid to be mobilized may be members of the same replication incompatibility group and so do not stably coexist in the same cell. Also, the self-transmissible plasmid may express its *tra* genes only for a short time after entering a recipient cell so that transfer is inefficient.

Triparental matings help overcome some of the barriers to efficient plasmid mobilization. Figure 5.11 illustrates the general method. As the name implies, three bacterial strains participate in the mating mixture. The first strain contains a self-transmissible plasmid, the second contains the plasmid to be mobilized, and the third is the eventual recipient. After the cells are mixed, some of the self-transmissible plasmids in the first strain are transferred into the strain carrying a plasmid that is capable of being mobilized (strain II in Figure 5.11). Because it is fertile when it first enters the cell, the self-transmissible plasmid quickly spreads through the population of cells containing the mobilizable plasmid. It is then able to mobilize the mobilizable plasmid into the third strain with high efficiency because new transconjugants retain their ability to transfer for at least six generations. Contrast this to a mating involving only two strains, one of which contains both the self-transmissible

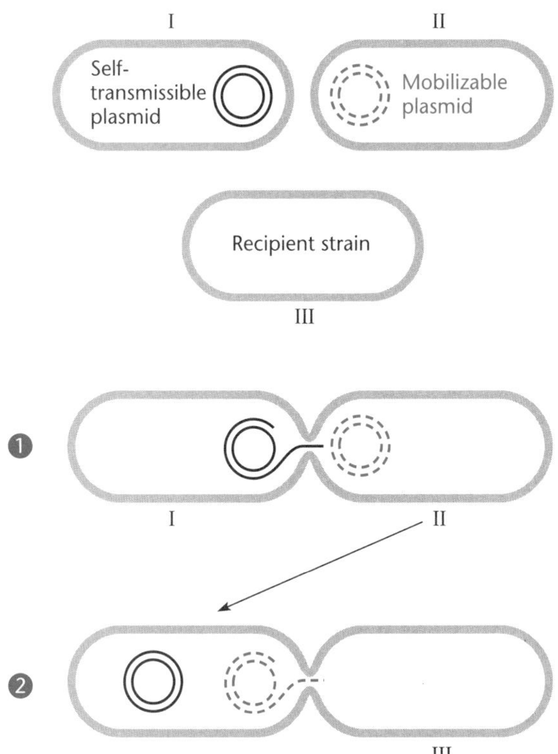

Figure 5.11 Triparental mating. In step 1, a self-transmissible plasmid from parent I transfers into parent II. In step 2, the self-transmissible plasmid transfers the mobilizable plasmid into parent III. This method works even if the self-transmissible plasmid and the mobilizable plasmid are members of the same Inc group (see the text) and if the self-transmissible plasmid cannot replicate in parent II.
doi:10.1128/9781555817169.ch5.f5.11

plasmid and the mobilizable plasmid. Only a small fraction of the cells are fertile and can mobilize the mobilizable plasmid into the recipient strain. Also, in a triparental mating, even if the two plasmids are members of the same replication incompatibility group, they coexist long enough for the mobilization to occur.

Chromosome Transfer by Plasmids

Usually during conjugation only a plasmid is transferred to another cell. However, plasmids sometimes transfer the chromosomal DNA of their bacterial host, a fact that has been put to good use in bacterial genetics. Without the transfer of genes, bacterial genetics is not possible, and conjugation is one of only three ways in which chromosomal and plasmid genes can be exchanged among bacteria (transduction and transformation are the others). In chapter 3, we discussed how these ways of exchanging genes between bacterial strains can be used to map genetic markers. In this chapter, we go into more detail about how plasmids transfer chromosomal DNA.

Formation of Hfr Strains

Sometimes plasmids integrate into chromosomes, and when such plasmids attempt to transfer, they take the chromosome with them. Bacteria that can transfer their chromosomes because of an integrated plasmid are called **Hfr strains**, where Hfr stands for high-frequency recombination. The name derives from the fact that many recombinants can appear when such a strain is mixed with another strain of the same bacterium, as we will discuss.

The integration of plasmids into the chromosome can occur by several different mechanisms, including recombination between sequences on the plasmid and sequences on the chromosome. For normal recombination to occur, the two DNAs must share a sequence (see chapter 10). Most plasmid sequences are unique to the plasmid, but sometimes the plasmid and the chromosome share an **insertion element**, which is a common source of insertion mutations (see chapter 3). These small transposons often exist in several copies in the chromosome and may also appear in plasmids; recombination between these common sequences can result in integration of the plasmid.

Figure 5.12 shows how recombination between the IS2 element in the F plasmid and an IS2 element in the chromosome of *E. coli* can lead to integration of the F plasmid by homologous recombination or site-specific recombination. Once integrated, the F plasmid is bracketed by two copies of the IS2 element. The bacterium is now an Hfr strain. The *E. coli* chromosome contains 20 sites for IS-mediated Hfr formation by the F plasmid,

Figure 5.12 Integration of the F plasmid by recombination between IS*2* elements in the plasmid and in the chromosome, forming an Hfr cell. Integration can also occur through recombination with one of the two IS*3* elements or Tn*1000* sequences on the F plasmid and in the chromosome (see Figure 5.2).
doi:10.1128/9781555817169.ch5.f5.12

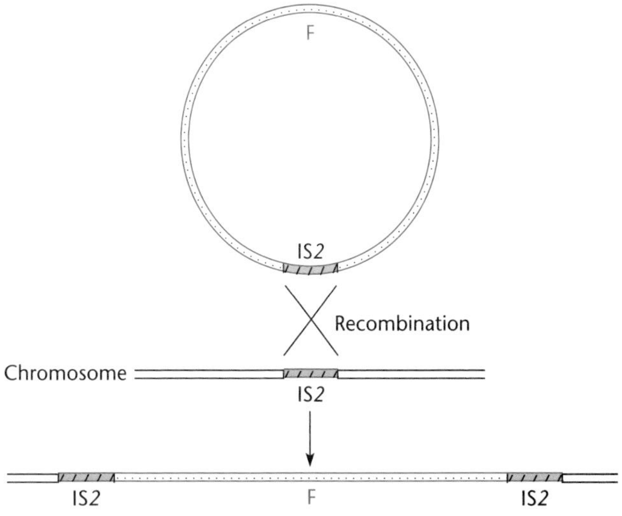

marking where IS elements are shared by the plasmid and the chromosome.

Transfer of Chromosomal DNA by Integrated Plasmids

We mentioned at the beginning of the chapter that self-transmissible plasmids were first detected in 1947 by Lederberg and Tatum, even though these investigators did not know what they were at the time. They observed recombinant types after mixing different strains of *E. coli*. Recombinant types differ from the two original, or parental, strains (see chapter 3) and in this case resulted from the transfer of chromosomal DNA from one strain to another by an F plasmid that had intermittently integrated into the chromosome and was transferring the chromosome to the other strain. In retrospect, it was fortuitous that the strains used by Lederberg and Tatum included an F plasmid and that the plasmid was constitutive for transfer.

Figure 3.35 gives an overview of the process by which chromosomal DNA is transferred in an Hfr strain. The process of initiating the transfer is the same as the process by which the transfer of the plasmid itself is initiated (Figure 5.5). The plasmid expresses its *tra* genes, even though it is integrated into the chromosome, and a pilus is synthesized. On contact with a recipient cell, the coupling protein communicates with the relaxase, and the integrated plasmid DNA is transferred starting at the *oriT* site. One strand is displaced into the recipient cell, while the other strand is replicated. Now, however, after transfer of the portion of the *oriT* sequence and plasmid on one side of the nick, chromosomal DNA is also transferred into the new cell. If the transfer continued long enough (in *E. coli*, approximately 100 min at 37°C), the entire bacterial chromosome would eventually be transferred, ending with the remaining plasmid *oriT* sequences. However, transfer of the entire chromosome (and thus the whole integrated plasmid) is rare, because the union between the cells can be broken, and the transferred strand likely contains nicks that would halt the process. This fact is exploited for genetic mapping by Hfr crosses in a method called gradient of transfer (see chapter 3). Also, because the remainder of the plasmid is seldom transferred, the recipient cell will not itself become a male cell.

Chromosome Mobilization

The Tra functions of self-transmissible plasmids can also mobilize the chromosome, provided that a mobilizable plasmid has integrated into the chromosome or that the chromosome contains the *oriT* sequence of the plasmid. Chromosome mobilization also begins at the *oriT* sequence in the chromosome. This has allowed the mapping of genes by gradient of transfer in many genera of bacteria by introducing the *oriT* site of a mobilizable plasmid into the chromosome on a transposon.

Prime Factors

Through the integration and excision of conjugal plasmids, chromosomal genes can get crossed onto the plasmid. These conjugal plasmids that now contain a portion of the host chromosome are called **prime factors**. When a prime factor moves to a new host, so will those chromosomal genes that now reside in the vector. Prime factors are usually designated by the name of the plasmid followed by a prime symbol, for example, F′ factor. An R plasmid, such as RK2, carrying bacterial chromosomal DNA is an R′ factor.

CREATION OF PRIME FACTORS

Prime factors are created through the same recombination processes that create Hfr strains. After an Hfr is formed, the plasmid can be restored if the same genetic process that formed the Hfr occurs in reverse. Occasionally, however, instead of the same insertion sequence being used as the site of recombination, a different insertion sequence that resides on both the plasmid and the chromosome is used. As shown in Figure 5.13, the result of this recombination event involves a portion of the chromosome adjacent to the Hfr being crossed onto the plasmid, forming an F′ factor. The F′ factor carries the chromosomal DNA that lies between the recombining DNA sequences. A prime factor can form from recombination between any repeated sequences, including identical IS elements or genes for rRNA, which often exist in more than one copy in bacteria.

Note that a deletion forms in the chromosome when the prime factor loops out. Some of the genes deleted from the chromosome may have been essential for the growth of the bacterium. Nevertheless, the cells do not die, because the prime factor still contains the essential genes, which should be passed on to daughter cells when the plasmid replicates. However, cells that lose the prime factor die.

Prime factors can be very large, almost as large as the chromosome itself. In general, the larger a prime factor is, the less stable it is. Maintaining large prime factors in the laboratory requires selection procedures designed so that cells die if they lose some or all of the prime factor. However, most prime factors are small enough to be transferred in their entirety. Because prime factors contain an entire self-transmissible plasmid, a cell receiving a prime factor becomes a donor and can transfer this DNA into other bacteria. Moreover, because prime factors are replicons with their own plasmid origin of replication, they can replicate in any new bacterium that falls within the plasmid host range (see chapter 4). The formation of F prime factors is a rare event; however, the products of this event can be selected. For example, gene products adjacent to an Hfr can be selected in recipient cells that are incapable of homologous recombination.

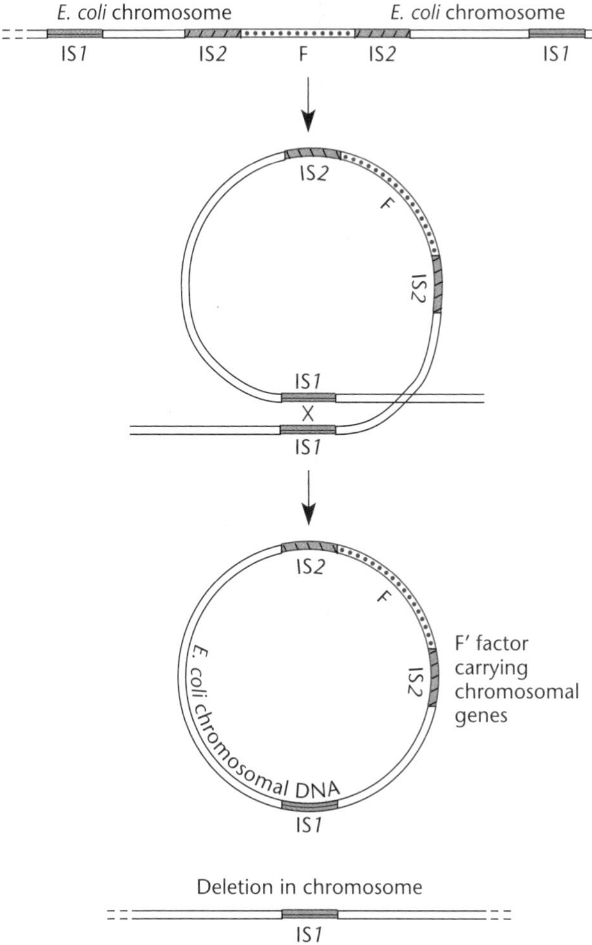

Figure 5.13 Creation of a prime factor by recombination. Recombination may occur between homologous sequences, such as IS sequences, in the chromosomal DNA outside the F factor. The F factor then contains chromosomal sequences, and the chromosome carries a deletion.
doi:10.1128/9781555817169.ch5.f5.13

COMPLEMENTATION TESTS USING PRIME FACTORS

Complementation tests can reveal whether two mutations are in the same or different genes and how many genes are represented by a collection of mutations that give the same phenotype (see chapter 3). Complementation tests can also provide needed information about the type of mutation being studied, whether the mutations are dominant or recessive, and whether they affect a *trans*-acting function or a *cis*-acting site on the DNA. However, complementation tests require that two different alleles of the same gene be introduced into the same cell, and bacteria are normally haploid, with only one allele for each gene in the cell.

Because prime factors contain a region of the chromosome, they can be used to create cells that are stable diploids for the region they carry. However, they contain only a short region of the chromosome and so are diploid for only part of the chromosome. Organisms diploid for only a region of their chromosome are called **partial diploids** or **merodiploids**.

ROLE OF PRIME FACTORS IN EVOLUTION

Prime factors formed with promiscuous, broad-host-range plasmids probably play an important role in bacterial evolution. Once chromosomal genes are on a broad-host-range promiscuous plasmid, they can be transferred into distantly related bacteria, where they are maintained as part of the broad-host-range plasmid replicon. They may then become integrated into the chromosome through various types of recombination.

Transfer Systems of Gram-Positive Bacteria

Self-transmissible plasmids have also been found in many types of gram-positive bacteria, including species of *Bacillus*, *Enterococcus*, *Streptococcus*, *Staphylococcus*, and *Streptomyces*. With the exception of some *Streptomyces* elements, these plasmids are known to transfer by systems that are similar to the gram-negative transfer systems discussed above. They are transferred complete with their own relaxases and *oriT* sequences. In fact, the *oriT* sequences of gram-positive plasmids are sometimes closely related to those of gram-negative bacteria and vegetative replication systems that utilize rolling-circle DNA replication (see chapter 4 and Kurenbach et al., Suggested Reading). As in the case of conjugal plasmids from gram-negative bacteria, the transmissible plasmids of gram-positive bacteria frequently have postsegregational killing systems and partitioning systems that stabilize the plasmid in the population. They also often encode functions such as virulence determinants and carry genes encoding antibiotic resistance that can aid the host. Transposable elements are also commonly found on conjugal plasmids in gram-positive bacteria.

The major differences between plasmids from the two bacterial groups are in the Mpf systems, which can be simpler in gram-positive bacteria because of the lack of an outer membrane. We discuss only plasmids from *Enterococcus* in this section because of their interesting method of attracting mating cells and their medical importance (for a review, see Clewell and Francia, Suggested Reading).

Plasmid-Attracting Pheromones

Some strains of *Enterococcus faecalis* excrete pheromone-like compounds that can stimulate mating with donor cells. These pheromones are small peptides, each of which stimulates mating with cells containing a particular plasmid. The pheromone-like peptides act by specifically stimulating the expression of *tra* genes in the plasmids of neighboring bacteria, thereby inducing aggregation and mating. Once a cell has acquired a plasmid, it reduces production of the cognate peptide pheromone, but it continues to excrete other peptides that stimulate mating with cells containing other plasmid types. As explained above, inducing Tra functions only when a potential recipient is nearby has many benefits.

The mechanism of pheromone sensing of recipient cells is outlined in Figure 5.14. All of the genes that encode the peptide pheromones are located on the *Enterococcus* chromosome and encode secreted lipoproteins (see De Boever et al., Suggested Reading). After the signal sequences are cut from the lipoprotein during export, the active pheromones are produced by proteolysis of the C-terminal 7 or 8 amino acids of the signal peptide sequence. Processing occurs as the pheromone is excreted from the cell. By cutting the pheromones from the signal sequences of many different lipoproteins, each cell is able to excrete a number of different pheromones, allowing specificity in interaction with different plasmids.

The genes encoding pheromone-sensing proteins are also conserved in the various plasmid families. Sensing of the pheromones requires several specific proteins located on the cell surface and in the cytoplasm. Each type of transmissible plasmid expresses proteins specific for one type of pheromone, and the pheromone is named after the plasmid it attracts. For example, in the well-studied plasmid pAD1, the pheromone is called cAD1, and the plasmid-encoded pheromone-binding protein on the cell surface is called TraC. TraC passes the bound cAD1 to the host oligopeptide permease system, and the imported peptide binds to TraA and abolishes the ability of TraA to repress transcription of the genes encoding the conjugation machinery of the donor cell. In this way, the pheromone can induce the production of products involved in plasmid transfer, including aggregation substance. This protein coats the donor cell surface and initiates contact with the recipient cell. Following cell-cell contact, the plasmid transfers through a mating channel, much like plasmid transfer in gram-negative bacteria.

Once the plasmid has entered the recipient, the new transconjugant no longer functions as a recipient and is unable to acquire additional plasmids of the same family. The limitation of plasmid uptake by a donor cell involves three mechanisms. The first mechanism is the expression of surface exclusion proteins that function much like the entry exclusion *eex* systems of gram-negative bacteria. The second mechanism involves the shutdown of pheromone sensing due to synthesis of an inhibitor encoded by a gene on the plasmid. The inhibitor gene product is a peptide of 7 or 8 amino acids, much like the pheromone itself, but it differs from the pheromone by only a few amino acids, which allows it to competitively inhibit the activity of the pheromone and prevents autoinduction (i.e., self-induction) of its own mating system. The third system involves a shutdown protein called TraB in pAD1 and PrgY in a related plasmid, pCF10, which has a similar method of pheromone sensing (Figure 5.15) (see Chandler et al., Suggested Reading). These proteins are located on the outer surface of the membrane and bind nascent pheromone molecules being exported from donor cells. This serves to reduce the production of active pheromone by cells that carry the plasmid to further prevent autoinduction and to avoid attracting other donors (see Chandler and Dunny, Suggested Reading).

The enterococci and their plasmids are of special importance, because these organisms are important hospital-acquired (nosocomial) pathogens. Their diverse plasmids carry both genes that enhance virulence and genes that confer resistance to multiple antibiotics. The enterococcal plasmids can also transfer their genes to other gram-positive bacteria, including the very dangerous pathogen *Staphylococcus aureus*. This is because *S. aureus* can produce pheromones to attract enterococcal plasmids (see De Boever et al., Suggested Reading). This type of transfer is of particular medical importance because enterococcal plasmids can confer resistance to vancomycin, which is often a "last-resort" antibiotic in the treatment of *S. aureus* infections.

Figure 5.14 Role of pheromones in plasmid transfer. **(A)** The recipient cell. The pheromone genes are located on the enterococcal chromosome; several examples are shown. The propheromone peptides cAD1 and cCF10 are processed from the signal sequences cut off of normal cellular proteins. The pheromones are processed from the propheromones Pro-cCF10 and Pro-cAD1 when exported. **(B)** The donor cell. The plasmid-carrying cell expresses TraA, which represses transcription of the other *tra* genes except *traC*, which encodes a cell surface protein that can sense the pheromone. Also shown is TraB, which is discussed in the legend to panel E. **(C)** Mating induction. The pheromone corresponding to the plasmid, in this case cAD1, binds TraC on the cell surface of a donor cell in close proximity and enters the cell via the oligopeptide permease system (Opp). The pheromone binds to the repressor TraA, releasing it from the DNA and derepressing the synthesis of TraE, which activates the expression of the *tra* genes, including the gene encoding the aggregation substance (Asa). **(D)** Plasmid transfer. The donor cell establishes contact with the recipient cell, and the plasmid transfers, producing a transconjugant. **(E)** Pheromone shutdown in the transconjugant. Once the cell has become a transconjugant, the inhibitor peptide iAD1 binds to TraC and prevents autoinduction or pheromone stimulation of mating with other donor cells. Also, TraB is an inhibitor protein that somehow functions to shut down induction by preventing the excretion of pheromone cAD1, but pheromone cCF10 continues to be expressed.
doi:10.1128/9781555817169.ch5.f5.14

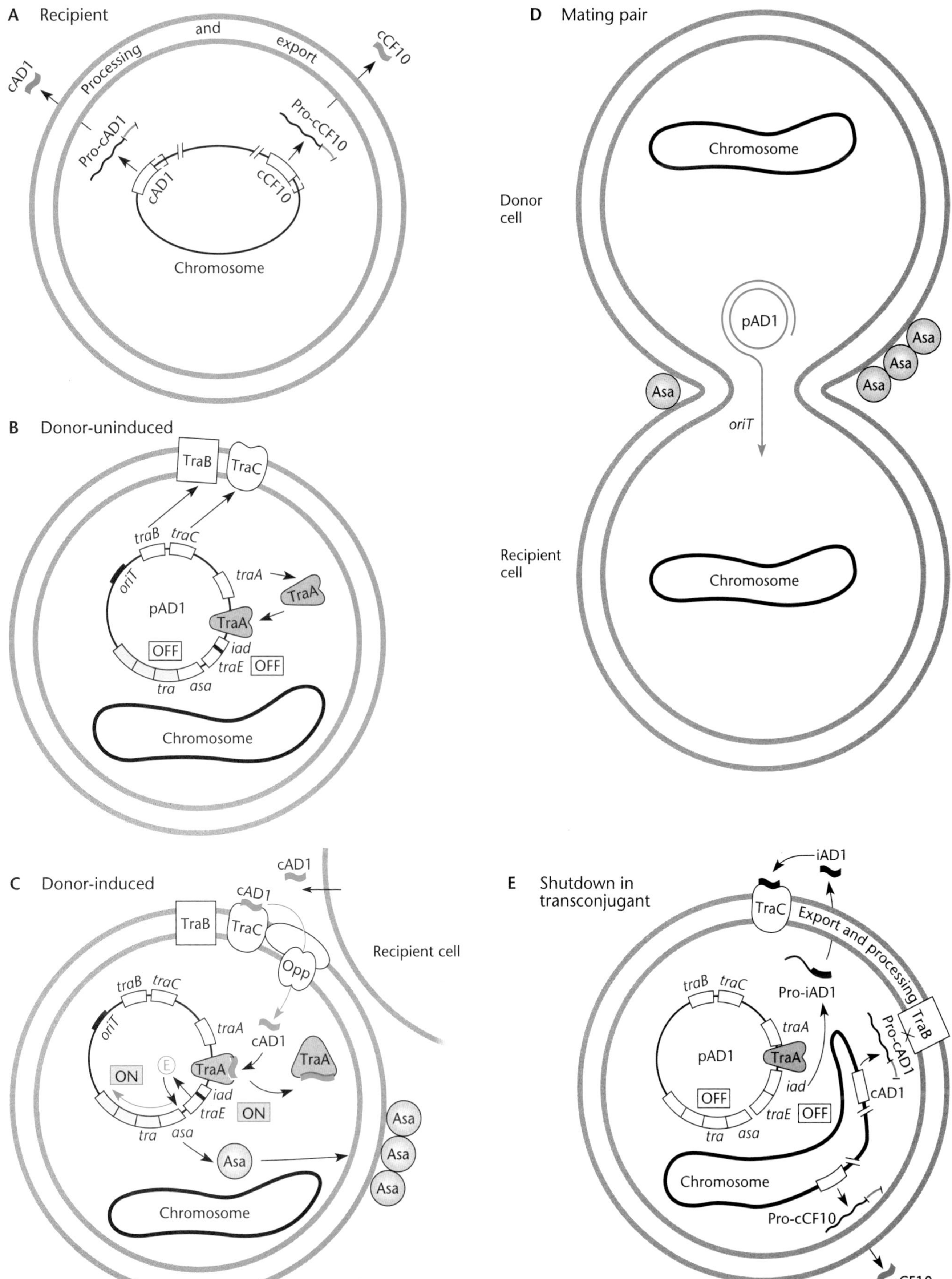
A Recipient
Processing and export
cAD1
cCF10
Pro-cAD1
Pro-cCF10
cAD1
cCF10
Chromosome
B Donor-uninduced
TraB
TraC
traB
traC
oriT
traA
TraA
pAD1
TraA
OFF
iad
traE
OFF
tra
asa
Chromosome
C Donor-induced
cAD1
cAD1
TraB
TraC
Opp
Recipient cell
traB
traC
oriT
traA
cAD1
TraA
E
ON
TraA
iad
traE
ON
tra
asa
Asa
Asa
Asa
Asa
Chromosome
D Mating pair
Chromosome
Donor cell
pAD1
Asa
Asa
Asa
Asa
oriT
Recipient cell
Chromosome
E Shutdown in transconjugant
iAD1
TraC
Export and processing
traB
traC
Pro-iAD1
TraB
traA
Pro-cAD1
pAD1
TraA
iad
cAD1
OFF
traE
OFF
tra
asa
Chromosome
Pro-cCF10
cCF10

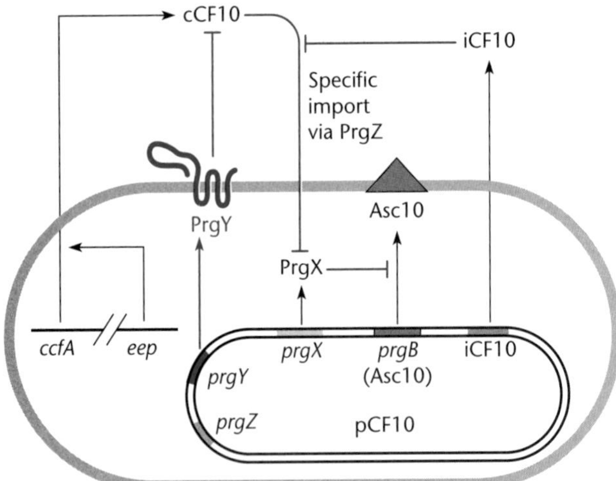

Figure 5.15 Model of pCF10 pheromone production, control, and response. *ccfA*, which encodes the pheromone of cCF10, and *eep* are both expressed from the chromosome. Mature cCF10 is processed from the signal sequence of the lipoprotein CcfA by the membrane protease Eep as the lipoprotein exits the cell. PrgY, PrgX, PrgZ, Asc10 (aggregation substance, expressed from *prgB*), and iCF10 are all encoded within pCF10. In the absence of PrgY or the inhibitor peptide, iCF10, endogenous mature cCF10 can internalize via PrgZ into donor cells to continually induce pCF10 transfer proteins. PrgY (blue) and the inhibitor peptide iCF10 prevent this self-induction by endogenous cCF10. iCF10 neutralizes endogenous cCF10 in the medium, and PrgY sequesters or blocks the activity of endogenous cCF10. PrgX functions as the on-off switch for induction and, on interaction with pheromone, releases the repression of Asc10 expression. Asc10 mediates aggregation of the cells, which is necessary for efficient transfer of pCF10 to recipient cells. doi:10.1128/9781555817169.ch5.f5.15

Integrating Conjugative Elements

Plasmids are not the only DNA elements in bacteria that are capable of transferring themselves by conjugation. ICE are normally integrated into the chromosome and do not exist autonomously, like plasmids. Nevertheless, they often encode Tra functions and can transfer themselves or mobilize other elements into recipient cells. ICE are very common vectors for the formation of islands of foreign DNA in bacterial chromosomes. When these elements were originally discovered, they were believed to be large transposons and were subsequently named with a "Tn" designation, for transposon. Later, it was discovered that they integrate into the host chromosome using a phage-like integrase rather than by transposition. These elements are now categorized with an ICE designation and a designation for the host where they were first identified. For example, it was recently shown that the model gram-positive bacterium *Bacillus subtilis* used by most laboratories has an active ICE in its chromosome that is now called ICE*Bs1*. A micrograph of a *B. subtilis* strain with the ICE*Bs1* element, where the ConE protein is fused to green fluorescent protein, is shown on the front cover of this book (see Berkmen et. al., Suggested Reading). ConE belongs to the same superfamily of ATPases as the conjugative plasmid protein VirB4 (Figure 5.8), which functions at translocation channels. The host proteins FtsK and SpoIIIE, which are important DNA-pumping enzymes for the chromosome, are also members of this superfamily (see chapter 1). The ConE protein is concentrated at the cell poles, but there is additional protein localized around the entire cell. This distribution of ConE (and presumably the other, unknown constituents of the pore complex) would enable the cells to transfer ICE*Bs1* side by side, but the high concentration at the poles allows mating to occur very efficiently in cell chains from pole to pole (see Babic et al., 2011, Suggested Reading).

A small and well-studied ICE is Tn*916* from *E. faecalis* (this element may be renamed under the ICE designation) (Figure 5.16). Like other ICE, the element is not capable of autonomous replication, but it can transfer itself from the chromosome of one bacterium to the chromosome of another bacterium without transferring chromosomal genes. It is transiently excised from one DNA, transfers itself into another cell, and then integrates into the DNA of the recipient bacterium (see Marra et al., Suggested Reading).

Tn*916* and its relatives are known to be promiscuous, and they transfer into many types of gram-positive bacteria and even into some gram-negative bacteria. The antibiotic resistance gene they carry, *tetM*, has also been found in many types of gram-positive and gram-negative bacteria. It is tempting to speculate that elements like Tn*916* are responsible for the widespread dissemination of the *tetM* gene. Other ICE, such as CTnDOT, which also carries tetracycline resistance, have been found in *Bacteroides* species. These *Bacteroides* ICE also mobilize other, smaller elements in the chromosome of *Bacteroides*.

To move from the DNA of one cell to the DNA of another cell, an ICE must first be excised from the DNA of the cell in which it resides, be transferred into the other cell, and then integrate into the DNA of the second cell. This process has been studied extensively for Tn*916*. Like phage λ (see chapter 8), Tn*916* requires two proteins, Int and Xis, to be excised. Excision of the element requires cutting the DNA next to the ends of the element. The integrase first makes a staggered break in the donor host DNA near the ends of the transposon to leave single-stranded ends 6 nucleotides long. The flanking sequences shown in the figure are arbitrary because they differ depending on where the transposon was inserted, although the element has a tendency to integrate next to a sequence similar to the sequence on one side of where it integrated previously. To the extent that they

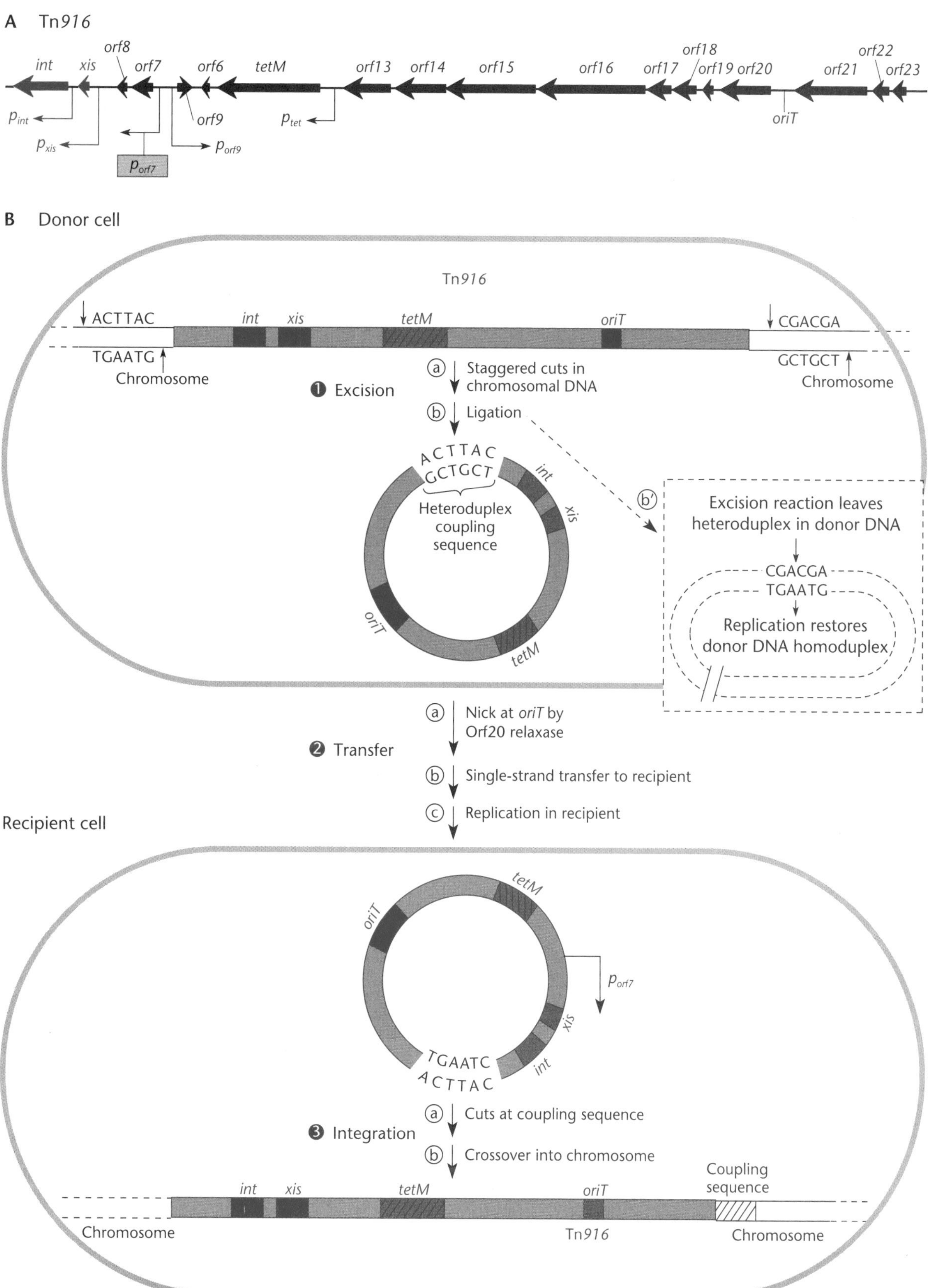

Figure 5.16 Genetic map and diagram of the integration and excision process of the ICE Tn*916*. **(A)** Genetic map of Tn*916*. **(B) (Top)** Mechanism used for excision of the element from the host chromosome; **(bottom)** mechanism used by Tn*916* to insert into the host chromosome following conjugation into a new host. doi:10.1128/9781555817169.ch5.f5.16

are random, these single-stranded ends (called the coupling sequences) are not complementary to each other. Nevertheless, the ends seem to pair to form a circular intermediate of the element, including the unpaired region formed by the ends of the element, which are thought to form a heteroduplex sequence called the coupling sequence. It has proven difficult to confirm various aspects of this model because of the possibility of directed mismatch repair in the coupling sequence before or after transfer.

The circular intermediate cannot replicate but can be transferred into another cell by a process that is much like plasmid transfer during conjugation. The element has its own *oriT* sequence and *tra* genes (Figure 5.16). In fact, this *oriT* sequence is related to those of the self-transmissible plasmids RP4 and F. The *tra* genes are *orf23* to *orf13*, and the *oriT* sequence lies between *orf20* and *orf21*. Orf20 functions as the Tn*916* relaxase (see Rocco and Churchward, Suggested Reading). To initiate the transfer, a single-strand break is made in the *oriT* sequence of the excised circular element, and a single strand of the element is transferred to the recipient cell by the element-encoded Tra functions. Once in the recipient cell, the ends rejoin and a complementary strand is synthesized to make another double-stranded circular element. The Int protein of the element alone then integrates the element into the DNA of the recipient cell; this is similar to the way the Int protein, but not Xis, is required to integrate the λ phage genome into the chromosome (see chapter 8).

It would seem that a bacterial strain harboring an ICE in its chromosome would be an Hfr strain, capable of transferring chromosomal DNA into recipient cells. However, this does not generally happen, because known ICE, including Tn*916*, can transfer only after they have been excised, not while they are still integrated in the chromosome. Tn*916* uses a clever mechanism to ensure that it will be transferred only after it is excised. The *tra* genes are arranged so that they can be transcribed only after the element has been excised and has formed a circle. It accomplishes this by positioning the promoter, called p_{orf7} (Figure 5.16), so that the *tra* genes (*orf23* to *orf13*) are transcribed only when the element has been excised and has formed a circle. The Int protein may also bind to the *oriT* sequence, while the ICE is inserted in the chromosome, blocking access to the *oriT* sequence by the Tra functions and precluding transfer. The advantage to the element of thus regulating its transfer seems obvious. If they could transfer themselves while still integrated in the chromosome, they would transfer the chromosome, much as self-transmissible plasmids transfer the chromosome in Hfr strains. However, as occurs during *Hfr* transfer, only the part of the element on one side of the *oriT* site would enter the recipient, and the element would essentially decapitate itself. There are special cases where cryptic *oriT* genes in a host chromosome can function as the starting point for Hfr-like transfer, and this type of system can be set up synthetically for genetic engineering.

The process of transfer of the *Bacteroides* conjugative elements, such as CTnDOT, may be similar, but there are important differences (see Cheng et al., Suggested Reading). The CTnDoT element and other ICE seem to be more restricted in target site selection than Tn*916*, preferring sites with a 10-bp sequence that is similar to a sequence on the element. The transfer of the CTnDOT element is also induced by tetracycline in the medium in much the same way that opines induce transfer of the Ti plasmid and unlike the transfer of Tn*916*, which is not induced by tetracycline.

SUMMARY

1. Self-transmissible plasmids can transfer themselves to other bacterial cells, a process called conjugation. Some plasmids can transfer themselves into a wide variety of bacteria from different genera. Such plasmids are said to be promiscuous.

2. The plasmid genes whose products are involved in transfer are called the *tra* genes. The site on the plasmid DNA at which transfer initiates is called the origin of transfer (*oriT*). The *tra* genes can be divided into two groups: those whose products are involved in mating pair formation (Mpf) and those whose products are involved in processing the plasmid DNA for transfer (Dtr).

3. The Mpf component includes a sex pilus that extrudes from the cell and holds mating cells together. The pilus is the site to which male-specific phages adsorb. The Mpf system also includes the channel in the membrane through which DNA and proteins pass, as well as a coupling protein that lies on the channel, docks the relaxase of the Dtr component, and translocates DNA through the channel.

4. The Dtr component includes the relaxase, which makes a nick within the *oriT* sequence and rejoins the ends of the plasmid in the recipient cell. The relaxase also often contains a helicase activity, which separates the strands of DNA during transfer. The Dtr component also includes proteins

SUMMARY (continued)

that bind to the *oriT* sequence to form the multiprotein complex called the relaxosome and a primase that primes replication in the recipient cell and is sometimes transferred along with the DNA.

5. Most plasmids transiently express their Mpf *tra* genes immediately after transfer to a recipient cell and only intermittently thereafter.

6. Mobilizable plasmids cannot transfer themselves but can be transferred by other plasmids. Mobilizable plasmids encode only a Dtr component; they lack genes to encode an Mpf component. In the context of a mobilizable plasmid, the genes that encode the Dtr component are called the *mob* genes. A mobilizable plasmid can be mobilized by a self-transmissible plasmid only if the coupling protein of the self-transmissible plasmid can dock the relaxase of the mobilizable plasmid. Because they lack an Mpf component, mobilizable plasmids can be much smaller than self-transmissible plasmids, which makes them very useful in molecular genetics and biotechnology.

7. Hfr strains of bacteria have a self-transmissible plasmid integrated into their chromosomes. Hfr strains have historically been useful for genetic mapping in bacteria because they transfer chromosomal DNA in a gradient, beginning at the site of integration of the plasmid. Hfr crosses were an important early tool for ordering genetic markers on the entire genome.

8. Prime factors are self-transmissible plasmids that have picked up part of the bacterial chromosome. They can be used to make partial diploids for complementation tests. If a prime factor is transferred into a cell, the cell will be a partial diploid for the region of the chromosome carried on the prime factor, making it useful for complementation experiments.

9. Self-transmissible plasmids also exist in gram-positive bacteria. However, these plasmids do not encode a sex pilus. Some gram-positive bacteria have the interesting property of excreting small pheromone-like compounds that stimulate mating with certain plasmids. The existence of such systems emphasizes the importance of plasmid exchange to bacteria.

10. Integrating elements that are themselves self-transmissible also exist. These ICE can be excised and transfer themselves into other cells even though they cannot replicate and therefore are not replicons.

QUESTIONS FOR THOUGHT

1. Why are the *tra* genes whose products are directly involved in DNA transfer usually adjacent to the *oriT* site?

2. Why do you suppose plasmids with a certain *mob* site are mobilized by only certain types of self-transmissible plasmids?

3. Why do self-transmissible plasmids usually encode their own primase function?

4. What do you think is different about the cell surfaces of gram-positive and gram-negative bacteria that causes only the self-transmissible plasmids of gram-negative bacteria to encode a pilus?

5. Why do so many types of phages use the sex pilus of plasmids as their adsorption site?

6. Why are so many plasmids either self-transmissible or mobilizable? Why are so many promiscuous?

7. F primes can be selected by mating an Hfr into a strain that is incapable of doing homologous recombination. Explain how selecting a late marker early using an Hfr donor and recipient that are proficient at homologous recombination could accomplish the same thing.

PROBLEMS

1. After mixing two strains of a bacterial species, you observe some recombinant types that are unlike either parent. These recombinant types seem to be the result of conjugation, because they appear only if the cells are in contact with each other. How would you determine which is the donor strain and which is the recipient? How would you determine whether the transfer is due to an Hfr strain or to a prime factor?

2. How would you determine which of the *tra* genes of a self-transmissible plasmid encodes the pilin protein? The site-specific DNA endonuclease that cuts at *oriT*? The helicase?

3. How would you show that only one strand of the plasmid DNA enters the recipient cell during plasmid transfer?

4. You have discovered that tetracycline resistance can be transferred from one strain of a bacterial species to another. How would you determine whether the tetracycline resistance gene being transferred is on a self-transmissible plasmid or on a conjugative transposon?

5. Can a male-specific phage infect bacteria containing only a mobilizable plasmid? Why or why not?

6. Outline how you would determine which of the open reading frames in the *tra* region of a plasmid encodes the Eex (entry exclusion) protein.

7. Outline how you would use mobilizable plasmids to determine if a plasmid indigenous to a wild-type bacterium you have isolated is self-transmissible if the plasmid does not have any selectable gene that you know of.

8. Why do you suppose a pheromone-responsive plasmid, such as pAD1, encodes both an inhibitory peptide, iAD1, and an inner membrane protein that interferes with pheromone processing?

SUGGESTED READING

Babic, A., A. B. Lindner, M. Vulic, E. J. Stewart, and M. Radman. 2008. Direct visualization of horizontal gene transfer. *Science* **319:**1533–1536.

Babic, A., M. B. Berkmen, C. A. Lee, and A. D. Grossman. 2011. Efficient gene transfer in bacterial cell chains. *mBio* **2:**e00027-11.

Bagdasarian, M., R. Lurz, B. Rukert, F. C. H. Franklin, M. M. Bagdasarian, J. Frey, and K. N. Timmis. 1981. Specific-purpose plasmid cloning vectors. II. Broad host range, high copy number, RSF1010-derived vectors, and a host-vector system for gene cloning in *Pseudomonas*. *Gene* **16:**237–247.

Berkmen, M. B., C. A. Lee, E. K. Loveday, and A. D. Grossman. 2010. Polar positioning of a conjugation protein from the integrative and conjugative element ICE*Bs1* of *Bacillus subtilis*. *J Bacteriol. 192:*38–45.

Chandler, J. R., and G. M. Dunny. 2004. Enterococcal peptide sex pheromones: synthesis and control of biological activity. *Peptides* **25:**1377–1388.

Chandler, J. R., A. R. Flynn, E. M. Bryan, and G. M. Dunny. 2005. Specific control of endogenous cCF10 pheromone by a conserved domain of the pCF10-encoded regulatory protein PrgY in *Entercoccus faecalis*. *J. Bacteriol.* **187:**4830–4843.

Cheng, Q., B. J. Paszkiet, N. B. Shoemaker, J. F. Gardner, and A. A. Salyers. 2000. Integration and excisions of a *Bacteroides* conjugative transposon, CTnDOT. *J. Bacteriol.* **182:**4035–4043.

Christie, P. J., K. Atmakuri, V. Krishnamoorthy, S. Jakubowski, and E. Cascales. 2005. Biogenesis, architecture, and function of bacterial type IV secretion systems. *Annu. Rev. Microbiol.* **59:**451–485.

Clewell, D. B., and M. V. Francia. 2004. Conjugation in gram-positive bacteria, p. 227–256. *In* B. E. Funnell and G. J. Phillips (ed.), *Plasmid Biology*. ASM Press, Washington, DC.

Covacci, A., J. L. Telford, G. Del Giudice, J. Parsonnet, and R. Rappuoli. 1999. *Helicobacter pylori* virulence and genetic geography. *Science* **284:**1328–1333.

D'Costa, V. M., K. M. McGrann, D. W. Hughes, and G. D. Wright. 2006. Sampling the antibiotic resistome. Science **311:** 374–377.

De Boever, E. H., D. B. Clewell, and C. M. Fraser. 2000. *Enterococcus faecalis* conjugative plasmid pAM373: complete nucleotide sequence and genetic analyses of sex pheromone response. *Mol. Microbiol.* **37:**1327–1341.

Derbyshire, K. M., G. Hatfull, and N. Willets. 1987. Mobilization of the nonconjugative plasmid RSF1010: a genetic and DNA sequence analysis of the mobilization region. *Mol. Gen. Genet.* **206:**161–168.

Eisenbrandt, R., M. Kalkum, R. Lurz, and E. Lanka. 2000. Maturation of IncP pilin precursors resembles the catalytic dyad-like mechanism of leader peptidases. *J. Bacteriol.* **182:**6751–6761.

Firth, N., K. Ippen-Ihler, and R. A. Skurray. 1996. Structure and function of the F factor and mechanism of conjugation, p. 2377–2401. *In* F. C. Neidhardt, R. Curtiss III, J. L. Ingraham, E. C. C. Lin, K. B. Low, B. Magasanik, W. S. Reznikoff, M. Riley, M. Schaechter, and H. E. Umbarger (ed.), Escherichia coli *and* Salmonella: *Cellular and Molecular Biology*, 2nd ed. ASM Press, Washington, DC.

Gilmour, M. W., T. D. Lawley, and D. E. Taylor. 15 November 2004, posting date. The cytology of bacterial conjugation. *In* A. Böck et al. (ed.), *EcoSal*—Escherichia coli *and* Salmonella: *Cellular and Molecular Biology*. ASM Press, Washington, DC. http://www.ecosal.org.

Grahn, A. M., J. Haase, D. H. Bamford, and E. Lanka. 2000. Components of the RP4 conjugative transfer apparatus form an envelope structure bridging inner and outer membranes of donor cells: implications for related macromolecule transport systems. *J. Bacteriol.* **182:**1564–1574.

Guglielmini, J., L. Quintais, M. P. Garcillan-Barcia, F. de la Cruz, and E. P. C. Rocha. 2011. The repertoire of ICE in prokaryotes underscores the unity, diversity, and ubiquity of conjugation. *PLoS Genet.* **7:**e1002222

Hinerfeld, D., and G. Churchward. 2001. Specific binding of integrase to the origin of transfer (*oriT*) of the conjugative transposon Tn*916*. *J. Bacteriol.* **183:**2947–2951.

Kurenbach, B., D. Grothe, M. E. Farias, U. Szewzyk, and E. Grohmann. 2002. The *tra* region of the conjugative plasmid pIP501 is organized in an operon with the first gene encoding the relaxase. *J. Bacteriol.* **184:**1801–1805.

Lawley, T., B. M. Wilkins, and L. S. Frost. 2004. Bacterial conjugation in gram-negative bacteria, p. 203–226. *In* B. E. Funnell and G. J. Phillips (ed.), *Plasmid Biology*. ASM Press, Washington, DC.

Lederberg, J., and E. L. Tatum. 1946. Gene recombination in *Escherichia coli*. *Nature* (London) **158:**558.

Low, K. B. 1968. Formation of merodiploids in matings with a class of Rec[2] recipient strains of *E. coli* K12. *Proc. Natl. Acad. Sci. USA* **60:**160–167.

Manchak, J., K. G. Anthony, and L. S. Frost. 2002. Mutational analysis of F-pilin reveals domains for pilus assembly, phage infection, and DNA transfer. *Mol. Microbiol.* **43:**195–205.

Marra, D., B. Pethel, G. G. Churchward, and J. R. Scott. 1999. The frequency of conjugative transposition of Tn*916* is not determined by the frequency of excision. *J. Bacteriol.* **181:**5414–5418.

Marra, D., and J. R. Scott. 1999. Regulation of excision of the conjugative transposon Tn*916*. *Mol. Microbiol.* **31:**609–621.

Matson, S. W., J. K. Sampson, and D. R. N. Byrd. 2001. F plasmid conjugative DNA transfer: the TraI helicase activity is essential for DNA strand transfer. *J. Biol. Chem.* **276:**2372–2379.

Parker, C., and R. J. Meyer. 2007. The R1162 relaxase/primase contains two type IV transport signals that require the small plasmid protein MobB. *Mol. Microbiol.* **66:**252–261.

Rees, C. E. D., and B. M. Wilkins. 1990. Protein transfer into the recipient cell during bacterial conjugation: studies with F and RP4. *Mol. Microbiol.* **4:**1199–1205.

Rocco, J. M., and G. Churchward. 2006. The integrase of the conjugative transposon Tn*916* directs strand- and sequence-specific cleavage of the origin of conjugal transfer, *oriT*, by the endonuclease Orf20. *J. Bacteriol.* **188:**2207–2213.

Samuels, A. L., E. Lanka, and J. E. Davies. 2000. Conjugative junctions in RP4-mediated mating of *Escherichia coli*. *J. Bacteriol.* **182:**2709–2715.

Schmidt, J. W., L. Rajeev, A. A. Salyers, and J. F. Gardner. 2006. NBU1 integrase: evidence for an altered recombination mechanism. *Mol. Microbiol.* **60:**152–164.

Schroder, G., and E. Lanka. 2005. The mating pair formation system of conjugative plasmids—a versatile secretion machinery for transfer of proteins and DNA. *Plasmid* **54:**1–25.

Senghas, E., J. M. Jones, M. Yamamoto, C. Gawron-Burke, and D. B. Clewell. 1988. Genetic organization of the bacterial conjugative transposon, Tn*916*. *J. Bacteriol.* **170:**245–249.

Vogel, J. P., H. L. Andrews, S. K. Wong, and R. R. Isberg. 1998. Conjugative transfer by the virulence system of *Legionella pneumophila*. *Science* **279:**873–876.

Watanabe, T., and T. Fukasawa. 1961. Episome-mediated transfer of drug resistance in *Enterobacteriaceae*. 1. Transfer of resistance factors by conjugation. *J. Bacteriol.* **81:**669–678.

Wilkins, B. M., and A. T. Thomas. 2000. DNA transfer independent transport of plasmid primase protein between bacteria by the I1 conjugation system. *Mol. Microbiol.* **38:**650–657.

Winans, S. C., and G. C. Walker. 1985. Conjugal transfer system of the IncN plasmid pKM101. *J. Bacteriol.* **161:**402–410.

CHAPTER 6

Transformation

DNA CAN BE EXCHANGED AMONG BACTERIA IN THREE WAYS: conjugation, transduction, and transformation. Chapter 5 covers the mechanism of conjugation, in which a plasmid or other self-transmissible DNA element transfers itself, and sometimes other DNA, into another bacterial cell. In transduction, which is discussed in detail in chapter 7, a phage carries DNA from one bacterium to another. In this chapter, we discuss transformation, a process in which cells take up free DNA directly from their environment.

Transformation is one of the cornerstones of molecular genetics because it is often the best way to reintroduce experimentally altered DNA into cells. Transformation was first discovered in bacteria, but methods have been devised to transform many types of animal and plant cells, as well.

The terminology of genetic analysis by transformation is similar to that of conjugation and transduction (see chapter 3). DNA is derived from a **donor bacterium** and taken up by a **recipient bacterium.** If the incoming DNA recombines with resident DNA in the cell, such as the chromosome, recombinant types can form; the cell that has taken up the incoming DNA is referred to as a **transformant.** The frequency of recombinant types for various genetic markers can be used for genetic analysis. Such genetic data obtained by transformation are analyzed similarly to those obtained by transduction. If the regions of two markers can be carried on the same piece of transforming DNA, the two markers are said to be **cotransformable.** The higher the **cotransformation frequency,** the more closely linked are the two markers on the DNA. The principles of using transduction and transformation for genetic mapping are outlined in chapter 3. In this chapter, we concentrate on the mechanism of transformation in various bacteria and its relationship to other biological phenomena.

doi:10.1128/9781555817169.ch6

Natural Transformation

Most types of cells cannot take up DNA efficiently unless they have been exposed to special chemical or electrical treatments to make them more permeable. However, some types of bacteria are **naturally transformable**, which means that they can take up DNA from their environment without requiring such treatments. Even naturally transformable bacteria are not always capable of taking up DNA but do so only at certain stages in their life cycle or under certain growth conditions. Bacteria at the stage in which they can take up DNA are said to be **competent**, and bacteria that are naturally capable of reaching this state are said to be **naturally competent**. Naturally competent transformable bacteria are found in several genera, including both gram-positive bacteria, such as *Bacillus subtilis*, a soil bacterium, and *Streptococcus pneumoniae*, which causes throat infections, and gram-negative bacteria, such as *Haemophilus influenzae*, a causative agent of pneumonia and spinal meningitis, *Neisseria gonorrhoeae*, which causes gonorrhea, and *Helicobacter pylori*, a stomach pathogen. *Acinetobacter baylyi*, another soil bacterium, is very highly transformable, as are some species of marine cyanobacteria, including *Synechococcus*. *Thermus thermophilus*, an extreme thermophile, and *Deinococcus radiodurans*, an organism resistant to high levels of radiation, are also naturally competent. Genome sequencing has revealed that many other organisms that have not been demonstrated to be naturally transformable contain some of the genes known to be involved in competence in other species, suggesting that some of these organisms may be transformable under certain conditions or that they have lost this property (see Johnsberg et al., Suggested Reading).

Discovery of Transformation

Transformation was the first mechanism of bacterial gene exchange to be discovered. In 1928, Fred Griffith found that one form of the pathogenic pneumococci (now called *S. pneumoniae*) could be mysteriously "transformed" into another form. Griffith's experiments were based on the fact that *S. pneumoniae* makes two types of colonies with different appearances, one type made by pathogenic bacteria and the other type made by bacteria that are incapable of causing infections (i.e., they are nonpathogenic). The colonies made by the pathogenic strains appear smooth on agar plates, because the bacteria excrete a polysaccharide capsule. The capsule apparently protects them and allows them to survive in vertebrate hosts, including mice, which they can infect and kill. However, rough-colony-forming mutants that cannot make the capsule sometimes arise from the smooth-colony formers, and these mutants are nonpathogenic in mice.

In his experiment, Griffith mixed dead *S. pneumoniae* cells that made smooth colonies with live nonpathogenic cells that made only rough colonies and injected the mixture into mice (Figure 6.1). Mice given injections of only the rough-colony-forming bacteria survived, but mice that received a mixture of dead smooth-colony formers and live rough-colony formers died. Furthermore, Griffith isolated live smooth-colony-forming bacteria from the blood of the dead mice. Concluding that the dead pathogenic bacteria gave off a "transforming principle" that changed the live nonpathogenic rough-colony-forming bacteria into the pathogenic smooth-colony form, he speculated that this transforming principle was the polysaccharide itself. Later, other researchers did an experiment in which they transformed rough-colony formers into smooth-colony formers by mixing the rough forms with extracts of the smooth-colony formers in a test tube. Then, about 16 years after Griffith did his experiments with mice, Oswald Avery and his collaborators purified the "transforming principle" from the extracts of the smooth-colony formers and showed that it is DNA (see Avery et al., Suggested Reading). Thus, Avery and colleagues were the first to demonstrate that DNA, and not protein or other factors in the cell, is the hereditary material (see the introductory chapter).

Competence

As mentioned above, the term "competence" refers to the state that some bacteria can enter in which they can take up naked DNA from their environment. Natural competence is genetically programmed, and the process

Figure 6.1 The Griffith experiment. **(A)** Type R nonencapsulated bacteria are nonpathogenic and do not survive in the host. **(B)** Type S encapsulated bacteria are pathogenic and are recovered from the host. **(C)** Heat-killed type S bacteria fail to kill the host and cannot be recovered. **(D)** Mixing heat-killed type S bacteria with live type R bacteria can convert the type R bacteria to the pathogenic capsulated form. doi:10.1128/9781555817169.ch6.f6.1

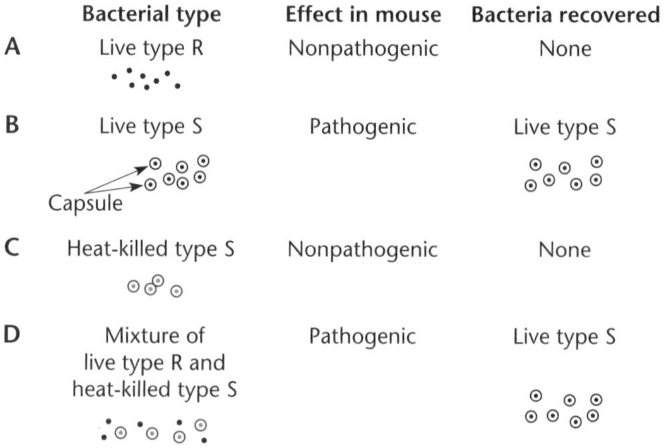

	Bacterial type	Effect in mouse	Bacteria recovered
A	Live type R	Nonpathogenic	None
B	Live type S (Capsule)	Pathogenic	Live type S
C	Heat-killed type S	Nonpathogenic	None
D	Mixture of live type R and heat-killed type S	Pathogenic	Live type S

of DNA uptake is often called "natural transformation" to distinguish it from transformation induced by laboratory treatments, such as electroporation, heat shock, Ca^{2+} treatment of cells, or removal of the cell wall to generate protoplasts. The genetic programming of competence is widespread but not universal. Generally, more than a dozen genes are involved, encoding both regulatory and structural components of the transformation process.

The general steps that occur in natural transformation differ somewhat in different systems. The best-characterized model systems are *B. subtilis* and *S. pneumoniae* (gram positive) and *H. influenzae* and *N. gonorrhoeae* (gram negative). The process of DNA uptake depends on whether the bacteria are gram negative or gram positive, due to the presence of the outer membrane in gram-negative bacteria. In gram-negative bacteria, the basic steps are (i) binding of double-stranded DNA to the outer cell surface of the bacterium, (ii) movement of the double-stranded DNA across the outer membrane and cell wall, (iii) degradation of one of the DNA strands, and (iv) translocation of the remaining single strand of DNA into the cytoplasm of the cell across the inner membrane. Once in the cell, the single-stranded transforming DNA might stably integrate into the genome by homologous recombination of the translocated single strand into the chromosome or other recipient DNA, reestablish itself as a plasmid after synthesis of the complementary strand and recyclization using recombination, or be degraded. In a gram-positive organism that lacks an outer membrane, double-stranded DNA binds to the outside surface, one strand is degraded, and the other strand is transported through the cell wall and membrane. The uptake of DNA in both gram-positive and gram-negative bacteria is discussed in more detail below. While the DNA uptake systems of gram-positive and gram-negative bacteria have features in common, they differ in certain important respects; therefore, they are discussed separately.

DNA UPTAKE IN GRAM-POSITIVE BACTERIA

Two gram-positive species in which transformation has been particularly well studied are *B. subtilis* and *S. pneumoniae*. These organisms will bind and take up DNA from any organism, although the fate of the DNA within the cell depends on whether the incoming DNA can participate in recombination with DNA already present in the cell or can establish independent replication (e.g., plasmid DNA). The proteins involved in transformation in these bacteria were discovered on the basis of isolation of mutants that lack the ability to take up DNA. The genes affected in the mutants were named *com* (for *com*petence defective).

In *B. subtilis*, the *com* genes are organized into several operons. The products of several of these, including the *comA* and *comK* operons, are involved in regulation of competence (see below). Others, including the products of genes in the *comE*, *comF*, and *comG* operons, become part of the competence machinery in the membrane that transports DNA into the bacterium. For historical reasons, the genes in these operons use a nonstandard type of nomenclature in which they are given two letters, the first for the operon and the second for the position of the gene in the operon. For example, *comFA* is the first gene of the *comF* operon, while *comEC* is the third gene of the *comE* operon. The corresponding protein products of the genes have the same name with the first letter capitalized, e.g., ComFA and ComEC, respectively.

The role played by some of the Com proteins in the competence machinery of *B. subtilis* is diagrammed in Figure 6.2A (see Chen and Dubnau and Claverys et al., Suggested Reading). The proteins encoded in the *comG* operon form a **pseudopilus** that resembles type IV pili that are characteristic of type II secretion systems and are similar to those used in some plasmid conjugation systems (see chapters 5 and 14). ComGC constitutes the major pilin protein; pilin processing is dependent on the ComC endopeptidase, and pseudopilus assembly requires the ComGB and ComGA proteins. Double-stranded DNA interacts directly with the pseudopilus, probably by nonspecific electrostatic interactions between the negatively charged DNA backbone and the positively charged amino acids in the pilin proteins. Double-stranded DNA at the cell surface is cleaved by the NucA endonuclease into segments that are approximately 6 kb in length in *S. pneumoniae* and 15 kb in *B. subtilis*, and the segments are then brought through the cell wall to the cell membrane by retraction of the pseudopilus, which probably occurs by disassembly of the pilus subunits, by analogy with retraction of some pili used in conjugation (see below and chapter 5).

The first gene of the *comE* operon, *comEA*, encodes the protein that directly binds extracellular double-stranded DNA at the outer surface of the membrane. One strand of the DNA is then degraded, and the other strand is transported through the membrane and into the cell through the ComEC channel, using ComFA as an ATP-dependent DNA translocase, possibly as part of an ATP-binding cassette (ABC) transporter, in conjunction with ComEA and ComEC. The EndA nuclease is responsible for generation of single-stranded DNA in *S. pneumoniae*, but the corresponding activity in *B. subtilis* has not been identified. Single-stranded DNA is transported into the cell at the rate of 80 to 100 nucleotides per second, and in *S. pneumoniae*, the DNA has been shown to enter with 3′-to-5′ polarity; at that rate, a 10-kilobase (kb) segment of DNA would be transported in about 2 minutes.

Transformation in *S. pneumoniae* utilizes proteins and mechanisms similar to those of transformation in *B.*

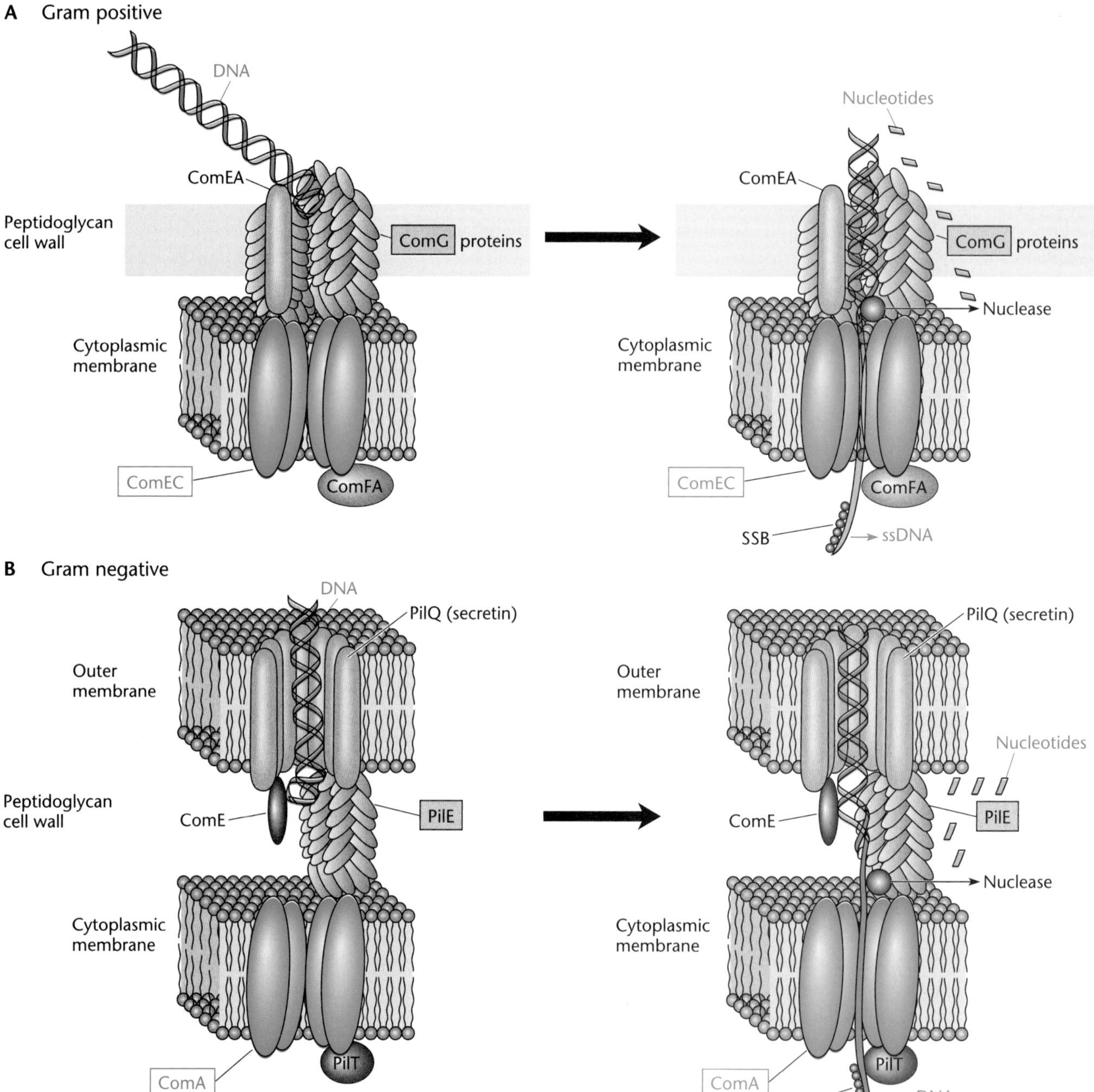

Figure 6.2 Structure of DNA uptake competence systems in gram-positive **(A)** and gram-negative **(B)** bacteria. Shown are some of the proteins involved and the channels they form. The nomenclature in panel A is based on *B. subtilis*, and that in panel B is based on *N. gonorrhoeae*. ComEA is a double-stranded DNA receptor protein in *B. subtilis*. Some of the *B. subtilis* ComG proteins are analogous to the *Neisseria* PilE protein. The endonuclease that generates single-stranded DNA is EndA in *S. pneumoniae*; the corresponding protein in *B. subtilis* has not been identified. The *B. subtilis* ComEC protein is analogous to the *Neisseria* ComA protein. ComEC forms a transmembrane channel through which the single-stranded DNA passes. Energy is provided by an ATPase (ComFA in *B. subtilis*; PilT in *Neisseria*). ss, single stranded. The DNA is shown running through the cell wall alongside the pseudopilus (ComG proteins in *B. subtilis*; PilE in *Neisseria*) but may be brought into the cell by retraction (disassembly) of the pseudopilus. Competence systems in both gram-positive and gram-negative organisms are related to type II protein secretion systems, which are discussed in chapter 14.
doi:10.1128/9781555817169.ch6.f6.2

subtilis, although the names of the *com* genes are often different (see Claverys et al., Suggested Reading).

DNA UPTAKE IN GRAM-NEGATIVE BACTERIA

As mentioned above, a variety of gram-negative bacteria are also capable of acquiring natural competence. Some examples are the bacterium *A. baylyi*, as well as the pathogens *H. pylori*, *Neisseria* spp., and *Haemophilus* spp. In the last two, specific uptake sequences are required for the binding of DNA, so these species usually take up DNA only of the same or closely related species (see below). This differs from the gram-positive bacteria and also from many other gram-negative bacteria, which do not have specific uptake sequences.

Transformation Systems Based on Type II Secretion Systems

Gram-negative bacteria utilize one of two fundamentally different types of DNA uptake systems. Most use a system related to type II secretion systems, which are used to assemble type IV pili on the cell surface and are similar to the pseudopilus system used in the gram-positive bacteria discussed above (Figure 6.2B). The major difference between the gram-negative bacterial competence systems and the gram-positive systems is necessitated by the presence of an outer membrane in the gram-negative bacteria. In gram-negative bacteria, the water-soluble (hydrophilic) DNA must first pass through this hydrophobic outer membrane before it can pass through the cytoplasmic membrane into the cytoplasm of the cell. To facilitate DNA transfer through the outer membrane, the competence systems of gram-negative bacteria also have a pore through the outer membrane, made up of 12 to 14 copies of a secretin protein (called PilQ in *Neisseria*). This pore has a hydrophilic aqueous channel through which the double-stranded DNA can pass. One strand of the DNA is then degraded as it passes through a second channel in the inner membrane; this channel is formed by a protein called ComA in *Neisseria*, which has sequences in common with the ComEC protein that forms a similar channel in *B. subtilis*.

Type IV pili are long, thin, hairlike appendages that stick out from the cell and are used to attach cells to solid surfaces, such as the surfaces of eukaryotic cells; they are often involved in pathogenicity. Type IV pilus systems involved in DNA transformation appear to work by a similar mechanism, binding to DNA on the cell surface and then retracting, pulling the DNA into the cell. The specific pilin proteins used to produce the pilus may be responsible for specific utilization in DNA uptake. As noted above, a dedicated competence-specific pilin (ComGC) is used in *B. subtilis*. In contrast, *Neisseria* utilizes the same pilin (PilE) for both competence and type II secretion. The association of PilE with different minor pilus proteins results in production of a pseudopilus that does not actually extrude from the cell but rather remains closely associated with the cell surface. In type II secretion systems, this pilus may grow and push proteins out of the cell through the secretin channel in the outer membrane (see chapter 14). In the case of transformation, the pseudopilus may do the reverse; it may bind DNA and retract, pulling the DNA into the cell through the secretin channel in the outer membrane.

Certain gram-negative bacteria, including *H. influenzae*, initially capture transforming DNA in membrane-associated vesicles called **transformasomes** (Figure 6.3). These vesicles contain the double-stranded DNA that has been transported through the outer membrane and are likely to represent the site at which the DNA is processed into single-stranded form and transported through the inner membrane.

Competence Systems Based on Type IV Secretion Systems

As mentioned above, most competent bacteria have competence systems based on type II secretion systems.

Figure 6.3 DNA uptake in *H. influenzae*. Double-stranded DNA is first taken up into transformasomes. One strand is degraded, and the other strand crosses the membrane and enters the cytoplasm. If the DNA is homologous to a region of the chromosome, the single-stranded transforming DNA invades the chromosome, displacing one chromosome strand, to form a heteroduplex, where one strand is from the donor and one strand is from the recipient.
doi:10.1128/9781555817169.ch6.f6.3

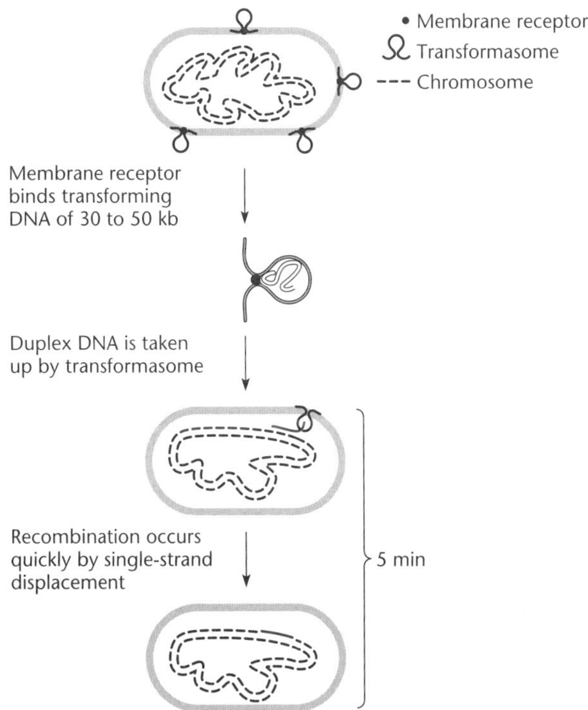
• Membrane receptor
Transformasome
--- Chromosome
Membrane receptor binds transforming DNA of 30 to 50 kb
Duplex DNA is taken up by transformasome
Recombination occurs quickly by single-strand displacement
5 min

To date, the only known exception is *H. pylori*, which has a system based on type IV secretion-conjugation systems, discussed in chapter 5. *H. pylori* is an opportunistic pathogen involved in gastrointestinal diseases. The similarity between the competence system of *H. pylori* and type IV secretion-conjugation systems was discovered because of the similarity between the proteins in this system and the VirB conjugation proteins in *Agrobacterium tumefaciens*, a plant pathogen, that transfer T-DNA from the Ti plasmid into plants (see Box 5.1). The Com proteins of *H. pylori*, a human pathogen, were therefore given letters and numbers corresponding to their orthologs (see Box 2.5 for definitions) in the T-DNA transfer system of the Ti plasmid in *A. tumefaciens*. Table 6.1 lists these Com proteins and their orthologs in the *Agrobacterium* Ti plasmid (see Karnholz et al., Suggested Reading). Apparently, type IV secretion pathway systems can function as two-way DNA transfer systems, capable of moving DNA both into and out of the cell. Interestingly, in addition to its transformation system, *H. pylori* has a bona fide type IV secretion system that secretes proteins directly into eukaryotic cells. However, even though the two systems are related, they function independently of each other and have no proteins in common.

The nomenclature of competence and secretion systems is admittedly very confusing. To reiterate, type IV pili and type IV secretion systems are not related to each other. Type IV secretion systems actually have type II pili, and type IV pili are assembled on the cell surface by systems related to type II secretion systems! Some conjugation systems have type IV pili in addition to their type II pili. The type IV pili help to hold the cells together during DNA transfer. There is also evidence that a type IV secretion system in *N. gonorrhoeae* is involved in releasing DNA into the environment, where it can be taken up by a transformation system related to type II secretion systems (see above). When the secretion systems, transformation systems, and pili were being named, no one could have predicted their relationships to each other; this unfortunate confusion is the result.

DNA Processing after Uptake

Single-stranded DNA is highly susceptible to degradation. For protection, it is rapidly covered with single-stranded-DNA-binding protein (SSB), using the normal SSB that functions during DNA replication and, in some systems, a competence-induced SSB. Single-stranded DNA is also a substrate for binding of the RecA recombination protein, which functions in the recombination of transforming DNA with the chromosome, as well as generally in recombination (see chapter 10).

If the transformed DNA is highly similar in sequence to DNA that is resident in the cell, the RecA protein will mediate homologous recombination. The transforming DNA integrates into the cellular DNA in a homologous region by means of strand displacement, a process discussed in detail in chapter 10. If the donor DNA and recipient DNA sequences differ slightly in this region, recombinant types can appear. The lengths of single-stranded DNA incorporated into the recipient chromosome are about 8.5 to 12 kb, as shown by cotransformation of genetic markers; the incorporation takes only a few minutes to be completed.

Experimental Evidence for Models of Natural Transformation

The above-mentioned models for natural transformation are based on experiments with a number of different systems. Experiments directed toward an understanding of DNA uptake during natural transformation have sought to answer three obvious questions: (i) how efficient is DNA uptake, (ii) can only DNA of the same species enter a given cell, and (iii) are both of the complementary DNA strands taken up and incorporated into the cellular DNA?

EFFICIENCY OF DNA UPTAKE

The efficiency of uptake is fairly easy to measure biochemically. Figure 6.4 shows an experiment based on the fact that transport of free DNA into the cell makes the DNA insensitive to deoxyribonucleases (DNases),

Table 6.1 Orthologous Com proteins of *H. pylori* and Vir proteins in *A. tumefaciens*

Helicobacter protein	Function	*Agrobacterium* ortholog
ComB2	Pseudopilus?	VirB2
ComB3	Unknown	VirB3
ComB4	ATPase	VirB4
ComB6	DNA binding?	VirB6
ComB7	Channel	VirB7
ComB8	Channel	VirB8
ComB9	Channel	VirB9
ComB10	Channel	VirB10

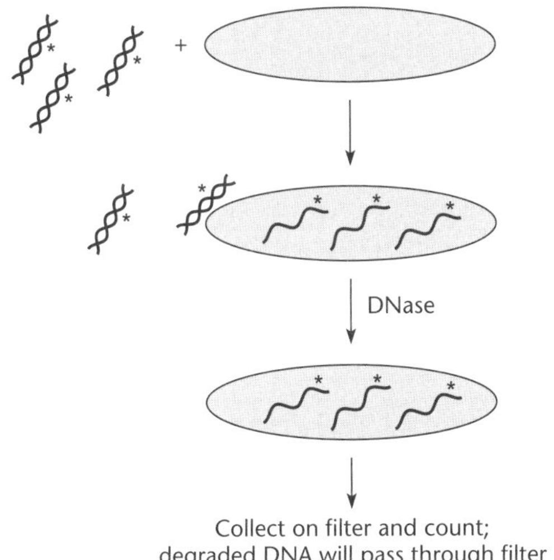

Figure 6.4 Determining the efficiency of DNA uptake during transformation. DNA in the cell is insensitive to DNase. DNA that did not enter the cell is degraded and passes through a filter. The asterisks indicate radioactively labeled DNA.
doi:10.1128/9781555817169.ch6.f6.4

which cannot enter the cell because competent cells take up only DNA and not proteins. Donor DNA is radioactively labeled by growing the cells in medium in which the phosphorus has been replaced with phosphorus-32, the radioisotope of phosphorus. The radioactive DNA is mixed with competent cells, and the mixture is treated with DNase at various times. Any DNA that is not degraded and survives intact must have been taken up by the cells, where it is protected from the DNase. The medium containing the cells is precipitated with acid and collected on a filter, and the radioactivity on the filter (which is due to undegraded DNA that must have been taken up by the cells) is counted and compared with the total radioactivity of the DNA that was added to the cells. Calculation of the percentage of DNA that is taken up gives the efficiency of DNA uptake. Experiments such as these have shown that some competent bacteria take up DNA very efficiently.

SPECIFICITY OF DNA UPTAKE

The second question, i.e., whether DNA from only the same species is taken up, is also fairly easy to answer. By using the same assay of resistance to DNases, it has been determined that some types of bacteria take up DNA from only their own species, whereas others can take up DNA from any source. The first group includes *N. gonorrhoeae* and *H. influenzae*.

Bacteria that preferentially take up the DNA of their own species do so because their DNA contains specific **uptake sequences**. Figure 6.5 shows the minimal uptake sequences for *H. influenzae* and *N. gonorrhoeae*. Uptake sequences are long enough that they almost never occur

Haemophilus influenzae	5′ AAGTGCGGT 3′
Neisseria gonorrhoeae	5′ GCCGTCTGAA 3′

Figure 6.5 Uptake sequences on DNA for some types of bacteria. Only DNA with these sequences is taken up by the bacteria indicated. One strand of the DNA is shown.
doi:10.1128/9781555817169.ch6.f6.5

by chance in other DNAs. In contrast, bacteria such as *B. subtilis* take up any DNA. Possible reasons why some bacteria preferentially take up DNA from their own species while others take up any DNA are subjects of speculation and are discussed below. Cells that take up DNA from different species can incorporate that DNA into their genomes only if the incoming DNA has high enough sequence similarity to resident DNA in the cell to promote homologous recombination or if the incoming DNA has the ability to promote independent replication (e.g., plasmid or bacteriophage DNA [see below]). The ability of DNA from a different species to be incorporated by recombination can be used as a measure of the close relationship between the species, at least in the genetic region that is tested for transformation activity.

GENETIC EVIDENCE FOR SINGLE-STRANDED DNA UPTAKE

Genetic experiments that take advantage of the molecular requirements for transformation can be used to study the molecular basis for transformation; in other words, transformation can be used to study itself. Evidence that DNA has transformed cells is usually based on the appearance of recombinant types after transformation. A recombinant type can form only if the donor and recipient bacteria differ in their genotypes and if the incoming DNA from the donor bacterium changes the genetic composition of the recipient bacterium. The chromosome of a recombinant type has the DNA sequence of the donor bacterium in the region of the transforming DNA.

Only double-stranded DNA can bind to the specific receptors on the cell surface, so double-stranded, but not single-stranded, DNA can transform cells and yield recombinant types. The fact that the DNA is converted into a single-stranded form in the course of the transformation process is also demonstrated by the observation that transforming DNA enters a phase during which it cannot be reisolated in a form that is active for transformation; this is termed the **eclipse phase**. For example, in the experiment shown in Figure 6.6, an Arg^- mutant that requires arginine for growth is used as the recipient strain and the corresponding Arg^+ prototroph is the source of donor DNA. At various times after the donor DNA has been mixed with the recipient cells, the recipients are treated with DNase, which cannot enter the cells but destroys any DNA remaining in the medium. The DNase is removed and the surviving DNA in the recipient cells is then

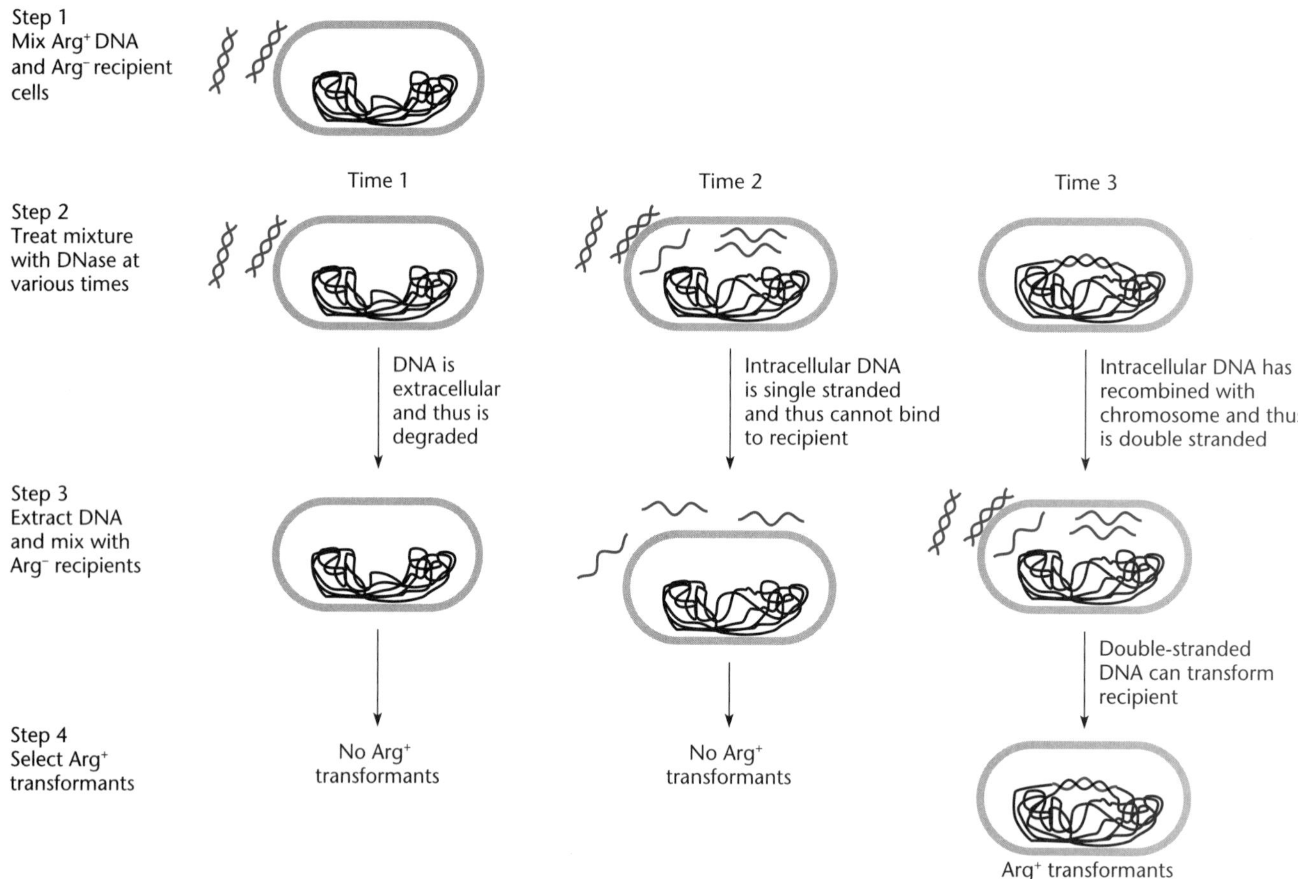
Step 1
Mix Arg⁺ DNA and Arg⁻ recipient cells
Time 1
Time 2
Time 3
Step 2
Treat mixture with DNase at various times
DNA is extracellular and thus is degraded
Intracellular DNA is single stranded and thus cannot bind to recipient
Intracellular DNA has recombined with chromosome and thus is double stranded
Step 3
Extract DNA and mix with Arg⁻ recipients
Double-stranded DNA can transform recipient
Step 4
Select Arg⁺ transformants
No Arg⁺ transformants
No Arg⁺ transformants
Arg⁺ transformants

Figure 6.6 Genetic assay for the state of DNA during transformation. Only double-stranded DNA binds to the cell to initiate transformation. After transformation is initiated, the donor DNA is converted to the single-stranded form, and reisolation of the DNA from the transformants does not allow transformation of new recipient cells until after the donor DNA has been incorporated into the chromosome by recombination. The appearance of transformants in step 4 indicates that the transforming DNA was intracellular and double stranded at the time of DNase treatment. doi:10.1128/9781555817169.ch6.f6.6

extracted and used for retransformation of new auxotrophic recipients, and Arg$^+$ transformants are selected on agar plates without the growth supplement arginine. Any Arg$^+$ transformants that arise in the second transformation experiment must have been due to double-stranded Arg$^+$ donor DNA in the recipient cells.

Whether transformants are observed depends on the time point at which the DNA was extracted from the cells. When the DNA is extracted at time 1 in Figure 6.6, while it is still outside the cells and accessible to the DNase, no Arg$^+$ transformants are observed because the Arg$^+$ donor DNA is all destroyed by the DNase. At time 2, some of the DNA is now inside the cells, where it cannot be degraded by the DNase, but this DNA is single stranded. It has not yet recombined with the chromosome, so Arg$^+$ transformants are still not observed in the retransformation experiment. Only at time 3, when some of the donor DNA has recombined with the chromosomal DNA and so is again double stranded, do Arg$^+$ transformants appear in the retransformation experiment. Thus, the transforming DNA enters the eclipse period for a short time after it is added to competent cells, as expected if it enters the cell in a single-stranded state.

Plasmid Transformation and Phage Transfection of Naturally Competent Bacteria

Chromosomal DNA can efficiently transform any bacterial cells from the same species that are naturally competent. However, neither plasmids nor phage DNAs can be efficiently introduced into naturally competent cells for two reasons. First, they must be double stranded to replicate. Natural transformation requires breakage of the double-stranded DNA and degradation of one of the two strands so that a linear single strand can enter the cell. Second, they must recyclize. However, pieces of plasmid or phage DNAs cannot recyclize if there are no repeated or complementary sequences at their ends.

Introduction of multiple copies of the same plasmid DNA into a single cell can regenerate an intact plasmid molecule if each double-stranded plasmid molecule is cleaved randomly during the initial binding step and plasmid strands are selected randomly for entry into the cell. Overlapping single-stranded plasmid molecules can hybridize to regenerate a partially double-stranded molecule (Figure 6.7), and these molecules are substrates for cellular repair systems that fill in the missing DNA and ligate the ends to regenerate a double-stranded circular plasmid.

Regulation of Natural Competence

Most naturally transformable bacteria express their competence genes only under certain growth conditions or in specific stages in the growth cycle. *S. pneumoniae* is maximally competent in the early exponential growth phase, and *B. subtilis* and *H. influenzae* become competent when nutrients are limited. In contrast, *N. gonorrhoeae* appears to express its competence genes under all conditions. Each of the regulated systems uses different mechanisms to monitor its nutritional state and regulate competence gene expression.

Figure 6.7 Transformation by plasmid DNA. If multiple plasmid molecules enter the same cell, linear single-stranded pieces of plasmid DNA anneal, the remaining gaps are filled in by cellular DNA repair systems, and nicks are sealed by DNA ligase. doi:10.1128/9781555817169.ch6.f6.7

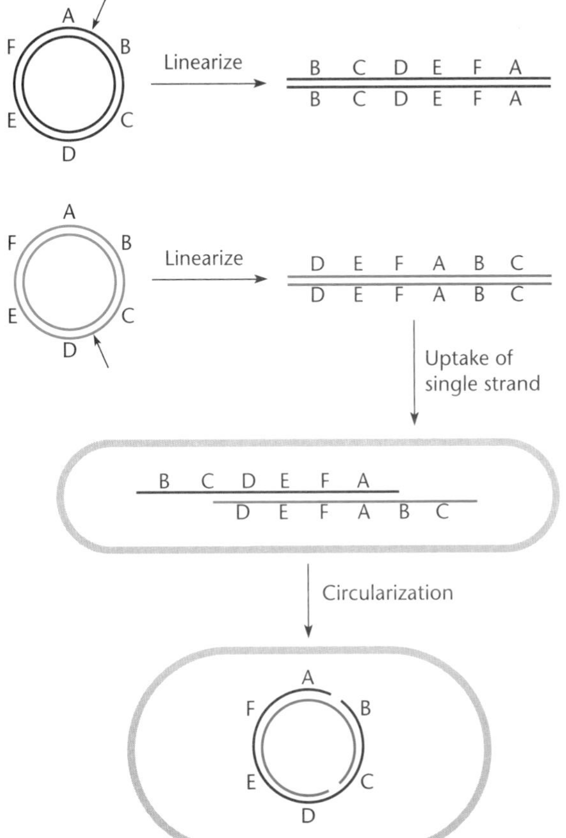

COMPETENCE REGULATION IN *B. SUBTILIS*

The *B. subtilis comE*, *comF*, and *comG* operons are all under the transcriptional control of ComK, a transcription factor that is itself regulated by the ComP-ComA **two-component regulatory system**, analogous to those used to regulate many other systems in bacteria (see chapter 13). In systems of this type, one protein senses a signal and transmits that signal to a second protein via phosphorylation events. Information that the cell is running out of nutrients and the population is reaching a high density is registered by ComP, a **sensor protein** in the membrane (Figure 6.8A). The signal for high cell density causes the sensor kinase protein to phosphorylate itself, i.e., to transfer a phosphate from ATP to a specific histidine residue within the ComP protein. The phosphate is then transferred from ComP to a specific aspartate residue in ComA, a **response regulator protein** in the cytoplasm. In the phosphorylated state, the ComA protein is a transcriptional activator (see chapters 12 and 13) for several genes, some of which are required for competence. One of the regulatory targets of phosphorylated ComA is another transcriptional activator, ComK, as well as ComS, which is a regulator of ComK. ComS protects ComK from proteolytic degradation, which allows ComK to accumulate to high enough levels to activate the transcription of other *com* genes, including those that form the transformation machinery illustrated in Figure 6.2A. This is an example of regulated proteolysis (see chapter 12).

Competence Pheromones

How does the cell know that other *B. subtilis* cells are nearby and that it should induce competence? High cell density is signaled through small peptides called **competence pheromones** that are excreted by the bacteria as they multiply (see Bongiorni et al. and Potahill and Lazazzera, Suggested Reading). Cells become competent only in the presence of high concentrations of these peptides, and the concentration of the peptides in the medium is high only when the concentration of cells giving them off is high. The requirement for competence pheromones ensures that cells are able to take up DNA only when other *B. subtilis* cells are nearby and giving off DNA to be taken up. This is one example of a phenomenon called **quorum sensing**, by which small molecules given off by cells send signals to other cells in the population that the cell density is high. Many such small molecules are known, including homoserine

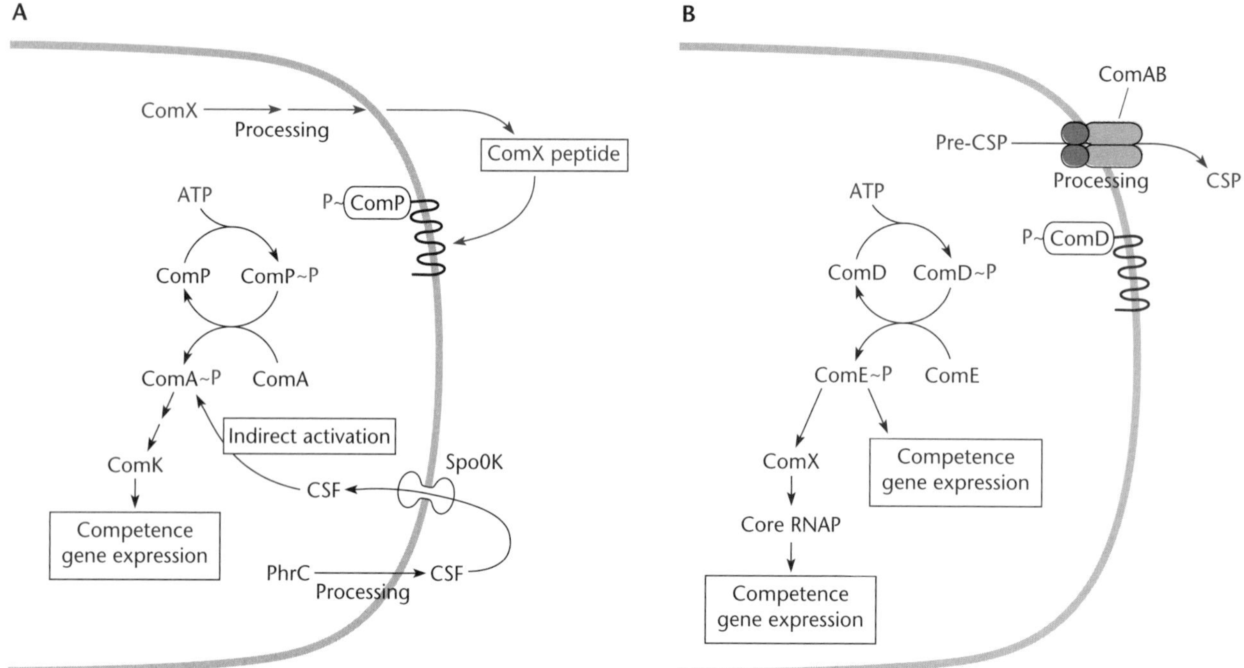

Figure 6.8 Regulation of competence development by quorum sensing. **(A)** In *B. subtilis,* the ComP protein in the membrane senses a high concentration of the ComX peptide (shown in blue) and phosphorylates itself by transferring a phosphate from ATP. The phosphate is then transferred to ComA, which allows the transcription of many genes, including those encoding ComK, the activator of the late *com* genes, and ComS, which prevents proteolytic degradation of ComK. In a separate pathway, the Phr peptide, processed from the signal sequence of another protein (PhrC), is imported into the cell by the Spo0K oligopeptide permease and indirectly activates phosphorylated ComA (ComA~P) by inactivating the RapC phosphatase, which otherwise will dephosphorylate ComA~P (not shown). **(B)** In *S. pneumoniae,* a peptide signal (CSP) activates the ComD-ComE two-component system, which directs expression of an initial early set of *com* genes that includes the *comX* gene. ComX is an alternate sigma factor that binds to RNA polymerase and directs transcription of the late *com* genes. doi:10.1128/9781555817169.ch6.f6.8

lactones that signal cell density in some gram-negative bacteria. Other examples of quorum sensing are discussed in chapter 13.

A second question is how the cell knows that the competence pheromone peptide came from other cells and was not produced internally. It does this by cutting the peptide out of a larger protein as the larger protein passes through the membrane of the cell in which it is synthesized. Once outside the cell, the peptide is diluted by the surrounding medium, and it can achieve high enough concentrations to induce competence only if the cell density is high and many surrounding cells are also giving off the peptide. In *B. subtilis*, the major competence pheromone peptide is called ComX and the longer polypeptide it is cut out of is the product of the *comX* gene. Another gene, *comQ*, which is immediately upstream of *comX*, is also required for synthesis of the competence pheromone because its product is the protease enzyme that cuts the competence pheromone out of the longer polypeptide. Once the peptide has been cut out of the longer molecule and is released, it can bind to the ComP protein in the membranes of nearby cells and trigger ComP autophosphorylation.

At best, only about 10% of *B. subtilis* cells ever become competent, no matter how favorable the conditions or how high the cell density. The advantages of this to the bacteria are obvious: if all the cells were competent, which cells would be giving off DNA to be taken up by the competent cells? This has been called a **bistable state** and seems to be determined by autoregulation of the ComK activator protein. Bistable states are common biological phenomena, and competence in *B. subtilis* has been used as an experimental model for such phenomena (see Maamer and Dubnau, Suggested Reading).

Relationship between Competence, Sporulation, and Other Cellular States

At about the same time that *B. subtilis* reaches the stationary phase, some cells acquire competence and other cells sporulate (see chapter 14). Sporulation, a developmental process found in some groups of bacteria, allows a bacterium to enter a dormant state and survive adverse conditions, such as starvation, irradiation, and heat. During sporulation, the bacterial chromosome is packaged into a resistant spore, where it remains viable until conditions improve and the spore can germinate into an actively growing bacterium. To coordinate sporulation and competence, *B. subtilis* cells produce other regulatory peptides (see Bongiorni et al., Suggested Reading). There are at least two such peptides (called Phr) that regulate ComA indirectly by inhibiting proteins called Rap proteins that dephosphorylate phosphorylated ComA and prevent it from binding to DNA and activating transcription. Like ComX, the Phr peptides are processed from the signal sequences of longer polypeptides, the products of the *phr* genes; unlike ComX, which interacts with ComP at the cell surface, the Phr peptides are transported into the cell by the oligopeptide permease Spo0K (Figure 6.8A).

The *spo0K* gene is an example of a regulatory gene that is required for both sporulation and the development of competence. The gene was first discovered because of its role in sporulation. A *spo0K* mutant is blocked in the first stage, the "0" stage, of sporulation. The *K* means that it was the 11th gene (as K is the 11th letter in the alphabet) involved in sporulation to be discovered in that collection.

REGULATION OF COMPETENCE IN *S. PNEUMONIAE*

As in *B. subtilis*, competence regulation in *S. pneumoniae* utilizes cell-cell signaling by a small peptide called competence-stimulating peptide (CSP). Binding of CSP to the membrane-bound ComD sensor kinase results in phosphorylation of its partner response regulator, ComE (Figure 6.8B); these are functionally analogous to ComP-ComA in *B. subtilis*. Phosphorylated ComE activates the transcription of 20 competence genes, one of which encodes a new RNA polymerase sigma factor (see chapters 2 and 12). Interaction of this sigma factor with core RNA polymerase directs the RNA polymerase to recognize a new set of promoter sequences that are found upstream of approximately 60 additional competence genes. These "late" competence genes encode most of the DNA-binding and transport machinery shown in Figure 6.2A. Both *B. subtilis* and *S. pneumoniae* use peptide signaling to monitor the presence of nearby related organisms, and both use two-component regulatory systems to transmit information about the extracellular peptide concentration to the gene expression machinery. They differ, however, in how they then control the expression of the late competence genes.

IDENTIFICATION OF COMPETENCE IN OTHER ORGANISMS

As noted above, genome sequencing has revealed genes related to known competence genes in organisms in which natural competence has not been demonstrated. Whereas in some cases these genes may have alternative functions (see Palchevskiy and Finkel, Suggested Reading), in other cases it is likely that the conditions for competence have not been identified. An interesting example is provided by studies in *Legionella pneumophila*, a pathogenic organism that causes pneumonia. Only certain strains were observed to be competent, and selection for "hypercompetent" strains, based on efficient incorporation of selectable antibiotic resistance markers, revealed mutations in genes that normally repress competence gene expression (see Sexton and Vogel, Suggested Reading).

Role of Natural Transformation

The fact that so many gene products play a direct role in competence indicates that the ability to take up DNA from the environment is advantageous. Below, we discuss three possible advantages and the arguments for and against them.

NUTRITION

Organisms may take up DNA for use as a carbon and nitrogen source (see Redfield, Suggested Reading). One argument against this hypothesis is that taking up whole DNA strands for degradation inside the cell may be more difficult than degrading the DNA outside the cell and then taking up the nucleotides. In fact, noncompetent *B. subtilis* cells excrete a DNase that degrades extracellular DNA so that the nucleotides can be taken up more easily. The major argument against this hypothesis as a general explanation for transformation in all bacteria is that some bacteria take up only DNA of their own species; if DNA is used only for nutrition, there would be no reason to selectively take up only certain DNA, since DNA from other organisms should offer the same nutritional benefits. Moreover, the fact that competence develops only in a minority of the population, at least in *B. subtilis*, argues against the nutrition hypothesis, since all the bacteria in the population would presumably need the nutrients.

These arguments are attractive but do not disprove the nutrition hypothesis, at least for all bacteria. The bacteria may consume DNA of only their own species because of the danger inherent in taking up foreign DNAs, which might contain prophages, transposons, or other elements that could become parasites of the organism. Furthermore, consumption of DNA from the same species may be a normal part of colony development; cell death and cannibalism are thought to be part of some prokaryotic developmental processes (see chapter 14). These processes would require that only some of the cells in the population become DNA consumers while the others become the "sacrificial lambs." The existence of specific cell-killing mechanisms that kill some cells in the population as *B. subtilis* enters the stationary phase lends support to such interpretations. Similarly, competent *S. pneumoniae* cells secrete a cell wall hydrolase that triggers lysis only of noncompetent cells, resulting in release of DNA that can be taken up by the competent cells. It is interesting that although natural transformation has never been observed in *Escherichia coli*, homologs of competence genes have been identified, and these genes are important for survival of the cells during stationary phase and for use of DNA as a nutritional source (see Palchevskiy and Finkel, Suggested Reading).

REPAIR

Cells may take up DNA from other cells to repair damage to their own DNA (see Mongold, Suggested Reading). Figure 6.9 illustrates this hypothesis, in which a population of cells is exposed to ultraviolet (UV) irradiation. The radiation damages the DNA, causing pyrimidine dimers and other lesions to form (see chapter 11). DNA leaks out of some of the dead cells and enters other bacteria. Because the damage to the DNA has not occurred at exactly the same places, undamaged incoming DNA sequences can replace the damaged regions in the recipient, allowing at least some of the bacteria to survive. This scenario explains why some bacteria take up DNA of only the same species, since in general, this is the only DNA that can recombine and thereby participate in the repair.

If natural transformation helps in DNA repair, we might expect that repair genes would be induced in response to developing competence and that competence would develop in response to UV irradiation or other types of DNA damage. In fact, in some bacteria, including *B. subtilis* and *S. pneumoniae*, the *recA* gene required for recombination repair is induced in response to the development of competence (see Haijema et al. and Raymond-Denise and Guillen, Suggested Reading). However, in other bacteria, such as *H. influenzae*, the *recA* gene is not induced in response to competence. There is also no evidence that competence genes are induced in response to DNA damage. Nevertheless, the need for DNA repair is an attractive explanation for why at least some types of bacteria develop competence.

Figure 6.9 Repair of DNA damage by transforming DNA. A region containing a thymine dimer (TT) induced by UV irradiation is replaced by the same, but undamaged, sequence from the DNA of a neighbor killed by the radiation.
doi:10.1128/9781555817169.ch6.f6.9

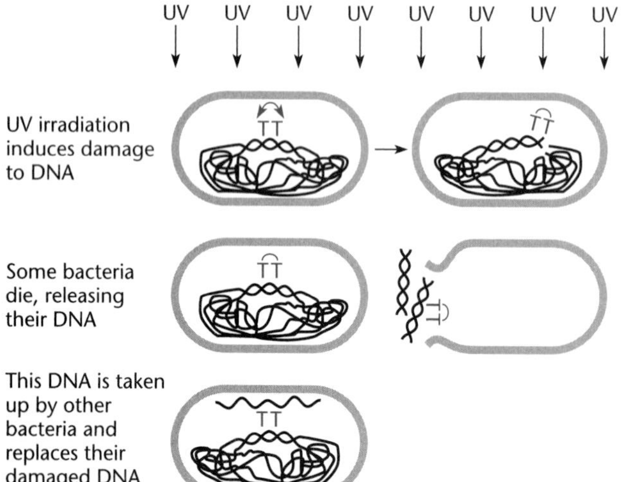

RECOMBINATION

The possibility that transformation allows recombination between individual members of a species is also an attractive hypothesis but is difficult to prove. According to this hypothesis, transformation serves the same function that sex serves in higher organisms: it allows the assembly of new combinations of genes and thereby increases diversity and speeds up evolution. Bacteria do not have an obligatory sexual cycle; therefore, without some means of genetic exchange, any genetic changes that a bacterium accumulates during its lifetime are not necessarily exchanged with other members of the species.

The gene exchange function of transformation is supported by the fact that cells of some naturally transformable bacteria leak DNA as they grow. Some *Neisseria* species have a type IV secretion system that appears to be dedicated to active export of DNA from the cell into the medium. It is hard to imagine what function this DNA export could perform unless the exported DNA is taken up by other bacteria to promote gene transfer to neighboring cells.

In several *Neisseria* species, including *N. gonorrhoeae*, transformation may enhance antigenic variability, allowing the organism to avoid the host immune system (Box 6.1). In mixed laboratory cultures, transformation does contribute substantially to the antigenic diversity in the species. However, under natural conditions, it is debatable whether most of this antigenic diversity results from recombination between DNAs brought together by transformation or simply from recombination between sequences within the chromosomal DNA of the bacterium itself.

We still do not know why some types of bacteria are naturally transformable and others are not. It seems possible that most types of bacteria are naturally transformable at low levels and that we have not identified the appropriate laboratory conditions to induce competence for some of these organisms. As noted above, many organisms for which natural transformation has not been observed contain competence genes in their genomes,

BOX 6.1

Antigenic Variation in *Neisseria gonorrhoeae*

Many types of pathogenic microorganisms avoid the host immune system by changing the antigens on their cell surfaces. Well-studied examples include trypanosomes, which cause sleeping sickness, and *N. gonorrhoeae*, which causes a sexually transmitted disease.

The pili of *N. gonorrhoeae* are involved in attaching the bacteria to the host epithelial cells. These pili are highly antigenic, and can undergo spontaneous alterations that can change the specificity of binding and confound the host immune system. *N. gonorrhoeae* appears to be capable of making millions of different pili.

The mechanism of pilin variation in *N. gonorrhoeae* is understood in some detail. The major protein subunit of the pilus is encoded by the *pilE* gene. In addition, silent copies of *pilE*, called *pilS*, lack promoters or have various parts deleted. These silent copies share some conserved sequences with each other and with *pilE* but differ in the so-called variable regions. Pilin protein is usually not expressed from these silent copies. However, recombination between a *pilS* gene and *pilE* can change the *pilE* gene and result in a somewhat different pilin protein. This recombination is a type of gene conversion, because reciprocal recombinants are not formed (see chapter 10). Interestingly, the availability of iron affects the frequency of antigenic variation. Many bacteria use the availability of iron as an indicator that they are in a eukaryotic host or in a particular tissue of that host, suggesting that the variation becomes activated in certain tissues.

Because *N. gonorrhoeae* is naturally transformable, not only could recombination occur between a *pilS* gene and the *pilE* gene in the same organism, but transformation could also allow even more variation through the exchange of *pilS* genes with other strains. Experiments indicate that pilin variation is affected by the presence of DNase, suggesting that transformation between individuals contributes to the variation. The recombination seems to utilize mostly the RecFOR pathway, rather than the RecBCD pathway, which is expected, since only a single strand of DNA enters during transformation and the RecBCD complex does not recognize single-stranded DNA (see chapter 10). Also, experiments with marked *pil* genes indicate that transformation can result in the exchange of *pil* genes between bacteria. These experiments suggest, but do not prove, that transformation plays an important role in pilin variation during infection. Proof of this would require experiments in the infected host, and humans are the only known host for *N. gonorrhoeae*.

References

Sechman, E. V., M. S. Rohrer, and H. S. Seifert. 2005. A genetic screen identifies genes and sites involved in pilin antigenic variation in *Neisseria gonorrhoeae*. *Mol. Microbiol.* **57:**468–483.

Seifert, H. S., R. S. Ajioka, C. Marchal, P. F. Sparling, and M. So. 1988. DNA transformation leads to pilin antigenic variation in *Neisseria gonorrhoeae*. *Nature* **336:**392–395.

which suggests that they have the ability to become competent under some conditions or that the competence genes serve another function (as suggested for *E. coli*). Transformation may serve different purposes in different organisms. Perhaps transformation is used for DNA repair in soil bacteria, such as *B. subtilis*, but is used to increase genetic variability in obligate parasites, such as *N. gonorrhoeae*.

Importance of Natural Transformation for Forward and Reverse Genetics

Whatever its purpose for individual bacterial species, natural transformation has many uses in molecular genetics. Transformation has been used in many bacteria to map genetic markers in chromosomes and to reintroduce DNA into cells after the DNA has been manipulated in the test tube. This has made naturally competent organisms, like *B. subtilis*, ideal model systems for molecular genetic studies. As mentioned, the interpretation of genetic data obtained by transformation is similar to the interpretation of data obtained by transduction (see chapter 3). In the bacterium *A. baylyi*, transformation is so efficient that it can occur after simply spotting DNA restriction fragments or PCR-amplified DNA fragments onto streaks of recipient bacteria on plates. This offers opportunities to construct many different types of mutations in genes, including **loss-**, **gain-**, or **change-of-function mutations** (see Young et al., Suggested Reading). Such manipulations are more difficult in bacteria that do not have efficient natural competence systems.

Congression

The presence of multiple DNA-binding sites on a single competent cell (approximately 20/cell for *B. subtilis*) allows the possibility of import of multiple DNA segments into a cell. This has important consequences for genetic mapping using transformation, as well as for the efficiency of transformation with plasmid DNA. In the case of genetic mapping, the conclusion that cotransformation of two markers is an indication of close physical linkage depends on the assumption that the two markers entered the cell on the same piece of DNA. Entry of markers on separate DNA segments is referred to as **congression**, and this occurs when cells are exposed to high concentrations of DNA so that multiple binding sites on a cell are simultaneously engaged in DNA uptake. For genetic mapping experiments, congression can be avoided by using subsaturating concentrations of DNA (Figure 6.10). As increasing amounts of DNA are added to a transformation mixture, the number of transformants for a particular chromosomal marker (Arg$^+$ in the example) will increase. Simultaneously, the number of transformants that have also obtained an unlinked genetic marker (Met$^+$) will begin to increase, but only after the DNA concentration has reached a level at which each cell is likely to have taken up both the *arg* and *met* DNA fragments. Note that the number of Met$^+$ transformants does not exceed approximately 5% of the Arg$^+$ transformants, because once all of the DNA-binding sites are saturated, additional DNA has no effect.

Congression is useful for introduction of a nonselectable marker into a cell, in the absence of a linked selectable marker. For example, if the donor cells are Arg$^+$ and Trp$^-$ and the recipient cells are Arg$^-$ and Trp$^+$, use of saturating concentrations of DNA and selection for Arg$^+$ will result in 5% Arg$^+$ Trp$^-$ transformants; in this case, recovery of the Trp$^-$ transformants requires that the transformation mixture be plated on medium lacking arginine (to select for Arg$^+$) and containing tryptophan (so that the Trp$^-$ transformants can grow).

Figure 6.10 Import of multiple DNA fragments into a single cell by congression. The donor strain is Arg$^+$ Met$^+$, the recipient strain is Arg$^-$ Met$^-$, and the transformation mixture was plated on medium lacking arginine but containing methionine; Arg$^+$ colonies were then tested for the ability to grow in the absence of methionine. Addition of increasing concentrations of DNA results in increasing numbers of Arg$^+$ transformants, until the DNA concentration reaches a level where the DNA-binding sites on the competent cells are saturated, at which point additional DNA has no effect on the number of transformants obtained. At low DNA concentrations, no transformants for the unlinked and nonselected Met$^+$ marker are observed. At high DNA concentrations, cells take up multiple DNA molecules, and incorporation of the unlinked Met$^+$ marker is observed, yielding Arg$^+$ Met$^+$ transformants.
doi:10.1128/9781555817169.ch6.f6.10

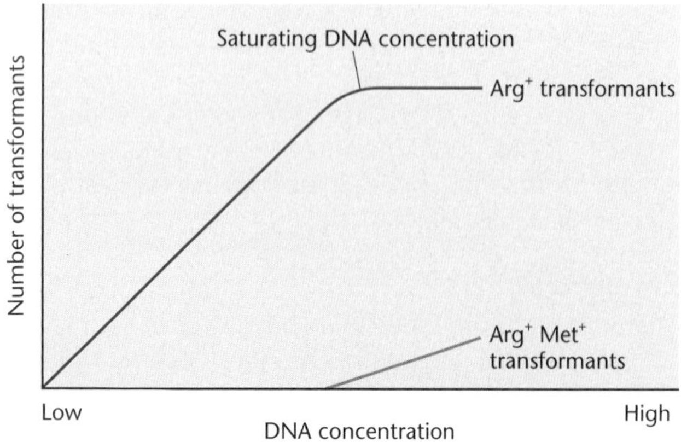

As noted above, transformation of plasmid DNA into naturally competent cells results in import of single-stranded DNA. Binding of multiple double-stranded plasmid molecules, and simultaneous uptake of the opposite strand from a different molecule through congression, increases the probability that the single strands will hybridize in the cell and regenerate an intact plasmid after repair of any gaps and nicks in the DNA.

Artificially Induced Competence

Most types of bacteria are not naturally transformable, at least not at easily detectable levels. Left to their own devices, these bacteria do not take up DNA from the environment. However, even these bacteria can sometimes be made competent by certain chemical treatments, or DNA can be forced into them by a strong electric field in a process called electroporation.

Chemical Induction

Treatment with calcium ions (see Cohen et al., Suggested Reading) or related chemicals, such as rubidium ions, can make some bacteria competent; examples include *E. coli* and *Salmonella* spp., as well as some *Pseudomonas* spp. The reason for this is not fully understood, but it is likely that the calcium ions perturb the cell surface, allowing DNA to leak inside.

Chemically induced transformation is usually inefficient, and only a small percentage of the cells are ever transformed. Accordingly, the cells must be plated under conditions selective for the transformed cells. Therefore, normally, the DNA used for the transformation should contain a selectable gene, such as one encoding resistance to an antibiotic, or another genetic trick must be used (see chapter 3).

TRANSFORMATION BY PLASMIDS

In contrast to naturally competent cells, cells made permeable to DNA by chemical treatment will take up both single-stranded and double-stranded DNA. Therefore, both linear and double-stranded circular plasmid DNAs can be efficiently introduced into chemically treated cells. This fact has made calcium ion-induced competence very useful for cloning and other applications that require the introduction of plasmid and phage

DNAs into cells. Transformation by circular DNAs is generally more efficient, as the DNA is less susceptible to degradation.

TRANSFECTION BY PHAGE DNA

In addition to plasmid DNAs, viral genomic DNAs or RNAs can often be introduced into cells by transformation, thereby initiating a viral infection. This process is called **transfection** rather than transformation, although the principle is the same. To detect transfection, the potentially transfected cells are usually mixed with indicator bacteria and plated (see the introduction). If the transfection is successful, a plaque forms where the transfected cells had produced phage, which then infected the indicator bacteria.

Some viral infections cannot be initiated merely by transfection with the viral DNA. These viruses cannot transfect cells, because in a natural infection, proteins in the viral head are normally injected along with the DNA, and these proteins are required to initiate the infection. For example, the *E. coli* phage N4 carries a phage-specific RNA polymerase in its head that is injected with the DNA and used to transcribe the early genes (see chapter 7). Transfection with the purified phage DNA does not initiate an infection, because the early genes are not transcribed without this phage-encoded RNA polymerase. Another example of a phage in which the infection cannot be initiated by the nucleic acid alone is phage ϕ6 (see Box 7.2). This phage has RNA instead of DNA in the phage head and must inject an RNA replicase to initiate the infection, so the cells cannot be transfected by the RNA alone. Such examples of phages that inject required proteins are rare; for most phages, the infection can be initiated by transfection.

TRANSFORMATION OF CELLS WITH CHROMOSOMAL GENES

Transformation with linear DNA is one method used to replace endogenous genes with genes altered in vitro. However, most types of bacteria made competent by calcium ion treatment are transformed poorly by chromosomal DNA because the linear pieces of double-stranded DNA entering the cell are degraded by an enzyme called the RecBCD nuclease (or a related enzyme). This nuclease degrades DNA from the ends; therefore, it does not degrade circular plasmid and phage DNAs.

Nevertheless, methods have been devised to transform competent *E. coli* with linear DNA. One way is to use a mutant *E. coli* strain lacking the D subunit of the RecBCD nuclease. These *recD* mutants are still capable of recombination, but because they lack the nuclease activity that degrades linear double-stranded DNA, they can be transformed with linear double-stranded DNAs. Other methods, sometimes called **recombineering**, use the recombination systems of phages, such as λ phage, instead of the host recombination functions. Double-stranded DNA can be introduced into cells expressing the λ recombination functions because of the ability of a λ protein named γ (gamma) to inhibit RecBCD. The method can also be used with single-stranded DNAs, such as PCR primers, which are not degraded by RecBCD. Methods of this type are discussed further in chapter 10.

Electroporation

Another way in which DNA can be introduced into bacterial cells is by **electroporation**. In the electroporation process, the bacteria are mixed with DNA and briefly exposed to a strong electric field using special equipment. It is important that the recipient cells first be washed extensively in buffer with very low ionic strength. The buffer also usually contains a nonionic solute, such as glycerol, to prevent osmotic shock. The brief electric fields across the cellular membranes might create artificial pores of H_2O-lined phospholipid head groups. DNA can pass through these temporary hydrophilic pores (see Tieleman, Suggested Reading). Electroporation works with most types of cells, including most bacteria, unlike the methods mentioned above, which are very specific for certain species. Also, electroporation can be used to introduce linear chromosomal and circular plasmid DNAs into cells.

Protoplast Transformation

Disruption of the bacterial cell wall results in production of osmotically sensitive protoplast forms of cells. Protoplasts are highly susceptible to lysis but can be maintained by incubation in a solution of the appropriate high osmolarity to balance the osmolarity of the cytoplasm. Exposure of protoplasts to DNA in the presence of polyethylene glycol triggers membrane fusion events that can trap the DNA within the cytoplasm, resulting in transformation. The protoplast forms of most bacteria are unable to divide, and the cell wall must be regenerated for transformants to be recovered. This is generally an inefficient process and must be optimized for each organism, which can be tedious. Protoplast transformation is most efficient for circular plasmid DNAs that are stable during the uptake process and can establish independent replication in the transformants. Despite its limitations, protoplast transformation represents a useful tool for introducing plasmid DNA into cells that are not easily transformed by other techniques.

SUMMARY

1. In transformation, DNA is taken up directly by cells. Transformation was the first form of genetic exchange to be discovered in bacteria, and the demonstration that DNA is the transforming principle was the first direct evidence that DNA is the hereditary material. The bacteria from which the DNA was taken are called the donors, and the bacteria to which the DNA has been added are called the recipients. Bacteria that have taken up DNA and are able to maintain the new DNA in a heritable form (for example, by recombination with the genome) are called transformants.

2. Bacteria that are capable of taking up DNA are said to be competent.

3. Some types of bacteria become competent naturally during part of their life cycle. A number of genes whose products form the competence machinery have been identified. Some of these encode proteins related to type II secretion systems, which form type IV pili, or to type IV secretion-conjugation systems.

4. The fate of the DNA during natural transformation is fairly well understood. The double-stranded DNA first binds to the cell surface and then is broken into smaller pieces by endonucleases. Then, one strand of the DNA is degraded by an exonuclease. The single-stranded pieces of DNA enter the cell and then invade the chromosome in homologous regions, displacing one strand of the chromosome at these sites. Through repair or subsequent replication, the sequence of the incoming DNA may replace the original chromosomal sequence in these regions.

5. Naturally competent cells can be transformed with linear chromosomal DNA but are not as efficiently transformed with monomeric circular plasmid or phage DNAs. Transformation by plasmid DNA usually occurs by simultaneous uptake of multiple molecules of the plasmid DNA that can hybridize to each other and recyclize.

6. Some types of bacteria, including *H. influenzae* and *N. gonorrhoeae*, take up DNA of only the same species. Their DNA contains short uptake sequences that are required for import of DNA into the cells. Other types of bacteria, including *B. subtilis* and *S. pneumoniae*, seem to be capable of taking up any DNA.

7. There are three possible roles for natural competence: a nutritional function allowing competent cells to use DNA as a carbon, energy, and nitrogen source; a repair function in which cells use DNA from neighboring bacteria to repair damage to their own chromosomes, thus ensuring survival of the species; and a recombination function in which bacteria exchange genetic material among members of their species, increasing diversity and accelerating evolution. Different types of bacteria may maintain transformation for different purposes, and individual types may maintain the process for multiple reasons.

8. Some types of bacteria that do not show natural competence can nevertheless be transformed after some types of chemical treatment or by generation of protoplasts, which lack their cell wall. The standard method for making *E. coli* permeable to DNA involves treatment with calcium ions. Cells made competent by calcium treatment can be transformed with plasmid and phage DNAs, making this method one of the cornerstones of molecular genetics.

9. If the cell is transformed with viral DNA to initiate an infection, the process is called transfection.

10. Brief exposure of cells to an electric field also allows cells to take up DNA, a process called electroporation.

QUESTIONS FOR THOUGHT

1. Why do you think some types of bacteria are capable of developing competence? What is the "real" function of competence?

2. How would you determine if the competence genes of *B. subtilis* are turned on by UV irradiation and other types of DNA damage?

3. How would you determine whether antigenic variation in *N. gonorrhoeae* is due to transformation between different bacteria or to recombination within the same bacterium?

PROBLEMS

1. How would you determine if a type of bacterium you have isolated is naturally competent? Outline the steps you would use.

2. You have isolated a naturally competent bacterium from the soil. Outline how you would isolate mutants of your bacterium that are defective in transformation. Distinguish

those that are defective in recombination from those that are defective in the uptake of DNA.

3. How would you determine if a naturally transformable bacterium can take up DNA of only its own species or can take up any DNA?

4. How would you determine if a piece of DNA contains the uptake sequence for that species?

5. How would you determine if the DNA of a phage can be used to transfect *E. coli*?

SUGGESTED READING

Avery, O. T., C. M. MacLeod, and M. McCarty. 1944. Studies on the chemical nature of the substance inducing transformation of pneumococcal types. Induction of transformation by a deoxyribonucleic acid fraction isolated from pneumococcus type III. *J. Exp. Med.* **79:**137–159.

Bongiorni, C., S. Ishikawa, S. Stephenson, N. Ogasawara, and M. Perego. 2005. Synergistic regulation of competence development in *Bacillus subtilis* by two Rap-Phr systems. *J. Bacteriol.* 187:4353–4361.

Chen, I., and D. Dubnau. 2004. DNA uptake during bacterial transformation. *Nat. Rev. Microbiol.* **2:**241–249.

Claverys, J.-P., B. Martin, and P. Polard. 2009. The genetic transformation machinery: composition, localization, and mechanism. *FEMS Microbiol. Rev. 33:*643–656.

Clerico, E. M., J. L. Ditty, and S. F. Golden. 2006. Specialized techniques for site directed mutagenesis in cyanobacteria, p. 155–171. *In* E. Rosato (ed.), *Methods in Molecular Biology.* Humana Press, Totowa, NJ.

Cohen, S. N., A. C. Y. Chang, and L. Hsu. 1972. Nonchromosomal antibiotic resistance in bacteria: genetic transformation of *Escherichia coli* by R-factor DNA. *Proc. Natl. Acad. Sci. USA* **69:**2110–2114.

Cornella, N., and A. D. Grossman. 2005. Conservation of genes and processes controlled by the quorum response in bacteria: characterization of genes controlled by the quorum-sensing transcription factor ComA in *Bacillus subtilis. Mol. Microbiol.* 57:1159–1174.

Haijema, B. J., D. van Sinderen, K. Winterling, J. Kooistra, G. Venema, and L. W. Hamoen. 1996. Regulated expression of the *dinR* and *recA* genes during competence development and SOS induction in *Bacillus subtilis. Mol. Microbiol.* **22:**75–86.

Johnsberg, O., V. Eldholm, and L. S. Haverstein. 2007. Genetic transformation: prevalence, mechanisms and function. *Res. Microbiol. 158:*767–778.

Karnholz, A., C. Hoefler, S. Odenbreit, W. Fischer, D. Hofreuter, and R. Haas. 2006. Functional and topological characterization of novel components of the *comB* DNA transformation system in *Helicobacter pylori. J. Bacteriol.* **188:** 882–893.

Maamer, H., and D. Dubnau. 2005. Bistability in the *Bacillus subtilis* K-state (competence) system requires a positive feedback loop. *Mol. Microbiol.* **56:**615–624.

Mongold, J. A. 1992. DNA repair and the evolution of competence in *Haemophilus influenzae. Genetics* **132:**893–898.

Palchevskiy, V., and S. E. Finkel. 2006. *Escherichia coli* competence gene homologs are essential for competitive fitness and the use of DNA as a nutrient. *J. Bacteriol.* **188:**3902–3910.

Peterson, S. N., C. K. Sung, R. Cline, B. V. Desai, E. C. Snesrud, P. Luo, J. Walling, H. Li, M. Mintz, G. Tsegaye, P. C. Burr, Y. Do, S. Ahn, J. Gilbert, R. D. Fleischmann, and D. A. Morrison. 2004. Identification of competence pheromone responsive genes in *Streptococcus pneumoniae* by use of DNA microarrays. *Mol. Microbiol.* **51:**1051–1070.

Potahill, M., and B. A. Lazazzera. 2003. The extracellular Phr-Rap phosphatase circuit of *Bacillus subtilis. Front. Biosci. 8:*d32–d45.

Provedi, R., I. Chen, and D. Dubnau. 2001. NucA is required for DNA cleavage during transformation of *Bacillus subtilis. Mol. Microbiol.* **40:**634–644.

Raymond-Denise, A., and N. Guillen. 1992. Expression of the *Bacillus subtilis dinR* and *recA* genes after DNA damage and during competence. *J. Bacteriol.* **174:**3171–3176.

Redfield, R. J. 1993. Genes for breakfast: the have your cake and eat it too of bacterial transformation. *J. Hered.* **84:**400–404.

Sexton, J. A., and J. P. Vogel. 2004. Regulation of hypercompetence in *Legionella pneumophila. J. Bacteriol. 186:*3814–3825.

Tieleman, D. P. 2004. The molecular basis of electroporation. *BMC Biochem.* **5:**10.

Young, D. M., D. Parke, and L. N. Ornston. 2005. Opportunities for genetic investigation afforded by *Acinetobacter baylyi*, a nutritionally versatile bacterial species that is highly competent for natural transformation. *Annu. Rev. Microbiol.* **59:**519–551.

CHAPTER 7

Bacteriophages: Lytic Development, Genetics, and Generalized Transduction

PROBABLY ALL ORGANISMS ON EARTH are parasitized by viruses, and bacteria are no exception. For purely historical reasons, viruses that infect bacteria are usually not called viruses but are called **bacteriophages** (**phages** for short), even though they are also viruses and have lifestyles similar to those of plant and animal viruses. Phages and other viruses are also probably very ancient, coexisting with cellular organisms since the earliest life on earth and probably influencing evolution in very substantial ways (see the introduction). As mentioned in the introduction, the name *phage* derives from the Greek verb "to eat," and it describes the eaten-out places, or **plaques,** that are formed on bacterial lawns. The plural of phage is also phage when referring to a number of phage of the same type, but we add an "s" (phages) when we are discussing more than one type of phage.

Phages are no biological sideshow. They are probably the most numerous biological entities, with an estimated 10^{31} phages on Earth. These phages infect an estimated 10^{23} bacteria per second. By their massive predation on cyanobacteria in the oceans alone, they play a major role in the ecology of the earth. They are also one of the most diverse biological entities, which has made them a source of interesting and useful functions whose future applications we can only imagine. The smallest phage known is the *Leucomotor* phage, with a genome of 2,435 base pairs (bp), and the largest, a *Pseudomonas* phage with a genome of 316,674 bp, is almost 10% as long as the genome of it host. However, most phages are much smaller, with genomes on the order of 40,000 bp.

Like all viruses, phages are so small that they can be seen only under an electron microscope. As shown in Figure 7.1A, phages are often spectacular in appearance, with **capsids**, or icosahedral heads, and elaborate tail structures that make them resemble lunar landing modules. The tail structures

doi:10.1128/9781555817169.ch7

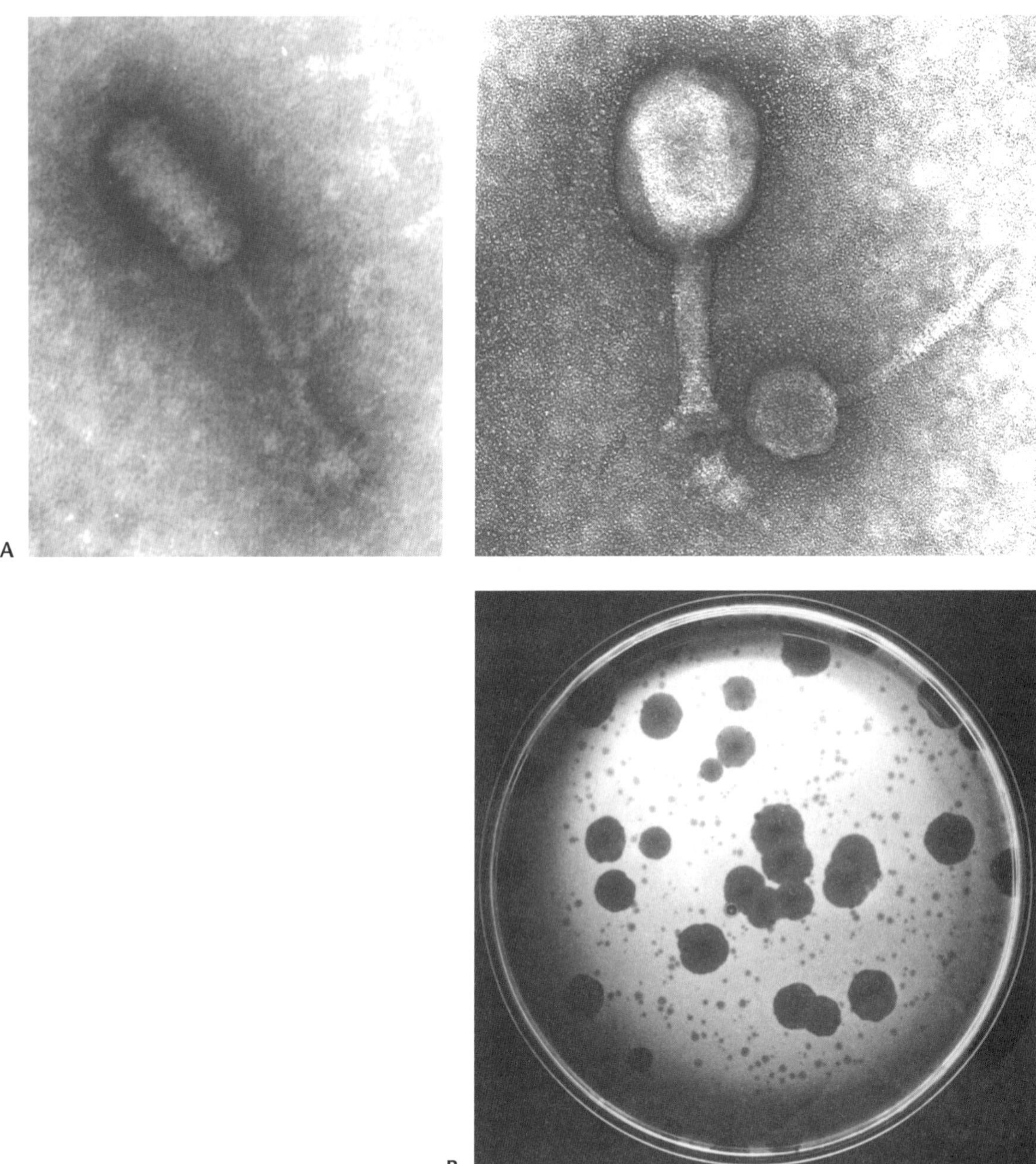

Figure 7.1 Electron micrographs and plaques of some bacteriophages. **(A) (Left)** A phage of *Enterococcus*. **(Right)** Electron micrograph of phages T4 and λ (left and right, respectively). **(B)** Plaques of *E. coli* phages M13 (smaller plaques) and T3, a relative of phage T7 (larger plaques). doi:10.1128/9781555817169.ch7.f7.1

allow them to penetrate bacterial membranes and cell walls to inject their DNA into the cell. Animal and plant viruses have much simpler shapes because they do not need such elaborate tail structures. They either are engulfed by the cell, in the case of animal viruses, or enter through wounds, in the case of plant viruses. Nevertheless, there are some types of phages that do not have tails or even protein capsids (Table 7.1). These are usually phages that infect bacteria that lack cell walls, such as mycoplasmas, or are taken up by the cell through its own secretion systems, for example, the filamentous phages.

Phages differ greatly in complexity, depending on their size. Smaller phages, such as MS2, usually have no tail, and their heads may consist of as few as two different types of proteins. The heads of some of the larger phages, such as T4, have up to 10 different proteins, each of which can exist in as few as 1 copy to almost

Table 7.1 Some families of phages by morphotype

Family	Example(s)	Characteristic(s)[a]
Siphoviridae	λ, N15, T1, T5	Linear dsDNA, long tail
Podoviridae	T3, T7, P22, N4, ϕ29	Linear dsDNA, short tail
Myoviridae	T4, P1, P2, Mu, SPO1	Linear dsDNA, large phage, contractile tail
Microviridae	ϕX174	Circular ssDNA, spherical capsid
Inoviridae	M13, f2, CTXϕ	Circular ssDNA, filamentous
Leviviridae	Qβ, R17, MS2	ssRNA, spherical
Cystoviridae	ϕ6	dsRNA, segmented, envelope
Corticoviridae	PM2	Internal membrane
Plasmaviridae	L2 (*Mycoplasma*)	No capsid, lipid envelope
Lipothrixviridae	TTV-1[b] (Archaea)	Filamentous, tail fibers, dsDNA

[a]ss, single stranded; ds, double stranded.
[b]TTV-1, *Thermoproteus tenax* virus 1.

1,000 copies depending on the structural role they play in the phage particle. The tails of phages can also contain many different proteins and have complicated structures, including base plates from which emanate tail fibers that help them bind to specific sites on the surfaces of their bacterial hosts. Often, the largest protein of the tail is the "tape measure protein" that extends the entire length of the tail and determines its length. These proteins are particularly long in phages of the family *Siphoviridae*, with a long and flexible tail (see below). The tails of phages allow them to infect specific bacterial hosts, and different phages have very different host ranges, as discussed below.

Phages fall into a relatively small number of families that can be composed of a large number of types that infect different bacterial hosts (Table 7.1). Phages within a family are closely related to each other, and their sequences bear little relationship to those of phages of other families or to their host bacteria, showing that they have evolved separately from their hosts. Also, recent work comparing the genomic sequences of phages within these families has revealed their "mosaic" nature, in which different phages of the same family seem to be assembled from shared "mosaic tiles" composed of individual genes or modules composed of groups of genes for the same function, for example, genes for DNA replication or for formation of the phage head (Box 7.1). The assumption is that these genes have been obtained either individually or as an interacting group from different phages of the same family at different evolutionary times. This suggests that phages do not just evolve with their host but evolve as a family of phages by somehow exchanging building modules with phages that are from the same family but that now infect different hosts.

Like all viruses, phages are not live organisms capable of living independently but merely a nucleic acid—either DNA or RNA, depending on the type of phage—wrapped in a protein and/or membrane coat for protection. This coat is lost during infection, although sometimes one or more different types of proteins that are encapsulated with the nucleic acid also enter the host during infection. The nucleic acid in the capsid carries genes that direct the synthesis of more phage. Depending on the type of phage, either DNA or RNA is carried in the head and is called the **phage genome**. These genome molecules can be very long, because the genome must be long enough to have at least one copy of each of the phage genes. They can also be in more than one segment, at least in the case of some phages with RNA genomes. The length of the DNA or RNA genome therefore reflects the size and complexity of the phage. For instance, the *Leviviridae* phage MS2 has only four genes and a rather small RNA genome, whereas the *Myoviridae* phage T4 has more than 200 genes and a DNA genome that is almost 10 μm long, 1/10 as long as that of its host. Long phage genomes, which can be as much as 1,000 times longer than the phage head, must be very tightly packed into the head of the phage, while in filamentous phages, the phage is as long as the genome DNA.

Because phages are so small, they are usually detected only by the plaques they form on **lawns** of susceptible host bacteria (Figure 7.1B) (see the introductory chapter). Each type of phage makes plaques on only certain host bacteria, which define its **host range**. Mutations in the phage DNA can alter the host range of a phage or the conditions under which the phage can form a plaque, which is often how mutations are detected. In this chapter, we discuss what is known about how some representative phages control their gene expression and replicate their genomes, as well as how they exit the cell after new phage have formed, summarizing some of the genetic and biochemical experiments that have contributed to this knowledge. We also discuss some important

BOX 7.1

Phage Genomics

Because of their relatively small size, phage genomes were the first to be sequenced, and at the time of writing, about 600 phage genomes have been sequenced, almost as many as for bacteria. Most of the genome sequences in data banks like those kept by the National Center for Biotechnology Information are composed of bacterial sequences, since they are larger. Nevertheless, a considerable portion of the diversity comes from phages, with many unique genes with unknown functions.

The sequencing of phage genomes has yielded a number of interesting observations. Much of this has come from the phages of *Mycobacterium*, 70 of which have been sequenced so far (see Hatfull, References). One observation is that, while the sizes of the mycobacterial phage genomes are widely distributed, their sizes seem to be concentrated within certain ranges, represented by the common families of phages (Table 7.1). The largest groupings are from about 30 kilobase pairs (kbp) to about 50 kbp, which includes the families *Siphoviridae* and *Podoviridae*, and another grouping around 120 kbp, which includes the family *Myoviridae*. Another striking feature revealed by comparing the sequences of a number of different phages is the mosaic nature of many phage genomes, especially within the *Siphoviridae*. The genomes were obviously assembled from genes obtained from other *Siphoviridae* as a unit at different times. This is illustrated in the figure. In panel A is shown the map of one mycophage (a phage named TM4 that infects *Mycobacterium*). Some genes in this phage are compared to the corresponding genes in 60 other *Siphoviridae* mycophages. In this representation, 60 sequenced *Siphoviridae* phages are displayed around a circle, with closely related phages adjacent to each other. The reference phage, TM4, was placed at about 59 min on this circle. Three adjacent genes or ORFs, *83*, *84*, and *85*, belonging to phage gene families (phamilies, or Phams) 114, 279, and 963, respectively (shown in blue on the map of TM4 in panel A of the figure), were compared to orthologs from other phages, and an arc was drawn to phages that had orthologs of these genes. Note that even though the genes are adjacent on the TM4 genome, they apparently came from widely different sources, as evidenced by the fact that the arc from each ORF connects TM4 to very different phages. Also, they came in at different times, since the arcs are different colors, with blue being more closely related and gray less closely, indicating how long they have been separated evolutionarily. Also, when a phage shares two or more of them, their map positions are often different in different phages, as indicated in panel C of the figure, which shows the relative positions of orthologs of 84 and 85 in two different phages, Che12 and Chah. This is further indication that they were acquired independently.

While some genes have been acquired independently, others have obviously been acquired as a unit or module of genes, particularly genes with products with the same function. For example, the tail genes of the phage have obviously been acquired as a module containing all the tail genes, as has the module containing the head genes (compare the map of phage TM4 in the figure to those of λ and other *Siphoviridae* phages in Figures 8.2, 8.19, and 8.20). In other words, if one tail gene shows considerable sequence homology with the same tail gene from another phage, all of the tail genes will show the same degree of sequence homology, indicating they were acquired at the same time from a common origin. However, the head genes will all show a different degree of homology from the tail genes, indicating that, like individual genes, the two modules have been acquired from different sources at different times.

Not only do genes within a module all share sequence homology, they also show considerable **synteny** (see "The Bacterial Genome" in chapter 1), meaning the genes in each module are in the same order as they are in λ and many other *Siphoviridae*. For example, the approximately 12 tail genes are almost always in the same order, with the largest being the gene for the tape measure protein that determines the length of the tail, so its product must be at least as long as the tail. Also, in the *Siphoviridae*, the head genes are almost always 5′ of and in the same orientation relative to the tail genes as they are in λ and other *Siphoviridae*. This synteny exists even if the sequences of the genes from different phages themselves show only very little homology, indicating that they were acquired from a common ancestor a long time ago. Why such synteny should have been preserved more faithfully throughout evolution than the sequences of the genes themselves is something of a mystery. Possibly, it confers some slight advantage for the assembly of the phage particle. Genes and larger modules apparently have not been exchanged nearly as frequently between different families of phages, since *Siphoviridae* modules have not yet been found in the *Myoviridae* phages or vice versa, even when they infect the same hosts. This indicates that the families of phages have evolved independently of each other, even though they may share the same hosts.

Another unknown is the nature of the events that gave rise to the mosaic structure of these phages. There are two hypotheses. One is that sequences flanking the modules promote the recombination that exchanges one copy of the module for another. However, in most cases, there is no evidence for such flanking sequences. A more

BOX 7.1 (continued)

Phage Genomics

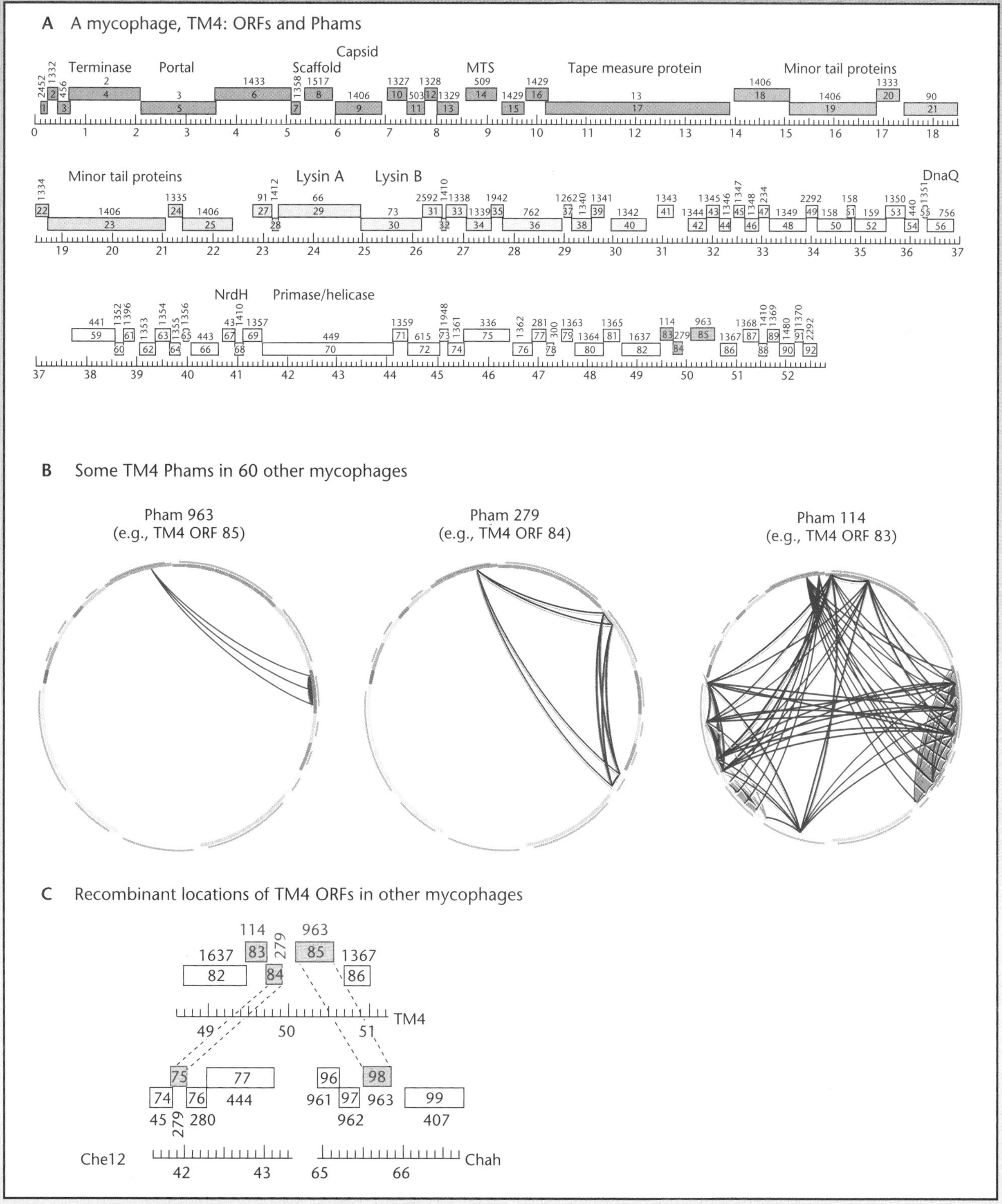

A A mycophage, TM4: ORFs and Phams

Terminase Portal Capsid Scaffold MTS Tape measure protein Minor tail proteins

Minor tail proteins Lysin A Lysin B DnaQ

NrdH Primase/helicase

B Some TM4 Phams in 60 other mycophages

Pham 963 (e.g., TM4 ORF 85)

Pham 279 (e.g., TM4 ORF 84)

Pham 114 (e.g., TM4 ORF 83)

C Recombinant locations of TM4 ORFs in other mycophages

TM4

Che12

Chah

doi:10.1128/9781555817169.ch7.Box7.1.f

(continued)

BOX 7.1 (continued)

Phage Genomics

attractive hypothesis is that random recombination followed by selection is sufficient to establish the modular exchange. According to this hypothesis, recombination occurs more or less randomly between two very different members of the same family of phages, depending on where sequence homology happens to occur. However, if the product of each of the genes within a module must physically interact with at least one other gene product from that module to make a particular structure, for example, the phage tail, they might not be able to functionally interact with proteins encoded by the other phage if its proteins are too different. Only phage that get an entire module of genes of interacting functions from one phage are viable and will be selected.

Another feature of phage genomes that is emerging is that they share very few genes with their bacterial hosts (see Kristensen et al., References). Apparently, phages evolve as a family quite separately from the hosts they infect. They also have fewer paralogs—genes with different functions that have arisen as the result of duplication—than bacterial genomes, and proteins seem less apt to have been assembled by interchange of different domains.

References

Hatfull, G. F. 2010. Bacteriophage research: gateway to learning science. *Microbe* **5:**243–250.

Kristensen, D. M., X. Cai , and A. Mushegian. 2011. Evolutionarily conserved orthologous families in phages are relatively rare in their prokaryotic hosts. *J. Bacteriol.* **193:**1806–1814.

new technologies that have been developed using phages and some mechanisms that cells use to defend themselves against phages, as well as why some phages, but not others, can be used for genetic mapping and strain construction in bacteria in a process called transduction. First, however, we review some general features of phage development. All the phages discussed in detail in this chapter use DNA as their genomes; the properties of some RNA phages are briefly described in Box 7.2.

Regulation of Gene Expression during Lytic Development

Because phages are viruses and are essentially genes wrapped in a protein or membrane coat, they cannot multiply without benefit of a host cell. The virus injects its genes into a cell, and the cell furnishes some or all of the means to express those genes and make more viruses.

Figure 7.2 illustrates the multiplication process for a typical large DNA phage. To start the infection, a phage adsorbs to an actively growing bacterial cell by binding to a specific receptor on the cell surface. In the next step, the phage injects its entire DNA into the cell, where transcription of RNA, usually by the host RNA polymerase, begins almost immediately. However, not all the genes of a phage are transcribed into mRNA when the DNA first enters the cell. The ones that are transcribed first usually have promoters that mimic those of the host cell DNA and so are recognized by the host RNA polymerase. Those transcribed immediately after infection are called the **early genes** of the phage and encode mostly enzymes involved in DNA synthesis, such as DNA polymerase, primase, DNA ligase, and helicase. With the help of these enzymes, the phage DNA begins to replicate, and many copies accumulate in the cell.

Next, mRNA is transcribed from the rest of the phage genes, the **late genes**, which may or may not be intermingled with the early genes in the phage DNA, depending on the phage. These genes have promoters that are unlike those of the host cell and so are not recognized by the host RNA polymerase alone or are recognized only by a phage-encoded RNA polymerase. Many of these genes encode proteins involved in assembly of the head and tail of the phage and lysis of the cell. After the phage particle is completed, the DNA is taken up (encapsulated) by the heads, and the tails are attached. Finally, the cells break open, or **lyse**, and the new phage are released to infect other sensitive cells. This whole process, known as the **lytic cycle**, takes less than 1 hour for many phages, and hundreds of progeny phage can be produced from a single infecting phage.

Actual phage development is usually much more complex than this basic process, proceeding through several intermediate stages in which the expression of different genes is regulated by specific mechanisms. Most of the regulation is achieved by transcribing different classes of genes into mRNA only at certain times; this type of regulation is called **transcriptional regulation** (see chapter 2). However, some genes undergo **posttranscriptional regulation**, which occurs after the mRNA has been made. For example, regulation may operate at the level of determining whether the mRNAs are translated; this is known as **translational regulation**. Other types of posttranscriptional regulation involve the stability of

BOX 7.2

RNA Phages

The capsids (i.e., heads or coats) of many animal and plant viruses contain RNA as their genome instead of DNA. Some of these viruses, the so-called retroviruses, use enzymes called reverse transcriptases to transcribe the RNA into DNA, and these enzymes, because they are essentially DNA polymerases, need primers. In contrast, other RNA-containing animal viruses, such as the influenza viruses, which cause flu, and the reoviruses, which cause colds, replicate their RNA by using RNA replicases and need no DNA intermediate. Because these RNA replicases have no need for primers, the genomes of RNA viruses can be linear without repeated ends. As we might expect, RNA viruses seem to have higher spontaneous mutation rates during replication, probably because their RNA replicases have no editing functions.

Some phages also have RNA as their genomes. Examples are Qβ, MS2, R17, f2, and ϕ6. The *E. coli* RNA phages Qβ, MS2, R17, and f2 are similar to each other. All have a single-stranded RNA genome that encodes only four proteins: a replicase, two head proteins, and a single protein lysin (see the text). Immediately after the genome RNA enters the cell, it serves as an mRNA and is translated into the replicase. This enzyme replicates the RNA, first making complementary **minus strands** and then using them as a template to synthesize more **plus strands**. The phage genomic RNA must serve as an mRNA to synthesize the replicase, because no such replicase enzyme exists in *E. coli* to synthesize RNA from an RNA template. Interestingly, the phage Qβ replicase has four subunits, only one of which is encoded by the phage. The other three are components of the host translational machinery: two of the elongation factors for translation, EF-Tu and EF-Ts, and a ribosomal protein, S1. They also need another protein, Hfq (host factor for Qβ), that is required to replicate their genomes. This protein, first discovered in the Qβ replicase, is used for regulation by small noncoding RNAs and is discussed extensively in chapter 13. Because the genomes of these RNA phages also function as mRNAs, they have served as a convenient source of a single species of mRNA in studies of translation, including the first sequences of translation initiation regions.

Another RNA phage, ϕ6, was isolated from the bean pathogen *Pseudomonas syringae* subsp. *phaseolicola*. The RNA genome of this phage is double stranded and exists in three segments in the phage capsid, much like the reoviruses of mammals. These three segments are called the S, M, and L segments, for small, medium, and large. Also like animal viruses, the phage is surrounded by membrane material derived from the host cell, i.e., an envelope, and the phage enters its host cells in much the same way that animal viruses enter their hosts. However, unlike most animal viruses, ϕ6 is released by lysis.

The replication, transcription, and translation of the double-stranded RNA of a virus such as ϕ6 present special problems. Not only must the phage replicate its double-stranded RNA, it also must transcribe it into single-stranded mRNA, since double-stranded RNA cannot be translated. The uninfected host cell contains neither of the enzymes required for these functions, which therefore must be virus encoded and packaged into the phage head so that they enter the cell with the RNA. Otherwise, neither the transcriptase nor the replicase could be made. Another interesting question with this phage is how three separate RNAs are encapsidated in the phage head. The three segments of the genome are transcribed into plus strand transcripts that are then packaged into a preformed head in sequential order, with the S segment first, the M segment second, and the L segment last. Packaging is initiated at 200-base *pac* sequences that seem to share little sequence or structural similarity among the three RNAs. Only after the single plus strand enters the head is its minus strand complement synthesized to make the double-stranded genomic RNA. This expands the prohead into its mature spherical form. The entire process can be performed in the test tube using phage proteins synthesized in *E. coli*, which spontaneously assemble into heads and the separate genomic RNAs that are taken up in the heads.

References

Blumenthal, T., and G. G. Carmichael. 1979. RNA replication: function and structure of Qβ replicase. *Annu. Rev. Biochem.* **48:**525–548.

Qiao, X., J. Qiao, and L. Mindich. 2003. Analysis of specific binding involved in genomic packaging of the double-stranded RNA bacteriophage ϕ6. *J. Bacteriol.* **185:**6409–6414.

certain RNAs that quickly degrade unless they are synthesized at the right stage of development.

Figure 7.3 shows the basic process of phage gene transcriptional regulation, in which one or more of the gene products synthesized during each stage of development turns on the transcription of the genes in the next stage of development. The gene products synthesized during each stage can also be responsible for turning off the transcription of genes expressed in the preceding stage. Genes whose products are responsible for regulating the transcription of other genes are called **regulatory genes**, and this type of regulation is called a **regulatory cascade**,

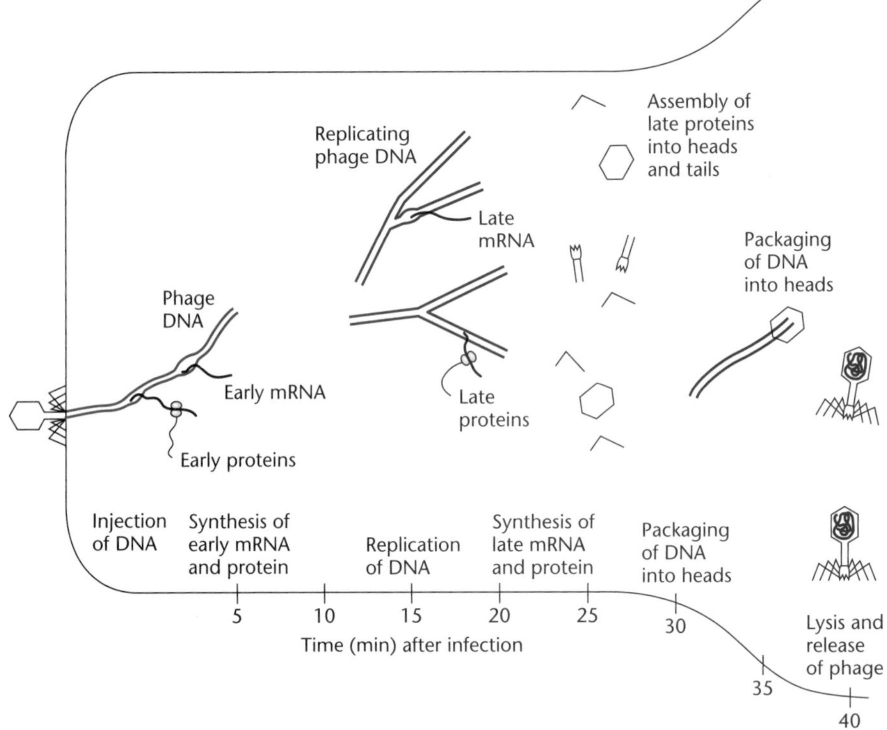

Figure 7.2 A typical bacteriophage multiplication cycle. After the phage injects its DNA, the early genes, most of which encode products involved in DNA replication, are transcribed and translated. Then, DNA replication begins, and the late genes are transcribed and translated to form the head and tail of the phage. The DNA is packaged into the heads, the tails are attached, and the cells lyse, releasing the phage to infect other cells. The phage DNA is shown in blue. doi:10.1128/9781555817169.ch7.f7.2

because each step triggers the next step and stops the preceding step. By having such a cascade of gene expression, all the information for the step-by-step development of the phage can be preprogrammed into the DNA of the phage.

Regulatory genes can usually be easily identified by mutations. Mutations in most genes affect only the product of the mutated gene. However, mutations in regulatory genes can affect the expression of many other genes. This fact has been used to identify the regulatory genes of many phages. Below, we discuss some of these genes and their functions, selecting our examples either because they are the basis for cloning and other technologies or because of the impact they have had on our understanding of regulatory mechanisms in general.

Figure 7.3 Transcriptional regulation by a regulatory cascade during development of a typical large DNA phage. The blue arrows indicate activation of gene expression; the black bars indicate repression of gene expression. doi:10.1128/9781555817169.ch7.f7.3

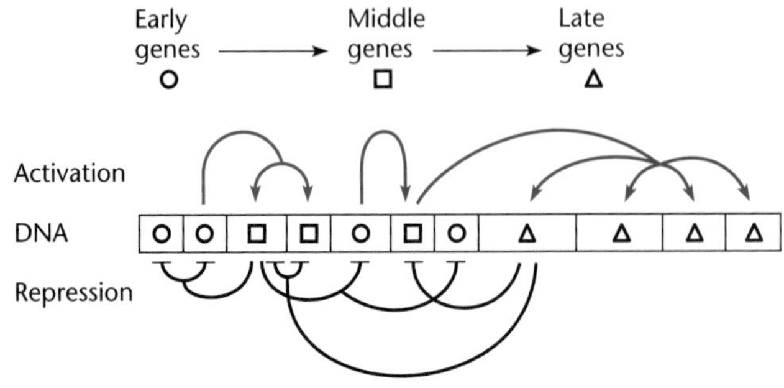

Phages That Encode Their Own RNA Polymerases

Some phages control their developmental cycle by encoding their own RNA polymerases. These RNA polymerases use only promoters on the phage DNA, which allows them to transcribe their own DNA in the presence of much larger amounts of host DNA, although at what stages in their development they use their own phage-encoded RNA polymerases depends on the type of phage.

T7: A NEW RNA POLYMERASE FOR THE LATE GENES

Compared with some of the larger phages, the typical short-tailed *Podoviridae* phage T7 has a relatively simple program of gene expression after infection, with only two major classes of genes, the early and late genes. The phage has about 50 genes, many of which are shown on the T7 genome map in Figure 7.4. After infection, expression of the T7 genes proceeds from left to right, with the genes on the extreme left of the genetic map, up to and including gene *1.3*, expressed first. These are the early genes. The genes to the right of gene *1.3*—the middle

DNA metabolism and late phage assembly genes—are transcribed after a few minutes delay.

Nonsense and temperature-sensitive mutations were used to identify which of the early-gene products are responsible for turning on the late genes. Under nonpermissive conditions, in which the mutated genes were inoperable, amber and temperature-sensitive mutations in gene *1* prevented transcription of the late genes, so gene *1* was a candidate for a regulatory gene. Later work showed that the product of gene *1* is an RNA polymerase that recognizes the promoters used to transcribe the late genes. The sequences of these promoters differ greatly from those recognized by bacterial RNA polymerases, so these phage late promoters are recognized only by the T7-specific RNA polymerase. In fact, transcription of these late genes by the gene *1* product helps pull the DNA containing the later genes into the cell, causing sequential gene expression based on the order of genes in the genome. The specificity of phage T7 RNA polymerase and other polymerases of phages in this family for their own promoters has been exploited in many applications in molecular genetics, including making riboprobes and high-level expression vectors, and some of these applications are discussed below and in other chapters.

T7 Phage-Based Expression Vectors

Some of the most widely used expression vectors use the T7 RNA polymerase to express foreign genes in *Escherichia coli*. The pET vectors (for plasmid expression T7) are a family of plasmid expression vectors that use the T7 phage RNA polymerase and T7 gene *10* promoter to express foreign genes in *E. coli*. The promoter that is used to transcribe the head protein gene (gene *10*) of T7 is a very strong promoter, since hundreds of thousands of copies of the T7 head protein must be synthesized within a few minutes after infection, making this one of the strongest known promoters. Downstream of the T7 promoter in the pET vector are a number of restriction sites into which foreign genes can be cloned. Any foreign gene cloned downstream of the T7 promoter is transcribed at very high rates by the T7 RNA polymerase. A number of variations of pET vectors have been designed. Some of the pET vectors have strong translational initiation regions (see chapter 2) immediately upstream of sequences for making translational fusions to affinity tags, such as a His tag, which makes the protein easy to detect and purify on nickel columns (see chapter 2 for a discussion of translational fusions and affinity tags). The T7 promoter can also be made inducible by providing the T7 RNA polymerase only when the foreign gene is to be expressed. This is important in cases where the fusion protein is toxic to the cell.

One commonly used strategy for making the T7 RNA polymerase inducible and thereby allowing the synthesis of a toxic gene product from a pET vector is illustrated in Figure 7.5. To provide a source of inducible phage T7 RNA polymerase, *E. coli* strains have been constructed in which phage gene *1* for RNA polymerase is cloned downstream of the inducible *lac* promoter and integrated into their chromosomes. In these strains, which often have the (DE3) suffix, as in *E. coli* JM109(DE3), the phage polymerase gene is transcribed only from the *lac* promoter, so that the T7 RNA polymerase is synthesized only if an inducer of the *lac* promoter, such as isopropyl-β-D-thiogalactopyranoside (IPTG), is added. The cells can be grown without inducer so that less of the gene product will be synthesized. When IPTG is added, newly synthesized T7 RNA polymerase then makes large amounts of mRNA on the foreign gene cloned into the pET plasmid and large amounts of its protein product.

Making Riboprobes and RNA-Processing Substrates

Another application of phage T7 and RNA polymerases from related phages is in making specific RNAs in vitro. RNAs made from a single gene are useful as probes for hybridization experiments (**riboprobes**) or as RNA substrates for processing reactions, such as splicing. This technology is also based on the fact that the phage RNA polymerases transcribe RNA only from their own promoters. For example, pBAC vectors, which were used in the Human Genome Project and other large genome

Figure 7.4 Genetic map of phage T7. The genes for the RNA polymerase used for expression vectors and the major capsid protein used for phage display are indicated. The gene for the major head protein used for phage display is highlighted in blue.
doi:10.1128/9781555817169.ch7.f7.4

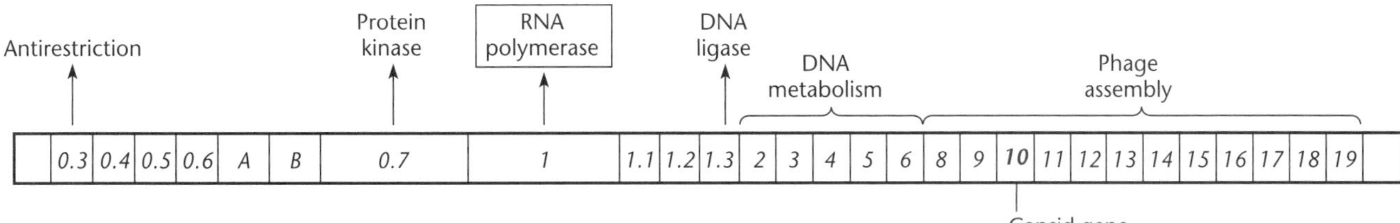

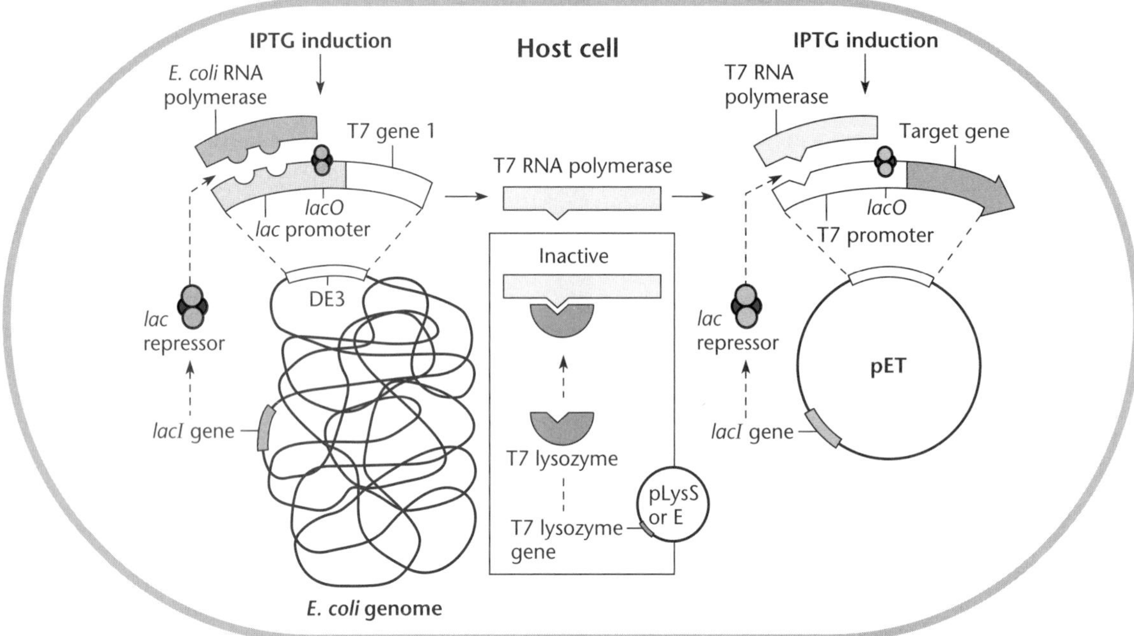

Figure 7.5 Strategy for regulating the expression of genes cloned into a pET vector. The gene for T7 RNA polymerase (gene *1*) is inserted into the chromosome of *E. coli* and transcribed from the *lac* promoter; therefore, it is expressed only if the inducer IPTG is added. The T7 RNA polymerase then transcribes the gene cloned into the pET vector downstream of the T7 late promoter on the pET cloning vector. If the protein product of the cloned gene is toxic, it may be necessary to further reduce the transcription of the cloned gene before induction. The T7 lysozyme encoded by a compatible plasmid, pLysS, binds to any residual T7 RNA polymerase made in the absence of induction and inactivates it. Also, the presence of *lac* operators between the T7 promoter and the cloned gene further reduces transcription of the cloned gene in the absence of IPTG. doi:10.1128/9781555817169.ch7.f7.5

projects, use phage promoters to make RNAs as hybridization probes to identify clones of neighboring sequences (see Figure 4.18). In this and other applications, the vectors are designed with multiple restriction sites bracketed by promoters for phage RNA polymerases. The gene on which RNA is to be made is cloned into one of the restriction sites on the vector, and the vector DNA is purified and cut with a restriction endonuclease on the other side of the cloned gene from the phage promoter. When purified phage RNA polymerase is added along with the other ingredients, including the ribonucleoside triphosphates needed for RNA synthesis, the only RNA that is made is complementary to the transcribed strand of the cloned gene. To make RNA complementary to the other strand, the gene can be cloned in the opposite orientation in the cloning vector or a special vector can be used that has the promoter for another phage on the other side of the cloning site. For example, in one such vector, the cloning site is bracketed by a T7 promoter on one side and an Sp6 promoter on the other. One strand of the cloned gene DNA is transcribed into RNA if purified T7 RNA polymerase is added, but the other strand is transcribed into RNA if Sp6 RNA polymerase is added. Purified RNA polymerases of T7 and some other phages are available from biochemical supply companies.

N4 PHAGE; TWO RNA POLYMERASES, ONE PACKAGED IN THE HEAD

The *E. coli* phage N4 also encodes its own RNA polymerase; in fact, it encodes two of them. One of these RNA polymerases is packaged in the phage head, as with many animal viruses (see Choi et al., Suggested Reading). Four copies of this polymerase are in each phage N4 head, and they are injected along with the DNA of the phage. They may help draw the DNA into the cell while they transcribe the earliest genes of the phage. The capsid N4 RNA polymerase has a number of interesting properties, including the ability to initiate transcription from hairpin-shaped promoters in the DNA. These hairpins can form in supercoiled N4 DNA with the help of the *E. coli* SSB (single-stranded-DNA-binding protein [see chapter 1]). One of the products made by genes transcribed by this RNA polymerase is a second RNA polymerase that specifically transcribes the middle genes. This RNA polymerase is unusual in that it can initiate transcription from single-stranded templates, but only with the help of the phage's own SSB protein. Only late in infection is the *E. coli* RNA polymerase used to transcribe the late genes. Therefore, N4 uses its own RNA polymerases and the host RNA polymerase in the reverse order in which they are used by T7 and most other *Podoviridae*. Note that if a phage-encoded RNA polymerase is required to transcribe the earliest genes, it must be encapsidated and injected with the phage DNA; otherwise, it could not be made.

Phage T4: Transcriptional Activators, a New Sigma Factor, and Replication-Coupled Transcription

Bacteriophage T4 is one of the largest known viruses, with a complex structure, making it a popular cover for biology textbooks (Figure 7.1A). Experiments with this phage have been so important in the development of molecular genetics that the phage deserves equal status with Mendel's peas in the history of genetics (see the introduction). The function of ribosomes, the existence of mRNA, the nature of the genetic code, the confirmation of codon assignments, and many other basic insights originally came from studies with this phage.

Phage T4 and other members of the family *Myoviridae* are much larger than phages related to T7 and other *Podoviridae*, with over 200 genes (the T4 genome map is shown in Figure 7.6). Many of these genes substitute for host genes, allowing the phage to multiply in many different hosts and under many different conditions. For example, it has genes to enhance its rate of DNA replication by encoding its own enzymes to make the precursors for DNA replication. It has enzymes to make the unusual

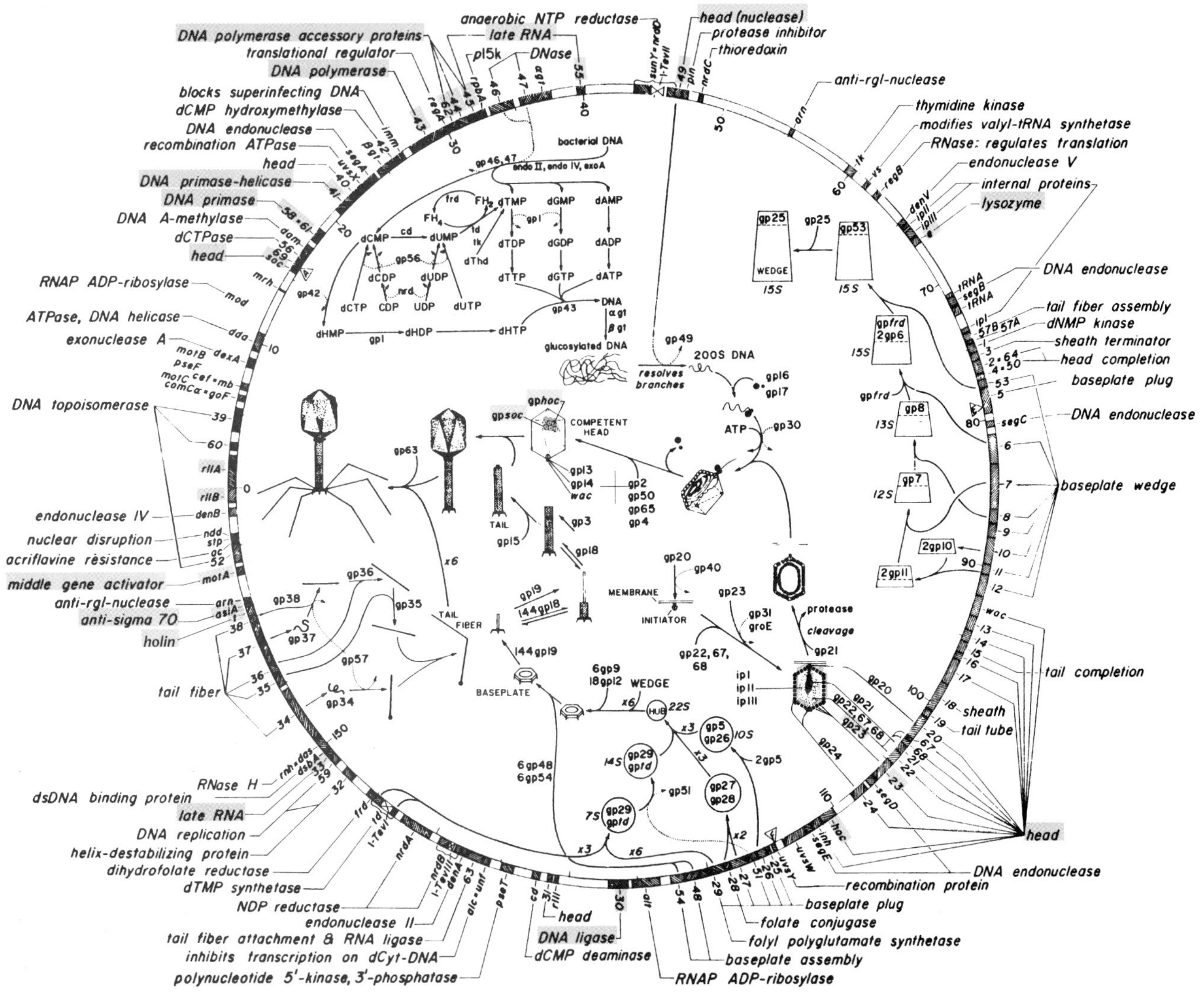

Figure 7.6 Genomic map of phage T4. Some genes whose products are discussed in the text are highlighted in blue. doi:10.1128/9781555817169.ch7.f7.6

nucleoside hydroxymethylcytosine and to attach glucose to it, which protects its DNA against restriction endonucleases. It has enzymes to enhance its multiplication when its host, *E. coli*, is under anaerobic conditions (without oxygen) in its normal environment, the vertebrate intestine. It encodes some of its own tRNAs to enhance its rate of translation, since its codon usage is somewhat different from that of its host, as its DNA is very AT rich. It encodes its own enzymes for DNA replication, which has made it a rich source of useful enzymes for DNA technology, including the T4 DNA ligase and polynucleotide kinase, which are extensively used in biotechnology. Because T4 has so many genes, the regulation of its gene expression is predictably complex. In fact, T4 uses many of the known mechanisms of regulation of gene expression at some stage of its life cycle, including making new sigma factors, antitermination, translational regulation, and replication-coupled transcription. The only thing it does not do is make its own RNA polymerase.

Figure 7.7 shows the time course of T4 protein synthesis after infection. Each band in the figure is the polypeptide product of a single phage T4 gene, and some of the gene products are identified by the gene that encodes them (details on how the bands were obtained are given in the legend to Figure 7.7). For example, gp37 is the *p*roduct of gene *37* (Figure 7.6). Because of the way the polypeptides were labeled, the time at which a band first appears is the time at which that gene begins to be expressed, and the time a band disappears is the time that gene is shut off. Clearly, some genes of T4 are expressed immediately after infection. They are called the immediate-early genes. Other genes, called the delayed-early and middle genes, are expressed only a few minutes after infection. This is followed by expression of the true-late genes, so called to distinguish them from some of the delayed-early and middle genes that continue to be expressed throughout infection. The assignments of the polypeptide products to genes were made by using amber mutations in the genes and observing which band was missing after infection of a nonsuppressor host. In cases where two or more bands were missing, the gene might encode a regulatory protein or the product of the gene might be processed after infection, such as the cleavage of the head protein gp23 into the smaller form gp23* during maturation of the head. The locations where these late-gene products fit into the T4 particle are shown in Figure 7.8.

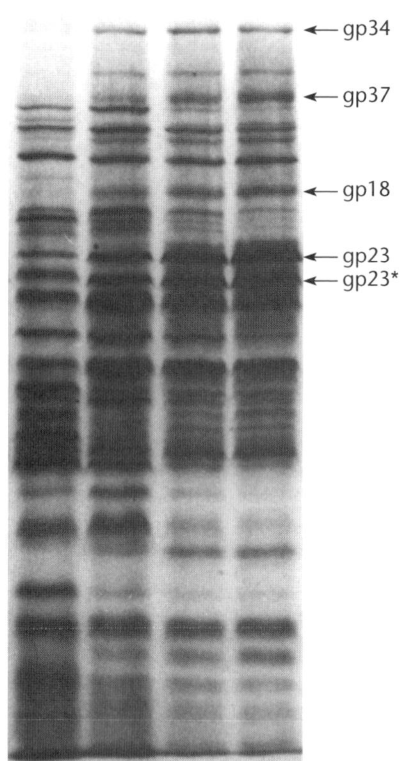

Figure 7.7 Polyacrylamide gel electrophoresis of proteins synthesized during the development of phage T4. The proteins were labeled by adding radioactive amino acids at various times after the phage were added to the bacteria. Because phage T4 stops host protein synthesis after infection, radioactive amino acids are incorporated only into phage proteins, and therefore only the phage proteins become radioactive. Moreover, only the phage proteins being made when the radioactive amino acids were added are labeled. Hence, we can tell which phage proteins are made at any given time. The proteins were denatured with sodium dodecyl sulfate (SDS) and run on an SDS-polyacrylamide gel electrophoresis (PAGE) gel. The gel was then exposed to autoradiography. Shown are proteins synthesized between 5 and 10 min (lane 1 from the left), 10 and 15 min (lane 2), 15 and 20 min (lane 3), and 30 and 35 min (lane 4) after infection.
doi:10.1128/9781555817169.ch7.f7.7

REGULATION OF GENE EXPRESSION DURING T4 DEVELOPMENT

Experiments like those described in the legend to Figure 7.7 were used to identify the regulatory genes of a phage. Mutations in these genes prevent the expression of many other genes and thus cause the disappearance of many bands from the gel. In T4, mutations in genes named *motA* and *asiA* prevent the appearance of the middle-gene products. Mutations in genes *33* and *55*, as well as mutations in many of the genes whose products are required for T4 DNA replication, prevent the appearance of the true-late-gene products. Therefore, *motA*, *asiA*, *33*, and *55*, as well as some genes whose products are required for DNA replication, are predicted to be regulatory genes. Many genetic and biochemical experiments have been directed toward understanding how these T4 regulatory gene products turn on the synthesis of other proteins.

T4 Early Transcription

Even the early genes of this complex phage are subject to regulation, causing them to be expressed sequentially after infection.

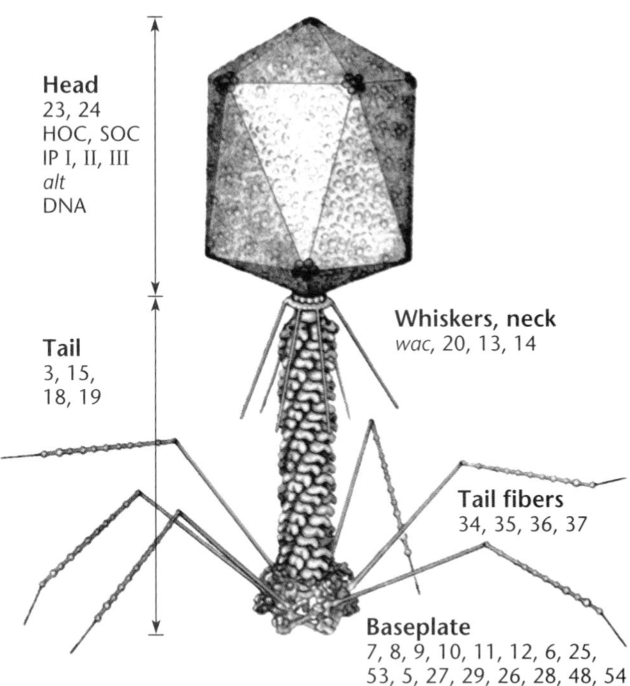

Figure 7.8 Structural components of the T4 particle. The features of the particle have been resolved to about 3 nm. The positions of the baseplate and tail fiber proteins are indicated. The HOC and SOC proteins used for phage display coat the head after it has assembled (see the text).
doi:10.1128/9781555817169.ch7.f7.8

Immediate-early genes. Some of the genes of T4 are turned on immediately after infection. They are called the immediate-early genes and are transcribed from σ^{70} promoters (see chapter 2) by the unmodified host RNA polymerase immediately after the DNA enters the cell. Some of the immediate-early-gene products then allow the synthesis of other early genes, after a few minutes delay.

Delayed-early genes. The delayed-early genes of T4 are transcribed from the same *E. coli*-type σ^{70} promoters as the immediate-early genes but are regulated by an antitermination mechanism. Without some T4 regulatory gene products, the RNA polymerase, which has initiated at an immediate-early promoter, would stop before it gets to the delayed-early genes. Therefore, the transcription of these genes must await the synthesis of antitermination factors encoded by immediate-early genes. Regulation by antitermination of transcription in the case of phage λ, where the mechanism is better understood, is discussed in chapter 8.

Middle genes. Regulation of the middle-gene products is still being actively studied because of what it reveals about the structure of RNA polymerase and its contact with DNA sequences at promoters. The middle genes are transcribed from their own promoters: the so-called "middle-mode" promoters, which look somewhat different from normal σ^{70} promoters in that their −35 sequence is replaced by a sequence centered at −30 called a Mot box (Figure 7.9). Because they are somewhat different, transcription from these promoters requires the phage-encoded MotA and AsiA proteins, themselves the products of delayed-early genes. Rather than functioning as σ factors, these two proteins act as coactivators that remodel the σ^{70} on RNA polymerase so that the RNA polymerase now recognizes promoters with the Mot box centered at −30 rather than with the consensus σ^{70} −35 sequence. The MotA protein is a sequence-specific DNA-binding protein that, in conjunction with region 4 of σ^{70}, recognizes the Mot box. AsiA, by binding to σ^{70} on RNA polymerase, releases region 4 of σ^{70} from the β flap of RNA polymerase, allowing region 4 to be repositioned to help MotA bind to the Mot box rather than to the −35 sequence of σ^{70} promoters. Additional stabilization is provided by a direct protein-protein contact between MotA and AsiA (see Yuan and Hochschild, Suggested Reading).

Because it repositions region 4, binding of AsiA inhibits the ability of σ^{70} RNA polymerase to recognize or initiate transcription from standard *E. coli* promoters and T4 early promoters, leading AsiA to be called an antisigma factor, the first antisigma protein to be discovered. However, AsiA is unusual in that it does not bind to σ^{70} directly and inactivate it, like other antisigma factors. We discuss the role of antisigma factors in bacterial gene regulation in subsequent chapters.

True-Late Transcription

The last genes of T4 to be transcribed are the true-late genes, named to distinguish them from those early genes that continue to be transcribed late into infection. The principal products of these genes are mostly the head, tail, and tail fiber components of the phage particle itself. The initiation of transcription of the true-late genes of T4 is of particular interest because it is coupled to the replication of the phage DNA. It also led to the discovery of the first alternative sigma factor.

Figure 7.9 Sequence of T4 middle-mode and late promoters. Only the sequences important for recognition by RNA polymerase are shown.
doi:10.1128/9781555817169.ch7.f7.9

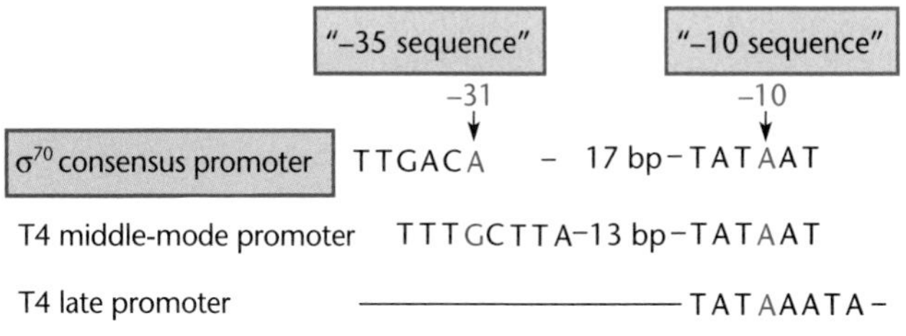

	"−35 sequence" (−31)		"−10 sequence" (−10)
σ^{70} consensus promoter	TTGACA	− 17 bp −	TATAAT
T4 middle-mode promoter	TTTGCTTA	−13 bp −	TATAAT
T4 late promoter		————	TATAAATA−

T4 alternative sigma factor. Like the middle genes, the true-late genes of T4 are transcribed from promoters that are quite different from those of the host σ^{70} promoters (Figure 7.9) (see chapter 2). They have the longer −10 sequence TATAAATA rather than the shorter −10 sequence of a typical bacterial σ^{70} promoter, and they lack a −35 sequence or any other defined sequence upstream of the −10 sequence. Because of this difference, the host RNA polymerase containing σ^{70} does not normally recognize the T4 true-late promoters. However, rather than being remodeled, the host σ^{70}, is displaced from the RNA polymerase altogether, and two new T4-encoded proteins, gp55 and gp33, bind in its place., This allows the RNA polymerase to recognize only the promoters for the T4 true-late genes. In a sense, these two proteins constitute a "split" sigma factor, although they have little sequence homology to other sigma factors. The gp55 portion has an identifiable region 2 that recognizes the altered −10 sequence of the T4 true-late promoters, and gp33 contributes some of the functions of region 4, in that it binds to the RNA polymerase β flap, where region 4 of σ^{70} normally binds, although gp33 does not recognize any specific upstream DNA sequences (see Geiduschek and Kassavetis, Suggested Reading). It is not clear why the coding sequences for various functions of this apparent sigma are split (map in Fig. 7.6), but it might reflect another role for gp55 in packaging phage DNA into heads (see Malys et al., Box 7.4 references).

Phage T4 and its close relatives are not the only phages to use alternative sigma factors to activate the transcription of their late genes. For example, the *B. subtilis* phage SPO1, another large member of the *Myoviridae* (Table 7.1), also uses this developmental mechanism. The SPO1 late promoters are more like the normal bacterial promoters, with recognizable −35 and −10 sequences, but are recognized only by *B. subtilis* RNA polymerase with the phage-encoded sigma factor attached.

The general strategy of changing which genes are transcribed by changing sigma factors on the RNA polymerase is also used in uninfected bacteria. Bacterial RNA polymerases can be adapted to recognize a wide variety of promoter sequences merely through the attachment of an alternative sigma factor, and sigma factors direct the transcription of specific genes following stresses, such as heat shock, and during developmental processes, such as sporulation, as discussed in chapters 13 and 14.

Replication-coupled transcription. The gp33 and gp55 proteins recognize the true-late promoters, but they cannot efficiently form an open complex and activate true-late transcription by themselves (in fact, gp33 inhibits it). They can only open the promoter if they bind to gp45, which contacts both gp33 and gp55. The gp45 protein is the T4 DNA polymerase accessory protein that acts like the β clamp in *E. coli* DNA replication and wraps around the DNA to form a "sliding clamp" (Table 7.2) (see

Table 7.2 T4 gene products involved in replication and their homologs in *E. coli* and eukaryotes

T4 gene product	*E. coli* function	Eukaryotic function[a]
Origin-specific replication		
gp43	PolIII α and ε	DNA Pol α, β, γ
gp45	β sliding clamp	PCNA
gp44, gp62	Clamp loader (γ complex)	RFC
gp41	Replicative helicase (DnaB)	Mcm complex
gp61	Primase (DnaG)	Pol α
gp39, gp52, gp60	GyrAB, topoisomerase IV	Topoisomerase II
gp30	DNA ligase	—[b]
Rnh	RNase H	—
RDR		
UvsW	RecG, RuvAB	—
UvsX	RecA	Rad51 (yeast)
UvsY	RecFOR	Rad52 (yeast)
gp46, gp47	RecBCD, SbcCD	Rad50, Mrell (yeast)
gp32	SSB	RPA
gp59	PriABC, DnaT, DnaC	—
gp49	RuvC	—

[a]PCNA, proliferating-cell nuclear antigen; RPA, replication protein A; RFC, replication factor C; Mrell, double-strand break processing.

[b]—, not identified.

"Phage T4: Another Phage That Forms Concatemers" below). It then moves with the DNA polymerase and helps prevent it from falling off the DNA during replication (Figure 7.10) (see chapter 1 for a more detailed description of replication). The gp44 and gp62 proteins are subunits of the clamp loader, analogous to the γ complex of *E. coli*, and load the gp45 clamp onto the DNA during replication. Once loaded on the DNA, this sliding clamp interacts with the gp43 DNA polymerase to keep it on the DNA, but it periodically cycles off the replication apparatus. It can then slide around on the DNA by itself and can also bind to gp33 and gp55 on RNA polymerase to activate true-late transcription. It is interesting that gp43, gp33, and gp55 all share the same short gp45-interacting sequence in their C termini through which they can all bind to the same regions on one side of the sliding clamp. In fact, that region of gp43 can be substituted for the corresponding regions in gp33 and gp55, and the sliding clamp can still activate transcription through them.

From the simple depiction in Figure 7.10, it is not obvious how the gp45 ring could contact the C termini of both gp55 and gp33 on the RNA polymerase, but like other such sliding clamps, the gp45 clamp has long protuberances that can extend into the RNA polymerase, as well as other proteins. It is also possible that gp55 extends further toward the back end of the RNA polymerase than is shown in the figure. This ability of sliding clamps to switch their binding from one protein to another has occasioned their comparison to "sliding tool belts" binding different "tools," or proteins, as needed. Recall from chapter 1 how such switching helps delay reinitiation of DNA replication and allows switching from replicating DNA polymerase to translesion DNA polymerases at damage in the DNA. The requirement for T4 late transcription for continuous T4 DNA replication can also be explained. The gp45 clamps periodically fall off the DNA and are depleted on the DNA unless they are continuously being reloaded at replication forks. The clamp loader can also load the gp45 clamp onto DNA at nicks, partially obviating the need for other replication proteins if the DNA is heavily nicked (Fig. 7.10B).

Coupling gene expression to the replication of the DNA would have obvious advantages in other systems, such as in coordinating gene expression and cell division with replication during the normal cell cycle. Other such systems are known, but so far, none of them are known to use the same mechanism as the DNA replication-transcription coupling of the true-late transcription of T4 and its relatives. Nevertheless, understanding the mechanism it uses has contributed to our overall understanding of the activation of transcription in general and has broadened our knowledge of the possible ways in which the replication apparatus can interact with other cellular functions.

Phage DNA Genome Replication and Packaging

Unlike the replication of chromosomal or plasmid DNAs, which must be coordinated with cell division, phage DNA replication is governed by only one purpose: to make the greatest number of copies of the phage genome in the shortest possible time. Phage DNA replication can be truly impressive. A single phage DNA molecule initially entering the cell can replicate to make hundreds or even thousands of copies to be packaged into phage heads in as little as 20 or 30 minutes (min). This unchecked replication often makes phage DNA replication easier to study than replication in other systems. Nevertheless, phage replication shares many of the features of cellular replication found in all types of living organisms, and phage DNA replication has served as a model system to understand DNA replication in bacteria and even in humans.

Phages with Single-Stranded Circular DNA

The genomes of some small phages consist of circular single-stranded DNA. The small *E. coli* phages that fit into this category can be separated into two groups. The representative phage of one group, ϕX174, has a spherical capsid with spikes sticking out, resembling the ball portion of the medieval weapon known as a "morning star." These phages are called icosahedral phages because their capsids are icosahedrons, like a geodesic dome with mostly six-sided building blocks and an occasional five-sided block. In the other group, represented by M13 and f1 and often referred to as **filamentous phages**, the phages have a single layer of protein covering the extended DNA molecule, making the phages filamentous in appearance.

The different shapes of phages of these two families determine how they enter and infect cells and then leave the infected cells. The icosahedral phages enter and leave cells much like other phages. They bind to the cell surface, and only the DNA enters the cell; they then lyse, or break open, the cell to exit after they have developed (see "Phage Lysis" below). In contrast, M13 and other filamentous phages are called male-specific phages because they specifically adsorb to the sex pilus encoded by certain plasmids and so infect only "male," or donor, strains of bacteria (see chapter 5). Unlike most other phages, filamentous phages do not inject their DNA. Instead, the entire phage is ingested by the cell, and the protein coat is removed from the DNA as the phage passes through the inner cytoplasmic membrane of the bacterium. After the phage DNA has replicated, it is again coated with protein as it leaks back out through the cytoplasmic membrane. Since these phages do not lyse infected cells and leak out only slowly, the cells they

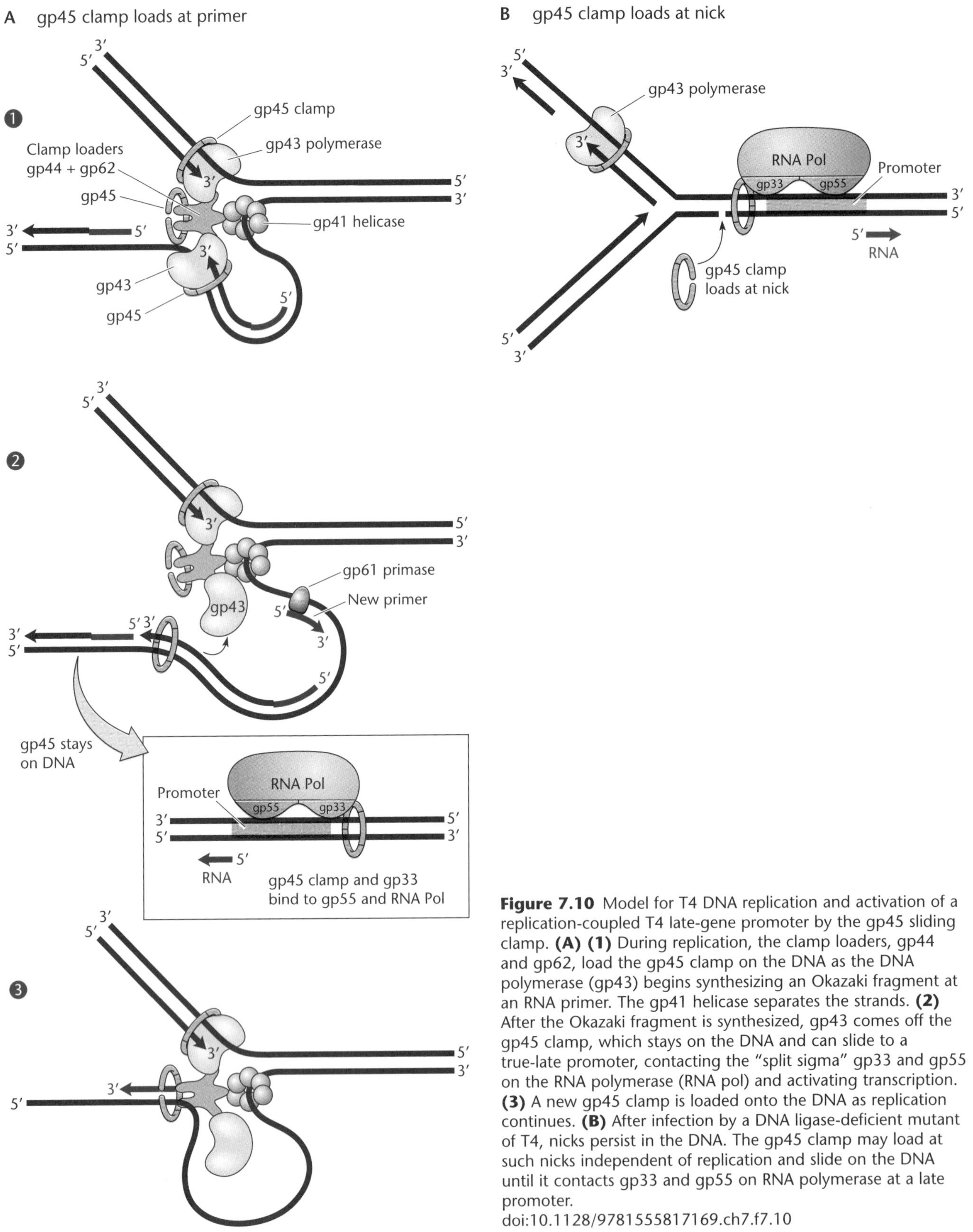

A gp45 clamp loads at primer

1

Clamp loaders gp44 + gp62

gp45 clamp

gp43 polymerase

gp45

gp41 helicase

gp43

gp45

2

gp61 primase

New primer

gp43

gp45 stays on DNA

Promoter

RNA Pol

gp55

gp33

RNA

gp45 clamp and gp33 bind to gp55 and RNA Pol

3

B gp45 clamp loads at nick

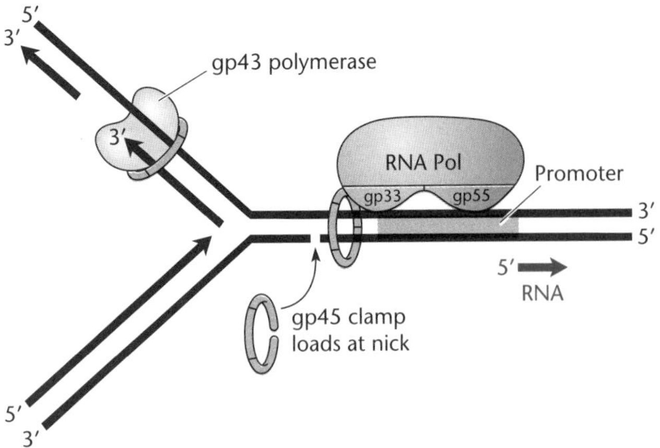

Figure 7.10 Model for T4 DNA replication and activation of a replication-coupled T4 late-gene promoter by the gp45 sliding clamp. **(A) (1)** During replication, the clamp loaders, gp44 and gp62, load the gp45 clamp on the DNA as the DNA polymerase (gp43) begins synthesizing an Okazaki fragment at an RNA primer. The gp41 helicase separates the strands. **(2)** After the Okazaki fragment is synthesized, gp43 comes off the gp45 clamp, which stays on the DNA and can slide to a true-late promoter, contacting the "split sigma" gp33 and gp55 on the RNA polymerase (RNA pol) and activating transcription. **(3)** A new gp45 clamp is loaded onto the DNA as replication continues. **(B)** After infection by a DNA ligase-deficient mutant of T4, nicks persist in the DNA. The gp45 clamp may load at such nicks independent of replication and slide on the DNA until it contacts gp33 and gp55 on RNA polymerase at a late promoter.
doi:10.1128/9781555817169.ch7.f7.10

infect are "chronically" infected rather than "acutely" infected. Nevertheless, the filamentous phages form visible plaques, because the infected cells grow more slowly than the uninfected lawn. Because filamentous phages continue to play a major role in biotechnology, we spend some time on the details of their multiplication.

The process of infection of the cell by a filamentous single-stranded DNA phage and its release from the cell has been studied in some detail because it serves as a model system for the ability of a large particle like a virus to get through a membrane (see Rakonjac et al., Suggested Reading). Most of these studies have been performed with phage f1, but related phages probably use similar mechanisms.

The phage f1 particle has only five proteins, one of which, the major head protein, pVIII, exists in about 2,700 copies and coats the DNA. The other four proteins are on the ends of the phage and exist in only four or five copies per phage particle. Two of these proteins, pVII and pIX, are located on one end of the phage, and the other two, pVI and pIII, are on the other end. To start the infection, the pIII protein on one end of the phage first makes contact with the end of the sex pilus. The sex pilus makes a good first contact point because it sticks out of the cell and hence is very accessible. The pilus retracts when the phage binds to it, drawing the phage to the cell surface. A different region of the same pIII protein then contacts a host inner membrane protein called TolA. How this contact is made is somewhat unclear. The TolA protein sticks into the periplasmic space and might make contact with the outer membrane, since it is part of a larger structure whose role has something to do with keeping the outer membrane intact and coordinating the division of the inner and outer membranes during cell division (see chapter 14). The phage DNA then enters the cytoplasm, while the major coat protein is stripped off into the host inner membrane.

Release of the phage from the cell uses a different process. Unlike the spherical single-stranded DNA phages that make a single protein that lyses the cell (see "Phage Lysis" below), the filamentous phages leak out of the cell. This process is illustrated in Figure 7.11. Unlike injection of the phage DNA, which must rely exclusively on host proteins, secretion of the phages from infected cells can use phage proteins newly synthesized during the course of the infection. After the phage DNA has replicated a few times to produce the replicative-form DNA (see below), it enters the rolling-circle stage of replication. As it rolls off the circle, the newly synthesized single-stranded DNA is coated by another protein, pV. The proteins that make up the phage coat have been synthesized and are waiting in the membrane, and the major head protein, pVIII, replaces the pV protein on the DNA as it enters the membrane. Only DNA containing the ***pac* site** of the phage is packaged. The other phage proteins are then added to the particle. Meanwhile, the phage-encoded

Figure 7.11 Infection cycle of the single-stranded DNA phage f1. Steps 1 through 7 show the encapsidation of phage DNA as it is secreted through the membrane pore formed by the pIV secretin to release the phage and infect a new cell. Details are given in the text. ssDNA, single-stranded DNA. doi:10.1128/9781555817169.ch7.f7.11

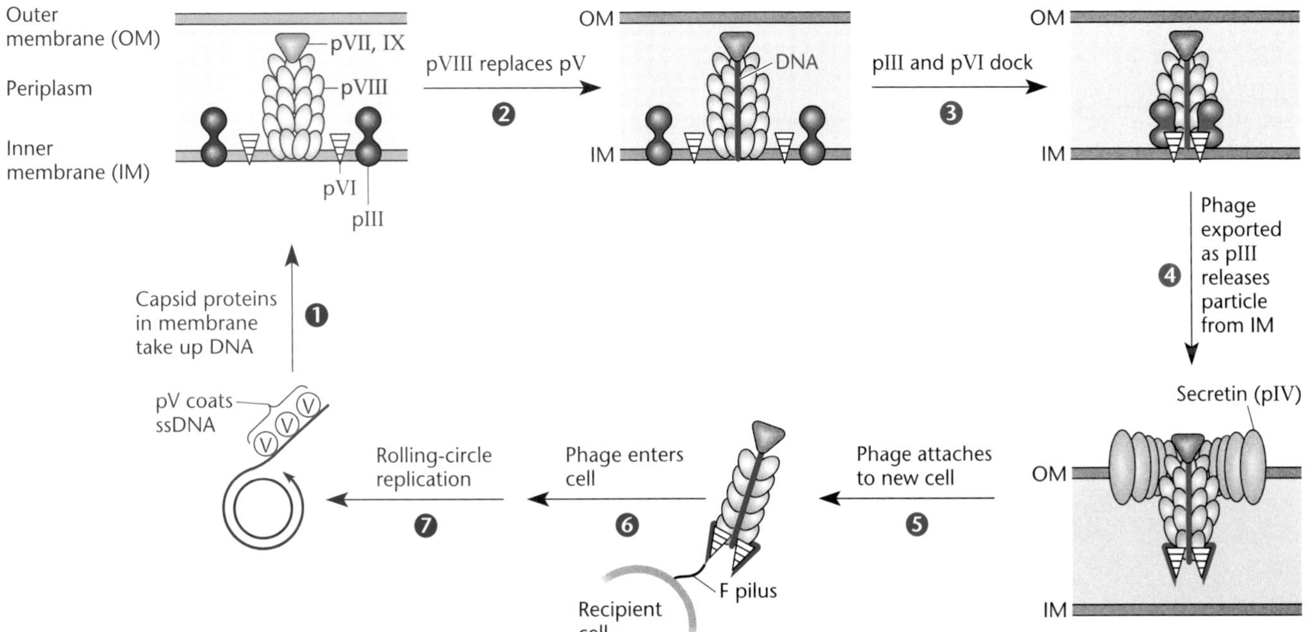

secretin protein, pIV, has formed a channel in the outer membrane through which the assembled phage can pass. This channel is related to the channels formed by type II secretion systems to assemble pili on the cell surface and by competence systems to allow DNA into the cell (see chapters 6 and 14).

REPLICATION OF SINGLE-STRANDED PHAGE DNA

Studies of the replication of single-stranded phage DNA have contributed much to our understanding of replication in general. It was with these phages that rolling-circle replication was discovered, as well as many proteins required for host DNA replication, including the proteins PriA, PriB, PriC, and DnaT, which are now known to be involved in restarting chromosomal replication forks after they have dissociated upon encountering DNA damage (see chapter 10). Many of the genes for these host proteins were found in searches for host mutants that cannot support the development of the phages and by reconstituting replication systems in vitro by adding host DNA replication proteins to phage DNA until replication was achieved.

The groups working on the secretion of single-stranded DNA phages from the cell have been different from those working on phage DNA replication, and the groups have used different phages. Much of the work on phage DNA replication has been done with M13 and ϕX174, and the genes of these phages are named somewhat differently from those of phage f1.

First, we talk about the replication of the DNA of the filamentous phage M13, and then we compare it to the replication of the icosahedral phage ϕX174, which turns out to be surprisingly different.

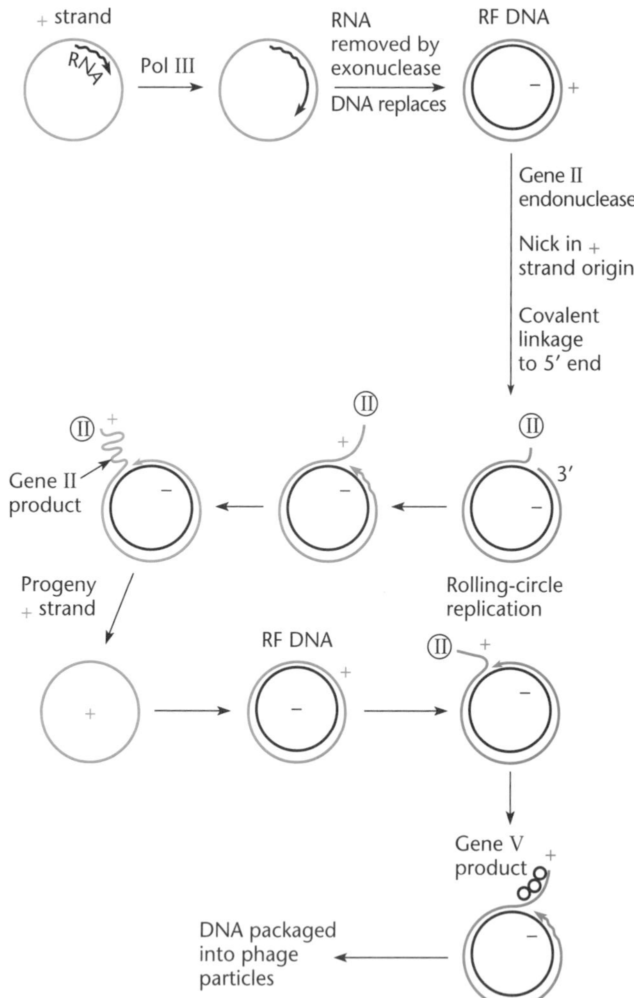

Figure 7.12 Replication of the circular single-stranded DNA phage M13. First, an RNA primer is used to synthesize the complementary minus strand (in black) to form double-stranded RF DNA. The product of gene II, an endonuclease, nicks the plus strand of the RF and remains attached to the 5′ phosphate at the nick. Then, more plus strands are synthesized via rolling-circle replication, and their minus strands are synthesized to make more RFs. Later, the gene V product binds to the plus strands as they are synthesized, preventing them from being used as templates for more RF synthesis and helping package them into phage heads. PolIII, DNA polymerase III. doi:10.1128/9781555817169.ch7.f7.12

Synthesis of the Complementary Strand To Form the First RF

The steps in the replication of phage M13 DNA are outlined in Figure 7.12. The DNA strand of the phage encapsidated in the phage head is called the plus strand. Immediately after the single-stranded plus strand DNA enters the cell, a complementary minus strand is synthesized on the plus strand to form a double-stranded DNA called the **replicative form** (**RF**). The formation of this first RF is dependent entirely on host functions, as it must be, since no phage proteins enter the cytoplasm with the phage DNA and phage proteins cannot be synthesized from single-stranded DNA. The synthesis of the complementary strand is primed by an RNA made by the normal host RNA polymerase. Normally the host RNA polymerase recognizes only double-stranded DNA, but the single-stranded phage DNA forms a hairpin at the origin of replication, making it double-stranded in this region. Once the RNA primer is synthesized, the DNA polymerase III and accessory proteins load on the DNA and synthesize the complementary strand until they have come full circle on the DNA and encounter the RNA primer. The 5′ exonuclease activity of DNA polymerase I then removes the RNA primer, its DNA polymerase activity fills the gap with DNA, and the nick is sealed by DNA ligase to leave a double-stranded covalently closed RF, which can then be supercoiled (see chapter 1). The replication also occurs on a specific site on the membrane of the bacterium, sometimes called the reduction sequence.

This binding may direct the phage DNA to sites where the replication machinery of the host is located.

The icosahedral phages, such as ϕX174, use a much more complicated mechanism to initiate synthesis of the first RF. Rather than using just the host RNA polymerase, they assemble a large primosome at a unique site on the single-stranded phage DNA. This primosome is composed of many copies of seven different proteins, including DnaB (the replicative helicase), DnaC (which loads the helicase on the DNA), and DnaG (the primase). As described in chapter 1, most of these proteins are required to initiate replication at the chromosomal origin, *oriC*, and they have been enlisted by the phage for the same purpose. However, the primosome also includes other proteins, PriA, PriB, PriC, and DnaT, which are not required for initiation of chromosomal replication at *oriC* but, rather, are required to restart chromosomal replication after it has been blocked due to encountering damage in the DNA template (see chapter 10). The role of the PriA, PriB, PriC, and DnaT proteins in the initiation of replication of ϕX174 DNA is not yet clear. Some of these proteins are helicases and may be required to open up a hairpin at the unique origin of replication of the single-stranded DNA and to allow the replication apparatus to load on the DNA and the primosome to move on the DNA.

The discovery that the Pri proteins play a role in ϕX174 DNA synthesis is an interesting lesson in history with important general ramifications in science and medicine. At first, it was puzzling why these proteins would be required for the initiation of replication of phage ϕX174 DNA, since they were not known to be required for replication of the host DNA. At the *oriC* site, where chromosomal replication normally initiates, only DnaC is required to load the replicative helicase, DnaB, and initiate replication. Later, it was shown that the Pri proteins and the DnaT protein were required to restart chromosomal replication forks after they had collapsed on encountering damage to the DNA. If the replication fork encounters damage in the template DNA and collapses, the recombination functions can promote the formation of a recombinational intermediate, and the Pri proteins and DnaT are required to load DnaB helicase at such a structure and to reinitiate replication. This is analogous to the recombination-dependent replication (RDR) of phage T4 (see below). The ϕX174 phage enlists the host functions to initiate its own RF replication, presumably by having an origin of replication that mimics the recombinational intermediate. Presumably, eukaryotes, including humans, use a similar mechanism to restart replication forks, and this plays an important role in preventing DNA damage and therefore cancer. This is yet another case where detailed studies of a relatively simple phage system have led unexpectedly to the discovery of universal phenomena applicable to all organisms.

Synthesis of More RFs and Phage DNA

The subsequent steps in replication are probably similar in all single-stranded DNA phages but are best understood in M13. Once the first RF of M13 is synthesized, more RFs are made by semiconservative replication. This process requires phage proteins that are synthesized from the first RF. The two strands of the RF are replicated separately and by very different mechanisms. The plus strand is replicated by rolling-circle replication from a different origin by a process similar to the replication of DNA during transfer of a plasmid by conjugation (see chapter 5). First, a nick is made in the RF, at the origin of plus strand synthesis, by a specific endonuclease, the product of gene II in M13. A host protein called Rep, a helicase, helps unwind the DNA at the nick. The gene II protein remains attached through one of its tyrosines to the 5′ end of the DNA at the nick, and the DNA polymerase III, with its accessory proteins, extends the 3′ end to synthesize more plus strand, displacing the old plus strand. The gene II protein bound to the 5′ end of the old, displaced strand then reseals the ends of the displaced strand by a transesterification reaction in which the phosphate attached to its tyrosine is passed back to the free 3′ end of the old strand. This recyclizes the old plus strand, which can then serve again as the template for minus strand synthesis to create another RF. Such transesterification reactions use little energy and are also used to recyclize plasmids after conjugation (see chapter 5) and in site-specific recombinases and some transposases (see chapter 9).

This process of accumulating RFs continues until the product of phage gene V begins to accumulate. This protein coats the single-stranded plus strand of DNA, probably with the help of the attached gene II product, and prevents the synthesis of more RF by a complicated process that is only incompletely understood. The single-stranded viral DNA is then encapsulated in the head, and the cell is lysed (for icosahedral phages, such as ϕX174) or transferred to the assembling viral particle in the membrane and leaked out of the cell (for filamentous phages, such as f1 and M13), as described above.

M13 CLONING VECTORS

Because M13 and related phages encapsulate only one of the two DNA strands in their heads, these phages have provided a convenient vehicle for cloned DNA to use in other applications that involve single-stranded DNA. Also, because filamentous phages, such as M13, have no fixed length and the phage particle is as long as its DNA, foreign DNAs of different lengths can be cloned into the phage DNA, producing a molecule that is longer than normal without disrupting the functionality of the phage.

Figure 7.13 shows one such phage-derived cloning vector, the M13 cloning vector M13mp18 (see Yanisch-Perron et al., Suggested Reading). Like pUC plasmid vectors (see Figure 4.17), the mp series of M13 phage vectors contain the α-fragment-coding portion of the *E. coli lacZ* gene into which some convenient restriction sites (shown as the polylinker cloning site in Figure 7.13, with the multiple restriction sites shown at the bottom of the figure) have been introduced. Phages with a foreign DNA insert in one of these sites can be identified easily by insertional inactivation because they make colorless instead of blue plaques on plates containing X-Gal (5-bromo-4-chloro-3-indolyl-β-D-galactopyranoside).

To use a single-stranded DNA phage vector, the double-stranded RF must be isolated from infected cells, since most restriction endonucleases and DNA ligase require double-stranded DNA. A piece of foreign DNA is cloned into the RF by using restriction endonucleases, and the recombinant DNA is used to transfect competent bacterial cells. The term "transfection" refers to the artificial initiation of a viral infection by viral DNA (see chapter 6). When the RF containing the clone replicates to form single-stranded progeny DNA, a single strand of the cloned DNA is packaged into the phage particle containing five different phage proteins. The structure of the assembled M13 phage particle is shown in Figure 7.14. The phage plaques obtained when these phage are plated are a convenient source of one strand of the cloned DNA. Which strand of the cloned DNA is represented in the single-stranded DNA will depend on the orientation of the cloned DNA in the cloning vector. If it is cloned in one orientation, one of the strands is obtained; if it is cloned in the other orientation, the other strand is obtained.

Some plasmid cloning vectors have also been engineered to contain the *pac* site of a single-stranded DNA phage, which allows it to be packaged in a phage particle (see "Replication and DNA Packaging: Linear Genomes" below). They are called **phasmids** because they are mostly plasmids but have a phage *pac* site. Phasmids are easier than phage RFs to isolate and use for cloning, since they replicate as a double-stranded plasmid. If cells

Figure 7.13 Map of the M13mp18 cloning vector. The positions of the genes of M13 and the polylinker cloning site containing multiple restriction sites are shown below the map. Cloning into one of these sites inactivates the portion of the *lacZ* gene on the cloning vector, a process called insertional inactivation. The cloning vector also contains the *lacI* gene, whose product represses transcription from the p_{lacZ} promoter (see chapter 11).
doi:10.1128/9781555817169.ch7.f7.13

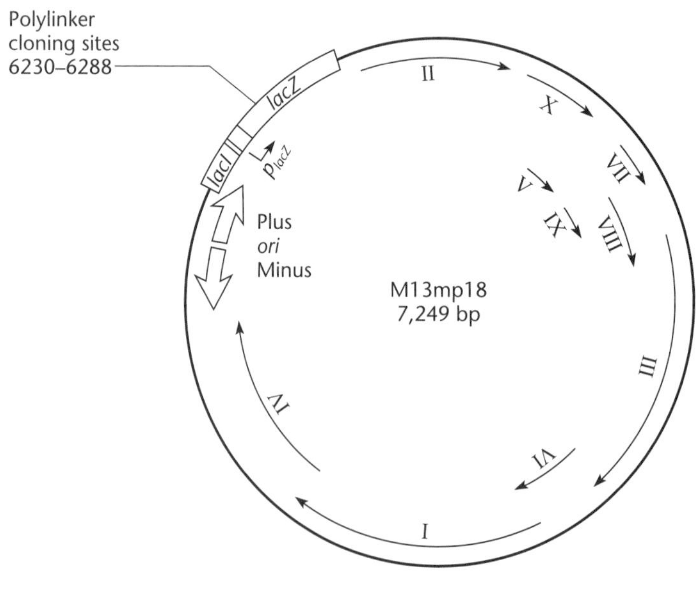

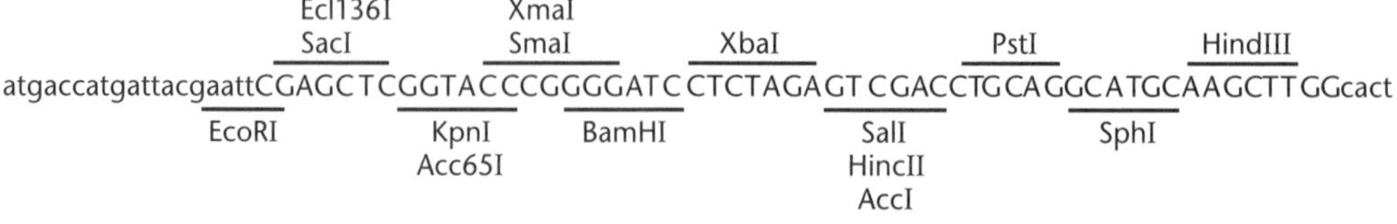

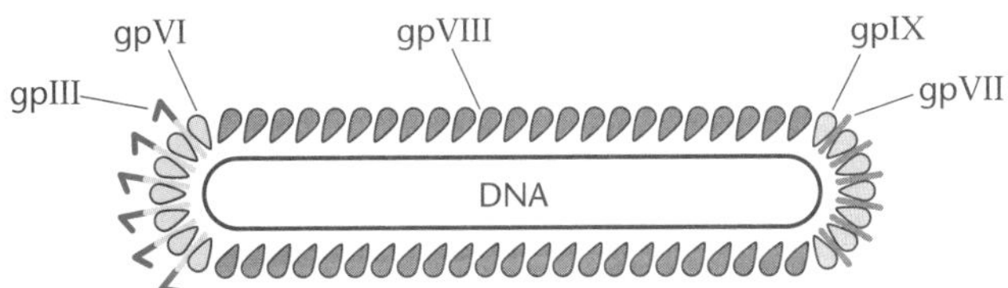

Figure 7.14 Schematic representation of the filamentous bacteriophage M13. The single-stranded circular DNA is coated with five viral proteins. The schematic locations of the different proteins are shown. The gpVIII protein is present at about 2,700 copies, while gpIII, gpVI, gpVII, and gpIX are present at about 5 copies each. All of the coat proteins can be used as platforms for protein display. With the exception of gpIII, the capsid proteins are small, with 33 to 112 amino acids. doi:10.1128/9781555817169.ch7.f7.14

containing such a phasmid are infected with the phage, called the **helper phage**, the phasmid is packaged into the helper phage head in a single-stranded form. Which strand is packaged depends on the orientation of the *pac* site in the phasmid. Another form of the cloning vectors, called **phagemids** because they are more phage and less plasmid, also have a plasmid replication origin but encode all of the proteins required to make a phage, so they do not need a helper phage to be packaged. Phasmids and phagemids have many uses in molecular genetics and biotechnology, and we discuss some of these in this and other chapters.

SITE-SPECIFIC MUTAGENESIS OF M13 CLONES

Because single-stranded DNA phages, such as M13, offer a convenient source of only one of the two strands of a cloned DNA, they were used in the first applications of site-specific mutagenesis and for DNA sequencing, which originally required a single-stranded template. However, such methods have now been supplanted by other methods, including various types of PCR mutagenesis and recombineering. We discuss these methods in chapter 1 and in subsequent chapters.

Replication and DNA Packaging: Linear Genomes

While many of the smaller phages have circular DNA, most of the larger phages have linear DNA, at least as it is packaged into the phage head. This creates special problems for replicating the DNA genome and packaging it into phage heads, as we discuss next.

THE PRIMER PROBLEM

Many of the strategies used by phages with linear DNA genomes to replicate their genomes can be understood in terms of their need to solve the primer problem. The primer problem exists because all known DNA polymerases cannot initiate the synthesis of new DNA but can only add to a preexisting primer (see chapters 1 and 4). This property of all DNA polymerases presents special problems for replication to the ends of any linear DNA, whether phage or human. When the DNA replication fork gets to the end of a linear template molecule, there is no way to make a DNA primer upstream from which to synthesize the last fragment on the lagging strand, because this would require the DNA polymerase to initiate synthesis of a new strand. RNA polymerases can initiate the synthesis of a new strand of RNA, but even if the primers for the 5′ ends are synthesized as RNA, once the RNA primer is removed, there is no DNA upstream to serve as a primer to replace the RNA primer. Because of this priming problem, a linear DNA molecule would get smaller each time it replicated until essential genes were lost. Eukaryotic chromosomes are linear but have throwaway lengths of DNA at their ends, called telomeres, which are dispensable and are enzymatically synthesized from the shortened end by telomerases after each replication, using an RNA in the telomerase enzyme as the template. However, no known phages with linear genomes have telomeres, and they must solve this primer problem in other ways. One way is to circularize the phage genome after infection. Some phages with linear genomes, for example, λ phage, have complementary *cos* sites on their ends that allow them to form circles after infection (see chapter 8). Some phages use protein primers rather than RNA primers (Box 7.3). Other phages have repeated sequences at the ends of their genomic DNA called terminal redundancies, which allow them to form concatemers, as discussed below. Still others have hairpin ends like some linear plasmids, which, in the prophage state as a plasmid, allow them to replicate around the ends and form dimeric circles that can then be resolved by telomere resolvase (see chapter 4). However, they only replicate this way as a linear plasmid prophage and switch to one of the other mechanisms when the prophage is induced to form new phage, so we will not discuss them here. In this section, we discuss in more detail how some phages with linear DNA solve the primer problem.

Phage T7: Linear DNA That Forms Concatemers

Some phages, including T7, never cyclize their DNA but form **concatemers** composed of individual genome-length DNAs linked end to end. The phage DNA can then be cut out of these concatemers so that no information is lost when the phage DNA is packaged.

As shown in Figure 7.15, T7 DNA replication begins at a unique *ori* site and proceeds toward both ends of the molecule, leaving the 3′ ends single stranded because there is no way to prime replication at these ends. However, because T7 has the same sequence at both ends, called a terminal redundancy, these single strands are

BOX 7.3

Protein Priming

Some viruses with linear genomes, including the animal adenoviruses and the *Bacillus subtilis* phage ϕ29, have solved the primer problem by using proteins, rather than RNA, to prime their DNA replication. In the ϕ29 phage head, a protein is covalently attached to the 5′ end of the virus DNA for protection. After infection, another copy of this protein binds to the attached protein to form a dimer, and this second copy serves as a primer for the synthesis of a new DNA strand, with the first nucleotide attached to a specific serine on the protein. By using the protein as a primer, the virus DNA does not need to form circles or concatemers. The phage DNA polymerase uses this protein to prime its replication by an unusual "sliding back" mechanism. First, the DNA polymerase adds a dAMP to the hydroxyl group of a specific serine on the protein. The incorporation of this dAMP is directed by a T in the template DNA. However, the T used as a template is the second nucleotide from the 3′ end of the template, not the first deoxynucleotide. The DNA polymerase then backs up to recapture the information in the 3′ deoxynucleotide before replication continues. The extra dAMP that is added to the end is removed later. After replication, the protein is transferred to the 5′ end of the newly replicated strand, and replication continues. In this way, no information is lost during replication.

Phage ϕ29 has also been an important model system with which to study phage maturation, because the phage DNA can be packaged very efficiently into phage heads in a test tube. Interestingly, its packaging motor contains an RNA, six copies of which are joined to form a ring around the entering DNA. This RNA ring binds ATP and might somehow rotate to help pump the DNA into the phage head, although this is difficult to prove, and RNA motors to pump DNA have not yet been identified in other systems.

References

Escarmis, C., D. Guirao, and M. Salas. 1989. Replication of recombinant ϕ29 DNA molecules in *Bacillus subtilis* protoplasts. *Virology* **169:**152–160.

Lee, T.-J., and P. Guo. 2006. Interaction of gp16 with pRNA and DNA for genome packaging by motor of bacterial virus ϕ29. *J. Mol. Biol.* **356:**589–599.

Meijer, W. J., J. A. Horcajadas, and M. Salas. 2001. ϕ29 family of phages. *Microbiol. Mol. Biol. Rev.* **65:**261–287.

complementary to each other and so can pair, forming a concatemer with the genomes linked end to end. Consequently, the information missing as a result of incomplete replication of the 3′ ends is provided by the complete information at the 5′ end of the other daughter DNA molecule. Individual molecules are then cut out of the concatemers at the unique *pac* sites at the ends of the T7 DNA and packaged into phage heads. It is not clear how the terminal redundancies are re-formed in the mature phage DNAs, but it might be done by making staggered breaks and then filling them in with DNA polymerase or merely by discarding every other genome in the concatemer, which seems wasteful.

GENETIC REQUIREMENTS OF T7 DNA REPLICATION

In contrast to single-stranded DNA phages, which encode only two of their own replication proteins (the products of genes II and V in M13) and otherwise depend on the host replication machinery, T7 encodes many of its own replication functions, including DNA polymerase, DNA ligase, DNA helicase, and primase. The phage T7 RNA polymerase is also required to synthesize the initial primer for phage T7 DNA synthesis. In addition to these proteins, the phage encodes a DNA endonuclease and an exonuclease that degrade host DNA to mononucleotides, thereby providing another source of deoxynucleotides for phage DNA replication. Analogous host enzymes can substitute for some of these T7-encoded gene products, so they are not absolutely required for T7 DNA replication. For example, the T7 DNA ligase is not required because the host ligase can act in its stead. Nevertheless, T7 DNA replication is a remarkably simple process that requires fewer gene products overall than the replication of bacterial chromosomes and of many other large DNA phages.

Phage T4: Another Phage That Forms Concatemers

In its head, phage T4 also has linear DNA that never cyclizes. It forms concatemers like T7, except that it forms them by recombination rather than by pairing between complementary single-stranded ends. However, T4 and T7 differ greatly in how the DNA replicates and is packaged. Also, befitting its larger size, T4 has many more gene products involved in replication than does T7. As many as 30 T4 gene products participate in replication, as shown in the map in Figure 7.6. In fact, one of the advantages of studying replication with T4 is that it encodes many of its own replication proteins rather than just using those of its host. We have already mentioned some of these. It encodes its own DNA polymerase,

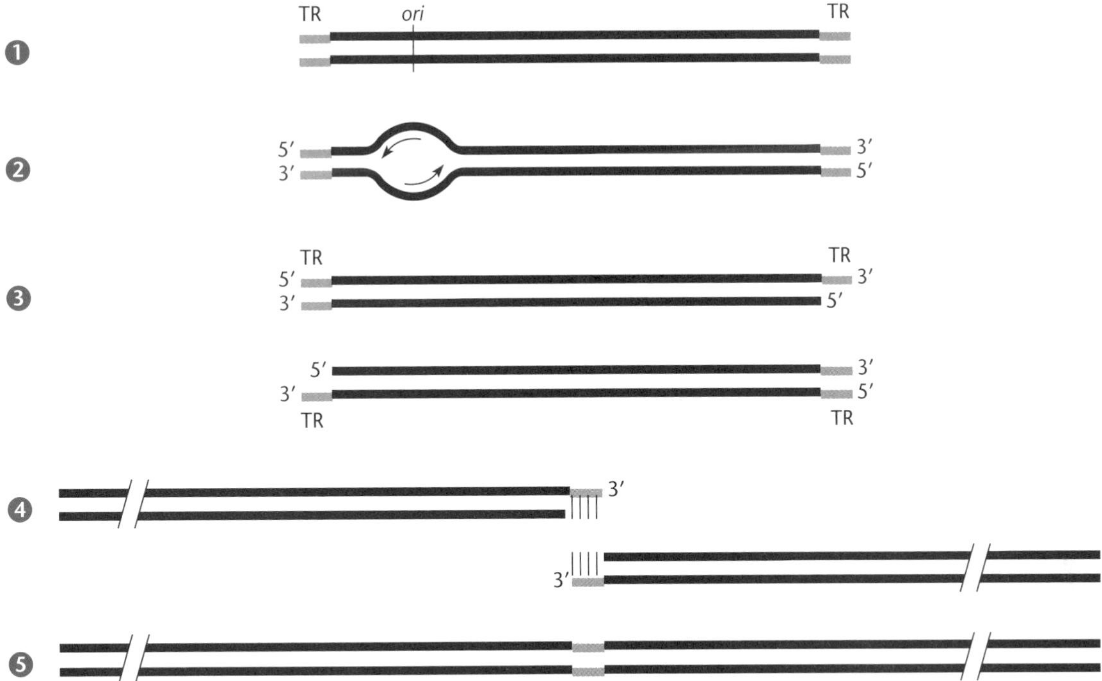

Figure 7.15 Replication of phage T7 DNA. Replication is initiated bidirectionally at the origin (*ori*). The replicated DNAs could pair at their terminally repeated ends (TR) to give long concatemers, as shown. doi:10.1128/9781555817169.ch7.f7.15

sliding clamp, clamp-loading proteins, primase, replicative helicase, DNA ligase, etc. Table 7.2 lists some of these functions and their functional equivalents in *E. coli* and in eukaryotes, where known.

OVERVIEW OF T4 PHAGE DNA REPLICATION AND PACKAGING

Phage T4 replication occurs in two stages, which are illustrated in Figure 7.16. In the first stage, T4 DNA replicates from a number of well-defined origins around the DNA. This type of replication is analogous to the replication of bacterial chromosomes and leads to the accumulation of single-genome length molecules. However, these two daughter molecules have single-stranded 3′ ends because of the inability of T4 DNA polymerase to completely replicate the ends—the primer problem again. They lose no information, however, because the sequences at the ends of T4 DNA are repeated, i.e., the DNAs are terminally redundant, as they are in T7. Somewhat later, this type of replication ceases and an entirely new type of replication ensues. The single-stranded repeated sequences at the **terminally redundant** ends of the genome length molecules can invade the same sequence in other daughter DNAs, forming D loops, which prime replication to form large branched concatemers. This RDR (see "Details of Stage 2 RDR" below), is analogous to the **replication restarts** discussed in chapter 10 and, while first discovered with this phage, is now known to occur in all organisms. The two stages of T4 DNA replication are discussed in more detail below (see Mosig, Suggested Reading).

Like T7, T4 DNA is packaged into the phage head from concatemers. Periodic cycles of RDR lead to the synthesis of very large branched concatemers from which individual genome length DNAs are packaged into phage

Figure 7.16 Initiation of replication of phage T4 DNA. In stage 1, replication initiates at specific origins, using RNA primers. In stage 2, recombinational intermediates furnish the primers for initiation. In both cases, once initiated, replication proceeds as in Figure 7.10. doi:10.1128/9781555817169.ch7.f7.16

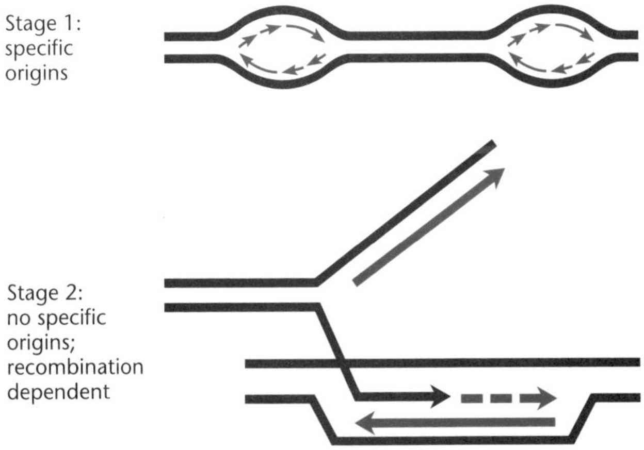

Stage 1: specific origins

Stage 2: no specific origins; recombination dependent

heads (see Black and Rao, Suggested Reading). A pentameric (five-sided) ring around the entrance to the phage head forms a "motor" that sucks DNA into the head at a rate of about 2,000 bp per second, cleaving ATP for energy to drive the motor. When the head is full, the DNA is cut so that the head has somewhat more than a genome length of DNA accounting for the terminal redundancy, as illustrated in Figure 7.17. This type of packaging is called **headful packaging**, because a length of DNA sufficient to fill the head is taken up. It is like sucking a very long strand of spaghetti into your mouth until your mouth is full and then biting it off—not very polite, but effective. If the packaging motor encounters one of the many branches in the concatemer, the branch is cut by gp49, a Holliday junction resolvase (X-phile nuclease [see chapter 10]), which is also involved in recombination.

Because they cut off more than a genome length of DNA, the genomes of the phages that come out of each infected cell are cyclic permutations of each other. The mathematical definition of a cyclic permutation is a permutation that shifts all elements of a set by a fixed offset, with the elements shifted off the end inserted back at the beginning. This explains why the genetic map of T4 is circular (Figure 7.6), even though T4 DNA itself never forms a circle. The way in which different modes of replication and packaging give rise to the various different genetic linkage maps of phages is discussed below (see "Genetic Analysis of Phages" below).

Details of Stage I Replication from Defined Origins

As mentioned above, the first stage of T4 DNA replication, from defined origins, is analogous to chromosome replication from unique origins in cellular organisms, bacteria and eukaryotes. Consistent with its large size, and also like chromosome replication in eukaryotes, T4 has a number of defined origins around the chromosome that can be used to initiate replication. This is unlike most bacteria and other phages, including T7, which usually use only one unique origin to initiate replication. However, T4 most often uses only one of these origins, *oriE*, to replicate each chromosome.

The first step in initiating replication from a T4 origin is to synthesize RNA primers on the origin, using the host RNA polymerase. These primer RNAs also sometimes double as mRNAs for the synthesis of middle-mode proteins and are made from middle-mode-type promoters. These promoters are first turned on a few minutes after infection and require an RNA polymerase whose σ^{70} has been remodeled by binding the MotA and AsiA proteins (see above). In their role as primers, these short RNAs invade the double-stranded DNA at the origin and hybridize to the strand of the DNA to which they are complementary, displacing the other strand to create a structure called an R loop. The invading RNA can then prime the leading strand of DNA replication from the origin from a replication apparatus made up mostly of T4-encoded proteins. Some of the T4 gene products involved in T4 DNA replication are listed in Table 7.2, and a schematic of how they are involved in replication is shown in Figure 7.10. The gp41 replicative helicase, which plays the role of DnaB in uninfected *E. coli*, is loaded on the DNA. The gp59 helicase-loading protein seems to assist in this but is not absolutely required. Other helicase-loading proteins may assist at particular promoters. Once the gp41 helicase is loaded, replication is under way. The gp41 helicase is associated with the lagging-strand primase (gp61), which primes replication of the lagging strand, similar to the role of the primase DnaG in *E. coli* DNA replication. Once replication is under way, the DNA polymerase (gp43) is held on the DNA by a sliding clamp (gp45), which has been loaded on the DNA by the clamp loader, comprised of proteins gp44 and gp62. Once synthesis of Okazaki fragments is complete, a T4-encoded DNA ligase (gp30) joins the pieces together, although the host DNA ligase can substitute to some extent for this function. The T4 DNA replication fork moves along the DNA like a trombone, much like the *E. coli* replication fork. In fact, the best evidence for

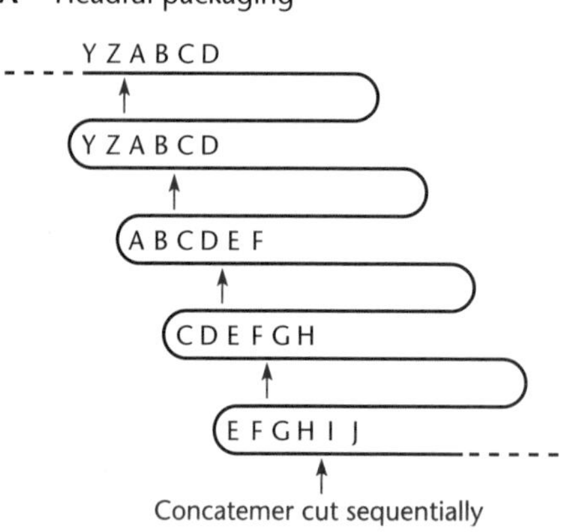

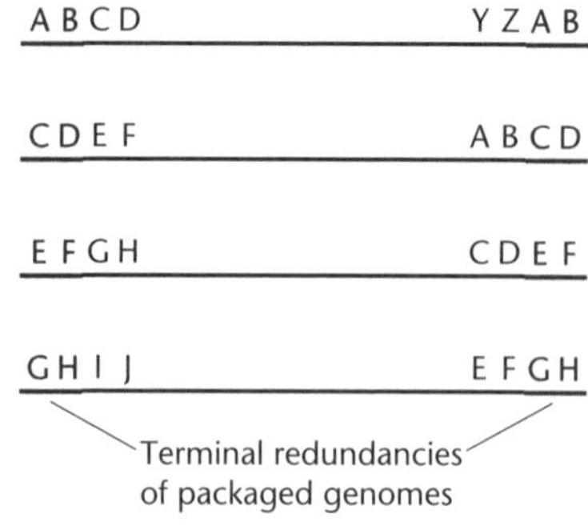

Figure 7.17 T4 DNA headful packaging. Packaging of DNA longer than a single genome equivalent gives rise to repeated terminally redundant ends and cyclically permuted genomes. **(A)** Headfuls of DNA are packaged sequentially from concatemers. The vertical arrows indicate the sites of cleavage during packaging. **(B)** Each packaged genome is a different cyclic permutation with different terminal redundancies.
doi:10.1128/9781555817169.ch7.f7.17

the trombone model has come from this phage. After one or very few copies of the DNA have been made from defined origins, a helicase called UvsW can displace these R loops, suppressing origin-specific replication in favor of the next mode of replication, RDR.

Details of Stage 2 RDR

In the second stage of replication, the leading strand of T4 replication is primed by recombination intermediates rather than by primer RNAs synthesized by RNA polymerase (see Mosig, Suggested Reading). This T4 RDR is similar to replication restarts in uninfected cells (see chapter 10) and uses similar recombination functions, some of which are listed in Table 7.2.

The first step in RDR is the invasion of a complementary double-stranded DNA by a single-stranded 3′ end to form a three-stranded **D loop** (Figure 7.16). This invading single-stranded 3′ end is created during an earlier round of origin-specific replication by the inability of the DNA polymerase to replicate to the end of the molecule and could be extended by the actions of exonucleases, such as gp46 and gp47, on the ends of the molecule. In this respect, gp46 and gp47 are analogous to the RecBCD proteins of *E. coli* (Table 7.2) (see chapter 10). If the cell has been infected by more than one T4 phage particle, the complementary sequence that the free 3′ end invades could be anywhere in the DNA of a coinfecting phage, since T4 DNAs are **cyclically permuted** (see above). However, if the cell has been infected by a single phage, the newly replicated phage DNAs have the same sequences at their ends, and the single-strand invasion would be into the terminal redundancy of the other daughter DNA. This pairing of the invading strand with the complementary strand in the invaded DNA is promoted by the T4 *uvsX* gene product (Table 7.2), which is analogous in function to the RecA protein, the *E. coli* function that promotes single-strand invasion in uninfected cells (see chapter 10). Normally, single-stranded T4 DNA is coated with the T4-coded single-stranded-DNA-binding protein gp32, and the UvsX protein might need the help of another T4 protein, UvsY, to displace the gp32 protein, much as RecFOR proteins displace the *E. coli* single-stranded-DNA-binding protein SSB in uninfected cells (see chapter 10). Once the D loop has formed, the replicative helicase gp41 is loaded on the DNA by gp59 in a process that seems similar to the loading on of the DnaB helicase by the Pri proteins in uninfected cells during replication restarts. The invading 3′ end can then serve as a primer for new leading-strand DNA replication, and the primase gp62 can be loaded on the displaced strand for lagging-strand replication. Later, when replication from defined origins has ceased and there are no double-stranded ends, recombination is initiated by other proteins, gp17 and gp49, that break the DNA and create ends for single-strand invasion, as described above, and the process continues. Repeated rounds of strand invasion and replication lead to very long branched concatemers, which are then packaged into phage heads This is a simplified version of what must be a much more complicated mechanism of RDR. It ignores some known features of RDR, such as its bidirectionality, as well as the roles of some of the helicases and exonucleases, among other enzymes, known to be required for the process. The details of RDR in this and other systems are still being uncovered.

The RDR of T4 was discovered because, when cells are infected by T4 with mutations in any of the phage recombination function genes, genes *46, 47, 49, 59, uvsX*, etc. (see chapter 10), replication begins normally but soon ceases (see Mosig, Suggested Reading). Only relatively recently has it come to be recognized that ubiquitous phenomena, such as double-strand break repair, replication restart, and intron and intein mobility—processes common to all organisms—use basically the same mechanism as RDR (see Kreuzer, Suggested Reading). Developments in these fields are covered in chapters 10 and 11.

Phage Lysis

Once the phage DNA has replicated, the phage particles assemble and are released from the infected cell to infect other cells. As mentioned above, some phages, such as the small filamentous DNA phages, including M13 and f1, assemble in the membrane and then leak out of the cell, using a modified type II secretion system. Type II systems have already been mentioned in chapter 6 because of their relationship to some competence systems for DNA transformation and because they are used to secrete the pilin proteins of type IV pili, but these phages have adapted them to their own use by encoding their own secretin proteins that form a channel through the outer membrane through which the assembled phage can pass.

Phages that leak out of the infected cell without killing them could be said to cause a chronic infection because the host is not killed and continues to produce phage. However, most phages cause an acute infection, in that a cell infected by one of these phages begins to round up later in infection and then suddenly breaks open or lyses, releasing the phage. Besides their scientific interest, phage lysis systems have potential as antibacterial agents as antibiotics lose their effectiveness due to resistance.

Single-Protein Lysis

In some of the simplest phages, including the spherical single-stranded DNA phage ϕX174 and the RNA phage Qβ, a single protein causes cell lysis. These lysis proteins inhibit enzymes that make precursors of the bacterial cell wall by binding directly to them. When the cell begins to divide, the resultant shortage of cell wall precursors causes the cell to lyse, releasing the phage. While the system adopted by these small phages requires very few phage-encoded gene products, it is inefficient, because the number of phage that are produced from an infected cell depends on the stage in the cell cycle at which the cell was infected. If the cell is infected by a phage shortly before it is to divide, very few phage are produced.

Timed Lysis

The larger double-stranded DNA phages have systems to more precisely time the lysis so that an optimum number of phage can be produced. If lysis occurs too early, phage will not yet have assembled, and the infection will not be productive. If it occurs too late, the infection process will take longer than necessary, and the phage may lose the race to infect new hosts.

The timing of cell lysis by double-stranded DNA phages is a complicated process that requires many proteins. Phages of gram-negative bacteria that are released from the cell by lysis usually encode at least five such proteins. One of these is the endolysin that breaks bonds in the cell wall. The cell wall is a rigid sheet around the cell that gives the cell its structure and integrity and is composed of a sheet of peptidylglycan that must be broken to release phage from the cell (see chapter 14). Other proteins required for lysis, at least of gram-negative bacteria, are the **spanins**, which are composed of a small lipoprotein and a protein that scans the periplasm, hence the name. It is not clear exactly how spanins work, but they may play a role in separating the outer membrane from the cell wall, which may be required for release of the phage under some conditions. In some systems, the spanins are required for lysis only in the presence of divalent cations, such as Mg^{2+}.

Timing of Lysis by Holins

The endolysins are made early in infection but are not active until later in infection, when they are suddenly activated by the action of **holins**. These are not enzymes but rather are membrane proteins that form holes, or pores, in the inner membrane. Still other proteins involved in lysis are **antiholins**, which inhibit holins until it is time for lysis. Understanding the timing of lysis depends upon learning how holins and antiholins work and how they are activated at the right time for lysis.

An early idea was that holins form holes in the inner membrane that allow the lysozyme to pass through the inner membrane to reach and degrade the peptidylglycan cell wall in the periplasmic space—hence the name holin. Some holins may in fact work this way, since they have a number of transmembrane domains (TMDs) and are known to form pores in the membrane large enough for the endolysin to pass through. Others may form only very small pores that allow the passage of protons through the membrane and destroy the membrane potential. In fact, it is known that in many systems, destroying the membrane potential causes endolysins to lyse the cell even before lysis normally occurs.

T4 PHAGE LYSIS

The lysis system of the T4 phage may be one of one of those in which the holin channel allows passage of the endolysin; it is illustrated in Figure 7.18A. In this phage, the endolysin, the product of gene *e*, is made in the active form in the cytoplasm relatively early in infection but cannot degrade the cell wall until later, because it cannot pass through the inner membrane. The holin, the product of the gene *t*, gp*t*, is in the membrane but is somehow inactive. At the time for lysis, the *t* gene product becomes active and allows endolysin through the inner membrane, which, with the help of two spanins, causes lysis of the cell. T4 has an antiholin, the product of the *r*1 gene, gp*r*1, but it is very unstable, so very little accumulates during phage infection. Its job seems to be to delay lysis if there are other T4 phage bound to the outside of the cell, in a process called lysis inhibition. These external phage can inject their DNA into the cell, but it becomes trapped in the periplasmic space rather than entering the cytoplasm. Ordinarily, the puncturing of the membrane by the T4 syringe would activate the holin and lyse the cell. Somehow, this periplasmic DNA stabilizes the gp*r*1 antiholin, which then prevents activation of the gp*t* holin by binding to its periplasmic domain, thereby inhibiting cell lysis. Lysis inhibition may ensure that the cells do not lyse when there are already plenty of T4 phage around that have been released from other infected cells nearby, and it may be better to stay in the infected cell and keep multiplying.

The phenotypes of mutations in the *t* and *r*1 genes have been known for a long time. Mutations in the *t* gene were known to delay lysis, and mutations in the *r*1 gene were known to prevent lysis inhibition. The name *r*1 means rapid lysis gene 1, because it is one of the four genes (the *r*II genes are two others) in which mutations prevent lysis inhibition, making the plaques of mutants clearer. We discuss the importance of the *r*II genes in the history of molecular genetics below. It was only relatively recently that the roles of the products of the *t* and *r*1 genes as holin and antiholin were realized.

λ PHAGE LYSIS

The action of the antiholin of the phage λ in timing lysis may be better understood (Figure 7.18B) (see White et al., Suggested Reading). As shown in the figure, two copies of the holin, named S105 because it is a product

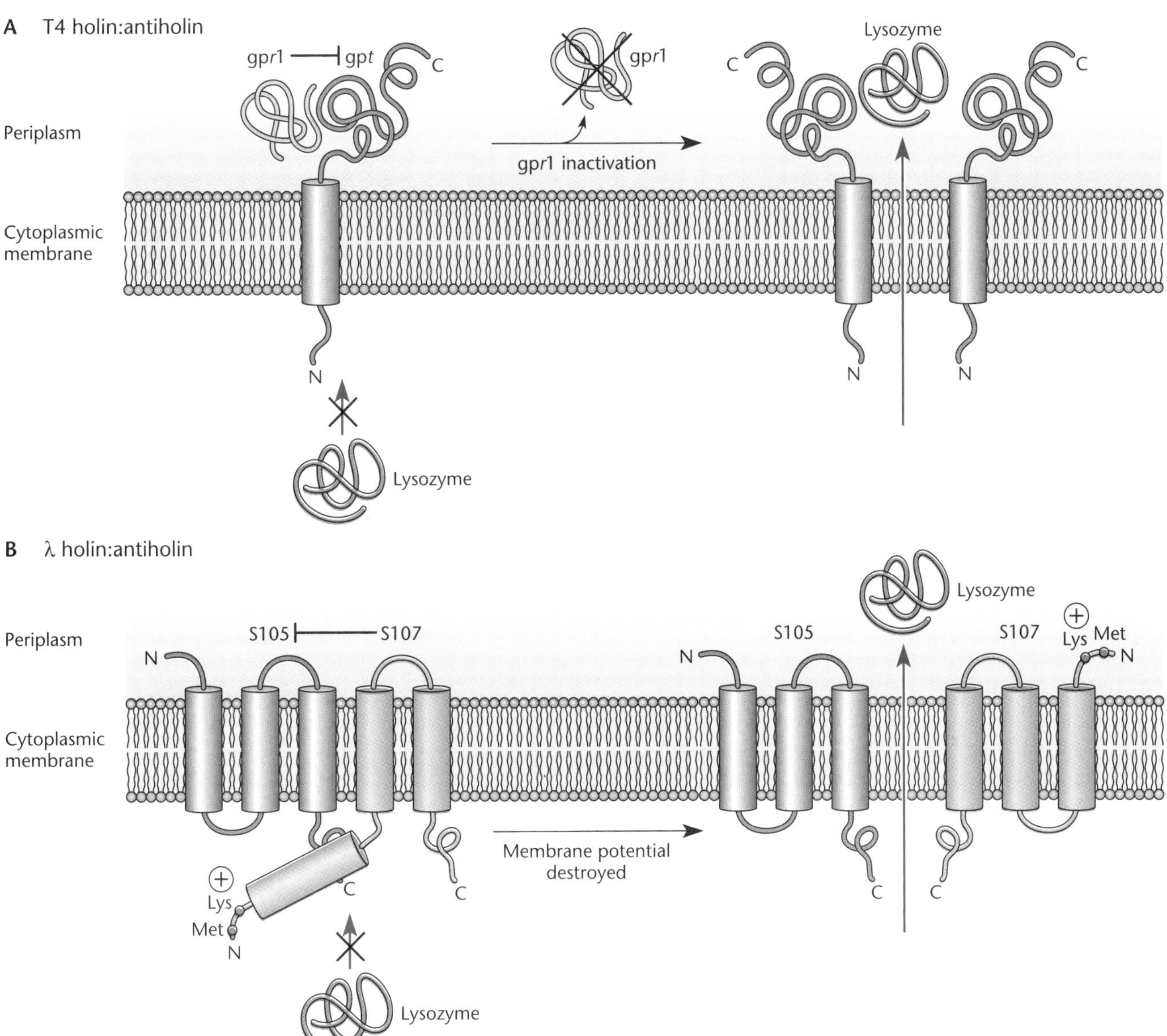

Figure 7.18 Timing of phage lysis by activation of holins. The antiholin keeps the holin inactive until the time of lysis. The holin then becomes active, forming a pore that allows the lysozyme (blue) to traverse the membrane and then to degrade the cell wall and lyse the cell. **(A)** Model for lysis inhibition in T4 phage. If phage are bound to the outside of the cell, the antiholin (gp*r*1) binds to the periplasmic domain of the holin (gp*t*) , which prevents it from forming pores. Much later, the antiholin somehow becomes inactive, allowing the holin to form a pore in the membrane through which the lysozyme can pass. **(B)** Model for timing of lysis by λ phage. The antiholin (S107) and holin (S105) differ only in that the antiholin has an extra 2 amino acids, Lys-Met, at its N terminus due to a second upstream translation initiation region (TIR). These extra amino acids prevent the first TMD of S107 from entering the membrane because they add a positive charge that cannot move against the membrane potential. This makes the S107 antiholin inactive as a holin, but it still binds to the S105 holin, interfering with its ability to form pores. At the time of lysis, the membrane loses its potential, allowing TMD1 of S107 to enter the membrane and participate with S105 in the formation of pores through which lysozyme can pass to reach the cell wall. The antiholin S107 is shown in blue. doi:10.1128/9781555817169.ch7.f7.18

of the *S* gene and has 105 amino acids, dimerize to form a pore through which the endolysin can pass. To form a pore, the holin, which has three TMDs, must traverse the membrane three times, so that its N terminus ends up in the periplasm and its C terminus ends up in the cytoplasm, as shown. The antiholin (S107) is 2 amino acids longer than the holin because, even though it is translated from the same open reading frame (ORF), its translation starts two codons upstream from a different AUG codon, adding an extra methionine and lysine at its N terminus. These two extra amino acids prevent the first TMD from entering the membrane, perhaps because the positively charged lysine cannot enter the membrane against the membrane potential. With only two TMDs in the membrane, S107 cannot form a holin, but it can still dimerize with an S105 holin. However, these hybrid structures are not active as holins and do not form pores. At the appropriate time for lysis, the membrane loses its potential by an unknown mechanism. This may allow the first TMD of the antiholin to enter the membrane and participate with the shorter form in the formation of active pores, which allows access of the endolysin, the product of gene *R*, to the cell wall and, with the help of the spanins Rz and Rz1, causes lysis.

ACTIVATION OF SAR ENDOLYSINS

Some endolysins, particularly in phages of gram-positive bacteria, have long hydrophobic TMDs at their N termini that function as signal sequences and direct the protein through the inner membrane by the Sec system (see chapter 14). The endolysin of phage P1 is one example. Since these endolysins are transported by the Sec system, they have no need for porin pores to get through the membrane. Nevertheless they do also use holins. Such endolysins, called SAR (signal anchor release) endolysins, are inactive after transport to the periplasm because the N-terminal TMD is not cleaved off as they pass through the SecYEG channel and remains anchored in the inner membrane. At the correct time for lysis, the holin releases the trapped domain, activating the endolysin. How the holin does this is not clear, but it might destroy the membrane potential, perhaps by forming pores in the membrane, and without a membrane potential, the N terminus of the endolysin is released. In fact, SAR endolysins are often paired with holins called pinholins, which make very small pores that may destroy the membrane potential by allowing the passage of protons but are not large enough to allow the passage of proteins, including the endolysin.

How release of the signal sequence from the membrane activates a SAR endolysin varies but often involves the shuffling, or **isomerization**, of disulfide bonds between cysteine amino acids in the protein. Trapping of the N terminus of the endolysin in the membrane allows an inhibitory disulfide bond to form elsewhere in the protein. In some cases, one of the cysteines in the inhibitory disulfide bond is in the active center of the endolysin, making it unavailable for the reaction; in other cases, the critical cysteine is not actually in the active center, but the disulfide bond still influences the activity of the endolysin, either by "caging" the active center by surrounding it with other sequences so it is not available or by some other mechanism affecting the conformation of the enzyme. In all these cases, a cysteine residue in the SAR domain trapped in the inner membrane is unpaired because it is not in the periplasm, where disulfide bonds form (see chapter 14). After the SAR domain is released by the holin, Dbs oxidoreductases in the periplasm direct this cysteine to pair with another cysteine to form a new disulfide bond, and formation of this new bond frees the critical cysteine from the inhibitory disulfide bond and activates the enzyme. Not all SAR endolysins are activated by disulfide bond isomerization. In some cases, release of the trapped N-terminal domain is sufficient to cause proper folding and activation of the endolysin.

GENETIC STRUCTURE OF LYSIS GENES

As an interesting aside, the genetic structure of the lysis genes of phages is often unusual, presumably to save space so the genome can be smaller. For example the product of the *A2* gene of the small RNA phage Qβ also has another function. The lysis gene *E* of the small single-stranded DNA phage ϕX174 is read in the −1 frame of another gene, *D*, required for phage assembly. The spanin genes *Rz* and *Rz1* of λ are completely overlapping, with *Rz1* completely embedded in *Rz* but read in the +1 frame relative to *Rz*.

Phage Display

The relative ease with which phages can be manipulated and propagated has allowed the development of a powerful technology called **phage display**. The principle behind phage display is quite simple. Since most phages are just genes wrapped in a protein coat encoded by those genes, proteins can be expressed (displayed) on the surface of the phage if their coding sequence is translationally fused to the coding sequence for one of the coat proteins encoded by one of the coat protein genes. This is called **genotype-phenotype linkage**, because the DNA encoding the displayed peptide or protein is encapsidated in the same phage coat displaying the peptide or protein. If even a single phage particle displaying that particular peptide can be somehow separated from most of the other phages that are expressing different peptides, the phage can be propagated to make hundreds of millions of identical copies. This makes it easy to obtain plenty of DNA to determine the sequence of the DNA

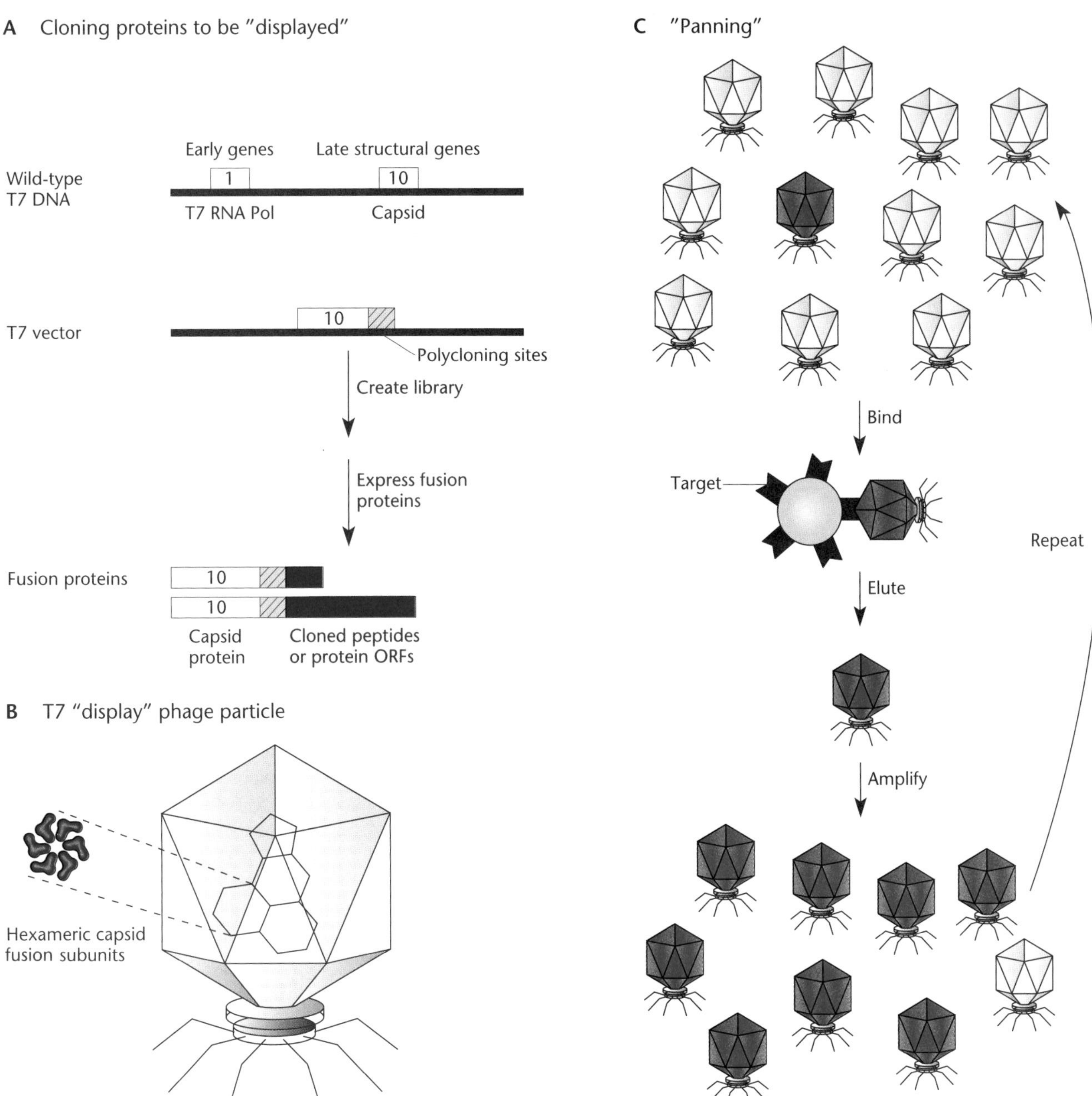

Figure 7.19 Use of phage T7 for phage display. **(A)** A randomized protein-coding sequence (shown in blue) is cloned into the polycloning site in the phage DNA to make translational fusions in which random peptides have been fused to the head protein-coding region, gene *10*, of the phage. The DNA is then introduced into cells by transfection or by packaging the DNA into a phage head and using the phage to infect cells. **(B)** When the phage multiply, the progeny of a phage displaying a particular peptide sequence all express that peptide on their surfaces fused to their head proteins. However, since a random collection of peptide-coding sequences has been fused to the gene *10* protein-coding sequence, the descendants of different phage in the population express different fusion peptides on their surfaces. Some of them (shown in blue) express a peptide that happens to bind to the target, while most (shown in gray) express peptides that do not bind. When this mixture of phage is exposed to the target and the target is washed, those expressing a peptide that binds are preferentially retained, and more of the others are washed away. The retained phage can then be eluted from the target and propagated. This process is repeated multiple times to further enrich for the phage that specifically bind. doi:10.1128/9781555817169.ch7.f7.19

and, therefore, the peptide displayed by that phage. If the displayed peptide binds to a particular ligand, it is possible to isolate a phage expressing that peptide. This is very useful, because the extreme complexity of protein-protein interactions makes it difficult or impossible to predict what peptide sequences will bind tightly to a particular **ligand**, such as a hormone or an antibody. With phage display, there is no need to make any such predictions. The ligand is presented with a huge variety of different peptide sequences and asked to choose among them.

Many phages have been adapted for phage display. The use of T7 phage is illustrated in Figure 7.19. A randomized peptide-coding sequence is translationally fused to the product of gene *10* so that different phages display different versions of the peptide on their surfaces. Phages that happen to display a version of the peptide that binds to the ligand molecule are then enriched from among the myriad of phages, the vast majority of which express nonbinding peptides, by a process called **biopanning**, which involves adding the mixture of phages to a solid surface to which the ligand is bound. Most of the phages are then washed off, and those that remain are enriched for those that have bound to the ligand, much as in panning for gold, where the heavier gold is left in a pan after lighter material is washed off. The bound phage are then eluted and propagated, and the process is repeated numerous times. Sequencing the variable regions of the DNAs of the phages that survive repeated cycles of biopanning reveals peptide sequences that may bind to the ligand molecule. However, their binding must still be tested directly, using other methods.

Of the different types of phages that have been engineered for phage display, each type has its advantages and disadvantages. Single-stranded DNA phages have the advantage that it is easier to clone the randomized sequences into the phage genome, and five different head proteins can be used to make the fusions (Figure 7.14). A number of phagemids have been designed for these phages that encode all of the proteins required to make a phage with convenient cloning sites next to their genes, as well as a *pac* site to allow them to be packaged in a phage head. However, the filamentous phages have the disadvantage that they must be secreted through the membrane, and generally, only very small peptides can be fused to the head proteins. The larger double-stranded phages that lyse the cell can generally accommodate larger fusions, but it is more difficult to clone into their larger genomes. Box 7.4 discusses the advantages and disadvantages of different types of phages for phage display and gives some recent examples of their use in more detail.

BOX 7.4

Phage Display

As discussed in the text, one of the most powerful current applications of phages is in phage display. This remarkable technology allows the selection of peptides that bind tightly to proteins and other **ligands** from among millions of nonbinders. The technology is currently being used to identify potential antigens, for example, in autoimmune diseases, such as multiple sclerosis; to purify antibodies against specific antigens from an artificial human antibody library; to identify drug targets; and to develop peptides that target chemotherapeutics to specific tissues, to list just a few examples.

Many different types of phages have been adapted for phage display, and each has its advantages and disadvantages. Which one to use depends on the specific application. In general, the more copies of the peptide that are expressed on the surface of the phage, the stronger the binding to the target. Some phage heads are composed of hundreds of copies of one head protein and one or very few copies of others, so for strong binding, you might choose to fuse the randomized peptide sequences to a head protein with lots of copies in the head. Strong binding is good, except that it can make it harder to elute the phage that have bound from the target, and the elution conditions must not inactivate the phage, since it must be propagated after elution. In order not to limit detection to more weakly binding peptides, it may be possible to "infect" the phage off the target by adding mid-log-phase bacteria that will be infected by the bound phage and produce more phage, leaving the bound phage protein behind, since it is the head that is bound, and the DNA is injected through the tail. It may also be possible to use CysDisplay to make the phage easier to elute, as we discuss below.

Another limitation is that the head proteins must be able to assemble into a phage head with the peptides attached. Some head proteins can tolerate longer peptides fused to them than others. Some that occur in many copies in a phage head can assemble into a phage head only if a minority of the copies in a head are fused to a peptide, while others can tolerate almost complete substitution. In general, those that can express longer polypeptides often require that only a small percentage of the head proteins be fused to a

BOX 7.4 (continued)

Phage Display

polypeptide; otherwise, the phage will be nonviable. This can be accomplished by coinfecting the cells with "helper phage" that express the normal head protein or by expressing the normal phage head protein from a plasmid clone.

Convenience for cloning and making gene fusions is also a factor. The cloning steps involved in making the gene fusions in the phage are usually easier with smaller phages than with larger ones because genotype-phenotype coupling requires that the peptide-coding sequence must be in the DNA that is packaged and that this packaged DNA must have most of the genes necessary to direct the synthesis of a new phage, which for large phages means a large, inconvenient cloning vector.

The use of T7 phage for phage display is outlined in the text. The gene *10* head protein of T7 can accommodate up to 50-amino-acid-long peptides on the carboxyl terminus of its head protein, the part exposed on the surface. Since there are 415 head proteins per phage, a large number of copies of a peptide can be expressed from each phage, enhancing binding. However, if longer peptides are required, only some copies of the head protein can be fused to a peptide, and most of the proteins must be normal or the phage are not viable. To incorporate normal copies of the head protein into the phage head, the normal gene *10* of T7 can be expressed from a plasmid in the *E. coli* strain used to propagate the phage. What fraction of the head protein in each phage head is fused to a peptide depends on the relative strengths of the promoters servicing the head gene in the phage and in the plasmid.

The phage T4 has also been adapted for phage display. In this system, the polypeptides to be biopanned for (see the text) can be fused to the HOC and/or SOC head protein of the phage (Figure 7.8 shows the structure of the phage). These proteins are not required for head formation, but between them more than 1,000 copies bind tightly to the outside of the head after it is assembled but before the DNA is encapsulated. To overcome the handicap of cloning into the very large phage genome, the coding sequence for the polypeptide to be biopanned is not cloned directly into the phage, as it usually is with T7, but is cloned into a plasmid that contains the HOC or SOC coding sequence to make the translational fusion. To achieve genotype-phenotype linkage, the cells containing the plasmid with the fusion are then infected by T4, and the fusion is crossed into the phage by recombination. The phage they are crossed into has a deletion mutation in its *e* gene, which encodes the lysozyme. This mutation prevents release of the phage unless external egg white lysozyme is added. Recombination with the cloned fusion gene integrates the *e* gene and allows phage release in the absence of added lysozyme, allowing the selection of phage that have integrated the fusion gene expressing the peptide into their genomes. The T4 system shows promise for displaying larger polypeptides, and even complete proteins of up to 35 kilodaltons have been displayed, but at a low copy number (see Ren et al., References).

Large phages, such as T4 and T7, have the advantage for phage display that they lyse the cell to release themselves rather than being secreted through the membranes like the smaller filamentous phages, such as M13. Larger, more hydrophilic peptides are difficult to secrete. Nevertheless, the first and most widely used phage vectors to be developed for phage display are based on single-stranded DNA phages. The small size of the phage genome makes cloning easier, and phagemids that can be packaged into the phage head have been designed to facilitate cloning of random coding sequences fused to the coding sequences for the head proteins. Any of the five head proteins of the phage (Figure 7.14) can in principle be used to make the translational fusion, but gpIII and gpVIII are most commonly used. While gpIII is present in many fewer copies than gpVIII, it is much larger and may tolerate longer peptide fusions. The random peptide sequence is inserted between the signal sequence required for secretion of the protein (see chapter 14) and the N terminus of the mature protein, where it will be displayed on the surface of the phage secreted from the cell. However, longer fused peptides interfere with the secretion of the head protein, and only very short peptides can be accommodated with 100% substitution. Helper phages that supply normal head protein are often used to reduce the frequency of the fusion proteins in the head and to increase phage viability. Also, mutations of the head proteins have been selected that show increased tolerance for longer fused peptides.

One clever way to achieve genotype-phenotype linkage of a peptide-coding sequence without translationally fusing it to a head protein is to fuse it to the head protein in the periplasm of the cell after it has been secreted. One such application, called CysDisplay (see Rothe et al., References), depends on the fact that disulfide bonds between cysteine amino acids in polypeptides form only in the oxidizing environment of the periplasmic space. If the head protein and the protein to be displayed are expressed from different genes on the phagemid and both have a signal sequence, they will be secreted separately into the periplasm. Oxidoreductases in the periplasm (see chapter 14) will form a disulfide bond between an unpaired cysteine in the head protein and an unpaired cysteine in the protein to be displayed, cross-linking the two

(continued)

BOX 7.4 (continued)

Phage Display

proteins and displaying the protein on the surface of the phage after it is assembled. The genotype-phenotype linkage is preserved because the cross-linking of the two proteins occurs before the phage leave the cell and the phagemid DNA that is packaged in the phage head encodes both of them. CysDisplay has a number of advantages. Because they are secreted separately from the head protein, larger polypeptides can be displayed in this way without interfering with phage assembly. Another major advantage is that the bound phage can be eluted merely by adding a reducing agent, such as dithiothreitol, that will break the disulfide bond and release the phage, leaving the protein bound to the surface. This is a big advantage, because the phage can be released no matter how tightly the protein binds to the ligand fixed to the surface, and the most strongly binding peptides are often the most desirable. Also, phage that are nonspecifically trapped on the surface are less apt to be released by the reducing agent than by strong denaturing conditions, theoretically making the enrichment for binding phage more effective.

To illustrate the power of this technology, we will discuss a recent application (see Rothe et al., References). This is illustrated in the figure. The goal of the Rothe et al. project was to make an artificial human antibody library with almost the same diversity as the human antibody repertoire and then use phage display to identify antibodies in the library that bind to a particular protein or other antigen. The structure of antibodies is shown in panel A. Antibodies are heterodimers consisting of two polypeptide chains, the longer "heavy" and shorter "light" chains. The N termini of the heavy and light chains cooperate to form an antigen-binding pocket (shown in blue) that is highly variable due largely to the presence of short complementarity-determining regions (CDRs) that vary greatly in sequence from one antibody molecule to another (hypervariable regions) and allow the specific binding of a vast array of antigens in the pocket. Antibodies consisting of just the N termini of the heavy and light chains, called Fab fragments, also shown in the figure, form the binding pocket and are useful for many applications, although they lack other functions of antibodies.

To construct a diverse collection of Fab fragments, Rothe et al. cloned random sequences sequentially into the coding regions for three CDRs in the heavy-chain and the three CDRs in the light-chain Fab fragments so that each sequence in one CDR in each light and heavy chain would be associated with all the other possible sequences in the other two CDRs in that chain to achieve diversity approximating that of the normal human antibody repertoire. They introduced random sequences for the CDRs by cloning randomly synthesized oligonucleotides that were synthesized using trinucleotides rather than mononucleotides to avoid nonsense codons. They also introduced an antibiotic resistance gene in frame downstream of the largest CDR in the heavy chain so that, by selecting only antibiotic-resistant transformants, they could eliminate frameshifts created during the cloning, which would limit the diversity. They then introduced these randomized light- and heavy-chain genes into an M13 phagemid (see the figure) in such a way that they each had their own signal sequence so they would be secreted independently into the periplasm. They also introduced cysteine codons into the heavy chain and head protein genes to introduce unpaired cysteines close to the C terminus of the heavy chain and close to the N terminus of the head protein. With this construct, the gpIII head protein of the phage was cross-linked in the periplasm to the heavy chain by disulfide bonds between the cysteines (S-S) in the two proteins. The heavy chain in turn bound to its natural partner, the light chain, to express the Fab fragment antigen-binding pocket on the surface of the phage, as shown in the insert in panel A of the figure. Panels B and C show the series of steps involved in isolating a phage expressing an antibody that binds to a particular antigen. In panel B, a random phage library is constructed in which each packaged phasmid encodes the random heavy and light chain expressed on its coat through linkage by a disulfide bond between its head protein and the heavy chain it encodes. In panel C, this random phage library is biopanned to isolate those phage displaying Fab fragments that bind specifically to a particular antigen. Phage expressing Fab fragments that do not bind to the antigen are preferentially washed off, and the specifically binding phage are eluted with dithiothreitol to break the disulfide bond joining the phage to the heavy chain. The phage are propagated, and the whole process is repeated numerous times to further enrich for phage expressing specifically binding Fab fragments. Phage that survive this process are then tested directly to determine if the Fab fragments they display bind to the antigen. For some applications that require more complete antibodies, or even bivalent antibodies, these can be constructed from the Fab fragments, once they are identified, by fusing their coding sequences to the coding sequences for the remainder of the antibody molecule. While it was a lot of work to make the library (note the number of coauthors of the paper), once constructed, a random phage library can be used over and over again to isolate antibodies to particular antigens, and if the library is depleted, it can be resynthesized just by propagating the phage library. This approach has the potential to supplant the making of natural human or mouse monoclonal antibodies for some applications because it is much cheaper

BOX 7.4 (continued)

Phage Display

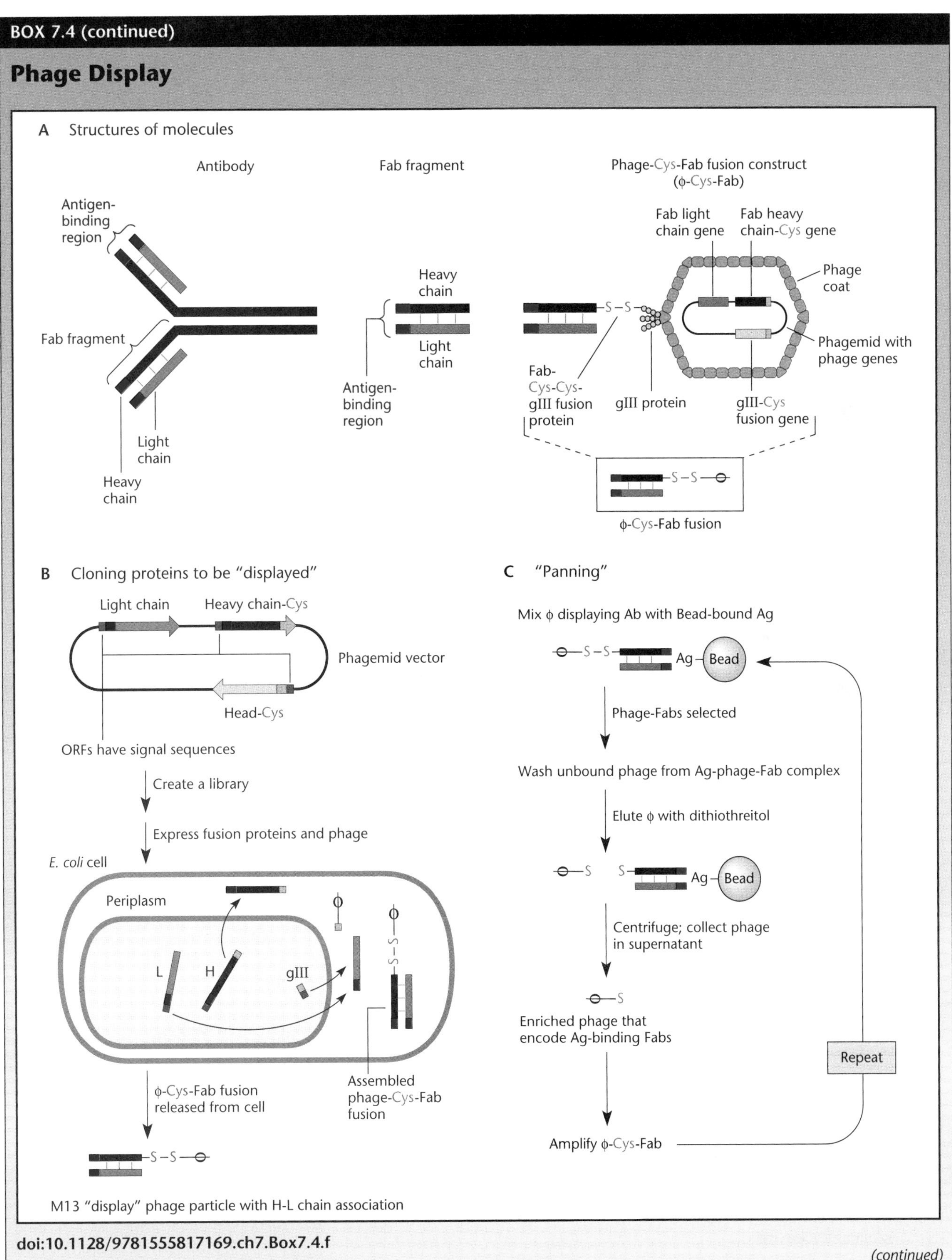

doi:10.1128/9781555817169.ch7.Box7.4.f

(continued)

BOX 7.4 (continued)

Phage Display

and lacks some of the limitations of such antibodies, which need to be remade each time and cannot be made against antigens that are highly toxic to mammalian cells or to self-antigens, which are naturally excluded from natural human antibody libraries.

References

Gupta, A., A. B. Oppenheim, and V. K. Chaudhary. 2005. Phage display: a molecular fashion show, p. 415–429. *In* M. K. Waldor, D. I. Friedman, and S. L. Adhya (ed.), *Phages: Their Role in Bacterial Pathogenesis and Biotechnology.* ASM Press, Washington, DC.

Malys, N., D.-Y. Chang, R. G. Baumann, D. Xie, and L. W. Black. 2002. A bipartite bacteriophage T4 SOC and HOC randomized peptide display library: detection and analysis of phage T4 terminase and late σ factor (gp55) interaction. *J. Mol. Biol.* **319:**289–304.

Ren, Z. J., C. J. Tian, Q. S. Zhu, M. Y. Zhao, A. G. Xin, W. X. Nie, S. R. Ling, M. W. Zhu, J. Y. Wu, H. Y. Lan, Y. C. Cao, and Y. Z. Bi. 2008. Orally delivered foot-and-mouth disease virus capsid protomer vaccine displayed on T4 bacteriophage surface: 100% protection from potency challenge in mice. *Vaccine* **26:**1471–1481.

Rothe, C., S. Urlinger, C. Lohning, J. Prassler, Y. Stark, U. Jager, B. Hubner, M. Badroff, I. Pradel, M. Boss, R. Bittlingmaier, T. Bataa, C. Frisch, B. Brocks, A. Honegger, and M. Urban. 2008. The human combinatorial antibody library HuCAL GOLD combines diversification of all six CDRs according to the natural immune system with a novel display method for efficient selection of high affinity antibodies. *J. Mol. Biol.* **366:**1182–1200.

Genetic Analysis of Phages

Phages are ideal for genetic analysis (see the introductory chapter). They have short generation times and are haploid. Mutant strains can be stored for long periods and resurrected only when needed. Also, phages multiply as clones in plaques, and large numbers can be propagated on plates or in small volumes of liquid media. Different phage mutants can be easily crossed with each other, and the progeny can be readily analyzed. Because of these advantages, phages were central to the development of molecular genetics, and important genetic principles, such as recombination, complementation, suppression, and *cis*- and *trans*-acting mutations, are most easily demonstrated with phages. In this section, we discuss the general principles of genetic analysis of phages. However, most of the genetic principles presented here are the same for all organisms, including humans. Only the details of how genetic experiments are performed differ from organism to organism.

Infection of Cells

The first step in doing a genetic analysis of phages, or any other virus for that matter, is to infect cells with different mutants of the phage. Phages can infect only cells that are sensitive or susceptible to them, and they can multiply only in cells that are permissive for their development. To multiply in a cell, not only must the phage adsorb to the cell surface of the bacterium and inject its nucleic acid, either DNA or RNA, but also, a **permissive host** cell must provide all of the functions needed for multiplication of the phage. Therefore, most phages can infect and multiply in only a very limited number of types of bacteria. The different types of bacterial cells in which a phage can multiply are called its host range. Sometimes, a normally permissive type of cell can become a **nonpermissive host** for the phage as a result of a single mutation or other genetic change. Alternatively, a mutant virus or phage may be able to multiply in a particular type of host cell under one set of conditions, for example, at lower temperatures, but not under a different set of conditions, for example, at higher temperatures. As in bacterial genetics, these are called **conditional-lethal mutations** and are very useful in phage genetics. The conditions under which a phage with a conditional-lethal mutation can multiply are **permissive conditions**, while the conditions under which it cannot multiply are **nonpermissive conditions.**

MULTIPLICITY OF INFECTION

Infecting permissive cells with a phage is simple enough in principle. The phage and potential bacterial host need only be mixed with each other, and some bacteria and phage will collide at random, leading to phage infection. However, the percentage of the cells that are infected depends on the concentrations of phage and bacteria. If the phage and bacteria are very concentrated, they collide with each other to initiate an infection more often than if they are more dilute.

The efficiency of infection is affected not only by the concentrations of phage and bacteria, but also by the ratio of phage to bacteria, the **multiplicity of infection (MOI)**. For example, if 2.5×10^9 phage are added to 5×10^8 bacteria, there are $2.5 \times 10^9/5 \times 10^8$, i.e., 5 phage for every cell, and the MOI is 5. If only 2.5×10^8 phage had been added to the same number of bacteria, the MOI would have been 0.5. If the number of phage exceeds the number of cells to infect, the cells are infected at a **high MOI**. Conversely, a **low MOI** indicates that the cells outnumber the phage. Whether a high or low MOI is used depends on the nature of the experiment. At a high enough MOI, most of the cells are infected by at least one phage; at a low MOI, many of the cells remain uninfected, but each infected cell is usually infected by only one phage.

Even at very high MOI, not all the cells are infected. There are two reasons for this. First, infection by phage is never 100% efficient. The surface of each cell may have only one or very few receptors for the phage, and a phage can infect a cell only if it happens to bind to one of these receptors. There is also the statistical variation in the number of phage that bind to each cell. Because the chance of each phage binding to a cell is random, the number of phage infecting each cell follows a normal distribution. At an MOI of 5, some cells are infected by five phage—the average MOI—but some are infected by six phage, some by four, some by three, and so on. Even at the highest MOIs, some cells by chance receive no phage and so remain uninfected.

The minimum fraction of cells that escape infection due to statistical variation can be calculated by using the Poisson distribution, which can be used to approximate the normal distribution in such situations. In chapter 3, we discuss how Luria and Delbrück used the Poisson distribution to estimate mutation rates. According to the Poisson distribution, the probability of a cell receiving no phage and remaining uninfected (P_0) is at least $e^{-\mathrm{MOI}}$, since the MOI is the average number of phage per cell. If the MOI is 5, then P_0 is equal to e^{-5} is equal to 0.0067; i.e., at least 0.67% of the cells remain uninfected. At an MOI of only 1, e^{-1}, or at least 37%, of the cells remain uninfected. In other words, at most ~63% of the cells are infected at an MOI of 1. Even this is an overestimation of the fraction of cells infected, since as mentioned, some of the viruses never actually infect a cell.

Phage Crosses

Crosses in phages and other viruses are easy; the cells are infected with different strains of the virus simultaneously. To be certain that many of the cells in a culture are simultaneously infected by both strains of a phage, we must use a high MOI of both phages. Again, the Poisson distribution can be used to calculate the maximum fraction of cells that will be infected by both mutant phages at a given MOI of each. If an MOI of 1 for each mutant phage is used for the infection, then at most 1 – 0.37, or 0.63 (63%), of the cells will be infected with each mutant strain of the phage. Since the chance of being infected with one strain is independent of the chance of being infected with the other strain, at most 0.63 × 0.63, or 0.40 (40%), of the cells will be infected by both phage strains at an MOI of 1. A high MOI of both mutant phages is required to infect most of the bacteria with both strains.

Recombination and Complementation Tests with Phages

As discussed in chapter 3, two basic concepts in classical genetic analysis are recombination and complementation. The types of information derived from these tests are completely different. In recombination, the DNAs of the two parent organisms are assembled in new combinations so that the progeny have DNA sequences from both parents. In complementation, the gene products synthesized from two different DNAs interact in the same cell to produce a phenotype. While the principles of recombination and complementation are the same in bacteria and phages, the ways in which data from recombination tests and complementation tests are obtained and interpreted are different, so we will emphasize the differences here.

RECOMBINATION TESTS

Recombination tests are easy to do with phages. Cells are infected with both mutants simultaneously at a high MOI. The cells also must be permissive for both mutants or the infection must be done under conditions that are permissive for both mutants. The two different DNAs then enter the same cell and recombine. Figure 7.20 gives a simplified view of what happens when two DNA molecules from different strains of the same phage recombine. In the example, the two mutant phage strains infecting the cell are almost identical, except that one has a mutation at one end of the DNA and the other has a mutation at the other end. The sequences of the two DNAs, therefore, differ only at the sites of the two mutations shown as differences in single base pairs at opposite ends of the molecules. Recombination occurs by means of a crossover between the two DNA molecules, and two new molecules are created, which are identical in sequence to the original molecules except that one part now comes from one of the original DNA molecules and the other part comes from the other. The effect, if any, of the crossover depends on where it occurs. If the crossover occurs between the sites of the two mutations, two new types of recombinant DNA molecules appear: one has neither mutation, and the other has both mutations. The DNAs that are produced, some of which have recombined while others have not, are then packaged into progeny phage particles. Progeny phage that have packaged the DNAs with these new DNA sequences are **recombinant types** because they are unlike either parent (see chapter 3). In the example, progeny phage that have packaged a DNA molecule with only one of the mutations are called **parental types** because they are like the original phage that infected the cell. The appearance of recombinant types tells us that recombination has occurred. Note that the decision about what is a recombinant type and what is a parental type depends on how the mutations were distributed between the parental phages. In Figure 7.20, one parental strain had one mutation and the other parent strain had the other mutation. However, one strain could have had both mutations and the other strain could have had neither. In that case, the two recombinant types would have had only one or the other of the two mutations and the two parental types would have had either both mutations or neither mutation.

Figure 7.20 Recombination between two phage mutations. The two different mutant parent phages infect the same permissive host cell, and their DNA replicates. Crossovers occur in the region between the two mutations, giving rise to recombinant types that are unlike either parent phage. Only the positions of the mutated base pairs are shown. The DNA of one parent phage is shown in black, and that of the other is shown in blue. doi:10.1128/9781555817169.ch7.f7.20

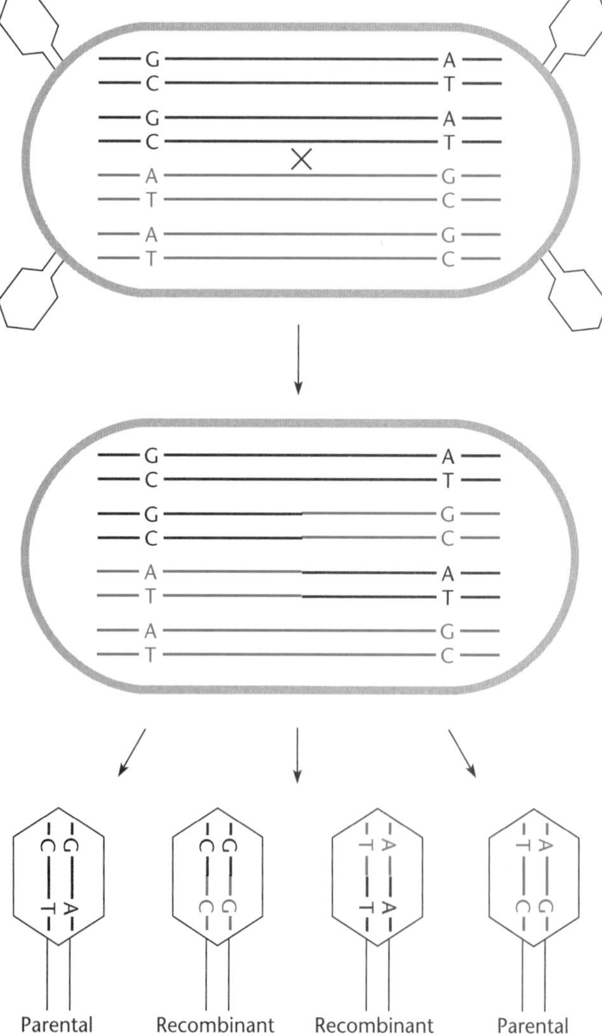

Recombination Frequency

Unlike in bacterial genetics, where mutations are mapped based on whether they are included in the same piece of donor DNA or on when they are transferred during Hfr crosses, mapping in phages (and most other organisms) is determined by recombination frequencies. The closer together the regions of sequence difference are to each other in the DNA, the less room there is between them for a crossover to occur. Therefore, the frequency of recombinant-type progeny is a measure of how far apart the mutations are in the DNA of the phage. This number is usually expressed as the **recombination frequency**. The general equation for recombination frequency used in all organisms from bacteria to fruit flies to humans is as follows: recombination frequency = number of recombinant progeny/total number of progeny. When the recombination frequency is expressed as a percentage, it is called the **map unit**. For example, the regions of two mutations in the DNA give a recombination frequency of 0.01 if 1 in 100 of the progeny is a recombinant type. The regions of the two mutations are then 0.01×100, or 1, map unit apart.

Different organisms differ greatly in their recombination efficiencies; therefore, map distance is only a relative measure, and a map unit represents different physical lengths of DNA for different organisms. Also, the recombination frequency can indicate the proximity of two mutated regions only when the mutations are not too far apart. If they are far apart, two or more crossovers often occur between them, reducing the apparent recombination frequency. Note that while one crossover between the regions of two mutations creates recombinant types, two crossovers recreate the parental types. In general, odd numbers of crossovers produce recombinant types and even numbers recreate parental types. There is also the technical difficulty of accurately measuring the number of parent phages of each type that are added. Unless they are added in equal numbers, the apparent recombination frequency will be lowered.

Complementation Tests

Complementation tests are easy to do with phages and other viruses, in fact, much easier than they are to do with bacteria. As with recombination tests, to perform a complementation test with phages, cells are infected simultaneously with different mutant strains of a particular phage. Now, however, rather than infecting permissive cells or infecting under conditions that are permissive for both mutations, we infect under conditions that are nonpermissive for both mutations. If the two mutations complement each other, both phages will multiply and form plaques. The phage in these plaques will contain both parental types, as well as any recombinant types. If the complementation occurs, the two mutations are probably in different genes (see Table 3.4 for exceptions). Figure 7.21 illustrates complementation tests with phages.

Genetic Experiments with the *r*II Genes of Phage T4

Because of their historical importance and to include a little history, we shall illustrate the basic principles of phage genetics by using the *r*II genes of phage T4. Experiments with these genes were responsible for many early developments in molecular genetics, including the discovery of the three-letter nature of the code and some of its other properties, the fact that some codons are nonsense, and even the discovery that genes can be subdivided and are not, as was believed by many at the time, like beads on a string. Considering their historical importance, it is ironic that we still do not know what the *r*II gene products do for the phage (see "Phage Defense Mechanisms" below). However, as discussed in chapter 3, one of the advantages of genetic analysis is that one can perform a genetic analysis without knowing the functions of the genes involved.

The name *r*II stands for rapid-lysis mutants type II. Phages with an *r*-type mutation cause the infected cells to lyse more quickly than the normal (r^+) phage, a property that *r*II mutant phages share with the other rapid-lysis mutant types, *r*I and *r*III. Recall from "Phage Lysis" above that the product of *r*I is an antiholin. The function of the *r*III product is unknown. Phages with a rapid-lysis mutation can be distinguished by the appearance of their plaques on *E. coli* B indicator bacteria. The plaques formed by wild-type r^+ phages have fuzzy edges because of lysis inhibition (see "Phage Release" above). However *r*-type mutants do not show lysis inhibition, which causes them to form hard-edged, clear plaques. The hard-edged, clear-plaque phenotype makes it easy to distinguish rapid-lysis mutants from the wild type (Figure 7.22).

A phage mutant that makes clear plaques on *E. coli* B strains could have an *r*I, *r*II, or *r*III mutation. The property of *r*II mutants that distinguishes them from the other types of rapid-lysis mutants is that they cannot multiply in strains of *E. coli* that are lysogenic for the λ prophage; these strains are designated *E. coli* K-12λ, or Kλ, to indicate that they harbor the λ prophage in their chromosomes. Lysogeny is discussed in chapter 8. Whatever the reason for the inability of *r*II mutants to multiply in λ lysogens, it greatly facilitates complementation

Figure 7.21 Tests of complementation between phage mutations. Phages with different mutations infect the same host cell, in which neither mutant phage can multiply. **(Left)** The mutations, represented by the minus signs, in different genes (*M* and *N*). Each mutant phage synthesizes the gene product that the other one cannot make; complementation occurs, and new phage are produced. **(Right)** Both mutations (minus signs) prevent the synthesis of the *M* gene product. There is no complementation, the mutants cannot help each other multiply, and no phage are produced.
doi:10.1128/9781555817169.ch7.f7.21

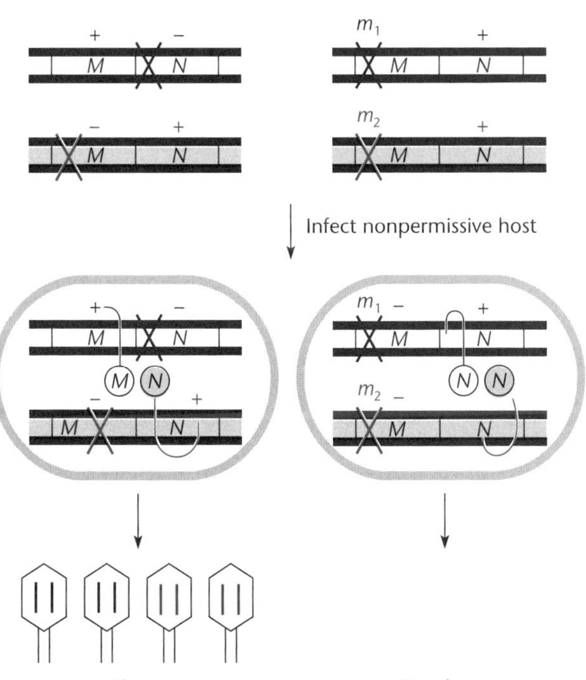

Figure 7.22 Plaques of phage T4. Most plaques are fuzzy edged, but some, due to *r*II or other *r*-type mutants, have hard edges because of rapid lysis of the host cells.
doi:10.1128/9781555817169.ch7.f7.22

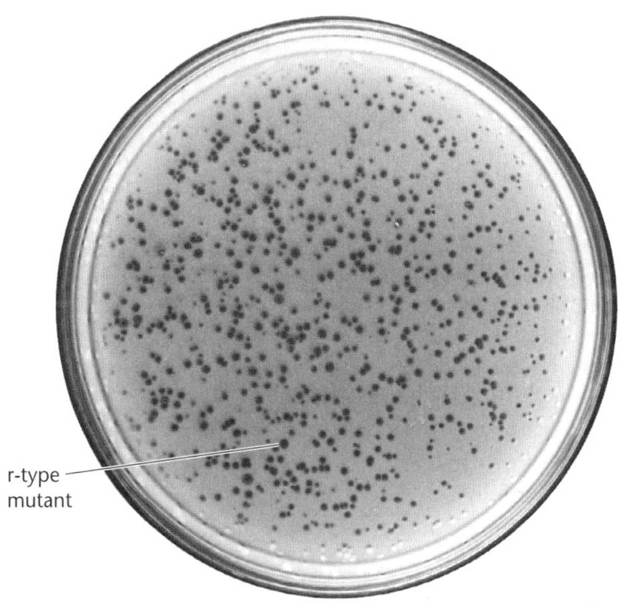

tests with *r*II mutants and makes possible the detection of even very rare recombinant types.

COMPLEMENTATION TESTS WITH *r*II MUTANTS
In about 1950, Seymour Benzer and others realized the potential of using the *r*II genes of T4 to determine the detailed structure of genes. The first question asked was how many genes, or **complementation groups**, are represented by *r*II mutations? To obtain an answer, numerous mutants of the phage with *r*II mutations were isolated, and pairwise complementation tests were performed in which two different *r*II mutants infected cells of the nonpermissive host, *E. coli* Kλ, simultaneously. If the two *r*II mutations complement each other, phage are produced. These complementation tests revealed that all *r*II mutations could be sorted into two complementation groups, or genes, which were named *r*IIA and *r*IIB. The investigators concluded that the *r*IIA and *r*IIB genes encode different polypeptides, both of which are required to make normal-appearing plaques and for multiplication in *E. coli* Kλ.

RECOMBINATION TESTS WITH *r*II MUTANTS
The next step in the genetic analysis was to perform recombination tests between *r*II mutants to determine the locations of the *r*IIA and *r*IIB genes with respect to each other and to order the mutations within the genes. Recombination between *r*II mutants was measured by infecting permissive *E. coli* B with two different *r*II mutants and allowing the phage to multiply. The progeny phage were then plated to measure the frequency of recombinant types.

If recombination can occur between the two *r*II mutants, two different recombinant-type progeny would appear: double mutants with both *r*II mutations and wild-type, or r^+, recombinants with neither mutation. The recombinant types with both mutations are difficult to distinguish from the parental types. However, the r^+ recombinants with neither *r*II mutation are easy to detect because they can multiply and form plaques on *E. coli* Kλ. Therefore, when the progeny of the cross are plated on *E. coli* Kλ, any plaques that appear are due to r^+ recombinants. As discussed above, the recombination frequency equals the total number of recombinant types divided by the total number of progeny of the cross. If we assume that about half of the recombinants are not being detected because they are double-mutant recombinants, the total number of recombinant types is the number of r^+ recombinants multiplied by 2. All the progeny should form plaques on *E. coli* B, so this is a measure of the total progeny. The phage to be plated on the indicator bacteria must be diluted by different amounts before plating because there are many fewer r^+ recombinant progeny than total progeny, and the calculation of the recombination frequency must take these differences in dilution into account.

ORDERING *r*IIA MUTATIONS BY DELETION MAPPING
One of the earliest contributions of *r*II genetics was the demonstration by Benzer that genes are composed of a large number of mutable elements and of the ordering of large numbers of mutations in the *r*II genes, both spontaneous mutations and mutations induced by mutagens (see Benzer, Suggested Reading). As part of his genetic analysis of the structure of the *r*II genes, Benzer wanted to determine how many sites there are for mutations in *r*IIA and *r*IIB and to determine whether all sites within these genes are equally mutable or whether some are preferred. To find these answers, he turned to **deletion mapping**. The principle behind this method is that if a phage with a point mutation is crossed with another phage with a deletion mutation, no wild-type recombinants appear when the point mutation lies within the deleted region. It is much easier to determine whether there are any r^+ recombinants at all than it is to carefully measure recombination frequencies. Therefore, deletion mapping offers a convenient way to map large numbers of mutations.

For this approach, Benzer needed deletion mutations extending for known distances into the *r*II genes. He mapped the endpoint of a number of deletions by crossing them against point mutations in *r*IIA that he had mapped previously. Figure 7.23 shows a set of Benzer deletions that are particularly useful for mapping *r*IIA mutations. These lengthy deletions begin at various places somewhere outside of *r*IIB and remove all of that gene, extending for various distances into *r*IIA. One deletion, *r1272*, extends through the entire *r*II region, completely removing both *r*IIA and *r*IIB.

Figure 7.23 Some of Benzer's *r*II deletions in phage T4. The deletions remove all of *r*IIB and extend various distances into *r*IIA. The bars shaded blue show the region deleted in each of the mutations.
doi:10.1128/9781555817169.ch7.f7.23

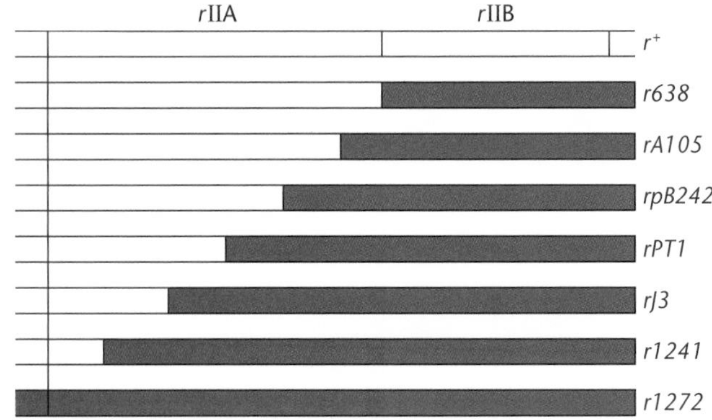

Armed with such deletions with known endpoints, Benzer was able to quickly localize the position of any new *r*IIA point mutation by crossing the mutant phage separately with phage containing each of these deletions. For example, if a point mutation gives r^+ recombinants when crossed with the deletion *rA105* but not when crossed with *rpB242*, the point mutation must lie in the short region between the endpoint of *rA105* and the endpoint of *rpB242*. Therefore, with no more than seven crosses, a mutation could be localized to one of seven segments of the *r*IIA gene. The position of the mutation could be located more precisely through additional crosses with other mutations located within this smaller segment.

MUTATIONAL SPECTRA

The numerous point mutations within the *r*IIA and *r*IIB genes that Benzer found by deletion mapping included spontaneous, as well as induced, mutations. Figure 7.24 illustrates the map locations of some of the spontaneous mutations. Spontaneous mutations can occur everywhere in the *r*II genes, but Benzer noted that some sites are "hot spots," in which many more mutations occur than at other sites. Mutagen-induced mutations also have hot spots, which differ depending on the mutagen and from spontaneous mutations (data not shown).

The tendency of different mutagens to mutate some sites much more frequently than others has practical consequences in genetic analysis (see chapter 3). It is apparent from Figure 7.24 that if Benzer had studied only spontaneous mutations in the *r*II genes, many of them, in fact, almost 30%, would have been at one site in A6c, a major hot spot for spontaneous mutations. Therefore, to obtain a random collection of mutations in a gene requires isolating not only spontaneous mutations, but also mutations induced with different mutagens. Methods are now available that allow essentially random mutagenesis of selected regions of DNA with some organisms. These methods involve the use of special oligonucleotide primers with random changes for site-specific mutagenesis and PCR mutagenesis. Some of these methods are discussed in chapters 1 and 3.

THE *r*II GENES AND THE NATURE OF THE GENETIC CODE

Of all the early experiments with the T4 *r*II genes, some of the most elegant were those by Francis Crick and his collaborators that revealed the nature of the genetic code (see Crick et al., Suggested Reading). These experiments not only have great historical importance, but also are a good illustration of classical genetic principles, and especially suppressor analysis.

At the time Crick and his collaborators began these experiments, he and James Watson had combined the X-ray diffraction data of Rosalind Franklin and Maurice Wilkins and the biochemical data of Erwin Chargaff and others to solve the structure of DNA (see the introduction). This structure indicated that the sequence of bases in DNA determines the sequence of amino acids in a protein, since proteins were known to consist of a chain of amino acids and there were known to be many thousands of different proteins in a cell that did most of the work of the cell. However, the question of how the sequence of bases is read remained. For example, how many bases in DNA encode each amino acid? Does every possible sequence of bases encode an amino acid? Is the code "punctuated," with each code word demarcated, or does the cell merely begin reading at the beginning of a gene and continue to the end, reading a certain number of bases each time? The ease with which the *r*II genes of T4 could be manipulated made the system the obvious choice for use in experiments to answer these questions.

The experiments of Crick et al. were successful for two reasons. First, the extreme N-terminal region of the *r*IIB polypeptide, the so-called B1 region, is nonessential for the activity of the *r*IIB protein. The B1 region can be deleted, or all the amino acids it encodes can be changed, without affecting the activity of the polypeptide. Note that this is not normal for proteins.

Figure 7.24 Mutational spectrum for spontaneous mutations in a short region of the *r*II genes. Each small box represents one mutation observed at that site. Large numbers of boxes at a site indicate hot spots, where spontaneous mutations often occur.
doi:10.1128/9781555817169.ch7.f7.24

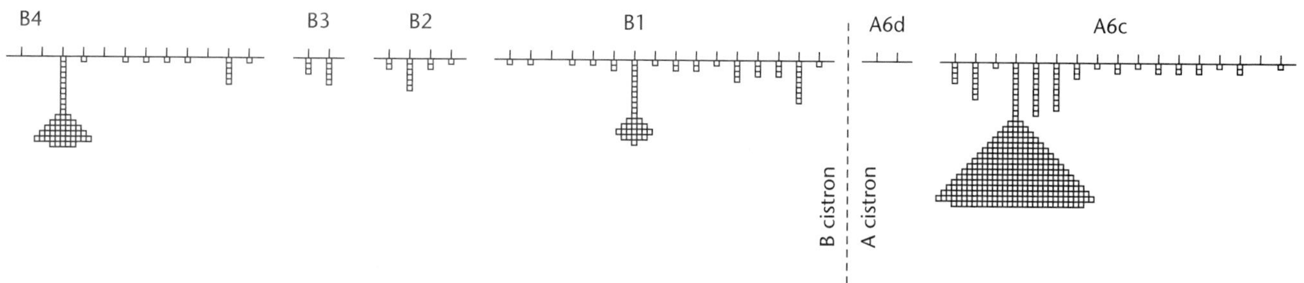

Most proteins cannot tolerate such extensive amino acid changes in any region.

The second reason for the success of these experiments was that the experimenters guessed correctly that acridine dyes specifically induce frameshift mutations by increasing the chance of the removal or addition of a base pair in DNA. This conclusion required a leap of faith at the time. The mutations caused by acridine dyes usually are not leaky but are obviously not deletions, because they map as point mutations and revert. Also, the frequency of revertants of mutations due to acridine dyes increases when the mutant phage are again propagated in the presence of acridine dyes, but not when they are propagated in the presence of base analogs, which even at the time were suspected to cause only base pair changes. It was reasoned that if acridine dye-induced mutations could not be reverted by base pair changes, the mutations induced by acridine dyes themselves could not be base pair changes. This evidence that acridine dyes cause frameshift mutations may seem flimsy in retrospect, yet it was convincing enough to Crick et al. that they proceeded with their experiments on the nature of the genetic code.

Isolating the FC0 Frameshift Mutation in *r*IIB1

The first step in their analysis was to induce a frameshift mutation in the *r*IIB1 region by propagating cells infected with the phage in the presence of the acridine dye proflavin and isolating a mutant with a mutation that prevented multiplication on Kλ and also mapped to the B1 region of *r*IIB. Crick and his colleagues named their first *r*II mutation FC0 for Francis Crick zero. The FC0 mutation prevents T4 multiplication in *E. coli* Kλ because it inactivates the *r*IIB polypeptide. The fact that an acridine-induced mutation in the region encoding the nonessential B1 portion of the B polypeptide can inactivate the *r*IIB polypeptide in itself suggests that the mutations are frameshifts and already says something about how the genetic code is read, since, as mentioned above, merely changing an amino acid in the B1 region should not inactivate the gene product, but a frameshift should change the reading frame of the entire gene if the B1 region is the first to be translated.

Selecting Intragenic Suppressor Mutations of FC0

The next step in the Crick et al. analysis was to select intragenic suppressor mutations of FC0. As discussed in chapter 3, an intragenic suppressor mutation restores the wild-type phenotype by altering the DNA sequence somewhere other than the site of the original mutation but in the same gene. To select suppressors of FC0, Crick et al. merely needed to plate large numbers of FC0 mutant phage on *E. coli* Kλ. A few plaques due to phenotypically r^+ phage appeared. These phage either could have been revertants of the original FC0 mutation or could have had two mutations, the original FC0 mutation plus a suppressor.

To determine which of the r^+ apparent revertants instead had intragenic suppressor mutations, Crick et al. applied the classic genetic test for suppression (see chapter 3 and Figure 7.25). In such a test, an apparent revertant is crossed with the wild type. If any of the progeny are recombinant types with the *r*II mutant phenotype and do not form plaques on Kλ, the interpretation is that the mutation had not reverted but had been suppressed by a second-site mutation that restored the wild-type phenotype. A very low recombination frequency also indicates the suppressor mutation is in the *r*IIB1 region. In the test used by Crick et al., most of the apparent r^+ revertants of FC0 produced a few *r*II mutant recombinants when crossed with the wild type; therefore, the FC0 mutation in these apparent revertants was being suppressed by a closely linked suppressor mutation rather than reverted.

Isolating the Suppressor Mutations

A double mutant with both the FC0 mutation and the suppressor mutation is r^+ and multiplies in *E. coli* Kλ, but would a mutant with a suppressor mutation alone be phenotypically an *r*II mutant or r^+? If the suppressor mutations by themselves produce a phenotypically *r*II mutant, presumably some of the *r*II mutant recombinants obtained by crossing the suppressed FC0 mutant with wild-type T4 would be single-mutant recombinants with only the suppressor mutation. This can be easily tested. If a recombinant is phenotypically *r*II$^+$ because it has

Figure 7.25 Classical genetic test for suppression. Phage that have apparently reverted to wild type (r^+) are crossed with wild-type r^+ phage. **(A)** If the mutation has reverted, all of the progeny will be r^+. **(B)** If the mutation has been suppressed by another mutation, x, there will be some *r*II mutant recombinants among the progeny. The plus sign indicates the wild-type sequence.
doi:10.1128/9781555817169.ch7.f7.25

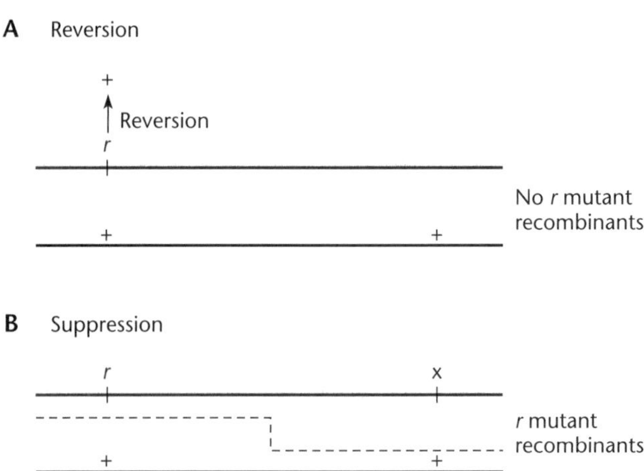

the suppressor mutation rather than the FC0 mutation, it should produce some r^+ recombinants when crossed with the FC0 **single mutant**. Some of the *r*II mutant recombinant phages did produce some r^+ recombinants when crossed with FC0 mutants, indicating that these phages have an *r*II mutation different from the FC0 mutation, presumably the suppressor mutation by itself.

Selecting Suppressor-of-Suppressor Mutations

Because the suppressor mutations of FC0 by themselves make the phage *r*IIB and prevent multiplication on *E. coli* Kλ, the next question was whether the suppressor mutations of FC0 could be suppressed by "suppressor-of-suppressor" mutations. As with the original FC0 mutation, when Crick et al. plated large numbers of T4 with a suppressor mutation on Kλ, they observed a few plaques. Most of them resulted from second-site suppressors. Moreover, these suppressor-of-suppressor mutations were phenotypically *r*IIB mutants when isolated by themselves. This process could be continued indefinitely.

Implications for the Genetic Code

To explain these results, Crick and his collaborators proposed the model shown in Figure 7.26. They proposed that FC0 is a frameshift mutation that alters the reading frame of the *r*IIB gene by adding or removing a base pair so that all the amino acids inserted in the protein from that point on are wrong. This explained how the FC0 mutation can inactivate the *r*IIB polypeptide, even though it occurs in the nonessential N-terminus-encoding B1 region of the gene.

The suppressors of FC0 are also frameshift mutations in the *r*IIB1 region. The suppressors either remove or add a base pair, depending on whether FC0 adds or removes a base pair, respectively. As long as the other mutation has an effect opposite that of FC0, either deleting a base pair if FC0 added one or adding one if FC0 deleted one, an active *r*IIB polypeptide is often synthesized. The results of these experiments had several implications for the genetic code.

The code is unpunctuated. At the time, it was not known if something demarcated the point where a code word in the DNA begins and ends. Consider a language in which all the words have the same number of letters. If there were spaces between the words, we could always read the words of a sentence correctly because we would know where one word ended and the next word began. However, if the words did not have spaces between them, the only way we would know where the words began and ended would be to count the letters. If a letter were left out or added to a word, we would read all the following words wrong. This is what happens when a base pair is added to or deleted from a gene. The remainder of the gene is read wrong; therefore, the code must be unpunctuated.

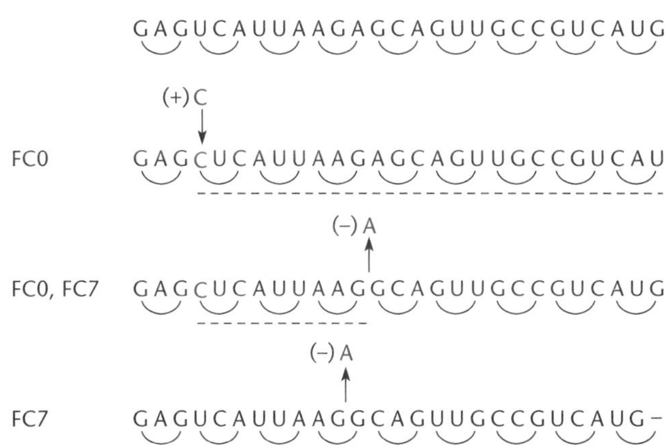

Figure 7.26 Frameshift mutations and suppression. The FC0 frameshift is assumed to be caused by the addition of 1 bp, which alters the reading frame and makes an *r*II mutant phenotype. FC0 can be suppressed by another mutation, FC7, which deletes 1 bp and restores the proper reading frame and the r^+ phenotype. The FC7 mutation by itself confers an *r*II mutant phenotype. Regions translated in the wrong frame are underlined, those in the –1 frame with dashed underlining and those in the +1 frame with solid underlining.
doi:10.1128/9781555817169.ch7.f7.26

The code is three lettered. The experiments of Crick et al. also answered the question of how many letters are in each word of the language, i.e., how many bases in DNA are being read for each amino acid inserted in the protein. At the time, there were theoretical reasons to believe that the number is larger than 2. Since DNA contains four "letters," or bases (A, G, T, and C), only 4 × 4, or 16, possible amino acid code words could be made out of only two of these letters. However, it was known that proteins contain at least 20 different amino acids (the known number is now 22 if you do not count selenocysteine, which is made on the tRNA [see Box 2.3]), so a two-letter code would not yield enough code words for all the amino acids. However, 3 bases per code word results in 4 × 4 × 4, or 64, possible code words, plenty to encode all 20 amino acids.

They could test the assumption that the code is three lettered. The reading frame of a three-letter code would not be altered if 3 bp were added to or removed from the *r*IIB1 region. Continuing with the letter analogy, an extra word would be put in or left out, but all the other words would be read correctly. Suppressors of FC0 have either all added or all removed a base pair, while suppressors of suppressors have all done the opposite. Therefore, if three suppressors of FC0 or three suppressors of suppressors were combined in the *r*IIB1 region, an amino acid would be added to or subtracted from the protein product of the *r*IIB gene and the correct reading frame would be restored, so all the amino acids after that would be correct. Thus, the phage should be r^+ and should multiply in *E. coli* Kλ. When they put three suppressors or suppressors of suppressors together, the phenotype was again usually r^+, indicating that 3 bp in DNA encode each amino acid inserted into a protein.

The code is redundant. The results of these experiments also indicated that the code is redundant, that is, more than one word codes for each amino acid. Crick et al. reasoned that if the code were not redundant, most of the code words, i.e., 64 – 20, or 44, would not encode an amino acid; then, a ribosome translating in the wrong frame would almost immediately encounter a code word that does not encode an amino acid, and translation would cease. The fact that most combinations of suppressors of FC0 and suppressors of suppressors restored the r^+ phenotype indicated that most of the possible code words do encode an amino acid.

Some code words are nonsense and terminate translation. Although their evidence indicated that most code words encode an amino acid, it also indicated that not all of them do. If all possible words signified an amino acid, the entire *r*IIB1 region should be translatable in any frame and a functional polypeptide would result, provided that the correct frame was restored before the translation mechanism entered the remainder of the *r*IIB gene. However, if not all the words encode an amino acid, a "forbidden" code word that does not encode an amino acid might be encountered during translation in the wrong frame. In this situation, not all combinations of suppressors and suppressors of suppressors would restore the r^+ phenotype. Crick et al. observed that some combinations of suppressors and suppressors of suppressors did cause "forbidden" code words to be encountered in the *r*IIB1 region. However, other combinations in the same region resulted in a functional *r*IIB polypeptide. For example, in Figure 7.27, a nonsense codon (UAA) is encountered when a region is translated in the +1 frame because 1 bp was removed. However, no nonsense codons are encountered when 1 bp is added and the same region is translated in the –1 frame.

Postscript on the Crick et al. Experiments

The experiments of Crick et al. laid the groundwork for the subsequent deciphering of the genetic code by Marshall Nirenberg and his colleagues, who assigned an amino acid to each of the 61 3-base sense codons. Other researchers later used reversion and suppression studies to determine that the nonsense codons are UAG, UAA, and UGA, at least in *E. coli* and most other organisms.

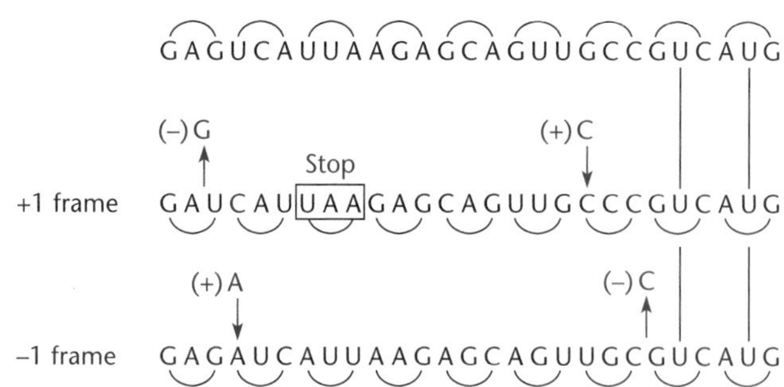

Figure 7.27 Frameshift suppression and nonsense codons. The nonsense codon UAA is encountered in the +1 frame because of a deletion of 1 bp in the DNA. Translation terminates even if the correct translational frame is restored farther downstream. In the –1 frame, due to addition of 1 bp, no nonsense codons are encountered. A downstream deletion of 1 bp restores the correct reading frame, and the active polypeptide is translated. doi:10.1128/9781555817169.ch7.f7.27

DETERMINING THE STRUCTURE OF DUPLICATION MUTATIONS OF THE *r*II REGION

Our final example of the genetic manipulation of the *r*II genes of T4 is the determination of the structure of tandem-duplication mutations in the region (see, for example, Symonds et al., Suggested Reading). These experiments help contrast the differences between complementation and recombination in phages and also illustrate some of the genetic properties of tandem-duplication mutations. In some ways, this analysis is similar to the analysis of *his* duplications presented in chapter 3, but the researchers were able to use the power of phage genetics to determine the structures of some *r*II duplications, including estimates of where the ectopic recombination that formed some of the duplications occurred.

Their isolation of tandem-duplication mutations of the *r*II region depended on the properties of two deletions in the *r*II region, the *r1589* deletion and the aforementioned *r638* deletion (Figure 7.28). The *r1589*

Figure 7.28 The *r*II deletions *r638* and *r1589.* An *r*IIAB fusion protein that has *r*IIB activity is made in a strain with *r1589.* See the text for details. doi:10.1128/9781555817169.ch7.f7.28

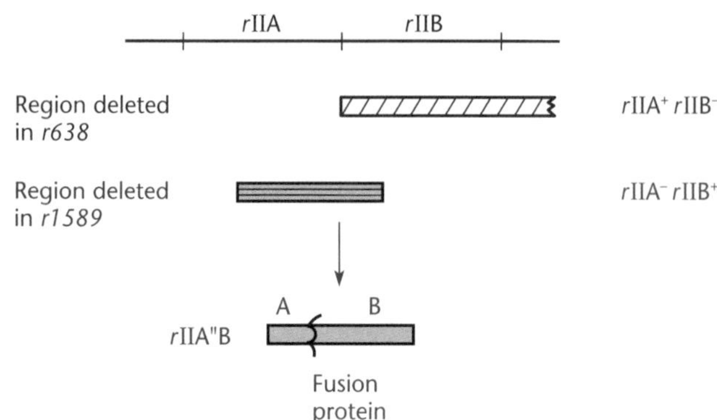

deletion removes the N-terminus-coding B1 region of the *r*IIB gene, as well as the C-terminus-coding part of *r*IIA. Phages with this deletion make a fusion protein with the N terminus of the *r*IIA protein fused to most of *r*IIB and are phenotypically *r*IIA$^-$ *r*IIB$^+$. The deletion mutation *r638* deletes all of *r*IIB but does not enter *r*IIA, so phages with this deletion are *r*IIA$^+$ *r*IIB$^-$. Because one deleted DNA makes the product of the *r*IIA gene and the other makes the product of the *r*IIB gene, the two deletion mutations can complement each other. However, they cannot recombine to give r^+ recombinants, because they overlap, both deleting the B1 region of the *r*IIB gene.

Selecting rII Duplications

Even though recombination should not occur between the two deletions to give r^+ recombinants, when *E. coli* B is infected simultaneously with the two deletion mutants and the progeny are plated on *E. coli* Kλ, a few rare plaques due to phenotypically r^+ phages arise. Rather than being r^+ recombinants, these phenotypically r^+ phages have tandem duplications of the *r*II region (Figure 7.29A). Each copy of the duplicated *r*II region has a different deletion mutation, and these mutations complement each other to give the r^+ phenotype. As is characteristic of duplications, they are very unstable and spontaneously segregate haploid *r*II$^-$ progeny, some of which have one deletion mutation while the others have the other deletion mutation.

Structure of the tandem duplications. The relative frequencies with which the two haploid segregants arise allowed the researchers to determine the structures of some of these duplications. This is illustrated in Figure 7.29B. In the example, the *r638* deletion is in the upstream copy and *r1589* in the downstream copy, and the distance between the upstream endpoint of *r1589* and the upstream site at which the ectopic recombination occurred is "*x*" while the distance from the downstream endpoint of *r638* and the downstream site is "*y*." A crossover between either of the duplicated segments, *x* or *y*, causes the intervening sequences to be deleted and one copy of the duplication to be lost. If the crossover occurs between the two "*x*" segments, the upstream copy is deleted and the copy with the *r1589* deletion remains, while if it occurs between the two "*y*" segments, the upstream copy with the *r638* deletion remains. The relative recombination frequency of each haploid type depends on the relative lengths of "*x*" and "*y*." Such considerations also allowed the researchers to determine which deletion was upstream in the duplication. However, these results can only be considered tentative, since recombination frequency is not an absolutely reliable measure of length on DNA.

Constructing the Genetic-Linkage Map of a Phage

Before the advent of DNA sequencing, the genetic maps of a number of phages were laboriously constructed. A large number of conditional-lethal mutations, both temperature-sensitive and nonsense mutations, as well as other types of mutations, were isolated and assigned to genes by complementation tests. One or more mutations in each gene were then placed on the expanding genetic map by ordering them using recombination frequencies and three-factor crosses (see chapter 3). A picture that shows many of the genes of an organism and how they are ordered with respect to each other is known as the **genetic linkage map** of the organism, so named because it shows the proximity or linkage of the genes to each other. Later, DNA sequencing and other physical methods for mapping DNA, discussed in chapter 1, gave rise to a physical genetic map, which could then be correlated with the genetic linkage map. Nowadays, the ease of sequencing the relatively small DNAs of phages makes it unlikely that the laborious task of constructing a genetic linkage map using phage crosses will ever again be performed for any phage.

FEATURES OF THE GENETIC MAPS OF SOME PHAGES

Using methods such as those described above, physical and genetic linkage maps have been determined for several commonly used phages. Some of these maps appear throughout this chapter and the next, along with the functions of some of the gene products, where known (examples are given in Figures 7.4, 7.6, and 7.13).

The genetic maps of phages have a number of obvious features. One noticeable feature of most phage genetic maps is that genes whose products must physically interact, such as the products involved in head or tail formation, tend to be clustered. In fact, if a gene is not in the same region as other genes whose products are involved in the same function, it is often an indication that the gene has a second, unknown function unrelated to the first. The argument is that the clustering of genes whose products physically interact may allow recombination between closely related phages without disruption of their shared function (Box 7.1). Another striking feature of phage genetic maps is that some are linear while others are circular (compare the T7 map in Figure 7.4 with the T4 map in Figure 7.6). That some phage genetic maps are circular came as a surprise in the early days of mapping phage genes. When the genes of the phage were being ordered from left to right by genetic crosses, researchers discovered that they had come full circle and the first gene was now the next one on the right. Also, the form of the genetic map does not necessarily

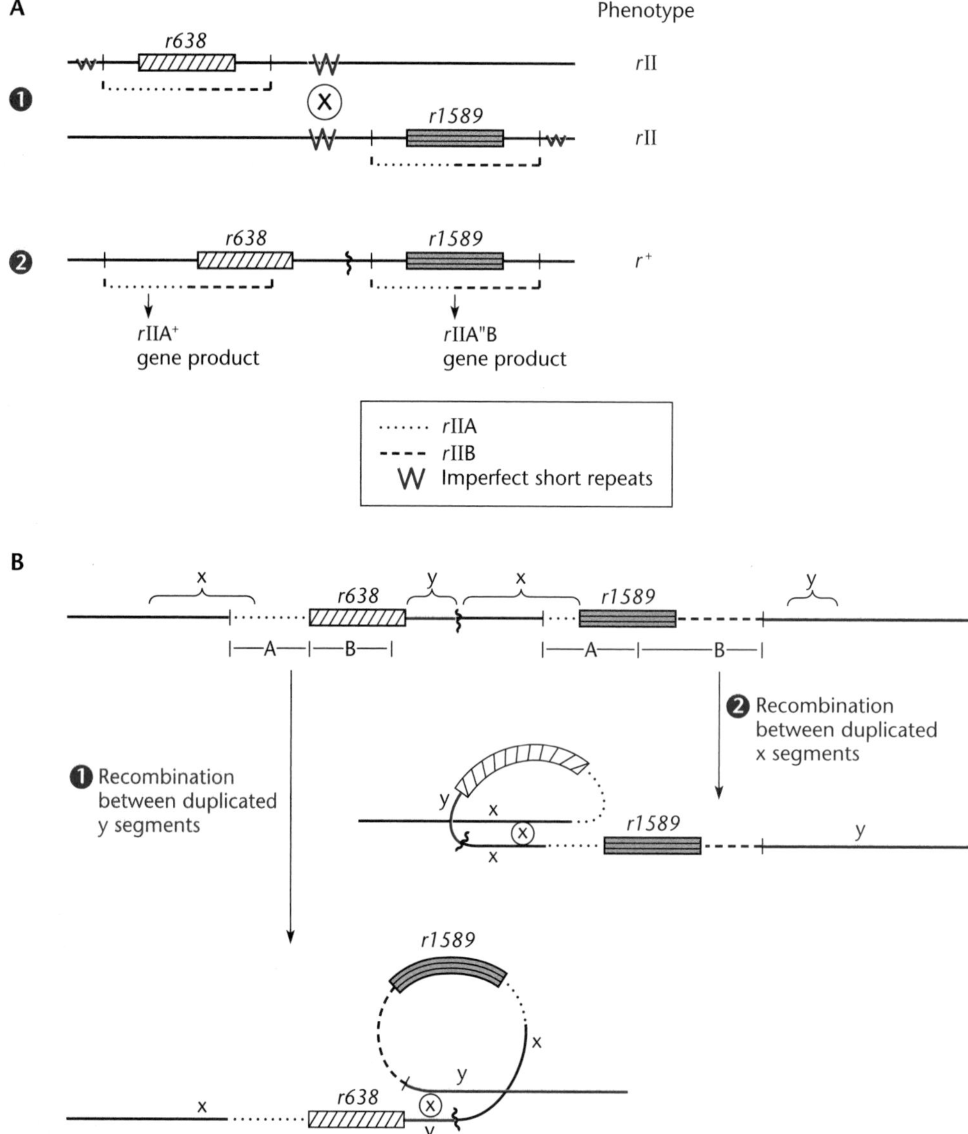

Figure 7.29 Using segregation frequencies to determine the structure of *r*II duplications. **(A)** If a cell is infected by two different mutant phages, one with the *r638* deletion mutation and the other with the *r1589* deletion, ectopic recombination between short repeated sequences flanking the *r*II region (W) may occur at low frequency, giving rise to a duplication of the *r*II region where one copy has one deletion and the other has the other deletion. The two deletions can complement each other, giving the r^+ phenotype, which allows the easy selection of phage with the duplication. **(B)** Recombination between the duplicated regions can destroy the duplication. Which deletion mutation remains in the haploid segregant depends on where the recombination occurs. In the example shown, recombination between the downstream "y" regions in the duplicated segments gives rise to a haploid segregant with only the *r638* deletion **(1)**, while recombination between the upstream "x" regions in the duplicated segments gives rise to a haploid segregant with only the *r1589* deletion **(2)**. The frequency of each of these segregants depends on the relative lengths of the x and y regions. In this example, *r1589* haploid segregants are about three times more frequent than *r638* segregants because x is approximately three times as long as y. A circled x indicates a crossover. doi:10.1128/9781555817169.ch7.f7.29

correlate with the linearity or circularity of the phage DNA itself. Some phages with circular DNA in the cell, such as λ during the early stages of replication (see chapter 8), have a linear linkage map (see Figure 8.2), while some with linear DNA that never circularizes, such as T4, have a circular map (Figure 7.6). To understand how the various genetic maps arise, we need to review how some phages replicate their DNA and how it is packaged into phage heads.

Phage λ

Phage λ has a linear genetic map, even though the DNA forms a circle after it enters the cell. Phage λ has a linear map because its concatemers are cleaved at unique *cos* sites before being packaged into the phage head. The positions of these *cos* sites determine the end of the linear genetic map. As an illustration, consider a cross between two phages with mutations in the *A* and *R* genes at opposite ends of the phage DNA (see the λ map in Figure 8.2). Even though different parental alleles of the *A* and *R* genes can be next to each other in the concatemers prior to packaging, these alleles are separated when the DNA is cut at the *cos* site during packaging of the DNA. Therefore, the *A* and *R* genes appear to be far apart and essentially unlinked when one measures recombination frequencies in genetic crosses, and the genetic map is linear, with the *A* and *R* genes at opposite ends. For this reason, all types of phages that package DNA from unique *pac* or *cos* sites have linear genetic linkage maps with the ends defined by the position of the *pac* or *cos* site, whether or not the DNA ever cyclizes.

Phage T4

In contrast, phage T4 has a circular genetic map, even though its DNA never forms a circle. The reason for its circular map is that T4 has at most only weak *pac* sites and the DNA is packaged by a headful mechanism from long concatemers (see Figure 7.17). Consequently, any two T4 phage DNAs in different phage heads from the same infection do not have the same ends but are cyclic permutations of each other. Therefore, genes that are next to each other in the concatemers will still be together in most of the phage heads unless they happen to be on the terminal redundancy, which is only 3% of the genome, and so will appear linked in crosses or when the DNA, which contains a mixture of molecules with different endpoints, is sequenced, producing a circular map.

Phage P22

Phage P22 is a phage of *Salmonella* that we have discussed in connection with transduction. Although it has a tail structure different from that of λ and thus is in a different morphological family (Table 7.1), it replicates by a similar mechanism. However, unlike λ, it has a circular linkage map. The difference is that P22 begins packaging at a unique *pac* site or *cos* site, like λ, but then packages a few genomes by a processive headful mechanism, like T4, giving rise to a circular genetic map and making it a good transducing phage, as discussed below.

Phage P1

In its head, phage P1 has linear DNA that forms a circle by recombination between terminally repeated sequences at its ends after infection. The DNA then replicates as a circle and forms concatemers from which the DNA is packaged by a headful mechanism, which makes it a good transducing phage. However, unlike most phages that package DNA by a headful mechanism, the genetic map of P1 is linear, because it has a very active site-specific recombination system called *cre-lox*, which promotes recombination at a particular site in the DNA. Because recombination at this site is so frequent, genetic markers on either side of the site appear to be unlinked, giving rise to a linear map terminating at the *cre-lox* site. The function of this site-specific recombination system in phage development is unknown, but it may resolve dimeric circles in the prophage state, where P1 DNA exists as a circular plasmid (see chapter 8). Because of its stability and specificity, the *cre-lox* recombination system of P1 has many uses. For example, it was used to show that some proteins encoded by the Ti plasmid enter the plant cell nucleus along with the T-DNA during the formation of crown gall tumors. It also has served as a model system for site-specific recombinases, as discussed in chapter 9.

Phage Defense Mechanisms

Bacteria live in a veritable sea of phages, and their survival as species depends upon defending themselves from them. The battle between bacteria and their phages is a continuous "arms race" in which the bacteria develop a defense against the phages and the phages develop a way to circumvent it. Whether a particular phage defense mechanism works against a particular phage depends on what stage of the arms race the two are in. Also, many phage defense mechanisms may have other purposes in the cell. They may also defend against other incoming DNA elements, such as plasmids, or may play a role in developmental processes, such as killing some cells in a biofilm to aid dispersal from the overgrown biofilm. Many phage defense mechanisms are not a normal part of the chromosome but are encoded on exchangeable DNA elements, such as prophages or plasmids, so they can be readily exchanged among various bacteria in the population. We mention some defenses against phages in other chapters. For example, probably one of the reasons that the Mpf systems of self-transmissible plasmids are often inducible is so that male-specific phages that use the pilus as an adsorption site are able to adsorb to only a small percentage of a particular bacterial population at any one time, preventing contagion by the phage.

Restriction-Modification Systems

Restriction-modification systems are the best-known type of phage defense mechanism. In chapter 1, we discussed the uses of restriction-modification systems in cloning and other manipulations with DNA. These systems degrade DNA by cutting at DNA sequences in which neighboring bases have not been properly modified. This allows the cell to distinguish its own DNA from incoming foreign DNAs, such as phage DNAs. In some cases, it can be the presence rather than the absence of a modified base that distinguishes a foreign DNA from the bacterium's own DNA. For example, T4 and many other T-even phages of *E. coli* and other gram-negative bacteria have the unusual base hydroxymethylcytosine in their DNA instead of the usual cytosine, and the presence of this unusual base makes them immune to many restriction endonucleases. However, it makes them susceptible to other restriction endonucleases, which cut only DNA in which the cytosines have been modified. In turn, many T-even phages attach glucose to the hydroxymethyl groups on their DNA, which then renders them immune to these restriction systems, and the arms race continues. Incidentally, the specificity of these restriction endonucleases for DNA with modified cytosines suggests that their major purpose in nature is to defend against this ubiquitous family of phages with hydroxymethylcytosine-containing DNA.

The roles of other restriction-modification systems may be more varied. Some may also serve as plasmid addiction systems. If the restriction system is encoded on a plasmid and the modification part of the system is less stable than the restriction part, then if a cell is cured of the plasmid and neither part continues to be synthesized, the restriction will outlast the modification of the DNA and the cell's own DNA will be killed. This may explain the existence of "8-hitter" restriction systems that recognize a specific 8-bp sequence. Their function was always something of a mystery, since by chance, most phage or plasmid DNAs do not contain even one such site.

Abi Systems

Some phage defense systems kill the cell in response to phage infection. These systems are sometimes called Abi systems for *ab*ortive *i*nfection because the infection begins normally but then something happens to the cell that causes it to die and the infection to be aborted (see Labrie et al., Suggested Reading). This has been compared to apoptosis in multicellular eukaryotes, where cells undergo programmed death if they are infected by a virus to prevent spread of the virus through the cells of the organism. There is an obvious advantage in a multicellular organism to such self-inflicted killing of a virus-infected cell, since it blocks multiplication of the virus before it can spread throughout the other cells of the organism. But such suicidal systems seem to make less sense for a single-celled bacterium. By killing themselves, they seem to be acting "altruistically" in helping other bacteria, even though they are doing nothing for themselves. However, even bacteria do not usually live as free single cells but are grouped together in structures called biofilms or other gatherings of bacteria, where they are surrounded by others of their kind. Preventing the spread of the phage through their relatives amounts to a kind of sibling selection, where the survival of other bacteria with identical genetic makeup is promoted by the sacrifice of the one.

The classic example of an Abi phage defense mechanism is the Rex exclusion system, encoded by λ phage. The Rex system was mentioned earlier in the chapter, since it was used in the classic analyses of the *r*II genes of bacteriophage T4. Nevertheless, how the system works to kill the T4-infected cell is still not completely understood. The Rex system consists of two proteins, the products of two λ genes, *rexA* and *rexB*, that are among the few λ genes expressed in a cell that is lysogenic for λ (see chapter 8). Somehow, these gene products are activated after infection by a wide variety of phages and destroy the membrane potential of the cell, thereby reversing the ATP synthase in the membrane and depleting the cell of ATP. The development of the infecting phage is thus aborted. It is not clear what feature of a phage-infected cell activates the Rex gene products, but it might be some sort of recombination intermediate that is common to the replication of many types of phages and is lacking or very infrequent in the uninfected cell.

Some other types of Abi systems are better understood. Two of them are directed against T4 and other T-even phages and stop the development of the infecting phage by cleaving a component of the translation apparatus after infection by the phage, thereby blocking all host cell translation and the multiplication of the infecting phage. One such Abi system is a protease encoded by the defective prophage e14 in *E. coli* K-12 strains. The protease is activated after infection by the binding of a short region of the newly synthesized major head protein, gp23, to elongation factor Tu (EF-Tu). The protease will cleave EF-Tu only when it is bound to that region of the phage head protein, so it is not active in an uninfected cell. Cleaving EF-Tu blocks translation and prevents multiplication of the phage. The function of binding the region of the head protein to EF-Tu may be to delay translation to allow time for the N terminus of the head protein to be taken up by the chaperonin,

GroEL, before translation of the head protein continues. However, this binding signals the Abi system that the cell has been infected by a T-even phage and that it should be sacrificed to prevent spread of the phage.

Another, better-understood Abi system is the Prr system, a lysine tRNA-specific ribonuclease (RNase) encoded by a prophage related to P1. This RNase is normally "masked" by binding to a restriction-modification system. The RNase is activated after T4 infection by a T4-encoded antirestriction peptide that dissociates the restriction-modification system, releasing the RNase, as well as an increase in cellular dTTP levels required for phage DNA replication. The activated RNase then cleaves the host lysine tRNA, blocking translation and multiplication of the phage. By encoding the antirestriction peptide, the phage plants the "seed of its own destruction." When it attempts to inactivate the restriction-modification system, it activates the RNase instead, blocking cellular translation and stopping its own development.

CRISPR Loci

A very different type of phage defense system is represented by CRISPR loci. This is an adaptive immune system reminiscent of our own adaptive immune system in that it can establish immunity to a heretofore unknown phage or transmissible plasmid after being exposed to it. The acronym CRISPR stands for cluster of regularly interspaced short palindromic repeats, and their striking structure caused them to be noticed in bacterial genome sequences long before their role in phage defense was realized. CRISPR loci are comprised of directly repeated identical sequences, often about 30 bp long, separated by apparently random "spacer" sequences, also often about 30 bp long, although the lengths of both vary significantly from one type of CRISPR locus to another. The number of repeats also varies greatly, with some types of CRISPR loci having only a few repeats and others as many as 500. They exist in most bacterial and archaeal genomes, and often more than one type of CRISPR locus is found in the same genome.

The clue that CRISPRs are a type of phage and plasmid defense system came when it was noticed that the spacer sequences often perfectly match a sequence from a phage or plasmid DNA found in that species of bacterium. An exact match of this length could not be due to chance, and the spacer must somehow have been copied directly from the DNA or RNA of the phage or plasmid. Different spacers come from the genomes of different phages or plasmids and from different sequences within the genomes of these phages and plasmids, explaining why the spacer sequences in a CRISPR are different from each other. The sequence in the plasmid or phage that is identical to a spacer was given the name protospacer because it gave rise to the spacer in the CRISPR locus (the prefix proto is from the Greek for first or original).

Elegant experiments with *Streptococcus thermophilus* confirmed that a CRISPR could acquire a spacer from a phage grown on it and that the spacer could confer resistance to that phage, presumably by directing the CRISPR to attack the protospacer sequence in the DNA or RNA of the phage (see Barrangou et al., Suggested Reading). The match has to be perfect, because mutant phages selected for the ability to form plaques on bacteria containing that spacer in a CRISPR locus often have a single base pair change in the corresponding protospacer. Others showed that the target of at least some CRISPR loci was the DNA rather than the RNA. They had found that a spacer sequence in a CRISPR locus in a *Staphylococcus* strain matched a sequence in a plasmid found commonly in that strain and that it prevented the transfer of the plasmid into the strain. To show whether the CRISPR was attacking the DNA or mRNA made in that region of the plasmid, they introduced a self-splicing intron into the protospacer sequence in the plasmid (see Box 2.4). The transfer of the plasmid was no longer prevented by the spacer in the CRISPR. Apparently, the mRNA could not be the target, since the intron was precisely spliced out of it. Only the DNA still contained the self-splicing intron. Also, the spacer sequence in the plasmid could be inverted, along with flanking sequences on either side of it, and the transfer was still prevented, even though the same sequence would not be transcribed into mRNA on the other strand. Not only is the DNA, rather than the RNA, the target of the CRISPR, at least in this system, but DNA also seems to be the source of the sequence in the CRISPR spacers (the adaptive phase), since spacers with both polarities from the same region of a phage can be found, even if only one strand from that region is transcribed into mRNA.

The molecular details of how CRISPR loci pick up sequences from infecting phage and plasmid DNAs (the adaptive phase) and how they then use them to attack the complementary sequence in the protospacer of incoming DNA (the immunity phase) are being uncovered. The model in Figure 7.30 summarizes the current understanding of this process. A number of conserved proteins, encoded by *cas* (CRISPR-*as*sociated) genes that are adjacent to the locus, are obviously involved in these processes. The roles of some of these proteins can be guessed from their biochemical activities and the relatedness of their sequences to the sequences of other proteins with known functions. Also, some of them can be assigned roles in either the adaptive phase or the immunity phase based on whether mutations in the particular *cas* gene can prevent the cell from acquiring resistance to the phage (the adaptive phase) or make an already resistant

A CRISPR-associated elements

❶ Bacterial chromosome

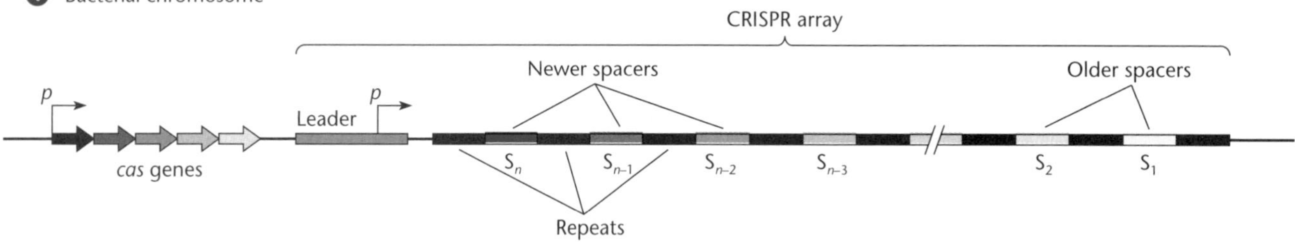

CRISPR (clustered regularly interspaced short palindromic repeats)
CRISPR array: one to hundreds of repeats

Repeat: 21–47 bp (CAGTTCCCCGCGCCAGCGGGGATAAACCG)

Spacer sample: ~30 bp $(\text{CTTTCGCAGACGCGCGGCGATACGCTCACGCA})_n$

Leader: Several hundred noncoding base pairs

cas genes (*cas*cade): *cas* complex for CRISPR-associated antivirus defense; acquisition of new spacers; immunity, e.g., processing pre-crRNAs

❷ Phage or plasmid DNA

ps_n

PAM

PAM: protospacer-adjacent motif (example) (CT/AT)

ps_n: proto spacer $(\text{CTTTCGCAGACGCGCGGCGATACGCTCACGCA})_n$

B Construction of CRISPR array: adaptive phase

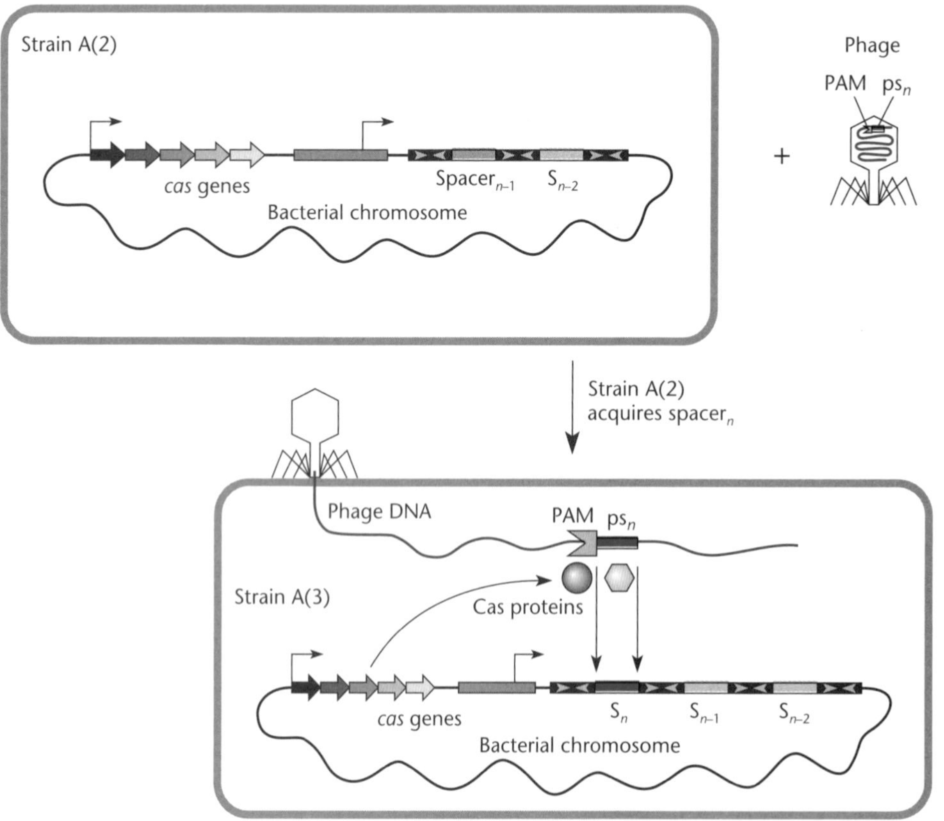

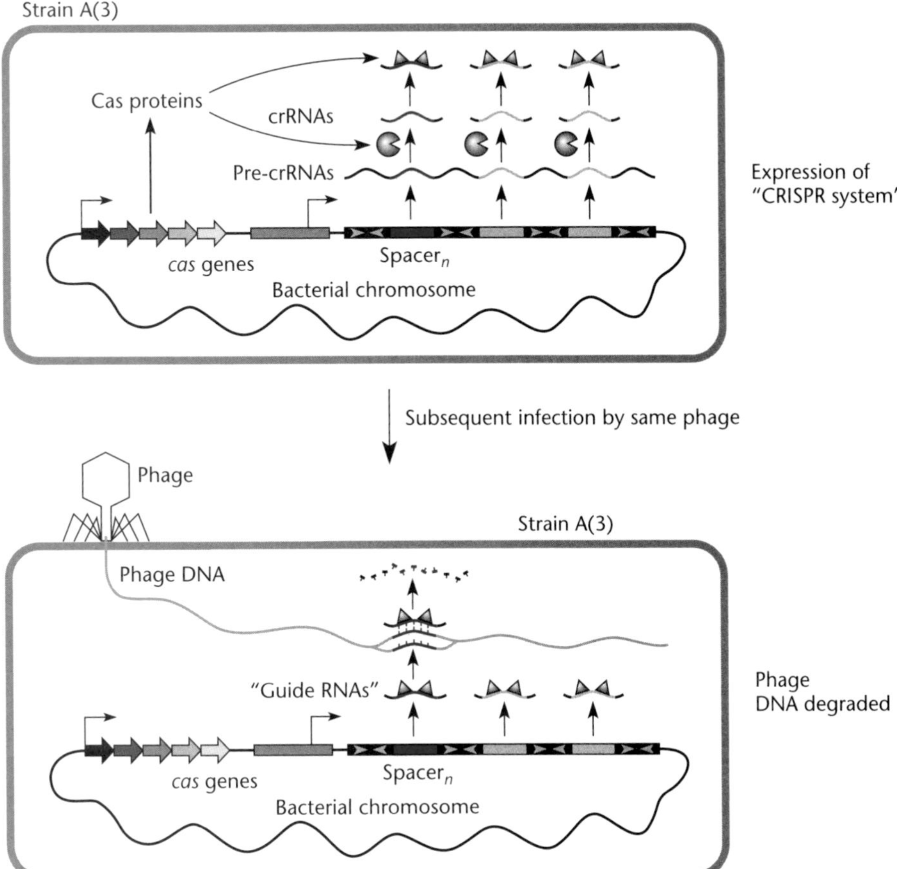

Figure 7.30 A model for how new CRISPR spacers are generated and how they direct the attack on incoming phage DNA. **(A) (1)** Typical CRISPR array. A number of repeated sequences interspersed with spacers s_n to s_1 (blue) of about the same length but with different sequences. A leader region array contains a promoter (p) from which the CRISPR array is transcribed and CRISPR-associated *cas* genes whose products are involved in taking up sequences and targeting protospacer sequences. The spacers closest to the leader sequence are the ones that have been taken up most recently. **(2)** A protospacer (ps_n) from a phage or plasmid with exactly the same sequence as one of the spacers (s_n) in the CRISPR. Adjacent to the protospacer is a protospacer-adjacent motif (PAM) sequence, which identifies it as a sequence to be taken up by the CRISPR. **(B)** The adaptive phase: picking up a protospacer sequence. One or more Cas proteins recognize a PAM sequence in incoming phage DNA and integrate the adjacent ps sequence (ps_n) as a new spacer (s_n) into the CRISPR array on the end closest to the leader. All the other spacers move down, with the one on the far end being discarded. **(C)** The immunity phase: targeting the incoming protospacer sequence on the phage DNA. The CRISPR array is transcribed into a single long RNA, which is cut in the repeat sequences to make "guide" *CR*ISPR RNAs (crRNAs), each with the sequence of one of the spacers. When the same type of phage infects the cell, the pairing of a crRNA with the s_n sequence with the identical protospacer sequence (ps_n) in the incoming phage DNA directs one or more of the Cas proteins to cut the protospacer DNA, inactivating the phage.
doi:10.1128/9781555817169.ch7.f7.30

cell lose its resistance (the immune phase). In the adaptive phase, Cas proteins incorporate new spacers next to the end of the CRISPR locus that contains a leader sequence and a promoter, duplicating the repeat sequence in the process. As new spacers arrive, they displace the old spacers, which move down the CRISPR locus, leaving a fossil trail of old resistance, until the oldest spacers eventually disappear from the other end.

More is understood about how the CRISPR locus attacks incoming DNA, i.e., the immunity phase. The CRISPR locus is transcribed from the promoter in the leader into a long RNA, out of which smaller RNAs are cut by cutting at the same site in each of the repeat sequences. Each of these small RNAs then contains a spacer sequence, which can serve as guide RNA to direct the DNA endonuclease activity of one of the Cas proteins to its complementary sequence in the protospacer in the incoming phage or plasmid DNA. They also cut only DNA that has a particular sequence next to it called the PAM (protospacer adjacent motif) sequence. In the case of CRISPR1 of *S. thermophilus*, this PAM sequence seems to be two purines followed by AGAA and then a pyrimidine. As evidence, this sequence is next to most protospacers in the phage DNA, and mutations in the sequence can also prevent the cutting (see Barrangou et al., Suggested Reading). Apparently, this PAM sequence is recognized by one or more of the Cas proteins as a requirement both for taking up the adjacent protospacer sequence and for cutting it. The requirement for an adjacent PAM sequence to cut a protospacer is sufficient to explain why the same spacer in the CRISPR locus itself is not cut, causing autoimmunity, since the PAM sequence is not included in the spacer. However, sequences within the repeats on either side of the spacer also inhibit cutting. These flanking repeat sequences are in the guide RNA, indicating that the homology with the spacer must be exact but must not extend into the repeat sequences for it to be cut.

A number of questions remain about CRISPR immunity. One big question is whether or how CRISPR systems avoid picking up chromosomal DNA. If they picked up chromosomal sequences, they would cleave the cell's own DNA and kill the cell. They take up DNA only from a protospacer that is next to a particular PAM sequence, but the known PAM sequences are so short they should occur by chance many times in the chromosomal DNA. Perhaps CRISPRs do often pick up chromosomal DNA, but then the cells are selected against, so those CRISPR loci that survive contain only plasmid or phage sequences as spacers. In fact, it has been shown that a CRISPR engineered to contain a spacer corresponding to λ phage sequences will cleave λ prophage in the chromosome, killing the cell (see Edgar and Qimron, Suggested Reading). Another interesting question is how the CRISPR immunity acquired by an infected cell can be passed on to subsequent generations. Either the original infected bacterium was not killed by the infection that led to immunity or it passed it on in spite of being killed. The cell may not have been killed if the phage is not always lytic and sometimes lysogenizes its host, or the immunity could be acquired and act before the phage has killed the cell, which seems less likely, at least in all cases. Even if the originally infected cell was killed, it might have passed its CRISPR locus with the newly acquired spacer to another cell in the population by transformation, or, ironically, by transduction by the phage itself after the bacterium had lysed.

Generalized Transduction

As discussed in the introduction and chapter 3, bacteriophages not only infect and kill cells, but also sometimes transfer bacterial DNA from one cell to another in the process called **transduction**. There are two types of transduction in bacteria: **generalized transduction,** in which essentially any region of the bacterial DNA can be transferred from one bacterium to another, and **specialized transduction,** in which only certain genes, those close to the attachment site of the prophage, can be transferred (discussed in connection with lysogeny in chapter 8). In this section, we discuss only the mechanism of generalized transduction and refer to it simply as transduction. Furthermore, since the analysis of genetic data obtained by transduction is discussed in chapter 3, we restrict ourselves to the mechanism of transduction and the properties a phage must have to be a transducing phage.

Figure 7.31 gives an overview of the process of transduction. While phages are packaging their own DNA, they sometimes mistakenly package the DNA of the bacterial host instead. These phages are still capable of infecting other cells, but progeny phages are not produced. What happens to the DNA after it enters the cell depends on the source of the bacterial DNA. If the bacterial DNA is a piece of the bacterial chromosome of the same species, it usually has extensive sequence homology to the chromosome and may recombine with the host chromosome to form recombinants. If the piece of DNA that was picked up and injected is a plasmid, it may replicate after it enters the cell and thus be maintained. If the incoming DNA contains a transposon, the transposon may hop, or insert itself, into a host plasmid or chromosome, even if the remainder of the DNA contains no sequence in common with the DNA of the bacterium it entered (see chapter 9).

The nomenclature of transduction is much like that of transformation and conjugation. Phages capable of

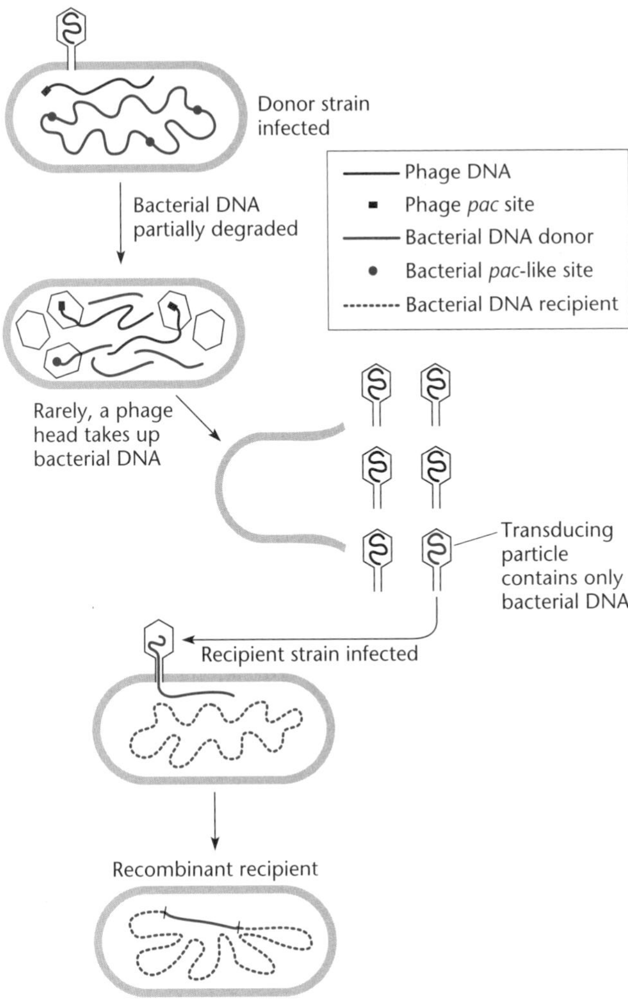

Figure 7.31 Generalized transduction. A phage infects one bacterium, and in the course of packaging DNA into heads, the phage mistakenly packages some bacterial DNA instead of its own DNA into a head because it mistakenly recognizes a *pac* site in the bacterial DNA. In the next infection, this transducing particle is different from most of the phage in the population in that it injects the bacterial DNA instead of phage DNA into the recipient bacterium. If the bacterial DNA that was packaged is a plasmid, it might replicate in the recipient cell, or if it is chromosomal DNA, it might recombine with the chromosome and form a recombinant type.
doi:10.1128/9781555817169.ch7.f7.31

transduction are called **transducing phages**. A phage that has picked up bacterial DNA is called a **transducing particle**. The original bacterial strain in which the transducing particle multiplied and picked up host DNA is called the **donor strain**. The bacterial strain it infects is called the **recipient strain**. Cells that have received DNA from another bacterium by transduction are called **transductants**.

Usually, transductants arise very rarely for a number of reasons. First, mistaken packaging of host DNA is itself rare, and transduced DNA must survive in the recipient cell to form a stable transductant. Since each of these steps has a limited probability of success, transduction can usually be detected only by powerful selection techniques.

What Makes a Transducing Phage?

Not all phages can transduce. To be a generalized transducing phage, the phage must have a number of characteristics. It must not degrade the host DNA completely after infection, or no host DNA will be available to be packaged into phage heads when packaging begins. It also must not be too particular about what DNA it packages. As discussed above, most phages package DNA from sites on the DNA called *pac* or *cos* sites. If a DNA lacks such specific sites, it is usually not packaged. The **packaging sites**, or *pac* sites, of transducing phages must not be so specific that such sequences do not occur by chance in host DNA (Figure 7.31). It also helps if the phage is not too particular about the length of the DNA it packages, so that even relatively nonspecific *pac* sites do not have to be properly spaced.

Some of the most useful transducing phages have a broad host range for adsorption. Then, they might be used to introduce DNA into a broad variety of bacteria. It is important to realize that to transduce, a phage does not actually have to be capable of multiplying in the recipient host, only of adsorbing to it and injecting its DNA. Transduction from other hosts using a broad-host-range phage sometimes offers a way of introducing plasmids or other DNA elements into hosts for which other means of gene exchange are not available. Even if a type of phage does not have all of the above characteristics, it may be possible to engineer it into a good transducer, and this has been done for some phages.

Table 7.3 compares two good transducing phages, P1 and P22. Phage P1, which infects gram-negative bacteria, is a good transducer because it has less *pac* site specificity than most phages and packages DNA via a headful mechanism; therefore, it efficiently packages host DNA. About 1 in 10^6 phage P1 particles transduces a particular marker. It also has a very broad host range for adsorption and can transduce DNA from *E. coli* into a wide variety of other gram-negative bacteria, including members of the genera *Klebsiella* and *Myxococcus*. It cannot multiply in hosts other than *E. coli*, but it can transfer plasmids, etc., from *E. coli*, on which it can be propagated, into these other hosts.

The *Salmonella enterica* serovar Typhimurium phage P22 is also a very good transducer and, in fact, was the first transducing phage to be discovered (see Zinder and Lederberg, Suggested Reading). Like P1, P22 has *pac* sites that are not too specific and packages DNA by a headful mechanism. From a single *pac*-like site, about

Table 7.3 Characteristics of generalized transducing phages P1 and P22

Characteristic	Value in phage: P1	P22
Length (kb) of DNA packaged	100	44
Length (%) of chromosome transduced	2	1
Packaging mechanism	Sequential headful	Sequential headful
Specificity of markers transduced	Almost none	Some markers transduced at low frequency
Packaging of host DNA	Packaged from ends	Packaged from *pac*-like sequences
% of transducing particles in lysate	1	2
% of transduced DNA recombined into chromosome	1–2	1–2

10 headfuls of DNA can be packaged. Because of even this limited *pac* site specificity, however, some regions of *Salmonella* DNA are transduced by P22 at a much higher frequency than others.

As mentioned above, other phages are not normally transducing phages but can be converted into them by special treatments. For example, T4 normally degrades the host DNA after infection but works extremely well as a transducing phage if its genes for the degradation of host DNA have been mutated. Because phage T4 DNA packaging does not require very specific *pac* sites, it packages any DNA, including the host DNA, with almost equal efficiency. It also packages by a headful mechanism (Figure 7.17), so even these very nonspecific *pac* sites do not have to be evenly spaced.

In contrast, phage λ does not work well for generalized transduction because it normally packages DNA between two *cos* sites rather than by a headful mechanism. It very infrequently picks up host DNA by mistake, but then it does not cut the DNA properly when the head is filled unless another *cos*-like sequence happens to lie at a genome length distance along the DNA. Thus, potential transducing particles usually have DNA hanging out of them that must be removed with DNase before the tails can be added. Even with these and other manipulations, λ works poorly as a generalized transducer. However, because of the efficiency with which it packages cosmids of the correct length containing *cos* sites, it is very useful for making cosmid libraries, in vitro packaging, and introducing DNA into *E. coli* cells.

Transducing phages have been isolated for a wide variety of bacteria and have greatly aided genetic analysis of these bacteria. Transduction is particularly useful for moving alleles into different strains of bacteria and making isogenic strains that differ only in a small region of their chromosomes. This makes transduction very useful for strain construction and gene knockouts for functional genomics (see chapter 3). However, if no transducing phage is known for a particular strain of bacterium, finding one can be very time-consuming. Therefore, for such bacteria, transduction needs to be replaced by other methods.

Shuttle Phasmids

One variation on transduction involves **shuttle phasmids**, which have been constructed and used effectively in bacterial strains of the gram-positive genera *Mycobacterium* and *Streptomyces*. As mentioned above, a phasmid is part phage and part plasmid. In a sense, this is a form of restricted transduction, because only some genes are transduced, those that have been cloned into the phasmid. Shuttle phasmids are useful for studying bacteria not closely related to *E. coli* or are constructed by combining parts of a phage of the bacterium being studied with a plasmid from *E. coli*. In fact, problems can arise if attempts are made to construct a shuttle phasmid for a bacterium closely related to *E. coli*, because some of the phage genes may be expressed in *E. coli*. Many phage gene products are toxic to bacteria and may kill the cell.

Because the DNA is introduced into the bacterium by infection with one of its own phages, shuttle phasmids allow the very efficient transfer of DNA into bacteria for which no other efficient DNA transfer system is available. Many applications in molecular genetics require efficient DNA transfer, including transposon mutagenesis and allele replacement.

In one procedure for constructing a shuttle phasmid (see Bardarov et al., Suggested Reading), a phage that can form plaques on a lawn of the bacterium is isolated. Little need be known about the phage except that it should have a genome size similar to that of λ (about 40 to 50 kilobases [kb]), which is a common size for phages. Also, like λ, it should have sticky cohesive ends that pair with each other; this is also a common characteristic of many phages, especially the *Siphoviridae*. The phage is purified, and the DNA is extracted. The DNAs are then ligated to each other to form a large concatemer. The concatemers are partially digested with a four-hitter restriction endonuclease, such as Sau3A, to leave pieces of about 40 to 45 kb, about the genome length of the phage but with random ends. The pieces are then ligated to the halves of a cosmid. A cosmid is a plasmid that contains the *pac* (*cos*) site of λ phage, a plasmid origin of replication that functions in *E. coli*, and a selectable antibiotic resistance gene in *E. coli*, such as for ampicillin. Cosmids are discussed in chapter 8. After ligation, long concatemers form in which phage genomes are sometimes bracketed by the halves of the cosmid. The cosmids can be packaged in λ heads by mixing them with λ heads and tails in the test tube, a process called in vitro packaging. These λ phage particles are then used to infect *E. coli*, selecting for the antibiotic resistance gene on the cosmid. The antibiotic-resistant cells contain the phage genome with the cosmid inserted at various places, rather than just the cosmid, which is not long enough to be packaged into a λ phage head, or just the DNA of the phage of the bacterium being studied, which would not be packaged in λ heads or replicate in *E. coli* even if it did get in. Moreover, even if it could replicate, it would not confer resistance to the antibiotic on *E. coli*.

To determine which of these antibiotic-resistant *E. coli* cells contains a usable shuttle phasmid, they are pooled and the DNA is electroporated into the bacterium being studied. These cells are then mixed with indicator bacteria of the same type and incubated to allow plaques to form. Unlike in *E. coli*, the phage DNA can replicate and direct the synthesis of new phage gene products in this host, provided none of its own essential genes have been inactivated by insertion of the cosmid. Any plaques that form may contain phage that have packaged a useful phasmid. However, they should be tested by repeating the process of using their DNA to form concatemers, packaging them into λ phage heads, and reinfecting *E. coli*, selecting the antibiotic resistance on the cosmid.

Once a useful phasmid has been obtained for the bacterium being studied, any DNA can be introduced into it by replacing the cosmid in the phasmid with a cosmid containing a cloned piece of DNA from the bacterium being studied. Then, the DNA can be reintroduced into the bacterium with high efficiency merely by infecting the bacterium with the phasmid. In some of these applications, such as transposon mutagenesis or allele replacement, it may be necessary to make a suicide vector from the phasmid. This can be done by isolating mutants of the phasmid that cannot replicate under some conditions, for example, at high temperatures. The use of phage suicide vectors for transposon mutagenesis is discussed in chapter 9, and the process is similar for allele replacement.

One promising recent application of shuttle phasmids is in phage typing and testing for antibiotic susceptibility (see, for example, Banaiee et al., Suggested Reading). Standard techniques for testing bacteria responsible for an infection involve culturing the bacteria on plates to identify their serotypes and to test their susceptibilities to various antibiotics. Culturing bacteria takes time, especially since some types of pathogenic bacteria grow slowly. Often, an antibiotic is administered to the patient before susceptibility tests are completed; if the bacterium turns out to be insensitive to the antibiotic, the patient must be given a different antibiotic, and valuable time is lost. Shuttle phasmids containing a reporter gene, such as a luciferase or green fluorescent protein gene, offer the opportunity to perform diagnostic tests more quickly. The reporter gene is cloned into the cosmid part of a series of shuttle phasmids that infect the target bacterium. The reporter gene must be cloned in such a way that it is expressed in the bacterium being tested. If the phage can infect the bacterium, then the reporter gene is expressed and can be readily detected—by seeing the cells "light up" with a luminometer if using a luciferase reporter or by microscopically watching cells fluoresce if using green fluorescent protein. Also, if the bacterium has been pretreated with an antibiotic to which it is susceptible, particularly one that blocks translation, as most of them do (see chapter 2), the reporter gene is not expressed after infection and the cells do not light up or fluoresce. The antibiotic can then be used to treat the infection without the need to wait for the results of other antibiotic sensitivity tests, which generally take longer.

Role of Transduction in Bacterial Evolution

Phages may play an important role in evolution by promoting the horizontal transfer of genes between individual members of a species, as well as between distantly related bacteria. The DNA in phage heads is usually more stable than naked DNA and so may persist longer in the environment. Also, many phages have a broad host range for adsorption. We have cited the example of phage P1, which infects and multiplies in *E. coli* but also injects its DNA into a number of other gram-negative bacterial species, including *Myxococcus xanthus*. The host range of P1 is partially affected by the orientation of an invertible DNA segment encoding the tail fibers (see chapter 9). Although incoming DNA from one species does not recombine with the chromosome of a different species if they share no sequences, stable transduction of genes between distantly related bacteria becomes possible when the transduced DNA is a broad-host-range plasmid that can replicate in the recipient strain or contains a broad-host-range transposon that can hop into the DNA of the recipient cell.

SUMMARY

1. Viruses that infect bacteria are called bacteriophages, or phages for short. The plural of phage is also phage if they are all the same type, while phages refers to more than one type.

2. The productive developmental cycle of a phage is called the lytic cycle. The larger DNA phages undergo a complex program of gene expression during development.

3. The products of phage regulatory genes regulate the expression of other phage genes during development. One or more regulatory genes at each stage of development turn on the genes in the following stage and turn off the genes in the preceding stage, creating a regulatory cascade. In this way, all the information for the stepwise development of the phage can be preprogrammed into the phage DNA.

4. Phage T7 encodes an RNA polymerase that specifically recognizes the promoters for the late genes of the phage. Because of its specificity and because transcription from the T7 promoters is so strong, this system, and others like it, have been used to make cloning vectors for expressing large amounts of protein and riboprobes from cloned genes. Phage N4 encodes two RNA polymerases, one that is encapsulated in the phage head with the DNA and another that transcribes the middle genes. Only the late genes are transcribed by the host RNA polymerase.

5. All the genes of phage T4 are transcribed by the host RNA polymerase, which undergoes many changes in the course of infection. Phage T4 goes through a number of steps of gene regulation in its development, and the genes are named based on when their products appear in the infected cell. The immediate-early genes are transcribed immediately after infection from σ^{70}-type promoters. The delayed-early genes are transcribed through antitermination of transcription from immediate-early genes. The transcription of the middle genes is from phage-specific middle-mode promoters and requires transcriptional activators, MotA and AsiA, that remodel the σ^{70} RNA polymerase so it can recognize middle promoters. The true-late genes are also transcribed from phage-specific promoters that require the binding of two proteins, gp55 and gp33, to the RNA polymerase. These proteins constitute a "split sigma" that partially substitutes for regions 2 and 4 of a sigma factor, respectively; gp55 substitutes for region 2 by recognizing the −10 sequence of the promoter, and gp33 binds to the β flap on RNA polymerase, where region 4 binds. However, they cannot efficiently activate transcription by themselves and must bind to gp45, the T4 DNA replication sliding clamp, which serves as a sort of mobile activator. This explains the coupling of T4 true-late transcription to the replication of the T4 DNA.

6. All phages encode at least some of the proteins required to replicate their nucleic acids. They borrow others from their hosts. Making their own replication proteins may allow them to broaden their host range. Often, these phage-encoded replication proteins are more closely related to those of eukaryotes than to those of their own host.

7. The requirement of all known DNA polymerases for a primer creates obstacles to the replication of the extreme 5′ ends of linear DNAs. Phages solve this replication primer problem in different ways. Some replicate as circular DNA. Others form long concatemers by rolling-circle replication, by linking single DNA molecules end to end by recombination, or by pairing through complementary ends. They can then package genome length DNAs from these concatemers by making staggered breaks in the DNA and resynthesizing the ends or package by a headful mechanism, leaving terminally redundant ends.

8. Phage endolysins break the bacterial cell wall, releasing the phages. In gram-negative bacteria, these endolysins are made early in infection in an active form but are not active until they pass through the inner membrane. Some holins form pores in the inner membrane that help endolysins pass through the inner membrane. They are inhibited by antiholins until it is time for lysis. Other endolysins are transported to the periplasm by the Sec system and so have no need for holins to transport them through the membrane. However, they remain tethered to the membrane and inactive until it is time for lysis, when they are released by the holin, which probably acts by allowing the passage of protons and destroying the membrane potential.

9. Bacteriophages are ideal for illustrating the basic principles of classical genetic analysis, including recombination and complementation. To perform recombination tests with phages, cells are infected by two different mutants or strains of phage, and the progeny are allowed to develop. The progeny are then tested for recombinant types. To perform complementation tests with phages, the cells are infected by two different strains under conditions that are nonpermissive for both mutants. If the mutations complement each other, the phages will multiply. Recombination tests can be used to order mutations with respect to each other. Complementation tests can be used to determine whether two mutations are in the same functional unit or gene.

10. Nonsense and temperature-sensitive mutations are very useful for identifying the genes of a phage whose products are essential for multiplication and for constructing genetic linkage maps of a phage.

11. The genetic linkage map of a phage shows the relative positions of all of known genes of the phage with respect to each other. Whether the linkage map of a phage is circular or linear depends on how the DNA of the phage replicates, how it is packaged into phage heads, and whether it contains an active site-specific recombination system.

12. Bacterial cells protect themselves from phages in a number of ways. Restriction-modification systems degrade

SUMMARY (continued)

incoming phage DNA. Abi (aborted-infection) systems kill the infected cell, preventing multiplication of the phage and its spread to other bacteria in the population. CRISPR loci are a type of adaptive immune system. They retain phage DNA sequences from previous infections and use them to make guide RNAs to target incoming phage DNA containing the sequences for cleavage.

13. Transduction occurs when a phage accidentally packages bacterial DNA into a head and carries it from one host to another. In generalized transduction, essentially any region of bacterial DNA can be carried.

14. Not all types of phages make good transducing phages. To be a good transducing phage, the phage must not degrade the host DNA after infection, must not have very specific *pac* sites, and must accept different-length DNAs or package DNA by a headful mechanism.

15. Shuttle phasmids can be made from the phages of bacteria for which no transducing phage is available. They are made by cloning cosmids into the phage so it can be packaged into λ phage and be transferred into and replicate as a plasmid in *E. coli*. After transfection into the bacterium being studied, any phage plaques might contain a phasmid that can be used to transduce DNA into the bacterium. The DNA to be transduced is cloned into the cosmid portion of the phasmid.

QUESTIONS FOR THOUGHT

1. Why do you suppose phages regulate their gene expression during development so that genes whose products are involved in DNA replication are transcribed before genes whose products become part of the phage particle? What do you suppose would happen if they did not do this?

2. What do you suppose the advantages are of a phage encoding its own RNA polymerase instead of merely changing the host RNA polymerase? What are the advantages of using the host RNA polymerase?

3. Why do phages often encode their own replicative machinery rather than depending on that of the host?

4. Why do some single-stranded DNA phages, such as ϕX174, use a complicated mechanism to initiate replication that uses the cellular proteins PriA, PriB, PriC, and DnaT, which are normally used by the cell to reinitiate replication at blocked forks, while others, such as f1, use a much simpler process involving only the host RNA polymerase to prime replication?

5. Phages often change the cell to prevent subsequent infection by other phages of the same type once the infection is under way. This is called superinfection exclusion. What do you suppose would be the consequences of another phage of the same type superinfecting a cell that is already in the late state of development of the phage if this mechanism did not exist?

6. How could CRISPR avoid killing the host cell through autoimmunity by not picking up chromosomal bacterial DNA sequences in the adaptive phase? What kinds of mechanisms can you propose?

PROBLEMS

1. You have precisely determined the titer or concentration of a stock of virus by counting the viruses in a particular volume under the electron microscope. How would you determine the effective MOI of the virus (i.e., the fraction of viruses that actually infect a cell) under a given set of conditions?

2. Phage components (e.g., heads and tails) can often be seen in lysates by electron microscopy even before they are assembled into phage particles. In studying a newly discovered phage of *Pseudomonas putida*, you have observed that amber mutations in gene *C* of the phage prevent the appearance of phage tails in a nonsuppressor host. Similarly, mutations in gene *T* prevent the appearance of heads. However, mutations in gene *M* prevent the appearance of either heads or tails. Which gene, *C*, *T*, or *M*, is most likely to be a regulatory gene? Why?

3. Would you expect to be able to isolate amber mutations in the *ori* sequence of a phage? Why or why not?

4. The *r1589* deletion in the *r*II genes of T4 begins in *r*IIA and extends into *r*IIB, deleting only the nonessential *r*IIB1 region. It is phenotypically *r*IIA$^-$ *r*IIB$^+$ and complements *r*IIA mutations. How would you isolate mutations in the portion of *r*IIA remaining in the *r1589* deletion? What types of mutations in the remaining *r*IIA sequences in *r1589* would you expect to make it *r*IIB$^-$ and not complement *r*IIA mutations?

5. To order the three genes *A*, *M*, and *Q* in a previously uncharacterized phage you have isolated, you cross a double mutant having an amber mutation in gene *A* and a temperature-sensitive mutation in gene *M* with a single mutant having an amber mutation in gene *Q*. About 90%

of the Am+ recombinants that can form plaques on the nonsuppressor host are temperature sensitive. Is the order *Q-A-M* or *A-Q-M*? Why? (Hint: draw a picture of the cross).

6. Phage T1 packages DNA from concatemers beginning at a unique *pac* site and then packaging by a processive headful mechanism, cutting about 6% longer than a genome length each time. However, it packages a maximum of only three headfuls from each concatemer. Would you expect T1 to have a linear or a circular genetic map? Draw a hypothetical T1 map.

7. Most types of phages encode endolysin enzymes that break the cell wall of the infected cell late in infection to release the phage. How would you determine which of the genes in the linkage map of a phage you have isolated encodes the endolysin? Hint: you can purchase egg white lysozyme from biochemical supply companies.

8. Chloroform ($CHCl_3$) dissolves the membranes of bacteria but does not affect the cell wall. What would be the effect of adding $CHCl_3$ to an *r*I (antiholin) mutant of T4? To a *t* (holin) mutant? To an *e* (lysozyme) mutant?

9. How would you use recombination and complementation tests to show that the spanin *Rz1* gene of λ is completely embedded in the *Rz* gene?

10. Why is it important to propagate phage for phage display at a low MOI?

SUGGESTED READING

Banaiee, N., M. Bobadilla-del-Valle, P. F. Riska, S. Bardarov, Jr., P. M. Small, A. Ponce-de-Leon, W. R. Jacobs, Jr., G. F. Hatfull, and J. Sifuentes-Osornio. 2003. Rapid identification and susceptibility testing of *Mycobacterium tuberculosis* from MGIT cultures with luciferase reporter mycobacteriophages. *J. Med. Microbiol.* **52:**557–561.

Bardarov, S., S. Bardarov, Jr., M. S. Pavelka, Jr., V. Sambandamurthy, M. Larsen, J. Tufariello, J. Chan, G. Hatfull, and W. R. Jacobs, Jr. 2002. Specialized transduction: an efficient method for generating marked and unmarked targeted gene disruptions in *Mycobacterium tuberculosis*, *M. bovis* and *M. smegmatis*. *Microbiology* **148:**3007–3017.

Barrangou, R., C. Fremaux, H. Deveau, M. Richards, P. Boyaval, S. Moineau, D. A. Romero and P. Horvath. 2007. CRISPR provides acquired resistance against viruses in prokayotes. *Science* **315:**1709–1712.

Benzer, S. 1961. On the topography of genetic fine structure. *Proc. Natl. Acad. Sci. USA* **47:**403–415.

Black, L. W., and V. B. Rao. 2012. Structure, assembly, and DNA packaging of the bacteriophage T4 head. *Adv. Virus Res.* **82:**119–153.

Brenner, S., A. O. W. Stretton, and S. Kaplan. 1965. Genetic code: the nonsense triplets for chain termination and their suppression. *Nature* (London) **206:**994–998.

Calender, R. (ed.). 2005. *The Bacteriophages*, 2nd ed. Oxford University Press, New York, NY.

Chastian, P. D., II, A. M. Makhov, N. G. Nossal, and J. Griffith. 2003. Architecture of the replication complex and DNA loops at the fork generated by the bacteriophage T4 proteins. *J. Biol. Chem.* **278:**21276–21285.

Choi, K. H., J. McPartland, I. Kaganman, V. D. Bowman, L. B. Rothman-Denes, and M. Rossmann. 2008. Insights into DNA and protein transport in double-stranded DNA viruses: the structure of bacteriophage N4. *J. Mol. Biol.* **378:**726–736.

Crick, F. H. C., L. Barnett, S. Brenner, and R. J. Watts-Tobin. 1961. General nature of the genetic code for proteins. *Nature* (London) **192:**1227–1232.

Edgar, R., and U. Qimron. 2010. *Escherichia coli* CRISPR system protects from λ lysogenization, lysogens, and prophage induction. *J. Bacteriol.* **192:**6291–6294.

Geiduschek, E. P., and G. A. Kassavetis. 2010. Transcription of the T4 late genes. *Virol. J.* 7:288.

Hendrix, R. W. 2005. Bacteriophage evolution and the role of phages in host evolution, p. 55–65. *In* M. K. Waldor, D. I. Friedman, and S. L. Adhya (ed.), *Phages: Their Role in Bacterial Pathogenesis and Biotechnology*. ASM Press, Washington, DC.

Herendeen, D. R., G. A. Kassavetis, and E. P. Geiduschek. 1992. A transcriptional enhancer whose function imposes a requirement that proteins track along the DNA. *Science* **256:**1298–1303.

Kreuzer, K. N. 2005. Interplay between replication and recombination in prokaryotes. *Annu. Rev. Microbiol.* **59:**43–67.

Labrie, S. J., J. E. Samson, and S. Molineau. 2010. Bacteriophage resistance mechanisms. *Nat. Rev. Microbiol. 8:*317–327.

Miller, E. S., E. Kutter, G. Mosig, F. Arisaka, T. Kunisawa, and W. Ruger. 2003. Bacteriophage T4 genome. *Microbiol. Mol. Biol. Rev.* **67:**86–156.

Mosig, G. 1998. Recombination and recombination-dependent replication in bacteriophage T4. *Annu. Rev. Genet.* **32:**379–413.

Nomura, M., and S. Benzer. 1961. The nature of deletion mutants in the *r*II region of phage T4. *J. Mol. Biol.* **3:**684–692.

Piuri, M., W. R. Jacobs, Jr., and G. F. Hatfull. 2009. Fluoromycobacteriophages for rapid, specific, and sensitive antibiotic susceptibility testing of *Mycobacterium tuberculosis*. *PLoS ONE* **4:**e4870.

Rakonjac, J., J. Feng, and P. Model. 1999. Filamentous phage are released from the bacterial membrane by a two-step mechanism involving a short C-terminal fragment of pIII. *J. Mol. Biol.* **289:**1253–1265.

Studier, F. W. 1969. The genetics and physiology of bacteriophage T7. *Virology* **39:**562–574.

Studier, F. W., A. H. Rosenberg, J. J. Dunn, and J. W. Dubendoff. 1990. Use of T7 RNA polymerase to direct expression of cloned genes. *Methods Enzymol.* **185:**60–89.

Symonds, N., P. Vander Ende, A. Dunston, and P. White. 1972. The structure of *r*II diploids of phage T4. *Mol. Gen. Genet.* **116:**223–238.

Van der Ost, J., M. M. Jore, E. R. Westra, M. Lundgreen, and S. J. J. Brouns. 2009. CRISPR-based adaptive and heritable immunity in prokaryotes. *Trends Biochem. Sci.* **34:**401–407.

White, R., T. A. Tran, C. A. Dankenbring, J. Deaton, and R. Young. 2010. The N terminal transmembrane domain of lambda S is required for holin but not antiholin function. *J. Bacteriol.* **192:**725–733.

Yanisch-Perron, C., J. Vieira, and J. Messing. 1985. Improved M13 phage cloning vectors and host strains: nucleotide sequences of the M13mp18 and pUC19 vectors. *Gene* **33:**103–119.

Yuan, A. H., and A. Hochschild. 2009. Direct activator/co-activator interaction is essential for bacteriophage T4 middle gene expression. *Mol. Microbiol.* **74:**1018–1030

Zinder, N. D., and J. Lederberg. 1952. Genetic exchange in *Salmonella. J. Bacteriol.* **64:**679–699.

CHAPTER 8

Lysogeny: the λ Paradigm and the Role of Lysogenic Conversion in Bacterial Pathogenesis

IN CHAPTER 7, we reviewed the lytic development of some representative phages. During lytic development, the phage infects a cell and multiplies, producing more phage that can then infect other cells. However, this is not the only lifestyle of which phages are capable. Some phages are able to maintain a stable relationship with the host cell in which they neither multiply nor are lost from the cell. Such a phage is sometimes called a lysogen-forming or **temperate phage**, although not everyone likes this dichotomy. Even temperate phages are capable of lytic development, and who is to say that any particular phage is not capable of lysogeny in some host? In the lysogenic state, the phage DNA either is integrated into the host chromosome or replicates as a plasmid. The phage DNA in the lysogenic state is called a **prophage**, and the bacterium harboring a prophage is a **lysogen** for that phage. Thus, a bacterium harboring the prophage P2 would be a P2 lysogen. Lysogeny was discovered almost at the same time as phages themselves.

In a lysogen, the prophage acts like any good parasite and does not place too great a burden on its host. The prophage DNA is mostly quiescent; most of the genes expressed are those required to maintain the lysogenic state, and most of the others are turned off. Often the only indication that the host cell carries a prophage is that the cell is immune to superinfection by another phage of the same type. The prophage state can continue almost indefinitely unless the host cell suffers potentially lethal damage to its chromosomal DNA or is infected by another phage of the same or a related type. Then, like a rat leaving a sinking ship, the phage can be induced and enter lytic development, producing more phage. The released phage can then infect other cells and develop lytically to produce more phage or lysogenize the new bacterial cell, hopefully one with a brighter future than its original host.

doi:10.1128/9781555817169.ch8

Studies of phage lysogeny were important in forming concepts of how viruses can remain dormant in their hosts and how they might convert cells into cancer cells. Also, many genes of benefit to the host bacterium are expressed from prophages, including phage defense systems against unrelated phages, as discussed in chapter 7, and virulence genes, whose products are required to make bacteria pathogenic, some examples of which are covered later in this chapter. In addition to prophages that can be induced under some circumstances to make more phage, bacteria carry many **defective prophages** that no longer can form infective phage because they have lost some of their essential genes. These DNA elements are suspected of being defective prophages rather than normal parts of the chromosome because they are not common to all the strains of a species of bacterium and they often carry phage-like genes. Defective prophages may be important in evolution because they eventually lose their identity and become part of the normal chromosome; this is one of the few ways bacteria can acquire new genes that can evolve to perform new functions. Prophages, including defective prophages, often compose a large percentage of the total chromosome of a bacterium and often contribute much of the diversity between closely related strains. In this chapter, we present what has been learned about the prototype lysogen-forming phage λ (lambda). We have reserved a discussion of the λ lytic cycle until this chapter because of its intimate relationship with its **lysogenic cycle**. We also discuss some other examples of lysogen-forming phages and the various means by which the lysogenic state is achieved and maintained. Some of the ways prophages contribute to the pathogenicity of their hosts are also covered. One striking insight to come from these studies is the extent to which bacteria and their prophages depend on each other, to the point that the distinction between the parasitized bacterial host and the parasitic virus begins to blur.

Phage λ

Phage λ is the lysogen-forming phage that has been studied most extensively and the one to which all others are compared. Although lysogeny was suspected as early as the 1920s, the first convincing demonstration that bacterial cells could carry phages in a quiescent state was made with λ in about 1950 (see Lwoff, Suggested Reading). In this experiment, apparently uninfected *Escherichia coli* cells could be made to produce λ by being irradiated with ultraviolet (UV) light. Since then, phage λ has played a central role in the development of the science of molecular genetics. A large number of very clever and industrious researchers have collaborated to make the interaction of λ with its host, *E. coli*, our best-understood biological system (see Gottesman, Suggested Reading). This research has revealed not only the complexity and subtlety of biological systems, but also their utility, robustness, and beauty.

Figure 8.1 gives an overview of the two life cycles of which λ is capable and the fate of the DNA in each cycle, while Figure 8.2 gives a more detailed map of the phage genome for reference. The phage DNA is linear in the phage head, and the map shows how it exists in the head. Immediately after the DNA is injected into the cell to initiate the infection, the DNA cyclizes, that is, forms a circular molecule, by pairing between the *cos* sites at the ends (Figure 8.3). This brings the lysis genes (*S* and *R*) and the head and tail genes (*A* to *J*) of the phage together and allows them all to be transcribed from the late promoter $p_{R'}$, as discussed below. This circular DNA can then either integrate into the host chromosome (lysogenic cycle) or replicate and be packaged into phage heads to form more phage (lytic cycle). Which decision is made depends on the physiological state of the cell, as we discuss below. Later, the integrated DNA in a lysogen can also be excised, replicate, and form more phage (induction). We first discuss in detail how λ replicates lytically and then describe how it can form a lysogen and, finally, how these two states are coordinated.

λ Lytic Development

Phage λ is fairly large, with a genome intermediate in size between those of T7 and T4. Some of the gene products and sites encoded by λ that are required for transcriptional regulation are listed in Tables 8.1 and 8.2, respectively. Phage λ goes through three major stages of gene expression during development. The first λ genes to be expressed after infection are *N* and *cro*. Most of the genes expressed next play a role in replication and recombination. Finally, the late genes of the phage, encoding the head and tail proteins of the phage particle and enzymes involved in cell lysis, are expressed.

PROCESSIVE TRANSCRIPTION ANTITERMINATION

Many mechanisms of gene regulation now known to be universal were first discovered in phage λ. One of these is regulation by **transcription antitermination** (see chapter 2). In regulation by transcription antitermination, transcription begins at the promoter but then terminates until certain conditions are met. Then, the transcription continues, hence the name "antitermination," because the mechanism works against, or "anti," termination. Antitermination can occur through a number of mechanisms. In some cases, specific proteins or other molecules bind to the nascent mRNA, allowing the formation of hairpins in the mRNA that compete with terminator hairpins for secondary structure, preventing formation of the terminator hairpin and allowing transcriptional readthrough. This type of transcription antitermination acts by preventing

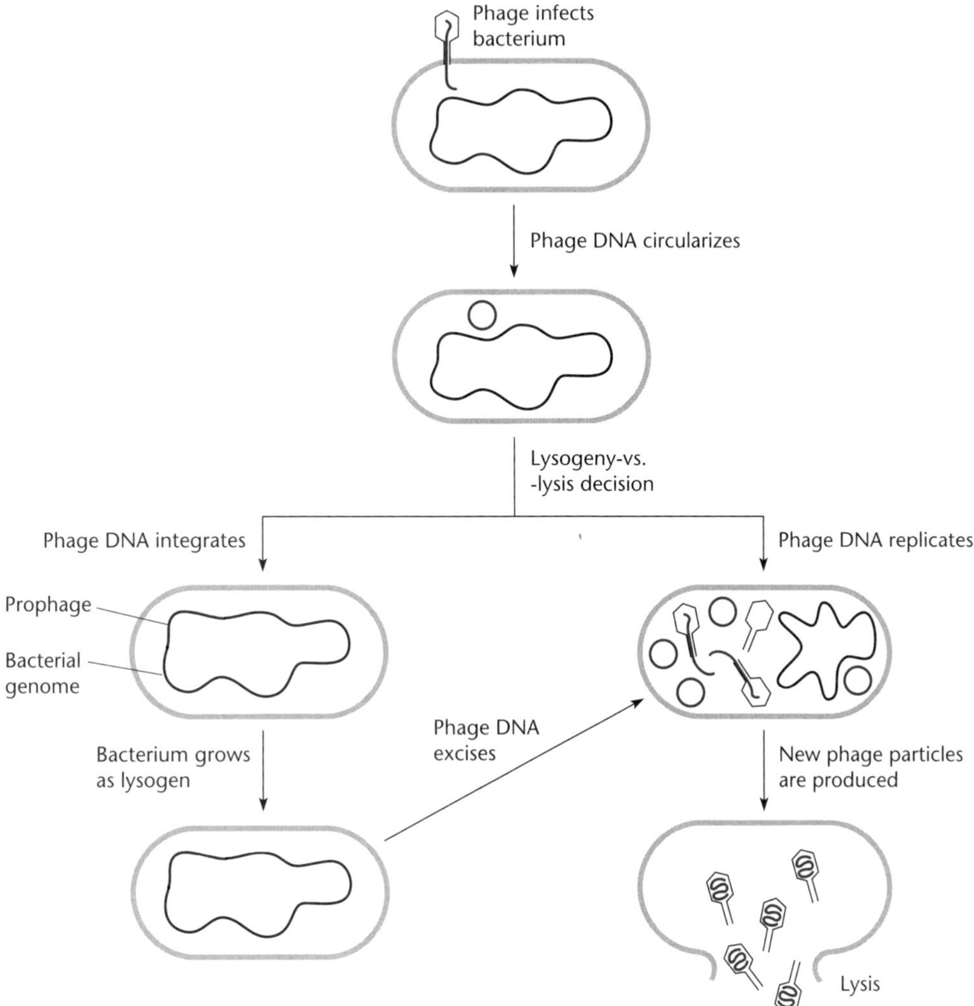

Figure 8.1 Overview of the fate of λ DNA in the lytic and lysogenic pathways. Phage DNA is shown in blue. doi:10.1128/9781555817169.ch8.f8.1

formation of only one transcription terminator and therefore affects only one terminator site. We discuss some of these regulatory mechanisms in chapters 12 and 13. In the case of λ and some other known antitermination mechanisms, specific proteins or sites on the nascent RNA bind to the RNA polymerase, often in conjunction with antiterminator proteins, and travel with the RNA polymerase, rendering it incapable of termination at many downstream sites. This is called **processive antitermination** to distinguish it from antitermination that often acts at only a single terminator.

Phage λ uses processive antitermination in a regulatory cascade to turn on different genes at different stages in its development. At an early stage, it uses antitermination protein N to turn on the synthesis of its recombination and replication functions and also to turn on the gene for another antitermination protein, Q. At a later stage, it uses the antitermination protein Q to turn on the synthesis of its late proteins, including the head and tail proteins.

The N Protein

The N protein is responsible for the first stage of λ antitermination, as illustrated in Figure 8.4. When the λ DNA first enters the cell, transcription immediately begins from two promoters, p_L and p_R, that face outward from the immunity region. This leads to the synthesis of two RNAs, one leftward on the genome and the other rightward. However, the synthesis of both RNAs terminates at transcription termination sites, $t_L{}^1$ and $t_R{}^1$, after only short RNAs are synthesized (Figure 8.4A). One of these short RNAs encodes the Cro protein, which is an inhibitor of repressor synthesis, as discussed in "Lysogeny by Phage λ" below. The other encodes the N protein, the antitermination factor that permits the RNA

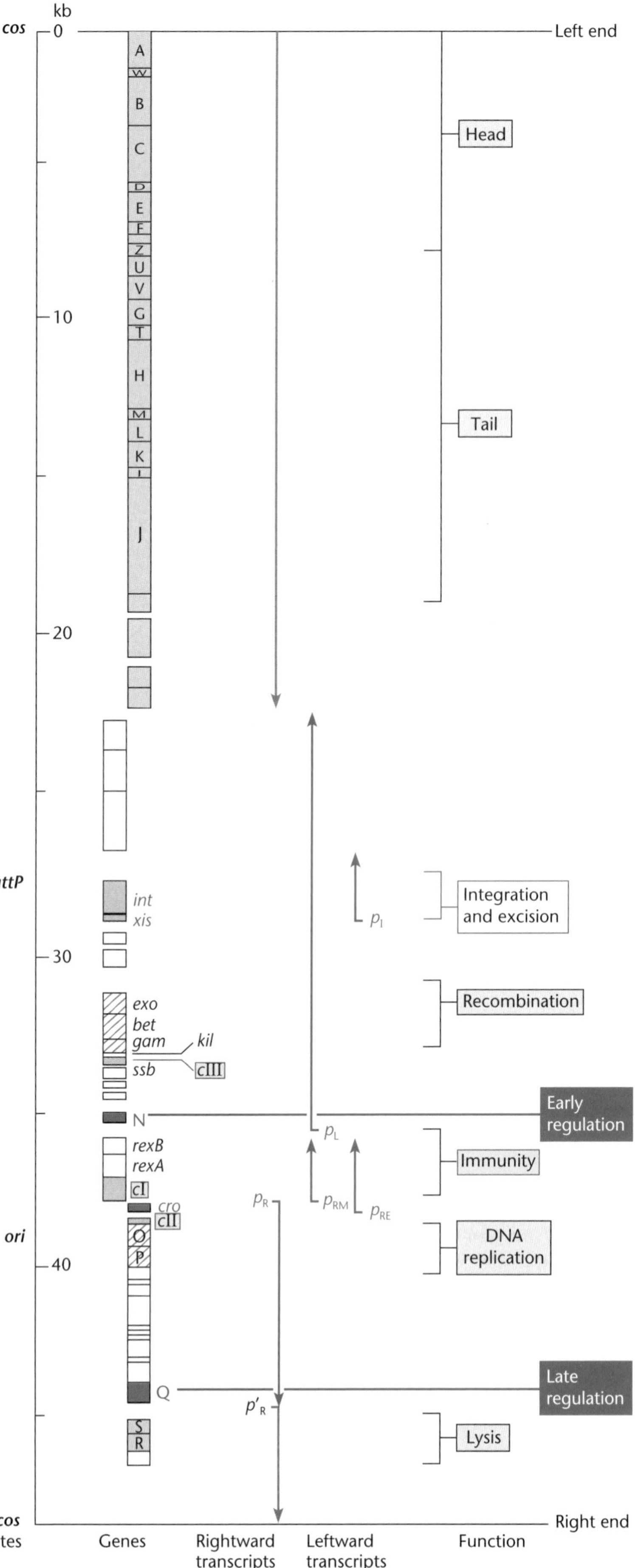

Figure 8.2 Genetic map of phage λ. Regulatory genes for the lytic pathway are shown in dark blue. Additional genes emphasized in the text are shown in light blue. Recombination and replication genes are indicated by hatched boxes. The structural genes for the phage particle and lysis genes are in gray. Promoters and transcripts discussed in the text are also indicated. The GenBank accession number for the λ genome is NC_001416. doi:10.1128/9781555817169.ch8.f8.2

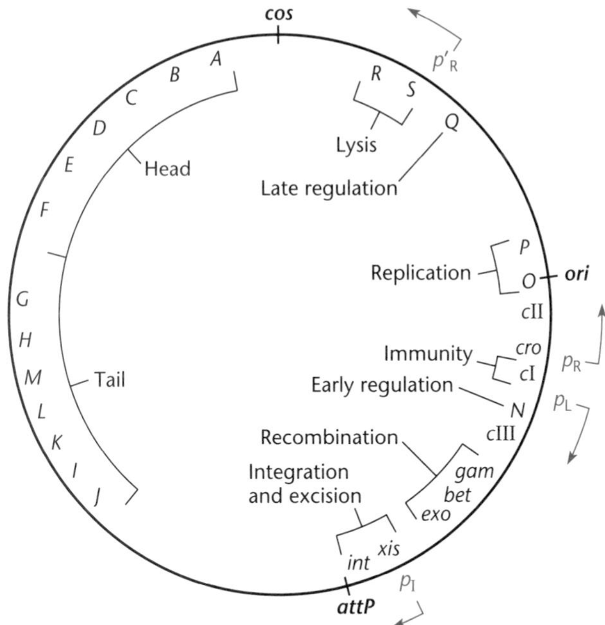

Figure 8.3 Genetic map of λ cyclized by pairing at the *cos* sites, shown at the top. The positions and directions of transcription from some promoters are shown in blue.
doi:10.1128/9781555817169.ch8.f8.3

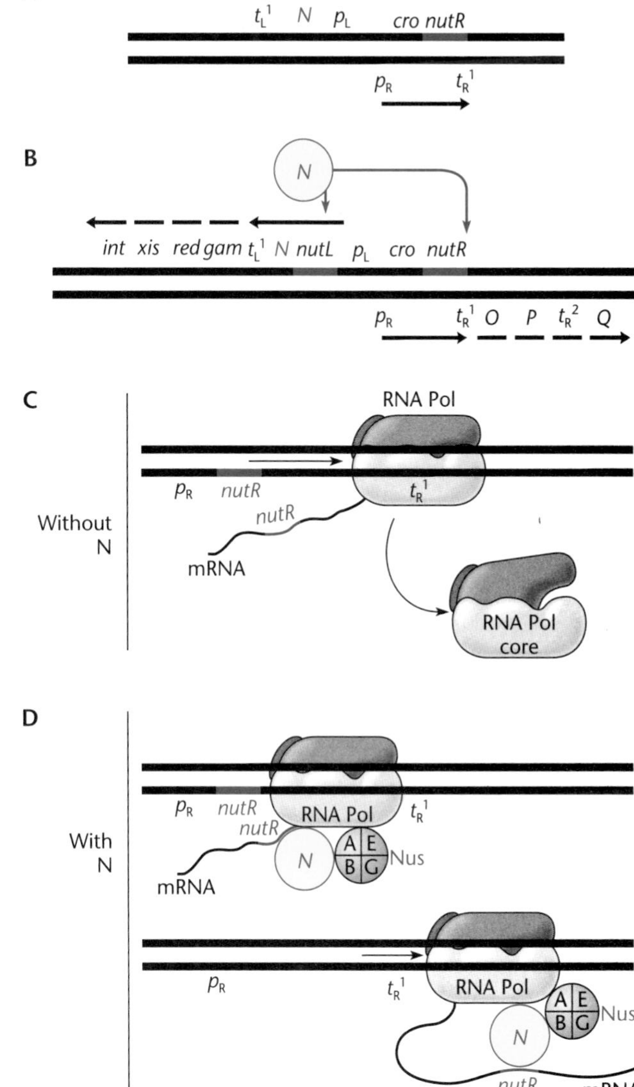

Figure 8.4 Antitermination of transcription in phage λ. **(A)** Before the N protein is synthesized, transcription starts at promoters p_L and p_R and stops at transcription terminators t_L^1 and t_R^1. **(B)** The N protein causes transcription to continue past t_L^1 and t_R^1 into *gam-red-xis-int* and *O-P-Q*, respectively. **(C and D)** Mechanism of antitermination by N, showing rightward transcription only. **(C)** In the absence of N, transcription initiated at p_R terminates at the terminator t_R^1. **(D)** If N has been made, it binds to RNA polymerase (RNA Pol) as the polymerase transcribes the *nutR* site. A conformational change occurs in N when it binds to the *nutR* sequence in the RNA, and this change is required before N can bind to the RNA polymerase. The binding of N to RNA polymerase is facilitated by the host Nus proteins NusA, -B, -E, and -G. The antitermination complex, composed of RNA polymerase, N, *nutR*, and NusABEG, then transcribes past transcriptional terminator t_R^1 plus any other transcriptional terminators downstream. The sites and λ gene products shown here are defined in Tables 8.1 and 8.2.
doi:10.1128/9781555817169.ch8.f8.4

polymerase to bypass the transcription termination sites and continue along the DNA, as shown in Figure 8.4B.

Figure 8.4C and D outline how N protein antiterminates, showing only rightward transcription. Initially, transcription initiated at the rightward p_R promoter terminates at the transcription terminator designated t_R^1. One of the sequences transcribed into RNA is *nutR* (for *N utilization rightward*), which binds to the RNA polymerase. In the meantime, the N protein is being translated from the leftward RNA. The N protein can bind to the RNA polymerase, but only after the *nutR* RNA is bound. Other host proteins, called the Nus proteins, help it bind (see below), and together, they all form an antitermination complex that travels with the RNA polymerase, making it resistant to t_R^1 and other transcription termination signals. When this antitermination complex transcribes through t_R^1, it reaches the genes *O* and *P*, which encode the replication proteins, and replication begins. The gene for the other termination protein, Q, is also transcribed, and this protein turns on the transcription of the late genes (see below).

Similar events occur during leftward transcription. When the *nut* site (for *N utilization site*) on the other side, called *nutL* (for *N utilization leftward*), is transcribed, it and the N protein bind to the RNA polymerase and prevent any further termination on the left, allowing transcription to continue into other genes on the left side, including *gam* and *red*, which are λ recombination

Table 8.1 Some λ gene products and their functions

Gene product	Function
N	Antitermination protein; loads on RNA polymerase at *nutL* and *nutR* sites and prevents termination at downstream terminators, including t_L^1, t_R^1, and t_R^2
O, P	Initiation of λ DNA replication
Q	Antitermination protein; loads on RNA polymerase at *qut* site and prevents termination at downstream terminators, including $t_{R'}$
CI	Repressor, activator; binds to o_L and o_R, preventing transcription from p_L and p_R and activating transcription from p_{RM}
CII	Activator of transcription of *cI* from p_{RE} and *int* from p_I
CIII	Stabilizes CII by inhibiting a cellular protease
Cro	Repressor of CI synthesis from p_{RM} after infection
Gam	Protein required for rolling-circle replication; inhibits RecBC
Red	Composed of two proteins, β and Exo, involved in λ recombination
Int	Integrase; site-specific recombinase required for recombination between *attP* and *attB* to integrate prophage
Xis	Excisase; directionality factor; works with Int to promote recombination between hybrid sites BOP′ and POB′ to excise prophage

functions. The RNA polymerase also continues into *int* and *xis*, genes involved in integrating and excising the prophage, although the mRNA from these genes is quickly degraded, as explained below. The important sequences in the *nutL* site, the box A and box B sequences, are identical to those in the *nutR* site (Fig. 8.5).

This model of how N regulates the expression of the early genes by antiterminating transcription was first proposed on the basis of indirect genetic experiments. First, it was shown that transcription initiated at the λ promoters p_R and p_L (see the genetic map in Figure 8.2) soon terminates unless the N protein is present. This led to the conclusion that N was acting as an antiterminator by allowing transcription through downstream transcription terminators. However, surprisingly, N could antiterminate only if transcription had initiated at the p_R and p_L promoters and not if transcription had initiated from other promoters closer to the terminators. This led to the conclusion that N does not act only at the terminators to prevent termination but that some sites upstream in or close to the p_R and p_L promoters are required for N action. Then, it was shown that the sites required for N antitermination were actually not the p_R and p_L promoters themselves but nearby sites somewhat downstream of the promoters. It was hypothesized that N must bind to these sites to allow transcription through the downstream termination sites, and they were named the N utilization (*nut*) sites.

Table 8.2 Some sites involved in phage λ transcription and replication

Site(s)	Function(s)
p_L	Left promoter
p_R, $p_{R'}$	Right promoters
o_L	Operator for leftward transcription; binding sites for CI and Cro repressors
o_R	Operator for rightward transcription; binding sites for CI and Cro repressors
t_L^1, t_L^2	Termination sites of leftward transcription
t_R^1, t_R^{234}, $t_{R'}$	Termination sites of rightward transcription
nutL	N utilization site for leftward-transcribing RNA Pol. Sequence in the mRNA binds to N, allowing it to bind to RNA Pol and resist termination.
nutR	N utilization site for rightward-transcribing RNA Pol
qut	Q utilization site for antitermination of transcription from $p_{R'}$ (on DNA); overlaps promoter
p_{RE}	Promoter for repressor establishment; activated by CII
p_{RM}	Promoter for repressor maintenance; activated by CI
p_I	Promoter for *int* transcription; activated by CII
POP′	Phage attachment site (*attP*)
cos	Cohesive ends of λ genome (complementary 12-bp single-stranded ends in linear genome anneal to form a circular genome after infection)

Identifying the hypothetical *nut* sites on mRNA involved some clever selectional genetics (see "Genetic Experiments with Phage λ" below and Salstrom and Szybalski, Suggested Reading). The investigators were able to isolate *nutL* mutations that had all the predicted characteristics of *nut* mutations. These mutations prevent the expression of genes downstream of the terminator site; hence, they no longer allow antitermination, and they are *cis*-acting and affect only mRNA from the DNA that has the mutation. They also map in approximately the right place for a predicted *nutL* mutation, just downstream of the p_L promoter and just upstream of the *N* gene. Once the site of *nutL* mutations was identified by DNA sequencing, a region containing some of the same sequences was found just to the right of the *cro* gene and was assumed to be *nutR*. The location of the *nut* sites supported another element of the model, i.e., that the N protein binds to the *nut* site sequence in the mRNA rather than in the DNA, since they are in regions transcribed into RNA and change their polarity with respect to the orientation of transcription. However, because of their location between genes, the *nut* sequences are not normally translated into proteins. In fact, translating the *nut* sites interferes with antitermination. Apparently, ribosomes translating the mRNA can interfere with the binding of the *nut* site to RNA polymerase or dislodge the antitermination complex from the RNA polymerase and thereby interfere with antitermination.

Figure 8.5 shows the sequences of the *nut* sites of λ. The *nut* sites consist of a sequence, called box B, that forms a "hairpin" secondary structure in the mRNA because it is encoded by a region of the DNA with twofold rotational symmetry (see chapter 2). The original *nutL* mutations all change bases in box B and disrupt the twofold symmetry of the sequence, preventing formation of the hairpin in the mRNA. Thus, apparently, the formation of the box B hairpin in the mRNA is important for binding of the N protein. In fact, structural studies performed more recently have indicated that the N protein can bind to RNA with just the box B hairpin and that both the N protein and the box B secondary structure change as a result of this binding. This supports the idea that the N protein changes its conformation on binding to the box B sequence and that only N in the changed conformation can bind to RNA polymerase and prevent termination.

Host Nus Proteins

The functions of the box A and box C sequences in *nut* sites are more obscure and have to do with host proteins that participate in antitermination. These host proteins are called Nus proteins, for N utilization substances, and were found by using an elegant selection approach to find host mutations that interfere with N antitermination (see Friedman et al., Suggested Reading, and "Genetic Experiments with Phage λ" below). At least four of these proteins, NusA, -B, -E, and -G, are known to be involved in transcriptional processes in uninfected *E. coli* and have been commandeered by λ to help with antitermination. One of the proteins, NusA, binds to the box B sequence after N has bound and helps stabilize the complex. The box A sequence binds a complex of NusB and NusE, and then all the proteins travel with the antitermination complex. Whatever their detailed functions, the box A and box C sequences are common to the *nut* sites of all λ-related phages, and box A-like sequences occur in some bacterial genes, including the genes for rRNA, where they play an important role in preventing premature termination of transcription (see below and chapter 13). Adding to the mystery is the fact that point mutations in box A of a phage λ *nut* site can prevent antitermination, but deletion of the entire box A sequence does not. No one has found a function for box C. It could be that these sites are required to regulate antitermination in subtle ways not easily detectable in a laboratory situation.

Surprisingly, *nusE* mutations are in a gene for a ribosomal protein, S10. This is surprising, because translation is not thought to be required for antitermination and, in fact, inhibits it (see above). Perhaps the S10 protein plays a dual role in the cell, one in translation and another in transcription antitermination. Other *nus* mutations have a less direct effect on antitermination. For example, *nusD* mutations affect the host ρ factor, which is required for transcription termination at ρ-dependent termination sites (see chapter 2). These mutations may affect N antitermination by causing stronger termination that cannot be overcome by N. Other *nus* mutations,

Figure 8.5 Sequences of the *nutL* and *nutR* regions of bacteriophage λ. Box A, box B, and box C are underlined in blue. The twofold rotational symmetry in box B that causes a hairpin to form in the mRNA is shown as arrows of opposite polarity in blue.
doi:10.1128/9781555817169.ch8.f8.5

Box A — Box B — Box C

λ*nutL* ATGAAGGTGACGCTCTTAAAAATTAAGCCCTGAAGAAGGGCAGCATTCAAAGCAGAAGGCTTTGGGGTGTGTGATAC

λ*nutR* TAAATAACCCCGCTCTTACACATTCCAGCCCTGAAAAAGGGCATCAAATTAAACCACACCTATGGTGTATGCATTTAT

called *nusC*, are located in the RNA polymerase β subunit. They may alter the binding of the antitermination complex to the RNA polymerase and thereby reduce antitermination.

The Q Antitermination Protein

As mentioned above, one of the genes under the control of the N antiterminator is gene *Q*, whose product is required for the transcription of the late genes of λ, including the head and tail genes. Thus, λ marches through a regulatory cascade, with one of the earliest gene products (N) directing the synthesis of early gene products, including another regulatory gene product (Q), which in turn directs the transcription of the late genes. Like N, the Q protein of λ is a processive antiterminator that binds to the RNA polymerase, rendering it incapable of termination and allowing transcription from the late promoter $p_{R'}$ to proceed through terminators into downstream genes. One of the host Nus proteins, NusA, also travels with this complex. However, the mechanism by which Q loads on the RNA polymerase is very different from that of N. Like N, the Q protein loads on RNA polymerase in response to a sequence located close to the promoter, in this case called *qut* (for *Q utilization site*) (Figure 8.6 and Table 8.2). However, the *qut* site is not in the mRNA but, rather, in the DNA. The required *qut* sequences are not even transcribed into mRNA, being upstream of the start site of transcription at p'_R.

The details of Q protein antitermination have been studied in some detail (Figure 8.6) (see deHaseth and Gott , Suggested Reading, for an overview). As a prelude to antitermination, the RNA polymerase transcribes a short RNA only 16 to 17 nucleotides long from the late promoter $p_{R'}$ before it pauses. As the short RNA exits the RNA polymerase, it displaces region 3.2 and region 4 of σ^{70} from the RNA polymerase, leaving only region 2, which normally contacts the −10 region at the promoter, to contact the DNA. The RNA polymerase pauses 16 base pairs (bp) downstream of the start site of transcription because region 2 recognizes a sequence that is similar to the −10 sequences of σ^{70} promoters (Figure 8.6 and chapter 2). Because the single-stranded DNA at the transcription bubble is more flexible, it can "scrunch," allowing region 2 to contact the −10-like sequence even though the spacing on the template is less than this. The Q protein then loads on the *qut* sequence and contacts the RNA polymerase, stabilizing the interaction between region 4 of σ^{70} and a −35-like sequence that is only 1 bp upstream of the −10-like sequence instead of the usual 17 bp at normal σ^{70} promoters. The displacement of the region 3.2 linker from the RNA polymerase must make σ^{70} flexible enough so that region 4 can double back and contact a −35 region that is much closer to the −10 region than it is at normal promoters. The Q protein then loads on the paused RNA polymerase, allowing the transcript to exit through the exit pore and displace σ^{70} to allow the RNA polymerase to escape from the pause site. Once bound, Q and NusA travel with the RNA polymerase, making it oblivious to further transcription termination and pause sites and allowing it to proceed rapidly through the late genes of the phage.

Apparently, the Q protein can load on RNA polymerase only if the RNA polymerase is positioned

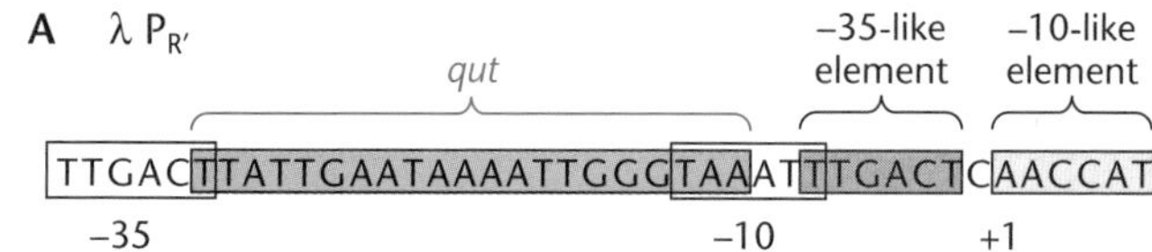

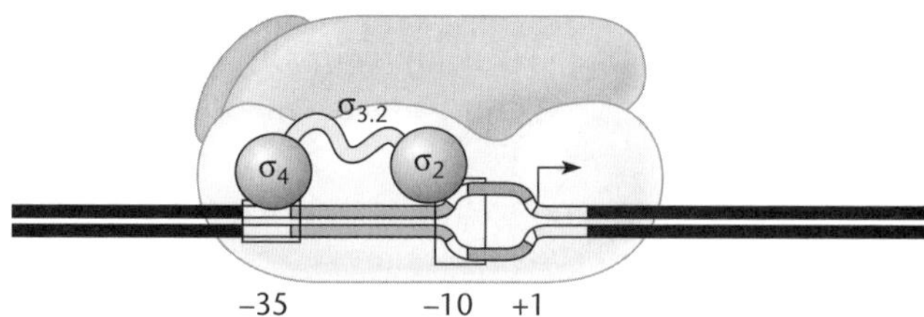

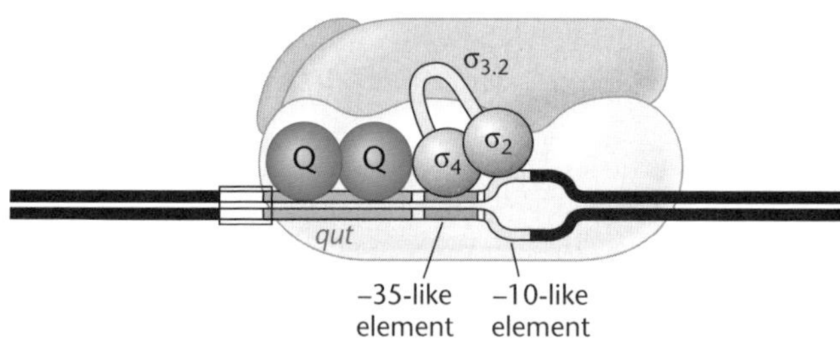

Figure 8.6 Formation of the Q protein antitermination complex at the $p_{R'}$ promoter. **(A)** The *qut* site (shown in gray) overlaps the −35 and −10 sites (boxed) of the σ^{70} promoter. Sequences just downstream, identified as −35-like and −10-like (boxed in dark and light blue, respectively) function as −10 and −35 sequences. **(B)** Transcription initiates as at other promoters, with region 2 of sigma contacting −10 and opening the helix and region 4 contacting −35 until the incipient 16-base mRNA displaces region 3.2 from the RNA polymerase and region 4 from the DNA and RNA polymerase. **(C)** As the RNA polymerase moves down the template, sigma region 2 scans the DNA until it recognizes the −10-like sequence, and the RNA polymerase pauses. This allows sigma region 4 to bind to the −35-like sequence with the help of Q bound at the *qut* site. The Q protein then loads on the RNA polymerase and moves with it down the DNA into the late genes, ignoring transcription termination signals. The "scrunching" of the DNA template allows sigma region 2 to contact the −10-like element 12 bp downstream of the −10 sequence even though the RNA polymerase has moved 16 bp on the DNA template.
doi:10.1128/9781555817169.ch8.f8.6

correctly at the pause site. Sometimes, the RNA polymerase overshoots the pause site and makes an RNA 17 nucleotides or more in length. Then, the RNA polymerase must backtrack to 16 nucleotides while the GreA and GreB proteins cleave the extra RNA that extrudes from the RNA polymerase as a result of the backtracking (see chapter 2). Only then can the Q protein load on the RNA polymerase and send it on its way at last.

PROCESSIVE ANTITERMINATION IN OTHER SYSTEMS

Processive antitermination was first discovered in phage λ but is now known to operate in many other phages, as well as systems other than phages. The HK022 phage and some other *Siphoviridae* phages related to λ also use progressive antitermination to regulate their gene expression (see King et al., Suggested Reading). However, antitermination in these phages only requires the transcription of *cis*-acting sites on the DNA called *put* sites, transcribed from the early promoters, and not a protein like N. The small *put* RNAs bind to the RNA polymerase as they are transcribed and make it resistant to termination at sites further downstream. Because they bind to the RNA polymerase as they are made, the *put* RNAs are *cis* acting, like the *nut* sites of λ. In fact, the two sites *putL* and *putR* are located in the HK022 genome approximately where the *nutL* and *nutR* sites are located in λ, just downstream of the p_L promoter and just downstream of the *cro* gene, respectively. However, these RNAs bind to RNA polymerase independent of an antitermination protein and do not even require any of the Nus proteins of the host, at least not absolutely, since some antitermination occurs in purified transcription systems with just RNA polymerase and DNA and none of the other factors. Why these phages forego the need for an antitermination protein and simply use the *cis*-acting RNAs for antitermination is a mystery. Perhaps the requirement for the N antitermination protein delays the onset of antitermination in λ while the protein is being translated, and it may reflect differences in the lifestyles of these phages.

Examples of regulation by processive antitermination are also accumulating in uninfected bacteria. For example, the transcription of the rRNA genes of *E. coli*, and presumably other bacteria, is enhanced by processive antitermination. After they are transcribed, box A-like sequences just downstream of the promoter bind to the RNA polymerase, along with many of the same Nus proteins that help N antitermination. This makes rRNA transcription insensitive to ρ-dependent termination signals, which could cause transcription to terminate, since rRNA is not translated (see chapter 2 for the mechanism of ρ termination). The RNA polymerase with the box A RNA and Nus proteins bound also ignores other pause signals, which allows very high rates of transcription of the rRNA genes.

Other cases are more like Q-dependent antitermination in that sites on the RNA are not involved, but rather, a protein binds to the RNA polymerase at certain sites in the DNA and travels with the RNA polymerase. One known case is the RfaH antitermination protein of *E. coli* and *Salmonella*, which is known to antiterminate transcription of *tra* genes in plasmids, some virulence genes, and genes for lipopolysaccharide transferases. This protein binds to RNA polymerase at the *ops* site in the DNA just downstream of the promoter and then travels with the RNA polymerase, preventing ρ-dependent termination. The regulatory strategy that is used by Q antitermination, i.e., having pause sites for RNA polymerase just downstream of promoters, is also used for many bacterial operons, although the purposes may vary. As with pausing downstream of the $p_{R'}$ promoter of λ, this pausing is due to σ^{70} remaining at least partially attached to the RNA polymerase and recognizing promoter-like sequences downstream of the promoter. We discuss these and other examples of processive antitermination again in chapters 12 and 13.

Replication of λ DNA

The replication of λ DNA has also been studied extensively and has been one of the major model systems for understanding replication in general. The λ DNA is linear in the phage head but cyclizes after it enters the cell through pairing between its cohesive ends, or ***cos* sites** (Figure 8.3). These ends are single stranded and complementary to each other for 12 bases and so can join by complementary base pairing, which makes them cohesive, or "sticky." Once the cohesive ends are paired, DNA ligase can join the two ends to form covalently closed circular λ DNA molecules. These circular DNA molecules can replicate in their entirety because there is always DNA upstream to serve as a primer (see chapter 7).

CIRCLE-TO-CIRCLE, OR θ, REPLICATION OF λ DNA

Once circular λ DNA molecules have formed in the cell, they replicate by a mechanism similar to the θ replication described for the chromosome in chapter 1 and for plasmids in chapter 4. Replication initiates at the *ori* site in gene *O* (λ map in Figure 8.2) and proceeds in both directions, with both leading- and lagging-strand synthesis in the replication fork (Figure 8.7). When the two replication forks meet somewhere on the other side of the circle, the two daughter molecules separate, and each can serve as the template to make another circular DNA molecule.

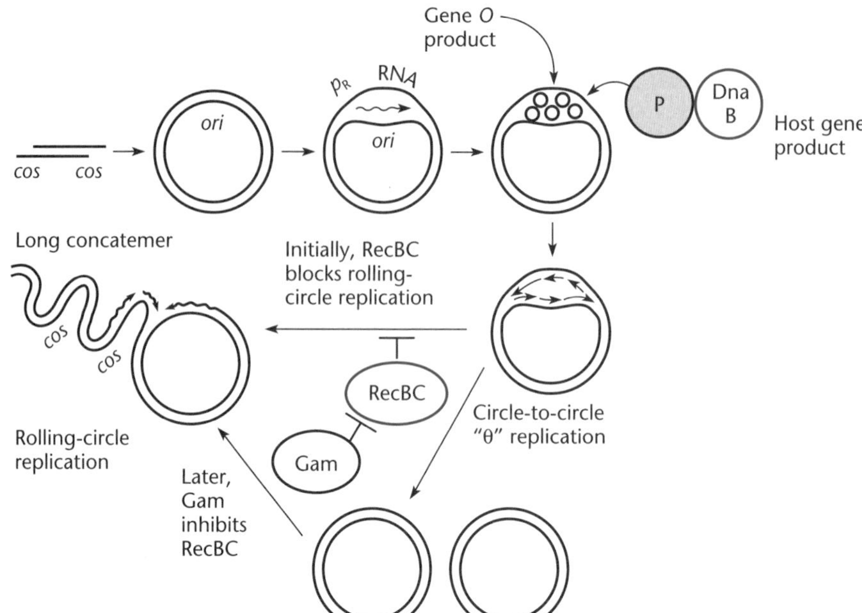

Figure 8.7 Overview of replication of phage λ. The *E. coli* gene products involved are outlined in blue.
doi:10.1128/9781555817169.ch8.f8.7

ROLLING-CIRCLE REPLICATION OF λ DNA

After a few circular λ DNA molecules have accumulated in the cell by θ replication, the rolling-circle type of replication ensues. The initiation of λ rolling-circle replication is similar to that of M13 in that one strand of the circular DNA is cut and the free 3′ end serves as a primer to initiate the synthesis of a new strand of DNA that displaces the old strand. DNA complementary to the displaced strand is also synthesized to make a new double-stranded DNA. The λ process differs in that the displaced individual single-stranded molecules are not released when replication around the circle is completed. Rather, the circle keeps rolling, giving rise to long tandem repeats of individual λ DNA molecules linked end to end, called **concatemers** (see chapter 7 and Figure 8.7). The formation of end-to-end concatemers by rolling-circle replication can be compared to what happens when an engraved ring, dipped in ink, is rolled across a piece of paper: the pattern on the ring is repeated over and over again on the paper.

In the final step, the long concatemers are cut at the *cos* sites into λ genome-length pieces as they are packaged into phage heads. Phage λ can package DNA only from concatemers, and at least two λ genomes must be linked end to end in a concatemer, because the packaging system in the λ head recognizes one *cos* site on the concatemeric DNA and takes up DNA until it arrives at the next *cos* site, which it cleaves to complete the packaging. The dependence on concatemeric DNA for λ DNA packaging was important in the discovery of *chi* (χ) sites and their role in RecBCD recombination, as we discuss in chapter 10.

GENETIC REQUIREMENTS FOR λ DNA REPLICATION

Unlike T4 and T7, which encode many proteins for replication, the products of only two λ genes, *O* and *P*, are required for λ DNA replication. As Figure 8.7 illustrates, both of these proteins are required for loading the host DnaB helicase and initiating DNA replication at the *ori* site. The O protein is thought to bend and open the DNA at this site by binding to repeated sequences, similar to the mechanism by which the DnaA protein bends and opens the DNA at the *oriC* site to initiate chromosome replication (see chapter 1) The P protein binds to the O protein and to the replicative helicase DnaB of the host replication machinery, thus commandeering DnaB for λ DNA replication; it therefore acts like DnaC. Appropriately, the P of the P gene product stands for "pirate."

RNA synthesis by the host RNA polymerase must also occur in the *ori* region for λ replication to initiate. It normally initiates at the p_R promoter and may be required to separate the DNA strands at the origin and/or serve as a primer for rightward replication.

A third λ protein, the product of the *gam* gene, is required for the shift to the rolling-circle type of replication, albeit indirectly. The RecBCD nuclease (an enzyme that facilitates recombination of *E. coli*) somehow inhibits the switch to rolling-circle replication, perhaps by degrading the free 3′ end that forms, but Gam inhibits the RecBCD nuclease. Therefore, a *gam* mutant of λ is restricted to the θ mode of replication, and concatemers can form from individual circular λ DNA molecules only by recombination. Because λ requires concatemers for packaging, a *gam* mutant of λ cannot multiply in the absence of a functional recombination system, either its own or the RecBCD pathway of its host. This fact was also important in the detection of *chi* sequences, as discussed further in chapter 10.

PHAGE λ CLONING VECTORS

The many cloning vectors derived from phage λ offer numerous advantages. The phage multiply to a high copy number, allowing the synthesis of large amounts of DNA and protein. Also, it matters less if the cloned gene encodes a toxic protein than with plasmid cloning vectors. The toxic protein is not synthesized until the phage infects the cell, and the infected cell is destined to die anyway. It is also relatively easy to store libraries in the stable λ phage head.

Cosmids

Some of the most widely used cloning vectors are cosmids based on phage λ. We discussed shuttle phasmids based on cosmids in chapter 7, and they are still in common use.

As mentioned above, λ packages DNA into its head by recognizing *cos* sites in concatemeric DNA; therefore, any DNA containing two *cos* sequences is packaged into phage heads if the two *cos* sequences are about 50 kilobases (kb) apart. In particular, plasmids containing *cos* sites can be packaged into λ phage heads. Such plasmid cloning vectors, called **cosmids**, also offer many advantages for genetic engineering, including **in vitro packaging**. In this procedure, cosmid DNA is mixed with commercially available packaging extracts, which are extracts of λ-infected cells containing heads and tails of the phage. The cosmid DNA is taken up by the heads, and because λ particles self-assemble in the test tube, the tails are attached to the heads to make infectious λ particles, which can then be used to introduce the cosmid into cells very efficiently by infection, which is a much more efficient way of introducing DNA into bacteria than either transfection, electroporation, or transformation.

Another major advantage of cosmids is that the size of the cloned DNA is limited by the size of the phage head. If the piece of DNA cloned into a cosmid is too large, the *cos* sites will be too far apart, and the DNA will be too long to fit into a phage head. However, if the cloned DNA is too small, the *cos* sites will be too close to each other, and the phage heads will have too little DNA and be unstable. Therefore, the use of cosmids ensures that the pieces of DNA cloned into a vector will all be approximately the same size, which is sometimes important for making DNA libraries (see chapter 1).

Lysogeny by Phage λ

Besides lytic development, phage λ has another lifestyle choice. It can form a lysogen. Phage λ is the classical example of a phage that can form lysogens. In the lysogenic state, very few λ genes are expressed, and essentially the only evidence that the cell harbors a prophage is that the lysogenic cells are immune to **superinfection** by more λ phage. The growth of the immune lysogens in the plaque is what gives the λ plaque its characteristic "fried-egg" appearance, with the lysogens forming the "yolk" in the middle of the plaque (Figure 8.8).

Some phage λ mutants form plaques that are clear because they do not contain immune lysogens. These phage have mutations in either the *c*I, *c*II, or *c*III gene, where the "*c*" stands for *c*lear plaque. These mutations prevent the formation of lysogens. Understanding the regulation of the λ lysogenic pathway and the functions of the *c*I, *c*II, and *c*III gene products in forming lysogens required the concerted efforts of many people. Their findings illustrate the complexity and subtlety of biological regulatory pathways and serve as a model for other systems (see Gottesman, Suggested Reading).

The Lytic-versus-Lysogen Decision: the Roles of cI, cII, and cIII Gene Products

Figure 8.9 illustrates the process of forming a lysogen after λ infection, how the *c*I, *c*II, and *c*III gene products are involved, and the central role of the CII protein. After λ infects a cell, the decision about whether the phage enters the lytic cycle and makes more phage or forms a lysogen

Figure 8.8 Phage λ plaques with typical cloudy centers, giving them a fried-egg appearance. The arrow points to a plaque formed by a clear-plaque mutant.
doi:10.1128/9781555817169.ch8.f8.8

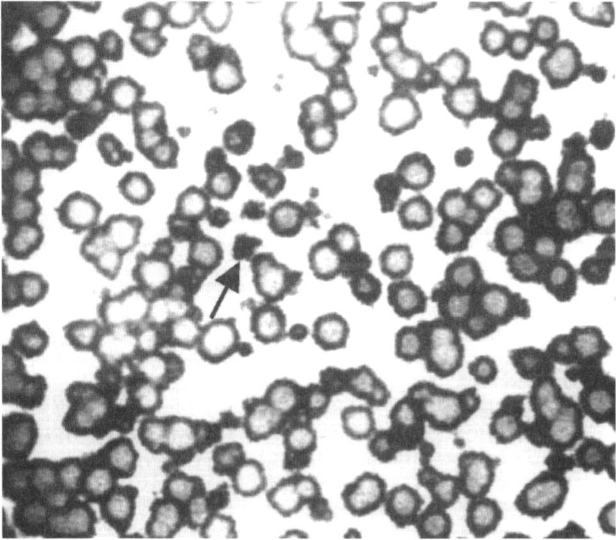

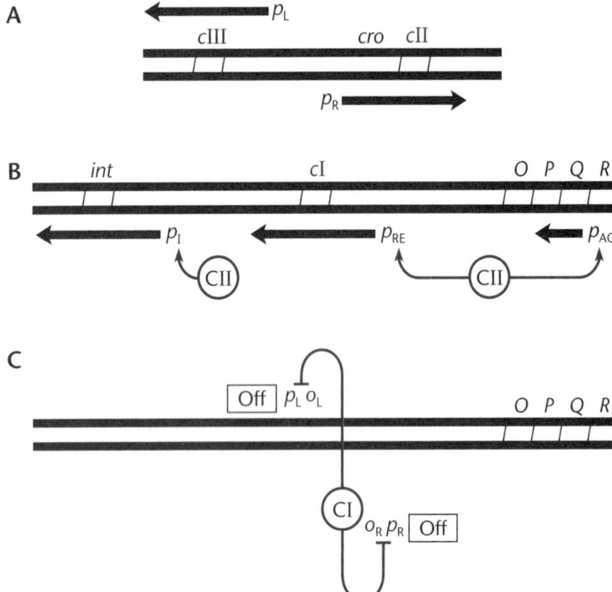

Figure 8.9 Formation of lysogens after λ infection. **(A)** The cII and cIII genes are transcribed from promoters p_R and p_L, respectively. **(B)** CII activates transcription from promoters p_{RE} and p_I, leading to the synthesis of CI repressor and the integrase Int, respectively. It also activates transcription from p_{AQ}, probably inhibiting synthesis of Q. **(C)** The repressor shuts off transcription from p_L and p_R by binding to o_R and o_L. Finally, the Int protein integrates the λ DNA into the chromosome (Figure 8.10).
doi:10.1128/9781555817169.ch8.f8.9

depends on the outcome of a competition between the product of the *c*II gene, which acts to form lysogens, and the products of the *cro* gene and of genes in the lytic cycle that replicate the DNA and make more phage particles. Which pathway wins most often depends on the conditions of infection. At a **low multiplicity of infection** (MOI), the lytic cycle usually wins, and in as many as 99% of the infected cells, the λ DNA replicates, and more phage are produced. However, for reasons we explain below, at a **high MOI**, the CII protein wins more often, and as many as 50% of the infected cells can form lysogens. The richness of the medium also plays a role. The reason is that cells that are growing very fast in rich medium have more RNase III than if they are growing more slowly, and more RNase III means more N protein, which favors lytic development (see Court et al, Suggested Reading). The reason they have more N protein is that the leftward transcript from p_L contains the *nutL* site just upstream of the translational initiation region (TIR) for the *N* gene (Figure 8.4). The Nus factors and N protein bound to the *nutL* site inhibit N translation from the nearby TIR for the *N* gene, so less N protein is made. There is a cleavage site for RNase III between the *nutL* site and the TIR for gene *N*, and the higher concentrations of RNase III when the cells are growing rapidly cleave the mRNA at a hairpin between the *nutL* site and the TIR for gene *N*, separating the *nutL* site from the *N* gene so more N is made. Note that it seems to make strategic sense to enter the lytic cycle when the MOI is low or when the cells are growing rapidly. If there are many more uninfected cells than phage, as there are if the MOI is low, it offers the opportunity to make many more phage, particularly if the growth conditions are good. If most of the cells are already infected, as they are if the MOI is high, it seems to make more sense to form a lysogen and wait until conditions improve, especially if growth conditions are poor at the time.

ROLE OF THE CII PROTEIN, A TRANSCRIPTION ACTIVATOR

Once the the CII protein is made from the p_R promoter, it promotes lysogeny by activating the RNA polymerase to begin transcribing at three promoters, which are otherwise inactive (Figure 8.9). Proteins that enable RNA polymerase to begin transcription at certain promoters are called **transcriptional activators** (see chapter 2). One of the promoters activated by CII is p_{RE}, which allows transcription of the *c*I gene. The product of this gene, the **CI repressor**, prevents transcription from the promoters p_R and p_L, which service many of the remaining genes of λ, including genes for replication of the DNA. The CI repressor is discussed in more detail below. Another promoter activated by the CII protein, p_I, allows transcription of the integrase (*int*) gene and synthesis of the Int protein. The Int enzyme integrates the λ DNA into the bacterial DNA to form the lysogen. Even though the *int* gene is transcribed from the p_L promoter, the Int protein cannot be made from the transcript for reasons that are explained below. The third promoter is the *pAQ* promoter, which directs the synthesis of an antisense RNA in the Q gene, thereby delaying the synthesis or activity of the Q protein, which would kill the cell.

ROLE OF THE CIII PROTEIN: A PROTEASE INHIBITOR

The role of the *c*III gene product in lysogeny is less direct. CIII inhibits a cellular protease, HflB, the product of the *ftsH* gene, that degrades CII. Therefore, in the absence of CIII, the CII protein is rapidly degraded and no lysogens form. Incidentally, this explains why more lysogens form when cells are infected at higher MOIs (see above). More CIII protein is made at higher MOIs and more CIII protein inhibits more HflB protease, leading to more CII activator protein and more lysogens.

Phage λ Integration

As discussed above, as soon as λ DNA enters the cell, it forms a circle by pairing between the complementary single-stranded *cos* sequences at its ends. The Int protein can then promote the integration of the circular λ DNA into the chromosome, as illustrated in Figure 8.10A. Int is a site-specific recombinase, a member of the tyrosine (Y) family of recombinases, that specifically promotes recombination between the attachment sequence (called *attP*, for *att*achment *p*hage) on the phage DNA and a site on the bacterial DNA (called *attB*, for *att*achment *b*acteria) that lies between the galactose (*gal*) and biotin (*bio*) operons in the chromosome of *E. coli*. This is a nonessential region of the *E. coli* chromosome, so integration of λ at this site causes no observable phenotype. Some phages do integrate their DNA into essential genes of the bacterium, which requires special adaptations (Box 8.1). Because the Int-promoted recombination does not occur at the ends of λ DNA but rather at the internal *attP* site, the prophage map is a cyclic permutation of the map of DNA found in the phage head with different genes at the ends. In the phage head, the λ DNA has the *A* gene at one end and the *R* gene at the other end (Figure 8.2, λ map). In contrast, in the prophage, the *int* gene is at one end and the *J* gene is at the other (Figure 8.10B). It was the difference between the phage genetic map and the prophage map that led to this model of integration, which is sometimes called the **Campbell model** after the person who first proposed it.

The recombination promoted by Int is called **site-specific recombination** because it occurs between specific sites, one on the bacterial DNA and another on the phage DNA. This site-specific recombination is not normal homologous recombination but rather a type of nonhomologous recombination, because the sequences of the phage and bacterial *att* sites are mostly dissimilar. They have a common core sequence, O, of only 15 bp—GCTT T(TTTATAC)TAA—flanked by two dissimilar sequences, B and B′ in *attB* and P and P′ in *attP* (Figure 8.10, inset). The recombination always occurs within the bracketed 7-bp sequence. Because the region of homology is so short, this recombination would not occur without the Int protein, which recognizes both *attP* and *attB* and promotes recombination between them. The mechanism of action of Y recombinases and other types of site-specific recombinases is discussed in chapter 9.

Figure 8.10 Integration of λ DNA into the chromosome of *E. coli*. **(A)** The Int protein promotes recombination between the *attP* sequence in the λ DNA and the *attB* sequence in the chromosome. The inset **(A′)** shows the region in more detail, with sequences POP′ and BOB′. The common core sequence of the two sites is shown in black. **(B)** Gene order in the prophage. The *cos* site is the location where the λ DNA is cut for packaging and recircularization after infection. The locations of the *int, xis, A,* and *J* genes in the prophage are shown (refer to the λ map in Figure 8.2). The *E. coli gal* and *bio* operons are on either side of the prophage DNA in the chromosome. Only the ends of the prophage are shown.
doi:10.1128/9781555817169.ch8.f8.10

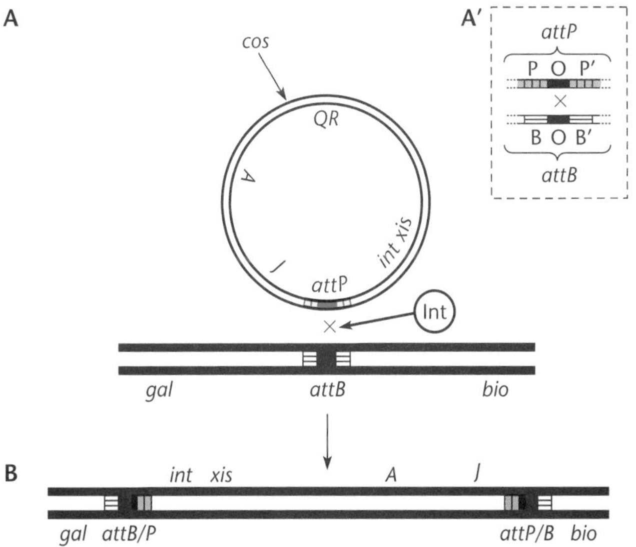

Maintenance of λ Lysogeny

After the lysogen has formed, the *c*I repressor gene is one of the few λ genes to be transcribed, and it maintains the lysogenic state by binding to two regions, called **operators**. These operators, o_R and o_L, are close to promoters p_R and p_L, respectively, and by binding the repressor, they prevent transcription from p_R and p_L and therefore transcription of most of the genes of λ, either directly or indirectly. Repressors and operators are discussed further in chapter 12. In the prophage state, the *c*I gene is transcribed from a different promoter, the p_{RM} promoter (for repression maintenance), which is immediately upstream of the *c*I gene, rather than from the p_{RE} promoter used immediately after infection (Figures 8.9 and 8.11). The p_{RM} promoter is not used immediately after infection because its activation requires the CI repressor, as we discuss below.

THE CI REPRESSOR

It is important to know the structure of the CI repressor to understand how it can regulate its own synthesis, as well as that of the other λ genes. Each CI polypeptide consists of two separable parts, or **domains**. It is generally drawn as a dumbbell to illustrate this domain structure (Figure 8.11). One of the domains of the CI polypeptide promotes the formation of dimers and tetramers by binding to the corresponding domain on other CI polypeptides, shown as two dumbbells binding to each other. For the CI repressor to function, two of these dumbbells must bind to each other through their dimerization domains to form a dimer made up of two copies of the polypeptide. In turn, a tetramer forms when two of these dimer dumbbells bind to each other through

BOX 8.1

Effects of Prophage Insertion on the Host

The insertion of the prophage into the host cell DNA can have many effects on the host. For example, the prophage may encode virulence proteins that increase the pathogenicity of the host in a process called lysogenic conversion. However, sometimes the insertion of the prophage causes phenotypes by itself, by disrupting genes of the host. For example, it has been known for decades that some phages of *S. aureus* integrate into the gene for beta hemolysin (β), inactivating the gene. However, this enzyme seems to be nonessential for pathogenicity in humans, and most strains that cause disease in humans seem to lack it, often because they have one of the prophages integrated in the gene. However, many of these same phages carry a gene for staphylokinase (SAK), a virulence determinant, so most pathogenic strains are β^- SAK$^+$. In some bacterial pathogens, disruption of a gene by insertion of a prophage may actually contribute to the pathogenicity of the bacterium in a case where the product of the disrupted gene interferes with pathogenicity (see Al Mamun et al., References). Some phages, including Mu, integrate almost randomly into the chromosome. Such phages are mutagenic because they inactivate any gene they happen to integrate into, hence the name Mu (for mutagenic phage). This property has made them useful for transposon mutagenesis (see chapter 9).

Surprisingly, most phages do not cause mutant phenotypes when they integrate into the chromosome for a number of reasons. The archetypal lysogenic phage λ avoids causing phenotypes by integrating into a nonessential region between the *gal* and *bio* operons of *E. coli* (see the text). However, some phages and other DNA elements, such as pathogenicity islands (see chapter 9), integrate directly into genes, often into genes for transfer (tRNAs). Examples include the *Salmonella* phage P22, which integrates into a threonine tRNA gene; the *E. coli* phage P4, which integrates into a leucine tRNA gene; the *Haemophilus influenzae* phage HPc1, which also integrates into a leucine tRNA gene; and the virus-like element SSV1 of the archaeon *Sulfolobus* sp., which integrates into an arginine tRNA gene. The genes for tRNAs may be chosen as integration sites because tRNA genes are relatively highly conserved in evolution, so the phage could find its *attB* site and form a lysogen in many types of bacteria. Another possible explanation is that phages seem to prefer sequences with twofold rotational symmetry for their attachment sites. The sequences of tRNA genes have such symmetry, since the tRNA products of the genes can form hairpin loops, and in fact, most phages seem to integrate into the region of the tRNA gene that encodes the anticodon loop.

Other phages integrate into essential protein-coding genes. For example, phage ϕ21, a close relative of λ, and phage e14, a defective prophage, both integrate into the isocitrate dehydrogenase (*icd*) gene of *E. coli*. The product of the *icd* gene is an enzyme of the tricarboxylic acid cycle and is required for optimal utilization of most energy sources, as well as for the production of precursors for some biosynthetic reactions. Inactivation of the *icd* gene would cause the cell to grow poorly on most carbon sources.

How can some phages integrate into essential genes and not kill the host? In some cases, the phages may disrupt the gene into which they integrate, for example, a tRNA gene, but they carry genes for tRNAs that may substitute for the tRNA gene they disrupt (see Ventura et al., below). However, there are other cases where insertion of the prophage does not inactivate the gene because it duplicates part of the gene next to its *attP* site. The 3′ end of the gene is repeated, with very few neutral changes next to the phage *attP* site, so that when the phage integrates, the normal 3′ end of the gene is replaced by the very similar phage-encoded sequence. This is true both of phages like ϕ21 that integrate into a protein-coding sequence and of some phages that integrate into an essential tRNA gene. It is an interesting question in evolution how the bacterial sequence could have arisen in the phage. Perhaps the phage first arose as a specialized transducing particle, which then somehow adapted to using the substituted bacterial genes as their normal attachment site in the chromosome.

References

Al Mamun, A., A. Tominaga, and M. Enomoto. 1997. Cloning and characterization of the region III flagellar operons of the four *Shigella* subgroups: genetic defects that cause loss of flagella of *Shigella boydii* and *Shigella sonnei*. *J. Bacteriol.* **179:**4493–4500.

Hill, C. W., J. A. Gray, and H. Brody. 1989. Use of the isocitrate dehydrogenase structural gene for attachment of e14 in *Escherichia coli* K-12. *J. Bacteriol.* **171:**4083–4084.

Ventura, M., C. Canchaya, D. Pridmore, B. Berger, and H. Brüssow. 2003. Integration and distribution of *Lactobacillus johnsonii* prophages. *J. Bacteriol.* **185:**4603–4608.

their tetramerization regions in the same domain. At very low concentrations of CI polypeptide, the dimers do not form and the repressor is not active. At somewhat higher concentrations, the dimers form and the repressor is active. The other domain on each polypeptide specifically binds to an operator sequence on the DNA or, more specifically, to subsequences within these operators, as we discuss below.

Regulation of Repressor Synthesis

It is important to maintain the level of repressor in the cell within normal limits, even after a lysogen has

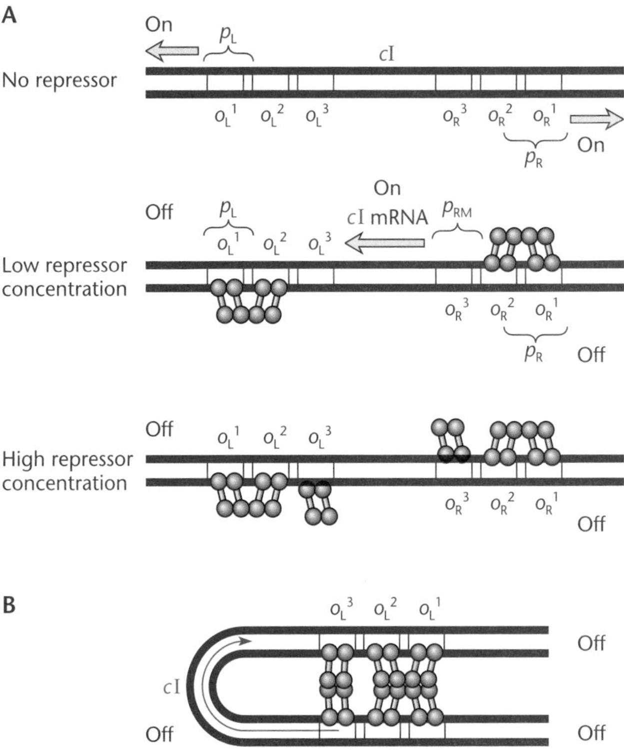

Figure 8.11 Regulation of repressor synthesis in the lysogenic state. The dumbbell shape represents the two domains of the repressor. **(A)** The dimeric repressor, shown as two dumbbells, binds cooperatively to o_R^1 and o_R^2 (and o_L^1 and o_L^2), repressing transcription from p_R (and p_L) and activating transcription from p_{RM}. At higher repressor concentrations, it also binds to o_R^3 and o_L^3, repressing transcription from p_{RM}. **(B)** Still higher concentrations cause the formation of tetramers that bend the DNA, further repressing transcription from p_{RM}. The relative affinities of the repressor for the sites is as follows: $o_R^1 > o_R^2 > o_R^3$ and $o_L^1 > o_L^2 > o_L^3$.
doi:10.1128/9781555817169.ch8.f8.11

formed. If the amount of repressor drops below a certain level, transcription of the lytic genes begins and the prophage is induced to produce phage. However, if the amount of repressor increases beyond optimal levels, cellular energy is wasted in making excess repressor, and it might be too difficult to induce the prophage should the need arise. The mechanism of regulation of repressor synthesis in lysogenic cells is well understood and has served as a model for gene regulation in other systems (see Ptashne, Suggested Reading). Figure 8.11 illustrates the regulation of CI repressor synthesis.

The repressor dimer regulates its own synthesis, as well as that of other λ gene products, by binding to the operator sequences, one on the right of the CI gene, called o_R, and the other on the left of the CI gene, called o_L. The CI protein can be either a repressor or an activator of transcription, depending on how many copies of it are bound to these operators. We first discuss binding to the o_R operator on the right of the repressor gene, since it is the most important operator for regulating repressor synthesis, although both operators have the same structure. The operator o_R can be divided into three repressor-binding sites, o_R^1, o_R^2, and o_R^3. If the concentration of repressor is low, only the o_R^1 site is occupied, which is sufficient to repress transcription from the promoter p_R, which overlaps the operator site (Figure 8.11). This prevents transcription of the replication genes O and P. However, as the repressor concentration increases, eventually o_R^2 also becomes occupied, because tetramers can form between the dimers bound at o_R^1 and o_R^2 (Figure 8.11). The formation of a tetramer stabilizes the binding of a CI dimer to o_R^2. This is called **cooperative binding** and is seen with many DNA-binding proteins. Repressor bound at o_R^2 is required to activate transcription from the promoter p_{RM}, which is why p_{RM} is used to transcribe the repressor gene in the lysogen only when there is some repressor in the cell. Since o_R and o_L have the same structure, o_L^1 and o_L^2 are occupied by repressor at the same repressor concentrations as the corresponding sites in o_R. Repressor dimer bound at o_L^1 blocks leftward transcription, and repressor dimers bound at o_L^1 and o_L^2 can bind to those at o_R^1 and o_R^2 to form an octamer, which bends the DNA and helps stabilize their binding and more completely represses transcription from p_L and p_R (Figure 8.11B). Other repressors are known to regulate transcription from a promoter by bending the DNA at the promoter, and we discuss some examples of this in chapter 12.

At very high concentrations, CI repressor dimers also cooperatively bind to o_L^3 and o_R^3, where they also form a tetramer and stabilize each other's binding. The repressor dimer bound at o_R^3 interferes with RNA polymerase binding to the p_{RM} promoter, reducing the synthesis of more repressor (Figure 8.11B). This complex regulation allows the prophage to synthesize more repressor when there is less in the cell, and vice versa, so that the cell maintains the levels of repressor within narrow limits. The synthesis of repressor also responds quickly to perturbations in the cell, which explains why λ lysogens are very stable and usually release phage only under unusual circumstances. The term **robust regulation** has been used to describe such interactive regulatory systems that are capable of responding to varying conditions.

Immunity to Superinfection

The CI repressor in the cell of a lysogen prevents not only the transcription of the other prophage genes by binding to operators o_L and o_R, but also the transcription of the genes of any other λ phage infecting the lysogenic cell by binding to the operators of that phage. Thus, bacteria lysogenic for λ are immune to λ superinfection. However, λ lysogens can still be lytically infected by any relative of λ phage that has different operator sequences to which the λ CI repressor cannot bind. Any two phages that differ in their operator sequences are said to be **heteroimmune phages**. If they have the same operator sequences, they can inhibit each other's transcription and are said to be **homoimmune phages**, no matter how different they are in their other genes. Many hybrid phages have been made between λ and some of its relatives that have λ genes but the repressor gene and operators of another phage, for example, phage 434. Such a phage is then called λ i434 to indicate that it is λ phage with the immunity of phage 434, so it will multiply in a λ lysogen but not a 434 lysogen.

THE CRO PROTEIN

Another protein involved in the regulation of repressor synthesis is Cro. This is one of the first proteins made after infection and inhibits the synthesis of repressor and thereby helps commit the phage to the lytic cycle. Cro does this by binding to the operator sequences, although in reverse order of repressor binding, as illustrated in Figure 8.12. Cro binds first to the o_R^3 site, occluding the p_{RM} promoter, and then to the o_R^2 site, thereby preventing the CI repressor from binding to the o_R^2 site and activating its own synthesis from the p_{RM} promoter. This prevents interference from the lysogenic cycle once the phage is in the lytic cycle. It seems that Cro could also play a role in induction of the prophage, committing the phage to the lytic cycle after induction (see below). However, a role for Cro in the induction of the prophage is still controversial.

Figure 8.12 Cro prevents repressor binding and synthesis by binding to the operator sites in reverse order from the repressor. By binding to o_R^3, Cro prevents repressor activation of transcription from p_{RM} while allowing transcription from p_R. Eventually, Cro accumulates to the point where it binds to o_R^1 and o_R^2 and blocks transcription of early RNA.
doi:10.1128/9781555817169.ch8.f8.12

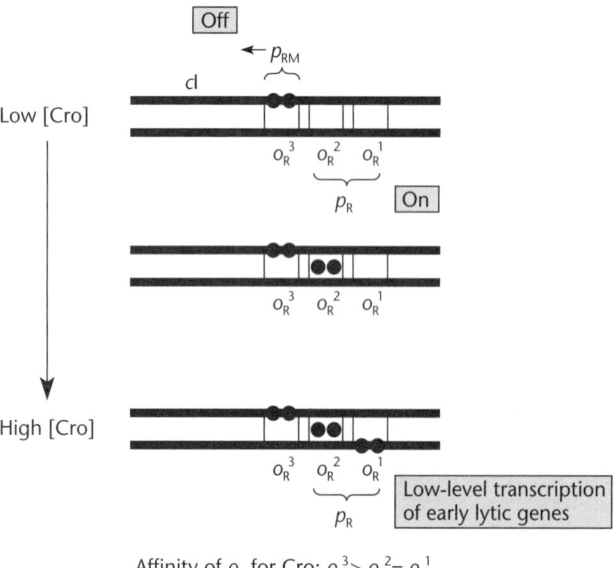

Induction of λ

Phage λ remains in the prophage state until the host cell DNA is severely damaged by irradiation or some types of chemicals that cause extensive damage to the DNA. The prophage is then induced to go through its lytic cycle. Figure 8.13 outlines the process of induction of λ. When the cell attempts to repair the damage to its DNA, short pieces of single-stranded DNA accumulate and bind to the RecA protein of the host, which reflects the role RecA plays in recombination (see chapter 10). This binding activates the **coprotease** activity of RecA, and the activated RecA protein with single-stranded DNA attached then binds to the λ CI repressor, causing the repressor to cleave itself (**autocleavage**). The autocleavage separates the DNA-binding domain in the CI polypeptide from the domain involved in dimer formation. Without the dimerization domain, the CI repressor can no longer form dimers and the DNA-binding domains can no longer bind to the operators. As the repressors

Figure 8.13 Induction of λ. Accumulation of single-stranded DNA (ssDNA) due to damage to the DNA results in activation of the RecA protein, which promotes the autocleavage of the CI repressor protein, separating the dimerization domain of the protein from the DNA-binding domain so that the repressor can no longer form dimers and bind to DNA. Transcription of *int-xis* and *cro*, *O*, and *P* ensues, and the phage DNA is excised from the chromosome and replicates.
doi:10.1128/9781555817169.ch8.f8.13

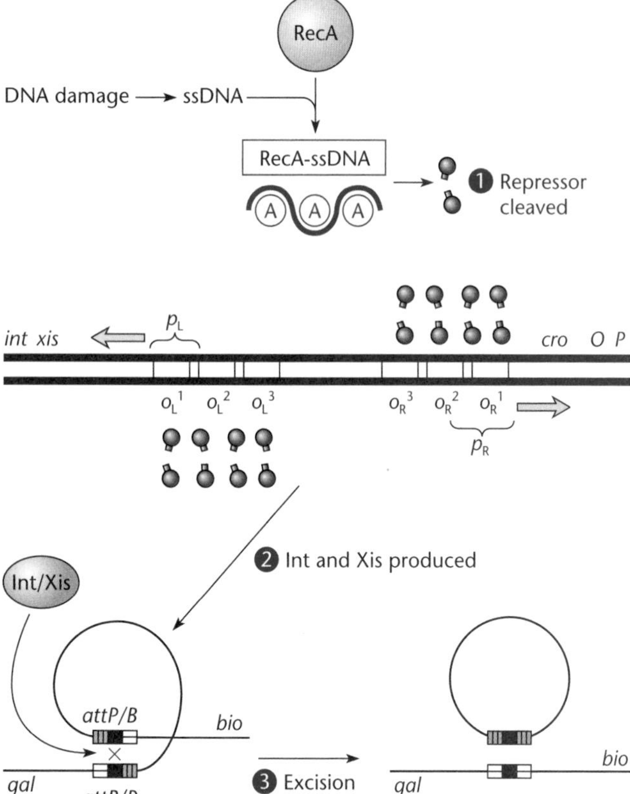

drop off the operators, transcription initiates from the promoters p_R and p_L, and the lytic cycle begins.

The induction of λ by cleavage of its repressor is yet another case in which studies with λ led to the discovery of more universal phenomena, in this case, SOS induction. The LexA repressor of *E. coli* and other bacteria also cleaves itself when bound by RecA and single-stranded DNA that accumulates following DNA damage. Autocleavage of LexA leads to the induction of many repair genes under its control, including those encoding RecA protein and mutagenic bypass polymerases. Thus, λ relies on the preexisting SOS system of *E. coli* to induce itself after DNA damage. Other lysogenic phages, including the *Vibrio cholerae* phage that makes cholera toxin (CT), "skip the middleman" and use LexA directly to regulate their transcription, so they can be induced following DNA damage by autocleavage of LexA (see below). We discuss SOS induction and mutagenic bypass polymerases in chapters 1 and 11.

EXCISION

Once the repressor is out of the way, transcription from p_L and p_R can begin in earnest. Some of the genes transcribed from p_L are required to excise the λ DNA from the chromosome. Excision requires site-specific recombination between hybrid *attP-attB* sequences that exist at the junctions between the prophage DNA and the chromosomal DNA (Figure 8.13). These hybrid sequences are different from either *attB* or *attP* and contain sequences from both; therefore, Int alone is not capable of recognizing and promoting recombination between them to excise the prophage. Another protein, called excisase (Xis), is also required to allow the Int protein to recognize these hybrid sequences. Proteins such as Xis are often called **directionality factors** because they promote recombination in only one direction, excision, unlike Int, which is required for both excision and integration. Accordingly, unlike after infection, when only Int is synthesized, after induction, both the Int and Xis proteins are synthesized (see below). In fact, it is necessary that only Int be made after infection, because if both Int and Xis were synthesized after infection, the λ prophage would be excised as soon as it integrates, and lysogens could not form.

Why only Int is made after infection but both Int and Xis are made after induction was a puzzle (Figure 8.14). After infection, Int is synthesized from the p_L promoter, which is in the *xis* gene, so Xis obviously cannot be synthesized from the p_I promoter; but why are both Xis and Int not also synthesized from the p_I promoter? After antitermination by N, the transcription from p_L should continue through both *int* and *xis*. To achieve this differential gene expression, λ takes advantage of the fact that the ends of the λ prophage are different from the ends of

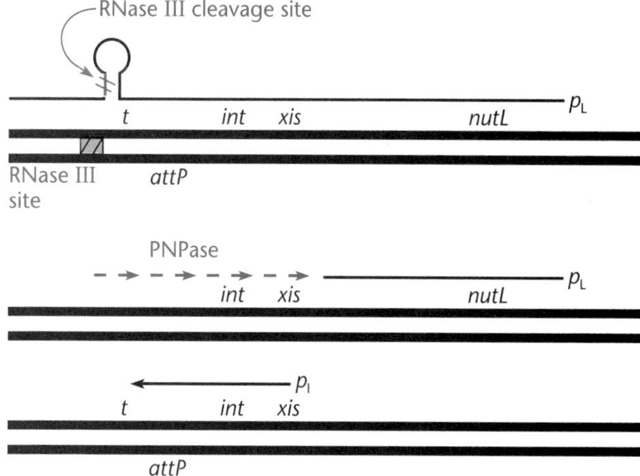

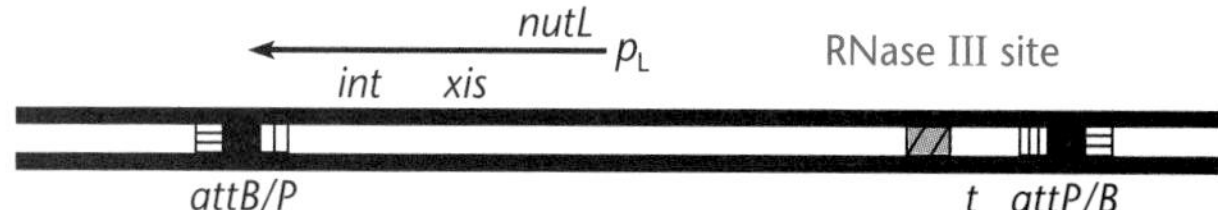

Figure 8.14 Retroregulation. After infection, the *xis* and *int* genes cannot be expressed from the p_L promoter, even though transcription extends that far. Because of N, transcription from p_L continues past the terminator, *t*, into an RNase III cleavage site on the other side of *attP*. The RNA is cleaved and digested back into *xis* and *int* by PNPase, removing the coding sequences for Int and Xis from the RNA. Xis also cannot be expressed from p_I because the p_I promoter is in the *xis* gene and Int cannot excise the DNA by itself. The RNA from p_I is stable, however, because the transcript does not contain the *nutL* site and so does not continue through *t* to the RNase III cleavage site beyond. The blue region indicates the location of the coding information for the RNase III site in the DNA, but RNase III cleaves only the mRNA transcript. **(B)** When the prophage is first induced, however, and before it is excised, the sequence encoding the RNase III cleavage site is separated from the *xis-int* coding sequence, so that the RNA made from p_L is stable and both Int and Xis are made. **(A)** Early after infection; **(B)** early after induction.
doi:10.1128/9781555817169.ch8.f8.14

the DNA in the phage head. Briefly, after infection, both the *xis* and *int* genes are transcribed from the p_L promoter, but because of antitermination, the transcription proceeds past *xis* and *int* into sequences on the other side of *attP*, as shown in the figure. One of these sequences on the other side of *attP* forms a hairpin that is cleaved by RNase III if it appears in an RNA (see Table 2.1 for a listing of *E. coli* RNases). Cleavage by RNase III at this site sets up the RNA to be degraded by a 3′-5′ exonuclease, PNPase, before either Int or Xis can be translated from the RNA. In contrast, the RNA made from the p_I promoter is stable and can synthesize Int because it does not include *nutL*, so it is not antiterminated and usually terminates at *t* before it transcribes the RNase III cleavage site, as shown in the figure.

The situation is very different immediately after induction, when the λ prophage DNA is still integrated into the chromosome. Integration of the phage DNA splits the RNase III cleavage site from promoter p_L, and it is now at the other end of the integrated prophage DNA (Figure 8.14B). Now, the mRNA transcribed from p_L is stable, and both Int and Xis can be translated from the mRNA. Regulation through splitting genes from promoters during integration is also used by other elements, including integrating conjugative elements (see chapter 5).

While the Int and Xis proteins are excising and cyclizing λ DNA from the chromosome, the *O* and *P* genes are being transcribed from p_R. These proteins promote replication of the excised λ DNA. The late genes, now joined to the early genes by cyclization of the DNA, are transcribed, and phage particles are produced. In about 1 h after the cellular DNA has been damaged, depending on the medium and the temperature, the cell lyses, spilling about 100 phage into the medium from a cell that, an hour before, showed few signs of harboring the phage.

Summary of the Lytic and Lysogenic Cycles

The lytic and lysogenic cycles, including the induction of prophages, involve so many interacting pathways that it is worth reviewing them. Figure 8.15 and Table 8.3 review the competition for entry into the lysogenic cycle versus the lytic cycle. After infection, when there is no CI repressor in the cell, the *N* and *cro* genes are transcribed. As discussed above, the *N* gene product acts as an antiterminator and allows the transcription of many genes, including *c*II and *c*III, as well as the genes encoding the replication proteins O and P, while the Cro protein inhibits repressor synthesis during the lytic cycle.

Whether the phage enters the lytic or the lysogenic cycle depends on the fate of the CII activator protein, which is determined by the MOI and the metabolic state of the infected cell. At higher MOIs, more CIII protein is made to inhibit the FtsH protease and more CII protein accumulates. More CII means more CI repressor is then made from the p_{RE} promoter, more Int is made from p_I, and less Q is made because of interference from an antisense RNA made from p_{AQ}. Consequently, more of the cells survive and enter the lysogenic phase. In the meantime, the Cro protein is trying to tip the phage in the direction of lytic development. By binding to $o_R{}^3$ it occludes the p_{RM} promoter, so less repressor is made and more of the cells enter the lytic phase. At high enough concentrations, it also binds to $o_R{}^2$ and $o_R{}^1$, displacing the C1 repressor and allowing transcription of the O and P genes. If the cells are growing in rich medium, more RNase III is made, which separates the *N* gene TIR from the *nutL* site on the same mRNA, allowing more N protein to be made, and more N protein also favors lytic development. Eventually, there is too much DNA for the repressor to bind to all of it, and transcription of *O* and *P* increases further, followed by yet more DNA replication. The Q protein accumulates, allowing transcription of the head, tail, and lysis genes.

The induction of a prophage is normally caused by damage to the DNA, which causes the CI repressor to cleave itself. Transcription then begins from promoters p_L and p_R, as after infection, and N antiterminates transcription from these promoters. Now, however, because the prophage DNA ends at the *attP* site, the leftward transcript is stable, and both Int and Xis are made by transcription from the p_L promoter. The two proteins promote recombination between the hybrid *attP-attB* sites at the ends of the prophage, and the phage DNA excises as a circle from the chromosome. Lytic development then proceeds much as after infection, although the Cro protein may not play as much of a role.

Specialized Transduction

In generalized transduction, phages carry or transduce host DNA, instead of their own DNA, from one cell to another (see chapter 7). This is called **generalized transduction** because almost any region of the chromosome can be transduced. Some phages that integrate their DNA into the chromosome during lysogeny are also capable of another type of transduction, called **specialized** or **restricted transduction**, because in this type of transduction,

Table 8.3 Steps leading to lytic growth and lysogeny

Step	Lytic growth	Lysogeny
1	Transcription from p_L and p_R	Same as lytic
2	Synthesis of N and Cro	Same as lytic
3	N allows expression of O and P.	N allows expression of CII.
4	Cro blocks CI synthesis from p_{RM}.	CII activates CI synthesis from p_{RE}, Int synthesis from p_I.
5	Low MOI; most CII degraded	High MOI; CII stabilized
6	Fast growth; more RNase III, more N	Slow growth; less RNase III, less N
7	More O, P, Q; lytic growth	More CI; lysogeny

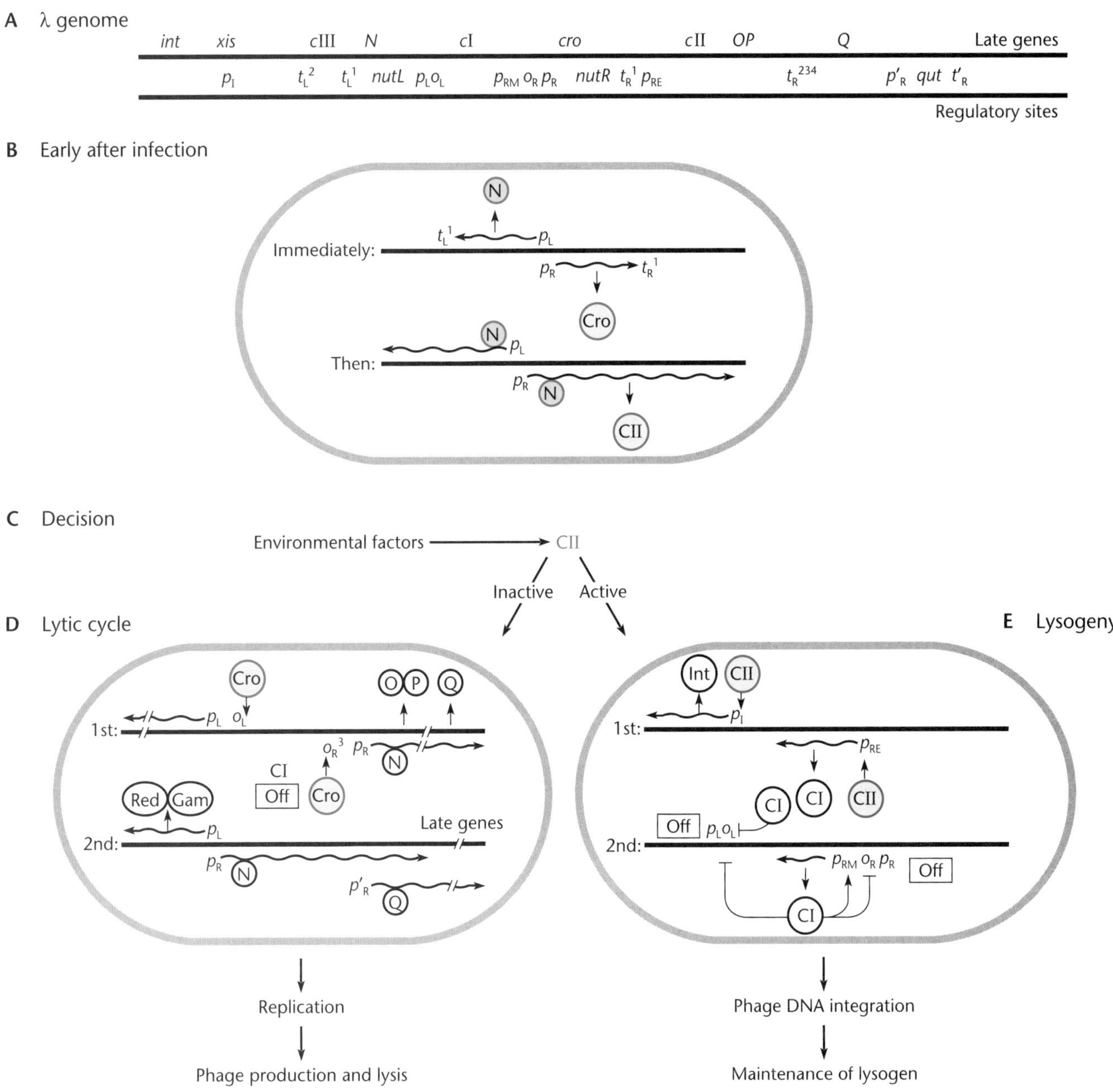

Figure 8.15 Competition determining whether phage will enter the lytic or lysogenic cycle. **(A)** Key genes (top line) and sites (bottom line). **(B)** Gene expression early after infection. **(C)** The abundance of active CII protein determines whether the phage enters the lytic or lysogenic cycle. **(D)** The synthesis of Cro promotes lytic development by repressing the synthesis of CI repressor. Once O and P are synthesized, the replication of λ DNA dilutes out the CI repressor. **(E)** Synthesis of CII promotes lysogeny. doi:10.1128/9781555817169.ch8.f8.15

only bacterial genes close to the attachment site of the prophage can be transduced. The specialized transducing phage particle carries both bacterial DNA and phage DNA instead of only bacterial DNA, reflecting very different mechanisms for the two types of transduction.

Figure 8.16 illustrates how a λ phage particle capable of specialized transduction of *gal* markers arises. The λ prophage is integrated between the closely linked *gal* and *bio* genes in the chromosome. The *gal* gene products degrade galactose for use as a carbon and energy source, while the *bio* gene products make the vitamin biotin.

Specialized transduction can occur when a phage picks up neighboring bacterial genes during induction of the prophage. As shown in Figure 8.16, a specialized transducing phage carrying the *gal* genes, called λd*gal*, where the "d" stands for defective, forms as the result of

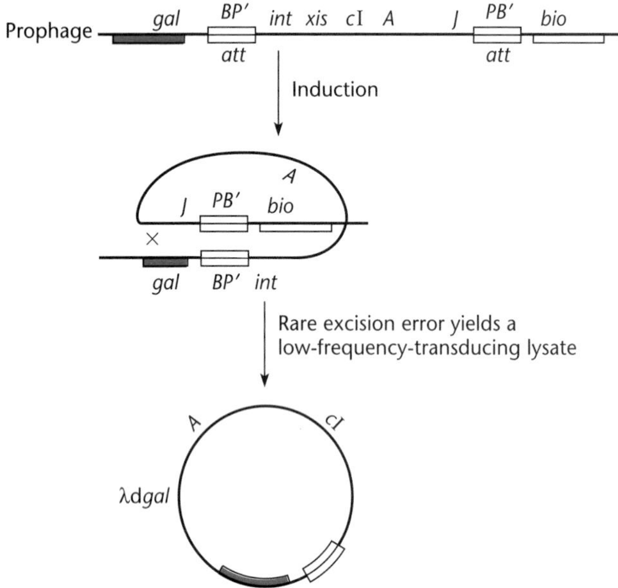

Figure 8.16 Formation of a λ*dgal* transducing particle. A rare mistake in recombination between a site in the prophage DNA (in this case between *A* and *J*) and a bacterial site on the left of the prophage in the *gal* operon results in excision of a DNA particle in which some bacterial DNA, including *gal*, has replaced phage DNA.
doi:10.1128/9781555817169.ch8.f8.16

a rare mistake during the excision recombination. Normally, the prophage DNA is excised from the bacterial DNA by recombination between the hybrid *attP-attB* sites at the junctions between the prophage and host DNA. However, if the recombination occurs between a site internal to the prophage DNA and a neighboring site in the bacterial DNA, in this case in the *gal* operon, the DNA later packaged into the head includes some bacterial sequences from the *gal* operon, as shown. Such transducing phages are very rare, because the erroneous recombination that gives rise to them is extremely infrequent, occurring at one-millionth the frequency of normal excision. Furthermore, the recombination must occur by chance between two sites that are approximately a λ genome length apart, with one site in the prophage DNA, or the excised DNA will not contain a *cos* site and be packaged in a phage λ head.

Selection of HFT Particles

Because of the rarity of these transducing phages, powerful selection techniques are required to detect them. In this case, rare galactose-positive (Gal$^+$) transductants are selected on plates with galactose as the sole carbon source.

In the rare Gal$^+$ transductants, a λ phage carrying *gal* genes has usually integrated into the chromosome, providing, by complementation, the *gal* gene product that the mutant lacks. If such a Gal$^+$ lysogen is colony purified and the prophage is induced from it, a high percentage of the resultant phage progeny will carry the *gal* genes. Such a lysogenic strain produces an **HFT lysate** (for high-frequency transduction), because it produces phage that can transduce bacterial genes at a very high frequency.

Normally, not all of the induced phage particles in a *gal*-transducing HFT lysate produced in this way carry the *gal* genes, and the lysate contains a mixture of transducing phage particles and wild-type λ. The reason for this is also apparent from Figure 8.16. Because the λ phage head can hold DNA of only a certain length, the transducing particles have of necessity lost some phage genes to make room for the bacterial genes. The properties of the transducing particle are determined by which phage genes are lost. For example, the λ*dgal* phage shown in Figure 8.16 lacks essential head and tail genes, beginning with the *J* gene, and so cannot multiply without a wild-type λ **helper phage** to provide the missing head and tail proteins. These phage particles are thus called λ*dgal* and usually can be produced only by induction from **dilysogens** that contain the λ*dgal* prophage and a wild-type phage integrated next to each other in tandem, as shown in Figure 8.17.

The molecular details of how these phage are induced and whether it requires Int and Xis or just Int depends

Figure 8.17 Induction of the λ*dgal* phage from a dilysogen containing both λ*dgal* and a wild-type λ in tandem. Recombination between the hybrid *PB′* and *BP′* sites at the ends excises both phages. The wild-type "helper" phage helps the λ*dgal* phage to form phage particles, and both are packaged from repeated *cos* sites in long concatemers. See the text for details.
doi:10.1128/9781555817169.ch8.f8.17

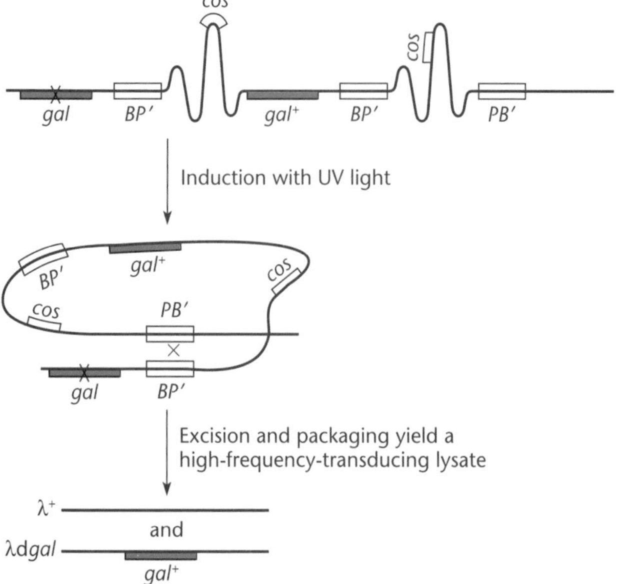

on how the dilysogen formed. In the example, when the λd*gal* integrated, it could have done so by homologous recombination between λ phage DNA sequences carried by both the λd*gal* DNA and the wild-type λ DNA, creating a structure like that shown in Figure 8.17. Then, the two could be excised together by Int and Xis as shown and then packaged independently from *cos* sites on the concatemers that form during DNA replication. Alternatively, the λd*gal* phage might have integrated next to a preexisting wild-type λ prophage, or the two might have integrated together as a dimeric circle. All of these cases require different combinations of Int and Xis for excision, and we leave this subject for the questions at the end of the chapter. However they have formed and been induced, the wild-type phage DNA must provide the head and tail genes that the λd*gal* DNA has lost. The two can then multiply together, so that roughly half the phage produced are now λd*gal*; the other half are wild-type λ.

The situation is very different if the HFT transducing phage are created by a mistaken recombination on the other side, replacing the *int* and *xis* genes with *bio* genes. These phage are able to multiply, since the genes on this side of *attP* are not required for multiplication. However, they cannot form a lysogen or be induced without the help of a wild-type phage because the *bio* genes of the host have replaced their *attP* site and their *int* and *xis* genes. Because they can multiply and form plaques, *bio*-transducing phages are called λp*bio*, in which the "p" stands for plaque forming.

Before the advent of DNA cloning, HFT phage particles played a major role in the development of microbial molecular genetics, including the first isolation of single genes and the discovery of insertion elements in bacteria. They have also been used to map phage λ genes and sites (see below). Although their general use has been largely supplanted by recombinant DNA techniques, they continue to have special applications.

Other Lysogen-Forming Phages

Phage λ was the first lysogen-forming phage to be extensively studied and thus serves as the archetypal lysogen-forming phage. However, many other types of lysogen-forming phages are known. Many use somewhat different strategies to achieve and maintain the prophage state. Some of them are described briefly here.

Phage P2

Phage P2 is another lysogen-forming phage of *E. coli*. The phage DNA is linear in the phage head but has cohesive ends, like λ, which cause the DNA to cyclize immediately after infection. The phage replicates as a circle, and the DNA is packaged from these circles instead of from concatemers as λ does normally. Also, like λ, the genetic map of P2 phage is linear, because it has a unique *cos* site at which the circles are cut during packaging.

One way in which P2 differs significantly from λ, which almost always integrates into a single site in the *E. coli* chromosome, is that P2 can integrate into many sites in the bacterial DNA, although it uses some sites more than others. Like λ, P2 requires one gene product to integrate and two gene products to be excised. P2 prophage is much more difficult to induce than λ, however. It is not inducible by UV light, and even temperature-sensitive repressor mutations cannot efficiently induce it. The only known way to induce it efficiently is to infect the lysogen with another P2 (or P4) phage (see below).

Phage P4: a Satellite Virus

Even viruses can have parasites! Phage P4 is a parasite that depends on phage P2 for its lytic development (see Kahn et al., Suggested Reading). Thus, it is a representative of a group called **satellite viruses**, which need other viruses to multiply. Phage P4 does not encode its own head and tail proteins but rather uses those of P2. Thus, P4 can multiply only in a cell that is lysogenic for P2 or that has been simultaneously infected with a P2 phage. When P4 multiplies in bacteria lysogenic for P2, it induces transcription of the head and tail genes of the P2 prophage, which are normally not transcribed in the P2 lysogen. This is illustrated in Figure 8.18A. P4 uses two mechanisms to induce transcription of the late genes of P2. It induces the P2 lysogen because it makes an inhibitor of the P2 repressor protein, which binds to the P2 repressor, inactivating it and inducing P2 to enter the lytic cycle. However, even though the P2 DNA replicates after induction by P4 and all the P2 proteins are made, most of the phage that are made contain P4 DNA genomes rather than P2 genomes. This is because P4 makes a protein called Sid, which causes the P2 proteins to assemble into heads that are smaller than normal, with only one-third the volume of a normal P2 head. These heads are too small to hold P2 DNA but large enough to hold P4 DNA, which is only about one-third the length of P2 DNA, so that the heads are filled with P4 DNA instead. Nevertheless, a P4 *sid* mutant, which cannot make the Sid protein, can still multiply in a P2 lysogen. Now, the heads in the lysate, which are the larger P2 size, contain either P2 DNA or P4 DNA. However, those that contain P4 DNA have two or three copies of the P4 DNA to fill the larger heads.

P4 can still multiply in P2 lysogens, even if it cannot induce the P2 prophage, which remains in the chromosome. At first, it was not obvious how P4 could induce the transcription of the late genes of P2, since, like T4 phage, the transcription of the late genes of P2 is enhanced by replication and the P2 DNA does not replicate if the prophage is not induced. P4 accomplishes this

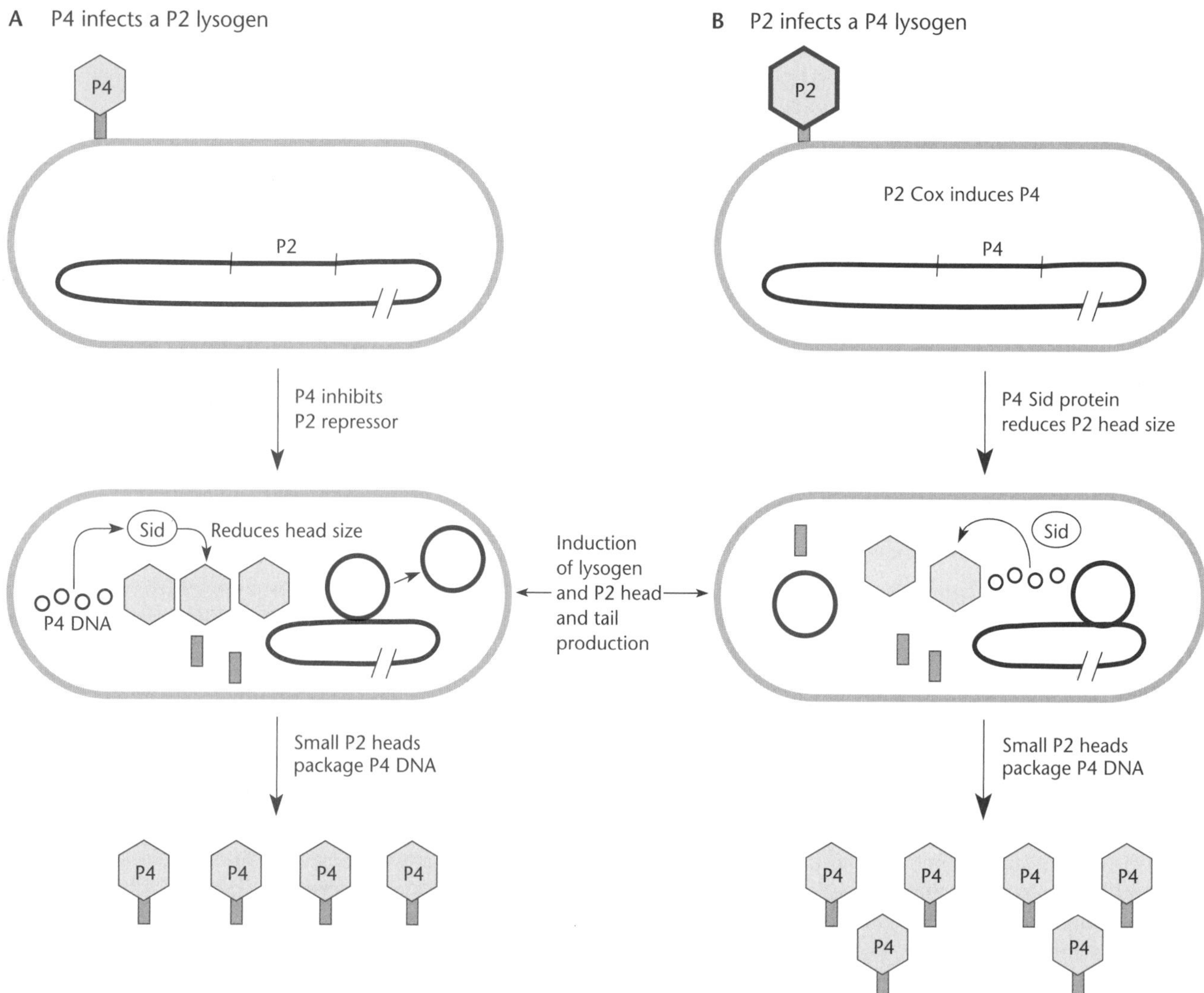

Figure 8.18 P2 can't win for losing. **(A)** A P2 lysogen is infected with P4. P4 makes a protein that binds to and inhibits the P2 repressor, inducing the P2 prophage. The P4 protein Sid makes the P2 head proteins form a smaller head, which packages the shorter P4 DNA rather than P2 DNA. Therefore, mostly P4 phage particles are released when the cells lyse. **(B)** A P4 lysogen is infected with P2. Now the P4 prophage is induced, and its replicating DNA is packaged by head proteins made by the infecting P2. Again, mostly P4 phage particles are released from the lysed cell, even though it was a P2 phage that infected the cell. Details are given in the text. doi:10.1128/9781555817169.ch8.f8.18

by *trans*-activating the transcription of the head and tail genes of P2 via synthesis of a protein called δ, which activates the transcription of the P2 late genes without P2 replication, even though the transcription seems to occur from the same promoters as the normal P2 replication-enhanced transcription.

Because it wears the protein coat of P2, the phage P4 particle looks similar to P2, except that it is smaller to accommodate the shorter DNA. While the DNAs of P2 and P4 have otherwise very different sequences, the *cos* sites at the ends of the DNA are the same, so that the head proteins of P2 can package either DNA.

Phage P4 can also form a lysogen; when it does so, it usually integrates into a unique site on the chromosome. Not only can P4 infection induce a P2 prophage, but P2 infection can also induce a P4 prophage. It does this inadvertently, by making a protein called Cox, which induces the P4 prophage (Figure 8.18B). Apparently, P4, which cannot multiply by itself, does not want to be caught sleeping as a prophage if the cell happens to be infected by P2. Not only would it die along with its host, it would also miss the opportunity to multiply and infect new hosts. Again, at least some of the phage that emerge from the infection after P4 is induced have P4 DNA wrapped in a smaller-than-normal P2 coat, even though it was a P2 phage that infected the cell. One phage enters the cell and emerges as a different phage. No matter who initiates the infection, P2 comes out the loser.

Not only can phage P4 integrate into the chromosome, it also can replicate autonomously as a circle in the prophage state, as does P1 (see below). Because of this ability to maintain itself as a circle, phage P4 has been engineered for use as a cloning vector.

Phages P2 and P4, as well as their many relatives, have a very broad host range and infect many members of the *Enterobacteriaceae*, including *Salmonella* and *Klebsiella* spp., as well as some *Pseudomonas* spp. They are also related to phage P1, although their lifestyles and strategies for lytic development and lysogeny are very different.

As is often the case, once the interaction of P2 and P4 had been discovered and characterized, other examples of DNA elements that parasitize phages to move between cells were discovered. We talk about one of these examples, the parasitization of a *Staphylococcus aureus* phage by pathogenicity islands carrying the genes for toxic shock syndrome, below under "Lysogenic Conversion and Bacterial Pathogenesis." Whether a DNA element that moves this way is a genetic island or a satellite virus becomes a matter of semantics.

Prophages That Replicate as Plasmids

Not all prophages integrate into the chromosome of the host to form a lysogen. Some form a prophage that replicates autonomously as a plasmid. These plasmid prophages can be very stable, with efficient and robust copy number control mechanisms and partitioning systems that make them resemble other plasmids, except that they also encode phage proteins. The prophages can also be either circular or linear plasmids. Other phages are known to sometimes exist as plasmids in the prophage state, including P4 (see above) and some mutants of λ. In these cases, partial repression of gene expression of the phage limits replication and keeps the copy number of the plasmid low but very variable, suggesting that this is not their normal state.

P1 PHAGE

P1 is the best-studied phage whose prophage replicates as a circular plasmid. The P1 plasmid prophage is a low-copy-number plasmid that maintains a copy number of only 1 and combines many of the other features of true plasmids, including a partitioning system and even a plasmid addiction system called PhD-Doc (see Box 4.2). Because it is a true plasmid, combined with the convenience of having a lytic phage cycle, which, among other advantages, facilitates the isolation of plasmid DNA, plasmid P1 is one of the major model systems for studying plasmid copy number control, segregation, and partitioning. Another interesting aspect of this phage is that it has an invertible segment (see chapters 7 and 9). A region of the phage DNA encoding the tail fibers frequently inverts, and the host range of the phage depends on the orientation of this invertible segment. The invertible segment thereby contributes to the very broad host range of P1. It also has a site-specific Y recombinase, Cre, that acts on the *lox* site to resolve plasmid dimers and prevent curing of the prophage. This very active *lox* site forms the end of the linear genetic map, as discussed in chapter 7. The Cre-*lox* system is the best understood of the Y recombinases and has been put to many uses in molecular genetics (see chapter 9).

N15 PHAGE

Another *E. coli* phage, N15, has a plasmid as its prophage, but this plasmid is linear rather than circular. It has served as a model system for how some types of linear plasmids replicate. It has hairpin ends with the 3′ and 5′ ends joined to each other and replicates around the ends from an internal origin to yield a dimeric circle. The dimeric circles are then resolved by a telomere resolvase. The replication of N15 prophage and other linear plasmids is discussed in Box 4.1.

Phage Mu: a Transposon Masquerading as a Phage

Another well-studied phage that forms lysogens is Mu, which integrates randomly into the chromosome. Because it integrates randomly, it often integrates into genes and causes random insertion mutations, hence its name, Mu (for *mu*tator phage). This phage is essentially a transposon wrapped in a phage coat, and it integrates and replicates by transposition. For this reason, the discussion of phage Mu is deferred to chapter 9.

Lysogenic Conversion and Bacterial Pathogenesis

In a surprising number of instances, prophages carry genes for virulence factors or toxins required for virulence by the pathogenic bacteria they lysogenize. This was discovered for diphtheria in 1952, almost as early as lysogeny itself. These genes that confer virulence on their hosts are sometimes called **morons** (for "more DNA") and are not found in all the phages of that type, suggesting that they were recently acquired (see Box 7.1). They also are often expressed from their own promoters and regulated by bacterial regulators that are active when the lysogen is in the eukaryote, so they are expressed in the lysogen in the eukaryotic host, where other prophage genes are usually repressed. Some examples of bacteria carrying prophages with morons that contribute to the diseases they cause are the bacteria that cause diphtheria, bacterial dysentery, scarlet fever, botulism, tetanus, toxic shock syndrome, and cholera. Even λ phage carries genes that confer on its *E. coli* host serum resistance and the ability to survive in macrophages. As mentioned above, the process by which a prophage converts a nonpathogenic bacterium to a pathogen is an example of **lysogenic conversion.**

E. coli and Dysentery: Shiga Toxins

Pathogenic strains of *E. coli* are prime examples of bacteria that are not pathogenic unless they harbor prophages or other DNA elements carrying virulence genes. These bacteria are part of the normal intestinal flora unless they carry certain DNA elements. Then, they can cause severe diseases, including bacterial dysentery, with symptoms such as bloody diarrhea. In fact, bacterial dysentery due to these bacteria is the major cause of infant mortality worldwide. The infamous *E. coli* strain O157:H7, which has caused many outbreaks of bacterial dysentery worldwide, is one example of such a lysogenic *E. coli* strain. It harbors more than 18 prophage-like elements, which account for more than 50% of the DNA that is not a normal part of the *E. coli* chromosome.

In one particularly clear example, a group of *Siphoviridae* prophages very closely related to λ can make *E. coli* pathogenic by encoding toxins called Shiga toxins (Stx), so named because they were first discovered in *Shigella dysenteriae*, which is so closely related to *E. coli* that it has recently been moved into the same genus. Like CT and many other toxins, the Shiga toxin is composed of two subunits, A and B. The B subunit helps the A subunit enter an endothelial cell of the host by binding to a specific receptor on the cell surfaces of some tissues. The A subunit is a very specific *N*-glycosylase that removes only a certain adenine base from the 28S rRNA. Removal of this adenine from the 28S rRNA in a ribosome blocks translation by interfering with binding of the translation factor elongation factor 1((EF-1α), the eukaryotic equivalent of EF-Tu, to the ribosome. Interestingly, this adenine in the 28S rRNA seems to be the "Achilles heel" of the ribosome and is a popular target of translation-blocking systems. The rRNAs of all organisms are highly conserved, and an adenine occurs in this position in the rRNAs of all eukaryotes. Plant enzymes, such as the herbicide ricin, that protect the plant by blocking translation in cells infected by virus are also *N*-glycosylases that remove the same adenine from their own 28S rRNA, killing the cell and preventing multiplication of the virus. Yeasts also make an *N*-glycosylase named saracin, which has the same target, although the function of the enzyme is unknown.

Shiga toxins can be divided into two groups based on their amino acid sequences: the Stx1 group, encoded by *E. coli* prophages, which also includes the toxin encoded in the chromosome of *S. dysenteriae* and the ricin toxin in plants, and the Stx2 group, which so far has been found only in prophages carried by *E. coli*. Usually, the bacteria responsible for the most serious human diseases carry the Stx1 type. Apparently, expression of the toxin is required to convert the disease from just watery diarrhea to hemolytic-uremic syndrome (HUS), which is the leading cause of kidney failure in children.

The regulation and secretion of the Shiga toxin from these prophages present intriguing clues about the etiology of the diseases caused by these bacteria. The *stx* genes in prophages are usually in the same place, downstream of $p_{R'}$ in the phage genome. A prophage genetic map of phage ϕ361, encoding Stx2, is shown in Figure 8.19. Note the remarkable similarity between the genetic map of this prophage and the genetic map of its close relative, λ prophage, shown in Figure 8.15.

Because the toxin genes *stx2A* and *stx2B* lie just downstream of the *Q* gene and upstream of the lysis genes *S* and *R*, they are expressed from the phage $p_{R'}$ promoter only late in induction. However, the toxin

Figure 8.19 Close relatives of λ encode Shiga toxins. Shown is the prophage genetic map of phage ϕ361, with the positions of the toxin genes indicated. The blue shading indicates that the repressor and toxin genes are expressed in the lysogen. Details are given in the text. doi:10.1128/9781555817169.ch8.f8.19

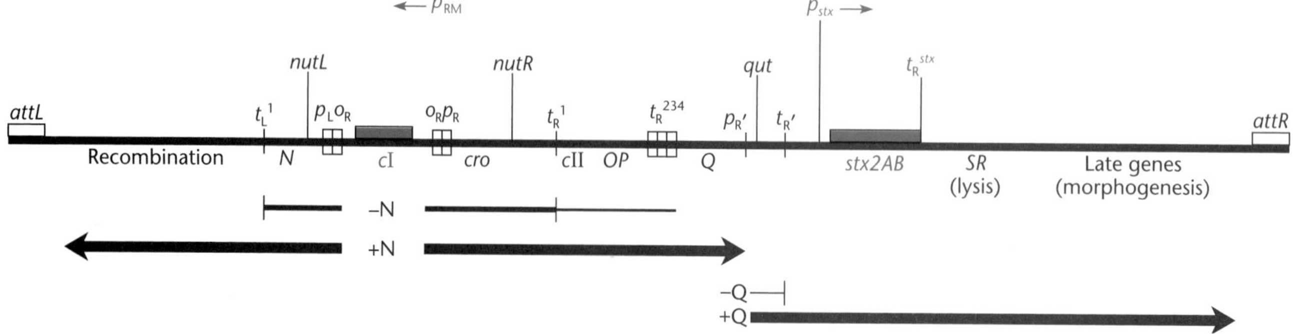

genes in ϕ361 and many of these phages have their own weak promoter, p_{stx}, so they are weakly expressed even in the uninduced lysogen (see Wagner et al., Suggested Reading).

Interestingly, this promoter is sometimes regulated by the presence of iron, as is an *stx1* gene found in the chromosome of a *Shigella* strain. Iron deficiency is often used as a sensor of the eukaryotic environment so that the gene is turned on only in the eukaryotic intestine (see chapter 13 for a discussion of iron regulation). However, these promoters are very weak, and not enough toxin would be expected to be made from them to account for the illness they cause. Moreover, even if the toxin protein is made at low levels, it is not clear how it would get out of the bacterial cell, which apparently does not harbor another type of secretion system capable of secreting the toxin into the extracellular environment of the intestine. Therefore, we surmise that the only way sufficient amounts of toxin can be made and get out of the bacterial cell is if the prophage is induced, makes large amounts of toxin from $p_{R'}$, and then lyses the cell. The phage could be induced by our own immune systems, since neutrophils release H_2O_2, which induces the SOS system in the bacteria and therefore induces the phage. This may happen, but it kills the pathogenic bacterium and thus seems counterproductive. One scenario is that some of the bacteria lyse and release the phage, which then infect and lysogenize nonpathogenic *E. coli* strains that are part of the normal bacterial flora. If the phage in these normally nonpathogenic bacteria are then induced, they kill the nonpathogenic bacteria that harbored the prophage, but the released Shiga toxin can participate in pathogenesis by the pathogenic strain, thereby multiplying the effects of the toxin and perhaps leading to HUS and other severe diseases related to the infection. This raises questions about the effects of some types of antibiotics, such as ciprofloxacin (see chapter 1), that damage DNA and therefore induce phages. They may help to spread these phages and increase, rather than decrease, the severity of the disease. In fact, there is some evidence that the use of ciprofloxacin and similar antibiotics to treat people with bacterial dysentery due to *E. coli* can increase the chance of the disease developing into the more serious HUS.

Diphtheria

Diphtheria is the classic example of a disease caused in part by the product of a gene carried by a phage. As early as 1918, the microbiologist Félix d'Herelle demonstrated plaque formation by phages on the bacterium that causes diphtheria, and in the late 1940s, pathogenic strains of *Corynebacterium diphtheriae* were being shown to differ from nonpathogenic strains in that they are lysogenic for a phage β or a closely related *Siphoviridae* phage. These prophages carry a gene, *tox*, for the diphtheria toxin (Figure 8.20), which is largely responsible for the effects of the disease and is the target of the vaccine against it. The diphtheria toxin is another typical AB toxin with two subunits, one that transposes the toxin into the eukaryotic cell while the other, the toxin, is an enzyme that kills eukaryotic cells by ADP ribosylating (attaching ADP to) the EF-2 translation factor, thereby inactivating it and blocking translation in the cell. The *tox* gene of the β prophage is transcribed only when *C. diphtheriae* infects its eukaryotic host or under low iron concentrations, a condition that mimics this environment. Even though the *tox* gene is on the prophage, it is regulated by the products of chromosomal genes involved in iron regulation, illustrating the close relationship between these bacteria and their phages. The diphtheria toxin has been extensively studied, and the mechanism of action of the diphtheria toxin and the regulation of the *tox* gene are discussed in chapter 13 under "Regulation of Virulence Genes in Pathogenic Bacteria."

Cholera

Another striking example of toxin genes carried by a phage is CT, encoded by the phage CTXϕ (for *CT* phage) of *Vibrio cholerae*, the bacterium that causes cholera (see Waldor and Mekalanos, Suggested Reading). It came as a surprise that the phage carrying the CT genes is a filamentous phage with single-stranded DNA. Even though

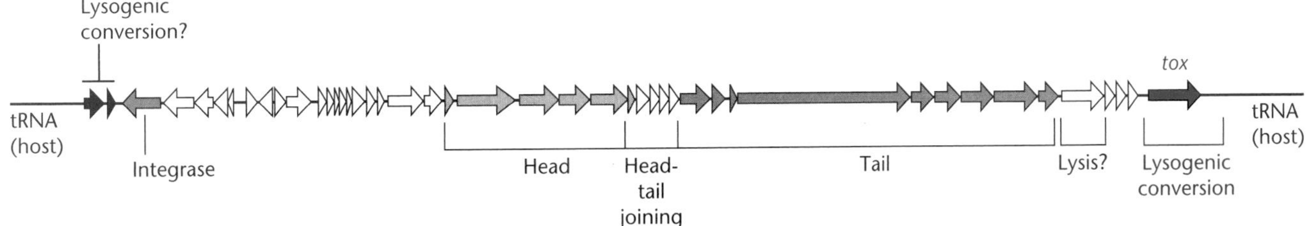

Figure 8.20 Genome map of a *C. diphtheriae* prophage containing the diphtheria toxin gene (*tox*). Selected genes are annotated. The prophage is bracketed by tRNA genes of the host. Other genes involved in lysogenic conversion may be on the other end of the prophage.
doi:10.1128/9781555817169.ch8.f8.20

the phage genome is single stranded, it forms a double-stranded replicative form (RF) after infection (see chapter 7), and this double-stranded RF can integrate into a specific site in the chromosome to form a prophage. If its *attB* site in the chromosome is deleted, it can form a lysogen by replicating as a double-stranded circular plasmid prophage, although it is fairly unstable.

In many ways, this phage is like fd and other filamentous single-stranded DNA phages of *E. coli* that we discussed in chapter 7. It infects the cell by attaching to a pilus, in this case the toxin-coregulated pilus, and then enters the cell through the TolA, TolQ, and TolR channel. It replicates its DNA through a double-stranded replicative intermediate and leaks from the infected cell through a type II-related secretion system.

One striking feature of the CTXϕ phage that distinguishes it from most other phages we have discussed is the extent to which it lives in harmony with its host and does not substantially affect the growth rate of the host even when induced (see McLeod et al., Suggested Reading). It also uses many host functions instead of encoding its own. For example, the CTXϕ phage does not have its own integrase but uses the **XerCD recombinase** of the host cell, which in the host cell resolves chromosomal dimers during replication by recombination at the *dif* site close to the terminus of chromosome replication (see chapter 1). Because of its specificity, the XerCD recombinase integrates the phage into the *dif* site of the chromosome, the site close to the terminus of chromosome replication normally recognized by XerCD. However, even with the prophage integrated, the *dif* site is still usable, and the cells can still resolve chromosome dimers and do not filament.

Another interesting feature of the phage is how it is maintained as a prophage and how it is induced following damage to cellular DNA. In some ways, it uses the same basic strategy as λ phage, but with significant differences that reflect its more modest lifestyle. Like λ, it has a repressor, which can also function as an activator, but this repressor/activator is not made in the prophage state because its promoter is repressed by a host protein, LexA. The function of this protein in the host is to repress repair genes that are part of the SOS system and are only induced following severe DNA damage (see above and chapter 11). The other phage genes (with the exception of those for CT [see below]) are also transcribed from the same promoter, so almost no phage genes are transcribed in the prophage state. However, if the DNA is damaged, for example, by UV irradiation, the RecA protein of the host promotes the self-cleavage of the LexA repressor and the repressor/activator gene and most of the other phage genes are transcribed. Once it is made, the repressor/activator can modulate its own transcription and the transcription of the other phage genes from the same promoter so that only a few phage are ever produced and the growth of the bacterial cell is not seriously affected.

The CTXϕ phage also uses some host functions to release itself from the cell. Rather than encoding its own complete type II secretion system, like the filamentous coliphages, it uses the secretin part of a host type II secretion system combined with some of its own proteins to secrete itself from the cell. Incidentally, this is the same type II secretion system that secretes CT from the cell (see below). We discuss the different types of secretion systems in chapter 14.

Another difference from most types of known phages capable of lysogeny is the way in which the CTXϕ prophage DNA escapes from the bacterial chromosome after induction. Rather than being excised from the chromosome, the prophage DNA replicates **in situ** (in place) while remaining in the chromosome. However, to do this, it must exist as a dilysogen, with two or more adjacent prophage genomes linked head to foot, or as commonly happens, a truncated form of the prophage is integrated immediately downstream of the prophage. In a process similar to λ rolling-circle replication and plasmid transfer during conjugation (see chapters 4 and 5) and to some types of transposition (see chapter 9), a nickase cleaves one strand of the DNA at the origin of replication of the upstream prophage and the free 3′ end primes replication that extends all the way to the origin of replication of the downstream prophage, displacing the old plus strand. This is why the phage can be induced only in a dilysogen; otherwise, only the phage genome on one side of the origin of replication would be replicated. The nickase presumably remains attached to the 5′ end of the displaced old plus strand, so it can cyclize it by a transesterification reaction, as in plasmid transfer. The cyclized plus strand can then direct the synthesis of the first RF, and the replication and DNA packaging proceed as with other single-stranded filamentous phages.

Because the sole phage promoter is repressed in the prophage state, the only phage genes transcribed are the CT genes, *ctxA* and *ctxB*, which are transcribed from their own promoter. The CT is another AB-type toxin in which the B subunit helps the A subunit toxin enter the eukaryotic cell. The A subunit is an enzyme that ADP ribosylates a mucosal membrane protein, causing cyclic AMP levels to rise and water to be excreted from the cells, leading to severe dehydration. The toxin subunits are secreted from the *V. cholerae* cell by a type II secretion system, the same one whose secretin subunit is used to transport the phage from the cell (see above). It is also interesting that the CT genes on the prophage are regulated by a transcriptional activator, ToxR, the product of a chromosomal gene, which acts through another transcriptional activator, ToxT, that also regulates the synthesis of the pili that serve as the receptor sites for the phage. This is another example of the close coordination between the phage and host functions involved in pathogenesis. These pili are also important virulence determinants, because they enable the bacteria to adhere to the intestinal mucosa. The regulation by ToxR of many genes involved in pathogenesis, including the *ctx* genes, is a good example of global gene regulation and is discussed in chapter 13.

S. aureus and Toxic Shock Syndrome

Yet another example of a pathogenic bacterium that depends on phages for its virulence is *S. aureus*. Not only are many of the virulence determinants of *S. aureus* carried on prophages, but those that are carried on DNA elements called **pathogenicity islands** often depend on phages for their mobility. Pathogenicity islands are a type of **genetic island** found in pathogenic bacteria and carry genes required for virulence. Many types of bacteria harbor genetic islands, often quite large, that are integrated into their chromosomes and encode functions that broaden the physiological capacity of the cells and therefore the number of ecosystems they can inhabit. These genetic islands are not found in all strains of a species and presumably have been acquired fairly recently by horizontal transfer from other bacteria. They are unable to replicate autonomously but usually encode an integrase that allows them to integrate into a specific site in the chromosome. However, in most cases, it is not clear how a genetic island has moved from one bacterium to another, since most of them cannot transfer in a laboratory setting and they encode no functions known to promote lateral transfer between bacteria.

S. aureus is a normal inhabitant of the skin of most people, but some strains can cause serious illnesses. One of the most serious illnesses caused by *S. aureus* is toxic shock syndrome. Toxins that elicit toxic shock syndrome, such as TSST-1 (toxic shock syndrome toxin 1), are superantigens that bind to nonvariable regions of antibodies and cause a severe immune response. In those strains of *S. aureus* that cause toxic shock syndrome, the toxin is often encoded on a small pathogenicity island called a ***Staphylococcus aureus* pathogenicity island (SaPI)**. While not technically prophages because they encode no recognizable phage proteins themselves, these SaPIs could be considered satellite phages, like P4, in that they are induced following infection by a specific type of phage. After they are induced, they replicate, probably with the help of phage proteins, and their DNA is packaged in a shortened version of the phage head, also like P4. The phage can then infect another strain of *S. aureus*, and the SaPI integrates into the chromosome using its own Int protein. This allows the SaPI to move very efficiently between strains of *S. aureus*, increasing the frequency of infection.

A number of different SaPIs exist, and they each have a specific relationship with their own type of *Siphoviridae* helper phage related to λ. For example, the *S. aureus* phage 80α is able to induce and transfer a number of different SaPIs, while a closely related phage, ϕ11, cannot induce any of these but is able to induce and transfer a different one. The SaPIs normally remain in a dormant state in the chromosome because their transcription is repressed by a repressor, Stl, which binds to a region between two divergent promoters that service most of the genes of the pathogenicity island, simultaneously preventing transcription from both promoters. Phage infection inactivates the Stl repressor, allowing transcription from the promoters and inducing the SaPI. The Stl repressors are not closely related, even those that are induced by the same helper phage, which raises the question of how a phage like 80α can induce so many different SaPIs.

A genetic analysis revealed what gene products of the phage induce the pathogenicity islands (see Tormos-Mas et al., Suggested Reading). This genetic analysis was made possible because the phage cannot form plaques if the bacterial indicator contains a SaPI it induces. When the phage induces an SaPI, most of the phage particles encapsulate the SaPI, and the yield of phage drops by 1 to 2 orders of magnitude, preventing the phage from forming a plaque. Any phage mutants that can multiply to form a plaque should have mutations in genes whose products are required to induce the SaPI. Some of the mutants they selected had nonsense mutations, indicating not only that the product of the mutated gene was required for induction of the SaPI, but also that the product of the gene was probably not essential for multiplication of the phage, since nonsense mutations are usually null mutations. When the location of the mutations was determined by DNA sequencing, it was discovered that induction of different SaPIs by the same phage required different phage proteins. In one case, the nonessential phage Dut protein, the uracil-*N*-glycosylase that removes uracil from the DNA (see chapter 1), induces one SaPI, while induction of a different SaPI requires a different nonessential protein. Apparently, each of these nonessential phage proteins binds specifically to the Stl repressor protein of a particular SAPI and inactivates it, inducing that particular SaPI. If a type of phage cannot induce a particular type of SaPI, quite often it lacks the gene to encode the nonessential protein required to induce that SaPI. This raises an interesting possibility for the evolution of the induction system. The SaPI repressor evolves to bind to and be inactivated by a particular phage protein, but if the phage protein mutates or the gene for it is deleted to avoid inducing the SaPI, the SaPI switches to a different protein to induce it. However, why the SaPI seems to adopt only nonessential phage proteins for this purpose is something of a mystery, since it seems the phage would have more difficulty changing a protein if it is essential for the development of the phage.

The only major difference between a satellite phage like P4 and an SaPI pathogenicity island is that P4 encodes all the proteins to replicate its own DNA while the SaPIs seem to depend on the helper phage for at least some of their replication proteins. Maybe P4 should be called a genetic island rather than a satellite phage, or vice versa, and you should never judge a DNA element by its coat alone.

As with the Shiga toxins, the mobilization of pathogenicity islands by phages raises special questions concerning the use of antibiotics. Some antibiotics, particularly those that inhibit the DNA gyrase, such as the fluoroquinolones, including ciprofloxacin (see chapter 1), cause DNA damage that can induce prophages. If the chromosomes of the cells in which the prophages exist also contain pathogenicity islands, these pathogenicity islands may be induced and encapsulated by the phage and moved into other, previously harmless strains.

Synopsis

More and more examples of genes that help make bacteria pathogenic being carried on prophages rather than being normal chromosomal genes are continuously being discovered. Detailed studies of lysogeny by λ and other phages laid the groundwork for understanding these systems and help in the search for more effective treatment of diseases, as well as in identifying pathogenic strains. We have listed only a few examples here; however, many other classical diseases that have plagued humankind at least throughout recorded history are turning out to have prophage involvement. It has been estimated that prophages make up as much as 10 to 20% of the DNA of some types of bacteria. They also are often the major contributors to strain diversity within a species. Within the same species, bacterial strains that cause very different diseases are often found to differ mostly in the prophages they harbor; but why are toxin genes and other virulence factor genes often carried by prophages instead of being normal genes in the chromosome? The argument is similar to that used to explain why genes are carried on plasmids (see chapter 4). Having toxin and other virulence genes on a movable DNA element like a phage may allow the bacterium to adapt to being pathogenic without all the members of the population having to carry extra genes. Furthermore, many virulence proteins are also strong antigens. If the virulence protein is encoded on a prophage, a nonlysogenic bacterium can colonize the host without alerting the host immune system. Once established, the strain can become pathogenic after being infected by the phage. We have also discussed possible complications of treating bacterial infections caused by virulence genes carried on prophages. Antibiotics that damage DNA and induce a prophage carrying a virulence gene might lead to infection of other, nonpathogenic strains, increasing the severity of the infection.

Uses of Lysogeny in Genetic Analysis and Biotechnology

Because prophages integrate in single copies in the chromosome but can be induced to produce many copies, the lysogenic cycles of phages have many uses in genetics and biotechnology and in genetic analysis in bacteria.

Complementation and Gene Expression Studies

Some of the major uses of lysogeny in basic science are in complementation and gene expression studies with cloned genes. Such studies include altering cloned genes in the test tube and then reintroducing them into the cell to study the effects of the changes on their ability to complement other alleles or on their expression or regulation. Ideally, such tests should be done with the cloned gene in only one copy; otherwise, multicopy suppression might be mistaken for real complementation (see chapter 3), and having multiple copies of a gene can affect its regulation. Some low-copy-number plasmid cloning vectors have been constructed for these studies, but such plasmids are difficult to isolate precisely because they exist in so few copies. Also, the expression of a gene in a plasmid can be affected by other factors, such as different degrees of supercoiling of plasmids relative to the chromosome or other features unique to plasmids. Expressing a cloned gene from a prophage avoids many of these complications. Since the prophage usually integrates in a single copy into the chromosome of the host, after the lysogen has formed, any cloned gene in the phage exists in at most only two copies, one in the prophage integration site and one at the normal site if the gene has been introduced into the host of origin. Obtaining large amounts of the cloned DNA to manipulate in vitro is also relatively easy, since the prophage can be induced to multiply as a phage, After the gene has been manipulated in vitro, it can be reintroduced into the cell in the single-copy prophage state to study the effects of the changes on its regulation or on its ability to complement another copy elsewhere in the chromosome.

Use of Phage Display and Frequency of Mixed Dilysogens To Detect Protein-Protein Interactions

Different technologies have been developed to detect protein-protein interactions, including the yeast two-hybrid system and the reconstruction of active adenyl cyclase from inactive fragments in the BACTH system (see chapter 13). However, most of these methods can detect only in vivo interactions, usually in a yeast or bacterial cell, and cannot detect binding outside the cell, although the adenyl cyclase reconstitution method has the advantage that it can detect the interaction between two membrane proteins. The 2λ system is a recent application using lysogeny to detect protein-protein interactions outside the cell (see Bair et al., Suggested Reading). The principle behind the 2λ system is illustrated in Figure 8.21. It uses the fact that λ phage often forms dilysogens, with two copies of the prophage integrated in tandem head to tail at the *attB* site in the chromosome. A dilysogen can form either by integration of one prophage next to a preexisting prophage in the chromosome or by integration of a dimerized λ chromosome after infection. If the cells have been simultaneously infected by two different genetic constructs of the phage, a mixed dilysogen can form, with one copy of the prophage having one construct and the other having the other construct, either due to recombination between the phages forming mixed dimerized circular DNA before they integrate (Figure 8.21B) or due to integration of a prophage with one construct next to another that has already integrated. These mixed dilysogens can be detected by introducing genes for resistance to different antibiotics into the two phage genomes and selecting cells that have become resistant to both antibiotics, as shown in Figure 8.21A. Such mixed dilysogens normally occur at a reasonable frequency only if the cells have been infected by one or both phages at a high MOI. At a low MOI of both phages, most of the cells, if they are infected at all, will have received only one of the two different constructs of the phage genome, and mixed dilysogens will be very rare. However, if the phages carrying the different antibiotic resistance genes somehow bind to each other prior to infection, they will infect the same cell, greatly increasing the frequency of mixed dilysogens, even at overall low MOIs. There is the additional advantage that the cells infected with such conjoined phages are infected at the relatively high effective MOI of 2 while the other cells are being infected at most at an MOI of 1. Since higher MOIs favor lysogen formation (see "The Lytic-versus-Lysogen Decision: the Roles of *c*I, *c*II, and *c*III Gene Products" above), the potential for mixed dilysogens among the lysogenic cells that survive the infection becomes even higher.

One way the two λ phages expressing different antibiotic resistance genes can bind to each other prior to infection is if the two different phages display peptides on their surfaces that bind to each other (see chapter 7 for a description of phage display). In λ, peptides can be displayed by translationally fusing their coding sequences to the coding sequence for the C terminus of protein D, one of the major head proteins of the phage. Controls showed that the frequency of formation of dilysogens resistant to both antibiotics after infection at low MOIs is much higher if they express peptides that are known to bind to each other and much lower if the phages do not display interacting peptides or proteins. Also, the phages expressing interacting peptides seem to be able to find and bind to each other and infect the same cell even in the presence of many other phages that are not expressing interacting peptides; this is an important attribute if this method is to be used to "biopan" for interacting peptides in a random library consisting mostly of peptides that do not bind (see Box 7.4).

Genetic Experiments with Phage λ

Above, we discussed how studies of the interaction of phage λ with its host have made major contributions to our present understanding of how cells function at the molecular level, but we have not gone into detail about the types of experiments that contributed to these concepts. As we go along, we will point out some of the conceptual advances that have come from these experiments.

Genetic Analysis of λ Lysogen Formation

As with many biological phenomena, the framework for understanding how λ forms lysogens was first developed by a genetic analysis. The first step in any such analysis is to determine how many gene products are required and at what stage each acts. Because phage λ is capable of lysogeny, the plaques of λ are cloudy in the middle due to the growth of immune lysogens in the plaque (Figure 8.8). Mutants of λ that cannot form lysogens do not contain these immune lysogens and so are easily identified by their clear plaques. These "clear-plaque mutants," called C-type mutants, have mutations in λ genes whose products are required for the phage to form lysogens.

Complementation tests have revealed how many genes are represented by clear-plaque mutants. Now, however, rather than asking whether two mutants can help each other to multiply, the question is whether two mutants can help each other to form a lysogen, since this is the function of the genes represented by clear-plaque mutants. Cells are infected by two different clear-plaque mutants simultaneously, and the appearance of lysogens is monitored. Lysogens can be recognized by their immunity to superinfection by the phage, which allows a lysogen to form colonies in the presence of the phage. An efficient way to perform this test is to mix one of the mutant phages with the bacteria and streak the mixture on a plate. The other mutant phage is then immediately streaked at right angles to the first streak. Very

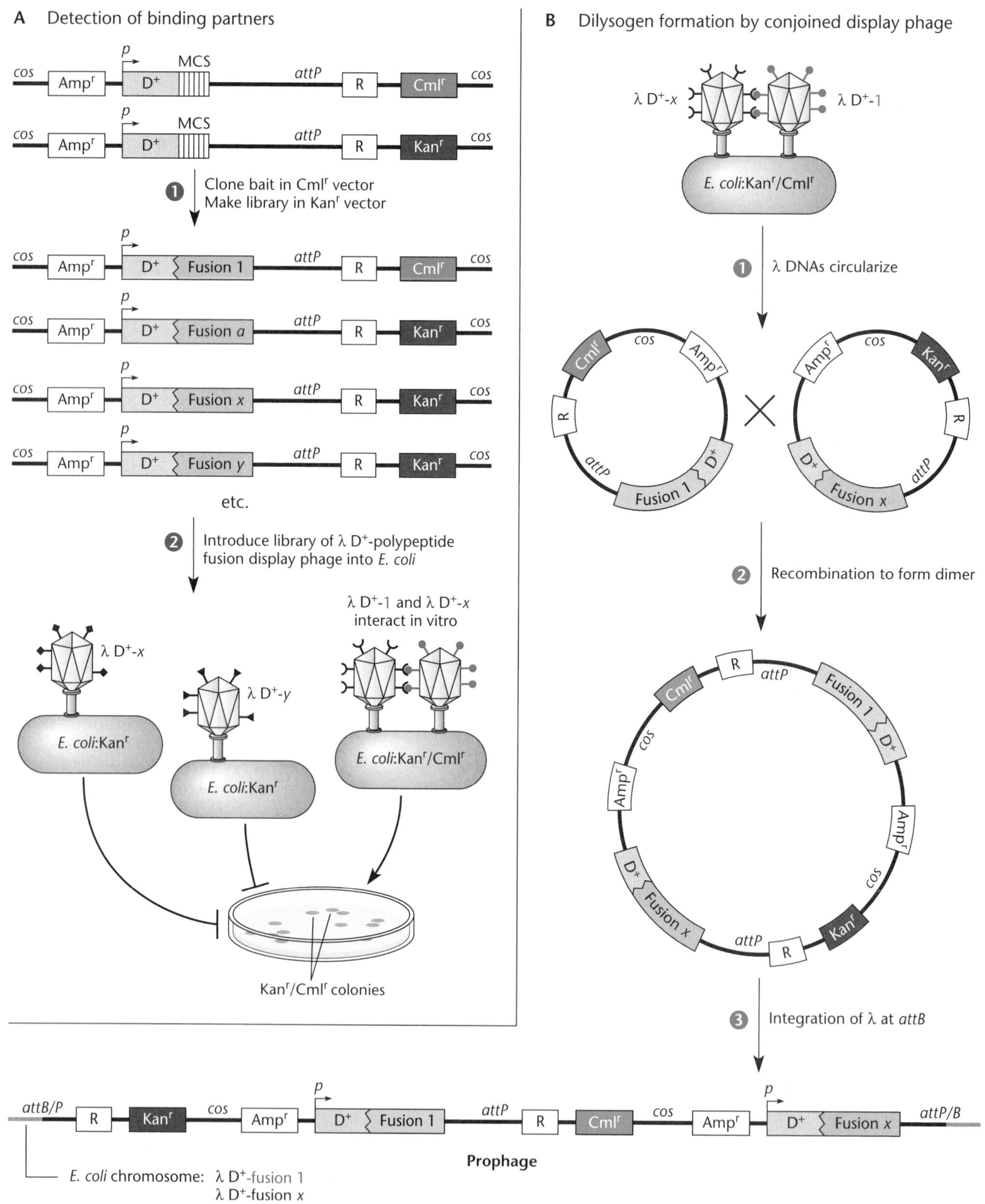

Figure 8.21 Using phage display and dilysogens to detect protein-protein interactions. **(A)** In the example, two λ phage vectors, one with a gene for chloramphenicol resistance (Cmlʳ) and another with a gene for kanamycin resistance (Kanʳ), have multiple cloning sites (MCS) that can be used to clone polypeptide-coding sequences of up to 3 kb to express different proteins fused

few bacterial colonies will grow in the original streak, because the individual mutants cannot form lysogens. However, if bacterial colonies grow in the region where the two streaks cross, immune lysogens form due to infection of some cells by both mutant phages, and complementation between the two mutations allows the phages to form a lysogen. Such crosswise complementation tests revealed three complementation groups of genes to which the clear-plaque mutations of λ belonged: *c*I, *c*II, and *c*III. In addition to mutations in the clear-plaque genes, mutations in the *int* gene can prevent the formation of stable lysogens, although Int-negative (Int$^-$) mutants make somewhat cloudy plaques. In this case, the λ DNA, while not integrated into the chromosome, may make the cells transiently immune due to the formation of unstable circular plasmid prophages.

Further genetic tests revealed the stages at which CI, CII, and CIII act in forming a lysogen. The question is whether each of these gene products is only required to form a lysogen or whether it is also required to maintain the lysogen once it has formed. Mutations in the *c*II and *c*III genes can be complemented to form lysogens, and these lysogens can harbor a single prophage with either a *c*II or a *c*III mutation. Therefore, the CII and CIII proteins are only required to form a lysogen, and neither of these gene products is required once it has formed. However, lysogens harboring a single prophage with a *c*I mutation are never seen. Apparently, the *c*I mutation in the phage must be complemented by another mutation to maintain the phage in the lysogenic state. Further studies showed, as we have described above, that the CI protein is the repressor that maintains the prophage in the dormant state while CII and CIII are required only after infection to activate the transcription of the genes to make the CI repressor and the other functions required to establish the lysogen.

Because they can be complemented, the researchers knew the *c*I, *c*II, and *c*III mutations affect *trans*-acting functions, either proteins or RNAs, required to form lysogens. Another set of mutations, called the ***vir* mutations**, also prevent lysogeny and cause clear-plaque formation, but they cannot be complemented and so are *cis* acting (see chapter 3). Unlike clear mutants, phages with *vir* mutations allow the mutant phage to multiply and form clear plaques even on λ lysogens. Later, DNA sequencing revealed that phages with *vir* mutations are multiple mutants with mutations in the o_R^1 and o_R^2 sequences, as well as o_L^1. These mutations change the operators so that they can no longer bind the CI repressor, thereby preventing lysogeny and allowing a *vir* mutant to multiply in a lysogen (see "Regulation of Repressor Synthesis" above). These experiments laid the groundwork for the concept of cooperative binding of proteins to DNA.

Genetics of the CI Repressor: Evidence for the Domain Structure of Proteins

Many proteins are now known to be assembled from "modules" or domains with separable functions, and one of the goals of modern proteomics is to identify the domains in proteins in an attempt to guess the overall function of the protein. The λ repressor product of the *c*I gene was the first protein shown to have separable domains. One domain of the CI protein binds to the operator sequences on the DNA, and the other domain binds to another repressor monomer to form the active repressor dimer.

The first indication that the CI repressor has separable domains came from genetic experiments that demonstrated intragenic complementation between temperature-sensitive mutations in the *c*I gene (see Lieb, Suggested Reading). As discussed in chapter 3, complementation usually occurs only between mutations in different genes, and intragenic complementation is possible only if the protein product of the gene is a homodimer (or higher multimer) composed of more than one identical polypeptide encoded by the same gene. Then, mutations in different domains of the protein can complement each other by forming active mixed dimers with one polypeptide chain in the dimer having one mutation and the other having the other mutation. Sometimes the mixed dimer is at least partially active even if dimers composed of one mutant polypeptide or the other are inactive.

Figure 8.22 illustrates the experiments that demonstrated intragenic complementation by some temperature-sensitive mutations in the *c*I gene. Lysogenic

to the head protein, D (step 1). For technical reasons, the gene *D* fusion is expressed from the *lac* promoter rather than being expressed with the other late genes, which are not shown. In the example, only the polypeptide in fusion *x* interacts with that in fusion *1* and not those in fusions *a*, *b*, *c*, *y*, *z*, etc. In step 2, the phage with construct fusion *1* and the random library are packaged separately and combined before being used to infect *E. coli* at a low MOI. If the two expressed proteins bind to each other, the conjoined phage can infect the same cell, forming a dilysogen expressing both antibiotic resistances. **(B)** How the dilysogens probably form. After infection of the same cell, the two DNAs circularize **(1)** and then recombine **(2)** to form a dimeric circular DNA that integrates to form the mixed dilysogen expressing both Cml^r and Kan^r **(3)**. **(C)** The two prophages form a tandem dilysogen, with one phage expressing Cml^r and the other expressing Kan^r. doi:10.1128/9781555817169.ch8.f8.21

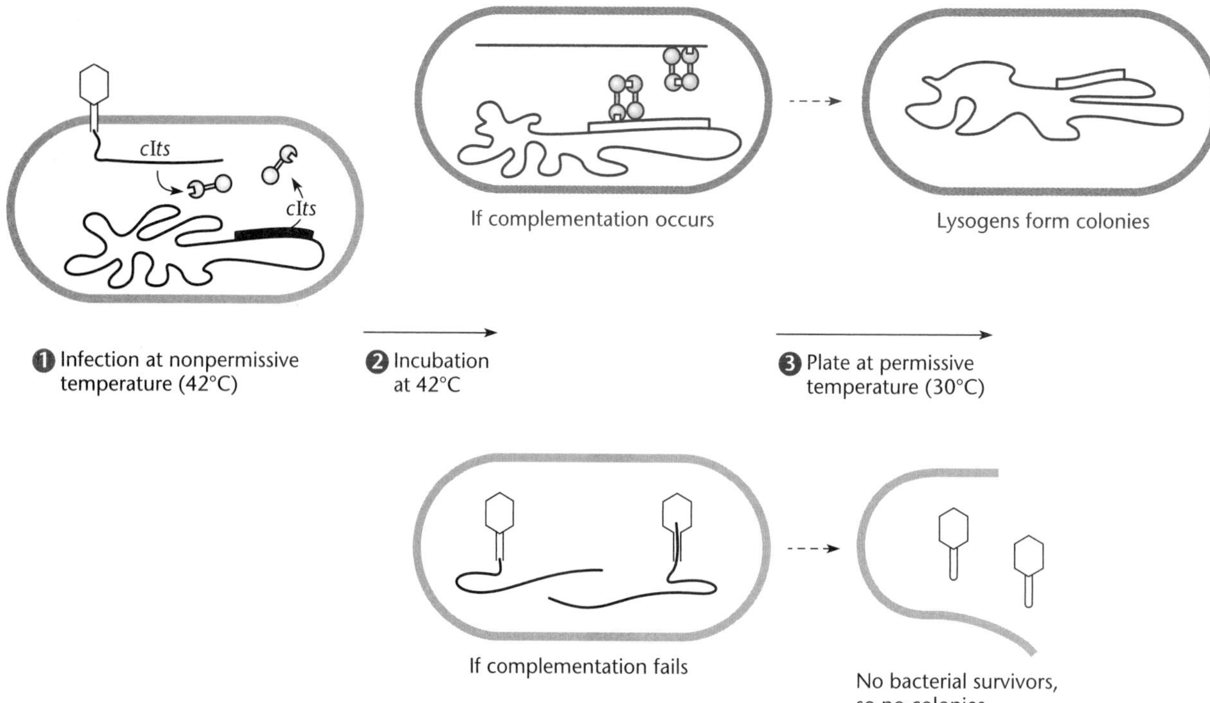
cIts
cIts
If complementation occurs
Lysogens form colonies
1 Infection at nonpermissive temperature (42°C)
2 Incubation at 42°C
3 Plate at permissive temperature (30°C)
If complementation fails
No bacterial survivors, so no colonies

Figure 8.22 An experiment to show intragenic complementation in the cI gene of λ. See the text for details. doi:10.1128/9781555817169.ch8.f8.22

cells containing a prophage with one *c*I temperature-sensitive mutation were heated to the nonpermissive temperature and infected with a phage carrying a different *c*I temperature-sensitive mutation. At this temperature, infection by one or the other mutant phage alone invariably kills the cell because the repressor is inactivated so that λ cannot form a lysogen. However, if the two mutations complement each other to form an active repressor, a few cells may become lysogens and survive.

The results clearly demonstrated intragenic complementation between some of the mutations in the *c*I gene. In particular, some mutations in the amino (N)-terminal part of the polypeptide, which we now know is involved in DNA binding, complement some mutations in the carboxyl (C)-terminal part, which we now know is involved in dimer formation. Apparently, in some cases, dimers can form if only one of the two polypeptides has a mutation in the C-terminal dimerization domain and the other is active for dimer formation. Furthermore, the mixed dimer can also sometimes bind to DNA if only one of the two polypeptides in the dimer has a mutation in the N-terminal DNA-binding domain.

Identification of λ *nut* Sites Involved in Progressive Transcription Antitermination

Some of the most elegant early genetic experiments with phage λ involved the isolation and mapping of mutations in the λ *nut* sites (see Salstrom and Szybalski, Suggested Reading). These experiments allowed the identification of the *nut* sites and made possible biochemical experiments to prove the model for progressive antitermination of transcription, whose counterparts have now been found in other systems. As discussed above, in λ transcriptional regulation, the *nut* sites are the sites on the mRNA to which N must bind before it can bind to the RNA polymerase and allow it to proceed through downstream transcription termination sites (Figure 8.4), thereby allowing transcription to proceed into the downstream genes. Therefore, mutations in the DNA coding sequence for one of these *nut* sites could prevent the binding of the N protein to the mRNA, thereby causing transcription to stop at the next transcription termination site and preventing transcription of the downstream genes.

The first decision was whether to isolate *nutL* or *nutR* mutants. The obvious choice is *nutL* mutants, because *nutR* mutations should be lethal. While *nutL* mutations should prevent the transcription of genes on the left of *c*I, including the *gam* and *red* genes, all of which are nonessential, *nutR* mutations should prevent the transcription of essential genes on the right of the *c*I gene, including the *O* and *P* genes required for replication (Figure 8.2, λ map). Then, a selection for *nutL* mutations was needed, since they might be very rare. Not only might *nut* sequences in DNA be very short, but it was possible that only a few base pair changes in a short sequence might inactivate the site. Designing a selection meant finding conditions under which phages with a mutation that inactivates the *nutL* site can form plaques whereas wild-type λ cannot.

The selection that made it possible to isolate *nutL* mutations is illustrated in Figure 8.23. The selection is based on the observation that, for unknown reasons, wild-type λ cannot multiply in *E. coli* lysogenized by another phage, P2, because either one of the products of the *gam* and *red* genes (the *red* gene later turned out to be two genes, *exo* and *bet*, used for recombineering [see chapter 10]) of the infecting λ interact somehow with the *old* gene product expressed from the P2 prophage and kill the cell. Phage with *nutL* mutations should fail to make both Gam and Red and so should form plaques on a P2 lysogen. Unfortunately, *nutL* mutants are not the only type of mutant that can form plaques under these conditions. As shown in the figure, double mutants of λ with point mutations in both the *gam* and *red* genes or with a deletion mutation that simultaneously inactivates both the *red* and *gam* genes also multiply and form plaques on a P2 lysogen. Fortunately, double mutants should not be much more common than *nutL* single mutants, since the chance of getting two independent mutations is the product of the chances of getting either single mutation. On the other hand, spontaneous deletion mutations that include both the *red* and *gam* genes could be much more frequent than *nutL* mutations. However, by inducing mutations using a specific mutagen that causes only point mutations, it is possible to lower the percentage of all the mutations that are deletions (see chapters 3 and 7). Nevertheless, in spite of all precautions, some of the λ mutants that multiply on P2 lysogens have both *gam* and *red* mutations, and these had to be distinguished from any *nutL* mutants.

Figure 8.23 Selection for λ *nutL* mutations. See the text for details. RP, RNA polymerase.
doi:10.1128/9781555817169.ch8.f8.23

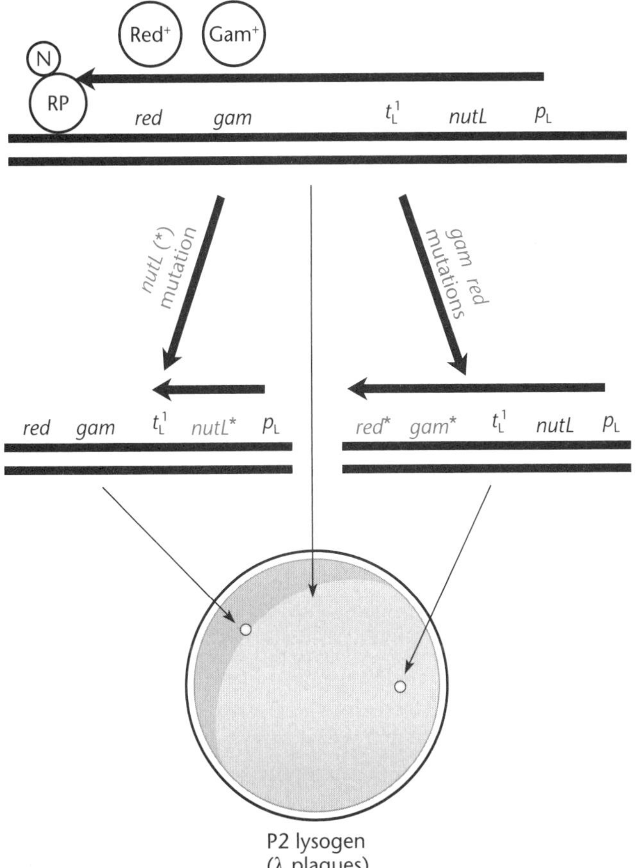

One way to distinguish *nutL* mutants from the other types of mutants that can form plaques on P2 lysogens is by genetic mapping, since *gam* and *red* mutations should map on the left of the t_L^1 terminator while *nutL* mutations should map on its right (Figure 8.2). Genetic mapping in this region of phage λ is facilitated by the collection of λ*pbio* specialized transducing phage in which *E. coli* genes have replaced some of the λ genes and the endpoints of the substitutions have been precisely mapped. Deletion mapping is not convenient with phage λ, because the phage package between two *cos* sites and phage with longer deletions are not viable. However, mapping using substitutions is like mapping using deletions in that if any recombinants occur, the mutation must lie outside the substituted region.

The way in which a λ*pbio* substitution can be used to map *red* and *gam* mutations that make the phage Red$^-$ Gam$^-$ is illustrated in Figure 8.24. In the example, a mutant of λ that is phenotypically Red$^-$ Gam$^-$ is crossed with a phage λ*pbio* that is also Red$^-$ Gam$^-$ because the *bio* substitution includes the *red* and *gam* genes. The appearance of Red$^+$ Gam$^+$ recombinants capable of forming plaques on a P2 lysogen indicates that the mutation that makes the phage Red$^-$ Gam$^-$ must lie outside the substituted region and therefore not in *red* and *gam*. The researchers could detect even very rare Red$^+$ Gam$^+$ recombinants because, as discussed in chapter 10 in connection with the discovery of *chi* sites, only Red$^+$ Gam$^+$ recombinants of λ can form concatemers and hence form plaques on RecA$^-$ *E. coli*.

Even if the mutation that allows the phage to multiply in a P2 lysogen maps on the right of the t_L^1 terminator, it is not necessarily a *nutL* mutation. Other types of mutations have the potential to reduce *gam* and *red* transcription. For example, leaky *N* mutations might reduce the transcription of *gam* and *red* enough to allow plaques to form on a P2 lysogen but allow sufficient *O* and *P* transcription for a plaque to form. However, such *N* mutations, as well as many other such types of mutations, can be distinguished from *nutL* mutations because they are *trans* acting rather than *cis* acting. To determine

A

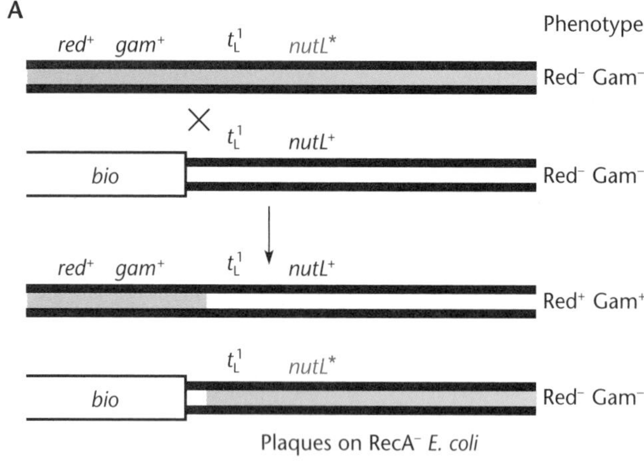

B

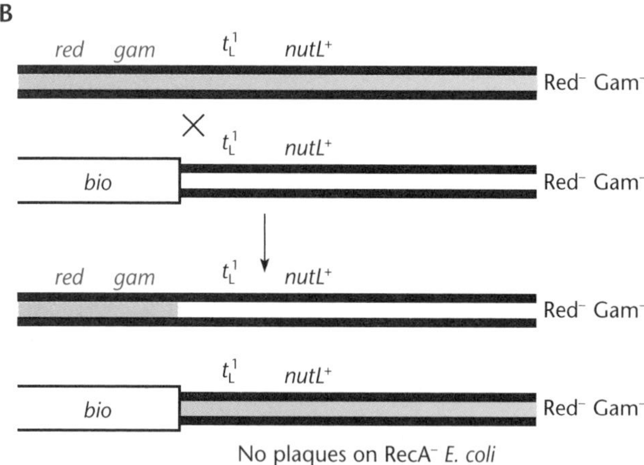

Figure 8.24 Use of λ*pbio*-substituted phage to map *nutL* mutations. **(A)** If the mutation that makes λ phenotypically Gam⁻ Red⁻ maps outside the substituted region, e.g., a *nutL* mutation, Gam⁺ Red⁺ recombinants arise that can multiply in RecA⁻ *E. coli.* **(B)** If the mutations lie within the substituted region, e.g., a *red gam* double mutant, no Red⁺ Gam⁺ recombinants should arise. Only the *nutL* region and the *gam* and *red* genes are shown.
doi:10.1128/9781555817169.ch8.f8.24

if the mutation is *cis* acting, a P2 lysogen is infected simultaneously with two different mutant λ phages, one of which has one of the potential *nutL* mutations while the other has a *gam red* deletion mutation. If the potential *nutL* mutation is *cis* acting, the phage should still multiply on the P2 lysogen, since the *nutL* mutant phage cannot make the *gam* and *red* gene products even if it is furnished with all the other λ gene products in *trans*. Note that *nut* mutations should behave as though they are *cis* acting, even though they affect a site on the RNA (a diffusible molecule) rather than a site on the DNA, because the mutation affects only transcription termination by RNA polymerase making RNA from the same DNA.

As mentioned above, once the *nutL* mutations had been genetically mapped, investigators could locate the base pair change in the mutation through DNA sequencing and comparison of the sequence of the mutant DNA to the known sequence of wild-type λ DNA in the region. They then found a site with identical sequences to the right of the *c*I gene and assumed that this was *nutR* (Figure 8.5).

Isolation of Host *nus* Mutations: *E. coli* Functions Involved in Transcription Elongation-Termination

The isolation of host *nus* mutations also illustrates some additional important concepts in selectional genetics. As mentioned above, host *nus* mutations affect host gene products that are required for N antitermination (see Friedman et al., Suggested Reading). The question was whether the λ *N* gene product acts alone or whether host proteins are also required for efficient N antitermination of transcription. Such host functions might be involved in similar phenomena in the uninfected cell. A host chromosomal mutation that affects one of these proteins may reduce antitermination by the λ *N* gene product. However, host *nus* mutations might be expected to be rare, especially if they are in genes whose products are essential for growth of the cell, again necessitating a selection.

The selection for *nus* mutations was based on the observation that, while induction of a wild-type λ prophage invariably kills the host, the cell can survive induction of an N⁻ mutant prophage because antitermination by N is required to synthesize the *P* gene product and other λ gene products that kill the cell. Since a *nus* mutation of the host should also allow a lysogen to survive induction of the prophage by preventing N antitermination, *nus* mutants should also survive. However, other types of mutants should also survive induction, and some of them are expected to be much more frequent than *nus* mutants. For example, cells cured of the prophage would also survive induction. To reduce the frequency of cured cells, the investigators used a deleted prophage that contained the *P* gene and that could kill the cell but could not be induced to be excised from the chromosome. There are also a myriad of mutations in the prophage itself that allow the cell to survive induction, for example, *N* mutations. The frequency of surviving mutants with *N* mutations or other types of mutations in the prophage could be reduced by using a double lysogen with two copies of the prophage in the chromosome. The investigators reasoned that a mutation in just one of the two copies of the prophage would not save the cell and that two mutations in the same gene, one in each copy of the prophage, would be required, greatly reducing the frequency of this type of survivor among the mutants.

Once surviving mutants were selected, they were mapped by Hfr crosses and transduction, using methods described in chapter 3. The investigators were not interested in mutations that mapped to where the prophage was inserted in the chromosome and that were presumably in the prophage; they were interested only in mutations elsewhere in the *E. coli* genome, and these mutations defined the chromosomal *nus* genes. In this way, they found the *E. coli* genes *nusA*, *nusB*, etc. The products of many of these were subsequently shown to travel with the phage antitermination complex and had been commandeered by the λ phage to promote its own transcriptional antitermination by N.

SUMMARY

1. Some phages are capable of lysogeny, in which they persist in the cell as prophages. A bacterium harboring a prophage is called a lysogen. In the lysogen, most of the phage gene products made are involved in maintaining the prophage state. The prophage DNA can either be integrated into the chromosome or replicate autonomously as a plasmid.

2. The *E. coli* phage λ is the prototype of a lysogen-forming phage. It was the first such phage to be discovered and is the one to which all others are compared.

3. Phage λ regulates its transcription through antitermination proteins N and Q, which allow transcription to continue into the early and late genes, respectively. These proteins bind to the RNA polymerase and allow it to transcribe through transcription termination sites. However, the conditions under which N and Q bind to the RNA polymerase differ greatly.

4. The N protein must first bind to a sequence, *nut*, in the mRNA before it can bind to the RNA polymerase. This ensures that N can bind only to RNA polymerase being transcribed from the p_R and p_L promoters. Other host proteins, called the Nus proteins, help N bind, and some of them travel with it on the RNA polymerase. The RNA polymerase with N, NusA, NusB, NusE, and NusG bound can then transcribe through transcription terminators into the *O*, *P*, and *Q* genes on the right and the *red* (*bet* and *exo*), *gam*, and *int* genes on the left. At least some of the Nus proteins are involved in transcription termination and antitermination in the host.

5. The Q protein must first bind to a sequence, *qut*, that overlaps the promoter before it can bind to the RNA polymerase. The RNA polymerase first makes a short RNA from the late $p_{R'}$ promoter and then stops at a sequence that resembles a σ^{70} promoter but with the −10 and −35 sequences separated by only 1 bp. The *Q* gene product then binds to a *qut* sequence in the DNA before it can bind to the stalled RNA polymerase. It and the NusA protein then travel with the RNA polymerase to allow it to transcribe through terminators into the late genes, including the head, tail, and lysis genes, of the phage.

6. Phage λ DNA is linear in the phage head, with short complementary single-stranded 5′ ends called the *cos* (for *co*hesive) ends. Because they have complementary sequences, the *cos* ends can base pair with each other to form a circle after the DNA enters the cell. The phage DNA then replicates as a circle a few times before it enters the rolling-circle mode of replication, which leads to the formation of long concatemers in which many genome length DNAs are linked end to end.

7. Phage λ DNA can be packaged only from concatemers and not from unit length genomes. This is because λ begins filling the head at one *cos* site and stops only when it gets to the next *cos* site in the concatemer. The concatemers from which λ DNA is packaged can be formed either by rolling-circle replication or by recombination between single-length circles.

8. Whether λ enters the lysogenic state depends on the outcome of a race between the *c*II gene product and the *cro* gene product of the phage. The product of the *c*II gene is a transcriptional activator that activates the transcription of the *c*I and *int* genes after infection. The *c*I gene product is the repressor that blocks transcription of most of the genes of λ in the prophage state, and the *int* gene product is a site-specific recombinase that integrates λ DNA into the bacterial chromosome by promoting recombination between the *attP* site on the phage DNA and the *attB* site in the chromosome. The *cro* gene product binds to the operators and prevents synthesis of the CI repressor from the p_{RM} promoter, enhancing lytic development.

9. The frequency with which lysogens form is a function of the MOI and the growth rate of the cells being infected. At a high MOI, as many as 50% of the infections can result in lysogeny because the excess CIII protein inhibits the protease that degrades CII, stabilizing CII. At high growth rates, increased concentrations of RNase III favor lytic development by separating the *nutL* site from the TIR for gene *N*, which increases the rate of translation of N.

10. The CI repressor protein of λ is a homodimer made up of two identical polypeptides encoded by the *c*I gene and the prototype of a protein made up of separable domains. The repressor blocks transcription by binding to operators o_R and o_L on both sides of the *c*I gene, preventing the utilization of the two promoters, p_R and p_L, that are responsible for the transcription of genes on the right and on the left of the *c*I gene, respectively. It is also a transcriptional activator that activates its own transcription in the lysogen by binding to o_R^2 and activating the p_{RM} promoter. The CI polypeptide is made up of separable domains, a domain that

(continued)

SUMMARY (continued)

binds to the operator sequences on DNA (the DNA-binding domain) and another domain that binds to another repressor monomer to make a dimer (the dimerization domain).

11. The repressor regulates its own synthesis in the lysogenic state through its cooperative binding to three repressor-binding sites within o_R. These sites are named o_R^1, o_R^2, and o_R^3 in the order of their affinity for the repressor. Repressor bound at o_R^1 blocks transcription from p_R. At higher concentrations, repressor also binds at o_R^2 and activates transcription of the *c*1 gene from the p_{RM} promoter. At still higher concentrations, repressor binds to o_R^3 and forms a tetramer with repressor bound at o_L^3, bending the DNA and preventing synthesis of more repressor. The repressors bound at o_L^1 and o_L^2 may also interact with the ones bound at o_R^1 and o_R^2 to stabilize the bend. The ability of a gene product to regulate its own synthesis is called autoregulation.

12. Damage to the DNA of its host can cause the λ prophage to be induced and produce phage. Single-stranded DNA that accumulates after DNA damage binds to host RecA protein, activating its coprotease activity. The activated RecA protein of the host causes the λ repressor protein to cleave itself between the DNA-binding and dimerization domains so that it can no longer form dimers and be active. The process of excision of λ DNA is essentially the reverse of integration, except that excision requires both the *int* and *xis* gene products of λ because it requires recombination between the hybrid *attP-attB* sites flanking the prophage.

13. Very rarely, when λ DNA is excised, it picks up neighboring bacterial DNA and becomes a transducing particle. This type of transduction is called specialized transduction, because only bacterial genes close to the insertion site of the prophage can be transduced.

14. Lysogen-forming phages are often useful as cloning vectors. A bacterial gene cloned into a prophage exists in only two copies, one in its normal site in the chromosome and another in the prophage at its *attB* site. This can be important in complementation tests or in other applications, such as gene expression studies, which sometimes require that a gene be expressed in a single copy from the chromosome. When the prophage is induced, the cloned DNA can be recovered in large amounts.

15. Prophages often carry genes for bacterial toxins. Examples are the toxins that cause HUS, diphtheria, botulism, scarlet fever, and cholera. The toxins expressed by some strains of *S. aureus* that cause toxic shock syndrome are often encoded by pathogenicity islands that are induced and packaged by phages, allowing them to move to other cells.

QUESTIONS FOR THOUGHT

1. Why do you suppose the λ prophage uses different promoters to transcribe the *c*I repressor gene immediately after infection and in the lysogenic state?

2. Why is only one protein, Int, required to integrate the phage DNA into the chromosome while two proteins, Int and Xis, are required to excise it? Why not just make one different Int-like protein that excises the prophage?

3. How do you suppose morons containing toxin and other virulence genes move onto a phage? What is the selective pressure for a phage to pick up a moron? Where do morons come from?

4. Why do you suppose some types of prophage can be induced only if another phage of the same type infects the lysogenic cell containing them? What purpose does this serve?

5. Is P4 a satellite phage or a genetic island? What distinguishes these two types of DNA elements?

6. Why do you suppose SaPIs use only nonessential gene products of the phage to inactivate their repressors and induce them?

PROBLEMS

1. λ *vir* mutations cause clear plaques because they change the operator sequences so that they no longer bind the repressor. How would you determine if a clear-plaque mutant you have isolated has a *vir* mutation instead of a mutation in any one of the three genes *c*I, *c*II, and *c*III?

2. λ integrates into the bacterial chromosome in the region between the galactose utilization (*gal*) genes and the biotin biosynthetic (*bio*) genes on the other side. Outline how you would isolate an HFT strain carrying the *bio* operon of *E. coli*. Would you expect your transducing phage to form plaques? Why or why not?

3. A *vir* mutation changes the operator sequences. Would you expect λ with *vir* mutations in the o_R^1 and o_L^1 sites to form plaques on λ lysogens? Why or why not?

4. The DNA of a λ*dgal* specialized transducing particle can be induced to make phage only if it is integrated next to a preexisting wild-type prophage. This integration could occur by integration of the λ*dgal* next to the preexisting prophage or by recombination between complementary sequences in the two phages either before either has integrated or after one has integrated. Draw the structures you would expect from these different mechanisms of integration. Make sure you show the structures of the *att* sites at the ends of both phages and consider whether you would need both Int and Xis to excise the phages in both structures. Also, what kinds of experiments could you do to determine which kind of structure has formed in a particular dilysogen?

5. Would you use a high or a low MOI to use phage display and the formation of dilysogens in the 2λ system to determine if two peptides bind to each other? Why?

6. What controls would you do using the 2λ system to measure the frequency of dilysogens to determine if two peptides interact?

7. How would you determine whether N antitermination in a Shiga toxin-encoding phage from *E. coli* O157:H7 requires the same Nus factors as λ phage? Hint: the phage forms plaques on the *E. coli* laboratory strain used to isolate Nus mutants.

8. You can sometimes get intragenic complementation between two temperature-sensitive mutations in the *c*I gene of λ phage. Would you expect ever to get intragenic complementation between two amber (UAG) mutations?

9. In the intragenic complementation tests that demonstrated the domain structure of the CI repressor, would you expect the lysogens that form after complementation of two temperature-sensitive mutations in the *c*I gene to be single lysogens or dilysogens with two different mutant prophages inserted in tandem? Why?

10. The λ CI repressor must dimerize to function. Outline how you would use this fact to identify the regions of another protein, e.g., LacZ, required for its dimerization.

11. You have isolated a relative of P2 phage from sewage, using *E. coli* as the indicator bacterium. How would you determine if P4 phage can parasitize (i.e., can be a satellite virus of) your P2-like phage?

SUGGESTED READING

Bair, C. L., A. Oppenheim, A. Trostel, and S. Adhya. 2008. A phage display system designed to detect and study protein-protein interactions. *Mol. Microbiol.* **67:**719–728

Brüssow, H., C. Canchaya, and W.-D. Hardt. 2004. Phages and the evolution of bacterial pathogens: from genomic rearrangements to lysogenic conversion. *Microbiol. Mol. Biol. Rev.* **68:**560–602.

Cairns, J., M. Delbrück, G. S. Stent, and J. D. Watson. 1966. *Phage and the Origins of Molecular Biology.* Cold Spring Harbor Laboratory Press, Cold Spring Harbor, NY.

Calendar, R. L. (ed.). 2005. *The Bacteriophages*, 2nd ed. Oxford University Press, New York, NY.

Court, D. L., A. B. Oppenheim, and S. L. Adhya. 2007. A new look at bacteriophage λ genetic networks. *J. Bacteriol.* *189:*298–304.

deHaseth, P. L., and J. M. Gott. 2010. Conformational flexibility of σ^{70} in antiterminator loading. *Mol. Microbiol.* **75:**543–546.

Echols, H., and H. Murialdo. 1978. Genetic map of bacteriophage lambda. *Microbiol. Rev.* **42:**577–591.

Friedman, D. I., M. F. Baumann, and L. S. Baron. 1976. Cooperative effects of bacterial mutations affecting λ N gene expression. *Virology* **73:**119–127.

Gottesman, M. 1999. Bacteriophage λ: the untold story. *J. Mol. Biol.* **293:**177–180.

Hendrix, R. W., J. W. Roberts, F. W. Stahl, and R. A. Weisberg (ed.). 1983. *Lambda II.* Cold Spring Harbor Laboratory Press, Cold Spring Harbor, NY.

Kahn, M. L., R. Ziermann, G. Deho, D. W. Ow, M. G. Sunshine, and R. Calendar. 1991. Bacteriophage P2 and P4. *Methods Enzymol.* **204:**264–280.

King, R. A., A. Wright, C. Miles, C. S. Pendleton, A. Ebelhar, S. Lane, and P. T. Parthasarathy. 2011. Newly discovered antiterminator RNAs in bacteriophage. *J. Bacteriol.* **193:**5784–5792.

Lieb, M. 1976. Lambda *c*I mutants: intragenic complementation and complementation with a *c*I promoter mutant. *Mol. Gen. Genet.* **146:**291–297.

Livny, J., and D. I. Friedman. 2004. Characterizing spontaneous induction of Stx encoding phages using a selectable reporter system. *Mol. Microbiol.* **51:**1691–1704.

Lwoff, A. 1953. Lysogeny. *Bacteriol. Rev.* **17:**269–337.

McLeod, S. M., H. H. Kimsey, B. M. Davis, and M. K. Waldor. 2005. CTXϕ and *Vibrio cholerae*: exploring a newly recognized type of phage-host cell relationship. *Mol. Microbiol.* **57:**347–356.

Ptashne, M. 2004. *A Genetic Switch: Phage Lambda Revisited*, 3rd ed. Cold Spring Harbor Laboratory Press, Cold Spring Harbor, NY.

Salstrom, J. S., and W. Szybalski. 1978. Coliphage λ *nutL*: a unique class of mutations defective in the site of N product utilization for antitermination of leftward transcription. *J. Mol. Biol.* **124:**195–222.

Sunagawa, H., T. Ohyama, T. Watanabe, and K. Inoue. 1992. The complete amino acid sequence of the *Clostridium botulinum* type D neurotoxin, deduced by nucleotide sequence analysis of the encoding phage d-16ϕ genome. *J. Vet. Med. Sci.* **54:**905–913.

Tormo-Más, M. A., I. Mir, A. Shrestha, S. M. Tallent, S. Campoy, Í. Lasa, J. Barbé, R. P. Novick, G. E. Christie, and J. R. Penadés. 2010. Moonlighting bacteriophage proteins derepress staphylococcal pathogenicity islands. *Nature* **465:**779–782.

Wagner, P. L., M. N. Neely, X. Zhang, D. W. K. Acheson, M. K. Waldor, and D. L. Friedman. 2001. Role for a phage promoter in Shiga toxin 2 expression from a pathogenic *Escherichia coli* strain. *J. Bacteriol.* **183:**2081–2085.

Waldor, M. K., D. I. Friedman, and S. L. Adhya (ed.). 2005. *Phages: Their Role in Bacterial Pathogenesis and Biotechnology.* ASM Press, Washington, DC.

Waldor, M. K., and J. J. Mekalanos. 1996. Lysogenic conversion by a filamentous phage encoding cholera toxin. *Science* **272:**1910–1914.

CHAPTER 9

Transposition, Site-Specific Recombination, and Families of Recombinases

RECOMBINATION IS THE BREAKING AND REJOINING of DNA in new combinations. In homologous recombination, which accounts for most recombination in the cell, the breaking and rejoining occur only between regions of two DNA molecules that have similar or identical sequences. Homologous recombination requires that the two DNAs pair through complementary base pairing, for which the two DNAs must have the same sequence (see chapter 10). However, other types of recombination, known as **nonhomologous recombination**, also occur in cells. As the name implies, these types of recombination do not depend on extensive homology between the two DNA sequences involved. Some types of nonhomologous recombination are due to the mistaken breaking and rejoining of DNA by enzymes such as topoisomerases (see chapter 1). Other types are not mistakes and have specific purposes in the cell. These types depend on specific enzymes that promote recombination between different regions in DNA, which may or may not have sequences in common. This chapter addresses some of these examples of nonhomologous recombination in bacteria and the mechanisms involved, including transposition by transposons, the site-specific recombination that occurs during the integration and excision of prophages and other DNA elements, the inversion of invertible sequences, and the resolution of cointegrates by resolvases. Evidence shows that the enzymes that perform these various functions have much in common across all three domains of life.

Transposition

Transposons are elements that can hop, or transpose, from one place in DNA to another. Transposable DNA elements were first discovered in corn by Barbara McClintock in the early 1950s and about 20 years later in

doi:10.1128/9781555817169.ch9

bacteria by others. Transposons are now known to exist in all organisms on Earth, including humans. In fact, from the Human Genome Project, it is apparent that almost half of our DNA may be transposons! The process by which a transposon moves is called **transposition**, and the enzymes that promote transposition are a type of recombinase called **transposases.** The transposon itself usually encodes its own transposases, so that it carries with it the ability to hop each time it moves. For this reason, transposons have been called "jumping genes."

Not all elements that can move are called transposons. The word transposon typically refers to DNA transposons that always move as DNA. There are other types of elements that utilize an RNA intermediate. One type of element that moves in this fashion is mobile group II introns, which cleave themselves out of an mRNA and move to a new insertion site as RNA. Insertion of the group II mobile intron occurs by a process that is typified by elements that are common in the human genome, where the RNA breaks the chromosome during the joining event to provide a 3′ end to be used to prime its own reverse transcription. DNA transposons should also be distinguished from **retrotransposons,** so named because they behave like RNA retroviruses with a DNA intermediate. An RNA copy of a region is made and then copied into DNA by a reverse transcriptase. The DNA intermediate then integrates elsewhere by various mechanisms that may or may not be analogous to transposition. Although only a few examples of retrotransposons are known in bacteria, such elements are well known in eukaryotes. Ty elements are common, well-studied retrotransposons in yeast.

Although transposons appear to exist in all organisms on Earth, they are best understood in bacteria, where they play an important role in evolution. While transposons are true parasites of the host, they also can encode beneficial gene products. As more DNA sequence information accumulates, it is becoming clear that transposons move among different genera of bacteria with some regularity. As mentioned in previous chapters, transposons may enter other genera of bacteria during transfer of promiscuous plasmids or via transducing phages.

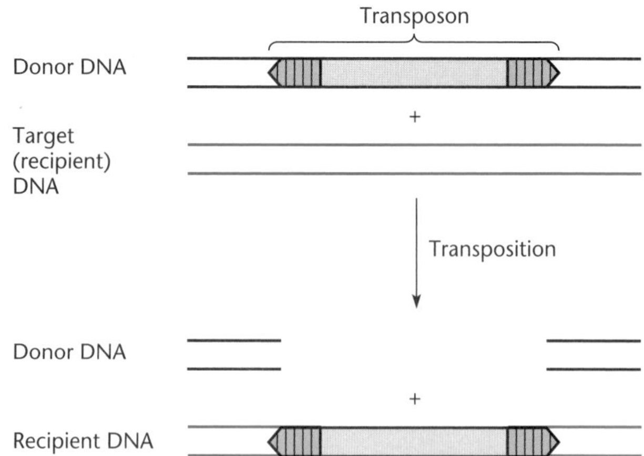

Figure 9.1 Overview of transposition. See the text for details. doi:10.1128/9781555817169.ch9.f9.1

Overview of Transposition

The net result of transposition is that the transposon now resides at a new place in the genome where it was not originally found. Many transposons are essentially "cut out" of one DNA and inserted into a new position (Figure 9.1), whereas other transposons are copied and then inserted elsewhere. Regardless of the type of transposon, however, the DNA from which the transposon originated is called the **donor DNA** and the DNA into which it hops is called the **target** or **recipient DNA.**

In the classically defined transposition events, the transposase enzyme breaks the donor DNA at the ends of the transposon and then inserts the transposon into the target DNA. However, the details of the mechanism can vary. Some types of transposons may exist free of other DNA during the act of transposition, but many transposons, before and after they move, remain contiguous with other, flanking DNA molecules. Later in this chapter, we discuss more detailed models for the various types of transposition.

Transposition must be tightly regulated and occur only rarely; otherwise, the cellular DNA would become riddled with the transposon, which would have many deleterious effects. Transposons have evolved elaborate mechanisms to regulate their transposition so that they move very infrequently and do not often kill the host cell. We mention some of these mechanisms later in this chapter, when we discuss individual examples of transposons. The frequency of transposition varies from about once in every 10^3 cell divisions to about once in every 10^8 cell divisions, depending on the type of transposon. Thus, the chance of a particular transposon hopping into a gene and inactivating it is sometimes not much higher than the chance that a gene will be inactivated by other types of mutations (see chapter 3).

Structure of Bacterial Transposons

There are many different types of bacterial transposons. Some of the smaller elements are about 1,000 bp long and carry only the gene for the transposase that promotes their movement. Larger transposons may also contain one or more other genes that are responsible for regulating transposition or for specialized targeting functions. In other cases, genes beneficial to the host may also be carried on the element, such as those for resistance to an antibiotic.

Classically defined transposons also have *cis*-acting sites at the ends of the element that are recognized by the transposase. These *cis*-acting sites are in opposite orientation and hence are referred to as **inverted repeats** (IRs) (Figure 9.2). As discussed in chapter 1, two regions of DNA are IRs if the sequence of nucleotides on one strand in one region, when read in the 5′-to-3′ direction, is the same or almost the same as the 5′-to-3′ sequence of the opposite strand in the other region. The transposition reaction does not occur until a **synapse** structure is formed, where the transposase proteins bound to each of the IR sequences also bind to one another (Figure 9.2). IRs are one of the features that help to identify transposons in a genome.

Another feature associated with DNA transposons is short **direct repeats**. Direct repeats have the same 5′-to-3′ sequence of nucleotides on the same strand. The direct repeats are a by-product of the way the element is joined into the target DNA (Figure 9.3). As shown in Figure 9.3, the target DNA originally contains only one copy of the sequence at the place where a transposon inserts.

Figure 9.2 Steps in transposon excision. IRs (shown in blue) that are recognized by the element-encoded transposase characterize the ends of the transposon. Genes carried by the transposon are not shown but reside in the region shaded grey. A synapse is formed when the transposase binds the IRs and the ends are paired with the transposase. This signals the transposase to carry out the cleavage and joining events that underlie the process of transposition. A completely excised element is not found for all transposition processes, and not all transposons leave a double-strand break in the donor DNA. doi:10.1128/9781555817169.ch9.f9.2

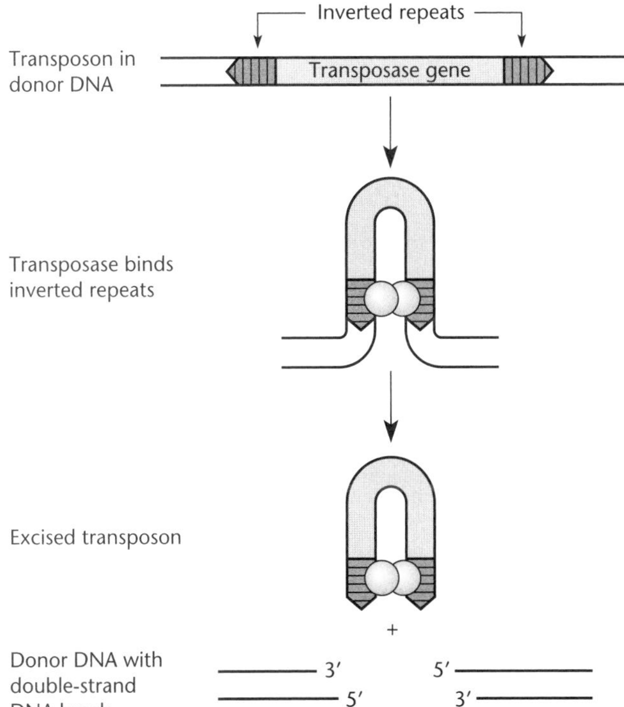

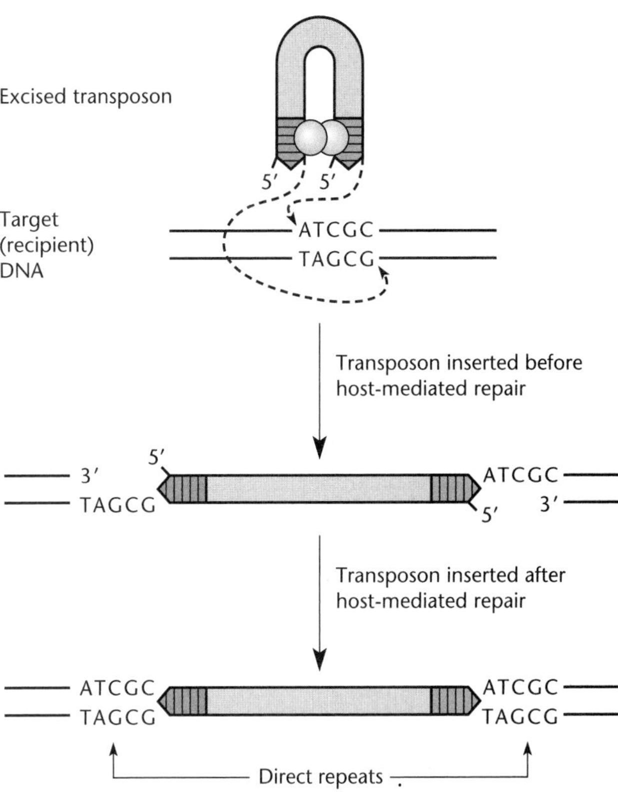

Figure 9.3 Steps in transposon insertion. The transposon inserts into a target DNA using staggered joins to the top and bottom strands (an excised transposon species found with many elements is shown). The process of inserting with staggered joins into the target DNA leaves gaps flanking the transposon that must be filled by a host DNA polymerase. Often, a few bases of sequence from the donor DNA also remain bound to the 5′ ends of the element, which must also be processed by host enzymes. The effect of staggered joining events and the subsequent host repair leads to the direct repeats that are indicative of transposition. doi:10.1128/9781555817169.ch9.f9.3

When the 3′ ends of the element are joined to offset positions (i.e., not directly across the DNA backbone) in a target DNA, single-stranded gaps are left behind. These gaps are repaired by replication by DNA polymerase I, which leads to the directly repeated sequence flanking the element. DNA polymerase I also processes a few bases that are often carried over from the donor DNA. Most transposons can insert into many places in DNA and have little or only limited target specificity. Thus, the duplicated sequence depends on the sequence at the site in the target DNA into which the transposon inserted. However, even though the duplicated sequences differ, the number of duplicated base pairs is characteristic of each transposon. Some duplicate as few as 2 bp, and others duplicate as many as 9 bp.

Types of Bacterial Transposons

In this section, we describe some of the common types of DNA transposons.

INSERTION SEQUENCE ELEMENTS

The smallest bacterial transposons are called **insertion sequence (IS) elements.** These transposons are usually only about 750 to 2,000 bp long and encode little more than the transposase enzymes that promote their transposition.

Because IS elements carry no selectable genes, they were discovered only because they inactivate a gene if they happen to insert into it. The first IS elements were detected as a type of *gal* mutation that was unlike any other known mutation. This type of mutation resembled deletion mutations in that it was nonleaky; however, unlike deletion mutations, it could revert, albeit at a much lower frequency than base pair changes or frameshifts. Such anomalous *gal* mutations were also very polar and could prevent the transcription of downstream genes (see chapter 2). Later work showed that these mutations resulted from insertion of about 1,000 bp of DNA into a *gal* gene. Moreover, they were due to insertion of not just any piece of DNA, but one of very few sequences.

Originally, four different IS elements were found in *Escherichia coli*: IS*1*, IS2, IS*3*, and IS*4*. Most strains of *E. coli* K-12 contain approximately six copies of IS*1*, seven copies of IS2, and fewer copies of the others. Although almost all bacteria carry IS elements, IS elements tend to have a level of species specificity where the most closely related bacteria have more closely related IS elements. To date, thousands of different IS elements have been found in bacteria. Plasmids also often carry IS elements, which are important in the assembly of the plasmid itself (see below) and in the formation of Hfr strains (see chapter 5). Although the original IS elements were discovered only because they had inserted into a gene, causing a detectable phenotype, IS elements are now more often discovered during genomic sequencing. In addition, DNAs that are introduced into *E. coli* by DNA cloning can also be a target for endogenous IS elements. There are many examples where the IS element sequences inadvertently show up in DNA that was passaged though *E. coli* before being subjected to sequencing. An interesting strategy for getting around this problem involved the construction of an *E. coli* K-12 derivative from which all of the IS elements were removed (see Pósfai, Suggested Reading).

COMPOSITE TRANSPOSONS

Sometimes two IS elements of the same type form a larger transposon, called a **composite transposon,** by bracketing other genes. Figure 9.4 shows the structures of three composite transposons, Tn*5*, Tn*9*, and Tn*10*. Tn*5* consists of genes for kanamycin resistance (Kan^r) and streptomycin resistance (Str^r) bracketed by copies of an IS element called IS*50*. Tn*9* has two copies of IS*1* bracketing a chloramphenicol resistance (Cam^r) gene. In Tn*10*, two copies of IS*10* flank a gene for tetracycline resistance (Tet^r). Some composite transposons, such as Tn*9*, have the bracketing IS elements in the same orientation, whereas others, including Tn*5* and Tn*10*, have them in opposite orientations.

Figure 9.5 illustrates transposition of a composite transposon. Each IS element can transpose independently as long as the transposase acts on both of its ends. However, because all the ends of the IS elements

Figure 9.4 Structures of some composite transposons. The left (L) and right (R) ends of the elements are indicated. The commonly used Kan^r genes and the Cam^r gene come from Tn*5* and Tn*9*, respectively. The active transposase gene is in one of the two IS elements. Note that the IS elements can be in either the same or opposite orientation. Str^r, gene encoding streptomycin resistance; Tet^r, gene encoding tetracycline resistance; Ble^r, gene encoding bleomycin resistance. doi:10.1128/9781555817169.ch9.f9.4

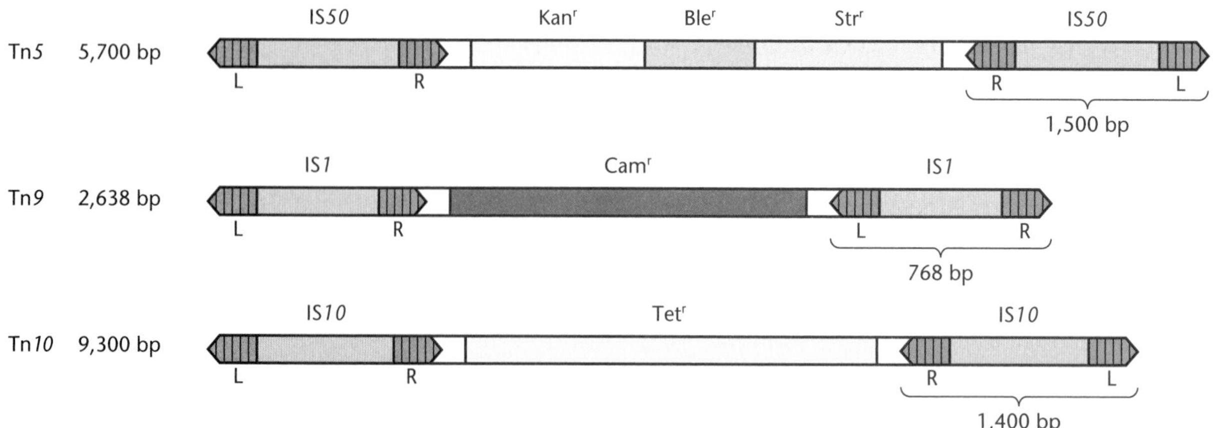

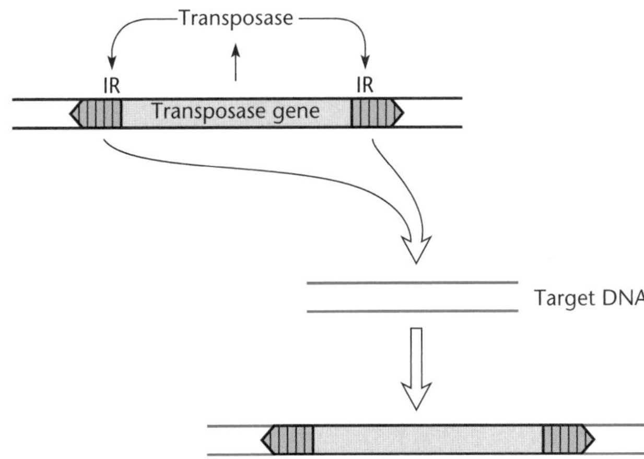

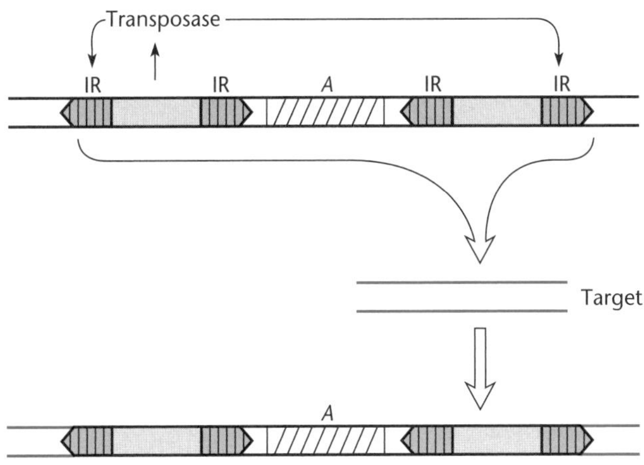

Figure 9.5 Two IS elements can transpose any DNA between them. **(A)** Action of the transposase at the ends of an isolated IS element causes it to transpose. **(B)** Two IS elements of the same type are close to each other in the DNA. Action of the transposase on their outside ends causes them to transpose together, carrying along the DNA between them. *A* denotes the DNA between the IS elements (diagonally hatched), and the arrows indicate the IR sequences at the ends of the IS elements. The blue lines represent the target DNA.
doi:10.1128/9781555817169.ch9.f9.5

in a composite transposon are the same, a transposase encoded by one of the IS elements can recognize the ends of either IS element. When such a transposase acts on the IRs at the farthest ends of a composite transposon, the two IS elements transpose as a unit, bringing along the genes between them. These two IRs are called the "outside ends" of the two IS elements because they are the farthest from each other.

The two IS elements that form composite transposons are often not completely autonomous because of mutations in the transposase gene of one of the elements. Thus, only one of the IS elements encodes an active transposase. However, this transposase can act on the outside ends to promote transposition of the composite transposon. Mutations in one or both of the "inside" IRs also enhance transposition of the composite transposon.

Assembly of Plasmids by IS Elements

Any time two IS elements of the same type happen to insert close to each other on the same DNA, a composite transposon is born. These transposons have not yet evolved a defined structure like the named transposons (e.g., Tn*10*) described above. Nevertheless, the two IS elements can transpose any DNA between them. In this way, "cassettes" of genes bracketed by IS elements can be moved from one DNA molecule to another.

Many plasmids seem to have been assembled from such cassettes. Figure 9.6 shows a naturally occurring plasmid that carries genes for resistance to many antibiotics. Such plasmids are historically called R factors, because they confer resistance to so many different antibiotics (see chapter 4). Notice that many of the resistance genes on the plasmid are bracketed by the same IS elements. IS*3* flanks the tetracycline resistance gene, and IS*1* brackets the genes for resistance to many other antibiotics. Apparently, the plasmid was assembled in nature by insertion of resistance genes onto the plasmid from some other DNA via the bracketing IS elements. In principle, any two transposons of the same type can move other DNA lying between them by a similar mechanism, but because IS elements are the most common transposons and often exist in more than one copy per

Figure 9.6 R factors, or plasmids containing many resistance genes, may have been assembled, in part, by IS elements. The Tet^r gene is bracketed by IS*3* elements, and the region containing the other resistance genes (the r determinant is outlined in blue) is bracketed by IS*1* elements.
doi:10.1128/9781555817169.ch9.f9.6

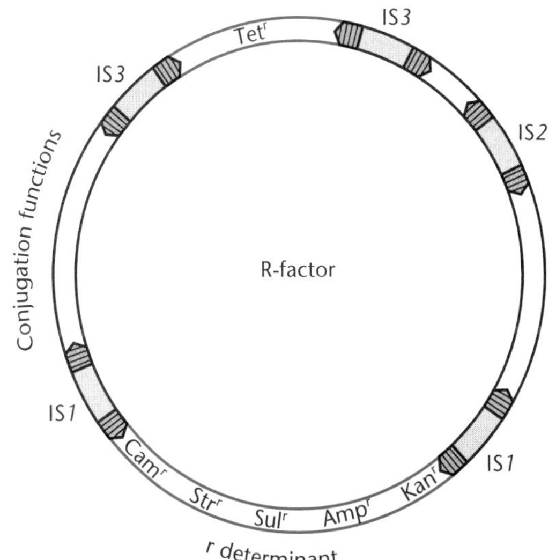

cell, they probably play the major role in the assembly of plasmids.

NONCOMPOSITE TRANSPOSONS

Composite transposons are not the only ones that carry resistance genes. Such genes can also be an integral part of transposons that are sometimes referred to as **noncomposite transposons** (Figure 9.7). They are bracketed by short IRs, but the resistance gene is part of the minimum transposable unit. Noncomposite transposons can also cause rearrangement of adjoining chromosomal DNA even though they do not have complete copies of IS elements at their ends and so do not have inside ends to participate in transposition events. This is because many of them transpose by a replicative mechanism (see below) in which the ends of the transposon remain attached to the flanking chromosomal DNA during the transposition event. Transposition by such a tranposon into a nearby site on the same DNA can then rearrange the DNA between the donor and target sites. This may be one reason why such transposons often exhibit target immunity (see below).

Noncomposite transposons seem to belong to a number of families in which the members are related to each other by sequence and structure. Interestingly, different members of transposon families, such as the Tn*21* family, often carry different resistance genes, even though they are almost identical otherwise (Figure 9.7). Often, this is the result of the resistance genes having integrated into the transposon as a cassette carrying one or more resistance genes. The cassettes, which existed elsewhere in the genome, had been excised to form a circle that then integrated into an *att* site on the transposon, using an integrase much like those used to integrate lysogenic phages (see "Integrases of Transposon Integrons" below). Integration of these cassettes into integrons provides one of the major ways by which bacteria can achieve resistance to a variety of antibiotics. In fact, the first known example of multiple drug resistance in pathogenic bacteria in Japan in the early 1950s was due to transposon Tn*21*, with multiple drug resistance cassettes acquired by integrons. This phenomenon of cassette insertion into transposons such as Tn*21* is part of a more general phenomenon in which numerous gene cassettes in superintegrons are stored in the chromosome, from which individual cassettes can then insert into mobile elements, such as plasmids (see "Integrases of Transposon Integrons" below).

Assays of Transposition

To study transposition, we must have assays to monitor the process. As mentioned above, insertion elements were discovered because they create mutations when they hop into a gene. However, this is usually not a convenient way to assay transposition, because transposition is infrequent, and it is laborious to distinguish insertion

Figure 9.7 Some examples of noncomposite transposons. The ORFs encoding the proteins are boxed. The terminal IR ends are shown in blue and hatched. *A* is the transposase, *R* is the resolvase and repressor of *A* transcription, *res* is the site at which resolvase acts, Mer^r is the mercury resistance region, and *merR* is the regulator of mercury resistance gene transcription. In*2* is an integron of the type shown in Figure 9.23. Tn*3* was originally found on the broad-host-range plasmid pR1*drd*19, Tn*501* was found on the *Pseudomonas* plasmid pUS1, Tn*21* was found on the *S. flexneri* plasmid R100, and γδ is found on the chromosome of *E. coli* and on the F plasmid. The arrows indicate the sites at which the proteins act.
doi:10.1128/9781555817169.ch9.f9.7

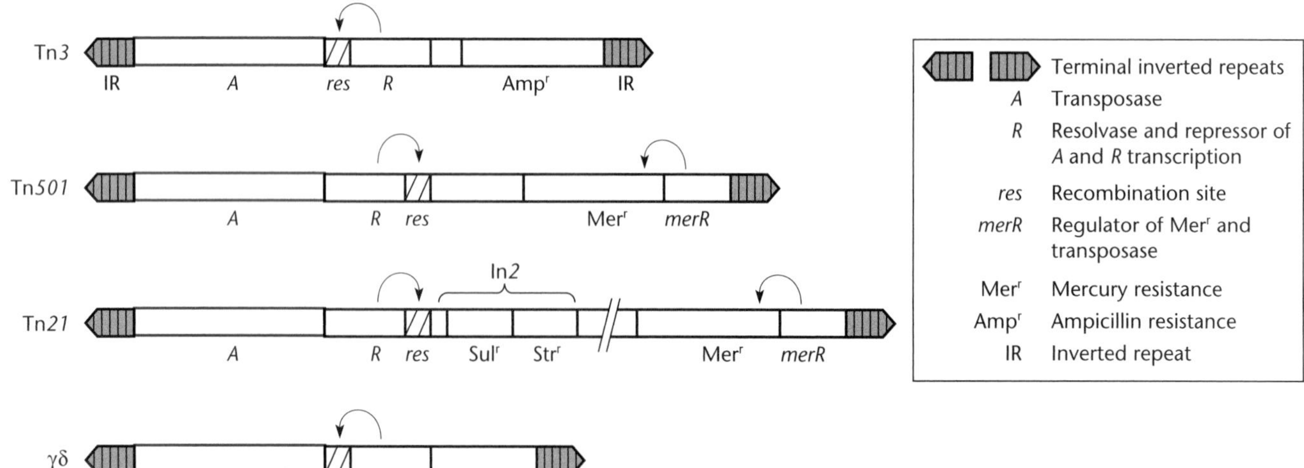

mutations from the myriad other mutations that can occur. If the transposon carries a resistance gene, the job of assaying transposition is easier, but how do we know if a transposon has moved in the cell? The cells are resistant to the antibiotic no matter where the transposon is inserted in the cellular DNA, so moving from one place to another makes no difference in the level of resistance of the cell. Obviously, detecting transposition requires special methods.

SUICIDE VECTORS

One way to assay transposition is with **suicide vectors**. Any DNA, including plasmid or phage DNA, that cannot replicate (i.e., is not a replicon) in a particular host can be used as a suicide vector. These DNAs are called suicide vectors because, by entering cells in which they cannot replicate, they essentially kill themselves. To assay transposition with a suicide vector, we use one to introduce a transposon carrying an antibiotic resistance gene into an appropriate host. The way in which the suicide vector itself is introduced into the cells depends on its source. If it is a phage, the cells could be infected with that phage. If it is a plasmid, it could be introduced into the cells through conjugation. However, whatever method is used, it should be very efficient, since transposition is a rare event.

Once in the cell, the suicide vector remains unreplicated and eventually is lost as the cell population grows. The only way the transposon can survive and confer antibiotic resistance on the cells is by moving to another DNA molecule that is capable of autonomous replication in those cells, for example, a replicating plasmid or the chromosome. Therefore, when the cells under study are plated on antibiotic-containing agar and incubated, the appearance of colonies, as a result of the multiplication of antibiotic-resistant bacteria, is evidence for transposition. These cells have been mutagenized by the transposon, since the transposon has moved into a cellular DNA molecule—either the chromosome or a plasmid—causing insertion mutations.

Phage Suicide Vectors

Some derivatives of phage λ have been modified to be used as suicide vectors in *E. coli*. These phages have been rendered incapable of replication in nonsuppressing hosts by the presence of nonsense codons in their replication genes *O* and *P* (see chapter 8). They have also been rendered incapable of integrating into the host DNA by deletion of their attachment region, *attP*. Such a λ phage can be propagated on an *E. coli* strain carrying a nonsense suppressor. However, in a nonsuppressor *E. coli*, it cannot replicate or integrate. Because of the narrow host range of λ, these suicide vectors can normally be used only in strains of *E. coli* K-12. Other bacteriophages have been adapted for delivery into other types of bacteria.

Plasmid Suicide Vectors

Plasmid cloning vectors can also be used as suicide vectors, provided that the plasmid cannot replicate in the cells in which transposition is occurring. The plasmid containing the transposon with a gene for antibiotic resistance can be propagated in a host in which it can replicate and is then introduced into a cell in which it cannot replicate. In principle, any plasmid with a conditional-lethal mutation (e.g., a nonsense or temperature-sensitive mutation) in a gene required for plasmid replication can be used as a suicide vector. The plasmid could be propagated in the permissive host or under permissive conditions and then introduced into a nonpermissive host or into the same host under nonpermissive conditions, depending on the type of mutation. Alternatively, a narrow-host-range plasmid could be used. It can be propagated in a host in which it can replicate and introduced into a different species in which it cannot replicate.

Many general methods for assaying transposition are based on promiscuous self-transmissible plasmids because the most efficient way to introduce a plasmid into cells is by conjugation, which can approach 100% under some conditions. If the plasmid containing the transposon contains a *mob* region, it can be mobilized into the recipient cell by using the Tra functions of a self-transmissible plasmid (see chapter 5). This technique is most highly developed for gram-negative bacteria. Taking advantage of the extreme promiscuity of some self-transmissible plasmids of gram-negative bacteria, the plasmid might be mobilized into almost any gram-negative bacterium. If the mobilizable plasmid has a narrow host range, it might not be able to replicate in the host into which it has been mobilized and will eventually be lost. ColE1-derived plasmids into which a *mob* site has been introduced are often the suicide vectors of choice in such applications because they can replicate only in some enteric bacteria, including *E. coli*, and so can be used as suicide vectors in any other gram-negative bacterium. The movement of the transposon can then be assayed if it carries a selectable gene, such as for antibiotic resistance, which is expressed in the recipient host. Some types of transposons are also very broad in their host range for transposition, so the transposon may hop in almost any recipient bacterium. Such methods are discussed in more detail later in this chapter (see "Transposon Mutagenesis").

THE MATING-OUT ASSAY FOR TRANSPOSITION

Transposition can also be assayed by using the "mating-out" assay, which is also based on conjugation. In this assay, a transposon in a nontransferable plasmid or the chromosome is not transferred into other cells unless it inserts into a plasmid that is transferable. Figure 9.8 shows a specific example of a mating-out assay with *E. coli*. In the example shown, transposon Tn*10* carrying tetracycline resistance has been inserted into a small plasmid that is neither self-transmissible nor mobilizable. This small plasmid is transferred by transformation (see chapter 6) into streptomycin-sensitive cells containing F, a larger, self-transmissible plasmid. While the cells are growing, the transposon may hop from the smaller plasmid into the F plasmid in a few of the cells. Later, when these cells are mixed with streptomycin-resistant recipient cells, any F plasmid into which the transposon hopped will carry the transposon when it transfers to a new cell, thus conferring tetracycline resistance on that cell. Transposition can be detected by plating the mating mixture on agar plates containing tetracycline and by including streptomycin to kill the donor cells (see chapter 3).

The appearance of antibiotic-resistant transconjugants in a mating-out assay is not by itself definitive proof of transposition. Some transconjugants could become antibiotic resistant by means other than transposition of the transposon into the larger, self-transmissible plasmid. The smaller plasmid containing the transposon could have been somehow mobilized by the larger plasmid, or the smaller plasmid could have been fused to the larger plasmid by recombination or by cointegrate formation (see below). A few representative transconjugants should be tested by DNA sequencing or other techniques to verify that they contain only the larger plasmid with the transposon inserted.

Figure 9.8 Example of a mating-out assay for transposition. See the text for details.
doi:10.1128/9781555817169.ch9.f9.8

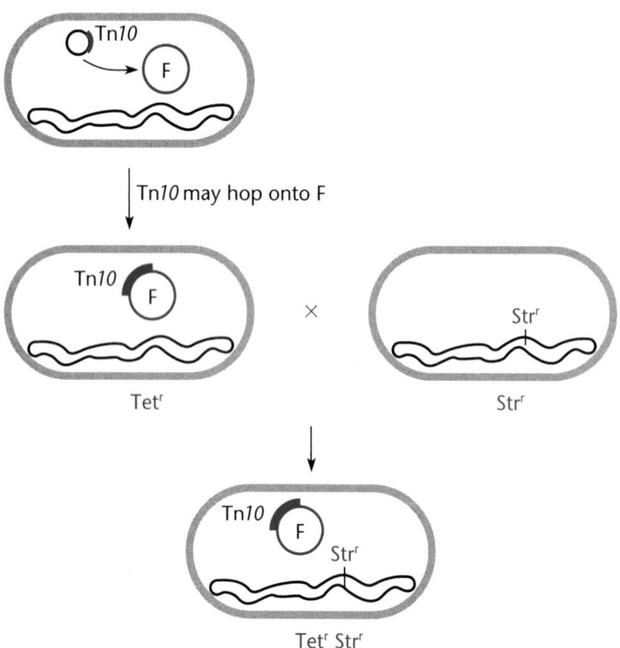

Mechanisms of Transposition

The process of figuring out how transposons move followed the usual course in molecular genetics. First, genetic analyses were done with certain selected transposons to identify the gene products and DNA sequences involved and to obtain an overview of the process. Then, the studies became more molecular, identifying the detailed molecular reactions required and determining the actual structures of the molecules involved and the ways in which the structures contribute to the transposition process. These studies of certain select transposons have revealed that transposons move by a number of different mechanisms, which are nevertheless conceptually related and often use related enzymes. We first review some of the earlier genetic studies on how some transposons move and then address the molecular details of these processes.

Genetic Requirements for Transposition of Tn*3*

The first analysis of the genetic requirements for transposition used the transposons Tn*3* and Mu, which happen to transpose by similar mechanisms. We will use Tn*3* as our primary example (see Gill et al., Suggested Reading; also see Figure 9.7 for a diagram of Tn*3*). The questions are essentially the same as for any other genetic analysis (see chapter 3). How many gene products are required for transposition of Tn*3*, and where are the sites at which they act? Where do the genes for these gene products lie on the transposon? Do any intermediates of transposition accumulate when one or more of these gene products are inactivated? Obtaining answers to these questions was the first step in developing a molecular model for transposition of Tn*3*.

ISOLATION OF MUTATIONS IN THE TRANSPOSON

As in any genetic analysis, the first step in analyzing the genetic requirements for transposition was to isolate mutations in the transposon. To facilitate the process, the element was mutagenized in a plasmid. These randomly placed mutations were tested for their effects on transposition by the mating-out assay. As illustrated in Figure 9.9, cells containing both a small nonmobilizable plasmid carrying the mutant Tn*3* and a larger, self-transmissible plasmid were mixed with recipient cells, and tranconjugants resistant to ampicillin were selected. Because the smaller plasmid was not mobilizable, ampicillin-resistant transconjugants could be produced only by donor cells in

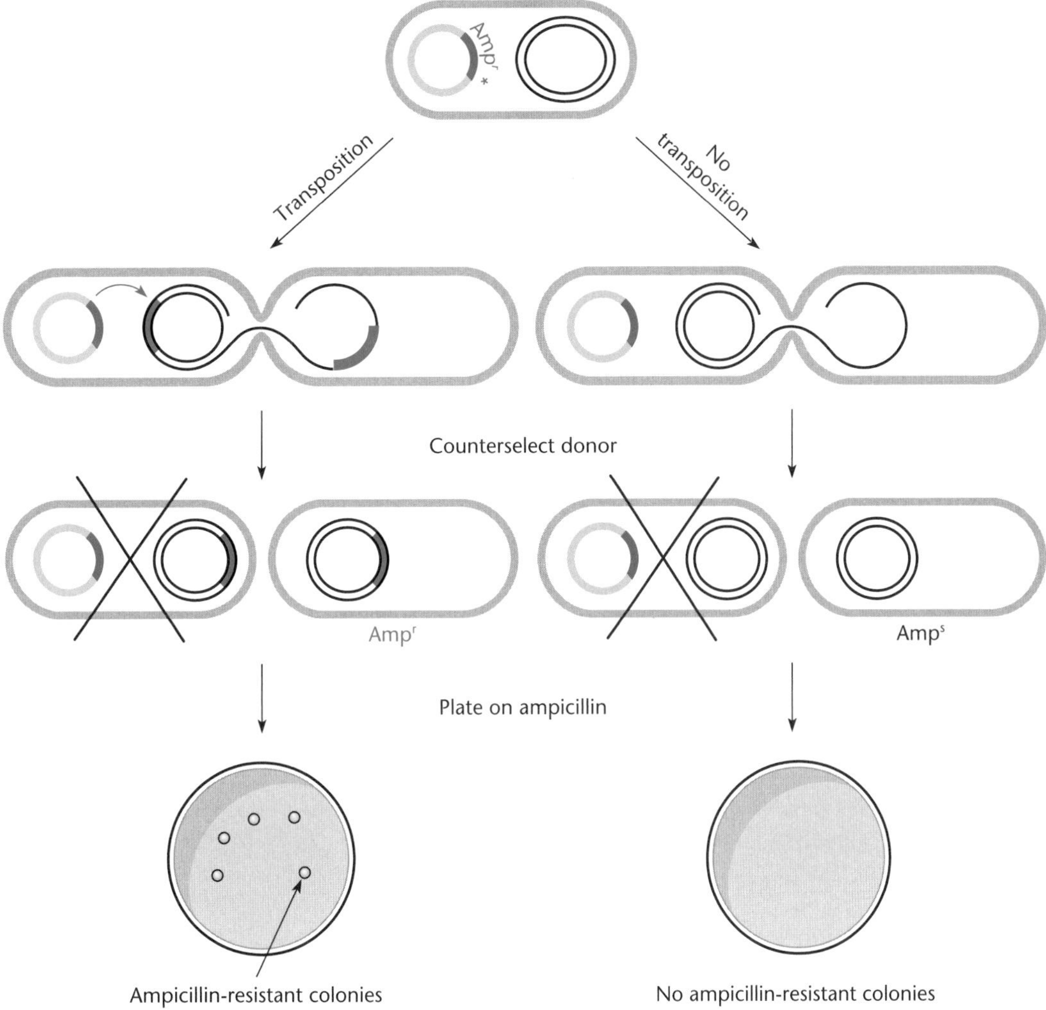

Figure 9.9 Molecular genetic analysis of transposition of the replicative transposon Tn*3*. The transposon is in blue. The asterisk marks the small nonmobilizable plasmid that was mutagenized. See the text for details. doi:10.1128/9781555817169.ch9.f9.9

which the transposon had hopped from the smaller plasmid into the larger, self-transmissible one, which was then transferred into the recipient cell. When no ampicillin-resistant transconjugants were observed, the mutation in the transposon must have prevented transposition into the self-transmissible plasmid. When larger-than-normal numbers of ampicillin-resistant transconjugants were observed, the mutation must have increased the frequency of transposition.

As expected, the effect of a mutation depended on its position in the transposon. Mutations in the IR sequences and mutations that disrupt the *tnpA* open reading frame (ORF) can totally prevent transposition (Figure 9.10). In contrast, mutations that disrupt the *tnpR* ORF result in higher-than-normal rates of transposition and the formation of cointegrates, in which the self-transmissible plasmid and the smaller plasmid, which originally contained the transposon, are now joined and are transferred together into the recipient strain. The cointegrate contains two copies of the transposon bracketing the smaller plasmid, as shown. Mutations in the short sequence called *res* (for *res*olution sequence) also give rise to cointegrates, but unlike *tnpR* mutations, they result in normal, not elevated, rates of transposition.

COMPLEMENTATION TESTS WITH TRANSPOSON MUTATIONS

The next step was to do complementation tests to determine which mutations in transposon Tn*3* disrupt *trans*-acting functions and which disrupt *cis*-acting sequences or sites (see chapter 3). The complementation tests used the same mating-out assay illustrated in Figure 9.9, except that the cell in which the transposition was to occur

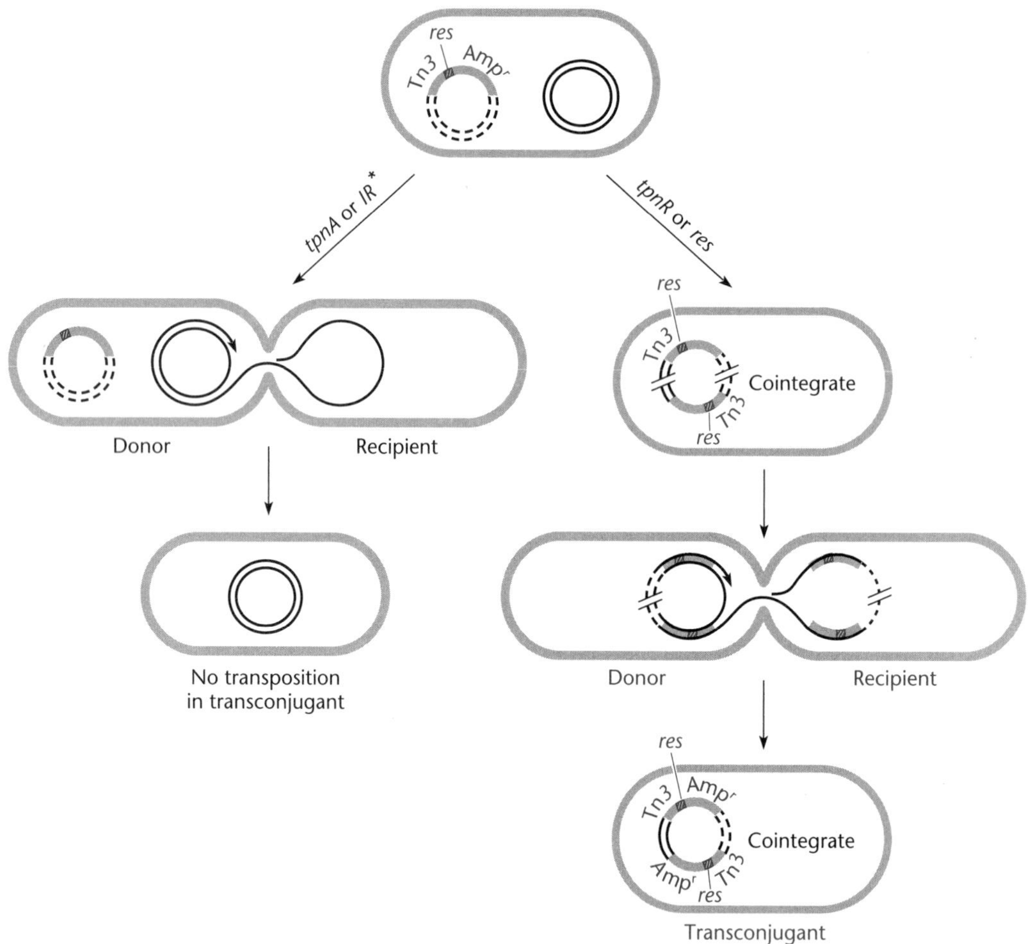

Figure 9.10 Effects of mutations in different genes required for transposition of Tn*3*. In the left-hand pathway, a *tpnA* or *IR* mutation prevents transposition, so no Amp^r transconjugants form. In the right-hand pathway, transposition by a *tnpR* or *res* mutant leads to the formation of Amp^r transconjugants that contain the mobilizable and self-transmissible plasmids fused to each other in a cointegrate. The asterisk indicates a mutation.
doi:10.1128/9781555817169.ch9.f9.10

also contained another Tn*3*-related transposon inserted into its chromosome (Figure 9.11). This other transposon was capable of transposition but lacked an ampicillin resistance (Amp^r) gene so that its own transposition did not create ampicillin-resistant transconjugants and confuse the analysis. The data are interpreted as in any other complementation test. If the mutation in the transposon in the plasmid inactivates a *trans*-acting function, it will be complemented by the corresponding gene in the transposon in the chromosome, and the transposon should be able to transpose. However, if the mutation in the plasmid transposon inactivates a *cis*-acting site, it will not be complemented and will not transpose properly, even in the presence of the chromosomal transposon, since mutations that inactivate *cis*-acting sites cannot be complemented. The complementation tests revealed that mutations that inactivate either the ORF called *A* or the ORF called *R* (Figure 9.7) could be complemented to give normal transposition. However, neither mutations in the IR sequences at the ends of the transposon nor those in the sequence called *res* could be complemented to give normal transposition. Mutations in an IR sequence prevented transposition altogether, even in the presence of the complementing copy of Tn*3*, while mutations in *res* permitted transposition but still gave rise to cointegrates. The investigators concluded that *tnpA* and *tnpR* encode *trans*-acting proteins while IR and *res* are *cis*-acting sites on the transposon DNA.

These genetic data prompted the formulation of a model for **replicative transposition** of Tn*3* and other Tn*3*-like transposons. Briefly, mutations in the *tnpA* gene prevent transposition because the *tnpA* gene encodes the transposase TnpA, which promotes transposition. Mutations in the IR elements at the ends of the

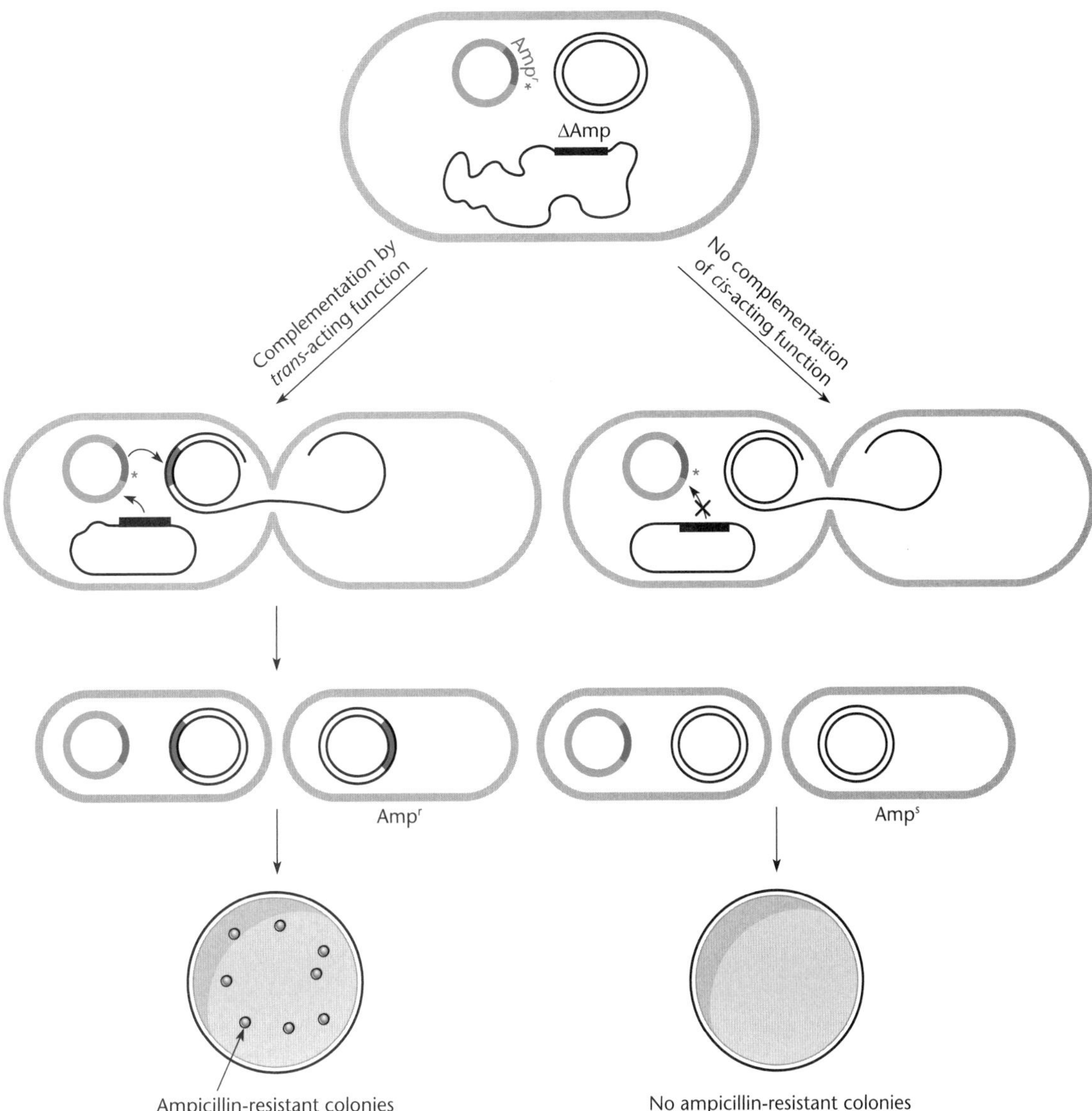

Figure 9.11 Complementation tests of transposition-defective Tn*3* mutations. The mutant Tn*3* transposon being complemented is in blue, and the asterisk indicates the mutation. The black bar indicates a Tn*3* element in the chromosome that is capable of producing proteins needed for transposition but lacks ampicillin resistance (ΔAmp).See the text for details. doi:10.1128/9781555817169.ch9.f9.11

transposon also prevent transposition, because these are the sites at which the TnpA transposase acts to promote transposition.

The behavior of mutations in *tnpR* and *res* was more difficult to explain. To reiterate, *tnpR* mutations are *trans* acting and not only cause higher-than-normal rates of transposition, but also cause the formation of cointegrates. Mutations in *res* also cause cointegrates to form but are *cis* acting and do not affect the frequency of transposition. To explain these results, the investigators proposed that *tnpR* encodes a protein with two functions. First, the TnpR protein acts as a repressor (see chapters 2 and 12) that represses the transcription of the *tnpA* gene for the transposase. By inactivating the repressor, *tnpR* mutations cause higher rates of transposition by allowing more TnpA synthesis. In addition to its role as a repressor, however, the TnpR protein acts as a recombinase that resolves cointegrates by promoting site-specific recombination between the *res* sequences in the two copies of the transposon in the cointegrate (Figure 9.12). This explains why both *tnpR* and *res* mutations cause the accumulation of cointegrates but only *tnpR* mutions can be complemented. Either type prevents the site-specific recombination that resolves the

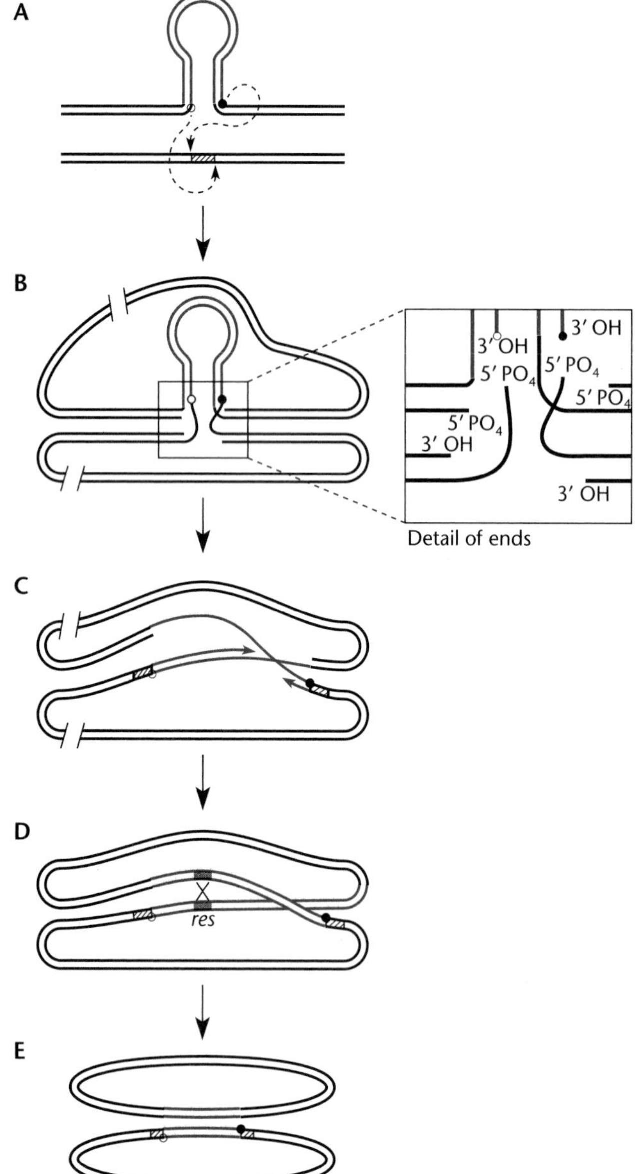

Figure 9.12 Replicative transposition of Tn*3* (blue outline) and formation and resolution of cointegrates. **(A)** Cleavage at the 3′ ends of the element occur in a concerted reaction with joins to staggered positions in the target DNA. **(B)** These joining events link the donor and target DNAs. The inset shows details of the boxed regions in panel B, with the 3′ OH and 5′ PO_4 that participate in the transactions carried out by the transposase. **(C)** The free 3′ ends are used to prime DNA replication that copies the top and bottom strands of the Tn*3* element (represented by the blue arrows). **(D)** The cointegrate is resolved by recombination promoted by the resolvase TnpR at the *res* sites (represented by the gray rectangles). **(E)** The final product is a Tn*3* element at a new position in a separate target DNA without loss of the original element in the donor DNA. The target site duplication that is formed in the reaction is indicated by a diagonally hatched rectangle. The solid and open circles indicate the original 3′ OHs at the transposon ends that were used in the joining reaction to the target DNA.
doi:10.1128/9781555817169.ch9.f9.12

cointegrates, causing cointegrates to accumulate, but only *tnpR* mutations can be complemented because only *tnpR* encodes a diffusible gene product.

A Molecular Model for Transposition of Tn*3*

The first detailed model to be developed for transposition attempted to explain all that was known about Tn*3* transposition. This model continues to be generally accepted for that type of transposon. The model incorporates the following observations, some of which have already been mentioned.

1. Whenever a transposon, such as Tn*3*, hops into a site, a short sequence of the target DNA is duplicated. The number of bases duplicated is characteristic of each transposon. (For Tn*3*, 5 bp is duplicated; for IS*1*, 9 bp is duplicated.)
2. The formation of a **cointegrate**, in which the donor and target DNAs have become fused and encode two copies of the transposon, is an intermediate step in the transposition process.
3. Once the cointegrate has formed, it can be resolved into separate donor and target DNA molecules either by the host recombination functions or by a transposon-encoded **resolvase** that promotes recombination at internal *res* sequences.
4. The donor DNA and target DNA molecules both have a copy of the transposon after resolution of the cointegrate. Therefore, the transposon does not actually move but duplicates itself, and a new copy appears somewhere else, hence the name "replicative transposition."
5. Neither transposition nor, for some transposons, resolution of cointegrates requires the normal recombination enzymes or extensive homology between the transposon and the target DNA. Special site-specific recombinases, such as TnpR, resolve the cointegrate by promoting recombination between specific sites in DNA, such as the *res* sites in Tn*3*.

Figure 9.12 shows the molecular details of the model for replicative transposition. The transposases used in the classically defined transposons (see "Details of Transposition by the DDE Transposons" below) are different from many other enzymes in that they break a DNA only through the process of joining it to something else. This fact is important when considering all of the DNA transactions that occur with replicative transposition (e.g., Tn*3*) and cut-and-paste transposition (e.g., Tn*5* and Tn*10*), as described below. In the first step of replicative transposition, the transposase catalyzes the strand transfer of the 3′ OH at the ends of the element to staggered positions in the target DNA, as indicated in Figure 9.12A and B. The inset indicates the actual 3′ OH and 5′ PO_4 that participate in the reactions. This reaction was introduced in Figure 9.3; however, in the case of replicative transposition, the 5′ ends of the element always remain attached to the donor DNA. In Figure 9.12, the positions of the two joining events are indicated by open and filled circles. After the 3′ ends of the element are joined to the staggered 5′ ends in the target DNA (Figure 9.12B), the 3′ OH ends in the target DNA are then available to initiate DNA replication. In some systems, this is known to require a special replication priming system normally involved in initiating DNA replication for repair (see chapter 10), and this is probably true for all replicative transposons. Replication then proceeds in both directions over the transposon (Figure 9.12C), and ligase is used at the 5′ ends of the donor DNA to complete the cointegrate. The last step, **resolution** of the cointegrate (Figure 9.12D), results from recombination between the two *res* sites in the cointegrate promoted by the resolvase of the transposon (see "S Recombinases: Mechanism" below). Resolution of the cointegrate gives rise to two copies of the transposon, one at the former (or donor) site and a new one at the target site (Figure 9.12E).

This model explained why cointegrates were obligate intermediates in replicative transposition. After replication has proceeded over the transposon in both directions, the donor DNA and the target DNA are fused to each other, separated by copies of the transposon, as shown. This model also explained why, after transposition, a short target DNA sequence of defined length had been duplicated at each end of the transposon, something that is true for all transposons.

Finally, this model explained why replicative transposition is independent of many host functions, including the normal recombination functions, such as RecA (see chapter 10). The normal recombination system is not needed to resolve the cointegrate into the original replicons, because the resolvase specifically promotes recombination between the *res* elements in the cointegrates. Although cointegrates can also be resolved by homologous recombination anywhere within the repeated copies of the transposon, the resolvase greatly increases the rate of resolution by actively promoting recombination between the *res* sequences.

Not all transposons that replicate by this mechanism resolve the cointegrates after they form. An interesting example of this involves the bacteriophage Mu. Phage Mu also provides a nice example of how evolution can mix and match different types of mobile DNA strategies to address the basic needs of a parasitic element (Figure 9.13). When the bacteriophage Mu replicates itself, it inserts itself around the chromosome of its bacterial host by a replicative mechanism using its transposase, the MuA protein, similar to the process carried out by TnA of Tn*3*. However, it does not resolve the cointegrates that form, and soon, the chromosome becomes loaded with Mu genomes. These genomes are then packaged directly from the chromosomal DNA into the phage head. Mu then goes though a typical lysis program to provide virulent particles to infect additional hosts. Interestingly, a slightly different process is carried out when the phage uses the same machinery to insert into the chromosome of a new host to form a lysogen (see below).

Transposition by Tn*10* and Tn*5*

Further evidence indicated that not all transposons transpose by the exact mechanism used by Tn*3* and Mu. Other transposons, such as the composite transposons, Tn*10* and Tn*5*, transpose by a **cut-and-paste** mechanism, in which the transposon is removed from one place and inserted into another, as illustrated in the simplified model in Figure 9.1. Cut-and-paste transposition is sometimes referred to as conservative transposition. In this simplified mechanism, the transposase makes double-strand breaks at the ends of the transposon, cutting it out of the donor DNA, and then pastes it into the target DNA at the site of a staggered break. The details of the mechanism with Tn*5* are indicated below, but as with all transposons, staggered joining events to the target DNA result in target site duplication (Figure 9.3). These elements leave a DNA double-strand break in the donor DNA, which is probably almost always repaired

Figure 9.13 Representation of bacteriophage Mu showing components from its bacteriophage lifestyle and its integration and replication strategy via transposition. See the text for details. doi:10.1128/9781555817169.ch9.f9.13

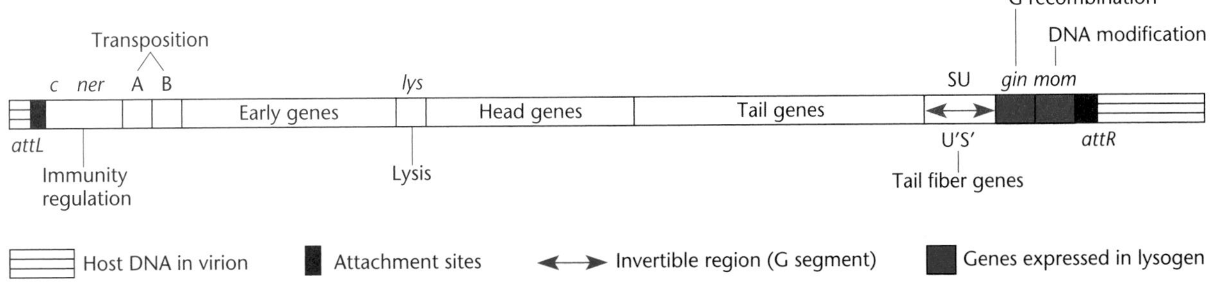

using homologous recombination with a sister chromosome. This is likely the major reason why Tn*5* and Tn*10* regulate transposition so that it is stimulated immediately following replication of the element (see below).

GENETIC EVIDENCE FOR CUT-AND-PASTE TRANSPOSITION

In the next few sections, we describe in detail some of the early evidence for cut-and-paste transposition by Tn*10* and how it can be differentiated from replicative transposition.

No Cointegrate Intermediate

In Tn*10* transposition, cointegrates do not form as a necessary intermediate, as they do in the replicative mechanism. This conclusion is supported by indirect evidence. For example, there are no mutants of transposon Tn*10* and Tn*5* that accumulate cointegrates, as there are for Tn*3*. Moreover, even if cointegrates are formed artificially by recombinant DNA techniques, there is no evidence that the cointegrates can be resolved except by the normal host recombination system. Therefore, these transposons do not seem to encode their own resolvases, which they would be likely to do if cointegrates were a normal intermediate in their transposition process.

Both Strands of the Transposon Transpose

The primary difference between replicative and cut-and-paste transposition is that in the latter, both strands (i.e., the top and bottom strand at each end) of the transposon move to the target DNA. The results of genetic experiments with Tn*10* (outlined in Figure 9.14) supported this conclusion (see Bender and Kleckner, Suggested Reading).

The first step in these experiments was to introduce different versions of transposon Tn*10* into a λ suicide vector. Both of the Tn*10* derivatives contained a copy of the *lacZ* gene, as well as the Tet^r gene usually carried by Tn*10*. However, one of the Tn*10* derivatives carried three missense mutations in the *lacZ* gene to inactivate it. The DNA of the two λ::Tn*10* derivatives was mixed, and the strands of the two λ DNAs were separated and reannealed. Some of the strands would reanneal with a strand of the DNA of the other derivative to make heteroduplex DNA, in which each of the strands came from a λ phage carrying a different derivative of Tn*10*. Consequently, these heteroduplex DNAs had one strand with a good copy of *lacZ* and another strand with the mutated copy of *lacZ*. In the next step, the heteroduplex DNA was packaged into λ heads in vitro (see chapter 8) and used to infect Lac$^-$ *E. coli* cells. Because this λ was a suicide vector, the only cells that became Tet^r were ones in which the Tn*10* derivatives had hopped into the chromosome. If the transposition had occurred by a replicative

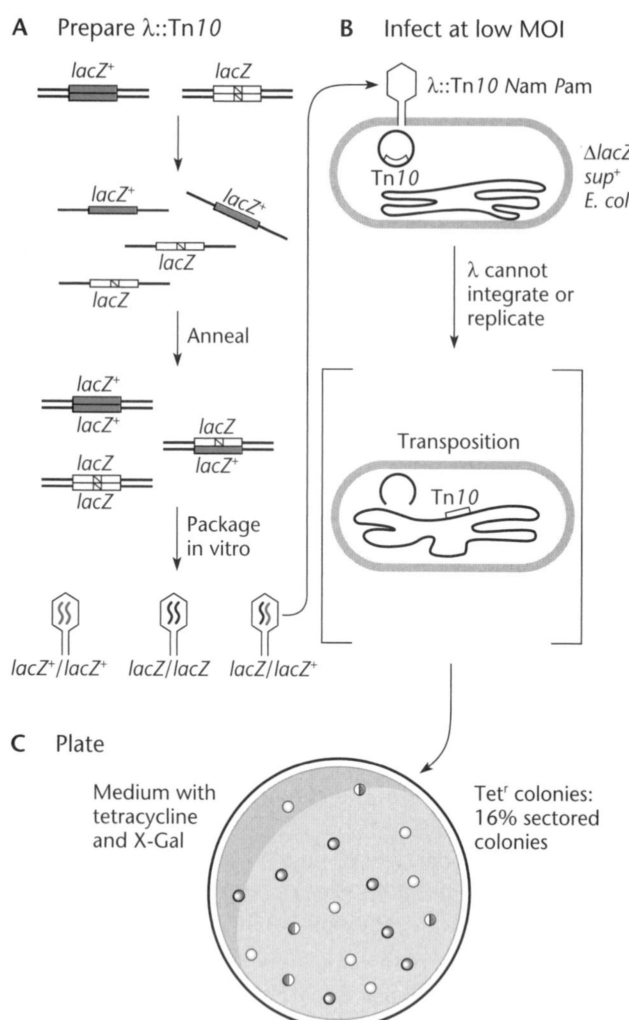

Figure 9.14 Genetic evidence for nonreplicative transposition by Tn*10*. **(A)** Preparation of λ::Tn*10* *lacZ*/*lacZ*$^+$ heteroduplex DNA. **(B)** λ::Tn*10* infects a nonsuppressor (sup$^+$) *E. coli* host. Because λ contains *Nam* and *Pam* mutations, it cannot integrate or replicate, and the cells become Tet^r only if the transposon hops. If the transposon replicates during transposition, the bacteria in the Tet^r colonies will get only one or the other strand of DNA in the heteroduplex, and the colonies will be all blue or all colorless. If both strands are transferred, some colonies will be sectored, part blue and part white. See the text for further details. MOI, multiplicity of infection.
doi:10.1128/9781555817169.ch9.f9.14

mechanism, the Tet^r colonies would have contained either Lac$^+$ or Lac$^-$ bacteria and not both (Figure 9.15), since the information in only one of the two strands could have been transferred. If, however, the transposition had occurred by a nonreplicative cut-and-paste mechanism, both strands of the Tn*10* from a heteroduplex would have hopped into the chromosome some of the time, so that one of the strands would have the good copy of the *lacZ* gene and the other would have the *lacZ* gene with the mutations. When these heteroduplex

DNAs replicated, they would give rise to both Lac$^+$ and Lac$^-$ bacteria in the same colony, making "sectored" blue-and-white colonies on plates containing 5-bromo-4-chloro-3-indolyl-β-D-galactopyranoside (X-Gal), as shown in Figure 9.14. In the experiment, about 16% of the colonies were sectored blue and white, supporting the conclusion that both strands were transferred.

Transposon Leaves the Donor DNA

A major difference between replicative and simple cut-and-paste transposition is the number of copies of the transposon after transposition. In the replicative mechanism, a new insertion occurs in the cell, but in addition, a copy of the element remains in the donor DNA, where the element was originally positioned. In the cut-and-paste mechanism, there is no increase in the number of transposons by the act of transposition itself. This is difficult to demonstrate genetically, because double-strand break repair (see chapter 10) is extremely efficient in bacteria, and the sister chromosome used for template repair contains the transposon, as mentioned above. However, this could be demonstrated by an increase in homologous recombination in the *lac* locus upon excision of a cut-and-paste element (see Hagemann and Craig, Suggested Reading). While most repair occurs via recombination, at a very low frequency, the site of excision can be repaired in a way that reestablishes the wild-type allele, something that must be selected for using reversion studies. For a transposon insertion mutation to revert, the transposon must be completely removed from the DNA in a process called **precise excision**. Not a trace of the transposon can remain, including the duplication of the short target sequence, or the gene would probably remain disrupted and nonfunctional.

If the transposon were precisely excised and the donor DNA were resealed every time a transposon moved by a cut-and-paste mechanism, transposon insertion mutations would revert every time the transposon moved. However, reversion of insertion mutations occurs at a much lower rate than does transposition itself, suggesting that the insertion mutation does not revert every time the transposon hops. Moreover, mutations in the transposon itself that inactivate the transposase and render the transposon incapable of transposition do not further lower the reversion frequency, as might be expected if the few revertants that are seen resulted from a transposition event. Presumably, the rare precise excisions that

Figure 9.15 Comparison of the results predicted in the experiment in Figure 9.14 if transposition of Tn*10* is by a cut-and-paste mechanism **(1)** or by a replicative mechanism **(2).** See the text for details. doi:10.1128/9781555817169.ch9.f9.15

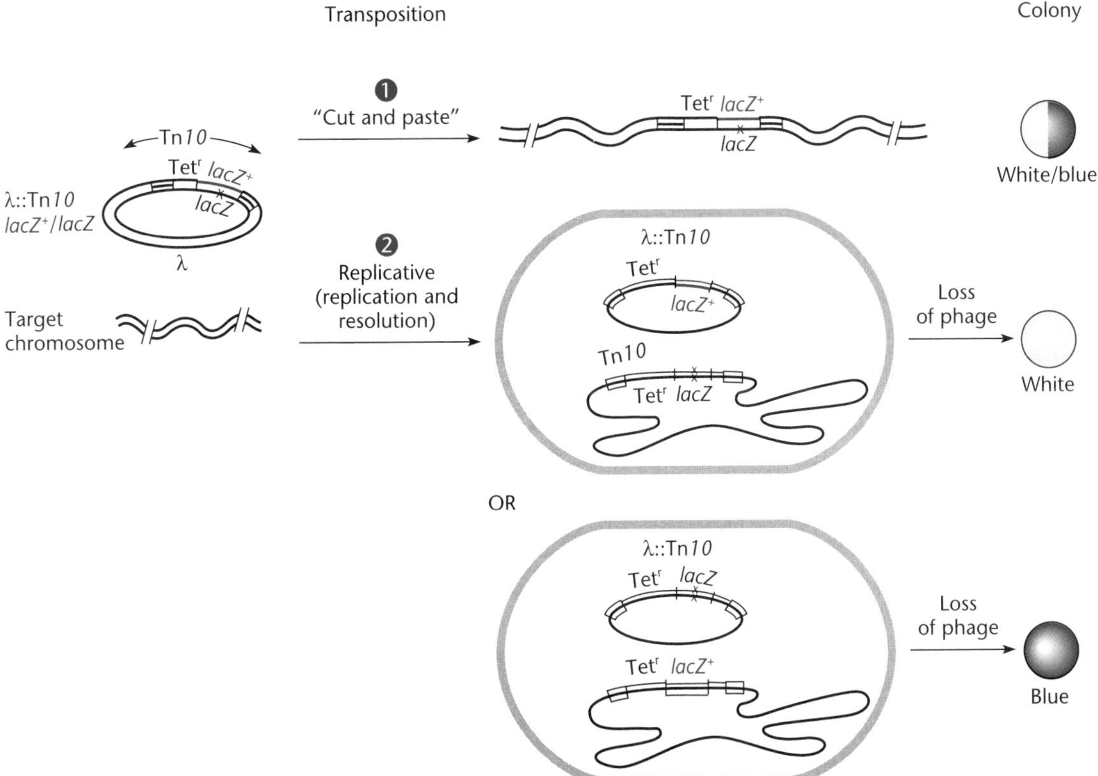

cause transposon insertion mutations to revert are due to rare recombination events using the microhomology between the short duplicated target sequences bracketing the transposon and are unrelated to transposition itself.

Details of Transposition by the DDE Transposons

All of the transposons we have discussed so far are considered **DDE transposons**, because their transposases all have two aspartates (D) and one glutamate (E) (see inside front cover) that are essential for their activity. These acidic amino acids are not next to each other in the polypeptide, but they are together in the active center when the protein is folded. Their job is to hold (by chelation) two magnesium ions (Mg^{2+}) that participate in the cleavage of phosphodiester bonds in the DNA during the transposition event. A similar structure is found for some other related enzymes, such as the human immunodeficiency virus integrase, the RAG-1 protein responsible for generating antibody diversity in vertebrates, and RuvC, the enzyme that cuts Holliday junctions during recombination (see chapter 10). However, the details of how many DNA strands are cut and the fates of the ends are different for the different enzymes.

Details of the Mechanism of Transposition by Tn*5* and Tn*7*

The mechanism of transposition of the DDE transposon Tn*5* has been studied extensively and is illustrated in Figure 9.16 (see Reznikoff, Suggested Reading). The first step is the binding of one copy (monomer) of the transposase to each of the ends of the transposon in the donor DNA. The two monomers then bind each other through dimerization domains in their carboxy termini to bring the two ends of the transposon together (synapsis). Then, the transposase bound to one end of the transposon breaks the DNA at the other end, and vice versa, to leave a reactive 3′ OH at each end of the transposon. These 3′ OH ends attack the phosphodiester bond on the other strand, forming 3′-5′ phosphodiester hairpins, as shown. This cuts the transposon out of the donor DNA. When the transposase binds to the target DNA, it breaks the two hairpin ends again, and the 3′ OH ends attack phosphodiester bonds 9 bp apart in the target DNA, cutting them and allowing the 5′ phosphate ends in the target DNA to join to the 3′ OH ends in the transposon, which inserts the transposon into the target DNA. The 9-bp single-stranded gaps on each side of the transposon are then filled in by DNA polymerase to make the 9-bp repeats in the target DNA characteristic of the Tn*5* transposon.

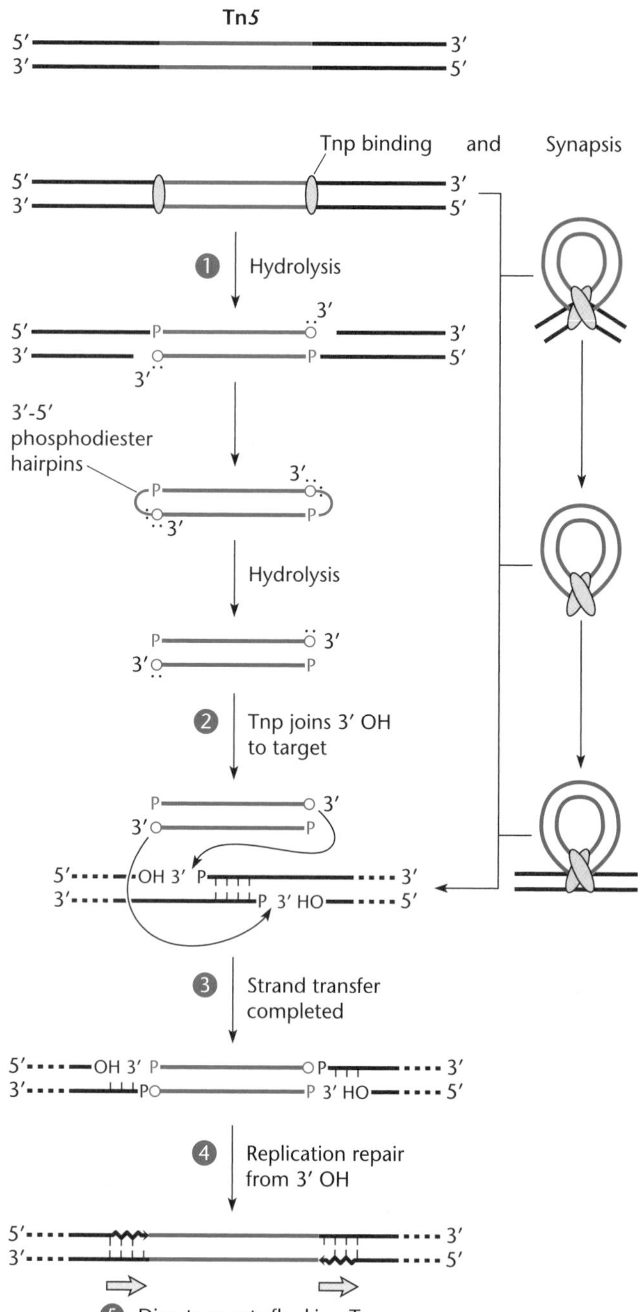

Figure 9.16 Mechanism of transposition by Tn*5*. Single copies of the transposase (Tnp) bind to each end of the transposon and then bind to each other, bringing the two ends of the transposon together (synapsis). The breaking-and-joining reaction cuts the transposon out of the donor DNA to form hairpins at the ends of the element. These hairpins are subsequently broken and joined at positions 9 bp apart in the target DNA. Replication completes the 9-bp target site duplications that flank the element. See the text for details.
doi:10.1128/9781555817169.ch9.f9.16

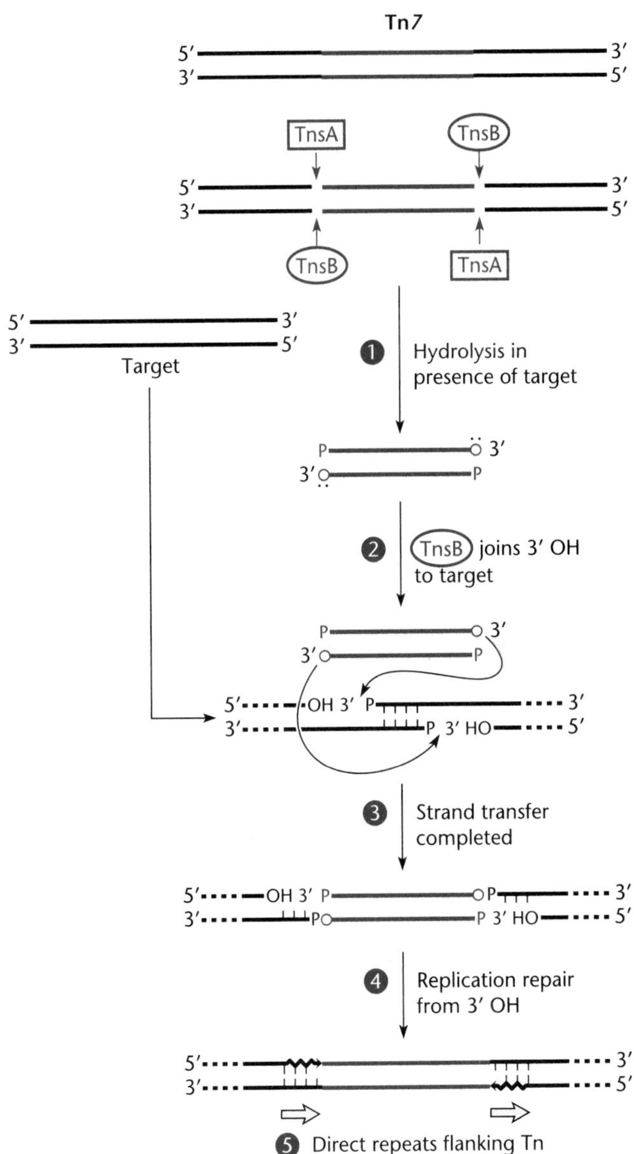

Figure 9.17 Transposition by Tn*7*. The TnsA and TnsB proteins are required for removing the element from the target DNA. TnsA nicks at the 5′ ends, and TnsB makes breaks at the 3′ ends of the element as they are joined at positions 5 bp apart in the target DNA. Replication completes the 5-bp target site duplications that flank the element. See the text for details. doi:10.1128/9781555817169.ch9.f9.17

RELATIONSHIP BETWEEN REPLICATIVE AND CUT-AND-PASTE TRANSPOSITION

As hinted at before, although replicative transposition and cut-and-paste transposition by DDE transposons seem different, they are actually mechanistically the same. Only the fates of the 5′ ends of the element are different in the two cases. In cut-and-paste transposition, both strands on each end of the element are released from the donor DNA. As shown in Figure 9.16, this occurs through joining events made to a nearby position across the DNA backbone, forming hairpins at the ends of the element. Insertion of the excised transposon occurs as a second event when the hairpins are opened as the element is joined to a target DNA (Figure 9.16). Therefore, by doing the breaking-and-joining reaction twice, transposase fully removes the element from the donor DNA. In replicative transposition, the joining event occurs directly to the target DNA with no intermediate step with the transposase but instead a separate series of processing events that involve extensive DNA replication and site-specific recombination (see below) to free the target and donor DNAs (Figure 9.12).

A dramatic confirmation of the similarity between the cut-and-paste and replicative mechanisms of transposition by DDE transposons came with the demonstration that the cut-and-paste transposon Tn*7* can be converted into a replicative transposon by a single amino acid change in one subunit of the transposase (see May and Craig, Suggested Reading). As defined genetically and biochemically, transposon Tn*7* normally transposes by a cut-and-paste mechanism in which different subunits of the transposase make the breaks in the opposite strands of DNA at the ends of the transposon, leaving a double-stranded DNA break in the donor DNA. At the most basic level, steps in Tn*7* transposition are most similar to Tn*3* and Mu transposition, where the transposon ends are joined directly to the target DNA (Figure 9.17). However, Tn*7* normally uses a second recombinase, an additional subunit of the transposase called TnsA, to make nicks at the 5′ ends of the element to prevent the formation of a cointegrate, as is found for Tn*3* transposition. Support for the unity of transposition came from studies where the TnsA active site was inactivated. While TnsA must be part of the transposase for any breaking-and-joining event to occur, it was found that mutating the active site in TnsA allows transposition, but the transposition events occur via a replicative mechanism involving extensive DNA replication, similar to what is found with Tn*3*.

Bacteriophage Mu also provides an example of the unity in the recombination processes handled by DDE transposases in the switch between cut-and-paste transposition and replicative transposition. As described above, Mu replicates in the host during its lytic cycle using replicative transposition. However, this process would not be useful when the bacteriophage first enters the host, when a cut-and-paste mechanism is needed to insert the phage into the host genome. Interestingly, the explanation for how this occurs shows similarities to what is known about the TnsA protein of Tn*7*. For bacteriophage Mu, it was found that a previously identified cryptic endonuclease activity in the Mu transposase is capable of cleaving off the DNA segment on the 5′ end of the element that is transferred from the donor site (in this case, from the previous host bacterium) (see Choi and Harshey, Suggested Reading). The endonuclease activity identified in the transposase is highly regulated and is the focus of ongoing investigation.

It is somewhat surprising that the same transposase enzymes can support both cut-and-paste and replicative transposition, given that different cellular replication machineries are needed for each. Replication of tens of thousands of base pairs during replicative transposition would be a much more involved process than replication of a few base pairs during cut-and-paste transposition.

Transposition Processes Not Involving the Canonical DDE Transposase Reaction

Other DDE transposons use a mechanism of transposition that cannot be described as either a strictly replicative or cut-and-paste mechanism. For example, the mechanism of transposition of the IS elements IS*2* and IS*3*, as well as IS*911*, has features of both replicative and cut-and-paste transposition. Basically, one strand of the transposon is cut out of the donor DNA. The ends of this strand are then joined to form a single-stranded circle (not illustrated), and the circular strand is replicated to form the double-stranded circular transposon. This double-stranded circular transposon is linearized at the transposon ends in the circular species, which can then be used in a more standard integration event. The strand cut out of the donor DNA is also replaced by replication, leaving a copy of the transposon in the donor DNA. Thus, this transposition is replicative because a copy of the transposon appears in the target DNA, but the donor DNA retains the transposon. However, a cointegrate does not form, and the transposon is, in a sense, cut out of the donor DNA and pasted into the target DNA.

Rolling-Circle Transposons

Not all transposons transpose by a strand exchange mechanism like that used by the DDE transposons. Other transposons, represented by IS*91*, are called **rolling-circle transposons** because they use a rolling-circle mechanism to transpose themselves into a target DNA. Rather than having the DDE motif, their transposase has two essential tyrosines in its active center, hence the alternate name **Y2 transposons**. We have encountered this form of replication in previous chapters as the mechanism of replication of some plasmids (see chapter 4) and phage DNAs (chapters 7 and 8) and the mechanism of strand displacement during DNA transfer in conjugation (see chapter 5). In all of these cases, the responsible protein has a tyrosine to which the 5′ phosphate at the end of the DNA is covalently joined during the replication or transfer process. Figure 9.18 illustrates the difference between DDE transposons and rolling-circle, or Y2, transposons.

The structure of Y2 transposons is also very different from that of DDE transposons, reflecting their very different mechanism of transposition. Y2 transposons do not have IR sequences at their ends, nor do they duplicate a target DNA sequence during integration. The details of how they transpose are not completely known, but basically, they cut one strand of the DNA close to one end of the transposon (called the *ori* end in analogy to the *ori* sequence of rolling-circle plasmids) and attach the 5′ phosphate at the cleavage site to one of the tyrosines in the active center of the transposase. The free 3′ OH end then serves as a primer to replicate over the transposon, ending at the other end of the transposon, called the *ter* end. The displaced old strand of the transposon enters the target DNA, and its complementary strand is synthesized in the target DNA, so that both the donor and target DNAs end up with a copy of the transposon. This is therefore a form of replicative transposition. It is not clear how integration into target DNA occurs in this process, nor are the exact roles of the two tyrosines known, since the other examples of rolling-circle replication mentioned above require only one tyrosine in the active center. The free 5′ phosphate ends created at the two ends of the transposon are presumably shuttled between the two tyrosines to allow replication back over the transposon to create two copies of the transposon.

It is becoming increasingly clear that a number of antibiotic resistance genes found on integrons are carried on Y2 transposons related to IS*91*. They were first identified because of common sequence elements, which led to them being called **IS*CR* elements**, for IS common regions. Therefore, rolling-circle transposons are another common way in which antibiotic genes move from one bacterium to another (see Toleman et al., Suggested Reading).

Y and S Transposons

Other transposon-like DNA elements exist that use neither a DDE transposase nor a rolling-circle transposase to transpose. They are sometimes called Y and S transposons because they have either a single essential tyrosine (Y) or serine (S) in the active center of their transposase. However, these transposases are more akin to integrases than they are to transposases, even though they often show less specificity in their integration sites. Accordingly, they are discussed along with integrases and other recombinases under "Site-Specific Recombination"

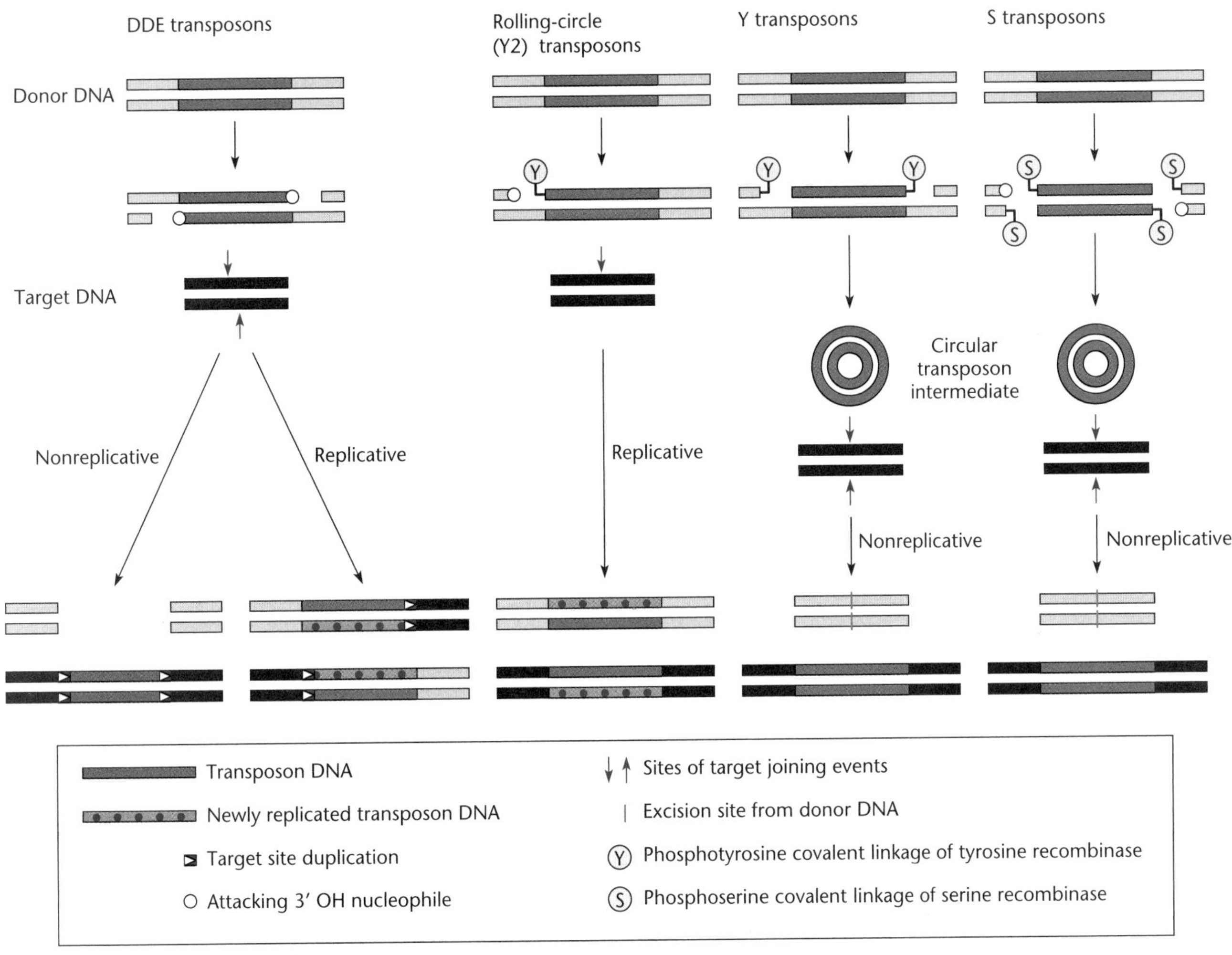

Figure 9.18 Comparison of multiple known mechanisms of transposition in bacteria. They differ in the initial strand cleavage, whether and how the transposon DNA is transiently attached to the transposase, the role of DNA replication, the existence of circular intermediates, whether a target site is duplicated, and the fate of the donor DNA. The Y transposons and S transposons use mechanisms more akin to Y and S recombinases.
doi:10.1128/9781555817169.ch9.f9.18

below. Figure 9.18 gives an overview of all of the known types of transposons, and Table 9.1 summarizes some of their distinctive properties.

General Properties of Transposons

There are some properties that are shared by many types of transposons, even if they differ in their mechanisms of transposition.

Target Site Specificity

While the transposition of some elements seems almost totally random, no transposable element inserts completely randomly into target DNA. Most transposable elements show some target specificity, inserting into some sites more often than into others. Even Tn*5* and Mu, which are famous for inserting almost at random, prefer some sites to others, although the preference is weak.

A bias for certain DNA sequences is not the only thing that affects where a transposon inserts in the bacterial genome. Transposons are adapted to their host and in a number of cases appear to avoid DNA involved in other cellular processes, such as highly transcribed genes. Transposons can also be attracted to DNA involved in certain cellular processes. As indicated below, Tn*7* preferentially inserts into DNAs undergoing active replication in some settings. Tn*917* shows a strong bias for the region where DNA replication terminates in many Gram-positive bacteria (see Shi et. al., Suggested Reading). The ability to target certain DNA processes presumably indicates an adaptation for the element, but on a practical level, these types of biases need to be

Table 9.1 Characteristics of transposon families

Family	Characteristic			Example(s)
	Active-site category	Protein-DNA covalent linkage	Target duplication	
DDE transposons	DDE	No	Yes	Tn*3*, Tn*5*, Tn*7*, Tn*10*, Mu
Rolling-circle/Y2 transposons	YY (related to ϕX174 A protein and to conjugative-plasmid relaxases)	Yes, to 5′ P	No	IS*91*
Y transposons	Y recombinase	Yes, to 3′ P	No, but flanking "coupling sequences" are transferred to one side of target	Tn*916* (see chapter 5)
S transposons	S recombinase	Yes, to 5′ P	No	IS*607* from *H. pylori*

taken into account when choosing an element as a random insertion mutagen (see below).

Tn7 is the extreme case of a transposon with target specificity and selectivity. Tn7 utilizes five proteins for transposition, TnsA, TnsB, TnsC, TnsD, and TnsE, which provide two pathways of transposition. One pathway recognizes a single site found in bacteria, called *attTn7*, and a second pathway recognizes actively conjugating plasmids. As described above, TnsA and TnsB make up the transposase that breaks and joins the DNA strands (Figure 9.17), while the other proteins play roles in regulation and targeting. The TnsD protein identifies *attTn7* as a target site where its binding induces a change in the structure of the DNA that helps to recruit the regulatory protein TnsC to establish the target complex. Along with host factors, the TnsD-*attTn7* complex is capable of conveying a signal to the TnsAB transposase to initiate Tn7 transposition. Insertion into the *attTn7* site may be advantageous, because it is a neutral position in *E. coli*. The neutrality of the site likely explains why this transposition pathway can occur at a frequency that is about 1,000-fold higher than that of other transposable elements. TnsD likely evolved to recognize this particular sequence because it is found in a highly conserved portion of an essential gene in bacteria. While an essential gene is recognized, the actual insertion event occurs 3′ to the gene in the transcriptional terminator for the operon, so insertion at that site has no effect on the function of the gene. Tn7 has a second targeting pathway for transposition that utilizes the Tn7 protein TnsE. This pathway is largely cryptic but is activated in the presence of actively conjugating DNA, where transposition events are directed into the conjugal plasmid. Activation involves an interaction with the β sliding-clamp subunit of DNA polymerase III (see Parks et. al., Suggested Reading), but how TnsC is recruited to the complex and how transposition shows such a high preference for replication found during conjugation are still being investigated. Having two complementary pathways of transposition appears to be beneficial, given that Tn7 elements are very broadly distributed across diverse bacteria in disparate environments around the world (see Parks and Peters, Suggested Reading). The existence of two pathways of transposition also allows Tn7-like elements to form genetic islands in bacteria, where horizontally transferred genes accumulate at a specific position in the genome. This is likely to be because Tn7 elements deliver genes to this single position in chromosomes, using the TnsD pathway in a process that is facilitated by the TnsE pathway, which preferentially targets mobile plasmids.

Effects on Genes Adjacent to the Insertion Site

Most insertion element and transposon insertions cause polar effects if they insert into a gene transcribed as a polycistronic mRNA (see chapter 2). The inserted element contains transcriptional stop signals and may also contain long stretches of sequence that are transcribed, but not translated. The latter may cause Rho-dependent transcriptional termination, which prevents transcription of genes located downstream of the insertion site in the target DNA.

In contrast to negative effects on the expression of downstream genes, some insertions may enhance the expression of a gene adjacent to the insertion site. This expression can result from transcription that originates within the transposon. For example, both Tn*5* and Tn*10* contain outward-facing promoters near their termini, and these promoters can initiate transcription into neighboring genes.

Regulation of Transposition

Transposition of most transposons occurs rarely, because transposons self-regulate their transposition activity (see Gueguen et al., Suggested Reading). The regulatory mechanisms used by various transposons differ greatly. In Tn*3*, the TnpR protein represses the transcription of the transposase gene. For some transposons, such as Tn*10*, transposition occurs very rarely and then primarily just after a replication fork has passed through the element. Newly replicated *E. coli* DNA is hemimethylated at GATC sites (see chapter 1), and hemimethylated DNA not only activates the promoter of the transposase gene of Tn*10*, but also increases the activity of the transposon ends. Also, the translation of the transposase gene of Tn*10* is repressed by an antisense RNA (see chapter 12). The transposase of Tn*5*, which already carries out only a low frequency of transposition, uses a truncated version of the transposase to further inhibit the active transposase. As illustrated in Figure 9.19, the translation of the truncated transposase is initiated at an internal translational initiation region (TIR), so it lacks the N terminus but has the C terminus involved in dimer formation. When this defective transposase pairs with the normal transposase, the hybrid transposase is inactive. Most transposons employ similar mechanisms to modulate the level of transposase transcription and/or translation, as well as the level of catalysis. Interestingly, in the case of Tn7, transcription and translation of the products required for transposition do not control the frequency or targeting of transposition; instead, the availability of the target site (*attTn7* and conjugal DNA replication) and various host proteins serves as the cue for the highly regulated transposition found with this element.

Regulating transposition with DNA replication is a common theme across all domains of life. For cut-and-paste transposons, one benefit of timing transposition to occur immediately after replication of the element is that the presence of a sister chromosome allows repair of the DNA double-strand break that is left in the donor DNA (Figure 9.20) (see chapter 10). A considerable additional advantage of this type of repair is that the allele in the sister chromosome also has a copy of the element, so there is a net gain in copy numbers even though the process of cut-and-paste transposition is itself a conservative process (Figure 9.20).

Figure 9.19 Regulation of Tn*5* transposition. Two similar IS*50* elements flank the antibiotic resistance genes. Only IS*50*R encodes the transposase Tnp and the inhibitor, which is an N terminally truncated version of itself. Also, Dam methylation of the inside ends (IE) of the IS*50* elements prevents the transposase from cutting these ends and transposing the individual IS*50* elements. See the text for details. OE, outer ends; Inh, inhibitor of transposase Tnp. The dashed lines indicate that Tnp and Inh made from IS*50*L are defective.
doi:10.1128/9781555817169.ch9.f9.19

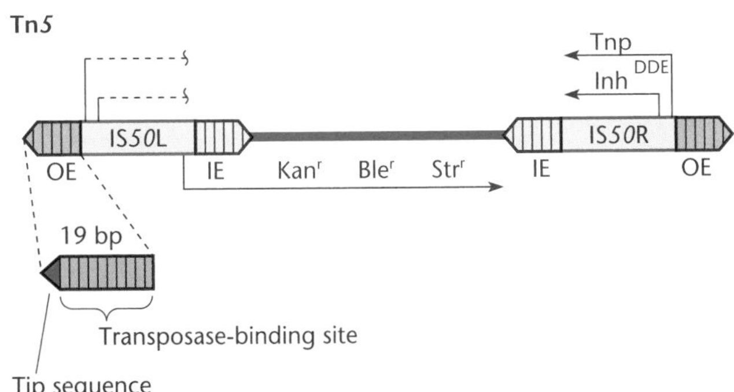

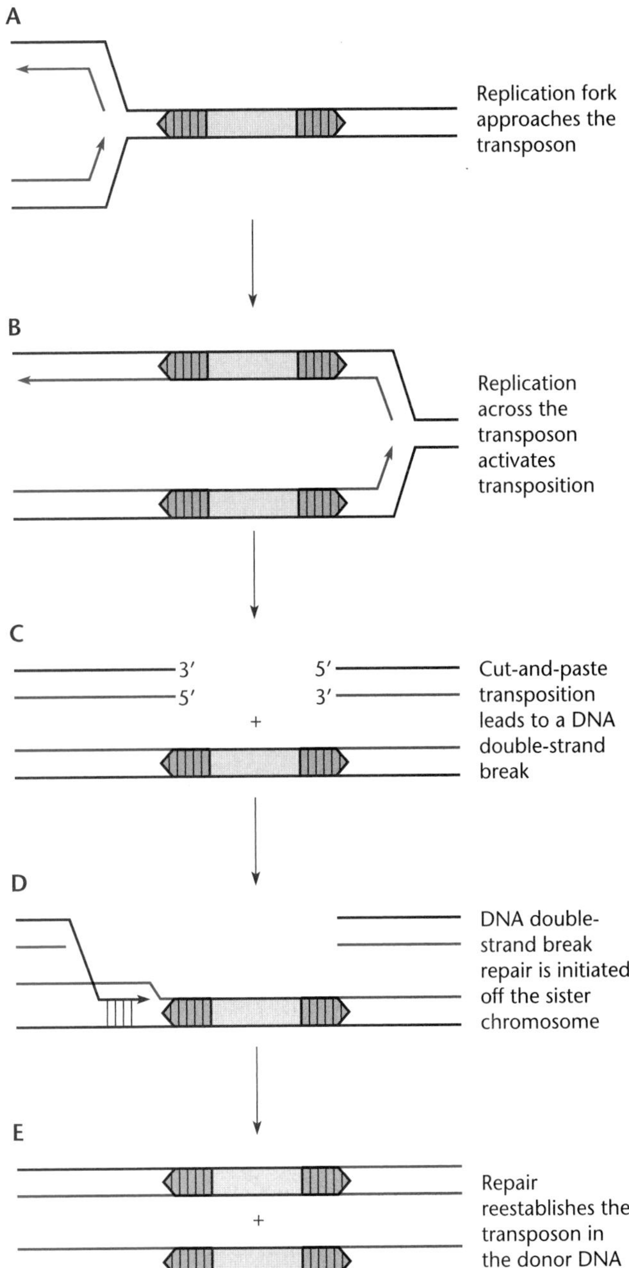

Figure 9.20 Transposition after DNA replication facilitates DNA repair. **(A and B)** For multiple transposons, DNA replication stimulates transposition of the element while producing a copy of the element. **(C and D)** When the element is lost from one sister chromosome through cut-and-paste transposition, repair of the DNA break can occur by recombination and DNA replication. **(E)** This process reestablishes the element in the donor DNA through the repair process.
doi:10.1128/9781555817169.ch9.f9.20

Target Immunity

Another feature of some transposons is that they prefer not to transpose to a site in the DNA close to another transposon of the same type. This is called **target site immunity**, because DNA sequences close to a transposon in the DNA are relatively immune to insertion of another copy of the same transposon. The immunity can extend over 100,000 bp of DNA, although its reach varies between transposons. There are likely many advantages for the host and transposon to having target site immunity. If two transposons were to insert close to each other, the resolution of the two copies by the transposon resolvase or homologous recombination between the two copies of the transposon would cause large deletions and often lead to death of the cell. One benefit of a larger transposon is that the process of immunity would discourage the insertion of the element into the "sister" element on the duplicated copy chromosome after DNA replication, which would likely destroy one, if not both, elements.

Target site immunity is limited to only some transposons of the types we have discussed, including the Mu, Tn*3* (Tn*21*), and Tn7 families of transposons. The mechanism of target site immunity has been addressed at the molecular level with Mu and Tn7, where certain themes may exist but the underlying mechanisms differ. A central paradox in target site immunity stems from the fact that the same proteins that are involved in carrying out transposition are also involved in an interaction that discourages nearby transposition. In the case of Tn7, an interaction between TnsB and TnsC discourages TnsC from establishing a transposition complex with TnsD or TnsE anywhere near the element in the genome. In the case of the TnsD-mediated pathway of transposition, target site immunity discourages multiple insertion events from occurring at the *attTn7* site from the same Tn7 element; this is likely to be essential because of the very high frequency of transposition found with this pathway. In the case of TnsE-mediated transposition, transposition is known to be discouraged over an entire F plasmid when a single Tn7 insertion resides in the plasmid. This could be important, because conjugal plasmids are preferentially targeted by TnsE-mediated transposition, but multiple Tn7 insertions could also destabilize these plasmids. The ability of TnsB to bind to the IR sequences is also critical for immunity (as is true for the MuA protein in Mu). TnsB binding to these sequences may cause a high concentration of TnsB around the element as it cycles on and off the DNA, which prevents TnsC from forming an active target complex with TnsD or TnsE near the element. An interaction between TnsB and the other subunit of the transposase, TnsA, seems to channel TnsC into an active target complex in regions where no Tn7 element resides (see Skelding et al., Suggested Reading).

Transposon Mutagenesis

One of the most important uses of transposons is in transposon mutagenesis. This is a particularly effective form of mutagenesis, because a gene that has been marked with a transposon is easy to map using arbitrary PCR or ligation-mediated PCR and is amenable to high-throughput DNA-sequencing strategies (Box 9.1). Furthermore, genes marked with a transposon are also relatively easy to clone by selecting for selectable genes carried on the transposon.

Transposon Mutagenesis In Vivo

Not all types of transposons are equally useful for mutagenesis. A transposon used for mutagenesis should have the following properties:

1. It should transpose at a fairly high frequency.
2. It should not be very selective in its target sequence.
3. It should carry an easily selectable gene, such as one for resistance to an antibiotic.
4. It should have a broad host range for transposition if it is to be used in several different kinds of bacteria.

Transposon Tn*5* is ideal for random mutagenesis of gram-negative bacteria because it embodies all of these features. Not only does Tn*5* transpose at a relatively high frequency, but also, it has almost no target specificity and transposes in essentially any gram-negative bacterium. It also carries a kanamycin resistance gene that is expressed in most gram-negative bacteria. Figure 9.21A illustrates a popular method for transposon mutagenesis of gram-negative bacteria other than *E. coli* (see Simon et al., 1983, Suggested Reading). In addition to the broad host range of Tn*5* and the promiscuity of plasmid RP4, this method takes advantage of the narrow host range of ColE1-derived plasmids, which replicate only in *E. coli* and a few other closely related species. Phage Mu is another transposon-like element that can transpose in many types of gram-negative bacteria and that shows little target specificity. Many of the original elements used in gram-positive bacteria (such as Tn*917*, Tn*916*, or Tn*10* derivatives) were plagued with targeting biases of one type or another. However, a *mariner*-type element called *Himar1* appears to provide the features elaborated in the four points mentioned above (see Rubin et. al. Suggested Reading). Somewhat remarkably, the *Himar1* element was originally obtained from the horn fly, *Haematobia irritans*, but has been adapted with delivery systems and antibiotic resistance genes that work in a variety of bacteria.

Transposon Mutagenesis In Vitro

While in vivo transposon mutagenesis is a very useful technology, it does have some limitations. One of the limitations is that it is necessary to introduce the transposon on a suicide vector, which may give some residual false-positive results for transposon insertion mutants if the suicide vector is capable of limited replication. Another limitation is that it is not very efficient and

BOX 9.1

Transposons and Genomics

For decades, transposons have been important tools for bacterial genetics. Transposons are a very useful type of insertion mutagen and offer the benefit of providing a selectable marker that is invaluable for genetic mapping. Historically, this has allowed the insertion to be mapped using Hfrs and cotransfer using P1 transduction (see chapters 3 and 5). A set of genetically mapped Tn*10* insertions at approximately 1-minute intervals across the entire *E. coli* genome was an important tool for mapping mutations in the chromosome. Transposon insertions can also be efficiently mapped by DNA sequencing. Early strategies involved cloning the transposon from the genome with some of its flanking DNA to allow determination of the DNA sequence from a plasmid. However, PCR sequencing strategies have all but replaced cloning. PCR involves using two short DNA oligonucleotide primers to amplify the intervening DNA sequences in sequential rounds of denaturation, annealing, and amplification with a thermostable DNA polymerase (see chapter 1). The difficulty with using PCR to map transposon insertions stems from the fact that, while the sequence of the transposon is known, by definition, the flanking DNA sequence is not known. There are various techniques for obtaining a place where a second primer can anneal for amplification of the transposon end, along with the flanking sequences, but they all involve somehow attaching a known DNA sequence to the flanking DNA sequence.

One efficient way to map insertions is arbitrary PCR, where one of the primers has a series of random nucleotides. In panel A of the figure, 10 random nucleotides (N_{10}) are integrated into the oligonucleotide, which anneals only at a fixed short sequence, GCTGG, that is found at the very 5′ end of the oligonucleotide. The positions of the four primers used in the experiment are shown. (B) In the first PCR, the arbitrary primer with an extra DNA sequence tag (primer 1) is used with a primer that is specific for the position inside the IR of the transposon (primer 2) (panel B). This step enriches only for amplified products that contain both the transposon end and a known DNA sequence tag. In the next step (panel C), a second round of PCR is done with a primer that recognizes the sequence tag (primer 3) and a primer that is specific for the transposon but closer to the outer end of the transposon (primer 4). This product is then subjected to DNA sequencing using primer 4 or a primer that anneals even closer to the end of the transposon (panel D). Part of the value of this procedure is that the template DNA can be whole cells that are added directly to the PCR mixture from a colony or an overnight culture. Another technique involves enzymatically digesting genomic DNA or physically shearing the DNA

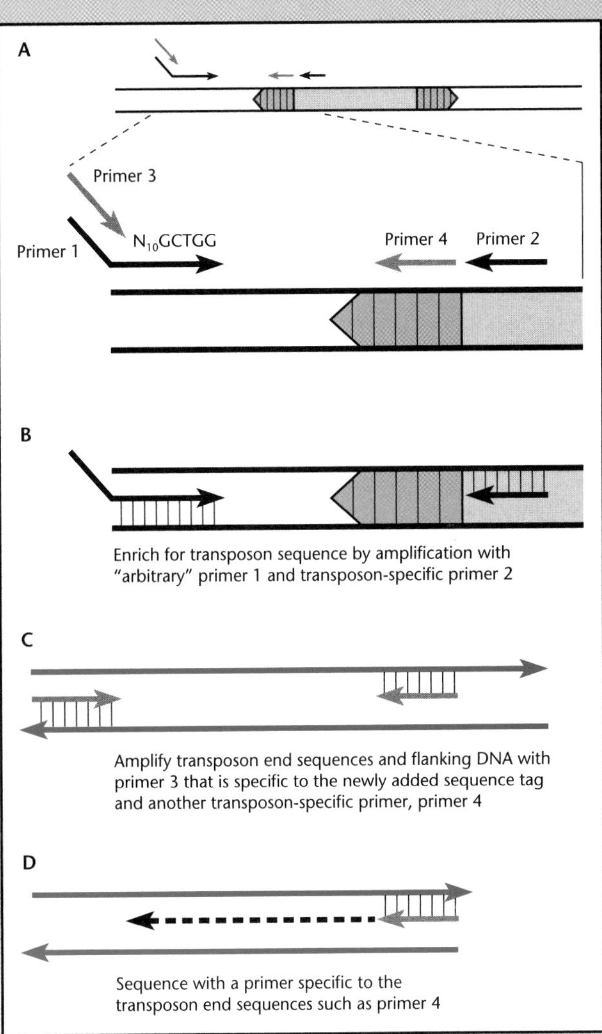
A
Primer 3
Primer 1
N10GCTGG
Primer 4
Primer 2
B
Enrich for transposon sequence by amplification with "arbitrary" primer 1 and transposon-specific primer 2
C
Amplify transposon end sequences and flanking DNA with primer 3 that is specific to the newly added sequence tag and another transposon-specific primer, primer 4
D
Sequence with a primer specific to the transposon end sequences such as primer 4

doi:10.1128/9781555817169.ch9.Box9.1.f

and directly ligating a known sequence onto the fragments to provide the sequence tag for amplifying the flanking DNA. The arrowheads indicate the 3′ ends of the DNAs.

High-throughput sequencing techniques are allowing large numbers of insertions to be mapped. They can be used to isolate transposon insertions in every nonessential gene in an organism for greatly facilitated genetics. In addition, tags can be added to allow next-generation sequencing technologies to be used to simultaneously map tens of thousands of insertions in a population of cells containing random insertions. These techniques can be used to identify genes that are likely to be essential because they will not tolerate transposon insertions, and transposons should not map to these

(continued)

BOX 9.1 (continued)

Transposons and Genomics

genes. Also, by mapping insertions in a population of cells with random insertions before and after growth under specialized conditions, one can determine the profile of genes that are required for growth under these conditions. For example, the cell population with random insertions could be constructed and grown first in rich medium. The distribution of transposons in this population of cells could be monitored before and after growth in minimal medium containing only salts and a carbon source. Presumably, this would identify all genes required for growth in minimal medium, because individuals in the population that had transposon insertions in genes that were required for growth in minimal medium would be lost from the population by dilution. This procedure could be adapted to many different growth conditions and environments (e.g., growth in a biofilm or in a model animal or plant system).

References

Liberati, N. T., J. M. Urbach, S. Miyata, D. G. Lee, E. Drenkard, G. Wu, J. Villanueva, T. Wei, and F. M. Ausubel. 2006. An ordered, nonredundant library of *Pseudomonas aeruginosa* strain PA14 transposon insertion mutants. *Proc. Natl. Acad. Sci. USA* **103:**2833–2838.

Singer, M., T. A. Baker, G. Schnitzler, S. M. Deischel, M. Goel, W. Dove, K. J. Jaacks, A. D. Grossman, J. W. Erickson, and C. A. Gross. 1989. A collection of strains containing genetically linked alternating antibiotic resistance elements for genetic mapping of *Escherichia coli*. *Microbiol. Rev.* **53:**1–24.

van Opijnen, T., K. L. Bodi, and A. Camilli. 2009. Tn-seq: high-throughput parallel sequencing for fitness and genetic interaction studies in microorganisms. *Nat. Methods* **6:**767–772.

requires powerful positive-selection techniques to isolate the mutants. Another limitation arises if a specific plasmid or other smaller DNA sequence is to be mutated. There is no target specificity to the insertion mutants, and so most of the time, the transposon hops into the chromosome; the few with transposon insertions in the smaller target DNA must be found among the overwhelming number of mutants in the chromosome. There is also the possibility of multiple transposition events.

In vitro transposon mutagenesis avoids many of these limitations. This technology is made possible by the fact that the transposase enzyme by itself performs the chemical reactions of the "cut-and-paste" transposition reaction. In the procedure, the target DNA is mixed with a donor DNA containing the transposon, and the purified transposase is added, allowing the transposon to insert into the target DNA in the test tube. Multiple transposases have been adapted for this process, and each has its own advantages. For example, mutants of the Tn*5* transposase are available that enhance the transposition frequency, which is necessary because the wild-type Tn*5* transposase is only weakly active. Also, only the sequences at the ends of the IRs of Tn*5* are needed, and they are only 19 bp long. One disadvantage of Tn*5* is that it is prone to multiple transposition events in a single DNA because the element lacks target site immunity. A mutant form of the Tn*7* regulator protein allows high-frequency transposition with the Tn*7* transposase, resulting in almost undetectable target site selectivity, with the added benefit of target site immunity, which greatly reduces, if not eliminates, multiple insertions (see Biery et. al., Suggested Reading). A drawback of Tn*7* is that the element has an array of IR sequences at each end, where the *cis*-acting sequences required for optimal transposition with the element are around 100 bp long. In these systems, the transposon that is used has been engineered to lack a transposition gene so that the element does not transpose in subsequent generations or cause genetic instability, such as deletions, once it is in the chromosome. In some of these systems controllable outward-facing promoters are engineered into the element to allow controllable expression of an adjacent gene or operon. Therefore, in addition to producing loss-of-activity mutations, one can also isolate mutants with altered expression (see Bordi et. al., Suggested Reading).

The target DNA can be either a replicon, such as a plasmid that replicates in the recipient cell, or random linear pieces of the chromosomal DNA of the recipient if it is being introduced into cells that can be transformed with linear DNA (see chapter 6). The linear pieces recombine with the chromosome and replace the chromosomal sequence with the sequence mutated with the transposon. This offers a method for doing random chromosomal transposon mutagenesis of bacteria for which no workable transposon mutagenesis system is available (for example, see Bordi et. al., Suggested Reading).

Another variation of this method for performing transposon mutagenesis, which can be applied to mutagenize the DNA of almost any bacterium, and even eukaryotic cells, where transformation works very well, is to use "transpososomes" (see Goryshin et al., Suggested Reading). A transpososome is a transposon to which the transposase protein is already attached so that it does not have to be made in the cell. This feature is important, because the transposase gene on the transposon might not be expressed in a distantly related bacterium and

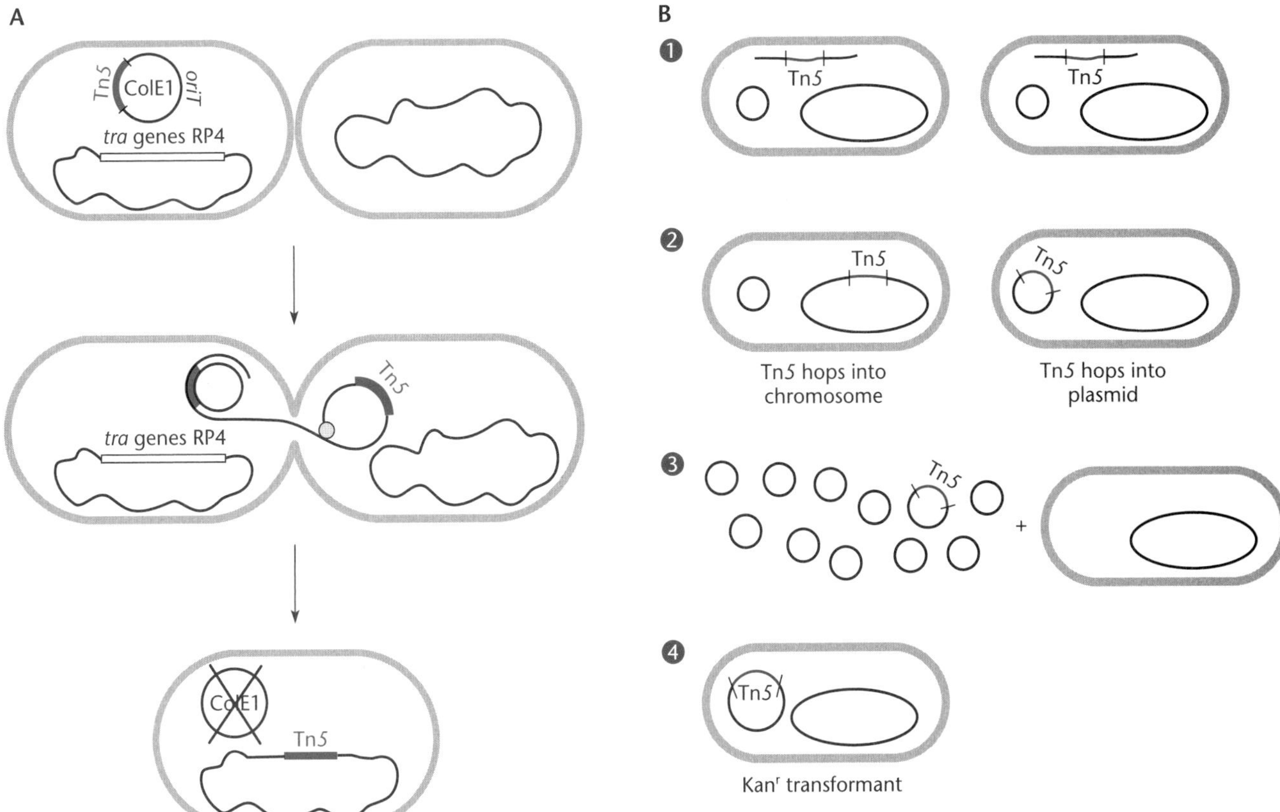

Figure 9.21 Transposon Tn*5* mutagenesis. **(A)** A standard protocol for transposon mutagenesis of gram-negative bacteria. A ColE1-derived suicide plasmid containing a *mob* site whose relaxase recognizes the coupling protein of the promiscuous plasmid RP4 and contains transposon Tn*5* is mobilized into the bacterium by the products of the RP4 transfer genes, which are inserted in the chromosome. The transposon hops into the chromosome of the recipient cell, and the ColE1 plasmid is lost because it cannot replicate. The Tn*5* transposon is shown in blue. **(B)** Random transposon mutagenesis of a plasmid. In step 1, transposon Tn*5* is introduced into cells on a suicide vector. In step 2, the culture is incubated, allowing the Tn*5* time to hop, either into the chromosome (large circle) or into a plasmid (small circle). Plating on kanamycin-containing medium results in the selection of cells in which a transposition has occurred. In step 3, plasmid DNA is prepared from Kan^r cells and used to transform a kanamycin-susceptible (Kan^s) recipient. In step 4, selection for Kan^r allows the identification of cells that have acquired a Tn*5*-carrying plasmid. doi:10.1128/9781555817169.ch9.f9.21

certainly not in a eukaryotic cell. As in other methods of transposon mutagenesis, the transposon should carry a selectable gene that is expressed in the cell to be mutagenized. Transpososomes based on Tn*5* are made by running the transposition reaction in vitro in the absence of magnesium ions. Under these conditions, the ends of the transposon in the donor DNA are not cut, but if the transposon has already been cut out of the donor DNA by some other process (for example, with restriction endonucleases), the transposon binds to the ends and remains attached, forming the transpososome. When the transpososome is introduced into the target cells by electroporation, the transposase attached to the transposon catalyzes the DNA strand exchanges required for transposition of the transposon into the chromosome or other cellular DNA. The transposase enzyme introduced with the transposon is quickly degraded, preventing further transposition.

Transposon Mutagenesis of Plasmids

One common use of transposon mutagenesis is to identify genes within large clones on a plasmid. If the transposon hops into a gene on the plasmid, it will disrupt the gene. DNA sequencing from the element will identify the gene of interest.

Figure 9.21B illustrates the steps in the selection of plasmids with transposon insertions in *E. coli*. A suicide vector containing the transposon (Tn*5* in the example) is introduced into cells harboring the plasmid. Cells in which the transposon has hopped into cellular DNA (either the plasmid or the chromosome) are then selected by plating them on medium containing the antibiotic to which a transposon gene confers resistance, in this case, kanamycin. Only the cells in which the transposon has hopped to another DNA become resistant to the antibiotic, since the transposons that remain in the suicide vector are lost with the suicide vector. In most antibiotic-resistant bacteria, the transposon will have hopped into the chromosome rather than into the plasmid simply because the chromosome is a larger target. The plasmids in these bacteria are normal. However, if the plasmid being mutagenized is self-transmissible, the plasmids can be isolated from the few bacteria in which the transposon has hopped into the plasmid by mating the plasmid into another *E. coli* strain and selecting the antibiotic resistance on the transposon. Alternatively, the antibiotic-resistant colonies that have the transposon either in the plasmid or in the chromosome can be pooled and the plasmids can be isolated. This mixture of plasmids, most of which are normal, is then used to transform another strain of *E. coli*, selecting for the antibiotic resistance gene on the transposon (the kanamycin resistance gene in the example). The antibiotic-resistant transformants should contain the plasmid with the transposon inserted somewhere in it. Voilà! In a few simple steps, plasmids with transposon insertion mutations have been isolated. This method can be used to randomly mutagenize a DNA segment cloned in a plasmid or to mutagenize the plasmid itself.

Transposon Mutagenesis of the Bacterial Chromosome

The same methods used to mutagenize a plasmid with a transposon can also be used to mutagenize the chromosome. A gene with a transposon insertion is much easier to map or clone than a gene with another type of mutation, making this a popular method for mutagenesis of chromosomal genes.

The major limitation of transposon mutagenesis is that transposon insertions usually inactivate a gene, a lethal event in a haploid bacterium if the gene is essential for growth. Therefore, this method can generally be used only to mutate genes that are nonessential or essential under only some conditions. However, it can still be used to map essential genes by isolating transposon insertion mutations that are not in the gene itself but close to it in the DNA. If the transposon is inserted close enough, it might be used to map or clone the gene. It is also important to remember that insertion of some transposons may increase the expression of genes near the insertion site. As described above, elements that contain controllable outward-facing promoters are also useful for isolating altered expression mutations. Transposon insertions are easily mapped using PCR, and procedures exist for mapping all of the random insertions in a pool of bacteria to identify essential genes or genes that are important or essential in certain environments (Box 9.1).

Transposon Mutagenesis of All Bacteria

One of the most useful features of transposon mutagenesis is that it can be applied to many types of bacteria, even ones that have not been extensively characterized. Methods have been developed to perform transposon mutagenesis of almost all gram-negative bacteria, as well as many other types of bacteria. All that is needed is a way of introducing a transposon into the bacterium, provided that the transposon can transpose in that bacterium. The transposon should also carry a gene that can be selected in the target organism. Some of these methods were mentioned above and are outlined in more detail below.

CLONING GENES MUTATED WITH A TRANSPOSON INSERTION

Genes that have been mutated by transposon insertion are usually relatively easy to identify by cloning a DNA segment that includes the easily identified antibiotic resistance gene in the transposon. Since some antibiotic genes, for example, the kanamycin resistance gene in Tn*5*, are expressed in many types of bacteria, this method can even be used to clone genes from one bacterium in a cloning vector from another. This is particularly desirable because most cloning vectors and recombinant DNA techniques have been designed for *E. coli*. To clone a gene mutated by a transposon from a bacterium distantly related to *E. coli*, the DNA from the mutagenized strain is cut with a restriction endonuclease that does not cut in the transposon and is ligated into an *E. coli* plasmid cloning vector cut with the same or a compatible enzyme. The ligation mixture is then introduced into *E. coli* by transformation, and the transformed cells are spread on a plate containing the antibiotic to which the transposon confers resistance. Only cells containing the mutated, cloned gene multiply to form a colony.

While DNA sequencing will identify the position of the transposon insertion, in a large unsequenced DNA fragment, it may be useful to be able to clone the subregion of interest that is responsible for a particular phenotype. Transposons can be engineered to make cloning of transposon insertions even more efficient by introducing an origin of replication into the transposon so that the DNA containing the transposon need only cyclize to replicate autonomously in *E. coli*. The use of such a transposon for cloning transposon insertions is illustrated in Figure 9.22. In the example, the transposon carrying a plasmid origin of replication has inserted into the gene to be cloned. The chromosomal DNA is isolated from the mutant cells and cut with a restriction endonuclease that does not cut in the transposon, EcoRI in the example. When the cut DNA is religated,

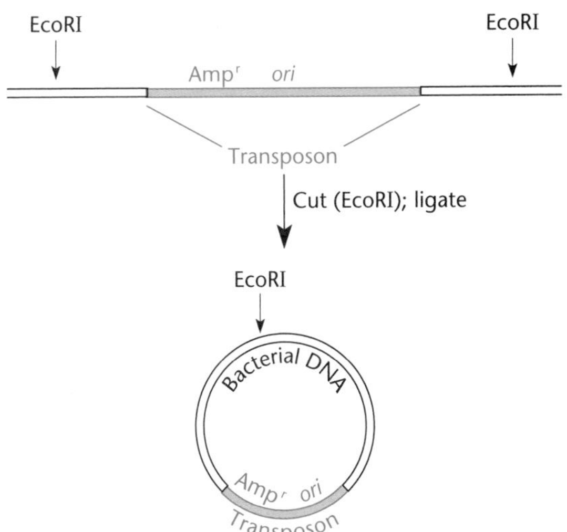

Figure 9.22 Cloning genes mutated by insertion of a transposon. A transposon used for mutagenesis of a chromosome contains a plasmid origin of replication (*ori*), and the chromosome is cut with the restriction endonuclease EcoRI and religated. If the ligation mixture is used to transform *E. coli*, the resulting plasmid in the Amp^r transformants will contain the sequences that flanked the transposon insertion in the chromosome. Chromosomal sequences are shown in black, and transposon sequences are shown in blue.
doi:10.1128/9781555817169.ch9.f9.22

the fragment containing the transposon becomes a circular replicon with the plasmid origin of replication. If the ligation mixture is used to transform *E. coli* and the ampicillin resistance gene on the transposon is selected, the chromosomal DNA surrounding the transposon has been cloned. Since the restriction endonuclease cuts outside the transposon, any clones of the transposon cut from the chromosome also include sequences from the gene of interest into which the transposon had inserted.

This method allows the cloning of genes about which nothing is known except the phenotype of mutations that inactivate the gene, and it can be easily adapted to clone genes from any bacterium in which the transposon can hop to create the original chromosomal mutation. Once mutants are obtained with the transposon inserted in the gene of interest, the remaining manipulations are performed in *E. coli*. This can be a particular advantage if the bacterium being studied is difficult to grow or maintain in a laboratory situation.

Using Transposon Mutagenesis To Make Random Gene Fusions

The ability of some transposons to hop randomly into DNA has made them very useful for making **random gene fusions** to reporter genes (see chapter 2). Fusing a gene to a reporter gene whose product (such as LacZ or green fluorescent protein) is easy to monitor can make regulation of the gene much easier to study or can be used to identify genes subject to a certain type of regulation or those localized to certain cellular compartments. Once a gene subject to a certain type of regulation has been identified in this way, it can be easily cloned and studied using methods such as those described above.

Transposons have been engineered to make either transcriptional or translational fusions. As discussed in chapter 2, in a transcriptional fusion, one gene is fused to the promoter for another gene so that the two genes are transcribed into an mRNA together. In a translational fusion, the ORFs for the two proteins are fused to each other in the same reading frame so that translation, initiated at the TIR for one protein, continues into the ORF for the other protein, making a fusion protein.

Site-Specific Recombination

Another type of nonhomologous recombination, **site-specific recombination**, occurs only between specific sequences or sites in DNA. It is promoted by enzymes called **site-specific recombinases**, which recognize two specific sites in DNA and promote recombination between them. Even though the two sites generally have short sequences in common, the regions of homology are usually too short for normal homologous recombination to occur efficiently. Therefore, efficient recombination between the two sites requires the presence of a specific recombinase enzyme. We have already mentioned some site-specific recombination systems in connection with the resolution of chromosome and ColE1 plasmid dimers by the XerCD site-specific recombinase (see chapters 1 and 4). The integrase of phage λ is another example, as are resolvases, such as TnpR of Tn*3*, that resolve cointegrates formed during replicative transposition. In this section, we discuss some other examples of site-specific recombination in bacteria and phages, which can all be placed into one of two groups, the S and the Y recombinases, based on their mechanism of action.

Integrases

Integrases are a type of site-specific recombinase. They recognize two sequences in DNA and promote recombination between them; therefore, they are no different in principle from the site-specific recombinases that resolve cointegrates. However, integrases act to integrate one DNA into another by promoting recombination between two sites on different DNAs.

PHAGE INTEGRASES

The best-known integrase is the Int enzyme of λ phage, which is responsible for the integration of circular phage DNA into the DNA of the host to form a prophage (see chapter 8). Briefly, the λ phage integrase specifically recognizes the *attP* site in the phage DNA and the *attB* site on the bacterial chromosome and promotes recombination between them. Usually, phage integrases are extremely specific. Only the *attP* and *attB* sites are recognized, so the DNA integrates only at one or at most a few places in the bacterial chromosome. Other integrases seem to be somewhat less specific, including the integrase of the integrating conjugative element Tn*916* (see chapter 5) and the integrases of integrons (see below), where there seems to be some flexibility in the sequence of the *attB* site. In a reversal of the reaction performed by the integrases, a combination of the integrase (Int) and another enzyme, often called the excisase (Xis), promotes recombination between the hybrid *attP-attB* sites flanking the integrated DNA to excise the integrated DNA, although the integrase is again the enzyme that performs the site-specific recombination.

Because of their specificity, phage integrases have a number of potential uses in molecular genetics. For example, the reaction performed by phage λ Int and Xis has been capitalized on as a cloning technology called the Gateway system, marketed by the biotechnology company Invitrogen. In this application, a PCR fragment containing the gene of interest is cloned into a plasmid vector, called the entry vector, so that the clone is flanked by one of the hybrid *attB-attP* sites which flank integrated prophage λ DNA in the chromosome. If the vector is mixed with a destination vector containing the other *attP-attB* hybrid site and λ integrase and excisase are added, site-specific recombination between the sites moves the cloned gene into the destination vector, where it becomes flanked by *attB* sites. While this technology does not remove the requirement for the initial cloning, which can be laborious, once a clone is made in the entry vector, it can be transferred quickly into a number of different destination vectors. This could be important if, for example, one wished to determine the effect on the solubility of a protein of fusing it to a number of different affinity tags which are encoded on a number of different destination vectors.

INTEGRASES OF TRANSPOSON INTEGRONS

Integrases are also important in the evolution of some types of transposons. The first clue was that transposons seemed to have picked up antibiotic resistance genes so that related members of some families of transposons, such as the Tn*21* family, have different resistance genes inserted in approximately the same place in the transposon. Figure 9.23 shows a more detailed structure of this region, called an **integron**, and how it presumably recruits antibiotic resistance genes. A basic integron consists of an integrase gene (called *intI1* in the figure) next to a site called *attI* (for *att*achment site *i*ntegron). A gene in the *attI* site is transcribed from the promoter p_c. The transposon originally had no antibiotic resistance gene inserted at the *attI* site. Elsewhere in the cell, there was an antibiotic resistance cassette that consisted of an antibiotic resistance gene and a site, *attC* (for *att*achment site *c*assette), recognized by the integrase. This cassette was excised and formed a circle, and the integrase then integrated it into the *attI* site on the integron by promoting site-specific recombination between the *attC* site on the cassette and the *attI* site in the integron. At another time, and in another cell, a similar cassette carrying a different resistance gene could integrate next to this one, adding another antibiotic resistance gene to the transposon. The advantages of this system are obvious. By carrying different resistance genes, the transposon allows the cell, and thereby itself, to survive in environments containing various toxic chemicals. However, where the resistance gene cassettes originally came from remains a mystery.

Integrons are not only found in transposons. Chromosomal **superintegrons** are found in a number of different types of bacteria (see Rowe-Magnus et al., Suggested Reading). The structure of one of them, from *Vibrio cholerae*, is shown in Figure 9.24. It consists of 179 cassettes carrying ORFs with largely unknown functions separated by partially conserved sequences that might be *attC* sites. Presumably, integrons in transposons will prove to be one example of a larger phenomenon in which useful genes can be recruited as needed from storage areas carrying large numbers of such cassettes.

GENETIC (PATHOGENICITY) ISLANDS

Integrases and nonhomologous recombination also play roles in the integration of at least some types of genetic islands into the chromosome. Like plasmids and prophages, genetic islands often carry clusters of genes that allow the bacterium to occupy specific environmental niches. Genetic islands can be hundreds of thousands of base pairs long and carry hundreds of genes. **Pathogenicity islands** (PAIs) are a type of genetic island that carries genes required for pathogenicity. PAIs carry genes for resistance to multiple antibiotics in *Shigella flexneri*, for alpha-hemolysin and fimbriae in pathogenic *E. coli*, for scavenging and storing iron in *Yersinia*, and for a type III secretion system in *Helicobacter pylori*, to give just some examples. We have already mentioned the PAI SaPI1, which carries the gene for the toxin involved in toxic shock syndrome in *Staphylococcus aureus* (see chapter 8). This PAI has its own integrase and so can integrate itself into the chromosomes of new strains of *S. aureus*, increasing their pathogenicity. It can also move from strain to strain of *S. aureus* by behaving like a satellite virus of phage 80α and allowing itself to be transduced by the phage. If a cell carrying the PAI is infected by phage 80α, the island will be excised and replicate

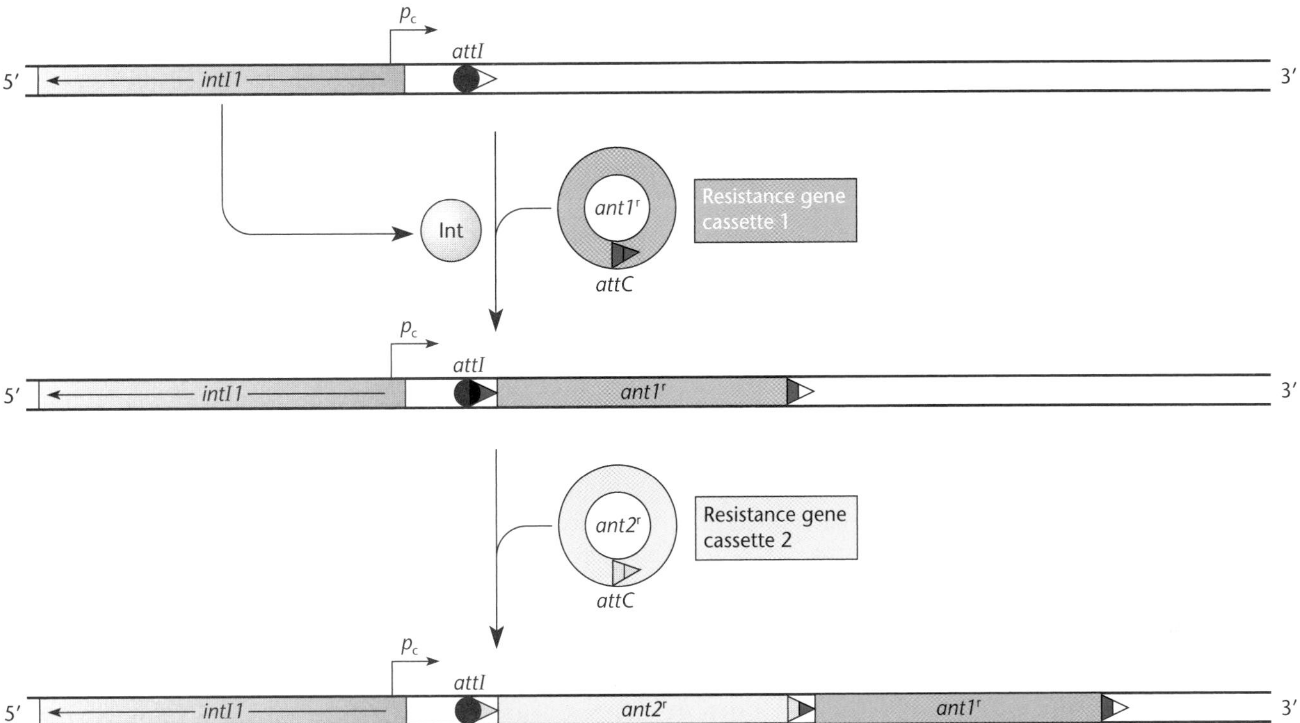

Figure 9.23 Assembly of integrons. The primary transposon carries an integron with a gene, *intI1*, encoding an integrase and a site, *attI*, transcribed by a strong promoter, p_c. A cassette carrying resistance to one antibiotic, *ant1ʳ*, has been excised from elsewhere and is integrated by the integrase by recombination between its *attC* site and the *attI* site on the integron. The antibiotic resistance gene is transcribed from the promoter on the integron. Later, the integrase integrates another cassette carrying a different antibiotic resistance gene, *ant2ʳ*, at the same place. In this way, a number of different antibiotic resistance genes can be assembled by the integron on the transposon. The *attC* sites, represented by triangles, contain conserved regions related to the *attC* sites between the cassettes of superintegrons shown in Figure 9.24. doi:10.1128/9781555817169.ch9.f9.23

with the help of phage replication proteins. It will then be packaged into phage heads, from where it can be injected into another bacterium that does not carry it and integrate into its chromosome, converting the new bacterium into a pathogen.

The demonstration that SaPI1 can move makes it the exception. Most PAIs have not been demonstrated to move from one bacterium to another or even to integrate into a DNA that lacks them. Often, the only evidence that they have moved recently is that they are not

Figure 9.24 Example of a superintegron from *V. cholerae*. More than 100 cassettes encoding resistance to different antibiotics and other functions are associated with partially homologous *attC* sites next to an integrase gene, *intIA*, and an *attI* attachment site. Regions between the cassettes corresponding to possible *attC* sites are shown as arrows. Regions of sequence conservation are shown as triangles. R, purine; Y, pyrimidine. doi:10.1128/9781555817169.ch9.f9.24

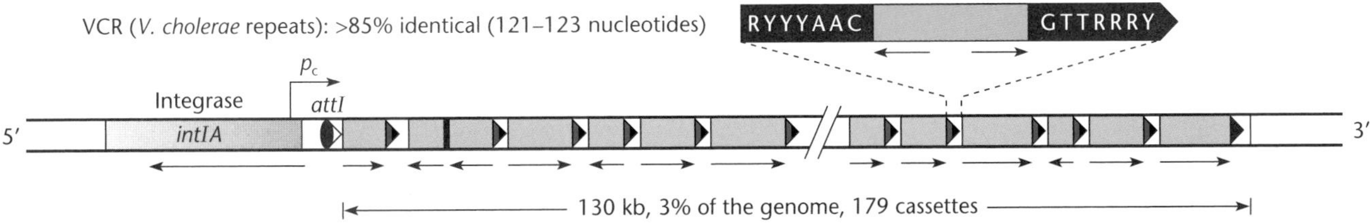

found in all the strains of a type of bacterium and that the base composition (G+C content) of their DNA and their codon usage are often different from those of the chromosomal DNA as a whole. These characteristics are often taken as evidence of recent horizontal transfer of genes from one strain of bacterium to another. PAIs also often carry pieces of integrase genes that are broken up by nonsense and other types of mutations, suggesting that they once encoded functional integrases that are no longer functional. The PAI elements are often flanked by short repeated sequences, either direct or inverted, that may be the sites at which the integrase acted to integrate the PAI. Apparently, the PAIs did move into the strain some time ago, perhaps on a promiscuous plasmid, and integrated into the chromosome, but their DNA has mutated over time so that they are no longer capable of moving. Interestingly, many PAIs are integrated into tRNA genes in the chromosome. Part of the tRNA gene is duplicated on the PAI, so the tRNA product of the gene is still functional (see Box 8.1). Chromosomal tRNA genes may be preferred sites of integration because they have almost the same sequence in different species of bacteria. By using a highly conserved tRNA gene as its bacterial attachment (*attB*) site, the PAI can integrate into the chromosome of any bacterial strain in which it finds itself.

Even though most PAIs cannot integrate, some researchers decided to make one that could do so (see Rakin et al., Suggested Reading). More accurately, they constructed a plasmid that could integrate using the integrase of a PAI. They accomplished this by using parts of a PAI from *Yersinia*. Different strains of *Yersinia* differ greatly in their pathogenicity depending on the DNA elements they carry. The most pathogenic species of *Yersinia* are *Yersinia pestis*, which causes bubonic plague, and *Y. enterocolitica* and *Y. pseudotuberculosis*, which cause mild intestinal upsets. These strains carry a PAI called HPI (for high-pathogenicity island). This 40-kb PAI is integrated into one of the asparagine tRNA ($tRNA^{Asn}$) genes and encodes enzymes to make small molecules called siderophores, which help scavenge for iron, which is in limited supply in the eukaryotic host. The authors reasoned that even though this PAI from different strains has not been demonstrated to move, perhaps the functional parts of each PAI could be assembled into a plasmid cloning vector of *E. coli*, which would then be able to integrate into a corresponding $tRNA^{Asn}$ gene of *E. coli*. As mentioned above, the tRNA genes are very similar in different species, and *E. coli* has four $tRNA^{Asn}$ genes, all of which are very similar in the region where the PAI is integrated in *Yersinia*.

In order to integrate, the PAI needs both a functional integrase and functional *attP* and *attB* sites, by analogy to phage λ. First, the investigators reasoned that the integrase itself from the PAI in *Y. pestis* might be functional, so they cloned the integrase gene into a plasmid cloning vector so that it would be transcribed at a high level from a phage T7 promoter (see the discussion of T7-based expression vectors in chapter 7). They also needed to reconstruct a site that would be recognized by the integrase, as the PAI itself might not have such a site. They guessed that the situation might be analogous to what happens when phage λ integrates (see chapter 8). The *attP* site on the phage recombines with the *attB* site in the chromosome, resulting in *attP-attB* hybrid sites flanking the integrated prophage, which are no longer recognized by the integrase alone. Such hybrid sites might exist at the ends of the PAI, and if so, they might also not be recognized by the integrase. To reconstruct the original site the integrase recognizes from the hybrid sites, the authors needed to know where in the hybrid sites the "*attP*" sequence ends and the "*attB*" sequence in the $tRNA^{Asn}$ gene begins. To determine this, they compared the sequences at the ends of the PAI to the sequence of $tRNA^{Asn}$ genes in strains that do not have the PAI integrated to determine which sequences are due to the *attB* sequence in the tRNA gene. They then constructed the original *attP* sequence by using PCR and cloned it into the plasmid that already contained the integrase. When the plasmid was introduced into *E. coli* and the integrase gene on the plasmid was induced, it integrated into a $tRNA^{Asn}$ gene of *E. coli*, showing that all the features needed to integrate the PAI were still present and active.

Resolvases

The resolvases of transposons such as the TnpR protein of Tn*3* are another type of site-specific recombinase. In fact, the resolvase of transposon γδ, a close relative of Tn*3*, is one of the best-studied site-specific recombinases and has been crystallized bound to its DNA substrate. These enzymes promote the resolution of cointegrates by recognizing the *res* sequences that occur in one copy in the transposon but in two copies in direct orientation in cointegrates. Recombination between the two *res* sequences in a cointegrate excises the DNA between them, resolving the cointegrate into the donor DNA and the target DNA, both containing the transposon.

Other resolvases already mentioned resolve dimers of plasmids. Dimer formation by plasmids reduces their stability, especially if they have a low copy number, because each dimer is treated as one plasmid molecule by the partitioning system and segregated to the same side of the dividing cell (see chapter 4). Because mutations in the resolvase gene can affect the segregation of the plasmid, some of these plasmid resolution systems were originally mistaken for partitioning systems and given the name Par (for partitioning). Examples of resolvases involved in resolving plasmid dimers are the Cre recombinase, which resolves dimers of the P1 prophage plasmid by promoting recombination between repeated *loxP* sites on the dimerized plasmid, and the XerCD recombinase, which resolves dimers of the ColE1 and pSC101 plasmids by promoting recombination between repeated *cer* and *psi* sites, respectively. The XerCD recombinases also resolve the chromosome dimers between repeated *dif* sites in the dimers during cell division (see chapter 1). Chromosome dimers can form between circular genomes when an uneven number of crossover events occur between two sister chromosomes, something that is common during recombination repair of stalled chromosome replication forks (see chapter 10). The XerCD system is also borrowed by bacteriophages, where it doubles as an integrase to integrate the cholera toxin-producing phage (see chapter 8), illustrating how the ability to promote recombination between specific sequences can be put to many uses.

DNA Invertases

The DNA invertases are like resolvases in that they promote site-specific recombination between two sites on the same DNA. The main difference between the reactions promoted by DNA invertases and those catalyzed by resolvases is that two sites recognized by invertases are in reverse orientation with respect to each other whereas the sites recognized by resolvases are in direct orientation. As discussed in chapter 3 in the section "Types of Mutations," recombination between two sequences that are in direct orientation deletes the DNA between the two sites, whereas recombination between two sites that are in inverse orientation with respect to each other inverts the intervening DNA.

The sequences acted on by DNA invertases are called **invertible sequences**. These short sequences may carry the gene for the invertase or may be adjacent to it. Therefore, the invertible sequence and its specific invertase form an inversion cassette that sometimes plays an important regulatory role in the cell, some examples of which follow.

PHASE VARIATION IN *SALMONELLA* SPECIES

The classic example of an invertible sequence is the one responsible for **phase variation** in some strains of *Salmonella*. Phase variation was discovered in the 1940s with the observation that some strains of *Salmonella* can change their surface antigens. They do this by shifting from making flagella composed of one flagellin protein, H1, to making flagella composed of a different flagellin protein, H2. The shift can also occur in reverse, i.e., from making H2-type flagella to making H1-type flagella. The flagellar proteins are the strongest antigens on the surfaces of many bacteria, and periodically changing their flagella may help these bacteria escape detection by the host immune system.

Two features of the *Salmonella* phase variation phenomenon suggested that the shift in flagellar type was not due to normal mutations. First, the shift occurs at a frequency of about 10^{-2} to 10^{-3} per cell, much higher than normal mutation rates. Second, both phenotypes are completely reversible—the cells switch back and forth, exhibiting first one type of flagella and then the other.

Figure 9.25 outlines the molecular basis for the *Salmonella* antigen phase variation (see Simon et al., 1980, Suggested Reading). As mentioned above, the two types of flagella are called H1 and H2. A DNA invertase called

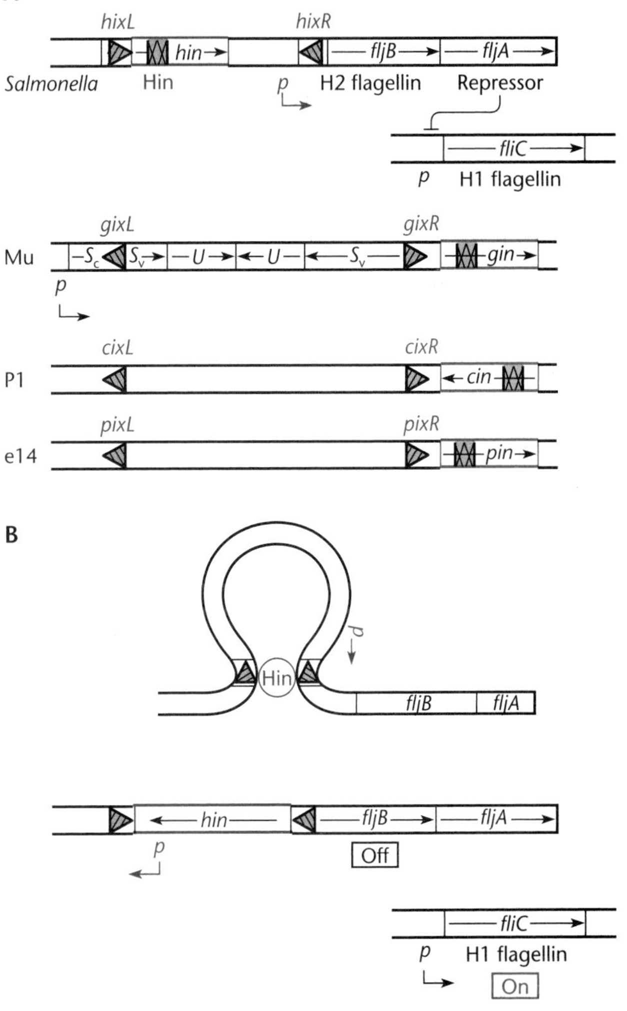

Figure 9.25 Regulation of *Salmonella* phase variation and some other members of the family of Hin invertases. **(A)** Invertible sequences, bordered by triangles, of *Salmonella* and several phages. The recombination sites are designated *hixL*, *hixR*, etc. **(B)** Hin-mediated inversion. In one orientation, the H2 flagellin gene, as well as the repressor gene, is transcribed from the promoter *p* (in blue). In the other orientation, neither of these genes is transcribed, and the H1 flagellin is synthesized instead. The invertase Hin is made constitutively from its own promoter. doi:10.1128/9781555817169.ch9.f9.25

the Hin invertase causes the phase variation by inverting an invertible sequence upstream of the gene for the H2 flagellin by promoting recombination between two sites, *hixL* and *hixR*. The invertible sequence contains the invertase gene itself and a promoter for two other genes: one, called *fljB*, encodes the H2 flagellin, and one, called *fljA*, encodes a repressor of H1 gene transcription. With the invertible sequence in one orientation, the promoter transcribes the H2 gene and the repressor gene, and only the H2-type flagellum is expressed on the cell surface. When the sequence is in the other orientation, neither the H2 gene nor the repressor gene is transcribed, because the promoter is facing backward. Now, however, without the repressor, the H1 gene, called *fliC*, can be transcribed; therefore, in this state, only the H1-type flagellum is expressed on the cell surface. Clearly, the Hin DNA invertase that is encoded in the invertible sequence is expressed in either orientation, or the inversion would not be reversible.

OTHER INVERTIBLE SEQUENCES

There are a few other known examples of regulation by invertible sequences in bacteria. For example, fimbria synthesis in some pathogenic strains of *E. coli* is regulated by an invertible sequence. Fimbriae are required for the attachment of the bacteria to the eukaryotic cell surface and may also be important targets of the host immune system. In an interesting case, a symbiont of nematodes, *Photorhabdus*, switches to a pathogenic state in insects by a single inversion that affects the expression of almost 10% of its genes.

Invertible sequences also exist in some phages. An example is the invertible G segment region of phage Mu (Figure 9.13). Both phage P1 and the defective prophage e14 also have invertible regions (shown in Figure 9.25). These phages use invertible sequences to change their tail fibers. The tail fibers made when the invertible sequences are in one orientation differ from those made when the sequences are in the other orientation, broadening the host range of the phage. In phage Mu, the tail fiber genes expressed when the invertible sequence is in one orientation allow the phage to adsorb to *E. coli* K-12, *Serratia marcescens*, and *Salmonella enterica* serovar Typhi. In the other orientation, the phage is able to adsorb to other strains of *E. coli*, *Citrobacter*, and *Shigella sonnei*.

Not only do these phage invertases perform similar functions, they are also closely related to each other. Dramatic evidence for their relationship came from the discovery that the Hin invertase inverts the Mu, P1, and e14 invertible sequences, and vice versa (see van de Putte et al., Suggested Reading). Apparently, inversion cassettes can be recruited for many purposes, much like antibiotic resistance cassettes are recruited by integrons.

Y and S Recombinases

As mentioned above, many site-specific recombinases, whether they are integrases, resolvases, or invertases, are closely related to each other. This is not surprising, considering that they all must perform the same basic reactions. First, they must cut a total of four strands of DNA—both strands in two recognition sequences—whether these recognition sequences are on the same DNA (resolvases and invertases) or on different DNAs (integrases). Then, they must join the cut end of each strand to the cut end of the corresponding strand from the other recognition sequence. We can anticipate some of the features that site-specific recombinases must have in order to perform these reactions. First, they must somehow hold the DNA ends after they cut them so that they are not free to flop around and join with the end of any strand they happen to encounter. Second, after the strands are cut, either the DNA or at least part of the recombinase must rotate in a defined way to place the cut ends from different strands in juxtaposition with each other so that the correct rejoining can occur. If they rejoin the ends of the same strands that were cut originally, there will be no recombination, and they will be back where they started.

All site-specific recombinases appear to fall into two unrelated families, the Y (tyrosine) family and the S (serine) family, based on which of these amino acids, called the catalytic amino acid, plays the crucial role in their active centers. The feature shared by these amino acids is a hydroxyl group in their side chains to which phosphates can be attached. After the DNA is cut, the hydroxyl group on the catalytic amino acid forms a covalent bond with the free phosphate end on the DNA. This protects the end of the strand and holds it while the recombinase moves it into position to be joined to the end of a different cut strand, the essence of recombination. However, when the strands are cut, the end of DNA that is joined to the catalytic amino acid and the structures that form after this cutting differ between the two groups of recombinases. Each of these pathways is outlined in the following sections.

Y Recombinases: Mechanism

The Y recombinases seem to be the most varied group and include recombinases that perform the most complex reactions. Some examples of Y recombinases are listed in Table 9.2, and the structures of some of them are shown in Figure 9.26. Some of them were mentioned in this and previous chapters; they include Cre recombinase, which resolves dimers of the P1 plasmid prophage by recombination between *loxP* sequences, and the XerCD recombinases, which resolve chromosome dimers by recombination between *dif* sites and resolve

Table 9.2 Examples of tyrosine (Y) recombinases

Source	Enzyme
Resolvases	
E. coli	Plasmid/phage P1 Cre
	Plasmid F ResD
	Phage N15 telomere resolvase
Borrelia spp.	Telomere resolvase
S. sonnei	Plasmid CoIIb-P9 shufflon
Bacteria	XerCD
Saccharomyces spp.	Flp
Integrases	
E. coli	Phage λ integrase
Bacteria	Integrase of integrons
Superfamily	
Eukaryotes	Topoisomerases
Viruses, e.g., vaccinia virus	Topoisomerase

plasmid ColE1 dimers by recombination between *cer* sites, as well as integrating the cholera toxin-producing phage into one of the *V. cholerae* chromosomes. The λ phage integrase discussed in chapter 8 is also a Y recombinase, as are the integrases of integrons and of the integrating conjugative elements (see chapter 5). Although the reactions they perform are somewhat different, the terminases that resolve the circular dimerized plasmids created during replication of linear plasmids, including those in *Borrelia* (see chapter 4), also belong to this family and use a similar mechanism. The Y recombinases are also not limited to eubacteria and include some resolvases of some eukaryotes, such as the Flp recombinase, which inverts a short sequence in the 2μm circle of yeast (Table 9.2), and they are related to some type I topoisomerases of eukaryotes, suggesting that they might have had a common origin.

Details of the molecular basis of recombination by the Y recombinases are outlined in Figure 9.27, and the structures of the sites recognized by some of them are shown in Figure 9.28. Much of what we know about how Y recombinases work comes from studies of the structure of the relatively simple Cre recombinase, which has been crystallized bound to various forms of its *loxP* DNA substrate. In the absence of evidence to the contrary, we may assume that at least most features of this reaction can be extrapolated to other Y recombinases, even ones that perform more complex reactions. The *loxP* site recognized by the Cre recombinase consists of a short sequence of 8 bp, where the crossover occurs. It also has two almost identical flanking sequences of 11 bp in inverse orientation that are recognized by the recombinase. In the first step, four copies of the Cre resolvase bind to two *loxP* recognition sites (two to each site) and hold them together in a large complex. Then, one strand of each of the recognition sequences is cut in the crossover region by an attack by the active-site tyrosine, creating 5′ OH ends. As they are cut, the 3′ phosphate ends are transferred to the side chain of the active-site tyrosine in two of the bound recombinase molecules to form tyrosyl-3′-phosphate bonds. This holds the 3′ phosphate ends and protects them. The free 5′ OH ends then

Figure 9.26 Domain structure of tyrosine recombinases (Cre, XerCD, etc.; λ Int; and Flp) and eukaryotic type IB topoisomerases. The conserved C-terminal catalytic domain of the proteins is shaded in blue. The brackets show the positions of three conserved regions of the catalytic domain: boxes I, II, and III. Residues of the catalytic signature of the family are indicated, and the tyrosine nucleophiles are circled. Other protein regions are shown in different shades of gray to indicate that they are structurally unrelated. Integrases, such as λ Int, have an additional DNA-binding domain at the N terminus to bind the Arm site sequences of the recombination site. In the human type IB topoisomerase core enzyme, residues 215 to 765, the catalytic domain is interrupted by a linker region spanning the region between the active-site histidine and the tyrosine nucleophile. doi:10.1128/9781555817169.ch9.f9.26

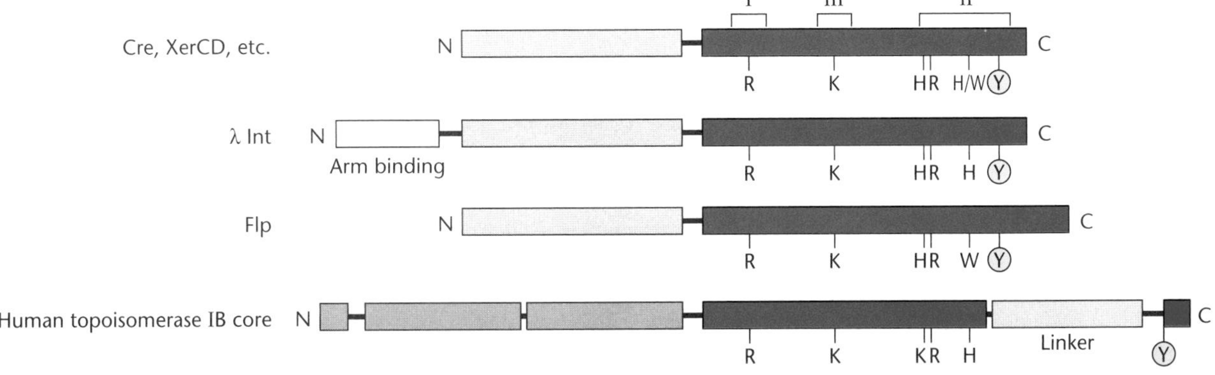

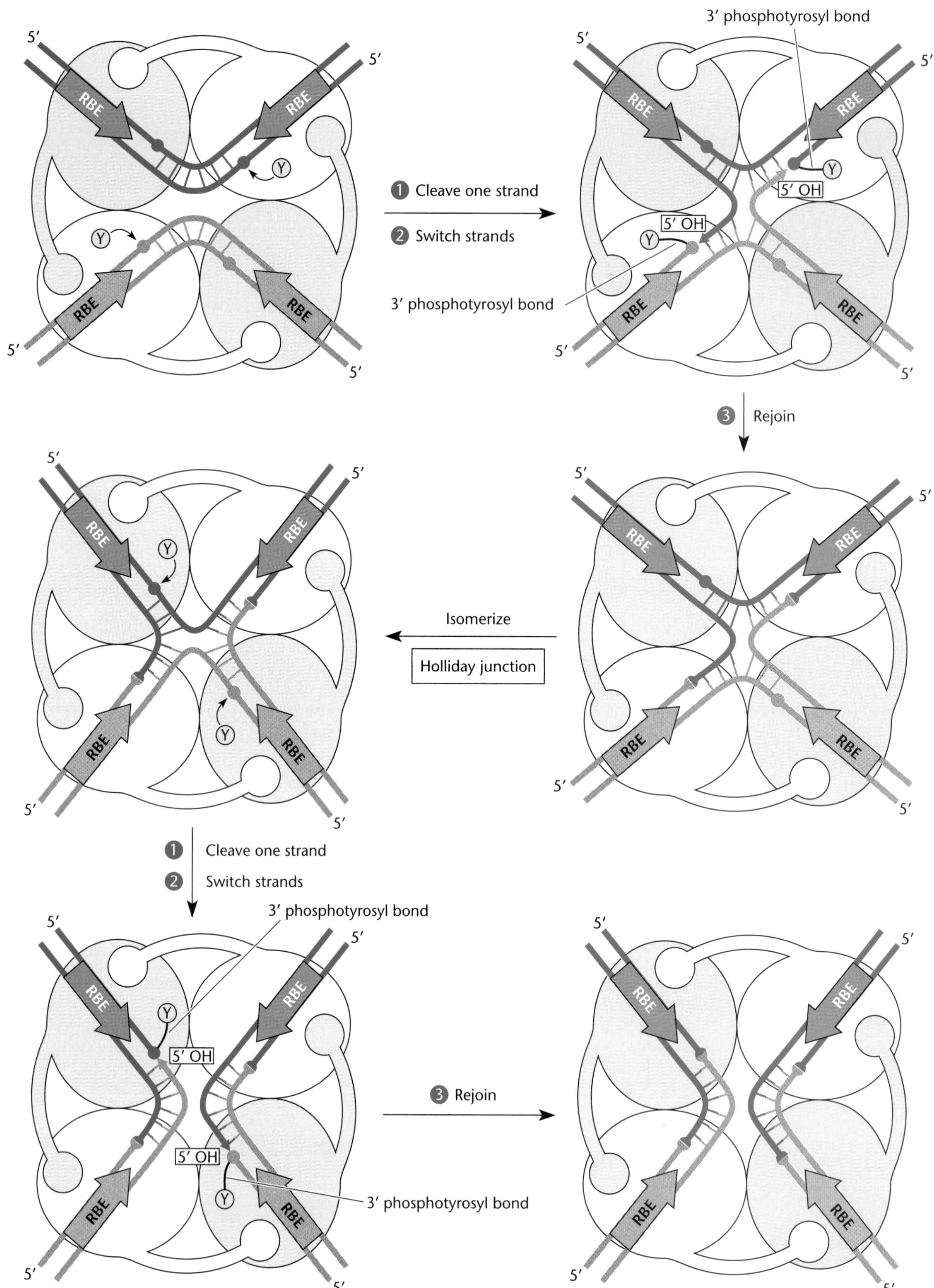

Figure 9.27 Model for the reaction promoted by the Cre tyrosine (Y) recombinase. Four Cre recombinase molecules bind two *loxP* sites, bringing them together. RBE, recombinase-binding element. **(1)** The active-site tyrosines in two of the Cre molecules, indicated by Y, cleave two of the strands in a phosphoryltransferase reaction that forms 3′ phosphotyrosyl bonds and 5′ OH

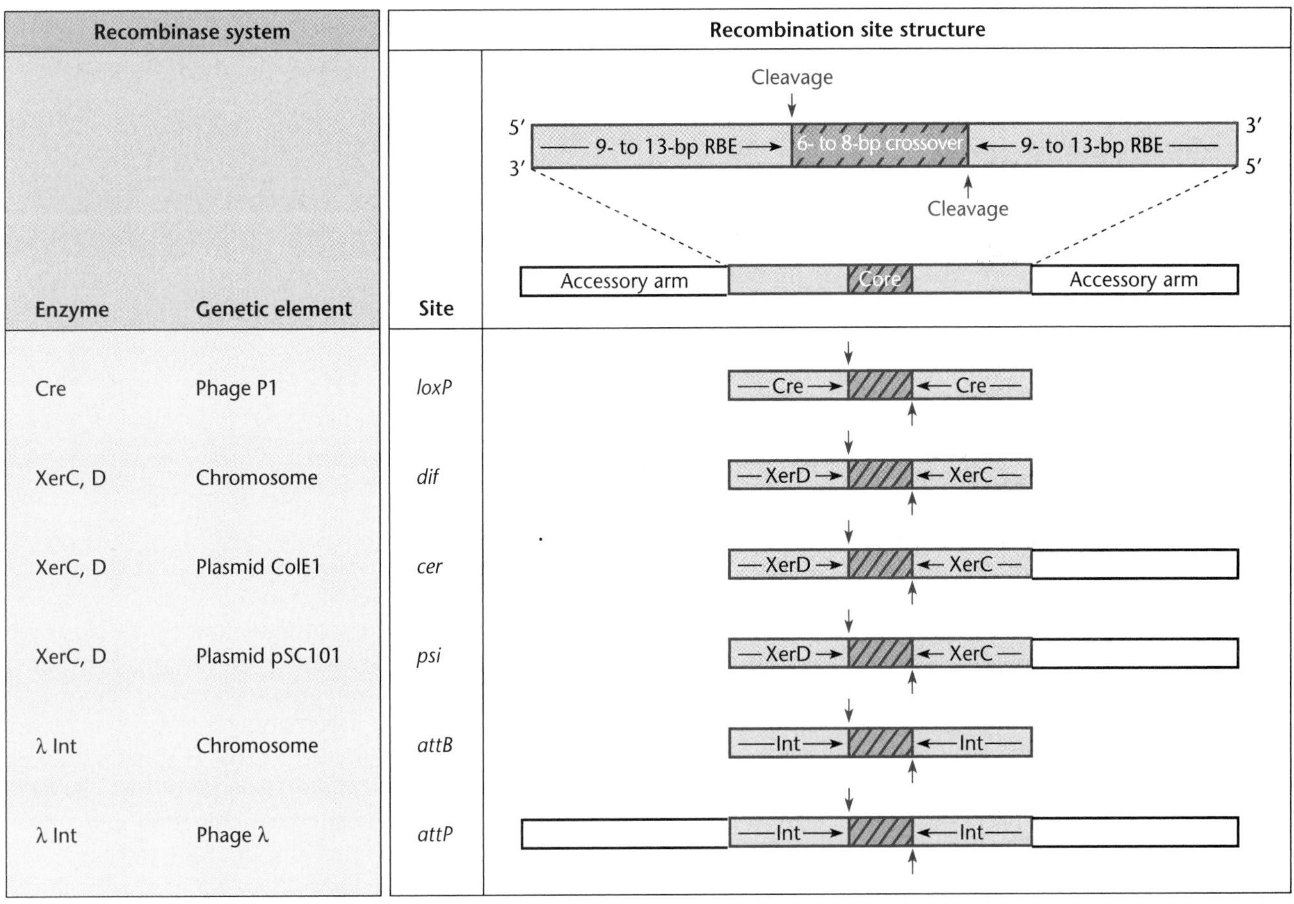

Figure 9.28 Structures of some sites recognized by tyrosine (Y) recombinases. The top part of the figure shows (in blue) the basic structure, with a core of a 6- to 8-bp crossover region flanked by two 9- to 13-bp sequences required for recombinase binding. These are referred to as recombinase-binding elements (RBE). Many sites also have flanking accessory protein-binding sites called the accessory arms. Proteins bind to these sites and help orient the recombinase and give it specificity. The bottom part of the figure shows the variations on the theme exhibited by some of the sites described in the text. The arrows indicate IR sequences.
doi:10.1128/9781555817169.ch9.f9.28

attack the 3′ tyrosyl phosphate bond in the other DNA, rejoining the 5′ OH ends to 3′ phosphate ends on the corresponding strand of the other DNA rather than on their own DNA. This causes two of the strands to cross over and hold the two DNAs together in what is called a Holliday junction. Holliday structures also form during homologous recombination; they are discussed in more detail in chapter 10. The crystal structure reveals that the Holliday junction is held very flat by the resolvase, with the four DNA branches coming out of the complex in the same plane.

What happens next is less clear and differs somewhat in different types of Y recombinases. To achieve recombination, the noncrossover strands must also be cut and joined to the corresponding strands on the other DNA. If the crossover strands are cut again and rejoined to their original strands, no recombination will occur. Therefore, the recombinase has to know

ends (arrows). **(2)** Each 5′ OH end then attacks the opposite 3′ phosphoryltyrosine bond, switching the strands and causing a 3- to 4-nucleotide swap in the complementary region. **(3)** The nucleophilic 5′ OH ends attack the phosphotyrosyl bonds, rejoining the ends to form a Holliday junction that isomerizes. The next steps are essentially a repeat of steps 1 to 3, but the other two Cre molecules cut the other two strands by a phosphotransferase reaction, and the strands are exchanged and rejoined to form the two recombinant DNA molecules.
doi:10.1128/9781555817169.ch9.f9.27

which strands to cut in the Holliday junction and which strands to join them to. Holliday junctions can do a number of wonderful things, as discussed in chapter 10. They can isomerize so that the crossed strands become the uncrossed strands, and vice versa. They can also migrate so that the position at which the strands are crossed over can move on the DNA, provided that the sequences of the two DNAs are almost the same in the region of migration. One possibility is that part of the resolvase rotates, forcing the isomerization of the Holliday junction so that the correct strands enter the active center of the other two copies of the resolvase to repeat the cutting and rejoining reaction. This seems unlikely, considering that major changes are not seen in the crystal structure of the complex in its various states. The other possibility is that the Holliday junction migrates, rotating the DNA so that the correct strands to be cut come in contact with the active centers of the other two copies of the resolvase and can be cut and rejoined. However, it is hard to imagine how the Holliday junction could migrate very far, since that would mean either drastically rotating the DNA arms that emerge from the complex or severely distorting the DNA in the complex.

Further complicating the models, some apparent Y resolvases, including the integrases of the integrating conjugative element, Tn*916*, and the integrases of integrons, do not require extensive homology in the crossover region of the two sites being recombined. The Tn*916* integrative element seems to integrate in many places in the chromosome, suggesting that it can use many different sequences as bacterial *att* sites, although it may prefer some sites over others. The *attC* sites in resistance cassettes seem to have very little homology either to each other or to the *intI* site on the integron. Extensive homology in the crossover region should be required for Holliday junction formation and for branch migration. Current research addresses these and other questions about the reactions performed by Y recombinases.

Table 9.3 Examples of serine (S) recombinases

Source	Enzyme
Resolvases	
E. coli	TnpR of Tn*3*
	Resolvase of $\gamma\delta$
	ParA of RP4/RK2
Enterococcus spp.	TnpR of Tn*917*
Invertases	
S. enterica serovar Typhimurium	Hin flagellar invertase
E. coli phage Mu	Gin tail fiber invertase
Superfamily	
Streptomyces coelicolor	Integrase of phage ϕC31
B. subtilis	SpoIVCA recombinase
Anabaena spp.	Heterocyst recombinase

Other proteins besides the Y recombinases themselves are often involved in the recombination reactions. These other proteins bind close to the core region, on what is referred to as the accessory arm regions, and help to stabilize the recombinase-DNA complexes and/or orient the recombinase proteins on the DNA (Figure 9.28). For example, the XerCD recombinase requires two proteins, ArgR and PepA, to promote recombination at *cer* sites in plasmid ColE1 and to resolve dimers (see chapter 4). These proteins also play other, very different roles in the cell: one is the repressor of the arginine biosynthetic operon, and the other is an aminopeptidase. It is not clear why the ColE1 plasmid recruits these particular proteins for dimer resolution, but they may happen to have structural and/or sequence features that make them easy to adapt to this role. Whatever the reason, such situations are a nightmare for geneticists trying to deduce the function of a gene product. For example, mutations in the *argR* gene cause constitutive synthesis of the arginine-biosynthetic enzymes but also reduce the stability of the ColE1 plasmid for a completely unrelated reason. The λ

Figure 9.29 Domain structure of serine (S) recombinases. The conserved catalytic domain (ca. 120 amino acids) is shown in blue. The conserved amino acids that play major roles in catalysis are indicated, and the active-site serines are circled.
doi:10.1128/9781555817169.ch9.f9.29

Resolvase/invertase subgroup (e.g., Tn*3* and $\gamma\delta$ resolvases and Gin and Hin invertases)

"Large" serine recombinase subgroup (e.g., ϕC31 integrase, Tn*4451* and Tn*5397* transposases, and SpoIVCA)

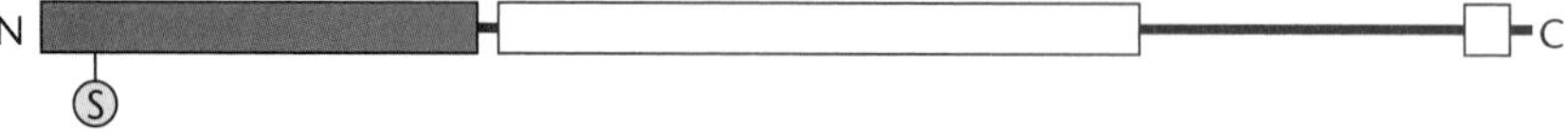

integrase also requires that other host proteins, including integration host factor (IHF), be bound close to the *attP* site for recombination to occur. This protein is bound at many places in the chromosomal DNA, where it plays multiple roles, making this less surprising.

S Recombinases: Mechanism

The S recombinases are also a large family, comprising many of the plasmid resolution and invertase systems in both gram-positive and gram-negative bacteria. Some S recombinases are listed in Table 9.3. The TnpR cointegrate resolution systems of Tn*3*-like transposons, including γδ, are members of this family, as are the dimer resolution systems of some plasmids, including the gram-negative promiscuous plasmids RK2 and RP4 and the gram-positive *Enterococcus faecalis* plasmid pAMβ1, to list some plasmids mentioned elsewhere in the book. The integrase of a *Streptomyces* phage, fC131, is an S recombinase, unlike the integrases of most phages, which are Y recombinases. The Hin invertase in *Salmonella* and its relatives that invert tail fiber genes in some phages are also members of this family, as are the excisases mentioned above, which remove the intervening sequences during sporulation in *Bacillus subtilis* and heterocyst formation in *Anabaena*. Some integrating conjugative elements use S integrases rather than Y integrases to integrate into the recipient cell DNA. So far, counterparts of the S recombinases have not been found in eukaryotes, with the possible exception of some in small plasmids in marine diatoms.

The domain structure of some S recombinases is shown in Figure 9.29, and a model of the reaction they perform is shown in Figure 9.30. Again, much of what we know comes from studies with one S recombinase, the resolvase of the transposon γδ, which has been crystallized with its DNA substrate. Superficially, the molecular mechanism of site-specific recombination by S recombinases is similar to that of recombination by Y recombinases. The recognition sites have a short crossover region bracketed by copies of recombinase-binding sites. Four copies of the recombinase bind to two

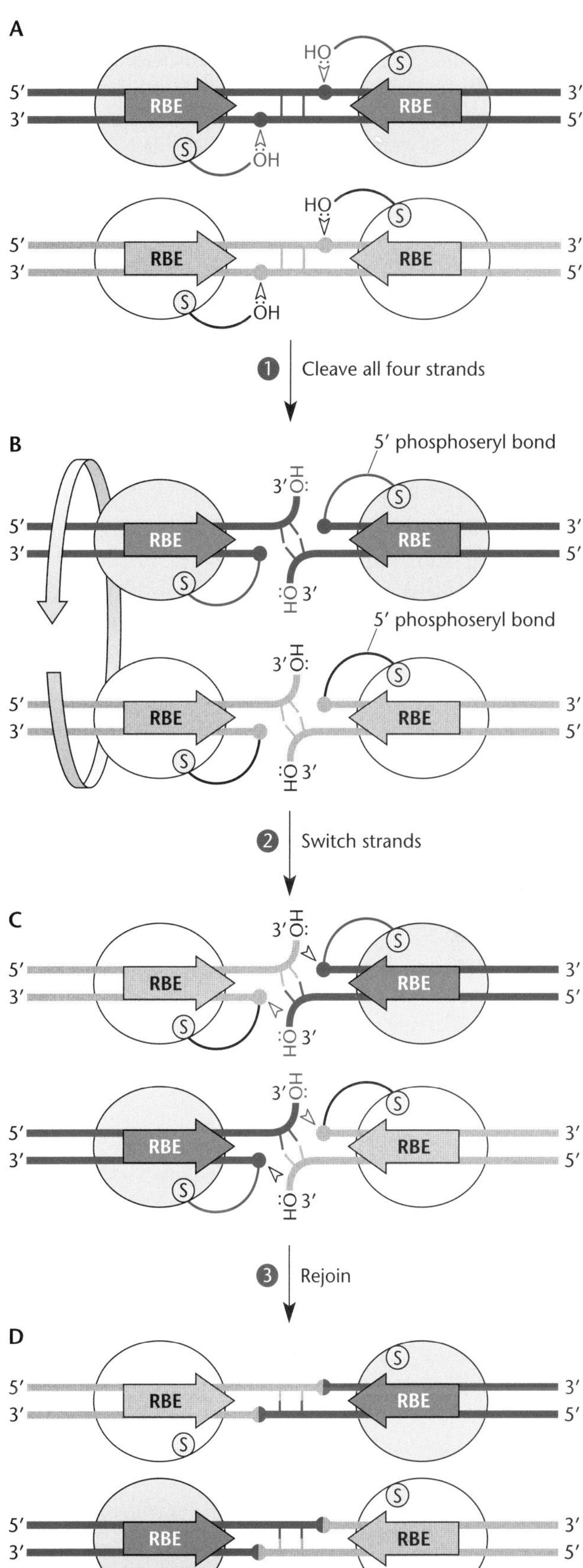

Figure 9.30 Model for the reaction promoted by the γδ recombinase. **(A)** Four recombinase molecules bind to two copies of the recombination sites, holding them together, and the active-site serines in all four recombinases attack phosphodiester bonds a few nucleotides apart to leave staggered breaks, with the 5′ phosphate overhangs forming phosphorylserine bonds with the active-site serines. **(B)** Rotation of a domain of the recombinase brings the corresponding ends of the two different recombination sites together. **(C)** The free 3′ OH ends then attack the 5′ phosphorylserine bond on the corresponding strand on the other DNA. **(D)** The nicks are sealed to form the recombinant product. RBE, recombinase-binding element.
doi:10.1128/9781555817169.ch9.f9.30

recognition sequences to form a complex in which the recombination occurs. This is where the similarities end, however. Rather than having the active-site tyrosines in two recombinase molecules make a nucleophilic attack on the same phosphodiester bond in the two DNAs, the active-site serines in all four recombinases make nucleophilic attacks a few nucleotides apart to create staggered breaks in both strands, for a total of four breaks. Also, the staggered breaks leave 5′ PO_4 and 3′ OH overhangs. The nucleophilic attacks leave the 5′ phosphate ends joined to the hydroxyl group in the side chain of the active-site serines on all four recombinase molecules to form 5′ phosphorylseryl bonds to the ends of the DNA, rather than 3′ phosphoryltyrosyl bonds, as in the Y recombinases; Figure 9.31 shows these nucleophilic attacks and how they leave 5′ phosphorylseryl bonds. The nucleophilic 3′ OH on each cut strand then attacks the phosphorylseryl bond in the corresponding strand of the other recognition sequence. Rejoining of the nicks leaves the recombinant product without the formation of a Holliday junction. A large 180° rotation in the complex appears to explain how the new configuration occurs to allow strand exchange. This was first suggested by the crystal structure, which revealed that the two dimers that are linked to the DNAs interface with a large flat surface (see Li et. al., Suggested Reading). This idea has received additional support recently from single-molecule work with a simpler recombinase.

Like the Y recombinases, the S recombinases often also use other proteins bound close to the recognition site to help stabilize the complex during recombination. In some cases, these are extra copies of the recombinase itself.

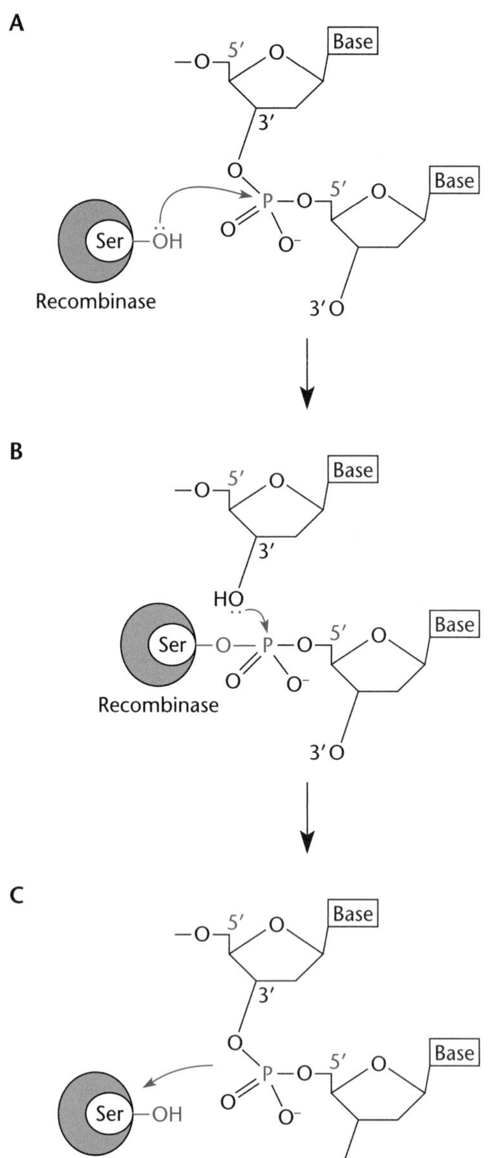

Figure 9.31 How successive attacks by nucleophilic hydroxyl groups of serine (S) recombinases can create a recombinant DNA product. **(A)** A nucleophilic attack by the hydroxyl group on the side chain of serine in the active center of a serine recombinase forms a 5′ phosphorylserine bond and breaks the phosphodiester bond in the DNA. **(B)** The free 3′ OH group can attack another 5′ phosphorylserine bond, breaking the bond. **(C)** This attack results in the re-formation of a phosphodiester bond to a different DNA strand.
doi:10.1128/9781555817169.ch9.f9.31

Importance of Transposition and Site-Specific Recombination in Bacterial Adaptation

One of the most important conclusions from studies of transposons and other types of moveable elements in bacteria is the contribution they make to bacterial adaptation. We have seen how integrons can integrate antibiotic resistance genes from large storage areas into transposons that can then transpose into other DNAs. Conjugative elements, including self-transmissible plasmids and integrating conjugative elements, can move these transposons from bacterium to bacterium and even into other genera of bacteria. This presumably allows bacterial genomes to remain small but still have access to many types of genes that increase their ability to adapt to different environments. One way this impacts humans directly is in the acquisition of antibiotic resistance by bacteria. Table 9.4 gives a summary of the moveable elements discussed so far that are known to carry antibiotic resistance genes in both gram-negative and gram-positive bacteria. If we are to continue to be able to treat bacterial infections effectively, we must have a clear idea of how these moveable elements can contribute to widespread antibiotic resistance and how to combat it.

Table 9.4 Characteristics of genetic elements involved in the spread of antibiotic resistance genes

Genetic element	Characteristics	Role in spread of resistance genes
Self-transmissible plasmid	Circular, autonomously replicating element; carries genes needed for conjugal DNA transfer	Transfers resistance genes; mobilizes other elements that carry resistance genes
Integrating conjugative element	Integrated element that can excise to form a non-replicating circular transfer intermediate; carries genes needed for conjugal DNA transfer	Same as self-transmissible plasmid: highly promiscuous, transferring between gram-positive and gram-negative genera and species
Mobilizable plasmid	Circular, autonomously replicating element; carries site and genes that allow it to use the conjugal apparatus provided by a self-transmissible plasmid	Transfer of resistance genes
Transposon	Moves from one DNA segment to another within the same cell	Carries resistance genes from chromosome to plasmid or vice versa
Gene cassette	Circular, nonreplicating DNA segment containing only ORFs; integrates into integrons	Carries resistance genes
Integron	Integrated DNA segment that contains an integrase, a promoter, and an integration site for gene cassettes	Forms clusters of resistance genes that are transcribed under control of the integron promoter

SUMMARY

1. Nonhomologous recombination is the recombination between specific sequences on DNA that occurs even if the sequences are mostly dissimilar.

2. Transposition is the movement of certain DNA sequences, called transposons, from one place in DNA to another. The smallest known bacterial transposons are IS elements, which contain only the genes required for their own transposition. Other transposons carry genes for resistance to substances such as antibiotics and heavy metals. Transposons have played an important role in evolution and are useful for mutagenesis, gene cloning, and random gene fusions.

3. Composite transposons are composed of DNA sequences bracketed by IS elements.

4. Most known transposons are DDE transposons, because their transposases have three amino acids, DDE (aspartate-aspartate-glutamate), which hold magnesium ions that play an essential role in transposition. They are characterized by having IR sequences at their ends and duplicating a short sequence in the target DNA on entry. Other transposons, called Y2 or rolling-circle transposons, essentially replicate themselves into a target DNA by using a free 3′ hydroxyl as a primer. They do not have IR sequences on their ends. Other transposons, called Y transposons and S transposons, are more like integrative elements and lysogenic phages in that they use integrases to integrate into the target DNA, although they generally have less target specificity than do integrases.

5. Bacterial DDE transposons transpose by two distinct mechanisms: replicative transposition and cut-and-paste, or conservative, transposition. In replicative transposition, which is used by such transposons as Tn*3* and Mu, the entire transposon is replicated, leading to formation of a cointegrate. In cut-and-paste transposition, which is used by many IS elements and composite transposons, the transposon is cut out of the donor DNA and inserted somewhere else. However, these two mechanisms are closely related and differ only in whether the donor DNA is cut in both strands at the ends of the transposon. The cut-and-paste DDE transposon Tn*7* has been converted into a replicative transposon by a single mutation in part of its transposition system so that it now cuts only one strand at the end of the transposon.

6. In transposon mutagenesis, a gene is disrupted by insertion of a transposon or by activation of adjacent genes. Transposons are mapped easily with DNA sequencing or facilitate cloning in unsequenced genomes.

7. Specially engineered transposons carrying reporter genes can be used to make random gene fusions. Insertion of the transposon into a gene can lead to expression of the reporter gene on the transposon from the promoter or TIR of the disrupted gene, depending on whether the fusion is transcriptional or translational.

8. Genes that have been mutated by insertion of a transposon are often easy to clone in *E. coli* if an antibiotic resistance gene is present on the transposon. Some transposons have been engineered to contain an *E. coli* plasmid origin of replication so that the restriction fragment containing the transposon need not be cloned in another cloning vector but need only be cyclized after ligation to form a replicon in *E. coli*.

9. Site-specific recombinases are enzymes that promote recombination between certain sites on the DNA. Examples

(continued)

SUMMARY (continued)

of site-specific recombinases are resolvases, integrases, and DNA invertases. The genes for many of these site-specific recombinases have sequences in common and appear to form two families with two distinct common ancestors.

10. Resolvases are site-specific recombinases encoded by replicative transposons that resolve cointegrates by promoting recombination between short *res* sequences in the copies of the transposon in the cointegrate.

11. Integrases promote nonhomologous recombination between specific sequences on a DNA element, such as a phage DNA and the chromosome, integrating the phage DNA into the chromosome to form lysogens. They also integrate antibiotic resistance gene cassettes into transposons. An integron consists of an integrase and an *att* site for integration of gene cassettes. Integrases also play a role in integrating PAIs and other types of genetic islands into the chromosome. Superintegrons are large DNA elements (50,000 to 100,000 bp) that carry genes, including genes for pathogenicity, that allow the bacterium to occupy unusual ecological niches.

12. DNA invertases promote nonhomologous recombination between short IRs, thereby changing the orientation of the DNA sequence between them. The sequences they invert, invertible sequences, are known to play an important role in changing the host range of phages and the bacterial surface antigens to avoid host immune defenses.

13. Recombinases can be divided into two types, Y (or tyrosine) recombinases and S (or serine) recombinases. Recombination by both types involves nucleolytic attacks by the hydroxyl group of the side chain of the amino acid on the phosphodiester bond in DNA, forming a phosphoryl bond to the amino acid. They differ in that Y transposons form a 3′ phosphoryltyrosine bond whereas S transposons form a 5′ phosphorylserine bond; other differences are that Y recombinases cut two strands simultaneously and form a Holliday junction, which can then isomerize, whereas S recombinases cut all four strands, not necessarily in any order, and do not form a Holliday junction. Rather, S recombinases appear to depend on rotation of part of the recombinase to bring the different strands into juxtaposition.

QUESTIONS FOR THOUGHT

1. For transposons that transpose replicatively (e.g., Tn*3*), why are there not multiple copies of the transposon around the genome?

2. How do you think that transposon Tn*3* and its relatives have spread throughout the bacterial kingdom?

3. While transposons are parasites of the host, they also mitigate their effects with beneficial functions. List some of the ways they help the host.

4. Where do you suppose the genes that were inserted into integrons in the evolution of transposons came from?

5. If the DNA invertase enzymes are made continuously, why do the invertible sequences invert so infrequently?

6. In the experiments of Bender and Kleckner on Tn*10* transposition, why were only 16% of the colonies sectored and not 50%?

PROBLEMS

1. Outline how you would use an HFT l strain to show that some *gal* mutations of *E. coli* are due to insertion of an IS element.

2. In the experiments shown in Figures 9.14 and 9.15, what would have been observed if Tn*10* transposed by a replicative mechanism?

3. List the advantages and disadvantages of transposon mutagenesis versus chemical mutagenesis in obtaining mutations.

4. Outline how you would use transposon mutagenesis to mutagenize plasmid pBR322 in *E. coli* with transposon Tn*5*.

5. How would you determine whether a new transposon you have discovered integrates randomly into DNA?

6. You have isolated a strain of *Pseudomonas putida* that can grow on the herbicide 2,4-dichlorophenoxyacetic acid (2,4-D) as the sole carbon and energy source. Outline how you would clone the genes for 2,4-D utilization (i) by transposon mutagenesis and (ii) by complementation in the original host.

7. The Int protein of the integrating conjugative elements, such as Tn*916*, must integrate the element into the chromosome of the recipient cell after transfer. How would you show whether the Int protein of the transposon must be synthesized in the recipient cell or can be transferred in with the transposon during conjugation, much like primases are transferred during some types of plasmid conjugation?

8. How would you show whether the G segment of Mu can invert while it is in the prophage state or whether it inverts only after the phage is induced?

9. The defective prophage e14 resides in the *E. coli* chromosome and has an invertible sequence. How would you determine if its invertase can invert the invertible sequences of Mu, P1, and *S. enterica* serovar Typhimurium?

10. The red color of *S. marcescens* is reversibly lost at a high frequency. Outline how you would attempt to determine if the change in pigment is due to an invertible sequence.

11. How would you prove that the Mu phage transposon probably does not encode a resolvase?

SUGGESTED READING

Bender, J., and N. Kleckner. 1986. Genetic evidence that Tn*10* transposes by a nonreplicative mechanism. *Cell* **45:**801–815.

Biery, M. C., F. J. Stewart, A. E. Stellwagen, E. A. Raleigh, and N. L. Craig. 2000. A simple *in vitro* Tn7-based transposition system with low target site selectivity for genome and gene analysis. *Nucleic Acids Res.* **28:**1067–1077.

Bordi, C., B. G. Butcher, Q. Shi, A.-B. Hachmann, J. E. Peters, and J. D. Helmann. 2008. In vitro mutagenesis of *Bacillus subtilis* by using a modified Tn7 transposon with an outward-facing inducible promoter. *Appl. Environ. Microbiol.* **74:**3419–3425.

Casadaban, M. J., and S. N. Cohen. 1979. Lactose genes fused to exogenous promoters in one step using a Mu-*lac* bacteriophage: *in vivo* probe for transcriptional control sequences. *Proc. Natl. Acad. Sci. USA* **76:**4530–4533.

Choi, W., and R. M. Harshey. 2010. DNA repair by the cryptic endonuclease activity of Mu transposase. *Proc. Natl. Acad. Sci. USA 107:*10014–10019.

Collis, C. M., G. D. Recchia, M.-J. Kim, H. W. Stokes, and R. M. Hall. 2001. Efficiency of recombination reactions catalyzed by class I integron integrase IntI1. *J. Bacteriol.* **183:**2535–2542.

Comfort, N. C. 2001. *The Tangled Field: Barbara McClintock's Search for the Patterns of Genetic Control.* Harvard University Press, Boston, MA.

Craig, N. L. 2002. Tn7, p. 423–456. *In* N. L. Craig, R. Craigie, M. Gellert, and A. M. Lambowitz (ed.), *Mobile DNA II.* ASM Press, Washington, DC.

Derbyshire, K. M., and N. D. F. Grindley. 2005. DNA transposons: different proteins and mechanisms but similar rearrangements, p. 467–497. *In* N. P. Higgins (ed.), *The Bacterial Chromosome.* ASM Press, Washington, DC.

Foster, T. J., M. A. Davis, D. E. Roberts, K. Takeshita, and N. Kleckner. 1981. Genetic organization of transposon Tn*10. Cell* **23:**201–213.

Gill, R., F. Heffron, G. Dougan, and S. Falkow. 1978. Analysis of sequences transposed by complementation of two classes of transposition-deficient mutants of Tn*3*. *J. Bacteriol.* **136:**742–756.

Golden, J. W., S. G. Robinson, and R. Haselkorn. 1985. Rearrangement of nitrogen fixation genes during heterocyst differentiation in the cyanobacterium *Anabaena. Nature* (London) **327:**419–423.

Goryshin, I. Y., J. Jendrisak, L. M. Hoffman, R. Mais, and W. S. Reznikoff. 2000. Insertional transposon mutagenesis by electroporation of released Tn*5* transposition complexes. *Nat. Biotechnol.* **18:**97–100.

Groisman, E. O., and M. J. Casadaban. 1986. Mini-Mu bacteriophage with plasmid replicons for in vivo cloning and *lac* gene fusing. *J. Bacteriol.* **168:**357–364.

Gueguen, E., P. Rousseau, G. Duval-Valentin, and M. Chandler. 2005. The transpososome: control of transposition at the level of catalysis. *Trends Microbiol.* **13:**543–549.

Hagemann, A. T., and N. L. Craig. 1993. Tn7 transposition creates a hotspot for homologous recombination at the transposon donor site. *Genetics* 133:9–16.

Hallet, B., V. Vanhooff, and F. Cornet. 2004. DNA site-specific resolution systems, p. 145–180. *In* B. E. Funnell and G. J. Phillips (ed.), *Plasmid Biology.* ASM Press, Washington, DC.

Hughes, K. Y., and S. M. Maloy (ed.). 2007. *Methods in Enzymology*, vol. 421. *Advanced Bacterial Genetics: Use of Transposons and Phage for Genomic Engineering.* Elsevier, London, United Kingdom.

Kenyon, C. J., and G. C. Walker. 1980. DNA-damaging agents stimulate gene expression at specific loci in *Escherichia coli. Proc. Natl. Acad. Sci. USA* 77:2819–2823.

Komano, T. 1999. Shufflons: multiple inversion systems and integrons. *Annu. Rev. Genet.* **33:**171–191.

Kunkel, B., R. Losick, and P. Stragier. 1990. The *Bacillus subtilis* gene for the developmental transcription factor s^K is generated by excision of a dispensable DNA element containing a sporulation recombinase gene. *Genes Dev.* **4:**525–535.

Li, W., S. Kamtekar, Y. Xiong, G. J. Sarkis, N. D. Grindley, and T. A. Steitz. 2005. Structure of a synaptic γδ resolvase tetramer covalently linked to two cleaved DNAs. *Science* **309:**1210–1215.

Martin, S. S., E. Pulido, V. C. Chu, T. S. Lechner, and E. P. Baldwin. 2002. The order of strand exchanges in Cre-LoxP recombination and its basis suggested by the crystal structure of the Cre-LoxP Holliday junction complex. *J. Mol. Biol.* **319:**107–127.

May, E. W., and N. L. Craig. 1996. Switching from cut-and-paste to replicative Tn7 transposition. *Science* **272:**401–404.

Parks, A. R., and J. E. Peters. 2009. Tn7 elements: engendering diversity from chromosomes to episomes. *Plasmid* **61:**1–14

Parks, A. R., Z. Li, Q. Shi, R. M. Owens, M. M. Jin, and J. E. Peters. 2009. Transposition into replicating DNA occurs through interaction with the processivity factor. *Cell* **138:**685–695.

Pósfai, G., G. Plunkett III, T. Fehér, D. Frisch, G. M. Keil, K. Umenhoffer, V. Kolisnychenko, B. Stahl, S. S. Sharma, M. de Arruda, V. Burland, S. W. Harcum, and F. R. Blattner. 2006.

Emergent properties of reduced-genome *Escherichia coli. Science* **312:**1044–1046.

Rakin, A., C. Noelting, P. Schropp, and J. Heesemann. 2001. Integrative module of the high-pathogenicity island of *Yersinia. Mol. Microbiol.* **39:**407–415.

Reznikoff, W. S. 2003. Tn*5* as a model for understanding DNA transposition. *Mol. Microbiol.* **47:**1199–1206.

Rowe-Magnus, D. A., A.-M. Guerout, P. Ploncard, B. Dychinco, J. Davies, and D. Mazel. 2001. The evolutionary history of chromosomal superintegrons provides an ancestry for multiresistant integrons. *Proc. Natl. Acad. Sci. USA* **98:**652–657.

Rubin, E. J., B. J. Akerley, V. N. Novik, D. J. Lampe, R. N. Husson, and J. J. Mekalanos. 1999. In vivo transposition of mariner-based elements in enteric bacteria and mycobacteria. *Proc. Natl. Acad. Sci. USA* **96:**1645–1650.

Shapiro, J. A. 1979. Molecular model for the transposition and replication of bacteriophage Mu and other transposable elements. *Proc. Natl. Acad. Sci. USA* **76:**1933–1937.

Shi, Q., J. C. Huguet-Tapia, and J. E. Peters. 2009. Tn*917* targets the region where DNA replication terminates in *Bacillus subtilis*, highlighting a difference in chromosome processing in the firmicutes. *J. Bacteriol.* **191:**7623–7627.

Siguier, P., J. Filée, and M. Chandler. 2006. Insertion sequences in prokaryotic genomes. *Curr. Opin. Microbiol. 9:*526–531.

Simon, M., J. Zeig, M. Silverman, G. Mandel, and R. Doolittle. 1980. Phase variation: evolution of a controlling element. *Science* **209:**1370–1374.

Simon, R., U. Preifer, and A. Puhler. 1983. A broad host range mobilization system for in vivo genetic engineering: transposon mutagenesis in gram negative bacteria. *Biotechnology* **1:**784–790.

Skelding, Z., J. Queen-Baker, and N. L. Craig. 2003. Alternative interactions between the Tn7 transposase and the Tn7 target DNA binding protein regulate target immunity and transposition. *EMBO J. 22:*5904–5917.

Toleman, M. A., P. M. Bennett, and T. R. Walsh. 2006. IS*CR* elements: novel gene-capturing systems of the 21st century. *Microbiol. Mol. Biol. Rev.* **70:**296–316.

van de Putte, P., R. Plasterk, and A. Kuijpers. 1984. A Mu *gin* complementing function and an invertible DNA region in *Escherichia coli* K-12 are situated on the genetic element e14. *J. Bacteriol.* **158:**517–522.

van Opijnen, T., K. L. Bodi, and A. Camilli. 2009. Tn-seq: high-throughput parallel sequencing for fitness and genetic interaction studies in microorganisms. *Nat. Methods* **6:**767–772.

CHAPTER 10

Molecular Mechanisms of Homologous Recombination

WHEN WE TALK ABOUT RECOMBINATION, WE are usually referring to homologous recombination, which occurs more generally than the other types of recombination discussed in chapter 9. Homologous recombination can occur between any two DNA sequences that are the same or very similar. Depending on the organism, homology-dependent recombination can occur between regions of homology that are as short as 23 bases, although longer homologies produce more frequent crossovers. All organisms on Earth probably have some mechanism of homologous recombination, suggesting that recombination is very important for species survival. It is well advertised that the new combinations of genes obtained through recombination allow the species to adapt more quickly to the environment and speed up the process of evolution. However, it has become increasingly clear that the most important role for homologous recombination is likely to involve the facilitation of DNA replication. This is especially clear in bacteria, which invariably have a single origin of DNA replication in their chromosome or in each of their chromosomes (this is also true for most plasmids and bacteriophages), making each DNA replication fork extremely important to maintain. As we will see below, nicks in either of the two template strands cause a DNA replication fork to collapse. A role for homologous recombination in restarting collapsed DNA replication forks almost certainly provided the original selective advantage for the evolution of homologous recombination because of the immediate advantage that it provides an organism. While the other benefits of recombination are certainly important, benefits allowing species to better adapt over evolutionary time are unlikely to have provided the selection that was needed for the evolution of these systems.

doi:10.1128/9781555817169.ch10

Because of its importance in genetics, homologous recombination has already been mentioned in previous chapters, for example, in discussions of deletion and inversion mutations and genetic analysis. Determination of recombination frequencies allows us to measure the distance between mutations and thus can be used to map mutations with respect to each other, as we discussed in chapters 3 and 7, among others. Moreover, clever use of recombination can take some of the hard work out of cloning genes and making DNA constructs, and we have already referred to some of these techniques. When we discussed the use of recombination for genetic mapping and many other types of applications in previous chapters, we used a simplified description of recombination: the strands of two DNA molecules were suggested to break at a place where they both have the same sequence of bases, and then the strands of the two DNA molecules join with each other to form a new molecule. In this chapter, we focus on the actual molecular mechanism of recombination—what actually happens to the DNA molecules involved. We will see that the process of homologous recombination is inextricably intertwined with the process of DNA replication. We also discuss the proteins involved in recombination, mostly in *Escherichia coli*, the bacterium for which recombination is best understood and which has served as the paradigm for recombination in all other organisms.

Homologous Recombination and DNA Replication in Bacteria

One of the most gratifying times in science comes when phenomena that were originally thought to be distinct are discovered to be different manifestations of the same process. Such a discovery is usually followed by rapid progress as the mass of information accumulated on the different phenomena is combined and reinterpreted. This is true of the fields of homologous recombination and DNA replication and the role they jointly play in DNA repair. Because of the connection between the replication, recombination, and repair processes, they are sometimes called the "three R's" in DNA metabolism. As discussed in chapter 1, all bacteria that have been examined to date appear to have a single origin of chromosomal replication. This can be gleaned from looking at the DNA sequence of a bacterial genome, because of many features in the sequence that indicate the direction of DNA replication (see Box 1.1).

While having a single *oriC* offers distinct advantages for regulating DNA replication and coordinating DNA replication with chromosome segregation, it also makes the process vulnerable to problems with the template strands. One such issue arises when there is a nick in the template strand. Even a single nick in the template strand will cause a DNA replication fork to collapse (Figure 10.1). Collapse of one arm of the DNA replication fork at a nick results in a broken end and the inability of the replication fork to continue on the contiguous strand. Presumably, without a way to deal with this event, a considerable amount of chromosomal DNA would need to be degraded and DNA replication would need to be started again at *oriC*. Bacterial cells likely could not tolerate this kind of loss. This is especially true because nicks in the template strands are very common; replication forks are believed to collapse nearly every generation at nicks generated as a by-product of the functioning of many enzymes that act on DNA, as discussed in chapters 1, 9, and 11. Homologous recombination provides a mechanism to restart DNA replication using the broken end of the chromosome (Figure 10.1). Viewing homologous recombination in this light provides an explanation for numerous diverse experimental observations and strongly suggests how homologous recombination occurs during P1 transduction and Hfr crosses and with linear DNA fragments that are transferred into the cell by a process that involves extensive DNA replication (see below).

Early Evidence for the Interdependence of Homologous Recombination and DNA Replication

Replication can be initiated in a number of different ways, many of which involve recombination. The process of completing homologous recombination itself very likely requires extensive DNA replication. However, it took a long time to recognize the interdependence of these processes, and they were studied in isolation for decades. Early on, it was known that some phages, such as T4, need recombination functions to initiate replication (see chapter 7 and Mosig, Suggested Reading). In these phages, recombination intermediates function to initiate DNA replication later in infection. However, this was thought to be unique to these phages.

Normally, initiation of chromosomal replication in bacteria does not require recombination functions. However, after extensive DNA damage due to irradiation or other agents, a new type of initiation, which does require recombination, comes into play. This type of initiation was originally called stable DNA replication because it continued even after protein synthesis stopped (see Kogoma, Suggested Reading). Initiation of DNA replication at the chromosomal *oriC* site normally requires new protein synthesis, and so, in the absence of protein synthesis, replication continues only until all the ongoing rounds of replication are completed and no new rounds are initiated (see chapter 1). However, after extensive DNA damage, such as irradiation, initiation of new rounds of replication occurs even in the absence of

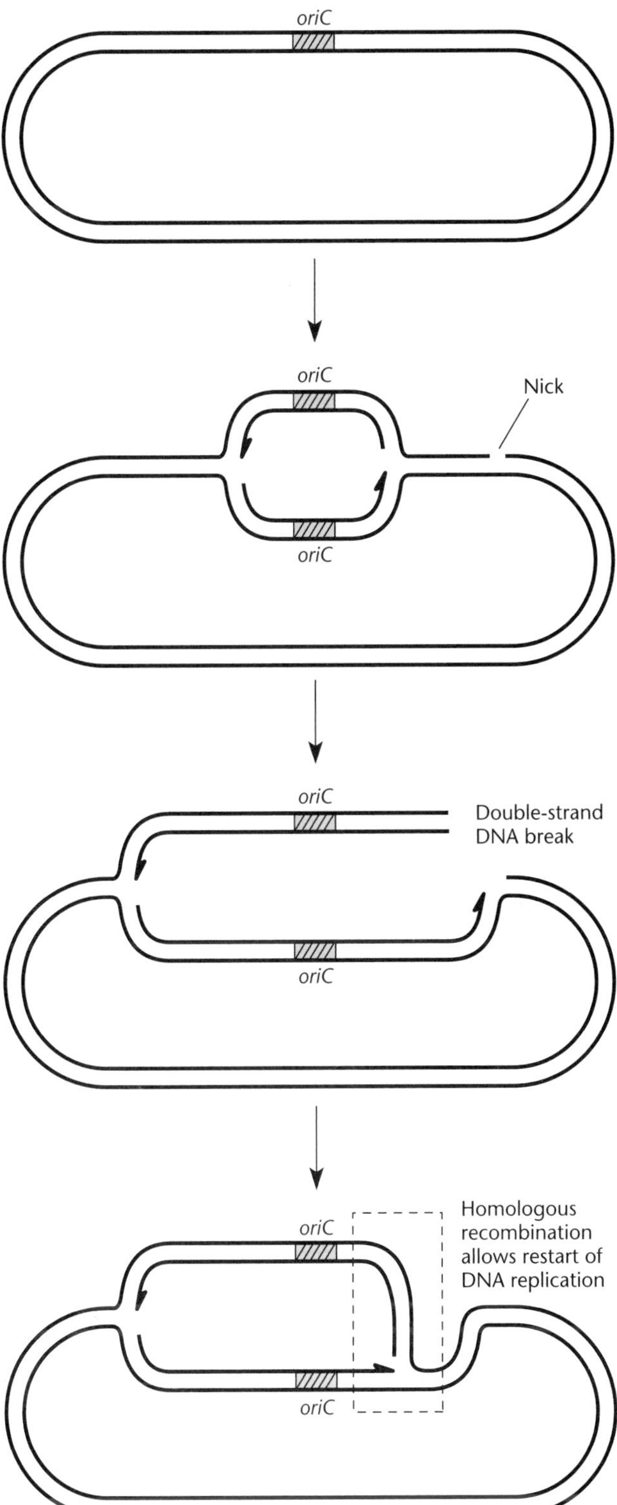

Figure 10.1 Replication forks initiated at *oriC* can collapse when there are nicks in the template strand. In the DNA strand with the nick, a double-strand broken end results, and the replication fork collapses. Recombination with the copy of the chromosome forms a structure that can reprime DNA replication to allow chromosomal replication to continue.
doi:10.1128/9781555817169.ch10.f10.1

protein synthesis. It is now believed that recombination is required for stable DNA replication because double-strand DNA breaks formed during irradiation are used to initiate DNA replication through the same process that reestablishes collapsed DNA replication forks (see Kuzminov and Stahl, Suggested Reading).

Homologous recombination allows the broken end that results from collapse of the DNA replication fork to interact with the homologous region that was just replicated (Figure 10.1). However, another process is required to allow a full DNA replication fork to be reestablished. This process is known to involve the Pri proteins, which also have a long and interesting history. There are three Pri proteins, PriA, PriB, and PriC, as well as another protein, DnaT, that help reload the replicative helicase DnaB through the use of DnaC. These proteins were first discovered because they are required for the initiation of replication of the DNA of some single-stranded DNA phages (see chapter 7), so it was assumed that they also play a role in *E. coli* DNA replication. However, mutations that inactivate the products of these genes are not lethal, although the mutants are defective in recombination and are more sensitive to DNA-damaging agents. Later work showed that different combinations of Pri proteins are responsible for acting on different types of substrates that can result when a DNA replication fork collapses (see Lopper et al., Suggested Reading). The findings that the initiation of DNA replication with the Pri proteins is actually required for homologous recombination and that a role for recombination is essential for restarting DNA replication after replication forks collapse provided some of the evidence for the unity of these processes.

The Molecular Basis for Recombination in *E. coli*

As with many cellular phenomena, much more is known about the molecular basis for recombination in *E. coli* than in any other organism. At least 25 proteins involved in recombination have been identified in *E. coli*, and specific roles have been assigned to many of these proteins (Table 10.1).

Chi (χ) Sites and the RecBCD Complex

The first analysis of the genetic requirements for recombination in *E. coli* used Hfr crosses (see Clark and Margulies, Suggested Reading, and chapter 3). The *recB* and *recC* genes were among the first *rec* genes found because their products are required for recombination after Hfr crosses (see Clark and Margulies, Suggested Reading, and "Genetic Analysis of Recombination in Bacteria" below). Their products were later shown to also be required for transductional crosses. The products of these

Table 10.1 Some genes encoding recombination functions in *E. coli*

Gene	Mutant phenotype	Enzymatic activity	Probable role in recombination
recA	Recombination deficient	Enhanced pairing of homologous DNAs	Synapse formation
recBC	Reduced recombination	Exonuclease, ATPase, helicase, χ-specific endonuclease	Initiates recombination by separating strands, degrading DNA up to a χ site, and loading RecA
recD	Rec⁺ χ independent	Stimulates exonuclease	Degrading 3′ends
recF	Reduced plasmid recombination	Binds ATP and single-stranded DNA	RecA loading at single-strand gaps
recJ	Reduced recombination in RecBC⁻	Single-stranded exonuclease	Processing at single-strand gaps
recN	Reduced recombination in RecBC⁻	ATP binding	RecA loading at single-strand gaps
recO	Reduced recombination in RecBC⁻	DNA binding and renaturation	RecA loading at single-strand gaps
recQ	Reduced recombination in RecBC–	DNA helicase	Processing at single-strand gaps
recR	Reduced recombination in RecBC⁻	Binds double-stranded DNA	RecA loading at single-strand gaps
recG	Reduced Rec in RuvA⁻ RuvB⁻ RuvC⁻	Branch-specific helicase	Migration of Holliday junctions, dismantling recombination structures
ruvA	Reduced recombination in RecG⁻	Binds to Holliday junctions	Migration of Holliday junctions
ruvB	Reduced recombination in RecG⁻	Holliday junction-specific helicase	Migration of Holliday junctions
ruvC	Reduced recombination in RecG⁻	Holliday junction-specific nuclease	Resolution of Holliday junctions
priA, priB, priC, dnaT	Reduced recombination	Helicase	Restart of replication forks

genes, together with the product of another gene, *recD*, form a heterotrimer, accordingly named the RecBCD nuclease. The *recD* gene product is not required for recombination after such crosses, so the gene had to be found in other ways (see below). We now know that the RecBCD enzyme is required for the recombination that occurs after conjugation and transduction because the small pieces of DNA transferred into cells by Hfr crosses or by transduction are the natural substrates for the RecBCD enzyme. The RecBCD enzyme processes the ends of these pieces to form single-stranded 3′ ends, which can then invade the chromosomal DNA to form recombinants (see below). Most importantly, RecBCD is of paramount importance in *E. coli* and many other bacteria for tailoring the double-strand break that is formed after a replication fork collapses at nicks in the template DNA.

HOW RecBCD WORKS

The RecBCD complex is a remarkable enzyme with many activities. It has single-stranded DNA endonuclease and exonuclease activities, as well as DNA helicase and DNA-dependent ATPase activities (see Taylor and Smith, Suggested Reading). To put all these activities in perspective, it is useful to think of the RecBCD complex as a DNA helicase with associated nuclease activities. In actuality, the RecBCD complex has two helicase activities of opposite polarity. When it moves down a duplex DNA, the two helicase motors work in tandem, because the strands are of opposite polarity to one another (Figure 10.2). Having two motors working together contributes to the speed and processivity of the enzyme, especially if there are gaps or other types of damage in the DNA strands. In its role of processing a double-strand broken end to prepare it for recombination, it is specifically capable of recognizing only broken ends where the top and bottom strands are flush or nearly completely flush (something that is important when we consider the two major "pathways" for recombination below). A 3′ or 5′ extension will prevent RecBCD from loading onto a DNA. As RecBCD unwinds the DNA, it loops out the top strand of the DNA as it goes (Figure 10.2B). The strand containing these loops is cut into small pieces by its 3′-to-5′ exonuclease activity, leaving the 5′-to-3′ strand mostly intact but vulnerable to occasional breaks caused by the same enzyme. This process continues for up to 30 kb or until the RecBCD protein encounters a sequence on the DNA called a chi (χ) site (for crossover hotspot instigator), which in *E. coli* has the sequence 5′-GCTGGTGG-3′ but in other types of bacteria has somewhat different sequences. These sites were first found because they stimulate recombination in phage λ (see "Discovery of χ Sites" below). When RecBCD encounters a χ site, its 3′-to-5′ exonuclease activity is inhibited but its 5′-to-3′ exonuclease activity is stimulated, leading to formation of the free 3′ single-stranded tail, as shown in Figure 10.2. Note that the χ sequence does not have twofold rotational symmetry like the sites recognized by many restriction endonucleases. This means that the sequence will be recognized on only one strand of the DNA and the RecBCD nuclease will pass over the sequence if it occurs on the other strand, making the orientation of the χ site extremely important, as we discuss below.

Figure 10.2 Model for promotion of recombination initiation at a χ site by the RecBCD complex. **(A)** RecBCD loads onto a double-stranded end. **(B)** Its helicase activity separates the strands, and its 3′-to-5′ nuclease activity degrades both strands until it encounters a χ site (χ). **(C)** The χ site signals the attenuation of the 3′-to-5′ nuclease and upregulation of the 5′-to-3′ nuclease. As a result of these changes in the activity of the RecBCD complex following interaction with the χ site, a single-stranded 3′ extension is created. Interaction with the χ site also signals the RecBCD complex to load RecA (blue circles) on the single-stranded DNA, forming it into an extended structure. **(D)** This RecA nucleoprotein filament can then invade a complementary double-stranded DNA, forming a D loop or a triple-stranded structure (see Figure 10.5).
doi:10.1128/9781555817169.ch10.f10.2

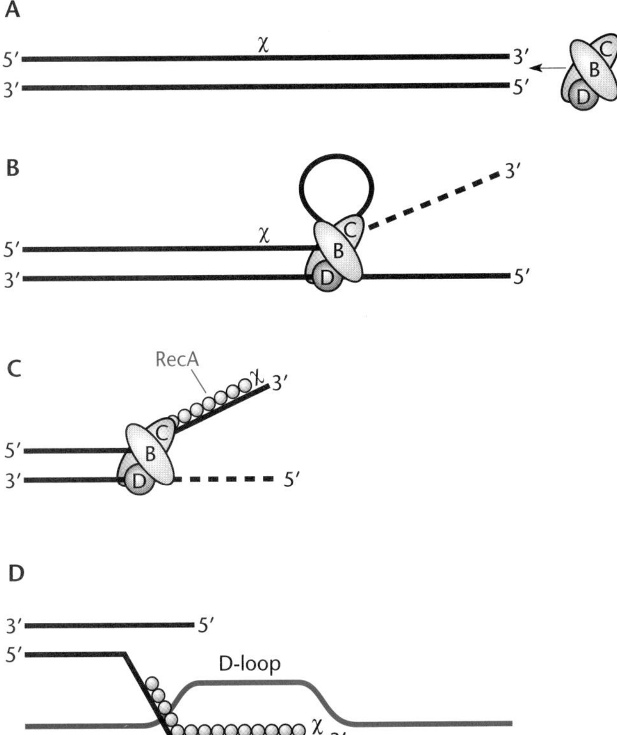

After a RecBC enzyme has formed a 3′ single-stranded tail on the DNA, it directs a RecA protein molecule to bind to the DNA next to where it is bound (see below). This is called cooperative binding, because one protein is helping another to bind. In more physicochemical terms, the incoming RecA protein makes contact both with the DNA and with the RecBC protein already on the DNA, which helps to stabilize the binding of RecA to the DNA. This cooperative binding is necessary, because another protein called single-stranded DNA-binding (SSB) binds single-stranded DNA with greater affinity than RecA. The RecD protein may inhibit RecA binding so that it does not bind until RecD has been inhibited by encountering a χ site. More RecA protein molecules then bind cooperatively to the first RecA protein to form a helical nucleoprotein filament to prepare for the next step in recombination, synapse formation, as discussed in more detail below. This dependence on RecBCD for loading RecA may help ensure that RecA does not bind to just any single-stranded DNA in the cell, only those that are destined for recombination.

The discovery of χ sites and their role in recombination came as a complete surprise, and it took many years and a lot of clever experimentation to figure out how they work (see "Discovery of χ Sites" below). A detailed model consistent with much of the available information has emerged from this work. According to this detailed model, the RecD subunit of the RecBCD enzyme does not itself have any exonuclease activity but instead stimulates the 3′-to-5′ exonuclease activity of the RecB subunit. The χ sites work by inhibiting the RecD subunit and thereby indirectly inhibiting the 3′-to-5′ exonuclease activity and stimulating the 5′-to-3′ nuclease activity of the remainder of the enzyme. Thus, as the RecBCD enzyme moves along the DNA, opening the strands, it largely degrades the 3′-to-5′ strand until it encounters the sequence 5′-GCTGGTGG-3′ (the χ sequence) on the strand being degraded. The DNA with this sequence can bind to the RecD subunit and inhibit it. The RecBCD enzyme continues to move on past the site, still degrading the 5′-to-3′ strand but leaving the 3′-to-5′ strand intact. The end result is a single-stranded 3′-ended tail that contains the χ site sequence at its end, as shown in Figure 10.2.

The evidence for this detailed model of χ site action is both biochemical and genetic. First, the RecBC enzyme does have some 3′-to-5′ exonuclease activity in the absence of the RecD subunit, but this activity is greatly stimulated by the RecD subunit. Because of this, a linear DNA can be introduced into a RecD⁻ mutant of *E. coli* by transformation (see chapter 6) and is not degraded. Another prediction of the model, which is fulfilled, is that RecD⁻ mutants are proficient for recombination, and this recombination does not require χ sites. This is due to the fact that, since the RecBC enzyme lacking the RecD subunit still has some of its helicase activity to separate the strands and these single-stranded ends are not degraded even if they do not contain a χ site, they are available to be loaded with RecA and to invade other DNAs and promote recombination. This property of RecD⁻ mutants of not degrading linear DNA but still being proficient for recombination is what makes RecD⁻ mutants of *E. coli* useful for gene replacements with linear DNA.

This model also explains why recombination is stimulated only on the 5′ side of a χ site. Until the RecBCD enzyme reaches a χ site, the displaced strand is degraded and so is not available for recombination. Only after the RecBCD protein passes completely through a χ site will the strand survive to invade another DNA, so that only DNA on the 5′ side of the site survives. This model is also supported by the known enzymatic activities of the RecBCD complex, as well as by electron microscopic visualization of the RecBCD complex acting on DNA. It has also received experimental support from the results of in vitro experiments with purified RecBCD enzyme and DNA containing a χ site (see Dixon and Kowalczykowski, Suggested Reading). Recombination on the 5′ side of χ is further stimulated because this is the region where RecBCD actively loads RecA.

WHY χ?

The hardest question to answer in biology is often "Why?" To answer this question with certainty, we must know everything about the organism and every situation in which it might find itself, both past and present. Nevertheless, it is tempting to ask why *E. coli* and many other bacteria use such a complicated mechanism involving χ sites for a major pathway of recombination. Adding to the mystery is the fact that they would not need χ sites at all if they were willing to dispense with the RecD subunit of the RecBCD nuclease. As mentioned above, in the absence of RecD, recombination proceeds just fine without χ sites, and the cells are viable.

One idea is that the self-inflicted dependency on χ sites allows the RecBCD nuclease to play a dual role in recombination and in defending against phages and other foreign DNAs. Small pieces of foreign DNA, such as a phage DNA entering the cell, are not likely to have a χ site, since 8-bp sequences like χ occur by chance only once in about 65 kb, longer than many phage DNAs. The RecBCD nuclease degrades a DNA until it encounters a χ site, and if it does not encounter a χ site, it degrades the entire DNA. *E. coli* DNA, in contrast, has many more of these sites than would be predicted by chance. In support of the idea that RecBCD is designed to help defend against phages is the extent to which phages go to avoid degradation by the enzyme. Many phages avoid degradation by RecBCD by attaching proteins to the ends of their DNA or by making proteins that inhibit RecBCD in more direct ways, such as the Gam protein of λ (see chapter 8 and below).

Another possible reason for having a complicated system that involves χ sites is that they control the direction of DNA replication following a double-strand break in the chromosome. Ionizing radiation, oxygen radicals, and other forms of DNA damage (see chapter 11), along with the action of restriction endonucleases, are just some of the ways that double-strand DNA breaks can occur in the chromosome. These breaks can be processed by the RecBCD complex to allow RecA to be loaded for the formation of a **D loop** that can be used for priming DNA replication for DNA repair on a copy of the chromosome during DNA replication (Figure 10.3). A D loop is found when a single strand of DNA invades a homologous duplex DNA, displacing one of the strands when it binds via hydrogen bonding to the other strand. The displaced strand of DNA forms the loop that gives the structure its name. However, unlike the case where a replication fork collapses at a nick in the template strand (Figure 10.1), there would be two ends where RecBCD could load and, with the help of RecA DNA replication forks, could be initiated in either direction (i.e., toward *oriC* or toward the terminus region) (Figure 10.3). However, the distribution of χ sites makes it very likely that replication will be directed toward the terminus region. As described in Box 1.1, χ sites are highly overrepresented in the *E. coli* chromosome, occurring on average about once every 4.5 kb. Even more significant, this enrichment for the polar χ sites is found for χ sites in only one orientation relative to the direction of DNA replication. Therefore, because of the overrepresentation of χ sequences in one orientation between *oriC* and the terminus region, the fates of two RecBCD complexes that load at a double-strand break will likely be very different (Figure 10.3B). The RecBCD complex that degrades DNA in the direction of *oriC* (1 in Figure 10.3B) will quickly encounter a χ site in the correct orientation and, with RecA, aid in the initiation of a DNA replication fork (Figure 10.3C). Therefore, replication in this scenario travels in the same direction as replication events that are normally initiated at *oriC*. However, a RecBCD complex that degrades DNA in the direction of the terminus (labeled 2 in Figure 10.3B) is much less likely reach a χ site that will attenuate its exonuclease activities. Given that RecBCD does not "see" every χ site, the few sites encountered in this direction may never be recognized, contributing to the degradation of the extra "arm" of the chromosome (Figure 10.3C). While homologous recombination initiated by the RecBCD complex is used for the repair of DNA double-strand breaks in *E. coli*, bacteria can have additional systems for the repair of these breaks (Box 10.1).

Whatever the purpose or purposes of χ sites, they seem to be very common among bacteria. Many bacteria from different branches (including *Bacillus subtilis*) contain an enzyme called AddAB (or RexAB) with a function similar to that of RecBCD. The *B. subtilis* enzyme is known to function similarly to RecBCD, except that the 5′-to-3′ strand might be degraded more extensively even before a χ site is encountered and the χ site has a different sequence in *B. subtilis*.

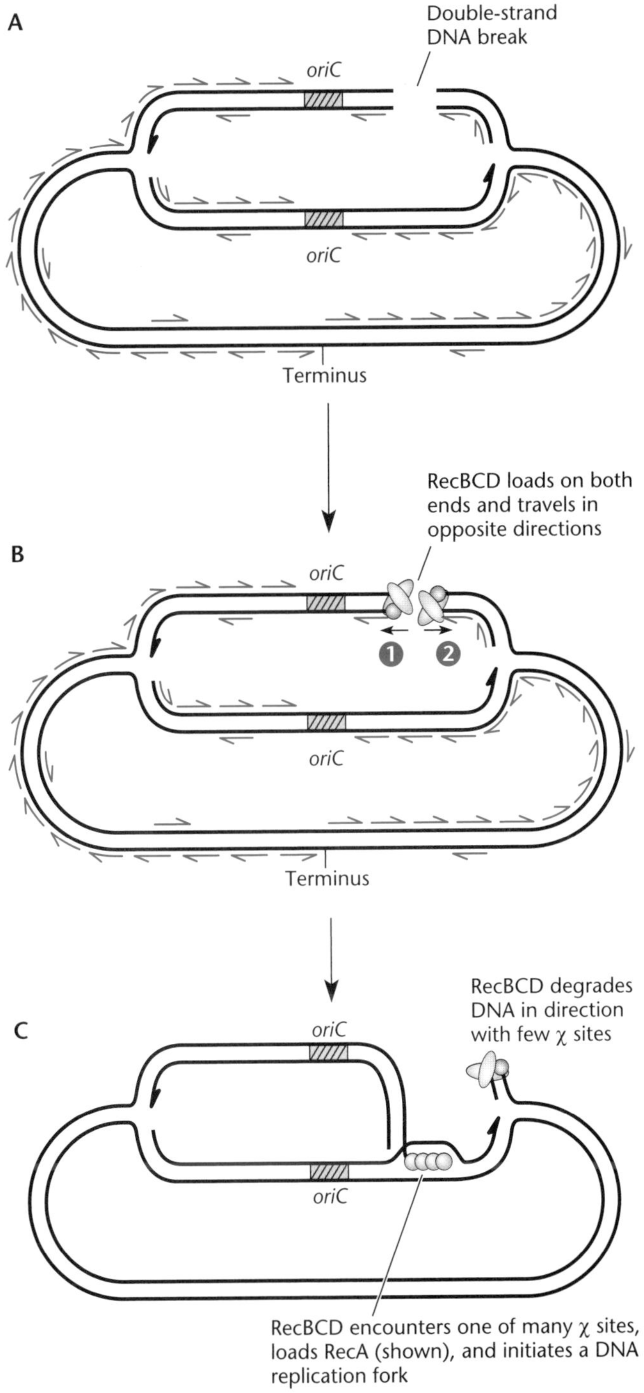

A

Double-strand DNA break

oriC

oriC

Terminus

B

RecBCD loads on both ends and travels in opposite directions

oriC

1 2

oriC

Terminus

C

RecBCD degrades DNA in direction with few χ sites

oriC

oriC

RecBCD encounters one of many χ sites, loads RecA (shown), and initiates a DNA replication fork

Figure 10.3 Model for how χ sites can help RecBCD load RecA to direct DNA replication forks toward the terminus during replication-mediated double-strand break repair. **(A)** A DNA double-strand break occurs during DNA replication. **(B)** Two ends with DNA double-strand breaks are available for processing by the RecBCD complex. One RecBCD complex, 1, progresses toward the origin, where many correctly oriented polar χ sites are encountered. A second RecBCD complex, 2, progresses toward the terminus, where there is little chance that it will encounter a χ site that will attenuate its nuclease activity. **(C)** The RecBCD complex progressing toward the origin quickly encounters a χ site, which attenuates its nuclease activity on the top strand and actively loads RecA protein that can produce a structure on the sister chromosome to initiate DNA replication toward the terminus. The RecBCD complex degrading toward the terminus is unlikely to encounter a χ site and continues degrading the arm of the chromosome, possibly providing another important role for removing this structure, which could stop subsequent DNA replication. The terminus region, where DNA replication terminates, and the origin of chromosomal replication (*oriC*) are shown in all panels. The χ sites are indicated in panels A and B by blue arrows with half heads.
doi:10.1128/9781555817169.ch10.f10.3

The RecF Pathway

The other major pathway used to prepare single-stranded DNA for invading another DNA in *E. coli* is the RecF pathway (Figure 10.4). This pathway is so named because it requires the products of the *recF* gene, as well as the *recO*, *recR*, *recQ*, and *recJ* genes. This pathway is used under different circumstances from the RecBCD pathway because it cannot prepare DNA ends for recombination and is used mostly to initiate recombination at single-stranded gaps in DNA, as shown in Figure 10.4. These gaps may be created during repair of DNA damage or by the movement of the replication fork past a lesion in the lagging strand of DNA, leaving a single-stranded gap (see chapter 1). The RecF proteins can then prepare the single-stranded DNA at the gap to invade a sister DNA. This structure can be used to restart replication, making the RecF pathway important in recombination repair of DNA damage (see chapter 11).

The RecF pathway does not have a "superstar" like RecBCD that can do it all, so it needs a number of proteins to perform the tasks that the RecBCD complex can perform alone (see Handa et. al., Suggested Reading). The RecQ protein is a helicase, like RecBCD, but it lacks a nuclease activity to degrade the strands it displaces. RecJ may provide the exonuclease activity that RecQ lacks, helping to extend the single-stranded gaps. The RecQ protein also lacks the ability to displace the SSB protein and therefore to load the RecA protein on the single-stranded DNA it creates with its helicase

BOX 10.1

Other Types of Double-Strand Break Repair in Bacteria

The RecBCD pathway and the related AddAB (and RexAB) system allow efficient repair of DNA double-strand breaks in bacteria. In many well-studied bacterial species, the RecBCD/AddAB systems process the broken ends to form a 3′ extension of DNA that is actively loaded with RecA for recombination on a "sister" chromosome that results from DNA replication (panel A of the figure; also see Figure 10.2). In other bacteria, the RecF pathway seems to participate in a similar reaction to allow the repair to use a sister chromosome as a template. Recombination-based systems are well represented in eukaryotes, but in many eukaryotes, there are also systems for directly joining breaks where the ends have no homology in a process called nonhomologous end joining (NHEJ) (panel B of the figure). In these systems, the DNA ends are directly joined together. The NHEJ system is inherently risky, because if there is any loss of sequence information before the ends are joined, there will be a permanent loss of genetic information. A second and more serious risk to genome stability comes from the chance of putting the wrong two ends together. One common mechanism of NHEJ in eukaryotes involves an end-binding protein; a heterodimer, Ku70/80; and a ligase that is responsible for working with the end-binding protein to seal the ends of the DNA break. Interestingly, it was shown that many bacteria (but not *E. coli*) have a protein that is homologous to the Ku70/80 proteins. Of additional interest, in many cases, the gene encoding this protein, called Ku, was adjacent to a predicted DNA ligase (e.g., LigD). Subsequent work has showed that the Ku and LigD systems are capable of repairing DNA double-strand breaks in a manner similar to NHEJ in eukaryotes. In addition to a ligase domain, LigD also has nuclease and polymerase domains, which can act as a template-independent terminal transferase on single-stranded or blunt-ended DNAs. The LigD ligase and other ligases not normally involved in sealing nicks during DNA synthesis and repair can work with Ku to allow NHEJ.

Why would some bacteria need an additional mechanism for DNA double-strand break repair beyond the efficient RecBCD/AddAB systems? Intriguingly, many of the types of bacteria that have Ku- and ligase-based systems are found in organisms that spend considerable amounts of time in a quiescent or nonreplicating state. For example, Ku and LigD are found in the spore-forming bacterium *B. subtilis*. Direct end joining may be required when a spore matures and DNA breaks need to be repaired in the absence of a second copy of the chromosome to act as a template for repair. *Mycobacterium tuberculosis*, the causative agent of tuberculosis, also has a Ku-based repair system, which may aid its DNA

doi:10.1128/9781555817169.ch10.Box10.1.f

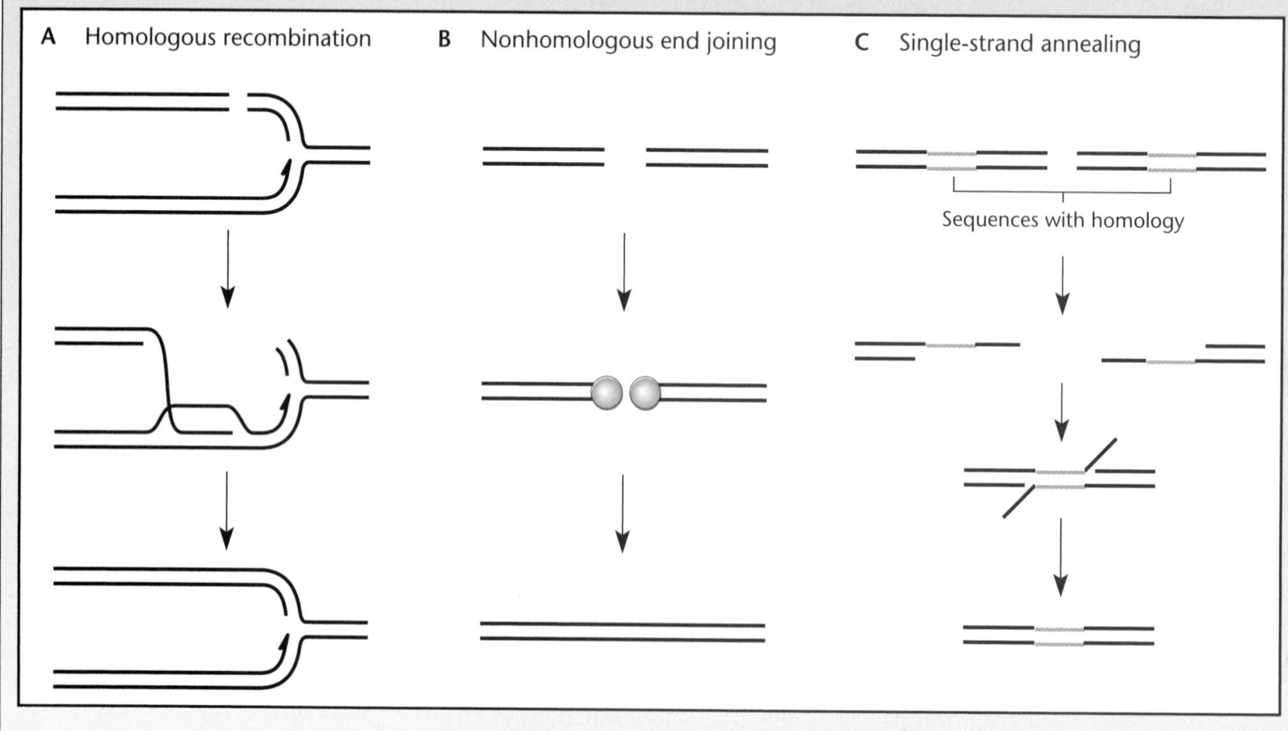

BOX 10.1 (continued)

Other Types of Double-Strand Break Repair in Bacteria

repair during long periods of inactivity in the latent stage of infection that is characteristic of the disease. Interestingly, eukaryotes seem to regulate their types of DNA repair according to the phase of growth; when cells are not replicating (i.e., in G_1), cells use NHEJ for DNA repair, but when they are in late S/G_2, where a second copy of the chromosome is found, they use homologous recombination for the repair of double-strand breaks. It is also thought that pathogens may benefit from the Ku- and ligase-based NHEJ system. This is because an important part of the defense response of animals involves mechanisms to damage the pathogen's DNA. *M. tuberculosis* actually appears to use three distinct types of double-strand break repair: a form of homology-based repair, as described for the RecBCD/AddAB systems; a Ku- and ligase-based system; and a third system based on single-strand annealing (SSA). SSA is also based on homology, but the homologies are shorter and are found in the same chromosome (panel C of the figure). Repair by SSA is highly error prone and results in deletions between the small regions of homology. Related forms of double-strand break repair involving portions of these repair pathways also exist.

References

Aravind, L., and E. V. Koonin. 2001. Prokaryotic homologs of the eukaryotic DNA-end-binding protein Ku, novel domains in the Ku protein and prediction of a prokaryotic double-strand break repair system. *Genome Res.* **11**:1365–1374.

Della, M., P. L. Palmbos, H. M. Tseng, L. M. Tonkin, J. M. Daley, L. M. Topper, R. S. Pitcher, A. E. Tomkinson, T. E. Wilson, and A. J. Doherty. Mycobacterial Ku and ligase proteins constitute a two-component NHEJ repair machine. *Science* **306**:683–685.

Gupta, R., D. Barkan, G. Redelman-Sidi, S. Shuman, and M. S. Glickman. 2010. Mycobacteria exploit three genetically distinct DNA double-strand break repair pathways. *Mol. Microbiol.* **79**:316–330.

activity. As we have seen, the RecBCD protein solves this problem by helping to load the first RecA protein onto the DNA. More RecA can then bind cooperatively, displacing SSB in the process. The RecF, RecO, and RecR proteins appear to help the RecA protein to bind to and coat the single-stranded DNA created by RecQ and RecJ. They do this by helping the first RecA protein bind to one end of the single-stranded gap and then displacing the SSB protein from the single-stranded DNA in the gap, as shown in Figure 10.4. The RecF proteins may also stop RecA from invading the neighboring double-stranded DNA before the synapse with another DNA occurs (see below).

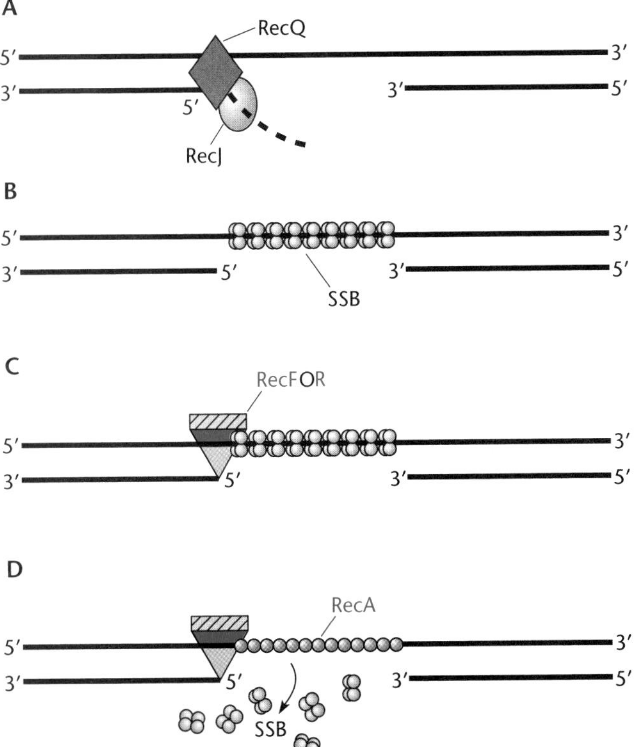

Figure 10.4 Model for recombination initiation by the RecF pathway. **(A)** RecQ, a helicase, and RecJ, an exonuclease, process gapped DNA. **(B)** SSB protein (gray circles) coats the gap. **(C)** RecF, RecO, and RecR bind to the SSB-coated gap. **(D)** The RecF complex nucleates the RecA nucleoprotein filament assembly (blue circles), displacing the SSB protein. The RecFOR proteins can then prepare the single-stranded DNA at the gap to invade a sister DNA (not shown).
doi:10.1128/9781555817169.ch10.f10.4

Synapse Formation and the RecA Protein

Once a single-stranded DNA is created by the RecBCD or RecF pathway, it must find and invade another DNA with the complementary sequence. The joining of two DNAs in this way is called a **synapse**, and the process by which an invading strand can replace one of the two strands in a double-stranded DNA is called **strand exchange**. This is a remarkable process. Somehow the single-stranded DNA must find its complementary sequence by scanning possibly all of the double-stranded

DNA in the cell, which even in a simple bacterium can be more than 1 mm long! How, then, could the single-stranded DNA know when the sequence is complementary, especially if it can only scan the outside of the DNA double helix and the bases of the double-stranded DNA are on the inside? Not only is synapse formation remarkably fast, it is also remarkably efficient. Once an incoming single-stranded DNA enters the cell, for example, during an Hfr cross, it finds and recombines with its complementary sequence in the chromosome almost 100% of the time.

Searching for complementary DNA is the job of the RecA protein, whose role in recombination is outlined in Figure 10.5. As the single-stranded DNA is created by RecBC or RecQ and RecJ, it is coated with RecA, as facilitated by RecBC or RecF, to form a nucleoprotein filament. As mentioned above and shown in Figures 10.2 and 10.4, the RecBCD and RecF proteins help load RecA onto single-stranded DNA in competition with SSB protein, which is a prerequisite to forming the filament. The DNA in the RecA filament is also helical but much extended relative to the normal helix in DNA; it takes about twice as many nucleotides to complete a helical turn. The helical nucleoprotein filament then scans the double-stranded DNA in the cell to find a homologous sequence. Some evidence suggests that it might scan double-stranded DNA through its major groove (see chapter 1) and that it can base pair with its complementary sequence in the major groove once it finds it without transiently disrupting the helix. Once the nucleoprotein filament finds its complementary sequence, it pairs with it. There is still some question of what happens next. Either it displaces one strand of the double-stranded DNA to form a D loop, as shown in Figure 10.2, or, as some evidence suggests, it actually forms a **triple-stranded structure**, as shown in Figure 10.5. Other evidence calls into question the formation of triple-stranded structures, and it has been proposed instead that RecA can approach DNA through its minor groove and somehow flip the bases out to test for complementarity. For our present purposes, we often use D loops to schematize strand displacement because they are easier to draw.

While the details of how a single-stranded DNA-RecA filament and a double-stranded DNA find each other and what kinds of structures are formed remain obscure, a number of observations have been made that may shed light on this process. One observation is that a single-stranded RecA nucleoprotein filament may somehow change, or "activate," double-stranded DNAs merely by binding to them, even if the two DNAs are not complementary. This activation presumably has something to do with the way in which the RecA nucleoprotein filament scans DNA looking for its complementary sequence. Once the strands of the double-stranded DNA are activated, even by a noncomplementary RecA nucleoprotein filament, a complementary single-stranded DNA not bound to RecA can invade it and exchange with one of its strands. This process was named *trans* activation because the RecA nucleoprotein filament that activated the DNA is not necessarily the one that invades it (see Mazin and Kowalczykowski, Suggested Reading). It is not clear what happens to a double-stranded DNA when it is activated by a RecA nucleoprotein filament, but the helix may be transiently extended and the strands may be partially separated to allow the single-stranded DNA to search for its complementary sequence. This is an area that needs further investigation.

Figure 10.5 Model for synapse formation and strand exchange between two homologous DNAs by RecA protein. The RecA protein (in blue), bound to a single-stranded end, formed as shown in Figures 10.2 and 10.4, and forced into an extended helical structure, pairs with a homologous double-stranded DNA in its major groove to form a three-stranded structure. RecA can drive this three-stranded structure into adjacent double-stranded DNA by a "spooling" mechanism, forming a four-stranded Holliday junction (not shown). doi:10.1128/9781555817169.ch10.f10.5

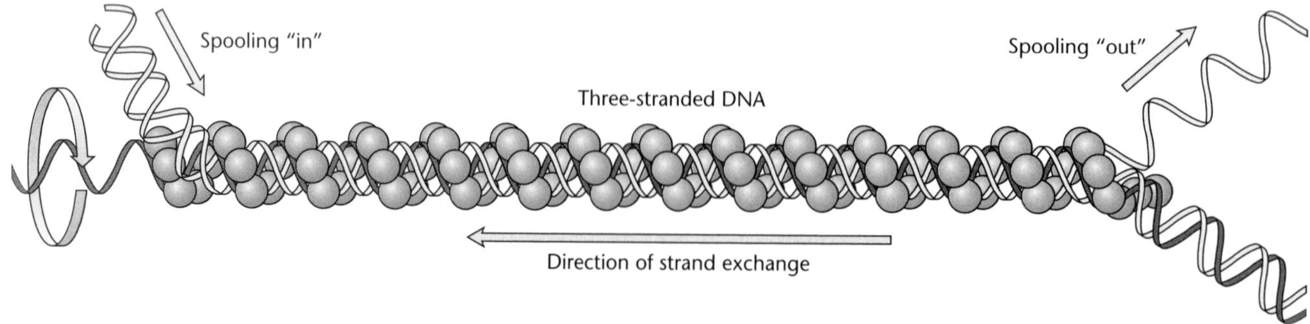

The RecA protein initially forms a nucleofilament only on single-stranded DNA, either an end or a gap, and does not invade the neighboring double-stranded DNA. However, after the nucleofilament has invaded a double-stranded DNA, the RecA protein can continue to polymerize on the same strand, extending the nucleofilament into the neighboring double-stranded DNA. While the filament grows in the 5′-to-3′ direction on the strand, eventually invading the neighboring double-stranded DNA in a process called "spooling" (Figure 10.5), it can also apparently depolymerize on the other end, leaving a single-stranded gap. The single-stranded gap can then be repaired by other cellular enzymes.

HOLLIDAY JUNCTIONS

The process by which RecA mediates the invasion of a single-stranded DNA into the homologous sequence in a double-stranded DNA that unwinds the adjacent DNA sequences as it grows creates a special type of junction called a **Holliday junction** (Figure 10.6). Holliday junctions were first proposed in 1964 as part of one of the early models of how homologous recombination occurs. The formation of Holliday junctions is central to many of the roles of recombination in bacterial genetics. Holliday junctions were introduced earlier because of their role during site-specific recombination by Cre and other Y recombinases (see chapter 9). While extension of the filament into double-stranded DNA requires energy in the form of ATP cleavage, in theory, Holliday junctions can migrate spontaneously along the DNA because as one bond is broken through movement of the junctions, another bond forms on the other side. However, in the cell, specialized enzymes move these junctions, creating what are called **heteroduplexes**. The regions of complementary base pairing between the two DNA molecules are called heteroduplexes because the strands in these regions come from different DNA molecules. The action of RecA and the movement of Holliday junctions both form heteroduplexes; however, filament formation with RecA is unidirectional while Holliday junctions are presumably free to move in either direction.

Once formed, Holliday junctions can undergo a rearrangement that changes the relationship of the strands to each other. This rearrangement is called an **isomerization** because no bonds are broken. As shown in Figure 10.6, isomerization causes the crossed strands in configuration I to uncross. A second isomerization occurs to create configuration II, where the ends of the two double-stranded DNA molecules are in the recombinant configuration with respect to each other. Notice that the strands that were not crossed before are now the crossed strands, and vice versa. It may seem surprising that the strands that have crossed over to hold the two DNA duplexes together can change places so readily, but experiments with models show that the two structures I and II are actually equivalent to each other and that the Holliday junction can change from one form to the other without breaking any hydrogen bonds between the bases. Hence, flipping from one configuration to the other requires no energy and should occur quickly, so that each configuration should be present approximately 50% of the time.

Figure 10.6 Holliday junctions can form through the action of RecA. The movement of Holliday junctions creates a heteroduplex region (shown as a region of blue paired with black). Holliday junctions can isomerize between forms I and II. Cutting and ligating resolve the Holliday junction. Depending on the conformation of the junction, the flanking markers A, B, a, and b recombine or remain in their original conformation. The product DNA molecules contain heteroduplex patches. doi:10.1128/9781555817169.ch10.f10.6

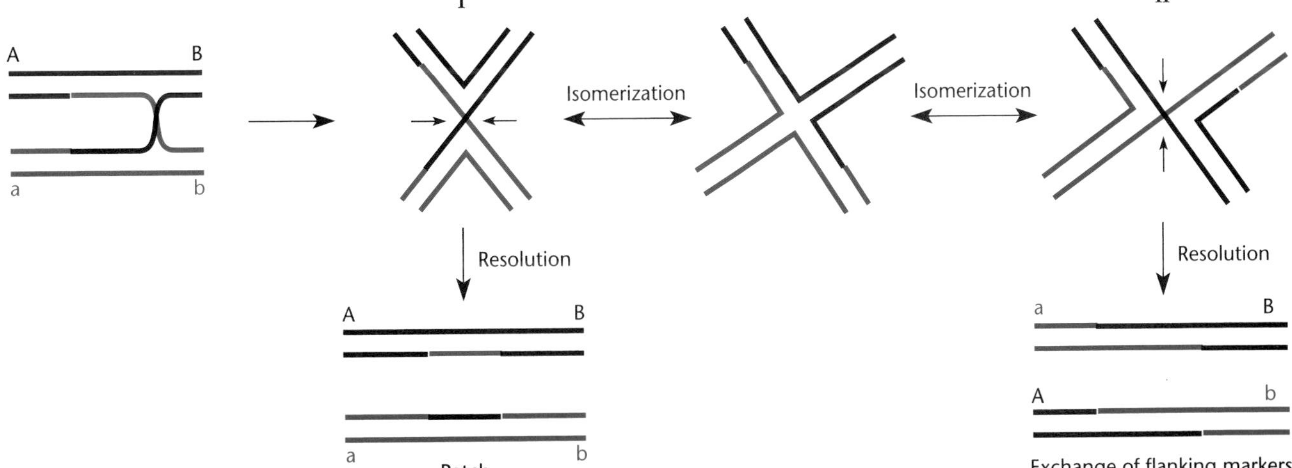

Once formed, the crossed strands in a Holliday junction can be cut and religated, or **resolved**, as shown in Figure 10.6 and discussed in more detail in the next section. Whether an exchange of markers occurs depends on the configuration of the junction at resolution, in other words, which of the strands are the crossed strands. If the Holliday junction is in configuration I when the crossed strands are cut, the flanking DNA sequences are not recombined and the two DNAs return to the way they were. However, if the Holliday junction is in configuration II when the crossed strands are cut, the flanking DNA sequences are exchanged between the two molecules and recombination occurs. This is indicated by recombination of the flanking markers shown in Figure 10.6.

The Ruv and RecG Proteins and the Migration and Cutting of Holliday Junctions

As discussed above, Holliday junctions are remarkable structures that can do many things. Once a Holliday junction has formed as the result of the concerted action of RecBCD and RecA or the RecF pathway and RecA, at least two different pathways can resolve the junctions to make recombinant products. One pathway uses the three Ruv proteins, RuvA, RuvB, and RuvC, which are encoded by adjacent genes. Another protein, called RecG, can move Holliday junctions, but its role during homologous recombination is unclear because it lacks its own host-encoded resolvase (see below).

RuvABC

The crystal structures of the Ruv proteins are interesting and give clues to how they function in the migration and **resolution of Holliday junctions** (Figure 10.7) (see West, Suggested Reading). The RuvA protein is a specific Holliday junction-binding protein whose role is to force the Holliday junction into a certain structure amenable to the subsequent steps of migration and resolution. Four copies of the RuvA protein come together to form a flat structure like a flower with four petals. The Holliday junction lies flat on the flower and thus is forced into a flat (planar) configuration. Binding of the RuvA tetramer also creates a short region in the middle of the Holliday junction where the strands are not base paired and the single strands form a sort of square. Another tetramer of RuvA may then bind to the first to form a sort of turtle shell, with the four arms of double-stranded DNA emerging from the "leg holes." The RuvB protein then forms a hexameric (six-member) ring encircling one arm of the DNA, as shown in Figure 10.7. The DNA is then pumped through the RuvB ring, using ATP cleavage to drive the pump, thereby forcing the Holliday junction to migrate.

After the RuvA and RuvB proteins have forced Holliday junctions to migrate, the junctions can be cut (resolved) by the RuvC protein. The RuvC protein cuts only Holliday junctions that are being held by RuvA and RuvB. The RuvC protein is a specialized DNA endonuclease that cuts the two crossed strands of a Holliday junction simultaneously. Such enzymes are often called **X-philes** because they cut the strands of DNA crossed in a Holliday junction or branch, which look like the letter X, while "phile" means "having an affinity for." Like many enzymes that make double-strand breaks in DNA, two identical polypeptides encoded by the *ruvC* gene come together to form a homodimer, which is the active form of the enzyme. Because the enzyme has two copies of the polypeptide, it has two DNA endonuclease active centers, each of which can cut one of the DNA strands to make a double-strand break.

The evidence that RuvC can cut only Holliday junctions that are bound to RuvA and RuvB is mostly genetic; mutants with either a *ruvA*, a *ruvB*, or a *ruvC* mutation are indistinguishable in that they are all defective in the resolution of Holliday junctions (see below). To a geneticist, this means that RuvC cannot act to resolve a Holliday junction without RuvA and RuvB being present and bound to the Holliday junction. However, this leads to an apparent conflict with the structural information about RuvA and RuvB discussed above, which indicates that RuvA forms a turtle shell-like structure over the Holliday junction. How could RuvC enter the turtle shell formed by RuvA to cleave the crossed DNA strands inside? One possibility is that a tetramer of RuvA is bound to only one face of the Holliday junction, leaving the other face open for RuvC to bind and cut the crossed strands. However, it seems unlikely that the Holliday junction could be held tightly enough in this way not to be dislodged by the RuvB pump. Another idea is that the RuvA shell opens up somehow to allow RuvC to enter and cut the crossed strands.

Support for this model of the concerted action of RuvA, RuvB, and RuvC has come from observations of purified Ruv proteins acting on artificially synthesized structures that resemble Holliday junctions (see Parsons et al., Suggested Reading). These junctions are constructed by annealing four synthetic single-stranded DNA chains that have pairwise complementarity to each other (Figure 10.8). These synthetic structures are not completely analogous to a real Holliday junction in that they are not made from two naturally occurring DNAs with the same sequence. Rather, four single strands that are complementary to each other in the regions shown and therefore form a pairwise cross are synthesized. A Holliday junction made in this way is much more stable than a natural Holliday junction because the branch cannot migrate spontaneously all the way to one end or the other. Real Holliday junctions are too unstable for these experiments; they quickly separate into two double-stranded DNA molecules.

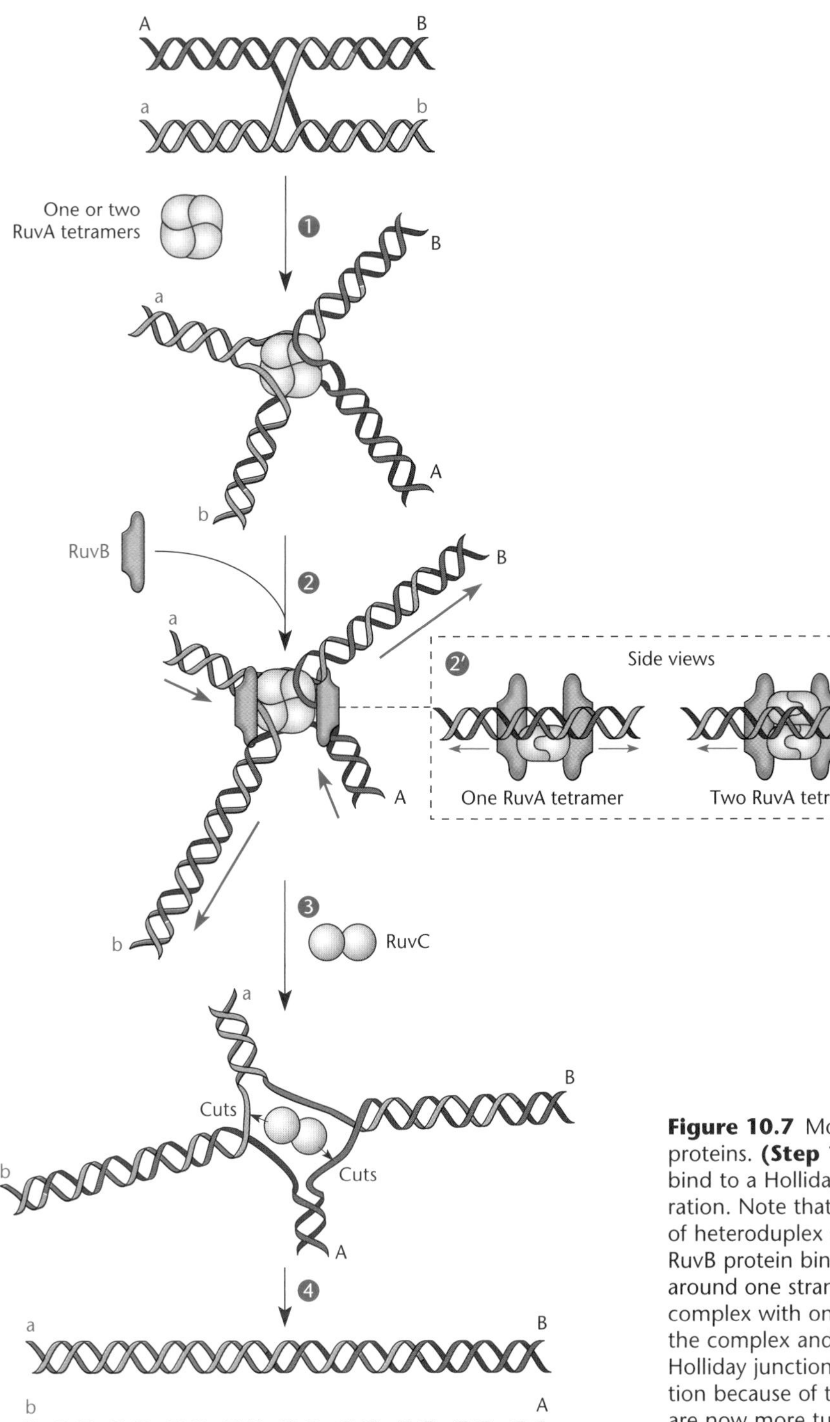

Figure 10.7 Model for the mechanism of action of the Ruv proteins. **(Step 1)** One or two tetramers of the RuvA protein bind to a Holliday junction, holding it in a planar (flat) configuration. Note that at the beginning, the DNA has only one turn of heteroduplex (blue-gray). **(Step 2)** Two hexamers of the RuvB protein bind to the RuvA complex, each forming a ring around one strand of the DNA. **(Step 2′)** Side view of the complex with one and two tetramers. **(Step 3)** RuvC binds to the complex and cuts two of the strands. **(Step 4)** The Holliday junction has been resolved into a different configuration because of the way the strands were cut. Note that there are now more turns of heteroduplex (blue-gray). doi:10.1128/9781555817169.ch10.f10.7

Experiments performed with such synthetic Holliday junctions indicated that purified Ruv proteins act sequentially on a synthetic Holliday junction in a manner consistent with the above-mentioned model. First, RuvA protein bound specifically to the synthetic Holliday junctions, and then a combination of RuvA and RuvB caused a disassociation of the synthetic Holliday junctions, simulating branch migration in a natural DNA molecule. The dissociation required the energy in ATP to break the hydrogen bonds holding the Holliday junction together, as predicted. The RuvC protein could also specifically cut the synthetic Holliday junction in two of the four strands.

RuvC has some sequence specificity at the site where it cuts. It always cuts just downstream of two Ts in the DNA preceded by either an A or a T and followed by either a G or a C (as shown in Figure 10.8). At first,

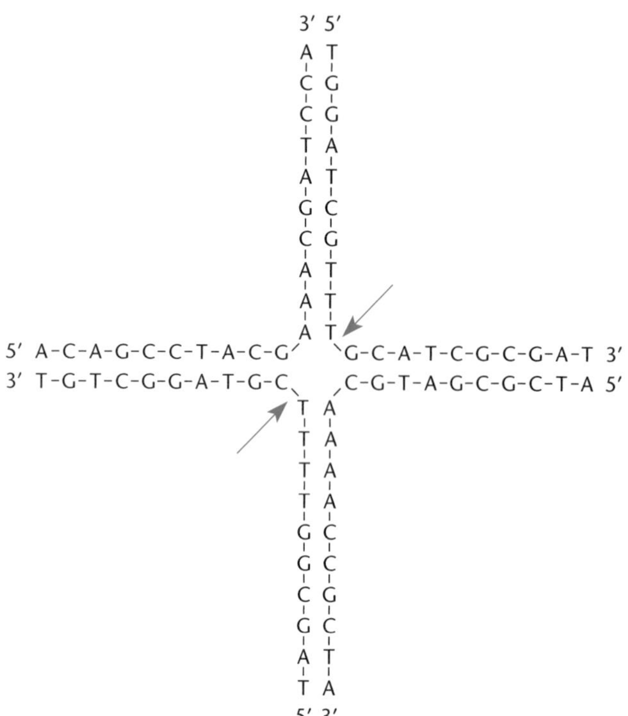

Figure 10.8 A synthetic Holliday junction with four complementary strands. The junction cannot migrate but can be disrupted by RuvA and RuvB. It can also be cut by RuvC (arrows) and other Holliday junction resolvases, such as the gene *49* and gene *3* products of phages T4 and T7, respectively.
doi:10.1128/9781555817169.ch10.f10.8

it seems impossible that two Ts could be opposite each other in the DNA until we remember that the crossed strands of DNA are not the complementary strands but the strands with the same sequence. If one strand has the sequence recognized by RuvC, 5′-(A/T)TT(G/C)-3′, the other strand has the same sequence at that position. Presumably, RuvA and RuvB cause the Holliday junction to migrate until this sequence is encountered and RuvC then can cut the crossed strands. The alternate resolvase in *E. coli*, RusA, which is encoded by a defective prophage, also has sequence specificity and cuts upstream of two Gs in the DNA. It has been speculated that Holliday junction resolvases may have sequence specificity to distinguish them from nucleases that can cut branches, such as those that arise from single-strand invasion of a double-stranded DNA. Holliday junction resolvases, such as RuvC and RusA, have been shown to cut such branches if the branch point has their recognition sequence. However, such branches cannot migrate, so the crossed-over strand always has the same sequence, which is not apt to be the sequence recognized by the Holliday junction resolvase; therefore, they are seldom cut by RuvC or RusA. Holliday junctions, on the other hand, can migrate, so eventually they migrate to the sequence recognized by the resolvase and are cut.

Low-G+C gram-positive bacteria, like *B. subtilis*, lack RuvC but possess a different protein, called RecU, which appears to act as the Holliday junction resolvase. The details of how RecU interfaces with the branch migration system are still being worked out.

RecG

Another helicase in *E. coli* that can help junctions to migrate is RecG (Table 10.1). For a long time, the function of this helicase has been debated. It has very little effect on recombination when RuvABC is present, which suggested that its role is redundant with respect to RuvABC. The idea was that RecG could play the role of RuvAB and promote the migration of Holliday junctions. Then, another resolvase could play the role of RuvC and resolve the Holliday junction, providing a backup for RuvABC. However, it now seems that the roles of these proteins are not redundant and that they each have their own unique role to play (see Bolt and Lloyd, Suggested Reading). For example, they move in opposite directions on single-stranded DNA. The RuvAB helicase moves in the 5′-to-3′ direction on single-stranded DNA, while the RecG helicase moves in the 3′-to-5′ direction. Also, unlike RuvAB, which seems to promote the migration of Holliday junctions only after they have formed, the RecG protein can turn blocked replication forks into a form of Holliday junction (see below) or bind to three-strand junctions, like those at a branch. The RecG protein may then cause a replication fork stalled at DNA damage or through another means to back up, much like a train backs up to allow the track to be repaired before it moves on. Backing up of the replication fork in this way could help with the formation of a type of Holliday junction called a "chicken foot" (see chapter 11). Following DNA repair or another type of processing event (as discussed in detail in chapter 11), replication could restart, in most cases with the help of the PriA proteins and DnaC to load the DnaB helicase back on the DNA. A more recent proposal for the function of RecG is not to facilitate the formation of substrates where the Pri proteins can initiate DNA replication. Instead, it is hypothesized that RecG could unwind these structures to prevent "pathological" DNA replication from being initiated, which would otherwise complicate future replication and recombination reactions. The role of RecG is still under active research.

Recombination between Different DNAs in Bacteria

Most of the earlier work on recombination in bacteria concerned the recombination between DNAs with different mutations, leading to the formation of recombinant types. This work led to the discovery of many of the gene products we have been discussing, and we review this genetics work at the end of the chapter. As discussed in chapter 3, recombination in bacteria has some differences from recombination in most other organisms. In bacteria, generally only small pieces of incoming donor DNA recombine with the chromosome, whereas in other organisms, two enormously long double-stranded DNA molecules of equal size usually recombine. Nevertheless, the requirements for recombination in bacteria are similar to those in other organisms. In fact, accumulating evidence supports the existence of proteins analogous to many of the recombination proteins of bacteria in both eukaryotes and archaea.

How Are Linear DNA Fragments Recombined into the *E. coli* Chromosome?

Throughout chapter 3 and elsewhere in the text, a simplified representation was shown, indicating how linear fragments are recombined into the host chromosome (or plasmid or bacteriophage). The requirements for integrating a linear DNA fragment into the chromosome are similar to those required to restart a collapsed replication fork, which is why they use the same functions. The ability of the RecBCD complex to efficiently process double-strand breaks to load the RecA protein to initiate DNA replication immediately suggests a model for how linear fragments could recombine into the chromosome. Figure 10.9 shows how recombination likely initiates DNA replication to copy an allele into the chromosome during DNA replication. The incoming DNA fragment is shown with blue dots to emphasize an allele on the DNA fragment, and the allele on the chromosome is shown with black dots. RecBCD processes the ends of the linear fragment to 3′ single-stranded overhangs that are loaded with the RecA protein, which can invade the chromosome. Replication that is initiated by the DNA fragment then continues to the terminus region (clockwise-traveling DNA replication fork) or is met by a replication fork that initiates from *oriC*. After chromosome segregation, each daughter cell will have one of the alleles, the allele from the parent (shown as black dots) or the allele that was recombined into the chromosome (shown as blue dots).

Phage Recombination Pathways

Many phages encode their own recombination functions, some of which can be important for the multiplication of the phages. As discussed in chapter 7, some phages use recombination to make primers for replication and concatemers for packaging. Also, phage recombination systems may be important for repairing damaged phage DNA and for exchanging DNA between related phages to increase diversity. Phages may encode their own recombination systems to avoid dependence on host systems for these important functions.

Many phage recombination functions are analogous to the recombination proteins of the host bacteria (Table 10.2), and in many cases, the phage recombination proteins were discovered before their host counterparts. As a result, studies of bacterial recombination systems have been heavily influenced by simultaneous studies of phage recombination systems.

Rec Proteins of Phages T4 and T7

Phages T4 and T7 depend on recombination for the formation of DNA concatemers after infection (see chapter 7). Therefore, recombination functions are essential for the multiplication of these phages. Many of the T4 and T7 Rec proteins are analogous to those of their hosts. For example, the gene *49* protein of T4 and the gene *3* protein of T7 are X-phile endonucleases that resolve Holliday junctions and are representative of phage proteins discovered before their host counterparts, in this case RuvC. The gene *46* and *47* products of phage T4 may perform a reaction similar to that of the RecBCD protein of the host, although no evidence indicates the presence of χ-like sites associated with this enzyme. The UvsX protein of T4 and the Bet protein of λ are analogous to the RecA protein of the host.

The RecE Pathway of the *rac* Prophage

Another classic example of a phage-encoded recombination pathway is the RecE pathway encoded by the *rac* prophage of *E. coli* K-12. The *rac* prophage is integrated at 29 min in the *E. coli* genetic map and is related to λ. This defective prophage cannot be induced to produce infective phage, since it lacks some essential functions for multiplication.

The RecE pathway was discovered by isolating suppressors of *recBCD* mutations, called *sbcA* mutations (for suppressor of *recB* and *recC* deletions), that restored recombination in conjugational crosses. The *sbcA*

Table 10.2 Analogy between phage and host recombination functions

Phage function	Analogous *E. coli* function
T4 UvsX	RecA
T4 gene *49*	RuvC
T7 gene *3*	RuvC
T4 genes *46* and *47*	RecBCD
λ ORF[a] in *nin* region	RecO, RecR, RecF
Rac *recE* gene	RecJ, RecQ
λ *gam*	Inhibits RecBCD
λ *exo*	RecBCD, RecJ
λ *bet*	RecA
rusA (DLP12 prophage)	RuvC

[a]ORF, open reading frame.

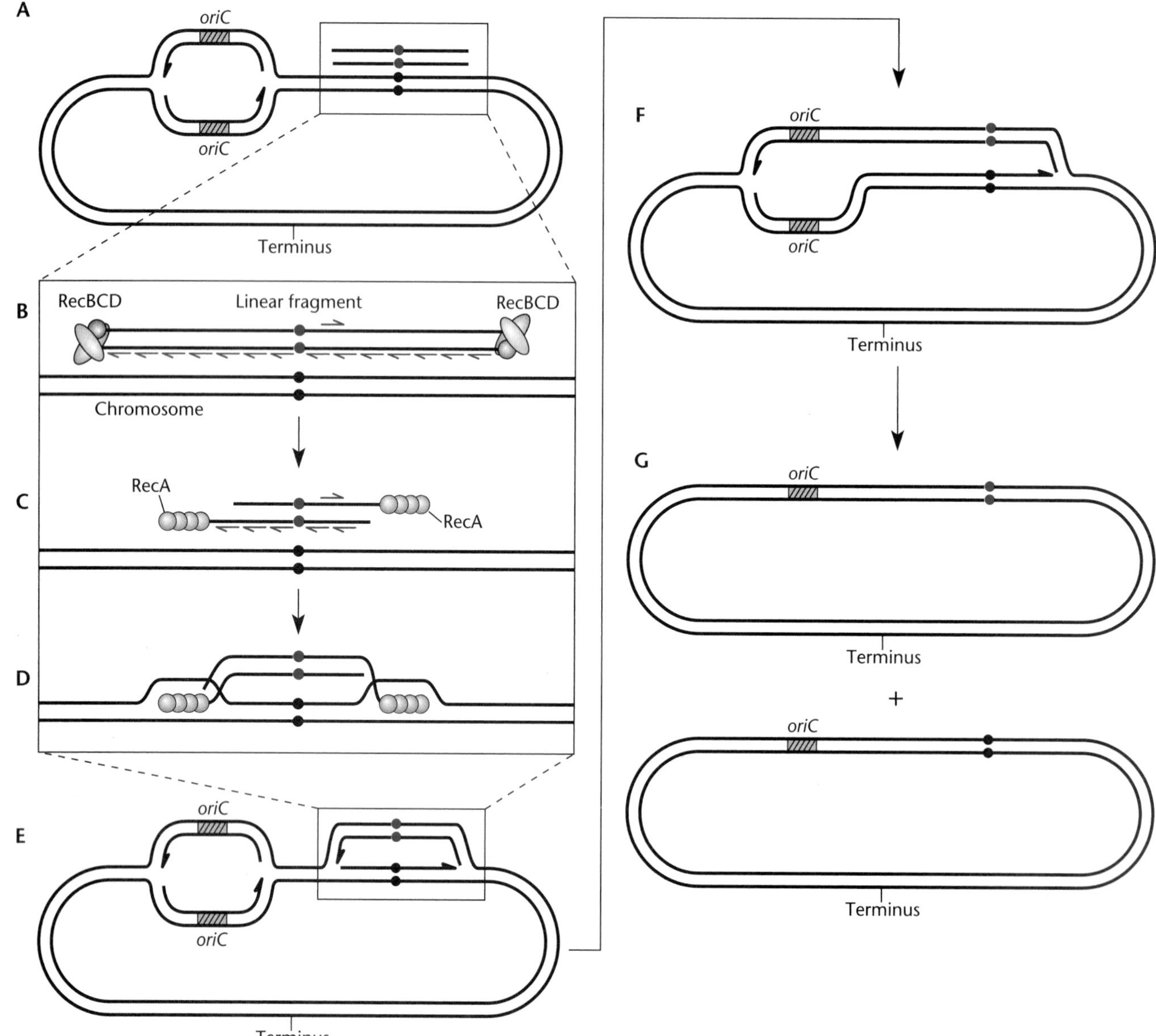

Figure 10.9 Model for how linear fragments are recombined into the chromosome using homologous recombination and DNA replication. **(A)** A homologous DNA fragment with a different allele than the chromosome (blue versus black dots). **(B to D)** Diagram of how the DNAs in the boxed region in panel A are converted to the structure shown in the boxed region in panel E. **(B)** The ends of the linear fragment are processed with RecBCD. The χ sites are indicated by blue arrows with half heads. **(C)** The ends are processed into 3′ single-strand extensions that end with a χ site and are coated with the RecA protein. **(D)** The single-stranded DNA coated with RecA invades the homologous DNA sequence and initiates DNA replication in the chromosome. **(E)** Replication that is initiated by the DNA fragment then continues in opposite directions. **(F)** The replication fork traveling in the clockwise direction progresses to the terminus region, while the counterclockwise replication fork is met by a replication fork that initiates from *oriC*. **(G)** After chromosome segregation, each daughter cell has one of the alleles, the allele from the parent (shown as black dots) or the allele that was recombined into the chromosome (shown as blue dots). doi:10.1128/9781555817169.ch10.f10.9

mutations were later found to activate a normally repressed recombination function of the defective prophage *rac*. Apparently, *sbcA* mutations inactivate a repressor that normally prevents the transcription of the *recE* and *recT* genes. When the repressor gene is inactivated, the RecE protein, as well as other prophage gene products that can substitute for the RecBCD exonuclease in recombination, is synthesized. These gene products, or homologous ones from other phages, can be used for recombineering, a powerful technique discussed below.

The Phage λ Red System

Phage λ also encodes recombination functions. The best characterized is the Red system, which requires the products of the adjacent λ genes *exo* and *bet*. The product of the *exo* gene is an exonuclease that degrades one strand of a double-stranded DNA from the 5′ end to leave a 3′ single-stranded tail. The *bet* gene product is known to help the renaturation of denatured DNA and to bind to the λ exonuclease. Unlike many of the other recombination systems that we have discussed, the λ Red recombination pathway does not require the RecA protein, since it has its own synapse-forming protein, Bet. The λ Red system was first used for recombineering (see below). Besides the Red system, phage λ encodes another recombination function that can substitute for components of the *E. coli* RecF pathway (see Sawitzke and Stahl, Suggested Reading, and Table 10.2). Apparently, phages can carry components of more than one recombination pathway.

Recombineering: Gene Replacements in *E. coli* with Phage λ Recombination Functions

One of the major advantages of using bacteria and other simple organisms for molecular genetic studies is the relative ease of doing gene replacements with some of these organisms (see the discussion of gene replacements in chapter 3). To perform a gene replacement, a piece of the DNA of an organism is manipulated in the test tube to change its sequence in some desired way. The DNA is then reintroduced into the cell, and the recombination systems of the cell cause the altered sequence of the reintroduced DNA to replace the normal sequence of the corresponding DNA in the chromosome. Because it depends on homologous recombination, gene replacement requires that the sequence of the reintroduced DNA be homologous to the sequence of the DNA it replaces. However, the homology need not be complete, and minor changes, such as base pair changes, can be introduced into the chromosome in this way as a type of site-specific mutagenesis. Also, the reintroduced DNA need not be homologous over its entire length; homology is needed only where the recombination occurs. This makes it possible to use gene replacement to make large alterations, such as construction of an in-frame deletion to avoid polarity effects (and insertion of an antibiotic resistance gene cassette into the chromosome [see Box 12.3]). If the sequences on both sides of the alteration (the flanking sequences) are homologous to sequences in the chromosome, recombination between these flanking sequences and the chromosome will insert the alteration. Methods for gene replacement in *E. coli* have usually relied on the RecBCD-RecA recombination pathway, since this is the major pathway for recombination in *E. coli*. We mention some of these methods in this chapter and chapter 3.

A newer and more useful method for manipulating DNA in *E. coli* is called recombineering. The term recombineering is used to describe various applications with bacteriophage recombination proteins that have been optimized for performing site-specific mutagenesis and gene replacements in *E. coli* (Table 10.2). One system in particular, the λ Red system, has many advantages over the RecBCD-RecA pathway for such manipulations. This method makes it possible to use single-stranded DNA oligonucleotides, possibly as short as 30 bases, although those 60 bases long or longer work better. This is important because the synthesis of single-stranded DNAs of these lengths has become routine for making PCR primers, and oligonucleotides with any desired sequence can be purchased for a reasonable cost. Other methods of site-specific mutagenesis for making specific changes in a sequence are more tedious and require a certain amount of technical skill. Probably most important, recombineering is very efficient. Minor changes, such as single-amino-acid changes in a protein, usually offer no positive selection, and most methods require the screening of thousands of individuals to find one with the replacement.

Figure 10.10 outlines the original procedure for using the λ Red system for gene replacements. Figure 10.10A shows the structure of the *E. coli* strain required. It carries a defective prophage in which most of the λ genes have been deleted, except the recombination (Red) genes *gam-bet-exo* (Table 10.2; see also Figure 8.2). Figure 10.10B shows the replacement of a sequence in the plasmid by the corresponding region on another plasmid, in which the sequence has been disrupted by introduction of an antibiotic resistance (Ab^r) cassette. This region of the plasmid has been amplified by PCR to produce a double-stranded DNA fragment carrying the antibiotic resistance cassette and some of the flanking sequences. First, the cells are briefly incubated at 42°C to inactivate the mutant λ repressor, inducing transcription of the Red genes of the prophage. Then, cells are made competent for electroporation, and the PCR fragment is

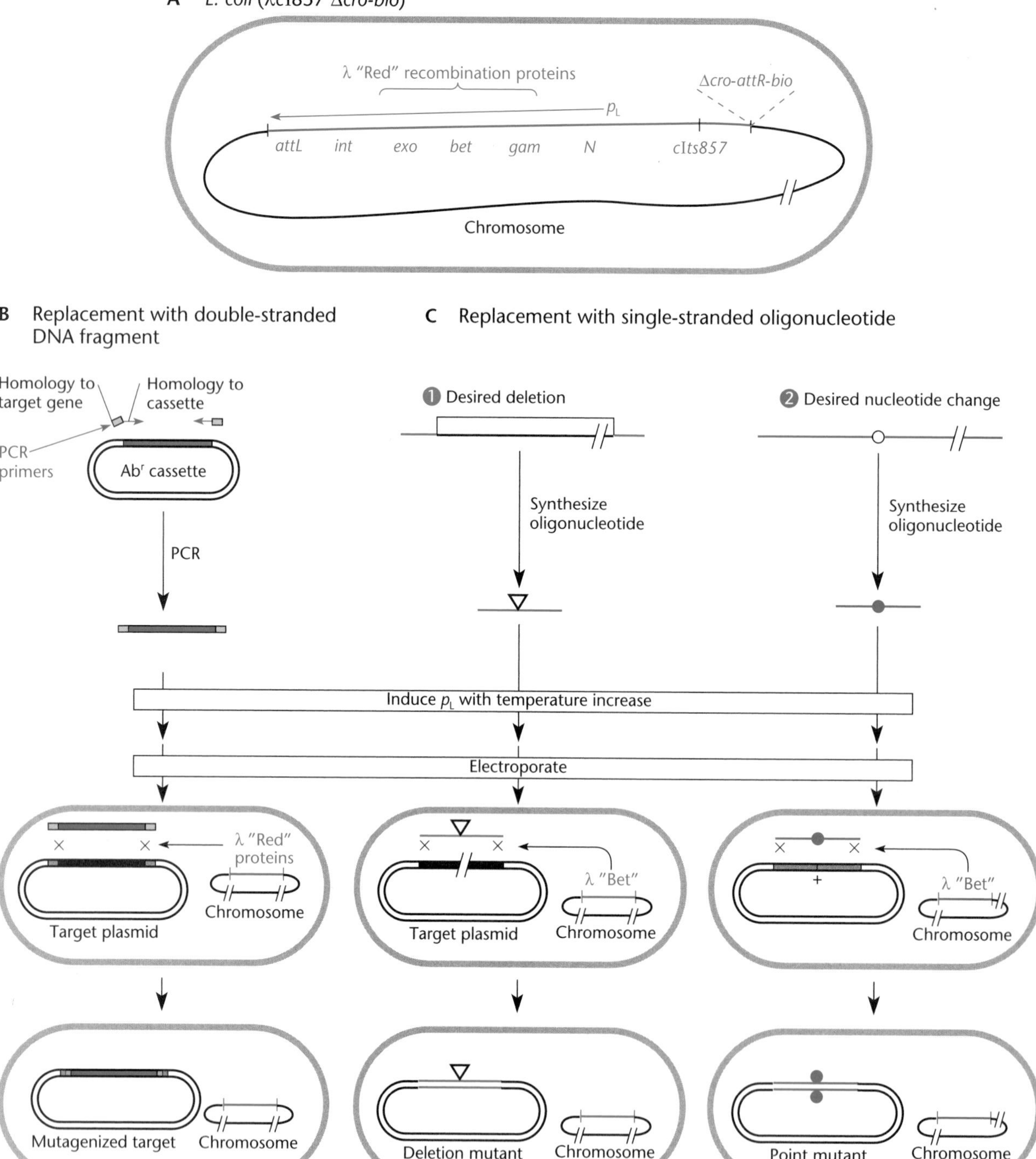

Figure 10.10 Recombineering: in vivo DNA modification in *E. coli* using λ phage-encoded proteins. **(A)** A deletion derivative of a λ lysogen with a temperature-sensitive repressor can be used to induce the λ Red recombination functions. **(B)** Double-stranded DNA cassettes can be amplified using primers with homology to the regions flanking the region to be replaced. Double-stranded DNA can be processed by the λ recombination proteins encoded by the *red* genes. **(C)** Oligonucleotides that contain a deletion compared to the target plasmid (indicated with a triangle), where only the Bet protein is needed, can be synthesized. **(D)** Oligonucleotides that contain a point mutation compared to the target plasmid (indicated with a small circle), where only the Bet protein is needed, can be synthesized.
doi:10.1128/9781555817169.ch10.f10.10

electroporated into them. The *gam* gene product, Gam, inhibits the RecBCD complex so that the linear DNA fragment is not degraded as soon as it enters the cell. The *exo* gene product, Exo, then processes the fragment for recombination. Exo is an exonuclease that plays the role of RecBCD, degrading one strand of a double-stranded DNA from the 5′ end, thereby exposing a **3′ overhang** single strand for strand invasion. The *bet* gene product, Bet, then plays the role of RecA, binding to the single-stranded DNA exposed by Exo and promoting synapse formation and strand exchange with a complementary DNA in the cell. The cells in which the PCR fragment has recombined with the cellular DNA so that the sequence containing the antibiotic resistance gene has replaced the corresponding sequence in the cellular DNA are then selected on plates containing the antibiotic.

To determine the effect on the cell of inactivating a gene product, it is best to delete the entire gene and replace it with an antibiotic resistance cassette. This can be accomplished by using PCR to amplify the cassette with primers whose 5′ sequences are complementary to sequences flanking the gene to be deleted. Recall that the 5′ sequences on a PCR primer need not be complementary to the sequences being amplified. When this amplified fragment is electroporated into the cells, the antibiotic resistance cassette replaces the entire gene.

Introducing an antibiotic resistance cassette into a gene simplifies the task of selecting the gene replacement and inactivating the gene. A variation of this procedure is to include sites recognized by a site-specific recombinase flanking the antibiotic resistance cassette. In this adaptation of the procedure, the recombinase can be expressed later to remove the cassette and leave an in-frame deletion, which is important to reduce the chance of polar effects (see chapter 2 and Datsenko and Wanner, Suggested Reading).

Sometimes, we want to introduce a small change into the gene for which there is no direct selection, for example, a specific change in one amino acid that we think may play an important role in the protein product of the gene. A variation on this method allows the selection of recombinants that have a single defined base pair change or some other small change. It depends on having a cassette that has both a gene that can be selected by positive selection and a gene whose product is toxic under some conditions. This allows us to select both for acquisition of the cassette and, later, for its loss. An example of such a system could involve insertion of the recognition site for the I-SceI homing endonuclease, which is long enough to not be found naturally in bacterial genomes. The I-SceI recognition site is toxic only when the I-SceI endonuclease is expressed in the same strain. First, a DNA cassette carrying both an antibiotic resistance gene and the cleavage site for the I-SceI homing endonuclease, flanked by sequences for the region of the gene to be replaced, is introduced into the cell by electroporation, selecting for the antibiotic resistance as described above. Then, another DNA fragment (or oligonucleotide), identical to the targeted region of the DNA but carrying the desired base pair change, is introduced. At the same time, a plasmid expressing the I-SceI endonuclease is also introduced by transformation and selected, using a plasmid-borne antibiotic resistance gene. Any DNA retaining the I-SceI recognition site will be destroyed by cleavage by the endonuclease, and only recombinants in which the corresponding region is replaced by sequence derived from the second DNA fragment will survive. Most of the surviving bacteria, therefore, have the sequence with the base pair change replacing the original sequence in the gene, and this can be verified by DNA sequencing. Various other techniques have been developed to make so-called "markerless" manipulations of the chromosome (see Box 12.3).

The utility of this method for making specific changes in a gene increased dramatically when it was discovered that single-stranded DNA can also be used for the electroporation (see Ellis et al., Suggested Reading). Single-stranded DNAs with a defined sequence are easily obtainable, as this is how DNA is chemically synthesized for PCR primers, etc. Using single-stranded DNAs also makes it possible to dispense with Gam and Exo, since single-stranded DNA is not degraded by RecBCD and the DNA does not need Exo to make it single stranded, since it is already single stranded. Figure 10.10C shows the replacement if a single-stranded oligonucleotide is used to introduce either an in-frame deletion or a single-base-pair change into the target DNA. The procedure is similar, except that only *bet* needs to be expressed from the prophage. The Bet protein promotes pairing between the introduced single-stranded DNA and the chromosomal DNA and strand exchange. Then, replication or repair replaces the normal sequence with the mutant sequence in both strands, as shown.

Interestingly, gene replacement with single-stranded DNA shows a strong strand bias, meaning that a single-stranded oligonucleotide complementary to one strand is more apt to replace the corresponding chromosomal sequence than is an oligonucleotide complementary to the other strand in any particular region. Which strand is preferred correlates with the direction of replication in the region. The *E. coli* chromosome replicates bidirectionally from the origin of replication (see chapter 1), so that on one side of the *oriC* region the replication fork moves in one direction while on the other side it moves in the opposite direction; the sequence to which it binds corresponds to the lagging strand. Presumably, the single-stranded gaps that are produced on the lagging strand at the replication fork are sites of strand invasion promoted by Bet, which, unlike RecA, is not able to help a single-stranded DNA invade a completely double-stranded DNA. Bet may need a single-stranded gap in the invaded DNA to pair with before it can get its "foot in the door" and promote the invasion of adjacent double-stranded DNA. Interestingly, there may be an inherent tendency for the lagging strand to be vulnerable to recombination. This stems from the surprising finding that several different species could naturally be manipulated with oligonucleotides without the benefit of discernible phage recombination genes in their genome sequences (see Swingle et al., Suggested Reading).The efficiency of recombineering is also increased in a strain that is deficient in mismatch repair or if a C-C mismatch, which will not be recognized by the mismatch repair system, is used. Other adaptations to the basic procedure that can increase the efficiency of Red-mediated recombination include using oligonucleotides that are modified in such a way that they are not easily recognized by host enzymes that could degrade them before recombination. All of these improvements greatly increase the frequency of progeny with the desired mutation. The λ Red recombination technique has also been shown to work with multiple oligonucleotides in the same transformation, where many sites can be modified at a single time (see Isaacs et al., Suggested Reading). Furthermore, specific changes need not always be targeted; instead, random changes can be made using oligonucleotides with random nucleotides included at some positions. For example, a collection of oligonucleotides that differed at a certain codon or codons could be transformed. Recombinants from this procedure could then be screened or subjected to genetic selection to identify useful mutations at one or many genetic positions in a bacterial genome.

The high-frequency λ-based method has been adapted to *E. coli* and many of its relatives, such as *Salmonella* and *Yersinia*. The number of bacteria in which it can be used is being expanded by identifying prophages with recombination functions in other bacteria through genome sequences for use with recombineering. In some of these, genes for the phage recombination enzymes are more homologous to *recET* genes of the *rac* prophage (see above) than they are to the genes of λ.

Genetic Analysis of Recombination in Bacteria

The major reason we understand so much more about recombination in *E. coli* than in most other organisms is because of the relative ease of doing genetic experiments with the organism. In this section, we discuss some of the genetic experiments that have led to our present picture of the mechanisms of recombination in *E. coli*.

Isolating Rec$^-$ Mutants of *E. coli*

As in any genetic analysis, the first step in studying recombination in *E. coli* was to isolate mutants defective in recombination. Such mutants are called Rec$^-$ mutants and have mutations in the *rec* genes, whose products are required for recombination. Two very different approaches were used in the first isolations of Rec$^-$ mutants of *E. coli*.

Some of the first Rec$^-$ mutants were selected directly, based on their inability to support recombination (see Clark and Margulies, Suggested Reading). The idea behind this selection was that an *E. coli* strain with a mutation that inactivates a required *rec* gene should not be able to produce recombinant types when crossed with another strain. In one experiment, a Leu$^-$ strain of *E. coli* was mutagenized with nitrosoguanidine. The mutagenized bacteria, some of which might now also have a *rec* mutation in addition to their *leu* mutation, were then crossed separately with a Leu$^+$ Hfr strain. A Rec$^-$ mutant should produce no Leu$^+$ recombinants when crossed with the Hfr strain.

To cross thousands of the mutagenized bacteria separately with the Hfr strain to find the few that had *rec* mutations and gave no recombinants would have been too laborious, so the investigators used replica plating to facilitate the identification of the mutants. When a plate containing colonies of individual mutagenized bacteria was replicated onto another plate lacking leucine on which an Hfr strain had been spread, the few Leu$^-$ colonies that arose within the replicated colony were due to recombinants. A few colonies produced no Leu$^-$ recombinants when crossed with the Hfr strain and were therefore candidates for Rec$^-$ mutants. We discuss bacterial genetic techniques, such as replica plating and Hfr crosses, in chapter 3.

However, just because the mutants produced no recombinants in a cross does not mean that they are necessarily Rec$^-$ mutants. For instance, the mutants might have been normal for recombination but defective in the uptake of DNA during conjugation. This possibility was ruled out by crossing the mutants with an F′ plasmid-containing strain instead of an Hfr strain. As discussed in chapter 5, apparent recombinant types can appear without recombination in an F′ cross because the F′ plasmid can replicate autonomously in the recipient cells; that is, it is a replicon. However, the DNA must still be taken up during transfer of the F′ plasmid, so if the mutants are defective in DNA uptake, no apparent recombination types would appear in the F′ cross. Normal frequencies of apparent recombinant types appeared when some of the mutants were crossed with F′ strains; therefore, these mutants were not defective in DNA uptake during conjugation, but rather, had defects in recombination. These and other criteria were used to isolate several recombination-deficient (Rec$^-$) mutants.

The approach used by others to isolate recombination-deficient mutants of *E. coli* was less direct (see Howard-Flanders and Theriot, Suggested Reading). Their isolation depended on the fact that some recombination functions are also involved in the repair of ultraviolet (UV)-damaged DNA. Therefore, using methods described in chapter 11, Howard-Flanders and Theriot isolated several repair-deficient mutants and tested them to determine if any were also deficient in recombination. Some, but not all, of these repair-deficient mutants could also be shown to be defective in recombination in crosses with Hfr strains.

COMPLEMENTATION TESTS WITH *rec* MUTATIONS

Once Rec⁻ mutants had been isolated, the number of *rec* genes could be determined by complementation tests (see chapter 3). The original *rec* mutations defined three genes of the bacterium: *recA*, *recB*, and *recC*. The *recB* and *recC* mutants were less defective in recombination and repair than were the *recA* mutants. In fact, the RecA function is the only gene product absolutely required for recombination in *E. coli* and many other bacteria.

MAPPING *rec* GENES

The next step was to map the *rec* genes. This task may seem impossible, since a *rec* mutation causes a deficiency in recombination, but it is the frequency of recombination with known markers that reveals the map position (see chapter 3). Crosses with Rec⁻ mutants can be successful only if the donor, but not the recipient, strain has the *rec* mutation. After DNA transfer, phenotypic lag ensures that even recipient cells that get the *rec* mutation by recombination, and which therefore eventually become Rec⁻, will retain recombination activity at least long enough for recombination to occur. Other markers are selected, and the presence or absence of the *rec* mutation is scored as an unselected marker either by the UV sensitivity of the recombinants or by some other phenotype due to the *rec* mutation.

Using crosses such as those described above, investigators found that *recA* mutations mapped at 51 min and *recB* and *recC* mutations mapped close to each other at 54 min on the *E. coli* genetic map. Later studies showed that the products of the *recB* and *recC* genes, as well as those of the adjacent *recD* gene, comprise the RecBCD nuclease, which initiates the major recombination pathway required after Hfr crosses (see above). The *recD* gene was not found in the original selection because its product is not essential for recombination under these conditions. In fact, as discussed above, mutations in the *recD* gene can stimulate recombination by preventing degradation of the displaced strand and making recombination independent of χ sites.

Isolating Mutants with Mutations in Other Recombination Genes

In addition to the *recA* and *recBCD* genes, other genes whose products participate in recombination in *E. coli* have been found (Table 10.1). Many of these genes were not found in the original selections because inactivating them alone does not sufficiently reduce recombination after Hfr or transductional crosses in wild-type *E. coli*.

THE RecF PATHWAY

As discussed above, the RecF pathway in *E. coli* involves the products of the *recF*, *recJ*, *recN*, *recO*, *recQ*, and *recR* genes. The genes of the RecF pathway of recombination were not discovered in the original selections of *rec* mutations because they are not normally required for recombination after Hfr crosses. By themselves, the *recB* and *recC* mutations reduce recombination after an Hfr cross to about 1% of its normal level. Mutations in any of the genes of the RecF pathway can prevent the residual recombination that occurs in a *recB* or *recC* cell, which suggested that these genes are responsible for only a minor pathway of recombination in *E. coli*. However, further evidence suggested that the RecF pathway was just as important as the RecBCD pathway but was initiated at single-stranded gaps in DNA rather than at free ends, such as occur during Hfr crosses and when replication forks collapse at nicks in the DNA templates.

The RecF pathway was discovered because mutations in two other genes suppressed the requirement for RecBC after Hfr crosses. These genes were therefore named *sbcB* and *sbcC* (because they suppress the recombination deficiency in *recB* and *recC* mutants). We now know that these suppressor mutations act by allowing the RecF pathway to function after Hfr crosses, because the extra recombination in a *recB sbcB* mutant was eliminated by a third mutation in a gene of the RecF pathway, for example, *recO*. In other words, a *recB sbcB recO* triple mutant was almost as defective in recombination as a *recA* mutant. More recently, the relationship between SbcB, SbcC, RecF, and RecBCD has become clearer. The RecF pathway for loading RecA actually appears to be more widespread than the RecBCD system (see Rocha et. al., Suggested Reading). The multiple single-stranded exonucleases, including SbcB and SbcC (and/or other exonucleases), appear to help to tailor double-stranded DNA breaks so that they are not recognized by the RecF machinery but instead are processed by RecBCD for loading RecA. This may be an adaptation that allows cells to use the RecBCD pathway for reestablishing collapsed replication forks and fixing DNA double-strand breaks while reserving the RecF pathway for repair at single-stranded gaps.

THE *ruvABC* AND *recG* GENES

As discussed above, the products of the *ruv* and *recG* genes are involved in the migration and cutting of Holliday junctions. There are three adjacent *ruv* genes: *ruvA*, *ruvB*, *and ruvC*. The *ruvA* and *ruvB* genes are transcribed into a polycistronic mRNA (see chapter 2), and the *ruvC* gene is adjacent but independently transcribed. The *recG* gene lies elsewhere in the genome.

The discovery of the role of the *ruv* genes in recombination involved some interesting genetics. The *recG* and *ruvABC* genes were not found in the original selections for recombination-deficient mutants because, by themselves, mutations in these genes do not significantly reduce recombination. The *ruv* genes were found only because mutations in them can increase the sensitivity of *E. coli* to killing by UV irradiation. The *recG* gene was found because double mutants with mutations in the *recG* gene and one of the *ruv* genes are severely deficient in recombination (see Lloyd, Suggested Reading). In genetic analysis, this is usually taken to mean that the RecG and Ruv proteins perform overlapping functions in recombination and so can substitute for each other. If any one of the *ruv* genes is inactive, the whole Ruv pathway is inactive, but the *recG* gene product is still available to help Holliday junctions migrate, and vice versa. The enzyme that resolves Holliday junctions in a *ruv* mutant is not clear. Redundancy of function is a common explanation in genetics for why some gene products are nonessential. We discuss other examples of redundant functions elsewhere in this book.

Once the Ruv proteins had been shown to be involved in recombination, it took some clever intuition to find that their role is in the migration and resolution of Holliday junctions (see Parsons et al., Suggested Reading). The Ruv proteins were first suspected to be involved in a late stage of recombination because of a puzzling observation: although *ruv* mutants produce normal numbers of transconjugants when crossed with an Hfr strain, they produce many fewer transconjugants when crossed with strains containing F′ plasmids. As mentioned above, mating with F′ plasmids should result in, if anything, more transconjugants than crosses with Hfr strains, because F′ plasmids are replicons that can multiply autonomously and do not rely on recombination for their maintenance.

One way to explain this observation is to propose that the Ruv proteins function late in recombination. If the Ruv proteins function late, recombinational intermediates might accumulate in cells with *ruv* mutations, thus having a deleterious effect on the cell. If so, *recA* mutations, which block an early step in recombination by preventing the formation of synapses, might suppress the deleterious effect of the *ruv* mutations on F′ crosses. This was found to be the case. A *ruv* mutation had no effect on the frequency of transconjugants in F′ crosses if the recipient cells also had a *recA* mutation. Once genetic evidence indicated a late role for the Ruv proteins in recombination, the process of Holliday junction resolution became the candidate for that role, since it is the last step in recombination. Biochemical experiments were then used to show that the Ruv proteins help in the migration and resolution of synthetic Holliday junctions, as described above.

DISCOVERY OF χ SITES

The discovery on DNA of χ sites that stimulate recombination by the RecBCD nuclease also required some interesting genetics and was a triumph of genetic reasoning. Many sites that are subject to single- or double-strand breaks are known to be "hot spots" for recombination. In some cases, such as recombination initiated by homing enzymes (Box 10.2), it is clear that breaks at specific sites in DNA can initiate recombination. In general, however, the frequency of recombination seems to correlate fairly well with physical distance on DNA, as though recombination occurs fairly uniformly throughout DNA molecules.

It came as a surprise, therefore, to discover that the major recombination pathway for Hfr crosses in *E. coli*, the RecBCD pathway, occurs through specific sites on the DNA—the χ sites. Like many important discoveries in science, the discovery of χ sites started with an astute observation. This observation was made during studies of host recombination functions using λ phage (see Stahl et al., Suggested Reading). The experiments were designed to analyze the recombinant types that formed when the phage Red recombination genes *exo* and *bet* were deleted. Without its own recombination functions, the phage requires the host RecBCD nuclease. In addition, if the phage is also a *gam* mutant, it does not form a plaque unless it can recombine. Therefore, plaque formation by a *gam* mutant is an indication that recombination has taken place.

The reason that *red gam* mutant λ phage cannot multiply to make a plaque without recombination is somewhat complicated. As discussed in chapter 8, λ phage cannot package DNA from genome length DNA molecules but only from concatemers in which the λ genomes are linked end to end. Normally, the phage makes concatemers by rolling-circle replication. However, if the phage is a *gam* mutant, it cannot switch to the rolling-circle mode of replication because the RecBCD nuclease, which is normally inhibited by Gam, somehow blocks the switch. Therefore, the only way a *gam* mutant λ phage can form concatemers is by recombination between the circular λ DNAs formed via θ replication (Figure 8.7). If the phage is also a *red* mutant (i.e., lacks its own recombination functions), the only way it can form concatemers is by RecBCD recombination, the major host pathway. Therefore, *red*

BOX 10.2

Breaking and Entering: Introns and Inteins Move by Double-Strand Break Repair or Retrohoming

Introns and inteins are parasitic DNA elements that are sometimes found in genes, both in eukaryotes and in prokaryotes (see Box 2.4). Like all good parasites, they have as little effect on the health of their host as possible. This is a matter of self-interest, because if their host dies, they die with it. When they move into a DNA, they could disrupt a gene, which could inactivate the gene product and be deleterious to their host. As discussed in Box 2.4, they can avoid inactivating the product of the gene in which they reside by splicing their sequences out of the mRNA before it is translated (in the case of introns) or out of the protein product of the gene after it is made (in the case of inteins). Sometimes this splicing requires other gene products of the intron or the host, and sometimes it occurs spontaneously, in a process called self-splicing. Self-splicing introns were one of the first known examples of RNA enzymes or ribozymes.

Many introns and inteins are able to move from one DNA to another. When they move, they usually move from one gene into exactly the same location in the same gene of a member of the same species that previously lacked them. In that way, they can move through a population until almost all of the individuals in the population have the intron in that location. Because they always return to the same site, this process is called homing. There is a good reason why they choose to move into exactly the same position in the same gene rather than into other places in the same gene or even other places in the genome. The sequences in the gene around the intron or intein, called exon or extein sequences, respectively, also participate in the splicing reaction; therefore, if they find themselves somewhere else where these flanking exon sequences are different, they will not be able to splice themselves out of the RNA or protein. Homing allows the intron or intein to spread through a population by parasitizing other DNAs that lack it but never disrupting the product of an essential gene and disabling or killing the new host as it moves.

There are two basic mechanisms by which introns home. Some introns, called group I introns, and inteins home by double-strand break repair. To move by double-strand break repair, the intron or intein first makes a double-strand break in the homing site of the target gene into which it must move. To accomplish this, the intron or intein encodes a specific DNA endonuclease called a homing nuclease, which makes a break only in this particular sequence. In group I introns, this homing endonuclease is usually encoded by an open reading frame on the intron. In inteins, the intein itself becomes the homing endonuclease after it is spliced out of the protein. After the double-strand cut is made by the specific endonuclease, the corresponding gene containing the intron repairs the cut by double-strand break repair, replacing the sequence without the intron with the corresponding sequence containing the intron by repair (Figure 10.9). After repair, both DNAs contain the intron in exactly the same position. Other DNA elements, including the mating-type loci of yeast, are known to move by a similar mechanism of double-strand break repair.

Other introns, called the group II introns, move by a process called retrohoming. These introns essentially splice themselves into one strand of the target DNA by a process analogous to splicing the intron out of the mRNA, but in reverse (see Box 2.4). The intron also encodes an endonuclease that makes a cut in the other strand, and the exposed 3′ hydroxyl end then primes an intron-encoded reverse transcriptase that makes a DNA copy of the intron, which is then joined to the target site DNA by host DNA repair enzymes. The intron is homed to its target site by homologous sequences in the intron and the target site. The most important sequence in the intron for this complementary base pairing with the homing site is only 15 bp long. Other, shorter complementary sites flanking this region are recognized by the nuclease that cuts the homing site, allowing integration of the intron.

Because group II introns recognize their homing site almost exclusively by complementary base pairing, it is possible to redirect the introns to other sites merely by changing the sequence of the 15-bp sequence on the intron so that it is complementary to a different region on the chromosome. The efficiency of insertion into the new target site increases the more the other DNA sequences flanking the new complementary sequence resemble the sequences recognized by the intron nuclease in the original homing site. The ability of group II introns to be redirected to new sites merely by changing the sequence of part of the intron has allowed their development as site-specific mutagenesis systems. This system, marketed as TargeTron by Sigma-Aldrich, in theory allows the insertion of the intron into essentially any gene in any organism. Basically, PCR is used to make a mutated version of the 15-bp region of the mobile element that is complementary to the sequence into which the intron is to be inserted. Software is provided to identify the best region in the gene into which to insert the intron, based on which region requires the fewest changes in the other flanking complementary sequences, which can then also be changed by using multiple PCR primers. Once this region

(continued)

BOX 10.2 (continued)

Breaking and Entering: Introns and Inteins Move by Double-Strand Break Repair or Retrohoming

of the transposon is PCR amplified, it is cloned into a vector containing the rest of the mobile element, plus a kanamycin resistance cassette that is expressed only if the intron has been excised from the vector. When the RNA is made on the intron by using the T7 phage RNA polymerase promoter and T7 RNA polymerase expressed from a different DNA, the RNA nuclease and reverse transcriptase encoded by the intron are expressed and the intron integrates itself specifically into the selected target site. Such integrations can be selected on kanamycin-containing plates. If another gene has been cloned into the intron on the plasmid, the other gene will be integrated at the new site. A variation of this method uses a mutant intron to introduce small changes, such as base pair changes, close to the new homing site. While less efficient than recombineering, this method has the advantage of being more easily adapted to bacteria other than *E. coli* and its close relatives by expressing the T7 RNA polymerase from a different vector in the bacterium to be mutagenized or by using a different promoter to transcribe the intron.

References

Belfort, M., M. E. Reaban, T. Coetzee, and J. Z. Dalgaard. 1995. Prokaryotic introns and inteins: a panoply of form and function. *J. Bacteriol.* **177:**3897–3903.

Karberg, M., H. Guo, J. Zhong, R. Coon, J. Perutka, and A. M. Lambowitz. 2001. Group II introns as controllable gene targeting vectors for genetic manipulation of bacteria. *Nat. Biotechnol.* **19:**1162–1167.

gam mutant λ phage requires RecBCD recombination to form plaques, and the formation of plaques can be used as a measure of RecBCD recombination under these conditions.

The first χ mutations were discovered when large numbers of *red gam* mutant λ phage were plated on RecBCD$^+$ *E. coli*. Very few phage were produced, and the plaques that formed were very tiny. Apparently, very little RecBCD recombination was occurring between the circular phage DNAs produced by θ replication. However, λ mutants that produced much larger plaques sometimes appeared. The circular λ DNAs in these mutants were apparently recombining at a much higher rate. The responsible mutations were named χ mutations because they increase the frequency of crossovers (a crossover in the chromosome is called a chiasma in eukaryotes). Once the mutations were mapped and the DNA was sequenced, χ mutations in λ were found to be base pair changes that created the sequence 5′-GCTGGTGG-3′ somewhere in the λ DNA. The presence of this sequence appears to stimulate recombination by the RecBCD pathway. Since wild-type λ does not need RecBCD and therefore has no such sequence anywhere in its DNA, recombination by RecBCD is very infrequent unless the χ sequence is created by a mutation.

Further experimentation with χ sites revealed several interesting properties. For example, they stimulate crossovers to only one side of themselves, the 5′ side. Very little stimulation of crossovers occurs on the 3′ side. Also, if only one of the two parental phage contains a χ site, most of the recombinant progeny will not have the χ site, so the χ site itself is preferentially lost during the recombination. These and other properties of χ sites eventually led to the model for χ site action presented earlier in this chapter.

Gene Conversion and Other Manifestations of Heteroduplex Formation during Recombination

As mentioned earlier in the chapter, Holliday junctions formed during recombination can migrate. There are consequences to the migration of Holliday junctions when the DNAs involved are not perfectly identical across the region of the heteroduplex (Figure 10.11). This is because mismatches result in a heteroduplex region that can be processed in different ways.

GENE CONVERSION

The first such evidence for heteroduplex formation during recombination came from studies of gene conversion in fungi. Understanding this process requires a little knowledge of the sexual cycles of fungi. Some fungi have long been favored organisms for the study of recombination because the spores that are the products of a single meiosis are often contained in the same bag, or ascus (see any general genetics textbook). When two haploid fungal cells mate, they fuse to form a diploid zygote. Then, the homologous chromosomes pair and replicate once to form four chromatids that recombine with each other before they are packaged into spores. Therefore, each ascus contains four spores (or eight in fungi such as *Neurospora crassa*, in which the chromatids replicate once more before spore packaging).

Since both haploid fungi contribute equal numbers of chromosomes to the zygote, their genes should show

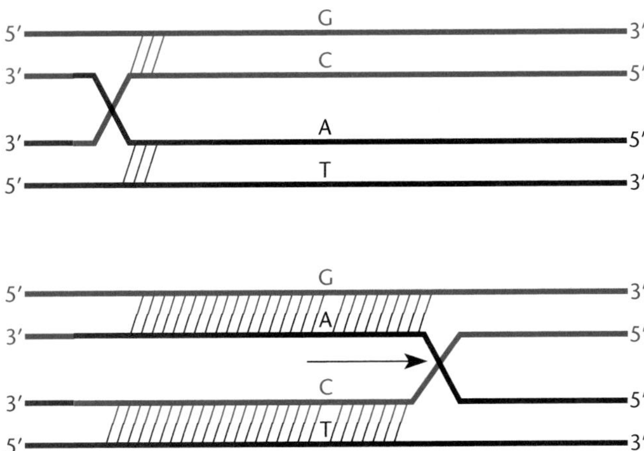

Figure 10.11 Migration of Holliday junctions. By breaking the hydrogen bonds holding the DNAs together in front of the branch and re-forming them behind, the junction migrates and extends the regions of pairing (i.e., the heteroduplexes) between the two DNAs. The heteroduplex region is hatched. In the example, two mismatches, GA and CT, form in the heteroduplex regions because one of the DNA molecules has a mutation in the region.
doi:10.1128/9781555817169.ch10.f10.11

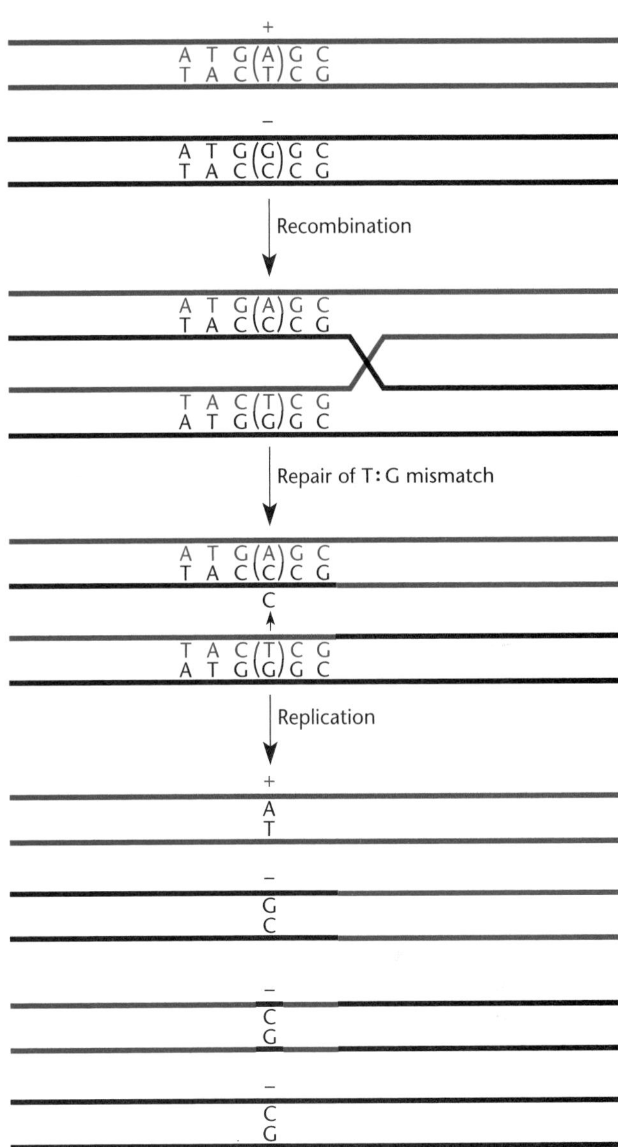

Figure 10.12 Repair of a mismatch in a heteroduplex region formed during recombination can cause gene conversion. A plus sign indicates the wild-type sequence, and a minus sign indicates the mutant sequence. See the text for details.
doi:10.1128/9781555817169.ch10.f10.12

up in equal numbers in the spores in the ascus. In other words, if the two haploid fungi have different alleles of the same gene, two of the four spores in each ascus should have the allele from one parent and the other two spores should have the allele from the other parent. This is called a 2:2 segregation. However, the two parental alleles sometimes do not appear in equal numbers in the spores. For example, three of the spores in a particular ascus might have the allele from one parent while the remaining spore has the allele from the other parent—a 3:1 segregation instead of the expected 2:2 segregation. In this case, an allele of one of the parents appears to have been converted into the allele of the other parent during meiosis, hence the name gene conversion.

Gene conversion is caused by repair of mismatches created on heteroduplexes during recombination, and Figure 10.12 shows how such mismatch repair can convert one allele into the other when the DNA molecules of the two parents recombine. In the illustration, the DNAs of the two parents are identical in the region of the recombination except that a mutation has changed a wild-type AT pair into a GC in one of the DNA molecules. Hence, the parents have different alleles of the gene. When the two individuals mate to form a diploid zygote and their DNAs recombine during meiosis, one strand of each DNA may pair with the complementary strand of the other DNA in this region. A mismatch will result, with a G opposite a T in one DNA and an A opposite a C in the other DNA (Figure 10.12). If a repair system changes the T opposite the G to a C in one of the DNAs, then after meiosis, three molecules will carry the mutant allele sequence, with GC at this position, but only one DNA will have the wild-type allele sequence, with AT at this position. Hence, one of the two wild-type alleles has been "converted" into the mutant allele.

MANIFESTATIONS OF MISMATCH REPAIR IN HETERODUPLEXES IN PHAGES AND BACTERIA

Gene conversion is more difficult to detect in crosses with bacteria and phages than with fungi, since the products of a single recombination event do not stay together in a bacterial or phage cross. However, the existence of heteroduplexes formed during recombination in bacteria and phages, as well as the repair of mismatches in these structures, is manifested in other ways.

Map Expansion

In prokaryotes, mismatch repair in heteroduplexes can increase the apparent recombination frequency between two closely linked markers, making the two markers seem farther apart than they really are. This manifestation of mismatch repair in heteroduplexes formed during recombination is called **map expansion** because the genetic map appears to increase in size.

Figure 10.13 shows how mismatch repair can affect the apparent recombination frequency between two markers. In the illustration, the two DNA molecules participating in the recombination have mutations that are very close to each other, so that crossovers between the two mutations to produce wild-type recombinants should be very rare. However, a Holliday junction occurs nearby, and the region of one of the two mutations is included in the heteroduplex, creating mismatches that can be repaired. If the G in the GT mismatch in one of the DNA molecules is repaired to an A, the progeny with the DNA will appear to be a wild-type recombinant. Therefore, even though the potential crossover that caused the formation of heteroduplexes did not occur in the region between the two mutations, apparent wild-type recombinants resulted. The apparent recombination due to mismatch repair might occur even when the Holliday junction is resolved so that the flanking sequences are returned to their original configuration; thus, a true crossover does not result. Therefore, although gene conversion and other manifestations of mismatch repair are generally associated with recombinant DNA molecules, they do not appear only in DNA molecules with crossovers.

Figure 10.13 Repair of mismatches can give rise to recombinant types between two mutations. A plus sign indicates the wild-type sequence, and a minus sign indicates the mutant sequence. The positions of the mutations are shown in parentheses. See the text for details.
doi:10.1128/9781555817169.ch10.f10.13

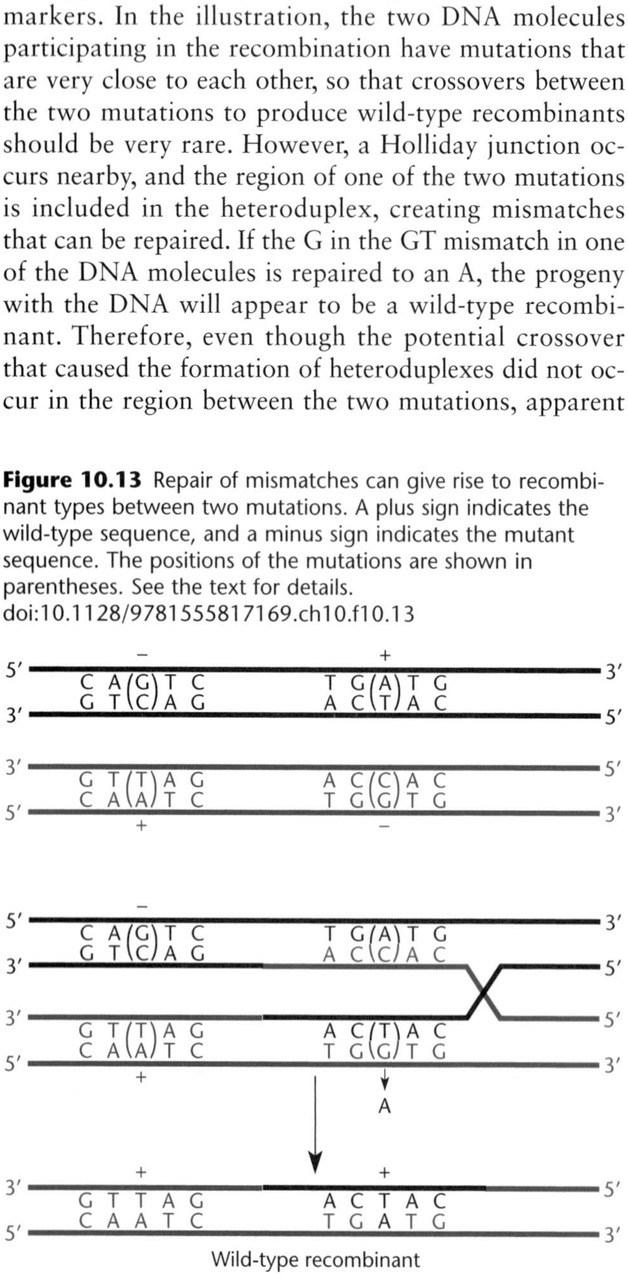

Marker Effects

Mismatch repair of heteroduplexes can also cause marker effects, phenomena in which two different markers at exactly the same locus show different recombination frequencies when crossed with the same nearby marker. For example, two different transversion mutations might change a UAC codon into UAA and UAG codons in different strains. However, when these two strains are crossed with another strain with a third nearby mutation, the recombination frequency between the UAA mutation and the third mutation might appear to be much higher than the recombination frequency between the UAG mutation and the third mutation, even though the UAG and UAA mutations are exactly the same distance on the DNA from the third mutation. Such a difference between the two recombination frequencies can be explained because the UAG and UAA mutations are causing different mismatches to form during recombination, and one of these may be recognized and repaired more readily by the mismatch repair system than the other. In *E. coli* at least, CC mismatches are not repaired by the mismatch repair system. Note that one of the heteroduplexes that the original UAC sequence forms with the UAG mutation at this site contains a CC mismatch but the heteroduplex formed with the UAA mutation does not.

Marker effects also occur because the lengths of single DNA strands removed and resynthesized by different mismatch repair systems vary (see chapter 11), and the chance that mismatch repair will lead to apparent recombination depends on the lengths of these sequences, or patches. As is apparent in Figure 10.13, a wild-type recombinant occurs only if mismatches due to both mutations are not removed on the same repair patch. If the patch that is removed in repairing one mismatch also removes the other mismatch, one of the parental DNA sequences will be restored and no apparent recombination will occur. In *E. coli*, mismatches due to deamination of certain methylated cytosines are repaired by very short patches (VSP repair [see chapter 11]).

High Negative Interference

Another manifestation of mismatch repair in heteroduplexes is **high negative interference**. This has the reverse effect of interference in eukaryotes, a phenomenon in which a molecular process helps regulate the number and spacing of crossover events that occur in each chromosome. In high negative interference, one crossover greatly increases the apparent frequency of another crossover nearby.

High negative interference is often detected during three-factor crosses with closely linked markers. In chapters 3 and 7, we discussed how three-factor crosses can be used to order three closely linked mutations. Briefly, if one parent has two mutations and the other parent has a third mutation, the frequency of the different types of recombinants after the cross will depend on the order of the sites of the three mutations in the DNA. Twice as many apparent crossovers are required to produce the rarest recombinant type as are needed for the more frequent recombinant types.

However, owing to mismatch repair of heteroduplexes, the rarest recombinant type can occur much more frequently than expected. Figure 10.14 shows a three-factor cross between markers that are created by three mutations. Wild-type sequences are marked with a plus, and mutant sequences are marked with a minus. With the molecules pictured, the formation of a wild-type recombinant should require two crossovers, one between mutations 1 and 2 and another between mutations 2 and 3. Theoretically, if the two crossovers were truly independent, the frequency of wild-type recombinants in the three-factor cross should equal the frequency when mutation 1 is crossed with mutation 2 times the frequency when mutation 2 is crossed with mutation 3 in separate crosses. Instead, the frequency of the wild-type recombinants in the three-factor cross is often much higher than this product. As shown in Figure 10.14, if a crossover occurs between two of the mutations, a heteroduplex formed by the Holliday junction might include the region of the third mutation. Then, repair of the third mutant site mismatch in the heteroduplex would give the appearance of a second nearby crossover, greatly increasing the apparent frequency of that crossover.

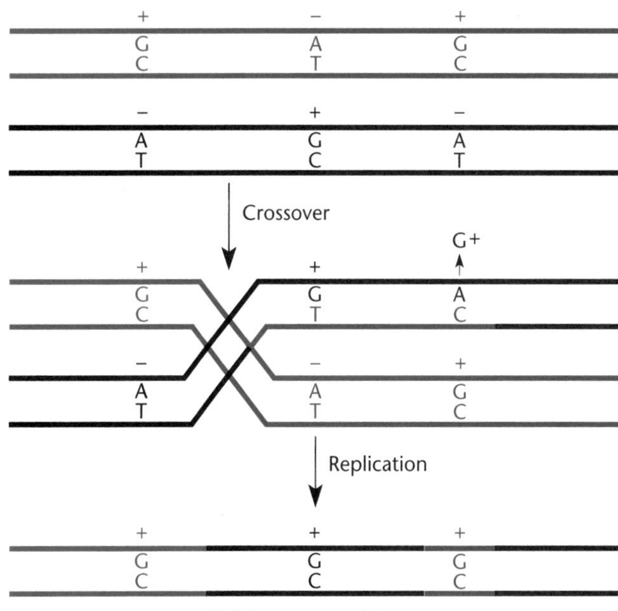

Figure 10.14 High negative interference due to mismatch repair. Inclusion of the region of the third mutation in the heteroduplex followed by repair of the mismatch can give the appearance of a second crossover.
doi:10.1128/9781555817169.ch10.f10.14

It is important to realize that repair of mismatches can occur wherever two DNAs are recombining. However, only if two markers are close enough together that normal crossovers are infrequent will mismatch repair in the heteroduplexes contribute significantly to apparent recombination frequencies and cause effects such as map expansion, high negative interference, and marker effects.

SUMMARY

1. Recombination is the joining of DNA strands in new combinations. Homologous recombination occurs only between two DNA molecules that have the same sequence in the region of the crossover.

2. All recombination models involve the formation of Holliday junctions. The Holliday junctions can migrate and isomerize so that the crossed strands uncross and then recross in a different orientation. Holliday junctions can then be resolved by specific DNA endonucleases to give recombinant DNA products.

3. The region over which two strands originating from different DNA molecules are paired in a Holliday junction is called a heteroduplex.

4. Repair of mismatches on the heteroduplex DNA molecules formed as intermediates in recombination can give rise to such phenomena as gene conversion, map expansion, high negative interference, and marker effects.

(continued)

SUMMARY (continued)

5. In *E. coli*, the major pathway for recombination during conjugation and transduction is the RecBCD pathway. The RecBCD protein loads on the DNA at a double-strand break and moves along the DNA, looping out a single strand and degrading one strand from the 3′ end. If a sequence in the DNA called the χ sequence is encountered as the protein moves along the DNA, the 3′-to-5′ nuclease activity on RecBCD is inhibited and 5′-to-3′ activity is stimulated, leaving a free 3′ end that can invade other double-stranded DNAs.

6. The RecF pathway is another recombination pathway in *E. coli*. It is required for recombination between single-stranded gaps in DNA and double-stranded DNA. The RecF pathway requires the products of the *recF*, *recJ*, *recN*, *recO*, *recQ*, and *recR* genes. Because in *E. coli* it functions only at single-stranded gaps in the DNA, this pathway can function in Hfr crosses only if the *sbcB*, *sbcC*, and *sbcD* genes have been inactivated or mutated. The SbcB, SbcC, and other enzymes destroy intermediates that would normally allow recombination by the RecF pathway and channel recombination to the RecBCD pathway.

7. The RecA protein promotes synapse formation and strand displacement and is required for recombination by both the RecBCD and RecF pathways. The RecA protein forces a single-stranded DNA into a helical nucleoprotein filament, which can then scan double-stranded DNAs looking for its complementary sequence. If it finds its complementary sequence, it invades it, forming a D loop or a three-strand structure, which can then migrate to form a Holliday junction.

8. Holliday junctions can migrate by at least two separate pathways in *E. coli*, the RuvABC pathway and the RecG pathway. In the first pathway, the RuvA protein binds to Holliday junctions, and then the RuvB protein binds to RuvA and promotes branch migration with the energy derived from cleaving ATP. The RuvC protein is a Holliday junction-specific endonuclease that cleaves Holliday junctions to resolve recombinant products. The RecG protein is also a Holliday junction-specific helicase, but it may be more important in reversing DNA replication forks and removing structures that could otherwise serve as sites for "pathological" initiation of DNA replication.

9. Many phages also encode their own recombination systems. Sometimes, phage recombination functions are analogous to host recombination functions. The gene *49* product of the T4 phage and the gene *3* product of T7 resolve Holliday junctions. A phage λ recombination system, the Red system, is encoded by two genes, *exo* and *bet*. The *exo* gene product is analogous to RecBCD in that it degrades one strand of a double-stranded DNA to make a single-stranded DNA for strand invasion. The *bet* gene product is analogous to RecA in that it promotes synapse formation between two DNAs. These λ genes have become the basis for a very useful way of doing site-specific mutagenesis and gene replacements, sometimes called recombineering.

QUESTIONS FOR THOUGHT

1. Why do you suppose that essentially all organisms have recombination systems?

2. Why do you suppose the RecBCD protein promotes recombination through such a complicated process?

3. Why are there overlapping pathways of recombination that can substitute for each other?

4. Why do you think the cell encodes the *sbcB* and *sbcC* gene products that interfere with the RecF pathway?

5. Why do some phages encode their own recombination systems? Why not rely exclusively on the host pathways?

6. Propose a model for how the RecG protein could substitute for RuvABC when it has only a helicase activity and not an X-phile resolvase activity.

PROBLEMS

1. Outline how you would determine if recombinants in an Hfr cross have a *recA* mutation. Note: *recA* mutations make the cells very sensitive to mitomycin and UV irradiation.

2. How would you determine whether the products of other genes participate in the RecG pathway of migration and resolution of Holliday junctions? How would you find such genes?

3. Describe the recombination promoted by homing double-stranded nucleases to insert an intron by using the double-strand break repair model.

4. Design an experiment to determine whether recombination due to the RecBC nuclease without the RecD subunit is still stimulated by χ.

SUGGESTED READING

Baharoglu, Z., M. Petranovic, M.-J. Flores, and B. Michel. 2006. RuvAB is essential for replication forks reversal in certain replication mutants. *EMBO J.* **25:**596–604.

Bolt, E. L., and R. G. Lloyd. 2002. Substrate specificity of RusA resolvase reveals the DNA structures targeted by RuvAB and RecG *in vivo*. *Mol. Cell* **10:**187–198.

Clark, A. J., and A. D. Margulies. 1965. Isolation and characterization of recombination-deficient mutants of *Escherichia coli* K12. *Proc. Natl. Acad. Sci. USA* **62:**451–459.

Cox, M. M. 2003. The bacterial RecA protein as a motor protein. *Annu. Rev. Microbiol.* **57:**551–577.

Datsenko, K. A., and B. L. Wanner. 2000. One-step inactivation of chromosomal genes in *Escherichia coli* K-12 using PCR products. *Proc. Natl. Acad. Sci. USA* **97:**6640–6645.

Dillingham, M. S., M. Spies, and S. C. Kowalczykowski. 2003. RecBCD enzyme is a bipolar DNA helicase. *Nature* **423:**893.

Dixon, D. A., and S. C. Kowalczykowski. 1993. The recombination hotspot Chi is a regulatory sequence that acts by attenuating the nuclease activity of the *E. coli* RecBCD enzyme. *Cell* **73:**87–96.

Ellis, H. M., D. Yu, T. DiTizio, and D. L. Court. 2001. High efficiency mutagenesis, repair and engineering of chromosomal DNA using single-stranded oligonucleotides. *Proc. Natl. Acad. Sci. USA 98:*6742–6746.

Handa, N., K. Morimatsu, S. T. Lovett, and S. C. Kowalczykowski. 2009. Reconstitution of initial steps of dsDNA break repair by the RecF pathway of *E. coli*. *Genes Dev.* **23:**1234–1245.

Holliday, R. 1964. A mechanism for gene conversion in fungi. *Genet. Res.* **5:**282–304.

Howard-Flanders, P., and L. Theriot. 1966. Mutants of *Escherichia coli* defective in DNA repair and in genetic recombination. *Genetics* **53:**1137–1150.

Isaacs, F. J., P. A. Carr, H. H. Wang, M. J. Lajoie, B. Sterling, L. Kraal, A. C. Tolonen, T. A. Gianoulis, D. B. Goodman, N. B. Reppas, C. J. Emig, D. Bang, S. J. Hwang, M. C. Jewett, J. M. Jacobson, and G. M. Church. 2011. Precise manipulation of chromosomes *in vivo* enables genome-wide codon replacement. *Science* **333:**348–353.

Jones, M., R. Wagner, and M. Radman. 1987. Mismatch repair and recombination in *E. coli*. *Cell* **50:**621–626.

Kogoma, T. 1997. Stable DNA replication: interplay between DNA replication, homologous recombination, and transcription. *Microbiol. Mol. Biol. Rev.* **61:**212–238.

Kuzminov, A., and F. W. Stahl. 1999. Double-strand end repair via the RecBCD pathway in *Escherichia coli* primes DNA replication. *Genes Dev.* **13:**345–356.

Lloyd, R. G. 1991. Conjugal recombination in resolvase-deficient *ruvC* mutants of *Escherichia coli* K-12 depends on *recG*. J. Bacteriol. **173:**5414–5418.

Lopper, M., R. Boonsombat, S. J. Sandler, and J. L. Keck. 2007. A hand-off mechanism for primosome assembly in replication restart. *Mol. Cell* **26:**781–793.

Lovett, S. T. 2005. Filling the gaps in replication restart pathways. *Mol. Cell* **17:**751–752.

Mazin, A. V., and S. C. Kowalczykowski. 1999. A novel property of the RecA nucleoprotein filament: activation of double-stranded DNA for strand exchange *in trans*. *Genes Dev.* **13:**2005–2016.

Meselson, M. S., and C. M. Radding. 1975. A general model for genetic recombination. *Proc. Natl. Acad. Sci. USA 72:*358–361.

Mosig, G. 1987. The essential role of recombination in T4 phage growth. *Annu. Rev. Genet.* **21:**347–371.

Parsons, C. A., I. Tsaneva, R. G. Lloyd, and S. C. West. 1992. Interaction of *Escherichia coli* RuvA and RuvB proteins with synthetic Holliday junctions. *Proc. Natl. Acad. Sci. USA* **89:**5452–5456.

Radding, C. M. 1991. Helical interactions in homologous pairing and strand exchange driven by the RecA protein. *J. Biol. Chem.* **266:**5355–5358.

Renzette, N., N. Gumlaw, and S. J. Sandler. 2007. DinI and RecX modulate RecA-DNA structures in *Escherichia coli* K-12. *Mol. Microbiol.* **63:**103–115.

Rocha, E. P., E. Cornet, and B. Michel. 2005. Comparative and evolutionary analysis of the bacterial homologous recombination systems. *PLoS Genet.* **1:**e15.

Sawitzke, J. A., and F. W. Stahl. 1992. Phage λ has an analog of *Escherichia coli recO, recR* and *recF* genes. *Genetics* **130:**7–16.

Sharples, G. J. 2001. The X philes: structure-specific endonucleases that resolve Holliday junctions. *Mol. Microbiol.* **39:**823–834.

Singleton, M. R., M. S. Dillingham, M. Gaudier, S. C. Kowalczykowski, and D. B. Wigley. 2004. Crystal structure of RecBCD enzyme reveals a machine for processing DNA breaks. *Nature* **432:**187–193.

Stahl, F. W., M. M. Stahl, R. E. Malone, and J. M. Crasemann. 1980. Directionality and nonreciprocality of Chi-stimulated recombination in phage λ. *Genetics* **94:**235–248.

Swingle, B., E. Markel, N. Costantino, M. G. Bubunenko, S. Cartinhour, and D. L. Court. 2010. Oligonucleotide recombination in Gram-negative bacteria. *Mol. Microbiol.* **75:**138–148.

Szostak, J. W., T. L. Orr-Weaver, R. J. Rothstein, and F. W. Stahl. 1983. The double-strand-break repair model for recombination. *Cell* **33:**25–35.

Taylor, A. F., and G. R. Smith. 2003. RecBCD enzyme is a DNA helicase with fast and slow motors of opposite polarity. *Nature* **423:**889–893.

West, S. C. 1998. RuvA gets X-rayed on Holliday. *Cell* **94:**699–701.

Zhang, J., A. A. Mahdi, G. S. Briggs, and R. G. Lloyd. 2010. Promoting and avoiding recombination: contrasting activities of the *Escherichia coli* RuvABC Holliday junction resolvase and RecG DNA translocase. *Genetics* **185:**23–37.

CHAPTER 11

DNA Repair and Mutagenesis

THE CONTINUITY OF SPECIES FROM ONE generation to the next is a tribute to the stability of DNA. If DNA were not so stable and were not reproduced so faithfully, there could be no species. Before DNA was known to be the hereditary material and its structure was determined, a lot of speculation centered on what types of materials would be stable enough to ensure the reliable transfer of genetic information over so many generations (see, for example, Schrodinger, Suggested Reading, in the introduction). Therefore, the discovery that the hereditary material is DNA—a chemical polymer no more stable than many other chemical polymers—came as a surprise.

Evolution has provided the selection for a DNA replication apparatus that minimizes mistakes (see chapter 1). However, mistakes during replication are not the only threats to DNA. Since DNA is a chemical, it is constantly damaged by chemical reactions. Many environmental factors can damage the molecule. Heat can speed up spontaneous chemical reactions, leading, for example, to the deamination of bases. Chemicals can react with DNA, adding groups to the bases or sugars, breaking the bonds of the DNA, or fusing parts of the molecule to each other. Irradiation at certain wavelengths can also chemically damage DNA, which can absorb the energy of the photons. Once the molecule is energized, bonds may be broken or parts may be erroneously fused. DNA damage can be very deleterious to cells because DNA polymerase may not be able to replicate over the damaged area, preventing the cells' multiplication. Even if the damage does not block replication, replicating over the damage can cause mutations, many of which may be deleterious or even lethal. Obviously, cells need mechanisms for DNA damage repair.

doi:10.1128/9781555817169.ch11

To describe DNA damage and its repair, we need first to define a few terms. Chemical damage in DNA is called a **lesion**. Chemical compounds or treatments that cause lesions in DNA can kill cells and can also increase the frequency of mutations in the DNA. Such treatments that generate mutations are called **mutagenic treatments** or **mutagens**. Some mutagens, known as **in vitro mutagens**, can be used to damage DNA in the test tube, which then produces mutations when the DNA is introduced into cells. Other mutagens damage DNA only in the cell, for example, by interfering with base pairing during replication. These are called **in vivo mutagens**.

In this chapter, we discuss the types of DNA damage, how each type of damage to DNA might cause mutations, and how bacterial cells repair the damage to their DNA. Many of these mechanisms seem to be universal, as related systems are found in other organisms, including humans.

Evidence for DNA Repair

Before discussing specific types of DNA damage, we should make some general comments on the outward manifestations of DNA damage and its repair. The first question is how we even know a cell has the means to repair a particular type of damage to its DNA. One way is to measure killing by exposure to chemicals or by irradiation. The chemical agents and radiation that damage DNA also often damage other cellular constituents, including RNA and proteins. Nevertheless, cells exposed to these agents usually die as a result of chemical damage to the DNA. The other components of the cell can usually be resynthesized and/or exist in many copies, so that even if some molecules are damaged, more of the same type of molecule will be there to substitute for the damaged ones. However, a single chemical change in the enormously long chromosomal DNA of a cell can prevent the replication of that molecule and subsequently cause cell death unless the damage is repaired.

To measure cell killing—and thereby demonstrate that a particular type of cell has DNA repair systems—we can compare the survival of the cells exposed intermittently to small doses of a DNA-damaging agent with that of cells that receive the same amount of treatment continuously. If the cells have DNA repair systems, more cells survive the short intervals of treatment because some of the damage is repaired between treatments. In contrast, if DNA is not repaired during the rest periods, it should make no difference whether the treatment occurs at intervals or continuously. The cells accumulate the same amount of damage regardless of the different treatments, and the same fraction of cells should survive both regimes.

Another indication of repair systems comes from the shape of the killing curves. A killing curve is a plot of the number of surviving cells versus the extent of treatment by an agent that damages DNA. The extent of treatment can refer to the length of time the cells are irradiated or exposed to a chemical that damages DNA or to the intensity of irradiation or the concentration of the damaging chemical.

The two curves in Figure 11.1 contrast the shapes of killing curves for cells with and without DNA repair systems. In the curve for cells without a repair system for the DNA-damaging treatment, the fraction of surviving cells drops exponentially, since the probability that each cell will be killed by a lethal "hit" to its DNA is the same during each time interval. This exponential decline gives rise to a straight line when plotted on a semilog scale, as shown (Figure 11.1).

The other curve shows what happens if the cell has DNA repair systems. Rather than dropping exponentially with increasing treatment, this curve extends horizontally first, forming a "shoulder." The shoulder appears because repair mechanisms can fix the lower levels of damage, allowing many of the cells to survive. Only with higher treatment levels, when the damage becomes so extensive that the repair systems can no longer cope with it, will the number of surviving cells drop exponentially with increasing levels of treatment.

Among the survivors of DNA-damaging agents, there may be many more mutants than before. However, it is very important to distinguish DNA damage

Figure 11.1 Survival of cells as a function of the time or extent of treatment with a DNA-damaging agent. The fraction of surviving cells is plotted against the duration of treatment. The shoulder on the survival curve indicates the presence of a repair mechanism.
doi:10.1128/9781555817169.ch11.f11.1

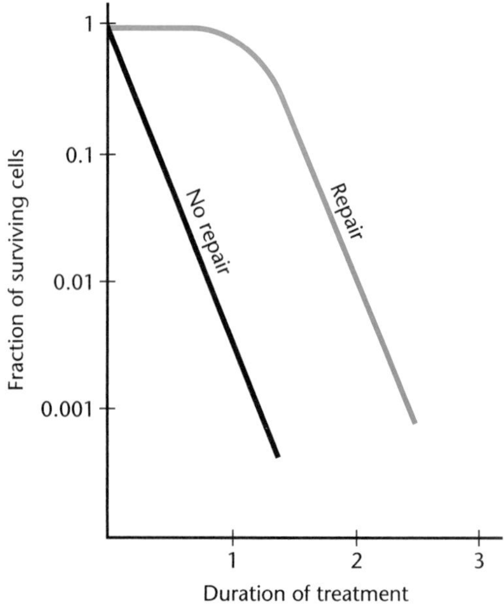

from mutagenesis. In particular types of cells, some types of DNA damage are mutagenic while others are not, independent of their effects on cell survival. Recall from chapter 3 that mutations are permanent heritable changes in the sequence of nucleotides in DNA. The damage to DNA caused by a chemical or by irradiation is not by itself a mutation because it is not heritable. A mutation might occur because the damage is not repaired and the replication apparatus must proceed over the damage, making mistakes because complementary base pairing does not occur properly at the site of the damage. Alternatively, mistakes might be made during attempts to repair the damage, causing changes in the sequence of nucleotides at the site of the damage. In the following sections, we discuss some types of DNA damage, how they might cause mutations, and the repair systems that can fix them. Most of what we describe is best known for *Escherichia coli*, for which these systems are best understood, but the universality of many of them is clear, and we share many of the same systems.

Specific Repair Pathways

Different agents damage DNA in different ways, and different repair pathways operate to repair the various forms of damage. Some of these pathways repair only a certain type of damage, whereas others are less specific and repair many types. We first discuss examples of damage repaired by specific repair pathways.

Deamination of Bases

One of the most common types of damage to DNA is the deamination of bases. Some of the amino groups in adenine, cytosine, and guanine are particularly vulnerable and can be removed spontaneously or by many chemical agents (Figure 11.2). When adenine is deaminated, it becomes **hypoxanthine**. When guanine is deaminated, it becomes **xanthine**. When cytosine is deaminated, it becomes **uracil**.

Deamination of DNA bases is mutagenic because it results in base mispairing. As shown in Figure 11.2, hypoxanthine derived from adenine pairs with the base cytosine instead of thymine, and uracil derived from the deamination of cytosine pairs with adenine instead of guanine.

The type of mutation caused by deamination depends on which base is altered. For example, the hypoxanthine that results from deamination of adenine pairs with cytosine during replication, incorporating C instead of T at that position. In a subsequent replication, the C pairs with the correct G, causing an AT-to-GC transition in the DNA. Similarly, a uracil resulting from the deamination of a cytosine pairs with an adenine during replication, causing a GC-to-AT transition type of mutation.

DEAMINATING AGENTS

Although deamination often occurs spontaneously, especially at higher temperatures, some types of chemicals react with DNA and remove amino groups from the bases. Treatment of cells or DNA with these chemicals, known as **deaminating agents**, can greatly increase the rate of mutation. Which deaminating agents are mutagenic in a particular situation depends on the properties of the chemical.

Hydroxylamine

Hydroxylamine specifically removes the amino group of cytosine and consequently causes only GC-to-AT transitions in the DNA. However, hydroxylamine, an in vitro mutagen, cannot enter cells, so it can be used only to mutagenize purified DNA or viruses. Mutagenesis by hydroxylamine is particularly effective when the treated DNA is introduced into cells deficient in repair by the uracil-*N*-glycosylase enzyme (see chapter 1) for reasons discussed below.

Nitrous Acid

Nitrous acid not only deaminates cytosines, but also removes the amino groups of adenine and guanine (Figure 11.2). It also causes other types of damage. Because it is less specific, nitrous acid can cause both GC-to-AT and AT-to-GC transitions, as well as deletions. Nitrous acid can enter some types of cells and so can be used as a mutagen both in vivo and in vitro.

REPAIR OF DEAMINATED BASES

Given that base deamination is potentially mutagenic, it is not surprising that special enzymes have evolved to remove deaminated bases from DNA. These enzymes, **DNA glycosylases**, break the glycosyl bond between the damaged base and the sugar in the nucleotide. These enzymes are highly specific, and a unique DNA glycosylase exists for each type of deaminated base and removes only that particular base. Specific DNA glycosylases remove bases damaged in other ways; glycosylases are discussed in later sections. There are at least a dozen specific *N*-glycosylases that remove damaged bases in *E. coli*, and they all work by basically the same mechanism.

Figure 11.3 illustrates the removal of damaged bases from DNA by DNA glycosylases. There are two types of glycosylases: those that remove just the base and others, called **AP lyases**, that both remove the base and cut the DNA backbone on the 3′ side of the damaged base. If just the base has been removed by the specific glycosylase, nucleases called **AP endonucleases** cut the sugar-phosphate backbone of the DNA on the 5′ side of the missing base, leaving a 3′ OH group. These enzymes can cut either next to the spot from which a pyrimidine (C or T) has been removed or next to where a purine (A or G) has been

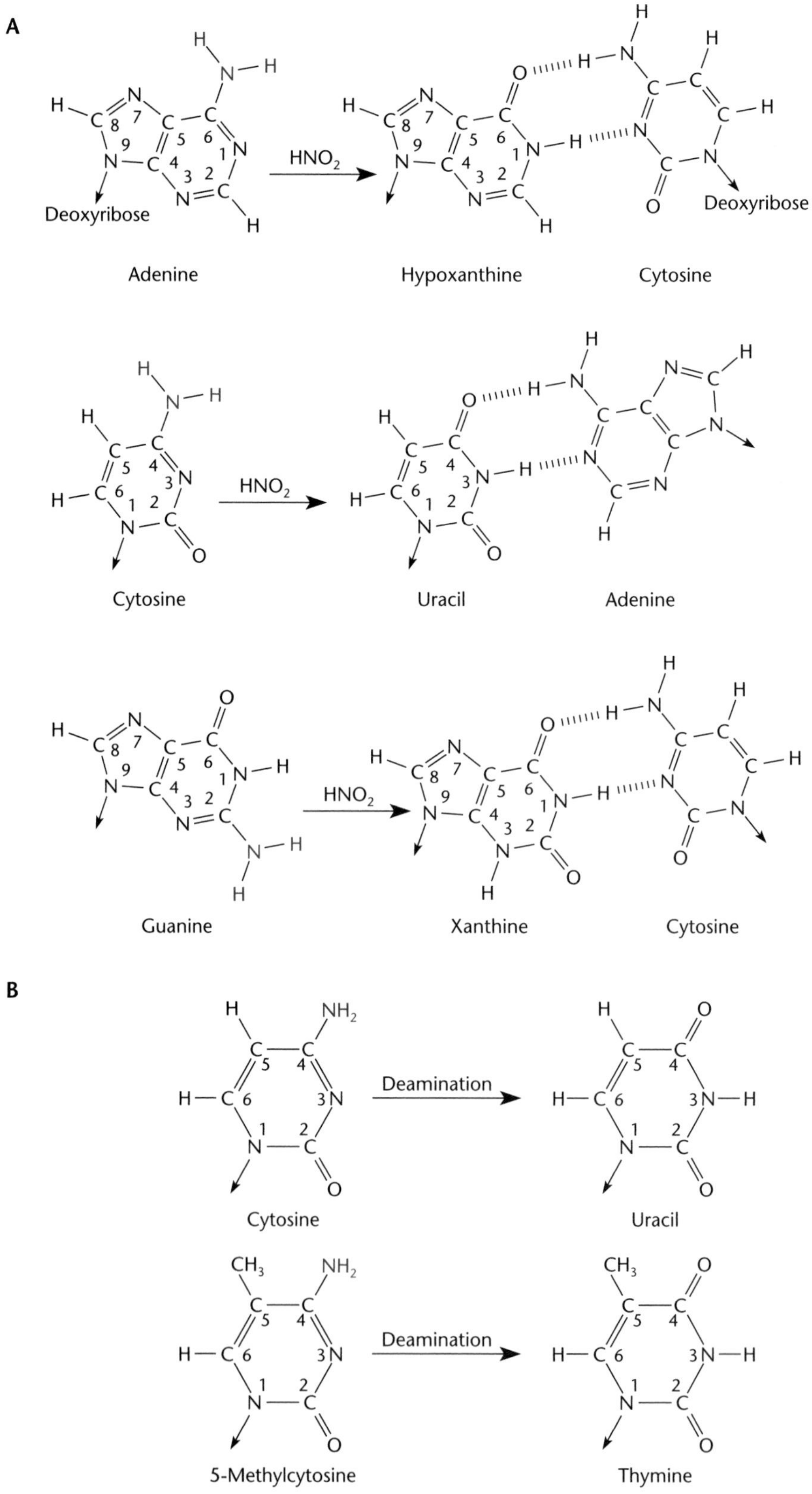

Figure 11.2 (A) Modified bases created by deaminating agents, such as nitrous acid (HNO_2). Some deaminated bases pair with the wrong base, causing mutations. **(B)** Spontaneous deaminations. Deamination of 5-methylcytosine produces a thymine that is not removable by the uracil-*N*-glycosylase.
doi:10.1128/9781555817169.ch11.f11.2

removed (apyrimidinic/apurinic sites). After the cut is made and processed, the free 3′ hydroxyl end is used as a primer by the repair DNA polymerase (DNA polymerase I [Pol I] in *E. coli*) to synthesize more DNA, while the 5′ exonuclease activity associated with the DNA polymerase degrades the strand ahead of the DNA polymerase. In this way, the entire region of the DNA strand around the deaminated base is resynthesized and the normal base is inserted in place of the damaged one.

VERY-SHORT-PATCH REPAIR OF DEAMINATED 5-METHYLCYTOSINE

Most organisms have some 5-methylcytosine bases instead of cytosines at specific sites in their DNA. These bases are cytosines with a methyl group at the 5 position on the pyrimidine ring instead of the usual hydrogen (Figure 11.2B). Specific enzymes called **methyltransferases** transfer the methyl group to this position after the DNA is synthesized, using *S*-adenosylmethionine as the methyl donor. The functions of these 5-methylcytosines are often obscure, but we know that they sometimes help protect DNA against cutting by restriction endonucleases in bacteria and play additional roles in eukaryotes.

The sites of 5-methylcytosine in DNA are often hot spots for mutagenesis, because deamination of 5-methylcytosine yields thymine rather than uracil (Figure 11.2B), and thymine in DNA is not recognized by the uracil-*N*-glycosylase, since it is a normal base in DNA. These thymines in DNA are located opposite guanines and so could in principle be repaired by the methyl-directed mismatch repair system (see below). However, in a GT mismatch created by a replication mistake, the mistakenly incorporated base can be identified because it is in a newly replicated strand, as yet unmethylated by Dam methylase (see chapter 1), whereas the GT mismatches created by the deamination of 5-methylcytosine are generally not found in newly synthesized DNA. Repairing the wrong strand causes a GC-to-AT transition in the DNA.

In *E. coli* K-12 and some other enteric bacteria, most of the 5-methylcytosine in the DNA occurs in the second C of the sequence 5′-CCWGG-3′/3′-GGWCC-5′, where the middle base pair (W) is generally either AT or TA. The second C in this sequence is methylated by an enzyme called DNA cytosine methylase (Dcm) to give $C^mCAGG/GGTC^mC$, where C^m indicates the position of the 5-methylcytosine. The mutation potential of this modification likely provided the selection for a special repair mechanism for deaminated 5-methylcytosines that occur in this sequence in *E. coli* K-12. This repair system specifically removes the thymine whenever it appears as a TG mismatch in this sequence.

Because a small region, or "patch," of the DNA strand containing the T is removed and resynthesized during the repair process, the mechanism is called **very-short-patch (VSP) repair**. In VSP repair, the Vsr endonuclease, the product of the *vsr* gene, binds to a TG mismatch in the $C^mCWGG/GGWTC$ sequence and makes a break next to the T. The T is then removed, and the strand is resynthesized by the repair DNA polymerase (DNA Pol I), which inserts the correct C.

The Vsr repair system is very specialized and usually repairs TG mismatches only in the sequence shown above. Therefore, this repair system would have only limited usefulness if methylation did not occur in the sequence. The

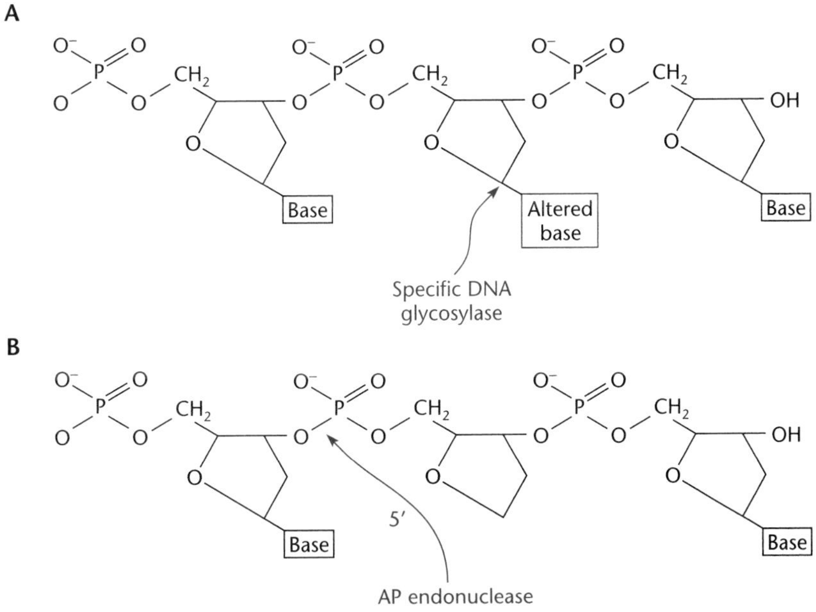

Figure 11.3 Repair of altered bases by DNA glycosylases. **(A)** The specific DNA glycosylase removes the altered base. **(B)** An apurinic, or AP, endonuclease cuts the DNA backbone on the 5′ side of the AP site. The strand is degraded and resynthesized, and the correct base is restored (not shown). doi:10.1128/9781555817169.ch11.f11.3

vsr gene is immediately downstream of the gene for the Dcm methylase, ensuring that cells that inherit the gene to methylate the C in C^{m}CWGG/GGWC^{m}C also usually inherit the ability to repair the mismatch correctly if it is deaminated. While only enteric bacteria like *E. coli* have been shown to have this particular repair system, many other organisms have 5-methylcytosine in their DNA, and we expect that similar repair systems will be found in these organisms. Interestingly, we will see that the Vsr repair system also interacts with the mismatch repair system (see Heinze et al., Suggested Reading).

Damage Due to Reactive Oxygen

Although molecular oxygen (O_2) is not very damaging to DNA and most other macromolecules, more reactive forms of oxygen (collectively called RO) are very damaging. These more reactive forms of oxygen include hydrogen peroxide, superoxide radicals, and hydroxyl free radicals, which are produced as byproducts by flavoenzymes in the presence of molecular oxygen (see Korshunov and Imlay, Suggested Reading). The molecular oxygen strips electrons from the flavin, converting itself into these more reactive forms. Of these, hydroxyl radicals are the most damaging to DNA. Hydrogen peroxide, which has many normal functions in cells, is also produced but has no unpaired electrons and is not particularly damaging to DNA; however, it can be converted into hydroxyl free radicals in the presence of iron (Fe^{2+}) atoms. Reactive forms of oxygen can also arise as a result of environmental factors, including ultraviolet (UV) irradiation and chemicals, such as the herbicide paraquat. Interestingly, most common antibiotics also appear to kill bacterial cells through the formation of reactive oxygen species (See Kohanski et al., Suggested Reading). Presumably by blocking translation, these antibiotics may change the relative concentrations of proteins in the respiratory chains, causing them to produce more RO.

Because the reactive forms of oxygen appear in cells growing in the presence of oxygen, all aerobic organisms must contend with the resulting DNA damage and have evolved elaborate mechanisms to remove these chemicals from the cellular environment. In bacteria, some of these systems are induced by the presence of the reactive forms of oxygen (Box 11.1), and these genes encode enzymes such as superoxide dismutases, catalases, and peroxide reductases that help destroy or scavenge the reactive forms. The same systems that induce enzymes to scavenge reactive forms of oxygen also induce the expression of proteins involved in exporting antibiotics or preventing the uptake of antibiotics, consistent with the idea that the lethal effects of antibiotics often occur though oxidative damage, so the two are inextricably linked. Other genes induced by reactive oxygen species include genes that encode repair enzymes that help repair the oxidative DNA damage. The accumulation of this type of damage stemming from reactive oxygen species may be responsible for the increase in cancer rates with age and for many age-related degenerative diseases (Box 11.1).

8-oxoG

One of the most mutagenic lesions in DNA caused by reactive oxygen is the oxidized base **7,8-dihydro-8-oxoguanine** (**8-oxoG**, or **GO**) (Figure 11.4). This base appears frequently in DNA because of damage caused by internally produced free radicals of oxygen, and unless repair systems deal with the damage, DNA Pol III often mispairs it with adenine, causing spontaneous GC-to-AT transition mutations. Because of the mutagenic potential of 8-oxoG, *E. coli* has evolved many mechanisms for avoiding the resultant mutations, and we discuss these below.

MutM, MutY, and MutT

The *mut* genes of an organism were so named because they were identified by their ability to reduce the normal rates of spontaneous mutagenesis (see "Isolation of *mut* Mutants" below). Organisms with a mutation that inactivates the product of a *mut* gene will suffer higher-than-normal rates of spontaneous mutagenesis. We have already discussed one of the *mut* genes of *E. coli*, which was selected as a mutation that increases the spontaneous mutation rate, the *dnaQ* (*mutD*) gene encoding the editing function ε (see chapter 1). Some of the other *mut* genes of *E. coli* are *mutM*, *mutY*, and *mutT*. The products of these *mut* genes are exclusively devoted to preventing mutations due to 8-oxoG. The generally high rate of spontaneous mutagenesis in *mutM*, *mutT*, and *mutY* mutants is testimony to the fact that internal oxidation of DNA is an important source of spontaneous mutations and that 8-oxoG, in particular, is a very mutagenic form of damage to DNA (Box 11.1).

The discovery that these three *mut* genes are dedicated to relieving the mutagenic effects of 8-oxoG in DNA, as well as the role played by each of them, was the result of some clever genetic experiments (see Michaels et al., Suggested Reading). First, there was the evidence that the functions of the three *mut* genes are additive to reduce spontaneous mutations, since the rate of spontaneous mutations is higher if two or all three of the *mut* genes are mutated than if only one of them is mutated. There was also evidence that mutations in each of the *mut* genes increased the frequency of some types of spontaneous mutations but not others. Below, we discuss the roles of the products of each of these *mut* genes and then discuss how the genetic evidence is consistent with each of these roles.

BOX 11.1

The Role of Reactive Oxygen Species in Cancer and Degenerative Diseases

To respire, all aerobic organisms, including humans, must take up molecular oxygen (O_2). At normal temperatures, molecular oxygen reacts with very few molecules. However, some of it is converted into more reactive forms, such as superoxide radicals (O_2^-), hydrogen peroxide (H_2O_2), and hydroxyl radicals. Hydrogen peroxide is also produced deliberately in the liver to help detoxify recalcitrant molecules and by lysozomes to kill invading bacteria. Iron in the cell can catalyze the conversion of hydrogen peroxide to hydroxyl radicals ($^{\cdot}OH$), which may be the form in which oxygen is most damaging to DNA (see the text).

As described in the text, many enzymes have evolved to help reduce this damage. In *E. coli*, the major mechanisms for signaling the presence of reactive oxygen species is through OxyR. Interestingly, it is the direct effect of reactive oxygen species on the OxyR protein that causes a specific disulfide bond to form between cysteine residues 199 and 208 within the protein, which changes OxyR from a repressor to a transcriptional activator of the target genes (see Lee et al., 2004, References). While the actual protein involved in signaling oxidative stress is completely different in humans, we employ the same strategy of using disulfide bond formation to detect reactive oxygen species (see Guo et al., References). In this case, human cells use one of the major DNA damage-signaling proteins, ATM. ATM becomes activated to signal DNA repair when a disulfide bond forms a covalent link between dimers of the protein at residue Cys 2991 in the presence of reactive oxygen species.

Accumulation of DNA damage due to RO has been linked to many degenerative diseases, such as cancer, arthritis, cataracts, and cardiovascular disease. Overexpression of catalase in the energy-producing mitochondria of mice leads to increases in the life span (see Schriner et al., References). This increase in life span may result from partial mitigation of the age-associated decline in the mitochondria, along with other health benefits, like a decrease in the incidence of insulin resistance, a precondition associated with type 2 diabetes (see Lee et al., 2010, References).

The importance of internally generated RO in cancer has received dramatic confirmation (see Chmiel et al., References). The authors report that a genetic disease characterized by increased rates of colon cancer is due to mutations in the human repair gene *MYH*, which is analogous to the *mutY* gene of *E. coli*. Siblings who have inherited this predisposition to cancer, called familial adenomatous polyposis, are heterozygous for different mutant alleles of the *MYH* gene. The *mutY* gene product of *E. coli* is a specific *N*-glycosylase that removes adenine bases that have mistakenly paired with 8-oxoG in the DNA (see the text), and the human enzyme is known to have a similar activity. Also, mice that have had their *Ogg-1* and *Myh* genes inactivated (so-called knockout mice) are much more prone to lung and ovarian tumors, as well as lymphomas. The Ogg-1 gene product of mice is functionally analogous to MutM of *E. coli*. Furthermore, the primary types of mutations in these knockout mice are GC-to-TA transversions, as they are in *E. coli* (see the text and Xie et al., References).

Obviously, any mechanism for reducing the levels of these active forms of oxygen should increase longevity and reduce the frequency of many degenerative diseases. Fruits and vegetables produce antioxidants, including ascorbic acid (vitamin C); tocopherol (vitamin E); carotenes, such as beta-carotene (in carrots and many other vegetables), lutein (in green leafy vegetables), and lycopene (in tomatoes and other fruits); and multiple forms of vitamin A, that destroy these molecules and thereby protect the DNA in their seeds and the photosynthetic apparatus in their leaves from damage due to oxygen free radicals produced by UV irradiation. Evidence suggests that consumption of the many fruits and vegetables that contain these compounds reduces the rates of cancer and degenerative diseases.

References

Chmiel, N. H., A. L. Livingston, and S. S. David. 2003. Insight into the functional consequences of inherited variants of the *hMYH* adenine glycosylase associated with colorectal cancer: complementation assays with *hMYH* variants and pre-steady state kinetics of the corresponding mutated *E. coli* enzymes. *J. Mol. Biol.* **327:**431–443.

Guo. Z., S. Kozlov, M. F. Lavin, M. D. Person, and T. T. Paull. 2010. ATM activation by oxidative stress. *Science* **330:**517–521.

Lee, C., S. M. Lee, P. Mukhopadhyay, S. J. Kim, S. C. Lee, W. S. Ahn, M. H. Yu, G. Storz, and S. E. Ryu. 2004. Redox regulation of OxyR requires specific disulfide bond formation involving a rapid kinetic reaction path. *Nat. Struct. Mol. Biol.* **11:**1179–1185.

Lee, H. Y., C. S. Choi, A. L. Birkenfeld, T. C. Alves, F. R. Jornayvaz, M. J. Jurczak, D. Zhang, D. K. Woo, G. S. Shadel, W. Ladiges, P. S. Rabinovitch, J. H. Santos, K. F. Petersen, V. T. Samuel, and G. I. Shulman. 2010. Targeted expression of catalase to mitochondria prevents age-associated reductions in mitochondrial function and insulin resistance. *Cell Metab.* **12:**668–674.

Schriner, S. E., N. J. Linford, G. M. Martin, P. Treuting, C. E. Ogburn, M. Emond, P. E. Coskun, W. Ladiges, N. Wolf, H. Van Remmen, D. C. Wallace, and P. S. Rabinovitch. 2005. Extension of murine life span by overexpression of catalase targeted to mitochondria. *Science* **308:**1909–1911.

Xie, Y., H. Yang, C. Cunanan, K. Okamoto, D. Shibata, J. Pan, D. E. Barnes, T. Lindahl, M. McIlhatton, R. Fishel, and J. H. Miller. 2004. Deficiencies in mouse *Myh* and *Ogg1* result in tumor predisposition and G to T mutations in codon 12 of the K-ras oncogene in lung tumors. *Cancer Res.* **64:**3096–3102.

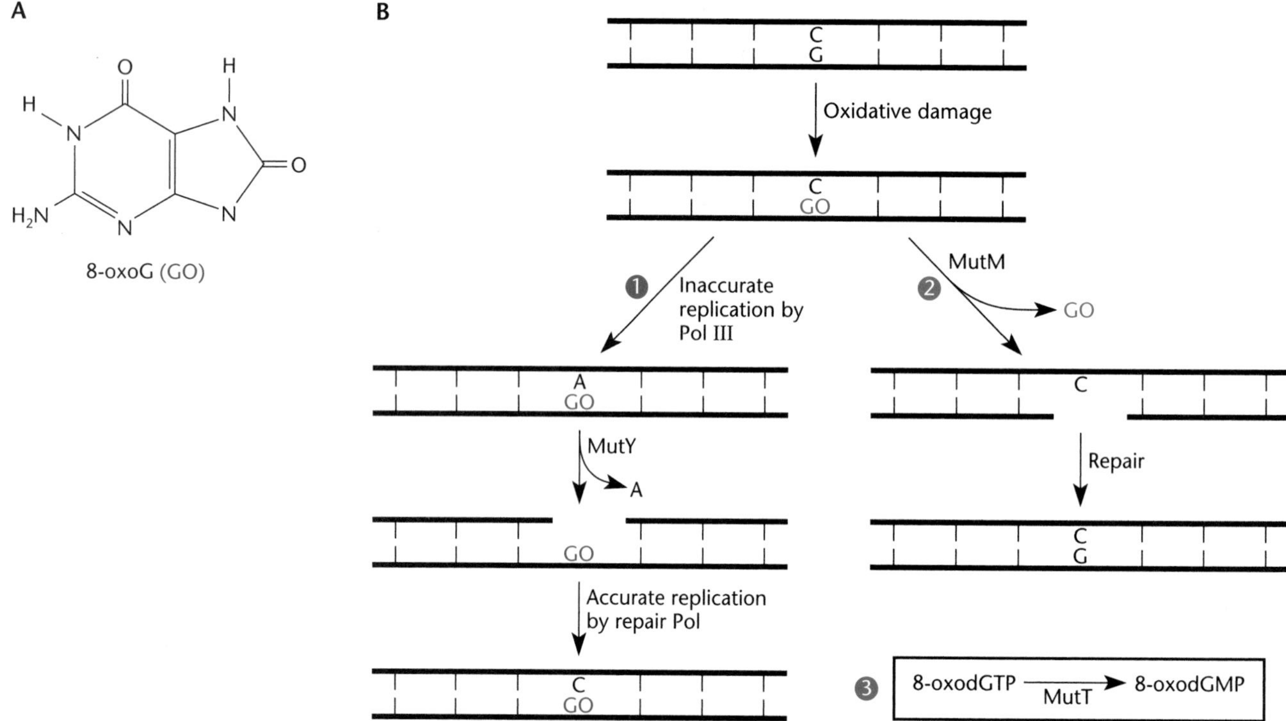

Figure 11.4 **(A)** Structure of 8-oxoG. **(B)** Mechanisms for avoiding mutagenesis due to 8-oxoG (GO). In pathway 1, an A mistakenly incorporated opposite 8-oxoG is removed by a specific glycosylase (MutY), and the strand is degraded and resynthesized with the correct C. In pathway 2, the 8-oxoG is itself removed by a specific glycosylase (MutM), and the strand is degraded and resynthesized with a normal G. In a third pathway, the 8-oxoG is prevented from entering the DNA by a specific phosphatase (MutT) that degrades the triphosphate 8-oxodGTP to the monophosphate 8-oxodGMP. doi:10.1128/9781555817169.ch11.f11.4

MutM

The MutM enzyme is an *N*-glycosylase that specifically removes the 8-oxoG base from the deoxyribose sugar in DNA (Figure 11.4). This repair pathway functions like other *N*-glycosylase repair pathways discussed in this book, except that the depurinated strand is cut by the AP endonuclease activity of MutM itself, degraded by an exonuclease, and resynthesized by DNA Pol I (Figure 11.3). The MutM protein is present in larger amounts in cells that have accumulated reactive oxygen, because the *mutM* gene is part of a regulon induced in response to oxidative stress. We discuss regulons in more detail in chapter 13.

MutY

The MutY enzyme is also a specific *N*-glycosylase. However, rather than removing 8-oxoG directly, the MutY *N*-glycosylase specifically removes adenine bases that have been mistakenly incorporated opposite an 8-oxoG in DNA (Figure 11.4). Repair synthesis by DNA Pol I then usually introduces the correct C to prevent a mutation, as with other *N*-glycosylase-initiated repair pathways.

The MutY enzyme also apparently recognizes a mismatch that results from accidental incorporation of an A opposite a normal G and removes the A. However, its major role in avoiding mutagenesis in the cell seems to be to prevent mutations due to 8-oxoG. As evidence, mutations that cause the overproduction of MutM completely suppress the mutator phenotype of *mutY* mutants. The interpretation of this result is as follows. If a significant proportion of all spontaneous mutations in a *mutY* mutant resulted from misincorporation of A's opposite normal Gs, excess MutM should have no effect on the mutation rate, because removal of 8-oxoG should not affect this type of mispairing. However, the fact that excess MutM almost completely suppresses the extra mutagenesis in *mutY* mutants suggests that very few of the extra spontaneous mutations in a *mutY* mutant are due to A-G mispairs and that most are due to A–8-oxoG mispairs.

MutT

The MutT enzyme operates by a very different mechanism (Figure 11.4): it prevents 8-oxoG from entering the DNA in the first place. The reactive forms of oxygen can oxidize not only guanine in DNA to 8-oxoG, but also the base in dGTP to form 8-oxodGTP. Without MutT, 8-oxodGTP is incorporated into DNA primarily by

DNA Pol IV, which cannot distinguish 8-oxodGTP from normal dGTP (see Foti et al., Suggested Reading). The MutT enzyme is a phosphatase that specifically degrades 8-oxodGTP to 8-oxodGMP so that it cannot be used in DNA synthesis.

GENETICS OF 8-OxoG MUTAGENESIS

There are a number of ways in which the genetic evidence obtained with *mutM*, *mutT*, and *mutY* mutants is consistent with these functions for the products of the genes. First, these activities explain why the effects of mutations in the genes are additive. If *mutT* is mutated, more 8-oxodGTP will be present in the cell to be incorporated into DNA, increasing the spontaneous-mutation rate. If MutM does not remove these 8-oxoGs from the DNA, spontaneous-mutation rates will increase even further. If MutY does not remove some of the A's that mistakenly pair with the 8-oxoGs, the spontaneous-mutation rate will be higher yet.

Mutations in the *mutM*, *mutY*, and *mutT* genes also increase the frequency of only some types of mutations, which again can be explained by the activities of these enzymes. For example, only the frequency of GC-to-TA transversion mutations is increased in *mutM* and *mutY* mutants. This is meaningful because, in general, transversions are less common than transition mutations (see chapter 3). The fact that mutations in both genes increase the frequency of the same type of relatively uncommon mutation first suggested that they function in the same pathway and also makes sense considering the functions of the gene products. If MutM does not remove 8-oxoG from DNA, mispairing of the 8-oxoG with A can lead to GC-to-TA transversions. Moreover, GC-to-TA transversions will occur if MutY does not remove the mispaired A's opposite the 8-oxoGs in the DNA. In contrast, while *mutT* mutations can increase the frequency of relatively uncommon GC-to-TA transversion mutations, they can also increase the frequency of TA-to-GC transversions. This is possible because an 8-oxodGTP molecule, which owes its existence to the lack of MutT to degrade it, may cause mutations in two different ways. It may incorrectly enter the DNA by pairing with an A and then, once in the DNA, correctly pair with a C to result in an AT-to-CG transversion, or it can enter the DNA correctly as a G by pairing with a C but then, once in the DNA, pair incorrectly with an A to result in a GC-TA transversion.

Damage Due to Alkylating Agents

Alkylation is another common type of damage to DNA. Both the bases and the phosphates in DNA can be alkylated. The responsible chemicals, known as **alkylating agents**, usually add alkyl groups (CH_3, CH_3CH_2, etc.) to the bases or phosphates in DNA, although any electrophilic reagent that reacts with DNA could be considered an alkylating agent. For example, the anticancer drug *cis*-diamminedichloroplatinum (cisplatin) is an alkylating agent that reacts with guanines in the DNA. Other examples of alkylating agents are ethyl methanesulfonate (EMS), nitrogen mustard gas, methyl methanesulfonate, and *N*-methyl-*N′*-nitro-*N*-nitrosoguanidine (nitrosoguanidine, NTG, or MNNG). Some of these alkylate DNA directly, whereas others react with cellular constituents, such as gluthionine, that are supposed to inactivate them but instead convert them into alkylating agents for DNA and worsen their effects. Many alkylating agents are known mutagens and carcinogens, and some are used as chemotherapeutic agents for the treatment of cancer. Not all alkylating agents are artificially synthesized; some are produced normally in cells or in the environment. For example, methylchloride, produced in large quantities by marine algae, is a DNA-alkylating agent, as are *S*-adenosylmethionine and methylurea, produced as normal cellular metabolites. Obviously, the cell needs repair systems to deal with alkylation damage to DNA.

Many reactive groups of the bases can be attacked by alkylating agents. The most reactive are N^7 of guanine and N^3 of adenine. These nitrogens can be alkylated by EMS or methyl methanesulfonate to yield methylated or ethylated bases, such as N^7-methylguanine and N^3-methyladenine, respectively. Alkylation of the bases at these positions can severely alter their pairing with other bases, causing major distortions in the helix.

Some alkylating agents, such as nitrosoguanidine, can also attack other atoms in the rings, including the O^6 of guanine and the O^4 of thymine. The addition of a methyl group to these atoms makes O^6-methylguanine (Figure 11.5) and O^4-methylthymine, respectively. Altered bases with an alkyl group at these positions are particularly mutagenic because the helix is not significantly distorted, so the lesions cannot be repaired by the more general repair systems discussed below. However, the altered base often mispairs, producing a mutation. In this section, we discuss the repair systems specific to these types of alkylated bases.

SPECIFIC *N*-GLYCOSYLASES

Some types of alkylated bases can be removed by specific *N*-glycosylases. The repair pathways involving these enzymes work in the same way as other *N*-glycosylase repair pathways in that the alkylated base is first removed by the specific *N*-glycosylase, and then the apurinic, or AP, DNA strand is cut by an AP endonuclease. Exonucleases degrade a portion of the cut strand, which is then resynthesized by DNA Pol I. In *E. coli*, two *N*-glycosylases that remove methylated and ethylated bases from the DNA have been identified. One of these, TagA (for *t*hree methyl*a*denine *g*lycosylase *A*) removes the base 3-methyladenine and some related methylated and ethylated bases. Another, AlkA, is less specific. It not only removes

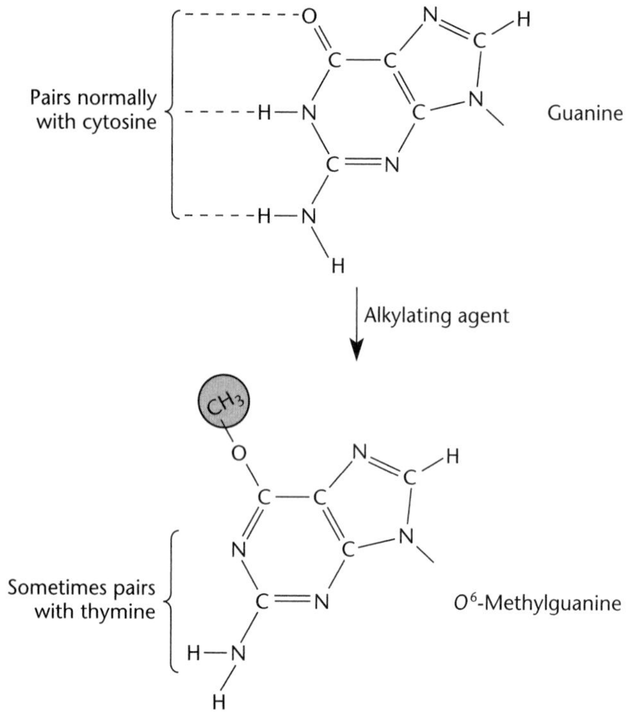

Figure 11.5 Alkylation of guanine to produce O^6-methylguanine. The altered base sometimes pairs with thymine, causing mutations.
doi:10.1128/9781555817169.ch11.f11.5

3-methyladenine from the DNA, but also removes many other alkylated bases, including 3-methylguanine and 7-methylguanine. This enzyme, which is encoded by the *alkA* gene, is induced as part of the adaptive response (see below).

METHYLTRANSFERASES

Other repair systems for alkylated bases act by repairing the damaged base rather than removing it and resynthesizing the DNA. These proteins, called methyltransferases, directly remove the alkyl group from the base by transferring the methyl or other alkyl group from the altered base in the DNA to themselves. They are not true enzymes, because they do not catalyze the reaction but rather are consumed during it. Once they have transferred a methyl or other alkyl group to themselves, they become inactive and are eventually degraded. The two major methyltransferases in *E. coli* are Ada and Ogt, sometimes called alkyltransferases I and II, respectively. Both of these proteins repair bases damaged from alkylation of the O^6 carbon of guanine and the O^4 carbon of thymine. Ogt plays the major role when the cells are growing actively, but when the cells reach stationary phase and stop growing, or if the cell is exposed to an external methylating agent, Ada is induced as part of the adaptive response and then becomes the major methyltransferase (see below). The fact that the cell is willing to sacrifice an entire protein molecule to repair a single O^6-methylguanine or O^4-methylthymine lesion is a tribute to the mutagenic potential of these lesions.

AlkB AND AidB

Two other enzymes that repair damage induced by alkylating agents are AlkB and AidB. The enzyme AlkB basically oxidizes the methyl groups on 1-methyladenine and 3-methylcytosine to formaldehyde, releasing them and restoring the normal base. More precisely, it is an α-ketoglutarate-dependent dioxygenase that couples the decarboxylation of α-ketoglutarate to the hydroxylation of the methyl group on 1-methyladenine or 3-methylcytosine to release formaldehyde (see Trewick et al., Suggested Reading). Its cofactor, α-ketoglutarate, is an intermediate in the tricarboxylic acid cycle, with many uses in nitrogen assimilation, etc., and therefore is always available in large quantities. The function of AidB is not proven; nevertheless, the structure of the protein suggests it is not involved in the direct repair of DNA but instead may have a role in deactivating alkylating agents that would otherwise damage DNA (see Bowles et al., Suggested Reading)

THE ADAPTIVE RESPONSE

The genes whose products repair alkylation damage in *E. coli* are part of the **adaptive response**, which includes the specific *N*-glycosylases and methyltransferases. The products of these genes are normally synthesized in small amounts, but they are produced in much greater amounts if the cells are exposed to an alkylating agent. The name "adaptive response" comes from early evidence suggesting that *E. coli* "adapted" to damage caused by alkylating agents. If *E. coli* cells are briefly treated with an alkylating agent, such as nitrosoguanidine, they will be better able to survive subsequent treatments with this and other alkylating agents. We now know that the cell adapts to the alkylating agents by inducing the expression of a number of genes whose products are involved in repairing alkylation damage to DNA. The adaptive-response genes seem to be the most important for conferring resistance to alkylating agents that transfer methyl (CH_3) groups to DNA. Resistance to alkylating agents that transfer longer groups, such as ethyl (CH_3CH_2) groups, to DNA seems to be due mostly to excision repair (see below).

Regulation of the Adaptive Response

Treatment of *E. coli* cells with an alkylating agent causes the concentration of some of the proteins involved in repairing alkylation damage to increase from a few to many thousands of copies. The genes induced as part of the adaptive response include *ada*, *aidB*, *alkA*, and *alkB*, discussed above. The regulation is achieved through the state of methylation of one of the alkylation-repairing proteins, the Ada protein (Figure 11.6). The *ada* gene is part of an operon with *alkB*, while the *aidB* and *alkA* genes are separately transcribed, as shown in the figure. The Ada protein can regulate the transcription of itself, as well as the other genes under its control, because, in addition to its role in repairing alkylation damage to DNA, it is a transcriptional activator. However, the Ada protein becomes a transcriptional activator only if the alkylation damage is quite extensive. It can discern the level of damage in the DNA by having two amino acids to which methyl groups can be transferred, cysteine-321 (the 321st amino acid from its N terminus) and cysteine-38 (the 38th amino acid from its N terminus).

Figure 11.6 (A) The adaptive response. **(B)** Regulation of the adaptive response. Only a few copies of the Ada protein normally exist in the cell. After damage due to alkylation, the Ada protein, a methyltransferase, transfers alkyl groups from methylated DNA phosphates to an amino acid in the N terminus (N) of itself, converting itself into a transcriptional activator (1), or from a methylated base to an amino acid in its C terminus (C), inactivating itself (2). See the text for details. doi:10.1128/9781555817169.ch11.f11.6

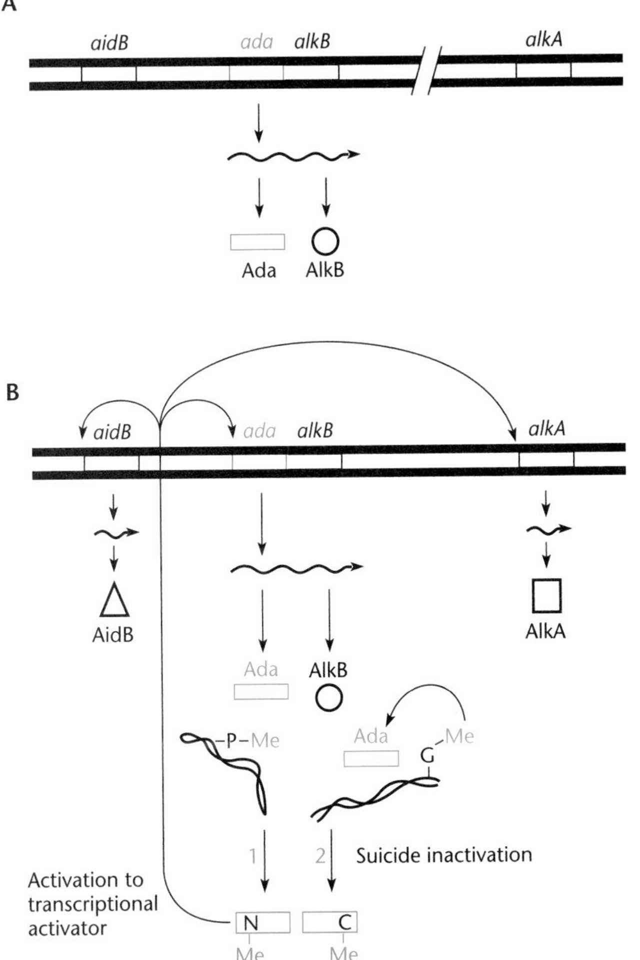

After mild alkylation damage has occurred, most of the methylation is confined to the bases; these methyl groups can be transferred from either O^6-methylguanine or O^4-methylthymine to cysteine-321, close to the C terminus. This removes the methyl groups from the damaged DNA bases but inactivates the Ada protein as far as transfer of more methyl groups from the bases is concerned. At higher levels of damage, some of the phosphates in the DNA background also become methylated in the form of phosphomethyltriesters, and these methyl groups can be transferred only to cysteine-38, close to the N terminus. The presence of a methyl group on cysteine-38 converts Ada into a transcriptional activator; modification at cysteine-38 alleviates a repulsive DNA-protein charge that allows the protein to interact with DNA, permitting the modified protein to activate transcription of its own gene, as well as the other genes under its control (see He, Suggested Reading). Transcriptional activators are discussed in detail in chapter 12.

Damage Due to UV Irradiation

One of the major sources of natural damage to DNA is UV irradiation due to sun exposure. Every organism that is exposed to sunlight must have mechanisms to repair UV damage to its DNA. The conjugated-ring structure of the bases in DNA causes them to strongly absorb light in the UV wavelengths. The absorbed photons energize the bases, causing their double bonds to react with other nearby atoms and hence to form additional chemical bonds. These chemical bonds result in abnormal linkages between the bases in the DNA and other bases or between bases and the sugars of the nucleotides.

One common type of UV irradiation damage is the **pyrimidine dimer**, in which the rings of two adjacent pyrimidines become fused (Figure 11.7). In one of the two possible dimers, the carbon atoms at positions 5 and 6 of two adjacent pyrimidines are joined to form a **cyclobutane ring.** In the other type of dimer, the carbon at position 6 of one pyrimidine is joined to the carbon at position 4 of an adjacent pyrimidine to form a **6-4 lesion.**

PHOTOREACTIVATION OF CYCLOBUTANE DIMERS

A special type of repair system called **photoreactivation** provides an efficient way to repair cyclobutane-type pyrimidine dimers due to UV irradiation. The photoreactivation repair systems separate the fused bases of the cyclobutane pyrimidine dimers rather than replacing them. This mechanism is named photoreactivation because this type of repair occurs only in the presence of visible light (and so used to be called light repair).

In fact, photoreactivation was the first DNA repair system to be discovered. In the 1940s, it was observed

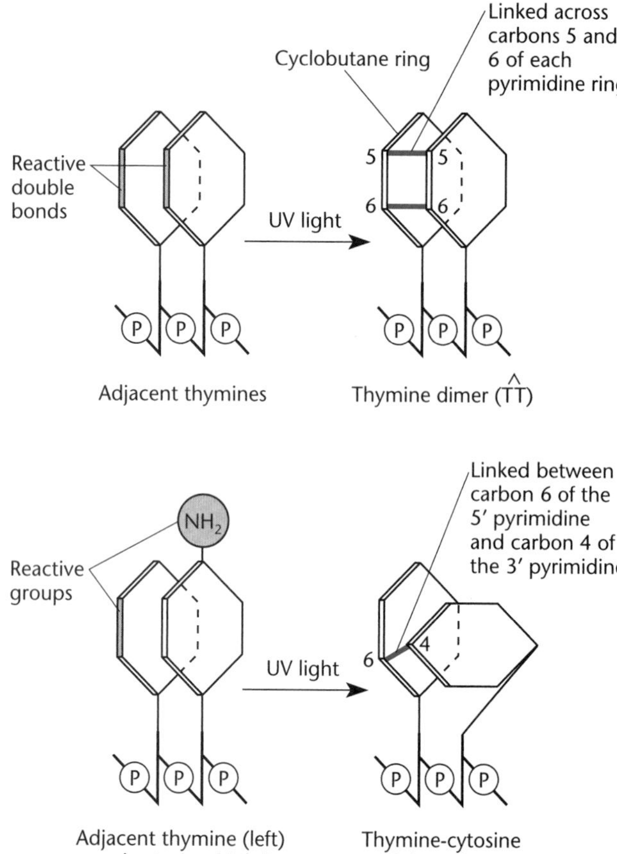

Figure 11.7 Two common types of pyrimidine dimers caused by UV irradiation. In the top diagram, two adjacent thymines are linked through the 5- and 6-carbons of their rings to form a cyclobutane ring. In the bottom diagram, a 6-4 dimer is formed between the 6-carbon of a cytosine and the 4-carbon of a thymine 3′ to it.
doi:10.1128/9781555817169.ch11.f11.7

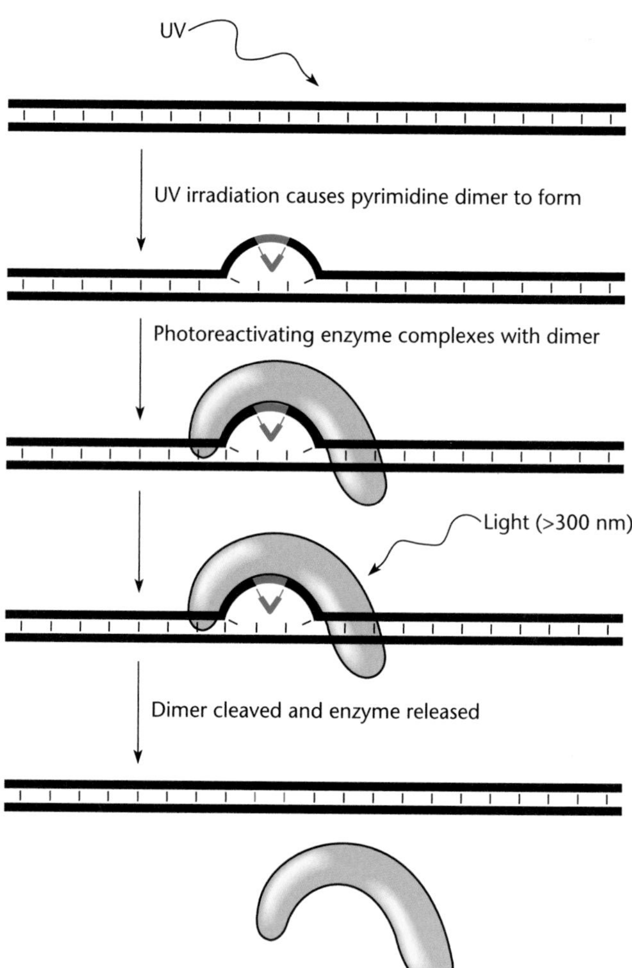

Figure 11.8 Photoreactivation. The photoreactivating enzyme (photolyase) binds to cyclobutane pyrimidine dimers (blue), even in the dark. Absorption of light by the photolyase causes it to cleave the bond between the two pyrimidines, restoring the bases to their original form.
doi:10.1128/9781555817169.ch11.f11.8

that the bacterium *Streptomyces griseus* was more likely to survive UV irradiation in the light than in the dark. Photoreactivation is now known to exist in most organisms on Earth that are exposed to UV irradiation, with the important exception of placental mammals, such as humans. Humans (and other placental mammals) instead appear to rely more heavily on nucleotide excision repair (see below).

The mechanism of action of photoreactivation is shown in Figure 11.8. The enzyme responsible for the repair is called **photolyase.** This enzyme, which contains a reduced flavin adenine dinucleotide group that absorbs light with wavelengths between 350 and 500 nm, binds to the fused bases. Absorption of light then gives photolyase the energy it needs to separate the fused bases. A different but related enzyme has evolved to repair 6-4 lesions in some eukaryotes.

There is some evidence that the photoreactivating system may also help to repair pyrimidine dimers, even in the dark, by cooperating with the excision repair system. By binding to pyrimidine dimers, it may make them more recognizable by the nucleotide excision repair system discussed below.

N-GLYCOSYLASES SPECIFIC TO PYRIMIDINE DIMERS

There are also specific *N*-glycosylases that recognize and remove pyrimidine dimers. This repair mechanism is similar to the mechanisms for deaminated and alkylated bases discussed above and involves AP endonucleases or lyases and the removal and resynthesis of strands of DNA containing the dimers.

General Repair Mechanisms

As mentioned above, not all repair mechanisms in cells are specific for a certain type of damage to DNA. Some types of repair systems can repair many different types of damage. Rather than recognizing the damage itself, these repair systems recognize distortions in the DNA structure caused by improper base pairing and repair them, independent of the type of damage that caused the distortion. There are a number of factors that can cause slight distortions in DNA, including 8-oxoG, incorporation of base analogs, frameshifts that occur by polymerase slippage, some types of alkylation damage, and mismatches. In general, DNA damage that causes only a minor distortion of the helix is repaired by the methyl-directed mismatch repair system, whereas other pathways, including nucleotide excision (see below), repair damage that causes more significant distortions.

Base Analogs

Base analogs are chemicals that resemble the normal bases in DNA. Because they resemble the normal bases, these analogs are sometimes converted into a deoxynucleoside triphosphate and enter DNA. Incorporation of a base analog can be mutagenic, because the analog often pairs with the wrong base, leading to base pair changes in the DNA. Figure 11.9 shows two base analogs, 2-aminopurine (2-AP) and 5-bromouracil (5-BU).

Figure 11.9 Base analogs 2-AP and 5-BU. The amino groups that are at different positions in adenine (A) and 2-AP are circled in blue, as are the methyl group in thymine (T) and the bromine group in 5-BU.
doi:10.1128/9781555817169.ch11.f11.9

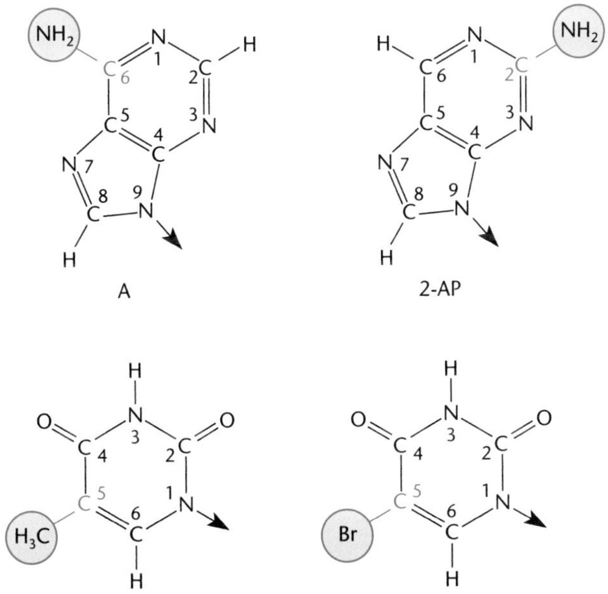

2-AP resembles adenine, except that it has the amino group at the 2 position rather than at the 6 position. The other base analog pictured, 5-BU, resembles thymine, except for a bromine atom instead of a methyl group at the 5 position.

Figure 11.10 shows how mispairing by a base analog might cause a mutation. In the illustration, the base analog 2-AP has entered a cell and has been converted into the nucleoside triphosphate. The deoxyribose 2-AP triphosphate is then incorporated during synthesis of DNA, sometimes pairing with cytosine in error. Which type of mutation occurs depends on when the 2-AP mistakenly pairs with cytosine. If the 2-AP enters the DNA by mistakenly pairing with C instead of T, in subsequent replications it usually pairs correctly with T. This causes a GC-to-AT transition mutation. However, if it is incorporated properly by pairing with T but pairs mistakenly with C in subsequent replications, an AT-to-GC transition mutation is the result. Similar arguments can be made for 5-BU, which sometimes mistakenly pairs with cytosine instead of adenine.

Frameshift Mutagens

Another type of damage repaired by the methyl-directed mismatch repair system is the incorporation of frameshift mutagens, which are usually planar molecules of the acridine dye family. These chemicals are mutagenic because they intercalate between bases in the same strand of the DNA, increasing the distance between the bases and preventing them from aligning properly with bases on the other strand. The frameshift mutagens include acridine dyes, such as 9-aminoacridine, proflavine, and ethidium bromide, as well as some aflatoxins made by fungi.

Figure 11.11 illustrates a model for mutagenesis by a frameshift mutagen. Intercalation of the dye forces two of the bases apart, causing the two strands to slip with respect to each other. One base is thus paired with the base next to the one with which it previously paired. This slippage is most likely to occur where a base pair in the DNA is repeated, for example, in a string of AT or GC base pairs. Whether a deletion or addition of a base pair occurs depends on which strand slips, as shown in Figure 11.11. If the dye is intercalated into the template DNA prior to replication, the newly synthesized strand might slip and incorporate an extra nucleotide. However, if it is incorporated into the newly synthesized strand, the strand might slip backward, leaving out a base pair in subsequent replication.

Methyl-Directed Mismatch Repair

One of the major pathways for avoiding mutations in *E. coli* is the **methyl-directed mismatch repair system.** Methyl-directed mismatch repair systems are most

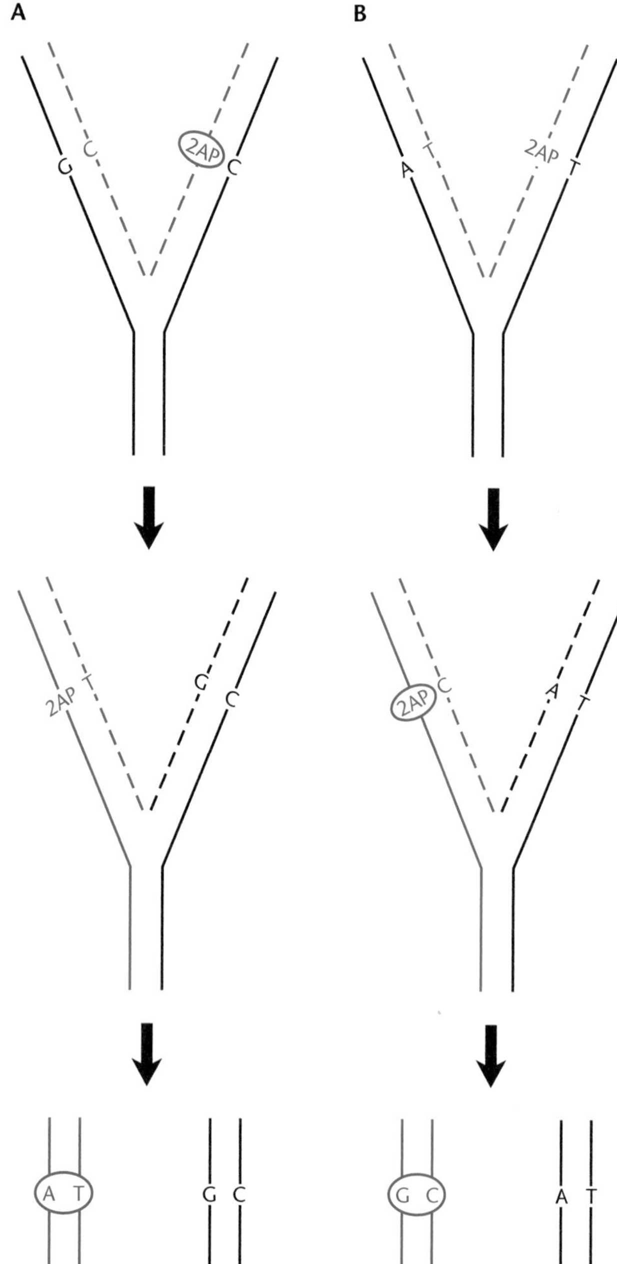

Figure 11.10 Mutagenesis by incorporation of the adenine analog 2-AP into DNA. The 2-AP is first converted to the deoxyribose nucleoside triphosphate and then inserted into the DNA. **(A)** The analog incorrectly pairs with a C during its incorporation into the DNA strand. **(B)** The analog is incorporated correctly opposite a T but mispairs with C during subsequent replication. The mutation is circled in both panels. doi:10.1128/9781555817169.ch11.f11.10

important immediately following DNA replication. This is because, in spite of the vigilance of the editing functions, sometimes the wrong base pair is inserted into DNA (see chapter 1). Mismatch repair allows the cell another chance to prevent a permanent mistake or mutation. This system recognizes the mismatch and removes it, as well as DNA in the same strand around the mismatch, leaving a gap in the DNA that is filled in by the action of DNA polymerase, which inserts the correct nucleotide.

The mismatch repair system is very effective at removing mismatches from DNA. However, by itself, it would not significantly lower the rate of spontaneous mutagenesis unless it repaired the correct strand of DNA at the mismatch. In the example shown in Figure 11.12, a T was mistakenly incorporated opposite a G. If the mismatch repair system changes the T to the correct C in the newly replicated DNA, a GC base pair will be restored at this position and no change in the sequence or mutation will occur. However, if it repairs the G in the old DNA in the mismatch to an A, the mismatch will have been removed and replaced by an AT base pair with correct pairing, but a GC-to-AT change would have occurred in the DNA sequence at the site of the mismatch, creating a mutation. To prevent mutations, the mismatch repair system must have some way of distinguishing the

Figure 11.11 Mutagenesis by a frameshift mutagen. Intercalation of a planar acridine dye molecule between two bases in a repeated sequence in DNA forces the bases apart and can lead to slippage. **(A)** The dye inserts itself into the new strand, resulting in deletion of a base pair (–). **(B)** The dye comes into the old strand, adding a base pair (+). doi:10.1128/9781555817169.ch11.f11.11

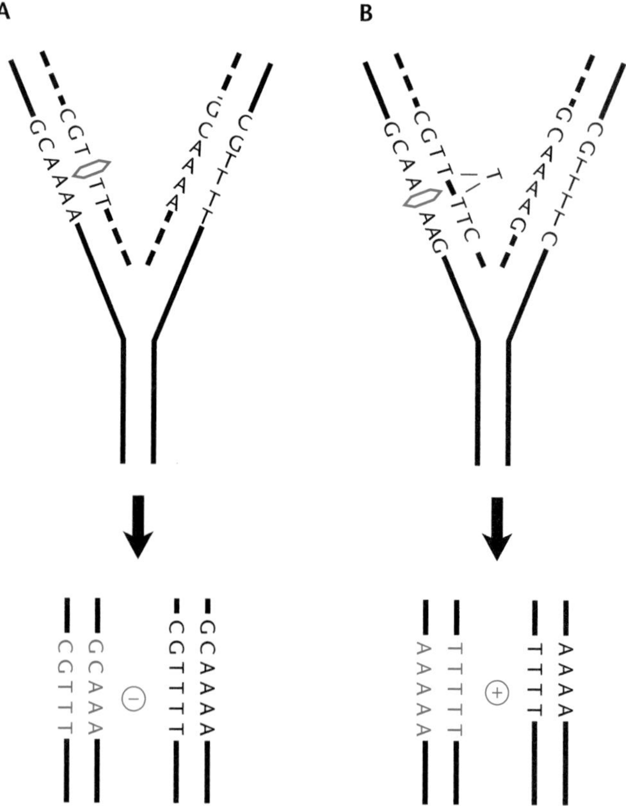

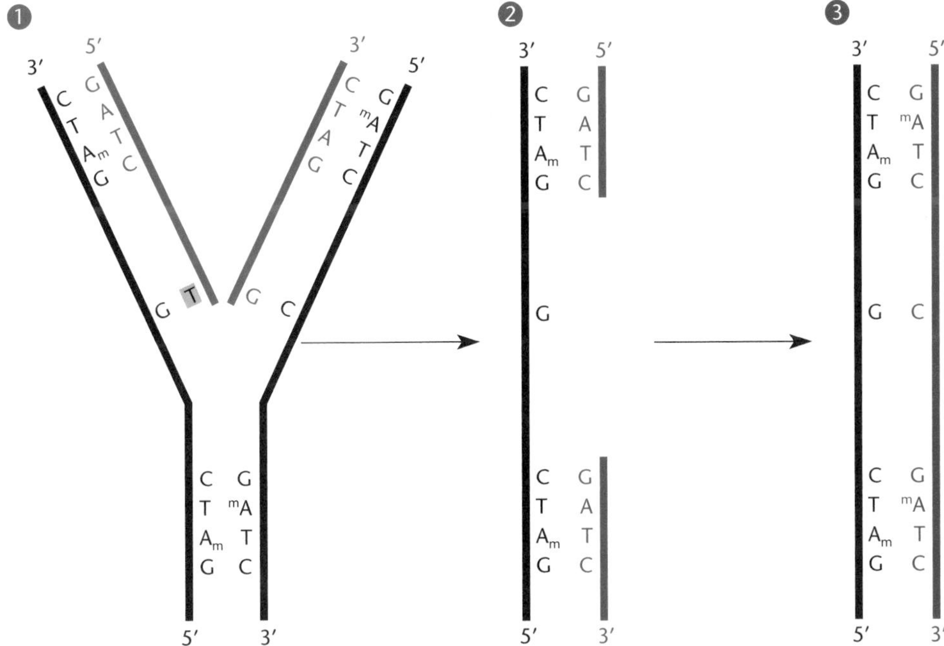

Figure 11.12 The methyl-directed mismatch repair system. The newly replicated DNA contains a GT mismatch **(1)**. The newly synthesized strand is recognized because it is not methylated at the nearby GATC sequence, and the T in the mismatch is removed, along with neighboring sequences **(2)**. The sequence is resynthesized, replacing the T with the correct C. The neighboring GATC sequence is then methylated by the Dam methylase **(3)**. The newly synthesized strand is shown in blue. doi:10.1128/9781555817169.ch11.f11.12

newly synthesized strand from the old strand so that it can replace the base in the right DNA strand.

Different organisms seem to use different mechanisms to distinguish the new and old strands for mismatch repair. In *E. coli*, it is the state of methylation of the DNA strands that allows the mismatch repair system to distinguish the new strand from the old strand after replication. In *E. coli* and some other members of the gammaproteobacteria, the A's in the **symmetric sequence** GATC/CTAG are methylated at the 6′ position of the larger of the two rings of the adenine base. These methyl groups are added to the bases by the enzyme deoxyadenosine methylase (Dam methylase) (see chapter 1), but this occurs only after the nucleotides have been incorporated into the DNA. Since DNA replicates by a semiconservative mechanism, the A in the GATC/CTAG sequence in the newly synthesized strand remains temporarily unmethylated after replication of a region containing the sequence. The DNA at this site is said to be hemimethylated if the bases on only one strand are methylated. Figure 11.12 shows that a hemimethylated GATC/CTAG sequence indicates to the mismatch repair system which strand is newly synthesized and should be repaired. For this reason, the repair system is called the methyl-directed mismatch repair system. The use of Dam hemimethylation to direct the mismatch repair system to the newly synthesized strand seems to be restricted to certain groups in the gammaproteobacteria, since most bacteria and eukaryotes do not have a methylation-sensing protein found in *E. coli* or the Dam methylase (see below). Nevertheless, nearly all organisms do possess a mismatch repair system, indicating that other mechanisms exist to distinguish the new from the old strand immediately following DNA replication.

The methyl-directed mismatch repair system in *E. coli* requires the products of the *mutS*, *mutL*, and *mutH* genes. Like the *mut* genes whose products are involved in avoiding mutagenesis due to 8-oxoG, these *mut* genes were found in a search for mutations that increase spontaneous-mutation rates (see below). It is somewhat mysterious how the mismatch repair system can use the state of methylation of the GATC/CTAG sequence to direct itself to the newly synthesized strand, even though the nearest GATC/CTAG sequence is probably some distance from the alteration. Figure 11.13 presents a model that has been used to explain this mechanism. First, a dimer of the MutS protein binds to the alteration in the DNA that is causing a minor distortion in the helix (marked with an X in the figure). A nearby GATC sequence is still unmethylated on the newly synthesized strand. Then, two copies of MutL bind to MutS, and a copy of MutH binds to the MutS-MutL complex. This binding activates the MutH nuclease to cut the nearest hemimethylated GATC sequence in the newly synthesized unmethylated strand, and exonucleases degrade the DNA past the site of the original mismatch. DNA Pol III then fills the gap, and ligase seals the break. This model is supported by experiments showing that the mismatch repair system preferentially repairs mismatches on hemimethylated DNA by correcting the sequence on the unmethylated strand of DNA to match the sequence on the methylated strand (see below and Pukkila et al., Suggested Reading). How, then, does

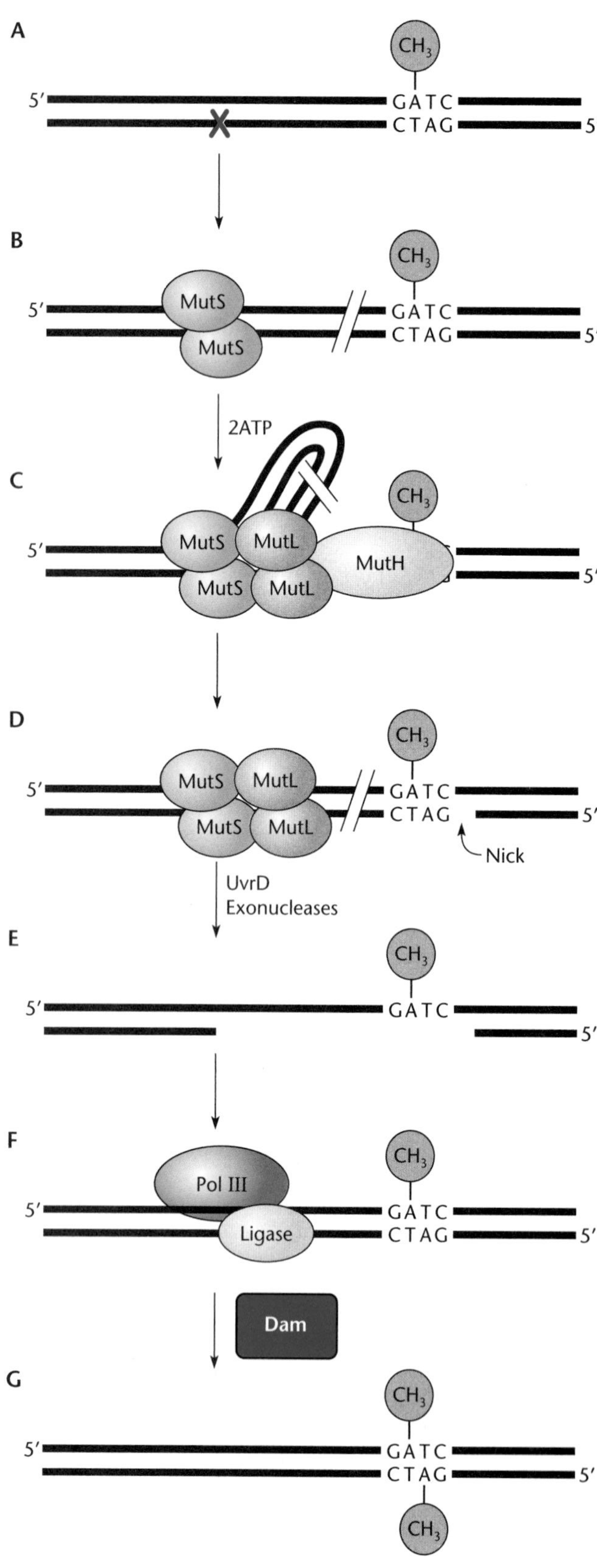

the cell know in what direction to degrade the DNA? The DNA is degraded in either the 3′-to-5′ or 5′-to-3′ direction depending on which side of the nearest GATC sequence the mismatch has occurred. According to the model, *E. coli* solves this problem by using four different exonucleases, two of which, ExoVII and RecJ, can degrade only in the 5′-to-3′ direction and two of which, ExoI and ExoX, can degrade only in the 3′-to-5′ direction. Also, these exonucleases can degrade only single-stranded DNA, so a helicase is needed to separate the strands. This is the job of the UvrD helicase, which is a general helicase also used by the excision repair system (see below). After the strand containing the mismatch has been degraded, DNA Pol III resynthesizes a new strand, using the other strand as the template and removing the cause of the distortion.

While this model more or less explains the genetic and biochemical evidence concerning the mismatch repair system (see below), evidence is accumulating that it may not be the whole story. For one thing, it is somewhat surprising that DNA Pol III, the replication DNA polymerase, is used for this repair. Most other repair reactions use DNA Pol I, the normal repair enzyme. How does DNA Pol III load itself on the DNA at the gap created by the exonucleases? It normally loads on DNA only at the origin of replication or with the help of Pri proteins at recombinational intermediates at blocked replication forks. Suspicions are also raised by the fact that other organisms seem to be able to identify the new strand for mismatch repair without the help of Dam methylation. Mismatch repair systems are universal, as is the ability to distinguish the newly synthesized strand from the old strand, even though most organisms, even most types of bacteria, do not have a Dam methylase and so could not use hemimethylation to identify the new strand. There is also evidence that the mismatch repair system is in much closer contact with the replicating polymerase than is indicated by the model. MutS may

Figure 11.13 MutSLH DNA repair in *E. coli.* **(A)** One arm of a replication fork is shown at the top of the figure, with methylated and unmethylated GATC (*dam*) sequences and a replication mistake (X) generating a base-base or deletion/insertion mismatch. **(B)** The mismatch is bound by MutS. **(C)** In an ATP-dependent reaction, a ternary complex is formed with MutS, MutL, and MutH proteins. **(D)** Incision by activated MutH occurs in the newly synthesized strand at the unmethylated GATC sequence. **(E)** The nick is extended into a gap by excision in either the 3′-to-5′ direction or the 5′-to-3′ direction. Only the 5′-to-3′ direction is depicted. The gap is formed by the actions of exonucleases, including exonuclease I (ExoI), ExoVII, ExoX, and RecJ, and the direction of excision is determined by the UvrD helicase. **(F)** Resynthesis is accomplished by the DNA Pol III holoenzyme, and the nick is sealed by DNA ligase. **(G)** Subsequent methylation by Dam completes the process.
doi:10.1128/9781555817169.ch11.f11.13

bind to the β clamp that holds the replicating DNA polymerase on the DNA, and MSH, the eukaryotic equivalent of MutS (Box 11.2), binds to the proliferating-cell nuclear antigen (PCNA) protein, which is the eukaryotic equivalent of the β clamp protein in bacteria. The MutL protein in *Bacillus subtilis* interacts with the β clamp, and this interaction regulates an endonuclease activity in the protein, likely providing nuclease activity similar to that provided by MutH in *E. coli* (see Pillon et al., Suggested Reading). It will be interesting to see how new evidence obtained with the bacterial systems furthers our understanding of this most interesting and important of DNA repair mechanisms, especially considering its role in preventing cancer in humans (Box 11.2).

VSP REPAIR

In addition to their role in general mismatch repair, the MutS and MutL proteins participate in VSP repair, which

BOX 11.2

DNA Repair and Cancer

Cancer is a multistep process that is initiated by mutations in oncogenes and other tumor-suppressing genes. It is not surprising, therefore, that DNA repair systems form an important line of defense against cancer. All organisms probably have mismatch repair systems that help reduce mutagenesis. However, unlike bacteria, which have only one or two different MutS proteins and one MutL protein, humans have at least five MutS analogs and four MutL homologs (see Kang et al., References). Attention was focused on the role of the mismatch repair system in cancer with the discovery that people with a mutation in a mismatch repair gene are much more likely to develop some types of cancer, including cancers of the colon, ovary, uterus, and kidney. One such genetic predisposition, called Lynch syndrome, results from mutations in a gene called *hMSH2* (for human *mutS* homolog 2) and leads to an increase in the incidence of certain types of colon cancer. This gene was first suspected to be involved in mismatch repair because people who inherited the mutant gene showed a higher frequency of short insertions and deletions that should normally be repaired by the mismatch repair system (see the text). People with this hereditary condition were found to have inherited a mutant form of a gene homologous to the *mutS* gene of *E. coli*. Moreover, the *hMSH2* gene can cause increased spontaneous mutations when expressed in *E. coli*, probably because it interferes with the normal mismatch repair system. Like its analog, MutS, the *hMSH2* gene product may bind to mismatches but then does not interact properly with MutL and MutH.

Another reason why the mismatch repair system is relevant to cancer research in humans is the role it plays in making cells sensitive to cancer therapeutic agents. Apparently, much of the toxicity of some antitumor agents, such as cisplatin and other alkylating agents, is due to the mismatch repair system. When tumor cells become resistant to the drug, it is often because they have acquired a mutation in a human *mut* gene (see Karran, References). Interestingly, tumor cells that are resistant to cisplatin can be resensitized to the drug by suppressing a polymerase normally induced by DNA damage (like the *E. coli* Pol IV and V) in human cells (see Doles et al., References). This indicates that translesion polymerases may be a target to make existing cancer therapies more effective.

Many other examples exist where processes involved in various forms of DNA repair discovered in bacteria are being found to play important roles in DNA repair in humans, and when defective, they lead to cancer. Mutations found in genes encoding the proteins BRCA1 and BRCA2, which are involved in repair involving homologous recombination, are associated with increases in the incidence of breast and ovarian cancer. The condition xeroderma pigmentosum stems from mutations in the nucleotide excision repair system in humans, resulting in extreme UV light sensitivity and skin cancer predisposition. Glioblastoma, a severe type of brain cancer, is associated with downregulation of the MGMT enzyme that repairs damage due to alkylating agents. These and numerous other examples indicate how findings from highly tractable bacterial systems will continue to help elucidate the link between DNA repair and cancer and cancer treatment.

References

Doles, J., T. G. Oliver, E. R. Cameron, G. Hsu, T. Jacks, G. C. Walker, and M. T. Hemann. 2010. Suppression of Rev3, the catalytic subunit of Polζ, sensitizes drug-resistant lung tumors to chemotherapy. *Proc. Natl. Acad. Sci. USA* **107:**20786–20791.

Fishel, R., M. K. Lescoe, M. R. S. Rao, N. G. Copeland, N. A. Jenkins, J. Garber, M. Kane, and R. Kolodner. 1993. The human mutator gene homolog MSH2 and its association with hereditary nonpolyposis colon cancer. *Cell* **75:**1027–1038.

Gradia, S., D. Subramanian, T. Wilson, S. Acharya, A. Makhov, J. Griffith, and R. Fishel. 1999. hMSH2-hMSH6 forms a hydrolysis-independent sliding clamp on mismatched DNA. *Mol. Cell* **3:**255–261.

Kang, J., S. Huang, and M. J. Blaser. 2005. Structural and functional divergence of MutS2 from bacterial MutS1 and eukaryotic MSH4-MSH5 homologs. *J. Bacteriol.* **187:**3528–3537.

Karran, P. 2001. Mechanisms of tolerance to DNA damaging therapeutic drugs. *Carcinogenesis* **22:**1931–1937.

occurs at the site of methylated cytosines in *E. coli* (see above and Lieb et al., Suggested Reading). The MutS protein may bind to the T in the T-G mismatch created by the deamination of the methylated C at this position, thereby attracting the attention of the Vsr endonuclease to the mismatch. The MutL protein may then recruit the UvrD helicase and exonucleases to degrade the strand causing the mismatch, consistent with the roles of these proteins in general mismatch repair.

GENETIC EVIDENCE FOR METHYL-DIRECTED MISMATCH REPAIR

Models like the one presented in Figure 11.12 are based on biochemical and genetic evidence. The results of these experiments support the conclusion that the state of methylation of GATC sequences in the DNA help direct the mismatch repair system to the newly synthesized strand of DNA. Some of this evidence is briefly reviewed in this section.

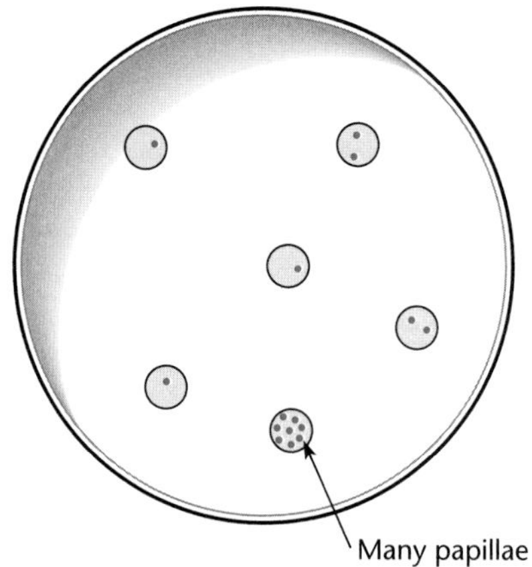

Figure 11.14 Colonies due to *mut* mutants have more papillae. A *lacZ* mutant was plated on X-Gal-containing medium. Revertants of the *lacZ* mutation produced blue sectors, or papillae (shown in blue). A *mut* mutant produced more blue papillae than normal owing to increased spontaneous-mutation frequencies (arrow). doi:10.1128/9781555817169.ch11.f11.14

Isolation of *mut* Mutants

The *mut* genes of *E. coli* were discovered because mutations in these genes increase spontaneous-mutation rates. The resulting phenotype is often referred to as the "mutator phenotype." Other mutations with this phenotype, including the *mutT*, *mutY*, and *mutM* mutations in the genes for reducing mutations due to oxygen damage to DNA and the *mutD* mutations in the gene for the editing function of DNA polymerase, are discussed in other sections. Mutations in the *uvrD*, *vsp*, and *dam* genes of *E. coli* also increase the rates of at least some spontaneous mutations, although these genes were found in other ways and so were not named *mut*.

A common method for detecting mutants with abnormally high mutation rates is colony papillation. This scheme is based on the fact that all of the descendants of a bacterium with a particular mutation are mutants of the same type. Colonies grow from the middle out, so if a mutation occurs in a growing colony, the mutant descendants of the original mutant stay together to form a sector, or **papilla**, as the colony grows. A growing colony contains many papillae composed of mutant bacteria of various types. A *mut* mutation increases the frequency of papillation for many types of mutants, so the phenotype used in a colony papillation test is purely a matter of convenience.

Lac$^+$ revertants of *lac* mutations in *E. coli* are an obvious choice for papillation tests, because the revertants form conspicuous blue papillae on 5-bromo-4-chloro-3-indolyl-β-D-galactopyranoside (X-Gal) plates. Figure 11.14 illustrates a papillation test using reversion of a *lac* mutation as an indicator of mutator activity. If the bacteria forming a colony are *mut* mutants with a higher-than-normal spontaneous-mutation frequency, the colony will have more blue papillae than normal. This is how many of the *mut* genes we have discussed were detected.

Once a number of *mut* genes were detected, they were classified into complementation groups and arbitrarily assigned letters (see chapter 3). The mutations were also combined in all of the available permutations to see how they interacted. The products of three of the *mut* genes, *mutS*, *mutL*, and *mutH*, were predicted to participate in the same repair pathway because of the observation that double mutants did not exhibit higher spontaneous-mutation rates than cells with mutations in each of the single genes alone. In other words, the effect of inactivating two of these *mut* genes by mutation is not additive. Generally, mutations that affect steps of the same pathway do not have additive effects.

Other experiments established the role of Dam methylation in mismatch repair. Probably the most convincing evidence for the role of methylation in directing the repair system came from experiments with heteroduplexes of λ DNA (see Pukkila et al., Suggested Reading). We discuss the synthesis and experimental use of heteroduplex DNAs in other chapters. For example, in chapter 9, we show how they can be used to distinguish replicative from cut-and-paste transposition.

To determine the role of methylation in directing the mismatch repair system, heteroduplex λ DNAs in which the two strands came from phages with different mutations were prepared. We shall call them phage λ mutants 1 and 2. The DNA from phage mutant 2 was unmethylated because it was propagated on *E. coli* with a *dam* mutation. After the DNA from both mutants was purified, the DNA strands of both were melted apart by heating, and the two complementary strands were selectively separated by a technique that made one strand denser on CsCl equilibrium gradients. The purified complementary strands from each of the phages were then mixed and rehybridized to create a mismatch at the site of the mutation, with one strand methylated at its GATC sequences and the other strand unmethylated. These heteroduplex DNAs were then transfected into cells, the phage progeny were plated, and their genotypes were tested to see which mutation they had inherited and which of the two strands had been preferentially repaired. The progeny phage exhibited predominantly the genotype of mutant λ, which had the methylated DNA. In the reverse experiment, mutant λ was propagated on the *dam* mutant so that it would have the unmethylated DNA. In this case, the progeny phage that arose from the heteroduplex DNA had the genotype of mutant 2; therefore, once again, the methylated DNA sequence was preferentially preserved. These results indicate that the sequence of the unmethylated strand is usually the one that is repaired to match the sequence of the methylated strand.

Other evidence supporting the role of methylation in directing the mismatch repair system came from genetic studies of 2-AP sensitivity in *E. coli* (see Glickman and Radman, Suggested Reading). These experiments were based on the observation that a *dam* mutation makes *E. coli* particularly sensitive to killing by 2-AP. This is expected from the model, since as mentioned above, cells incorporate 2-AP indiscriminately into their DNA because they cannot distinguish it from the normal base adenine. The cellular mismatch repair system repairs DNA containing 2-AP because the incorporated 2-AP causes a slight distortion in the helix. Since the 2-AP is incorporated into the newly synthesized strand during the replication of the DNA, the strand containing the 2-AP is normally transiently unmethylated, so that the 2-AP-containing strand is repaired, removing the 2-AP. In a *dam* mutant, however, neither strand is methylated, so the mismatch repair system cannot tell which strand was newly synthesized and may try to repair both strands. The rationale was that if two 2-APs are incorporated close enough to each other, the mismatch repair system may try to simultaneously remove the two mismatches by repairing the opposite strands. However, under this scenario, cutting two strands at sites opposite each other could cause extensive double-strand breaks in the DNA, which may kill the cell.

One prediction based on this proposed mechanism for the sensitivity of *dam* mutants to 2-AP is that the toxicity of 2-AP in *dam* mutants should be reduced if the cells also have a *mutL*, *mutS*, or *mutH* mutation. Without the products of the mismatch repair enzymes, the DNA is not cut on either strand, much less on both strands simultaneously. The cells may suffer higher rates of mutation, but at least they will survive. This prediction was fulfilled. Double mutants that have both a *dam* mutation and a mutation in one of the three *mut* genes were much less sensitive to 2-AP than were mutants with a *dam* mutation alone. Furthermore, *mutL*, *mutH*, and *mutS* mutations can be isolated as suppressors of *dam* mutations on media containing 2-AP. In other words, if large numbers of *dam* mutant *E. coli* cells are plated on medium containing 2-AP, the bacteria that survive are often double mutants with the original *dam* mutation and a spontaneous *mutL*, *mutS*, or *mutH* mutation. A similar situation seems to prevail in human cells (Box 11.2).

ROLE OF THE MISMATCH REPAIR SYSTEM IN PREVENTING HOMEOLOGOUS AND ECTOPIC RECOMBINATION

As discussed in chapter 3, some DNA rearrangements, such as deletions and inversions, are caused by recombination between similar sequences in different places in the DNA. This is called **ectopic recombination**, or "out-of-place" recombination. Many sequences at which ectopic recombination occurs are similar but not identical. Also, recombination between DNAs from different species often occurs between sequences that are similar but not identical. In general, recombination between similar but not identical sequences is called **homeologous recombination.** Recall that recombination between identical sequences is called homologous recombination.

The mismatch repair system helps reduce ectopic and other types of homeologous recombination. As evidence, recombination between similar but unrelated bacteria, such as *E. coli* and *Salmonella enterica* serovar Typhimurium, is greatly enhanced if the recipient cell has a *mutL*, *mutH*, or *mutS* mutation. Also, the frequency of deletions and other types of DNA rearrangements is enhanced among bacteria with a *mutL*, *mutH*, or *mutS* mutation, since these rearrangements often depend on recombination between similar but not identical sequences. The frequency of mutations engineered using recombineering is also enhanced by *mutS* mutations (see chapter 10), since this also depends on homeologous recombination between the DNA electroporated into the cell and the endogenous DNA. Evidence suggests that the mismatch repair system may inhibit homeologous and ectopic recombination by interfering with the ability of RecA to form synapses where there are extensive mismatches (see Worth et al., Suggested Reading). Because the sequences at which the homeologous recombination occurs are not identical, the heteroduplexes formed during recombination between different regions in the DNA or the same region from different species often contain many mismatches that bind MutS and the other proteins of the mismatch repair system and interfere with the binding of RecA.

Nucleotide Excision Repair

One of the most important general repair systems in cells is **nucleotide excision repair**, so named because the entire damaged nucleotides are cut out of the DNA and replaced. This type of repair is very efficient and seems to be common to most types of organisms on Earth. It is also relatively nonspecific and is responsible for repairing many different types of damage. Because of its efficiency and relative lack of specificity, the nucleotide excision repair system is very important to the ability of the cell to survive damage to its DNA.

The nucleotide excision repair system is relatively nonspecific because, like the mismatch repair system, it recognizes the distortions in the normal DNA helix that result from damage, rather than the chemical structure of the damage itself. This makes it capable of recognizing and repairing damage to DNA as diverse as most types of alkylation and almost all the types of damage caused by UV irradiation, including cyclobutane dimers, 6-4 lesions, and base-sugar cross-links. Nucleotide excision repair also collaborates with **recombination repair** to remove cross-links formed between the two strands by some chemical agents, such as psoralens, cisplatin, and mitomycin (see below). However, because nucleotide excision repair recognizes only major distortions in the helix, it does not repair lesions such as base mismatches, O^6-methylguanine, O^4-methylthymine, 8-oxoG, or base analogs, all of which result in only minor distortions and must be repaired by other repair systems.

MECHANISM OF NUCLEOTIDE EXCISION REPAIR

Because nucleotide excision repair is such an important line of defense against some types of DNA-damaging agents, including UV irradiation, mutations in the genes whose products are required for this type of repair can make cells much more sensitive to these agents. In fact, mutants defective in excision repair were identified because they are killed by much lower doses of irradiation than is the wild type. Table 11.1 lists the *E. coli* genes whose products are required for nucleotide excision repair. Comparative genomic analysis has found UvrA, UvrB, and UvrC orthologs across bacterial species, as well as in some members of the *Archaea*. The products of some of these genes, such as *uvrA*, *uvrB*, and *uvrC*, are involved only in excision repair, while the products of others, including the *polA* and *uvrD* genes, are also required for other types of repair.

How these gene products participate in excision repair is illustrated in Figure 11.15. The products of the *uvrA*, *uvrB*, and *uvrC* genes interact to form what is called the **UvrABC endonuclease**. The function of these gene products is to make a nick close to the damaged nucleotide, causing it to be excised. In more detail, two copies of the UvrA protein and one copy of the UvrB protein form a complex that binds nonspecifically to DNA, even if it is not damaged. This complex then migrates up and down the DNA until it hits a place where the helix is distorted because of DNA damage (in the figure, because of a thymine dimer). The complex then stops, the UvrB protein binds to the damage, and the UvrA protein leaves and is replaced by UvrC. The binding of the UvrC protein to UvrB causes UvrB to cut the DNA about 4 nucleotides 3′ of the damage. The UvrC protein then cuts the DNA 7 nucleotides 5′ of the damage. Once the DNA is cut, the UvrD helicase removes the oligonucleotide containing the damage and DNA Pol I resynthesizes the strand that was removed, using the complementary strand as a template.

Table 11.1 Genes involved in the UvrABC endonuclease repair pathway

Gene	Function of gene product
uvrA	DNA-binding protein
uvrB	Loaded by UvrA to form a DNA complex; nicks DNA 3′ of lesion
uvrC	Binds to UvrB-DNA complex; nicks DNA 5′ of lesion
uvrD	Helicase II; helps remove damage-containing oligonucleotide
polA	Pol I; fills in single-strand gap
lig	Ligase; seals single-strand nick

Figure 11.15 Model for nucleotide excision repair by the UvrABC endonuclease. See the text for details. A, UvrA; B, UvrB; C, UvrC; D, UvrD; I, DNA Pol I.
doi:10.1128/9781555817169.ch11.f11.15

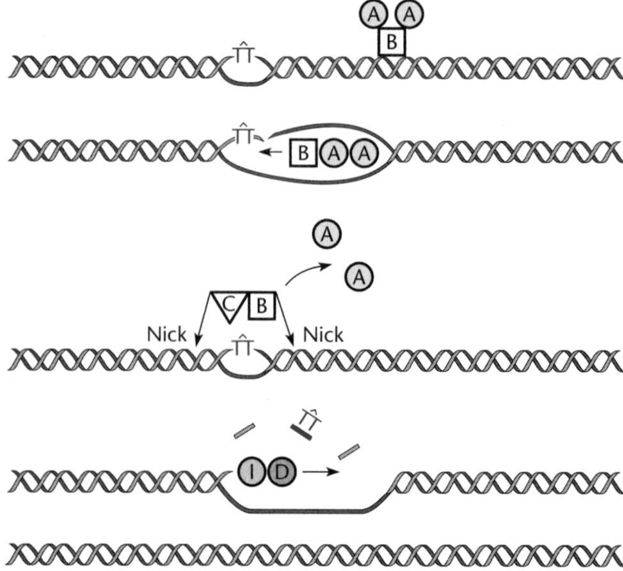

TRANSCRIPTION-COUPLED REPAIR

Damage in some regions of the DNA presents a more immediate problem for the cell than does damage in other regions. For example, pyrimidine dimers in transcribed regions of DNA block not only replication of the DNA, but also transcription of RNA from the DNA, when the RNA polymerase stalls at the damage. It makes sense for the cell to first repair the damage that occurs in transcribed genes so that these genes can be transcribed and translated into proteins. RNA polymerase that stalls during transcription can also stall translation because transcription and translation are coupled (see chapter 2). Stalled transcription/translation complexes would consequently stall DNA replication if not removed, thereby corrupting three of the most important processes in the cell. These risks likely explain why there is a special mechanism to deal with RNA polymerase that has stalled at damage to the DNA. This is called transcription-coupled repair, and the factor involved in bacteria is called the Mfd protein (for mutation frequency decline). The gene for this protein, *mfd*, was discovered more than 50 years ago by Evelyn Witkin by isolation of mutations that prevent a decrease in mutations if protein synthesis is inhibited immediately following DNA damage (see Witkin, Suggested Reading). Similar systems exist in eukaryotes, where they are called Rad26 in yeast and CSB in humans. One of the early pieces of evidence suggesting the existence of transcription-coupled repair was the fact that DNA damage that occurs within transcribed regions of the DNA and in the transcribed strand is repaired preferentially by the nucleotide excision repair system. Also, mutations occur more frequently when damage occurs in the nontranscribed strand of DNA, as would be expected if damage in this strand were not repaired by the relatively mistake-free excision repair system. It was some time before the Mfd protein was linked to these phenomena.

RNA polymerase stalled at damage in the DNA creates two potential problems for the cell. One is that the stalled RNA polymerase can block access of the nucleotide excision repair system to the damage and thereby prevent its repair. Another is that the stalled RNA polymerase can block the passage of replication forks. The Mfd protein overcomes these potential problems through its translocase activity. When RNA polymerase stalls on the DNA, it backtracks, shoving the 3′ hydroxyl end of the growing RNA out of the active center of the enzyme into the secondary channel (see chapter 2). Presumably, it would remain in this conformation indefinitely unless something came along that could force the 3′ hydroxyl end of the RNA back into position. The Gre factors do this by cleaving the 3′ end of the RNA (see chapter 2). The Mfd protein accomplishes it by binding to the DNA behind the stalled RNA polymerase and translocating (moving) itself forward, pushing the RNA polymerase ahead of it, like pushing your manual transmission car to "bump" or "push" start it when the starter or battery is dead (see Park et al., Suggested Reading). The action of Mfd can have one of two effects. If the damage is still there, the RNA polymerase cannot move forward and the forward movement will disrupt the RNA-DNA transcription bubble that was otherwise holding the RNA polymerase on the DNA. This causes the RNA polymerase to be released, getting it out of the way of repair and replication. If the RNA polymerase can move forward, the block is relieved and the RNA polymerase can continue to make the RNA. Probably of greater significance, the Mfd protein is also capable of recruiting the nucleotide excision repair system to allow repair. Mfd has a region that is homologous to the UvrB protein. This region of Mfd accounts for its ability to recruit the nucleotide excision repair system; deletion of the domain results in a protein that is capable of displacing RNA polymerase but that is unable to carry out transcription-coupled repair (see Manelyte et al., Suggested Reading). Therefore, the Mfd protein provides two important genome stability functions, first, by clearing backtracked RNA polymerase from transcripts (a function also carried out by the Gre factors) and, second, by facilitating repair by "favoring" actively transcribed regions of the genome that would be most sensitive to the effects of DNA damage.

INDUCTION OF NUCLEOTIDE EXCISION REPAIR

Although the genes of the excision repair system are almost always expressed at low levels, *uvrA*, *uvrB*, and *uvrD* are expressed at much higher levels after DNA has been damaged. This is a survival mechanism that ensures that larger amounts of the repair proteins are synthesized when they are needed. The *uvr* genes, along with many other proteins involved in DNA repair, are induced by DNA damage. The *uvr* genes are induced following exposure to DNA-damaging agents or irradiation and are members of the *din* genes (for "*d*amage *in*ducible"), which include *recF*, *recA*, *umuC*, and *umuD*. Many *din* genes, including *uvrA*, *uvrB*, and *uvrD*, are induced because they are part of the SOS regulon (see below).

DNA Damage Tolerance Mechanisms

In all of the repair systems discussed above, the cell removes the damage from the DNA, often using the information in the complementary strand to restore the correct DNA sequence. In each of these systems, one might imagine that the hope is that the damage is repaired before the replication apparatus arrives on the scene and tries to replicate over the damage. However, what happens if the damage is not repaired before a replication fork arrives? In all the instances where the damage is not repaired, the cell has no choice but to tolerate it by replicating over the damage. Mechanisms that allow the cell to tolerate DNA damage without repairing it at that moment are called **damage tolerance mechanisms**. Genetically, mutants in these systems show increased sensitivity to DNA damage with or without a change in the frequency of mutations. While some of these systems were presented in chapters 1 and 10, we now have enough background to go into more detail on the varied ways that DNA recombination allows bacteria to tolerate DNA damage to facilitate DNA repair and replication.

Homologous Recombination and DNA Replication

Homologous recombination is a significant mechanism for damage tolerance. In chapter 10, we learned how homologous recombination is important for reestablishing a replication fork that collapses at a nick in one of the template strands. Homologous recombination also plays an important role in allowing a DNA replication fork to **bypass** damage in DNA. After a replication fork has moved on, the damage remains in the chromosome, where it can be repaired. Such damage could involve various types of alterations to the bases so that they cannot base pair properly or alterations that make them too bulky move through the DNA polymerase involved in normal DNA replication. There are also structures that can form during DNA replication that provoke endogenous nucleases to make double-strand breaks in the chromosome. These breaks will be fixed using recombination to initiate replication using the same pathway that repairs collapsed DNA replication forks.

LAGGING-STRAND DAMAGE

The consequences of encountering damage in the template strand undergoing lagging-strand replication (the strand running in the 5′-to-3′ direction) are expected to be very different from the consequences of encountering it in the leading strand (running in the 3′-to-5′ direction). If damage is on the lagging strand, the replication

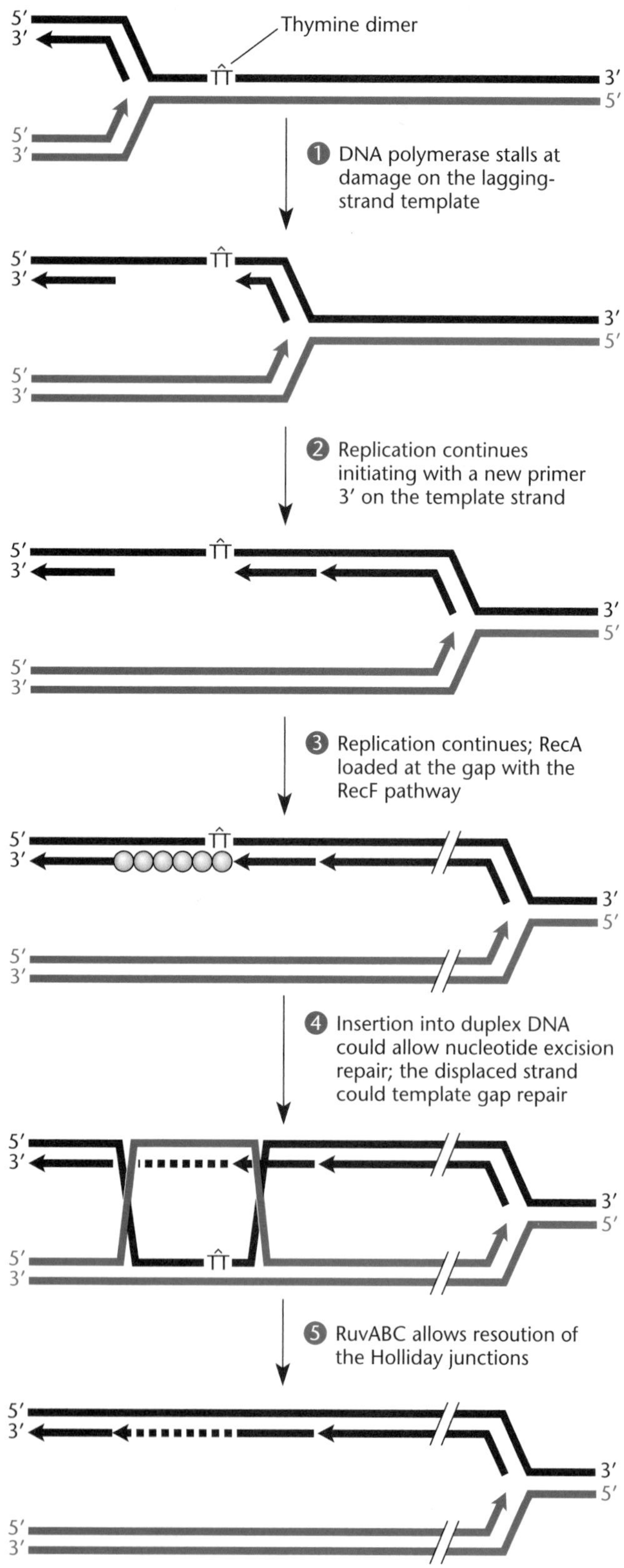

5′
3′
Thymine dimer
TT
3′
5′
5′
3′
1 DNA polymerase stalls at damage on the lagging-strand template
2 Replication continues initiating with a new primer 3′ on the template strand
3 Replication continues; RecA loaded at the gap with the RecF pathway
4 Insertion into duplex DNA could allow nucleotide excision repair; the displaced strand could template gap repair
5 RuvABC allows resoution of the Holliday junctions

Figure 11.16 Model for recombination-mediated bypass of DNA damage in the DNA strand undergoing lagging-strand replication. DNA Pol III stalls at the thymine dimer **(1)**, but DNA replication can continue **(2)**. The RecF pathway loads RecA onto the single-stranded DNA at the gap **(3)**, and the RecA nucleoprotein filament invades the sister DNA **(4)**. The gap is now opposite a good strand and can be filled in, leaving two Holliday junctions, which are resolved by RuvABC **(5)**. The DNA in which the thymine dimer remains depends on how the Holliday junctions are resolved.
doi:10.1128/9781555817169.ch11.f11.16

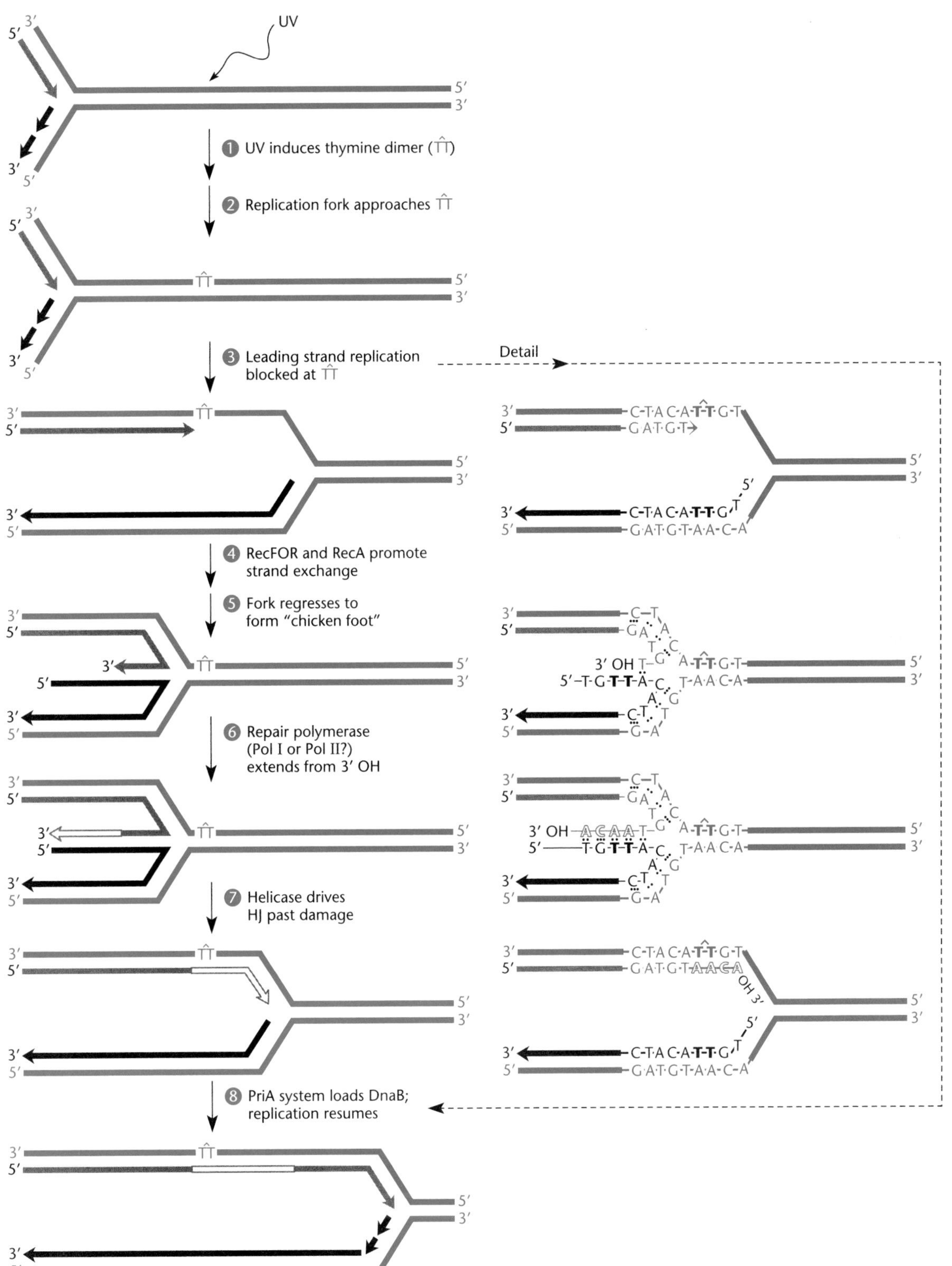

Figure 11.17 Fork regression model for recombination-mediated replicative bypass of a thymine dimer in DNA when the damage is on the leading strand **(1 and 2)**. The leading-strand replicating polymerase could stall, but the lagging-strand replicating polymerase keeps going, making good double-stranded DNA opposite the damage **(3)**. The fork could then back up (regress) to form a Holliday junction (HJ)-like "chicken foot" in which the newly synthesized strands have paired with each other, perhaps with the help of the RecFOR pathway and RecA **(4 and 5)**. The free 3′ end due to truncated synthesis of the leading strand could then serve as a primer for the synthesis of DNA past the original site of the damage **(6)**. This is illustrated on the right. The junction could then migrate back over the damage **(7)**, and the replication fork machinery could be reloaded by the PriA pathway **(8)**. The site of the thymine dimer is in boldface. doi:10.1128/9781555817169.ch11.f11.17

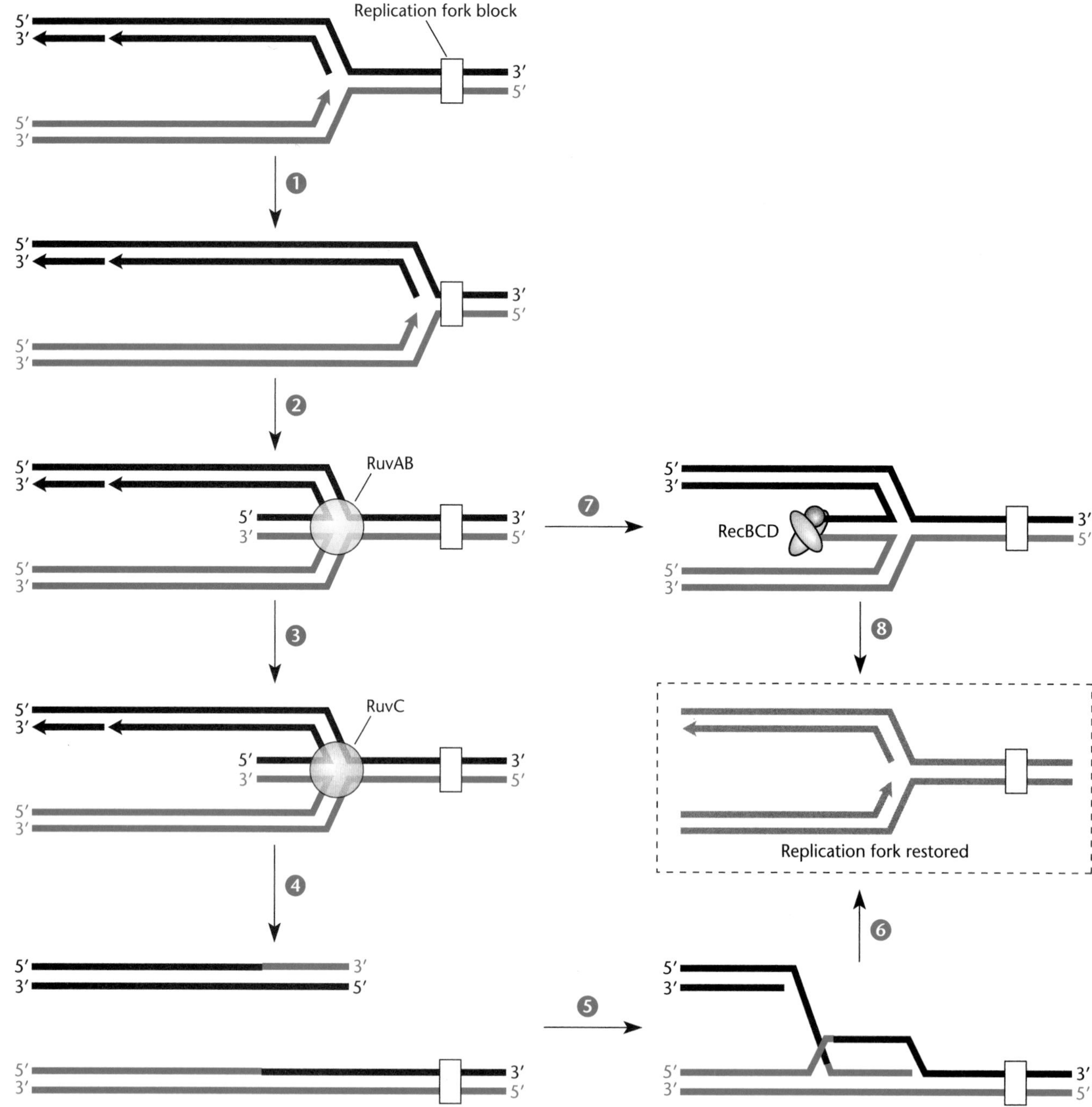

Figure 11.18 Models for how regressed replication forks can be repaired by multiple mechanisms. **(1)** DNA replication can stall due to various features (indicated by a rectangle). **(2 and 3)** Annealing of the newly synthesized strands allows the replication fork to reverse or regress (Figure 11.17), providing a Holliday junction structure that can be recognized by RuvAB. **(4)** Processing by RuvC, in collaboration with RuvAB, results in a break in the chromosome. **(5 and 6)** Processing of the break by RecBCD and loading of RecA (not shown) allows strand invasion and the initiation of replication with a sister chromosome. **(7)** Alternatively, the arm of the regressed replication fork that is formed by annealing of the newly synthesized strands can be processed by RecBCD. **(8)** This structure can be converted without the formation of a double-strand break in the chromosome that can be restarted by the Pri proteins. doi:10.1128/9781555817169.ch11.f11.18

fork has the opportunity to pass right over the damage provided that the DnaB helicase, which encircles the lagging strand (see Figure 1.10), can proceed over the damage. While DNA damage is likely to stall DNA replication by DNA Pol III (in the example, the block is due to a cyclobutane thymine dimer) (Figure 11.16), DNA replication on this template strand is naturally discontinuous, and replication can initiate at a new primer made by DnaG as a normal part of lagging-strand DNA replication (see Figure 1.13A). As indicated in Figure 11.16, this leaves a gap in the template strand that must be repaired. The gap structure that results is recognized by the RecF pathway (see Figure 10.4). The RecF pathway can direct loading of RecA at the gap, which would allow the damaged strand to invade the other daughter chromosome formed by replication (Figure 11.16). In this model, invasion into the undamaged chromosome provides a template to allow nucleotide excision repair to recognize the damage and for DNA polymerase to repair the damaged strand.

LEADING-STRAND DAMAGE

The consequences of encountering damage on the leading strand have historically been assumed to be more severe than those for encountering damage on the lagging-strand template. However, there is evidence for multiple types of repair that could help deal with damage to the leading strand. As described in chapter 1, it has been shown that the DnaG primase can prime DNA replication past DNA damage on the leading-strand template when a block is found on the template strand (see Figure 1.13B), This process is dependent on the PriC pathway for reinitiated DNA replication (see Heller and Marians, Suggested Reading). Presumably, the gap that was left behind under this scenario would be repaired using the same series of events modeled in Figure 11.16 for DNA damage on the DNA template replicated by lagging-strand replication.

Replication that is stalled on the DNA template replicated by leading-strand replication may have an additional method for facilitating DNA repair. In this model for repair of the thymine dimer, the replication fork can regress, as shown in Figure 11.17. The model starts out the same as the one discussed above, in that single-stranded DNA forms when replication of the leading strand is blocked at the damage. The newly synthesized strands might then pair with each other in a strand exchange reaction promoted by the RecFOR and RecA proteins. This creates a branch that can migrate backward (fork regression) to form a Holliday junction in what is sometimes called a "chicken foot" structure, in which the two new strands are hybridized to each other. The shorter new strand that results from blocked replication of the leading strand could then furnish a 3′ hydroxyl primer for a repair polymerase using an undefined DNA polymerase to extend the shorter strand until the two strands are the same length. On the right side of the figure, some arbitrary base pairs are shown to illustrate how this allows replication past the site of the thymine dimer. The chicken foot Holliday junction could then migrate forward, possibly with the help of RecG or some other helicase, past the site of the damage. We then have a bona fide replication fork on the other side of the damage with a free 3′ hydroxyl group to prime replication of the leading strand. The DnaB helicase and other replication proteins could then be reloaded, probably by the PriA-PriB-DnaT-DnaC pathway, and replication is again under way.

These are just some scenarios to explain the genetic evidence for how the recombination functions can help the cell to tolerate damage to the DNA template. Final proof of any model requires more detailed studies of what happens when the extremely complex replication fork encounters damage on the DNA.

BREAKING THE CHROMOSOME TO REPAIR THE CHROMOSOME

Interestingly, there appear to be scenarios where repair enzymes actually break the chromosome, possibly as a tool to facilitate DNA repair. An impressive early example of evidence that regressed forks do occur in living cells and that they could be recognized by a nuclease was based on the observation that double-strand breaks occur in the chromosome in strains in which the Rep helicase was inactivated (Seigneur et al., Suggested Reading). The DnaB helicase travels on the 5′-to-3′ template strand and unwinds DNA ahead of the replication fork, and the Rep helicase interacts with DnaB but travels on the 3′-to-5′ template strand to help displace proteins ahead of DNA replication (see chapter 1). In a *rep* strain background, the RecBCD complex is essential; in a *rep* strain with temperature-sensitive (Ts) mutations in *recB* and *recC*, cell death occurs at the nonpermissive temperature through the accumulation of double-strand DNA breaks that are not repaired. In the *rep recB*(Ts) *recC*(Ts) strain, mutations in *ruvA* and *ruvB* restore viability at the otherwise nonpermissive temperature. In other experiments, it was found that in a *dnaB*(Ts) mutant, and even in a wild-type background, replication forks likely reverse using RuvAB branch migration and become targets for various processing enzymes (Figure 11.18). Replication forks stall at various impediments in the chromosome, which is exacerbated in *rep* and *dnaB* helicase mutants (Figure 11.18, step 1). A replication fork may reverse though the action of RuvAB (possibly aided by RecG) (step 2). The Holliday junction resolvase can cleave the Holliday junction formed by RuvAB (step 3). This resolution of the Holliday junction forms a double-strand break in the chromosome (step 4). Processing of the broken end by RecBCD allows RecA-mediated strand invasion (step 5). The PriA primosome can then restart DNA replication, allowing the replication fork to be restored. Alternatively, the extruded DNA formed as the replication fork reverses could be recognized by RecBCD (step 7), and the PriA primosome could then restart DNA replication, allowing the replication fork to be restored (step 8).

In *E. coli*, self-inflicted double-strand breaks are also found through the action of the SbcCD nuclease. The gene encoding the SbcCD nuclease was originally identified based on its ability to suppress mutations in *recB* and *recC* (see chapter 10). The SbcCD nuclease cleaves preferentially at large palindromes that are capable of forming hairpin structures. These structures appear to form only following DNA replication and may occur in the transient single-stranded DNA that occurs following separation of the strands. What benefit could the cell derive from self-cleavage of the chromosome? One possibility is that the palindromes themselves lead to genome instability, and by forming breaks at these structures, the enzyme may facilitate their removal. Interestingly, in the absence of SbcBC, large experimentally introduced palindromes can lead to chromosomal inversions, which can be highly detrimental to the cell (see Darmon et al., Suggested Reading).

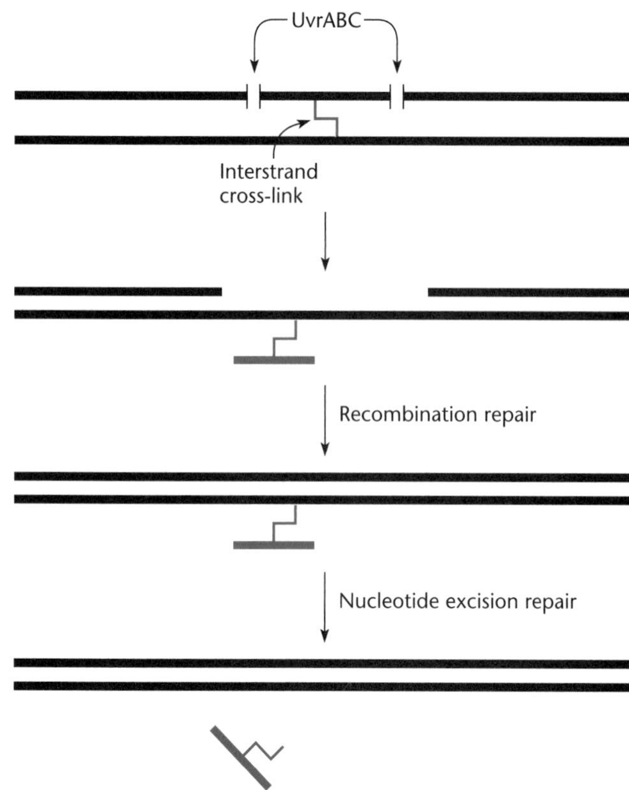

Figure 11.19 Repair of a DNA interstrand cross-link through the combined action of nucleotide excision repair and recombination repair. See the text for details.
doi:10.1128/9781555817169.ch11.f11.19

REPAIR OF INTERSTRAND CROSS-LINKS IN DNA

There are other situations where recombination functions may also collaborate with other repair functions to repair damage to DNA. One example might be in the repair of chemical cross-links in the DNA (Figure 11.19). Many chemicals, such as light-activated psoralens, mitomycin, cisplatin, and EMS, can form **interstrand cross-links** in the DNA, in which two bases in the opposite strands of the DNA are covalently joined to each other (hence the prefix "inter," or "between"). Interstrand cross-links present special problems and cannot be repaired by either nucleotide excision repair or recombination repair alone. Cutting both strands of the DNA with the UvrABC endonuclease would cause double-strand breakage that would be difficult to repair because recombination functions could not separate the strands.

Although DNA cross-links cannot be repaired by either excision or recombination repair alone, they can be repaired by a combination of recombination repair and nucleotide excision repair, as shown in Figure 11.19. In the first step, the UvrABC endonuclease makes nicks in one strand on either side of the interstrand cross-link, as though it were repairing any other type of damage. This leaves a gap opposite the DNA damage, as shown in the figure. In a second step, recombination repair replaces the gap with an undamaged strand from the other daughter DNA in the cell (see above). The DNA damage is then confined to only one strand and is opposite an undamaged strand; therefore, it can be repaired by the nucleotide excision repair pathway. Like the other recombination-based strategies for repair, this repair is possible only when the DNA has already replicated and there are already multiple copies of the DNA in the cell.

SOS-Inducible Repair

As mentioned above, DNA damage leads to the induction of genes whose products are required for DNA repair. In this way, the cell is better able to repair the damage and survive. The first indication that repair systems for UV damage are inducible came from the work of Jean Weigle on the reactivation of UV-irradiated λ phage (see Weigle, Suggested Reading). He irradiated phage λ and its hosts and tested the survival of the phage by their ability to form plaques when they were plated on *E. coli*. He observed that more irradiated phage survived when plated on *E. coli* cells that had themselves been irradiated than when plated on unirradiated *E. coli* cells. Because the phage were restored to viability, or reactivated, by being plated on preirradiated *E. coli* cells, this phenomenon was named **Weigle reactivation**, or **W reactivation.** Apparently, in the irradiated cells, a repair system that could repair the damage to the λ DNA when the DNA entered the cell was being induced.

THE SOS RESPONSE

Earlier in this chapter, we mentioned a class of genes induced after DNA damage, called the *din* genes, which includes genes that encode products that are part of the excision repair and recombination repair pathways. The products of other *din* genes help the cell survive DNA damage in other ways. For example, some *din* gene products transiently delay cell division until the damage can be repaired, and some allow the cell to replicate past the DNA damage (see below).

Many *din* genes are regulated by the **SOS response,** so named because the mechanism rescues cells that have suffered severe DNA damage. Genes under this type of control are called **SOS genes.** Originally, classical genetic analysis uncovered some 30 SOS genes. More recent microarray analyses have found a few more, not all of which may be directly regulated by the SOS system. For example, some of the new genes are on cryptic prophages that are induced by DNA damage, indirectly increasing the expression of the gene (see below and Courcelle et al., Suggested Reading).

Figure 11.20 illustrates how genes under SOS control are induced. The SOS genes are normally repressed by a protein called the LexA repressor, which binds to sequences called the **SOS box** upstream of the SOS genes and prevents their transcription. The SOS box is the operator sequence that binds the LexA repressor by analogy to other operator sites, such as *lac*O and the operator sites that bind the λ repressor (see chapters 8 and 12). Any gene directly regulated by LexA would therefore have one of these SOS boxes close to its promoter. Quite often, genes in the same regulon have a common upstream sequence that binds the regulatory protein, and these are often referred to as boxes, with the name of the regulon. In fact, this is often how the genes of a regulon are first identified, by the presence of one of these boxes close to their promoter. Examples of regulons and their boxes are discussed in chapter 13.

The LexA repressor remains bound to the SOS boxes upstream of the promoters for the genes, repressing their transcription, until the DNA is extensively damaged by UV irradiation or other DNA-damaging agents. This causes the LexA repressor to cleave itself, an action known as **autocleavage,** thereby inactivating itself and allowing transcription of the SOS genes. This is reminiscent of the cleavage of the λ repressor, which also cleaves itself following DNA damage during induction of phage λ, as described in chapter 8. The two mechanisms are in fact remarkably similar, as discussed below.

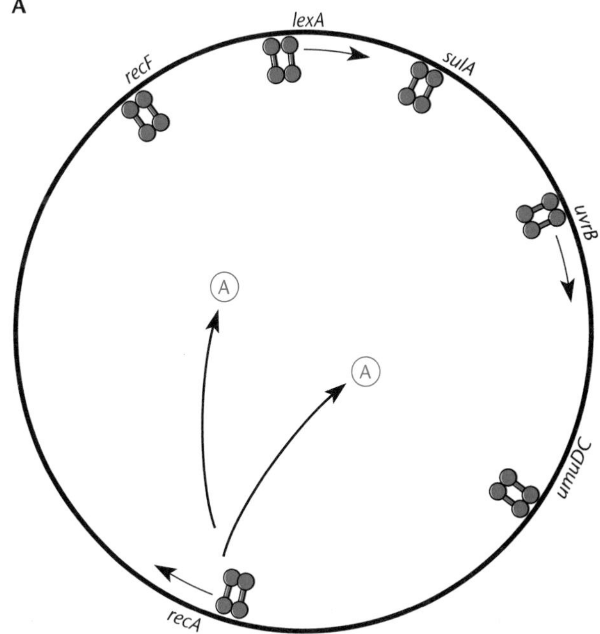

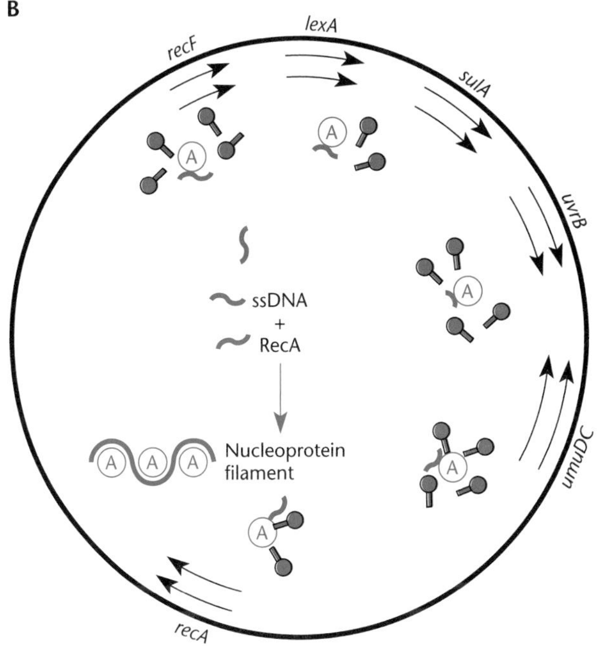

Figure 11.20 Regulation of the SOS response regulon in *E. coli.* **(A)** About 30 genes around the *E. coli* chromosome are normally repressed by the binding of a LexA dimer (barbell structure) to their operators; only a few of these genes are shown. Some SOS genes are expressed at low levels, as indicated by single arrows. **(B)** After DNA damage, the single-stranded DNA (ssDNA) that accumulates in the cell binds to RecA (circled A), forming a RecA nucleoprotein filament, which binds to LexA, causing LexA to cleave itself. The cleaved repressor can no longer bind to the operators of the genes, and the genes are induced, as indicated by two arrows. The approximate positions of some of the genes of the SOS regulon are shown.
doi:10.1128/9781555817169.ch11.f11.20

The reason why the LexA repressor no longer binds to DNA after it is cleaved is also well understood. Each LexA polypeptide has two separable domains. One, the **dimerization domain**, binds to another LexA polypeptide to form a dimer (hence its name), and the other, the **DNA-binding domain**, binds to the DNA operators upstream of the SOS genes and blocks their transcription. The LexA protein binds to DNA only if it is in the dimer state. The point of cleavage of the LexA polypeptide is between the two domains, and autocleavage separates the DNA-binding domain from the dimerization domain. The DNA-binding domain cannot dimerize and so by itself cannot bind to DNA and block transcription of the SOS genes.

Figure 11.20 also illustrates the answer to the next question: why does the LexA repressor cleave itself only after DNA damage? After DNA damage, single-stranded DNA appears in the cell, probably due to blockage of replication forks at the damaged sites. This single-stranded DNA is recognized and bound by the RecA protein to form RecA nucleoprotein filaments, which then bind to LexA and cause it to autocleave. This feature of the model—that LexA cleaves itself in response to RecA binding rather than being cleaved by RecA—is supported by experimental evidence. When heated under certain conditions, even in the absence of RecA, LexA eventually cleaves itself. This result indicates that RecA acts as a **coprotease** to facilitate LexA autocleavage, rather than doing the cleaving itself like a standard protease.

The RecA protein thus plays a central role in the induction of the SOS response. It senses the single-stranded DNA that accumulates in the cell as a by-product of attempts to repair damage to the DNA and then causes LexA to cleave itself, inducing the SOS genes. We have already discussed another activity of RecA, the recombination function involved in synapse formation; below, we discuss yet another in **translesion synthesis (TLS)**.

Another level of control in this system comes from the relative affinity of LexA for the SOS boxes. While the SOS boxes recognized by LexA are very similar, they are not identical. These differences result in the need for different amounts of LexA to repress transcription from different SOS genes. These differences in the operators allow a tuneable effect, where the SOS boxes that are bound weakly cause the gene they control to be turned on with a small decrease in LexA (i.e., these genes are turned on almost immediately after SOS induction). In contrast, genes that are controlled by SOS boxes that are bound very tightly by LexA remain off unless the level of LexA falls very low (i.e., these genes are turned on only after prolonged DNA damage). Some SOS boxes allow some leaky expression even without SOS induction to provide a basal level of activity in the cell (Figure 11.20).

The RecA protein therefore plays two roles, one in recombination and one in inducing the SOS response. We can speculate why the RecA protein might play these two different roles. It must bind to single-stranded DNA to promote synapse formation during recombination, so it can easily serve as a sensor of single-stranded DNA in the cell. Also, even though it is itself encoded by an SOS gene and is induced following DNA damage, it is always present in large enough amounts to bind to all of the LexA repressor and to quickly promote autocleavage of all of the repressor after DNA damage.

As mentioned above, the regulation of the SOS response through cleavage of the LexA repressor is strikingly similar to the induction of λ through autocleavage of the CI repressor (see chapter 8). Like the LexA repressor, the λ repressor must be a dimer to bind to the operator sequences in DNA and repress λ transcription. Each λ repressor polypeptide consists of an N-terminal DNA-binding domain and a C-terminal dimerization domain. Autocleavage of the λ repressor, stimulated by the activated RecA nucleoprotein filaments, also separates the DNA-binding domain and the dimerization domain, preventing the binding of the λ repressor to the operators and allowing transcription of λ lytic genes. The sequences of amino acids around the sites at which the LexA and λ repressors are cleaved are also similar. Evolution of the phage λ repressor, therefore, appears to have taken advantage of the endogenous SOS regulatory system of the host to induce its genes following DNA damage by mimicking LexA, thereby allowing λ to escape a doomed host cell. The activated RecA coprotease also promotes the autocleavage of the UmuD protein involved in SOS mutagenesis (see below).

GENETICS OF SOS-INDUCIBLE MUTAGENESIS

It is well known that many types of DNA damage, including UV irradiation, are mutagenic and increase the number of mutations in the surviving cells. This is true of all organisms, from bacteria to humans (Box 11.2). This implies that one or more of the repair mechanisms used to repair damage are mistake prone. Early evidence suggested that at least one of the repair systems encoded by the SOS genes appears to be very mistake prone. As we show, this system, known as TLS, allows the replication fork to proceed over damaged DNA so that the molecule can be replicated and the cell can survive. This mechanism seems to be a last resort that operates only when the DNA damage is so extensive that it cannot be repaired by other, less mistake-prone mechanisms.

The first indications that a mistake-prone pathway for UV damage repair is inducible in *E. coli* came from the same studies that showed that repair itself is inducible (see Weigle, Suggested Reading). In addition to measuring the survival of UV-irradiated λ plated on

UV-irradiated *E. coli*, Weigle counted the clear-plaque mutants among the surviving phage. (Recall from chapter 8 that lysogens form in the centers of wild-type phage λ plaques, making the plaques cloudy, but mutants that cannot lysogenize form clear plaques that can be easily identified.) There were more clear-plaque mutants among the surviving phage if the bacterial hosts had been UV irradiated prior to infection than if they had not been irradiated. Therefore, the increased mutagenesis was due to induction of a mutagenic repair system after DNA damage. This inducible mutagenesis was named **Weigle mutagenesis**, or just **W mutagenesis**. Later studies showed that the inducible mutagenesis results from induction of one or more of the SOS genes; it does not occur without RecA and cleavage of the LexA repressor. Thus, the inducible mutagenesis Weigle observed is now often called **SOS mutagenesis**.

Determining Which Repair Pathway Is Mutagenic

Although Weigle's experiments showed that one of the inducible UV damage repair systems in *E. coli* is mistake prone and causes mutations, they did not indicate which system was responsible. A genetic approach was used to answer this question (see Kato et al., Suggested Reading). To detect UV-induced mutations, the experimenters used the reversion of a *his* mutation. Their basic approach was to make double mutants with both a *his* mutation and a mutation in one or more of the genes of each of the repair pathways. The repair-deficient mutants were then irradiated with UV light and plated on medium containing limiting amounts of histidine so that only His$^+$ revertants could multiply to form a colony. Under these conditions, each reversion to *his*$^+$ results in only one colony, making it possible to measure directly the number of *his*$^+$ reversions that have occurred. This number, divided by the total number of surviving bacteria, gives an estimation of the susceptibility of the cells to mutagenesis by UV light.

The results of these experiments led to the conclusion that recombination repair of blocked replication forks does not seem to be mistake prone. While *recB* and *recF* mutations reduced the survival of the cells exposed to UV light, the number of *his*$^+$ reversions per surviving cell was no different from that of cells lacking mutations in their *rec* genes. Also, nucleotide excision repair does not seem to be mistake prone. Addition of a *uvrB* mutation to the *recB* and *recF* mutations made the cells even more susceptible to killing by UV light but also did not reduce the number of *his*$^+$ reversions per surviving cell.

However, *recA* mutations did seem to prevent UV mutagenesis. While these mutations made the cells extremely sensitive to killing by UV light, the few survivors did not contain additional mutations. We now know that two genes, *umuC* and *umuD*, must be induced for mutagenic repair and that *recA* mutations prevent their induction. Thus, the UmuC and UmuD proteins act in the opposite way to most repair systems. Rather than repairing the damage before mistakes in the form of mutations are made, the UmuC and UmuD proteins actually make the mistakes themselves. If they were not present, the cells that survive UV irradiation and some other types of DNA damage would have fewer mutations. The payoff, however, is that presumably, more cells survive. Besides its role in inducing the SOS genes, the RecA protein is directly involved in SOS mutagenesis (see below).

Isolation of *umuC* and *umuD* Mutants

Once it was established that most of the mutagenesis after UV irradiation can be attributed to the products of SOS-inducible genes and that these gene products are not the known ones involved in recombination and nucleotide excision repair, the next step was to identify the unknown genes. Note that the products of these genes seem to work in the opposite way to most repair pathways. Mutations that inactivate a *mut* gene or another repair pathway gene increase the rate of spontaneous mutations because normally, the gene products repair damage before it can cause mistakes in replication. However, as noted above, mutations that inactivate gene products of a mutagenic or mistake-prone repair pathway have the opposite effect, decreasing the rates of at least some induced mutations. Because the repair pathway itself is mutagenic, mutations should be less frequent if the repair pathway does not exist than if it does exist. Cells with a mutation in a gene of the mistake-prone repair pathway may be less likely to survive DNA damage, but there should be a lower percentage of newly generated mutants among the survivors.

Mutations that reduce the frequency of mutants after UV irradiation were found to fall in two genes, named *umuC* and *umuD* (for *u*ltraviolet-induced *mu*tations C and *D*). The first *umuC* and *umuD* mutants were isolated in two different laboratories by essentially the same method—reversion of a *his* mutation that makes cell growth require histidine to measure mutation rates—but we describe only the one used by Kato and Shinoura (see Suggested Reading). The basic strategy was to treat the *his* mutant with a mutagen that induces DNA damage similar to that caused by UV irradiation and to identify mutant bacteria in which fewer *his*$^+$ revertants occurred. These mutants would have a second mutation that inactivated the mutagenic repair pathway and reduced the rate of reversion of the *his* mutation.

Figure 11.21 illustrates the selection in more detail. The *his* mutant of *E. coli* was heavily mutagenized so that some of the bacteria would have mutations in the putative mutagenic repair genes. Individual colonies of the bacteria were then patched with a loop onto plates

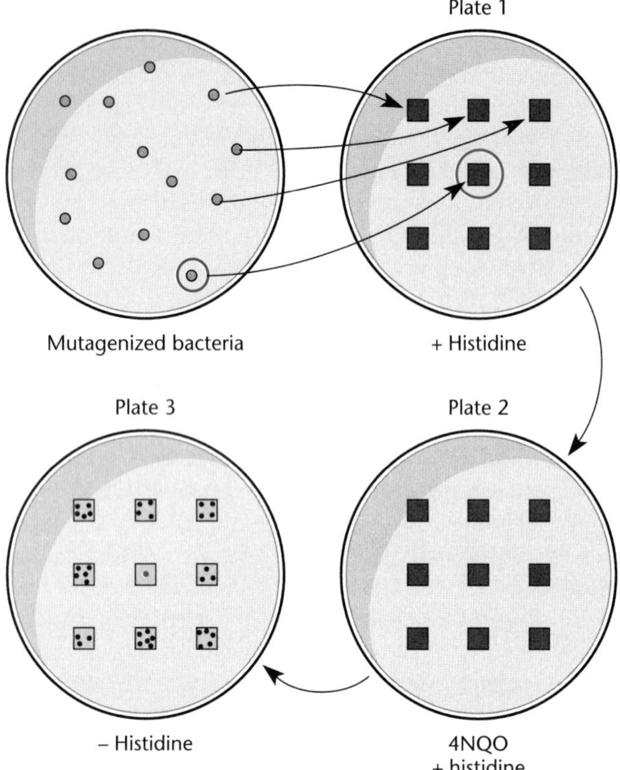

Figure 11.21 Detection of a mutant defective in mutagenic repair. Colonies of mutagenized *his* bacteria are picked individually from a plate and patched onto a new plate containing histidine. This plate (plate 1) is then replicated onto a plate containing 4NQO (plate 2) to induce DNA damage similar to that induced by UV irradiation. The 4NQO-containing plate is then replicated onto another plate with limiting amounts of histidine (plate 3). After incubation, mottling of a patch caused by many His+ revertants indicates that the bacterium that made the colony on the original plate was capable of mutagenic repair. The colony circled in blue on the original plate may have arisen from a mutant deficient in mutagenic repair, because it produces fewer His+ revertants when replicated onto plate 3.
doi:10.1128/9781555817169.ch11.f11.21

with medium containing histidine, and the plates were incubated until patches due to bacterial growth first appeared. Each plate was then replicated onto another plate containing 4-nitroquinoline-1-oxide (4NQO), which causes DNA damage similar to that caused by UV irradiation. After the patches developed, this plate was replicated onto a third plate containing limiting amounts of histidine. Most bacteria formed patches with a few regions of heavier growth due to His+ revertants, indicating that they were being mutagenized by the 4NQO. However, a few bacteria formed patches with very few areas of heavier growth and therefore had fewer His+ revertants. The bacteria in these patches were candidates for descendants of mutants that could not be mutagenized by 4NQO or, presumably, by UV irradiation, since, as mentioned above, the types of damage caused by the two mutagens are similar.

The next step was to map the mutations in some of the mutants. Some of the mutations that prevented mutagenesis by UV irradiation mapped to either *recA* or *lexA*. These mutations could have been anticipated, because mutations in these genes could prevent the induction of all the SOS genes, including those for mutagenic repair. The *recA* mutations presumably inactivate the coprotease activity of the RecA protein, preventing it from causing autocleavage of the LexA repressor and thereby preventing induction of the SOS genes. The *lexA* mutations in all probability change the LexA repressor protein so that it cannot be cleaved, presumably because one of the amino acids around the site of cleavage has been altered. This is a special type of *lexA* mutation called an Ind− mutation. If the LexA protein is not cleaved following UV irradiation of the cells, the SOS genes, including the genes for mutagenic repair, will not be induced.

Some of the mutations that prevented UV mutagenesis mapped to a previously unknown locus distant from *recA* or *lexA* on the *E. coli* map. Complementation tests between mutations at this locus revealed two genes at the site required for UV mutagenesis, named *umuC* and *umuD*. Later experiments also showed that these genes are transcribed into the same mRNA and so form an operon, *umuDC*, in which the *umuD* gene is transcribed first. Experiments with *lacZ* fusions revealed that the *umuDC* operon is inducible by UV light and is an SOS operon, since it is under the control of the LexA repressor (see Bagg et al., Suggested Reading).

Experiments Showing that Only *umuC* and *umuD* Must Be Induced for SOS Mutagenesis

The fact that *umuC* and *umuD* are inducible by DNA damage and are required for SOS mutagenesis does not mean that they are the only genes that must be induced for this pathway. Other genes that must be induced for SOS mutagenesis might have been missed in the mutant selections. Some investigators sought an answer to this question (see Sommer et al., Suggested Reading). Their experiments used a *lexA*(Ind−) mutant, which, as mentioned above, should permanently repress all the SOS genes. They also mutated the operator site of the *umuDC* operon so that these genes would be expressed constitutively and would no longer be under the control of the LexA repressor. Under these conditions, the only SOS gene products that should be present are those of the *umuC* and *umuD* genes, since all other such SOS genes are permanently repressed by the LexA(Ind−) repressor. In addition, a shortened form of the *umuD* gene was used. This altered gene synthesizes only the carboxyl-terminal UmuD′ fragment that is the active form for SOS mutagenesis (see below). With the altered *umuD* gene, the RecA coprotease is not required for UmuD to be autocleaved to the active form.

The experiments showed that UV irradiation is mutagenic if UmuC and UmuD′ are expressed constitutively, even if the cells are *lexA*(Ind⁻) mutants, indicating that *umuC* and *umuD* are the only genes that need to be induced by UV irradiation for SOS mutagenesis. However, this result does not entirely eliminate the possibility that other SOS gene products are involved. As discussed below, the RecA protein is also directly required for UV mutagenesis. The *recA* gene is induced to higher levels following UV irradiation but apparently is present in large enough amounts for UV mutagenesis even without induction. The GroEL and GroES proteins are also required for UV mutagenesis, presumably because they help fold mutagenic repair proteins, but the *groEL* and *groES* genes are not under the control of the LexA repressor and so are expressed even in the *lexA*(Ind⁻) mutants.

Experiments Show that RecA Has a Role in UV Mutagenesis in Addition to Its Role as a Coprotease

Similar experiments were performed to show that RecA has a required role in UV mutagenesis in addition to its role in promoting the autocleavage of LexA and UmuD. A sufficient explanation for why mutations in the *recA* gene can prevent UV mutagenesis was that they prevent the induction of all the SOS genes, including *umuC* and *umuD*, and that they also prevent the autocleavage of UmuD to UmuD′, which is required for TLS (see below). However, this did not eliminate the possibility that RecA plays another role in TLS besides its role as a LexA and UmuD coprotease. The mutants that express UmuC and the cleaved form of UmuD′ constitutively could also be used to answer this question. If the coprotease activity of RecA alone is required for TLS, *recA* mutations should not affect UV mutagenesis in this genetic background. However, it was found that *recA* mutations still prevented UV mutagenesis, even if UmuC and UmuD′ were made constitutively, indicating that RecA also participates directly in mutagenesis. This inspired new models in which UmuCD-based mutagenic polymerase needed a RecA nucleoprotein filament for a separate role beyond inducing the autocleavage of UmuD (see below).

Mechanism of TLS by the Pol V Mutasome

Dramatic progress has been made in understanding how the UmuC and UmuD proteins promote mutagenesis and allow an *E. coli* cell to tolerate damage to its DNA. As is often the case, these discoveries have implications far beyond the UmuC and UmuD proteins and DNA repair in *E. coli* (Box 11.2). The UmuC protein was found to be a DNA polymerase that, in contrast to the normal replicative DNA polymerase, can replicate right over some types of damage to the DNA, including the thymine cyclobutane dimers and cytosine-thymine 6-4 dimers induced by UV irradiation. It can also replicate over abasic sites in which the base has been removed by a glycosylase (see above); obviously, therefore, it does not require correct base pairing before it can move on. Accordingly, UmuC was renamed **DNA Pol V**; because it is capable of replicating over DNA damage or lesions in the template, it was said to be capable of TLS. This suggested an explanation of why UmuC could be mutagenic. Because thymine dimers and other types of damaged bases cannot pair properly, DNA Pol V must incorporate bases almost at random opposite the damaged bases, causing mutations.

UmuC is almost undetectable during normal growth, but with SOS induction, it can go as high as 200 molecules per cell. While at first pass this might seem to explain why Pol V is only active with SOS induction, in actuality, the regulation of UV mutagenesis is much more complicated and refined than this, presumably because it is in the best interest of the cell not to induce this mutagenic system unless the damage cannot be repaired in other ways. Figure 11.22 illustrates why SOS mutagenesis occurs only after extensive damage to the DNA. The first factor that limits SOS mutagenesis only to situations where severe DNA damage is found involves the amount of LexA needed to regulate the *umuDC* operon; even a small amount of LexA will prevent transcription of these genes, and it is estimated that prolonged exposure to RecA-inducing conditions (>30 minutes) is required for the genes to be expressed. Therefore, these proteins will not even be available until very late in SOS induction (Figure 11.22A and B). Once expressed, the newly synthesized UmuC and UmuD proteins come together to form a heterotrimer complex with two copies of the UmuD protein and one copy of the UmuC protein ($UmuD_2C$). This complex does not have DNA polymerase activity, although it might bind to DNA Pol III and temporarily arrest replication to create a "checkpoint" (see Sutton et al., Suggested Reading), while $UmuD_2$ appears to play an additional regulatory role (see below). However, as the single-stranded DNA-RecA nucleoprotein filaments accumulate, they also cause UmuD to cleave itself (to form UmuD′), in much the same way that they cause LexA and the λ repressor to cleave themselves. The autocleavage of UmuD to UmuD′ requires a higher concentration of RecA nucleoprotein filaments than does the autocleavage of LexA; therefore, rather than happening immediately, it occurs only if the damage is so extensive that it cannot be immediately repaired by other SOS functions (Figure 11.22C). Once UmuD is cleaved to UmuD′, the UmuC in the $UmuD'_2C$ complex is ready for activation using a separate

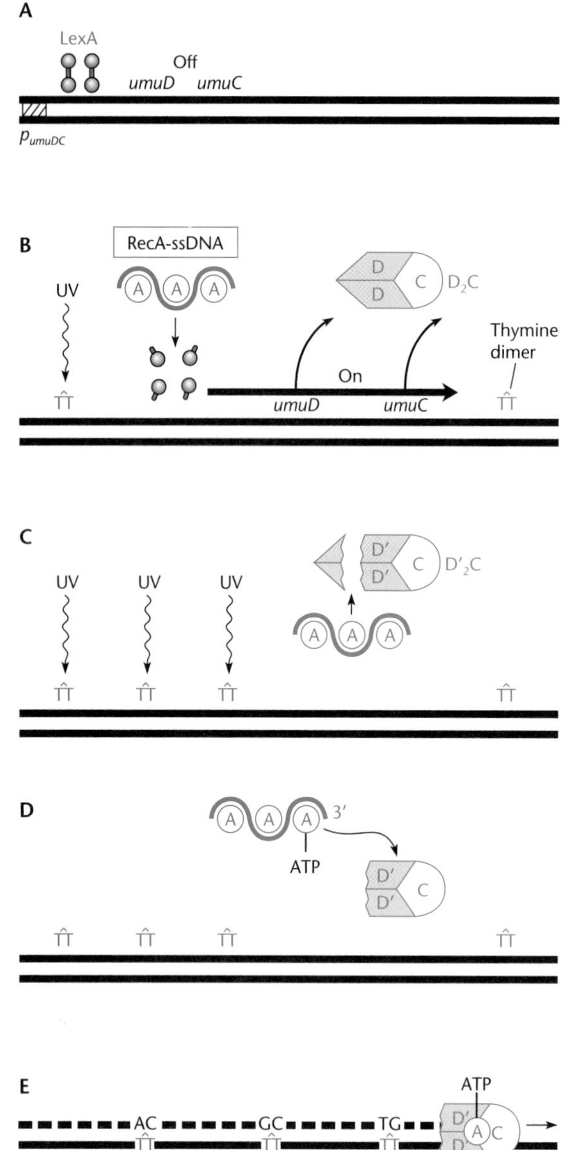

Figure 11.22 Regulation of SOS mutagenesis in *E. coli*. **(A)** Before DNA damage occurs, the LexA protein represses the transcription of SOS genes, including the *umuDC* operon. **(B)** After DNA damage, the RecA protein binds to the single-stranded DNA (ssDNA), which accumulates, forming RecA nucleoprotein filaments. These filaments bind to LexA, promoting its autocleavage and inducing the SOS genes. One of the operons induced late in SOS induction is the *umuDC* operon. The UmuC and UmuD proteins form a heterotrimer composed of two copies of UmuD and one copy of UmuC ($UmuD_2C$). **(C)** More damage causes more RecA nucleoprotein filaments to accumulate, eventually promoting the autocleavage of UmuD to form $UmuD'_2C$. **(D)** A RecA monomer that is bound to ATP can be delivered to the $UmcD'_2C$ complex from the very 3′ end of a RecA nucleoprotein filament, allowing the formation of the Pol V mutasome. **(E)** The Pol V mutasome can replicate right over the damage, often making mistakes in the process; some wrong bases are shown mistakenly incorporated opposite thymine dimers.
doi:10.1128/9781555817169.ch11.f11.22

additional property of activated RecA that is distinct from changes in expression through LexA or induced cleavage of UmuD. As described below, the final step in making the complex capable of TLS appears to involve the formation of an $UmuD'_2C$-ATP-RecA complex.

Figure 11.22 shows a recent model for how the $UmuD'_2C$ complex performs TLS and causes mutations. This model integrates the earlier observations and a more recent finding that activated RecA bound to ATP must be integrated with $UmuD'_2C$ to make an error-prone DNA Pol V referred to as the Pol V mutasome, $UmuD_2C$-RecA-ATP (see Jiang et al., Suggested Reading). The required addition of RecA-ATP is carried out with a nucleoprotein filament separate from the DNA that is actually subject to replication. In the earlier biochemical work, it was impossible to know if RecA was acting in *trans* or in *cis* to the DNA that was to be replicated. More recently, it was shown biochemically that a RecA monomer that was bound to ATP could be delivered to the $UmcD'_2C$ complex from the very 3′ end of a RecA nucleoprotein filament, allowing the formation of the Pol V mutasome (Figure 11.22D and E). Probably as a safeguard for genome stability, the Pol V mutasome shows very poor processivity and is capable of staying on DNA for only a short distance. After its short span of DNA polymerization, it is also rapidly deactivated when it dissociates from the template DNA. These characteristics ensure that the Pol V mutasome will only operate over a short distance and will only work once, thereby limiting mutations.

There are still questions to be answered about how TLS functions. The Pol V mutasome must function with the β clamp processivity factor, which is very likely coopted from DNA Pol III. However, the regulation that allows the exchange of polymerase with the β clamp is still a mystery. There are five separate DNA polymerases in *E. coli* (additional TLS polymerases are also found on some plasmids). Given that every polymerase likely works with a β clamp, there must be a molecular mechanism that helps coordinate these transfers at the DNA replication fork. The idea of polymerase switching is relevant to our own human genome; currently, it is believed that there are at least 15 different DNA polymerases that are also subject to similar processivity clamp exchanges.

Other Specialized Polymerases and Their Regulation

The UmuC polymerase is a member of a large group of polymerases called the **Y-family polymerases**. Structural studies have shown that the active centers of Y polymerases are more open than those of the replicative polymerases and have fewer contacts between the polymerase and the DNA template, allowing the Y-family polymerases to replicate over many types of lesions in the DNA. They also lack editing functions that would cause them to stall at mismatches. Most types of bacteria, archaea, and eukaryotes are known to have Y-family DNA polymerases related to UmuC. In addition, some naturally occurring plasmids carry analogs of the *umuC* and *umuD* genes. The best studied of these genes are the *mucA* and *mucB* genes of plasmid R46. The products of these plasmid genes can substitute for UmuC and UmuD′ in mutagenic repair of *E. coli*, and *Salmonella* spp. carrying the R46 derivative plasmid pKM101 are more sensitive to mutagenesis by many types of DNA-damaging agents. Because of the sensitivity that the *mucA* and *mucB* genes impose on a bacterial strain with the pKM101 plasmid, it has been introduced into the *Salmonella* strains used for the Ames test (Box 11.3).

DNA Pol II

Besides Pol V, *E. coli* has two other DNA polymerases that are induced as part of the SOS response, DNA Pol II and DNA Pol IV. DNA Pol II is a B-family polymerase that seems to help restart replication after UV irradiation, possibly by extending 3′ OH ends that the Pol III holoenzyme cannot extend. While DNA Pol II is present at 50 or more molecules per cell during normal growth, it is induced about 10-fold upon SOS induction. DNA Pol II is capable of doing some TLS. However, unlike Pol V, but typical of B-family polymerases, DNA Pol II has exonuclease activity for proofreading and is processive for DNA replication.

The crystal structure of DNA Pol II has shed light on how the polymerase can replicate both normal DNA and damaged DNA (see Wang and Yang, Suggested Reading). The structure revealed small protein cavities outside the active site of the enzyme that may allow the template DNA to loop out of the enzyme to facilitate primer extension past DNA damage or possibly allow it to skip over DNA damage that would otherwise halt the polymerase. DNA Pol II is expressed early during SOS induction. Early expression of the polymerase after DNA damage may be beneficial, because the enzyme could allow replication past lesions that stall DNA Pol III without the risk of mutagenesis that goes along with expression of the other SOS-induced DNA polymerases. For example, if DNA Pol II is induced early and can replicate past the lesion that stalled DNA Pol III, it will prevent single-stranded DNA from lingering long enough to allow the error-prone polymerases to be expressed and activated.

DNA Pol IV

DNA Pol IV (also known as DinB) is a Y family DNA polymerase, and its participation in TLS has been known for some time. Unlike Pol V, it causes spontaneous mutations when it is overproduced, even without DNA damage. Many of the mutations found with Pol IV are

BOX 11.3

The Ames Test

It is now well established that cancer is initiated by mutations in genes, including oncogenes and tumor-suppressing genes. Therefore, chemicals that cause mutations are often carcinogenic for humans. Many new chemicals are being used as food additives or otherwise come into contact with humans, and each of these chemicals must be tested for its carcinogenic potential. However, such testing in animal models is expensive and time-consuming. Because many carcinogenic chemicals damage DNA, they are mutagenic for bacteria, as well as humans. Therefore, bacteria can be used in initial tests to determine if chemicals are apt to be carcinogenic. The most widely used of these tests is the Ames test, developed by Bruce Ames and his collaborators. This test uses revertants of *his* mutations of *Salmonella* spp. to detect mutations. The chemical is placed on a plate lacking histidine on which has been spread a His$^-$ mutant of *Salmonella*. If the chemical can revert the *his* mutation, a ring of His$^+$ revertant colonies will appear around the chemical on the plate. A number of different *his* mutations must be used, because different mutagens cause different types of mutations, and they all have preferred sites of mutagenesis (hot spots). Also, the test is made more sensitive by eliminating the nonmutagenic nucleotide excision repair system with a *uvrA* mutation and introducing a plasmid containing the *mucA* and *mucB* genes. These genes are analogs of *umuC* and *umuD*, so they increase mutagenesis (see the text). Some chemicals are not mutagenic themselves but can be converted into mutagens by enzymes in the mammalian liver. To detect these precursors of mutagens, we can add a liver extract from rats to the plates and spot the chemical on the extract. If the extract converts the chemical into a mutagen, His$^+$ revertants will appear on the plate.

References

McCann, J., and B. N. Ames. 1976. Detection of carcinogens as mutagens in the *Salmonella*/microsome test assay of 300 chemicals: discussion. *Proc. Natl. Acad. Sci. USA* **73:**950–954.

frameshifts, but only −1 frameshifts. While it is induced 10-fold following DNA damage, it is always present in the cell at a fairly high level of about 150 to 250 copies per cell. DinB is largely responsible for the spontaneous mutations that occur during special types of long-term selection experiments, when the cells are not growing. More recently, it was shown that the mutation potential (and TLS activity) of DinB appears to be regulated by its interaction with other important proteins that have been discussed in this chapter, $UmuD_2$ and RecA (see Godoy et al., Suggested Reading). The three proteins interact physically and functionally, and structural modeling suggests that the DinB interaction with $UmuD_2$ and RecA helps enclose the relatively open active site of DinB, thereby preventing the template bulging responsible for −1 frameshifts. While the finding that a subunit of DNA Pol V (i.e., UmuD) can function with DNA Pol IV (i.e., DinB) blurs the historical nomenclature designation, it also reveals an exciting and unexpected regulatory option for cells for DNA replication. Presumably, in the earliest stages of SOS induction (<30 minutes), DinB interacts with $UmuD_2$ and RecA to allow replication that can bypass some lesions without the formation of −1 frameshifts. However, in the presence of higher levels of DNA damage and prolonged SOS induction (>30 minutes), the mutagenic TLS activity of DinB could allow replication over DNA with different bulky adducts or (controversially) could actually provoke mutations that might help the population overcome stressful conditions. DinB is the only Y polymerase that is conserved across all three domains of life. It is therefore of additional interest that amino acids in the proposed region of interaction between UmuD, RecA, and DinB covary throughout evolution, suggesting that this regulatory activity for mutation potential could be conserved.

Eukaryotes have a large number of different mutagenic polymerases related to UmuC and Pol IV. Each type of mutagenic DNA polymerase may be required to replicate over a particular type of damage to DNA and thereby may play a role in avoiding mutations due to a particular type of DNA damage. Mutations in some of these genes have been implicated in increased cancer risk. In the absence of the specialized Y-type polymerase for a particular type of lesion in DNA, another one will take over, causing the mutations that lead to increased risk of cancer.

Summary of Repair Pathways in *E. coli*

Table 11.2 shows the repair pathways that have been discussed and some of the genes whose products participate in each pathway. Some pathways, such as photoreactivation and most base excision pathways, have evolved to repair specific types of damage to DNA. Some, such as VSP repair, mend damage only in certain sequences. Others, such as mismatch repair and nucleotide excision repair, are much more general and repair essentially any damage to DNA, provided that it causes a distortion in the DNA structure.

The separation of repair functions into different pathways is in some cases artificial. Some repair genes are inducible, and the repair enzymes themselves can play a role in their induction, as well as in the induction of genes in other pathways. For example, the RecBCD nuclease is involved in recombination repair but can also help make the single-stranded DNA that activates the RecA coprotease activity after DNA damage to induce SOS functions. The RecA protein is required for both recombination repair and induction of the SOS functions, and it is directly involved in SOS mutagenesis. The proposed role of UmuD in the Pol V mutasome and with DinB/Pol IV provides another example of such an artificial grouping. Needless to say, the overlap of the functions of the repair gene products in different repair pathways has complicated the assignment of roles in these pathways.

Bacteriophage Repair Pathways

The DNA genomes of bacteriophages are also subject to DNA damage, either when the DNA is in the phage particle or when it is replicating in a host cell. Not surprisingly, many phages encode their own DNA repair enzymes. In fact, the discovery of some phage repair pathways preceded and anticipated the discovery of the corresponding bacterial repair pathways. By encoding their own repair pathways, phages avoid dependence on host pathways, and repair proceeds more efficiently than it might with the host pathways alone.

The repair pathways of phage T4 are perhaps the best understood. This large phage encodes at least seven different repair enzymes that help repair DNA damage due to UV irradiation, and some of these enzymes are also involved in recombination (see chapter 10). Table 11.3 lists the functions of some T4 gene products and their homologous bacterial enzymes. The phage also uses some of the corresponding host enzymes to repair damage to its DNA.

One of the most important functions for repairing UV damage in phage T4 is the product of the *denV* gene, which is an AP lyase (see above) having both *N*-glycosylase and DNA endonuclease activities (see Dodson and Lloyd, Suggested Reading). The *N*-glycosylase activity specifically breaks the bond holding one of the pyrimidines to its sugar in pyrimidine cyclobutane dimers of the *cis-syn* type (Figure 11.7). The endonuclease

Table 11.2 Genetic pathways for damage repair and tolerance

Repair mechanism	Genetic locus	Function
Methyl-directed mismatch repair	*dam*	DNA adenine methylase
	mutS	Mismatch recognition
	mutH	Endonuclease that cuts at hemimethylated sites
	mutL	Interacts with MutS and MutH
	uvrD	Helicase
VSP repair	*dcm*	DNA cytosine methylase
	vsr	Endonuclease that cuts at 5′ side of T in TG mismatch
GO (guanine oxidations)	*mutM*	Glycosylase that acts on GO
	mutY	Glycosylase that removes A from A-GO mismatch
	mutT	8-OxodGTP phosphatase
Alkylation/adaptive response	*ada*	Alkyltransferase and transcriptional activator
	alkA	Glycosylase for alkylpurines
	alkB	α-Ketoglutarate-dependent dioxygenase
Nucleotide excision	*uvrA*	Component of UvrABC
	uvrB	Component of UvrABC
	uvrC	Component of UvrABC
	uvrD	Helicase
	polA	Repair synthesis
Base excision	*xthA*	AP endonuclease
	nfo	AP endonuclease
Photoreactivation	*phr*	Photolyase
Recombination repair	*recA*	Strand exchange
	recBCD	DNA processing and RecA loading at double-strand breaks
	recFOR	RecA loading at single-strand gaps
	ssb	Single-stranded DNA-binding protein
SOS system	*recA*	Coprotease and recombinase
	lexA	Repressor
	umuDC	DNA Pol V
	dinB	DNA Pol IV
	polB	DNA Pol II

activity then cuts the DNA just 3′ of the pyrimidine dimer, and the dimer is removed by the exonuclease activity of the host cell DNA Pol I. As mentioned above, AP lyases thus work very differently from the UvrABC endonuclease of *E. coli* and, in a sense, combine both the *N*-glycosylase and AP endonuclease activities of some other repair pathways. The bacterium *Micrococcus luteus* has a similar enzyme. The endonuclease activity of the DenV protein also functions independently of the *N*-glycosylase activity and cuts apurinic sites in DNA, as well as heteroduplex loops caused by short insertions or deletions. Because it cuts next to pyrimidine dimers in DNA, the purified DenV endonuclease of T4 is often used to determine the persistence of pyrimidine dimers after UV irradiation.

Table 11.3 Bacteriophage T4 repair enzymes

Repair enzyme	Host analog
DenV	AP lyase (*M. luteus*)
UvsX	RecA
UvsY	RecFOR
UvsW	RecG
gp46/gp47 exonuclease	RecBCD
gp49 resolvase	RuvABC
gp59	PriA

SUMMARY

1. All organisms on Earth probably have mechanisms to repair damage to their DNA. Some of these repair systems are specific to certain types of damage, while others are more general and repair any damage that makes a significant distortion in the DNA helix.

2. Specific DNA glycosylases remove some types of damaged bases from DNA. Specific DNA glycosylases that remove uracil, hypoxanthine, some types of alkylated bases, 8-oxoG, and any A mistakenly incorporated opposite 8-oxoG in DNA are known. After the damaged base is removed by the specific glycosylase, the DNA is cut by an AP endonuclease, and the strand is degraded and resynthesized to restore the correct base.

3. The positions of 5-methylcytosine in DNA are particularly susceptible to mutagenesis, because deamination of 5-methylcytosine produces thymine instead of uracil, and thymine is not removed by the uracil-*N*-glycosylase. *E. coli* has a special repair system, called VSP repair, that recognizes the thymine in the thymine-guanine mismatch at the site of 5-methylcytosine and removes it, preventing mutagenesis. The products of the *mutS* and *mutL* genes of the mismatch repair system also make this pathway more efficient, perhaps by helping attract the Vsr endonuclease to the mismatch.

4. The photoreactivation system uses an enzyme called photolyase to specifically separate the bases of one type of pyrimidine dimer created during UV irradiation. The photolyase binds to pyrimidine dimers in the dark but requires visible light to separate the fused bases.

5. Methyltransferase enzymes remove the methyl group from certain alkylated bases and phosphates and transfer it to themselves. Others are dioxygenases that oxidize the methyl group, converting it into formaldehyde and removing it. In *E. coli*, some of these alkylation defense proteins are inducible as part of the adaptive response. Methylation of the Ada protein converts it into a transcriptional activator for its own gene, as well as for some other repair genes involved in repairing alkylation damage.

6. The methyl-directed mismatch repair system recognizes mismatches in DNA and removes and resynthesizes one of the two strands, restoring the correct pairing. The products of the *mutL*, *mutS*, and *mutH* genes participate in this pathway in *E. coli*. The Dam methylase product of the *dam* gene helps mark the strand to be degraded. The region including the mismatch in the newly synthesized strand is degraded and resynthesized because the A in neighboring GATC sequences on that strand has not yet been methylated by the Dam methylase.

7. The nucleotide excision repair pathway encoded by the *uvr* genes of *E. coli* removes many types of DNA damage that cause gross distortions in the DNA helix. The UvrABC endonuclease cuts on both sides of the DNA damage, and the entire oligonucleotide including the damage is removed and resynthesized.

8. Recombination repair is important for the repair of double-strand DNA breaks and gaps. Recombination repair does not actually repair damage, as described in this chapter, but instead helps the cell tolerate it. Lesions on both strands will stop DNA polymerases, resulting in gaps in the template strands as replication is initiated ahead of the damage, leaving a gap opposite the damaged bases. Recombination with the other strand can put a good strand opposite the damage, and the replication can continue. This type of repair in *E. coli* requires the recombination functions RecBCD, RecFOR, RecA, RecG, and RuvABC, as well as PriA, PriB, PriC, DnaC, and DnaT, to restart replication forks.

9. A combination of excision and recombination repair may remove interstrand cross-links in DNA. The excision repair system cuts one strand, and the single-strand break is enlarged by exonucleases to leave a gap opposite the damage. Recombination repair can then transfer a good strand to a position opposite the damage, and excision repair can remove the damage, since it is now confined to one strand. Interstrand cross-links can be repaired only if there are two or more copies of that region of the chromosome in the cell.

10. The SOS regulon includes many genes that are induced following DNA damage. The genes are normally repressed by the LexA repressor, which cleaves itself (autocleavage) after extensive DNA damage. The autocleavage is triggered by single-stranded DNA-RecA nucleoprotein filaments that accumulate following DNA damage.

11. SOS mutagenesis is due to the products of the genes *umuC* and *umuD*, which are induced following DNA damage. After induction, these proteins form a heterotrimer, $UmuD_2C$, which is not active. The UmuD and UmuC proteins can form a mutagenic polymerase called a Pol V mutasome with the aid of RecA. A RecA single-stranded DNA nucleoprotein filament promotes the autocleavage of UmuD to UmuD′, and the nucleoprotein filament also delivers RecA bound to ATP to activate the mutasome.

12. The Pol V mutasome can replicate over abasic sites and the two forms of pyrimidine dimers formed by UV irradiation, as well as some other types of DNA damage, inserting bases randomly opposite the damage and causing mutations. This polymerase is a member of a large family of translesion DNA polymerases called the Y-family polymerases, which are found in all the domains of life.

13. In *E. coli*, three polymerases are induced during the SOS response at different times. These different polymerases may provide multiple options to deal with various types of DNA damage. The regulation of this system likely allows the cell to balance mutagenic repair to enable replication of the chromosome while minimizing the accumulation of mutations.

QUESTIONS FOR THOUGHT

1. Some types of organisms, for example, yeasts, do not have methylated bases in their DNA and so would not be expected to have a methyl-directed mismatch repair system. Can you think of any other ways besides methylation that a mismatch repair system could be directed to the newly synthesized strand during replication?

2. How would you reconcile the results of Pukkila et al., using heteroduplex DNA to show that the unmethylated strand is preferentially repaired, with the speculation that GATC sequences stay in contact with the DNA Pol III holoenzyme after the replication fork has passed?

3. Why do you suppose so many pathways exist to repair some types of damage in DNA?

4. Why do you think the SOS mutagenesis pathway exists if it contributes so little to survival after DNA damage?

PROBLEMS

1. Outline how you would determine if a bacterium isolated from the gut of a marine organism at the bottom of the ocean has a photoreactivation-based DNA repair system.

2. How would you show in detail that the mismatch repair system preferentially repairs the base in a mismatch on the strand unmethylated by Dam methylase? Hint: make heteroduplex DNA of two mutants of your choice.

3. Outline how you would determine if the photoreactivating system is mutagenic.

4. Outline how you would determine if the nucleotide excision repair system of *E. coli* can repair damage due to aflatoxin B. Hint: use a *uvr* mutant.

5. How would you determine if the RecA protein plays a direct role in SOS mutagenesis independent of its role in cleaving LexA and UmuD?

6. Outline how you would determine if the *recN* gene is a member of the SOS regulon.

SUGGESTED READING

Bagg, A., C. J. Kenyon, and G. C. Walker. 1981. Inducibility of a gene product required for UV and chemical mutagenesis in *Escherichia coli*. *Proc. Natl. Acad. Sci. USA* **78**:5749–5753.

Bowles, T., A. H. Metz, J. O'Quin, Z. B. Wawrzak, and F. Eichman. 2008. Structure and DNA binding of alkylation response protein AidB. *Proc. Natl. Acad. Sci. USA* **105**:15299–15304.

Courcelle, J., A. Khodursky, B. Peter, P. O. Brown, and P. C. Hanawalt. 2001. Comparative gene expression profiles following UV exposure in wild-type and SOS-deficient *Escherichia coli* cells. *Genetics* **158**:41–64.

Darmon, E., J. K. Eykelenboom, F. Lincker, L. H. Jones, M. White, E. Okely, J. K. Blackwood, and D. R. Leach. 2010. *E. coli* SbcCD and RecA control chromosomal rearrangement induced by an interrupted palindrome. *Mol. Cell* **39**:59–70.

Deaconescu, A. M., A. L. Chambers, A. J. Smith, B. E. Nickels, A. Hochschild, N. J. Savery, and S. A. Darst. 2005. Structural basis for bacterial transcription-coupled DNA repair. *Cell* **124**:507–520.

Dodson, M. L., and R. S. Lloyd. 1989. Structure-function studies of the T4 endonuclease V repair enzyme. *Mutat. Res.* **218**:49–65.

Friedberg, E. C., G. C. Walker, W. Siede, R. D. Wood, R. A. Schultz, and T. Ellenberger. 2006. *DNA Repair and Mutagenesis*, 2nd ed. ASM Press, Washington, DC.

Foti, J. J., B. Devadoss, J. A. Winkler, J. J. Collins, and G. C. Walker. 2012. Oxidation of the guanine nucleotide pool underlies cell death by bactericidal antibiotics. *Science* **336**:315–319.

Friedberg, E. C., A. R. Lehmann, and R. P. P. Fuchs. 2005. Trading places: how do DNA polymerases switch during translesion DNA synthesis? *Mol. Cell* **18**:499–505.

Glickman, B. W., and M. Radman. 1980. *Escherichia coli* mutator mutants deficient in methylation instructed DNA mismatch correction. *Proc. Natl. Acad. Sci. USA* **77**:1063–1067.

Godoy, V. G., D. F. Jarosz, S. M. Simon, A. Abyzov, V. Ilyin, and G. C. Walker. 2007. UmuD and RecA directly modulate the mutagenic potential of the Y family DNA polymerase DinB. *Mol. Cell* **28**:1058–1070.

He, C., J. C. Hus, L. J. Sun, P. Zhou, D. P. Norman, V. Dotsch, H. Wei, J. D. Gross, W. S. Lane, G. Wagner, and G. L. Verdine. 2005. A methylation-dependent electrostatic switch controls DNA repair and transcriptional activation by *E. coli ada*. *Mol. Cell* **20**:117–129

Heinze, R. J., L. Giron-Monzon, A. Solovyova, S. L. Elliot, S. Geisler, C. G. Cupples, B. A. Connolly, and P. Friedhoff. 2009. Physical and functional interactions between *Escherichia coli* MutL and the Vsr repair endonuclease. Nucleic Acids Res. **37**:4453–4463.

Heller, R. C., and K. J. Marians. 2005. The disposition of nascent strands at stalled replication forks dictates the pathway of replisome loading during restart. *Mol. Cell* **17**:733–743.

Jiang, Q., K. Karata, R. Woodgate, M. M. Cox, and M. F. Goodman. 2009. The active form of DNA polymerase V is UmuD$'_2$C-RecA-ATP. *Nature* **460**:359–363.

Kato, T., R. H. Rothman, and A. J. Clark. 1977. Analysis of the role of recombination and repair in mutagenesis of *Escherichia coli* by UV irradiation. *Genetics* **87**:1–18.

Kato, T., and Y. Shinoura. 1977. Isolation and characterization of mutants of *Escherichia coli* deficient in induction of mutations by ultraviolet light. *Mol. Gen. Genet.* **156**:121–131.

Kohanski, M. A., D. J. Dwyer, B. Hayete, C. A. Lawrence, and J. J. Collins. 2007. A common mechanism of cellular death induced by bactericidal antibiotics. *Cell* **130:**797–810.

Korshunov, S., and J. A. Imlay. 2010. Two sources of endogenous hydrogen peroxide in *Escherichia coli. Mol. Microbiol.* **75:**1389–1401.

Landini, P., and M. R. Volkert. 2000. Regulatory responses of the adaptive response to alkylation damage: a simple regulon with complex regulatory features. *J. Bacteriol.* **182:**6543–6549.

Lieb, M., S. Rehmat, and A. S. Bhagwat. 2001. Interaction of MutS and Vsr: some dominant-negative *mutS* mutations that disable methyladenine-directed mismatch repair are active in very-short-patch repair. *J. Bacteriol.* **183:**6487–6490.

Manelyte, L, Y. I. Kim, A. J. Smith, R. M. Smith, and N. J. Savery. 2010. Regulation and rate enhancement during transcription-coupled DNA repair. *Mol. Cell* **40:**714–724.

Michaels, M. L., C. Cruz, A. P. Grollman, and J. H. Miller. 1992. Evidence that MutY and MutM combine to prevent mutations by an oxidatively damaged form of guanine in DNA. *Proc. Natl. Acad. Sci. USA* **89:**7022–7025.

Nohni, T., J. R. Battista, L. A. Dodson, and G. C. Walker. 1988. RecA-mediated cleavage activates UmuD for mutagenesis: mechanistic relationship between transcriptional derepression and posttranslational activation. *Proc. Natl. Acad. Sci. USA* **85:**1816–1820.

Pandya, G. A., I. Y. Yang, A. P. Grollman, and M. Moriya. 2000. *Escherichia coli* responses to a single DNA adduct. *J. Bacteriol.* **182:**6598–6604.

Park, J. S., M. T. Marr, and J. W. Roberts. 2002. *E. coli* transcription repair coupling factor (Mfd protein) rescues arrested complexes by promoting forward translocation. *Cell* **109:**757–767.

Pillon, M. C., J. J. Lorenowicz, K. Uckelmann, A. D. Klocko, R. R. Mitchell, Y. S. Chung, P. Modrich, G. C. Walker, L. A. Simmons, P. Friedhoff, A. Guarné. 2010. Structure of the endonuclease domain of MutL: unlicensed to cut. *Mol. Cell* **39:**145–151.

Pukkila, P. J., J. Peterson, G. Herman, P. Modrich, and M. Meselson. 1983. Effects of high levels of adenine methylation on methyl-directed mismatch repair in *Escherichia coli. Genetics* **104:**571–582.

Rupp, W. D., C. E. I. Wilde, D. L. Reno, and P. Howard-Flanders. 1971. Exchanges between DNA strands in ultraviolet irradiated *E. coli. J. Mol. Biol.* **61:**25–44.

Seigneur, M., V. Bidnenko, S. D. Ehrlich, and B. Michel. 1998. RuvAB acts at arrested replication forks. *Cell* **95:**419–430.

Sommer, S., K. Knezevic, A. Bailone, and R. Devoret. 1993. Induction of only one SOS operon, *umuDC*, is required for SOS mutagenesis in *E. coli. Mol. Gen. Genet.* **239:**137–144.

Sutton, M. D., M. F. Farrow, B. M. Burton, and G. C. Walker. 2001. Genetic interactions between the *Escherichia coli umuDC* gene products and the beta processivity clamp of the replicative DNA polymerase. *J. Bacteriol.* **183:**2897–2909.

Tang, M., X. Shen, E. G. Frank, M. O'Donnell, R. Woodgate, and M. F. Goodman. 1999. UmuD'$_2$C is an error-prone DNA polymerase *Escherichia coli* Pol V. *Proc. Natl. Acad. Sci. USA* **96:**8919–8924.

Teo, I., B. Sedgwick, M. W. Kilpatrick, T. V. McCarthy, and T. Lindahl. 1986. The intracellular signal for induction of resistance to alkylating agents in *E. coli. Cell* **45:**315–324.

Trewick, S. C., T. F. Henshaw, R. P. Hausinger, T. Lindahl, and B. Sedgwick. 2002. Oxidative demethylation by *Escherichia coli* AlkB directly reverts DNA base damage. *Nature* **419:**174–178.

Wang, F., and W. Yang. 2009. Structural insight into translesion synthesis by DNA Pol II. *Cell* **139:**1279–1289.

Weigle, J. J. 1953. Induction of mutation in a bacterial virus. *Proc. Natl. Acad. Sci. USA* **39:**628–636.

Witkin, E. M. 1956. Time, temperature and protein synthesis: a study of ultraviolet-induced mutation in bacteria. *Cold Spring Harbor Symp. Quant. Biol.* **21:**123–140.

Worth, L., Jr., S. Clark, M. Radman, and P. Modrich. 1994. Mismatch repair proteins MutS and MutL inhibit RecA-catalyzed strand transfer between diverged DNAs. *Proc. Natl. Acad. Sci. USA* **91:**3238–3241.

CHAPTER 12

Regulation of Gene Expression: Genes and Operons

THE DNA OF A CELL contains thousands to hundreds of thousands of genes, depending on whether the organism is a relatively simple single-celled bacterium or a complex multicellular eukaryote, like a human. All of the features of the organism are due, either directly or indirectly, to the products of these genes. However, all the cells of an organism do not always look or act the same, even though they share essentially the same genes. Even the cells of a single-celled bacterium can look or act differently depending on the conditions under which the cells find themselves, because the genes of a cell are not always expressed at the same levels. The process by which the expression of genes is turned on and off at different times and under different conditions is called the **regulation of gene expression.**

Cells regulate the expression of their genes for many reasons. A cell may express only the genes that it needs in a particular environment so that it does not waste energy making RNAs and proteins that are not needed at that time, or the cell may turn off genes whose products might interfere with other processes going on in the cell at the time. Cells also regulate their genes as part of developmental processes, such as sporulation. Genes can be expressed independently, as monocistronic units, or their expression can be coordinated through cotranscription in a polycistronic unit, or operon (see chapter 2).

As described in chapter 2, the expression of a gene moves through many stages, each of which offers an opportunity for regulation. First, the information in the nucleotide sequence in the DNA is copied into RNA. Even if RNA is the final gene product, as is the case for rRNAs and tRNAs, the transcript may require further processing for activity. If the final product of the gene is a protein, the mRNA synthesized from the gene might

doi:10.1128/9781555817169.ch12

have to be processed before it can be translated into protein. Protein synthesis provides another opportunity for regulatory input. The protein might have to be further processed or transported to its final location to be active. Even after the gene product is synthesized in its final form, its activity might be modulated under certain environmental conditions. The rate of degradation of the RNA or protein can also dramatically affect gene expression.

By far the most common type of regulation occurs at the first stage, when RNA is made. Genes that are regulated at this level are said to be **transcriptionally regulated**. This form of gene regulation is the most efficient, since synthesizing mRNA that will not be translated seems wasteful. However, not all genes are transcriptionally regulated, at least not exclusively. Examples in which the expression of a gene is regulated after RNA synthesis abound.

Any regulation that occurs after the gene is transcribed into mRNA is called **posttranscriptional regulation**. There are many types of posttranscriptional regulation; the most common is **translational regulation**. If a gene is translationally regulated, the mRNA may be continuously transcribed from the gene, but its translation is sometimes inhibited. Despite the apparent inefficiency of producing an mRNA that is not translated, it has advantages for certain classes of genes. Examples of a variety of mechanisms that can be used to regulate individual genes or operons are discussed in this chapter. Expression of multiple genes and operons can be coordinated in more complicated global regulatory systems, which are discussed in chapter 13.

Transcriptional Regulation in Bacteria

Thanks to the relative ease of doing genetics with bacteria, transcriptional regulation in these organisms is better understood than in other systems and has served as a framework for understanding transcriptional regulation in eukaryotic organisms. There are important differences between the mechanisms of transcriptional regulation in bacteria and eukaryotes, many of which are related to the presence of a nuclear membrane in eukaryotes and the complexity of the transcriptional machinery. Archaea are prokaryotes, like bacteria, but use a more complex transcriptional apparatus that more closely resembles that found in eukaryotes. Despite these differences, many of the strategies used are similar throughout the biological world, and many general principles have been uncovered through studies of bacterial transcriptional regulation.

As discussed in chapter 2, most transcriptional regulation occurs at the level of transcription initiation at the promoter. Transcriptional regulation occurs primarily through proteins called **transcriptional regulators**, which usually bind to DNA, often with helix-turn-helix (HTH) motifs (Box 12.1). Regulation of transcription initiation can be either negative or positive, which is defined by whether the regulatory protein turns gene expression off or on (Table 12.1). If regulation is negative, expression is controlled by a **repressor** that binds to a repressor-binding site or operator sequence in the DNA and prevents initiation of transcription by RNA polymerase. If the regulation is positive, initiation of transcription is controlled by an **activator** that is required for efficient initiation by RNA polymerase at the promoter. The activity of a regulatory protein can also be changed in response to a physiological signal, such as the binding of a small molecule. A signal that increases gene expression (either by inactivating a repressor or by activating an activator) is called an **inducer**, while a signal that decreases gene expression (by activating a repressor or inactivating an activator) is called a **corepressor**. Inducers are often used for catabolic systems, where expression of genes involved in degradation of a substrate is induced only when the substrate is available (an **inducible** system). Expression of genes encoding biosynthetic pathways is often repressed by the end product of the pathway, so that the cell expresses these genes only when the cell needs more of the product of the pathway (a repressible system). Activators, repressors, inducers, and corepressors can be used in all four combinations, as shown in Table 12.1. Examples of each of these types of systems are described below.

Some regulatory proteins can be either repressors or activators, depending on where they bind to the promoter region. If the binding site overlaps the binding site for RNA polymerase to the promoter, the protein might sterically inhibit (i.e., physically get in the way of) the binding of RNA polymerase to the promoter and repress

Table 12.1 Transcriptional regulatory systems

Effector molecule	Regulatory protein	
	Negative (repressor)	Positive (activator)
Inducer	Negative inducible (*E. coli lac* operon; LacI)	Positive inducible (*E. coli ara* operon; AraC)
Corepressor	Negative repressible (*E. coli trp* operon; TrpR)	Positive repressible (*E. coli fab* operon; FadR)

BOX 12.1

The Helix-Turn-Helix Motif of DNA-Binding Proteins

Proteins that bind to DNA, including repressors and activators, often share similar structural motifs determined by the interaction between the protein and the DNA helix. One such motif is the HTH motif. A region of approximately 7 to 9 amino acids forms an α-helical structure called helix 1. This region is separated by about 4 amino acids from another α-helical region of 7 to 9 amino acids called helix 2. The two helices are at approximately right angles to each other, hence the name HTH. When the protein binds to DNA, helix 2 lies in the major groove of the DNA double helix while helix 1 lies crosswise to the DNA, as shown in the figure. Because they lie in the major groove of the DNA double helix, the amino acids in helix 2 can contact and form hydrogen bonds with specific bases in the DNA. Thus, a DNA-binding protein containing an HTH motif recognizes and binds to specific regions on the DNA. Many DNA-binding proteins exist as dimers and bind to inverted-repeat DNA sequences. In such cases, the two polypeptides in the dimer are arranged head to tail so that the amino acids in helix 2 of each polypeptide can make contact with the same bases in the inverted repeats.

In the absence of structural information, the existence of an HTH motif in a protein can often be predicted from the amino acid sequence, since some sequences of amino acids cause the polypeptide to assume an α-helical form and the bent region between the two helices usually contains a glycine. The presence of an HTH motif in a protein helps to identify it as a DNA-binding protein.

A variant on the HTH domain is the winged HTH domain, in which the "winged turn" is 10 amino acids or more in length, longer than the 3 or 4 amino acids of the "turns" of other HTH domains (see Kenney, References).

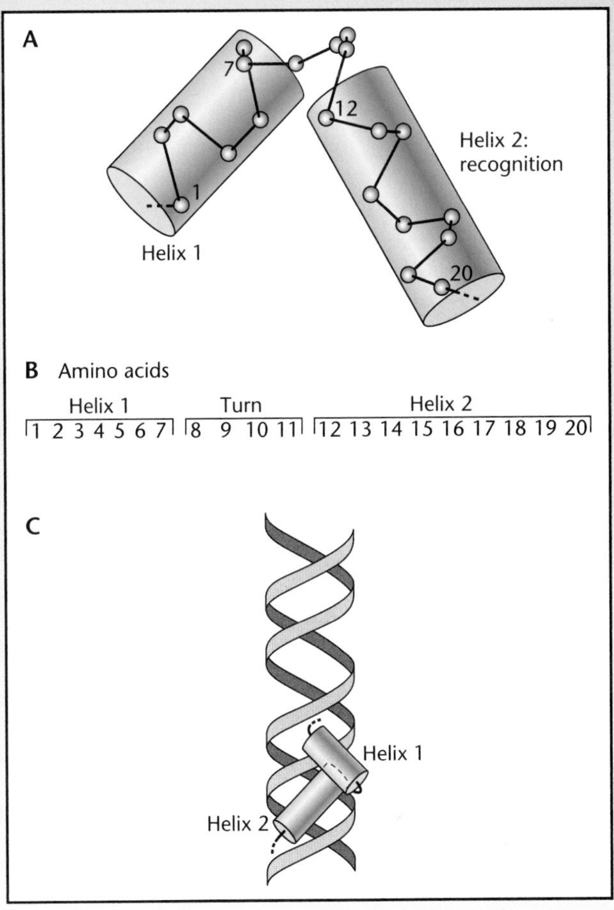

Figure 1 The HTH motif of DNA-binding proteins. (A) The structure of an HTH domain. (B) Number of amino acids in the HTH domain of the CAP protein. (C) Interactions of helix 1 and helix 2 with double-stranded DNA.
doi:10.1128/9781555817169.ch12.Box12.1.f

References

Kenney, L. J. 2002. Structure/function relationships in OmpR and other winged-helix transcription factors. *Curr. Opin. Microbiol.* **5**:135–141.

Steitz, T. A., D. H. Ohlendorf, D. B. McKay, W. F. Anderson, and B. W. Matthews. 1982. Structural similarity in the DNA-binding domains of catabolite gene activator and *cro* repressor proteins. *Proc. Natl. Acad. Sci. USA* **79**:3097–3100.

transcription. However, if it binds further upstream, it might make contacts with RNA polymerase that stabilize the binding of RNA polymerase to the promoter and activate transcription. Binding of a regulatory protein can have different effects on different promoters, depending on the characteristics of the promoter and its interactions with RNA polymerase.

While transcription initiation is a common level of gene regulation, transcription can also be regulated after RNA polymerase leaves the promoter. This can occur by premature termination of transcription, or **transcription attenuation.** Changes in the processivity of RNA polymerase can also be used to regulate transcription elongation, for example, by affecting the probability of termination at Rho-dependent terminators. As is the case for regulation of transcription initiation, transcription elongation and termination effects can be mediated by regulatory proteins and other factors that interact with the nascent RNA transcript. These factors can act either negatively (to repress gene expression) or positively (to activate gene expression). One of the most important initial steps in investigating the mechanism of gene regulation is to determine whether regulation is positive or negative (or even involves a combination of mechanisms).

Genetic Evidence for Negative and Positive Regulation

Negatively and positively regulated genes or operons behave very differently in genetic tests. One difference is in the effect of mutations that inactivate the regulatory gene. In **negative regulation**, a mutation that inactivates the regulatory gene allows transcription of the target genes, even in the absence of whatever physiological signal normally turns on expression. If the regulation is positive, mutations that inactivate the regulatory gene prevent transcription of the target genes, even in the presence of the regulatory signal, giving an **uninducible phenotype**. A mutant in which the target genes are always transcribed, even in the absence of the inducing signal, is called a **constitutive mutant**. Constitutive mutations are much more common with negatively regulated systems than with positively regulated systems because any mutation that inactivates the repressor will result in the constitutive phenotype. With positively regulated systems, a constitutive phenotype can be caused only by mutational changes that do not inactivate the activator protein but alter it so that it can activate transcription without the inducing signal. Such changes tend to be rare and may not be possible.

Complementation tests reveal another difference between negatively and positively regulated systems. Constitutive mutations of a negatively regulated system are often recessive to the wild type (i.e., the phenotype of the mutant is masked if the cell contains a second wild-type copy of the regulatory gene [see chapter 3 for genetic definitions]). This is because any normal repressor protein in the cell encoded by a wild-type copy of the gene binds to the operator and blocks transcription, even if the repressor encoded by the mutant copy of the gene in the same cell is inactive. In contrast, constitutive mutations in a positively regulated operon are often dominant to the wild type. A mutant activator protein that is active without the inducing signal might activate transcription even in the presence of a wild-type activator protein. In the following sections, we describe some examples of transcriptional regulatory systems and how genetic evidence contributed to our understanding of these systems.

Negative Regulation of Transcription Initiation

By definition, a negative regulatory system uses a repressor to turn off gene expression. A repressor can prevent initiation of transcription in a number of ways. The operator sequence may overlap with the promoter sequence so that binding of the repressor prevents binding of the RNA polymerase to the promoter. Alternatively, or in addition, the repressor might bend the promoter DNA so that RNA polymerase can no longer bind. The repressor might also hold the RNA polymerase on the promoter so that it cannot leave as it attempts to make RNA (see Rojo, Suggested Reading). The activity of the repressor can in turn be controlled by a physiological signal, often a small molecule. If the signal inactivates the repressor and therefore turns on expression of the system, then the system is inducible. If the signal activates the repressing activity of the repressor and turns off gene expression, the system is repressible.

Negative Inducible Systems

Negative inducible systems are systems in which expression of the regulated genes is turned off by the action of a repressor protein. The effector turns on gene expression (and is therefore an inducer) by inactivating the repressor. This type of regulatory mechanism is common for catabolic operons, where the inducer is often the substrate for the operon. The cell therefore needs the regulated gene products, which are involved in utilization of the substrate, only when the substrate is available. Negative inducible systems were the first type of regulatory system to be characterized and serve as a general paradigm for gene regulation. We will use the *Escherichia coli lac* and *gal* operons as examples.

THE *E. coli lac* OPERON

The classic example of a negative inducible system is provided by the *E. coli lac* operon, which encodes the enzymes responsible for the utilization of the sugar lactose. The experiments of François Jacob, Jacques Monod, and their collaborators on the regulation of the *E. coli lac* genes are excellent examples of the genetic analysis of a biological phenomenon in bacteria (see Jacob and Monod, Suggested Reading). Although these experiments were performed in the late 1950s, only shortly after the discovery of the structure of DNA and the existence of mRNA, they still stand as the paradigm to which all other studies of gene regulation are compared. It is also important to note that while we refer to the structure of the *lac* operon throughout this section, Jacob and Monod had no such information availalable to guide their work.

Mutations of the *lac* Operon

When Jacob and Monod began their classic work, it was known that the enzymes of lactose metabolism are inducible in that they are expressed only when the sugar lactose is present in the medium. If no lactose is present, the enzymes are not made. From the standpoint of the cell, this is a sensible strategy, since there is no point in making the enzymes for lactose utilization unless lactose is available for use as a carbon and energy source.

To understand the regulation of the lactose genes, Jacob and Monod performed a genetic analysis (see chapter 3). First, they isolated many mutants in which lactose metabolism and regulation were affected. These mutants fell into two fundamentally different groups. Some mutants were unable to grow with lactose as the sole carbon and energy source and so were called Lac$^-$ mutants. Other mutants made the lactose-metabolizing enzymes whether or not lactose was present in the medium and so were called constitutive mutants. Note that the fact that many constitutive mutants were isolated provided the first clue that regulation is in fact negative.

Complementation Tests with *lac* Mutations

Jacob and Monod needed to know which of the mutations affected *trans*-acting gene products—either protein or RNA—involved in the regulation and how many different genes these mutations represented. They also wished to know if any of the mutations were *cis* acting (affecting sites on the DNA involved in regulation).

To answer these questions, they first determined whether a particular *lac* mutation is dominant or recessive, which required that the organisms be diploid, with two copies of the genes being tested. Bacteria are normally haploid, with only one copy of each of their genes, but are "partial diploids" for any genes carried on an introduced F′ plasmid. Recall that an F′ plasmid is a plasmid into which some of the bacterial chromosomal genes have been inserted (see chapter 5). They introduced an F′ plasmid carrying the wild-type *lac* region into a strain with the *lac* mutation in the chromosome. If the partial diploid bacteria are Lac$^+$ and can multiply to form colonies on minimal plates with lactose as the sole carbon and energy source, the *lac* mutation is recessive. If the partial diploid cells are Lac$^-$ and cannot form colonies on lactose minimal plates, the *lac* mutation is dominant. Jacob and Monod discovered that most *lac* mutations are recessive to the wild type and so presumably inactivate genes whose products are required for lactose utilization.

The question of how many genes are represented by recessive *lac* mutations could be answered by performing pairwise complementation tests between different *lac* mutations. F′ plasmids carrying the *lac* region with one *lac* mutation were introduced into a mutant strain with another *lac* mutation in the chromosome (Figure 12.1). In this kind of experiment, if the partial diploid cells are Lac$^+$, the two recessive mutations can complement each other and are members of different complementation groups or genes. If the partial diploid cells are Lac$^-$, the two mutations cannot complement each other and are members of the same complementation group or gene. Jacob and Monod found that most of the *lac* mutations sorted into two different complementation groups, which they named *lacZ* and *lacY*. We now know of a third gene, *lacA*, which was not discovered in their original selections because its product is not required for growth on lactose or its regulation.

cis-acting lac *mutations.* Not all Lac– mutants have *lac* mutations that affect diffusible gene products and can be complemented. Immediately adjacent to the *lacZ* mutations are other *lac* mutations that are much rarer and have radically different properties. These mutations cannot be complemented to allow the expression of the *lac* genes on the same DNA, even in the presence of wild-type copies of the *lac* genes. Mutations that cannot be complemented are *cis* acting and presumably affect a site on DNA rather than a diffusible gene product like an RNA or protein (see chapter 3).

To show that a *lac* mutation is *cis* acting, i.e., affects only the expression of genes on the same DNA where it occurs, we could introduce an F′ plasmid containing the potential *cis*-acting *lac* mutation into cells containing either a *lacZ* or *lacY* mutation in the chromosome (Figure 12.2). Any *trans*-acting gene products encoded by the *lacZ* or *lacY* gene on the F′ plasmid would complement the chromosomal *lacY* or *lacZ* mutation, respectively. However, if a Lac$^-$ phenotype resulted, the *lac* mutation in the F′ plasmid must prevent synthesis of both LacZ and LacY proteins from the F′ plasmid. The mutation in the F′ plasmid is therefore *cis* acting.

As discussed below, Jacob and Monod named one type of the *cis*-acting *lac* mutations "*lacp* mutations" and hypothesized that they affect the binding of RNA polymerase to the beginning of the gene, i.e., are mutations

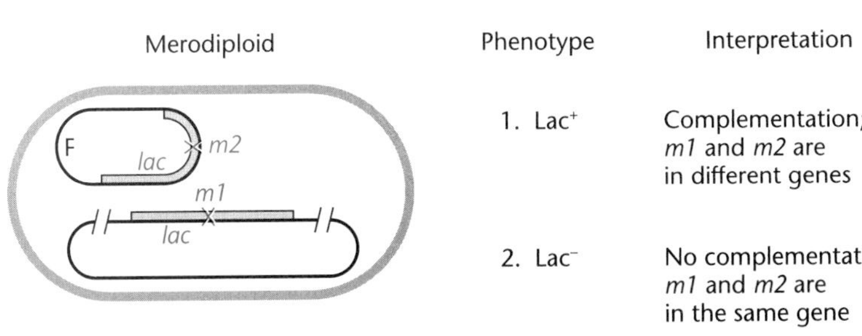
Merodiploid
Phenotype
Interpretation
F
lac
m2
m1
lac
1. Lac+
Complementation; m1 and m2 are in different genes
2. Lac−
No complementation; m1 and m2 are in the same gene

Figure 12.1 Complementation of two recessive *lac* mutations. One mutation (*m1*) is in the chromosome, and the other (*m2*) is in an F′ plasmid. If the two mutations complement each other, the cells will be Lac$^+$ and will grow with lactose as the sole carbon and energy source. The mutations will not complement each other if they are in the same gene or if one affects a regulatory site or is polar. See the text for more details.
doi:10.1128/9781555817169.ch12.f12.1

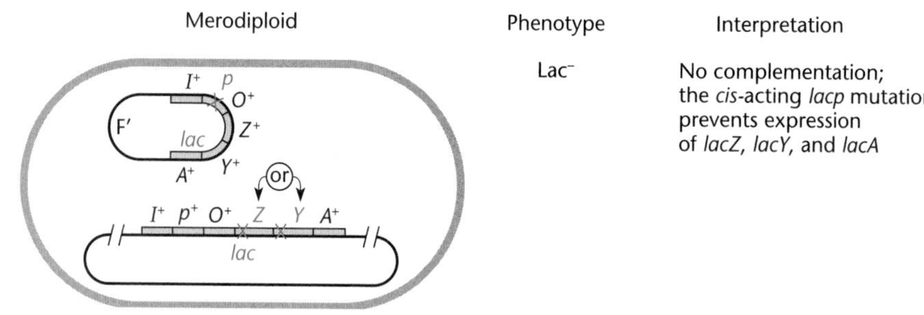

Figure 12.2 The *lacp* mutations cannot be complemented and are *cis* acting. A *lacp* mutation in the F′ plasmid prevents the expression of any of the other *lac* genes on F′, so a *lac* mutation in the chromosome will not be complemented. Partial diploid cells will be Lac$^-$. Mutated *lac* regions are shown in blue.
doi:10.1128/9781555817169.ch12.f12.2

in the promoter region. We now know that many of these mutations are actually strong polar mutations in the beginning of the *lacZ* gene that also prevent the transcription of the downstream *lacY* gene (see below).

Lac$^-$ mutants with dominant mutations. Some Lac$^-$ mutants have mutations that affect diffusible gene products but are dominant rather than recessive. A dominant *lac* mutation makes the cell Lac$^-$ and unable to use lactose even if there is another good copy of the *lac* operon in the cell, either in the chromosome or in the F′ plasmid. These dominant *lac* mutations are called *lacI*^{s} mutations, for "superrepressor mutations." As shown below, these mutants have mutations that change the repressor so that it can no longer bind the inducer.

Complementation tests with constitutive mutations. As mentioned above, some *lac* mutations do not make the cells Lac$^-$, but rather, make them constitutive, so that they express the *lacZ* and *lacY* genes even in the absence of the inducer lactose. In complementation tests between constitutive mutations, partial diploids are made in which either the chromosome or the F′ plasmid, or both, carry a constitutive mutation. The partial diploid cells are then tested to determine whether they express the *lac* genes constitutively or only in the presence of the inducer. If the partial diploid cells express the *lac* gene in the absence of the inducer, the constitutive mutation is dominant. However, if the partial diploid cells express the *lac* genes only in the presence of the inducer, the mutations are recessive. Using this test, Jacob and Monod found that some of the constitutive mutations, which could be recessive or dominant, affect a *trans*-acting gene product, either protein or RNA. Complementation between the recessive constitutive mutations revealed that they are all in the same gene, which these investigators named *lacI* (Figure 12.3A).

cis-acting lacO^c mutations. A rarer constitutive mutation in this and related regulatory systems is *cis* acting, allowing constitutive expression of the *lacZ* and *lacY* genes from the DNA that has the mutation, even in the

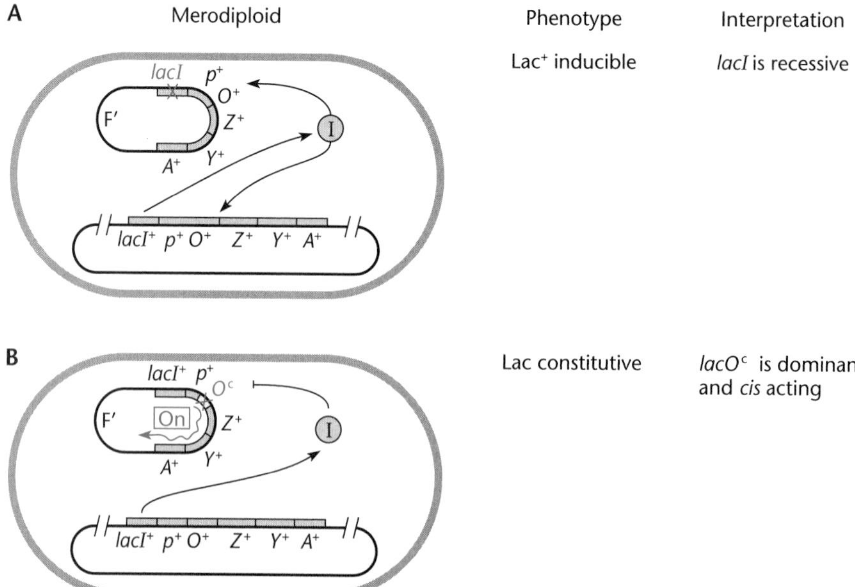

Figure 12.3 Complementation with two types of constitutive mutations. **(A)** The *lacI* mutation can be complemented, so other genes on the F′ plasmid are inducible in the presence of a wild-type copy of the *lacI* gene. **(B)** In contrast, *lacO*^{c} mutations cannot be complemented by a wild-type *lacO* region in the chromosome and so are *cis* acting. Mutated regions are shown in blue.
doi:10.1128/9781555817169.ch12.f12.3

Table 12.2 *lac* operon mutations

Mutation	Function affected	Phenotype
lacZ	β-Galactosidase	Lac⁻, recessive, *trans* acting
lacI	Operator binding	Constitutive, recessive, *trans* acting
lacI^{s}	Inducer binding	Uninducible, dominant, *trans* acting
lacI^{-d}	Operator binding	Constitutive, dominant, *trans* acting
lacI^{q}	Quantity of LacI	Inducible, dominant, *trans* acting
lacO^{c}	Repressor binding	Constitutive, dominant, *cis* acting
lacP	RNA polymerase binding	Uninducible, recessive, *cis* acting

presence of a wild-type copy of the *lac* DNA. Jacob and Monod named these *cis*-acting constitutive mutations *lacO*^{c} mutations, for *lac* operator-constitutive mutations. Figure 12.3B shows the partial diploid cells used in these complementation tests.

trans-acting dominant constitutive mutations. Some *lacI* mutations, called *lacI*$^{-d}$ mutations, are *trans* dominant, making the cell constitutive for expression of the *lac* operon even in the presence of a good copy of the *lacI* gene. These *lacI*$^{-d}$ mutations are possible because the LacI polypeptides form a homotetramer (a complex of four copies of the LacI protein). A mixture of normal and defective subunits can be nonfunctional, causing the constitutive LacI^{c} phenotype. Hence, the *lacI*$^{-d}$ mutations are *trans* dominant, because one defective subunit in the tetramer results in an inactive complex. Table 12.2 summarizes the behavior of the various *lac* mutations in complementation tests.

The Jacob and Monod Operon Model

On the basis of this genetic analysis, Jacob and Monod proposed their **operon model** for *lac* gene regulation (Figure 12.4). The *lac* operon includes the genes *lacZ* and *lacY* (remember that the third gene, *lacA*, was identified later). These genes, known as the **structural genes** of the operon, encode the enzymes required for lactose utilization. The *lacZ* gene product is a β-galactosidase that cleaves lactose to form glucose and galactose, which can then be used by other pathways. The *lacY* gene product is a permease that allows lactose into the cell. The *lacA* gene product is a transacetylase whose function is still unclear.

The operon model explained why the structural genes are expressed only in the presence of lactose. The product of the *lacI* gene is a repressor protein. In the absence of lactose, this repressor binds to the operator sequence (*lacO*) close to the promoter and thereby prevents binding of RNA polymerase to the promoter and blocks the transcription of the structural genes. In contrast, when lactose is available, the inducer binds to the repressor and changes its conformation so that it can no longer bind to the operator sequence. RNA polymerase can then bind to *lacp* and transcribe the *lacZ*, *lacY*, and *lacA* genes. The LacI repressor is very effective at blocking transcription of the structural genes of the operon. Transcription is about 1,000 times more active in the absence of repressor than in its presence.

It is worth emphasizing how the Jacob and Monod operon model explains the behavior of mutations that affect the regulation of the *lac* enzymes. Mutants with *lacZ* and *lacY* mutations are Lac⁻ because they do not make an active β-galactosidase or permease, respectively, both of which are required for lactose utilization. These mutations are clearly *trans* acting, because they are recessive and can be complemented. An active β-galactosidase or permease made from another DNA in the same cell can provide the missing enzyme and allow lactose utilization.

Figure 12.4 The Jacob and Monod model for negative regulation of the *lac* operon. In the absence of the inducer, lactose, the LacI repressor binds to the operator region, preventing transcription of the other genes of the operon by RNA polymerase (RNA Pol). In the presence of lactose, the repressor can no longer bind to the operator, allowing transcription of the *lacZ, lacY,* and *lacA* genes. It was later determined that the true inducer is allolactose, a metabolite of lactose, rather than lactose.
doi:10.1128/9781555817169.ch12.f12.4

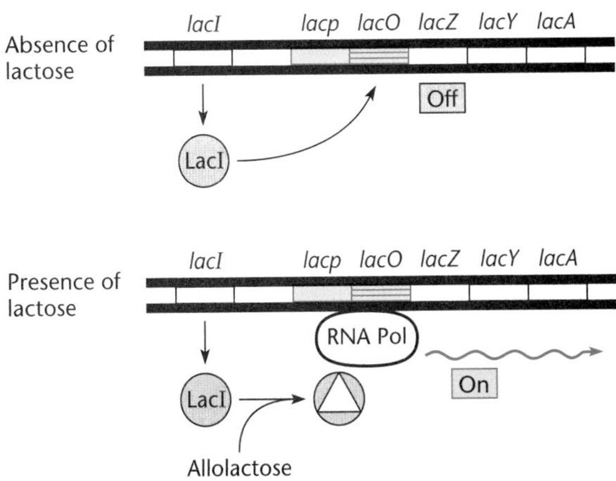

The behavior of *lacp* mutations is also explained by the model. Jacob and Monod proposed that *lacp* mutations change the sequence on the DNA to which the RNA polymerase binds. This explains why *lacp* mutations are *cis* acting; if the site on DNA at which RNA polymerase initiates transcription is changed by a mutation so that it no longer binds RNA polymerase, the *lacZ*, *lacY*, and *lacA* genes on that DNA are not transcribed into mRNA, even in the presence of a good copy of the *lacp* region elsewhere in the cell.

Their model also explains the behavior of the two constitutive mutations: *lacI* (*trans* acting) and *lacO*^{c} (*cis* acting). The *lacI* mutations affect a *trans*-acting function because they inactivate the repressor protein that binds to the operator and prevents transcription. The LacI repressor made from a functional copy of the *lacI* gene anywhere in the cell can bind to the operator sequence and block transcription in *trans*. However, the *lacO*^{c} mutations change the sequence on DNA to which the LacI repressor binds to block transcription. The LacI repressor cannot bind to this altered *lacO* sequence, even in the absence of lactose. Therefore, RNA polymerase is free to bind to the promoter and transcribe the structural genes. The *lacO*^{c} mutations are *cis* acting because they allow constitutive expression of the *lacZ*, *lacY*, and *lacA* genes from the same DNA, even in the presence of a good copy of the *lac* operon elsewhere in the cell.

The existence of superrepressor *lacI*^{s} mutations is also explained by their model. These are mutations that change the repressor molecule so that it can no longer bind the inducer. The mutated repressor binds to the operator even in the presence of inducer, making the cells permanently repressed and phenotypically Lac$^-$. The fact that this type of mutation is dominant over the wild type is also explained. The mutated repressors repress the transcription of any *lac* operon in the same cell, even in the presence of inducer, so they make the cell Lac$^-$ even in the presence of a good *lacI* gene, either in the chromosome or on an F′ plasmid.

The *lac* genes provide a good example of what is meant by "operon." An operon includes all the genes that are transcribed into the same mRNA plus any adjacent *cis*-acting sites that are involved in the transcription or regulation of transcription of the genes. The *lac* operon of *E. coli* consists of the three structural genes, *lacZ*, *lacY*, and *lacA*, which are transcribed into the same mRNA, as well as the *lac* promoter from which these genes are transcribed. It also includes the *lac* operator, since it is a *cis*-acting regulatory sequence involved in regulating the transcription of the structural genes. However, the *lac* operon does not include the gene for the repressor, *lacI*. The *lacI* gene is adjacent to the *lacZ*, *lacY*, and *lacA* genes and regulates their transcription, but it is not transcribed onto the same mRNA as the structural genes. Moreover, its product is *trans* acting rather than *cis* acting.

Refinements to the Regulation of the *lac* Operon

The operon model of Jacob and Monod has survived the passage of time. In 1965, it earned them a Nobel Prize, which they shared with André Lwoff. Because of its elegant simplicity, the operon model for the regulation of the *lac* genes of *E. coli* serves as the paradigm for understanding gene regulation in other organisms. The *lac* genes and *cis*-acting sites also have many uses in the molecular genetics of all organisms. They have been introduced into many other types of organisms, where they are used to study many aspects of gene regulation and developmental and cell biology.

While still largely intact, the operon model has undergone a few refinements over the years. As mentioned above, Jacob and Monod did not know of the existence of the *lacA* gene and thought that *lacY* encoded the transacetylase rather than the permease, which was unknown at the time. Also, most of the mutations that Jacob and Monod defined as *lacp* were not in fact promoter mutations but instead were strong polar mutations in *lacZ* that prevent the transcription of all three structural genes (*lacZ*, *lacY*, and *lacA*). Later studies also revealed that the true inducer that binds to the LacI repressor is allolactose, a metabolite of lactose, rather than lactose. In most experiments, an analog of allolactose called isopropyl-β-D-thiogalactopyranoside (IPTG) is used as the inducer because it is not metabolized by the cells.

The most significant refinement of the Jacob and Monod operon model came from the discovery that the LacI repressor can bind to not just one but three operators, called o_1, o_2, and o_3 (Figure 12.5). The operator closest to the promoter, o_1, seems to be the most important for repressing the transcription of *lac* expression and acts by sterically interfering with the binding of RNA polymerase to the promoter, because the sequence is centered at position +11 relative to the transcription start point (+1) and therefore overlaps with the position occupied by RNA polymerase during transcription initiation. Deletion of either o_2 or o_3 has little effect on repression. However, deleting both o_2 and o_3 diminishes repression as much as 50-fold.

Why does the *lac* operon have more than one operator, especially since these operators are so far from the RNA polymerase binding site that it seems unlikely that they could block binding of the RNA polymerase to the promoter? The purpose of having more than one operator is so that the same LacI repressor molecule can bind to two operators simultaneously, bending the DNA—and the promoter—between them to form a DNA loop (Figure 12.5B). This DNA loop stabilizes the interaction of LacI with the DNA. As discussed below, many other operons,

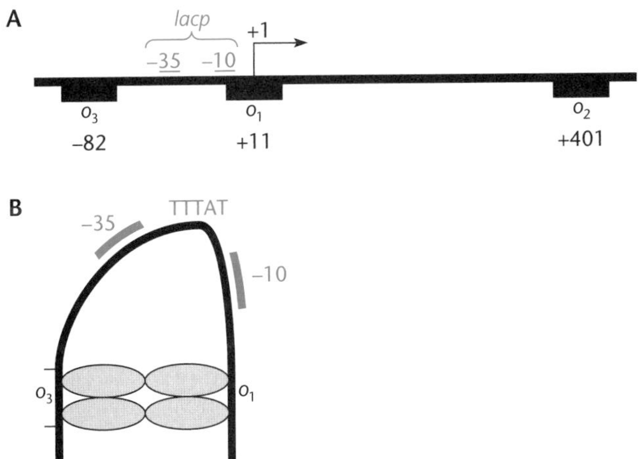

Figure 12.5 Locations of the three operators in the *lac* operon **(A)** and a model for how binding of the tetrameric LacI repressor to two of these operators may help repress the operon **(B)**. Repressor (solid ellipses) bound to o_1 and o_3, or o_1 and o_2 (not shown), can bend the DNA in the promoter region and stabilize binding of LacI to the DNA to help prevent binding of RNA polymerase to the promoter. The AT-rich region may facilitate bending.
doi:10.1128/9781555817169.ch12.f12.5

including the *gal* and *ara* operons, also contain multiple operators that promote DNA looping in the promoter region. In the *lac* system, formation of the DNA loop assists in repression but is not required for regulation to occur. New technologies, including single-molecule studies that allow visualization of interactions of LacI and RNA polymerase on an individual template DNA molecule, are likely to provide new insights into the molecular mechanisms that underlie transcriptional regulation in this system (see Sanchez et al., Suggested Reading).

Catabolite Regulation of the *lac* Operon

In addition to being under the control of its own specific repressor, the *lac* operon is regulated in response to the availability of other carbon sources through **catabolite repression**. Catabolic pathways are used to break down substrates to yield carbon or energy. The catabolite repression system ensures that the genes for lactose utilization are not expressed if a better carbon and energy source, such as glucose, is available. The name "catabolite repression" is a misnomer, at least in *E. coli*, since the expression of operons under catabolite control in *E. coli* requires a transcriptional activator, the catabolite activator protein (CAP, historically called Crp, for catabolite repression protein), and the small-molecule effector cyclic AMP (cAMP). Many operons are under the control of CAP-cAMP, and we defer a detailed discussion of the mechanism of **catabolite regulation** until chapter 13, since it is a type of global regulation.

Structure of the *lac* Control Region

Figure 12.6 illustrates the structure of the *lac* control region in detail, showing the nucleotide sequences of the *lac* promoter and one of the operators (o_1), as well as the region to which CAP binds. The *lac* promoter is a typical σ^{70} bacterial promoter with characteristic –10 and –35 regions (see chapter 2). One of the operators to which the LacI repressor binds (o_1) overlaps the transcription start site (+1 in Figure 12.6A) for transcription of *lacZ*, *lacY*, and *lacA*. The other *lac* operator sequences are positioned upstream and downstream of the promoter, and they are shown in Figure 12.6B. Each symmetrical half of an operator binds a LacI monomer. Simultaneous binding of the LacI tetramer to two operator sites (e.g., o_1 and o_3, as shown in Figure 12.5B) allows each of the four subunits of the tetramer to interact with an operator half-site. The CAP-binding site, which enhances binding of RNA polymerase in the absence of glucose, is just upstream of the promoter, as shown.

Locations of *lacI* Mutations in the Three-Dimensional Structure of the LacI Repressor

When the *lacI* gene was sequenced, the exact amino acid changes in the *lac* repressor due to some of the mutations described above (and listed in Table 12.2) could be identified, which provided information about the regions of the LacI protein that are involved in the various functions of the repressor. For example, the *lacI*^{s} mutations should be largely confined to the site on the protein where the inducer binds, allowing the site to be identified. Likewise, *lacI*$^{-d}$ mutations should not inactivate the regions involved in dimer or tetramer formation, since the mutant proteins must form mixed dimers or tetramers with the wild-type polypeptide to be dominant. We would expect that constitutive *lacI*$^{-}$ mutations would be scattered around the protein, since they could affect almost any activity of the protein, including its ability to bind to DNA and its ability to fold into its final conformation. However, it was found that all of the various mutation types are often scattered around the gene and are not always concentrated in a certain region. Many years later, when there was a three-dimensional structure of the protein, the distribution of the mutations was easier to understand, since amino acids that are not in the same region in the linear polypeptide might be close to each other in the folded protein.

The LacI protein was crystallized, and its three-dimensional structure was determined by X-ray diffraction (see Lewis et al., Suggested Reading). Nuclear magnetic resonance spectroscopy was also used to determine its interaction with the *lac* operators and how this structure changes when the inducer IPTG is bound. Figure 12.7 shows the structure of the LacI repressor and the regions affected by amino acid changes due to some

A The *lac* operon regulatory region

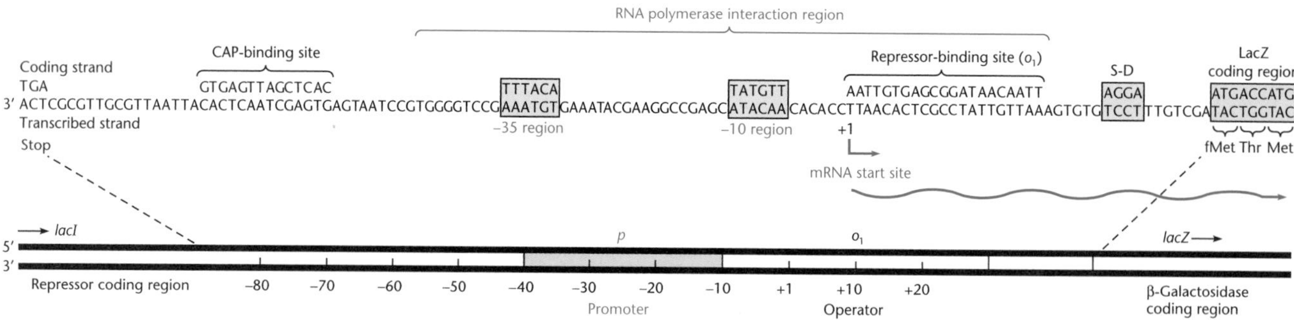

B The *lac* operator sequences

o_1 5′ AATTGTGAGCGGATAACAATT 3′
o_2 5′ AAaTGTGAGCGAGTAACAAcc 3′
o_3 5′ ggcaGTGAGCGcAacgCAATT 3′

Symmetrical operator halves

Figure 12.6 (A) DNA sequence of the promoter and operator regions of the *lac* operon. The entire region is only 100 bp long. Only the position of the o_1 operator sequence is shown. o_2 is centered at +401, in the LacZ coding region; o_3 is centered at position −82, upstream of the CAP-binding site. **(B)** Alignment of the three natural *lac* operator sequences. Nucleotides of o_3 and o_2 that do not match o_1 are shown in lowercase. doi:10.1128/9781555817169.ch12.f12.6

types of mutations. For example, the amino acid changes due to *lacI*[s] mutations are often in the inducer-binding pocket (see Pace et al., Suggested Reading). The spacing between the DNA-binding N-terminal domains changes when inducer is bound, preventing its binding to the operator. An interaction in which inducer binding in one region of the protein can promote changes in another region is called an **allosteric interaction.** Some *lacI*[s] mutations also change amino acids in the allosteric signaling region, the region that signals that inducer is bound to the DNA-binding domains in the inducer-binding pocket. This region doubles as the dimerization domain, which helps hold two monomers together, and changing the orientation of the two monomers may be part of the signal. The way in which some other repressors interact with their operators and how this interaction changes with the binding of inducers are discussed later in this chapter in the sections on the *gal* and *trp* repressors (see "The *E. coli gal* Operon" and "The *E. coli trp* Operon").

Experimental Uses of the *lac* Operon

The *lac* genes and regulatory regions have found many uses in molecular genetics (see Shuman and Silhavy, Suggested Reading). For example, the *lacZ* gene is probably the most widely used reporter gene (see chapter 2) and has been introduced into a wide variety of different organisms ranging from bacteria to fruit flies to human cells. It is so popular because its product, β-galactosidase, is easily detected by colorimetric assays using substrates such as X-Gal (5-bromo-4-chloro-3-indolyl-β-D-galactopyranoside) and ONPG (*o*-nitrophenyl-β-D-galactopyranoside). Also, the N-terminal portion of the protein is nonessential for activity, making it easier to make translation fusions. The only disadvantage of *lacZ* as a reporter gene is that its polypeptide product is very large, which can be a disadvantage in some types of expression systems.

The *lac* promoter or its derivatives are used in many expression vectors (see chapter 4). This promoter offers many advantages in these expression vectors. It is fairly strong, allowing high levels of transcription of a cloned gene. It is also inducible, which makes it possible to clone genes whose products are toxic to the cell. The cells can be grown in the absence of the inducer IPTG so that the cloned gene is not transcribed. When the cells reach a high density, the inducer IPTG is added and the cloned gene is expressed. Even if the protein is toxic and kills the cell when made in such large amounts, enough of the protein is usually synthesized before the cell dies.

There are many derivatives of the *lac* promoter in use. These derivatives retain some of the desirable properties of the wild-type *lac* promoter but have additional features. For example, the mutated *lac* promoter *lacUV5* is no longer sensitive to catabolite repression and so is active even if glucose is present in the medium (see chapter 13). A hybrid *trp-lac* promoter called the *tac* promoter

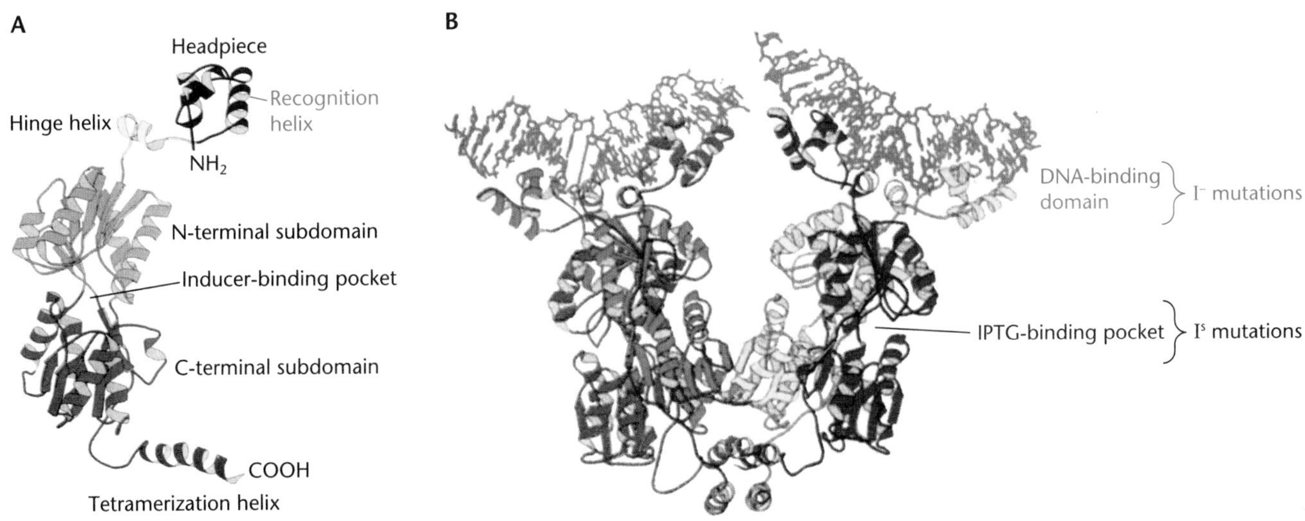

Figure 12.7 Three-dimensional structure of the LacI protein, showing the monomer and some regions devoted to its various functions, as well as the tetramer formed by two dimers, and how it interacts with *lac* operators on the DNA. Also shown are the sites of some types of mutational changes mentioned in the text. doi:10.1128/9781555817169.ch12.f12.7

has also been widely used. The *tac* promoter has the advantages that it is stronger than the *lac* promoter and is insensitive to catabolite repression but still retains its inducibility by IPTG.

Because it binds so tightly to its operator sequences, the LacI repressor protein also has many uses. One current use is to locate regions of DNA in the cell in bacterial cell biology. In one application, the LacI protein is translationally fused to green fluorescent protein, which can be detected in the cell by fluorescence microscopy (see chapter 14). If this fusion protein is expressed in a cell in which multiple copies of the operator sequence have been introduced into a region of the chromosome, for example, close to the origin of replication, the fusion protein binds in multiple copies to the origin region of the chromosome. The location of the origin region of the chromosome can then be tracked in the cell as it goes through its cell cycle by monitoring the fluorescence given off by the fusion protein. The same system can be used to experimentally stall DNA replication forks to analyze mechanisms for chromosome organization (see Possoz et al., Suggested Reading).

THE *E. COLI gal* OPERON

The *gal* operon of *E. coli*, which is involved in the utilization of the sugar galactose, is another classic example of negative regulation. Figure 12.8 shows the organization of the genes in this operon. Transcription initiates from two promoters, p_{G1} and p_{G2}. The products of three structural genes, *galE*, *galT*, and *galK*, are required for the conversion of galactose into glucose, which can then enter the glycolysis pathway.

The specific reactions catalyzed by each of the enzymes of the *gal* pathway appear in Figure 12.9. The *galK* gene product is a kinase that phosphorylates galactose to make galactose-1-phosphate. The product of the *galT* gene is a transferase that transfers the galactose-1-phosphate to UDP-glucose, displacing the glucose to make UDP-galactose. The released glucose can then be used as a carbon and energy source. The GalE gene product is an epimerase that converts UDP-galactose to UDP-glucose to continue the cycle. It is not clear why *E. coli* cells use this seemingly convoluted pathway to convert

Figure 12.8 Structure of the galactose operon of *E. coli.* The *galE, galT,* and *galK* genes are transcribed from two promoters, p_{G1} and p_{G2}. CAP with cAMP bound turns on p_{G1} and turns off p_{G2}, as shown. There are also two operators, o_E and o_I. The repressor genes are some distance away, as indicated by the broken line. Only the *galR* repressor gene is shown. doi:10.1128/9781555817169.ch12.f12.8

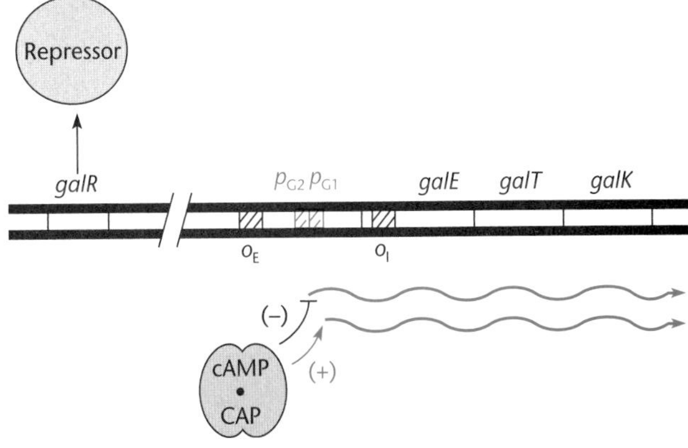

Galactose + ATP —GalK→ Galactose-1-PO_4 + ADP

Galactose-1-PO_4 + UDP-glucose —GalT→ UDP-galactose + glucose-1-PO_4

UDP-galactose —GalE→ UDP-glucose

UTP + glucose-1-PO_4 —GalU→ UDP-glucose

Figure 12.9 Pathway for galactose utilization in *E. coli.*
doi:10.1128/9781555817169.ch12.f12.9

galactose to glucose so that the latter can be used as a carbon and energy source. However, many organisms, including both plants and animals, use this pathway.

Unlike the genes for lactose utilization, not all the genes for galactose utilization are closely linked in the *E. coli* chromosome. The *galU* gene, whose product synthesizes UDP-glucose, is located in a different region of the chromosome. Also, the genes for the galactose permeases, which are responsible for transporting galactose into the cell, are not part of the *gal* operon. Another difference from the *lac* operon is that not only are there two repressor genes, *galR* and *galS*, but also they are nowhere near the operon, unlike the *lacI* gene, which is adjacent to the *lac* operon. This scattering of the genes for galactose metabolism reflects the fact that galactose not only serves as a carbon and energy source, but also plays other roles. For example, the UDP-galactose synthesized by the *gal* operon donates galactose to make polysaccharides for lipopolysaccharide and capsular synthesis.

Two *gal* Repressors: GalR and GalS

The two repressors that control the *gal* operon are GalR and GalS, encoded by the *galR* and *galS* genes, respectively. GalR was discovered first because mutations in *galR* cause constitutive expression of the *gal* operon. However, it was apparent that the *gal* operon is also subject to other regulation. If the GalR repressor were solely responsible for regulating the *gal* operon, mutations that inactivate the *galR* gene should result in the same level of *gal* expression whether galactose is present or not. However, some regulation of the *gal* operon could be observed, even in *galR* mutants. When galactose was added to the medium in which *galR* mutant cells were growing, more of the enzymes of the *gal* operon was made than if the cells were growing in the absence of galactose. The product of another gene, *galS*, was responsible for the residual regulation. Double mutants with mutations that inactivate both *galR* and *galS* are fully constitutive.

The GalS and GalR repressor proteins are closely related, and they both bind the inducer galactose. Even so, they may play somewhat different roles. The GalR repressor is responsible for most of the repression of the *gal* operon in the absence of galactose. The GalS repressor plays only a minor role in regulating the *gal* operon but solely controls the genes of the galactose transport system, which transports galactose into the cell. The reason for this two-tier regulation is unclear but also may be related to the diverse roles of galactose in the cell.

Two *gal* Operators

There are also two operators in the *gal* operon. One is upstream of the promoters, and the other is internal to the first gene, *galE* (Figure 12.8). The two operators are named o_E and o_I for operator *e*xternal to the *galE* gene and operator *i*nternal to the *galE* gene, respectively. Both of the operator sequences are palindromic, similar in organization to the *lac* operator sequences.

The first mutant with an o_I mutation was isolated as part of a collection of constitutive mutants of the *gal* operon (see Irani et al., Suggested Reading). These mutants are easier to isolate in strains with superrepressor *galR*^{s} mutations than in wild-type *E. coli.* The *galR*^{s} mutations are analogous to *lacI*^{s} mutations. The superrepressor mutation makes a strain Gal$^-$ and uninducible because galactose cannot bind to the mutated repressor. Therefore, *E. coli* with a *galR*^{s} mutation cannot multiply to form colonies on plates containing only galactose as the carbon and energy source. However, a constitutive mutation that inactivates the GalR^s repressor or changes the operator sequence prevents the mutant repressor from binding to the operator and allows the cells to use the galactose and multiply to form a colony. Thus, if bacteria with a *galR*^{s} mutation are plated on medium with galactose as the sole carbon and energy source, only constitutive mutants multiply to form a colony. However, most of the constitutive mutants isolated in this way have mutations in the *galR* gene that inactivate the GalR^s repressor rather than operator mutations, since the operator is by far the smaller target. Many *galR* mutants would have to be screened before a single operator mutant was found. Therefore, to make this method practicable for isolating constitutive mutants with operator mutations, the frequency of *galR* mutants must be decreased until it is not too much higher than that of constitutive mutants with mutations in the operator sequences.

One way to reduce the frequency of *galR* mutants is to use a strain that is diploid for (has two copies of) the *galR*^{s} gene. Then, even if one *galR*^{s} gene is inactivated by a mutation, the other *galR*^{s} gene continues to make the GalR^s protein, making the cell phenotypically Gal$^-$. Two independent mutations, one in each *galR*^{s} gene, are required to make the cell constitutive. Since the frequency of two independent mutations is the product of the frequency of each of the single mutations, the presence of two independent *galR* mutations should be very rare, probably no more frequent than single operator mutations, making cells with operator mutations a significant fraction of the total constitutive mutants and easier to identify. Moreover, constitutive mutants with operator mutations can be distinguished from the double mutants with mutations in both *galR*^s genes by the locations where they map. Operator mutations map in the *gal* operon, unlike mutations in the *galR*^s genes, which map elsewhere in the genome.

Accordingly, a partial diploid that had one copy of the *galR*^s gene in the normal position and another copy in a specialized transducing λ phage integrated at the λ attachment site was constructed (see chapter 8). When this strain was plated on medium containing galactose as the sole carbon and energy source, a few Gal⁺ colonies arose due to constitutive mutants. The mutations in two of these constitutive mutants mapped in the region of the *gal* operon and so were presumed to be operator mutations. When the DNA of the two mutants was sequenced, it was discovered that one mutation had changed a base pair in the known operator region (o_E) just upstream of the promoters, as expected. However, the other operator mutation had changed a base pair downstream in the *galE* gene, suggesting that a sequence in that gene also functions as an operator. This operator was named o_I. Furthermore, the o_I mutation occurred in a sequence that was similar to the 15-bp known o_E operator. Moreover, the mutation in the *galE* gene was *cis* acting for constitutive expression of the *gal* operon, one of the criteria for an operator mutation.

If the o_I sequence in the *galE* gene truly is an operator, it should bind the GalR repressor protein. To test this, the experiment illustrated in Figure 12.10 was performed. First, DNA containing the o_I site was inserted into a multicopy plasmid. When this multicopy plasmid is introduced into a cell by transformation, that cell contains many copies of the o_I site gene. If this site binds the repressor, these extra copies should bind most of the GalR protein in the cell, leaving too little to completely repress expression of the *gal* operon. Thus, the cells would appear to be constitutive mutants. This general method is called **titration**, and the enzymes of the *gal* operon are synthesized through **escape synthesis**, so named because the operon is "escaping" the effects of the repressor. In the actual experiment, cells containing many copies of the o_I site did exhibit a partially constitutive phenotype. In contrast, multiple copies of mutant DNA with the putative operator mutation (i.e., o_I^C) do not cause escape synthesis of the Gal enzymes, presumably because GalR no longer binds to the mutated site. If the plasmid DNA contained both o_I and o_E, the chromosomal *gal* operon was fully induced (Figure 12.10C). This result was interpreted to confirm the presence in *galE* of a second binding site for the repressor and that both sites are important for regulation.

Figure 12.11 shows a model for how the two operators cooperate to block transcription in the absence of galactose. Repressor molecules bound to the two operators interact with each other to bend the DNA of the promoter that is between the two operators. The bent promoter does not bind RNA polymerase, so there is no initiation of transcription at the promoter. Repressor bound to only one of the two operators might still

Figure 12.10 Escape synthesis of the enzymes of the *gal* operon caused by additional copies of the operator regions. Clones of the operator regions in a multicopy plasmid dilute out the repressor, inducing the operon even in the absence of galactose. **(A)** The cell contains the normal number of operators, and the operon is not induced. **(B)** The multicopy plasmid contains only o_I, and the operon is only partially induced. **(C)** The multicopy plasmid contains both o_E and o_I, and the operon is fully induced. doi:10.1128/9781555817169.ch12.f12.10

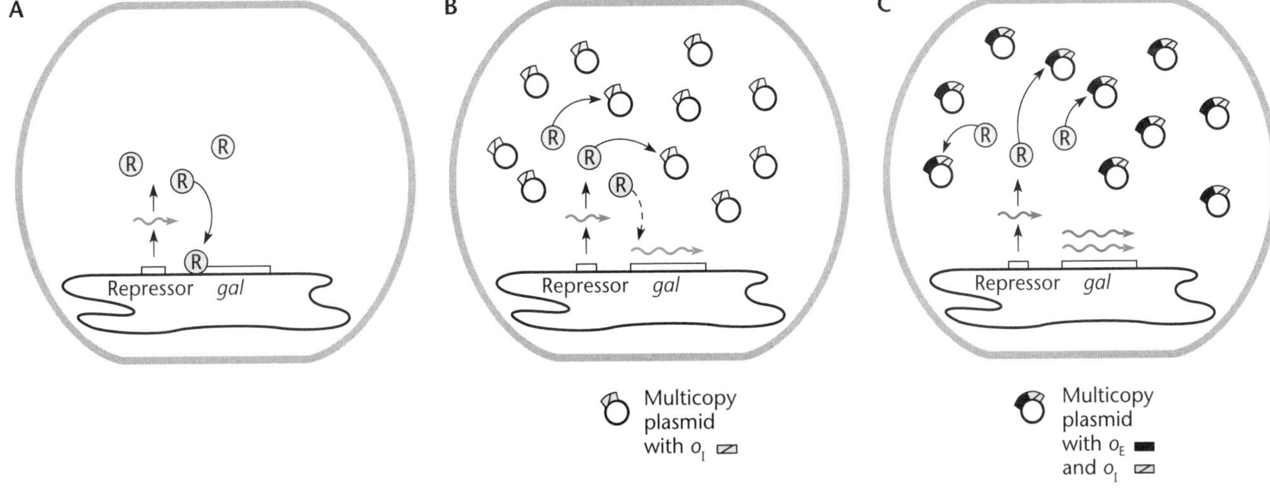

A

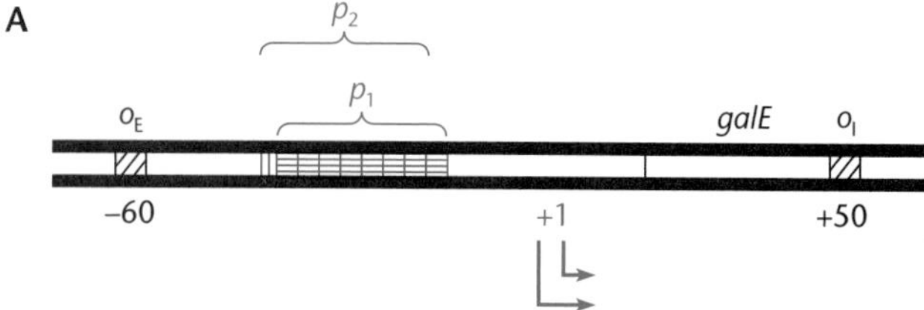

B

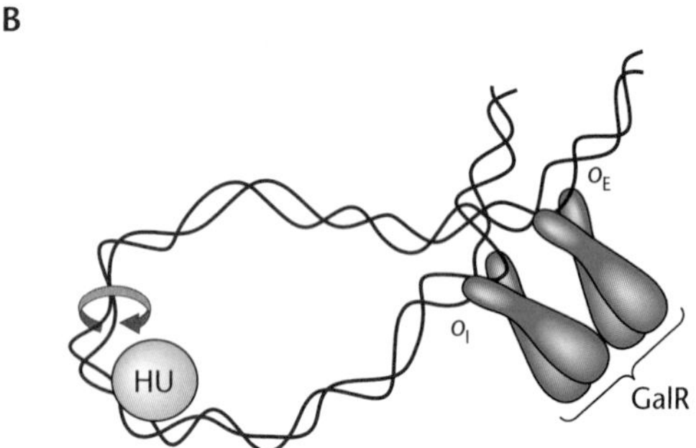

Figure 12.11 Formation of the *gal* operon repressosome. **(A)** Structure of the *gal* operon regulatory region. **(B)** Two dimers of the *galR* repressor gene product bind to each other and to the operators o_E and o_I to bend the DNA between the operators. A histone-like DNA-binding protein, HU, introduces a 180° twist in the DNA (shown by the arrow) so that the repressor can bind to the two operators in an antiparallel configuration. Bending of the DNA in the promoter regions inactivates the promoters.
doi:10.1128/9781555817169.ch12.f12.11

interfere with transcription to some extent, but the promoter would not be bent and the repression is much less severe.

Further studies showed that the positions of the two operator sites relative to each other are critical for regulation. In addition, a histone-like protein called HU binds between the two operators and introduces a 180° turn in the DNA that helps to position the operators so that GalR dimers bound at each of the two operator sites can interact with each other to form a stable DNA loop (see Geanacopoulos et al., Suggested Reading). The overall complex is called the repressosome (Figure 12.11). An additional requirement is that the regulation normally occurs only on supercoiled DNA; the supercoiling of the DNA is presumably required for accurate placement of the GalR dimers to form the DNA loop required for repression.

Two *gal* Promoters and Catabolite Regulation of the *gal* Operon

As mentioned, the *gal* operon also has two promoters called p_{G1} and p_{G2} (Figure 12.8). The *gal* operon may have two promoters because, unlike *lac*, the enzymes are needed even when a better carbon source is available, since they are involved in making polysaccharides, as well as in utilizing galactose (see above). One of these promoters is like the *lac* promoter in that it is poorly expressed if a better carbon source, such as glucose, is available. However, the other promoter is active even in the presence of glucose and continues to promote expression of the Gal enzymes so that other cellular constituents containing galactose can continue to be made. Catabolite regulation is discussed in more detail in chapter 13.

Negative Repressible Systems

The enzymes encoded by the *lac* and *gal* operons are involved in catabolism of specific sugars as sources of carbon and energy. Consequently, these operons are called **catabolic operons** or **degradative operons**. As described for the *lac* and *gal* systems, expression is induced only when the substrate of the enzymes encoded in the regulated operons is available to the cell, i.e., the appropriate sugar (lactose or galactose) acts as an inducer that inactivates the repressor. Not all operons are involved in degrading compounds, however. The enzymes encoded by some operons synthesize compounds needed by the cell, such as nucleotides, amino acids, and vitamins. These operons are called **biosynthetic operons**.

The regulation of a biosynthetic operon is essentially opposite to that of a degradative operon. The enzymes of a biosynthetic pathway should not be synthesized in the presence of the end product of the pathway, since the product is already available and energy should not be wasted in synthesizing more. However, the mechanisms by which degradative and biosynthetic operons are regulated are often similar. Biosynthetic operons can also be regulated negatively by repressors. If the genes of a biosynthetic operon are constitutively expressed in the absence of the regulatory gene product, even if the compound is present in the medium, the biosynthetic operon is negatively regulated.

The terminology used to describe the negative regulation of biosynthetic operons differs somewhat from that used for catabolic operons, despite shared principles. In a negative repressible system, the effector that binds to the repressor and allows it to bind to the operators is called the **corepressor**, because it functions together with the repressor. A repressor that negatively regulates a biosynthetic operon is not active in the absence of the corepressor and in this state is called the **aporepressor**. However, once the corepressor is bound, the protein is able to bind to the operator and so is now called the repressor.

THE *E. COLI trp* OPERON

The tryptophan (*trp*) operon of *E. coli* is the classic example of a negative repressible system (see Yanofsky and Crawford, Suggested Reading). The enzymes encoded by the *trp* operon (Figure 12.12) are responsible for synthesizing the amino acid L-tryptophan, which is a

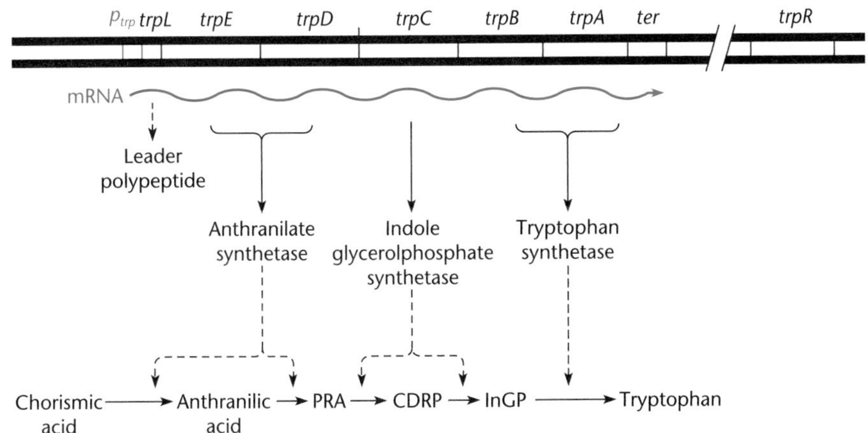

Figure 12.12 Structure of the tryptophan biosynthetic (*trp*) operon of *E. coli.* The structural genes *trpEDCBA* are transcribed from the promoter p_{trp}. Upstream of the structural genes is a short coding sequence for the leader peptide called *trpL.* The *trpR* repressor gene is unlinked, as shown by the broken line. PRA, phosphoribosyl anthranilate; CDRP, 1-(*o*-carboxyphenylamino)-1-deoxyribulose-5-phosphate; InGP, indole glycerolphosphate.
doi:10.1128/9781555817169.ch12.f12.12

constituent of most proteins and so must be synthesized if none is available in the medium. The products of five structural genes in the operon are required to make tryptophan from chorismic acid. These genes are transcribed from a single promoter, p_{trp}. The *trp* operon is negatively regulated by the TrpR repressor protein, whose gene, like the *gal* repressor gene, is unlinked to the rest of the operon. This may reflect the fact that TrpR regulates not only the *trp* operon, but also the *aroH* operon, which encodes enzymes involved in synthesis of chorismic acid, a precursor for other pathways.

Figure 12.13 shows the model for the regulation of the *trp* operon by the TrpR repressor. Binding of the TrpR repressor to the operator prevents transcription from the p_{trp} promoter. However, the TrpR repressor can bind to the operator only if the corepressor tryptophan is present in the cell. Tryptophan binds to the TrpR aporepressor protein and changes its conformation so that it can bind to the operator.

The TrpR repressor has been crystallized, and its structure has been determined in both the aporepressor form (when it cannot bind to DNA) and the repressor form with tryptophan bound (when it can bind to the operator). These structures have led to a satisfying explanation of how tryptophan binding changes the TrpR protein so that it can bind to the operator (Figure 12.14). The TrpR repressor is a dimer, and each copy of the TrpR polypeptide has α-helical structures (shown

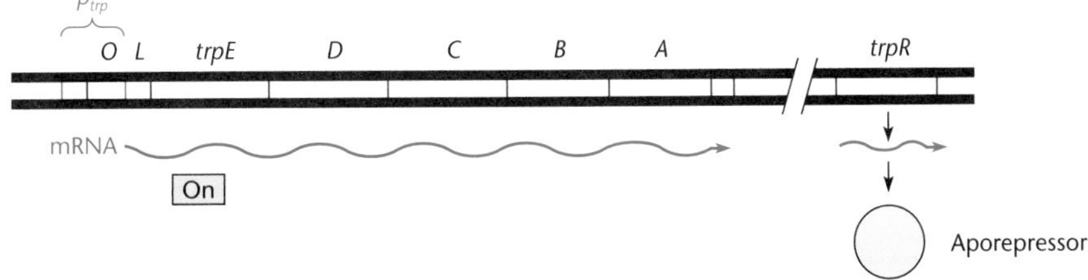

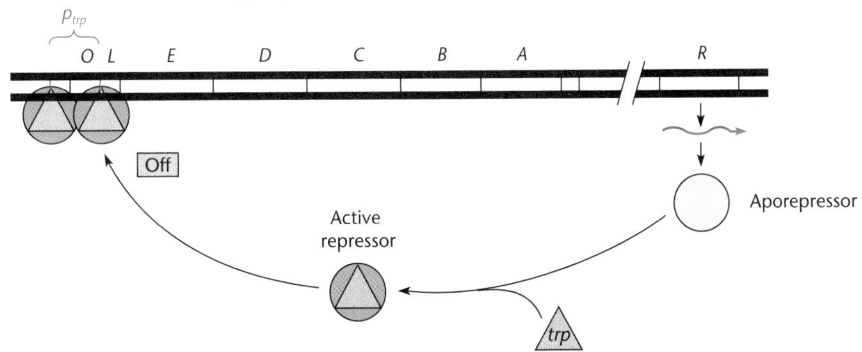

Figure 12.13 Negative regulation of the *trp* operon by the TrpR repressor. Binding of the corepressor tryptophan to the aporepressor converts it to the repressor confirmation, which is active in DNA binding.
doi:10.1128/9781555817169.ch12.f12.13

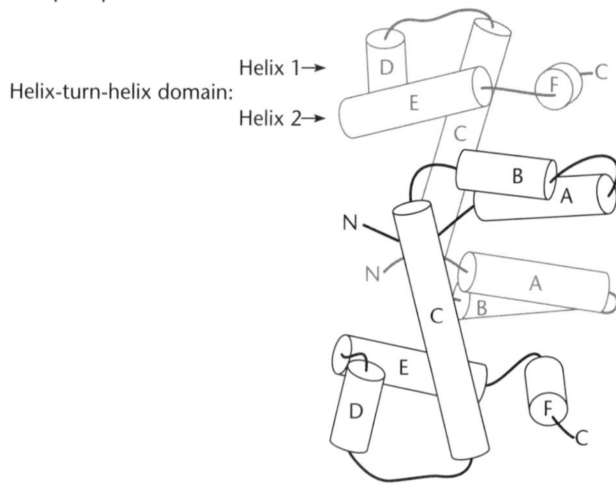

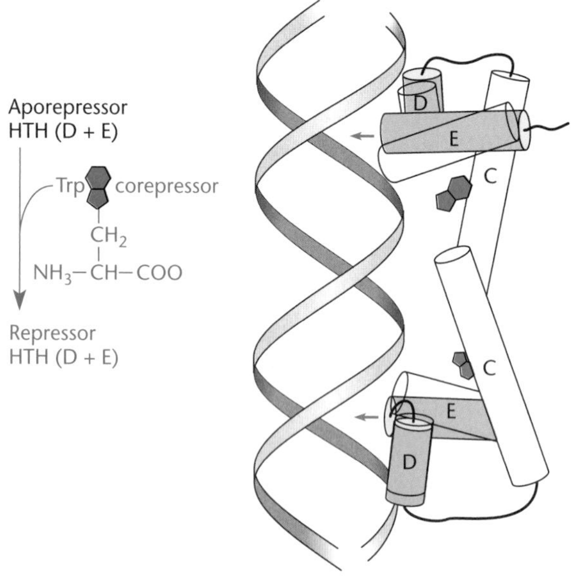

Figure 12.14 Structure of the TrpR repressor and an illustration of how tryptophan binding allows it to convert from the aporepressor to the repressor that binds to the operator. **(A)** The helices (shown as cylinders) of the aporepressor dimer in the inactive state with no tryptophan bound. **(B)** The blue cylinders represent the HTH domains of the active repressor with tryptophan bound.
doi:10.1128/9781555817169.ch12.f12.14

as cylinders). Helices D and E form the helix-turn-helix (HTH) DNA-binding domain (Box 12.1). Helix D corresponds to helix 1, the nonspecific DNA-binding helix, and helix E corresponds to helix 2, the DNA sequence-specific recognition helix. In the aporepressor state, the conformations of the two HTH domains in the dimer do not allow proper interactions with successive major grooves appearing on one side of the DNA helix. Binding of the typtophan corepressor alters the HTH conformations, allowing the repressor to bind to the operators.

Isolation of *trpR* Mutants

As is the case for other negatively regulated operons, constitutive mutations of the *trp* operon are quite common, and most map in *trpR*, inactivating the product of the gene or preventing binding of the corepressor. Mutants with constitutive expression of the *trp* operon can be obtained by selecting for mutants resistant to the tryptophan analog 5-methyltryptophan in the absence of tryptophan. This tryptophan analog binds to the TrpR repressor and acts as a corepressor. However, 5-methyltryptophan cannot be used in place of tryptophan in protein synthesis. Therefore, in the presence of the analog, the *trp* operon is not induced even in the absence of tryptophan, and the cells will starve for tryptophan. Only constitutive mutants that continue to express the genes of the operon in the presence of 5-methyltryptophan can multiply to form colonies on plates with this analog but without tryptophan.

Other Types of Regulation of the *E. coli trp* Operon

The *trp* operon is also subject to a second level of regulation called transcription attenuation. This type of regulation is discussed below. Also, as in many biosynthetic pathways, the first enzyme of the *trp* pathway is subject to **feedback inhibition** by the end product of the pathway, tryptophan. We also return to feedback inhibition below.

Molecular Mechanisms of Transcriptional Repression

Repression of transcription initiation can occur by a variety of mechanisms. The simplest mechanism is steric hindrance, where binding of the repressor protein to the DNA physically blocks access of RNA polymerase to the promoter site. The *E. coli* LacI protein provides an example of this mechanism. A second possibility is that binding of the repressor protein to the DNA results in a change in DNA structure, like a DNA loop, that in turn prevents binding of RNA polymerase to the promoter, as is the case for the *E. coli gal* operon. More complicated DNA structural changes also can occur, some of which are discussed in the context of global regulators (see chapter 13). Another possibility is that a repressor does not directly inhibit transcription, but instead prevents the positive activity of an activator that is required for efficient transcription. This mechanism, which is termed antiactivation, is discussed below in the context of transcriptional activation.

Positive Regulation of Transcription Initiation

The first part of this chapter covers the classic examples of negative regulation by repressors. However, many operons are regulated positively by activators. An operon under the control of an activator protein is transcribed only in the presence of that protein. As we saw in the discussion of negative regulation, the activity of the activator can be activated by an inducer (e.g., in the arabinose operon) or inactivated by a corepressor (e.g., in the fatty acid biosynthetic operon). Activators often work by increasing the tightness of binding of the RNA polymerase to the promoter, by allowing it to open the strands of DNA at the promoter, or by rotating and bending the promoter to bring the recognition sites together.

Positive Inducible Systems

Positive inducible systems utilize an activator to turn on gene expression, often by aiding in the recruitment of RNA polymerase. However, the activator protein itself is unable to bind to DNA to increase transcription in the absence of its effector molecule, which serves as an inducer, or can bind but not activate transcription without the effector. As is the case with other inducible systems, positive inducible systems are often involved in catabolism of the inducer or a related molecule, so that expression of the regulated genes is necessary only when the substrate of the pathway, often the inducer itself, is present.

THE *E. coli ara* OPERON

The *E. coli ara* operon was the first example of **positive regulation** in bacteria to be discovered (although it actually shows many levels of regulation [see below]). The *ara* operon is responsible for conversion of the five-carbon sugar L-arabinose into D-xylulose-5-phosphate, which can be used by other pathways. *E. coli* can also utilize D-arabinose, an isomer of L-arabinose, but the enzymes for D-arabinose utilization are encoded by a different operon, which lies elsewhere in the chromosome.

Figure 12.15A illustrates the structure of the *ara* operon. Three structural genes in the operon, *araB*, *araA*, and *araD*, are transcribed from a single promoter, p_{BAD}. Upstream of the promoter is the activator region, *araI*, where the activator protein AraC binds to activate transcription in the presence of L-arabinose, and the CAP site, at which CAP binds to mediate catabolite regulation (see chapter 13). There are also two operators, *ara*O_1 and *ara*O_2, at which the AraC protein binds to repress transcription (see below). The *araC* gene, which encodes

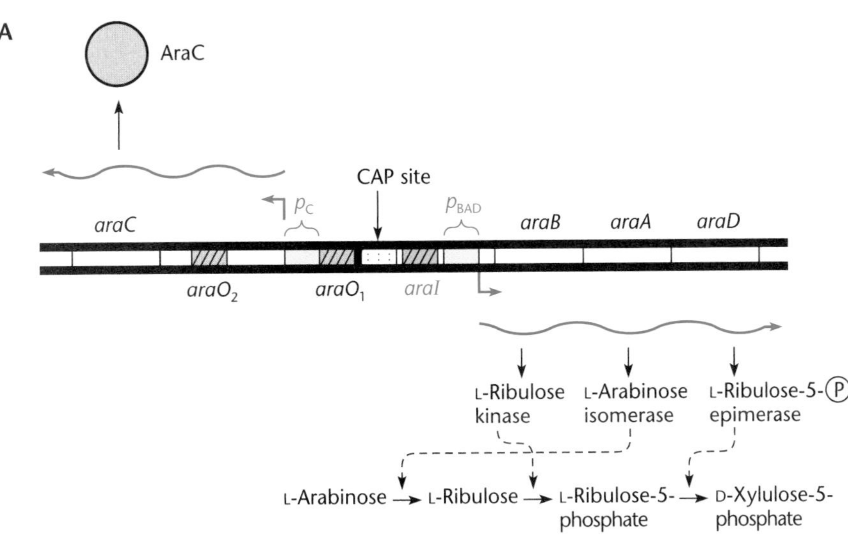

Figure 12.15 **(A)** Structure and function of of the L-arabinose operon of *E. coli.* **(B)** Binding of the inducer L-arabinose converts the AraC protein from an antiactivator P1 form to an activator P2 form. See the text for details.
doi:10.1128/9781555817169.ch12.f12.15

the regulatory protein, is also shown. This gene is transcribed from the promoter p_C in the opposite direction from *araBAD*, as shown by the arrows in the figure. As described below, the AraC protein is a positive activator of transcription. As such, it is a member of a large family of activator proteins (Box 12.2).

BOX 12.2

Families of Regulators

The techniques of comparative genomics have made it possible to identify repressor and activator genes in a wide variety of bacteria. These transcriptional regulatory proteins belong to a limited number of known families, based on sequence and structural conservation, even though they regulate operons with very different functions and respond to different effectors. Transcriptional regulators can be assigned to a family based on sequence and structural homology, the organization of their motifs, and whether they use the same motif to bind to DNA. Many of them use an HTH motif or a winged HTH motif (Box 12.1), or they can use a looped-hinge helix or a zinc finger motif, among others. There are at least 15 different families of transcriptional regulators. Some of these families are quite large, with many members. Some families consist only of repressors, others consist only of activators, and some consist of both repressors and activators. Most of the families are named after the first member of the family to be studied. For example, the LacI family includes GalR and consists mostly of repressors that regulate operons involved in carbon source utilization and generally also respond to CAP and cAMP. These repressors usually function as homotetramers and have an HTH DNA-binding motif to bind to DNA at their N termini, an effector-binding motif in the middle, and dimerization and tetramerization motifs at their C termini.

The TetR family of repressors, named after the repressor that regulates the tetracycline resistance gene in Tn*10*, is even larger and includes LuxR, which regulates light emission in chemiluminescent bacteria (see chapter 13). These family members are found in both gram-negative and gram-positive bacteria, and they repress operons involved in a variety of functions, including antibiotic resistance and synthesis, osmotic regulation, efflux pumps for multidrug resistance, and virulence genes in pathogenic bacteria. The TetR repressor itself is widely used to regulate gene expression in eukaryotic cells because it binds very tightly to its operators and because tetracycline readily diffuses into eukaryotic cells. TetR has an HTH motif that binds to DNA, but only when the repressor has not bound tetracycline, and partially unwinds the operator DNA through its major groove. Presumably, this can change the structure of the DNA at the promoter. Interestingly, a member of this family that regulates a multidrug efflux pump has a very broad effector-binding pocket, allowing it to bind a variety of antibiotics.

The AraC family of activators is also very large. These activators seem to fall into at least two subfamilies, those that regulate carbon source utilization, like AraC, and function as dimers and those that respond to stress responses, like SoxS, and function as monomers. A signature of this family is that the HTH motif that binds to DNA is in the C-terminal part of the protein, not in the N terminus, like many other activators.

Other large classes are the LysR activators and the NtrC activators. The NtrC activators are particularly interesting in that they activate transcription only from σ^N promoters (the nitrogen sigma factor [see chapter 13]). These activators are organized into distinguishable domains (see Box 13.3): the domain of the activator protein that either binds the inducer or is phosphorylated is located at the N terminus, and the DNA-binding domain is located at the C terminus. The middle region of the polypeptide contains a region that interacts with RNA polymerase and has an ATPase activity required for activation.

Experiments with hybrid activators, made by fusing the C-terminal DNA-binding domain of one activator protein to the N-terminal inducer-binding domain of another activator protein from the same family, provide a graphic demonstration that members of a family of regulators all use the same basic strategy to activate transcription of their respective operons (see Parek et al., References). Sometimes, such hybrid activators can still activate the transcription of an operon, but the operon that is activated by the hybrid activator depends on the source of the C-terminal DNA-binding domain, while the inducer that induces the operon depends on the source of the N-terminal inducer-binding domain. This leads to a situation where an operon is induced by the inducer of a different operon. It is intriguing to think that all activator proteins may have evolved from a single precursor protein through simple changes in its effector-binding and DNA-binding regions, yet they continue to activate the RNA polymerase by the same basic mechanism.

Regulatory proteins use almost every conceivable mechanism to regulate trancription. Repressors act on essentially every step required for initiation of transcription, although many affect more than one step. Some repressors act, at least in part, by preventing the binding of RNA polymerase to the promoter either by getting in the way (steric hindrance) or by bending the DNA at the promoter. They can also act by

BOX 12.2 (continued)

Families of Regulators

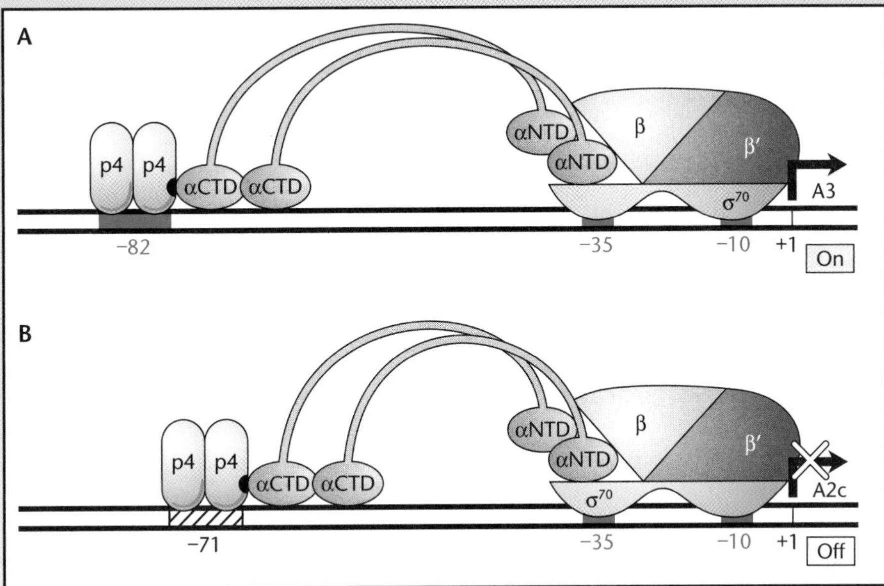

Figure 1 doi:10.1128/9781555817169.ch12.Box12.2.f1

preventing RNA polymerase from separating the strands of DNA at the promoter (open-complex formation) or even hindering the ability of RNA polymerase to move out of the promoter and begin making RNA (promoter escape). In one particularly illustrative case, shown in Figure 1, the repressor/activator protein p4 of a *B. subtilis* phage represses an already strong promoter by making it so strong that the RNA polymerase cannot escape it and begin transcription. Figure 1A shows how p4 binding to a sequence at −82 relative to the start site of the A3 promoter activates transcription from that promoter (shown as "On" in the figure), while p4 binding to a sequence at −71 relative to the start site of another promoter, A2c, inhibits transcription from that promoter (shown as "Off" in the figure). The crucial difference between these promoters is that the A3 promoter has weak affinity for RNA polymerase and therefore benefits from activation by p4, while the A2C promoter has strong affinity for RNA polymerase, so the extra affinity provided by the p4 protein locks RNA polymerase in place and prevents promoter escape.

Activators can also act at any of the steps of transcription initiation. Many of them "recruit" the RNA polymerase to the promoter by binding both to the DNA close to the promoter and to the RNA polymerase, thereby stabilizing the binding of the RNA polymerase to the promoter. Figure 2 gives some examples of how activators and sequences around a promoter can recruit RNA polymerase to promoters. Figure 2A shows the RNA polymerase binding to the −35 and −10 regions of a σ^{70} promoter. Figure 2B shows how binding of the αCTD domains (the C-terminal domains of the α subunits of RNA polymerase) to a sequence upstream of the promoter called an UP element can stabilize the binding. Figure 2C and D show how CAP bound at different sites upstream of a promoter can contact different regions of the RNA polymerase and stabilize its binding. In Figure 2C, CAP is bound further upstream and contacts one αCTD domain. In Figure 2D, it is bound closer to the start site and can contact one of the αNTDs (N-terminal domains of the α subunit), as well as the αCTDs. Figure 2E shows how the CI protein of phage λ (shown as dumbbells) can activate transcription from the p_{RM} promoter by binding cooperatively to the operators o_R^1 and o_R^2 (see chapter 8), from where they can contact both σ^{70} and the αCTDs. Some activators, such as SoxS, may even bind to the RNA polymerase before it binds to the promoter (Figure 3B). Only after the SoxS activator has bound to it can the RNA polymerase bind to the promoter. Others, such as the NtrC-type activators, do not recruit the RNA polymerase but allow an RNA polymerase already bound at the promoter to open the DNA to form an open complex (see chapter 13). Activators in the MerR family activate transcription by changing the conformation of the promoter DNA to increase binding of RNA polymerase (see Huffman and Brennan, References). Normally the spacing between the −35 sequence and the −10 sequence in a σ^{70}-type promoter is 17 bp, but this promoter has a spacer region that is 19 bp (Figure 4), which rotates the two

(continued)

BOX 12.2 (continued)

Families of Regulators

Figure 2 doi:10.1128/9781555817169.ch12.Box12.2.f2

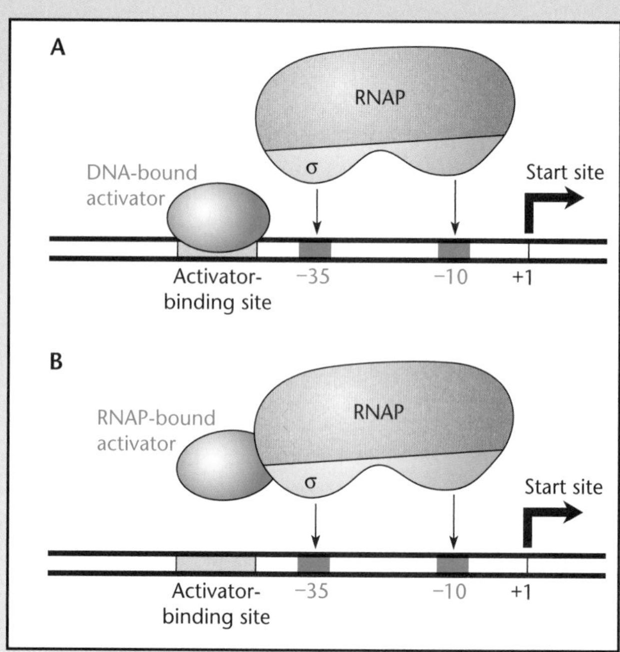

Figure 3 doi:10.1128/9781555817169.ch12.Box12.2.f3

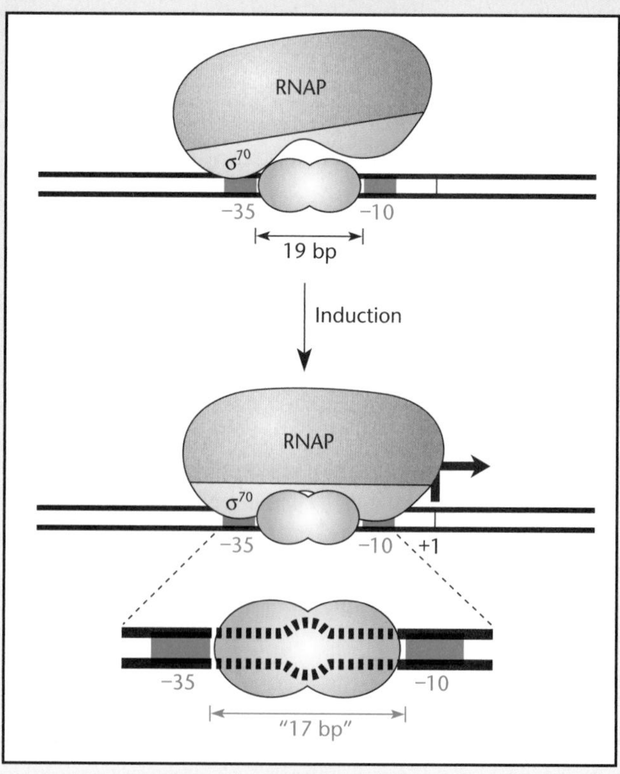

Figure 4 doi:10.1128/9781555817169.ch12.Box12.2.f4

BOX 12.2 (continued)

Families of Regulators

elements by about 70°, making it difficult for the σ^{70} subunit of the RNA polymerase to contact both of them (see chapter 2). The activator binds to the promoter even in the absence of inducer but fails to activate under these conditions. When the inducer binds, the conformation of the activator changes, and this bends and twists the DNA between the two promoter elements. This brings the –35 and –10 elements closer together and rotates them so that their orientation and spacing more closely resemble those in a normal σ^{70} promoter.

References

Dove, S. L., and A. Hochschild. 2005. How transcription initiation can be regulated in bacteria, p. 297–310. *In* N. P. Higgins (ed.), *The Bacterial Chromosome.* ASM Press, Washington, DC.

Egan, S. 2002. Growing repertoire of AraC/XylS activators. *J. Bacteriol.* **184:**5529–5532.

Huffman, J. L., and R. G. Brennan. 2002. Prokaryotic transcription regulators: more than just the helix-turn-helix motif. *Curr. Opin. Struct. Biol.* **12:**98–106.

Parek, M. R., S. M. McFall, D. L. Shinabarger, and A. M. Chakrabarty. 1994. Interaction of two LysR-type regulatory proteins CatR and ClcR with heterologous promoters: functional and evolutionary implications. *Proc. Natl. Acad. Sci. USA* **91:**12393–12397.

Ramos, J. L., M. Martinez-Bueno, A. J. Molina-Henares, W. Teran, K. Watanabe, X. Zhang, M. T. Gallegos, R. Brennan, and R. Tobes. 2005. The TetR family of transcriptional repressors. *Microbiol. Mol. Biol. Rev.* **69:**326–356.

Genetic Evidence for Positive Regulation of the *ara* Operon

Early genetic evidence indicated that the *lac* and *ara* operons are regulated by very different mechanisms (see Englesberg et al. and Schleif, Suggested Reading). One observation was that loss of the regulatory proteins results in very different phenotypes. For example, deletions and nonsense mutations in the *araC* gene—mutations that presumably inactivate the protein product of the gene—lead to an uninducible phenotype in which the genes of the *ara* operon are not expressed, even in the presence of the inducer arabinose. Recall that deletion or nonsense mutations in the regulatory gene of a negatively regulated operon such as *lac* result in a constitutive phenotype, not an uninducible phenotype. Another difference between *ara* and negatively regulated operons is in the frequency of constitutive mutants. Mutants that constitutively express a negatively regulated operon are relatively common, because any mutation that inactivates the repressor gene causes constitutive expression. However, mutants that constitutively express *ara* are very rare, which suggests that mutations that result in the constitutive phenotype do not merely inactivate AraC.

Isolating Constitutive Mutations of the *ara* Operon

Because constitutive mutations of the *ara* operon are so rare, special tricks are required to isolate them. One method for isolating rare constitutive mutations in *araC* uses the anti-inducer D-fucose. This anti-inducer binds to the AraC protein and prevents it from binding L-arabinose, thereby preventing induction of the operon. As a consequence, wild-type *E. coli* cannot multiply to form colonies on agar plates containing D-fucose with L-arabinose as the sole carbon and energy source. Only mutants that constitutively express the genes of the *ara* operon can form colonies under these conditions.

Another selection for constitutive mutations in the L-arabinose operon cleverly plays off the operon responsible for the utilization of its isomer, D-arabinose. The enzymes produced by the L-arabinose and D-arabinose operons cannot use the intermediate of the other operon, with one exception. The product of the *araB* gene (the ribulose kinase enzyme of the L-arabinose operon pathway, which phosphorylates L-ribulose as the second step of the pathway) can also phosphorylate D-ribulose, so that the L-arabinose kinase can substitute for that of the D-arabinose operon. Nevertheless, *E. coli* mutants that lack the D-arabinose kinase cannot multiply to form colonies on plates containing only D-arabinose as a carbon and energy source, because D-arabinose is not an inducer of the L-*ara* operon. Only constitutive mutants of the L-*ara* operon can grow if D-arabinose kinase-deficient mutants are plated on agar plates containing D-arabinose as the sole carbon and energy source.

A Model for Positive Regulation of the *ara* Operon

The contrast in phenotypes between mutations that inactivate the *lacI* and *araC* genes led to an early model for the regulation of the *ara* operon. According to this early model, the AraC protein can exist in two states, called P1 and P2 (Figue 12.15B). In the absence of the inducer, L-arabinose, the AraC protein is in the P1 state and inactive. If L-arabinose is present, it binds to AraC and changes the protein conformation to the P2 state. In this state, AraC binds to the DNA at the site called *araI* (Figure 12.15A) in the promoter region and activates transcription of the *araB*, *araA*, and *araD* genes.

This early model explained some, but not all, of the behavior of the *araC* mutations. It explained why mutations in *araC* that cause the constitutive phenotype are rare but do occur at a very low frequency. According to this model, these mutations, called *araC*^{c} mutations, change AraC so that it is permanently in the P2 state, even in the absence of L-arabinose, and thus the operon is always transcribed. Such mutations would be expected to be very rare because only a few amino acid changes in the AraC protein could specifically change the conformation of the AraC protein to the P2 state.

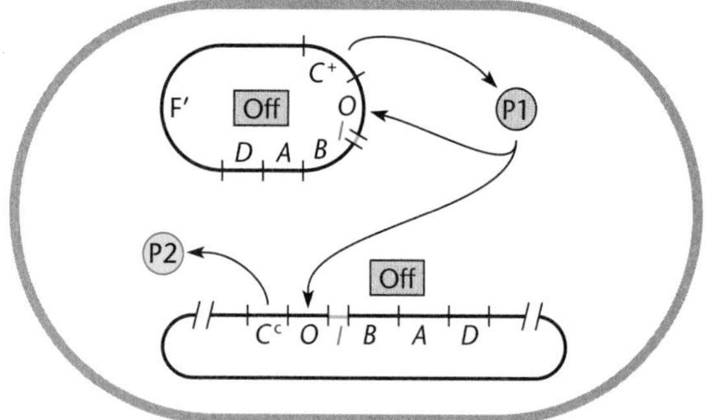

P1 form binds to operators of both operons, preventing activation of *araBAD* expression by the AraC in the P2 state

Figure 12.16 Recessiveness of *araC*^{c} mutations. The presence of a wild-type copy of the *araC* gene prevents activation of transcription of the operon by AraC^c. See the text for details.
doi:10.1128/9781555817169.ch12.f12.16

AraC Is Not Just an Activator

One prediction of this model for regulation of the *ara* operon is that *araC*^{c} mutations should be dominant over the wild-type allele in complementation tests. If AraC acts solely as an activator, partial diploid cells that have both an *araC*^{c} allele and the wild-type allele would be expected to constitutively express the *araB*, *araA*, and *araD* genes. That would make *araC*^{c} mutations dominant over the wild type, since the mutant AraC in the P2 state should activate transcription of *araBAD*, even in the presence of wild-type AraC protein in the P1 state.

The prediction of the model was tested with complementation. An F′ plasmid carrying the wild-type *ara* operon was introduced into cells with an *araC*^{c} mutation in the chromosome. Figure 12.16 illustrates that the partial diploid cells were inducible, not constitutive, indicating that *araC*^{c} mutations were recessive rather than dominant. This observation was contrary to the prediction of the model. Therefore, the model had to be changed.

Figure 12.17 illustrates a more detailed model to explain the recessiveness of *araC*^{c} mutations. In this model, transcriptional activation requires that the P2 form of AraC, which exists in the presence of arabinose, binds at both the *araI*$_1$ and *araI*$_2$ sites (which together make up the *araI* site in the original model [Figure 12.15]). The P1 form of the AraC protein that exists in the absence of arabinose is not simply inactive but takes on a new role as an **antiactivator** (Figure 12.17A). The P1 state is called an antiactivator rather than a repressor because it does not repress transcription like a classical repressor but instead acts to prevent activation by the P2 state of the protein. In the P1 state, the AraC protein preferentially binds to the operators *araO*$_2$ and *araI*$_1$. Binding of AraC to both *araO*$_2$ and *araI*$_1$ bends the DNA between the two sites in a manner similar to the DNA looping caused by the GalR repressor. Because AraC in the P1 form preferentially binds to *araO*$_2$, it cannot bind to *araI*$_2$ and activate transcription from the p_{BAD} promoter. In the presence of L-arabinose, however, the AraC protein changes to the P2 form and now preferentially binds to *araI*$_1$ and *araI*$_2$, activating transcription of the operon by RNA polymerase (Figure 12.17B).

This model explains why the *araC*^{c} mutations are recessive to the wild type in complementation tests, because AraC^c in the P2 form can no longer bind to *araI*$_1$ and *araI*$_2$ to activate transcription as long as wild-type AraC in the P1 state is already bound to *araO*$_2$ and *araI*$_1$.

Face-of-the-Helix Dependence

The *ara* operon provides a classic example of a regulatory system that exhibits **face-of-the-helix dependence**. In systems of this type, the binding site for a regulatory protein on the regulated gene is constrained by how a regulatory protein bound to that site will be positioned relative to some other site (for example, the binding site for another regulatory protein or RNA polymerase). This effect is identified by construction of mutants containing small insertions or deletions. In the case of the *ara* operon, repression by AraC in the absence of arabinose requires that AraC bound to *araI*$_1$ is able to interact with AraC bound to *araO*$_2$ to form the DNA loop. If 5 bp of DNA is inserted between the *araI*$_1$ and *araO*$_2$ sites, AraC bound to one of the sites will be unable to interact with AraC bound to the other site, because addition of 5 bp of DNA (half of a turn of the DNA helix [see chapter 1]) positions the second AraC not only further away in linear distance, but also on the opposite side of the DNA helix (Figure 12.18). Addition of 10 bp of DNA, which corresponds to a full turn of the DNA helix, extends the linear distance between

A Absence of L-arabinose

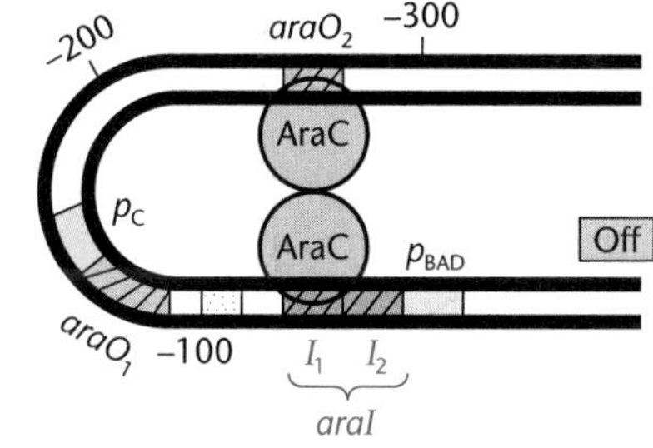

B Presence of L-arabinose

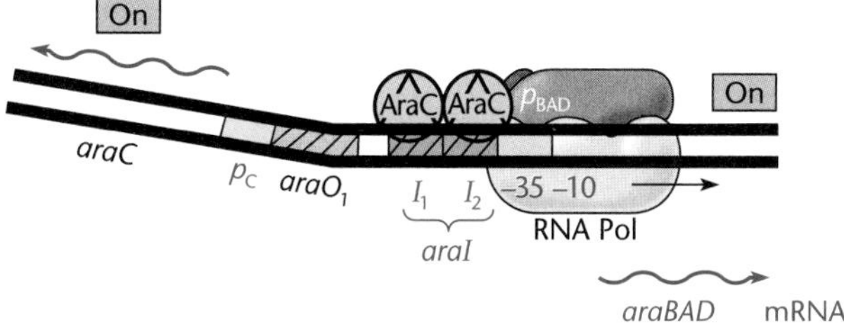

C Excess of AraC

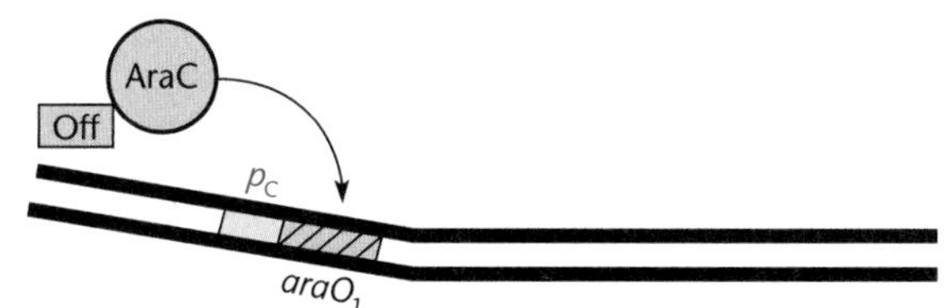

Figure 12.17 A model to explain how AraC can be a positive activator of the *ara* operon in the presence of L-arabinose and an antiactivator in the absence of L-arabinose, as well as how AraC can negatively autoregulate transcription of its own gene. **(A)** In the absence of arabinose, AraC molecules in the P1 state preferentially bind to *araI*$_1$ and *araO*$_2$, preventing any AraC in the P2 state from binding to *araI*$_2$. No transcription occurs, because AraC must bind to *araI*$_2$ to activate transcription from p_{BAD}. Bending of the DNA between the two sites may also inhibit transcription of the *araC* gene itself by inhibiting transcription from the p_C promoter. The bend in the DNA is also facilitated by the binding by the CAP protein (see chapter 13). **(B)** In the presence of arabinose, AraC shifts to the P2 state and preferentially binds to *araI*$_1$ and *araI*$_2$. AraC bound to *araI*$_2$ activates transcription from p_{BAD}. **(C)** If the AraC concentration becomes very high, it will also bind to *araO*$_1$, thereby repressing transcription from its own promoter, p_C.
doi:10.1128/9781555817169.ch12.f12.17

the binding sites, but more importantly, it repositions the second AraC so that it can now interact with the first AraC. Therefore, addition of 5 bp of DNA blocks repression, but addition of 10 bp does not. This pattern of response to insertion (or deletion) of partial or full turns of the DNA helix is the hallmark of face-of-the-helix dependence and is interpreted to indicate a requirement for an interaction between factors bound at the two sites. In the case of the *ara* operon, the key interaction is between the two molecules of AraC that must interact to form the DNA loop.

Autoregulation of *araC*

The AraC protein not only regulates the transcription of the *ara* operon, but also negatively autoregulates its own transcription. Like TrpR, the AraC protein represses its own synthesis, so less AraC protein is synthesized in the absence of arabinose than in its presence. However, if the concentration of AraC becomes too high, its synthesis will again be repressed.

Figure 12.17 also shows a model for the **autoregulation** of AraC synthesis. In the absence of arabinose, the interaction of two AraC monomers bound at *araO*$_2$ and *araI*$_1$ bends the DNA in the region of the *araC* promoter p_C, thereby inhibiting transcription from the promoter (Figure 12.17A). In the presence of arabinose, the AraC protein is no longer bound to *araO*$_2$, so the p_C promoter is no longer bent and transcription from p_C occurs. However, if the AraC concentration becomes too high, the excess AraC protein binds to the operator *araO*$_1$, preventing further transcription of *araC* from the p_C promoter (Figure 12.17C). A similar autoregulation of the λ *cI* gene by its product, the CI protein, was described in chapter 8.

Catabolite Regulation of the *ara* Operon

The *ara* operon is also regulated in response to the availabilty of other carbon sources, like glucose (see chapter 13). CAP, which regulates the transcription of genes subject to catabolite regulation, is a transcriptional activator, like the AraC protein; CAP activates gene expression only when the cell lacks a better carbon source. By binding to the CAP binding site shown in Figure 12.15A, CAP may help to open the DNA loop created when AraC binds to *araO*$_2$ and *araI*$_1$. Opening the loop may prevent binding of AraC to *araO*$_2$ and *araI*$_1$ and facilitate the binding of AraC to *araI*$_1$ and *araI*$_2$ and the activation of transcription from p_{BAD}. Thus, the absence of glucose or another carbon source better than arabinose enhances the transcription of the *ara* operon.

Uses of the *ara* Operon

Besides its historic importance in the pioneering studies of positive regulation, the *ara* operon has many uses in biotechnology. The p_{BAD} promoter, from which the genes of the *ara* operon are transcribed, is often used instead of the *lac* promoter in expression vectors because it is more tightly regulated than the *lac* promoter. Because of its combination of positive and negative regulation by the AraC activator, very little transcription occurs from the promoter unless L-arabinose is present in the medium. The p_{BAD} promoter is also more tightly regulated by catabolite repression than the *lac* promoter, making it more

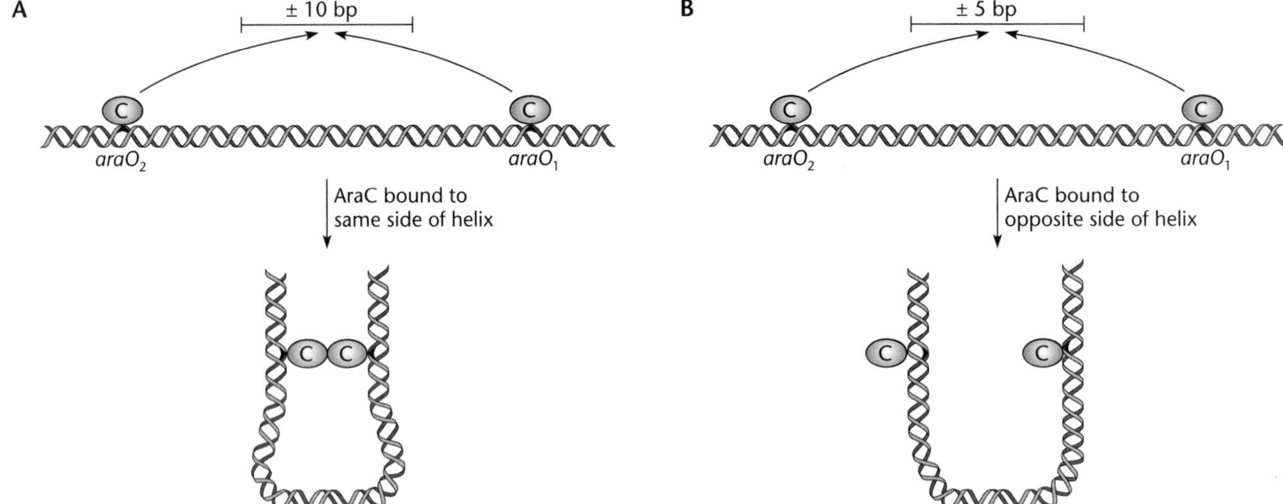

Figure 12.18 Face-of-the-helix dependence. **(A)** Molecules of AraC in the P1 state are bound to *araI*$_1$ and *araO*$_2$ on the same face of the DNA and can interact with each other to form a DNA loop. Insertion or deletion of 10 bp (a full turn of the DNA helix) between the operator sites is tolerated, as the protein molecules remain on the same side of the helix. **(B)** Insertion or deletion of 5 bp between the operator sites (half of a turn of the DNA helix) positions the protein on opposite sides of the helix, so the proteins are no longer able to interact with each other to form a DNA loop. doi:10.1128/9781555817169.ch12.f12.18

suitable for expression of large amounts of a toxic gene product by first growing the cells in medium containing glucose and then washing out the glucose and adding L-arabinose. A widely used series of *E. coli* expression vectors use the p_{BAD} promoter and have other desirable features (see Guzman et al., Suggested Reading). They have a variety of antibiotic resistance genes for selection, and some have origins of replication that allow them to coexist with the more standard cloning vectors that have the ColE1 origin of replication (see chapter 4).

THE *E. COLI* MALTOSE OPERONS

Other well-studied and heavily used positively regulated operons in bacteria include those for the utilization of the sugar maltose and polymers of maltose in *E. coli*, shown in Figure 12.19. Rather than being organized in one operon, the genes for maltose transport and metabolism are organized in four clusters at 36, 75, 80, and 91 minutes (min) on the *E. coli* genetic map. The operon at 75 min has two genes, *malQ* and *malP*, whose products are involved in converting maltose and polymers of maltose into glucose and glucose-1-phosphate. This cluster also includes the regulatory gene *malT*. The *malS* gene, at 80 min, encodes an enzyme that breaks down polymers of maltose, such as amylase. The other cluster, at 91 min, has two operons whose gene products can transport maltose into the cell. An operon at 36 min encodes enzymes that degrade polymers of maltose.

Although they allow the cell to use maltose as a carbon source, the more significant function of the products of these operons is probably to enable the cell to transport and degrade polymers of maltose called **maltodextrins**. These compounds are products of the breakdown of starch molecules, which are very long polysaccharides stored by cells to conserve energy. The sugar maltose is itself a disaccharide composed of two glucose residues with a 1-4 linkage, and the enzymes of the *malP-malQ* operon can break the maltodextrins down into maltose and then into glucose-1-phosphate, which can enter other pathways. Some bacteria, including species of *Klebsiella*, excrete extracellular enzymes that degrade long starch molecules and allow the bacteria to grow on starch as the sole carbon and energy source. *E. coli* lacks some of the genes needed to degrade starch to maltodextrins; therefore, in nature, it probably depends on neighboring microorganisms to break the starch down to the smaller maltodextrins that it can use.

The Maltose Transport System

Most of the protein products of the *mal* operons are involved in transporting maltodextrins and maltose through the outer and inner membranes into the cell (Figure 12.20). Five different proteins make up the transport system. The product of the *lamB* gene resides in the outer membrane, where it can bind maltodextrins in the medium. This protein forms a channel in the outer membrane through which the maltodextrins can pass. LamB is not required for growth on maltose, probably because maltose is small enough to pass through the outer membrane without its help. The LamB protein in

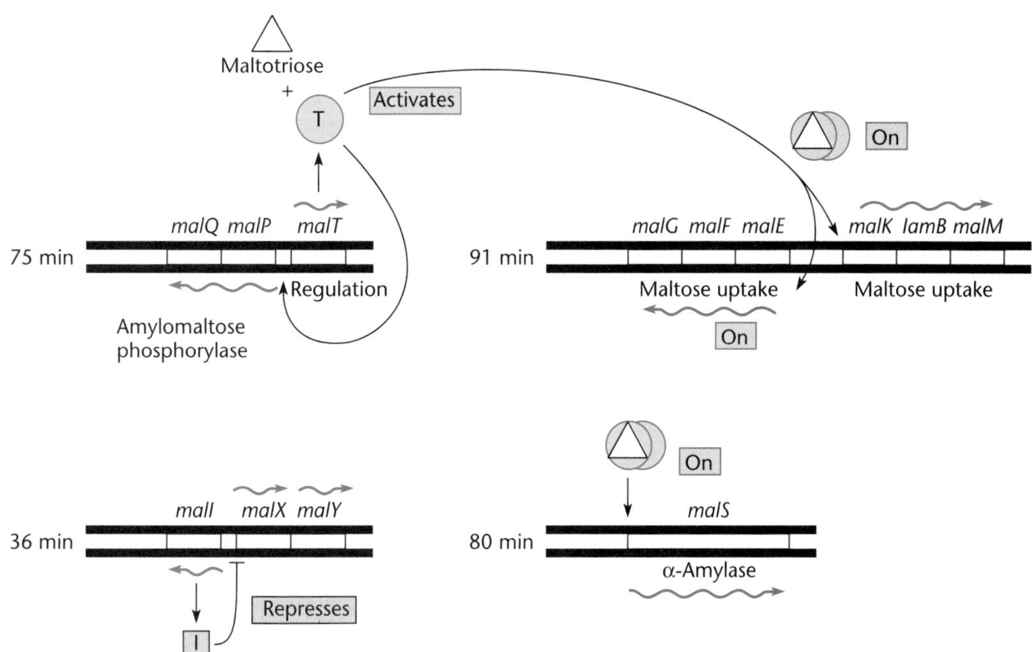

Figure 12.19 The maltose operons in *E. coli*. The MalT activator protein regulates both operons at 75 min and the operons at 80 and 91 min on the *E. coli* map. Another operon at 36 min is also induced by maltose. If maltose is not being transported, MalK, a part of the transport system, binds to MalT, inactivating it. doi:10.1128/9781555817169.ch12.f12.19

the outer membrane also serves as the cell surface receptor for phage λ, so the gene name is derived from the phage name (*lamB* from lambda). Mutants of *E. coli* resistant to λ have mutations in the *lamB* gene and lack the receptor for λ in the outer membrane.

Once maltodextrins are through the outer membrane, the MalS protein in the periplasm may degrade them into smaller polymers before they can be transported through the inner membrane. The smaller polymers of maltose bind to MalE in the periplasm between the outer and inner membranes. The MalF, MalG, and MalK proteins then transport the maltodextrin through the inner membrane. The MalF and MalG proteins are membrane-bound permeases, while the MalK protein is the ATPase that provides the energy. Altogether, these proteins form what is called a high-affinity ATP-binding cassette (ABC) transporter, based on the presence of an ATP-binding motif. There are many examples of related ABC transporters involved in transporting substances into the cell or that allow export as part of protein-secreting systems (see chapter 14).

Regulation of the *mal* Operons

The regulation of the *mal* operons is illustrated in Figure 12.19. The inducer of the *mal* operons is **maltotriose**, a trisaccharide composed of three molecules of glucose. Maltotriose can be synthesized from maltose by some of the enzymes encoded by the *mal* operons. Also, the cell normally contains polymers of maltose that were synthesized in the cell from glucose (and so do not need to be transported in) and can be broken down into maltotriose. The enzymes that degrade maltose polymers, therefore, play an indirect role in regulating the operons.

The genes in all three clusters are regulated by a single activator, encoded by the *malT* gene in the first cluster. The MalT activator is a member of a large family of activators that includes SoxS, which regulates genes involved in the response to oxidative stress (Box 12.2). The MalT activator protein specifically binds the inducer, maltotriose, and activates transcription of the operons. As with many catabolic operons, however, transcription of the *mal* operons occurs only if glucose, a better carbon source than maltose, is not present. Glucose regulates the ability of MalT to activate transcription of the *mal* operons in two ways, one through repressing the synthesis of MalT and the other through regulating its activity. Glucose represses the synthesis of the MalT activator by acting through a repressor protein, Mlc (for makes large colonies), which represses transcription of the *malT* gene, as well as many other genes involved in using alternative carbon sources. How Mlc represses the transcription of *malT* and other genes is interesting but complicated. Whether Mlc can repress transcription of the *malT* gene depends on the state of the system that transports glucose and other sugars into the cell; this system is called the phosphotransferase system. If glucose is present in the medium, the transporter of glucose

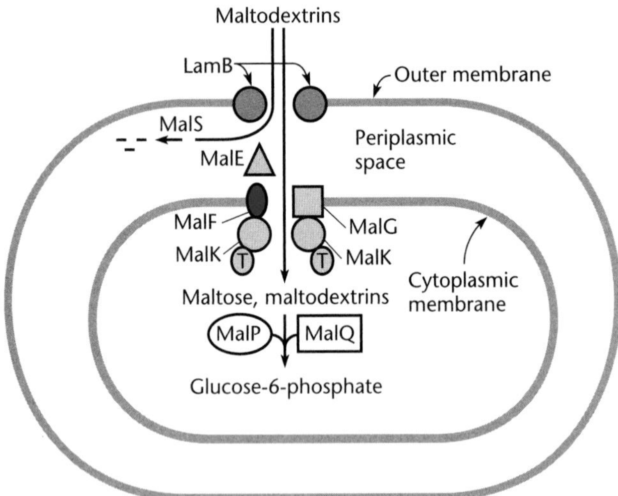

Figure 12.20 Functions of the genes of the maltose regulon in the transport and processing of maltodextrins in *E. coli.* The LamB protein binds maltodextrins and transports them across the outer membrane. The MalE protein in the periplasmic space then passes them through a pore in the cytoplasmic, or inner, membrane formed by the MalF, MalB, and MalK proteins, an ABC transporter. The MalK protein binds the MalT transcriptional activator and releases MalT only if maltose is being transported. Once in the cytoplasm, the maltodextrins and maltose are degraded by MalP and MalQ to glucose-1-phosphate and glucose, respectively. These compounds can then be converted into glucose-6-phosphate for use as energy and carbon sources.
doi:10.1128/9781555817169.ch12.f12.20

in the membrane, PtsG, transports glucose into the cell by transferring phosphate to glucose. PtsG, therefore, tends to be unphosphorylated if there is glucose; however, if there is no glucose, PtsG tends to be phosphorylated. Unphosphorylated PtsG binds specifically to Mlc, sequestering it so that it is unavailable to repress transcription of *malT* and the other genes under Mlc control. The other way glucose prevents MalT function is by acting through the catabolite regulation system via the CAP protein (see chapter 13). When glucose is low, CAP (in the presence of cAMP) binds upstream of the *malE* protein at a site adjacent to the MalT-binding site and repositions MalT on the DNA so that it can interact with RNA polymerase to activate transcription.

Positive Repressible Systems

The activity of an activator protein can be controlled either by the action of an inducer that activates it (like arabinose for AraC in the *ara* operon) or by an effector that inactivates it and therefore is formally a corepressor. As was discussed above in the context of negative repressible systems, expression of genes involved in biosynthesis of a cellular component is often turned off in the presence of high concentrations of the product of the pathway, for example, in the *E. coli trp* operon, where the TrpR repressor requires tryptophan for its DNA-binding activity. In the case of a positively regulated system that is repressible, the role of the effector molecule is to inactivate the activator, which has the same genetic effect as a corepressor that activates a repressor (as tryptophan does for TrpR). The common feature is that the effector molecule turns off gene expression, regardless of the mechanism.

THE *E. COLI fab* OPERON

The *E. coli fab* operon encodes genes involved in fatty acid biosynthesis. The FadR protein serves as a transcriptional activator, analogous to AraC or MalT. However, in this case, FadR binds to its target site upstream of the *fab* operon promoter in the absence of an effector molecule, so the default state of the system is for gene expression to be on. Fatty acids, the end product produced by the *fab* operon gene products, bind to FadR and prevent it from binding to the DNA. The products of the pathway signal the cell that the enzymes involved in producing them are no longer required. Since the role of the effector is to turn gene expression off, this represents a repressible system. The regulatory protein is an activator, which turns gene expression on, so this system is an example of a positive repressible system.

Molecular Mechanisms of Transcriptional Activation

Like transcriptional repression, activation can utilize a variety of mechanisms (see Browning and Busby, Suggested Reading, and Box 12.2). Promoters that are dependent upon the action of an activator protein generally have relatively low affinity for binding of RNA polymerase. This results in low levels of transcription in the absence of the activator, a condition that is necessary for transcription to be dependent upon the activator. The poor affinity for RNA polymerase is usually reflected by weak similarity to the consensus recognition sequence for RNA polymerase holoenzyme containing σ^{70} (see chapter 2 and Box 12.2). In many cases, the promoter contains a very poor −35 region, which results in loss of contact between region 4 of σ^{70} and the promoter DNA. Many transcriptional activators replace this protein-DNA contact between RNA polymerase and the promoter with a protein-protein contact with RNA polymerase by positioning of the activator on the DNA immediately upstream of the RNA polymerase-binding site.

A large class of activators function by interacting with the carboxyl-terminal domains of the RNA polymerase α subunits (αCTDs). As described in chapter 2, the αCTDs are connected to the rest of the RNA polymerase by a flexible linker. The αCTD can interact directly with the DNA upstream of the −35 region of the promoter if the DNA contains an UP element (see chapter 2 and Box 12.2). UP elements are found in promoters of rRNA operons and contribute to the very high transcriptional activity of these promoters (see chapter 13). Binding of certain transcriptional activators (including CAP [see chapter 13]) upstream of the –35 region allows a protein-protein interaction between the αCTD and the activator to mimic the protein-DNA interaction between αCTD and an UP element and activates transcription by inreasing the affinity of RNA polymerase for the promoter. Both UP elements and binding sites for activators that interact with αCTD generally exhibit face-of-the-helix dependence for their position relative to the –35 region of the promoter. Addition of 10 bp (a full turn of the DNA helix) between the –35 region and the activator-binding site (or UP element) is often tolerated, because the flexible linker allows αCTD to reach further upstream on the DNA. However, addition of 5 bp (half a turn of the DNA helix) usually disrupts activation, because αCTD is unable to reach around to the other side of the DNA helix.

Transcriptional activators can also interact with other parts of RNA polymerase to facilitate binding to the promoter. An example of this is provided by the phage λ CI protein, which is a central regulator of the phage λ gene expression program (see chapter 8). The CI protein acts primarily as a repressor, but also activates transcription of its own gene when levels of the protein are very low. Transcriptional activation requires binding of CI immediately upstream of the –35 region of the promoter, where the protein interacts with the σ subunit. Activators that interact with σ do not exhibit face-of-the-helix dependence of their binding site relative to the binding site for σ, because, unlike αCTD, σ is unable to "stretch" to reach a binding site that is further upstream on the DNA.

Activators that function to increase the affinity of RNA polymerase for a promoter can in some cases act as repressors if the promoter already exhibits high affinity for RNA polymerase. This occurs because RNA polymerase must release from the promoter DNA to move into the elongation phase of transcription (see chapter 2). This represents an example of how a regulatory protein can act as either an activator or a repressor, depending on either the position of its binding site relative to the promoter or features of the promoter itself (Box 12.2).

INTERACTIONS OF ACTIVATORS AND REPRESSORS

Individual genes or operons are often subject to regulation by multiple regulatory proteins that act together to modulate expression of the target genes in response to multiple physiological signals. In some cases, different regulatory proteins cooperate to allow activation, where either protein alone is insufficient. The *mal* operon, where the MalT activator must be repositioned by CAP plus cAMP (when glucose levels are low), allows the operon to be induced only when maltose is present and glucose is absent. Other systems coordinate multiple signals by using antiactivation. As described above for negative regulation, repressor proteins usually bind to the promoter region at a position that overlaps with the RNA polymerase-binding site and prevent transcription simply by preventing access of RNA polymerase to the promoter. In contrast, antiactivators bind to the DNA upstream of the promoter, at a site that overlaps the position where a required activator protein must bind. The presence of the repressor protein prevents binding of the activator and therefore inhibits transcription by preventing activation. These types of complexities illustrate how cells can use a fixed set of regulatory proteins in a variety of ways to integrate physiological signals of a variety of types.

Regulation by Transcription Attenuation

In the above examples, the transcription of a gene or operon is regulated through effects on the initiation of RNA synthesis. However, transcription can also be regulated after RNA polymerase leaves the promoter. One way is by transcription attenuation. Unlike repressors and activators, which turn transcription from the promoter on or off, the attenuation mechanism works by allowing transcription to begin constitutively at the promoter but then terminates it in the **leader region** of the transcript before the RNA polymerase reaches the first structural gene of the operon (Figure 12.21A). The termination event occurs only if the gene products of the operon are not needed (see Merino and Yanofsky, Suggested Reading). The classic example of regulation by transcriptional attenuation is the *trp* operon of *E. coli*, which uses translation of a short coding sequence to regulate transcription termination. The *Bacillus subtilis trp* operon and the *E. coli bgl* operon use RNA-binding proteins, and in riboswitch systems, the leader RNA directly monitors a regulatory signal. In all of these cases, physiological effects cause changes in the secondary structure of the leader RNA, which affects termination of transcription. The processivity of RNA polymerase can also be modified by proteins that associate with RNA polymease, as is the case for the phage λ N and *E. coli* RfaH proteins. Representative examples of these types of systems are discussed below.

Modulation of RNA Structure

Many transcription attenuation mechanisms operate by effects on the leader RNA secondary structure. Factor-independent transcriptional terminators are composed

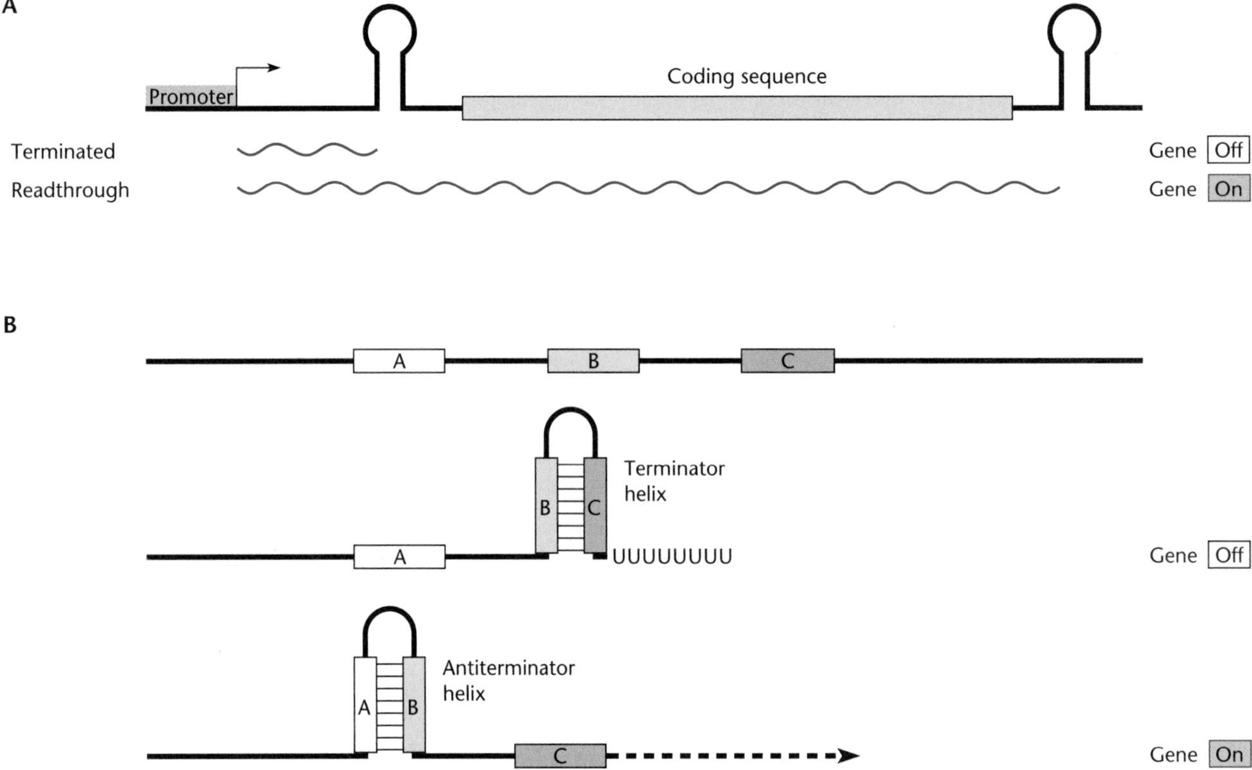

Figure 12.21 Transcription attenuation. **(A)** The presence of a transcription terminator in the leader region of a gene can lead to premature termination of transcription and synthesis of only a short RNA that does not include the downstream coding sequence. Readthrough of the termination site is required for expression of the downstream gene. **(B)** Alternate folding of the leader RNA into either the B-C terminator helix or the A-B antiterminator helix determines whether the short terminated RNA or the full-length RNA is made. doi:10.1128/9781555817169.ch12.f12.21

of a short helix immediately adjacent to a stretch of U residues in the newly synthesized (or nascent) RNA. As described in chapter 2, pausing of RNA polymerase during synthesis of the U residues allows the nascent RNA that is emerging from RNA polymerase to fold into the helix, which triggers RNA polymerase to terminate transcription. Termination can be prevented by folding of the nascent RNA into a different RNA structure that captures sequences that would otherwise form the terminator helix (Figure 12.21B). This alternate structure is called an **antiterminator**, because it competes with formation of the structure that acts as a factor-independent terminator. The choice between termination and antitermination is often mediated by interaction of regulatory molecules with the RNA, which determines which of the competing structures will form.

REGULATION OF THE *E. COLI trp* OPERON BY LEADER PEPTIDE TRANSLATION

The archetype of transcription attenuation control is the *trp* operon of *E. coli*. As discussed above, the *trp* operon is like the *lac* operon in that it is negatively regulated at the level of transcription initiation, in this case, by the TrpR repressor protein with tryptophan as the corepressor. However, early genetic evidence suggested that this is not the only type of regulation for the *trp* operon. If the *trp* operon were regulated solely by the TrpR repressor, the levels of the *trp* operon enzymes in a *trpR* mutant would be the same in the absence and the presence of tryptophan. However, even in a *trpR*-null mutant, the expression of these enzymes is higher in the absence of tryptophan than in its presence, indicating that the *trp* operon is subject to a regulatory system in addition to the TrpR repressor and that this second level also responds to tryptophan availability.

Early evidence suggested that $tRNA^{Trp}$ (the tRNA that is responsible for insertion of tryptophan at UGG tryptophan codons) plays a role in the regulation of the *trp* operon (see Morse and Morse, Suggested Reading). Mutations in the structural gene for $tRNA^{Trp}$ and in the gene encoding the aminoacyl-tRNA synthetase responsible for transferring tryptophan to $tRNA^{Trp}$, as well as mutations in genes whose products are responsible for modifying $tRNA^{Trp}$, increase the expression of the operon. All these mutations presumably lower the amount of tryptophanyl-tRNA^{Trp} in the cell, suggesting that this other regulatory mechanism senses the amount of tryptophan bound to tRNA^{Trp}.

Other evidence suggested that the region targeted by this other type of regulation is the leader region, or *trpL* (Figures 12.12 and 12.22). Deletions in this region, which lies between the promoter and *trpE*, the first gene of the operon, eliminate the regulation, so that double mutants, with both a deletion mutation of the leader region and a *trpR* mutation, are completely constitutive for expression of the *trp* operon. Deletions of the leader region are also *cis* acting and affect only the expression of the *trp* operon on the same DNA. Later evidence indicated that transcription terminated in this leader region in the presence of tryptophan because of an excess of aminoacylated tRNA^{Trp}. Because the regulation seemed to be able to stop, or attenuate, transcription that had already initiated at the promoter, it was called attenuation of transcription, in agreement with an analogous type of regulation observed in the *Salmonella his* operon.

Model for Regulation of the *E. coli trp* Operon by Attenuation

Figures 12.22 and 12.23 illustrate the current model for regulation of the *trp* operon by attenuation (see Oxender et al., Suggested Reading). According to this model, the percentage of the tRNA^{Trp} that is aminoacylated (i.e., has tryptophan attached) determines which of several alternative secondary-structure helices will form in the leader RNA. Recall from chapter 2 that formation of an RNA helix results from complementary pairing between the bases in RNA transcribed from inverted-repeat sequences in the DNA.

Whether transcription termination occurs depends on whether the attenuation mechanism senses relatively low or high levels of tryptophan. The *trpL* region, which contains two consecutive Trp codons, monitors the signal. If levels of tryptophan are low, the levels of tryptophanyl-tRNA^{Trp} will also be low. When a ribosome encounters one of the UGG Trp codons, it temporarily stalls, unable to insert a tryptophan. This stalled ribosome in the *trpL* region therefore communicates that the tryptophan concentration is low and that transcription should continue (Figure 12.23). This mechanism is dependent on the fact that transcription and translation are coupled in bacteria, and a ribosome can initiate translation of a transcript before synthesis of the transcript is complete.

Figures 12.22 and 12.23 show how the helices operate in attenuation. Four different regions in the *trpL* leader RNA, regions 1, 2, 3, and 4, can form three different helices (1:2, 2:3, and 3:4) by alternate pairing of region 2 with either region 1 or region 3 (1:2 versus 2:3) and alternate pairing of region 3 with either region 2 or region 4 (2:3 versus 3:4), as shown in Figure 12.22. The formation of helix 3:4 causes RNA polymerase to terminate transcription, because this helix is part of a factor-independent transcription termination signal (see chapter 2). Whether helix 3:4 forms is determined by the dynamic relationship between ribosomal translation

Figure 12.22 Structure and relevant features of the leader region of the *trp* operon involved in regulation by attentuation. UGGUGG (in blue) indicates the two Trp codons in the leader region. doi:10.1128/9781555817169.ch12.f12.22

A The *trpL* leader region

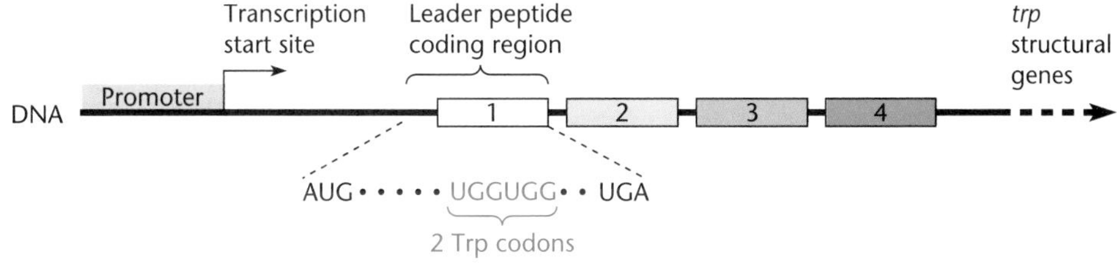

B Alternative pairing structures for *trpL* RNA

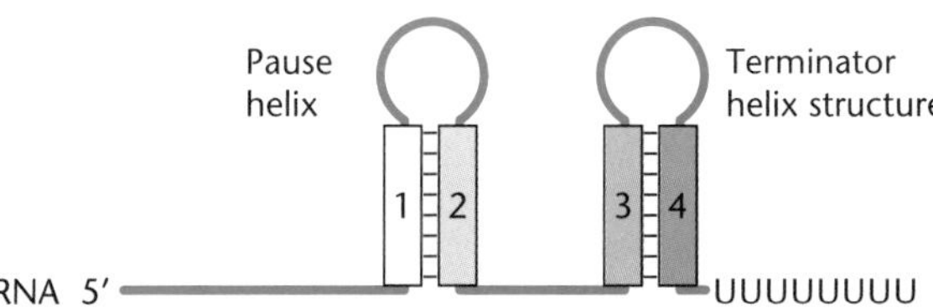

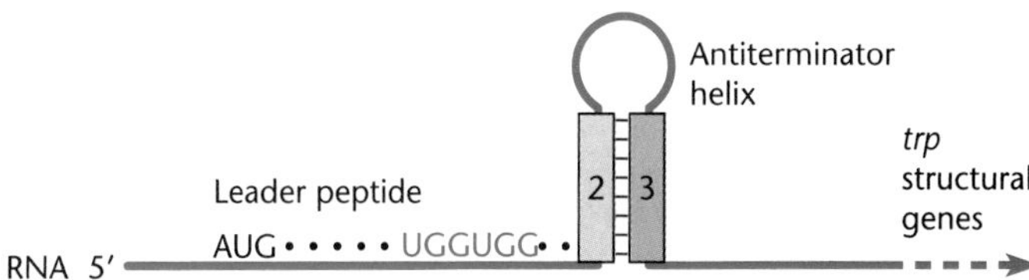

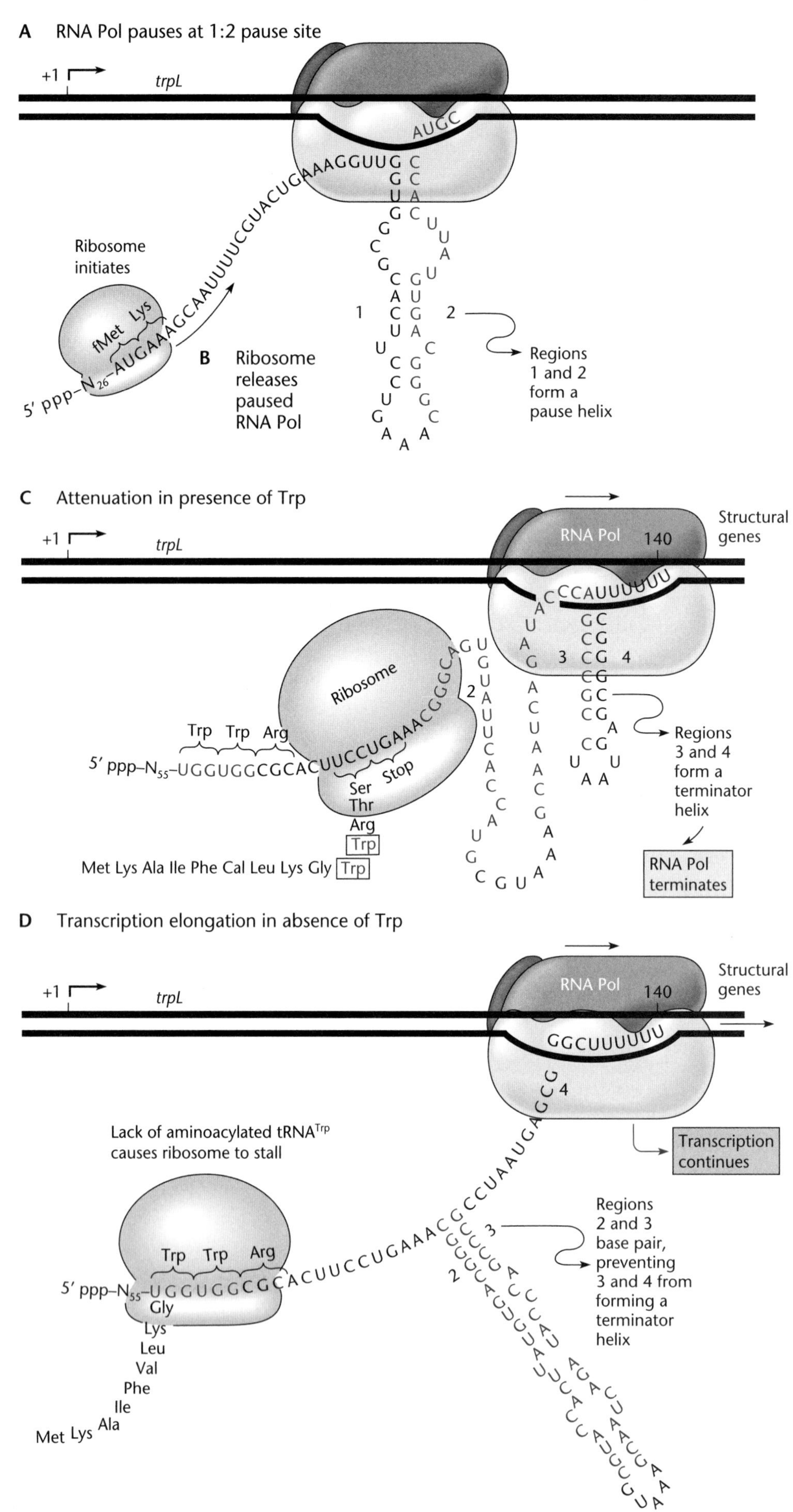
A RNA Pol pauses at 1:2 pause site
+1
trpL
Ribosome initiates
fMet Lys
5′ ppp–N26–AUGAAAGCAAUUUUCGUACUGAAAGGUUG
B Ribosome releases paused RNA Pol
1
2
Regions 1 and 2 form a pause helix
C Attenuation in presence of Trp
+1
trpL
RNA Pol
140
Structural genes
Ribosome
Trp Trp Arg
5′ ppp–N55–UGGUGGCGCACUUCCUGAAACGGGCA
Ser Stop
Thr
Arg
Trp
Met Lys Ala Ile Phe Cal Leu Lys Gly Trp
2
3
4
Regions 3 and 4 form a terminator helix
RNA Pol terminates
D Transcription elongation in absence of Trp
+1
trpL
RNA Pol
140
Structural genes
GGCUUUUUU
4
Transcription continues
Lack of aminoacylated tRNATrp causes ribosome to stall
Trp Trp Arg
5′ ppp–N55–UGGUGGCGCACUUCCUGAAACGC
Gly
Lys
Leu
Val
Phe
Ile
Met Lys Ala
2
3
Regions 2 and 3 base pair, preventing 3 and 4 from forming a terminator helix

of the Trp codons in the *trpL* region and the progress of RNA polymerase through the *trpL* region (also illustrated in Figure 12.23). After RNA polymerase initiates transcription at the promoter, it moves through the *trpL* region to a site located just after region 2, where it pauses. The helix formed by leader RNA regions 1 and 2 (1:2) is an important part of the signal to pause. The pause is short, but it ensures that a ribosome has time to load onto the leader RNA before the RNA polymerase proceeds to region 3. The moving ribosome may help to release the paused RNA polymerase by catching up and colliding with it.

The progress of ribosome translation through the Trp codons of *trpL*, then, determines whether helix 3:4 will form, causing termination, or whether region 2 will pair instead with region 3 to form the 2:3 helix. The 2:3 helix sequesters region 3, preventing formation of the 3:4 helix; helix 2:3 therefore acts as an antiterminator, because its formation prevents termination. Region 3 will pair with region 2 if the ribosome stalls at the Trp codons because of low tryptophan concentrations and therefore low availability of aminoacylated $tRNA^{Trp}$ (Figure 12.23D). If the ribosome does not stall at the Trp codons, it will continue until it reaches the UGA stop codon at the end of *trpL*. By remaining at the stop codon while region 4 is synthesized, the ribosome prevents formation of helix 2:3. Therefore, helix 3:4 can form and terminate transcription (Figure 12.23C).

Genetic Evidence for the *trp* Attenuation Model

The existence and in vivo function of helix 2:3 were supported by the phenotypes produced by mutation *trpL75*. This mutation, which changes one of the nucleotides and prevents pairing of two of the bases holding the helix together, should destabilize the helix. In the *trpL75* mutant, transcription terminates in the *trpL* region even in the absence of tryptophan, consistent with the model that formation of helix 2:3 normally prevents formation of helix 3:4.

The idea that translation of the leader peptide from the *trpL* region is essential to regulation is supported by the phenotypes of mutation *trpL29*, which changes the AUG start codon of the leader peptide to AUA, preventing initiation of translation. In *trpL29* mutants, termination occurs even in the absence of tryptophan. The model also explains this observation as long as we can assume that the RNA polymerase paused at helix 1:2 will eventually move on, even without a translating ribosome to nudge it, and will eventually transcribe the 3:4 region. Without a ribosome stalled at the Trp codons, however, helix 1:2 will persist, preventing the formation of helix 2:3. If helix 2:3 does not form, helix 3:4 will form and transcription will terminate.

One final prediction of the model is that stalling translation at codons other than those for tryptophan in *trpL* should also cause increased transcription through the termination site. Changing the tryptophan codons in the *trpL* region to codons for other amino acids results in increased readthrough in response to starvation for those amino acids, as predicted by the model. Several other amino acid-related operons in *E. coli* have similar leader peptide coding regions, with the codons appropriate for those operons (e.g., phenylalanine codons for a phenylalanine-related operon). This mechanism is easily adaptable to new types of genes during evolution simply by changing the identities of the codons within the leader peptide coding region.

REGULATION OF THE *B. SUBTILIS trp* OPERON BY AN RNA-BINDING PROTEIN

Comparing the regulation of the same operon in different types of bacteria often reveals that the same regulatory response can be achieved in different ways. The *trp* operon of *B. subtilis* consists of seven genes whose products are required to make tryptophan from chorismic acid. Interestingly, although *B. subtilis* uses different mechanisms than *E. coli* to regulate its *trp* operon, the result is the same: the operon can respond both to the amount of free tryptophan in the cell and to the aminoacylation state of $tRNA^{Trp}$.

Like the *E. coli trp* operon, the *B. subtilis trp* operon has a leader region that includes a factor-independent transcriptional terminator (Figure 12.24). However, rather than depending on pausing by the ribosome at tryptophan codons to sense limiting tryptophan and alter the secondary structure of the mRNA, the

Figure 12.23 Details of regulation by transcription attenuation in the *trp* operon of *E. coli.* **(A)** RNA polymerase pauses after transcribing regions 1 and 2. **(B)** A ribosome has time to load onto the mRNA and begins translating, eventually reaching the RNA polymerase and bumping it off the pause site. **(C)** In the presence of tryptophan, aminoacylated $tRNA^{Trp}$ is available and the ribosome translates through the Trp codons and prevents the formation of the 2:3 helix, thereby allowing the formation of helix 3:4, which is part of a transcription terminator. Transcription terminates. **(D)** In the absence of tryptophan, aminoacylated $tRNA^{Trp}$ is not available, the ribosome stalls at the Trp codons, and helix 2:3 forms, preventing the formation of helix 3:4 and allowing transcription to continue through the terminator site.
doi:10.1128/9781555817169.ch12.f12.23

A Transcription attenuation

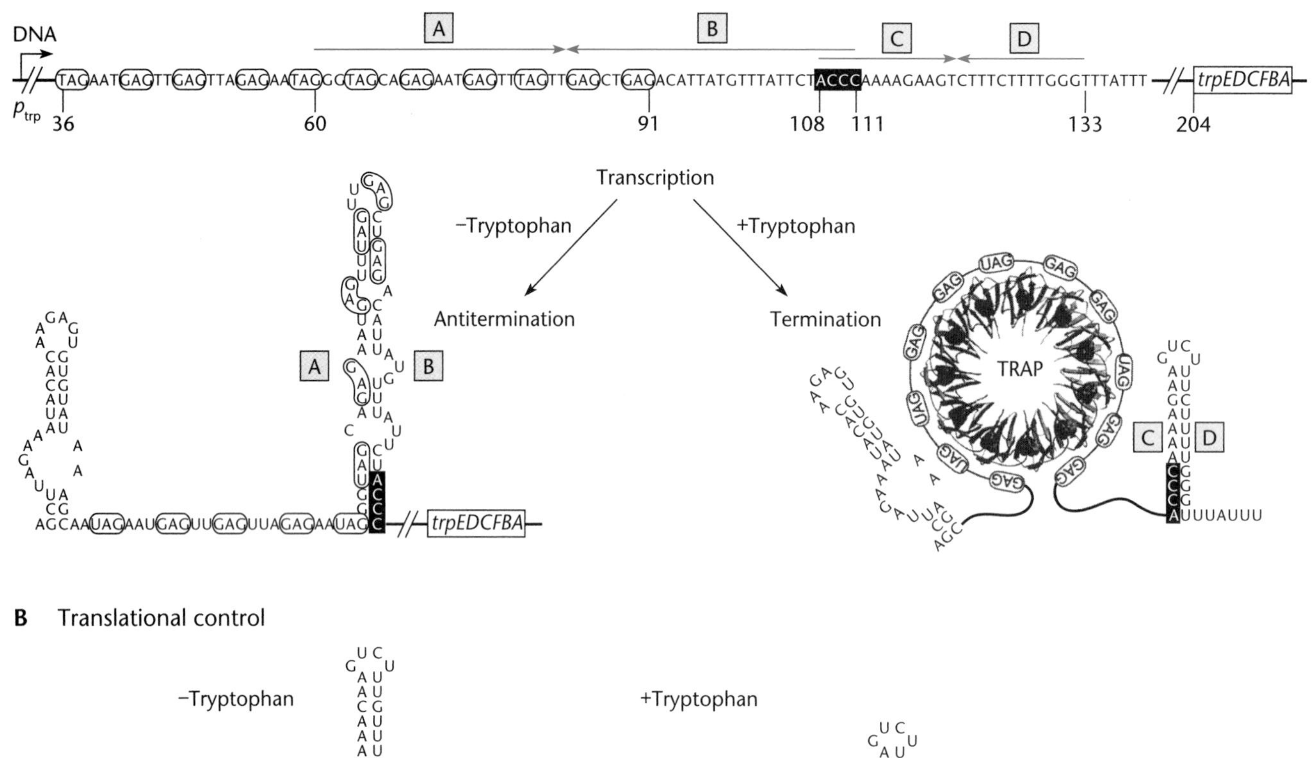

B Translational control

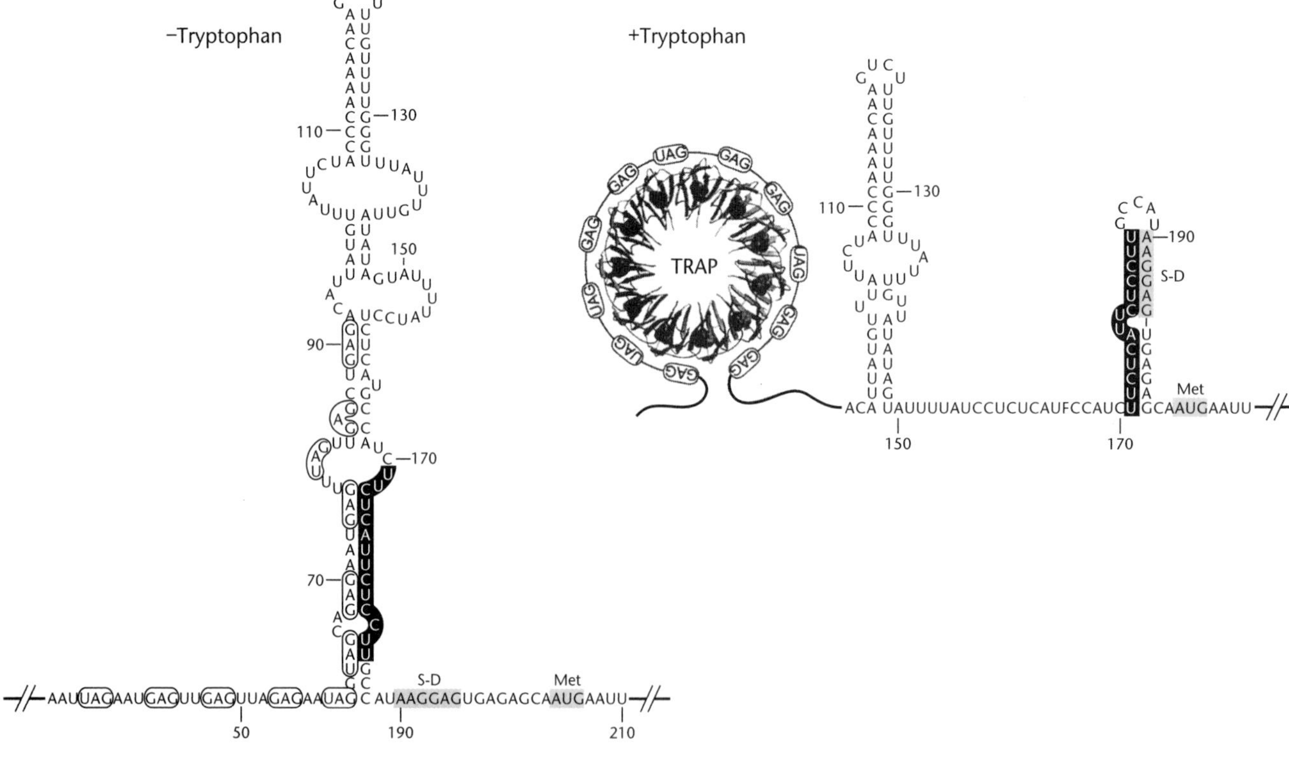

Figure 12.24 TRAP regulation of the *trp* operon in *B. subtilis.* **(A)** Model for transcription attenuation of the *trp* operon. When tryptophan is limiting (−Tryptophan), TRAP is not activated. During transcription, antiterminator formation (A-B helix) prevents formation of the terminator (C-D helix), which results in transcription of the *trp* operon structural genes. When tryptophan is in excess (+Tryptophan), TRAP is activated. Tryptophan-activated TRAP can bind to the (G/U)AG repeats and promote termination by preventing antiterminator formation. The overlap between the antiterminator and terminator structures is shown. **(B)** Translational control of *trpE* by TRAP. Under tryptophan-limiting conditions, TRAP is not activated and is unable to bind to the *trp* leader transcript. In this case, the *trp* leader RNA adopts a structure such that the *trpE* S-D sequence is single stranded and available for translation. When tryptophan levels are high, TRAP is activated and binds to the (G/U)AG repeats. As a consequence, a helix that sequesters the *trpE* S-D forms, which prevents ribosome binding and translation. The overlap between the two alternative structures is shown. Numbering is from the start of transcription. doi:10.1128/9781555817169.ch12.f12.24

B. subtilis trp operon uses an RNA-binding repressor protein called TRAP (for *trp* *R*NA-binding *a*ttenuation *p*rotein) (see Babitzke and Gollnick, Suggested Reading). This protein has 11 subunits, each of which can bind tryptophan. The subunits are arranged in a wheel, and each subunit binds to a repeated 3-base (triplet) sequence (either GAG or UAG) in the leader mRNA, but only if the TRAP subunit is bound to a tryptophan molecule. The optimal spacing between the triplets is 2 or 3 bp. This causes the leader RNA to wrap around TRAP, as shown, which sequesters sequences needed for formation of an antiterminator helix. If the antiterminator helix does not form, sequences needed for formation of a less stable downstream terminator helix are available, and transcription termination occurs. Therefore, transcription termination prevents expression of the operon only if tryptophan is present in the medium. The fact that there are 11 sites on TRAP for binding tryptophan may allow the regulation to be finely tuned in response to intermediate levels of tryptophan when only some of the 11 sites are occupied. TRAP is formally a repressor (like LacI or GalR), because mutations that inactivate the *mtrB* gene that encodes TRAP result in constitutive *trp* operon expression. Like the TrpR repressor, TRAP requires tryptophan as a corepressor.

TRAP, when bound to tryptophan, can also directly block translation of the first gene of the *trp* operon by binding to similar 3-base repeats just upstream of the Shine-Dalgarno (S-D) sequence of the *trpE* gene, as well as other tryptophan-related genes. TRAP binding close to the S-D sequence prevents the ribosome from binding and blocks translation of the gene (see below). Therefore, tryptophan in the medium can inhibit both the transcription and the translation of the genes of the *trp* operon. Yet another level of regulation that controls the activity of the TRAP RNA-binding protein in response to the aminoacylation state of $tRNA^{Trp}$ utilizes the T box riboswitch mechanism, which is discussed below.

REGULATION OF THE *E. COLI bgl* OPERON BY AN RNA-BINDING PROTEIN

TRAP acts a repressor by destabilizing an antiterminator helix and promoting termination. An RNA-binding protein can also act as an activator of gene expression by stabilizing an antiterminator helix and preventing termination. This is the strategy used by the *bgl* operon of *E. coli*. The *bgl* operon encodes proteins required for degradation of β-glucoside sugars for use as carbon and energy sources (see Fux et al., Suggested Reading). The BglG RNA-binding protein binds to and stabilizes an antiterminator helix that shares sequences with a terminator helix in the leader region of the *bgl* operon. If the BglG protein stabilizes the antiterminator helix, the terminator helix does not form, and transcription continues into the operon, which results in transcription of the genes for β-glucoside uptake and degradation.

As with other catabolic pathways, the *bgl* operon is induced only when β-glucosides are present in the medium. However, the β-glucoside inducer does not bind directly to the BglG protein. Instead, the activity of BglG is coupled to the transport of β-glucosides into the cell. The BglG protein is normally phosphorylated by the BglF protein, which is part of the transport system responsible for import of β-glucosides. When β-glucosides are transported by BglF, BglF transfers a phosphate group onto the sugar and does not phosphorylate BglG. If no β-glucoside transport is occurring, BglF instead phosphorylates the BglG protein. Phosphorylation of BglG prevents dimerization of BglG; since dimers are the active form of the BlgG protein that can bind to the antiterminator helix, phosphorylation inactivates BglG and results in transcription attenuation. This ensures that the genes for β-glucoside utilization are turned on only if β-glucosides are being transported into the cell. Additional interactions with the sugar transport system have also been demonstrated that suggest that BglG is captured by binding to BglF in the membrane and is released by sugar transport, which facilitates its activity in transcription antitermination (see Raveh et al., Suggested Reading).

The BglG protein is an activator, as inactivation of the *bglG* gene results in loss of *bgl* operon expression, even in the presence of the β-glucoside inducer. BglF is formally a repressor, since it inactivates the BglG activator. Inactivation of the *bglF* gene results in constitutive *bgl* operon expression, because BglG is not phosphorylated and is therefore always active. However, growth on β-glucoside sugars cannot occur in a *bglF* mutant because transport of the sugars across the membrane reguires the active BglF transporter.

RIBOSWITCH RNAs DIRECTLY SENSE METABOLIC SIGNALS

A **riboswitch** is an RNA element, usually found in leader RNAs, that controls expression of the downstream coding sequences by sensing a regulatory signal directly, without a requirement for translation (as in the *E. coli trp* operon) or a regulatory protein (as in the *B. subtilis trp* and *E. coli bgl* systems). These RNA elements usually fold into a complex three-dimensional structure that specifically recognizes their cognate ligand. Binding of the ligand results in a conformational change in the RNA, and this conformational change is responsible for the gene expression effect. Regulation can occur either at the level of transcription attenuation, by folding of the RNA into alternate terminator and antiterminator helices, or at the level of translation initiation, by folding of the RNA into a structure that sequesters the ribosome-binding site (see below). Riboswitches have been identified that respond to a variety of ligands, including tRNA and small molecules, such as nucleotides, cofactors, amino acids, and metal ions.

The T Box Mechanism: tRNA-Sensing Riboswitches

The first example of an effector of gene expression acting directly on the mRNA to be discovered was the regulation of the transcription of the genes for the aminoacyl-tRNA synthetase (aaRS) genes in *B. subtilis*. Depriving the cell of an amino acid causes an accumulation of the uncharged form of the tRNA for that amino acid. The bacteria respond by synthesizing higher levels of the aaRS for that tRNA, which allows more efficient attachment of the limiting amino acids to their cognate tRNAs.

The expression of the aaRS genes in *B. subtilis* is regulated through transcription attenuation. If the tRNA for that amino acid is mostly unaminoacylated (with no amino acid attached), transcription terminates less often in the leader sequence and more aaRS is made. If most of the tRNA for that amino acid has the amino acid attached (i.e., is aminoacylated), transcription of the aaRS gene often terminates in the leader region. Therefore, whether transcription terminates in the leader sequence of each gene is determined by the relative levels of the unaminoacylated and aminoacylated cognate tRNA for that aaRS.

The leader RNAs of genes in this family contain a series of conserved sequence and structural features, which include a terminator helix and a competing antiterminator helix (see Grundy and Henkin, Suggested Reading). An additional crucial element is a large helix that contains a small internal unpaired region called the specifier loop (Figure 12.25). This unpaired region includes a triplet sequence that matches a codon that corresponds to the amino acid specificity of the regulated aaRS (e.g., the tyrosyl-tRNA synthetase gene leader RNA contains a UAC tyrosine codon or a tryptophanyl-tRNA synthetase gene leader RNA contains a UGG tryptophan codon). A simple genetic experiment proved that this triplet codon is responsible for the specific response of each gene to the aminoacylation status of its cognate amino acid. Mutation of the UAC Tyr codon in the *tyrS* gene to a UUC Phe codon resulted in loss of response to tyrosine availability and a switch to a response to limitation of phenylalanine. Mutation of the triplet to a UAG nonsense codon resulted in constitutive termination and could be suppressed by introduction into the cell of a tRNA in which the anticodon was mutated to match the UAG codon (a nonsense suppressor tRNA) (see chapter 3).

These and other experiments suggested a model in which binding of a specific unaminoacylated tRNA to the leader RNA stabilizes the antiterminator, which sequesters sequences that would otherwise participate in formation of the terminator helix. The presence of the amino acid on the 3′ end of the aminoacylated tRNA prevents the interaction with the antiterminator, which allows the system to differentiate between aminoacylated and unaminoacylated tRNA. The triplet sequence, which was designated the specifier sequence, determines which tRNA a particular leader RNA will monitor.

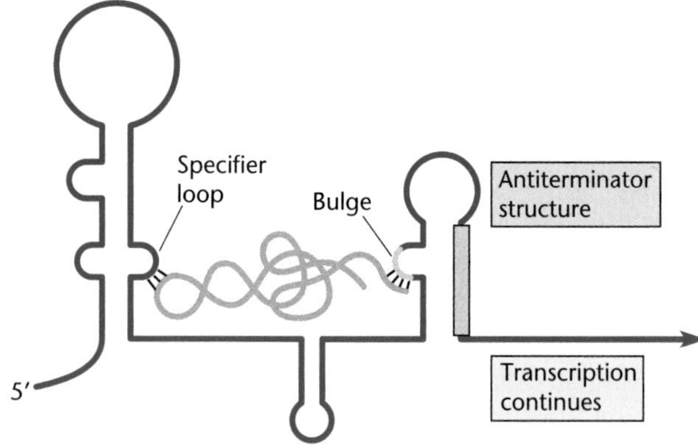

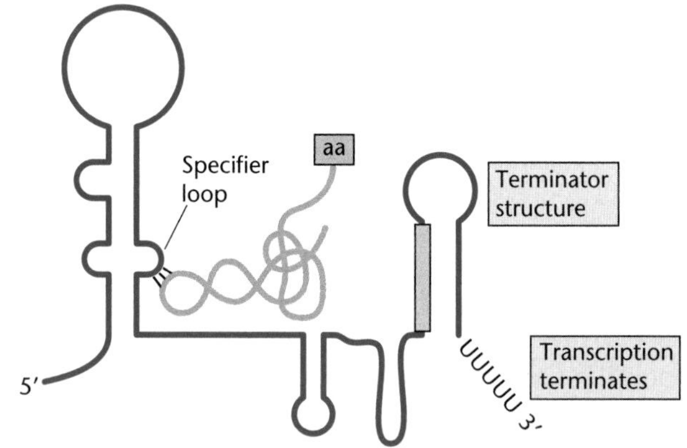

Figure 12.25 The tRNA-responsive T box riboswitch system. The leader RNAs for amino acid-related genes contain competing terminator and antiterminator helices. An upstream helix includes the specifier loop, which contains a triplet sequence corresponding to a codon of the same amino acid class as the regulated gene. A specific tRNA binds to the specifier loop through codon-anticodon base pairing. **(A)** If the tRNA is not aminoacylated, the 3′ end of the tRNA makes a second interaction with a bulge in the antiterminator helix, which stabilizes the antiterminator and prevents termination, allowing expression of the downstream gene. **(B)** If the tRNA is aminoacylated, it is unable to pair with the bulge. The antiterminator helix is not stabilized, and the terminator helix forms, which results in transcription termination. The sequence region shared by the antiterminator and terminator helices is boxed.
doi:10.1128/9781555817169.ch12.f12.25

In vitro experiments have confirmed many aspects of this model, including the fact that binding of tRNA to the leader RNA can affect its secondary structure and stabilize the antiterminator hairpin (see Yousef et al., Suggested Reading). Antitermination occurs when appropriate unaminoacylated tRNA is added to a purified system that has only the RNA polymerase and the DNA template, confirming that no other factors are required. Many different aaRS genes (as well as genes involved in amino acid biosynthesis, transport, and gene regulation) can be regulated by the same mechanism, but each gene responds only to levels of the cognate amino acid, monitored via the aminoacylation status of the corresponding tRNA.

As described above, the *B. subtilis trp* operon is regulated in response to tryptophan by the TRAP RNA-binding protein. However, like the *E. coli trp* operon, the *B. subtilis trp* operon also responds to the aminoacylation of $tRNA^{Trp}$. Another regulatory gene that contains a series of 3-base repeats characteristic of a TRAP-binding site in its leader RNA was identified. The leader sequence of this gene also contained a T box RNA with a UGG Trp specifier sequence, suggesting that its expression would be induced only when $tRNA^{Trp}$ is poorly aminoacylated. Overproduction of the regulatory gene resulted in constitutive expression of the *trp* operon, and the protein product was shown to bind directly to TRAP, even in the presence of tryptophan, blocking the ability of the TRAP-Trp complex to bind to the *trp* operon leader RNA. The protein was therefore named anti-TRAP. The transcription of the gene for anti-TRAP is induced when aminoacylation of $tRNA^{Trp}$ is low, resulting in increased *trp* operon expression. The *trp* operon of *B. subtilis* behaves much like the *trp* operon of *E. coli* in that expression is induced by both low tryptophan and low $tRNA^{Trp}$ aminoacylation, although the molecular mechanisms used are very different.

Metabolite-Binding Riboswitches

Riboswitch regulation can also occur by binding of small molecules directly to the leader RNA. Many of these riboswitches bind metabolites that are the end products of biosynthetic pathways and regulate genes that are involved with the pathways. As with the tRNA-binding T box riboswitch, binding of the ligand affects the secondary structure of the leader RNA, and the resulting structural change affects expression of the downstream genes. Examples of small molecules that are used in riboswitch regulation are amino acids (lysine and glycine), vitamins (B_{12} and thiamine pyrophosphate [B_1]), nucleic acid bases (guanine and adenine), cofactors (flavin mononucleotide and *S*-adenosylmethionine [SAM]), and metal ions (magnesium). The regulation is usually through transcription attenuation, as with the *B. subtilis* aaRS genes, but also can occur through translation by blocking the translation initiation region (TIR) on the mRNA (see below). A few examples have been found in eukaryotes, where binding of the ligand affects mRNA stability or RNA splicing.

One well-studied example of a metabolite-binding riboswitch is the S box riboswitch that regulates genes involved in methionine metabolism in *B. subtilis* and related organisms by transcription attenuation (see McDaniel et al., Suggested Reading). These genes contain leader RNA sequences with similar structural features that include a terminator and a competing antiterminator. If methionine is limiting, the leader RNA forms the more stable antiterminator helix, and the genes for methionine biosynthesis are expressed, whereas high methionine levels result in stabilization of the terminator helix and transcription attenuation (Figure 12.26). However, methionine does not act directly as the signal molecule. It plays a special role in the cell, in addition to its role in as a building block in protein synthesis. Methionine is also converted into SAM, which is the donor of methyl groups in many biochemical reactions in the cell. We mentioned some of these reactions in earlier chapters, including the methylation of DNA by restriction endonucleases, but there are many more, some of which are essential to the survival of the cell. The cell therefore determines its need for methionine by monitoring the level of SAM, and SAM is the effector that directly binds to the S box riboswitches upstream of methionine biosynthesis operons.

Binding of SAM to the S box riboswitch results in stabilization of a structural element in the RNA that includes sequences that would otherwise participate in formation of the stable antiterminator helix. The SAM-binding domain, therefore, serves as an antiantiterminator, as it prevents formation of the antiterminator. If the antiterminator is not stabilized by binding of SAM, the terminator can form, resulting in transcription attenuation only when SAM levels are high. Like other metabolite-binding riboswitches, the S box RNA binds very specifically to SAM and does not recognize related compounds, including *S*-adenosylhomocysteine, which is the product after SAM donates its methyl group in methyltransferase reactions. This specificity is very important, as it ensures that the genes for biosynthesis of methionine (and therefore SAM) are repressed only when SAM is abundant. As with the tRNA-responsive T box riboswitch, SAM-dependent termination can occur in a purified transcription system consisting only of RNA polymerase and template DNA, demonstrating that no other factors are required. The crystal structures of a number of riboswitches bound to their cognate ligand have been determined, and they provide insight into how each RNA specifically recognizes the appropriate metabolite.

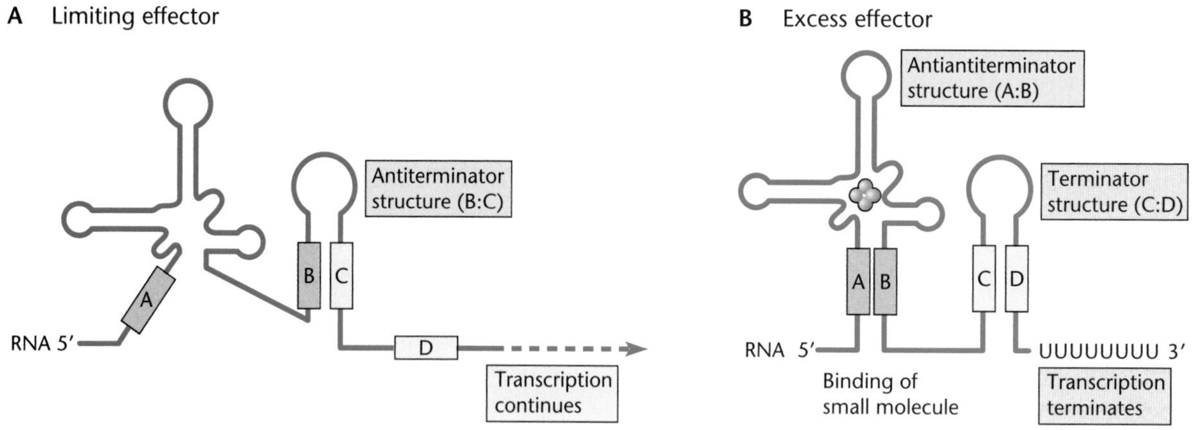

Figure 12.26 Metabolite-binding riboswitch regulation of transcription attenuation. Regions A, B, C, and D represent mRNA sequences that can bind to form alternative secondary structures. **(A)** Limiting effector. The antiterminator structure (B-C) forms, allowing transcription to continue. **(B)** High levels of effector. Binding of the effector molecule causes the antiantiterminator structure (A-B) to form. This sequesters sequences (B) that would otherwise participate in formation of the antiterminator. Consequently, the terminator helix (C-D) can form, and transcription terminates. doi:10.1128/9781555817169.ch12.f12.26

Changes in Processivity of RNA Polymerase

RNA structure can have a major effect on transcription termination, because of the dependence of factor-independent termination on formation of a helix in the nascent RNA immediately upstream of the run of U's in the RNA that form the U-A RNA-DNA hybrid within the transcription elongation complex (TEC) (see chapter 2). However, factors other than RNA structure can also affect transcription termination at both factor-independent and Rho-dependent terminators. They include factors that interact with RNA polymerase and affect its **processivity,** which means the efficiency with which it moves from nucleotide to nucleotide along the DNA template. Transcription termination requires that RNA polymerase pause at the termination site. In the case of factor-independent terminators, pausing allows the RNA to fold into the terminator helix before the RNA polymerase moves past the termination site. For Rho-dependent termination, pausing provides an opportunity for the Rho factor (which binds to the newly synthesized RNA and migrates in a 5′-to-3′ direction behind RNA polymerase) to catch up to the RNA polymerase and promote termination.

Factors that affect RNA polymerase processivity can target specific genes if those genes include sequences responsible for recruitment of the factors to the TEC. These factors often remain associated with the TEC and can therefore promote readthrough of termination sites encountered by the TEC over long distances. This phenomenon is called **processive antitermination** and results in generation of long transcripts. Processive antitermination is important both in regulatory systems, as described here, and in transcription of the long rRNA operons (see chapter 13).

PHAGE λ N AND Q PROTEINS

We have already discussed one example of processive antitermination. As described in chapter 8, phage λ shifts its gene expression into the lytic cycle by using the phage-encoded N protein to direct the TEC to ignore transcription termination sites, resulting in generation of longer transcripts that include genes required for phage DNA replication. N is an RNA-binding protein that binds to two specific sites, one on each of the two major transcripts that originate from the p_L and p_R promoters. These sites are called *nut* sites, for *N ut*ilization, and they are composed of a short helix. Binding of N to the nascent RNA transcript as it emerges from RNA polymerase results in recruitment of additional host cell-encoded factors, termed *nus* factors (for *N us*age), and the resulting complex has high processivity and ignores subsequent termination sites that it encounters. This results in expression of a new set of genes.

A similar antitermination event is responsible for transcription of genes required for the late stage of phage λ development. One of the proteins produced as a result of the activity of N is the Q protein, which promotes antitermination of transcripts that initiate at a different set of promoters. Unlike N, which binds to the newly synthesized RNA transcript, Q binds to the DNA at a specific site near the promoters for the genes on which it acts. Binding of Q modifies the TEC in a manner similar to the effect of N and results in processive antitermination and late gene expression.

BACTERIAL PROCESSIVE ANTITERMINATION SYSTEMS

Bacteria also use processive antitermination to increase expression of their genes. As noted above, transcription of very long operons (including rRNA operons) poses a special problem for the cell because of the high probability that RNA polymerase will terminate at some point during transcription. This is especially important in operons that are not efficiently translated or are not translated at all (like rRNA operons), because of the presence of sites at which Rho protein may bind to promote termination before the end of the operon. The cell has developed special antitermination systems to modify the TEC so that transcription becomes more processive and therefore less likely to terminate. These systems, which are especially important for ribosome biosynthesis (see chapter 13), likely represent the evolutionary basis for the presence of the host-encoded Nus factors that phage λ has borrowed for its own purposes.

The RfaH System

Transcription of several long operons in *E. coli*, including the genes for lipopolysaccharide production, certain genes involved in virulence, and plasmid-borne *tra* genes, relies on the RfaH protein. This protein acts in a manner similar to that described for the λ Q protein in that it binds to the template DNA at specific sites (called *ops* sites) at which RNA polymerase pauses during transcription of the operon. RfaH interacts with the nontemplate strand of the DNA in the transcription bubble, which is displaced to the outside of the TEC when RNA polymerase is transcribing the template strand of the same region. The protein remains associated with RNA polymerase once the complex leaves the *ops* site and promotes high processivity and resistance to Rho-dependent termination. The presence of systems like this in the host organism is likely to have provided the tools for an invading virus like λ to develop a related regulatory mechanism.

Regulation of mRNA Degradation

Synthesis of the mRNA is obviously a crucial step in gene expression. However, how efficiently an mRNA can be translated, and how many protein products can be generated per mRNA molecule, is dependent not only on the synthesis rate of the transcript (regulated by transcription initiation or elongation, as described above), but also on the length of time that mRNA persists in the cell. RNA stability is quantified by measurements of the persistence of an mRNA in the cell. The standard approach is to measure the **half-life** of the transcript, which is the time it takes for the amount of a particular transcript to be reduced to half of the amount observed at the start of the experiment. Most bacterial mRNAs have a relatively short half-life of 1 to 5 min; this contrasts with much greater mRNA stability in eukaryotic organisms, where transcripts can persist for hours (see chapter 2). Increasing the half-life of an individual transcript can result in higher gene expression, while decreasing the half-life results in lower gene expression.

RNAs are destroyed by endoribonucleases that cleave internally in the RNA chain and by exoribonucleases that degrade from the end(s) of the molecule (see chapter 2). The susceptibility of a particular RNA to degradation is dependent on the features of that RNA (for example, whether it is highly structured or single stranded) and can be modified by changes in the structure or by changes in the accessibility of sites at which the ribonucleases (RNases) can act. This can be exploited as a mechanism for regulation of gene expression. Regulation at the level of mRNA degradation is complicated by the fact that mRNAs that are efficiently translated (and are therefore densely covered with elongating ribosomes) are often more stable than mRNAs that are not efficiently translated. This means that regulatory mechanisms that directly affect translation (see below) may result in indirect effects on mRNA stability. Nevertheless, there are several systems in which it is clear that mRNA degradation is specifically controlled by a regulatory mechanism.

Protein-Dependent Effects on RNA Stability

RNA degradation usually requires binding of RNases to the transcript. The positioning of an RNA-binding protein on a transcript can affect the ability of an RNase to access its target site and therefore can affect the stability of the mRNA. The RNA-binding protein may hide a site at which the RNase would otherwise act, resulting in increased stability of the transcript. An RNA-binding protein can also recruit an RNase to a transcript that it otherwise fails to recognize, decreasing the stability of the transcript.

The simplest case of regulation of mRNA stability by an RNA-binding protein is provided by the *rne* gene, which encodes RNase E, one of the major endoribonucleases in *E. coli*. The *rne* gene includes a leader region that causes this transcript to be highly sensitive to cleavage by RNase E. When RNase E levels are high, the transcript is rapidly degraded, which prevents synthesis of additional RNase E protein. Low intracellular levels of RNase E allow the transcript to persist and to be translated, which results in an increase in the amount of the enzyme (see Schuck et al., Suggested Reading). This is another example of autoregulation, where a gene product controls the expression of its own gene. Autoregulation allows tight control of the amount of a gene product in the cell by direct measurement of the gene product itself. This is especially important in genes like *rne*, because the RNase E enzyme is essential for cell survival but too much of the enzyme could lead to inappropriate degradation of mRNAs that the cell needs.

The importance of the leader region as the target for regulation was established in this case by transplanting the *rne* leader region upstream of a different mRNA. This resulted in RNase E-dependent destabilization of this new mRNA, which demonstrated that the leader region alone was sufficient to confer the regulatory response. The role of RNase E as the enzyme responsible for degradation was more difficult to establish, because the *rne* gene is essential. It was therefore not possible to inactivate the gene to test the effect on gene expression. Instead, a temperature-sensitive conditional mutation (see chapter 3) in *rne* that resulted in reduced RNase E activity at high growth temperatures was used to allow comparison of expression levels in the presence of high versus low RNase E activity. This illustrates some of the complexities that are encountered during analysis of the regulation of cellular processes that are essential for survival.

RNA-Dependent Effects on RNA Stability

The stability of an mRNA can be affected by the action of a regulatory RNA. In most cases, this occurs by binding of a *trans*-acting small regulatory RNA (sRNA) that binds to the target mRNA, which changes its stability. A *cis*-acting regulatory element in the target mRNA itself can also affect mRNA stability. In both cases, the regulatory RNA acts by altering the susceptibility of the target mRNA to RNases, thereby changing its stability and the level of gene expression.

REGULATION OF mRNA DEGRADATION BY sRNAs

The role of sRNAs in regulation of gene expression has become increasingly evident in the last 10 years (see Storz et al., Suggested Reading). Most (but not all) sRNAs regulate multiple targets by base pairing between the sRNA and the mRNA target; as many of these sRNAs are part of global regulatory systems, they are discussed in more detail in chapter 13. The RNA-binding protein Hfq often facilitates the interaction between an sRNA and its target. Regulation usually occurs either at the level of translation initiation (by sequestration of the TIR [see below]) or at the level of mRNA degradation. As noted above, translational repression can have secondary effects on transcript stability, but there are several cases in which there appears to be a primary effect on the susceptibility of an mRNA to specific RNases.

sRNAs can affect the stability of their target mRNAs by a variety of mechanisms. In some cases, the sRNA (probably in conjunction with the Hfq protein) facilitates the recruitment of an RNase to the complex, which results in simultaneous degradation of both the sRNA and its target; this appears to be the case for the *E. coli* RyhB sRNA, which is involved in iron regulation. Degradation usually involves RNase E or RNase III (see chapter 2), both of which are endoribonucleases and presumably trigger attack by additional endonucleases. In some cases, a specific endonucleolytic cleavage can increase mRNA stability by removing a segment of the mRNA that otherwise leads to rapid degradation. Stabilization can also occur simply by formation of the sRNA-mRNA complex if the complex blocks the access of an endoribonuclease or the processivity of an exoribonuclease, resulting in protection of the mRNA. Analysis of the complexities of how different sRNAs act on their targets, and the mechanisms of their effects on gene expression, is a very active field of research.

THE *glmS* RIBOZYME

The *B. subtilis glmS* gene is an example of a gene in which expression is controlled by modulation of mRNA stability. This gene encodes the enzyme required for biosynthesis of glucosamine-6-phosphate (GlcN6P), a cell wall component. Like metabolite-binding riboswitches, the leader region of the *glmS* gene binds GlcN6P, and the binding results in reduced synthesis of the enzyme. However, in this case, binding of the signal molecule does not result in a change in the leader RNA structure, but instead activates the RNA to cleave itself at a specific position. The *glmS* leader RNA therefore acts as a metabolite-induced ribozyme, an RNA enzyme (see chapter 2). Cleavage of the RNA results in removal of the 5′ end of the transcript, and the mRNA now contains a 5′-hydroxyl end instead of the normal 5′ phosphate. The presence of the 5′ hydroxyl appears to target the mRNA for degradation by RNase J1, which is a 5′-3′ exoribonuclease (see chapter 2). Because the ribozyme activity is activated by GlcN6P, the end product of the pathway, the mRNA is stable when the supply of GlcN6P is low, and degradation of the mRNA (and repression of *glmS* gene expression) occurs when the cell has an adequate supply of GlcN6P (see Collins et al., Suggested Reading).

Regulation of Translation

Translation of an mRNA to generate a protein product requires that the 30S ribosomal subunit be able to access the TIR, positioning the initator methionyl tRNA (fMet-tRNA^{fMet}) at the AUG start codon, and that the resulting 70S translation elongation complex can move processively down the mRNA. Most known examples of translational regulation operate at the level of translation initiation, as that step is highly sensitive to the structure of the target mRNA (see Geissmann et al, Suggested

Reading), whereas translation elongation is usually (but not always) highly processive and relatively insensitive to the mRNA structure. Examples of both levels of regulation are described below. It is interesting that several of the regulatory mechanisms described above that affect transcription attenuation or mRNA stability can also be modified in simple ways to operate instead at the level of translation initiation.

Regulation of Translation Initiation

The first step of translation involves binding of the 30S initiation complex, which includes the 30S ribosomal subunit, the initiator fMet-tRNA^{fMet}, and initiation factors, to the TIR, which is composed of the S-D sequence and the initiator codon (usually AUG), as described in chapter 2. Anything that inhibits access of the initiation complex to the TIR, such as physical blocking of the TIR by binding of an RNA-binding protein to the mRNA or secondary structure in the mRNA that sequesters the TIR, inhibits translation initiation. This has been exploited by a number of regulatory mechanisms. Changes in the composition of the 30S initiation complex can also affect its ability to recognize TIRs with specific characteristics, resulting in effects on the efficiency of translation.

TRANSLATIONAL REGULATION BY RNA-BINDING PROTEINS

Binding of an RNA-binding protein to a TIR can sequester the TIR and block binding of the 30S initiation complex. As noted above, the TRAP transcription attenuation protein can also regulate *trp* gene expression at the translational level by binding to a site that overlaps the TIR. Other examples include many operons for the genes that encode ribosomal proteins. For each operon, a single ribosomal protein encoded in the operon acts as the regulator for the entire operon, usually at the level of translation. Binding of the regulatory protein to the mRNA, at a site that overlaps the TIR for the first gene in the operon, represses translation of the entire mRNA. The effect on the first gene is easy to understand, because the presence of the protein prevents binding of the 30S initiation complex. This is a form of autoregulation, as the protein represses its own expression, and is straightforward for a monocistronic operon. However, for a polycistronic operon, the normal expectation is that the other genes in the operon have their own independent TIR. Repression of the downstream genes is likely to occur by **translational coupling** (see chapter 2). A probable mechanism is that the TIRs of the downstream genes are sequestered in a secondary structure by pairing of the TIR with a complementary region in the 3′ part of the coding sequence of the upstream gene. If the upstream gene is being translated (because its TIR is not blocked by the ribosomal protein that acts as the repressor), the mRNA region that would otherwise sequester the TIR is occupied by translating ribosomes, and the TIR for the downstream gene remains accessible. If the upstream gene is not translated (because its TIR is blocked by the regulatory protein), the TIR for the downstream gene will be sequestered by base pairing, and its translation will also be inhibited. Regulation of ribosome biosynthesis is discussed in more detail in chapter 13.

TRANSLATIONAL REGULATION BY sRNAs

sRNAs can regulate target genes in a variety of ways and commonly operate by base pairing with a complementary region of the target mRNA (see Storz et al., Suggested Reading). Binding of the sRNA to its target can affect degradation of the target mRNA, as described above. In addition, if the sRNA binds to a site that overlaps the TIR, binding of the sRNA can inhibit translation of the mRNA. This is similar to the effect of an RNA-binding protein in that the major effect is that the TIR is made unavailable for binding of the 30S translation initiation complex. In many cases, interaction between the sRNA and the mRNA is dependent on the Hfq protein, which facilitates the pairing of the two RNAs. A number of sRNAs have been shown to function in this way, including the *E. coli* DsrA sRNA, which inhibits translation of multiple genes, including the *hns* gene, which encodes the H-NS global regulatory protein (see chapter 13). Interestingly, in addition to its role as a repressor of the *hns* gene, DsrA can also activate expression of another global regulator, the RpoS stationary-phase sigma factor, which is encoded by the *rpoS* gene. The *rpoS* mRNA contains a structure that sequesters its TIR, so that the transcript is inactive unless that structure is disrupted. Binding of the DsrA sRNA to the region of the mRNA that would otherwise pair with the TIR releases the TIR, which is now available for binding of the 30S initiation complex. Repression of *hns* and activation of *rpoS* utilize different regions of the DsrA sRNA, so DsrA has the ability to regulate multiple sets of targets. Since both RpoS and H-NS are global regulators (see chapter 13), DsrA impacts a large number of genes in the cell.

RNA THERMOSENSORS: REGULATION BY MELTING SECONDARY STRUCTURE IN THE mRNA

An mRNA leader region can affect the expression of a gene directly through the effects of temperature on its secondary structure. Base pairing between complementary sequences on the mRNA can cause secondary structures to form in the RNA in the form of helices and more complicated structures (see chapter 2). Secondary structures are less stable at higher temperatures because the base pairing that holds them together can melt at these temperatures. One way in which temperature can regulate the expression of a gene is if the secondary structures that have formed block access of the ribosome to the TIR of the mRNA, for example, if they include the S-D sequence and/or the initiator codon. At lower temperatures, the structure is stable, and translation initiation is inhibited. When the temperature rises, these secondary structures can melt, exposing the TIR so that the ribosomes can bind and initiate translation of the mRNA. RNAs of this type are called RNA **thermosensors**, as they have the ability to directly sense a change in temperature.

The *E. coli rpoH* Heat Shock Thermosensor

The first RNA thermosensor that was discovered was in the *E. coli rpoH* gene, which encodes the heat shock sigma factor, σ^H. The RNA polymerase holoenzyme containing σ^H instead of the normal σ^{70} recognizes a new set of promoters that are responsible for transcription of genes that help the cell to respond to an abrupt increase in temperature called a heat shock. The cell always contains a certain amount of *rpoH* mRNA, but this mRNA is usually inactive for translation because of secondary structure that sequesters the TIR (Figure 12.27). An abrupt increase in temperature causes the secondary structure in the mRNA to melt and allows translation of the transcript, which results in a rapid increase in the amount of σ^H in the cell. This allows the cell to respond very quickly to the temperature increase, because the *rpoH* mRNA is already present in the cell before the heat shock occurs. The heat shock response is a form of global regulation and is discussed in more detail in chapter 13.

RNA Thermosensors That Control Virulence

Some pathogenic bacteria use RNA thermosensors to regulate their virulence genes. Our body temperature, and that of most other warm-blooded hosts, is much higher than those of the outside environments usually inhabited by bacteria. Pathogenic bacteria often use temperature as one of the clues that they are in a mammalian host, and a rise in temperature tells them that it is time to turn on the virulence genes that allow them to survive and multiply in the host. Placement of an RNA thermosensor in the leader region of a gene that regulates virulence allows virulence gene expression to be repressed when the temperature is low and rapidly induced when the organism enters the host. As with the heat shock sigma factor discussed above, the response is rapid because the mRNA for the regulatory protein is already present in the cell prior to entry into the host, and melting of the secondary structure to allow protein synthesis to proceed is nearly instantaneous. One example of this type of temperature regulation is in the expression of the *lcrF* gene in *Yersinia pestis*, the bacterium that causes bubonic plague. The product of this gene is the transcriptional activator that turns on virulence genes in mammals. During growth at low temperature (e.g., in the flea, which is an intermediate host for this organism), an RNA thermosensor in the *lcrF* mRNA blocks the S-D sequence, preventing translation; this secondary structure melts at 37°C (the body temperature

Figure 12.27 Regulation of the *E. coli rpoH* gene by an RNA thermosensor. Transcription of the *rpoH* gene under normal growth conditions yields an mRNA in which the S-D sequence for translation of the *rpoH* coding region is sequestered into a helix. Exposure of the cells to heat shock results in melting of the helix and release of the S-D sequence. The mRNA is active for ribosome binding, and σ^H is synthesized. doi:10.1128/9781555817169.ch12.f12.27

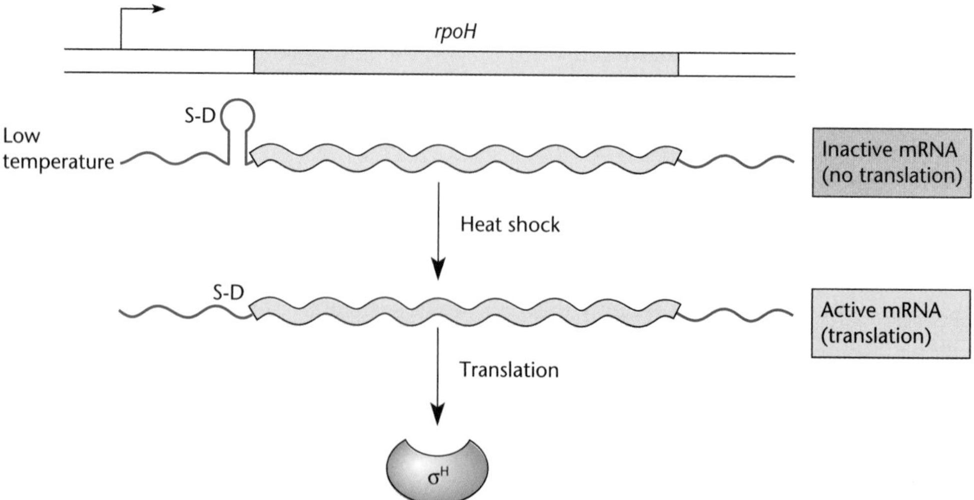

of the mammalian host), allowing translation of the mRNA and synthesis of the LcrF transcriptional regulator. This allows the cell to quickly activate transcription of the genes involved in virulence, so that the organism can multiply and kill the mammalian host.

RIBOSWITCH REGULATION OF TRANSLATION INITIATION

Many riboswitches operate at the level of transcription attenuation, as described above. However, the same types of ligand-dependent RNA rearrangements can also regulate gene expression at the level of translation initiation. The key difference is that in riboswitches of this type, the transcription terminator helix is replaced by a helix that sequesters the S-D sequence (by pairing of the S-D with a complementary anti-S-D [ASD] sequence), and the antiterminator helix is replaced by a helix that sequesters the ASD sequence (often called an ASD-sequestering helix).

The crucial difference between these two classes of riboswitches is that in transcriptional riboswitches, binding of the signal molecule determines whether the full-length transcript is synthesized, while in translational riboswitches, the transcript is always made, but binding of the ligand determines whether the mRNA will be translated. A second important distinction is that for riboswitches that operate at the level of transcription attenuation, the ligand must bind during transcription of the leader RNA, before RNA polymerase reaches the termination site. In contrast, for translational riboswitches, the ligand can bind either cotranscriptionally or after transcription is complete. A single class of riboswitches (e.g., the S box SAM-binding riboswitch described above) is often found in both transcriptional and translational forms, often in different groups of organisms. Riboswitches that function at the level of transcription attenuation tend to predominate in *B. subtilis* and related organisms, while translational control is found more frequently in gram-negative bacteria, for reasons that are unknown but that may reflect evolutionary history.

TRANSLATIONAL AUTOREGULATION OF INITIATION FACTOR IF3

All of the translational mechanisms described so far involve physical blocking of the TIR either by a protein or an sRNA or by secondary structure within the leader RNA. An alternative mechanism involves changes in the translation initiation complex itself, which can affect the ability of the complex to recognize specific mRNAs. This type of mechanism is used for the regulation of the *infC* gene, which encodes translation initiation factor IF3, a normal component of the 30S translation initiation complex.

As mentioned in chapter 2, the initiation codon for translation is usually AUG but can also be GUG or, more rarely, UUG. An AUU initiation codon is used only to initiate the translation for the *infC* gene. Use of an AUU initiation codon only for *infC* is conserved in many organisms, suggesting that it is an important feature of the gene. The AUU codon plays a crucial role in regulation of the *infC* gene so that the cell produces only as much IF3 as it needs. IF3 is responsible for directing the 30S translation initiation complex to authentic TIRs and for positioning the fMet-tRNA^{fMet} accurately at the initiation codon. Translation initiation complexes that contain IF3 will therefore discriminate against the *infC* mRNA because the mRNA has an unfavorable AUU initiation codon. If translation initiation complexes are saturated with IF3, the *infC* mRNA will not be translated and no additional IF3 will be synthesized. If there is insufficient IF3 in the cell, translation initiation complexes that lack IF3 will be present, and they will fail to discriminate against the AUU codon and IF3 synthesis will increase. This regulatory mechanism takes advantage of the function of the regulated gene product in monitoring the levels of that product. We will see a similar principle in the regulation of the gene that encodes the RF-2 translation termination factor (see "Regulation of Translation Termination" below).

Translational Regulation in the Exit Channel of the Ribosome

After the details of translation were worked out in the early 1960s, it was assumed that ribosomes translate an mRNA independent of its sequence, depending on the sequence of nucleotides in the mRNA only to direct the insertion of amino acids into the growing polypeptide. External features, such as the secondary structure of the mRNA and codon usage, were recognized to influence the initiation, termination, and rate of translation, but the ribosome itself did not appear to discriminate between different coding sequences. It came as a surprise to discover that ribosomes could interact differently with different polypeptides in the exit channel and that some polypeptides could cause translation to arrest.

As peptide bonds form in the peptidyl transfer center (PTC) of the ribosome, the growing peptide enters the exit channel in the large 50S subunit of the ribosome and emerges on the other end about 30 amino acids later. This exit channel is constructed mostly of the 23S rRNA, but some proteins, including L4 and L22, help form a constriction that narrows the channel, which the growing polypeptide enters some 9 amino acids from the PTC. The growing polypeptide is exposed both before it enters the narrower part of the channel and after its N terminus emerges from the ribosome, providing opportunities for external factors to influence the movement of the polypeptide through the channel and regulate the translation of the polypeptide. Specific polypeptide sequences called **stalling sequences** in growing polypeptides are now known to be capable of contacting specific regions in the walls of the exit channel, causing translation to arrest in the PTC unless certain conditions are met.

Two general types of such regulation have been described (see Ito et al., Suggested Reading). In both types, the specific stalling sequence can cause translation to arrest when it enters the narrow part of the exit channel. In some situations, the binding of a small effector to the stalling sequence is necessary for the arrest to occur, and in others, the arrest always occurs unless it is relieved by entry of the N terminus of the polypeptide into some cellular structure, such as the membrane, forcing coordinate translation of the polypeptide with its insertion into the structure. Interestingly, in all of the examples known so far, it is the translation of an upstream leader polypeptide that is arrested, rather than the translation of the gene itself. The translational arrest of this upstream leader polypeptide then regulates the expression of the downstream gene, either translationally, through translational coupling, or transcriptionally, through antitermination.

Examples of the first type, in which a small-molecule effector is required for regulation of a gene by translational arrest, are genes whose products confer resistance to the antibiotic erythromycin or chloramphenicol, both of which block translation by binding to the ribosome (see chapter 2). This regulation is illustrated for an erythromycin resistance gene in Figure 12.28A. In this case, sublethal concentrations of erythromycin bind to the ribosome and the stalling peptide as the peptide enters the narrow part of the exit channel, causing translation to arrest, even though this concentration of antibiotic would not normally block translation. Stalling of the ribosome affects the structure of the RNA and prevents the formation of a helix that would otherwise sequester the TIR of the downstream *ermC* gene, which encodes an enzyme that methylates a specific base in 23S rRNA. This methylation prevents binding of erythromycin to the ribosome and confers resistance to the antibiotic.

Another example of an effector-dependent translational arrest is the regulation of the tryptophan utilization *tna* operon of some enteric bacteria, including *E. coli*. In this case, binding of tryptophan to the stalling peptide in the ribosome arrests translation of the leader peptide. The stalled ribosome masks a Rho utilization site (*rut* site), thereby preventing Rho-dependent termination and allowing transcription of the downstream *tna* operon. In this way, the genes for using tryptophan as a carbon, nitrogen, and energy source are turned on only if tryptophan is present.

Other known examples of regulation in the ribosome exit channel are proteins involved in transporting other proteins into and through the inner membrane (see chapter 14). In one well-studied example in *E. coli*, translational arrest during synthesis of a leader polypeptide, SecM, after the N terminus of the protein has exited the ribosome prevents translation of the downstream *secA* gene, whose product is required for insertion of some proteins into the SecYEG translocon in the inner membrane (see Figure 12.28B and chapter 14). If the N-terminal region of the SecM protein is inserted into the SecYEG translocon, the arrest is relieved and the downstream *secA* gene is translated. It is not clear how the binding of SecM to the SecYEG translocon relieves binding of the stalling peptide to the channel. The SecM protein itself has no apparent function and is quickly degraded when it enters the periplasm. The translational arrest occurs only at very low temperatures or if the SecYEG channels are disrupted in some way. This mechanism may reduce the synthesis of SecA protein when the other translocation systems are inadequate and/or may have the effect of localizing the SecA protein to the SecYEG translocon in the membrane, where it performs its function.

Recent work has begun to clarify how these translation arrests might occur. Specific amino acids in the stalling peptide contact specific bases in the 23S rRNA and the L4 and L22 proteins in the channel. It has been proposed that binding of the stalling peptide to the channel, with or without the help of a small-molecule effector (depending on the system), causes a distortion in the orientation of the peptidyl tRNA in the P site of the ribosome and prevents formation of the next peptide bond. It matters which amino acid is attached to the aminoacyl-tRNA in the A site, and sometimes which amino acid in the growing peptide it has to bond to. The molecular basis for these effects is not yet understood.

Regulation of Translation Termination

As described above for transcription attenuation systems, it is possible to regulate expression of a gene by positioning a signal that should stop the gene expression machinery early in an mRNA, providing a regulatory mechanism that permits bypassing of that signal under certain circumstances to allow expression of the gene. The *E. coli prfB* gene, which encodes ribosome release factor 2 (RF2), provides an example of how that type of event can operate at the translational level.

RF2 recognizes UGA and UAA nonsense codons to terminate translation (see chapter 2). The *prfB* gene is very unusual in that it contains a UGA codon early in the coding sequence. Furthermore, the downstream part of the coding sequence is "out of frame" with the portion that is upstream of the UGA, so synthesis of full-length RF2 requires that the ribosome slip back on the mRNA by 1 nucleotide into the −1 frame. When the translating ribosome reaches the UGA codon in the

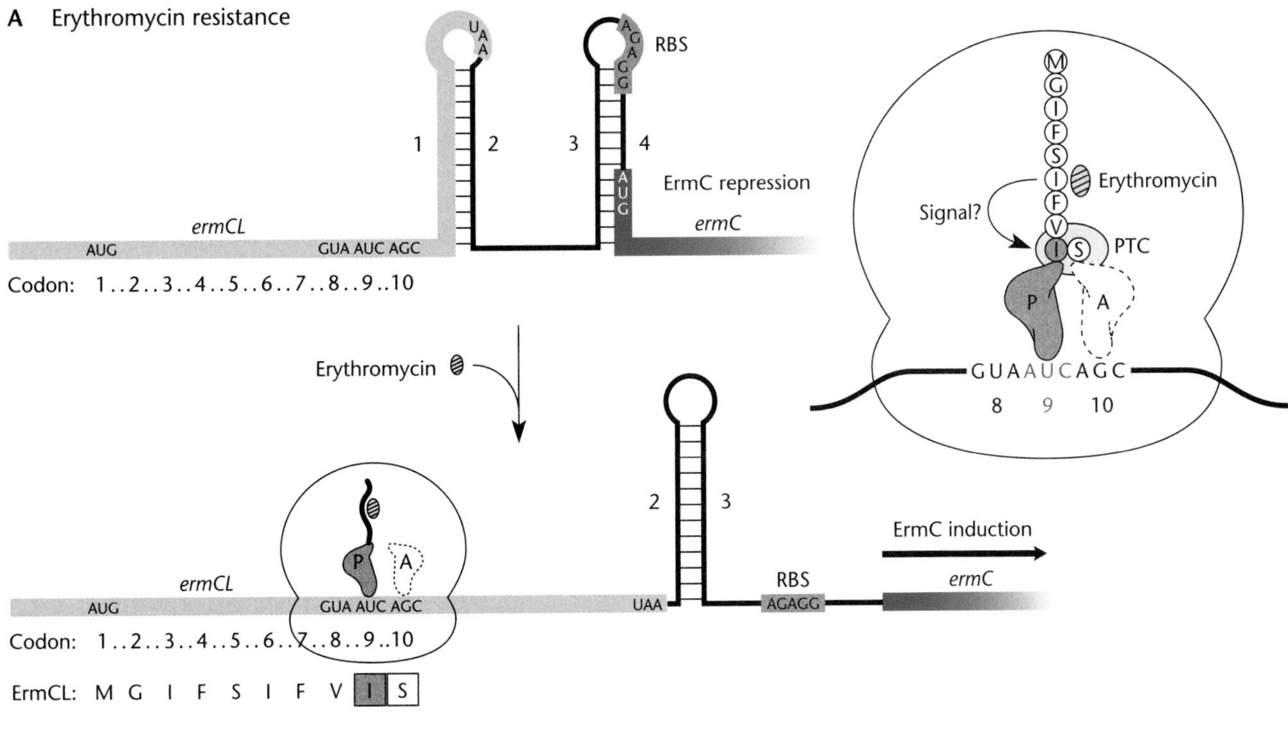

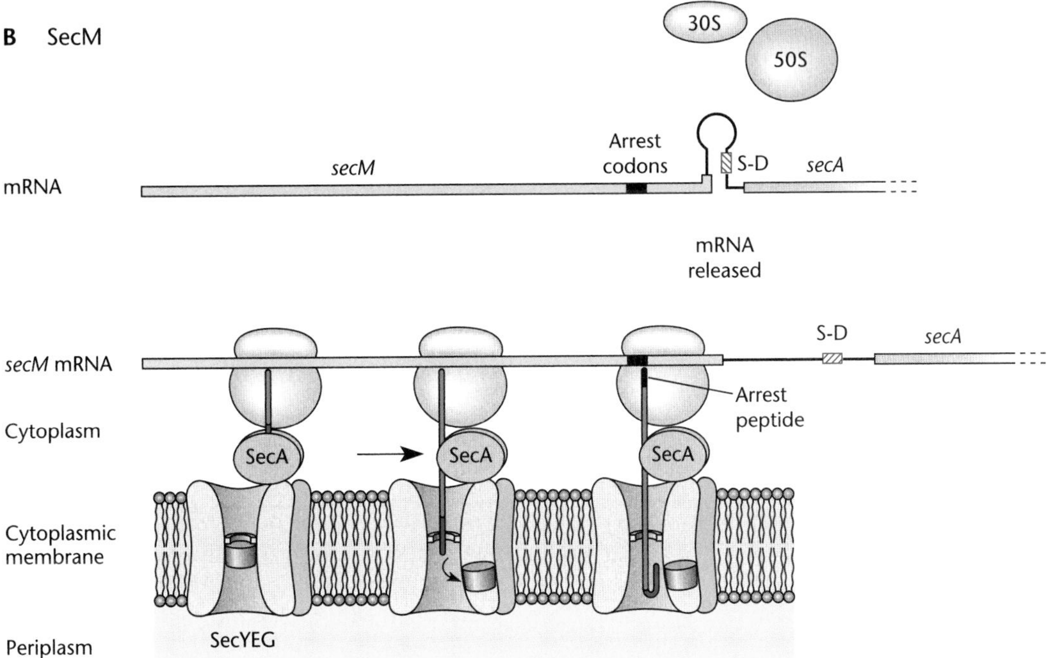

Figure 12.28 Regulation by translational arrest in the ribosome. **(A)** Regulation of the *ermC* gene by erythromycin. Binding of erythromycin to the stalling peptide in the exit channel causes the ribosome to stall on the mRNA, preventing formation of the 1:2 helix. This allows formation of the 2:3 helix and releases region 4 to allow binding of the ribosome to the TIR of the *ermC* gene. The diagram on the right shows how the relative orientations of the incoming amino acid in the A site (serine [S]) and the amino acid in the P site (isoleucine [I]) could be altered, preventing peptide bond formation. **(B)** Arrest of translation of the upstream coding sequence for SecM prevents translation of the downstream gene for SecA. If SecM enters the SecYEG channel in the membrane, the *secA* TIR is exposed and translation of SecA occurs. See chapter 14 for details of protein translocation through the membrane.
doi:10.1128/9781555817169.ch12.f12.28

mRNA, it terminates and is released from the mRNA only if RF2 levels are high enough to allow efficient recognition of the UGA codon. If RF2 levels are low, the translating ribosome pauses at the UGA codon, which results in a high frequency of the frameshift event that allows translation to continue in the −1 frame and synthesis of more RF2 protein. This is an example of **programmed frameshifting** (see Box 2.3) and, like the *infC* system described above, provides an example of translational autoregulation that exploits the biological function of the gene product (in this case, translation termination) as a key feature of the regulatory mechanism.

Posttranslational Regulation

We usually consider gene expression to be complete once a polypeptide has been released by the translating ribosome. However, other steps may be necessary to determine how much active protein product is in the cell. As described in chapter 2, the protein may need to fold correctly, to form higher order complexes, or to be modified to be fully active. The level of protein product in the cell also depends on how long each protein molecule persists in the cell. Finally, the activity of the protein can be affected by binding of small molecules, as in feedback inhibition of a biosynthetic enzyme by the end product of the pathway in which it participates. Each of these represents an opportunity for posttranslational regulation.

Posttranslational Protein Modification

Posttranslational regulation can occur by reversible modification of specific sites on a protein. These modifications, which in bacteria include phosphorylation, methylation, and acetylation, can change the activity of the protein, and changing the activity of the enzymes responsible for the modification can control their presence or absence.

One example of an important posttranslational modification is the adenoribosylation (addition of AMP) to the *E. coli* glutamine synthetase enzyme, which synthesizes glutamine from glutamate. At high concentrations of glutamine, this enzyme is adenoribosylated, which temporarily inhibits its activity until glutamine levels drop. The AMP groups are then removed, and the activity is restored (see chapter 13). Phosphorylation is another common modification that can change protein activity. This is used to control the activity of a large set of regulatory proteins through partnership of a phosphorylation enzyme (a "sensor kinase") and a DNA-binding protein (a "response regulator") in what are termed **two-component regulatory systems**, which are responsible for a wide range of global regulatory responses (see chapters 13 and 14). Other common posttranslational modifications include the methylations that are used in bacterial chemotaxis systems to allow the cell to monitor gradients of molecules in its external environment and to respond by moving toward attactants (compounds they want to use) and away from toxic compounds (see chapter 14). Protein acetylation has also emerged recently as a mechanism to regulate enzyme activity in response to cellular metabolism.

A common theme of these modification systems is that the modification is reversible, allowing the cell to continually sense whether conditions have changed. The cellular response is therefore dependent on the relative activities of the enzymes responsible for addition and removal of the modification, providing opportunities for tight control of the activity of the target protein.

Regulation of Protein Turnover

We discussed earlier that the level of an mRNA is determined not only by its synthesis rate, but also by its rate of degradation. This also applies to protein levels. A protein that is very stable can accumulate to high levels even if its synthesis rate is relatively low, while a protein that is very unstable will not accumulate to high levels even if its synthesis rate is high. The stability of a protein (measured by its half-life, which is the time it takes for the amount of protein present at an initial time point to be reduced by half) is determined by intrinsic properties of the protein (i.e., its susceptibility to cellular proteases [see chapter 2]) but can also be changed in response to changes in environmental conditions, such as heat shock (see chapter 13); this can occur because the stress condition is damaging to cellular proteins or can be targeted to specific proteins in a process called **regulated proteolysis.** Individual proteins can be targeted for destruction by proteins called **adaptors**, whose activity can in turn be controlled by other proteins, called **antiadaptors** because they prevent the activity of the adaptors, resulting in increased stability of the protein that would otherwise be recognized by the partner adaptor. Adaptors and antiadaptors are usually highly specific in both the proteins they regulate and the protease with which they interact.

REGULATION OF THE RpoS SIGMA FACTOR BY ADAPTORS AND ANTIADAPTORS

One of the best-characterized systems of regulated proteolysis is the *E. coli* RpoS protein, which is a sigma factor (σ^S) responsible for transcription of stress response genes during stationary phase and in response to certain stress conditions. Some of the σ^S protein is synthesized under nonstress conditions. However, the protein is rapidly degraded by the ClpXP protease (see chapter 2). Rapid degradation of σ^S requires the RssB protein, which acts as an adaptor protein to specifically bind σ^S and deliver it to ClpXP for destruction. This results in low levels of σ^S under conditions when transcription of the stress response genes is not required (Figure 12.29). When the cell encounters starvation or stress conditions, the RssB protein is inactivated by one of a set of antiadaptor proteins, the Ira (inhibitor of RssB activity) proteins, each of which is induced in response to a different stressful condition (e.g., starvation and low phosphate). Binding of an Ira protein to RssB prevents binding of RssB to σ^S and therefore results in increased stability of σ^S and transcription of the stress response genes (see Bougdour et al., Suggested Reading). This system allows the cell to be poised for response to a variety of environmental stresses by synthesis of σ^S before it is needed, but its activity is maintained at a low level by regulated proteolysis until the cell experiences a stress condition for which σ^S directs the response.

Feedback Inhibition of Enzyme Activity

Biosynthetic pathways are not regulated solely through transcriptional and translational regulation of their operons and by covalent modifications or degradation of their enzymes; they are also often regulated by feedback inhibition of the enzymes once they are made. In **feedback inhibition**, the end product of a pathway binds to the first enzyme of the pathway and inhibits its activity. Feedback inhibition is common to many types of biosynthetic pathways and is a more sensitive and rapid mechanism for modulating the amount of the end product than are transcriptional regulation and translational regulation, which respond more slowly to changes in the concentration of the end product of the pathway. Feedback inhibition is also easily reversible, because the enzymes remain in the cell, ready to resume activity if the level of the end product of the pathway drops as the cell utilizes the compound.

FEEDBACK INHIBITION OF THE *trp* OPERON

The tryptophan biosynthetic pathway of *E. coli* is subject to feedback inhibition. Tryptophan binds to the first enzyme of the tryptophan synthesis pathway, anthranilate synthetase, and inhibits it, thereby preventing the synthesis of more tryptophan. The tryptophan analog 5-methyltryptophan has been used to study this process. At high concentrations, 5-methyltryptophan binds to anthranilate synthetase in place of tryptophan and inhibits the activity of the enzyme, starving the cells for tryptophan. Only mutants defective in feedback inhibition because of a missense mutation in the *trpE* gene that prevents the binding of tryptophan (and 5-methyltryptophan) to the anthranilate synthetase enzyme can multiply to form a colony in the absence of tryptophan.

A similar method is described above for isolating constitutive mutants with mutations of the *trp* operon, but selection of constitutive mutants requires lower concentrations of 5-methyltryptophan. If the concentration of this analog is high enough, even constitutive mutants will be starved for tryptophan, because binding of the analog inactivates all of the anthranilate synthase that is synthesized.

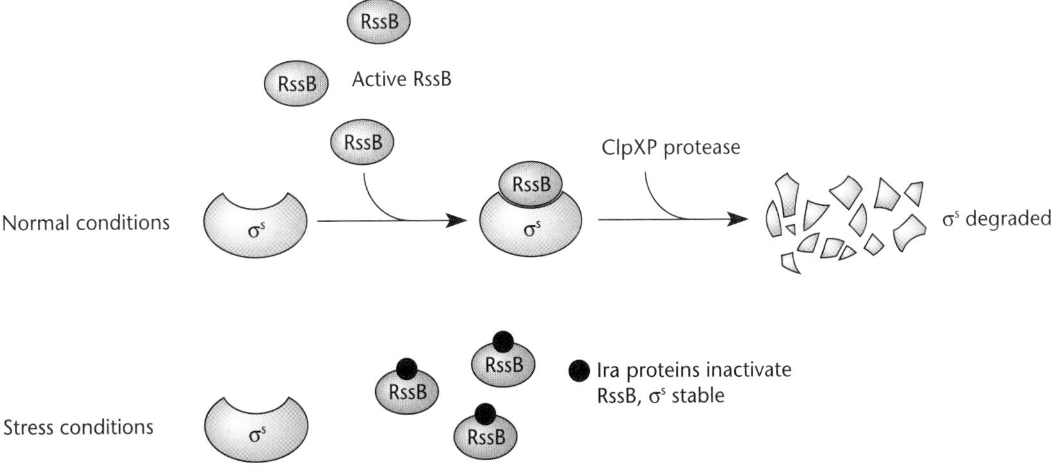

Figure 12.29 Regulated proteolysis of the σ^S factor by adaptors and antiadaptors. Under normal growth conditions, both σ^S and the RssB adaptor protein are present in the cell. RssB is active and binds to σ^S and delivers it to the ClpXP protease for degradation. When the cell enters stationary phase or is subjected to certain stressful conditions, the Ira antiadaptor proteins are synthesized. The Ira proteins bind to RssB and inactivate it. This prevents degradation of σ^S, which is now able to direct the transcription of genes required for response to the stressful conditions. doi:10.1128/9781555817169.ch12.f12.29

BOX 12.3

Special Problems in Genetic Analysis of Operons

Alleles of Operon Genes

Mutant alleles of structural genes may alter the function of a gene to eliminate, change, or increase its activity. A mutation that eliminates gene function creates a null allele. The term "loss-of-function allele" generally is interchangeable with "null allele." One mechanism for studying gene function is to isolate a transposon insertion in the gene. This is sometimes refered to as a "knockout" mutation. Caution is warranted with transposon insertions because, while this implies a null mutation, additional experimentation would need to be done, since some activity could remain, depending on where the transposon was inserted in the coding sequence. Although many valuable collections of mutants have been obtained by transposon mutagenesis, it should really be considered a method of identifying a candidate for follow-up experiments, because there is no assurance that all of these are null mutations. Further complicating the issue, such mutations can be polar and prevent expression of other genes downstream in the same operon. In some cases, they might even increase the expression of a downstream gene or, alternatively, provide promoter activity and express genes upstream that are transcribed in the direction opposite that of the gene in which the insertion occurred. Exciting strategies involving high-density transposon mutagenesis and high-throughput DNA-sequencing techniques (see Box 9.1) are getting around some of the issues with classical transposon mutagenesis experiments when looking over the entire geneome, but the analysis of individual insertions will still involve the same safeguards.

Using Reverse Genetics To Construct Null, Nonpolar Alleles

When the sequence of the DNA of a bacterium is known, it is often possible to make null mutations of genes using systematic methods that avoid many of the complications discussed above. Figure 1 illustrates one such method for *E. coli* (see Baba et al., References). This procedure is designed to delete the entire gene but leave the translation initiation region at the beginning of the gene and the terminator codon at the end of the gene plus some sequences upstream of the terminator codon in case the coding sequence for the gene being deleted includes the S-D sequence for the downstream gene. Overlap of the terminator codon for one gene with the TIR for the next gene in an operon often occurs and can cause translational coupling (see chapter 2). In this method, upstream and downstream PCR primers are designed to amplify an antibiotic resistance gene from a plasmid (Figure 1B). However, additional sequences are included on the 5′ ends of the primers (H1 and H2 in Figure 1B). They are not needed for the amplification part of PCR, but instead, these synthetic ends are added to the final product from PCR, which allows recombination with the chromosome (not shown). These sequences match and introduce sequences that provide homology to the N-terminal and C-terminal coding information in "gene *B*" (Figure 1A). This fragment is then introduced into an *E. coli* strain that expresses the Red functions of phage λ to promote recombineering between the amplified fragment and the gene in the chromosome (see chapter 10). Recombination between the sequences flanking the antibiotic resistance gene in the amplified fragment and the corresponding sequences in the gene in the chromosome deletes most of the gene, replacing it with the antibiotic resistance gene. As an additional feature, the antibiotic resistance gene can be removed later if the antibiotic resistance gene on the plasmid is flanked by sequences for a site-specific recombinase, for example, the Flp recombinase of yeast. When another plasmid expressing this recombinase is introduced into the *E. coli* strain, the antibiotic resistance gene is excised, leaving behind a scar. Encoded in this scar is a short polypeptide that contains too few sequences from the original gene to be active but whose expression prevents polarity or effects of translational coupling on the expression of the downstream gene. The Keio collection has gene knockouts for all of the nonessential genes in the MG1655 genome and is available through the National Institute of Genetics of Japan or the *E. coli* Genetic Stock Center in the United States.

The technique shown in Figure 1 is efficient, but the "scar" that is left behind can complicate procedures if it is necessary to knock out multiple genes with the Flp recombinase. Newer techniques allow a "scarless" method that depends on having both a gene that can be selected, e.g., for antibiotic resistance (Ab^r), and an element that can be counterselected. The homing endonuclease I-SceI provides an especially good tool for counterselction for multiple reasons. For one, the cut site is large enough (30 bp) that no bacterial genome studied to date has it by chance but small enough that it can easily be included with the positive-selection marker. Another big advantage comes from providing a double-strand break that allows loading of the Red recombination machinery. In the procedure, primers are designed to amplify a positive selection marker and the I-SceI cut site sequence out of a template. Like the procedure in Figure 1, additional sequences are included on the 5′ ends of the primers that have homology regions encoding the N-terminal and C-terminal regions of the protein (Figure 2A and B). The positions along the template DNA are indicated with numbers (1 to 10) to emphasize the regions that are deleted in the procedure. The PCR product is introduced by transformation into cells induced for the

BOX 12.3 (continued)

Special Problems in Genetic Analysis of Operons

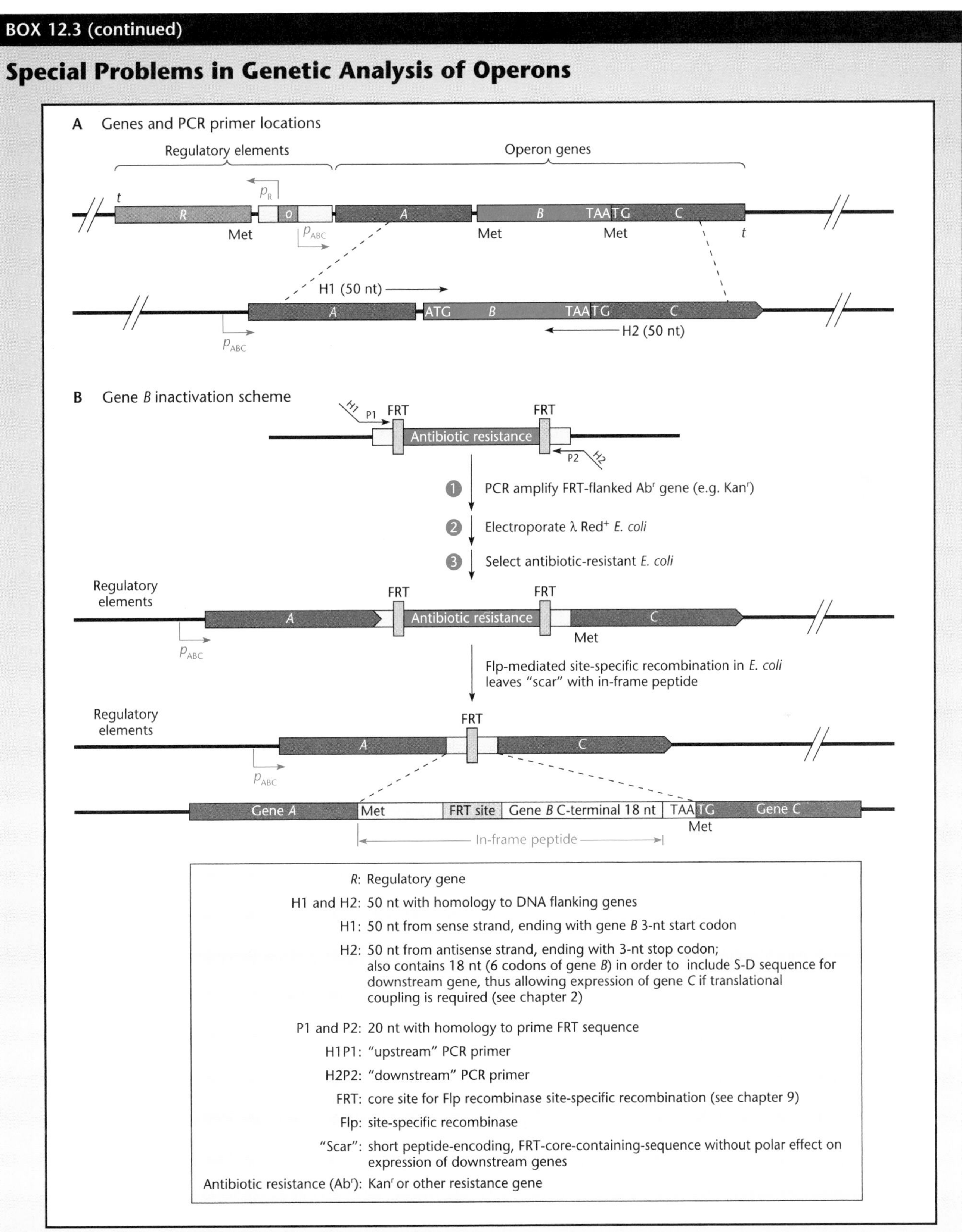

Figure 1 doi:10.1128/9781555817169.ch12.Box12.3.f1

(continued)

BOX 12.3 (continued)

Special Problems in Genetic Analysis of Operons

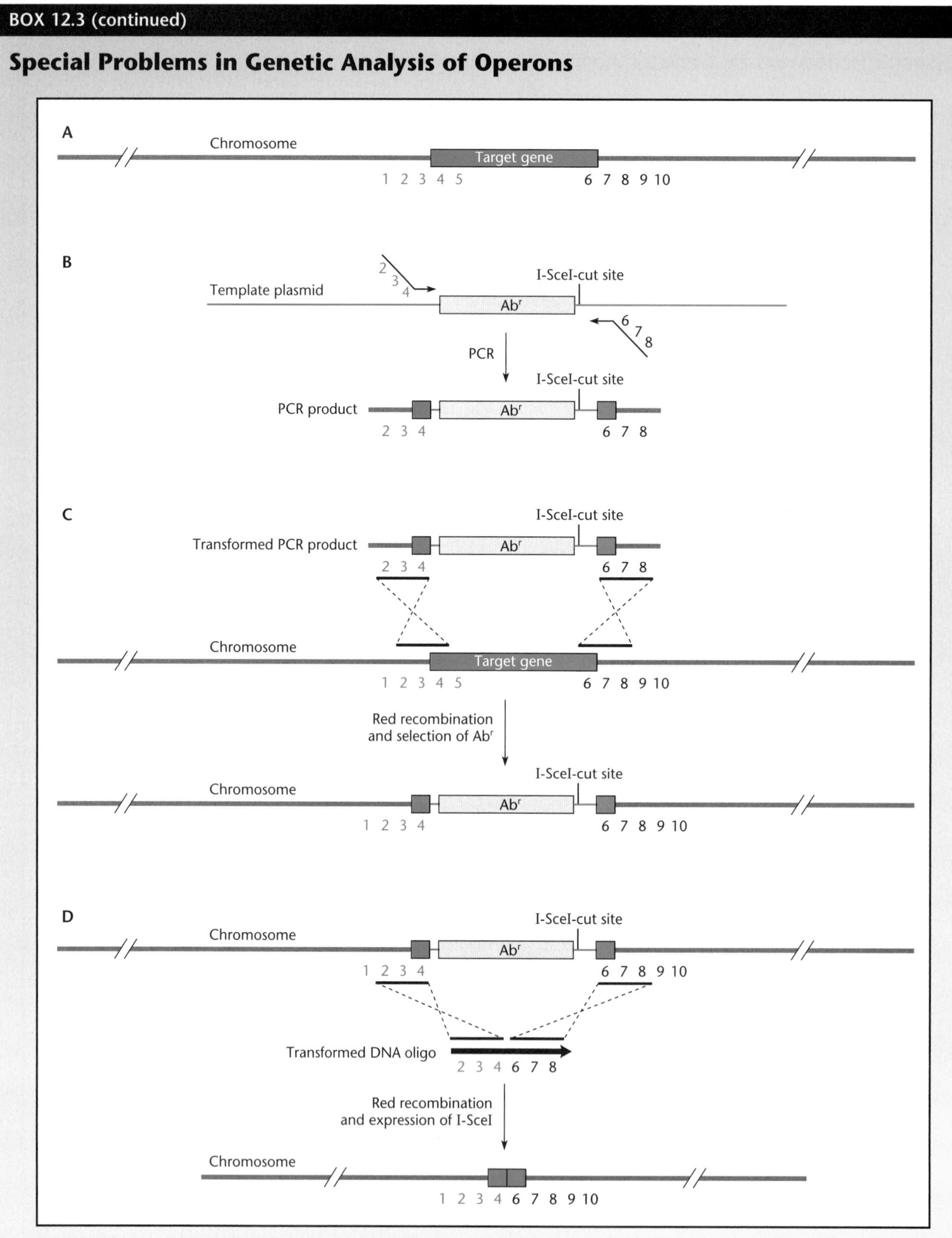

Figure 2 doi:10.1128/9781555817169.ch12.Box12.3.f2

BOX 12.3 (continued)

Special Problems in Genetic Analysis of Operons

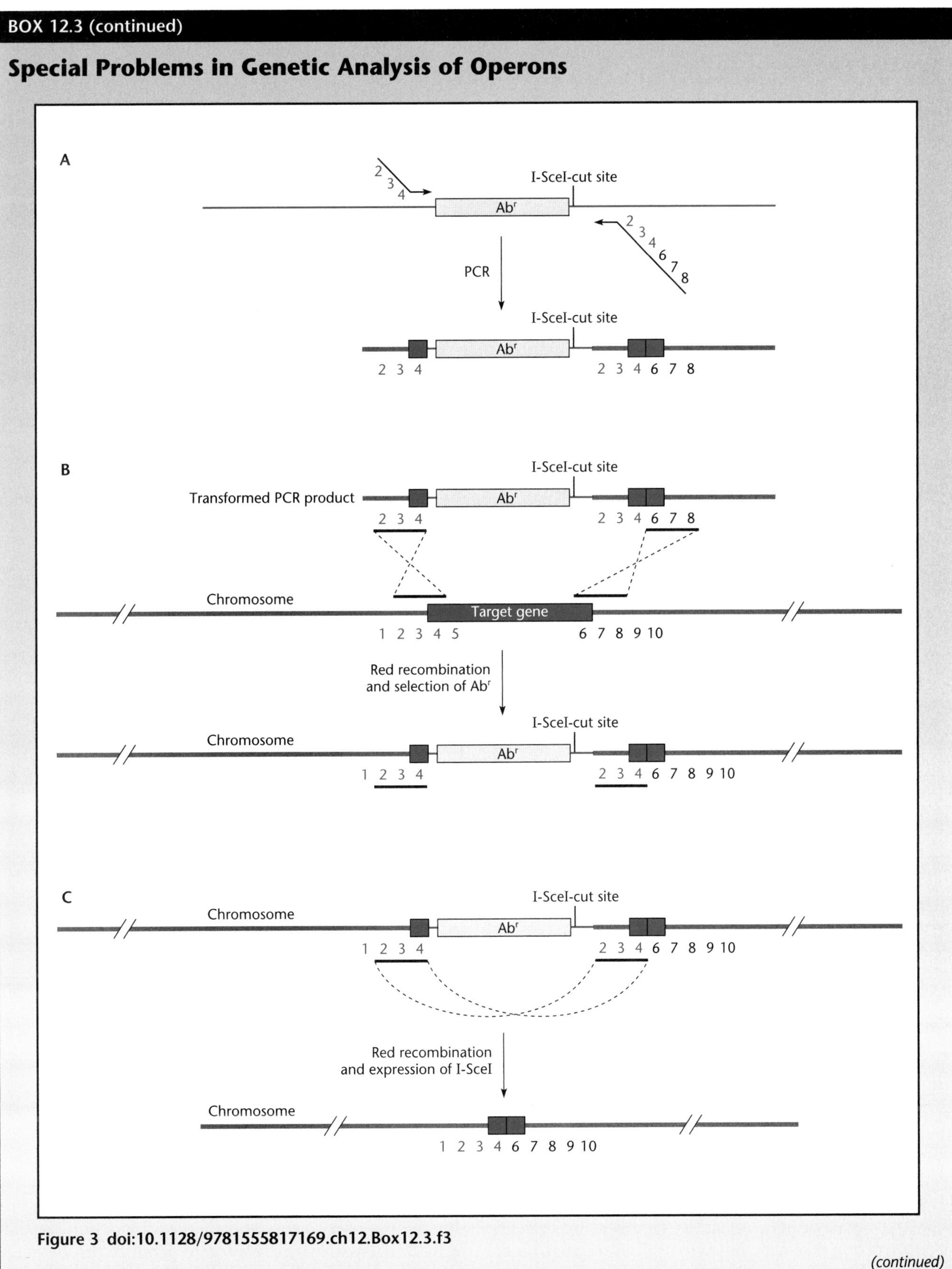

Figure 3 doi:10.1128/9781555817169.ch12.Box12.3.f3

(continued)

BOX 12.3 (continued)

Special Problems in Genetic Analysis of Operons

λ Red recombination system (Figure 2B). In the final step, a DNA oligonucleotide with the desired deletion is designed. It is introduced into the cell at the same time that the λ Red recombination system is induced and the I-SceI endonuclease is expressed. The efficient cleavage of the chromosome at the I-SceI cut site provides a strong **counterselection** for the desired recombination event (Figure 2D). Successful recombinants will not be sensitive to the positive-selection marker, but sequencing is used to confirm the modification.

In the systems where λ Red recombination is used, the limiting factor is always the efficiency of transformation. More advanced techniques involve harnessing endogenous DNA substrates for recombination to increase efficiency. One such technique is called *en passant* mutagenesis (named after an advanced move in chess involving pawns) (see Tischer et. al., References). The initial step in the procedure is very similar to the procedure explained in Figure 2. However, in this procedure, one of the primers includes a region of homology that allows the second recombination event (indicated with "2 3 4" in Figures 3A and B). The PCR product is introduced by transformation in cells induced for the λ Red recombination system. By selecting for antibiotic resistance, the only recombinants that are encouraged involve the sequences flanking the Ab^r maker (i.e., 2 3 4 and 6 7 8 in Figure 2B). The final step in the procedures is the most efficient because all cells already have the DNA substrates for recombination (Figure 3C). In the procedure, the double-strand break allows the entry of the λ Red recombination machinery and provides negative selection for recombinants that have the desired event (Figure 3C). This procedure can also be adapted for introducing point mutations, insertion events, etc. Another procedure that also capitalizes on having the plasmid already present in the cell involves including the I-SceI cut sites on the same DNA substrate that is to be recombined into the host chromosome. In this procedure, the plasmid-encoded DNA substrate is linearized in vivo via expression if the I-SceI enzyme, along with the λ Red recombination system (see Herring et. al., References).

Alleles of Regulatory Genes and Elements

The methods discussed above can also be applied to assess a gene to determine if it has a regulatory function. For example, a gene that is adjacent to an operon in a genome sequence, e.g., *lacI* or *araC*, is a candidate for encoding a regulator, especially if it contains an HTH motif. Determination of the null phenotype of the gene can show whether the gene product is a positive or a negative regulator. Generation of other allele types can add to an understanding of how the regulator functions. However, it is still very difficult to predict which amino acid changes will elicit a particular phenotype, for example, a superrepressor phenotype, even if the regulator falls into one of the families discussed above. Imagine how naive our view of regulation would be if some regulatory genes had not been intensively studied by using selectional genetics to identify the amino acids important for the various phenotypes.

References

Baba, T., T. Ara, M. Hasegawa, Y. Takai, Y. Okumura, M. Baba, K. A. Datsenko, M. Tomita, B. L. Wanner, and H. Mori. 2006. Construction of *Escherichia coli* K-12 in-frame, single-gene knockout mutants: the Keio collection. *Mol. Syst. Biol.* **2**:2006.0008.

Herring, C. D, J. D. Glasner, and F. R. Blattner. 2003. Gene replacement without selection: regulated suppression of amber mutations in *Escherichia coli*. *Gene* **311**:153–163.

Tischer, B. K., G. A. Smith, and N. Osterrieder. 2010. *En passant* mutagenesis: a two step markerless Red recombination system. *Methods Mol. Biol.* **634**:421–430.

Why Are There So Many Mechanisms of Gene Regulation?

As discussed elsewhere in this book, "why" questions are difficult if not impossible to answer definitively in biology. However, it is evident from the discussion in this chapter that gene expression can be regulated in a wide variety of ways and by a wide variety of mechanisms. It is relatively easy to understand why a catabolic system is induced by the substrate of the regulated pathway and a biosynthetic system is repressed by the end product of the pathway. But why isn't all regulation at the level of transcription initiation? It would appear to be wasteful to regulate gene expression at the level of transcript degradation or translation, since the cell is producing mRNAs that it may never use. In some cases, the level of gene expression at which regulation occurs is dictated by the function of the gene product. For example, the gene encoding IF3 or RF2 exploits the function of its product in its regulatory mechanism. Another advantage of posttranscriptional regulation is that the cell always contains an adequate supply of the mRNA for the regulated gene, and the decision to utilize that mRNA to synthesize the protein product can be made very rapidly in response to subtle physiological changes.

This is especially important in stress responses, such as the heat shock response, which in *E. coli* utilizes an RNA thermosensor to directly and rapidly sense an increase in temperature, enabling the cell to immediately begin synthesis of gene products that protect it from this stressful condition.

It is not always obvious why one set of genes in one set of organisms uses one mechanism while related genes in other organisms use a different mechanism. It is important to note that these mechanisms evolved in response to selective pressures faced by an ancestral cell, and we may not know what those selective pressures were. Nevertheless, the bacterial cells we see today must find that these regulatory solutions serve their current needs, and it is likely that regulatory patterns will shift as organisms face new sets of selective pressures.

Operon Analysis for Sequenced Genomes

Genetic tools have played a crucial role in the investigation of genetic regulatory mechanisms. The operons discussed in this chapter were all discovered in genetic analyses that used forward genetics; that is, they were discovered by isolating and analyzing mutations that changed the wild-type sequence to a mutant sequence, i.e., changed a wild-type allele to a mutant allele. Moreover, the mutations were found because mutant strains could be identified on the basis of phenotypic changes that were related to the biological role of the operon studied. For example, *lac* mutations could be isolated on color indicator plates, such as MacConkey agar plates, which use a color change to indicate a perturbation in β-galactoside metabolism. Also, both mutant selections and mutant screens could be used, and creative exploitation of the metabolic properties of the operon studied yielded robust collections of mutations. It is because of studies of the variety of mutant alleles in both the structural genes and the regulatory genes that we now know as much as we do about operons and their regulation.

The dramatic improvements in DNA-sequencing technologies have provided a wealth of information. It is therefore now possible to initiate an investigation of a new pathway, or a new organism, without first identifying mutations in genes involved in the pathway. The availability of mutations with known sequence effects, however, is extremely useful, and new approaches have been developed that allow introduction of specific mutations into the cell. These approaches are summarized in Box 12.3.

SUMMARY

1. Regulation of gene expression can occur at any stage in the expression of a gene. If the amount of mRNA synthesized from the gene differs under different conditions, the gene is transcriptionally regulated. If the regulation occurs after the mRNA is made, the gene is posttranscriptionally regulated. A gene is translationally regulated if the mRNA is made but not always translated at the same rate. The mRNA or protein product of a gene can also be stabilized or degraded, or the protein product can be modified.

2. In bacteria, more than one gene is sometimes transcribed into the same mRNA. Such a cluster of genes, along with the adjacent *cis*-acting regulatory sites, is called an operon.

3. The regulation of operon transcription can be negative, positive, or a combination of the two. If a protein blocks the transcription of the operon, the operon is negatively regulated and the regulatory protein is a repressor. If a protein is required for transcription of an operon, the operon is positively regulated and the regulatory protein is an activator.

4. If an operon is negatively regulated, mutations that inactivate the regulatory gene product result in constitutive mutants in which the operon genes are always expressed. If the operon is positively regulated, mutations that inactivate the regulatory protein cause permanent loss of expression of the operon. In general, because most mutations are inactivating mutations, constitutive mutations are much more common with negatively regulated operons than with positively regulated operons.

5. Sometimes the same protein can be both a repressor and an activator in different situations, which complicates the genetic analysis of the regulation.

6. The regulation of transcription of bacterial operons is often achieved through small molecules called effectors, which bind to the repressor or activator protein, changing its conformation. If the presence of the effector causes the operon to be transcribed, it is called an inducer; if its presence blocks trancription of the operon, it is called a corepressor. The substrates of catabolic operons are usually inducers, whereas the end products of biosynthetic pathways are usually corepressors.

7. The regions on DNA to which repressors bind are called operators. Some repressors act by physically interfering with the binding of the RNA polymerase to the promoter (preventing closed-complex formation). Others allow repressor binding but prevent opening of the DNA at the promoter (preventing open-complex formation). Yet others

(continued)

SUMMARY (continued)

prevent the RNA polymerase from escaping the promoter to begin RNA synthesis (preventing promoter clearance). Some repressors act by binding to two operators on either side of the promoter simultaneously, bending the DNA between them and inactivating the promoter.

8. The regions to which activator proteins bind are called activator sequences. Some activator proteins recruit RNA polymerase to the promoter by binding both to a region on the DNA close to the promoter and to an exposed region of the RNA polymerase, thereby stabilizing the binding of the RNA polymerase to the promoter. Others interact with RNA polymerase already at the promoter and allow it to form an open complex. Still others remodel the promoter by changing the conformation of the DNA, thereby optimizing the spacing and orientation of the $^{-}10$ and $^{-}35$ regions.

9. Binding sites for repressor proteins usually overlap with the RNA polymerase-binding site or are downstream. Binding sites for activators are usually upstream of the promoter so that the activator does not interfere with access of RNA polymerase to the promoter. Some regulatory proteins can bind to different regions on DNA, depending on the location of the binding site, and can act as both repressors and activators.

10. Some operons are transcriptionally regulated by a mechanism called attenuation. In operons regulated by attenuation, transcription begins on the operon but then terminates after a short leader sequence has been transcribed if the enzymes encoded by the operon are not needed.

11. Attenuation is sometimes determined by whether certain codons in the leader sequence are translated. Pausing of the ribosome at these codons can cause secondary-structure changes in the leader RNA, leading to termination of transcription by RNA polymerase before it reaches the first gene of the operon. In other operons, attenuation is mediated by RNA-binding proteins that either stabilize or destabilize leader RNA structural elements that determine whether transcription will terminate.

12. Leader RNA structural changes can affect both transcription attenuation and translation initiation. Riboswitches are leader RNA elements that directly sense a physiological signal. Binding of the signal changes the leader RNA secondary structure, affecting transcription attenuation or translation of the downstream gene. RNA-binding proteins can also affect translation initiation by sequestration of the TIR.

13. Gene expression can be regulated by affecting either the synthesis or the degradation of the mRNA and protein products. The half-life of the mRNA or protein product is the time required for the amount of the product to be reduced to half of the amout that was present at a given time point. Changes in half-life can have major effects on the amount of product present in the cell.

14. The activity of a protein can be regulated through reversible regulation of the activities of the enzymes of the pathway. This reversible regulation can occur by feedback inhibition, which results from binding of the end product of the biosynthetic pathway to the first enzyme of the pathway, or it can occur by reversible covalent modification of the protein.

QUESTIONS FOR THOUGHT

1. Why do you suppose both negative and positive mechanisms of transcriptional regulation are used to regulate bacterial operons?

2. Why are regulatory protein genes sometimes autoregulated?

3. Why do you suppose the genes for the biosynthesis of most amino acids, such as tryptophan, isoleucine-valine, and histidine, are arranged together in operons?

4. What advantages or disadvantages are there to regulation by attenuation? Would it not be less wasteful to regulate all operons through initiation of RNA synthesis at the promoter by repressors or activators?

5. Why does the stability of an mRNA transcript or protein product matter? Why is the synthesis rate not the only important parameter?

6. When genes are regulated through translational arrest in the exit channel of the ribosome, why is it that the translational arrest occurs in an upstream leader sequence rather than in the gene itself?

PROBLEMS

1. Outline how you would isolate a *lacI*^{s} mutant of *E. coli.*

2. Is the AraC protein in the P1 or P2 state with D-fucose bound?

3. What would the phenotype of the following merodiploid *E. coli* cells be with one form of the operon region in the F′ plasmid and the other in the chromosome?

   a. F′ *lac*$^{+}$/*lacI*^{s}

   b. F′ *lac*$^{+}$/*lacO*^{c} *lacZ*(Am) [the *lacZ*(Am) mutation is an N-terminal polar nonsense mutation]

   c. F′ *ara*$^{+}$/*araC* (the *araC* mutation is inactivating)

   d. F′ *ara*$^{+}$/*araI*

   e. F′ *ara*$^{+}$/*araP*$_{BAD}$

4. The *phoA* gene of *E. coli* is turned on only if phosphate is limiting in the medium. What kind of genetic experiments would you do to determine whether the *phoA* gene is positively or negatively regulated?

5. Outline how you would use 5-methyltryptophan to isolate constitutive mutants of the *trp* operon and then use them to isolate feedback inhibition mutants.

6. Would you expect BglG^{-} mutants to be constitutive or superrepressed? What about BglF^{-} mutants?

7. What experiment would you do to determine if codon discrimination in the A site of the ribosome leading to ribosome stalling by nascent polypeptides in the exit channel is due to the amino acid or to the tRNA in the incoming aminocylated tRNA? Hint: how do nonsense suppressor tRNAs work?

SUGGESTED READING

Babitzke, P., and P. Gollnick. 2001. Posttranscription initiation control of tryptophan metabolism in *Bacillus subtilis* by the *trp* RNA-binding attenuation protein (TRAP), anti-TRAP, and RNA structure. *J. Bacteriol. 183*:5795–5802.

Bougdour, A., C. Cunning, P. J. Baptiste, T. Elliott, and S. Gottesman. 2008. Multiple pathways for regulation of σ^S (RpoS) stability in *Eschericha coli* via the action of multiple anti-adaptors. *Mol. Microbiol.* **68**:298–313.

Browning, D. F., and S. Busby. 2004. The regulation of bacterial transcription initiation. *Nat. Rev. Microbiol. 2*:1–9.

Collins, J. A., I. Irnov, S. Baker, and W. C. Winkler. 2007. Mechanism of mRNA destabilization by the *glmS* ribozyme. *Genes Dev. 21*:3356–3368.

Englesberg, E., C. Squires, and F. Meronk. 1969. The arabinose operon in *Escherichia coli* B/r: a genetic demonstration of two functional states of the product of a regulator gene. *Proc. Natl. Acad. Sci. USA* **62**:1100–1107.

Fux, L., A. Nussbaum-Shochat, L. Lopian, and O. Amster-Choder. 2004. Modulation of monomer formation of the BglG transcriptional antiterminator from *Escherichia coli. J. Bacteriol.* **186**:6775–6781.

Geanacopoulos, M., G. Vasmatzis, V. B. Zhurkin, and S. Adhya. 2001. Gal repressosome contains an antiparallel DNA loop. *Nat. Struct. Biol.* **8**:432–436.

Geissmann, T., S. Marzi, and P. Romby. 2009. The role of mRNA structure in translational control in bacteria. *RNA Biol. 6*:153–160.

Grundy, F. J., and T. M. Henkin. 1993. tRNA-directed transcription antitermination. *Cell* **74**:475–482.

Guzman, L. M., D. Belin, M. J. Carson, and J. Beckwith. 1995. Tight regulation, modulation, and high-level expression by vectors containing the arabinose P$_{BAD}$ promoter. *J. Bacteriol.* **177**:4121–4130.

Irani, M. H., L. Orosz, and S. Adhya. 1983. A control element within a structural gene: the *gal* operon of *Escherichia coli. Cell* **32**:783–788.

Ito, K., S. Chiba, and K. Pogliano. 2010. Divergent stalling sequences sense and control cellular physiology. *Biochem. Biophys. Res. Commun.* **393**:1–5.

Jacob, F., and J. Monod. 1961. Genetic regulatory mechanisms in the synthesis of proteins. *J. Mol. Biol.* **3**:318–356.

Lewis, M., G. Chang, N. C. Horton, M. A. Kercher, H. C. Pace, M. A. Schumacher, R. G. Brennan, and P. Lu. 1996. Crystal structure of the lactose operon repressor and its complexes with DNA and inducer. *Science* **271**:1247–1254.

McDaniel, B. A., F. J. Grundy, and T. M. Henkin. 2005. A tertiary structural element in S box leader RNAs required for S-adenosylmethionine-directed transcription termination. *Mol. Microbiol.* **57**:1008–1021.

Merino, E., and C. Yanofsky. 2005. Transcription attenuation: a highly conserved regulatory strategy used by bacteria. *Trends Genet.* **21**:260–264.

Morse, D. E., and A. N. C. Morse. 1976. Dual control of the tryptophan operon is mediated by both tryptophanyl-tRNA synthetase and the repressor. *J. Mol. Biol.* **103**:209–226.

Oxender, D. L., G. Zurawski, and C. Yanofsky. 1979. Attenuation in the *Escherichia coli* tryptophan operon. Role of RNA secondary structure involving the tryptophan codon region. *Proc. Natl. Acad. Sci. USA* **76**:5524–5528.

Pace, H. C., M. A. Kercher, P. Lu, P. Markiewicz, J. H. Miller, G. Chang, and M. Lewis. 1997. Lac repressor genetic map in real space. *Trends Biochem. Sci.* **22**:334–339.

Possoz, C., S. R. Filipe, I. Grainge, and D. J. Sherratt. 2006. Tracking of controlled *Escherichia coli* replication fork stalling and restart at repressor-bound DNA *in vivo*. *EMBO J.* **25**:2596–2604.

Raveh, H., L. Lopian, A. Nussbaum-Shochat, A. Wright, and O. Amster-Choder. 2009. Modulation of transcription antitermination in the *bgl* operon by the PTS. *Proc. Natl. Acad. Sci. USA* **106**:13523–13528.

Rojo, F. 1999. Repression of transcription initiation in bacteria. *J. Bacteriol.* **181**:2987–2991.

Sanchez, A., M. L. Osborne, L. J. Friedman, J. Kondev, and J. Gelles. 2011. Mechanism of transcriptional repression at a bacterial promoter by analysis of single molecules. *EMBO J.* **30:**3940-3946.

Schleif, R. 2000. Regulation of the L-arabinose operon of *Escherichia coli. Trends Genet.* **16:**559–565.

Schuck, A., A. Diwa, and J. G. Belasco. 2009. RNase E autoregulates its synthesis in *Escherichia coli* by binding directly to a stem-loop in the *rne* 5′ untranslated region. *Mol. Microbiol.* **72:**470–478.

Shuman, H. A., and T. J. Silhavy. 2003. The art and design of genetic screens: *Escherichia coli. Nat. Rev. Genet.* **4:**419–431.

Storz, G., J. Vogel, and K. M. Wassarman. 2011. Regulation by small RNAs in bacteria: expanding frontiers. *Mol. Cell* **43:**880–891.

Yanofsky, C., and I. P. Crawford. 1987. The tryptophan operon, p. 1453–1472. *In* F. C. Neidhardt, J. L. Ingraham, K. B. Low, B. Magasanik, M. Schaechter, and H. E. Umbarger (ed.), Escherichia coli *and* Salmonella typhimurium: *Cellular and Molecular Biology*, vol. 2. American Society for Microbiology, Washington, DC.

Yousef, M. R., F. J. Grundy, and T. M. Henkin. 2005. Structural transitions induced by the interaction between $tRNA^{Gly}$ and the *Bacillus subtilis glyQS* T box leader RNA. *J. Mol. Biol.* **349:**273–287.

CHAPTER 13

Global Regulation: Regulons and Stimulons

BACTERIA MUST BE ABLE TO ADAPT to a wide range of environmental conditions to survive. Nutrients are usually limiting, so bacteria must be able to recognize the availability of nutrients and protect themselves against starvation until an adequate food source becomes available. Different environments also vary greatly in the amount of water or in the concentration of solutes, so bacteria must also be able to adjust to desiccation and differences in osmolarity, as well as other stresses. Temperature fluctuations are also a problem for bacteria. Unlike humans and other warm-blooded animals, bacteria cannot maintain their own cell temperature and so must be able to function over wide ranges of temperature. Pathogenic bacteria must be able to sense that they have entered the host environment and adapt to the new conditions there.

Survival alone is not enough for a species to prevail, however. The species also must compete effectively with other organisms in the environment. Competing effectively might mean being able to use scarce nutrients efficiently or taking advantage of plentiful ones to achieve higher growth rates and thereby become a higher percentage of the total population of organisms in the environment. Moreover, different compounds may be available for use as carbon and energy sources. The bacterium may need to choose the carbon and energy source it can use most efficiently and ignore the rest, so that it does not waste energy making extra enzymes.

Not only do conditions vary in the environment, but the changes also can be abrupt. The bacterium may have to adjust the rate of synthesis of its cellular constituents quickly in response to a change in growth conditions. For example, different carbon and energy sources allow different rates of bacterial growth. Different growth rates require different rates of synthesis of cellular macromolecules, such as DNA, RNA, and proteins, which in

doi:10.1128/9781555817169.ch13

turn require different concentrations of the components of the cellular macromolecular synthesis machinery, such as ribosomes, tRNA, and RNA polymerase. Moreover, the relative rates of synthesis of the different cellular components must be coordinated so that the cell does not accumulate more of some component than it needs.

Adjusting to major changes in the environment requires regulatory systems that simultaneously regulate numerous operons. These systems are called **global regulatory mechanisms**. Often in global regulation, a single regulatory protein (or RNA) controls a large number of genes and operons, which are then said to be members of the same **regulon**. Most genes are part of some regulon, and some regulons are very large. Individual genes or operons can also be members of multiple regulons, which allows a response to multiple input signals. Regulons often overlap in their responses to changing conditions. The collection of regulons that respond to the same set of environmental conditions is called a **stimulon**. Large-scale genomic analyses can be used to identify most of the genes of a regulon or stimulon. Such studies reveal that only seven regulators control almost half of all the genes of *Escherichia coli*.

Table 13.1 lists some global regulatory mechanisms known to exist in *E. coli*. If the genes are under the control of a single regulatory gene (and so are members of the same regulon), the regulatory gene is also listed. Some examples of regulons are discussed in previous chapters. For example, all the genes under the control of the TrpR repressor, including the *trpR* gene itself, are part of the TrpR regulon. The Ada regulon comprises the adaptive-response genes, including those encoding the methyltransferases that repair alkylation damage to DNA; all of these genes are under the control of the Ada protein. Similarly, the SOS genes that are induced after UV irradiation and some other types of DNA-damaging treatments are all under the control of the same protein, the LexA repressor, and so are part of the LexA regulon (see chapter 11). In other cases, the molecular basis of the global regulation is less well understood and may involve a complex interaction among several cellular signals or regulators.

In this chapter, we discuss what is known about how some global regulatory mechanisms operate on the molecular level and describe some of the genetic experiments that have contributed to this knowledge. The ongoing studies of the molecular basis of global regulatory mechanisms represent one of the most active areas of research involving bacterial molecular genetics.

Carbon Catabolite Regulation

One of the largest global regulatory systems in bacteria coordinates the expression of genes involved in carbon and energy source utilization. All cells must have access to high-energy, carbon-containing compounds, which they degrade to generate ATP for energy and smaller molecules needed as building blocks for cellular constituents. Smaller molecules resulting from the metabolic breakdown of larger molecules are called **catabolites**.

In nutrient-rich environments, bacterial cells may be growing in the presence of several different carbon and energy sources, some of which can be used more efficiently than others. Energy must be expended to synthesize the enzymes needed to metabolize the different carbon sources, and the utilization of some carbon compounds requires more enzymes, and yields less energy, than does the utilization of others. By making only the enzymes for utilization of the carbon and energy source that yields the highest return, the cell gets the most catabolites and energy, in the form of ATP, for the energy it expends. The mechanism for ensuring that the cell preferentially uses the best carbon and energy source available is called **catabolite regulation**. Operons that are subject to catabolite regulation are said to be catabolite sensitive. Historically, this regulatory response has been refered to as "catabolite repression" based on the fact that cells growing in better carbon sources, such as glucose, seem to repress the expression of operons for the utilization of poorer carbon sources. However, as we shall see, the name "catabolite repression" is often a misnomer, because in at least some of the regulatory systems in *E. coli*, the genes under catabolite control are activated when poorer carbon sources are the only ones available. Catabolite regulation is also sometimes called the **glucose effect** because glucose, which yields the highest return of ATP per unit of expended energy, usually strongly represses operons for other carbon sources. To use glucose, the cell need only convert it to glucose-6-phosphate, which can enter the glycolytic pathway. Thus, glucose is the preferred carbon and energy source for most (but not all) types of bacteria.

Figure 13.1 illustrates what happens when *E. coli* cells are growing in a mixture of glucose and galactose. The cells first use the glucose, and only after it is depleted do they begin to use the galactose. When the glucose is gone, the cells stop growing briefly while they synthesize the enzymes for galactose utilization. Following the appropriate regulatory changes in the cell, growth resumes, but at a slightly lower rate. This growth pattern is called **diauxie** and is commonly observed during growth in a mixture of carbon sources.

Catabolite Regulation in *E. coli*: Catabolite Activator Protein (CAP) and cAMP

Most bacteria and lower eukaryotes are known to have systems for catabolite regulation. The best understood is the **cyclic AMP** (**cAMP**)-dependent system of *E. coli* and other enteric bacteria. cAMP is similar to AMP (a

Table 13.1 A sampling of *E. coli* global regulatory systems

System	Response	Regulatory gene(s) (protein[s])	Category of mechanism	Some genes, operons, regulons, and stimulons
Nutrient limitation				
Carbon	Catabolite regulation	*crp* (CAP, also called CRP)	DNA-binding activator or repressor	*lac, ara, gal, mal,* and numerous other C source operons
	Control of fermentative vs. oxidative metabolism	*cra* (Cra)	DNA-binding activator or repressor	Enzymes of glycolysis, Krebs cycle
Nitrogen	Response to ammonia limitation	*rpoN*	Sigma factor (σ^N)	*glnA* (GS) and operons for amino acid degradation
		ntrBC (NtrBC)	Two-component system	
Phosphorus	Starvation for inorganic orthophosphate (P_i)	*phoBR* (PhoBR)	Two-component system	>38 genes, including *phoA* (bacterial alkaline phosphatase) and *pst* operon (P_i uptake)
Growth limitation				
Stringent response	Response to lack of sufficient aminoacylated-tRNAs for protein synthesis	*relA* (RelA), *spoT* (SpoT)	(p)ppGpp metabolism	rRNA, tRNA, ribosomal proteins, amino acid biosynthesis operons
Stationary phase	Switch to maintenance metabolism and stress protection	*rpoS* (RpoS), sigma factor (σ^S)	Many genes with σ^S promoters; complex effects on many operons	
Oxygen	Response to anaerobic environment	*fnr* (Fnr)	CAP family of DNA-binding proteins	>31 transcripts, including *narGHJI* (nitrate reductase)
	Response to presence of oxygen	*arcAB* (ArcAB)	Two-component system	>20 genes, including *cob* (cobalamin synthesis)
Stress				
Osmoregulation	Response to abrupt osmotic upshift	*kdpDE* (KdpD, KdpE)	Two-component system	*kdpFABC* (K^+ uptake system)
	Adjustment to osmotic environment	*envZ/ompR* (EnvZ/OmpR)	Two-component system	OmpC and OmpF outer membrane proteins
		micF	sRNA	*ompF* (porin)
Oxygen stress	Protection against reactive oxygen species	*soxS* (SoxS)	AraC family of DNA-binding proteins	Regulon, including *sodA* (superoxide dismutase) and *micF* (sRNA regulator of *ompF*)
		oxyR (OxyR)	LysR family of DNA-binding proteins	Regulon, including *katG* (catalase)
Heat shock	Tolerance of abrupt temperature increase	*rpoH* (RpoH)	Sigma factor (σ^H)	Stimulon; Hsps (heat shock proteins), including *dnaK, dnaJ,* and *grpE* (chaperones), and *lon, clpP, clpX,* and *hflB* (proteases)
Envelope stress	Misfolded Omp proteins	*rpoE* (RpoE)	Sigma factor (σ^E)	>10 genes, including *rpoH* (σ^H) and *degP* (encoding a periplasmic protease)
	Misfolded pilus	*cpxAR* (CpxAR)	Two-component system	Overlap with RpoE regulon
pH shock	Tolerance of acidic environment	Many	Many	Complex stimulon

constituent of an RNA chain; see chapter 2), with a single phosphate group on the ribose sugar; however, the phosphate is attached to both the 5′-hydroxyl and the 3′-hydroxyl groups of the sugar, thereby making a circle out of the phosphate and sugar. Only *E. coli* and other closely related enteric bacteria seem to use this cAMP-dependent system. Some bacteria have an entirely different system that does not involve cAMP (see below). Even *E. coli* has a second, cAMP-independent system for catabolite regulation, which is discussed in Box 13.1.

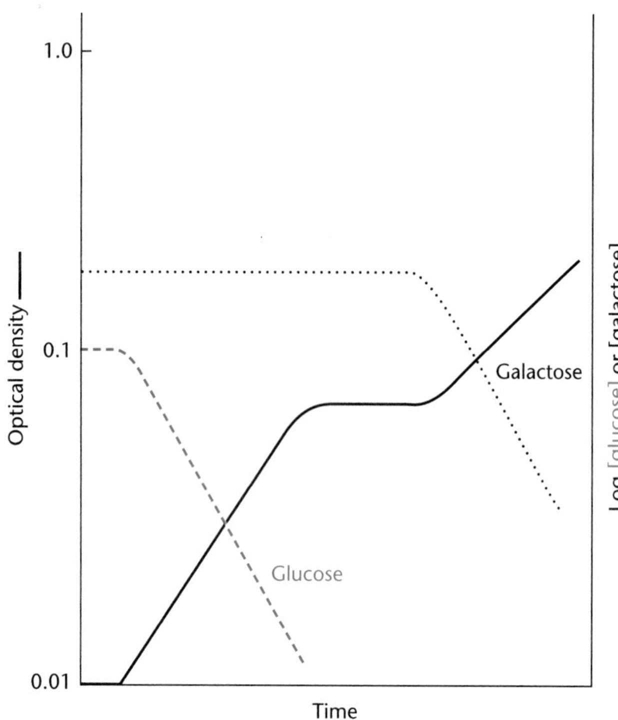

Figure 13.1 Diauxic growth of *E. coli* in a mixture of glucose and galactose. The concentrations of the sugars in the medium are shown as dashed and dotted lines. The optical density, a measure of cell growth, is shown as a solid line. The cells first deplete all the glucose and then show a short lag while they induce the *gal* operon (plateau in optical density). They then grow more slowly on the galactose.
doi:10.1128/9781555817169.ch13.f13.1

REGULATION OF cAMP SYNTHESIS

Catabolite regulation in *E. coli* is achieved through fluctuation in the levels of cAMP, which vary inversely with the availability of readily metabolizable carbon sources, such as glucose. In other words, cellular concentrations of cAMP are higher when levels of easily metabolized carbon sources are lower; this occurs when the bacteria are growing in a relatively poor source of carbon, such as lactose or maltose. The synthesis of cAMP is controlled through the regulation of the activity of adenylate cyclase. This enzyme, which makes cAMP from ATP, is more active when glucose is low and less active when glucose is high. The adenylate cyclase enzyme is associated with the inner membrane and is the product of the *cya* gene.

Figure 13.2 outlines the current picture of the regulation of adenylate cyclase activity. An important factor in the regulation is the phosphoenolpyruvate (PEP)-dependent sugar phosphotransferase system (PTS), which, as the name implies, is responsible for transporting certain sugars, including glucose, into the cell. We mentioned PTS systems in connection with the regulation of the *mal* and *bgl* operons in chapter 12. One of the protein components of the PTS, named IIA^{Glc}, can exist in either an unphosphorylated (IIA^{Glc}) or a phosphorylated (IIA^{Glc}~P) form. The IIA^{Glc}~P form activates adenylate cyclase to make cAMP. When levels of glucose or another sugar that IIA^{Glc} transports are high in the growth medium, most of the IIA^{Glc} is in the unphosphorylated form. As a result, little of the IIA^{Glc}~P form then exists to activate the adenylate cyclase, and cAMP levels drop.

The ratio of IIA^{Glc}~P to IIA^{Glc} is determined largely by the ratio of PEP to pyruvate in the cell. When a rapidly metabolizable substrate, such as glucose, is present in the medium, the PEP/pyruvate ratio is low; when only

Figure 13.2 Exogenous glucose inhibits both cAMP synthesis and the uptake of other sugars, such as lactose. **(A)** In the presence of glucose, the ratio of IIA^{Glc} to IIA^{Glc}~P is high, as glucose is phosphorylated when it is transported by the glucose transporter, II^{Glc}. Unphosphorylated IIA^{Glc} inhibits the lactose permease, resulting in "inducer exclusion." **(B)** In the absence of glucose, the IIA^{Glc}~P concentration is high, and it activates adenylate cyclase. Also, lactose transport is permitted.
doi:10.1128/9781555817169.ch13.f13.2

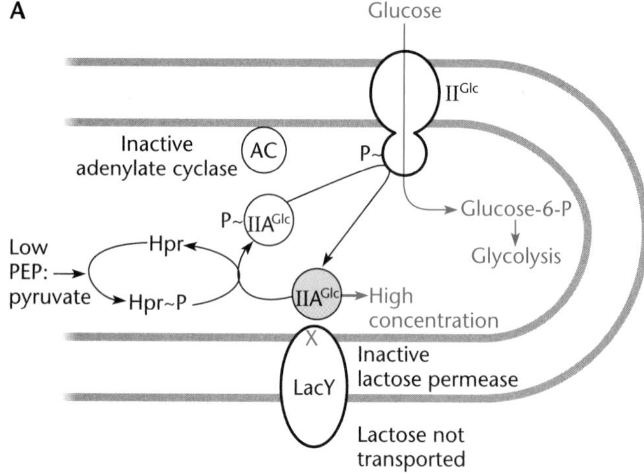

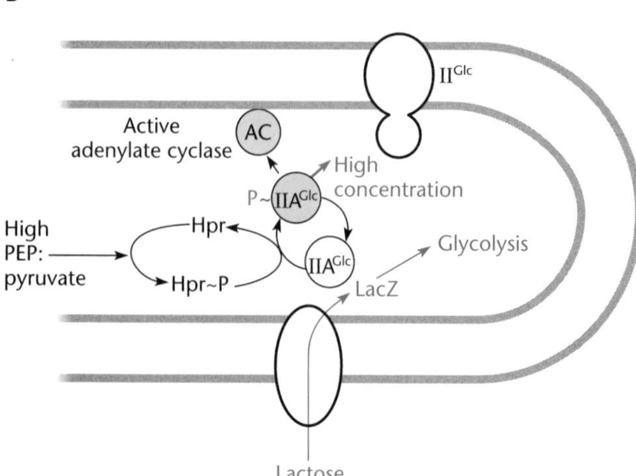

BOX 13.1

cAMP-Independent Carbon Catabolite Regulation in *E. coli*

Not all catabolite regulation in bacteria is mediated by cAMP. In fact, most gram-positive bacteria, including *B. subtilis*, do not even have cAMP and use a mechanism for catabolite regulation that is very different from the CAP-cAMP system. Even in *E. coli* and other enteric bacteria, there is a mechanism of catabolite repression that does not depend on cAMP. This mechanism involves the Cra protein, named for its function as a catabolite repressor/activator (originally called FruR, for *fru*ctose *r*epressor). The Cra protein is encoded by the *cra* gene and is a DNA-binding protein similar to LacI and GalR. Cra was discovered during the identification of mutations that suppress *ptsH* mutations of *E. coli* and *S. enterica* serovar Typhimurium. The *ptsH* gene encodes the Hpr protein, which phosphorylates many sugars during transport that can be transported by the PTS system, including glucose (Figure 13.2). Therefore, *ptsH* mutants cannot use these sugars, because phosphorylation is the first step in the glycolytic pathway. The *cra* mutations suppress *ptsH* mutations and allow growth on PTS sugars by allowing the constitutive expression of the fructose catabolic operon, which includes a gene that encodes a protein that can substitute for Hpr. The *cra* mutants were found to be pleiotropic in that they are unable to synthesize glucose from many substrates, including acetate, pyruvate, alanine, and citrate. They also demonstrate elevated expression of genes involved in glycolytic pathways.

The pleiotropic phenotype of *cra* mutants suggested that Cra functions as a global regulatory protein, activating the transcription of some genes and repressing the synthesis of others. As for other regulatory proteins, including *B. subtilis* CcpA (Figure 13.7), whether Cra activates or represses transcription depends on where it binds relative to the promoter of the regulated gene. If its binding site is upstream of the promoter, it activates transcription of the operon; if its binding site overlaps or is downstream of the promoter, it represses transcription. In either case, the DNA-binding activity of Cra depends on the presence of a signal molecule. In the absence of the signal molecules, Cra is active in DNA binding. Cra is inactivated by binding of fructose-1-phosphate (F1P) or fructose-1,6-bisphosphate (FBP), which are present at high concentrations during growth in the presence of sugars, such as glucose. The effect of this on the transcription of a particular operon depends on whether Cra functions as a repressor or an activator of that operon. If it functions as a repressor, the transcription of the operon increases when levels of glucose (and therefore F1P or FBP) are high; if it functions as an activator, the transcription of the operon decreases when levels of F1P or FBP are high. In general, the Cra protein represses operons whose products are involved in central pathways for sugar catabolism, such as the Embden-Meyerhof and Entner-Doudoroff pathways, so the transcription of these genes increases when glucose and other good carbon sources are available. In contrast, it usually activates operons whose products are involved in synthesizing glucose from pyruvate and other metabolites (gluconeogenesis), so it does not activate the transcription of these operons if glucose is present. The activity of Cra allows another layer of regulation of carbon metabolism independent of the activity of the sugar transport system, which is what is monitored (indirectly) by the CAP-cAMP system. Cra instead monitors internal pools of sugar metabolites.

Reference

Saier, M. H., Jr., and T. M. Ramseier. 1996. The catabolite repressor/activator (Cra) protein of enteric bacteria. *J. Bacteriol.* **178:**3411–3417.

poorer carbon sources are available, the ratio is high. The PEP transfers its phosphate to another protein called Hpr (for *h*istidine *pr*otein; histidine is the amino acid in the protein to which the phosphate is transferred to generate Hpr~P) and becomes pyruvate. The phosphate from Hpr~P is then transferred to IIA^{Glc} to make IIA^{Glc}~P. Therefore, the higher the PEP/pyruvate ratio, the higher the Hpr~P/Hpr ratio and the higher the IIA^{Glc}~P/IIA^{Glc} ratio. High IIA^{Glc}~P results in high cAMP, which serves as the signal that only poorer carbon sources are available. The transfer of phosphate from PEP to HPr to IIA^{Glc} is called a **phosphorylation cascade** because phosphates are transferred from one molecule to another, much like water is transferred down a cascade of waterfalls. We give other examples of phosphorylation cascades later in this chapter and in chapter 14.

The unphosphorylated form of IIA^{Glc} (which is present when glucose is high) also inhibits other sugar-specific permeases that transport sugars, such as lactose (Fig. 13.2). Therefore, less of these other sugars enters the cell if glucose or another, better carbon source is available, and less inducer is present to induce transcription of their respective operons (see chapter 12). This effect is called **inducer exclusion** (Figure 13.2). For systems like the *lac* and *gal* operons that require both activation when glucose is low and induction by their specific sugar substrate (lactose or galactose, respectively), it is often difficult to distinguish the effects of inducer exclusion on operon induction from the effects of cAMP on the promoter (see Inada et al., Suggested Reading, and below).

CATABOLITE ACTIVATOR PROTEIN

The mechanism by which cAMP turns on catabolite-sensitive operons in *E. coli* is well understood and has served as a model for transcriptional activation. The cAMP binds to the protein product of the *crp* gene, which is an activator of transcription of catabolite-sensitive operons. This activator protein goes by two names, CAP (for catabolite activator protein) and CRP (for cAMP receptor protein). We will use the CAP terminology, as it reflects the molecular mechanism by which the protein acts on transcription. The activator CAP with cAMP bound (CAP-cAMP) functions like other activator proteins discussed in chapter 12 in that it interacts with the RNA polymerase to activate transcription from promoters for operons under its control, including *lac*, *gal*, *ara*, and *mal*. These operons are all members of the **CAP regulon** or the catabolite-sensitive regulon (Table 13.1). However, the mechanism of CAP-cAMP regulation varies. CAP also can function not only as an activator, but as a repressor, depending on where it binds relative to the promoter (see below).

REGULATION BY CAP-cAMP

The mechanism by which CAP activates transcription varies from promoter to promoter. Some of these mechanisms are shown in Figure 13.3. Upstream of the promoter is a short sequence called the **CAP-binding site**, which is similar in all catabolite-sensitive operons and so can be easily identified. CAP is active as a DNA-binding protein only when it is bound to cAMP, so the site is occupied only when cAMP levels are high. CAP functions like many other activators to make contact with the RNA polymerase at the promoter and to stimulate one or more of the steps in the initiation of transcription (see Browning and Busby, Suggested Reading). Transcriptional activators are discussed in chapter 12. CAP can contact different regions of the RNA polymerase and stimulate different steps in initiation, depending on where it is bound relative to the promoter. This is illustrated in Figure 13.3B. At class I CAP-dependent promoters, such as the *lac* promoter, a dimer of CAP with cAMP binds upstream of the promoter and contacts the C-terminal end of the α subunit of RNA polymerase (α C-terminal domain [αCTD]) (see chapters 2 and 12). This contact strengthens the binding of RNA polymerase to the promoter (to form the closed complex). In class II CAP-dependent promoters, such as the *gal* promoter p_{G1}, the CAP dimer-binding site slightly overlaps that of RNA polymerase, and CAP contacts a region in the N terminus of the α subunit (α N-terminal domain [αNTD]). In this position, it stimulates the opening of the DNA at the promoter (open complex formation). There are even promoters in which more than one CAP dimer binds to stimulate both binding and open complex formation. CAP also can bend the DNA when it binds to the CAP-binding site, which can affect access to other regulatory proteins and to RNA polymerase.

A

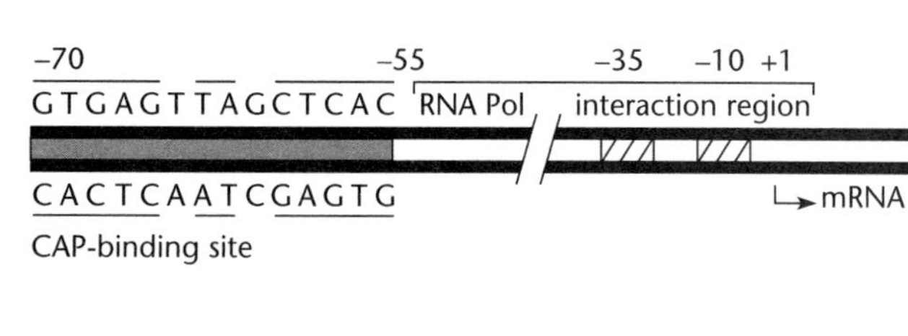

B

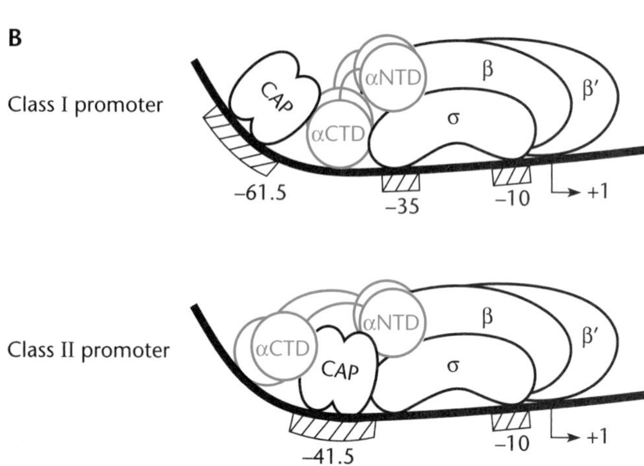

Figure 13.3 Model for CAP activation at class I and class II CAP-dependent promoters. **(A)** Sequence of the CAP-binding site upstream of the class I *lac* promoter. Pol, polymerase. **(B)** Binding and location interactions of CAP with the CTD and NTD of the α subunit with class I and class II promoters, respectively.
doi:10.1128/9781555817169.ch13.f13.3

The positioning of the CAP-binding sequence relative to the promoter can be very different from that of *lac* and *gal*. In the *ara* operon, the CAP-binding site is further upstream, with the AraC-binding site between it and the promoter (Figure 13.4). Nevertheless, it can still make contact with the αCTD of RNA polymerase, which can reach up along the DNA, as shown. The ability of the αCTD to interact with CAP at this site requires that the CAP-binding site is on the same face of the DNA helix as the promoter, so that the αCTD can reach upstream without reaching around to the other side of the DNA helix. This is an example of face-of-the-helix dependence (see chapter 12). CAP can also stimulate transcription by interacting with another activator or can stimulate transcription by preventing the binding of a repressor (see chapter 12).

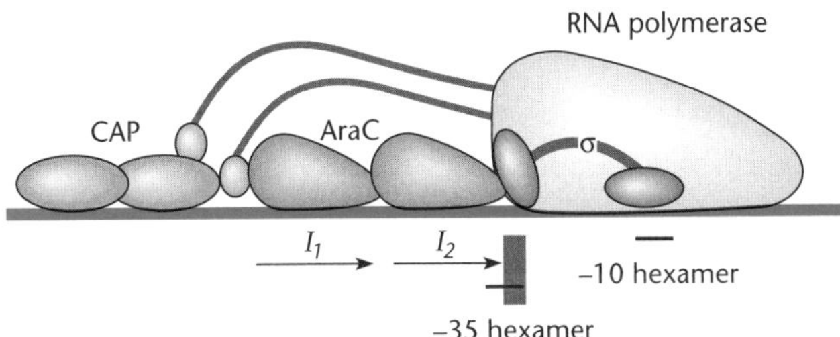

Figure 13.4 Summary of the RNA polymerase-promoter and activator-promoter interactions at the p_{BAD} promoter of the *ara* operon. The σ^{70} subunit of RNA polymerase contacts the –35 and –10 hexamers. Occupancy of the I_1 and I_2 half-sites by AraC activates transcription with the aid of CAP, utilizing the α subunit-activator interactions, as shown. The binding sites of σ^{70} and AraC overlap by 4 bp at p_{BAD}. The nucleotides in the –35 hexamer that lie outside the region of overlap are shaded. doi:10.1128/9781555817169.ch13.f13.4

Certain operons in the CAP regulon, such as *gal*, are less sensitive to catabolite repression than others. As discussed in chapter 12, some transcription of the *gal* operon is maintained when glucose is present in the medium because it has two promoters, p_{G1} and p_{G2} (see Figure 12.8). Transcription from p_{G2} does not require CAP-cAMP for its activation. The p_{G2} promoter permits some expression of the *gal* operon even in the presence of glucose. This low level of expression is necessary to allow the synthesis of cell wall components that include galactose, since the UDP-galactose synthesized by the operon serves as the donor of galactose in biosynthetic reactions. However, the level of expression of the *gal* operon from p_{G2} is not high enough for the cells to grow well on galactose as a carbon and energy source.

The dependence of transcriptional activation by CAP on cAMP has been exploited in a number of genetic tools. One of these, a **two-hybrid system** to test for protein-protein interactions using *E. coli* as a host, is described in Box 13.2.

RELATIONSHIP OF CATABOLITE REGULATION TO INDUCTION

An important point about CAP-dependent regulation of catabolite-sensitive operons is that it occurs in addition to any other regulation to which the operon is subject. Two conditions must be met before catabolite-sensitive operons can be transcribed: better carbon sources, such as glucose, must be absent, and the inducer of the operon must be present. Take the example of the *lac* operon (Figure 13.5). If a carbon source better than lactose is available, cAMP levels are low, and CAP-cAMP does not bind upstream of the *lac* promoter to activate transcription. Also, the *lac* transport system is inhibited, excluding the inducer from the cell. However, even at high cAMP levels, the *lac* operon is not transcribed unless the inducer, allolactose (a metabolite of lactose), is also present. In the absence of inducer, the LacI repressor is bound to the operator and prevents the RNA polymerase from binding to the promoter and transcribing the operon (see chapter 12).

GENETIC ANALYSIS OF CATABOLITE REGULATION IN *E. coli*

The above model for the regulation of catabolite-sensitive operons is supported by both genetic and biochemical analyses of catabolite regulation in *E. coli*. These analyses have involved the isolation of mutants

BOX 13.2

A Bacterial Two-Hybrid System Based on Adenylate Cyclase

Two-hybrid systems are experimental approaches for identifying proteins that interact with each other. These systems are based on using fragments of two proteins that exhibit an easily measurable property (like an effect on gene expression) when they are brought together in a complex. Fusion of each of these fragments to different proteins that are suspected to interact allows testing whether they interact. This is commonly done using a yeast system to identify partners for a test protein (the "bait"). A two-hybrid system named the BACTH (for bacterial adenylate cyclase two-hybrid) system that is based on interaction-mediated reconstitution of the cAMP signaling cascade has been developed for testing specific partner interactions in *E. coli*. The *B. pertussis* adenylate cyclase enzyme has been separated into two complementing fragments: T25, from amino acids 1 to 224, and T18, from amino acids 225 to 399. Functional complementation of these two fragments results in cAMP synthesis and so allows transcription of cAMP-dependent catabolic operons. Clones expressing the two *B. pertussis* fragments, T25 and T18, can be introduced into an *E. coli* mutant with an adenylate cyclase mutation, e.g., Δ*cya*, and the *Bordetella* adenylate cyclase complements the *E. coli* mutation provided the two fragments come together to form the functional adenylate cyclase. Functional complementation occurs only if the T25 and T18 gene fragments are fused to proteins that interact with each other, since the two fragments do not interact

(continued)

BOX 13.2 (continued)

A Bacterial Two-Hybrid System Based on Adenylate Cyclase

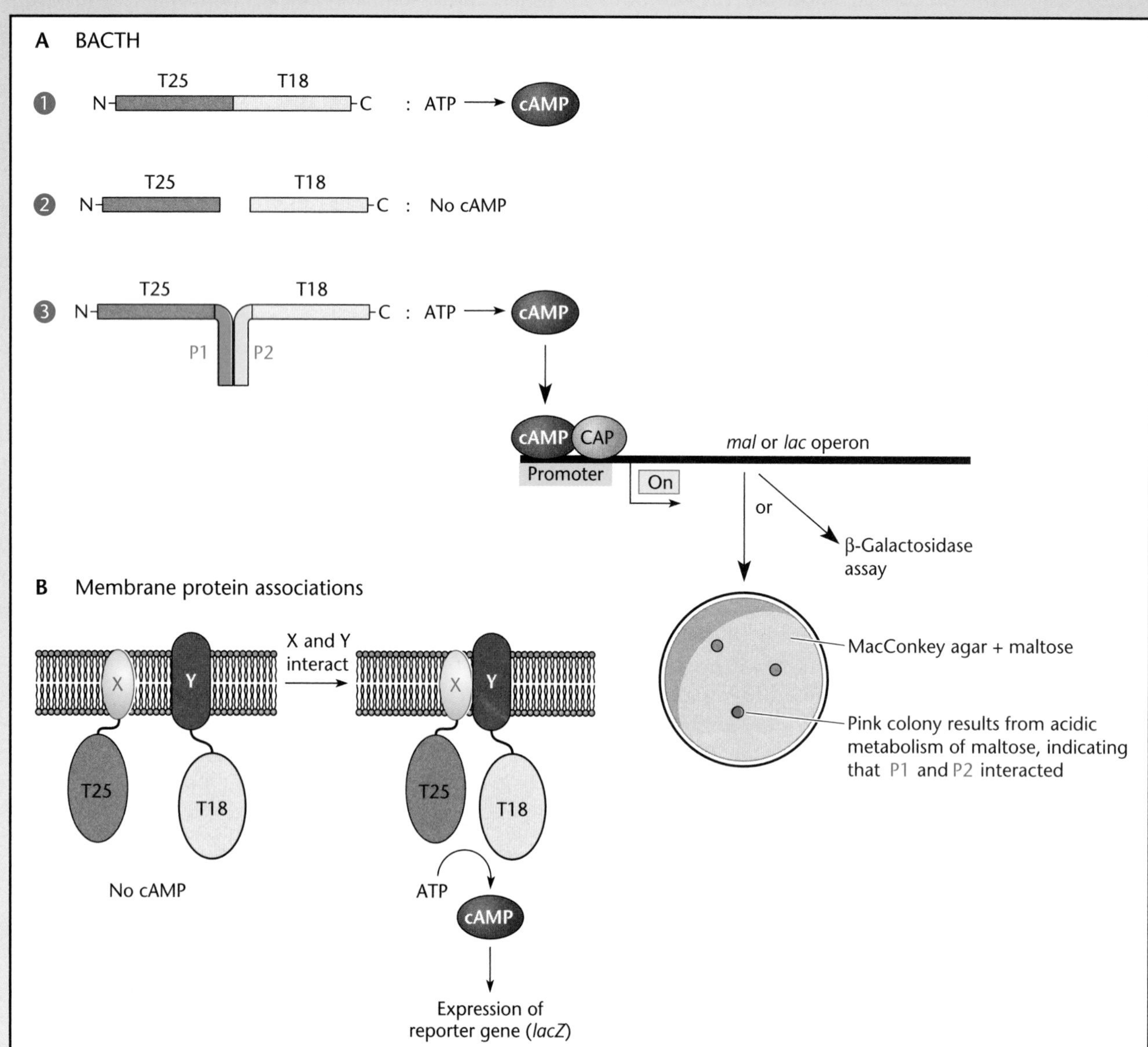

doi:10.1128/9781555817169.ch13.Box13.2.f

by themselves, and can be monitored by culture assay of β-galactosidase or by plating colonies on indicator plates.

A notable advantage of the BACTH system is that, unlike the yeast two-hybrid system, it does not require that interactions between hybrid proteins take place in association with the transcription complex. For this reason, it seems to work well for membrane proteins. For example, in one application, protein-protein interactions of the *E. coli* membrane-associated septum assembly proteins were investigated.

References

Karimova, G., J. Pidoux, A. Ullmann, and D. Ladant. 1998. A bacterial two-hybrid system based on a reconstituted signal transduction pathway. *Proc. Natl. Acad. Sci. USA* **95:**5752–5756.

Karimova, G., N. Dautin, and D. Ladant. 2005. Interaction network among *Escherichia coli* membrane proteins involved in cell division as revealed by bacterial two-hybrid analysis. *J. Bacteriol.* **187:**2233–2243.

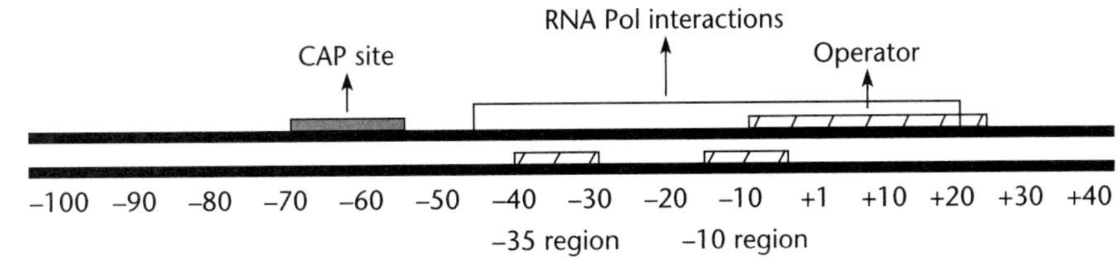

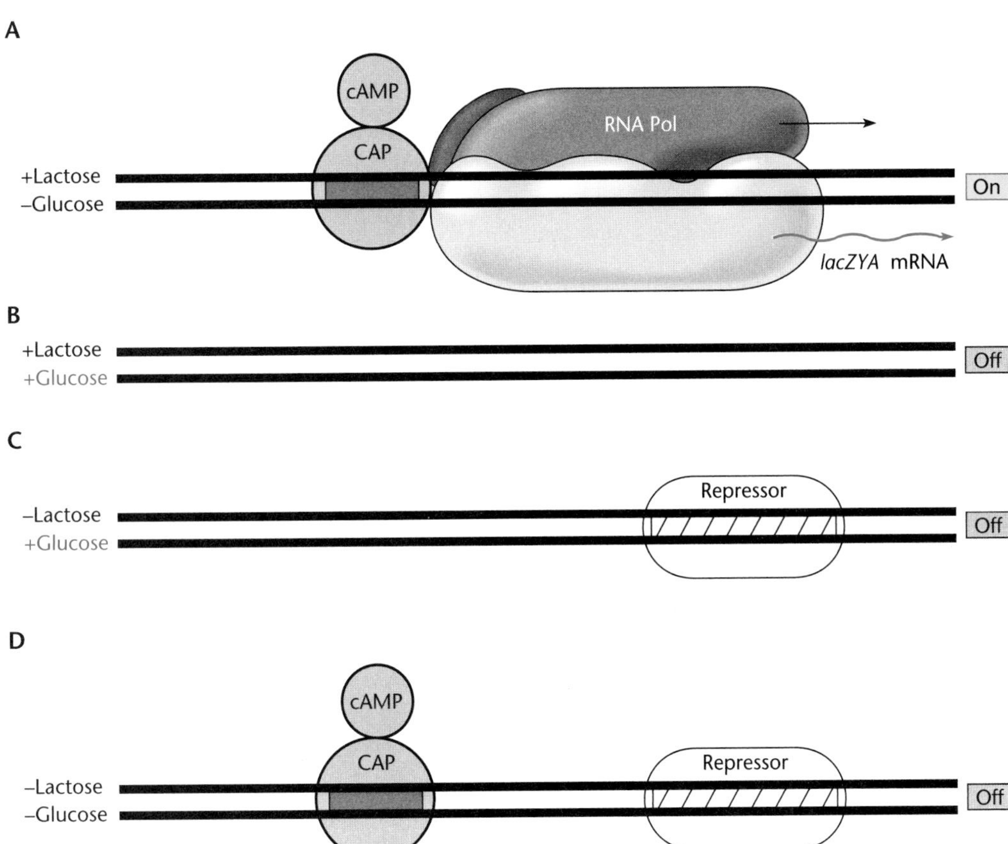

Figure 13.5 Regulation of the *lac* operon by both glucose and the inducer lactose. **(A)** The operon is on only in the absence of glucose and the presence of lactose, which is converted into the inducer allolactose. **(B and C)** The operon is off in the presence of glucose whether or not lactose is present, because the CAP-cAMP complex is not bound to the CAP site. **(D)** The operon is also off if lactose is not present, even if glucose is also not present, because the LacI repressor is bound to the operator. The relative positions of the CAP-binding site, operator, and promoter are shown. The entire regulatory region covers about 100 bp of DNA. doi:10.1128/9781555817169.ch13.f13.5

defective in the global regulation of all catabolite-sensitive operons, as well as mutants defective in the catabolite regulation of specific operons.

Isolation of *crp* and *cya* Mutations

According to the model presented above, mutations that inactivate the *cya* and *crp* genes for adenylate cyclase and CAP, respectively, should prevent transcription of all the catabolite-sensitive operons. In these mutants, there is no CAP with cAMP attached to bind to the promoters. In other words, *cya* and *crp* mutants should be Lac$^-$, Gal$^-$, Ara$^-$, Mal$^-$, and so on. In genetic terms, *cya* and *crp* mutations are **pleiotropic mutations**, because they cause many phenotypes, i.e., the inability to use many different sugars as carbon and energy sources.

The fact that *cya* and *crp* mutations should prevent cells from using several sugars was used in the first isolations of *crp* and *cya* mutants (see Schwartz and Beckwith, Suggested Reading). The identification of these mutants was based on the fact that colonies of bacteria turn tetrazolium salts red as they multiply, provided that the pH remains high. However, bacteria that are fermenting a carbon source give off organic acids, such as lactic acid, that lower the pH, preventing the conversion to red. As a consequence, wild-type *E. coli* cells growing on a fermentable carbon source form white colonies on tetrazolium-containing plates, whereas mutant bacteria that cannot use the fermentable carbon source utilize a different carbon source in the medium and so form red colonies. Some of these red-colony-forming mutants might have *crp* or *cya* mutations, although most would have mutations that inactivate a gene within the operon for the utilization of the fermentable carbon source. Thus, without a way to increase the frequency of *crp* and *cya* mutants among the red-colony-forming mutants, many red-colony-forming mutants would have to be tested to find any with mutations in either *cya* or *crp*.

For these experiments, the investigators reasoned that they could increase the frequency of *crp* and *cya* mutants by plating heavily mutagenized bacteria on tetrazolium agar containing two different fermentable sugars, for example, lactose plus galactose. Failure to utilize either of the two sugars would require either two mutations, one in each sugar utilization operon, or a single mutation in *cya* or *crp*. Since mutants with single mutations should be much more frequent than mutants with two independent mutations, the *crp* and *cya* mutants should be a much larger fraction of the total red-colony-forming mutants growing on two carbon and energy sources. Indeed, when the red-colony-forming mutants that could not use either of the two sugars provided were tested, most of them were found to be deficient in adenylate cyclase activity or to lack the protein, now named CAP, that was later shown to be required for the activation of the *lac* and *gal* promoters.

Promoter Mutations That Affect Activation by CAP-cAMP

Genetic experiments with the *lac* promoter also contributed to the models of CAP activation. Three classes of mutations have been isolated in the *lac* promoter. Those belonging to class I change the CAP-binding site so that CAP can no longer bind to it. The *lac* promoter mutation L8 is an example (Figure 13.6). By preventing the binding of CAP-cAMP upstream of the promoter, this mutation weakens the *lac* promoter. As a result, the *lac* operon is expressed poorly, as measured by β-galactosidase activity, even when cells are growing in lactose without glucose and cAMP levels are high. The low level of expression of the *lac* operon in the mutant is less strongly affected by the carbon source and is not reduced much more if glucose is added and cAMP levels drop, presumably because the remaining expression occurs without binding of CAP to the promoter and therefore fails to respond to cAMP levels.

Other promoter mutations, called class II mutations, change the −35 region of the RNA polymerase-binding site so that the promoter is less active even when cAMP levels are high. However, with this type of mutation, the residual expression of the *lac* operon is still sensitive to catabolite repression. Consequently, the amount of β-galactosidase synthesized when cells grow in the presence of lactose plus glucose, when cAMP levels are low, is less than the amount synthesized when the cells are growing in the presence of a poorer carbon source plus lactose, when cAMP levels are high.

A third, very useful mutated *lac* promoter type, termed class III, was found by isolating Lac$^+$ revertants of class I mutations, such as L8, or of *cya* or *crp* mutations. One such mutation is called *placUV5*. This mutant promoter is stronger than the wild-type *lac* promoter and no longer requires CAP-cAMP for activation. As shown in Figure 13.6, the *lacUV5* mutation changes 2 base pairs (bp) within the −10 region of the *lac* promoter so that the sequence reads TATAAT instead of TATGTT. This mutant −10 sequence perfectly matches the sequence of a consensus σ^{70} promoter (see chapter 2), which increases

Figure 13.6 Mutations in the *lac* regulatory region that affect activation by CAP. The class I mutation L8 changes the CAP-binding site so that CAP can no longer bind and the promoter cannot be turned on even in the absence of glucose (high cAMP). The class III mutation UV5 changes 2 bp in the −10 sequence of the promoter so that the promoter no longer requires activation by CAP and the operon can be induced even in the presence of glucose (low cAMP). No class II mutations are shown. The changes in the sequence in each mutation appear in blue. doi:10.1128/9781555817169.ch13.f13.6

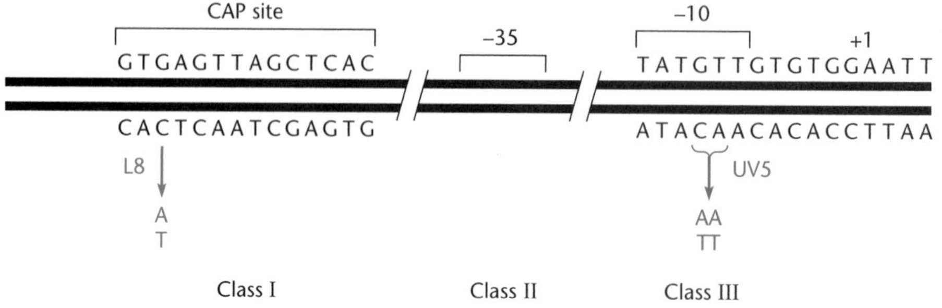

its affinity for RNA polymerase and results in loss of dependence on activation by CAP-cAMP. Some expression vectors use the *lacUV5* promoter rather than the wild-type *lac* promoter, so the promoter can be induced even if the bacteria are growing in glucose-containing medium.

Carbon Catabolite Regulation in *B. subtilis*: CcpA and Hpr

The use of cAMP as a signal for carbon catabolite regulation is not universal. *Bacillus subtilis* and its relatives use a completely different mechanism. Unlike *E. coli*, this bacterium does not have cAMP and so depends exclusively on cAMP-independent pathways to regulate its carbon source utilization pathways. In this system, catabolite repression is actually repression (in contrast to the use of a transcriptional activator in *E. coli*). The repressor protein, called CcpA (for catabolite control protein A), is a helix-turn-helix DNA-binding protein and is a member of the LacI/GalR family of regulators (see chapter 12). The CcpA repressor binds to operator sites called *cre* (for *c*atabolite *re*pressor) sites in the promoters of many catabolite-sensitive genes and represses the transcription of genes involved in carbon source utilization. Approximately 100 genes in *B. subtilis* are known to be under the control of CcpA.

CcpA also acts as an activator of transcription of genes that encode functions needed by the cell when carbon source availability is high. During growth in the presence of high glucose levels, *B. subtilis* produces large amounts of acetate, which is excreted from the cell. This results in a drop in the extracellular pH. To avoid too great a drop in pH, which is toxic, the cells can shift to production of a neutral compound called acetoin. The genes for production of both acetate and acetoin contain *cre* sites and are dependent on CcpA for transcriptional activation. The major difference between the genes CcpA activates and those that it represses is that the *cre* sites for the activated genes are positioned upstream of the promoter while the genes that are repressed by CcpA have *cre* sites that overlap the promoter, so that binding of CcpA to the *cre* site inhibits binding of RNA polymerase (Figure 13.7A).

How is CcpA activity controlled in response to carbon source availability? The DNA-binding activity of CcpA depends on the Hpr protein, which we discussed above in reference to the regulation of adenylate cyclase in *E. coli*. The phosphorylation state of Hpr also plays a crucial regulatory role in *B. subtilis*, but the mechanism is different. If cells are growing in a good carbon source, such as glucose, high levels of intermediates in the glycolytic pathway, including fructose-1,6-bisphosphate (FBP), accumulate. High levels of FBP stimulate the phosphorylation of Hpr on a specific serine residue in

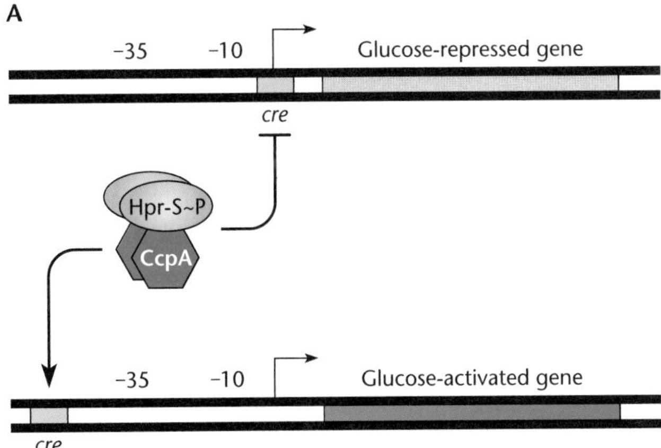

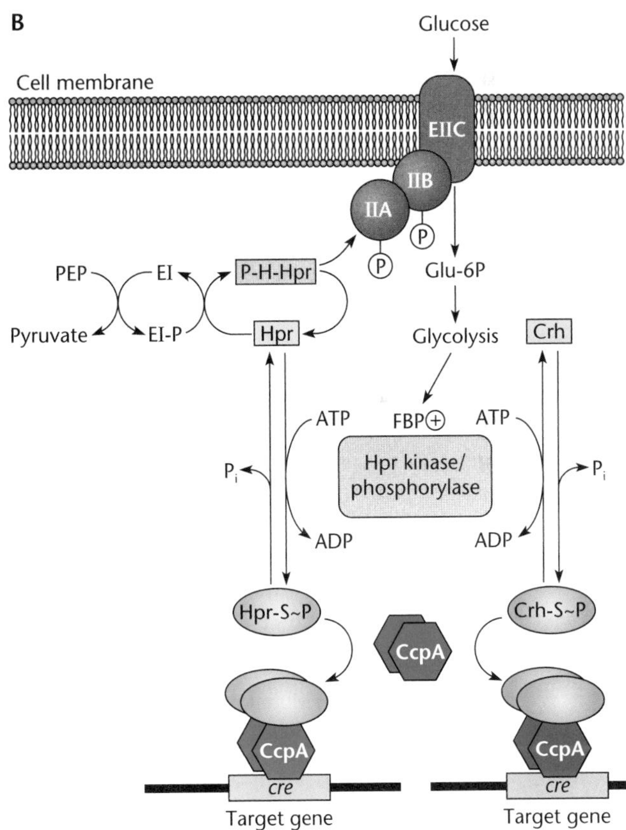

Figure 13.7 Carbon catabolite regulation in *B. subtilis.* **(A)** The CcpA regulatory protein represses genes for carbon source utilization pathways, which have *cre* sites within or downstream of the promoter, and activates genes for carbon excretion pathways, which have *cre* sites upstream of the promoter. **(B)** Binding of CcpA to *cre* sites requires a protein-protein interaction with the Hpr protein phosphorylated on a specific serine residue (Hpr-S~P) by Hpr kinase, which is activated by the glycolytic intermediate FBP. Crh-S~P can replace Hpr-S~P for this interaction.
doi:10.1128/9781555817169.ch13.f13.7

the protein. Phosphorylation of Hpr to generate Hpr-S~P utilizes another protein called Hpr kinase (HprK), the activity of which is stimulated by FBP. Hpr-S~P binds to CcpA, and the CcpA-Hpr~P complex binds to the *cre* sites (Figure 13.7B). HprK is also responsible for dephosphorylating Hpr-S~P when FBP is low.

Interestingly, the same Hpr protein that serves as a coregulator with CcpA when it is phosphorylated on the serine residue also serves as the phosphate donor in the PTS for sugar transport in a role similar to that described for the *E. coli* system. Sugar transport involves phosphorylation of a histidine residue in Hpr, and the phosphate is then donated to the PTS transport protein IIA^{Glc}. Phosphorylation of Hpr at the serine can inhibit phosphorylation at the histidine and therefore inhibits the transport of sugars that use the PTS system. This allows the close coordination of sugar transport and the regulation of catabolite-sensitive operons. A second protein, called Crh, is related to Hpr and can mimic the activity of Hpr to regulate CcpA activity. Generation of Crh-S~P also requires FBP-stimulated phosphorylation by HprK, but Crh does not act in the PTS system.

Regulation of Nitrogen Assimilation

Nitrogen is a component of many biological molecules, including nucleotides, amino acids, and vitamins. Thus, all organisms must have a source of nitrogen atoms for growth to occur. For most bacteria, possible sources include ammonia (NH_3) and nitrate (NO_3^-), as well as nitrogen-containing organic molecules, such as amino acids and the bases in nucleosides. Some bacteria can even use atmospheric nitrogen (N_2) as a nitrogen source. "Fixing" atmospheric nitrogen is a crucial step in the nitrogen cycle on Earth that few organisms can do, and the nitrogen cycle is one of the many cycles on the planet that are required for our existence (Box 13.3).

Whatever the source of nitrogen, all biosynthetic reactions that utilize it ultimately involve either the incorporation of nitrogen in the form of NH_3 or the transfer of nitrogen in the form of an NH_2 group from glutamate and glutamine, which in turn are synthesized by directly adding NH_3 to α-ketoglutarate and glutamate, respectively. Thus, because NH_3 is directly or indirectly the source of nitrogen in biosynthetic reactions, most other forms of nitrogen must be reduced to NH_3 before they can be used in these reactions. This process is called **assimilatory reduction** of the nitrogen-containing compounds, because the nitrogen-containing compound converted into NH_3 is introduced, or assimilated, into biological molecules. In another type of reduction, **dissimilatory reduction**, oxidized nitrogen-containing compounds, such as NO_3^-, are reduced when they serve as electron acceptors in anaerobic respiration (in the absence of oxygen). However, these compounds are generally not reduced all the way to NH_3 in this process, and the nitrogen is not assimilated into biological molecules and may be released as atmospheric N_2. Here, we discuss only the assimilatory uses of nitrogen-containing compounds. The genes whose products are required for anaerobic respiration in *E. coli* are members of a different regulon, the FNR regulon, which is turned on only in the absence of oxygen, when other, less efficient electron acceptors are required (Table 13.1).

Pathways for Nitrogen Assimilation

Enteric bacteria use different pathways to assimilate nitrogen depending on whether NH_3 concentrations are low or high (Figure 13.8). When NH_3 concentrations are low, for example, when the nitrogen sources are amino acids which must be degraded to release their NH_3, an enzyme named **glutamine synthetase**, the product of the *glnA* gene, adds the NH_3 directly to glutamate to make glutamine. About 75% of this glutamine is then converted to glutamate by another enzyme, **glutamate synthase**, sometimes called GOGAT, which removes an -NH_2 group from glutamine and adds it to α-ketoglutarate to make two glutamates. These glutamates can in turn be converted into glutamine by glutamine

Figure 13.8 Pathways for nitrogen assimilation in *E. coli* and other enteric bacteria. When NH_3 concentrations are low, the glutamine synthetase enzyme adds NH_3 directly to glutamate to make glutamine. Glutamate synthase (GOGAT) can then convert the glutamine plus α-ketoglutarate into two glutamates, which can reenter the cycle. In the presence of high NH_3 concentrations, the NH_3 is added directly to α-ketoglutarate by glutamate dehydrogenase to make glutamate, which can be subsequently converted to glutamine by glutamine synthetase.
doi:10.1128/9781555817169.ch13.f13.8

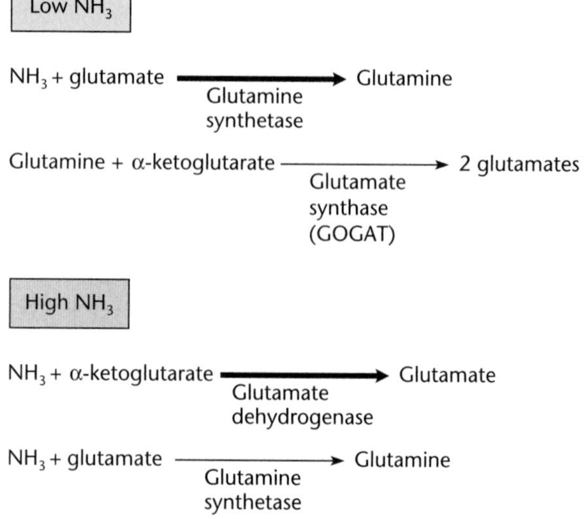

Low NH_3

NH_3 + glutamate → Glutamine (Glutamine synthetase)

Glutamine + α-ketoglutarate → 2 glutamates (Glutamate synthase (GOGAT))

High NH_3

NH_3 + α-ketoglutarate → Glutamate (Glutamate dehydrogenase)

NH_3 + glutamate → Glutamine (Glutamine synthetase)

BOX 13.3

Nitrogen Fixation

Some bacteria can use atmospheric nitrogen (N_2) as a nitrogen source by converting it to NH_3 in a process called nitrogen fixation, which appears to be unique to bacteria. However, N_2 is a very inconvenient source of nitrogen. The very stable bond holding the two nitrogen atoms together must be broken, and 16 moles (mol) of ATP must be cleaved to cleave 1 mol of dinitrogen. Bacteria that can fix nitrogen include members of the cyanobacteria and members of the genera *Klebsiella, Azotobacter, Rhizobium,* and *Azorhizobium.* These organisms play an important role in nitrogen cycles on Earth.

Some types of nitrogen-fixing bacteria, including members of the genera *Rhizobium* and *Azorhizobium,* are symbionts that fix N_2 in nodules on the roots or stems of plants and allow the plants to live in nitrogen-deficient soil. In return, the plant furnishes nutrients and an oxygen-free atmosphere in which the bacterium can fix N_2. This symbiosis therefore benefits both the bacterium and the plant. An active area of biotechnology is the use of N_2-fixing bacteria as a source of natural fertilizers.

The fixing of N_2 requires the products of many genes, called the *nif* genes. In free-living nitrogen-fixing bacteria, such as *Klebsiella* spp., there are about 20 *nif* genes arranged in eight adjacent operons. Some of the *nif* genes encode the nitrogenase enzymes directly responsible for fixing N_2. Others encode proteins involved in assembling the nitrogenase enzyme and in regulating the genes. Plant-symbiotic bacteria also require many other genes whose products produce the nodules on the plant (*nod* genes) and allow the bacterium to live and fix nitrogen in the nodules (*fix* genes).

Because nitrogen fixation requires a large investment of energy, the genes involved in N_2 fixation are part of the Ntr regulon and are under the control of the NtrC activator protein. In *K. pneumoniae,* in which the regulation of the *nif* genes has been studied most extensively, the phosphorylated form of NtrC (NtrC~P) does not directly activate all eight operons involved in N_2 fixation. Instead, NtrC~P activates the transcription of another activator gene, *nifA,* whose product is directly required for the activation of the eight *nif* operons. The nitrogenase enzymes are very sensitive to oxygen, and in the presence of oxygen, the *nif* operons are negatively regulated by the product of the *nifL* gene. The NifL protein is able to sense oxygen because it is a flavoprotein with a bound flavin adenine dinucleotide (FAD) group, which is oxidized in the presence of oxygen. The NifL protein then forms a stable complex with NifA and inactivates it so that the *nif* genes are not transcribed.

References

Margolin, W. 2000. Differentiation of free-living rhizobia into endosymbiotic bacteroids, p. 441–466. *In* Y. V. Brun and L. J. Shimkets (ed.), *Prokaryotic Development.* ASM Press, Washington, DC.

Martinez-Argudo, I., R. Little, N. Shearer, P. Johnson, and R. Dixon. 2004. The NifL-NifA system: a multidomain transcriptional regulatory complex that integrates environmental signals. *J. Bacteriol.* **186:**601–610.

synthetase. Because the NH_3 must all be routed by glutamine synthetase to glutamine when NH_3 concentrations are low, the cell needs a lot of the glutamine synthetase enzyme under these conditions. This pathway requires a lot of energy, but it is necessary if nitrogen availability is limited. The significance of this is addressed later.

If NH_3 concentrations are high because the medium contains NH_3 (usually in the form of NH_4OH) but carbon sources are limited, the nitrogen is assimilated through a very different pathway. This pathway requires less energy but is possible only if NH_3 concentrations are high. In this case, the enzyme **glutamate dehydrogenase** adds the NH_3 directly to α-ketoglutarate to make glutamate. Some of the glutamate is subsequently converted into glutamine by glutamine synthetase. Much less glutamine is required for protein synthesis and biosynthetic reactions than for assimilation of limiting nitrogen from the medium. Therefore, cells need much less glutamine synthetase when growing in high concentrations of NH_3 than when growing in low concentrations.

Regulation of Nitrogen Assimilation Pathways in *E. coli* by the Ntr System

The operons for nitrogen utilization in *E. coli* are part of the **Ntr system**, for *ni*trogen *r*egulated. Ntr regulation ensures that the cell does not waste energy making enzymes for the use of nitrogen sources such as amino acids or nitrate when NH_3 is available. Transport systems for alternative nitrogen sources are also part of this regulon. In this section, we discuss what is known about how the Ntr global regulatory system works. As usual, geneticists led the way by identifying the genes whose products are involved in the regulation, so that a role could eventually be assigned to each one. The Ntr regulatory systems in a variety of gram-negative bacteria, including *Escherichia*, *Salmonella*, *Klebsiella*, and

Rhizobium, are similar, but with important exceptions, some of which are pointed out here.

REGULATION OF THE *glnA-ntrB-ntrC* OPERON BY A SIGNAL TRANSDUCTION PATHWAY

Since cells need more glutamine synthetase when growing at low NH_3 concentrations than when growing at high NH_3 concentrations, the expression of the *glnA* gene, which encodes glutamine synthetase, must be regulated according to the nitrogen source that is available. This gene is part of an operon that includes three genes, *glnA*, *ntrB*, and *ntrC*. The products of the *ntrB* and *ntrC* genes are involved in regulating the operon. (These proteins are also called NR_{II} and NR_I, respectively, but we use the Ntr names in this chapter.) Because the *ntrB* and *ntrC* genes are part of the same operon as *glnA*, their genes are autoregulated, and their products are also synthesized at higher levels when NH_3 concentrations are low.

Figure 13.9 illustrates the regulation of the *glnA-ntrB-ntrC* operon and other Ntr genes. In addition to NtrB and NtrC, the proteins GlnD and P_{II} participate in the regulation of the operon. These four proteins form a **signal transduction pathway** (Box 13.4) in which information about nitrogen source availability is passed, or transduced, from one protein to another until it gets to its final destination, the transcriptional regulator, NtrC, which activates genes in the Ntr regulon.

The availability of nitrogen is sensed through the level of glutamine in the cell. If the cell is growing in a nitrogen-rich environment, the levels of glutamine are high, whereas if the cell is growing under limiting nitrogen conditions, the levels of glutamine are low. How the levels of glutamine affect the regulation of the Ntr genes involved in using alternative nitrogen sources is probably best explained by working backward from the last protein in the signal transduction pathway, NtrC. The NtrC protein can be phosphorylated to form NtrC~P, the form in which it is a transcriptional activator that activates transcription of the Ntr genes, which are turned on under limiting nitrogen (Figure 13.9). The penultimate protein in the pathway is NtrB. Together, NtrB and NtrC form a **two-component regulatory system** (Box 13.4) in which the NtrB protein is the **sensor kinase** and NtrC is the **response regulator**. NtrB is a protein kinase that can phosphorylate itself on a specific histidine residue to form NtrB~P; NtrB~P can then transfer this phosphate to NtrC to form NtrC~P, which is the active form that can activate transcription of the Ntr genes. The autophosphorylation activity of NtrB occurs only if nitrogen is limiting and depends on the state of modification of another regulatory protein called P_{II}. The P_{II} protein is modified, not by phosphorylation, but by having UMP attached to it (to form P_{II}~UMP). If nitrogen is limiting, making the glutamine level low, most of this protein exists as P_{II}~UMP. However, if nitrogen is in excess and the glutamine level is high, the glutamine stimulates an enzyme called GlnD to remove the UMP from P_{II}. The unmodified P_{II} protein binds to NtrB and inhibits its autokinase activity so that it cannot phosphorylate itself to form NtrB~P. If it cannot phosphorylate itself, it cannot transfer a phosphate to NtrC, and most of the NtrC remains in the unphosphorylated state that is unable to activate transcription of the Ntr genes involved in using alternate nitrogen sources.

Figure 13.9 Regulation of Ntr operons by a signal transduction pathway in response to NH_3 levels. At low NH_3 concentrations, the reactions shown in blue predominate. At high NH_3 concentrations, the reactions shown in black predominate. High glutamine (Gln) causes low levels of uridylylation of P_{II}, which in turn affects the activity of the NtrB sensor kinase. α-KG, α-ketoglutarate. The effect of α-ketoglutarate on P_{II} is indirect. See the text for details.
doi:10.1128/9781555817169.ch13.f13.9

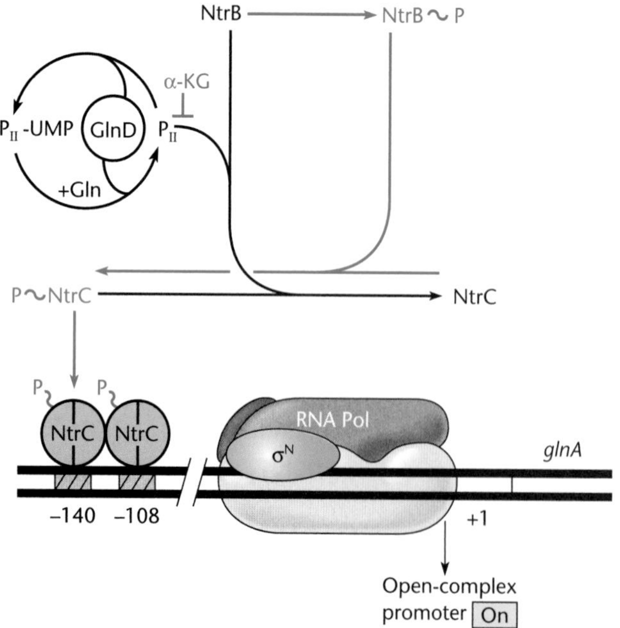

NtrB and NtrC: a Two-Component Sensor Kinase-Response Regulator System

As described above, the NtrB and NtrC proteins form a two-component system in which NtrB is a sensor kinase and NtrC is a response regulator. Such protein pairs are common in bacteria, and the corresponding members of such pairs are remarkably similar to each other (Box 13.4). Typically, one protein of the pair "senses" an environmental parameter and phosphorylates itself at a histidine. This phosphoryl group is then passed on to an aspartate in the second protein. The activity of the second protein, the response regulator, depends on whether it is phosphorylated. Many response regulators

BOX 13.4

Signal Transduction Systems in Bacteria

Many regulatory mechanisms require that the cell sense changes in the external environment and change the expression of its genes or the activities of its proteins accordingly. The sensors can be of many types (Figure 1). Some are serine-threonine-tyrosine kinases–phosphatases (STYK), such as the Rsb proteins that activate the stress sigma factor, σ^B, of *B. subtilis* and many other bacteria. Others already discussed are adenylate cyclases (ACyc), which make cAMP in enteric bacteria, such as *E. coli*, in response to nutritional conditions, such as a relatively poor carbon source. An interesting, recently discovered type of signaling molecule is cyclic diGMP (c-diGMP), which consists of two guanosine monophosphate (GMP) molecules linked to each other's 3′ carbons through their 5′ phosphates to form a sort of circle of phosphate-ribose sugar groups. Specific enzymes called diguanylate cyclases make this small-molecule effector, and specific phosphodiesterases destroy it. These enzymes were discovered primarily through genomic analysis because the cyclases have the domain GGDEF and the phosphodiesterases have the domain EAL (see front endsheet for amino acid assignments). Many signaling proteins in these pathways have both a diguanylate cyclase and a diguanylate phosphodiesterase domain. They are widespread, having been found in many types of bacteria, and play diverse roles in the attachment of bacteria to surfaces, in the formation of biofilms, in the regulation of photosynthesis, and in motility. However, how this effector acts is not well understood (see Povlotsky and Hengge, References).

Some of the most common and widely studied sensor systems are the so-called two-component signal transduction systems, which consist of a sensor kinase (SK) that autophosphorylates (transfers phosphates to itself) on a histidine residue and a response regulator (RR) that accepts the phosphate from the sensor kinase onto an aspartate residue and then performs a specific action in the cell. Removal of the phosphate from the response regulator is another important step, to allow the system to be "reset" to monitor another signal; this can occur spontaneously, through the action of a specific phosphatase, or through the phosphatase activity of the sensor kinase.

Two-component systems have been found in all bacteria and some plants but, at least in this form, are absent from animals. Bacteria with large genomes can have hundreds of these systems. As the name implies, they usually consist of two proteins, but in some cases, the sensor kinase and the response regulator activities are domains of the same protein. In only a few cases is the stimulus to which the sensor kinase responds known (see Gao and Stock, References). The output responses of the systems also vary (see Galperin, References). To name just a few, the response regulator is quite often a DNA-binding transcriptional regulator with a helix-turn-helix (HTH) domain, but it can also destabilize a protein by targeting it for proteolysis. The cellular functions that enlist two-component signal transduction systems also vary widely, including involvement in motility in response to chemical attractants, i.e., the methylated chemotaxis proteins (MCP

Figure 1 doi:10.1128/9781555817169.ch13.Box13.4.f1

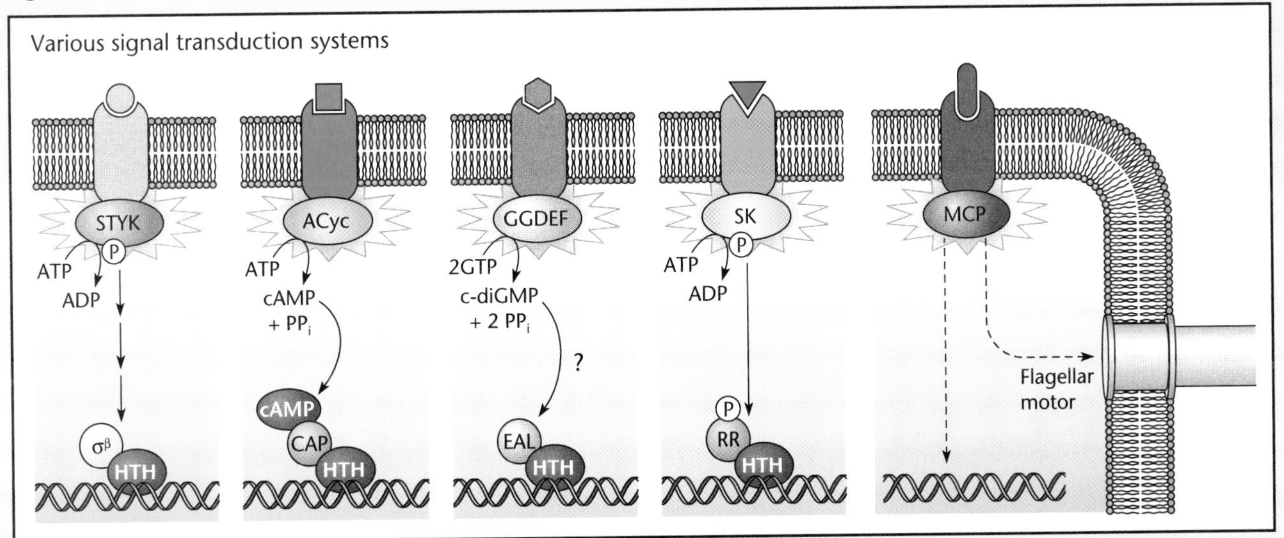

(continued)

BOX 13.4 (continued)

Signal Transduction Systems in Bacteria

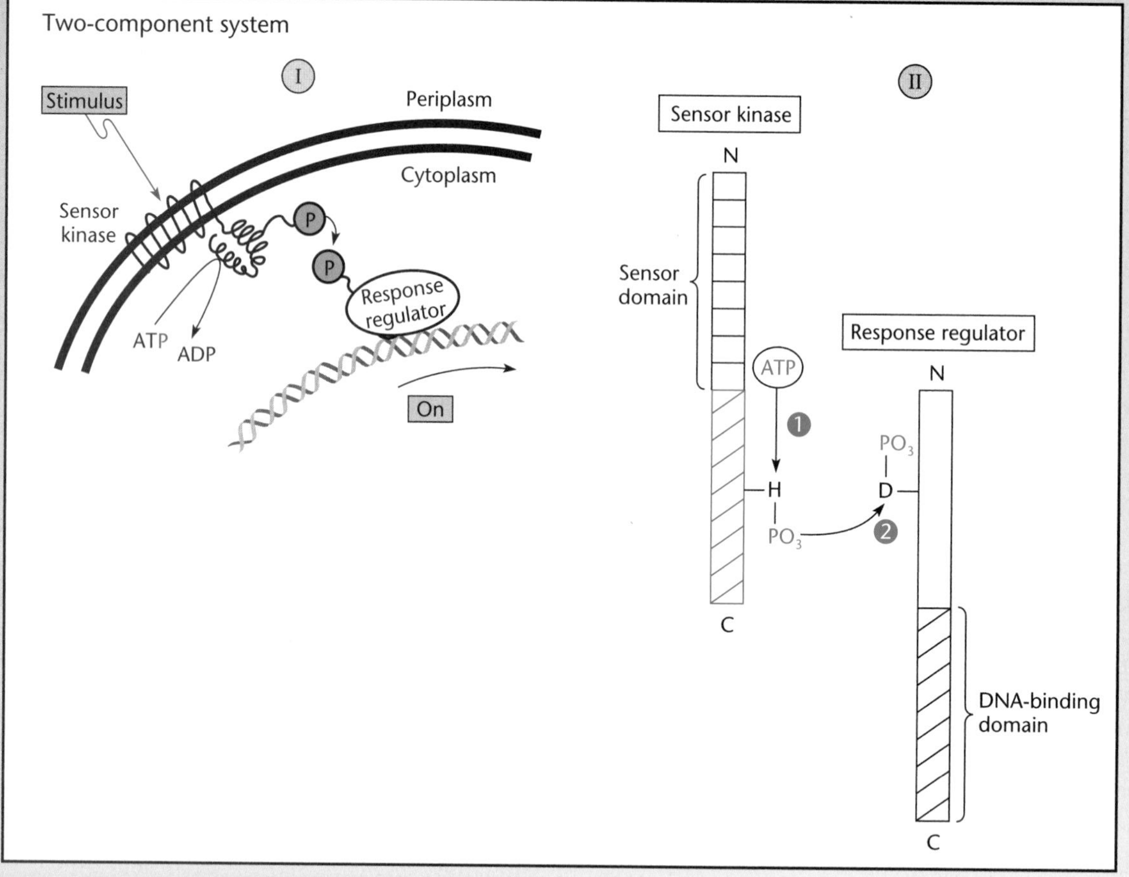

Figure 2 doi:10.1128/9781555817169.ch13.Box13.4.f2

in Figure 1) (see Box 14.1), the induction of pathogenesis operons after entry into a suitable host, and the activation of extracellular stress responses.

The way in which these two-component sensor kinase and response regulator systems operate in general is illustrated in Figure 2. Part I shows that sensor kinases are often integral membrane proteins responsive to external signals. Part II shows the functions of the protein domains. The CTD of a sensor kinase has the conserved histidine that is phosphorylated (step 1). The response regulators are similar in their N-terminal regions, which includes the phosphorylated aspartate (step 2). The remainder of the protein differs depending on its function, although different subfamilies of response regulators show regions of high homology in other parts of the protein, including the helix-turn-helix motif of many transcriptional regulators.

References

Galperin, M. Y. 2006. Structural classification of bacterial response regulators: diversity of output domains and domain combinations. *J. Bacteriol.* **188:**4169–4182.

Gao, R., and A. M. Stock. 2009. Biological insights from structures of two-component proteins. *Annu. Rev. Microbiol.* **63:**133–154.

Povlotsky, T. L., and R. Hengge. 2012. "Life-style" control networks in *Escherichia coli*: signaling by the second messenger c-di-GMP. *J. Biotechnol.* **160:**10–16.

are transcriptional regulators. In chapter 6, we discuss another example of a sensor kinase-response regulator pair, ComP and ComA, involved in the development of transformation competence in *B. subtilis*, and we describe other pairs later in this chapter and in chapter 14.

Regulation of Other Ntr Operons

The other operons, besides *glnA-ntrB-ntrC*, that are activated by NtrC~P depend on the type of bacteria and the other nitrogen sources they can use. In general, operons under the control of NtrC~P are those involved in using poorer nitrogen sources. For example, genes for the uptake of the amino acids glutamine in *E. coli* and histidine and arginine in *Salmonella enterica* serovar Typhimurium are under the control of NtrC~P. An operon for the utilization of nitrate as a nitrogen source in *Klebsiella pneumoniae* is activated by NtrC~P, but neither *E. coli* nor *S. enterica* serovar Typhimurium has such an operon.

In some bacteria, the Ntr genes are not regulated directly by NtrC~P but are under the control of another gene product whose transcription is activated by NtrC~P. For example, operons for amino acid degradative pathways in *Klebsiella aerogenes* do not require direct activation by NtrC~P. However, they are indirectly under the control of NtrC~P because transcription of the gene for their transcriptional activator, *nac*, is activated by NtrC~P. The nitrogen fixation genes of *K. pneumoniae* are similarly under the indirect control of NtrC~P because NtrC~P activates transcription of the gene for their activator protein, *nifA* (Box 13.3).

TRANSCRIPTION OF THE *glnA-ntrB-ntrC* OPERON

The *glnA-ntrB-ntrC* operon is transcribed from three promoters, only one of which is NtrC~P dependent. The positions of the three promoters and the RNAs made from each are shown in Figure 13.10. Of the three promoters, only the p_2 promoter is activated by NtrC~P and is responsible for the high levels of glutamine synthetase and NtrB and NtrC synthesis under conditions of low NH_3. The other two promoters, p_1 and p_3, are discussed below.

The p_2 promoter is immediately upstream of the *glnA* gene, and RNA synthesis initiated at the promoter continues through all three genes, as shown in Figure 13.10. However, some transcription terminates at a transcriptional terminator located between the *glnA* and *ntrB* genes, so that much less NtrB and NtrC than glutamine synthetase is made.

Figure 13.10 The *glnA-ntrB-ntrC* operon of *E. coli*. There are three promoters, and the arrows show the mRNAs that are made from each promoter. The blue arrow indicates the mRNA expressed from the nitrogen-regulated promoter, p_2. The thicknesses of the lines indicate how much RNA is made from each promoter in each region.
doi:10.1128/9781555817169.ch13.f13.10

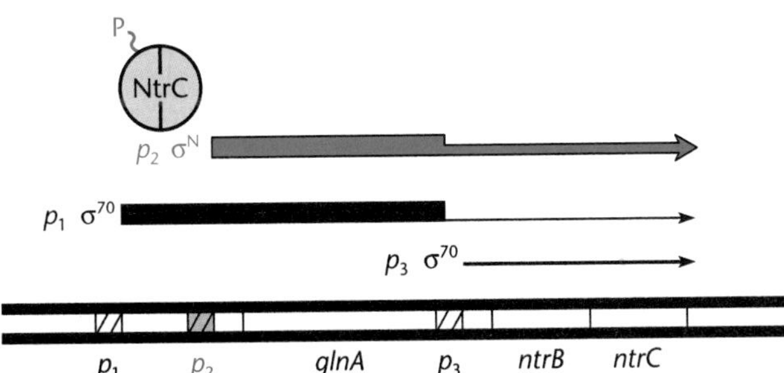

The Nitrogen Sigma Factor, σ^N

The p_2 promoter and other Ntr-type promoters activated by NtrC~P are unusual in terms of the RNA polymerase holoenzyme that recognizes them. Most promoters are recognized by the RNA polymerase holoenzyme with σ^{70} attached, but the Ntr-type promoters are recognized by a holoenzyme containing a special σ factor, designated σ^{54} or σ^N (Box 13.5). As shown in Figure 13.11, promoters recognized by the σ^N holoenzyme look very different from promoters recognized by the σ^{70} holoenzyme. Unlike the typical σ^{70} promoter, which has RNA polymerase-binding sequences centered at bp −35 and −10 relative to the transcription start site, the σ^N promoters have very different binding sequences centered at bp −24 and −12. Because promoters for the genes involved in Ntr regulation are recognized by RNA polymerase with σ^N, this sigma factor was named the nitrogen sigma factor and the gene was named *rpoN* (Table 13.1). However, σ^N-type promoters have been found in many operons unrelated to nitrogen utilization in other organisms, including in the flagellar genes of *Caulobacter* spp. and some promoters of the toluene-biodegradative operons of the *Pseudomonas putida* Tol plasmid. Interestingly, all of the known σ^N-type promoters require activation by an activator protein.

Figure 13.11 Sequence comparison of the promoters recognized by the RNA polymerase holoenzyme carrying the normal sigma factor (σ^{70}), the nitrogen sigma factor (σ^N), and the heat shock sigma factor (σ^H). Instead of consensus sequences centered at bp −10 and −35 with respect to the RNA start site, the σ^N promoter has consensus sequences at bp −12 and −24. The σ^H promoter has consensus sequences centered at approximately bp −10 and −35, but they are different from the consensus sequences of the σ^{70} promoter. X indicates that any base pair can be present at this position. +1 is the start site of transcription.
doi:10.1128/9781555817169.ch13.f13.11

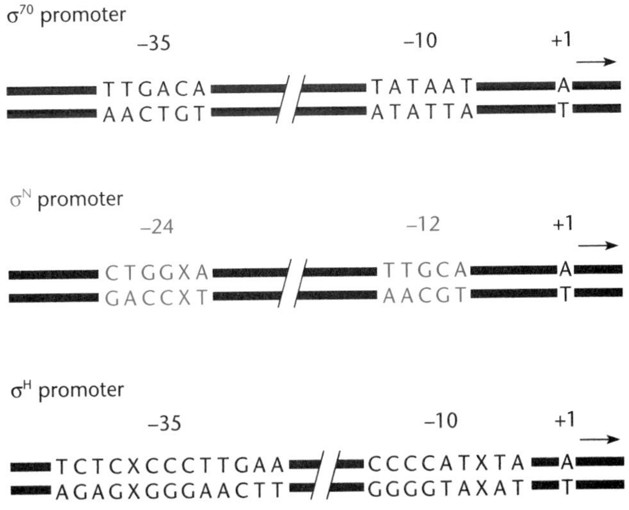

σ^{70} promoter
−35 −10 +1
TTGACA // TATAAT A
AACTGT // ATATTA T

σ^N promoter
−24 −12 +1
CTGGXA // TTGCA A
GACCXT // AACGT T

σ^H promoter
−35 −10 +1
TCTCXCCCTTGAA // CCCCATXTA A
AGAGXGGGAACTT // GGGGTAXAT T

BOX 13.5

Sigma Factors

Sigma (σ) factors seem to be unique to bacteria and their phages and are not found in eukaryotes. These proteins cycle on and off of RNA polymerase and help direct it to specific promoters (see chapter 2). They also help RNA polymerase separate the DNA strands at the promoter to initiate transcription, and they help in promoter clearance after initiation. They also may contain the contact points of activator proteins that help these proteins stabilize the RNA polymerase on the promoter and activate transcription (see chapter 12). Promoters are often identified by the σ factor they use; for example, a "σ^{70} promoter" is one that uses the RNA polymerase holoenzyme with σ^{70} attached, while a σ^{H} promoter uses RNA polymerase holoenzyme containing σ^{H}. A caveat: different σ factors are often given the same name in different bacteria. For example, σ^{E} in *E. coli* and *B. subtilis* refers to very different σ factors; the σ^{E} in *E. coli* is the extracytoplasmic stress sigma factor, while the σ^{E} in *B. subtilis* is involved in sporulation. The transcriptional activity of a σ is dependent on its ability to bind to core RNA polymerase, and competition for core binding influences the transcriptional pattern in the cell (see Mooney et al., References).

Sigma factors can be found in the genomic sequences of bacteria based on their sequence conservation. This has revealed that the number of different sigma factors varies widely from one bacterial type to another. The bacterium with the least known so far, *Helicobacter pylori*, has only 3 different types, while the current record holder, *Streptomyces coelicolor*, has 63. In general, bacteria that are free living have more sigma factors than do obligate parasites, probably reflecting the greater environmental challenges faced by free-living bacteria.

There are two major classes of sigma factors in bacteria: the σ^{70} class, which comprises most of the sigma factors discussed in this book, including σ^{S}, σ^{H}, σ^{B}, and σ^{E}, and another class, σ^{N}, which seems to form a class by itself. While all sigma factors in the σ^{70} class have some sequence and functional homology, there is no sequence similarity between members of this class and members of the σ^{N} class. The two classes also seem to differ fundamentally in their mechanisms of action (see below). Members of the σ^{70} class are found in all bacteria and play many diverse roles, some of which are discussed in previous chapters. The σ^{N}-type promoters are also widely distributed among both gram-positive and gram-negative bacteria, but they are not universal. This sigma factor was originally named the "nitrogen sigma factor," σ^{N}, because the promoters it uses were first found in the genes for Ntr regulation that are turned on during nitrogen-limited growth in *E. coli* (see the text). However, it is now known that there is no common theme for σ^{N}-expressed genes. In some soil

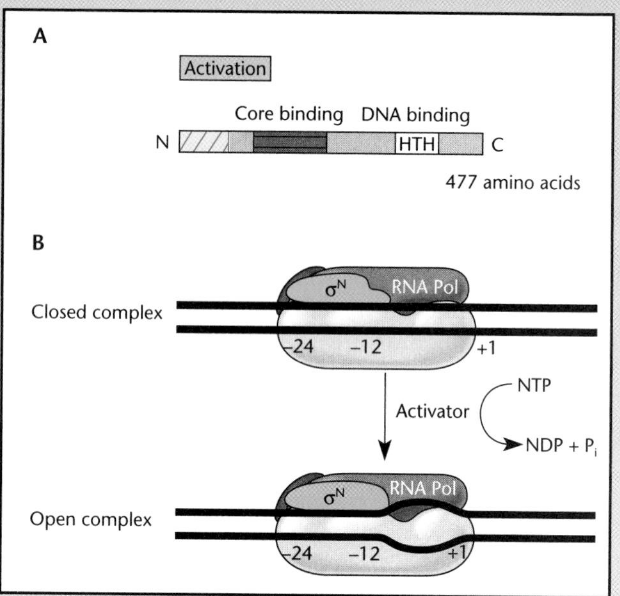

doi:10.1128/9781555817169.ch13.Box13.5.f

bacteria, this sigma factor is used to express biodegradative genes, for example, to degrade toluene, and in other species, including some pathogens, it is used by some of the flagellar genes to make components of type III secretion systems, as well as to make alginate in *P. aeruginosa* (see Kazmierczak et al., References).

One major distinction between the two classes of sigma factors is in the way their promoters are activated. For example, many promoters that use a sigma factor of the σ^{70} family can initiate transcription without the help of an activator protein. If a σ^{70} promoter requires an activator, it generally binds adjacent to the promoter and helps to recruit RNA polymerase to the promoter (see chapter 12). However, all σ^{N} promoters studied thus far absolutely require a specialized activator protein with ATPase activity, which binds to an enhancer sequence that can be hundreds of base pairs upstream of the promoter. In some ways, this makes σ^{N} promoters more like the RNA polymerase II promoters of eukaryotes. They also differ from σ^{70} promoters in their mechanism of activation. The activation of a σ^{N} promoter is illustrated in the figure. Panel A shows the functionally important domains of σ^{N} that allow binding to core RNA polymerase and DNA. The NTD allows σ^{N} to respond to activators. Panel B shows that the σ^{N}-RNA polymerase forms a stable but closed complex with the promoter, even in the absence of the activator bound to the upstream enhancer. The activator has a latent ATPase activity, which becomes activated by phosphorylation that is often passed down from a sensor kinase, either directly

BOX 13.5 (continued)

Sigma Factors

or through a phosphorylation cascade, with NtrC being the prototype (see the text). Phosphorylation of the N terminus of the activator alters its activity or its affinity for enhancer sites. Multimerization and formation of a DNA-bound complex activates the ATPase activity in its central domain. Once activated, the ATPase can cause the σ^N to undergo a conformational change that stimulates open-complex formation by the σ^N-RNA polymerase and allows initiation at the promoter.

References

Kazmierczak, M. J., M. Weidmann, and K. J. Boor. 2005. Alternate sigma factors and their roles in bacterial virulence. *Microbiol. Mol. Biol. Rev.* **69**:527–543.

Mooney, R. A., S. A. Darst, and R. Landick. 2005. Sigma and RNA polymerase: an on-again, off-again relationship? *Mol. Cell* **20**:335–345.

Osterberg, S., T. del Peso-Santos, and V. Shingler. 2011. Regulation of alternative sigma factor use. *Annu. Rev. Microbiol.* **65**:37–55.

Paget, M. S. B., and J. D. Helmann. 2003. The sigma 70 family of sigma factors. *Genome Biol.* **4**:203–215.

Wigneshweraraj, S., D. Bose, P. C. Burrows, N. Joly, J. Schumacher, M. Rappas, T. Pape, X. Zhang, P. Stockley, K. Severinov, and M. Buck. 2008. Modus operandi of the bacterial RNA polymerase containing the σ^{54} promoter specificity factor. *Mol. Microbiol.* **68**:538–546.

The Transcription Activator NtrC

The polypeptide chains of NtrC-type activators have the basic arrangement shown in Figure 13.12A. A DNA-binding domain that recognizes the σ^N-type promoter lies at the carboxyl-terminal end of the polypeptide. A regulatory domain that is phosphorylated (or in some cases, binds a regulatory factor) is present at the amino-terminal end. The region of the polypeptide responsible for transcriptional activation is in the middle. This region has an ATP-binding domain and an ATPase activity that cleaves ATP to ADP. The NTD somehow masks the middle domain for activation unless the NTD has been phosphorylated (or has bound its inducer, in some members of the NtrC family).

The mechanism of activation by NtrC-type activators has been studied in some detail. The NtrC-activated promoters, including p_2 of the *glnA-ntrB-ntrC* operon, are unusual in that NtrC binds to an **upstream activator sequence (UAS)**, which lies more than 100 bp upstream of the promoter. For most positively regulated promoters, the activator protein-binding sequences are adjacent to the site at which RNA polymerase binds (see chapter 12). Activation at a distance, such as occurs with the NtrC-activated promoters, is much more common in eukaryotes, where many examples are known.

Figure 13.12B shows a detailed model of how NtrC~P activates transcription from p_2 and other Ntr-activated promoters. The RNA polymerase holoenzyme can bind to the promoter even when nitrogen is not limiting and NtrC is not phosphorylated. Also, NtrC can bind to the UAS even if it is not phosphorylated, but no transcriptional activation occurs unless NtrC is phosphorylated. When nitrogen is limiting and NtrC becomes phosphorylated, oligomers of NtrC~P bound at the UAS activate transcription from the promoter, perhaps because the phosphorylated NtrC molecules interact more tightly with each other and this stimulates their ATPase activity. Activation of the promoter must involve bending of the DNA to bring the NtrC activators in contact with the RNA polymerase bound at the promoter, and for some NtrC-type activators, DNA bending is facilitated by an additional accessory protein (such as IHF, the integration host factor first identified by its role in phage λ integration into the chromosome [see chapter 8]). Activation of transcription at a σ^N promoter requires cleavage of ATP by NtrC~P for open-complex formation, unlike transcription from other promoters, where RNA polymerase can carry out open-complex formation in the absence of ATP cleavage (see chapter 2). The fact that the limiting step for transcription is open-complex formation rather than binding is consistent with the observation that binding of RNA polymerase containing σ^N can occur in the absence of activation. This may allow the system to be poised for an immediate response, as soon as NtrC becomes phosphorylated.

Functions of the Other Promoters of the *glnA-ntrB-ntrC* Operon

As mentioned above, the p_2 promoter is one of three promoters that are responsible for transcription of the *glnA-ntrB-ntrC* operon (Figure 13.10). The other promoters are p_1 and p_3. The p_1 promoter is further upstream of *glnA* than is p_2, and most of the transcription events that initiate at the p_1 promoter terminate at the transcription termination signal just downstream of *glnA*. The p_3 promoter is between the *glnA* and *ntrB* genes and allows transcription of the *ntrB* and *ntrC* genes.

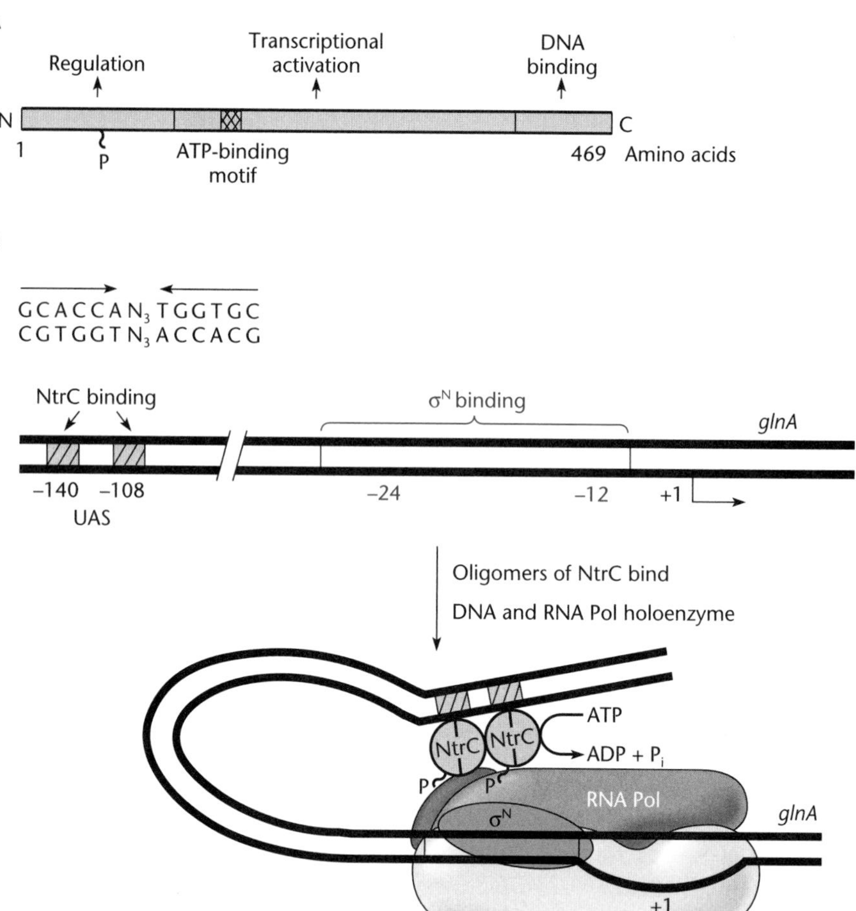

Figure 13.12 Model for the activation of the p_2 promoter by phosphorylated NtrC protein. **(A)** Functions of the various domains of NtrC. P denotes a phosphate. **(B)** Oligomers (shown as two dimers) of phosphorylated NtrC bind to the inverted repeats in the UAS. The DNA is bent between the UAS and the promoter, allowing contact between NtrC and the RNA polymerase containing σ^N bound at the promoter more than 100 bp downstream. Cleavage of ATP due to the NtrC ATPase activity is required for the RNA polymerase to form open complexes at the promoter.
doi:10.1128/9781555817169.ch13.f13.12

The p_1 and p_3 promoters use σ^{70} and so do not require NtrC~P for their activation. In fact, transcription initiation is repressed by NtrC~P, so transcription does not occur if NH_3 concentrations are low (conditions under which NtrC~P is high). The function of these promoters is presumably to ensure that the cell has some glutamine synthetase, NtrB, and NtrC even when NH_3 concentrations are high. Unless glutamine is provided in the medium, the cell must have glutamine synthetase to make glutamine for use in protein synthesis and as the -NH_2 group donor in some biosynthetic reactions. The cell must also have some NtrB and NtrC in case conditions change suddenly from a high-NH_3 to a low-NH_3 environment, in which the products of the Ntr genes are needed. Other genes of the Ntr regulon also have σ^S promoters, which allow them to be turned on in stationary phase and at times of stress (see below). This is one of the reasons for suspecting that the Ntr response is also a stress response.

ADENYLYLATION OF GLUTAMINE SYNTHETASE

Regulating the transcription of the *glnA* gene is not the only way that the activity of glutamine synthetase is regulated in the cell. The activity is also modulated by the adenylylation of (transfer of AMP to) a specific tyrosine in the glutamine synthetase enzyme by an adenylyltransferase enzyme when NH_3 concentrations are high. As mentioned in chapter 12, posttranslational modification of proteins can have a major effect on protein activity, and adenylylation of glutamine synthetase represents an example of this. The adenylylated form of the glutamine synthetase enzyme is less active and is also much more susceptible to feedback inhibition by glutamine than is the unadenylylated form (see chapter 12 for an explanation of feedback inhibition). This makes sense, considering that glutamine synthetase plays different roles when NH_3 concentrations are high and when they are low. When NH_3 concentrations are high, the primary role of glutamine synthetase is to make glutamine for protein synthesis, which requires less enzyme; the little enzyme activity that remains should be feedback inhibited by glutamine to ensure that the cells do not accumulate too much glutamine. When NH_3 concentrations are low, more enzyme is required. In this situation, the enzyme should not be feedback inhibited, because its major role is to assimilate nitrogen.

The state of adenylylation of the glutamine synthetase enzyme is also regulated by GlnD and the state of the P_{II} protein. When the cells are in low NH_3 concentrations so that the P_{II} protein has UMP attached (Figure 13.9), the adenylyltransferase removes AMP from glutamine synthetase. When the cells are in high NH_3 concentrations, so that the P_{II} protein does not have UMP attached, the P_{II} protein binds to and stimulates the adenylyltransferase to add more AMP to the glutamine synthetase.

The P_{II} protein is not be the only way to stimulate the deadenylylation of glutamine synthetase in *E. coli*, however. Another protein, GlnK, forms GlnK-UMP in response to ammonia deprivation. The GlnK-UMP protein can also stimulate the adenylyltransferase to adenylylate glutamine synthetase when NH_3 concentrations are high (see van Heeswijk et al., Suggested Reading).

COORDINATION OF CATABOLITE REPRESSION, THE Ntr SYSTEM, AND THE REGULATION OF AMINO ACID-DEGRADATIVE OPERONS

Not only must bacteria sometimes use one or more of the 20 amino acids as a nitrogen source, they also must sometimes use amino acids as carbon and energy sources. The types of amino acids that different bacteria can use vary. For example, *E. coli* can use almost any amino acid as a nitrogen source except tryptophan, histidine, and the branched-chain amino acids, such as valine. However, it can use only alanine, tryptophan, aspartate, asparagine, proline, and serine as carbon sources. *Salmonella* can also use many amino acids as nitrogen sources but can use only alanine, cysteine, proline, and serine as carbon sources. Like the sugar-utilizing and biosynthetic operons, the amino acid-utilizing operons not only have their own specific regulatory genes, so that they are transcribed only in the presence of their own inducer (usually the amino acid that the gene products can utilize), but often are also part of larger regulons. As discussed in this section, they are often under Ntr regulation and so are not induced in the presence of their inducer while a better nitrogen source, such as NH_3, is in the medium. In some bacteria, these operons are also under the control of the CAP catabolite regulation system and are not expressed in the presence of better carbon sources, such as glucose.

In addition to their potential as nitrogen, carbon, and energy sources, amino acids are necessary for other purposes. Most obviously, an amino acid is needed for protein synthesis if it is in short supply relative to the other aminos acids. However, other funtions include the use of proline in osmoregulation. Therefore, the use of amino acids as carbon and nitrogen sources can present strategic problems for the cell. The way in which all these potentially conflicting regulatory needs are resolved is often complicated (see Commichau et al., Suggested Reading).

GENETIC ANALYSIS OF NITROGEN REGULATION IN ENTERIC BACTERIA

The present picture of nitrogen regulation in bacteria began with genetic studies. Most of this work was first done with *K. aerogenes*, although some genes were first found in *S. enterica* serovar Typhimurium or *E. coli*.

The first indications of the central role of glutamine and glutamine synthetase in Ntr regulation came from the extraordinary number of genes that, when mutated, could affect the regulation of glutamine synthetase or give rise to an auxotrophic growth requirement for glutamine (see Magasanik, Suggested Reading). The *gln* genes were originally named *glnA*, *glnB*, *glnD*, *glnE*, *glnF*, *glnG*, and *glnL* (Table 13.2). These genes are not lettered consecutively because, as often happens in genetics, genes presumed to exist because of a certain phenotype were later found not to exist or to be the same as another gene, so their letters were retired. Sorting out

Table 13.2 Genes for nitrogen regulation

Gene	Alternate name	Product	Function
glnA		Glutamine synthetase	Synthesize glutamine
glnB		PII	Inhibit phosphatase of NtrB, activate adenylyltransferase
glnD		Uridylyltransferase/uridylyl-removing enzyme	Transfer UMP to and from P_{II}
glnE		Adenylyltransferase	Transfer AMP to glutamine synthetase
glnF	*rpoN*	σ^N	RNA polymerase recognition of promoters of Ntr operons
ntrC	*glnG*	NtrC	Activate promoters of Ntr operons
ntrB	*glnL*	NtrB	Autokinase, phosphatase; phosphate transferred to NtrC

the various contributions of the *gln* gene products to arrive at the model for nitrogen regulation outlined above was a remarkable achievement. It took many years and required the involvement of many people. In this section, we describe how the various *gln* genes were first discovered and how the phenotypes caused by mutations in the genes led to the model.

The *glnB* Gene

The *glnB* gene encodes the P_{II} protein. The first *glnB* mutations were found among a collection of *K. aerogenes* mutants with the Gln^- phenotype, the inability to multiply without glutamine in the medium. Mutations in *glnB* apparently can prevent the cell from making enough glutamine for growth. However, early genetic evidence indicated that these *glnB* mutations do not exert their Gln^- phenotype by inactivating the P_{II} protein. As evidence, transposon insertions and other mutations that should totally inactivate the *glnB* gene (null mutations) do not result in the Gln^- phenotype. In fact, null mutations in *glnB* are intragenic suppressors of the Gln^- phenotype of the original *glnB* mutations (see chapter 3 for a discussion of the different types of suppressors).

We now know that the original *glnB* mutations do not inactivate P_{II} but instead change the binding site for UMP so that UMP cannot be attached to it by GlnD. This should have two effects. P_{II} without UMP binds to NtrB~P and prevents phosphorylation of NtrC. Therefore, even under low NH_3 concentrations, expression of the *glnA* gene is low and little glutamine is synthesized. By itself, however, this effect does not explain the Gln^- phenotype, since the *glnA* gene can also be transcribed from the p_1 promoter, which does not require NtrC~P for activation (see above). It is the second function of P_{II}—the stimulation of the adenylyltransferase—that causes the Gln^- phenotype. In these *glnB* mutants, enough P_{II} without UMP attached accumulates to stimulate the adenylyltransferase to the extent that glutamine synthetase is too heavily adenylylated to synthesize enough glutamine for growth. This also explains why null mutations in *glnB* do not cause the Gln^- phenotype. By inactivating P_{II} completely, null mutations prevent the P_{II} protein from stimulating the adenylyltransferase, so that less glutamine synthetase is adenylylated and enough glutamine is synthesized for growth.

The *glnD* Gene

The *glnD* gene was also discovered in a collection of mutants with mutations that cause the Gln^- phenotype. However, in this case, null mutations in *glnD* cause the Gln^- phenotype. Furthermore, null mutations in *glnB* suppress the Gln^- phenotype of null mutations in *glnD*. These observations are consistent with the above model. Since the GlnD protein is the enzyme that transfers UMP to P_{II}, null mutations in *glnD* should behave like the original *glnB* mutations and prevent UMP attachment to P_{II} but leave the P_{II} protein intact. This makes the cells Gln^- for the reasons given above. Null mutations in *glnD* are suppressed by null mutations in *glnB* because the absence of the GlnD protein has no effect if the cell contains no P_{II} protein to bind to the adenylyltransferase and stimulate the attachment of AMP to glutamine synthetase. The glutamine synthetase without AMP attached remains active and synthesizes enough glutamine for growth.

The *ntrB* Gene

The *ntrB* gene was originally named *glnL*. Mutations in *ntrB* were first discovered as extragenic suppressors of *glnD* and *glnB* mutations. These mutations do not inactivate the kinase activity of NtrB; instead, they prevent the binding of P_{II} so that NtrB transfers its phosphate to NtrC regardless of the presence of P_{II}-UMP.

The *ntrC* Gene

The *ntrC* gene, originally called *glnG*, was discovered because mutations in it can suppress the Gln^- phenotype of *glnF* mutations. The *glnF* gene encodes the nitrogen sigma factor, σ^N. Further work showed that null mutations in *ntrC* suppress *glnF* mutations. Null mutations in *ntrC* do not cause the Gln^- phenotype, because the *glnA* gene can also be transcribed from the p_1 promoter, which does not require NtrC for activation. They do, however, prevent the expression of other Ntr operons that require NtrC~P for their activation, many of which do not have alternative promoters.

Some mutations in *ntrC* do cause the Gln^- phenotype, however. They presumably are mutations that change NtrC so that it can no longer activate transcription from p_2, but it retains the ability to repress transcription from p_1 because it can be phosphorylated.

The *glnF* Gene

As mentioned above, the *glnF* gene, now renamed *rpoN*, encodes the nitrogen sigma factor, σ^N. The gene was also discovered in a collection of Gln^- mutants. Without σ^N, the p_2 promoter cannot be used to transcribe the *glnA-ntrB-ntrC* operon. By itself, inactivation of p_2 would not be enough to cause the Gln^- phenotype, since the *glnA* gene can also be transcribed from the p_1 promoter, which does not require σ^N. However, the NtrC~P form of NtrC represses the p_1 promoter if the cells are growing in low NH_3 concentrations (see above). In high NH_3 concentrations, the p_1 promoter is not repressed, but the small amount of glutamine synthetase synthesized is heavily adenylylated, preventing the synthesis of sufficient glutamine for growth. This interpretation of the Gln^- phenotype of *rpoN* mutations is consistent with the fact that null mutations in *ntrC* suppress the Gln⁻ phenotype of *rpoN* mutations, as discussed above. Without NtrC~P present to repress the p_1 promoter, sufficient glutamine synthetase is synthesized from the p_1 promoter, even in low NH_3 concentrations.

Regulation of Nitrogen Assimilation in *B. subtilis*

As described above, nitrogen regulation in enteric bacteria relies heavily on posttranslational protein modification (e.g., phosphorylation of NtrC, uridylylation of P_{II}, and adenylylation of glutamine synthetase) and a dedicated sigma factor (σ^N). Regulation of the same pathway in *B. subtilis* depends primarily on protein-protein interactions, using a regulatory protein called TnrA, for *trans*-acting nitrogen regulation.

THE TnrA PROTEIN REGULATES NITROGEN METABOLISM IN *B. subtilis*

When glutamine is high, TnrA not only acts as a transcriptional repressor of the glutamine synthetase gene, but also forms a complex with existing glutamine synthetase and glutamine to inhibit TnrA activity (see Wray et al., Suggested Reading). When glutamine levels are low, the inhibitory activity of TnrA is blocked, and TnrA shifts its role to serve as an activator of transcription of genes for utilization of secondary nitrogen sources. The expression of the gene for GOGAT (see above), which is involved in glutamate biosynthesis, is repressed by TnrA when glutamine is high, but also is subject to regulation by the GltC transcriptional activator. GltC-dependent activation of GOGAT synthesis is increased if the cell has large amounts of α-ketoglutarate (the substrate for glutamate synthesis) and is decreased by large amounts of glutamate (the product of GOGAT activity). This allows the cell to produce the enzyme only when glutamate biosynthesis should occur.

The regulatory activity of TnrA is not modulated by posttranslational modification but instead is determined by formation of a complex with glutamine synthetase enzyme in the presence of glutamine. Binding of TnrA to glutamine synthetase and glutamine occurs only when glutamine is abundant and results in activation of TnrA as a repressor of *glnA* gene expression. Glutamine synthetase therefore autoregulates expression of its own gene, although repression requires both high levels of glutamine synthetase enzyme and the end product of the pathway.

THE CodY GLOBAL REGULATOR

The CodY protein is a global regulator that is conserved in many gram-positive bacteria. It serves as a general sensor of cell physiology and acts primarily as a repressor of genes whose products help the cell to adjust to nutrient limitation (see Sonenshein, Suggested Reading). The DNA-binding activity of CodY is activated by either high GTP levels or high concentrations of the branched-chain amino acids (leucine, isoleucine, and valine). These serve as useful signal molecules of metabolic activity, as their synthesis is dependent on the availability of nitrogen, carbon, and even phosphorus, so that high levels indicate that the cell is not starved for any of these crucial nutrients. High GTP levels also signal efficient energy production. CodY represses a number of genes involved in nitrogen metabolism, including the gene for GOGAT, thereby inhibiting synthesis of glutamate, and also represses genes for enzymes involved in the uptake and utilization of other amino acids that can be used to generate glutamate and glutamine. Interactions with CcpA, the central regulator of carbon metabolism in gram-positive bacteria that responds to the FBP levels (see above), further allow the cell to integrate information about general metabolism to control a large variety of pathways.

Regulation of Ribosome and tRNA Synthesis

To compete effectively in the environment, cells must make the most efficient use possible of the available energy. We have already talked about the competitive advantage of using the carbon source that allows the most efficient energy production first. However, a competitive edge can also be gained by economizing within the cell. One of the major ways in which cells conserve energy is by regulating the synthesis of their ribosomes and tRNAs so that they make only enough to meet their needs. More than half of the RNA made by the cell at any one time is rRNA and tRNA. Moreover, each ribosome is composed of about 50 different proteins, and there are about as many different tRNAs. Synthesis of the translational machinery, therefore, represents a major investment of cellular resources and requires careful regulation, not only to ensure the appropriate total quantity of the components, but also to maintain all of the components in the correct relative amounts so that the machinery can operate at maximum efficiency.

The numbers of ribosomes and tRNA molecules needed by the cell vary greatly, depending on the growth rate. Fast-growing cells require many ribosomes and tRNA molecules to maintain the high rates of protein synthesis required for rapid doubling of cell mass. Cells growing more slowly, either because they are using a relatively poor carbon and energy source or because some nutrient is limiting, need fewer ribosomes and tRNAs. As a consequence, a rapidly growing *E. coli* cell contains as many as 70,000 ribosomes, but a slow-growing cell has fewer than 20,000. As with most of the global regulatory systems we have discussed, the regulation of ribosome and tRNA synthesis is much better understood for

E. coli than for any other organism. In this section, we confine our discussion to *E. coli*, with occasional references to other bacteria when information is available.

Ribosomal Protein Gene Regulation

Ribosomes are composed of both proteins and RNAs (see chapter 2). A special nomenclature is used for the components of the ribosome. The ribosomal proteins are designated by the letter L or S, to indicate whether they are from the large (50S) or small (30S) subunit of the ribosome, respectively, followed by a number for the particular protein. Thus, protein L11 is protein number 11 from the large 50S subunit of the ribosome, whereas protein S12 is protein number 12 from the small 30S subunit. The gene names begin with *rp*, for ribosomal protein, followed by a lowercase *l* or *s* to indicate whether the protein product resides in the large or small subunit. Another capital letter designates the specific gene. For example, the gene *rplK* is ribosomal protein gene K encoding the L11 protein; note that *K* is the 11th letter of the alphabet. Similarly, *rpsL* encodes the S12 protein; *L* is the 12th letter of the alphabet.

MAPPING OF RIBOSOMAL PROTEIN GENES

A total of 54 different genes encode the 54 polypeptides that comprise the *E. coli* ribosome, and mapping these genes was a major undertaking. Some ribosomal protein genes were mapped by mapping mutations that caused resistance to antibiotics, such as streptomycin, which binds to the ribosomal protein S12, blocking the translation of other genes (see chapter 2). More complex techniques involving specialized transducing phages and DNA cloning (see chapters 1 and 8) were needed to map the other ribosomal protein genes. Often, clones containing these genes were identified because they direct synthesis of a particular ribosomal protein in coupled in vitro transcription-translation systems (see chapter 2). As these genes are highly conserved, identification in newly sequenced genomes is usually based on similarity to the related gene in *E. coli.*

The mapping results revealed some intriguing aspects of the organization of the ribosomal protein genes in the chromosome of *E. coli*. Rather than being randomly scattered around the chromosome, the 54 genes are organized into large clusters of operons, with the two largest clusters at 73 and 90 minutes in the *E. coli* genome. Furthermore, these operons also contain genes for other components of macromolecular synthesis, including subunits of RNA polymerase, tRNAs, and genes for proteins of the DNA replication apparatus. Clustering of this type is found in many bacterial genomes.

In the cluster shown in Figure 13.13, four genes for tRNAs, *thrU*, *tyrU*, *glyT*, and *thrT*, are followed by *tufB*, a gene for the translation elongation factor EF-Tu. These

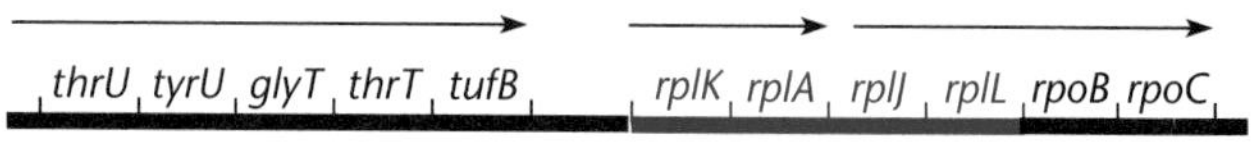

thrU *tyrU* *glyT* *thrT* *tufB* *rplK* *rplA* *rplJ* *rplL* *rpoB* *rpoC*

Figure 13.13 Arrangement of a gene cluster in *E. coli* encoding ribosomal proteins and other gene products involved in macromolecular synthesis. The cluster contains three operons, transcribed in the direction shown by the arrows. The *thrU, tyrU, glyT,* and *thrT* genes all encode tRNAs. The *tufB* gene encodes translation elongation factor EF-Tu. The *rplK, rplA, rplJ,* and *rplL* genes encode proteins of the large subunit of the ribosome. The *rpoB* and *rpoC* genes encode subunits of the RNA polymerase. doi:10.1128/9781555817169.ch13.f13.13

five genes constitute one operon; they are all cotranscribed into one long precursor RNA, from which the individual tRNAs are cut out later, leaving the mRNA for production of EF-Tu. The next operon in the cluster contains two genes for ribosomal proteins, *rplK* and *rplA*, and a third operon has four genes, *rplJ* and *rplL* (encoding two more ribosomal proteins) and *rpoB* and *rpoC* (encoding the β and β′ subunits of the RNA polymerase, respectively).

Several hypotheses have been proposed to explain why genes involved in macromolecular synthesis are clustered in the *E. coli* genome. First, the products of these genes must all be synthesized in large amounts to meet the cellular requirements. Some clusters are near the origin of replication, *oriC*, and cells growing at high growth rates have more than one copy of the genes near this site (see chapter 1), which allows higher rates of synthesis of the gene products. Other possible reasons have to do with the structure of the bacterial nucleoid (see chapter 1). Clustered genes are probably on the same loop of the nucleoid. A loop for the macromolecular synthesis genes might be relatively large and extend out from the core of the nucleoid to allow RNA polymerase and ribosomes to gain easier access to the genes. A third possible explanation is that the genes for macromolecular synthesis must be coordinately regulated with the growth rate, and their assembly in clusters of operons may facilitate their coordinate regulation.

REGULATION OF THE SYNTHESIS OF RIBOSOMAL PROTEINS

The regulation of the synthesis of ribosomal proteins and rRNA presents an interesting case of coordinate regulation. Even though the ribosomal proteins and rRNAs are synthesized independently and only later are assembled into mature ribosomes, there is never an excess of either free ribosomal proteins or free rRNA in the cell, suggesting that their synthesis is somehow coordinated. This indicates that either the rate of ribosomal protein synthesis is adjusted to match the rate of synthesis of the rRNA, or vice versa. As it turns out, the rate of ribosomal protein synthesis is adjusted to match the rate of rRNA synthesis, evidence for which is discussed below.

The regulation of ribosomal protein synthesis is best understood in *E. coli*. However, it is likely that the regulation is similar in all bacteria and probably, to some extent, in eukaryotes as well. The synthesis of many of the ribosomal proteins is **translationally autoregulated** (see chapter 12). The ribosomal proteins bind to translation initiation regions (TIRs) in their own mRNA and repress their own translation. However, rather than having each ribosomal gene of the operon translationally regulate itself independently, the protein product of only one of the genes of the operon is responsible for repressing the translation of all the ribosomal proteins encoded by the operon. Basically, this protein translationally represses the first gene in the operon, and the translation of the other protein-coding regions on the same mRNA is also translationally regulated, because their translation is coupled to the translation of the first protein (see chapter 2 for an explanation of translational coupling). The designated regulatory protein can also bind to free rRNA in the cell, which, as discussed below, is what coordinates the synthesis of the ribosomal proteins with the synthesis of rRNA.

Figure 13.14 illustrates this regulation with the relatively simple *rplK-rplA* operon, which encodes L11 and L1. The L1 protein serves as the regulator of the operon. The normal role of the L1 protein is to bind to free rRNA to assemble new ribosomes. L1 can also bind to the TIR for the *rplK* gene on the mRNA for the *rplK-rplA* operon. Binding of L1 to the *rplK* TIR directly inhibits translation of the *rplK* coding region and also inhibits translation of its own gene, because the *rplA* TIR is sequestered in the mRNA and is unavailable for translation unless *rplK* is actively translated, which releases the *rplA* TIR (see chapter 2). However, if the cell contains free rRNA, the L1 protein preferentially binds to it instead of binding to the *rplK* TIR, and translation of L11 (and L1) resumes. When there is no longer any free rRNA in the cell—because it has all been assembled into the ribosomes—the free L1 protein begins to accumulate, and it can bind to the TIR for the *rplK* gene and again repress the translation of the *rplK-rplA* mRNA. Note that this is a form of feedback repression, but in this case, the protein product of a gene directly represses its own synthesis.

A similar pattern is observed in other ribosomal protein gene operons. Their designated regulatory protein accumulates when rRNA is not freely available; the free designated regulatory ribosomal protein binds to the TIR for an early gene in the operon and represses the synthesis of all the ribosomal proteins encoded in the operon. The synthesis resumes when more free rRNA

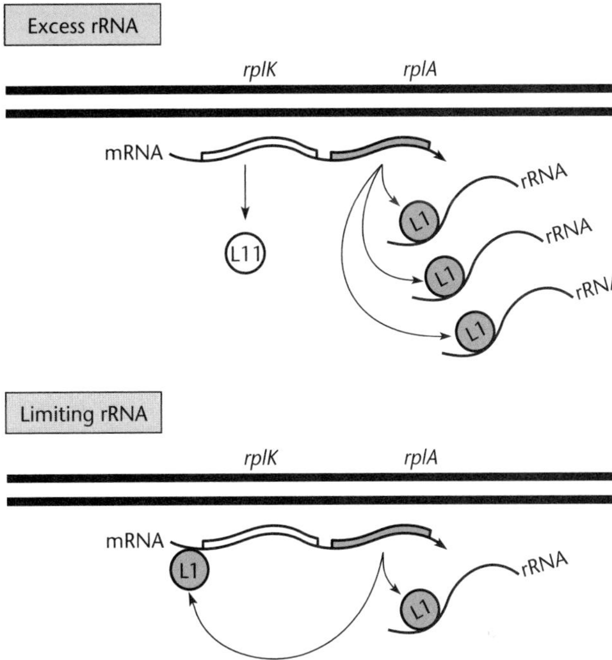

Figure 13.14 Translational autoregulation of ribosomal protein gene expression, as illustrated by the *rplK-rplA* operon shown in Figure 13.13. When rRNA is in excess, the regulatory ribosomal protein (L1 in the example) binds to rRNA to assemble more ribosomes, and the *rplK-rplA* mRNA is efficiently translated. When free rRNA is limiting, L1 binds instead to a target site on the mRNA that overlaps the TIR for *rplK*. This directly inhibits *rplK* translation by blocking binding of the ribosome and indirectly inhibits *rplA* translation by translational coupling. doi:10.1128/9781555817169.ch13.f13.14

accumulates. In this way, the cell ensures that there will be neither an excess of free ribosomal proteins nor an excess of free rRNA, and regulation of rRNA synthesis (see below) results in coordinate regulation of ribosomal protein synthesis.

The protein in each operon designated the regulator is usually a protein that normally binds to rRNA early during the assembly of the ribosome and so already has an RNA-binding activity. In at least some cases, the RNA structural element to which the designated protein binds in the mRNA is related to the binding site for that protein in the rRNA (see Nomura et al., Suggested Reading).

Experimental Support for Translational Autoregulation of Ribosomal Protein Genes

Evidence that the synthesis of ribosomal proteins is translationally autoregulated came from a series of experiments called **gene dosage experiments**, in which the number of copies, or dosage, of a gene in the cell is increased and the effect of this increase on the rate of synthesis of the protein product of the gene is determined (see Yates and Nomura, 1980 and 1981, Suggested

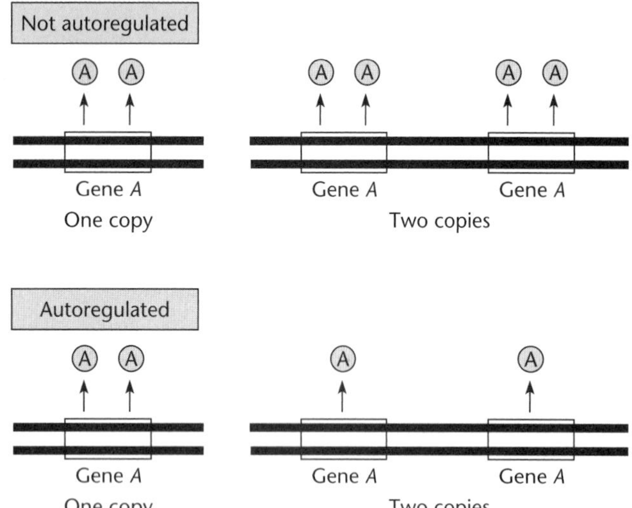

Figure 13.15 A gene dosage experiment to determine if a gene is autoregulated. The number of copies of the gene for the A protein is doubled. If the gene is not autoregulated, twice as much protein A is made. If gene *A* is autoregulated, the same amount of protein A is made.
doi:10.1128/9781555817169.ch13.f13.15

Reading). Figure 13.15 illustrates the principle behind a gene dosage experiment. If the gene is not autoregulated, the rate of synthesis of the gene product (if the gene is in the induced state) should be approximately proportional to the number of copies of the gene. However, if the gene is autoregulated, the product of the gene should repress its own synthesis, and the rate of synthesis of the gene product should not increase, regardless of the number of gene copies.

To introduce additional copies of the genes, the experiments were performed with cloning vectors based on phage λ, which is capable of lysogeny (see chapter 8). The phage carrying one or more ribosomal protein genes were then integrated at the normal phage attachment site. As a result, the cell contained two copies of these ribosomal protein genes, one at the normal position and another at the site of integration of the phage DNA. It was found that the same amounts of the ribosomal proteins were synthesized with two copies of the ribosomal protein genes as with one, indicating that their synthesis is autoregulated.

The next step was to determine if the autoregulation occurs at the level of transcription or translation. If the synthesis of the ribosomal proteins were transcriptionally autoregulated, the rate of transcription of the ribosomal genes would not increase when the number of copies of the genes increased. However, if the autoregulation occurred only at the level of translation, the rate of gene transcription would increase even though the rate of protein accumulation remained constant. The investigators found that the rate of transcription approximately doubled when the gene dosage was doubled, indicating that the autoregulation did not occur transcriptionally. This suggested that the genes must be translationally autoregulated. Therefore, the ribosomal proteins are capable of repressing their own translation.

The next step was to determine whether all the proteins in the operon independently repress their own translation or whether some of the proteins in each operon are responsible for regulating their own translation, as well as that of the others. To answer this question, the investigators systematically deleted genes for some of the proteins encoded by each operon in the phage and evaluated the effect of their absence on the synthesis of the other proteins. Deleting most of the genes in each operon had no effect on the synthesis of the proteins encoded by the other genes. However, when one particular gene in each operon was deleted, the rate of synthesis of the other proteins doubled. Therefore, the one protein represses the translation of itself and the other proteins encoded by the same operon.

Regulation of rRNA and tRNA Synthesis

If the translation of the ribosomal proteins is coupled to the synthesis of the rRNAs under different growth conditions, then how is the synthesis of rRNA regulated? As discussed in chapter 2, the 16S, 23S, and 5S rRNAs are synthesized together as a long precursor RNA, often with tRNA sequences positioned between the rRNA sequences. After synthesis, the long precursor RNA is processed into the individual rRNAs and tRNAs. Every ribosome contains one copy of each of the three types of rRNAs, and synthesizing all three rRNAs as part of the same precursor RNA ensures that all three are made in equal amounts.

REGULATION OF rRNA TRANSCRIPTION

Each cell has tens of thousands of ribosomes, requiring the synthesis of large amounts of rRNA. To meet this need, bacteria have evolved many ways to increase the output of their rRNA genes. For example, many bacteria have more than one copy of the genes for rRNA. For example, *E. coli* has 7 copies of the rRNA operons and *B. subtilis* has 10 copies. Many bacteria also have very strong promoters for their rRNA genes, with high affinity for RNA polymerase. The RNA polymerase molecules initiate transcription and start down the operon, one immediately after another, so that under conditions when rRNA synthesis is at its fastest, as many as 50 RNA polymerase molecules can be transcribing each rRNA operon simultaneously. The rRNA promoters are so strong because their sequences at the −10 and −35 elements closely match the preferred recognition sites for RNA polymerase holoenzyme containing σ^{70}, and they also have a sequence called the UP element upstream of the promoter (see Figure 2.13B). This sequence enhances initiation of transcription from the promoter by interacting with the α-CTD domain of RNA polymerase; the effect of this interaction with α-CTD is similar to that of CAP and some other activator proteins that use a similar interaction to increase the affinity of a promoter for RNA polymerase. The protein activators use protein-protein contacts with RNA polymerase, whereas the rRNA operons use DNA-protein contacts between the UP element sequence and α-CTD (see chapter 12).

An additional mechanism for increasing transcription from rRNA operon promoters utilizes the FIS protein. FIS is a DNA-binding protein that participates in a number of regulatory systems, as well as in certain site-specific recombination systems, including integration of the phage λ prophage during lysogeny (see chapter 8). FIS is more abundant in rapidly growing cells, and its levels drop as cells enter stationary phase. Binding of FIS upstream of rRNA operon promoters increases their transcriptional activity, and since FIS is more abundant during rapid growth, this allows increased rRNA synthesis under these conditions.

GROWTH RATE REGULATION OF rRNA AND tRNA TRANSCRIPTION

As mentioned above, cells growing quickly in rich medium have many more ribosomes and a higher concentration of tRNA than do cells growing slowly in poor medium. This regulation of rRNA and tRNA synthesis is called **growth rate regulation.**

An interesting model has been proposed to explain the growth rate dependence of rRNA synthesis (see Barker and Gourse, Suggested Reading). According to this model, rRNA (and tRNA) operon promoters are unusual in that the rate of initiation of transcription is very sensitive to the concentration of the initiating nucleotide. Like most transcription, rRNA synthesis begins with ATP or GTP, so one or the other of these initiates transcription of each of the rRNA genes. The rRNA promoters are unusual in that they have a high affinity for RNA polymerase but form very short-lived open complexes (see chapter 2 for a detailed discussion of closed and open complexes and transcription initiation). The high affinity for RNA polymerase allows them to form closed complexes at a high rate and compete very effectively for RNA polymerase. They then quickly form the open complex. However, because these open complexes are unstable, the transcription complex rapidly returns to the closed-complex state unless the initiating nucleotide immediately enters through the secondary channel and transcription initiates. The higher the concentration of the initiating nucleotide, the more likely this is to happen before the open complex reverts to the closed-complex state. When growth rates are high, the concentrations of ATP and GTP increase, leading to more frequent initiations at the rRNA promoters and therefore more synthesis of rRNAs. When growth rates are low, the concentrations of ATP and GTP are usually lower, leading to less frequent initiation of transcription from these promoters and lower synthesis of rRNA and tRNA. Since ribosomal protein synthesis is coupled to rRNA availability, changes in rRNA abundance lead to changes in ribosome production.

In addition to the response of rRNA operon promoters to the steady-state growth rate via monitoring of the availability of GTP or ATP, there is also a response to sudden changes in growth conditions, e.g., starvation for amino acids. This is called the stringent response and is discussed below.

ANTITERMINATION OF rRNA OPERONS

Another mechanism that promotes a high rate of rRNA gene transcription is the presence of antitermination sequences that are positioned just downstream of the promoter and in the spacer region between the 16S and 23S coding sequences. These antitermination sequences reduce pausing by RNA polymerase and prevent termination at sites that can promote binding of the ρ factor to cause ρ-dependent transcription termination (see Figure 2.20). Since rRNA is not translated, transcription of these long operons is particularly susceptible to ρ-dependent termination. The antitermination sequences, therefore, allow the synthesis of more full-length rRNAs and also allow them to be completed in a shorter time because RNA polymerase moves more rapidly along the DNA (see Condon et al., Suggested Reading).

The Stringent Response

In addition to the growth rate regulation discussed above, the rRNA and tRNA genes in *E. coli* are subject to another type of regulation called the **stringent response**, which causes rRNA and tRNA synthesis to cease when cells are starved. Both growth rate regulation and the stringent response involve the use of nucleotides as signals of the nutritional status of the cell.

Protein synthesis requires all 20 amino acids, which must be made by the cell or furnished in the medium. A cell is said to be **starved** for an amino acid when the amino acid is missing from the medium and the cell cannot make a sufficient amount. A translating ribosome stalls when it encounters a codon for the missing amino acid because the corresponding aminoacyl-tRNA is not available for insertion into the growing polypeptide chain.

In principle, rRNA synthesis can continue in cells starved for an amino acid, since RNA does not contain amino acids. However, in *E. coli* and many other types of cells, the synthesis of rRNA and tRNA ceases when an amino acid is lacking. This coupling of the synthesis of rRNA and tRNA to the synthesis of proteins is called the stringent response. Because ribosomal protein gene expression is coupled to rRNA availability, turning off rRNA synthesis also turns off ribosomal protein synthesis. The stringent response saves energy and cellular resources and conserves any remaining stores of the amino acid that is limiting for growth; there is no point in making ribosomes and tRNA if one of the amino acids is not available for protein synthesis.

SYNTHESIS OF ppGpp DURING THE STRINGENT RESPONSE

In *E. coli* and many other organisms, the reduction of rRNA and tRNA synthesis results from the accumulation of an unusual nucleotide, **guanosine tetraphosphate (ppGpp)**. This nucleotide is first made as **guanosine pentaphosphate (pppGpp)** by transferring two phosphates from ATP to the 3′ hydroxyl of GTP but is then quickly converted to ppGpp by a phosphatase. These nucleotides were originally called "magic spot" I and II (MSI and MSII) because they show up as distinct spots during some types of chromatography (see Cashel and Gallant, Suggested Reading).

Figure 13.16 shows a model for how amino acid starvation stimulates the synthesis of ppGpp. The nucleotide is made by an enzyme called RelA (for *re*laxed control gene *A*), which is bound to the ribosome. When *E. coli* cells are starved for an amino acid (lysine in the example), the tRNAs for that amino acid (e.g., $tRNA^{Lys}$) are uncharged. The low level of $tRNA^{Lys}$ with lysine attached (Lys-$tRNA^{Lys}$) results in failure of EF-Tu to bring Lys-$tRNA^{Lys}$ into the aminoacyl (A) site of the ribosome (see chapter 2). EF-Tu has very low affinity for uncharged $tRNA^{Lys}$, and therefore, when a ribosome moving along an mRNA encounters a codon for that amino acid (the codon AAA in the example), the ribosome stalls. If the ribosome stalls long enough, an uncharged tRNA ($tRNA^{Lys}$ in the example) may eventually enter the A site of the ribosome even though it is not bound to EF-Tu. Uncharged tRNA entering the A site stimulates the RelA protein, which is transiently associated with the ribosome, to synthesize pppGpp, which is then converted to ppGpp.

The intracellular levels of ppGpp during amino acid starvation are also regulated by the SpoT (for "magic spot") protein, which is the product of the *spoT* gene. The SpoT protein has pppGpp synthesis activity, like RelA, and also an activity that degrades ppGpp. The ppGpp degradation activity of SpoT is inhibited after amino acid starvation, leading to greater accumulation of ppGpp. Therefore, after amino acid starvation, the cellular concentration of ppGpp is determined both by the activation of the pppGpp synthesis activity of RelA and by the inhibition of the ppGpp-degrading activity of SpoT. The *spoT* gene product not only can degrade ppGpp, but also can synthesize pppGpp in response to other starvation conditions. While *E. coli* has separate RelA and SpoT proteins, many other types of bacteria,

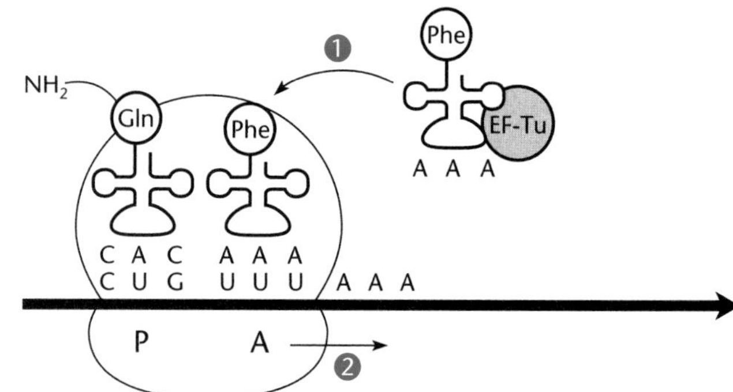

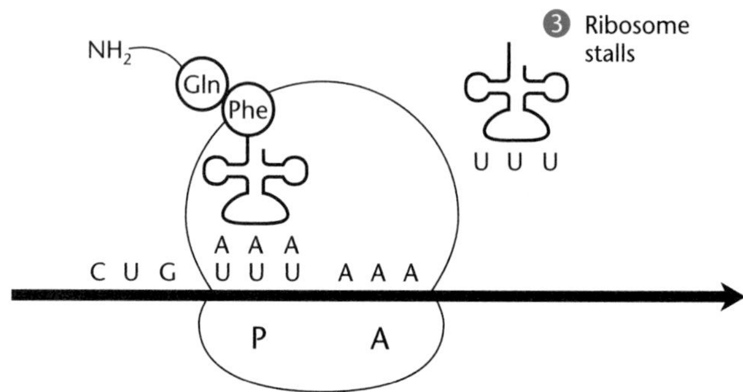

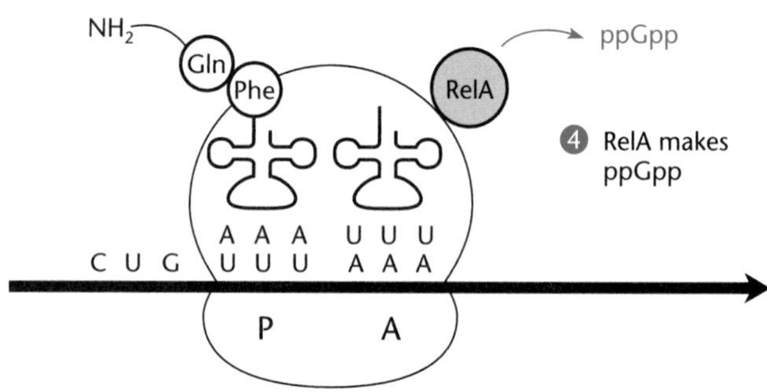

Figure 13.16 Model for synthesis of ppGpp after amino acid starvation. Cells are starved for the amino acid lysine. The $tRNA^{Lys}$ then has no lysine attached, and EF-Tu cannot bind to a tRNA that is not aminoacylated. A ribosome moving along the mRNA stops when it arrives at a codon for lysine (AAA) because it has no aminoacylated $tRNA^{Lys}$ to translate the codon. This results in a ribosome with a peptidyl-tRNA in the P site and an empty A site (see chapter 2). If the ribosome remains stalled at the codon long enough, an unaminoacylated $tRNA^{Lys}$ (anticodon UUU in the example) binds to the A site of the ribosome even though EF-Tu is not bound. This binding causes RelA to synthesize ppGpp.
doi:10.1128/9781555817169.ch13.f13.16

including *B. subtilis*, have only a single protein with both of these activities. This class of enzyme is designated Rsh (RelA-SpoT homolog) and seems to perform both roles.

ISOLATION OF *relA* MUTANTS

Evidence that the accumulation of ppGpp is responsible for the stringent response in *E. coli* came from the characterization of *relA* mutants that lack the RelA enzyme activity. These mutants do not accumulate ppGpp after amino acid starvation and also do not shut off rRNA and tRNA synthesis. Because rRNA synthesis and tRNA synthesis are not stringently coupled with protein synthesis in *relA* mutants, strains with *relA* mutations are called **relaxed strains**, which is the origin of the gene name.

As is often the case, the first *relA* mutant was isolated by chance. A mutant strain of *E. coli* was observed to have a difficult time recovering after amino acid starvation, whereas the wild-type parent could start growing almost immediately after the amino acid was restored. A later study showed that rRNA and tRNA synthesis continued after the mutant was starved for an amino acid, so the mutant was called a relaxed mutant. The growth phenotype was presumed to be due to the fact that the strain continued to produce ribosomes after it was starved, resulting in further depletion of cellular resources that made recovery less efficient once nutrients were restored.

The poor ability of the original *relA* mutant to recover from amino acid starvation suggested a way to enrich for *relA* mutants (see Fiil and Frieson, Suggested Reading). This procedure is based on the fact that growing cells are killed by ampicillin but cells that are not growing survive (see chapter 3 for a description of ampicillin enrichments). *E. coli* cells that were auxotrophic for an amino acid were mutagenized, washed, and resuspended in a medium without the amino acid so that they stopped growing. After a long period of incubation, the amino acid was added back to the medium and ampicillin was added. The ampicillin was removed shortly thereafter, and the process was repeated. After a few cycles of this treatment, the bacteria were plated. A high percentage of the bacteria that survived and could multiply to form colonies were *relA* mutants, as shown by the fact that most of these strains continued to synthesize rRNA and tRNA after amino acid starvation. Since ampicillin is toxic only to cells that are actively engaged in cell wall synthesis, the failure of the *relA* mutants to return to active growth rapidly after addition of the missing amino acid protected them from being killed by ampicillin, which selectively killed the wild-type cells, resulting in enrichment for *relA* mutants in the population of surviving cells.

ROLE OF ppGpp IN GROWTH RATE REGULATION, AFTER STRESS, AND IN STATIONARY PHASE

Not only are ppGpp levels higher when *E. coli* cells are growing more slowly in poorer medium, but also, they increase when cells run out of nutrients and begin to reach stationary phase. Many stress conditions can also cause ppGpp levels to increase. Under these conditions, the rates of rRNA and tRNA synthesis are reduced, but other genes whose products are required in stationary phase or for stress responses are turned on. This suggests that ppGpp is a sort of general "alarmone," signaling that major changes in gene expression will soon be required (see Magnusson et al., Suggested Reading). SpoT, which plays a role in degrading ppGpp in the stringent response, seems to play the major role in synthesizing it during during stress responses. However, mutations in both *relA* and *spoT* are required to completely block ppGpp synthesis, and a *relA spoT* double mutant completely lacks ppGpp (and is sometimes referred to as a $ppGpp^0$ strain).

Because *relA spoT* double mutants lack ppGpp, many experiments comparing these mutants with the wild type have been done to determine the effect of ppGpp on cells. However, in spite of many years of such experimentation, it is not completely clear how ppGpp affects transcription, and it is likely that more than one mechanism is involved. The ppGpp nucleotide seems to have both positive and negative effects on gene transcription. For example, in addition to the effect on rRNA and tRNA synthesis, *relA spoT* mutants do not grow in minimal medium without added amino acids, i.e., they are effectively auxotrophic for some amino acids. In *E. coli*, *relA spoT* mutants require nine different amino acids, suggesting that ppGpp serves as an inducer for the transcription of the operons to make these amino acids when they are lacking in the medium. Also, *relA spoT* double mutants show phenotypes similar to those of mutants that lack σ^S (the stationary-phase σ factor), which responds to many types of stress in the cell (see below). The activities of other alternate "stress" sigma factors, including σ^H (the heat shock sigma factor) and σ^E (the extracytoplasmic stress sigma factor), are also enhanced by ppGpp (see below).

One factor that complicates the interpretation of these experiments and necessitates the use of careful controls is the possibility of indirect effects due to competition for RNA polymerase. Genes that seem to be positively regulated by ppGpp might be those that do not bind RNA polymerase as tightly as the promoters for rRNA and tRNA genes; they would therefore require higher concentrations of free RNA polymerase to be active. During normal rapid growth, at least half of all the RNA polymerase in the cell is devoted to transcription of rRNA and tRNA genes. If the initiation of synthesis of the rRNAs and tRNAs is blocked by ppGpp, more RNA polymerase becomes available to transcribe other genes. However, in vitro experiments have indicated that competition for RNA polymerase is not the sole explanation for the apparent positive effect of ppGpp on the transcription of some genes (see below).

DksA: A PARTNER IN ppGpp ACTION

In *E. coli*, the effect of ppGpp on RNA polymerase is enhanced by a protein named DksA. As described for RelA, the gene for this protein was discovered by chance, and its role in ppGpp-mediated regulation was not suspected for some time. It was first found because overexpression of the *dksA* gene suppresses growth defects of *dnaK* mutants (the name DksA means *dnaK* suppressor A). It is not clear, even in retrospect, why excess DksA would have this effect. Recall from chapter 2 that DnaK is the bacterial Hsp70 protein chaperone that helps to fold some proteins as they emerge from the ribosome and also plays the role of the cellular thermometer that induces heat shock (see below). The *dnaK* null mutants are sick because DnaK normally binds to the heat shock sigma, σ^H, and targets it for degradation. In the absence of DnaK, σ^H accumulates, and the heat shock genes are expressed constitutively, slowing cell growth. Apparently, excess DksA relieves the toxicity of the constitutive induction of the heat shock response, but how it does this is not clear.

The role of DksA in ppGpp-mediated regulation was not discovered until later, when it was noticed that the phenotypes of deletion mutants of the *dksA* gene (Δ*dksA*) are similar to the phenotypes of *relA spoT* double mutants that do not make ppGpp. The Δ*dksA* mutants are effectively auxotrophic for some (but not all) of the same amino acids; they also show increased rates of rRNA transcription, and they seem to lack stringent control and growth rate regulation of rRNA synthesis. In vitro experiments that investigated the effect of adding ppGpp to RNA polymerase, with or without purified DksA, showed that DksA markedly enhances the ability of ppGpp to inhibit transcription from rRNA promoters (see Paul et al., Suggested Reading). It also greatly increases the dependence of initiation from rRNA promoters on the concentration of the initiating nucleotides (ATP or GTP) and seems to further shorten the half-life of the open complexes on these promoters. DksA also enhances the ppGpp stimulation of transcription from the promoters for some amino acid biosynthetic operons, indicating that this effect of ppGpp is not all due to competition for RNA polymerase.

Clues to how the DksA protein has these effects came from the observation that DksA bears a remarkable structural similiarity to the GreA protein (see Peredina et al., Suggested Reading). Recall from chapter 2 that GreA is a transcription factor that inserts into the secondary channel in RNA polymerase through which deoxynucleotide triphosphates enter (see Figure 2.18). It has a long extended coiled-coil probe that extends all of the way into the channel to the active center and has two acidic amino acids (aspartate and glutamate) on the end of the probe that may bind the Mg^{2+} in the active center, which plays a role in the polymerization reaction. From this position, GreA can degrade backtracked RNAs and release stalled transcription complexes. DksA also has an extended coiled-coil structure of about the same length, with two conserved acidic amino acids in the end (in this case, both aspartates), although there is little amino acid sequence similarity otherwise. This striking similarity in structure suggests that DksA may also enter RNA polymerase through the secondary channel, like GreA. The details of how DksA promotes the activity of ppGpp remain to be determined but are likely to provide important insight into how the activity of RNA polymerase can be controlled by both protein and nucleotide regulators.

STRINGENT RESPONSE IN OTHER BACTERIA

Homologs of RelA and/or SpoT are found in nearly all bacterial genomes. However, it is not clear that all bacteria respond similarly to ppGpp accumulation. There is no apparent homolog of DksA in many organisms, including *B. subtilis*. Furthermore, it appears that the crucial parameter that regulates rRNA operon transcription in *B. subtilis* is the GTP concentration, rather than ppGpp itself. Recall that ppGpp is synthesized from GTP. Therefore, activation of ppGpp synthesis results in a rapid drop in the GTP concentration. As in *E. coli*, the promoters for rRNA operons in *B. subtilis* are highly sensitive to the concentration of the initiating nucleoside triphosphate, and in this organism, nearly all of the rRNA and tRNA promoters use GTP as the initiating nucleotide (see Krasny and Gourse, Suggested Reading). It therefore may be the case in *B. subtilis* that the major effect of the stringent response (at least for rRNA transcription) is due to the effect on GTP pools rather than a direct effect of interaction of ppGpp with RNA polymerase, but additional studies will be necessary to clarify this issue.

Stress Responses in Bacteria

Many of the global regulons in bacteria are designed to deal with stress. In order to survive, all organisms must be able to deal with abrupt changes in the conditions in which they find themselves. The osmolarity, temperature, or pH of their surroundings might abruptly increase or decrease or they might be suddenly deprived of nutrients required for growth and have to enter a dormant state.

If they have invaded a eukaryotic host, they might suddenly be exposed to reactive forms of oxygen or nitric oxide as part of the host defense. Not only must they be able to respond quickly to such changes, but also, their response must be flexible enough to deal with a variety of different stresses, or even more than one stress at a time. Furthermore, they must be able to sense when the stress condition has ended so that the cell can return to a normal state. As expected, bacteria and other organisms have evolved complicated interactive pathways to deal with such changes, and this is a subject of active current research. In this section, we discuss what has been learned about some major pathways and how they interact.

Heat Shock Regulation

The regulation of gene expression following a heat shock is one of the most extensively studied and highly conserved global regulatory responses in bacteria and other organisms. One of the major challenges facing cells is to survive abrupt changes in temperature. To adjust to abrupt temperature increases, cells induce at least 30 different genes encoding proteins called the **heat shock proteins (Hsps)**. The concentrations of these proteins quickly increase in the cell after a temperature upshift and then slowly decline, a phenomenon known as the **heat shock response**. Besides being induced by abrupt increases in temperature, the heat shock genes are induced by other types of stress that damage proteins, such as the presence of ethanol and other organic solvents in the medium. Therefore, the heat shock response is more of a general stress response rather than a specific response to an abrupt increase in temperature, although there is a component of the regulatory mechanism that responds specifically to heat.

Unlike most shared cellular processes, the heat shock response was observed in cells of higher organisms long before it was seen in bacteria. Some of the Hsps are remarkably similar in all organisms and presumably play similar roles in protecting all cells against heat shock. Some of the mechanisms of regulation of the heat shock response may also be similar in organisms ranging from bacteria to higher eukaryotes.

HEAT SHOCK REGULATION IN *E. COLI*

The molecular basis for the heat shock response was first understood in *E. coli* and is illustrated in Figure 13.17. In this bacterium, about 30 genes encoding 30 different Hsps are turned on following a heat shock. The functions of many of these Hsps are known. Most of the Hsps play roles during the normal growth of the cell and so are always present at low concentrations, but after a heat shock, their rate of synthesis increases markedly and then slowly declines to normal levels.

Some Hsps, including GroEL, DnaK, DnaJ, and GrpE, are chaperones that direct the folding of newly synthesized proteins (see chapter 2). The names of these proteins do not reflect their functions, but rather, how they were orginally discovered. For example, DnaK and DnaJ were found because they affect the assembly of a protein complex required for phage λ DNA replication, but they are not themselves involved directly in replication. Chaperones help the cell survive a heat shock by binding to proteins denatured by the sudden rise in temperature and either helping them to refold properly or targeting them for destruction. As mentioned above, the chaperones are among the most highly conserved proteins in cells.

Other Hsps, including Lon and Clp, are proteases that degrade proteins that are so badly denatured by the heat shock that they are irreparable and so are best degraded before they poison the cell. Some other Hsps are proteins normally involved in protein synthesis. The function of this type of Hsp in protecting the cell after a temperature rise is not clear.

Knowing that many Hsps are involved in helping proteins to fold properly or in destroying denatured proteins helps to explain the transient nature of the heat shock response. Immediately after the temperature increases, the concentrations of salts and other cellular components that were adjusted for growth at lower temperatures are not appropriate for protein stability at the higher temperature; this leads to massive protein unfolding. Later, after the temperature has been elevated for some time, the internal conditions have had time to adjust, so the level of denatured proteins is lower and the increased amount of chaperones and other Hsps is no longer necessary. Hence, synthesis of the Hsps declines.

Genetic Analysis of Heat Shock in *E. coli*

As with other regulatory systems, the analysis of the heat shock response was greatly aided by the discovery of mutants with defective regulatory genes. The first such mutant was found in a collection of temperature-sensitive mutants and was shown to be unable to induce Hsp production after a shift to high temperature (see Zhou et al., Suggested Reading). This mutant, which failed to make the regulatory gene product, made it possible to clone the regulatory gene by complementation. A library of wild-type *E. coli* DNA was introduced into the mutant strain, and clones that permitted the cells to survive at high temperatures were isolated. When the sequence of the cloned gene for the regulatory Hsp was compared with those of genes encoding other proteins, the gene was found to encode a new type of sigma factor that was only 32 kilodaltons (kDa) in mass; therefore, it was named σ^{32}, heat shock sigma factor, or σ^H. The RNA polymerase holoenzyme with σ^H attached recognizes

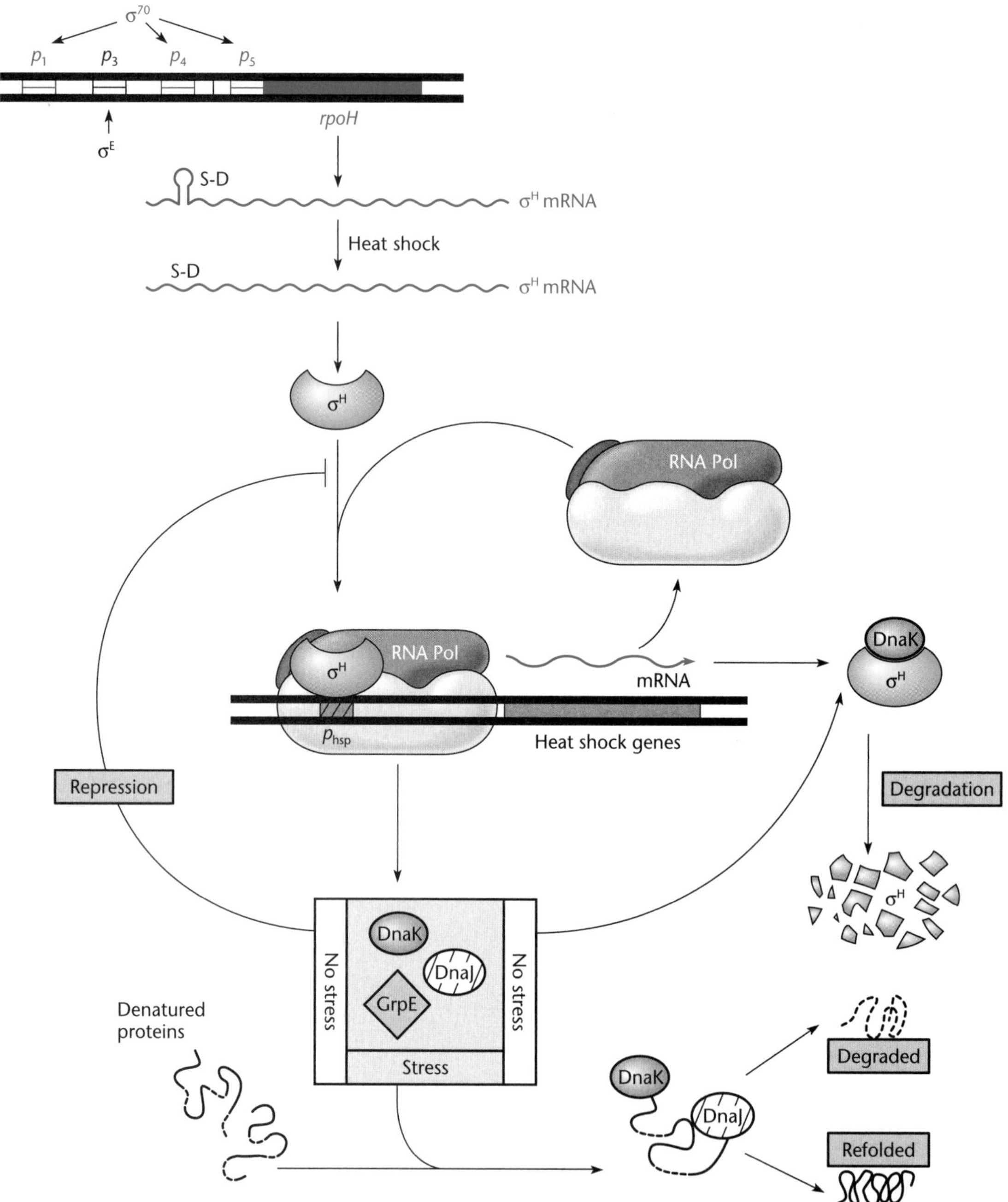

Figure 13.17 Induction of the heat shock response in *E. coli.* The *rpoH* mRNA is an RNA thermosensor, which is inactive for translation at low temperature and is unfolded and translated at high temperature. This results in rapid synthesis of σ^H after heat shock. The σ^H is quickly inactivated by binding to DnaK, which both makes it less active and targets it for degradation. After an abrupt increase in temperature, many other proteins are denatured, and DnaK, with the help of GrpE and DnaJ (see chapter 2), binds to them to help them refold or be degraded. Binding of DnaK to other protein substrates frees σ^H, stabilizing it and making it more active for transcription initiation at heat shock gene promoters (p_{hsp}). When the cell adjusts to the higher temperature and DnaK accumulates to the point where some is again available to bind to σ^H, the activity of σ^H in the cell again drops, and the transcription of the heat shock genes returns to basal levels. doi:10.1128/9781555817169.ch13.f13.17

promoters for the heat shock genes that are different from the promoters recognized by the normal σ^{70} factor and the nitrogen sigma factor, σ^N (Figure 13.11). The gene for the heat shock sigma factor was named *rpoH* (for *R*NA *po*lymerase subunit *h*eat shock).

Regulation of σ^H Synthesis

Normally, very few copies of σ^H exist in the cell. However, immediately after an increase in temperature from 30 to 42°C, the amount of σ^H in the cell increases 15-fold. This increase in concentration leads to a significant rise in the rate of transcription of the heat shock genes, since they are transcribed from σ^H-type promoters. Understanding how the heat shock genes are turned on after a heat shock requires an understanding of how this increase in the amount of σ^H occurs.

An abrupt increase in temperature might increase the amount of σ^H through several mechanisms. A crucial observation is that the amount of *rpoH* mRNA does not increase substantially after heat shock. This indicates that the increase in σ^H protein must occur at a posttranscriptional level. In fact, immediately after the temperature upshift, the translation rate of the *rpoH* mRNA increases 10-fold (i.e., there is translational regulation), and the half-life of the σ^H protein increases markedly (i.e., there is posttranslational regulation [see chapter 12]).

rpoH mRNA: an RNA Thermosensor

As described in chapter 12, RNA structure that sequesters the TIR of an mRNA can prevent binding of the 30S ribosomal subunit. RNA structure can be stabilized by binding of an RNA-binding protein or small molecule. Another possibility is that the RNA structure can be stable at low temperature and unstable at high temperature; an RNA element of this type is called an RNA thermosensor (see chapter 12) and allows translation initiation to be sensitive to temperature. This is the case for the *rpoH* mRNA, which is poorly translated during growth at low temperature but is efficiently translated after heat shock because the hydrogen bonds that stabilize the structure that hides the TIR are disrupted at high temperature (see Nagai et al., Suggested Reading). The *rpoH* mRNA, therefore, preexists in the cell under normal growth conditions but is inactive, and exposure to heat shock allows the mRNA to be translated. This type of mechanism is advantageous for a stress response, as the increase in σ^H protein is very rapid, allowing the cell to quickly induce synthesis of the Hsps.

DnaK: the *E. coli* Cellular Thermometer?

Posttranslational regulation of σ^H levels after a heat shock is due to the protein chaperone DnaK. Like *rpoH* mRNA, DnaK senses the change in temperature and promotes the heat shock response. However, unlike *rpoH* mRNA, which directly senses the temperature increase because of its effect on mRNA structure, DnaK senses temperature indirectly by measuring the cellular consequences of the temperature increase. It can play this role because it normally binds to nascent proteins in the process of being synthesized and helps them to fold properly. Under heat shock conditions, the DnaK chaperone can also bind to denatured proteins and help them refold.

Figure 13.17 shows a model for how the ability of the DnaK chaperone to bind unfolded proteins indirectly regulates the synthesis of the heat shock proteins. One of the proteins to which DnaK binds is σ^H (see Liberek et al., Suggested Reading). By binding to σ^H, the DnaK protein regulates the transcription of the heat shock genes in two ways. First, it affects the stability of σ^H because the σ^H protein with DnaK bound is more susceptible to a cellular protease called FtsH than is free σ^H (see Meyer and Baker, Suggested Reading). The σ^H bound to DnaK is rapidly degraded, so that any σ^H that is produced prior to heat shock is unable to direct the transcription of the heat shock genes. Second, the binding of DnaK inhibits the activity of σ^H so that the σ^H-DnaK complex is less active in transcription, which lowers the transcription of the heat shock genes even more.

How, then, does a sudden increase in temperature result in increased σ^H-dependent transcription? The answer lies in the chaperone role of DnaK. DnaK binds to denatured proteins to help them to refold properly. After heat shock, many denatured proteins appear in the cell, and most of the DnaK protein binds to these unfolded proteins. This leaves less DnaK available to bind to σ^H. The σ^H protein is then more stable and accumulates in the cell. It is also more active, increasing the transcription of the heat shock genes, including the *dnaK* gene itself. Synthesis of the σ^H protein also rises, due to the activation of translation of the *rpoH* transcript, resulting in rapid upregulation of the **heat shock regulon**.

This model also explains the transient nature of the heat shock response, in which the concentration of the Hsps increases sharply after an increase in temperature and then slowly declines. When enough DnaK has accumulated to bind to all the unfolded proteins and internal conditions have adjusted so that the proteins are more stable at the higher temperature, extra DnaK once again becomes available to bind to and inactivate σ^H, leading to the observed drop in the rate of synthesis of the Hsps. Furthermore, if the temperature returns to normal, the *rpoH* mRNA again folds into the repressed state, and additional synthesis of σ^H is inhibited.

Another alternative sigma factor, σ^E, is responsible for transcription of some heat shock genes at high temperature, including the *rpoH* gene encoding σ^H. Because σ^E is activated by damage to the outer membrane of the cell by heat and other agents, it is discussed below in connection with extracytoplasmic stress responses.

HEAT SHOCK REGULATION IN OTHER BACTERIA
Once heat shock regulation in *E. coli* was fairly well understood, it was of interest to see whether other bacteria use the same mechanism. Surprisingly, most other bacteria, including *B. subtilis*, in which it has been studied in the greatest detail, use a very different mechanism. Rather than using a heat shock sigma factor analogous to σ^H, *B. subtilis* and many other types of bacteria use the normal sigma factor and a repressor protein named HrcA to repress transcription from heat shock genes during growth at lower temperatures. The HrcA repressor binds to an operator sequence called CIRCE, which is highly conserved among bacteria as diverse as *B. subtilis* and cyanobacteria, suggesting that this type of regulation may be very ancient. Bacteria that use HrcA do use a chaperone as a cellular thermometer, but rather than using DnaK, they use the chaperonin GroEL (see chapter 2). GroEL may be required to fold HrcA; when the temperature increases abruptly and GroEL is recruited to help other proteins fold, HrcA may misfold. This then causes derepression of the heat shock genes HrcA would otherwise repress.

General Stress Response in Gram-Negative Bacteria

In addition to the heat shock sigma factor σ^H, which responds to an abrupt increase in temperature, *E. coli* has another sigma factor, called the stationary-phase sigma factor (σ^S), which is used to transcribe genes that are involved in the general stress response (Box 13.5). The gene for this sigma factor, *rpoS*, is turned on following many different types of stress, including nutritional deprivation, oxidative damage, and acidic conditions, and σ^S is active during stationary phase (as suggested by its name). σ^S is closely related to the normal vegetative sigma factor, σ^{70}, which transcribes most genes in *E. coli* and recognizes very similar promoters. The only difference is that the –10 sequence recognized by σ^S may be somewhat extended, and in fact, some promoters may be recognized by both σ^S and σ^{70}.

DNA microarray analysis (see below) has been used in attempts to identify all of the *E. coli* genes whose transcription is affected by RNA polymerase containing σ^S (see Weber et al., Suggested Reading). A total of 481 genes, more than 10% of the total number of genes in *E. coli*, are affected, either positively or negatively, by the absence of σ^S. Of these, 140 are affected under all the conditions tested in early stationary phase, while the remaining 341 are transcribed under only some conditions, such as low pH or high osmolarity in the medium. Many of these are regulatory genes that are activated only under a certain set of conditions. Besides genes whose products are obviously involved in stress responses, many genes involved in central energy metabolism, such as glycolysis, are also affected. The products of some of these genes probably play roles in switching the cell from aerobic metabolism to anaerobic metabolism as the cell enters stationary phase. Others are transporters that may help to remove toxic compounds from the cell or to scavenge for rare nutrients. The picture thus arises of σ^S as the master regulator at the top of a large regulatory pyramid, turning on a number of genes for more specialized activators that then respond to more individualized stress conditions.

With so much of the fate of *E. coli* in its hands, σ^S must be able to respond quickly to a number of different environmental signals. Transcriptional regulation is efficient but rather slow. While different stress conditions do affect *rpoS* transcription (see Magnusson et al., Suggested Reading), the levels of *rpoS* mRNA remain high throughout the exponential phase, indicating that most of the regulation occurs posttranscriptionally, either in the ability of the *rpoS* mRNA to be translated or in the stability of σ^S itself; we therefore concentrate on these.

One way in which translation of *rpoS* mRNA is regulated is through the action of a small RNA (sRNA) called DsrA, whose synthesis increases at low temperatures. DsrA stimulates *rpoS* mRNA translation by binding to the 5′ untranslated region of the *rpoS* mRNA with the help of a protein, Hfq (Figure 13.18). The *rpoS* mRNA normally contains a helix that sequesters the TIR (similar to the case for the *rpoH*) mRNA described above. Binding of DsrA to a region that would otherwise pair with the TIR opens up the mRNA secondary structure and exposes the TIR for translation initiation. Another sRNA, RprA, can substitute for DsrA under some conditions, suggesting redundancy in the regulation.

DsrA also acts as a regulator of other genes, including the *hns* gene, which encodes the H-NS global regulatory protein. H-NS plays a major role in repressing the expression of a variety of genes involved in growth and metabolism in *E. coli*, often by generating DNA loops that prevent transcription of its target genes. DsrA represses translation of the *hns* mRNA under the same conditions under which *rpoS* translation is enhanced, further amplifying the shift in gene expression under stress conditions. It is interesting to note that repression of *hns* mRNA translation utilizes a different region of the DsrA sRNA than that used to bind to and activate translation of the *rpoS* mRNA, so that DsrA acts as a compound sRNA with two different domains, each of which has different mRNA targets (Figure 13.18).

Levels of σ^S are also regulated through effects on the stability of the protein after it is made. While cells are growing exponentially, σ^S is being continuously made,

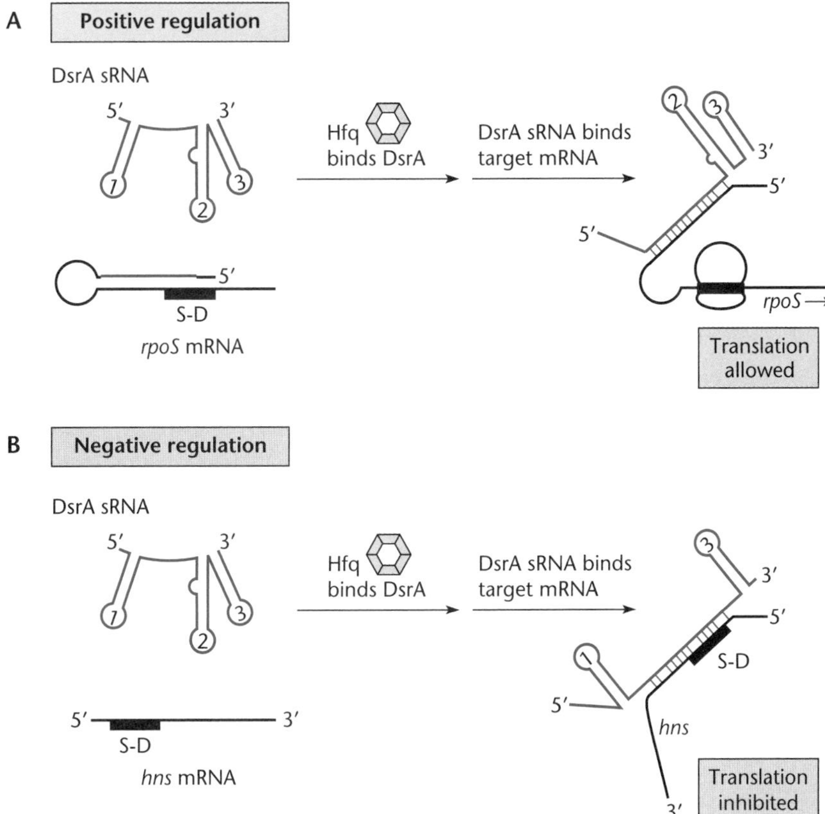

Figure 13.18 Repression and activation by the DsrA sRNA. **(A)** One domain (1) of the DsrA sRNA can bind to the 5′ region of the *rpoS* mRNA. This mRNA normally contains a structure that sequesters the Shine-Dalgarno (S-D) sequence and prevents *rpoS* translation. Binding of DsrA to the region of the RNA that would otherwise pair with the S-D sequence releases the S-D sequence and allows synthesis of σ^S. Binding requires the assistance of the Hfq protein. **(B)** A different domain (2) of DsrA binds to the *hns* mRNA (also with the help of Hfq) to repress translation by direct sequestration of the S-D sequence.
doi:10.1128/9781555817169.ch13.f13.18.

but very little of it accumulates, since it has a half-life of only 1 to 2 minutes (see chapters 2 and 12). However, when the cells run out of energy or are subjected to some other stress, the half-life of σ^S increases, leading to its rapid accumulation. The reason for this is well understood. During normal growth, the σ^S protein is degraded by the ClpXP protease, which consists of two proteins, a barrel-shaped chaperone made up of six copies of the ClpX protein that unfolds proteins, and the ClpP protease, which then degrades the unfolded protein (see chapter 2). However, ClpXP can degrade σ^S only if σ^S is bound to another protein, RssB. The RssB protein is therefore an adaptor protein that targets σ^S for degradation during normal growth. Exposure of cells to stress results in induction of the synthesis of the Ira antiadaptor proteins, which inactivate RssB and therefore allow σ^S to accumulate (see Figure 12.29). This is an example of regulated proteolysis, and in this case, even small changes in σ^S levels can have a dramatic effect on transcription.

Yet another posttranslational level of regulation is exerted by a different type of sRNA that, unlike DsrA, does not act by base pairing with its target (Box 13.6). The 6S RNA was discovered many years ago, but its function has been elucidated only recently. This RNA is a structural mimic of promoter DNA recognized by RNA polymerase containing σ^{70}, but it does not bind to RNA polymerase containing σ^S. 6S RNA accumulates as cells enter stationary phase and binds to and inactivates the normal RNA polymerase containing σ^{70} (see Wassarman, Suggested Reading). This then allows transcription by RNA polymerase containing σ^S to proceed more efficiently, as scarce resources are not being used in transcription of normal promoters. The combination of all of these effects results in an efficient reprogramming of transcription under stressful conditions, which enables the cell to respond most effectively to these conditions.

General Stress Response in Gram-Positive Bacteria

Many gram-negative bacteria are known to have a general stress response based on σ^S and probably similar to that of *E. coli*. However, gram-positive bacteria, including *B. subtilis*, use another sigma factor, σ^B, to transcribe stress-induced genes. This sigma factor is quite different from σ^S in sequence and is activated by a very different pathway. Its mechanism of activation has features in common with the activation of σ^E in *E. coli* (see below) and σ^F and σ^G in *B. subtilis* sporulation (see chapter 14) in that it depends on inactivation of an **anti-sigma factor**, a partner protein that otherwise binds to the sigma factor and prevents it from interacting with RNA polymerase to direct transcription. The signaling pathway

BOX 13.6

Regulatory RNAs

It is becoming increasingly evident that small regulatory RNAs (sRNAs) play a significant role in regulation in bacterial systems (see Storz et al., References). This mode of regulation is sometimes called "riboregulation" and can occur at many different levels. The majority of regulatory RNAs that have been identified function by base pairing with the RNA they regulate. This type of sRNA can be encoded on the strand of the DNA opposite that from which the target RNA is derived; these sRNAs are therefore *cis* encoded, because they come from the same region of the DNA, and their sequences are perfect matches to those of their RNA targets. sRNAs of this type can inhibit translation if they are complementary to the TIR of the target mRNA. Others can affect transcription termination, for example, in the pheromone-responsive plasmid transfer in *Enterococcus faecalis* (see Tomita and Clewell, References). Other *cis*-encoded sRNAs regulate plasmid replication, as discussed in chapter 4.

Most sRNAs are not encoded in the same DNA region as their targets. These sRNAs are referred to as *trans* encoded, and they are not perfectly complementary to their targets. In addition, they often have multiple target genes. Some of these sRNAs are mentioned in the text; they include the DsrA, MicF, and RyhB RNAs, which regulate *rpoS*, *ompF*, and the genes for iron-containing proteins, respectively. They are usually highly regulated and allow another level of regulation of the genes they control. By binding to their RNA targets, they can regulate gene expression in many ways. For example, they can inhibit translation by binding close to the TIR of an mRNA and blocking access by the ribosome to the TIR (e.g., MicF repression of *ompF*), or they can stimulate translation by binding close to the TIR and melting a secondary structure that includes the TIR (DsrA activation of *rpoS*). They can also target the mRNA for degradation by a cellular RNase (RyhB). For these *trans*-encoded sRNAs, the regions of complementarity are short and interrupted by mismatches. Because of such short interactions, a single antisense RNA may be able to regulate more than one target gene, with different regions of the sRNA base pairing with the various target sequences (e.g., DsrA regulation of *rpoS* and *hns*).

Many *trans*-encoded sRNAs require a protein, Hfq, to bind to their target RNAs (see Valentin-Hansen et al., References). This protein was first found as a host-encoded protein in the phage Qβ replicase that is required to replicate the genomic RNA of this small phage, hence its name, Hfq (for host factor Qβ). This protein is found in many bacteria and is homologous to the Sm RNA-binding proteins involved in

Figure 1 doi:10.1128/9781555817169.ch13.Box13.6.f1

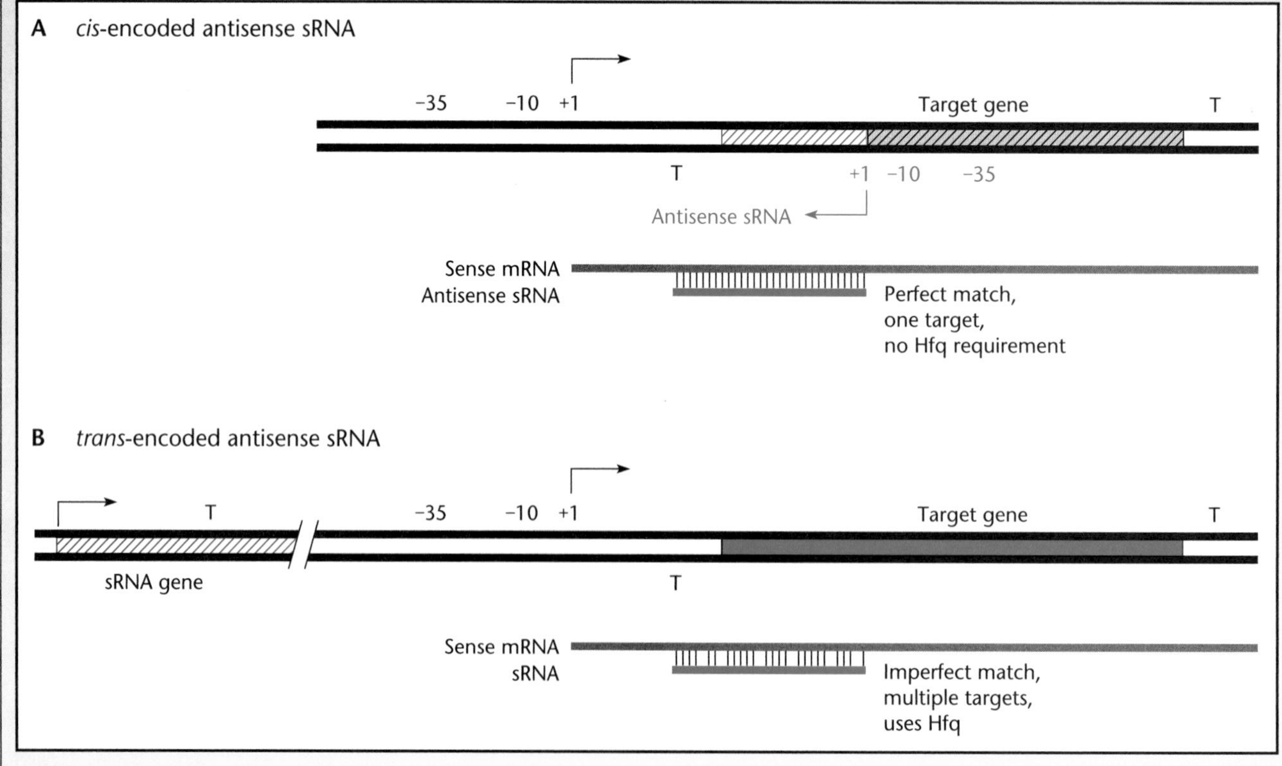

BOX 13.6 (continued)

Regulatory RNAs

RNA splicing in eukaryotes. Six polypeptide products of the *hfq* gene form a ring (a hexameric ring). The protein helps the sRNA bind to a specific region of the target RNA, even though there is very little complementary base pairing to hold them together. Hfq may also act as an RNA chaperone to hold the sRNA and mRNA in the proper conformation to allow them to interact with each other.

A very different class of sRNAs do not base pair with their regulatory targets. Instead, these sRNAs bind to proteins and regulate their activity. One example is provided by the CsrB family of sRNAs, which exhibit a repeated structure with multiple small helical domains. These domains mimic the binding site of an RNA-binding repressor protein called CsrA, which was first identified because of its role in the production of glycogen, which is made by *E. coli* as a carbon storage compound (see Babitzke and Romeo, References). The genes that CsrA controls are normally repressed by binding of CsrA to the target mRNA, which stabilizes a helical domain that sequesters the TIR of the mRNA. When the CsrB sRNAs are produced, they bind multiple copies of CsrA protein, titrating

Figure 2 doi:10.1128/9781555817169.ch13.Box13.6.f2

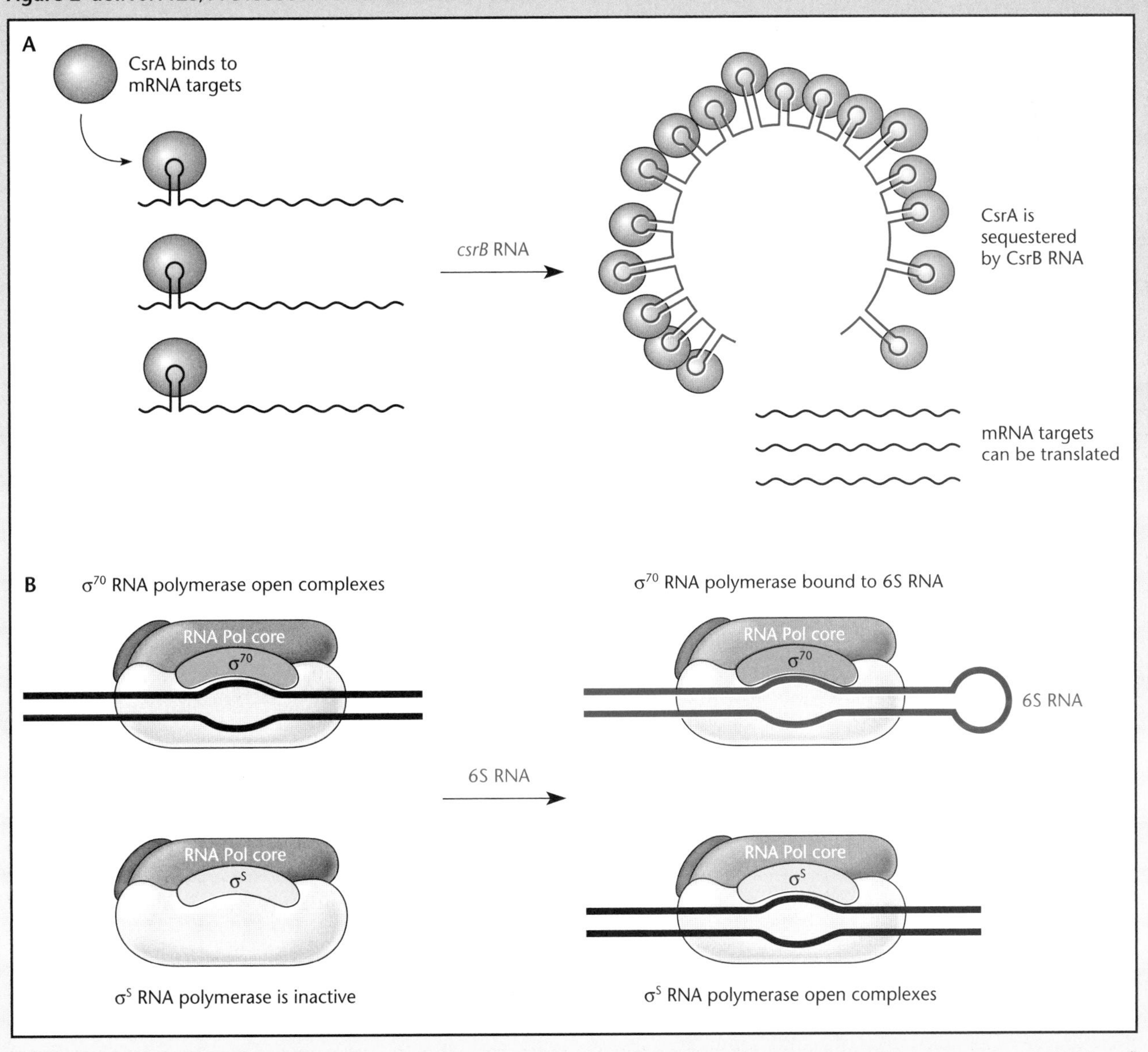

(continued)

BOX 13.6 (continued)

Regulatory RNAs

the protein so that the target mRNAs can be translated. Similar systems have now been uncovered in many bacteria (including *V. cholerae* [see the text]), where they control a variety of systems, including virulence and biofilm formation.

Another unusual sRNA is the 6S RNA, which binds to RNA polymerase containing σ^{70} in stationary phase and inhibits its activity. As described in the text, 6S RNA is a structural mimic of open-complex DNA, and inhibition of RNA polymerase containing σ^{70} enhances the transition to transcription by RNA polymerase containing σ^{S} during stationary phase (see Wassarman, References).

The first sRNAs were found by accident, for example, when they inhibited the synthesis of a gene product when they were overproduced from a multicopy plasmid. None of them were found in classical genetic analyses, perhaps because these genes are small and therefore are small targets for mutagenesis or because they often have redundant functions, so that inactivating mutations have no obvious phenotypes. Now, however, new sRNAs can be found by analysis of the genomic sequences of bacteria. Using such methods, over 100 sRNAs have been found in *E. coli*. The sRNAs are often encoded in intergenic regions (the regions between genes) and can sometimes be recognized because they have consensus promoters and transcription terminators but lack an obvious open reading frame encoding a protein between them. Alternatively, they might be recognized as conserved sequences in intergenic regions by a comparison of the genomes of closely related bacteria or by special types of genomics analyses, including tiling microarrays and high-throughput sequencing of cDNAs generated from total RNA isolated from the cell (a technique called RNASeq [Box 13.7]). For example, sRNAs have been found for *E. coli* by comparing the *E. coli* genome to the genomes of other closely related bacteria, *K. pneumoniae* and *Salmonella* spp., using a combined computational and microarray approach (see Wassarman et al., References). The reasoning is that the sequences of intergenic regions, which normally differ, are conserved if they encode important regulatory sRNAs. Another approach uses the Hfq protein to fish sRNAs out of the RNA pool from cells, since many of them bind to Hfq. Any approach that relies on isolation of the RNA pool from the cell is limited by the fact that only sRNAs that are expressed under the conditions under which the cells were grown will be present in the pool. These approaches are currently being applied to a growing number of bacterial genomes, and the interesting task is to identify the targets of these sRNAs and their regulatory roles.

References

Babitzke, P., and T. Romeo. 2007. CsrB sRNA family: sequestration of RNA-binding regulatory proteins. *Curr. Opin. Microbiol.* **10:**156–163.

Storz, G., J. Vogel, and K. M. Wassarman. 2011. Regulation by small RNAs in bacteria: an expanding frontier. *Mol. Cell* **43:**880–891.

Tomita, H., and D. B. Clewell. 2000. A pAD1-encoded small RNA molecule, mD, negatively regulates *Enterococcus faecalis* pheromone response by enhancing transcription termination. *J. Bacteriol.* **182:**1062–1073.

Valentin-Hansen, P., M. Eriksen, and C. Udesen. 2004. The bacterial Sm-like protein Hfq: a key player in RNA transactions. *Mol. Microbiol.* **51:**1525–1533.

Wassarman, K. M. 2007. 6S RNA: a small RNA regulator of transcription. *Curr. Opin. Microbiol.* **10:**164–168.

Wassarman, K. M., F. Repoila, C. Rosenow, G. Storz, and S. Gottesman. 2001. Identification of novel small RNAs using comparative genomics and microarrays. *Genes Dev.* **15:**1637–1651.

that activates σ^B is fairly long and complicated, using a phosphorelay system involving serine-threonine kinases and phosphatases that is more reminiscent of eukaryotes than it is of bacteria. Also, there seems to be one pathway to sense energy deficiency when the cells run out of a carbon and energy source and a different pathway to sense an environmental stress, such as heat shock or a pH change.

Recent research has concentrated on how environmental signals are communicated to the first step of the pathway (see Kim et al., Suggested Reading). A total of five periplasmic proteins called the Rsb proteins were known to be involved in the induction of the stress response. They all seem to be part of the same large complex (the stressosome) that senses external stresses and induces the stress response by activating the signal transduction system that activates σ^B. All five Rsb proteins are similar in their carboxyl termini, but one of them, RsbS, is shorter, consisting of only the shared carboxyl terminus. These proteins are all serine or threonine kinases, and phosphate groups are added to or removed from them in response to external stresses. Each of the longer Rsb proteins (RsbRA, RsbRB, RsbRC, and RsbRD) responds to a different but overlapping external stress, and they all phosphorylate RsbS in response to their particular stress. Phosphorylation of RsbS triggers a cascade of events that involve phosphorylation or dephosphorylation of a series of regulatory proteins. The final outcome of this regulatory cascade is the release of the σ^B sigma factor from an anti-sigma factor that holds it in an inactive state under normal growth conditions. RNA polymerase containing σ^B directs the transcription of genes whose products help the cell to recover from all of the various stress conditions to which the Rsb proteins respond.

Extracytoplasmic (Envelope) Stress Responses

The membranes of bacteria are the first line of defense against external stresses. They are also particularly sensitive to abrupt changes in osmolarity and damaging agents, such as hydrophobic toxins, heat shock, and pH changes. Not surprisingly, many stress responses are dedicated to preserving the integrity of the bacterial membranes. Responses of this type are often referred to as extracytoplasmic stress responses because they respond to changes outside of the cytoplasm.

REGULATION OF PORIN SYNTHESIS

One of the challenges often faced by bacteria is a change in osmolarity due to changing solute concentrations outside the cell. The osmotic pressure is normally higher inside the cell than outside it. This pressure would cause water to enter the cell and the bacterium to swell, but the rigid cell wall can help mitigate this by keeping the cell from expanding. However, even the cell wall is not invincible, and bacteria must keep the difference in osmotic pressure inside and outside the cell from becoming too great. The ability to monitor osmolarity can be important to bacteria for a second reason: bacteria also sometimes sense changes in their external environment by detecting changes in osmolarity. In fact, one way in which pathogenic bacteria sense that they are inside a host, and induce their virulence genes, is by the much higher osmolarity inside the host (see "Regulation of Virulence Genes in Pathogenic Bacteria" below). The systems by which bacterial cells sense these changes in osmolarity and adapt are global regulatory mechanisms, and many genes are involved.

Much is known about how bacteria respond to media with different osmolarities. One way they regulate the differences in osmotic pressure across the membrane is by excreting or accumulating K ions and other solutes, such as proline and glycine betaine. Gram-negative bacteria, such as *E. coli*, have the additional problem of maintaining osmotic pressure in the periplasm, as well as in the cytoplasm. They achieve this in part by synthesizing oligosaccharides in the periplasmic space to balance solutes in the external environment.

One of the major mechanisms by which gram-negative bacteria balance osmotic pressure across the outer membrane is by synthesizing pores to let solutes into and out of the periplasmic space. These pores are composed of outer membrane proteins called **porins**. To form pores, three of the polypeptide products of these genes come together (trimerize) in the outer membrane to form what are called β barrels with central channels that selectively allow hydrophilic molecules through the very hydrophobic outer membrane.

The two major porin proteins in *E. coli* are OmpC and OmpF. Pores composed of OmpC are smaller than those composed of OmpF, and the sizes of the pores can determine which solutes can pass through the pores and thus confer protection under some conditions. For example, the smaller pores, composed of OmpC, may prevent the passage of some toxins, such as the bile salts in the intestine. The larger pores, composed of OmpF, may allow more rapid passage of solutes and so confer an advantage in dilute aqueous environments. Accordingly, *E. coli* cells growing in a medium of high osmolarity, such as the human intestine, have more OmpC than OmpF, whereas *E. coli* cells growing in a medium of low osmolarity, such as dilute aqueous solutions, have less OmpC than OmpF.

Other environmental factors besides osmolarity can alter the ratio of OmpC to OmpF. This ratio increases at higher temperatures or pH or when the cell is under oxidative stress due to the accumulation of reactive forms of oxygen. The ratio also increases when the bacterium is growing in the presence of organic solvents, such as ethanol, or some antibiotics and other toxins. Presumably, the smaller size of OmpC pores limits the passage of many toxic chemicals into the cell. Many of these abrupt changes in porin proteins occur when the *E. coli* bacterium leaves the external environment and passes through the stomach into the intestine of a warm-blooded vertebrate host, its normal habitat. It then must synthesize mostly OmpC-containing pores to keep out toxic materials, such as bile salts, as mentioned above. Other conditions cause a decrease in both OmpC and OmpF concentrations. To respond to all of these other changes, the *ompC* and *ompF* genes are in a number of different regulons, which respond to different external stresses. We first discuss one of these pathways in *E. coli*, the regulation of *ompC* and *ompF* expression by EnvZ and OmpR. While not yet completely understood, this system has served as a model for two-component signal transduction systems that allow the cell to sense the external environment and adjust its gene expression accordingly; therefore, we discuss this subject in some detail.

GENETIC ANALYSIS OF PORIN REGULATION

As in the genetic analysis of any regulatory system, the first step in studying the osmotic regulation of porin synthesis in *E. coli* was to identify the genes whose products are involved in the regulation. The isolation of mutants defective in the regulation of porin synthesis was greatly aided by the fact that the some of the porin proteins also serve as receptors for phages and bacteriocins, so that mutants that lack a particular porin are resistant to a given phage or bacteriocin. This offers an easy selection for mutants defective in porin synthesis, as only mutants that lack a certain porin in the outer membrane are able to form colonies in the presence of the corresponding phage or bacteriocin.

Using such selections, investigators isolated mutants that had reduced amounts of the porin protein OmpF in their outer membranes. These mutants were found to have mutations in two different loci, which were named *ompF* and *ompB*. Mutations in the *ompF* locus can completely block OmpF synthesis, whereas mutations in *ompB* only partially prevent its synthesis. The quantitative difference in the effects of mutations in the two loci suggested that *ompF* is the structural gene for the OmpF protein and that the *ompB* locus is required for the expression of the *ompF* gene. Complementation experiments showed that the *ompB* locus actually consists of two genes, *envZ* and *ompR*. Using *lacZ* fusions to *ompF* to monitor the transcription of the *ompF* gene (see chapter 2), investigators confirmed that EnvZ and OmpR are required for optimal transcription of the *ompF* gene (see Hall and Silhavy, Suggested Reading).

EnvZ and OmpR: a Sensor Kinase and Response Regulator Partnership

The *envZ* and *ompR* genes were cloned and sequenced by methods such as those discussed in chapter 1. Similarities in amino acid sequence between EnvZ and OmpR and other sensor kinase and response regulator pairs, including NtrB and NtrC, suggested that these proteins are also a sensor kinase and response regulator pair of proteins. Like many (but not all) sensor proteins, EnvZ is an inner membrane protein, with its NTD in the periplasm and its CTD in the cytoplasm (Box 13.4). The NTD of EnvZ apparently senses an unknown signal in the periplasm that reflects the osmolarity and transfers this information to the cytoplasmic domain. The information is then transferred to the OmpR protein, a transcriptional regulator that regulates transcription of the porin genes. Like the NtrB protein, the EnvZ protein is autophosphorylated, and its phosphate is transferred to OmpR. This led to a model in which EnvZ is more heavily phosphorylated at high osmolarity and the phosphate is then transferred to OmpR. If levels of OmpR~P are high, transcription of *ompC* is activated; if levels of OmpR~P are low, transcription of *ompF* occurs.

Mutations in *envZ* and *ompR*

Table 13.3 lists several relevant phenotypes of *envZ* and *ompR* mutations. Mutations that totally inactivate the EnvZ protein (*envZ* null mutations) should completely prevent the phosphorylation of OmpR under any conditions. Therefore, *envZ* null mutations would be predicted to completely prevent the transcription of *ompC*. Both *ompR* and *envZ* null mutations prevent the transcription of *ompC* and also allow only very limited transcription of *ompF* (shown as OmpF$^{\pm}$ in Table 13.3; see Slauch et al., Suggested Reading). This indicates that the EnvZ protein, and presumably OmpR~P, is required to activate the transcription of both porin genes. Constitutive mutations in *ompR* proved more difficult to explain. One type of constitutive mutation, called *ompR2*(Con), prevents the expression of *ompC* but causes the constitutive expression of *ompF*, even when combined with a null mutation in *envZ*. A second type of constitutive mutation, called *ompR3*(Con), causes constitutive expression of *ompC* but prevents the expression of *ompF*. When the two mutations are combined in partial diploids, *ompC* is constitutively expressed but *ompF* is not expressed at all.

Table 13.3 Phenotypes of *envZ* and *ompR* mutations

Genotype	Phenotype
envZ$^+$ *ompR*$^+$	OmpC$^+$ OmpF$^+$
envZ$^+$ *ompR1*	OmpC$^-$ OmpF$^-$
envZ(null) *ompR*$^+$	OmpC$^-$ OmpF$^{\pm a}$
envZ$^+$ *ompR2*(Con)	OmpC$^-$ OmpF$^+$ (low osmolarity)
	OmpC$^-$ OmpF$^+$ (high osmolarity)
envZ(null) *ompR2*(Con)	OmpC$^-$ OmpF$^+$ (low osmolarity)
	OmpC$^-$ OmpF$^+$ (high osmolarity)
envZ$^+$ *ompR3*(Con)	OmpC$^+$ OmpF$^-$ (low osmolarity)
	OmpC$^+$ OmpF$^-$ (high osmolarity)
envZ$^+$ *ompR3*(Con)/*envZ*$^+$ *ompR*$^+$	OmpC$^+$ OmpF$^-$ (low osmolarity)
	OmpC$^+$ OmpF$^-$ (high osmolarity)

^{a}±indicates that OmpF levels are reduced but not eliminated.

The Affinity Model for Regulation of *ompC* and *ompF*

The phenotypes of the *envZ* and *ompR* mutants can be explained by a model that takes into account the concept that phosphorylation of OmpR is a reversible event and that the crucial parameter is the concentration of OmpR~P in the cell. Like many sensor kinases, the EnvZ protein is known to have both phosphotransferase and phosphatase activities, allowing it to both donate a phosphoryl group to and remove one from OmpR. Whether the phosphorylated or unphosphorylated form of OmpR predominates depends on whether the phosphotransferase or phosphatase activity of EnvZ is most active. Under conditions of high osmolarity, the phosphotransferase activity predominates and the levels of OmpR~P are high. Under conditions of low osmolarity, the phosphatase activity predominates, and most of the OmpR is unphosphorylated.

To explain how *envZ* null mutations can prevent the optimal transcription of both genes, we must propose that OmpR~P is required to activate transcription of both the *ompC* and *ompF* genes. However, how could higher levels of OmpR~P (under high-osmolarity conditions) favor the transcription of *ompC* while lower levels of OmpR~P (under low-osmolarity conditions) favor the transcription of *ompF* even though both require OmpR to be phosphorylated? One model is based on the existence of multiple binding sites for OmpR~P upstream of the promoters for the *ompF* and *ompC* genes (see Yoshida et al., Suggested Reading). Some of these sites bind OmpR~P more tightly than others; in other words, some are high-affinity sites, whereas others are low-affinity sites. By binding to these sites, the OmpR~P protein can either activate or repress transcription from the promoters, depending on the position of the binding site relative to the promoter (see chapters 2 and 12). The presence of low-affinity sites at an activating position upstream of the *ompC* promoter would result in transcription only when OmpR~P levels are high, which is what was observed. The dependence of *ompF* transcription on some amount of OmpR~P further suggests that the *ompF* gene should contain high-affinity sites at an activating position so that activation of the promoter occurs when levels of OmpR~P are low. However, the observation that *ompF* is not transcribed when levels of OmpR~P are high suggests the additional presence of low-affinity binding sites at a repressive position in the *ompF* promoter that would result in inhibition of transcription when OmpR~P levels are high, which fits the observed pattern of expression.

The prediction that high OmpR~P levels repress *ompF* transcription also explains the phenotype of the *ompR* constitutive mutations: the *ompR2*(Con) mutations, which result in constitutive *ompF* expression but no *ompC* expression, are predicted to result in constitutive low levels of OmpR~P; the *ompR3*(Con) mutations, which result in constitutive *ompC* expression, are predicted to result in constitutive high levels of OmpR~P; and the *ompR2*(Con) *ompR3*(Con) partial diploids behave like *ompR3*(Con) because the diploid has high levels of OmpR~P. Further studies have indicated that regulation of this system is even more complicated, as we will see below, but this model illustrates how a combination of binding sites with different affinities for a regulatory protein can be combined with differential positioning of the binding sites relative to the promoter to give different regulatory outcomes.

REGULATING OmpF BY THE MicF sRNA

As mentioned above, the OmpC/OmpF ratio increases, not only when the osmolarity increases, but also when the temperature or pH increases or when toxic chemicals, including active forms of oxygen, nitric oxide, or organic solvents, such as ethanol, are in the medium. In general, under conditions where levels of nutrients and toxins are high, such as in the vertebrate intestine, OmpC levels are high. Since OmpC forms narrower channels, fewer toxins can get in. Fewer nutrients can get in as well, but since their concentration in the intestine is high, this is not a problem. If nutrient levels are low, such as in water outside the vertebrate host, OmpF levels are higher, because allowing the available nutrients to get into the cell becomes more important than keeping toxins out. The rationale for regulating the porins by temperature or pH is less obvious. One possibility is that the bacterium uses temperature and pH as signals to indicate that the water in which it lives has just been drunk by a vertebrate and the bacterium is about to pass into the vertebrate intestine. The porins would then need to be regulated to combat the onslaught of toxins that will be faced by the bacterium, including oxidative bursts by macrophages and bile salts.

Much of the regulation of the OmpF porin in response to these other forms of stress is through the MicF sRNA. The mechanisms of action of regulatory sRNAs, including DsrA, the sRNA that regulates the translation of σ^S in *E. coli* (see above), are discussed in chapter 12 and Box 13.6. The MicF RNA was one of the first sRNAs to be discovered, and it was found by chance because its gene is adjacent to the *ompC* gene but is transcribed in the opposite direction (i.e., *ompC* and *micF* are divergently transcribed). When the region of the chromosome containing the *ompC* gene was cloned into a high-copy-number plasmid and introduced into *E. coli* cells, the synthesis of OmpF was inhibited. At first, it was assumed that the OmpC protein was somehow inhibiting the synthesis of the OmpF protein. However, it was observed that the OmpC coding region was not required for inhibition, which instead required an sRNA encoded just upstream of the *ompC* gene. The sRNA was named MicF for (*m*ulticopy *i*nhibitor of Omp*F*). Later, it was shown that part of MicF is complementary to the 5′ region of the mRNA for *ompF*, including the *ompF* TIR, and pairing of MicF with the *ompF* mRNA inhibits its translation. As is the case for many sRNAs that are only partially complementary to their mRNA targets, repression requires the Hfq RNA chaperone (Box 13.6). High concentrations of MicF sRNA therefore result in lower levels of OmpF synthesis.

The cellular levels of MicF sRNA increase under certain conditions because the promoter for the *micF* gene contains binding sites for many transcriptional activators. The activators seem to work independently, and each activates transcription from the *micF* promoter under its own particular set of conditions. For example, the transcriptional activator SoxS activates transcription from the *micF* promoter when the cell is under oxidative stress. Another activator, MarA, activates the transcription of *micF* when weak acids or some antibiotics are present. A third activator, Rob, induces transcription of *micF* in the presence of cationic peptide antibiotics. OmpR~P, which binds upstream of the *ompC* promoter, can also activate *micF* transcription; the dependence of this activation on high levels of OmpR~P (conditions under which *ompF* transcription is repressed) allows MicF to rapidly inactivate any *ompF* mRNA that already is present in the cell, resulting in faster shutoff of OmpF synthesis.

The MicF sRNA inhibits the translation of OmpF, but not OmpC. We might expect that a similar regulation would apply to OmpC so that less of it would be made under conditions that favor OmpF. In fact, OmpC translation is inhibited by its own sRNA, named MicC by analogy to the MicF sRNA. The MicC sRNA was discovered because it has some sequence complementarity to the 5′ region of *ompC* mRNA and overexpression of the *micC* gene was shown to inhibit OmpC trasnslation. Interestingly, the *micC* gene is also adjacent to a gene for a porin, in this case OmpN, which is very poorly expressed in *E. coli*, at least under laboratory conditions, but might be expected to form pores with sizes similar to those of OmpF. It is likely that OmpA, yet another porin, is also regulated by an sRNA. The regulation of porin synthesis by osmolarity and other environmental factors is obviously central to cell survival, which is why it is so complicated and involves so many interacting systems and regulatory molecules.

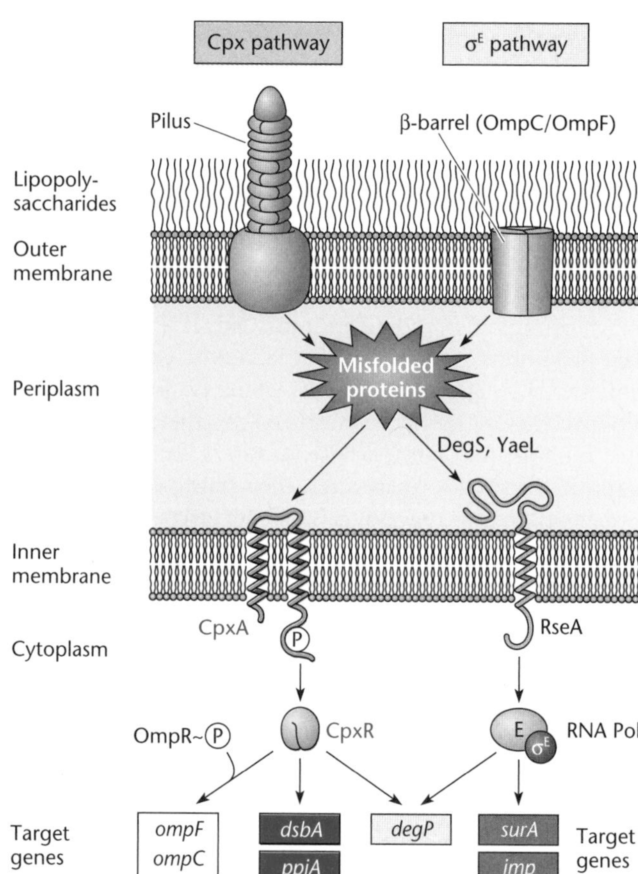

Figure 13.19 Two envelope stress responses in *E. coli* that respond to different stress signals in the periplasm. The Cpx pathway responds to the accumulation of pilin subunits in the periplasm, while the σ^E pathway responds to the accumulation of OmpA. DsbA is a periplasmic chaperone and oxidoreductase that makes disulfide bonds in exported proteins (see chapter 14), PpiA is a peptidyl-prolyl *cis-trans* isomerase, DegP is a protease and chaperone, SurA is a chaperone, and Imp is an usher protein for outer membrane proteins. doi:10.1128/9781555817169.ch13.f13.19

REGULATION OF THE ENVELOPE STRESS RESPONSE BY THE CpxA-CpxR TWO-COMPONENT SYSTEM

Another way that *E. coli* senses stress to the outer membrane is via the two-component system CpxA-CpxR (Figure 13.19). This two-component system works like many two-component systems (Box 13.4) in that CpxA is a sensor kinase that phosphorylates itself in response to a signal and then transfers the phosphate to the response regulator CpxR, a transcriptional activator that activates the transcription of more than 100 genes under its control. The CpxA sensor kinase phosphorylates itself when it senses that proteins to be secreted, such as pili or curli fibers that play a role in attaching the bacterium to surfaces such as eukaryotic cells, are piling up in the periplasm as a result of some defect in transport due to damage to the outer membrane (see Ruiz and Silhavy, Suggested Reading). The accumulation of these proteins in the periplasm is toxic, and many of the genes regulated by CpxR encode proteases and chaperones in the periplasm that help fold and degrade these proteins. The synthesis of the proteins that make up cellular appendages, such as pili and curli fibers, is also inhibited by phosphorylated CpxR, perhaps so that their intermediates will not accumulate in the periplasm.

The finding that the CpxA-CpxR two-component system also regulates the ratio of OmpC to OmpF came from a genetic screen for mutations that affect the ratio of OmpF to OmpC (see Batchelor et al., Suggested Reading). Since this analysis used some of the techniques described in previous chapters, it is discussed in some detail here. First, the investigators needed an easy way to determine if the ratio of OmpF to OmpC had been altered, since they would have to screen many mutants. To do this, they used transcriptional fusions that fused two different derivatives of the *gfp* reporter gene, called *yfp* and *cfp*, to the promoters for *ompF* and *ompC*. Fluoresence by the original green fluorescent protein (GFP) emits green light, while the mutant forms emit different colors, yellow and cyan (blue), respectively. If the *ompF* promoter is more active, the colonies fluoresce more yellow, whereas if the *ompC* promoter is more active, the colonies fluoresce more blue. The investigators constructed a strain containing both fusions by inserting the fusions into lysogenic phages and then integrating the phages into the chromosome so that the fusions would be present in only one copy.

The next step was to mutagenize their strain to try to obtain mutants that were altered in the relative expression of *ompF* and *ompC*. They used transposon mutagenesis because it was easier to characterize the resulting mutations (see chapter 9). About 5,000 transposon insertion mutants were plated, and the colors of their colonies were observed. One mutant in particular exhibited a dramatic decrease in *ompF* transcription and a modest increase in *ompC* transcription, especially in glucose minimal medium, and the transposon was determined to be in the *cpxA* gene.

To be certain that the transposon insertion was inactivating the *cpxA* gene product and that this was the basis for the relative effects on *ompC* and *ompF* expression, the investigators used recombineering (see chapters 10 and 12) to replace the entire *cpxA* gene in a different strain with a chloramphenicol resistance gene cassette and observed the same phenotypes with this strain. They also directly measured the amounts of the OmpF and OmpC proteins in the mutant cells to ensure that the amount of fluorescence accurately reflected the amount of protein product made from the native genes. It is easy to detect OmpF and OmpC on stained sodium dodecyl sulfate-polyacrylamide gels because they are two of the most abundant proteins in *E. coli*, making their bands clearly visible among all the other protein bands.

Since CpxA and CpxR were already known to act as partners in a two-component signal transduction system, the effect of inactivating both the *cpxR* gene and the *cpxA* gene in the same bacterium on *ompC* and *ompF* expression was investigated. We might expect that inactivating CpxR when CpxA had already been inactivated would have no effect, since the only known role of CpxA is to transfer its phosphate to CpxR. However, when both *cpxA* and *cpxR* were inactivated, OmpF and OmpC were made in normal amounts. One possible explanation is that the altered phenotypes in a *cpxA* mutant were due to a high level of phosphorylation of CpxR, rather than to no phosphorylation of CpxR. Like EnvZ and other sensor kinases, the CpxA protein is both a phosphotransferase, which attaches a phosphate to CpxR, and a phosphatase, which removes a phosphate from CpxR. In the absence of the phosphatase, phosphates might accumulate on CpxR as a result of another phosphorylating enzyme that transfers phosphates to many proteins. The investigators concluded that high levels of CpxR with phosphate attached (CpxR~P) repress *ompF* transcription but stimulate *ompC* transcription. The role of the CpxA-CpxR system is to detect damage in the periplasm that results in activation of CpxA and accumulation of CpxR~P. Repression of *ompF* and activation of *ompC* are likely to be mediated by the action of OmpR~P, as described above. The net result is that fewer toxins can get into the periplasm because more of the OmpC porin, which has a smaller pore than OmpF, is made. In this way, the composition of the porins can respond both to changes in osmolarity (via EnvZ and OmpR) and to the presence of toxins or proteins in the periplasm (via CpxA, CpxR, and OmpR). An open question is how activation of CpxA results in accumulation of OmpR~P and how inactivation of the *cpxA* gene results in constitutive accumulation of CpxR~P (and presumably OmpR~P).

THE EXTRACYTOPLASMIC SIGMA FACTOR σ^E

In *E. coli*, extracytoplasmic stress is also sensed by another system that uses a specialized sigma factor to transcribe stress response genes (see Alba and Gross, Suggested Reading). This alternative sigma factor is called the extracytoplasmic function (ECF) sigma factor, σ^E, because it is used mostly to express genes that function in the periplasm, such as proteases that degrade defective proteins in the periplasm and chaperones that help fold proteins as they pass through the periplasm. It was first discovered because one of the promoters that direct transcription of the *rpoH* gene for the heat shock sigma factor at high temperatures is recognized by this sigma factor. Sigma factors in this family are generally referred to as σ^{ECF}, and some bacteria have multiple σ^{ECF} family members.

The activation of σ^E in *E. coli* is similar to the activation of the σ^B general stress response sigma factor in *B. subtilis* in that it involves inactivation of an anti-sigma factor that holds σ^E in an inactive state in the absence of stress conditions. In this case, the activity of the σ^E anti-sigma factor is controlled by proteolytic degradation in response to envelope stress. This aspect of the induction of the envelope stress response is quite well understood and is illustrated in Figure 13.19. The anti-sigma factor RseA is an inner-membrane-spanning protein with domains in both the periplasm and the cytoplasm. The cytoplasmic domain binds σ^E, inactivating it and sequestering it to the membrane. When the outer membrane is damaged, Omp proteins, including OmpC, accumulate in the periplasm because they cannot be assembled into porins in the damaged outer membrane. The carboxyl-terminal domain of the Omp proteins that have accumulated in the periplasm binds to the carboxyl-terminal domain of a protease called DegS in the periplasm. The carboxyl terminus of DegS may normally inhibit its protease activity, and binding of the carboxyl terminus of an Omp protein to DegS activates the proteolytic activity of DegS, which then cleaves off the periplasmic domain of RseA. A second protease, named YaeL, then degrades the transmembrane domain of RseA, which releases σ^E from RseA and allows σ^E to bind RNA polymerase and transcribe genes in the envelope stress response.

There is experimental evidence for each of these steps. Mutations that inactivate the *rpoE* gene for σ^E are lethal at any temperature, showing that the σ^E protein is essential for viability at any temperature. It is not clear why σ^E is essential, even in the absence of envelope stress conditions, but some of the periplasmic proteins under its control, such as protein chaperones in the periplasm that help insert other proteins into the outer membrane, may be essential. This suggests that some level of σ^E-dependent transcription is likely to occur under nonstress conditions. Mutations that inactivate *rseA* cause constitutive induction of the σ^E response, as expected, since there is no anti-sigma factor to inhibit σ^E even in the absence of stress. Mutations that inactivate DegS and YaeL are lethal, but double mutants with both an *rseA* mutation and either a *degS* or *yaeL* mutation are viable. In genetic terms, *rseA* mutations are suppressors of *degS* and *yaeL* mutations. This shows that the only essential role for DegS and YaeL is to degrade RseA and, by extension, induce the σ^E stress response. Again, since inactivation of RseA is required for viability even under nonstress conditions, it is likely that some level of σ^E-dependent transcription is essential. The role of the carboxyl terminus of DegS in inhibiting its own protease activity was found when the region of the *degS* gene encoding the carboxyl terminus was deleted, which led to constitutive activation of σ^E. Finally, the role of the carboxyl termini of the Omp proteins in activating the protease activity of DegS was discovered when it was shown that overproducing the carboxyl termini of these proteins was sufficient to induce σ^E but did not increase the induction further if the carboxyl terminus of DegS had been deleted. It is important to note that activation of σ^E is reversible. When the stress responsible for activation of DegS has been resolved, newly synthesized RseA will not be cleaved by DegS and YaeL, and it will sequester and inactivate σ^E, resulting in turnoff of the transcription of σ^E-dependent genes.

The combination of the detection of environmental signals by the EnvZ-OmpR two-component system, the detection of defects in protein assembly into the outer membrane and/or periplasmic damage caused by toxins by the CpxA-CpxR two-component system, and the detection of accumulation of damaged or misplaced (notably Omp) proteins in the periplasm by the DegS protease, which in turn activates σ^E, allows the cell to detect a variety of effects that require modification of the outer cell surface. The resulting changes in gene expression include modification of the ratios of different porins (e.g., OmpF versus OmpC) that modulate the rate of transport into and out of the cell. Synthesis of chaperones and proteases that act specifically in the periplasm to facilitate outer membrane assembly is also induced (Figure 13.19), which allows the cell to repair the damage and restore outer membrane and periplasmic functions. The complexity and interconnection of these responses reflects the importance of the outer cell surface to cell integrity and function. The presence of multiple σ^{ECF} family members in some organisms further illustrates the importance of this mode of regulation.

Iron Regulation in *E. coli*

Iron is an important nutrient, both for bacteria and for humans (see Kadner, Suggested Reading). Many enzymes use iron as a catalyst in their active centers; many transcriptional regulators, such as FNR, which regulates genes for anaerobic metabolism, use it as a sensor of oxygen levels, and hemes use it as an oxygen carrier. However, too much iron can also be very damaging to cells and requires a stress response. Iron catalyzes the conversion of hydrogen peroxide and other reactive forms of oxygen to hydroxyl free radicals, the most mutagenic form of oxygen (see Box 11.1). It is therefore essential to maintain an adequate supply of iron without accumulation of excess levels.

Iron exists in two states in the environment, the ferric (Fe^{3+}) state, with a valence of three, and the ferrous (Fe^{2+}) state, with a valence of two. Iron in the ferric state forms largely insoluble compounds that are not easily used by bacteria and other organisms. Because oxygen quickly converts iron in the ferrous state to iron in the ferric state, most iron in aerobic environments exists in the insoluble ferric state. Accordingly, many bacteria secrete proteins called siderophores that bind ferric ions and transport them into the cell, where they can be converted to ferrous ions in the reducing atmosphere of the cytoplasm. These siderophores are made and secreted only if iron is limiting, to avoid the damaging effects of too much iron in the cell. Iron limitation is especially important for bacterial pathogens, as discussed below.

There are three basic mechanisms in *E. coli* and most other bacteria for regulating genes involved in iron metabolism: the Fur system, which uses a DNA-binding repressor protein that regulates transcription initiation; an sRNA called RyhB, which is regulated by Fur and promotes mRNA degradation; and the aconitase enzyme of the tricarboxylic acid (TCA) cycle, which is an iron-binding protein and doubles as a translational repressor. We discuss the Fur regulon first.

The Fur Regulon

The Fur repressor is a classical repressor with a helix-turn-helix DNA-binding domain in its amino terminus (see Box 12.1), a dimerization domain in its carboxyl terminus, and an effector-binding pocket in the middle. Figure 13.20 illustrates regulation by Fur, which is much like the regulation of the *trp* operon by the TrpR repressor (see chapter 12). When ferrous iron is in excess, it acts as a corepressor by binding to the Fur aporepressor, changing its conformation to the repressor form, which can bind to an operator sequence called a Fur box. By binding to the Fur box, which overlaps the −10 sequence of the σ^{70}-dependent promoters it regulates, the Fur protein blocks access of RNA polymerase to the promoters and represses transcription of the operons of the Fur regulon. If ferrous ions are in short supply, the repressor is in the aporepressor state and cannot bind to the operator sequences. The transcription of the genes under Fur control, including genes for the siderophores and iron transporters in the membrane (called iron assimilation proteins in Figure 13.20), is therefore high when ferrous iron is low and is repressed when ferrous iron is high. The Fur regulation system is highly conserved among gram-negative bacteria. Some gram-positive bacteria have a similar system based on repressors related to the DtxR repressor of *Corynebacterium diphtheriae* (see below), which has little sequence similarity to Fur but is structurally similar and probably works in similar ways.

The RyhB sRNA

While many genes are turned off when iron is in excess, other genes are turned on. They include the ferritin-like iron storage proteins and many other proteins that contain iron, including aconitase A (AcnA), an

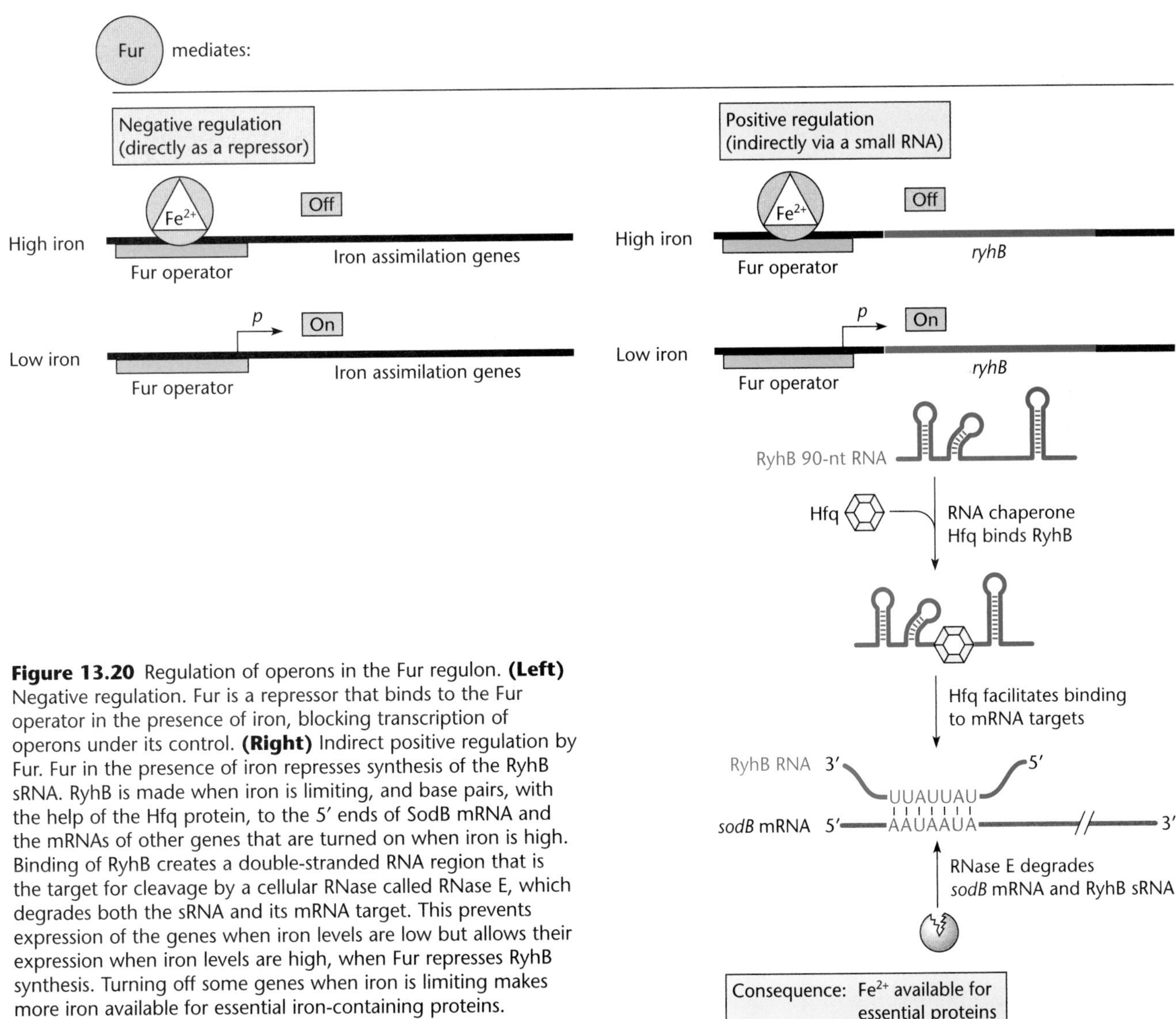

Figure 13.20 Regulation of operons in the Fur regulon. **(Left)** Negative regulation. Fur is a repressor that binds to the Fur operator in the presence of iron, blocking transcription of operons under its control. **(Right)** Indirect positive regulation by Fur. Fur in the presence of iron represses synthesis of the RyhB sRNA. RyhB is made when iron is limiting, and base pairs, with the help of the Hfq protein, to the 5′ ends of SodB mRNA and the mRNAs of other genes that are turned on when iron is high. Binding of RyhB creates a double-stranded RNA region that is the target for cleavage by a cellular RNase called RNase E, which degrades both the sRNA and its mRNA target. This prevents expression of the genes when iron levels are low but allows their expression when iron levels are high, when Fur represses RyhB synthesis. Turning off some genes when iron is limiting makes more iron available for essential iron-containing proteins.
doi:10.1128/9781555817169.ch13.f13.20

iron-containing enzyme of the TCA cycle, and an iron-containing superoxide dismutase that destroys peroxides in the cell before they can be converted into hydroxyl radicals by the iron and damage DNA and other cellular constituents. In the presence of excess iron, the concentration of these proteins increases rather than decreases, but only in Fur$^+$ cells. Very little of them is made in a Fur– mutant, suggesting that Fur activates their transcription. Microarray analysis (Box 13.7) comparing the levels of mRNA for all genes of *E. coli* in the presence and absence of iron revealed that the transcription of 53 genes increases when iron is in excess while that of 48 genes decreases. Accordingly, it was proposed that Fur could act either as a repressor or as an activator, like many other transcriptional regulators, repressing some genes in the presence of iron and activating others. However, activation by Fur does not occur by a direct effect on the target genes but instead occurs via an sRNA called RyhB. This sRNA inhibits the expression of iron-responsive genes under its control. In the presence of excess iron, Fur represses the synthesis of RyhB, and RyhB repression of the genes it regulates is relieved, causing an increase in expression of these iron-responsive genes.

The way in which the RyhB RNA regulates the genes under its control is shown in Figure 13.20. Like many such sRNAs, the RyhB RNA sequence is partially complementary to sequences close to the 5′ end of the mRNA of genes under its control. This complementarity allows RyhB to pair with the mRNA, with the help of the Hfq protein, in a manner similar to that described for DsrA and MicF (Box 13.6). However, rather than blocking translation of the mRNAs, like other sRNAs discussed so far, RyhB binding creates a partially double-stranded RNA that is a substrate of the RNase E enzyme, one of the major RNases in *E. coli* (see chapter 2). RNase E cleaves both the mRNA and the sRNA, resulting in degradation of both RNAs. Prevention of the synthesis of many iron-containing proteins by the RyhB sRNA when iron is limiting reserves the available iron for the most essential iron-containing enzymes in *E. coli*, including ribonucleotide reductase, which is required to make deoxynucleotides and hence DNA. When iron levels are high, the inhibition of RyhB synthesis by Fur allows these iron-containing enzymes to be produced.

The Aconitase Translational Repressor

The more we learn about cells, the more we discover about how many functions are coordinated. These discoveries are often serendipitous, as was the case when aconitase was discovered not only to function in metabolism, but also to play a role in iron regulation in bacteria and eukaryotes. This dual role was first discovered in eukaryotes during studies of proteins called iron-responsive proteins (IRPs). When iron is limiting, IRPs bind to the mRNAs for proteins involved in iron metabolism. They either can inhibit the translation of an mRNA by binding to the 5′ end of the transcript and preventing access of the ribosome to the initiator codon or they can increase expression by binding to the 3′ end of the mRNA and stabilizing the mRNA. In general, IRPs inhibit the translation of proteins such as ferritins that are needed only when iron is high and stimulate the synthesis of proteins such as iron transport proteins or transferrins that are needed when iron is limiting. The sequences to which they bind are highly conserved in evolution and are called iron-responsive elements.

It came as a complete surprise when IRPs were purified, partially sequenced, and discovered to be aconitases. Aconitases are enzymes that function in the TCA cycle to convert citrate into isocitrate. The TCA cycle produces essential carbon-containing compounds (including α-ketoglutarate) used in many biochemical reactions and generates reducing power in the form of reduced nicotinamide adenine dinucleotide (NADH) to feed electrons into the electron transport system to make ATP. Aconitases contain iron in the form of an iron-sulfur cluster, often written $[4Fe\text{-}4S]^{2+}$. Many iron-containing enzymes have the iron in this prosthetic group, and it usually makes them sensitive to oxygen.

Bacteria and mitochondria have aconitases related to the cytoplasmic aconitases of eukaryotes. *E. coli* has two aconitases, which differ in their regulation and their sensitivity to oxygen. AcnA is induced following stress and in the stationary phase and is an iron-containing protein whose expression is repressed by RyhB (and therefore is induced when iron levels are high). The other aconitase, aconitase B (AcnB), is the major aconitase synthesized during exponential growth and is more sensitive to oxygen. AcnB also regulates the translation and stability of mRNAs involved in iron metabolism in response to iron deficiency, much like the aconitases from eukaryotes. A similar dual role has been found for the aconitases of other bacteria, including *B. subtilis* and some pathogenic bacteria. Also, some pathogenic bacteria may use their aconitases to sense the availability of iron as part of the signal that they are inside a eukaryotic host and to adjust their physiology accordingly.

It is not clear why aconitase plays the dual role of sensing iron levels and performing an essential step in the TCA cycle. One possibility is related to the extreme toxicity and mutagenic properties of hydroxyl radicals (see Box 11.1). These are produced from hydrogen peroxide in the presence of high iron concentrations. If the cellular iron concentration is high, the TCA cycle may run at full capacity to increase the reducing power in the cell and to reduce the amount of such dangerous reactive oxygen species. If the cellular iron concentration is low, the TCA cycle can run at a lower rate, just fast enough to produce essential intermediates and electron donors for the electron transport system.

BOX 13.7

Tools for Studying Global Regulation

Each new genome sequence tempts us with a renewed challenge: to pursue the quest for complete understanding of cellular regulation. The ever-growing sequence databases allow the assignment of presumptive functions, enzymatic or structural, to most genes of an organism. It is important to remember, however, that the predicted gene identities are based on sequence similarities and are not always accurate. Regulation of these predicted genes poses an even greater challenge.

Transcriptome Analysis

RNA profiling can be carried out through a variety of techniques, including high-density DNA microarrays (see Rhodius and LaRossa, References). The technologies for sequencing and adaptations for microarray strategies are evolving very rapidly, and many tools and services are available at genomics centers at universities or commercially. The basic principle of microarray technologies is to have known segments of a genome positioned at known positions on a solid substrate (a "**gene chip**") and then isolating RNAs from cells grown under two different conditions, or from wild-type versus mutant strains, and using different fluorescent dyes to differentially label the cDNAs generated from those RNAs by reverse transcription. The two cDNA preparations are hybridized separately to the genomic arrays, and the intensities of the hybridization signals for the two cDNA preparations at each genomic position are compared to identify genes for which expression increases or decreases as a result of the change in growth conditions or introduction of a mutation in one of the strains.

A more recent approach to identifying the set of RNAs present in cells under a specific condition involves high-throughput sequencing of cDNA generated from the total pool of cellular RNA. This approach, commonly called RNASeq, allows quantitative measurements of even low-abundance RNAs that might have been missed in traditional microarray experiments (see Wang et al., References).

Proteome Analysis

The techniques of **proteomics** can be used to identify proteins in order to analyze the levels of proteins in cells, to determine relative changes in protein levels of regulons or stimulons, to evaluate protein-protein interactions, and to study subcellular localization of proteins (see Han and Lee, References). **Two-dimensional polyacrylamide gel protein electrophoresis (2D-PAGE)** can separate many of the proteins of the cell into individual spots on a membrane, and the spots can be further characterized. Because sample complexity limits the ability to identify individual proteins, nongel protein separation techniques, such as liquid chromatography are often used. This method separates the protein samples into subsamples that are less complex.

Mass spectrometry (MS) is an important tool of proteomics. It is used both to identify proteins and to quantitate protein expression. Protein MS involves the ionization of proteins and peptides and subsequent measurements of mass-to-charge (*m*/*z*) ratios. Once proteins have been obtained from a biological sample, they are fragmented into peptides, often by trypsin digestion, because it produces small peptides that are suitable for the first step in MS, which is ionization. Two popular ionization methods are **matrix-assisted laser desorption ionization (MALDI)** and **electrospray ionization (ESI)** (see Kolker et al., References).

In **tandem mass spectrometry (MS-MS)**, after a protein has been subjected to MS, a single peptide is isolated and shunted through a "collision" chamber in which nitrogen or argon gas breaks the peptide into subfragments. These fragments are then further analyzed so that the sequence of the peptide can be deduced. Computerized comparison to databases can identify the sequence of the peptide and, if the genome sequence is known, the protein and gene from which the peptide came.

References

Han, M.-J., and S. Y. Lee. 2006. The *Escherichia coli* proteome: past, present, and future prospects. *Microbiol. Mol. Biol. Rev.* **70**:362–439.

Kolker, E., R. Rigdon, and J. M. Hogan. 2006. Protein identification and expression analysis using mass spectrometry. *Trends Microbiol.* **14**:229–235.

Rhodius, V. A., and R. A. LaRossa. 2003. Uses and pitfalls of microarrays for studying transcriptional regulation. *Curr. Opin. Microbiol.* **6**:114–119.

Wang, Z., M. Gerstein, and M. Snyder. 2009. RNA-Seq: a revolutionary tool for transcriptomics. *Nat. Rev. Genet.* **10**:57–63.

Regulation of Virulence Genes in Pathogenic Bacteria

Many of the stress responses and other types of regulons discussed above are directly relevant to the ability of bacterial pathogens to survive in a eukaryotic host. The pathogen must recognize that it has entered a new environment, adapt its metabolism to allow it to exploit this new environment, and express virulence genes that allow the organism to survive in the host and cause disease. The virulence genes of pathogenic bacteria represent a type of global regulon. Most pathogenic bacteria express their virulence genes only in the eukaryotic host;

somehow, the conditions inside the host turn on the expression of these genes. Virulence genes can be identified because mutations that inactivate them render the bacterium nonpathogenic but do not affect its growth outside the host. Note that many of the basic regulatory responses of bacterial cells discussed above, such as catabolite regulation, the stringent response, and a variety of stress responses, are also very important for a bacterial pathogen to cause disease (see Eisenreich et al., Suggested Reading). Furthermore, interactions among bacteria can be crucial for disease, for example, in sensing the presence of other members of the same or different species or for formation of complex community structures, such as biofilms, which can have a major impact, not only on the disease process, but also on the susceptibility of the bacterial cells to antibiotics. In this section, we focus on examples of the regulation of genes that affect bacterial pathogenesis.

Diphtheria

Diphtheria is caused by the bacterium *C. diphtheriae*, a gram-positive bacterium that colonizes the human throat. It is spread from human to human through aerosols created by coughing or sneezing. The colonization of the throat by itself results in few symptoms. However, strains of *C. diphtheriae* that harbor a prophage named β (see chapter 8) produce diphtheria toxin, which is responsible for most of the disease symptoms. The toxin is excreted from the bacteria in the throat and enters the bloodstream, where it does its damage.

DIPHTHERIA TOXIN

Diphtheria toxin is a member of a large group of A-B toxins, so named because they have two subunits, A and B. In most A-B toxins, the A subunit is an enzyme that damages host cells and the B subunit helps the A subunit enter the host cell by binding to specific cell receptors. The two parts of the diphtheria toxin are first synthesized from the *tox* gene as a single polypeptide chain, which is cleaved into the A and B subunits as it is excreted from the bacterium. These two subunits are held together by a disulfide bond until they are translocated into the host cell, where the disulfide bond is reduced and broken, releasing the individual A subunit into the cell.

The action of the diphtheria toxin A subunit on eukaryotic cells is well understood and is widely used in studies of the eukaryotic cell. The A subunit enzyme specifically ADP-ribosylates (adds ADP-ribose to) a modified histidine amino acid of the translation elongation factor EF-2 (equivalent to EF-G in bacteria [see chapter 2]). The ADP-ribosylation of the translation factor blocks translation and kills the cell. The opportunistic pathogen *Pseudomonas aeruginosa* makes a toxin that is identical in action to the diphtheria toxin, although it has a somewhat different sequence.

Regulation of the *tox* Gene of *C. diphtheriae*

Iron limitation presents a problem for bacteria in general and for pathogenic bacteria in particular. All of the iron in the human body is tied up in other molecules, such as transferrins and hemoglobin. Thus, to multiply in a eukaryotic host, a pathogenic bacterium must extract the iron from the transferrins and other proteins to which it is bound and transport it into its own cell. For this purpose, *C. diphtheriae* and many other pathogenic bacteria synthesize very efficient siderophores, much like those of free-living bacteria (see above). These small siderophores are excreted from the bacterial cells into the host, where they bind Fe^{2+} more tightly than do other molecules and so can extract Fe^{2+} from them. The siderophore-Fe^{2+} complexes are transported back into the bacterial cell. As in free-living bacteria, the genes for making the siderophores and a high-efficiency transport system for iron are expressed only when iron is limiting.

Not only is iron an essential nutrient for bacteria, but also, its limitation is often used as a signal that the bacterium is in the eukaryotic host environment and the virulence genes should be turned on. In *C. diphtheriae*, the virulence genes, including the *tox* gene and iron uptake genes, are under the control of the same global regulator, DtxR (for *d*iphtheria *tox*in *r*egulator). The DtxR protein of *C. diphtheriae* is a repressor that functions similarly to the Fur repressor protein of gram-negative bacteria, including *E. coli* (Figure 13.21). Like Fur, the DtxR protein requires iron as a corepressor to bind to the operators of genes under its control. Interestingly, even though the *tox* gene encoding the toxin of *C. diphtheriae* is carried on the lysogenic phage, it is regulated by the DtxR repressor, which is encoded on the chromosome (Figure 13.21) (see Schmidt and Holmes, Suggested Reading). Most other genes controlled by DtxR are chromosomal genes. This is just one of many examples of the contribution of lysogenic phages to the pathogenicity of bacteria. A more detailed map of phage β is shown in Figure 8.20.

Cholera and Quorum Sensing

Cholera is another well-studied example of the global regulation of virulence genes. *Vibrio cholerae*, the causative agent, is a gram-negative bacterium that is spread through water contaminated with human feces. The disease continues to be a major health problem worldwide, with periodic outbreaks, especially in countries with poor sanitation. When ingested by a human, *V. cholerae* colonizes the small intestine, where it synthesizes cholera toxin, which acts on the mucosal cells to cause a severe form of diarrhea. Other virulence

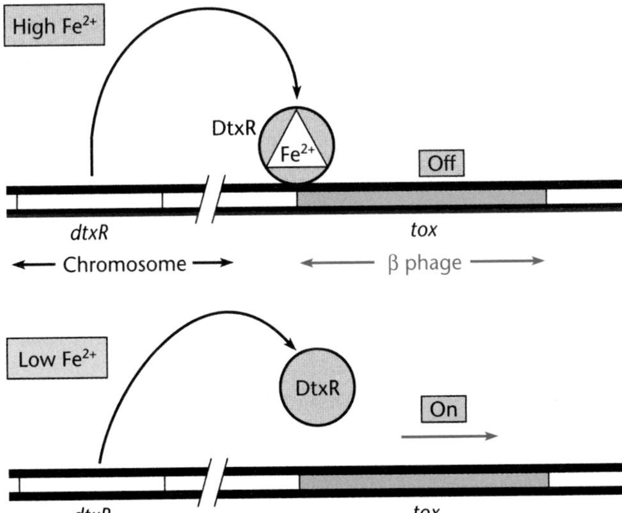

Figure 13.21 Regulation of the *C. diphtheriae tox* gene of prophage β. The DtxR repressor protein, which is encoded by the chromosomal *dtxR* gene, binds to the operator for the *tox* gene, which is encoded on the β prophage. DtxR repressor activity requires ferrous ions (Fe^{2+}).
doi:10.1128/9781555817169.ch13.f13.21

determinants are the flagellum, which allows the bacterium to move in the mucosal layer of the small intestine, and pili called toxin-coregulated pili (TCP) that allow it to stick to the mucosal surface. **Quorum sensing**, which allows the bacterium to sense when other *V. cholerae* cells are nearby (see below), and biofilm formation, in which the bacteria band together and surround themselves with an impregnable layer of polymers to keep from being washed out of the intestine and to resist host defense systems, also play important roles in the disease process.

CHOLERA TOXIN

The mechanism of action of cholera toxin has been the subject of intense investigation, in part because of what it reveals about the normal action of eukaryotic cells. The cholera toxin is composed of two subunits, CtxA and CtxB, which are exported from the bacterial cell by a type II secretion system (see chapter 14). The two subunits are secreted through the inner membrane by the SecYEG channel and then assemble in the periplasm before being released to the outside of the cell through a large structure called a secretin in the outer membrane. Once outside the cell, the CtxB subunit helps the CtxA subunit enter the eukaryotic cell. Like diphtheria toxin, the CtxA subunit of cholera toxin is an ADP-ribosylating enzyme. However, rather than ADP-ribosylating a translation elongation factor, CtxA ADP-ribosylates a mucosal cell membrane protein called Gs, which is part of a signal transduction pathway that regulates the activity of the adenylcyclase enzyme that makes cAMP. The ADP-ribosylation of Gs causes cAMP levels to rise and alters the activities of transport systems for sodium and chloride ions. This results in loss of sodium and chloride ions from the cells, and the change in osmotic pressure releases water from the cells, resulting in severe diarrhea and dehydration. This facilitates further spread of the disease to new hosts. The treatment is to aggressively administer water orally or intraveneously until the condition of the patient improves.

Regulation of the Synthesis of Cholera Toxin and Other Virulence Determinants

The *ctxA* and *ctxB* genes encoding the cholera toxin are part of a large regulon containing as many as 20 genes. In addition to the *ctx* genes, the genes of this regulon include those encoding pili, colonization factors, and outer membrane proteins (e.g., OmpT and OmpU) related to osmoregulation. Although some of these genes, including the *ctx* genes, are carried on a prophage, others, including the pilin genes, are carried on the bacterial chromosome. The transcription of the genes of this regulon is activated only under conditions of high osmolarity and in the presence of certain amino acids, conditions that may mimic those in the small intestine. Here, we describe the cascade of genes involved in the regulation of transcription of *ctx* and other virulence genes.

ToxR-ToxS. The cholera virulence regulon was first found to be under the control (either directly or indirectly) of the activator protein ToxR, the product of the *toxR* gene (see Yu and DiRita, Suggested Reading). The ToxR protein combines in a single polypeptide elements that are normally part of two different proteins of the two-component sensor and response regulator type of system (Box 13.4). The ToxR polypeptide spans the inner membrane so that the carboxyl-terminal part of the protein is in the periplasm, where it can sense the external environment. The amino-terminal part is in the cytoplasm, where it contains an OmpR-like DNA-binding domain that can activate the transcription of genes under its control.

While the ToxR protein resembles other response regulators in some respects, it is unlike most response regulator proteins in that it is not activated by phosphorylation. Also, it is not known to bind to any small-molecule effectors. A clue to its mechanism of activation could be found in another protein, ToxS. The *toxS* gene is immediately downstream of *toxR* in the same operon, and the gene was discovered because mutations that inactivate it also prevent expression of the genes of the ToxR regulon. Like ToxR, the ToxS protein is anchored in the inner membrane, but a large domain protrudes into the periplasm.

One model for how ToxS might activate ToxR came from experiments designed to investigate the membrane topology of ToxR. The purpose of these experiments was to determine if part of ToxR spans the inner membrane and extends into the periplasm. A method for determining the membrane topology of proteins in gram-negative bacteria uses translational fusions of various regions of the protein to PhoA (see chapter 2 for a general discussion of reporter gene fusions). The PhoA protein is an alkaline phosphatase enzyme that cleaves XP (5-bromo-4-chloro-3-indolylphosphate), which is like X-Gal (5-bromo-4-chloro-3-indolyl-β-D-galactopyranoside), the indicator for β-galactosidase activity, except that the dye is fused to phosphate instead of galactose. If PhoA cleaves the phosphate off of XP, the colonies turn blue. However, PhoA must be in the periplasm to be active, probably because it must form dimers that are held together by disulfide bonds between cysteines in its subunits, and these disulfide bonds form only in the periplasm and not in the cytoplasm. Therefore, if a region of a protein is fused to PhoA, the alkaline phosphatase will be active and the strain carrying that fusion will form blue colonies on XP plates only if that region of the protein is in the periplasm (see chapter 14). To use this method to determine which parts of ToxR are in the periplasm, fusion proteins composed of the PhoA reporter fused to various portions of ToxR were generated. Bacteria containing fusions to the amino-terminal portion of ToxR did not form blue colonies on XP plates, while strains containing fusions to the carboxyl-terminal portion did, suggesting that the carboxyl-terminal portion of the ToxR protein is in the periplasm but the amino-terminal portion is in the cytoplasm. A surprising result was that some of the ToxR-PhoA fusions functioned like wild-type ToxR in activation of transcription, even when *toxS* was inactivated by a mutation. One explanation for this result is that the PhoA part of these fusions was driving the dimerization of the ToxR portion in the periplasm and that dimerization was required for ToxR to activate transcription. This suggested that the normal function of ToxS is to dimerize ToxR, since ToxS could be dispensed with if PhoA promoted the dimerization.

Other evidence suggests that dimerization, while important, may not be all that is needed to activate the ToxR protein. These results suggested that some feature related to the membrane anchoring of ToxR may also be required. If dimerization were sufficient, attaching other dimerization domains, such as the ones from the CI repressor protein of λ phage (see chapter 8), to the cytoplasmic domain of the ToxR protein should cause ToxR to dimerize in the cytoplasm and activate transcription. However, fusion proteins composed of ToxR and other dimerization domains are still inactive unless the ToxR protein retains its transmembrane domain. This suggests that the ToxR protein must be at least partly in the membrane to be active and that ToxS may play an additional role in stabilizing ToxR. Anchoring ToxR in the membrane may allow ToxR to be activated directly by external signals, and ToxS may help in this activation. It is known that membrane-damaging agents, such as bile, can activate ToxR, consistent with the idea that ToxR senses properties of the membrane.

ToxT and TcpP. The regulation of virulence genes in *V. cholerae* is much more complicated than a single activator, ToxR, turning on virulence genes. A general outline of the various regulatory pathways used to turn on the genes required for *V. cholerae* pathogenicity is given in Figure 13.22. As mentioned above, *V. cholerae* has many virulence genes besides the toxin genes that are also considered part of the ToxR regulon, since *toxR* mutations prevent their expression. However, while ToxR can activate the transcription of the toxin genes directly, the ToxR protein does not directly control most of the other ToxR regulon genes. Instead, another transcriptional activator, the ToxT protein, which is a member of the AraC family of regulators (see Box 12.2), controls these genes. ToxR activates the transcription of the *toxT* gene, and ToxT then activates the transcription of the other genes. The ToxT-activated genes are therefore part of the ToxR regulon, as their expression is dependent on ToxR, and mutations in *toxR* result in loss of expression because of the loss of *toxT* expression. ToxT also activates the expression of its own gene, so that an initial signal from ToxR can be amplified to generate high levels of ToxT synthesis. This is an example of an **autoinduction loop**, where a regulatory gene activates its own expression, and can result in a very strong regulatory response.

Activation of the *toxT* gene also requires yet another activator, TcpP. The activity of the TcpP activator requires the activity of another protein, TcpH. Both TcpP and TcpH are inner membrane proteins, but it is not clear how TcpH regulates TcpP activity. Transcription of the *tcpP-tcpH* operon responds to environmental cues, but we do not understand how these two genes, together with the *toxR-toxS* genes, transduce environmental signals that the bacterium is in the intestine of its host into activation of ToxT. This type of serial activation, where one regulator activates another regulator, which activates another regulator, is a called a **regulatory cascade**. Obviously, much more needs to be done before we can begin to understand this important model system of bacterial pathogenicity.

Interestingly, the *toxT*, *tcpP*, and *tcpH* genes are all located on a DNA element in *V. cholerae* called VPI (*V.

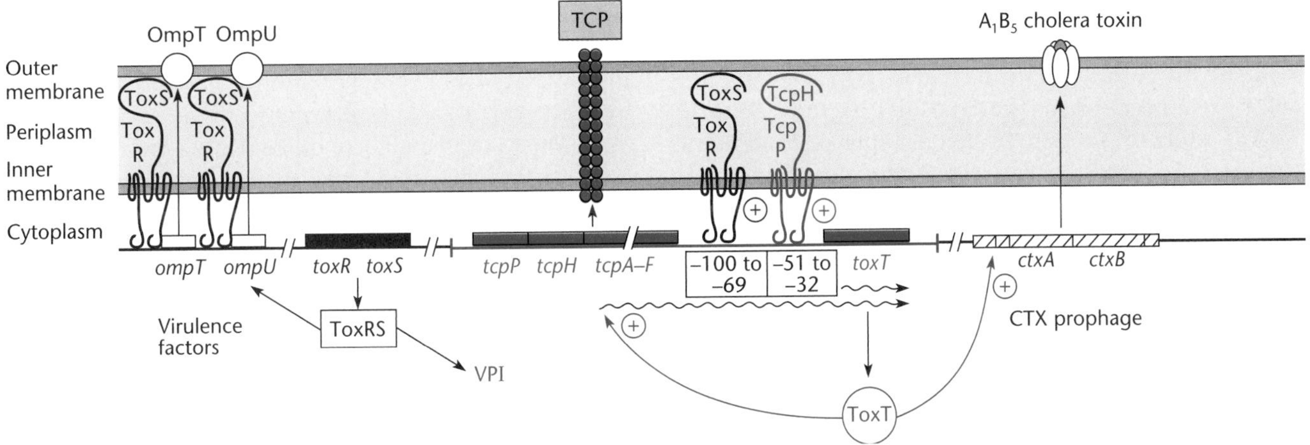

Figure 13.22 Regulatory cascade for *V. cholerae* virulence factors. The ToxR-ToxS proteins directly regulate *omp* virulence factors and the *toxT* regulatory gene located on a *V. cholerae* pathogenicity island (VPI) (indicated in blue). The VPI-encoded TcpP-TcpH proteins also regulate *toxT* transcription. ToxT activates the Ctx prophage-borne *ctxAB* toxin genes and the TCP genes. ToxT also positively regulates its own expression from the promoter for transcription of the *tcpA–F* operon. doi:10.1128/9781555817169.ch13.f13.22

cholerae pathogenicity island). A **pathogenicity island** is a segment of DNA that contains a cluster of genes involved in virulence, and this set of genes has usually been acquired by horizontal gene transfer from another organism (see chapter 1). The cholera toxin is also encoded by a horizontally acquired DNA element, in this case, a lysogenic phage (see chapter 8). These genes are regulated by ToxR, encoded by a chromosomal gene. This is yet another example of virulence traits in pathogenic bacteria that are encoded on exchangeable DNA elements but interact with the products of chromosomal genes, as well as another example of pathogenic bacteria that are derived from their free-living relatives by acquisition of virulence genes encoded on interchangeable DNA elements.

QUORUM SENSING

As mentioned above, quorum sensing is an important contributor to virulence in *V. cholerae*. For a long time, it was thought that single-celled bacteria, such as *V. cholerae*, lived as isolated cells, with no way of telling if other members of the same species were nearby. However, with the discovery of quorum sensing, it became apparent that many bacteria have ways of communicating with each other. They do this by releasing small molecules that can be taken up by other bacteria, usually members of the same species. When the concentration of bacteria is high, the concentration of these secreted small molecules also becomes high, signaling that the bacterium is in the presence of many other bacteria of the same type. In response, they induce certain types of genes to adapt them for a more communal existence, such as in a biofilm. We have discussed cell-cell signaling in reference to competence gene regulation in gram-positive bacteria, which use small peptides as their signals (see chapter 6).

Quorum sensing was first discovered in the marine bacterium *Photobacterium fischeri* (*Vibrio fischeri*), which is a facultative symbiont that can live either free in the ocean or in the light organs of some fishes and squids, where they give off light due to chemiluminescence. The light is due to induction of the *lux* genes, which are turned on only when a bacterium is close to other bacteria of the same type. The squid and fish with which these bacteria form a symbiosis live deep in the ocean, where there is little light. By emitting light when they are concentrated in the light organ, the bacteria help the marine organisms find each other. Recent work has concentrated on a free-living marine bacterium, *Vibrio harveyi*, which also gives off light when the bacteria are concentrated in certain regions of the ocean, but it is not known to be a symbiont of any marine animal. A likely explanation is that *V. harveyi* also forms such a symbiosis, but with an unknown marine organism. This explanation is supported by the fact that some of the genes induced along with *lux* form a type III secretion system. In other types of bacteria, such systems are used to inject proteins into eukaryotic cells and are important in avoiding host defenses and establishing infections or symbioses (see chapter 14).

Extensive research has revealed that the chemiluminescence of *V. harveyi* is induced because the bacteria give off two small molecules, called autoinducers AI-1 and AI-2. AI-1 is a homoserine lactone, and AI-2 is a furanosyl borate diester; the structures of these compounds are shown in Figure 13.23. Autoinducer AI-2 is made by the product of the *luxS* gene, which is found in many gram-negative and gram-positive bacteria. Because AI-2 is made by so many different types of bacteria, it has been proposed that it is a universal autoinducer allowing different types of bacteria to communicate with each other, for example, in the formation of biofilms.

Both AI-1 and AI-2 act through two-component sensor kinase and response regulator cascades that determine the state of phosphorylation of LuxO, a transcriptional regulator (Figure 13.23A). LuxO is a member of the NtrC family of activators and, like NtrC, is active only if it is phosphorylated. It also activates transcription only from σ^N-type promoters, like the other members of this family of regulators (Box 13.5). Genes under the control of the LuxO activator include four noncoding sRNAs. These sRNAs, with the help of the Hfq protein, inhibit the synthesis of an activator named LuxR by binding to and destabilizing its mRNA. If LuxR is not made, the *lux* operon is not transcribed, and the cells do not give off light. When the cell population density is low, the concentrations of AI-1 and AI-2 are low, and the kinase activity of the sensor kinases is high, which leads to the phosphorylation of LuxO, so that it is active as a transcriptional activator. The four sRNAs are made, LuxR is not made, the *lux* operon is not transcribed, and the cells do not give off light. At a high concentration of cells, and therefore at a high concentration of the autoinducers, the binding of the autoinducers to the sensor kinases activates the phosphatase activity of the sensor kinases, which results in removal of phosphate from the response regulator protein in the next step of the pathway until eventually LuxO loses its phosphate and is unphosphorylated. If LuxO is not phosphorylated, the sRNAs are not made; this allows synthesis of the LuxR activator, the *lux* operon is transcribed, and the cells give off light.

Quorum Sensing in *V. cholerae*

Analysis of the nonpathogenic *V. harveyi* provided crucial insights into mechanisms of virulence gene regulation in the pathogenic *V. cholerae*, which was also found to have quorum-sensing systems, as do many types of pathogenic bacteria, including plant pathogens, such as *Agrobacterium*. While it lacks the AI-1 system, *V. cholerae* has the AI-2 system and two other systems with unknown autoinducers named CA-1 and VarS-VarA, which also function through sensor kinases and response regulators to finally determine the state of phosphorylation of LuxO.

A comparison of quorum sensing in *V. cholerae* and *V. harveyi* is presented in Figure 13.23. A total of seven sRNAs are used by the pathways for quorum-sensing signaling in *V. cholerae* (see Lenz et al., 2005, Suggested Reading). Not only does *V. cholerae* have the four sRNAs whose transcription is activated by LuxO in *V. harveyi*, it also has three others that are used in the VarS-VarA pathway. These other sRNAs bind to and inhibit the activity of a regulatory protein named CsrA (Box 13.6), which in turn affects the state of phosphorylation of LuxO by an unknown mechanism. Instead of an activator, LuxR, which activates the transcription of the *lux* gene and other genes appropriate for the free-living *V. harveyi*, *V. cholerae* has HapR, which differentially regulates *lux* and virulence genes and genes involved in biofilm formation, etc., that are important for pathogenesis.

Genetic Experiments That Led to the Detection of the sRNAs That Regulate LuxR

It took many investigators a long time to unravel the signal transduction pathways of quorum sensing in *V. harveyi*, and these efforts are still under way. We discuss only one set of experiments here, those that led to the discovery of the sRNAs that bind to and destabilize the mRNA for the LuxR activator (see Lenz et al., 2004, Suggested Reading). These experiments illustrate some principles of genetics and some of the methods already described. They also illustrate the power of comparing two closely related bacteria to find mechanisms that are common to both.

It was hypothesized that in *V. harveyi* the response regulator LuxO does not act directly to activate the transcription of the *lux* operon and instead activates the transcription of a repressor of LuxR. This was based on the observation that *luxO* null mutations result in constitutive light production. If the LuxO activator is not made, the putative repressor would not be made, the LuxR activator would be made constitutively, and the cells should give off light, even at low concentrations. This suggested a way of isolating mutants with mutations in the putative repressor, since they should have the same phenotype as *luxO* mutations, i.e., constitutive light production even when the cells are at low concentrations. The problem is that it is difficult to see the light given off by individual cells when the cells are present at low concentrations. It is much easier to screen colonies for mutants, but the cells in a colony are effectively at high concentrations. To overcome this difficulty, the investigators introduced a mutated form of the *luxO* gene into the cells by site-specific mutagenesis. This mutated form of the gene produces a LuxO protein in which the aspartate to which the phosphate normally attaches is changed to a glutamate, whose side chain more closely resembles a phosphate group. The *V. harveyi* strain with this mutated form of LuxO should make the putative repressor constitutively, and as predicted, cells fail to

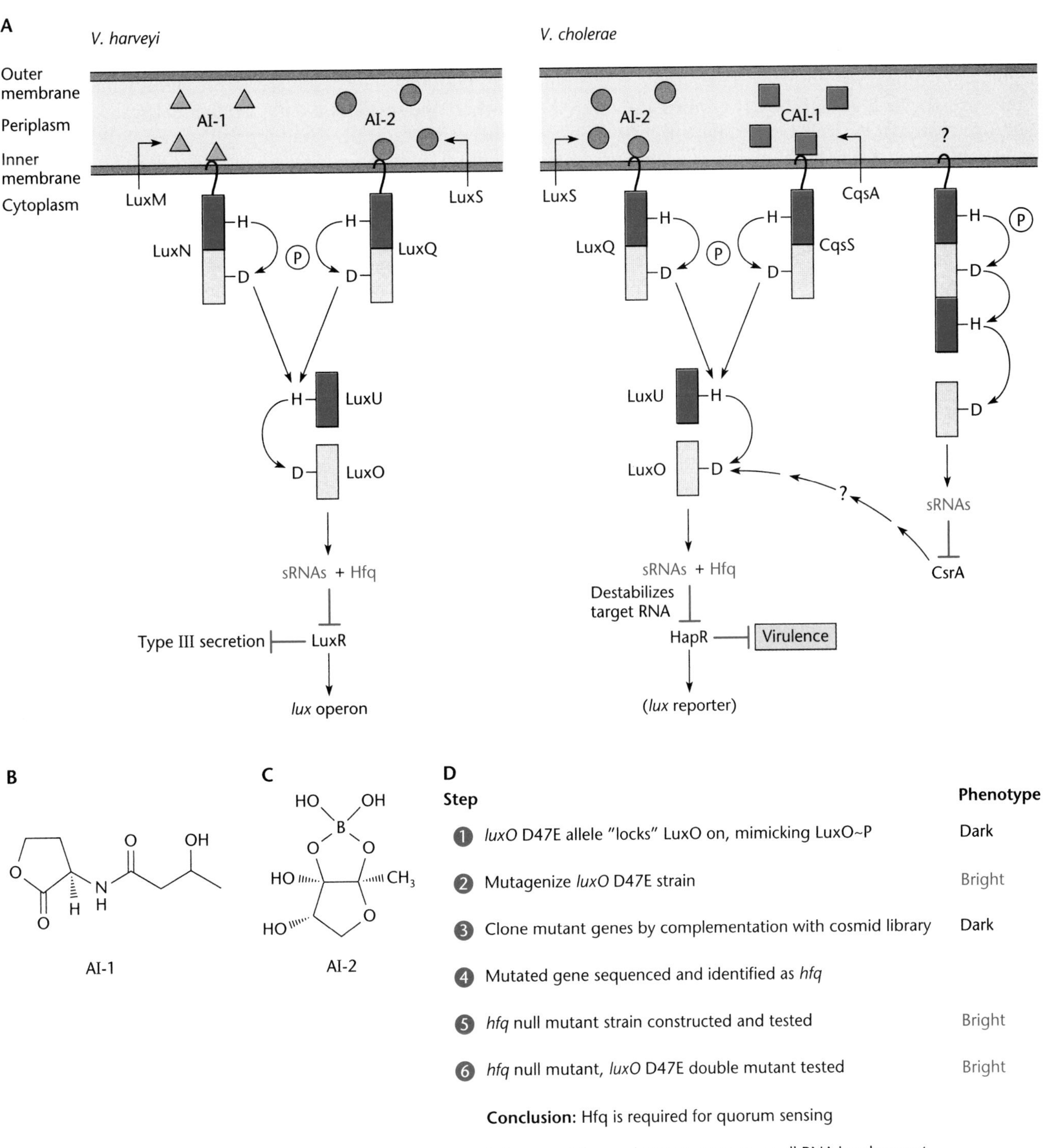

Figure 13.23 Quorum sensing in *V. harveyi* and *V. cholerae.* **(A)** *V. cholerae* shares the AI-2 quorum-sensing system with *V. harveyi* and contains two additional phosphorelay systems to regulate virulence. AI-1 (triangles), AI-2 (circles), and CAI-1 (squares) are small molecules. LuxS synthesizes AI-2, while LuxM and Cqs synthesize AI-1 and CAI-1, respectively. LuxN, LuxQ, LuxS, LuxU, and LuxO form a phosphorelay (see Box 14.4). The flow of phosphate depicts a low-cell-density state. High cell density would reverse the flow. **(B)** Structure of AI-1. **(C)** Structure of AI-2. **(D)** Steps in identifying quorum-sensing regulatory sRNAs.
doi:10.1128/9781555817169.ch13.f13.23

produce light even when the cells are at high concentration in colonies. The investigators then mutagenized these cells and looked for mutants that formed constitutive light-producing colonies. They found *luxO* mutants, as expected, since LuxO is required to make the putative repressor, as well as mutants with mutations in *rpoN*, the gene for the σ^N homolog that is required for LuxO to act. However, they also found mutants of *hfq*, the gene for the sRNA-binding protein that promotes the binding of many sRNAs to their targets (Box 13.6). This led them to suspect that the repressor was an sRNA rather than a protein and that the *hfq* mutant resulted in loss of sRNA-dependent repression. The reason that none of the constitutive mutants had mutations in the gene for this putative sRNA could merely be because the sRNA gene is small and small genes are difficult to inactivate by a mutation. Figure 13.23D shows the steps that implicated an sRNA in this regulation.

Because it is hard to identify genes for sRNAs by a classical genetic approach, the investigators searched for the sRNA genes directly, using the fact that the genome of *V. cholerae* had been sequenced, a prerequisite for such studies. They thought they could use *V. cholerae* for this search because the two species are very similar and *hfq* mutants were also constitutive for light production in *V. cholerae* into which the *lux* operon and the *luxR* gene had been introduced. In particular, they were interested in regions that contained a σ^N-type promoter (since LuxO activates promoters of this class) and a factor-independent transcriptional terminator not far downstream from the promoter, as is the pattern observed in sRNA genes. Four candidate sRNA genes were identified. The transcription of these RNAs was under the control of the LuxO activator. Mutations in one, two, or even three of these genes had no effect on the synthesis of LuxR or HapR. However, inactivation of all four of the sRNA genes caused constitutive expression of the *lux* operon. This indicates that the sRNAs are redundant in function and explains why mutations in these genes were not identified in the initial search for constitutive mutants, as single mutations in any of the sRNA genes have no effect. Further experiments showed that these sRNAs act by base pairing to the mRNA for LuxR (or HapR in *V. cholerae*) and create the double-stranded RNA substrate for a cellular RNase, which then degrades both the mRNA and the sRNA in a manner similar to that described for the RyhB iron-regulatory sRNA.

Whooping Cough

Another well-studied disease used to illustrate global regulation of virulence genes is whooping cough, caused by the gram-negative bacterium *Bordetella pertussis*. Whooping cough is mainly a childhood disease and is characterized by uncontrolled coughing, hence the name. The bacteria colonize the human throat and are spread through aerosols that result from the coughing. Effective vaccines have been developed, but the disease continues to kill thousands of children worldwide, mainly in areas where the vaccines are not available. Because outbreaks occur sporadically even in populations with high vaccination rates, an addtional vaccination in later years of life has become routine.

Despite their very different symptoms, the diseases caused by *V. cholerae* and *B. pertussis* have similar molecular mechanisms. *B. pertussis* makes a complex A-B toxin (pertussis toxin) that is in some ways similar to the cholera toxin. The pertussis toxin has six subunits, although only two of them are identical. One of the subunits (S1) is the active portion of the toxin, while the others are involved in adhesion to the mucosal surface of the throat. The toxin is first secreted through the outer membrane by the SecYEG translocase and is then exported with a type IV secretion system (see chapter 14). Once outside the bacterial cell, the B domains of the toxin bind to receptors on ciliated epithelial cells and transfer the A domains into the cells, where they ADP-ribosylatate the G protein in a signal transduction pathway involved in deactivating the adenylate cyclase, which leads to elevated cAMP levels. However, rather than causing a loss of water from the cells, as is the case for cholera toxin, the elevated cAMP levels in throat epithelial cells cause an increase in mucus production.

In addition to pertussis toxin, *B. pertussis* synthesizes a number of other toxins and other virulence proteins. They include an adenylate cyclase enzyme that enters host cells and presumably directly increases intracellular cAMP levels by synthesizing cAMP. This observation supports the importance of increased cAMP levels to the pathogenesis of the bacterium, although the contribution of this adenylate cyclase to the symptoms is unknown. Other known toxins include one that causes necrotic lesions on the skin of mice and a cytotoxin that is a peptidoglycan fragment that kills ciliated cells of the throat. Other virulence factors are involved in the adhesion of the bacterium to the mucosal layer and fimbriae and factors required to survive nutrient deprivation, motility, etc.

REGULATION OF PERTUSSIS VIRULENCE GENES

Like the virulence genes of *C. diphtheriae* and *V. cholerae*, many of the virulence genes of *B. pertussis* are expressed only when the bacterium enters the eukaryotic host at the same time that other genes that are expressed in the free-living state are repressed. The regulation of the virulence genes of *B. pertussis* is achieved by a sensor kinase and response regulator pair of proteins encoded by linked genes, *bvgA* and *bvgS* (for *Bordetella* virulence genes); *bvgS* encodes the sensor kinase, and *bvgA* encodes the transcriptional regulator.

The BvgS-BvgA system is similar to many other sensor kinase and response regulator pairs in that the BvgS protein is a transmembrane protein, with its N terminus in the periplasm and its C terminus in the cytoplasm, allowing it to communicate information from the external environment across the membrane to the inside of the cell (Box 13.4). It also exists as a dimer in the inner membrane. Furthermore, like many other sensor kinase proteins that work in two-component systems, the BvgS protein autophosphorylates in response to a signal from the external environment and donates this phosphate to the BvgA protein, which then regulates transcription of the virulence genes.

Attempts have been made to determine the signals to which BvgS responds to phosphorylate itself. In laboratory cultures, the signal transduction pathway necessary for transcription of the pertussis toxin gene and other virulence genes is normally constitutively active but is repressed by high nicotinamide and magnesium levels. Also, expression of the virulence genes is highest at 37°C, the temperature of the human body.

Once phosphorylated, the BvgA response regulator can be either an activator or a repressor. It activates the transcription of genes required in the host and represses genes that allow it to survive outside the host. This is particularly clear for *Bordetella bronchiseptica*, a close relative of *B. pertussis* that can infect other mammals besides humans and can survive for longer times outside the host. This more versatile relative has a BvgS-BvgA system very closely related to that of *B. pertussis*; under conditions that mimic those inside its hosts, the phosphorylated BvgA protein represses a number of genes whose products are required in the free-living state (see Cummings et al., Suggested Reading).

One difference between BvgS and many other sensor kinases is that it has more than one amino acid that can accept phosphates when it receives its signal. As with other sensor kinases, one histidine in the cytoplasmic N terminus of the protein phosphorylates itself in response to conditions that mimic entrance into the host bronchial tubes. This phosphate group can then be transferred to an aspartate closer to the C terminus of the same BvgS polypeptide and then transferred again to another histidine that is even closer to the C terminus before it is finally transferred to an aspartate in the BvgA response regulator. Thus, in successive steps, the phosphate is transferred closer and closer to the C terminus of the sensor kinase and therefore closer to the BvgA response regulator in the cytoplasm. Sensor kinases like BvgS, which transfer phosphates within themselves in a phosphorelay, have been called multidomain sensors because they transfer phosphates in a phosphorelay within the same polypeptide as opposed to phosphorelays from one protein to another (Box 13.4).

There is speculation about why *Bordetella* should use a multidomain sensor to signal that it is in a mammalian throat rather than just having a single site of phosphorylation in the sensor kinase or a multiprotein phosphorelay, as in many other signal transduction systems. This speculation centers on recent research indicating that changes in gene expression occur in more than one stage. When *Bordetella* encounters conditions like those in the host, the virulence genes are not just turned on, but rather, they proceed through a least one intermediate stage (see Cotter and Jones, Suggested Reading). The three stages have been named Bvg^- for outside the host, Bvg^+ for inside the host, and Bvg^i for an intermediate state when the bacterium has just entered the host and has not yet established an infection and is not in a state where it can be spread to another host in an aerosol through coughing. Genes expressed in the Bvg^- state include those that allow it to survive in a free-living state, such as genes for carbon source utilization and growth at low temperatures. Genes expressed in the Bvg^i state are those that promote attachment to the epithelial cells in the throat, and those expressed in the Bvg^+ state are the toxin genes and others that should be expressed only when the bacterium has already established an infection and is ready to spread to other hosts. The multidomain sensor of BvgS may facilitate sensing of multiple signals to modulate the final concentration of BvgA~P.

From Genes to Regulons to Networks

Analysis of large sets of genes that respond to multiple regulatory inputs is a complex process. As we have seen in this chapter, multiple regulators can affect an individual gene or operon, and genes affected by a global regulator may simultaneously respond to other regulators. There are a variety of tools available to study genetic regulons, which include whole-genome analyses. They include microarrays and high-throughput sequencing approaches (e.g., RNASeq) to identify all transcripts in a cell under a particular condition (referred to as transcriptomics) and proteomic analyses to identify all proteins produced under a particular condition (Box 13.7). It is important to note that transcriptomic analyses may miss regulation that occurs at a posttranscriptional level, such as translational regulation or regulated proteolysis (see chapter 12), which can be detected by proteomics.

Transcriptomic and proteomic analyses are most useful if they are seen as hypothesis-generating tools rather than as an end in themselves. Regulon and stimulon components that are identified by whole-genome or high-throughput technologies can be studied further by the use of genetic techniques. Gene knockout experiments, performed so as to avoid polar effects (see chapter 12), are necessary to determine the loss-of-function phenotypes of individual genes. However, as we can see in many of the examples of genetic analysis discussed in this textbook, an understanding of gene function often requires the study of a variety of allelic forms that result from base pair changes. With knowledge of genome sequences, we can better direct base pair changes to specific regions of proteins. Regions of proteins or RNAs with presumptive functions or predicted regulatory sequences can be altered at will by using recently developed methods of site-specific mutagenesis, including methods that involve the phage λ Red system (see chapter 10). This type of reverse genetic approach is not applicable to all bacteria. One limitation is that the genetic constructs are usually made in *E. coli*, and electroporation or transformation of bacteria that are not naturally transformable can also be a limiting factor. It is clear, however, that the rapid development of new tools and techniques enhances the power of genetic approaches in analyzing complex systems and how they interact.

SUMMARY

1. The coordinated regulation of a large number of genes is called global regulation. Operons that are regulated by the same regulatory protein are part of the same regulon.

2. In catabolite regulation, the operons for the use of alternate carbon sources cannot be induced when a better carbon and energy source, such as glucose, is present. In *E. coli* and other enteric bacteria, this catabolite regulation is achieved, in part, by cAMP, which is made by adenylate cyclase, the product of the *cya* gene. When the bacteria are growing in a poor carbon source, such as lactose or galactose, the adenylate cyclase is activated and cAMP levels are high. When the bacteria are growing in a good carbon source, such as glucose, cAMP levels are low. The cAMP acts through a protein called CAP (also called CRP), the product of the *crp* gene. CAP is a transcriptional activator, which, with cAMP bound, activates the transcription of catabolite-sensitive operons, such as *lac* and *gal*.

3. Gram-positive bacteria, like *B. subtilis*, use a regulatory protein called CcpA to repress genes for carbon source utilization pathways and activate genes for carbon excretion pathways when glucose is high. CcpA activity is controlled by protein-protein interactions with a protein called Hpr that is part of the sugar transport system.

4. Bacterial cells induce different genes depending on the nitrogen sources available. Genes that are regulated through the nitrogen source are called Ntr genes. Most bacteria, including *E. coli*, prefer NH_3 as a nitrogen source and do not transcribe genes for using other nitrogen sources when growing in NH_3. Glutamine concentrations are low when NH_3 concentrations are low. A signal transduction pathway is then activated, culminating in the phosphorylation of NtrC. This signal transduction pathway begins with the GlnD protein, a uridylyltransferase, which is the sensor of the glutamine concentration in the cell. At low concentrations of glutamine, GlnD transfers UMP to the P_{II} protein, inactivating it. However, at high concentrations of glutamine, the GlnD protein removes UMP from P_{II}. The P_{II} protein without UMP attached can bind to NtrB, somehow preventing the transfer of phosphate to NtrC and causing the removal of phosphates from NtrC. The phosphorylated NtrC protein activates the transcription of the *glnA* gene, the gene for glutamine synthetase, as well as the *ntrB* gene and its own gene, *ntrC*, since they are part of the same operon as *glnA*. It also activates the trancription of operons for using other nitrogen sources.

5. The NtrB and NtrC proteins form a sensor and response regulator pair and are highly homologous to other sensor kinase and response regulator pairs in bacteria.

6. NtrC-regulated promoters of *E. coli* and other enterics require a special sigma factor called σ^N. Sigma factors related to σ^N are used to transcribe the flagellar genes and some biodegradative operons in other types of bacteria.

7. The cell also regulates the activity of glutamine synthetase by adenylating the glutamine synthetase enzyme. The enzyme is highly adenylated at high glutamine–to–α -ketoglutarate ratios, which makes it less active and subject to feedback inhibition.

8. The genes encoding the ribosomal proteins, rRNAs, and tRNAs are part of a large regulon in bacteria, with hundreds of genes that are coordinately regulated. A large proportion of the cellular energy goes into making the rRNAs, tRNAs, and ribosomal proteins; therefore, regulating the expression of these genes saves the cell considerable energy.

9. The synthesis of ribosomal proteins is coordinated by coupling the translation of the ribosomal protein genes to the amount of free rRNA that is not yet in a ribosome. The ribosomal protein genes are organized into operons, and one ribosomal protein of each operon plays the role of translational repressor. The same protein also binds to free rRNA, so that when there is excess rRNA in the cell, all of the repressor protein binds to the free rRNA, and none is available to repress translation.

10. The synthesis of rRNA and tRNA following amino acid starvation is inhibited by ppGpp, synthesized by an enzyme associated with the ribosome called RelA. All types of bacteria contain ppGpp, so the regulation may be universal. However, it is not yet clear how higher levels of ppGpp inhibit transcription of the genes for rRNA and tRNA and stimulate the transcription of others. A protein named DksA may enter the secondary channel of RNA polymerase and help ppGpp regulate transcription in *E. coli*.

SUMMARY (continued)

11. Cells contain fewer ribosomes when they are growing more slowly in poorer media. This is called growth rate control and may be due to the lower concentration of the initiating ribonucleosides, ATP and GTP, in slower-growing cells. RNA polymerase forms short-lived open complexes on the promoters for the rRNA genes, and these may have to be stabilized by immediate initiation of transcription with high concentrations of ATP and GTP. ppGpp also plays a role in growth rate control, perhaps by reducing the concentration of ATP and GTP or by competing with ATP and GTP for the initiating complex.

12. Bacteria induce a set of proteins called the heat shock proteins in response to an abrupt increase in temperature. Some of the heat shock proteins are chaperones, which assist in the refolding of denatured proteins; others are proteases, which degrade denatured proteins. The heat shock response is common to all organisms, and some of the heat shock proteins have been highly conserved throughout evolution.

13. In *E. coli*, the promoters of the heat shock genes are recognized by RNA polymerase holoenzyme with an alternative sigma factor called the heat shock sigma factor, or σ^H. The amount of this sigma factor markedly increases following heat shock, leading to increased transcription of the heat shock genes. The increase in σ^H following heat shock involves an RNA thermosensor in the mRNA encoding σ^H and DnaK, a chaperone that is one of the heat shock proteins. The DnaK protein normally binds to σ^H, targeting the sigma factor for degradation. Immediately after a heat shock, DnaK binds to other denatured proteins, making less DnaK available to bind to σ^H so that the sigma factor is stabilized and more of it accumulates.

14. In addition to the heat shock sigma factor, bacteria have other stress sigma factors that are activated by a wide variety of different stresses.

15. Bacteria also have ways of detecting stress to their membranes, including osmotic stress and damage to the outer membrane. These are called extracytoplasmic stresses.

16. One of the ways that bacteria adjust to changes in the osmolarity of the medium is by changing the ratio of their porin proteins, which form pores in the outer membrane through which solutes can pass to equalize the osmotic pressure on both sides of the membrane. The major porins of *E. coli* are OmpC and OmpF, which make pores of different sizes, thereby allowing the passage of different-size solutes. The relative amounts of OmpC and OmpF change in response to changes in the osmolarity of the medium. The *ompC* and *ompF* genes in *E. coli* are regulated by a sensor and response regulator pair of proteins, EnvZ and OmpR, which are similar to NtrB and NtrC. The EnvZ protein is an inner membrane protein with both kinase and phosphatase activities that, in response to a change in osmolarity, can transfer a phosphoryl group to or remove one from OmpR, a transcriptional activator. The state of phosphorylation of OmpR affects the relative rates of transcription of the *ompC* and *ompF* genes.

17. The ratio of OmpF to OmpC porin proteins is also affected by an antisense RNA named MicF. A region of the MicF RNA can base pair with the TIR of the OmpF mRNA and block access by ribosomes, thereby inhibiting OmpF translation. The *micF* gene is regulated by a number of transcriptional regulatory proteins, including SoxS, which induces the oxidative stress regulon.

18. Bacteria detect damage to their outer membrane by detecting the accumulation of outer membrane proteins in the periplasm. The two systems in *E. coli* are Cpx and σ^E, which respond to the accumulation in the periplasm of pilin subunits and Omp proteins, respectively.

19. The virulence genes of pathogenic bacteria can also be members of global regulons and are normally transcribed only when the bacterium is in its host.

20. The diphtheria toxin gene, *tox*, encoded by a prophage of *C. diphtheriae*, is turned on only when iron is limiting, a condition mimicking that in the host. The *tox* gene is regulated by a chromosomally encoded repressor protein, DtxR, which is similar to the Fur protein involved in regulating the genes of iron availability pathways in *E. coli* and other enteric bacteria.

21. The toxin genes of *V. cholerae* are also carried on a prophage and are regulated by a regulatory cascade that begins with a transcriptional activator, ToxR. The ToxR protein traverses the inner membrane and is activated by a second protein, ToxS. ToxR and ToxS act in concert with another gene pair, TcpP-TcpH, to activate the transcription of *toxT*, whose gene product in turn activates the transcription of virulence genes.

22. The virulence genes of *B. pertussis* are regulated by a sensor and response regulator pair of proteins, BvgS and BvgA. The regulation goes through multiple stages as the bacterium enters its host.

23. Many sRNAs play important roles in gene regulation in bacteria. Most sRNAs function by pairing with complementary sequences in mRNA, often with the help of the Hfq RNA-binding protein, and block translation or target the mRNA for degradation by RNases. Other sRNAs function by direct interaction with RNA polymerase or by titration of a regulatory protein.

24. Techniques such as microarrays and other high-throughput approaches to transcript analysis have made it possible to monitor the expression of many genes simultaneously. This has led to the discovery of many more genes belonging to the same regulon. In the techniques of proteomics, proteins can be isolated and identified by tandem mass spectrometry. If the genome sequence of the organism is known, the gene and protein can be identified.

QUESTIONS FOR THOUGHT

1. Why do you suppose that proteins involved in gene expression (i.e., transcription and translation) are among the heat shock proteins?

2. Why do you think genes for the utilization of amino acids as a nitrogen source are not under Ntr regulation in *Salmonella* spp. but are under Ntr regulation in *Klebsiella* spp.?

3. Why are the corresponding sensor kinase and response regulator genes of the various two-component systems so similar to each other?

4. Why is the enzyme responsible for ppGpp synthesis in response to amino acid starvation different from the one responsible for ppGpp synthesis during growth rate control? Why might SpoT be used to degrade ppGpp made by RelA after amino acid starvation but be used to synthesize it during growth rate control?

5. Why do bacteria use small molecules to sense other bacteria in their environment? What benefit might this have to the bacteria?

PROBLEMS

1. You have isolated a mutant of *E. coli* that cannot use either maltose or arabinose as a carbon and energy source. How would you determine if your mutant has a *cya* or *crp* mutation or whether it is a double mutant with mutations in both the *ara* operon and a *mal* operon?

2. What would you expect the phenotypes of the following mutations to be?

   a. A *glnA* (glutamine synthetase) null mutation
   b. An *ntrB* null mutation
   c. An *ntrC* null mutation
   d. A *glnD* null mutation that inactivates the UTase so that P_{II} has no UMP attached
   e. A constitutive *ntrC* mutation that changes the NtrC protein so that it no longer needs to be phosphorylated to be active
   f. A *dnaK* null mutation
   g. A *dtrR* null mutation of *C. diphtheriae*
   h. A *relA spoT* double-null mutant

3. How would you show that the toxin gene of a pathogenic bacterium is not a normal chromosomal gene but is carried on a prophage not common to all the bacteria of the species?

4. Explain how you would use gene dosage experiments to prove that the heat shock sigma factor (σ^H) gene is not transcriptionally autoregulated.

5. Explain how you would show which of the ribosomal proteins in the *rplJ-rplL* operon is the translational repressor.

SUGGESTED READING

Alba, B. M., and C. A. Gross. 2004. Regulation of the *Escherichia coli* σ^E-dependent envelope stress response. *Mol. Microbiol.* **52:**613–620.

Barker, M. M., and R. L. Gourse. 2001. Regulation of rRNA transcription correlates with nucleoside triphosphate sensing. *J. Bacteriol.* **183:**6315–6323.

Batchelor, E., D. Walthers, L. J. Kenney, and M. Goulian. 2005. The *Escherichia coli* CpxA-CpxR envelope stress response system regulates expression of the porins OmpF and OmpC. *J. Bacteriol.* **187:**5723–5731.

Browning, D. F., and S. Busby. 2004. The regulation of bacterial transcription initiation. *Nat. Rev. Microbiol.* 2:57–65.

Cashel, M., and J. Gallant. 1969. Two compounds implicated in the function of the RC gene in *Escherichia coli. Nature* **221:**838–841.

Commichau, F. M., K. Forchhammer, and J. Stulke. 2006. Regulatory links between carbon and nitrogen metabolism. *Curr. Opin. Microbiol.* 9:167–172.

Condon, C., C. Squires, and C. L. Squires. 1995. Control of rRNA transcription in *Escherichia coli. Microbiol. Rev.* **59:**623–645.

Cotter, P. A., and A. M. Jones. 2003. Phosphorelay control of virulence gene expression in *Bordetella. Trends Microbiol.* **11:**367–373.

Cummings, C. A., H. J. Bootsma, D. A. Relman, and J. F. Miller. 2006. Species- and strain-specific control of a complex flexible regulon by *Bordetella* BvgAS. *J. Bacteriol.* **188:**1775–1785.

Eisenreich, W., T. Dandekar, J. Heesemann, and W. Goebel. 2010. Carbon metabolism of intracellular pathogens and possible links to virulence. *Nat. Rev. Microbiol.* 8:401–412.

Fiil, N., and J. D. Friesen. 1968. Isolation of relaxed mutants of *Escherichia coli. J. Bacteriol.* **95:**729–731.

Hall, M. N., and T. J. Silhavy. 1981. Genetic analysis of the *ompB* locus in *Escherichia coli* K-12. *J. Mol. Biol.* **151:**1–15.

Inada, T., K. Kimada, and H. Aiba. 1996. Mechanisms responsible for glucose-lactose diauxie in *Escherichia coli*: challenge to the cAMP model. *Genes Cells* 1:293–301.

Kadner, R. J. 2005. Regulation by iron: RNA rules the rust. *J. Bacteriol.* **187:**6870–6873.

Kim, T.-J., T. A. Gaidenko, and C. W. Price. 2004. A multicomponent protein complex mediates environmental stress signaling in *Bacillus subtilis. J. Mol. Biol.* **341:**135–150.

Krasny, L., and R. L. Gourse. 2004. An alternative strategy for bacterial ribosome synthesis: *Bacillus subtilis* rRNA transcription. *EMBO J. 23*:4473–4483.

Lenz, D. H., M. B. Miller, J. Zhu, R. V. Kulkarni, and B. L. Bassler. 2005. CsrA and three redundant small RNAs regulate quorum sensing in *Vibrio cholerae*. *Mol. Microbiol.* **58**:1186–1202.

Lenz, D. H., K. C. Mok, B. N. Lilley, R. V. Kulkarni, N. S. Wingreen, and B. L. Bassler. 2004. The small RNA chaperone Hfq and multiple small RNAs control quorum sensing in *Vibrio harveyi* and *Vibrio cholerae*. *Cell* **118**:69–82.

Liberek, K., T. P. Galitski, M. Zyliez, and C. Georgopoulos. 1992. The DnaK chaperon modulates the heat shock response of *E. coli* by binding to the σ^{32} transcription factor. *Proc. Natl. Acad. Sci. USA* **89**:3516–3520.

Magasanik, B. 1982. Genetic control of nitrogen assimilation in bacteria. *Annu. Rev. Genet.* **16**:135–168.

Magnusson, L. U., A. Farewell, and T. Nystrom. 2005. ppGpp: a global regulator in *Escherichia coli. Trends Microbiol.* 13:236–242.

Meyer, A. S., and T. A. Baker. 2011. Proteolysis in the *Escherichia coli* heat shock response: a player at many levels. *Curr. Opin. Microbiol. 14*:194–199.

Nagai, H., H. Yuzawa, and T. Yura. 1991. Interplay of two *cis*-acting mRNA regions in translational control of σ^{32} synthesis during the heat shock response of *Escherichia coli. Proc. Natl. Acad. Sci. USA* **88**:10515–10519.

Nomura, M., J. L. Yates, D. Dean, and L. E. Post. 1980. Feedback regulation of ribosomal protein gene expression in *Escherichia coli*: structural homology of ribosomal RNA and ribosomal protein mRNA. *Proc. Natl. Acad. Sci. USA 77*:7084–7088.

Paul, B. J., M. M. Barker, W. Ross, D. A. Schneider, C. Webb, J. W. Foster, and R. L. Gourse. 2004. DksA: a critical component of the transcription initiation machinery that potentiates the regulation of rRNA promoters by ppGpp and the initiating NTP. *Cell* **118**:311–322.

Peredina, A., V. Setlov, M. N. Vassylyeva, T. H. Tahirov, S. Yokoyama, I. Artsimovitch, and D. G. Vassylyev. 2004. Regulation through the secondary-channel structural framework for ppGpp-DksA synergism during transcription. *Cell* **118**:297–309.

Ruiz, N., and T. J. Silhavy. 2005. Sensing external stress: watchdogs of the *Escherichia coli* cell envelope. *Curr. Opin. Microbiol. 8*:122–126.

Schmidt, M., and R. K. Holmes. 1993. Analysis of diphtheria toxin repressor-operator interactions and the characterization of mutant repressor with decreasing binding activity for divalent metals. *Mol. Microbiol.* **9**:173–181.

Schwartz, D., and J. R. Beckwith. 1970. Mutants missing a factor necessary for the expression of catabolite-sensitive operons in *E. coli*, p. 417–422. *In* J. R. Beckwith and D. Zipser (ed.), *The Lactose Operon*. Cold Spring Harbor Laboratory Press, Cold Spring Harbor, NY.

Slauch, J. M., S. Garrett, D. E. Jackson, and T. J. Silhavy. 1988. EnvZ functions through OmpR to control porin gene expression in *Escherichia coli* K-12. *J. Bacteriol.* **170**:439–441.

Sonenshein, A. L. 2007. Control of key metabolic intersections in *Bacillus subtilis. Nat. Rev. Microbiol. 5*:917–927.

van Heeswijk, W. C., S. Hoving, D. Molenaar, B. Stegeman, D. Kahn, and H. V. Westerhoff. 1996. An alternative P_{II} protein in the regulation of glutamine synthetase in *Escherichia coli. Mol. Microbiol.* **21**:133–146.

Wassarman, K. M. 2007. 6S RNA: a regulator of transcription. *Mol. Microbiol. 65*:1425–1431.

Weber, H., T. Polen, J. Heuveling, V. F. Wendisch, and R. Hengge. 2005. Genome-wide analysis of the general stress response network in *Escherichia coli*: σ^S-dependent genes, promoters, and sigma factor selectivity. *J. Bacteriol.* **187**:1591–1603.

Wray, L. V., J. M. Zalieckas, and S. H. Fisher. 2001. *Bacillus subtilis* glutamine synthetase controls gene expression through a protein-protein interaction with transcription factor TnrA. *Cell* **107**:427–435.

Yates, J. L., and M. Nomura. 1980. *E. coli* ribosomal protein L4 is a feedback regulatory protein. *Cell* **21**:517–522.

Yates, J. L., and M. Nomura. 1981. Localization of the mRNA binding sites for ribosomal proteins. *Cell* **24**:243–249.

Yoshida. T., L. Qin, L. A. Egger, and M. Inouye. 2006. Transcription regulation of *ompF* and *ompC* by a single transcription factor, OmpR. *J. Biol. Chem. 28*:17114–17123.

Yu, R. R., and V. J. DiRita. 2002. Regulation of gene expression in *Vibrio cholerae* by ToxT involves both antirepression and RNA polymerase stimulation. *Mol. Microbiol.* **43**:119–134.

Zhou, Y. N., N. Kusukawa, J. W. Erickson, C. A. Gross, and T. Yura. 1988. Isolation and characterization of *Escherichia coli* mutants that lack the heat shock sigma factor σ^{32}. *J. Bacteriol.* **170**:3640–3649.

CHAPTER 14

Bacterial Cell Biology and Development

IN EARLIER CHAPTERS, WE MOSTLY DISCUSSED processes that occur in only one of the compartments of the bacterial cell, the cytoplasmic compartment. However, bacteria also have other compartments with which the cytoplasmic compartment must communicate. For their functions, proteins must be transported into these various components and even outside the cell. Also, bacterial cells must grow and divide during their cell cycle and coordinate all of the activities in the various compartments with the time in the cell cycle. Some bacteria even go through developmental processes much like those of eukaryotic cells, and some form multicellular structures. These developmental processes are rigorously programmed and require communication between different cells in the developing structure as development proceeds.

In this chapter, we discuss a selection of the best-studied examples of cellular compartmentalization, communication, and development in bacteria and explain how molecular genetic analysis has contributed to our understanding of these phenomena. Some of these topics were touched on in earlier chapters; in this chapter, we provide more details about the processes involved and the experiments that contributed to our understanding. These are certainly some of the most important current areas of research in biological science and, besides being important in their own right, have helped inform our understanding of related phenomena in eukaryotic organisms.

Membrane Proteins and Protein Export

About one-fifth of the proteins made in a bacterium do not remain in the cytoplasm but are transported or exported into or through the surrounding membranes. The terminology is often used loosely, but we refer to proteins

doi:10.1128/9781555817169.ch14

that leave the cytoplasm as being transported. The process of transferring them through one or both membranes is **secretion**. If they are transferred through both membranes to the exterior of the cell, they are exported. Correspondingly, proteins that remain in either the inner or outer membrane are **inner membrane proteins** or **outer membrane proteins**, while those that remain in the periplasmic space are **periplasmic proteins**. Proteins that were passed all the way out of the cell into the surrounding environment are **exported proteins**.

A number of transported proteins were discussed in other chapters. For example, the LamB protein resides in the outer membrane, where it binds polymers of maltose and serves as the receptor for phage λ. The β-lactamase enzyme that makes the cell resistant to penicillin resides in the periplasm and so must be secreted through the inner membrane. The disulfide isomerase proteins also reside in the periplasm and form disulfide linkages in some periplasmic or extracellular enzymes as these other proteins pass through the periplasm (see below). The maltose-binding protein MalE also resides in the periplasm, where it can help transport maltodextrins into the cell, while MalF is in the inner membrane and helps form the channel through which maltodextrins pass into the cytoplasm. The *tonB* gene product of *Escherichia coli* also must pass through the inner membrane to its final destination in the outer membrane, where it participates in transport processes and serves as a receptor for some phages and colicins.

By far the largest group of proteins that are exported from the cytoplasm are destined for the inner membrane. Inner membrane proteins often extend through the membrane a number of times and have stretches that are in the periplasm and other stretches that are in the cytoplasm. The stretches that traverse the membrane have mostly uncharged, nonpolar (hydrophobic [see the inner cover]) amino acids, which make them more soluble in the membranes. A stretch of about 20 mostly hydrophobic amino acids is long enough to extend from one side of the bipolar lipid membrane to the other, and such stretches in proteins are called the **transmembrane domains**. The less hydrophobic stretches between them are called the **cytoplasmic domains** and the **periplasmic domains**, depending on whether they extend into the cytoplasm on one side of the membrane or into the periplasm on the other side. Proteins with domains on both sides of the membrane are called **transmembrane proteins** and have many uses, because they allow communication from outside the cell to the cytoplasm. Some transmemembrane proteins that play such a communicating role are discussed in chapter 13.

The Translocase System

Transported proteins contain many amino acids that are either polar or charged (basic or acidic), which makes it difficult for them to pass through the membranes. They must be helped in their translocation through the membrane by other specialized proteins. Some of these proteins form a channel in the membrane. Some transported proteins make their own dedicated channel, but most use the more general channel called the **translocase**, so named because its function is to translocate proteins.

A current picture of the structure of the translocase that helps proteins pass through the inner membrane, as well as how it works, is outlined in Figure 14.1. We can predict some of the features this channel must have. It must have a relatively hydrophilic inner channel through which charged and polar amino acids can pass. It also must normally be closed and open only when a protein is passing through it; otherwise, other proteins and small molecules would leak in and out of the cell through the channel. Even the leakage of molecules as small as protons cannot be tolerated, as it would destroy the proton motive force (PMF). The channel is made up of one each of three proteins, SecY, SecE, and SecG, and is therefore called the **SecYEG channel** or **SecYEG translocase**. These three proteins form a heterotrimer made up of one each of the three different polypeptides. The SecY protein is by far the largest of the three proteins and forms the major part of the channel, while the other two, smaller proteins play more ancillary, albeit important, roles. One heterotrimer, made up of one copy of each of the proteins, can form a large enough channel to let an unfolded protein through (see van den Berg et al., Suggested Reading), but it seems likely that more than one of these heterotrimers (called protomers) is involved. Some structural studies support the idea that two heterotrimers are joined back to back and that one of them plays a regulatory role, binding SecA before transport (see below).

Besides forming the major part of the channel, a region of the SecY protein forms a hydrophobic "plug," which opens only when an exported protein is passing through (Figure 14.1). The binding of a signal sequence (see below) in a protein to be transported causes the plug to move over toward SecE on the side of the channel, opening the channel, as shown in the figure. This prevents the translocation of proteins that do not have a bona fide signal sequence. SecG (not shown in the figure) is not absolutely required for protein translocation but seems to stimulate the rate of protein movement through the channel. Two other nonessential proteins, called SecD and SecF, are also bound to the channel and are highly evolutionarily conserved, but their roles in transport are not known.

The Signal Sequence

As mentioned above, the defining feature of proteins that are to be transported into the inner membrane or beyond by the SecYEG channel is the presence at their N termini of a signal sequence. The nature and fate of this signal sequence depends upon the ultimate destination

A Export channel

Translocase
SecY monomer
Ring
Exported protein
Signal sequence
Cytoplasm
Inner membrane
E
E
Periplasm
Plug
SecY monomer

B Posttranslational export

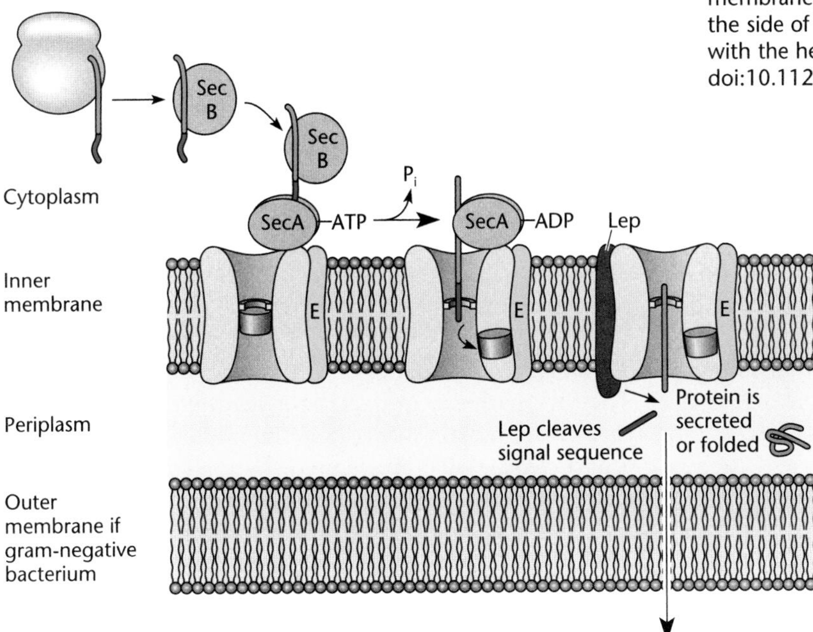

C Cotranslational transport

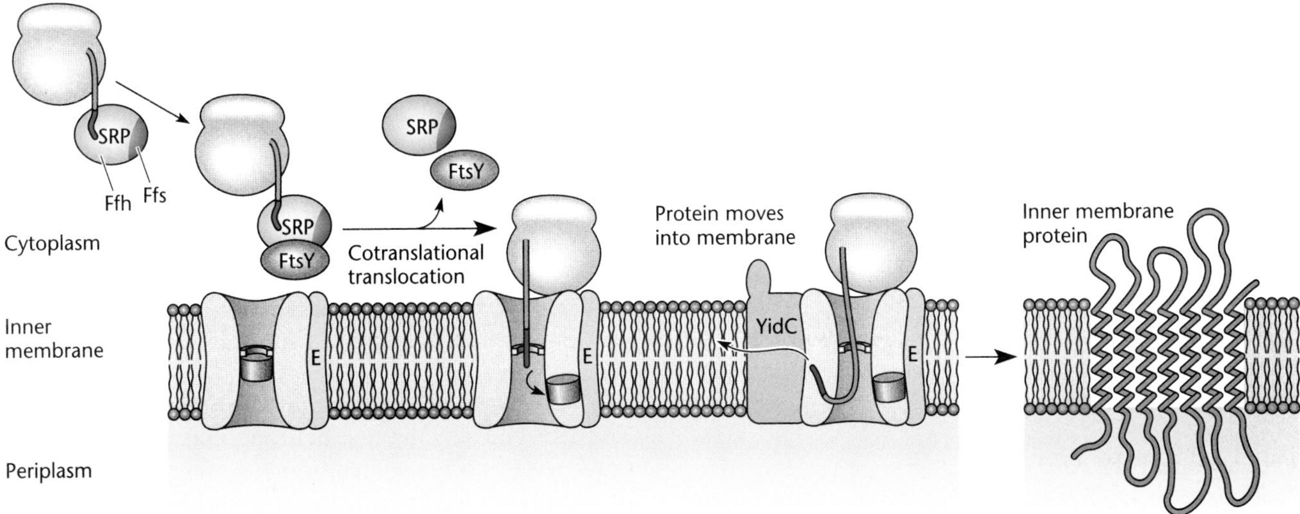

Figure 14.1 Protein transport. **(A)** Cutaway view of the secretion *sec* channel. SecY, SecE, and SecG (not shown) form the translocase. SecY forms the channel, ring, and plug. The signal sequence of the transported protein moves the plug toward SecE. **(B)** Posttranslational secretion by the SecB-SecA system. SecB keeps the protein unfolded until it binds to SecA, which interacts with SecY. The signal sequence is removed, in this case by Lep protease. The exported protein is folded in the periplasm or may be secreted across the outer membrane by one of the dedicated secretion systems in a gram-negative bacterium. **(C)** Cotranslational transport by the SRP system. SRP binds to the first transmembrane domain as it emerges from the ribosome and then binds to the FtsY docking protein, bringing the ribosome to interact with SecY. The protein is translated, driving it into the SecYEG channel. The transmembrane domains of the protein somehow escape through the side of the channel into the membrane, in some cases with the help of the YidC protein, as shown.
doi:10.1128/9781555817169.ch14.f14.1

of the transported protein. For proteins that are to be transported through the inner membrane into the periplasm and beyond, the signal sequence is approximately 20 amino acids long and consists of a basic region at the N terminus, called the n region, followed by a mostly hydrophobic region called the h region and then a region with some polar amino acids, called the c region. In contrast, most proteins whose final destination is the inner membrane merely use their first N-terminal transmembrane domain as a signal sequence.

Also, if the protein is to be secreted through the membrane, the signal sequence is removed by a protease as the protein passes through the SecYEG channel (Figure 14.1B). The most prevalent of the proteases that clip off signal sequences in *E. coli* is the **Lep protease** (for *le*ader *p*eptide protease), but there is at least one other, more specialized protease called LspA, which removes the leader sequence from some lipoproteins destined for the outer membrane. Proteins that are destined to be transported beyond the inner membrane but have just been synthesized and so still retain their signal sequences are called **presecretory proteins**. When the short signal sequence is removed in the SecYEG channel, the presecretory protein becomes somewhat shorter before it reaches its final destination in the periplasm or the outer membrane or outside the cell. This shortening of the protein after it is synthesized is easy to detect on sodium dodecyl sulfate-polyacrylamide gels and is often taken as evidence that the protein is secreted through the SecYEG channel.

The Targeting Factors

The targeting factors recognize proteins to be transported into or through the inner membrane and help target them to the membrane. Which type of signal sequence a protein has determines which of the targeting factors directs it to the SecYEG translocon. Enteric gram-negative bacteria like *E. coli* have at least two separate systems that target proteins to and through the membranes. One of these is the **SecB** system. This targeting system is dedicated to proteins that are directed through the inner membrane into the periplasm or exported from the cell. The other system is the **signal recognition particle (SRP)** system, which may exist in all organisms, including humans. In bacteria, this targeting system seems to be dedicated to proteins that are mostly destined to reside in the inner membrane. Another protein, SecA, participates in both pathways, at least for some proteins; it is found in all bacteria, but not in archaea or eukaryotes, although in eukaryotes other proteins may play a similar role.

THE SecB PATHWAY

Proteins that have a removable signal sequence and are transported through the inner membrane into the periplasm or beyond are most often targeted by the SecB system in *E. coli* and the other gram-negative bacteria that have it. The SecB protein is a specialized chaperone that binds to presecretory proteins either cotranslationally (e.g., as soon as the N-terminal region of the polypeptide emerges from the ribosome) or after they are completely synthesized (**posttranslational translocation**), thereby preventing them from folding prematurely and ensuring that the signal sequence is exposed. The SecB chaperone then passes the unfolded protein to SecA, which facilitates the association of the protein with the SecYEG channel, perhaps by binding simultaneously to the signal sequence and to a SecYEG heterotrimer, either the one forming the channel in the membrane, as in Figure 14.1B, or one associated with it (see above). Some bacteria have paralogs of SecA that may be dedicated to transporting only one or very few proteins. After SecA binds to the channel, the cleavage of ATP to ADP on SecA provides the energy to drive the protein into the channel, aided by the PMF of the membrane. As the protein passes through the channel, it loses its signal sequence, as shown in the figure. SecB is not an essential protein, and the cell can use DnaK or other general chaperones as substitutes for SecB to help transport some proteins.

THE SRP PATHWAY

The SRP pathway in bacteria generally targets proteins that are to remain in the inner membrane. It consists of a particle (the SRP) made up of both a small **4.5S RNA**, encoded by the *ffs* gene, and at least one protein, **Ffh**, as well as a specific receptor on the membrane, called **FtsY** in *E. coli*, to which the SRP binds. FtsY is sometimes referred to as the docking protein because it "docks" proteins targeted by the SRP pathway to the SecYEG channel in the membrane. Its name is a misnomer. The *ftsY* (filament temperature-sensitive Y) gene was originally identified through temperature-sensitive mutations that cause *E. coli* not to divide properly and to form long filaments of many cells linked end to end at higher temperatures. Such temperature-sensitive mutations have played an important role in understanding cell division in bacteria and are discussed later in the chapter. However, the role of FtsY in cell division is indirect. Apparently, FtsY is required to insert one or more inner membrane proteins required for cell division into the inner membrane, and these proteins are not inserted properly at higher temperatures due to the mutational defect in FtsY.

Figure 14.1C illustrates how the SRP system works. The SRP binds to the first hydrophobic transmembrane sequence of an inner membrane protein as this region of the protein emerges from the ribosome. It is debatable whether binding of the SRP stops translation of the emerging protein in bacteria as it does in eukaryotes (see below) or whether the particle binds quickly enough so that the complex has time to bind to the FtsY receptor in the membrane before translation continues. It is also not clear whether the FtsY receptor remains on the membrane or can bind to the SRP complex in the cytoplasm and then direct it to the membrane. In any case, once the complex has bound to the membrane, synthesis of the protein continues, feeding the protein directly into the SecYEG translocon as the protein emerges from the ribosome. The energy of translation due to cleavage of GTP to GDP drives the polypeptide out of the ribosome into the SecYEG translocon, obviating the need for SecA, at least in most cases, although the SecA protein might still be required for transmembrane proteins with long periplasmic domains.

The process of translating a protein as it is inserted into the translocon is called **cotranslational translocation.** In eukaryotes, it appears that all proteins are transported this way. There is a good reason why proteins destined for the inner membrane are cotranslated with their insertion into the translocon in the membrane while proteins targeted by the SecB pathway can first be translated in their entirety and then inserted into the translocon. Inner membrane proteins are much more hydrophobic than exported proteins and would form an insoluble aggregate in the aqueous cytoplasm if they were translated in their entirety before being transported into the membrane (see Lee and Bernstein, Suggested Reading).

A Lateral Gate?

What happens after an inner membrane protein enters the SecYEG channel is less clear. The transmembrane domains of the protein must escape the SecYEG channel and enter the surrounding membrane, while the periplasmic and cytoplasmic domains must stay in the correct compartments. Presumably, the SecYEG channel has a lateral gate that opens and allows the transmembrane domains of the protein to escape into the membrane. Another inner membrane protein called **YidC** might help in this process (Figure 14.1C) (see Xie and Dalbey, Suggested Reading). The role of YidC in protein translocation is not clear, but it seems to be required for the lateral escape of some proteins, but not others. Some inner membrane proteins bypass SecYEG altogether and require only YidC to enter the inner membrane.

Sec Systems of Archaea and Eukaryotes

It is interesting to compare the **Sec systems** of *E. coli* with those of archaea and eukaryotes. Archaea and eukaryotes do not have SecB or SecA and use the SRP system to translocate all exported proteins through the translocon. Although they lack SecA, they may have other systems that help direct already translated proteins to the translocon. The translocon itself was first discovered in eukaryotes and is composed of three proteins that form similar structures in all three kingdoms of life. It is highly conserved evolutionarily, and the amino acid sequences of the SecY and SecE subunits are similar in all three kingdoms; only the sequence of the third subunit (SecG in bacteria) is very different in eukaryotes and archaea, where it may have different functions. While eukaryotes have other such channels, the translocase, which helps transported proteins to enter the endoplasmic reticulum of eukaryotic cells, is the one most similar to the SecYEG channel of bacteria. This may not be a coincidence, since the endoplasmic reticulum of eukaryotes plays a role in protein translocation similar to the role played by the inner membrane of bacteria.

The YidC protein might also be conserved in all three domains of life. Interestingly, mitochondria lack a SecYEG translocon and have only a YidC homolog, called Oxa1 in yeast. Oxa1 is required to transport electron transport components into the mitochondrial membrane, which is similar to the role YidC plays in *E. coli.* These components are some of the few proteins in bacteria known to require only YidC for insertion into the membrane. Chloroplasts have both a SecYEG translocon and YidC. Archaea have a translocon, and some may also have a YidC homolog, although a putative *yidC* gene has been identified only by sequence homology and has not actually been shown to play the same role.

The SRP system was also first described in eukaryotes, where it is much larger, consisting of a 300-nucleotide RNA and eight proteins, six in the SRP and two in the docking protein, called the SRP receptor (SR). However, some of the proteins in eukaryotes are very similar to those in bacteria, such as the Ffh protein (for fifty-four homolog) in the SRP of bacteria, which is similar to the 54-kDa SRP protein in eukaryotes. The SRP system of eukaryotes targets both membrane and presecretory proteins to the endoplasmic reticulum, which is, as mentioned above, the organelle that plays a role in protein transport analogous to the role played by the inner membrane of bacteria and which seems to recognize both proteins with a removable signal sequence and those that use the first transmembrane domain as their signal sequence. Furthermore, the SRP of eukaryotes is known to stop translation of both secreted and inner membrane proteins as the signal sequence emerges from the ribosome, enforcing cotranslational transport of both types of protein (see Wild et al., Suggested Reading). When the SRP binds to the signal sequence of a protein as the protein emerges from the exit channel, it is long enough to extend all the way to the A site on the ribosome, blocking the entrance of aminoacylated tRNAs into the A site and stopping translation. When the SRP-ribosome complex then binds to the SR docking or receptor protein, the SRP is removed and translation resumes, feeding the protein directly into the translocon channel. If binding of the SRP to the emerging signal sequence also arrests translation in bacteria, it must be by a different mechanism, since the smaller bacterial SRP could not extend all the way from the exit channel to the ribosome A site.

The Tat Secretion Pathway

Not all proteins that are transported into or through the cytoplasmic membrane use the SecYEG channel. Only proteins that have not yet folded and so are still long, flexible polypeptides can be transported by this narrow channel. Once folded into their final three-dimensional structure, proteins are much too wide to fit through the channel. However, some proteins must fold in the cytoplasm before they can be transported. Often, these are membrane proteins that contain redox factors, such as molybdopterin and FeS clusters, which are synthesized in the cytoplasm and can be inserted into the protein only after it has folded. These cofactors would not be available in the periplasm when the protein folded. Other examples of proteins that are transported after they are folded are some heterodimers in which only one member has a signal sequence, so the other partner would be left behind if the signal sequence-containing polypeptide is transported before the two polypeptides have combined and folded. Folded proteins and complexes are transported by the Tat system.

STRUCTURE OF THE Tat SYSTEM

The Tat system of *E. coli* has three subunits, TatA, TatB, and TatC, while that of *Bacillus subtilis* has only two, with TatA and TatB seemingly combined into one larger subunit. In *E. coli*, TatB and TatC might initially bind the signal peptide on the protein to be transported and then recruit TatA, which forms the channel in the membranes. In this way, the channel may form only when there is a protein to be transported. Unlike the Sec translocon, which uses both the energy of ATP cleavage by SecA and the PMF to drive the protein through the channel, the Tat system may use only the PMF.

The Tat Signal Sequence

The signal sequence recognized by the Tat system is structurally similar to the signal sequence recognized by the Sec system, with a positively charged n region followed by a longer hydrophobic h region and a polar c region, and is cleaved from the protein as it passes through the channel. However, it is somewhat longer, especially in the positively charged n region. Two of the positively charged amino acids at the junction of the n and h regions are usually arginines, which give the Tat system its name. Tat stands for twin-arginine transport. Characteristically, the arginines are in the motif S-R-R-, although the first of the twin arginines is sometimes a lysine (K). This sequence is followed by two hydrophobic amino acids, usually F and L, and then often by a K.

The presence of this particular signal sequence at the N terminus of a newly synthesized protein indicates that the protein is to be transported by the Tat system rather than by the SecYEG channel. This raises an interesting question. How does the system know the protein has already folded properly and contains all the needed cofactors, etc., so that it is time to transport it? If anything, the signal sequence should be more accessible in the polypeptide before rather than after it has folded. The Tat system needs a "quality control" system to ensure that it transports only properly folded proteins and does not transport proteins that are unfolded or only partially folded. This quality control system should also be specific for each protein to be transported, since each type of folded protein has a unique structure (see chapter 2).

The *E. coli* cell seems to solve this problem by encoding dedicated proteins that specifically bind to the Tat signal sequence of only one type of protein and come off only when that protein has folded properly (see Palmer et al., Suggested Reading). The HyaE protein may play this role for HyaA, the small subunit of hydrogenase 1, and a different protein, HybE, may interact with the signal sequence of the small subunit HybO of hydrogenase 2. Interestingly, in the latter case, the HybE protein may also interact with the large subunit of hydrogenase 2, HybC, which lacks a signal sequence. The hydrogenases are heterodimers composed of the two subunits and an NiFe cofactor. When the protein folds, the interaction with the newly acquired large subunit in the proper position may cause the HybE protein to come off the signal sequence, indicating that the two subunits of hydrogenase 2 have already come together, folded, and bound the NiFe cofactor and that hydrogenase 2 is ready to be transported.

Tat Systems in Other Organisms

At least most bacteria and archaea, as well as the chloroplasts of plants (remember that chloroplasts are descended from cyanobacteria [see the introduction]), have a Tat secretion system, although they might differ from that of *E. coli* in the number of subunits. Some bacteria, including *B. subtilis*, have two or more Tat systems. Some of them are dedicated to the transport of only one or very few proteins.

Disulfide Bonds

Another characteristic of proteins that are exported to the periplasm or secreted outside the cell is that many of them have disulfide bonds between cysteines (C) (see the inside front cover). In other words, two of the cysteines in the protein are held together by covalent bonds between their sulfides. The sulfur atom of a cysteine in a disulfide bond is in its oxidized form because one of its electrons is shared by the two sulfurs, while the sulfur atom of an unbound cysteine is in its reduced form because it has unshared electrons. These disulfide bonds can be either between two cysteines in the same polypeptide chain or between cysteines in different polypeptide chains. Exported proteins need the covalent disulfide bonds to hold them together in the harsh environments of the periplasm and outside the cell. Failure to form the correct disulfide bonds or formation of disulfide bonds between the wrong cysteines can result in inactivity of the protein.

The disulfide bonds in proteins are formed by enzymes called **disulfide oxidoreductases**, DsbA, DsbB, etc., as the proteins pass through the oxidizing environment in the periplasmic space between the inner and outer membranes of gram-negative bacteria. The oxidoreductases contain the motif C-X-X-C, two cysteines separated by two other amino acids, where X can be any amino acid. This is sometimes called the thioredoxin motif because it is also found in thioredoxin, which plays a role in reducing proteins inside the cytoplasm (see Tan et al., Suggested Reading). The way the oxidoreductases work to create disulfide bonds in the periplasm is illustrated in Figure 14.2. Basically, the disulfide bonds are formed by passing electrons to stronger oxidizers from weaker ones down a chain. The disufide bond in the exported protein forms when an extra electron is passed from a cysteine in the protein being exported to a stronger oxidizer, like DsbA, in the periplasm. This oxidizes the cysteine in the exported protein, causing it to form a disulfide bond with another cysteine somewhere else in the exported protein. The electron transferred to DsbA reduces the cysteines in its C-X-X-C (actually C-P-H-C) motif, destroying their disulfide bond. These cysteines in DsbA are in turn oxidized to form another disulfide bond by passing the extra electron to DsbB, reducing its C-X-X-C motif. DsbB in turn passes the electron to quinones in the membrane, whose job it is to pass electrons to electron acceptors, such as oxygen, during electron transport. The disulfide bond in the C-P-H-C motif in DsbA can then be used to form another disulfide bond between two cysteines in another exported protein in the periplasm.

Not only do proteins that are found inside the cell in the cytoplasm lack disulfide bonds, but this type of bond also cannot normally form in the cytoplasm. This is because of the "reducing atmosphere" inside the cytoplasm due to the presence of high concentrations of small reducing molecules, mostly glutathione and thioredoxin. In fact, the appearance of disulfide bonds in some cytoplasmic regulatory proteins is taken as a signal by the cell that oxidizing chemicals are accumulating in the cell and that proteins should be made to combat the potentially lethal oxidative chemical stress, as we have discussed in earlier chapters.

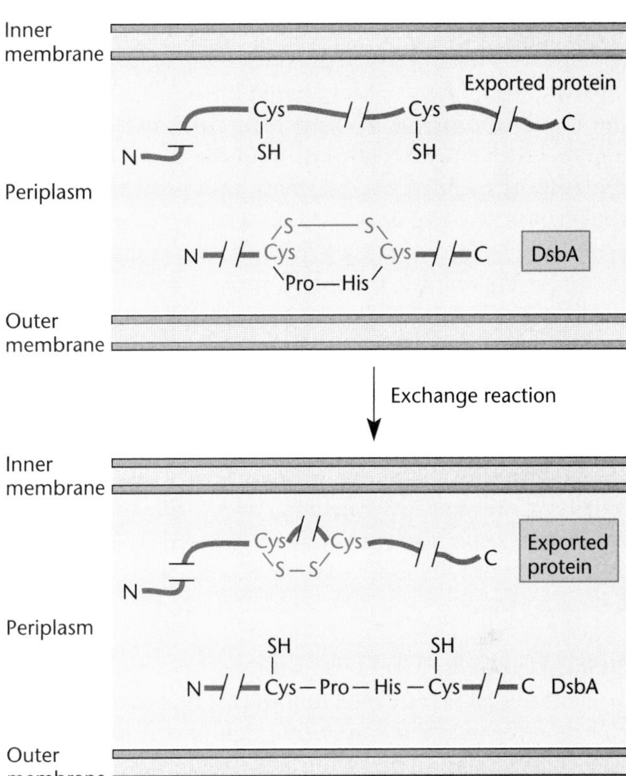

Figure 14.2 Disulfide bond formation in the periplasm. Oxidoreductases in the periplasm exchange disulfide bonds (in blue) with the protein as it enters the periplasm. Only DsbA is shown. The broken line indicates that the cysteines in the exported protein are different distances apart and can even be on different polypeptides. Details are given in the text. doi:10.1128/9781555817169.ch14.f14.2

Use of *mal-lac* Fusions To Study Protein Transport in *E. coli*

Identification of the genes and proteins of the protein transport systems in *E. coli* involved some very elegant genetic experiments, which also demonstrate the power of selectional genetics. Most of the sequences and genes involved in protein transport could be expected to be essential, so only relatively rare change-of-function base pair changes can be isolated in these genes. Therefore, it required powerful selections, or at least easily screened-for phenotypes. The selections were based on the ability or inability of *E. coli* mutants to transport a single protein into the inner membrane or beyond. The proteins chosen were those of the maltose transport system that transports maltose and maltodextrins, polymers of maltose, into the cell. To fulfil their roles, these proteins, which are the products of the *mal* operons, must be transported into or through the inner membrane and into the periplasm or outer membrane. We discussed the regulation of the *mal* operons in chapter 12, and Figure 12.20 shows the location in the membranes and the role of each of the proteins in transport. Which gene product to use depended on what types of mutants were being selected. For a protein that is transported to the outer membrane, the LamB protein was used; for a protein that is transported into the periplasm, the MalE protein was used; and for a protein that is transported into the inner membrane, the MalF protein was used. Since the transport of a protein is mostly determined by its N terminus, including the signal sequence, only the N terminus of the transported protein was needed. To make the mutants easier to select, the N terminus of the *mal* gene was translationally fused to a reporter gene, generally the *lacZ* gene of *E. coli*, which encodes β-galactosidase.

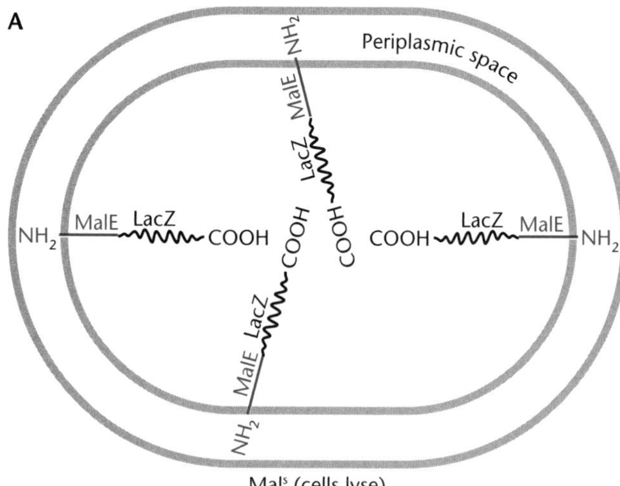

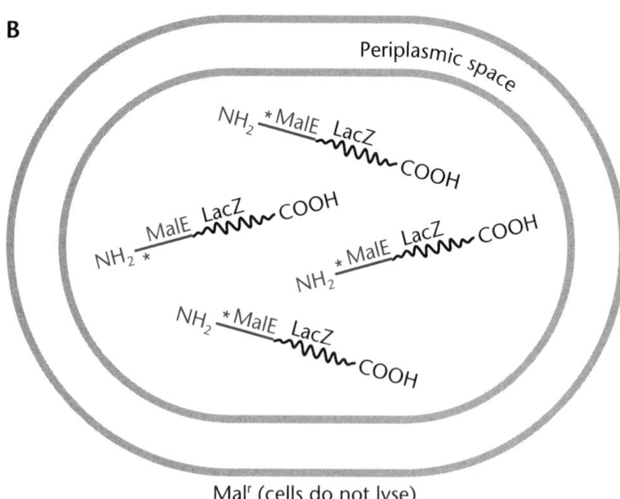

Figure 14.3 Model for the maltose sensitivity (Mal^s) of cells containing a *malE-lacZ* gene fusion. **(A)** In Mal^s cells, the presence of maltose induces the synthesis of the fusion protein, which cannot be transported completely through the membrane and so lodges in the membrane, causing the cell to lyse. **(B)** In Mal^r cells, a mutation in the region encoding the signal sequence (asterisks) prevents transport of the fusion protein into the membrane.
doi:10.1128/9781555817169.ch14.f14.3

SIGNAL SEQUENCE MUTATIONS

To isolate signal sequence mutations, the selection was based on the fact that a mutation that inactivates the signal sequence should prevent transfer of a MalE-LacZ fusion protein into the SecYEG channel. For reasons that are not completely understood, the transfer of the large fusion protein including LacZ kills the cell, perhaps because it jams the SecYEG channels. This offers positive selection for mutants with mutations that inactivate the signal sequence, as shown in Figure 14.3. Synthesis of the fusion protein is induced by maltose, so the cells are killed by maltose in the medium (Mal susceptible [Mal^s]). A signal sequence mutation that prevents transport of the fusion protein into the SecYEG channel will allow the cell to survive, so the cell will become maltose resistant (Mal^r) and make colonies in the presence of maltose. Of course, the investigators had to show that the responsible mutations mapped in the coding region for the N terminus of the MalE protein and did not affect the stability of the fusion protein.

On the basis of this selection, a number of signal sequence mutations were isolated and later sequenced to determine which amino acid changes could prevent the function of the signal sequence. Many of these were changes from hydrophobic to charged amino acids in the hydrophobic middle region of the signal peptide. These changes presumably interfere with the insertion of the signal sequence into the SecYEG channel in the hydrophobic membrane or its recognition by the SecB-SecA targeting system.

Sec MUTATIONS

It was possible to use a similar selection to isolate mutations in the *sec* genes encoding proteins required for protein transport. However, unlike signal sequence mutations, which affect only the transport of the fusion protein, a mutation in a *sec* gene should affect the transport of many proteins and is apt to be lethal if it completely inactivates the Sec protein. Therefore, a somewhat different, more sensitive selection was needed to select mutations in the *sec* genes.

Why the selection works is illustrated in Figure 14.4A for *secB* mutants and is based on the observation that cells containing a particular *malE-lacZ* fusion do make

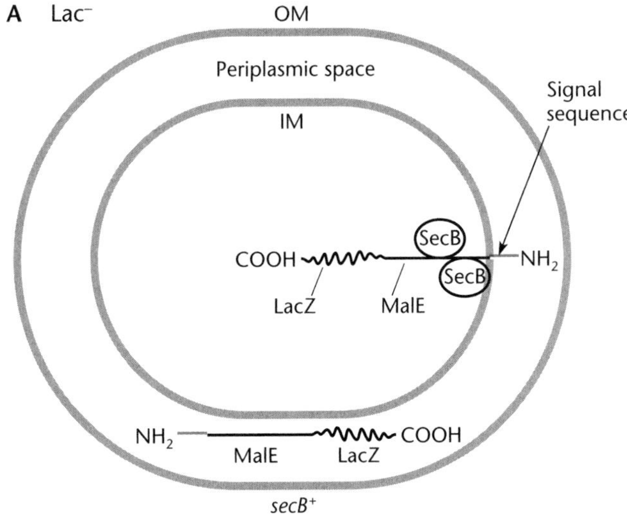

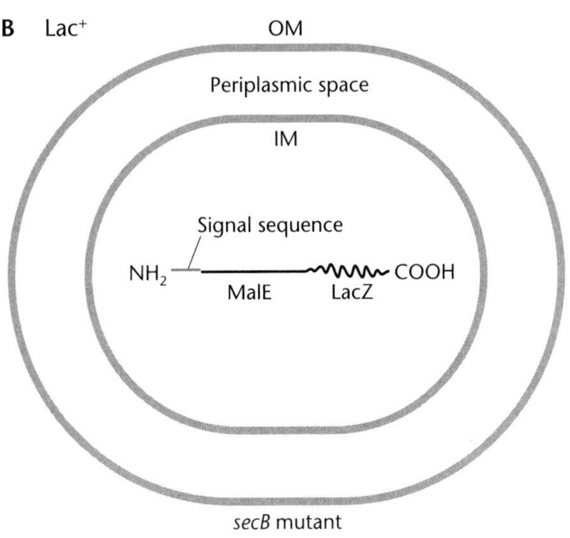

Figure 14.4 Function of SecB in protein transport and how *secB* mutations can be selected. **(A)** The SecB protein (black) prevents premature folding of a protein, keeping the signal sequence (blue) exposed so that it can enter the membrane after protein synthesis is complete. The fusion protein can then enter the periplasm, where the LacZ portion is not active. **(B)** In a *secB* mutant, the signal sequence folds into the interior of the protein, so the protein cannot be transported. The fusion protein stays in the cytoplasm, where the β-galactosidase portion of the fusion is active. OM, outer membrane; IM, inner membrane. doi:10.1128/9781555817169.ch14.f14.4

some MalE-LacZ fusion protein, even in the absence of maltose in the medium, but do not make enough to kill the cells. Nevertheless, even though Western blots indicated that the cells were making enough fusion protein so that they should be phenotypically Lac$^+$, the cells were instead Lac$^-$ and unable to multiply with lactose as the sole carbon and energy source, because what fusion protein was made was transported into the periplasm (Figure 14.4A) or, more likely, was being jammed in the SecYEG channel.

The fact that partial transport of the fusion protein inactivates its β-galactosidase activity, thereby preventing growth on lactose, offered a selection for *sec* mutations that prevent its transport. Any mutation that prevents transport of the fusion protein through the membrane, e.g., a *secB* mutation, as shown in the figure, should make the cells Lac$^+$ and able to form colonies on lactose minimal plates, because at least some of the MalE-LacZ fusion proteins should remain in the cytoplasm and retain their β-galactosidase activity (Figure 14.4B). Therefore, to isolate *sec* mutants, the investigators could merely plate cells containing the *malE-lacZ* fusion on minimal plates containing lactose but no maltose. They identified six different *sec* genes in this way and named them *secA*, *secB*, *secD*, *secE*, *secF*, and *secY*.

ISOLATION OF MUTATIONS IN THE SRP PATHWAY FOR INNER MEMBRANE PROTEINS

To isolate mutations in the SRP system, which mostly transports proteins that remain in the inner membrane, the investigators used the inner membrane protein MalF, which resides in the inner membrane with some domains in the periplasm and some in the cytoplasm and therefore presumably uses the SRP system for its transport into the membrane (Figure 12.20). They fused the N-terminal coding region for MalF, including the first transmembrane domain and the first periplasmic domain, to *lacZ* and introduced the gene fusion into a phage λ vector, which they then integrated into the chromosome. Now, however, rather than selecting for resistance to maltose or the ability to grow on lactose as a sole carbon source, they looked for blue colonies on 5-bromo-4-chloro-3-indolyl-β-D-galactopyranoside (X-Gal) plates. Cells containing this fusion normally make white colonies on X-Gal plates, perhaps because the MalF portion of the fusion protein is transported into the inner membrane, dragging the β-galactosidase portion of the fusion along with it, possibly as far as the periplasm. Whatever the reason, the transport of the β-galactosidase in the fusion protein somehow makes it inactive, so that it cannot cleave the X-Gal on the plates and turn the colonies blue, and therefore, they remain white. Any bacteria that form blue colonies are candidates for mutants that do not transport the MalF portion of the fusion protein into the inner membrane.

The mutants that were isolated in this way had mutations in the gene encoding the 4.5S RNA part of the SRP particle, in the *ffh* gene, and in the *ftsY* gene. Surprisingly, the only mutations found in a *sec* gene were in *secM*, the synthesis of whose product in the ribosome exit channel plays a role in regulating SecA (see Figure 12.28). Why they did not find mutations in the *secY* or *secE* genes is unclear, since we now know that the products of these genes form the channel that also plays a role in transporting proteins targeted by the SRP system into the inner membrane.

SUPPRESSORS OF SIGNAL SEQUENCE AND *sec* MUTATIONS

Just as the *mal* genes could be used to isolate signal sequence and *sec* mutations, they could be used to isolate suppressors of these mutations. However, the original selections would mean a screen for such suppressors, because mutants with the suppressor would not grow under the selective conditions. To convert what is a screen for suppressors of signal sequence mutations into a positive selection, the researchers reconstructed complete *malE* genes from the *malE-lacZ* fusion genes containing each of the signal sequence mutations. A strain containing a mutated *malE* gene cannot grow with maltose as the sole carbon and energy source. Any bacteria that form colonies on maltose minimal plates are candidates for mutants with mutations that suppress the signal sequence mutation in the *malE* gene. If they map to the *malE* gene, they are intragenic suppressors or may even be revertants of the original signal sequence mutation. If they map elsewhere, they are extragenic suppressors and are apt to be in one of the *sec* genes. Some extragenic suppressors were in *secY* and *secE* and seemed to open the SecYEG channel sufficiently so that a fully functional signal sequence was no longer required for transport of the protein. Such mutants, while sick (presumably because they leak other molecules through the SecYEG channel that should not be leaked), are still viable. If this is the case, then combining two such mutations, for example, one in *secY* and one in *secE*, in the same mutant strain might cause the channels to open even more and be lethal; such an outcome is called synthetic lethality. This test is possible because the *sec* mutations are dominant (see chapter 3). Each channel has only one copy of the protein, and even if some are constructed from a wild-type protein, those constructed from the mutant protein will be leaky, causing the phenotype. To do the test, one gene, say *secE*, with one of the mutations is expressed from an expression vector plasmid from an inducible promoter in a cell with *secY* in the chromosome with the other mutation. If adding the inducer kills the cell, the combination of the two mutations is lethal, and they have demonstrated conditional lethality. Such experiments have been very useful in interpreting the results of structural studies and have contributed to the model for transport through the SecYEG channel shown in Figure 14.1.

Genetic Analysis of Transmembrane Domains of Inner Membrane Proteins in Gram-Negative Bacteria

Rather than being completely buried in the membrane, most inner membrane proteins of gram-negative bacteria have regions exposed both to the cytoplasm and to the periplasm. Having exposed regions on both sides of the membrane allows the membrane protein to pass information from the external environment to the interior of the cell, or the reverse. As defined above, proteins that are exposed at both surfaces of a membrane are called transmembrane proteins, and the regions of the polypeptide that traverse the membrane from one surface to the other are called the transmembrane domains, which can often be distinguished by their chain of 20 or more mostly hydrophobic amino acids. Regions with more hydrophilic charged and polar amino acids define the cytoplasmic and periplasmic domains. Some transmembrane proteins traverse the membrane many times. The transmembrane domains alternate with the cytoplasmic and periplasmic domains (e.g., cytoplasmic-transmembrane-periplasmic-transmembrane-cytoplasmic, etc.), and once you know whether one of the hydrophilic domains is in the cytoplasm or the periplasm, you can count the transmembrane domains between them to assign the others to one compartment or the other. The term **membrane topology** refers to the way the different sections of the protein are distributed in the membrane and in the external and internal compartments.

Translational fusions to the alkaline phosphatase gene (*phoA*) of *E. coli* have been used to study the membrane topology of inner membrane proteins of *E. coli* and some closely related gram-negative bacteria (see, for example, San Milan et al., Suggested Reading). The alkaline phosphatase product of the *phoA* gene is a scavenger enzyme that cleaves phosphates off larger molecules so that the phosphates can be transported into the cell to be used in cellular reactions. PhoA is active only in the periplasmic compartment, because it must form a homodimer of two identical polypeptide products of the *phoA* gene held together by disulfide bonds that form only in the periplasm. The PhoA enzyme is also easy to assay, and bacteria that synthesize active PhoA make blue colonies on plates containing the chromogenic compound 5-bromo-4-chloro-3-indolylphosphate (XP), which turns blue when the phosphate is cleaved off by alkaline phosphatase.

The way *phoA* translational fusions can be used to determine the domains of a transmembrane protein that are in the periplasm is illustrated in Figure 14.5. Briefly, to determine if a hydrophilic domain of a membrane protein is in the periplasm, the carboxy-terminal coding region for PhoA, without its signal sequence, is translationally fused to the coding sequence for the domain. If the region of the protein to which PhoA is fused is in the periplasm, the cells will form blue colonies on XP plates; otherwise, they will be colorless.

Figure 14.5 Using *phoA* fusions to determine the membrane topology of a transmembrane protein. **(A)** Transmembrane protein, showing both periplasmic and cytoplasmic domains. **(B and C)** Fusion that joins the transmembrane protein to alkaline phosphatase (AP) at x or z leaves the alkaline phosphatase in the periplasm, where it is active. The bacteria form blue colonies on XP plates. **(D)** The transmembrane protein is fused at y to alkaline phosphatase, leaving the alkaline phosphatase in the cytoplasm, where it is inactive. The bacteria form colorless colonies on XP plates.
doi:10.1128/9781555817169.ch14.f14.5

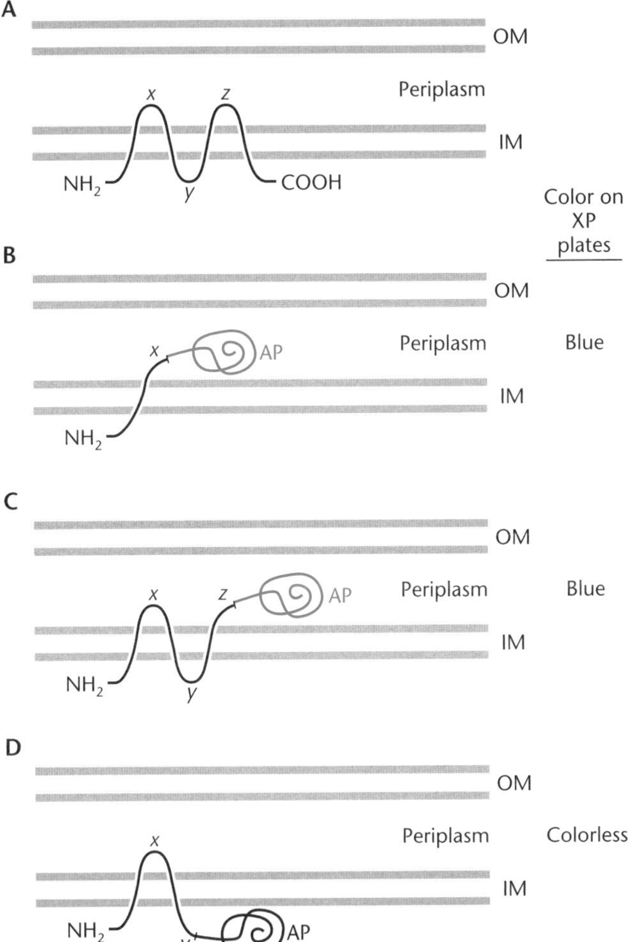

Identification of Genes for Inner Membrane Proteins by Random *phoA* Fusions

The fact that PhoA is active only in the periplasm also allows it to be used to identify inner membrane proteins in gram-negative bacteria. Some transposons have been engineered to generate random gene fusions by transposon mutagenesis (see chapter 9). These transposons contain a reporter gene that is expressed only if the transposon hops into an expressed gene in the correct orientation. One such transposon, Tn*phoA*, was developed to identify genes whose protein products are transported into or through the inner membrane (see Gutierrez et al., Suggested Reading). Tn*phoA* has the *phoA* reporter gene inserted so that it does not have its own promoter or translational initiation region and lacks its own signal-sequence-coding region. A fusion protein with PhoA fused to another protein is synthesized whenever the transposon hops into an expressed open reading frame (ORF) in such a way that PhoA is translated in the correct reading frame. The PhoA part of the fusion protein has alkaline phosphatase activity and turns colonies blue on XP plates if Tn*phoA* has integrated into a gene whose protein product is translocated and the *phoA* gene happens to be fused to a periplasmic domain of the protein. Obviously, such insertions are rare, but blue colonies can be easily spotted, even on plates with crowded colonies.

Protein Secretion

Some proteins are transported through the membranes to the outside of the cell, where they can remain attached, enter the surrounding medium, or even directly enter another cell. As mentioned above, these proteins are called exported proteins, and the structures that export them are called **protein secretion systems.** The process differs markedly between gram-negative and gram-positive bacteria, since the latter do not have an outer membrane. In gram-negative bacteria, once a protein has been translocated through the inner membrane, it is only in the periplasm and still faces the challenge of getting through the extremely hydrophobic outer membrane. Because of the additional challenge created by the outer membrane, gram-negative bacteria have developed elaborate specialized structures to export proteins. Many of these play important roles in bacterial pathogenesis, so they have attracted considerable attention. These structures are discussed in the next section.

Protein Secretion Systems in Gram-Negative Bacteria

The protein secretion systems of gram-negative bacteria come in at least six basic types, imaginatively named types I to VI. All of these secretion systems rely on channels in the outer membrane (called β-barrels or secretins) formed from β-sheets organized in a ring (see Figure 2.25 for an explanation of protein secondary and tertiary structures). The β-barrels are assembled so that the side chains of charged and polar amino acids tend to be in the center of the barrel, where they are in contact with hydrophilic proteins that are passing through, while the side chains of hydrophobic amino acids are on the outside of the barrel in contact with the very hydrophobic surrounding membrane. Assembly of the β barrels requires a complex of proteins called the Bam complex (BamA, -B, -C, etc.) and periplasmic chaperones, including Skp.

Having channels in the outer membrane presents some of the same problems associated with having channels in the cytoplasmic membrane, such as the SecYEG channel. For example, how do they select the proteins that are to go through without letting others through, and how do they keep smaller molecules from going in and out? This process is called **channel gating**; the gate is open only when the protein being exported passes through. However, they also have problems unique to them. Where does the energy come from to export a protein through the outer membrane? There is no ATP or GTP in the periplasmic space to provide energy, and the outer membrane is not known to have a proton gradient across it to create an electric field. Also, how do they themselves get through the inner membrane to reach the outer membrane, where the secretion system is assembled? Not all of these questions have been completely answered, but in this section, we try to address possible mechanisms used by the various secretion systems for solving these and other problems. We also mention some examples of proteins exported by each of the systems.

TYPE I SECRETION SYSTEMS

Type I secretion systems (T1SS) secrete a protein directly from the cytoplasm to the outside of the cell (Figure 14.6). They are different from members of the other types of secretion systems and more closely related to a large family of ATP-binding cassette (ABC) transporters that export small molecules, including antibiotics and toxins, from the cell. These ABC transporters tend to be more specialized, exporting only certain molecules from the cell. To get the protein through the inner membrane, T1SS use a dedicated system that consists of two

Figure 14.6 Schematic representation of the type I, II, III, and IV protein secretion systems. The examples shown are for type I (hemolysin A [HlyA] of *E. coli*), type II (pullulanase of *K. oxytoca*), type III (Yop of *Yersinia*), and type IV (*vir* of *A. tumefaciens*). EM, extracellular milieu; OM, outer membrane; Peri, periplasmic space; IM, inner membrane; Cyto, cytoplasm. The arrows indicate which pathways use the Sec and Tat pathways through the inner membrane. Blue indicates the secretin-formed channels. doi:10.1128/9781555817169.ch14.f14.6

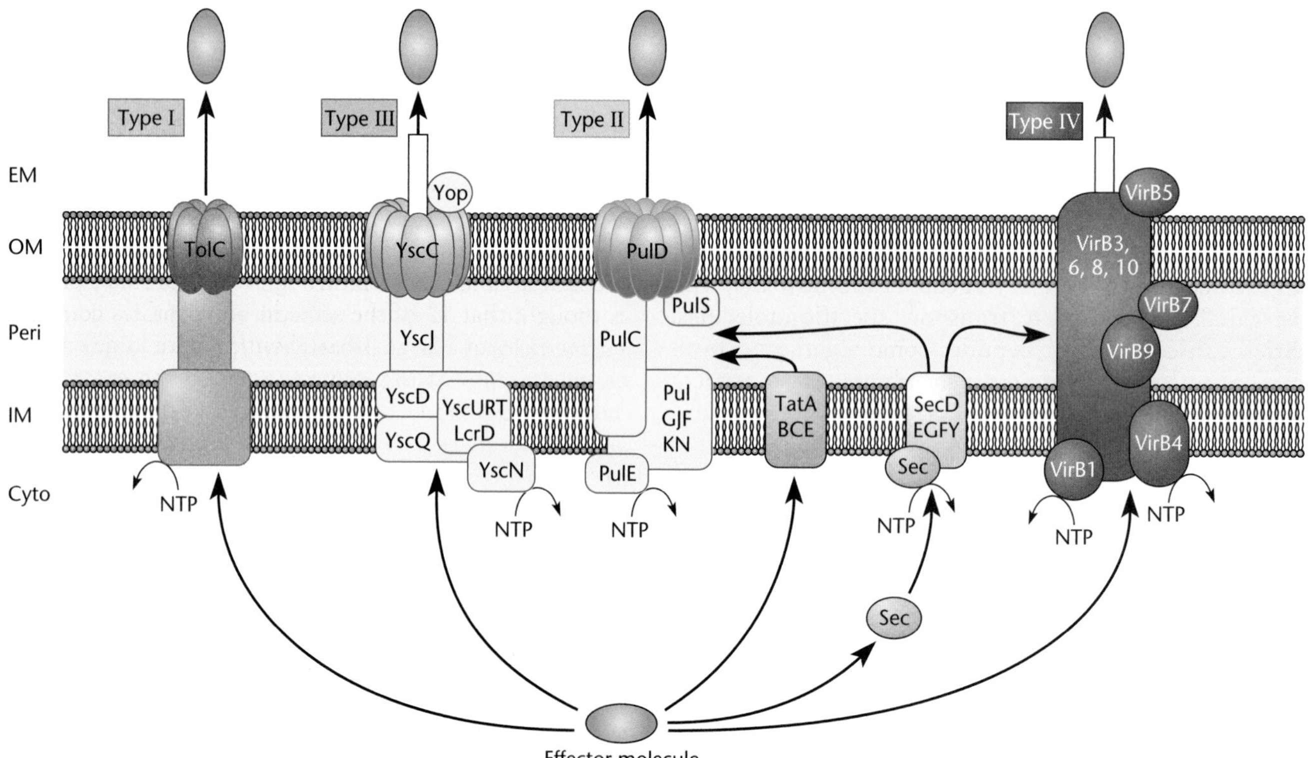

proteins: an ABC protein in the inner membrane and an integral membrane protein that bridges the inner and outer membranes. To get through the outer membrane, T1SS use a multiuse protein, TolC, that forms the β-barrel channel in the outer membrane. Because the TolC channel has other uses and also exports other molecules, including toxic compounds, from the cell, it is recruited to this system only when the specific protein is to be secreted. When the molecule to be secreted binds to the ABC protein, the integral membrane protein recruits TolC, which then forms the β-barrel in the outer membrane through which the molecule can pass. The cleavage of ATP by the ABC protein presumably provides the energy to push the molecule all the way through the TolC channel to the outside of the cell.

The classical example of a protein secreted by a T1SS is the HylA hemolysin protein of pathogenic *E. coli*. This toxin inserts itself into the plasma membrane of eukaryotic cells, creating pores that allow the contents to leak out. It also has its own dedicated T1SS composed of HylB (the ABC protein) and HylD (the integral membrane protein), which secretes it through the membranes. Because HylA is not transported through the inner membrane by either the SecYEG channel or the Tat system, it does not contain a cleavable N-terminal signal sequence. Instead, like all proteins secreted by T1SS, it has a sequence at its carboxyl terminus that is recognized by the ABC transporter but, unlike a signal sequence, is not cleaved off as the protein is exported.

Another well-studied protein secreted by a T1SS is the adenylate cyclase toxin of *Bordetella pertussis*. This toxin enters eukaryotic cells and makes cyclic AMP, thereby disrupting their signaling pathways. The use of the pertussis adenyl cyclase in bacterial two-hybrid selections (the BACTH system) is discussed in chapter 13.

The TolC channel has been crystallized and its structure determined (see Koronakis et al., Suggested Reading). This structure has provided interesting insights into the structure of β-barrels in general and how they can be gated and opened to transport specific molecules. Briefly, three TolC polypeptides come together to form the channel through the outer membrane. Each of these monomers contributes four transmembrane domains to form a β-barrel that is always open on one side of the outer membrane, the side on the outside of the cell. In addition, each monomer has four longer α-helical domains that are long enough to extend all the way across the periplasm. These four α-helical domains contribute to the formation of a second channel that is aligned with the first channel and traverses the periplasm. Because of these two channels, the secreted protein can be transported all the way from the inner membrane to the outside of the cell. In addition, the channel in the periplasm can open and close and therefore "gate" the channel. When a protein is being transported and the TolC channel is recruited, the α-helical domains of the periplasmic channel may rotate, which untwists them and opens the gate on the periplasmic side. The molecule is then secreted all the way through both channels to the outside of the cell.

TYPE II SECRETION SYSTEMS

We have already mentioned **type II secretion systems** (T2SS) because of their relationship to some competence systems (see chapter 6). They are also closely related to the systems that assemble type IV pili on the cell surface (see below). Type II secretion was originally called the main terminal branch of the Sec secretion pathway because it was thought that all gram-negative bacteria have the systems. However, it is now known that, even though very common, they are not universally shared among gram-negative bacteria.

Some examples of proteins secreted by T2SS are the pullulanase of *Klebsiella oxytoca* and the cholera toxin of *Vibrio cholerae*. The pullulanase degrades starch, and the cholera toxin is responsible for the watery diarrhea associated with the disease cholera (see chapter 13), illustrating the variety of proteins secreted by these systems. The T2SS can secrete already folded proteins. The cholera toxin is composed of two subunits, A and B, and after transport by the SecYEG channel, one of the A and five of the B subunits assemble in the periplasm, from where they are secreted through the secretin channel and into the intestine of the vertebrate host. The associated B subunit then assists the A subunit into mucosal cells, where it ADP-ribosylates, i.e., adds ADP, to a membrane protein that regulates the adenylate cyclase (see chapter 13). This disrupts the signaling pathways and causes diarrhea.

T2SS are very complex, consisting of as many as 15 different proteins (Figure 14.6). Most of these proteins are in the inner membrane and periplasm, and only one is in the outer membrane, where it forms a β-barrel secretin that is the channel through which proteins pass. It is thought that 12 of the secretin polypeptides come together to form a large β-barrel with a pore large enough to pass already folded proteins, such as the periplasmically folded cholera toxin. The formation of this channel is not spontaneous but requires the participation of normal cellular lipoproteins that might become part of the structure. The secretin protein has a long N terminus that might extend all the way through the periplasm to make contact with other proteins of the T2SS in the inner membrane. This periplasmic portion of the secretin may also gate the channel, as with the TolC channel.

Even though many of the components of the T2SS are in the inner membrane, they use either the SecYEG channel or the Tat pathway to get their substrates through the inner membrane. Therefore, proteins transported by this system have cleavable signal sequences at their N termini, either the Sec type or the Tat type. Once in the periplasm, the proteins usually fold, if they have not already, before they are transported through the outer membrane. Some of the periplasmic and inner membrane proteins of the secretion system are related to components of pili and have been called pseudopilin proteins, even though they do not normally appear outside the cell (see chapter 6). It has been proposed that the formation and retraction of these pseudopili works like a piston to push the protein through the secretin channel in the outer membrane to the outside of the cell. In this way, the energy for secretion could come from the inner membrane or the cytoplasm, as shown in the figure, since, as mentioned above, there is no source of energy in the periplasm. In support of this model, the pseudopili have been seen to produce pili outside an *E. coli* cell when the gene for the pilin-like protein was cloned and overproduced in *E. coli.*

TYPE III SECRETION SYSTEMS

The **type III secretion systems** (T3SS) are probably the most impressive of the secretion systems in gram-negative bacteria. They form a syringe-like structure composed of about 20 proteins, which takes up virulence proteins called effectors from the cytoplasm of the bacterium and injects them directly through both membranes into a eukaryotic cell (Figure 14.6). For this reason, they are sometimes called **injectisomes.** They exist in many gram-negative animal pathogens, including *Salmonella* and *Yersinia*, but are also found in many plant pathogens, including *Erwinia* and *Xanthomonas*. One striking feature of T3SS is how similar they are in both animal and plant pathogens. In all these bacteria, the parts of the secretion systems involved in getting the secreted protein through the bacterial membranes are very similar. Where they differ is in the protuberance, called the needle, that penetrates the eukaryotic cell wall to allow injection through the wall into the host cell cytoplasm. This difference is expected. Animal and plant cells are surrounded by very different cell walls, so the needle of a syringe that can penetrate the membrane of a mammalian cell would be expected to be very different from a needle that can penetrate a plant cell wall.

T3SS are usually encoded on pathogenicity islands, and their genes are induced only when the bacterium encounters its vertebrate host or if they are cultivated under conditions that are designed to mimic the host. They then induce the genes for the injectisome and assemble it in the cell membranes. The effector proteins they inject are also encoded by the same DNA element, and their genes are turned on at the same time. The part of the injectisome that traverses the outer membrane is composed of a secretin protein related to those of the T2SS. It also forms a β-barrel composed of about a dozen secretin subunits. Like the secretins of the T2SS, these might require normal bacterial lipoproteins to assemble the channel in the membrane, but these lipoproteins might not remain as part of the barrel as they do in T2SS. They might also require other components of the secretion machinery to assemble. The T3SS syringes are related to the flagellar motor that drives bacterial movement in liquid media (Box 14.1).

The identifying mark of effector proteins to be secreted by at least some T3SS is a short sequence located on the N terminus of the protein, as it is for the *sec* and Tat systems, but this signal is not cleaved off when the protein is injected as it is in those systems. Some of them may even use a sequence at the 5′ end of the mRNA encoding the protein to drag it to the injectisome to be secreted as it is translated, although this has not been proven unequivocally.

Many of the effector proteins secreted into eukaryotic cells by T3SS are involved in subverting the host defenses against infection by bacteria. This can be illustrated by *Yersinia pestis*, the bacterium that causes bubonic plague and in which T3SS were first discovered. In animals, one of the first lines of defense against infecting bacteria are the macrophages, phagocytic white blood cells that engulf invading bacteria and destroy them by emitting a burst of oxidizing compounds. However, when a macrophage binds to a *Yersinia* cell, the bacterium injects effectors called Yop proteins into the macrophage cell before it can be engulfed. Once in the eukaryotic cell, these effectors disarm the cell by interfering with its signaling systems and thus diverting the macrophage from its purpose of engulfing the bacterium. The effector proteins that are injected show a remarkable ability to interfere with these pathways. For example, one of the Yop proteins is a tyrosine phosphatase, which removes phosphates from proteins in a signal transduction system in the macrophage, blocking the signal to take up the bacterium and preventing the burst of oxidizing compounds. Some T3SS even inject proteins called intimins that provide receptors on the cell surface to which the bacterium can adsorb in order to enter the eukaryotic cell. This allows them to open the door into the cell from the inside, if the eukaryotic cell has not been nice enough to provide a receptor on its surface to which the bacterium can bind.

Plants use very different defense mechanisms against bacteria, so plant pathogens have to adapt their strategy accordingly. Plants defend themselves against infection by inducing necrosis (destruction of the infected tissue) and inducing phenolic compounds that destroy the bacterium. This is called a hypersensitive response, and it is induced by proteins called Avr (for *avir*ulence) proteins that are injected into the plant cell by a T3SS. In

BOX 14.1

Secretion Systems and Motility

T3SS are structurally related to the flagellar motors that provide motility to many bacteria in liquid environments (see Blocker et al., References). Not only are they structurally related, but also, some flagellar systems may play a role in secreting some virulence proteins. This makes sense, since the two systems are superficially similar. Flagellar motors have appendages that extrude from the cell surface, but in the form of flagella rather than the needle of the type III secretion apparatus. They also consist of a motor buried in the membranes that rotates the flagellum, and these proteins are structurally related to the syringe-forming proteins of the T3SS. However, the major functions of the systems are very different, as reflected in their structures. A number of flagella are clustered on one end of the cell. If they rotate in one direction, counterclockwise, the flagella wrap around each other (bundle), they all turn in the same direction, and the cell moves forward in a straight line. If they rotate in the other direction, clockwise, the flagella separate, and the cell moves in circles without any defined direction (tumbles). The direction in which the flagella rotate depends on the state of methylation of a number of protein receptors in the inner membrane, called MCPs (methyl-accepting chemotaxis proteins), which bind attractants, such as amino acids that might provide a food source (see Bray, References). The state of methylation of these receptors is determined by how much attractant is bound, and this level is adjusted every 3 or 4 seconds. If the level of methylation is not consistent with how much attractant is bound, it means that the cell is moving up or down a gradient of the attractant. This information is communicated down a phosphorelay system called the Che proteins to a component of the flagellar motor, which then continues rotating counterclockwise (if it is moving up a gradient and should continue in the same direction) or clockwise (if is moving down a gradient and should try another direction).

The relationship between flagella and the T3SS is not the only known example of a secretion system that has been adapted to provide motility to the cell; T2SS have also been been adapted to assemble type IV pili on the cell surface (see the text). These pili provide motility on solid surfaces to some bacteria, including *Myxococcus xanthus*, by extending and contracting and thus pulling the cell along. They are on the front end of the cell, where they can fulfill their pulling role, while flagella are on the rear end, where they can push. Thus, secretion systems, which have extensions that extend through the outer membrane, seem particularly adaptable to providing motility.

References

Blocker, A., K. Komoriya, and S. Aizawa. 2003. Type III secretion systems and bacterial flagella: insights into their function from structural similarities. *Proc. Natl. Acad. Sci. USA* **100:**3027–3030.

Bray, D. 2002. Bacterial chemotaxis and the question of gain. *Proc. Natl. Acad. Sci USA* **99:**7–9.

doi:10.1128/9781555817169.ch14.Box14.1.f

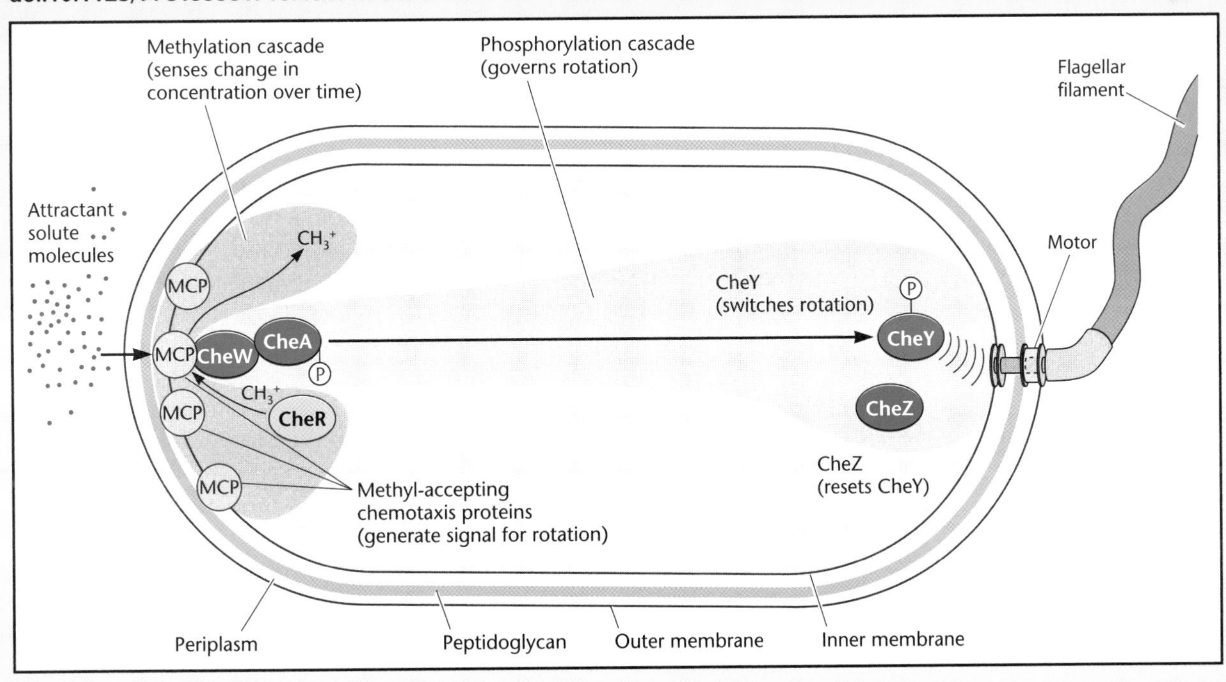

a susceptible plant, these Avr proteins do not elicit the hypersensitive response.

TYPE IV SECRETION SYSTEMS

Type IV secretion systems (T4SS) are discussed in chapters 5 and 6, because they are also involved in DNA transfer during conjugation and transformation. Of all the secretion systems they have the widest distribution, found also in gram-positive bacteria and even *Crenarchaeota*, a branch of the archaea where they are involved in plasmid transfer. However, so far, T4SS systems involved in secretion of protein effectors have been found only in gram-negative bacteria. Like T3SS, they can inject proteins through both membranes directly into eukaryotic cells, although there could be some exceptions that use the Sec system to get through the inner membrane (see below).

As discussed in chapter 5, T4SS are closely related to the conjugation systems of self-transmissible plasmids. In fact, the T-DNA transfer system of *Agrobacterium tumefaciens* has served as the prototype T4SS and is the one about which the most is known and to which all others are compared (Box 5.1 and Figure 14.6). Accordingly, the genes and proteins of other T4SS are numbered after their counterparts in the T-DNA transfer system, named the *vir* genes because of their virulence in plants. VirB9 is a secretin-like protein that forms a β-barrel channel in the outer membrane and extends into the periplasm, where it makes contact with proteins in the inner membrane. However, unlike true secretins, it seems to require another outer membrane protein, VirB7, to make a channel. The VirB9 protein is covalently attached to the VirB7 protein, which in turn is covalently attached to the lipid membrane, making the structure very stable. Some of the genes in the T-DNA are plant-like genes that encode plant hormones that cause growth of the plant cell, leading to the formation of tumors called crown galls. Others encode enzymes that make unusual compounds, called opines, that can be used by the bacterium as a carbon, nitrogen, and energy source (see Box 5.1). In addition to the T-DNA, this system directly injects proteins into the plant cell, which also makes it a bona fide protein secretion system.

T4SS work through a coupling protein, named VirD4 in the T-DNA transfer system, which binds proteins to be secreted. The coupling protein then allows them into the channel. Therefore, to be secreted, a protein must bind to this coupling protein, ensuring that only certain proteins are secreted. These proteins presumably have a short domain that specifically binds the coupling protein but that has been identified in only a few cases, including some relaxases and the VirB-VirD4 secretion system of *Bartonella* spp. (see below). The energy of secretion probably comes from the cleavage of ATP or GTP in the cytoplasm by some channel-associated proteins (Figure 14.6).

TYPE V SECRETION SYSTEMS: AUTOTRANSPORTERS

All of the secretion systems discussed above use some sort of structure formed of β-sheets assembled into a ring called a β-barrel to get them through the outer membrane. Some of them are part of the secretion apparatus itself, while some, like TolC, are recruited from other functions in the cell. However, some secreted proteins do not take it for granted that they will find a β-barrel in the outer membrane to allow them through to the outside of the cell when they get there. They carry their own β-barrel with them in the form of a domain of the protein that can create a β-barrel when it gets to the outer membrane. These proteins are called **autotransporters** because they transport themselves. The prototypical autotransporter is the immunoglobulin A protease of *Neisseria gonorrhoeae*. It is involved in evading the host immune system by cleaving antibodies. Most known autotransporters are large virulence proteins, such as toxins and intimins, that perform various roles in bacterial pathogenesis or in helping evade the host immune system. A particularly interesting autotransporter is the IcsA protein of *Shigella flexneri*, a cause of bacterial dysentery. It is localized to the outer membrane, where it recruits a host actin-regulating protein (N-WASP), which in turn recruits another host complex that polymerizes host actin into filaments, pushing the bacterium through the eukaryotic cell cytoplasm as part of the infection mechanism, a process called actin-based motility (see Box 14.3).

The mechanism used by autotransporters is illustrated in Figure 14.7, which also shows their basic structure. Most autotransporters consist of four domains, the translocator domain at the C terminus that forms a β-barrel in the outer membrane, an adjacent flexible linker domain (not shown) that may extend into the periplasm, a passenger domain that contains the functional part of the autotransported protein, and sometimes a protease domain that may cleave the passenger domain off of the translocator domain after it passes through the channel formed by the translocator domain.

Autotransporters are typically transported through the SecYEG channel, so they have a signal sequence that is cleaved off as they pass into the periplasm. Their translocator domain then enters the outer membrane, where it forms a 12-stranded β-barrel. This assembly does not occur by itself but requires the same accessory factors as the assembly of many secretins, including the periplasmic chaperone, Skp, and the Bam complex (see above). The flexible linker domain then guides the passenger domain into and through the channel to the outside of the

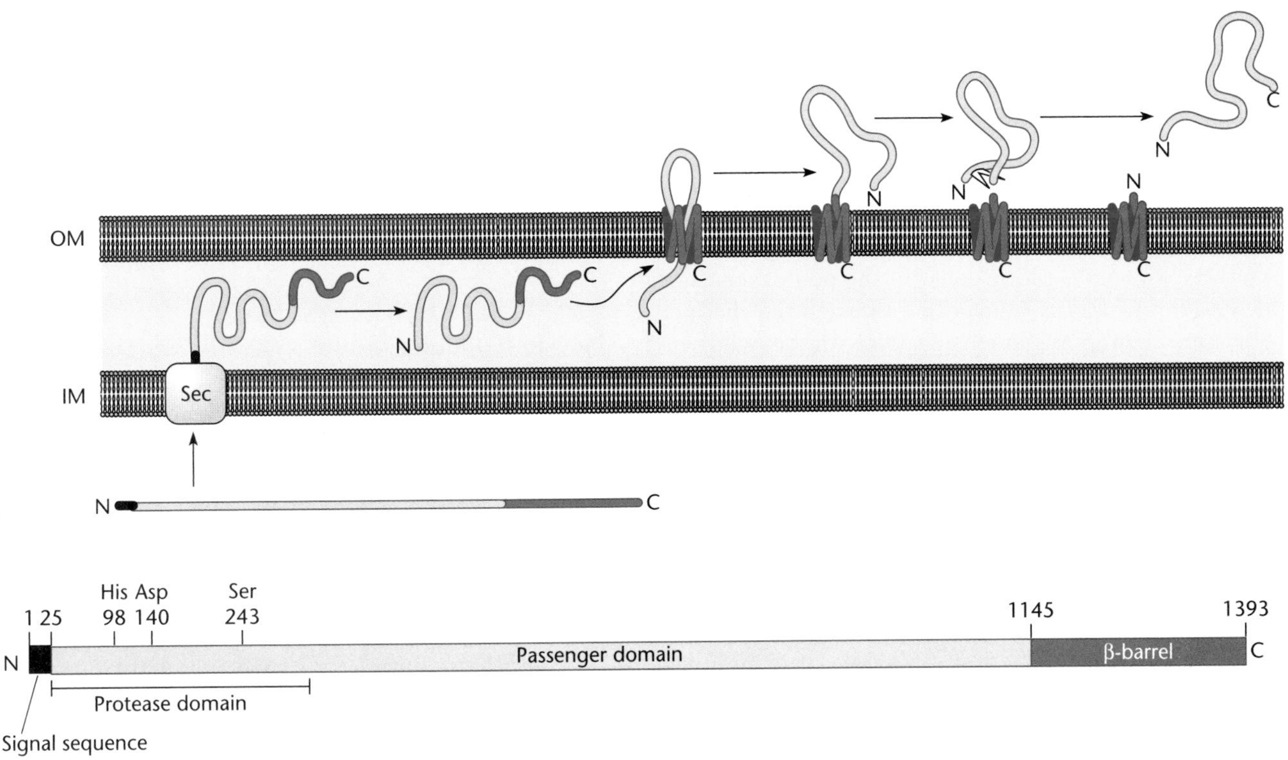

Figure 14.7 Structure and function of a typical autotransporter. Shown is a *Haemophilus influenzae* adhesin; the length in amino acids of each domain, where known, is indicated by the number above the structure, as are some of the important amino acids in the protease domain. The transporter domain at the C terminus that forms a β-barrel in the outer membrane is shown in dark blue; the passenger domain and the protease domain that cleaves the passenger domain off the transporter domain outside the cell are shown in light blue. The flexible linker domain is not indicated. The signal sequence that is cleaved off when the protein passes through the SecYEG (Sec) channel in the inner membrane is shown in black. OM, outer membrane; IM, inner membrane. doi:10.1128/9781555817169.ch14.f14.7

cell. The passenger domain can then be cleaved off by its own protease domain or remain attached to the translocator domain and protrude outside the cell, depending on the function of the passenger domain.

In spite of this simple picture, some questions about autotransporters remain. The first is the question of where the passenger domain folds. This question is related to the size of the pore formed by the transport domain; the channel formed by a single translocator domain would be too small to accommodate a folded passenger domain, much less the linker domain, if this is guiding it into the pore. Perhaps the passenger domain folds on the outside of the cell once it is exported. However, this would require that it have the capacity to fold spontaneously in the hostile environment outside the cell without the help of other cellular constituents. One way out of this dilemma is to propose that a number of translocator domains come together to form a larger pore, with each monomer contributing some β-structure domains to the larger β-barrel. Passenger domains could then pass through this shared larger β-barrel even if they had already folded. There is some structural evidence for such shared pores, at least for some autotransporters. Some autotransporters, called trimeric autotranporters, are known to cooperate to form the pore. They consist of three identical polypeptides, and each translocator domain contributes 4 strands to the 12-stranded pore (see Mikula et al., Suggested Reading).

There is also the question of where the energy for autotransportation comes from, since, as mentioned, there is no ATP or GTP in the periplasm and the outer membrane does not have a membrane potential. Perhaps the autotransporter arrives at the periplasm in a "cocked" or high-energy state that drives its own transport. Such an explanation has been proposed for the transport of some pilin proteins by a chaperone-usher system (see below).

Two-Partner Secretion

In a variation on autotransporters, sometimes the β-barrel-forming domains and the passenger domains are on different polypeptides. This has been called

two-partner secretion and is found in a large variety of gram-negative proteobacteria, where it is largely responsible for transporting large toxins, much like the autotransporter pathway. The two-partner polypeptides, called TpsA and TpsB, are transported separately through the inner membrane, and the TpsB protein forms a β-barrel in the outer membrane. The TpsA protein, the equivalent of the passenger domain in autotransporters, interacts with the TpsB protein on the periplasmic side of the channel and is transported through the channel, where it can either remain associated with the cell or enter the surrounding medium. The TpsB protein is highly specific for its TpsA partner and secretes no other partner proteins. It also contains motifs that may participate in the processing and folding of its partner protein. This is another case where the source of the energy for secretion is not clear.

Chaperone-Usher Secretion

Another type of secretion related to type V secretion is **chaperone-usher secretion**. This type of secretion is usually used to assemble some pilins on the cell surface, such as the P pilus of uropathogenic *E. coli*. The secretion system consists of three proteins, a β-barrel-forming protein in the outer membrane called the usher, a periplasmic protein called the chaperone, and the pilin subunit to be assembled on the cell surface. The pilin protein is transported through the inner membrane by the SecYEG channel and therefore has a cleavable signal sequence. Once in the periplasm, the pilin protein is bound by the dedicated periplasmic chaperone. However, rather than merely helping it fold like other chaperones (see chapter 2), this chaperone actually contributes a strand to the pilus protein that completes one of the folds of the pilin protein and makes it much more stable (see Waksman and Hultgren, Suggested Reading). This fold is called an immunoglobulin (Ig) fold because it was first found in immunoglobulins and contributes to their extreme stability. The complex of pilin protein and chaperone is then targeted to the usher channel in the outer membrane, where it is assembled into the growing pilus by a process called **donor strand exchange**. In this process, the chaperone strand that previously completed the Ig fold is replaced by a corresponding strand in the neighboring pilin subunit, joining the two subunits and making a very stable pilus. The chaperone that is released then helps the next pilin subunit add to the growing pilus. The pilus assembles from the end, with the adhesin at the far end of the pilus added first, followed by other subunits. Apparently, the usher knows what type of subunit to put on next, depending on where it is in the pilus. Also, two usher channels seem to cooperate with each other, forming twin channels that may alternate somehow in adding subunits. Again, there is the problem of where the energy for pilus assembly comes from, since the assembly of the pilus occurs at the inner face of the outer membrane after the pilin protein has been transported through the inner membrane and cytoplasm, which are the sources of energy. One idea is that the periplasmic chaperone holds the pilin protein in a high-energy state and its eventual folding at the usher drives the assembly process.

TYPE VI SECRETION SYSTEMS

The latest additions to the known secretion systems of gram-negative bacteria are the type VI secretion systems (T6SS). They were only officially discovered in 2006, although their existence was suspected for a decade or more. These systems are now known to be required for the secretion of proteins involved in pathogenesis and symbiosis, but they are also found in marine and soil bacteria, where they may have roles in biofilm formation and cell-to-cell communication. The T6SS are very large, with up to 21 proteins encoded within the gene cluster, 12 of which are found in all T6SS, where they are highly conserved and are thought to play structural roles in the secretion apparatus. At least some of the others unique to the system may encode effectors specifically transported by that system. An intriguing observation has been made that some of the highly conserved proteins structurally resemble components of phage tails. The highly conserved protein Hcp resembles gp19, the major tail protein of phage T4, and also forms hexameric (six-sided) rings, and another, VgrG, resembles the syringe of the baseplate of phage T4, consisting of gp27 and gp5 (see chapter 7). This has led to speculation that T6SS are inverted phage tails that now eject proteins from the bacterial cell instead of injecting proteins (and DNA) into the cell.

There can be little doubt of the importance of these systems, since mutations in their genes can have dramatic phenotypes, but many questions remain. So far, no one has been able to obtain a crystal structure for a T6SS, so the roles of the various proteins are purely speculative. Also, both Hcp and VgrG are themselves exported into the surrounding medium, and this export requires other components of the T6SS. If these phage-like proteins are structural components of the apparatus, why would they also be exported? They could double as effectors, but if so, they are common to all T6SS and not specific. So far, no specific functions have been attributed to effectors for any of these systems. We await further discoveries about these important secretion systems.

Protein Secretion in Gram-Positive Bacteria

So far, we have limited our discussion of protein transport to the mechanisms used by gram-negative bacteria. The protein secretion systems discussed above are, of necessity, restricted to gram-negative bacteria; gram-positive bacteria lack an outer lipid bilayer membrane and so have no need for them. However, some of our best friends (and worst enemies) are gram-positive bacteria, so we must not neglect them. To give some examples, the lactobacilli that are used to make food products, including yogurt and cheese, are gram positive, as are the biodegradable-insecticide-producing *Bacillus thuringiensis* and the *Streptomyces* species that make most of the known antibiotics. However, so are *Staphylococcus aureus*, the agent of many serious infections; *Bacillus anthracis*, the cause of anthrax; and *Streptococcus mutans*, which causes dental plaque. In this section, we discuss some features of secretion systems that are unique to gram-positive bacteria.

INJECTOSOMES OF GRAM-POSITIVE BACTERIA

While pathogenic gram-positive bacteria lack the T1SS, T3SS, and T4SS described above, they require mechanisms to translocate virulence effectors into eukaryotic cells. Some AB-type toxins, such as diphtheria toxin, and clostridial neurotoxins, such as botulinum toxin, are self-translocating. Once outside the cell, the B subunit binds to the surface receptors on eukaryotic cells and helps the A subunit toxin enter the cell. However, some gram-positive bacteria, such as *Streptococcus pyogenes*, inject a virulence effector into the mammalian target cell by a mechanism functionally analogous to the gram-negative T3SS (see Madden et al., Suggested Reading). This has been named an **injectosome** to distinguish it from the gram-negative type III injectisome (note the different spellings). As shown in Figure 14.8, the function of the injectosome requires translocation of the effector by the Sec-dependent secretion system across the bacterial membrane before it translocates the effector directly across the membrane of the eukaryotic target cell. The figure illustrates the functional analogy of the gram-positive injectosome and the gram-negative injectisome.

Sortases

Since gram-positive bacteria lack a bilipid outer membrane, their cell wall is not surrounded by a membrane and so is available to the external surface of the cell. The gram-positive bacteria take advantage of this by attaching some proteins directly to their cell wall so that the proteins will be exposed on the cell surface. In gram-positive bacteria, proteins destined for covalent attachment to the outer cell surface are the targets of a type of cell wall-sorting enzyme called a **sortase**. Gram-positive bacteria often have a number of such sortases. The general sortase, sortase A (StrA) sometimes called the housekeeping sortase, may attach as many as 40 different proteins to the cell wall. Others are more specific for certain proteins.

Figure 14.8 The gram-positive injectosome **(A)** compared to the gram-negative type III injectisome **(B)**. See the text for details. doi:10.1128/9781555817169.ch14.f14.8

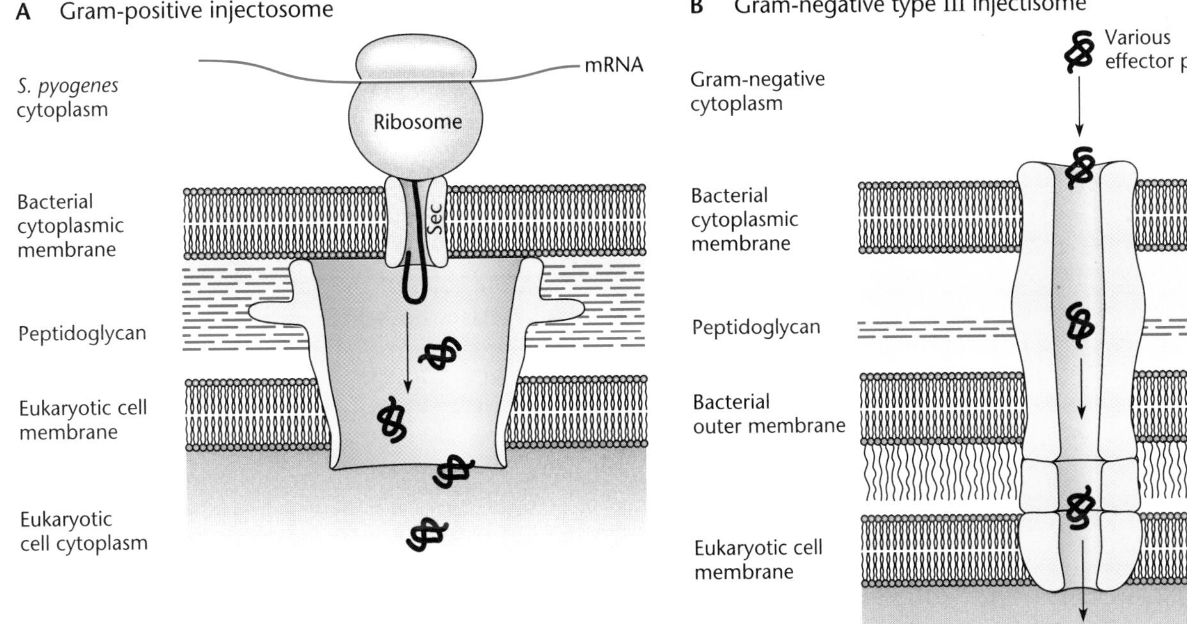

A sortase is able to create covalent attachments between peptides by catalyzing a transpeptidation reaction. Figure 14.9 illustrates a typical StrA pathway. Surface proteins that are StrA targets include an N-terminal signal peptide and a 30- to 40-residue C-terminal sorting signal, which is composed of a pentapeptide cleavage site, LPXTG, and a hydrophobic domain that together constitute the cell wall-sorting signal (Cws) (Figure 14.9A). The N-terminal signal sequence of the sortase target protein directs the protein to the membrane translocase, where the signal sequence is removed. After the protein has been translocated across the cytoplasmic membrane, the sorting signal is processed by the sortase. The sortase cuts between the threonine (T) and glycine (G) in the pentapeptide sorting signal and then covalently links the carboxyl group of the threonine to a specific cysteine in the sortase C terminus. The sortase then attaches it to lipid II, in the example, the terminal glycine in the five-glycine interlinking peptide that links two peptide cross-links in the *S. aureus* cell wall (see "Synthesis of the Cell Wall" below). When the MurNac in this lipid II is incorporated into the cell wall, the protein becomes covalently attached to the cell wall.

Five sortase subfamilies are currently defined, differing in their taxonomic distributions in the gram-positive genera, but also on the basis of differences in the sorting

Figure 14.9 The sortase A pathway. **(A)** Typical sortase substrate. The protein is composed of an N-terminal signal peptide and a C-terminal cell wall-sorting signal (Cws). The Cws contains a conserved LPXTG motif followed by a hydrophobic stretch of amino acids and positively charged residues at the C terminus. **(B)** Model for the cell wall sortase A pathway in *S. aureus*. **(1)** The full-length surface protein precursor is secreted through the cytoplasmic membrane via an N-terminal signal sequence. **(2)** A charged tail (+) at the C terminus of the protein may serve as a stop transfer signal. Following cleavage of this secretion signal, a sortase enzyme cleaves the protein between the threonine and glycine residues of the LPXTG motif, forming a thioacyl-enzyme intermediate to a specific cysteine in the sortase **(3).** It is then attached to the free amine of the five-glycine cross-bridge of lipid II **(4)** before transfer into the cell wall **(5).** The Pro-Gly-Ser-Thr region may help it through the thick cell wall so that it is expressed on the cell surface. PP is the site of MurNAc pentapeptide attachment to bactoprenyl (C_{55}) in the membrane in lipid II (Figure 14.11). doi:10.1128/9781555817169.ch14.f14.9

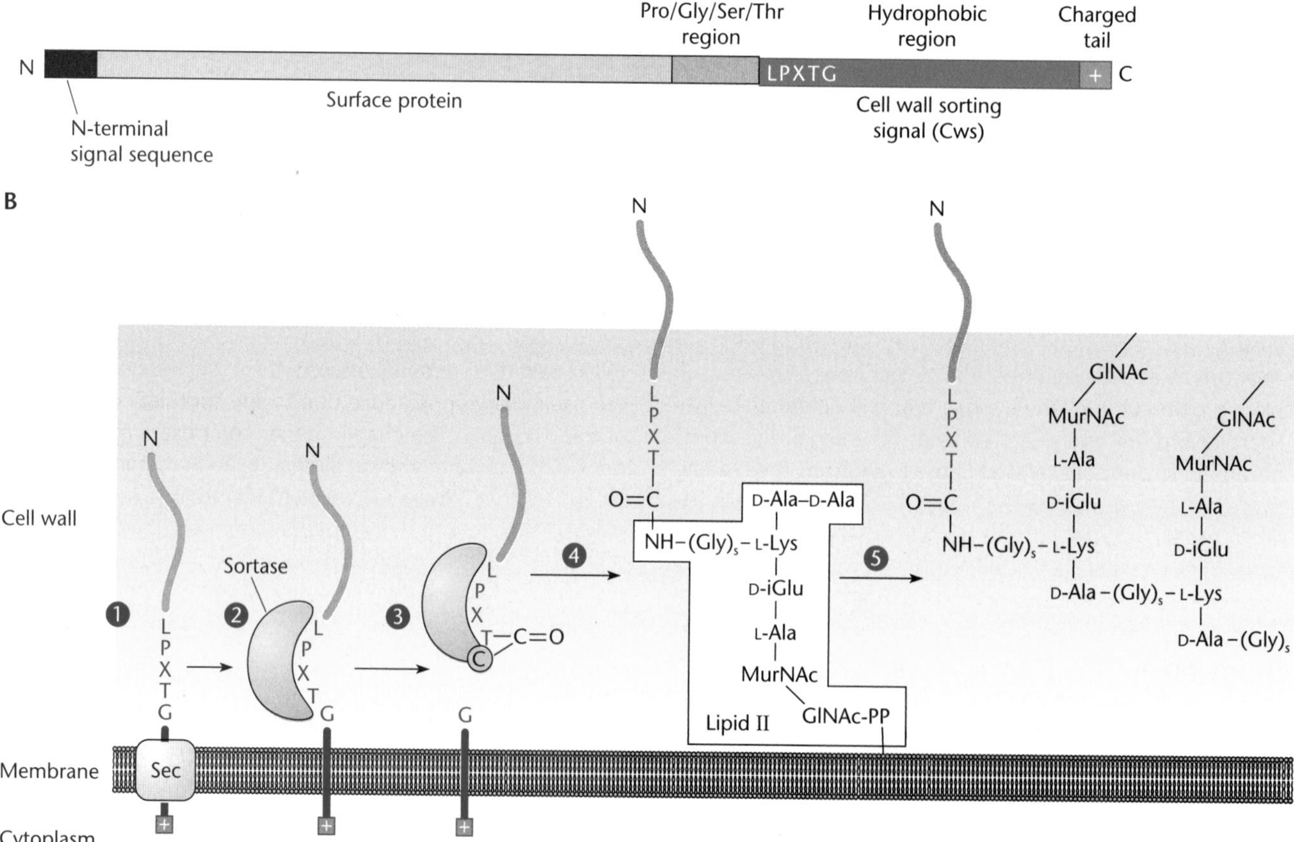

signal and the peptide to which it can be attached (see Hendrickx et al, Suggested Reading). A particularly interesting example is the assembly of pili from their subunits on the cell surfaces of gram-positive bacteria by sortases. Pili are composed of one major subunit, often called pilin, and one or more minor subunits. In the simplest example, pili composed of pilin and only one other subunit, the minor subunit at the tip is attached by a specialized sortase to the first pilin subunit, which is then attached to the next pilin subunit by the same specialized sortase. Finally, the last subunit is attached to the cell wall by the housekeeping sortase A, which attaches most other proteins to the cell wall. This makes for a very stable pilus and secures its attachment to the cell wall. Box 14.2 discusses another example of a sortase-dependent pathway in *Streptomyces coelicolor* that attaches a family of hydrophobic proteins to the cell walls of the aerial mycelia that make spores, making the mycelia more hydrophobic to help disperse the spores.

Bacterial Cell Biology and the Cell Cycle

One of the most amazing features of biological systems is the way cells grow and divide to create new organisms that remain virtually identical over thousands of generations. This seemingly magical process can be studied most easily in single-celled bacteria, with their short generation times, relatively simple cells, and more easily

BOX 14.2

Example of a Sortase-Dependent Pathway: Sporulation in *S. coelicolor*

Many proteins required for the virulence of pathogenic gram-positive bacteria have been shown to be exposed on the cell surface by sortases. However, gram-positive soil bacteria also use this mechanism. The common soil bacterium *S. coelicolor* uses sortase enzymes in a specific phase of growth: morphological differentiation. This gram-positive genus of bacteria and many of its relatives in the actinomycetes have been studied extensively, partly because of their complex fungus-like morphological development and because they produce many of the most useful antibiotics we have discussed in this book. Some of them are also responsible for the smell of fresh-turned soil in the spring (see Guest et al., References). *S. coelicolor* bacteria grow as filaments (hyphae), and morphological differentiation produces colonies that are reminiscent of filamentous fungi in that a feeding mycelium of hyphae "morphs" into a fluffy-surfaced colony covered with specialized spore-forming hyphae. The figure shows a cross section of a mature differentiated sporulating *S. coelicolor* colony as visualized by scanning electron microscopy. In plate-grown cultures, the hyphae that will differentiate into spore chains are often referred to as "the aerial mycelium" because the prespore hyphae grow away from the feeding mycelium, which means that they grow vertically into the air. In their natural soil habitat, growth of the sporulating hyphae from one soil particle-water droplet to another allows the dispersal of spores into new soil niches. This requires that the sporulating hyphae be very hydrophobic to escape from one water droplet to another. Their hydrophobicity is due, in part, to the production of a layer of hydrophobic proteins called **chaplins** (for coelicolor hydrophobic aerial protein).

doi:10.1128/9781555817169.ch14.Box14.2.f

The chaplin genes, *chp*, were discovered by using microarrays and comparing RNAs made by mutants that did not make sporulating hyphae to the wild type. These mutants are called bald (Bld) mutants because their colonies do not have the usual hairy appearance due to the sporulating hyphae of the wild type. The chaplin genes comprise a **multigene family** with eight members, all sharing an N-terminal hydrophobic domain. Some of them also have a cell wall-sorting signal, LAXTG, where X can be any amino acid, which allows them to be attached to the cell surface by a sortase cell wall-localizing pathway, such as the one shown in Figure 14.9.

Genetic analysis of the chaplin-encoding multigene family demonstrated that they have overlapping functions and can substitute for each other. This was accomplished by knocking out the genes progressively and showing that four or more of the eight chaplin genes had to be inactivated before the aerial mycelia were affected (see Elliot et al., References).

(continued)

BOX 14.2 (continued)

Example of a Sortase-Dependent Pathway: Sporulation in *S. coelicolor*

To do this, the investigators had to introduce each chaplin gene inactivated by insertion of an antibiotic insertion from *E. coli* by use of a promiscuous plasmid/suicide vector. They were able to construct this plasmid because they knew the sequence of the *Streptomyces* chromosomal DNA in and around the chaplin gene. Recombination between the flanking sequences of the inactivated gene on the incoming DNA and the same flanking sequences of the endogenous gene replaced the endogenous chaplin gene with the inactivated gene carrying antibiotic resistance, making the cells antibiotic resistant so that they could be selected. They then repeated the mating, this time introducing the gene with an in-frame deletion instead of the antibiotic resistance insertion and selecting for antibiotic sensitivity. The antibiotic-sensitive cells had replaced the antibiotic resistance gene with the deleted gene. They were then able to repeat the process with the next chaplin gene to obtain a strain in which two chaplin genes had been deleted, and so forth. Eventually, they were able to delete all eight chaplin genes and show that the presence of at least one of the chaplins on the colony surface is essential for the morphological-differentiation process.

In additional experiments, isolation of surface proteins and analysis with matrix-assisted laser desorption ionization–time-of-flight mass spectrometry confirmed that the chaplins could indeed be identified on the cell surface. Finally, transcriptional expression of the chaplin genes in the aerial mycelium could be seen by using green fluorescent protein transcriptional fusions.

References

Duong, A., D. S. Capstick, C. Di Berado, K. C. Findlay, A. Hesketh, H.-J. Hong, and M. A. Elliot. 2012. Aerial development in *Streptomyces coelicolor* requires sortase activity. *Mol. Microbiol.* **83:**992–1005.

Flardh, K. 2003. Growth polarity and cell division in *Streptomyces. Curr. Opin. Microbiol.* **6:**564–571.

Guest, B., G. L. Challis, K. Fowler, T. Kieser, and K. F. Chater. 2003. Gene replacement by PCR targeting in *Streptomyces* and its use to identify a protein domain involved in the biosynthesis of the sesquiterpene odor geosmin. *Proc. Natl. Acad. Sci. USA* **100:**1541–1546.

manipulated genetic systems. If history is any guide, the knowledge obtained from experiments with relatively simple bacteria should then serve as a basis for a much better understanding of the cell cycle and development in more complex multicellular organisms. Much of the work on the bacterial cell cycle is being performed with a few model organisms, including *E. coli*, *B. subtilis*, and *Caulobacter crescentus*, with contributions from a few other types of bacteria. While this work is in its early stages, it is becoming increasingly clear that different bacteria use somewhat different strategies to determine their cell shapes and to regulate their cell cycles but that they have many features in common.

The Bacterial Cell Wall

At the heart of generating a new bacterial cell is the synthesis of the cell wall. The cell wall of bacteria, sometimes called the **sacculus** or **murein layer**, is a unique structure in the biological world. It is one continuous large molecule completely surrounding the cell. It gives the cell its shape and rigidity and allows it to withstand the rigors of different environments, including large differences in osmotic pressure inside and outside the cell. The cell membranes do not have any rigidity and get their shape only through their association with the cell wall (see Fischer and Bremer, Suggested Reading).

Since the cell wall is one continuous molecule, there are no free ends on which to add new cell wall material, and new cell wall material must be inserted into the preexisting wall. This means the old wall must be temporarily broken before new material can be added, and it must be broken in such a way that the structure is not disrupted, even temporarily. Otherwise, osmotic pressure will lyse the cells and/or cellular materials will escape from the cell. Bacteria that are not spherical like *Staphylococcus* but are either rod-shaped, like *E. coli* and *B. subtilis*, or egg-shaped, like *Streptococcus*, have the additional requirement that cell wall synthesis also has to change direction at a carefully regulated time during the cell cycle in order to grow inward at the midcell to form the septum. Even after the cell wall has grown inward at the septum, it is still one continuous molecule that must be separated into two continuous cell walls, forming the two daughter sacculi, each with its own intact continuous cell wall, again without even temporary disruption of their structure. All of these steps require many enzymes and careful control of their activities, both spatial and temporal. They must also be coordinated with everything else that is happening in the cell, including the growth rate and the replication and segregation of the chromosome. In this section, we discuss what is known of how this control is achieved. In addition, because of the complexity and uniqueness of bacterial cell wall synthesis, many of the most useful antibiotics target the various steps of bacterial cell wall synthesis. We shall mention some of the most important of these antibiotics when we discuss the steps they target.

STRUCTURE OF THE CELL WALL

While the cell walls of different types of bacteria differ somewhat in composition, their basic structure is shared. This basic structure is illustrated in Figure 14.10. The bacterial cell wall is composed of peptidoglycan (PG), which consists of chains of sugars (glycans) held together by cross-links of short peptides (peptido). The glycan chains consist of repeated subunits in which each subunit is a disaccharide composed of the two sugars *N*-acetylglucosamine (GlcNAc) and *N*-acetylmuramic acid (MurNAc). The length of the glycan chains averages about 30 subunits in gram-negative bacteria but is much more variable in gram-positive bacteria, ranging from as short as an average of 10 disaccharide units in *Helicobacter pylori* to more than 100 in some of the bacilli.

The rigidity of the cell wall is achieved by having the long glycan chains cross-linked to each other through short peptide chains. Each MurNAc sugar in a glycan chain is attached to a 4-amino-acid peptide, and this peptide is often joined to a peptide chain attached to a MurNAc in another glycan chain. These short cross-linking peptides are composed of unusual amino acids that are normally not found in proteins, including, in many bacteria, *meso*-A_2pm (also called 2,6-diaminopimelic acid), an intermediate in the synthesis of L-lysine. In fact, some gram-positive bacteria use L-lysine instead. Also, often the cross-linking peptides have the D form of the amino acid instead of the L-amino acid found in proteins, and the amino acids in the peptide are not always joined by peptide bonds, as they are in proteins, but also can be joined by other types of bonds, some through the side chains of the amino acids. Typically, the sequence of 4 amino acids in the cross-linking peptide is L-Ala–D-Glu–*meso*-A_2pm–D-Ala. In the cross-links, the terminal D-Ala of one peptide is typically directly attached to the third amino acid of another peptide, usually *meso*-A_2pm, although in some types of bacteria, other short peptides intervene, such as short strings of glycines in *S. aureus*

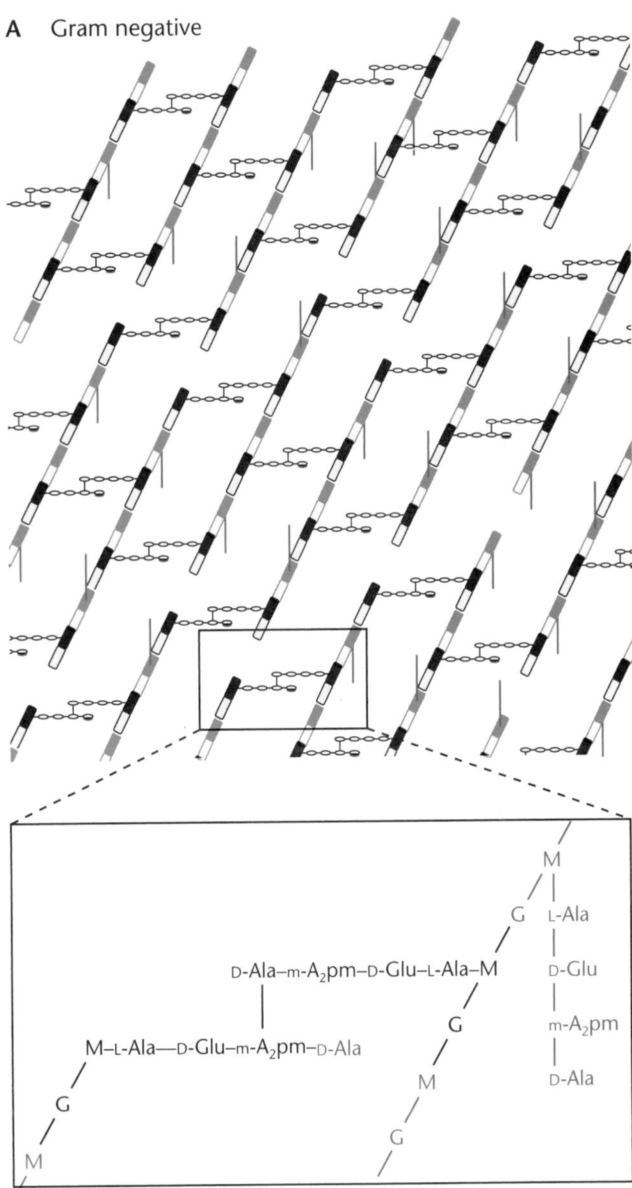

Figure 14.10 Cell wall PG. The glycan strands and peptide cross-links are shown. **(A)** In gram-negative bacteria, the alternating MurNAc (M) and GlycNAc (G) are shown as shaded and unshaded boxes, respectively. The cross-linked MurNAcs are shown in black, and the non-cross-linked MurNAcs are shown in blue. The non-cross-linked peptides are also shown as blue lines extending from a blue MurNAc. Some of these may be attached to a MurNAc in another layer. The inset shows details of the peptide cross-links. The invariant D-Ala is shown in blue. **(B)** Some differences in gram-positive bacteria. The third amino acid is often L-lysine and is attached to another peptide through a variable-length cross-link of amino acids (aa). m-A_2pm, *meso*-A_2pm.
doi:10.1128/9781555817169.ch14.f14.10

(for details and for all the variety that has been observed, see Vollmer et al., Suggested Reading).

Not all of the peptides in PG are cross-linked to another peptide, and the final cross-linked peptidoglycan structure, while rigid enough to provide the cell shape and prevent its rupture, is porous enough to allow larger molecules, including folded proteins, to pass through. Therefore, it does not provide a barrier to the movement of most molecules in and out of the cell; this role is played by the membranes. As discussed above, passage of most molecules through these membranes requires specialized gated channels, such as the SecYEG and Tat channels through the cytoplasmic membrane and the Omp protein channels in the form of β-barrels in the outer membranes of gram-negative bacteria, not to mention the six or more types of secretion systems in gram-negative bacteria that specifically secrete certain proteins through one or both membranes (see above). While the membranes are too flexible to provide much structure and are not strong enough to withstand large changes in osmotic pressure, some bacteria and all archaea lack PG cell walls. Even these cells often do maintain some shape, sometimes through layers of proteins called S layers.

Because the cell wall is essentially one enormous molecule surrounding the entire cell, it has been difficult to answer even some basic questions about its overall structure. For example, it is not known for certain whether the glycan strands run parallel or perpendicular to the long cell axis in rod-shaped bacteria like *E. coli* or *B. subtilis*, although recent evidence suggests that the glycan strands run perpendicular to the long axis in *E. coli* and *C. crescentus* (see Gan et al., Suggested Reading). It is not even known for certain how many layers of PG there are in different types of bacteria. The amount of cell wall material in at least some gram-negative bacteria is consistent with there being only one layer of PG in these bacteria, but it is possible that there is more than one, or even that the cell wall has different thicknesses in different regions. Gram-positive bacteria clearly have more than one layer of PG, since the cell wall is much thicker in these organisms.

SYNTHESIS OF THE CELL WALL

The cell wall exists just outside the cytoplasmic membrane, in the periplasmic space between the inner and outer membranes in gram-negative bacteria and on the outside of the cell in gram-positive bacteria, which lack an outer membrane. Therefore, in both types of bacteria, the precursors must be transported through the cytoplasmic membrane before they can be assembled into the continuous cell wall. How the subunits are synthesized and how they are assembled at their final destination is understood in some detail, but some questions still remain.

Synthesis of the Precursors

The precursors for cell wall synthesis are all made by enzymatic pathways in the cytoplasm, encoded by genes that are sufficiently conserved to be identifiable in genome sequences and often clustered together among other genes involved in cell division.

The structure and synthesis of the precursors of cell wall synthesis are shown in Figure 14.11. First, the sugar *N*-acetylglucosamine-1-PO_4 (GlcNAc-P) is made from fructose-1-PO_4 by the Glm enzymes, GlmS, -M, and -U. Then, UTP reacts with it to make the donor molecule UDP-GlcNAc, a reaction performed by a separate enzymatic activity of the same GlmU protein. Some of this UDP-GlcNAc is then converted into the donor molecule UDP-MurNAc by the Mur enzymes, MurA and MurB. To make what will be the cross-linking peptides in the cell wall, amino acids are then polymerized stepwise onto UDP-MurNAc, using the Mur enzymes (MurC to -F), known as the Mur ligases. If the bacterium uses the standard 4-amino-acid cross-linking peptide, first, L-Ala is attached to UDP-MurNAc. Then, D-Glu is attached to the L-Ala and *meso*-A_2pm to the D-Glu.

The last 2 amino acids in the 5-amino-acid peptide are both D-alanine (D-Ala) and they are added together as the dipeptide D-Ala–D-Ala. The amino acid D-Ala is unique to the cell wall in most bacteria and is made from L-alanine by a racemase. Incidentally, this racemase is inactivated by cycloserine, a useful antibiotic. Two D-Alas are then joined together by a specific D-Ala ligase to form the dipeptide before it is added to the growing peptide on UDP-MurNAc to make the 5-amino-acid peptide. As we discuss below, the last of these D-Alas will be lost during the cross-linking reaction when the PG is assembled from its subunits, leaving only a 4-amino-acid bridge to the cross-linked peptide.

Bactoprenyl: Flipping through the Cytoplasmic Membrane

The newly synthesized precursors in the cytoplasm must be transported, or "flipped," through the cytoplasmic membrane before they can be assembled into the cell wall outside the membrane. This is accomplished by first attaching the precursors sequentially to the cytoplasmic side of a long hydrophobic carrier lipid molecule, undecaprenyl phosphate (C_{55}-P), sometimes called bactoprenyl, made up of 11 (undeca) very hydrophobic five-carbon prenyl groups that make a chain long enough to insert through the cytoplasmic membrane. First, the MraY enzyme uses the donor UDP-MurNAc-peptide to transfer the precursor MurNAc-peptide to the phosphate end of the long carrier lipid molecule, with the release of UMP. The product, undecaprenyl-pyrophosphoryl-MurNAc-pentapeptide (C_{55}-P-P-MurNAc-pentapeptide), is called lipid I. Then, the MurG enzyme transfers the GlcNAc

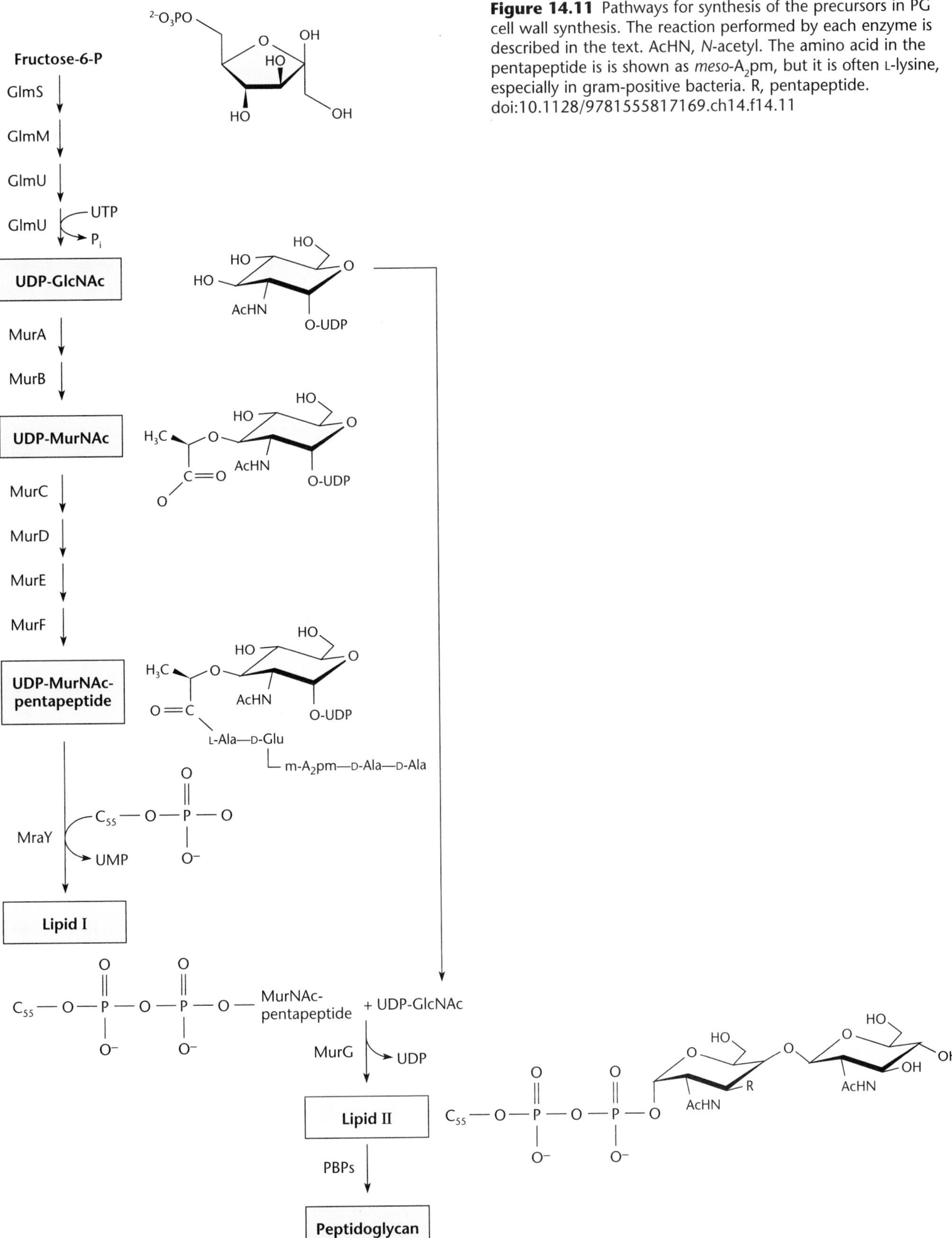

Figure 14.11 Pathways for synthesis of the precursors in PG cell wall synthesis. The reaction performed by each enzyme is described in the text. AcHN, *N*-acetyl. The amino acid in the pentapeptide is is shown as *meso*-A_2pm, but it is often L-lysine, especially in gram-positive bacteria. R, pentapeptide. doi:10.1128/9781555817169.ch14.f14.11

from the GlcNAc-UDP donor to lipid I, releasing its bound UDP to make C_{55}-P-P-MurNAc-pentapeptide-GlcNAc, called lipid II. Once the precursors are attached, the long lipid II molecule somehow "flips" in the membrane, moving the attached cell wall subunits to the outside of the membrane by an unknown mechanism. Another protein(s) might be involved in this flipping, but it has yet to be discovered.

As an aside, the undecaprenyl lipid (bactoprenyl) used to transport the precursors for cell wall synthesis across the inner membrane is not unique to cell wall synthesis and is also used to transport other cellular constituents through the membrane. The carrier lipid precursor, undecaprenylpyrophosphate (C_{55}-PP), is central to bacterial metabolism and is the target of the useful antibiotic bacitracin, which sequesters it, making it unavailable for biosynthetic reactions.

PG Synthesis

Once the disaccharide subunits have been flipped to the outside of the cytoplasmic membrane, they are polymerized by enzymes called the **penicillin-binding proteins (PBPs)**. These enzymes were discovered in the 1960s because they covalently bind penicillin and other β-lactam antibiotics, hence their name (see below). There are many different PBPs, and different types of bacteria have different numbers of these enzymes; for example, *E. coli* has 12 and *B. subtilis* has 16, mostly identified on the genome sequence by certain characteristic motifs. Some PBPs are larger and have two enzymatic domains, a glycosyltransferase activity responsible for making the long glycan strands and a transpeptidase activity responsible for cross-linking the peptide chains by attaching one peptide attached to a MurNAc to the third amino acid (often *meso*-A_2pm) of another peptide attached to a different MurNAc while removing the terminal D-Ala. Other, smaller enzymes may only have a D-Ala–D-Ala-cleaving peptidase domain. The role of these smaller enzymes may be to remove the terminal D-Ala from some of the peptides, preventing their use for cross-linking and increasing the pore size of the PG layer.

Antibiotics That Inhibit Cell Wall Synthesis

It is the transpeptidase domain of PBPs that is responsible for binding to penicillin and other β-lactam antibiotics. These antibiotics are mimics of the D-Ala–D-Ala dipeptide that forms the ends of the 5-amino-acid peptides that are the substrates of the transpeptidase domain (see above). The β-lactam antibiotic binds to the enzyme in lieu of the cross-linking peptide and becomes covalently bound to it, inactivating the enzyme and slowing PG synthesis. As we discuss below, this is a particular problem when the cell wall-synthesizing machinery is attempting to form a septum. Cells growing in the presence of one of these antibiotics first break at the site of septum formation, killing the cell and releasing the cell contents.

Other very useful antibiotics are vancomycin and other glycopeptides, some of the last defenses against methicillin-resistant *S. aureus*. These antibiotics bind to lipid II, more specifically, the D-Ala–D-Ala end of the peptide, and prevent its incorporation into PG. Vancomycin consists of a disaccharide with a 7-amino-acid peptide attached. This antibiotic cannot pass through membranes and so cannot enter the periplasmic space of gram-negative bacteria and only works against gram-positive bacteria, including *S. aureus*. In those gram-positive bacteria that are naturally resistant to vancomycin, often the cross-linking peptide ends in D-Ala–D-Lac instead of D-Ala–D-Ala and so does not bind vancomycin.

Interestingly, *S. aureus* and other bacteria do not seem to be able to acquire resistance to vancomycin through chromosomal mutations. When they become resistant to vancomycin, they usually have acquired resistance genes called *van* genes by lateral transfer from other naturally resistant bacteria. The products of the *van* resistance genes incorporate D-Ala–D-Lac into the cross-linking peptides attached to MurNac and also inhibit the enzyme that incorporates D-Ala–D-Ala into the growing peptide so that most of the cross-linking peptides in the strains that have acquired resistance now end in D-Ala–D-Lac, as in the naturally resistant bacteria.

The *van* resistance genes are normally repressed but can be induced by vancomycin. However, vancomycin does not enter the cytoplasm and so cannot directly serve as an inducer of the genes. Instead, the *van* genes are induced by a two-component system (see chapter 13) with a sensor kinase that binds vancomycin in the periplasm and communicates that information via phosphorylation to a response regulator in the cytoplasm that then induces the *van* genes. Other glycopeptide antibiotics that are more effective against vancomycin-resistant strains have been developed. Some of them have a short lipid attached and more effectively kill resistant strains because they can better mimic D-Ala–D-Lac, as well as D-Ala–D-Ala, or because they do not induce the *van* resistance genes through the two-component system.

Lytic Transglycolysases: Breaking and Entering

This simple view of cell wall synthesis is complicated by the aforementioned fact that the existing cell wall is a continuous molecule surrounding the entire cell. New material cannot be added to it without breaking and entering the preexisting cell wall. The new material must also be inserted in such a way that the cell is not left vulnerable to leakage or breakage due to osmotic and other forces. Breaking the preexisting cell wall is the job of a number of lytic transglycolysases or PG hydrolases that cleave bonds in the PG, allowing access to new subunits. There are at least 35 PG hydrolases in *E. coli*, which have been detected only by their enzymatic activities, and hydrolases that break each of the bonds holding the saccharides and amino acids together in PG are known. The genes for very few of them have been identified, and not much is known about the specific role of each of these enzymes. These hydrolase enzymes must be carefully controlled to prevent cell lysis, so the breaking of old bonds is tightly coupled to the formation of new ones. In the next section, we show that at least some of these enzymes are included in large complexes that make cell wall synthesis more efficient and help coordinate their various activities.

Cell Wall Synthesis Proteins Form a Complex

Because cell wall synthesis must move rapidly and efficiently and because the synthesis and breakage activities must be carefully coordinated, the enzymes involved in cell wall synthesis might be predicted to form a complex. Such complexes have been found in a number of bacteria, and the partial structure of the complex in *C. crescentus* is shown in Figure 14.12. The structure of this complex was assembled stepwise by determining which proteins bind to a particular protein, as determined by two-hybrid analyses, such as BACTH (see chapter 13); by biochemical experiments to identify proteins that bind to the purified protein; and from the way in which mutations in the genes for various proteins affect the distribution of other proteins in the cell (see White et al., Suggested Reading).

With the exception of the first enzyme of the pathway, MurA, the Mur enzymes involved in synthesizing the cytoplasmic precursors seem to form a complex in which the enzymes are colocalized at certain sites in the cell, which seem also to be the sites at which PG synthesis is occurring across the membrane in the periplasm. The advantages of such a complex seem obvious. By forming a complex, they can pass the intermediates in precursor synthesis from one enzyme to the next in the pathway more efficiently. The complex of Mur enzymes is bound to the inner face of the membrane, because the last of these enzymes, MurF, is bound to an inner membrane protein, MraY, the enzyme that forms lipid I by ligating MurNAc-pentapeptide to bactoprenyl lipid. MraY, in turn, binds to MurG, which forms lipid II by ligating GlcNAc to lipid I in the cytoplasm. An interesting protein called MreB helps organize this complex in the cytoplasm. It forms filaments and binds ATP and is thought to be distantly related to eukaryotic actin (Box 14.3). The filaments it forms are required for the proper assembly and movement of the PG-synthesizing complex, since the inhibition of MreB polymerization by the antibiotic A22 causes the complex to disperse and center elsewhere in the cell. Cell wall synthesis continues under these conditions, but the cells become spherical instead of rod shaped, as though MreB filaments are somehow required for lateral movement and orientation of the PG-synthesizing complex. Without MreB, PG synthesis may be in a constant state of septum formation, possibly because the PG-synthesizing complex is now organized by the FtsZ septal ring instead of by MreB (see

Figure 14.12 PG-synthesizing protein complex in *C. crescentus*. A representation of a cross section of the cell is shown. Proteins shown to be touching each other are hypothesized to bind to each other, as discussed in the text. Inner membrane proteins are drawn to reflect the relative number of transmembrane domains each has. See the text for details.
doi:10.1128/9781555817169.ch14.f14.12

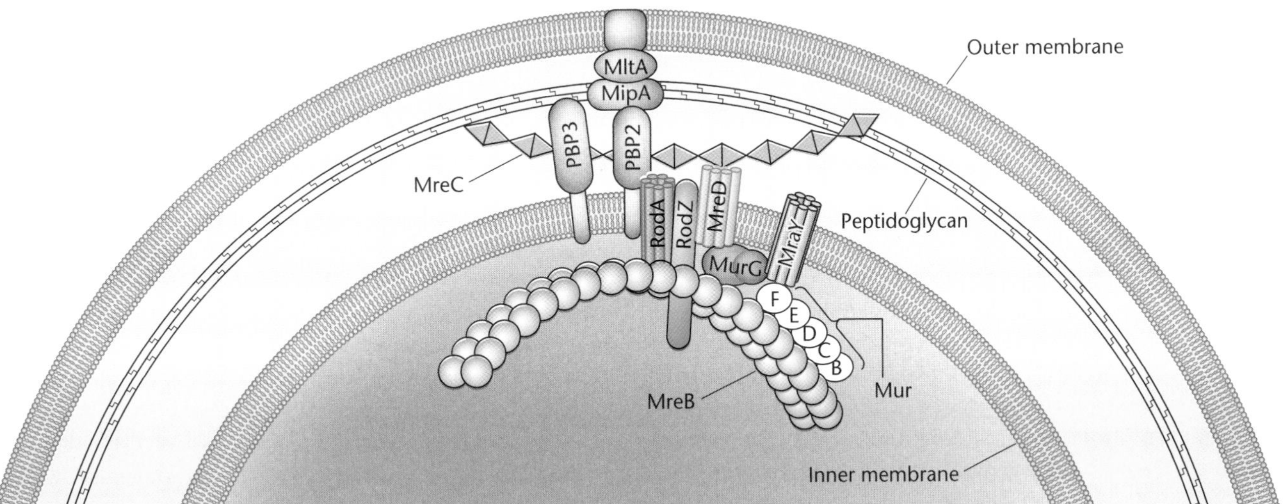

BOX 14.3

Evolutionary Origin of the Eukaryotic Cytoskeleton

Like many constituents of eukaryotic cells, the filamentous cytoskeleton of eukaryotes seems to have had its origin in an ancestor common to bacteria and archaea. Filaments related to all three types of filaments of eukaryotic cells—actin, tubulin, and intermediate filaments—have been found in bacteria and archaea, although they seem to play different roles in the different kingdoms of life. Some of them have been shown to be cytomotive and to move by polymerization on one end and depolymerization or contraction on the other, a process called treadmilling. Bacterial and archaeal cells are sufficiently small that diffusion may be rapid enough to move cellular constituents from one place to another, but eukaryotic cells, being generally much larger, can no longer depend solely on diffusion for such movement. It is as though the ability of these filaments to grow and contract by polymerization was put to good use in eukaryotic organisms to move cellular constituents around. Some filament-forming proteins from bacteria and eukaryotes are compared in the figure, with the bacterial protein shown in panel A opposite its related eukaryotic protein in panel B.

The first cytoskeletion filament discovered to have a counterpart in bacteria was tubulin, which forms microtubules in eukaryotic cells. Microtubules consist of about 13 protofilaments lying side by side to form a hollow tube (see the figure, panel B) and play various roles, including pushing chromosomes apart during mitosis (the mitotic spindle). The protofilaments are due to the polymerization of a heterodimer (see chapter 2 for a definition of heterodimer) made up of two different subunits, α- and β-tubulin. These subunits have an intrinsic GTPase activity (the active center of the GTPase is shown in blue). GTP bound to each of the newly added subunits encourages further polymerization, but after it hydrolyzes to GDP, it destabilizes the filaments and causes loss at the other end, leading to movement by treadmilling.

The FtsZ protein discussed in the text because of its central role in septum formation was the first protein in bacteria and archaea found to be homologous to tubulin. While FtsZ has little sequence similarity to tubulin and polymerizes as a monomer rather than a heterodimer, the structure of the FtsZ monomer is strikingly similar to that of both α- and β-tubulin, which are also similar to each other (see the figure, panel A). FtsZ monomers also bind GTP, which favors their polymerization into protofilaments, and in the protofilament they hydrolyze GTP to GDP, which favors depolymerization and dissociation (see the text). The active center for the GTPase is also shown in blue. The protofilaments associate side by side to form thicker filaments, which can be curved but are not known to form hollow tubes. The ubiquity of FtsZ in most bacteria and some archaea suggests that tubulin had its origin in a common precursor to all life on Earth.

After FtsZ was found to be a tubulin homolog, other bacterial proteins were found to be tubulin-like, including TubZ, the partitioning function of a *B. thuringiensis* plasmid (see the figure, panel A, and Ni et al., References). TubZ seems to have a function more similar to that of eukaryotic microtubules in that it has a role in segregating the plasmid that encodes it. It also has very little sequence similarity to tubulin but does seem to polymerize in vivo into double-stranded helical filaments that move by treadmilling and have a GTPase activity that aids their polymerization. However, these double-stranded filaments are composed of homodimers instead of heterodimers like tubulin (see the figure), and the protofilaments do not seem to interact laterally to form tubes.

The eukaryotic cytoskeletal protein F-actin seems also to have had its origin in an ancestor of bacteria and archaea. The various forms of actin play many roles in eukaryotes, but the cytoskeletal form, F-actin, forms double-stranded helical filaments that polymerize in the presence of ATP and hydrolyze ATP to ADP, with the active center shown in blue. The filaments move by treadmilling, with ATP-bound monomers being added at one end and ADP-bound monomers dissociating at the other end. If the dissociation is blocked, the filaments grow and can push cellular constituents from one place in the cell to another. Many proteins interact with actin in vivo to promote its instability and movement by binding to the monomers, cleaving the filaments, or capping one end to promote their contraction and expansion. Some pathogenic bacteria, including *Listeria*, take advantage of the treadmilling of actin filaments to move themselves within eukaryotic cells. F-actin is one of the most highly conserved proteins in cells, and its amino acid sequences are almost 90% identical from yeast to mammals.

The first bacterial protein discovered to be related to actin is MreB. This protein originally attracted interest because mutations in the *mreB* gene can cause normally cylindrical bacteria, like *E. coli*, to become spherical, as though the protein plays a role in directing lateral PG synthesis (see the text). Consistent with such a role, an *mreB* gene is found in rod- and egg-shaped bacteria but is missing from spherical bacteria, like *S. aureus*. While it shares little amino acid similarity to F-actin, the structures of their monomers are remarkably similar (compare panels A and B of the figure). Also, it binds ATP, has an intrinsic ATPase activity, and forms double-stranded filaments in vitro, although they are linear rather than helical like those of F-actin. It can also bind and hydrolyze GTP, as well as ATP, which distinguishes it from F-actin, which can only bind and hydrolyze ATP. This active center is also shown in blue.

BOX 14.3 (continued)

Evolutionary Origin of the Eukaryotic Cytoskeleton

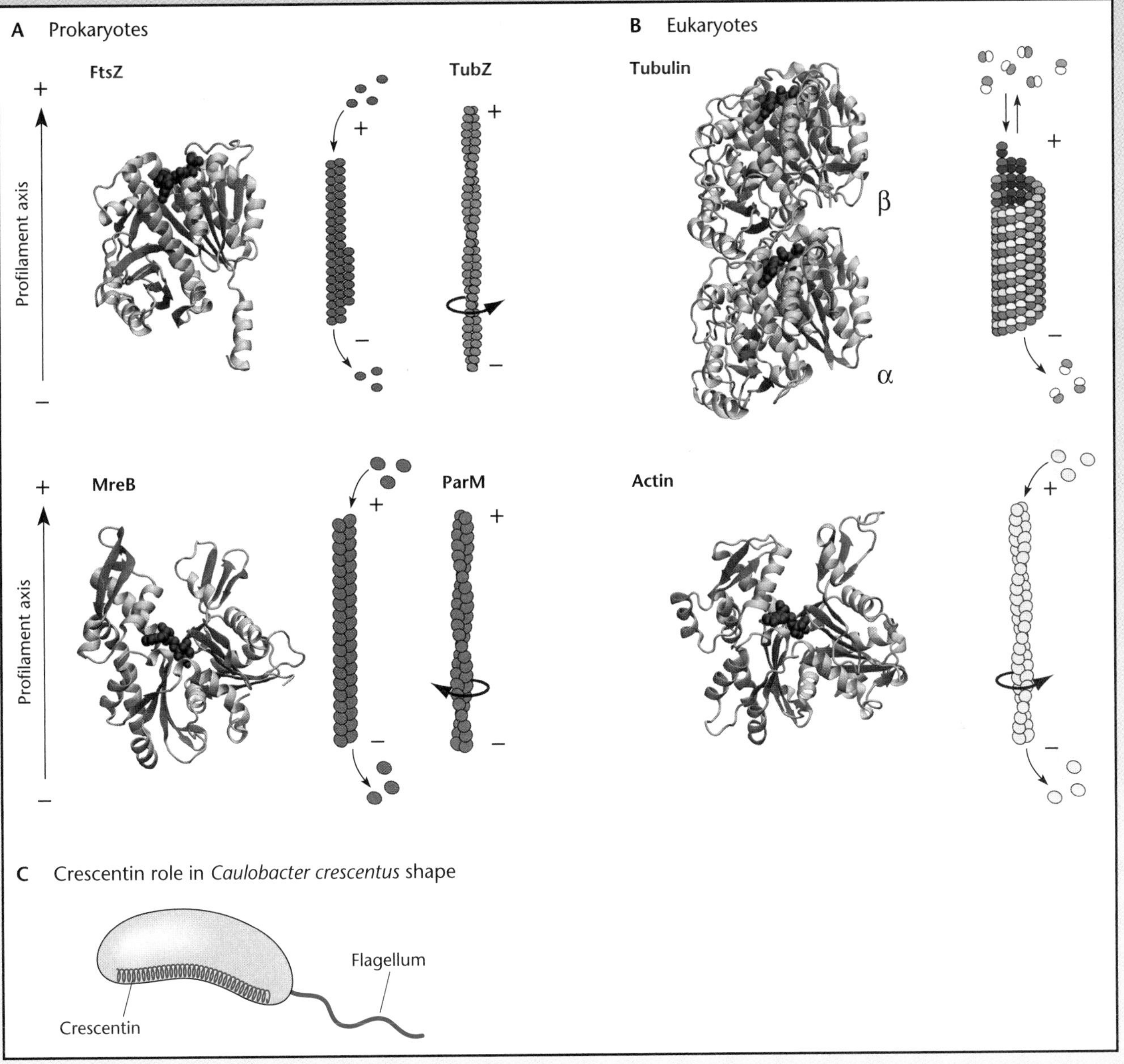

A Prokaryotes

FtsZ TubZ MreB ParM

Profilament axis

B Eukaryotes

Tubulin Actin

C Crescentin role in *Caulobacter crescentus* shape

Flagellum

Crescentin

doi:10.1128/9781555817169.ch14.Box14.3.f

As discussed in the text, in spite of extensive studies, it is still unclear what role MreB plays in organizing cell wall synthesis. It is known to move with the PG-synthesizing complex and to be required for its organization and orientation, but it is not clear how long or dynamic its filaments are in vivo and what role these filaments play in organizing the PG-synthesizing complex. Another difference from F-actin, at least in vitro, is that the MreB protein forms filaments in both the ATP- and ADP-bound forms. Also, the ATPase activity is not required for the lateral motion of the PG-synthesizing complex (see the text), suggesting that the hydrolysis of ATP by MreB is not driving the movement of the PG-synthesizing complex, although the ATPase activity must be required for some aspect(s) of MreB function in vivo, since mutations that specifically inactivate the ATPase activity render the protein nonfunctional.The FtsA protein involved in stabilizing FtsZ filaments is also related to actin, but even less is known of its role.

(continued)

BOX 14.3 (continued)

Evolutionary Origin of the Eukaryotic Cytoskeleton

The filament-forming functions of another bacterial plasmid actin-like protein, ParM, seem more like those of actin. The ATP-dependent filamentation and treadmilling of these proteins are required to segregate the plasmids that encode them and seem to push the plasmids apart (see chapter 4). They form double-stranded filaments in vitro, like actin, but these helices are left handed, unlike F-actin helical filaments, which are right handed (see the figure, panel B). They share very little amino acid sequence with either F-actin or MreB, but the structure of their monomers is similar to both.

So far, the winner of the look-alike contest between eukaryotic F-actin and bacterial and archaeal proteins is crenactin, a protein found in one of the major groups of archaea, the crenarchaeota. These single-celled organisms, originally thought to include only extremophiles that can live at very high temperatures in restricted environments, are now known to be one of the most common cellular organisms in the ocean, at least at lower depths. Unlike many other archaea, crenarchaea lack an *mreB* ortholog, but they have a gene for another protein, crenactin, that is much more closely related in its amino acid sequence to eukaryotic F-actin (see Ettema et al., References). The gene has been cloned from one such strain, and the protein has been purified. Like MreB, it has an intrinsic ATPase that also can use GTP. This protein may form filaments in cells, but the structure of these filaments is not known, and because of the difficulty in doing genetics with these organisms, it has not been possible to say with certainty that the filaments are involved in cell shape determination. Nevertheless, the discovery of this protein in a branch of the archaea increases the likelihood that actin had its origins in primitive microorganisms and is not unique to eukaryotes.

The third eukaryotic-type filament that may have had its origin in an ancestor shared with bacteria is the intermediate filament. These filaments play structural roles in eukaryotic cells, including giving cells some of their rigidity. The homologous protein found in bacteria is crescentin in *C. crescentus*, which shares little amino acid sequence with intermediate filaments but is structurally similar and behaves similarily in vitro. Crescentins are long proteins containing coiled-coiled domains in the middle. In vitro, they can be denatured in strong detergents, and when the detergent is removed, they reassemble spontaneously into long filaments without any requirement for energy or even divalent cations. The crescentin protein of *C. crescentus* forms helical bundles of filaments in the inside curvature of the cell that give the bacterium its characteristic crescent shape and its name (see the figure, panel C). Interestingly, the crescentin bundles seem to act by binding to and relieving stress on the cell wall, leading to a reduced rate of PG synthesis on the side they are on and the observed curvature in the cell (see Cabeen et al., References). This has fueled speculation that some of the energy for cell wall synthesis comes from the stress on the cell wall due to large differences in osmotic pressure inside and outside the cell, so that reducing the stress causes synthesis to slow down. However, it has been difficult to get detailed structural information about any of these proteins, making comparisons somewhat ambiguous. Also, there is still little evidence that orthologs of crescentin exist in bacteria outside of *C. crescentus*, so even if crescentin has a common origin with intermediate filaments, it is possible the gene transfer came by lateral transfer from a eukaryote to a bacterium rather than vertically by inheritance from a common ancestor.

References

Cabeen, M. T., G. Charbon. W. Vollmer, P. Born, N. Ausmees, D. B.Weil, and C. C. Jacobs-Wagner. 2009. Bacterial cell curvature through mechanical control of cell growth. *EMBO J.* **28:**1208–1219

Ettema, T. J. G., A.-C. Lindas, and R. Bernander. 2011. An actin-based cytoskeleton in archaea. *Mol. Microbiol.* **80:**1052–1061.

Ni, L., W. Xu, M. Kumara Swami, and M. A. Schumacher. 2010. Plasmid protein TubR uses a distinct mode of HTH DNA binding and recruits prokaryotic tubulin homolog TubZ to effect DNA partitioning. *Proc. Natl. Acad. Sci. USA* **107:**11763–11768.

Papp, D., and R. C. Robinson. 2011. Many ways to build an actin filament. *Mol. Microbiol.* **80:**300–308.

Wickstead, B., and K. Gull. 2011. The evolution of the cytoskeleton. *J. Cell Biol.* **194:**513–525.

below). This conclusion is supported by the observation that only rod- and egg-shaped bacteria seem to have a gene for MreB, which is not found in spherical bacteria, such as *S. aureus*. Two other inner membrane proteins of unknown function, RodA and RodZ, which are required to maintain the rod-like shape of the cell, are also part of the complex.

The complex in the cytoplasm that synthesizes the precursors is also connected to the PBP enzymes in the periplasm that insert them into the cell wall. The protein that makes this link is the inner membrane protein MreD, in at least some bacteria. This protein bridges the enzymes in the cytoplasm and in the periplasm by binding MreB in the cytoplasm, and also MreC in the periplasm. MreC is another actin-like filament-forming protein that may perform an organizing function in the periplasm similar to that performed by MreB in the cytoplasm, in this case binding PBP and lytic enzymes and organizing them into a PG-synthesizing complex. The organization of this complex allows the required tight coordination between the breaking of preexisting PG bonds and the insertion of new material. The locations of many of these proteins in the cell are mutually interdependent, and in *C. crescentus*, mutations in *mreD* cause MreB to mislocate, and vice versa, suggesting that there is not one master protein that directs the cellular location of the others and that their assembly is interdependent (see White et al., Suggested Reading).

While this is the general picture that is emerging, there seem to be some differences between species. For example, unlike in *E. coli* and *B. subtilis*, in *C. crescentus*, the MreC protein seems to be exclusively in the periplasm and not to have an inner membrane domain, and it has not been shown to bind directly to MreD. What seems clear, however, is that many of the proteins involved in the various stages of PG synthesis are assembled into a large complex spanning the membrane, and this large complex moves as it synthesizes the PG.

What Directs Cell Wall Synthesis?

The large PG-synthesizing complex moves through the cell wall, adding new cell wall material as it goes. Does something drive and direct it, or does it move itself, driven by the act of PG synthesis and directed by the preexisting cell wall? In principal, during lateral cell wall synthesis in rod-shaped bacteria, adding new material anywhere in the cylindrical sections of preexisting wall will cause the cell wall to grow laterally, provided there are some restrictions on its direction of growth. It has been known for a long time that lateral cell wall synthesis occurs in many places in the cell wall simultaneously, at least in *E. coli*, but it has not been known whether there is any pattern to this synthesis or whether it just occurs randomly. Recent advances in light microscopy have made it possible to focus on only one side of the cell and to locate and measure the direction and rate of movement of individual complexes of the cell wall synthesis machinery on this side of the cell (see Garner et al., Suggested Reading). The positions of the complexes and their movement can be followed by having one of the proteins in the complex fused to a fluorescent tag, for example, green fluorescent protein. These experiments suggest that new cell wall is synthesized simultaneously from a number of complexes that move independently of each other at right angels to the long axis of *B. subtilis* and other rod-shaped bacteria. There does not seem to be any direct communication between individual PG-synthesizing complexes, since even complexes that are close together can move in opposite directions. Thus, filament-forming proteins like MreB, which are required to assemble and orient the PG-synthesizing complexes, seem only to travel with them and not to play a role in positioning them relative to each other or in the cell as a whole. Furthermore, anything that slows PG synthesis, such as ampicillin, also slows the movement of the PG-synthesizing complexes, suggesting that they are propelled by PG synthesis itself and not, for example, by contraction and or growth of potential cell-spanning filaments, like MreB. However, we have known for a long time that a preexisting cell wall is not necessary to direct new cell wall synthesis, since spheroplasts that have had their cell walls substantially removed are still capable of regenerating their cell walls and returning to normal growth, provided they are maintained in a medium of high osmolarity so that they do not lyse. Moreover, MreB does seem to play an essential role in directing cell wall synthesis, since restoring MreB filaments in cells that have been treated with an inhibitor of MreB filamentation allows the renewed directed synthesis of PG. Thus, while MreB may not form a cytoskeleton, at least one that spans the cell, as was suggested by its relatedness to eukaryotic actin (see Box 4.3), its role is central to directing cell wall synthesis. In fact, in *C. crescentus*, MreB is required to identify the pole on which to synthesize the stalk (see Figure 14.14 and Wagner et al., Suggested Reading). The stalk is synthesized like the normal cell wall, except that it has a smaller diameter and almost always grows from only one pole. If MreB is inhibited, stalk synthesis is inhibited. When MreB inhibition is lifted, stalks will form more often on both ends and in the middle of the cell. In other alphaproteobacteria, including *Agrobacterium tumefaciens*, the cell wall grows from only one pole. When the cell has doubled its length, it divides in the middle, much like other types of bacteria (see "Septum Formation" below). These bacteria lack both MreB (and MreC and MreD) and DivIVA (see below), so they must use other systems to direct their cell wall synthesis (see Brown et al., Suggested Reading). The direction and control of cell wall synthesis are exciting areas of research and are getting at one of the most central questions of biology: how the overall structure of the cell is maintained from generation to generation.

Septum Formation

Not only must the cell wall grow, it must divide into two cell walls separating two daughter cells, a process called septum synthesis. Unlike spherical (coccoid) cells, such as *S. aureus*, which are in a continuous state of septum synthesis and division, rod-shaped cells, like *E. coli*, *B. subtilis*, and *C. crescentus*, and egg-shaped (ovoid) cells, like *Streptococcus pneumonia* and *Lactococcus lactis*, must have at least two states with different directions of overall PG synthesis: lateral synthesis in the direction of the long axis of the cell and septal synthesis, curving toward the middle of the cell so the walls can meet and divide into two cell walls. Unlike lateral synthesis, which can occur anywhere in the lateral part of the cell wall or even at the tips (see below), septal synthesis must occur only at the site of cell division and at the time in the cell cycle when the cells are to divide. In this section, we review what is understood about this process.

INVAGINATION OF THE MEMBRANES

Not only must the cell wall constrict and separate during cell division, but also, the membranes must do the same. The first stages of invagination of the inner membrane following the invagination of the cell wall may be subject to the same constrictive forces as invagination of the cell wall (see "The FtsZ Protein and the Septal Ring" below). However, the latest stages of the division of the cytoplasmic membrane seem to be independent of the division of the cell wall, since, in some mutants, the cytoplasmic membrane compartmentalizes into individual cells while the cell wall forms tubes. In the later stages of division, it is possible that the very hydrophobic cytoplasmic membrane may behave similarly to a large soap bubble. If a soap bubble is pressed down in the middle, it will spontaneously divide into two smaller soap bubbles.

The fate of the gram-negative outer membrane during cell division is more variable, depending on the species of bacterium. In some types of gram-negative bacteria, including *E. coli*, the outer membrane is attached to the cell wall through the Braun lipoprotein, which is embedded in the outer membrane and covalently attached to the cell wall. In these bacteria, the invagination of the outer membrane by necessity closely follows the invagination of the cell wall to which it is attached. In other gram-negative bacteria, including *C. crescentus*, there are no covalent attachments between proteins in the outer membrane and the cell wall. In these bacteria, the outer membrane invagination trails behind the division of the cell wall. In at least some of them, the Tol-Pal complex is largely responsible for outer membrane invagination. While not well understood, this complex of at least five proteins is highly conserved in gram-negative bacteria and spans the inner and outer membranes. Its role in communicating between the inner and outer membranes has many physiological manifestations, including being required for expressing surface lipoproteins and reducing sensitivity to detergents. This complex concentrates at the division plane of *C. crescentus* and other bacteria before division and seems to be required for the localization of some division proteins. In *C. crescentus*, it has been shown that mutations in the *pal* gene can cause division of the outer membrane to be disconnected from division of the cell wall, forming cylinders of outer membrane that divide only after a delay in cell wall division. The evidence from this and other gram-negative bacteria suggests that the division of the outer membrane is less carefully regulated than that of the other cellular envelope structures and seems to more or less come along for the ride.

The FtsZ Protein and the Septal Ring

In the 1970s, an extensive genetic analysis in *E. coli* revealed a number of genes whose products are required for cell division. A large number of temperature-sensitive mutants that could not grow to form colonies at a higher, nonpermissive temperature but could grow to form colonies at a lower, permissive temperature were screened visually to find ones that formed filaments of undivided cells at the higher, nonpermissive temperature. Presumably, such mutants had a reversible temperature-sensitive mutation in a gene whose protein product is required for cell division (see chapter 3). The genes in which such mutations could occur were named *ftsA*, *ftsB*, etc., for filamentous temperature sensitive. As more technology became available over the years, the products of the *fts* genes were assigned roles in the increasingly detailed picture of cell division in *E. coli*. One of the most central of these turned out to be *ftsZ*, so named because it was one of the last to be discovered, having for years been mistakenly thought to be part of one of the first such genes to be discovered, *ftsA*. Found in most types of bacteria, some archaea, and even chloroplasts and some mitochondria, FtsZ forms a ring at the site of future septum formation (Figure 14.13) and recruits at least 24 other proteins to the septal ring.

The FtsZ protein is fairly easy to purify and study in vitro, where it has many intriguing properties that are presumably related to its in vivo behavior. It has been compared to tubulin in eukaryotes, which it closely resembles in many of its properties (Box 14.3). FtsZ binds GTP and has an intrinsic GTPase activity that spontaneously degrades GTP to GDP. The GTP-bound monomers form filaments, which dissociate in the GDP-bound form. In fact, GTP can be degraded to GDP in the filaments only because the GTPase active center in FtsZ is divided between individual monomers in the filament, with the "head" of one monomer having part of the active center and the "tail" of the adjacent monomer having the other part, so that the active center for GTPase activity is reconstructed only in the filaments. The short filaments, called protofilaments, have a natural curvature and can be quite stiff, suggesting that they might be independently capable of forming rings and providing the force for membrane contraction during septum formation. Besides growing and contracting, the protofilaments can stack laterally by binding to each other randomly side by side to form sheets and even rings. However, it is always difficult to correlate the in vitro behavior of purified protein with what happens in the cell under different conditions and with the possible

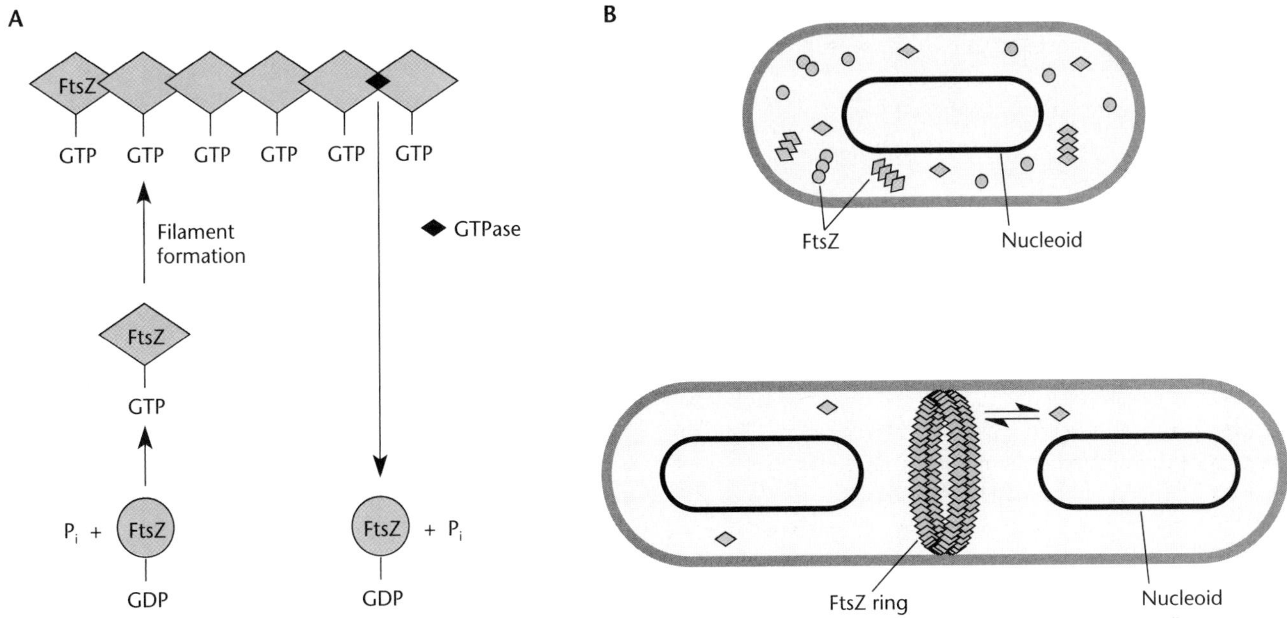

Figure 14.13 Formation of the FtsZ ring. **(A)** Binding of GTP to FtsZ favors filamentation. Hydrolysis of GTP to GDP plus PO_4 (inorganic phosphate [P_i]) in filaments favors dissociation into monomers. The GTPase active center formed by adjacent monomers is shown. **(B)** Early in the cell cycle, FtsZ is scattered throughout the cell as monomers and short protofilaments. After the nucleoid has segregated, the FtsZ monomers and protofilaments polymerize and associate to form the multistranded dynamic FtsZ ring. doi:10.1128/9781555817169.ch14.f14.13

involvement of other proteins and structures, making this is an ongoing but fascinating project.

SEPTAL-RING ASSEMBLY

Before cell division, the FtsZ protein exists as monomers and short protofilaments scattered throughout the cell. Its concentration remains largely unchanged during the cell cycle, but when the cell attains the correct size (depending on the growth conditions), the FtsZ monomers and short protofilaments assemble to form a preliminary ring or protoring in the center of the cell. Initially, this ring contains only FtsZ and two accessory proteins, FtsA and ZipA, which help the FtsZ ring bind to the membrane and help stabilize the FtsZ ring. The protoring persists for almost one-fifth of the cell cycle, at which time many other proteins are loaded on. Some of these newly added proteins traverse the membrane, and others assemble in the periplasmic space and organize and bind the various PBPs and amidases for septal-wall synthesis. Table 14.1 shows a partial list of such proteins and what is known of their functions. While some PBPs may be unique to septum synthesis, others are shared by lateral and septal syntheses. When they are finally assembled, the proteins of the septal ring play various roles in dividing the cell, including stabilizing FtsZ rings, invaginating the cytoplasmic membrane, and changing the direction of PG synthesis by remodeling the PG-synthesizing complex. Their addition is followed closely by the initiation of constriction of the cell wall and the membranes.

Some of the last proteins to be added to the septal ring are the amidases that divide the cell wall by cleaving the peptide cross-links that are holding the two walls together (see Peters et al., Suggested Reading). Their functions overlap somewhat, because mutants with a single mutation in any one of the amidase (*ami*) genes show few effects, but multiple mutants of *E. coli* with mutations in a number of these genes form long filaments of cells in which the inner membranes have constricted and separated but the cells are still attached through their cell walls. This is part of the evidence stated above that the final stages of membrane constriction and separation occur independently of cell wall separation. The separation of the cell walls by the amidases is a vulnerable time in cell wall synthesis, and the activities of the amidases must be carefully controlled to prevent lysis. Some of the amidases are not activated until they bind to other proteins previously bound to the septum, EnvC and NplD. In turn, these proteins require the previous binding of FtsE and FtsX, an ABC transporter, to the septum. The fact that ABC transporters require ATP cleavage to change their conformation raises the interesting possibility that the activation of the amidases depends on the levels of ATP and therefore on the energy state of the cell.

Table 14.1 Some septal ring proteins of *E. coli*

Protein(s)	Function
FtsZ	Spontaneously forms rings and may provide constrictive force
FtsA, ZipA	Stabilize FtsZ rings and attach them to membrane
FtsK	DNA translocase; activates site-specific recombinase to separate chromosome dimers
FtsL, FtsB, FtsQ	Trimeric complex in inner membrane; connect FtsZ ring in cytoplasm with membrane-associated and periplasmic proteins
FtsE, FtsX	Putative ABC transporters; required to bind EnvC and NplD to the septum
FtsW, FtsI	FtsW recruits FtsI (PBP3) to septal ring
EnvC, NlpD	Proteins with LytM domain in periplasmic septal ring; activate amidases A, B, and C (*N*-acetylmuramyl-L-alanine amidases) during cell separation
FtsN	Last protein added; binds to PBP1b transpeptidase involved in both lateral and septal PG synthesis and to septal PG through its SPOR domain
AmiA, AmiB, AmiC	Amidases that cleave peptide cross-links during division of the cell wall; require EnvC and NlpD bound to the septal ring for activation

The latest stage in cell division, separation of a single cell wall into two cell walls by amidases, is also the stage when the cell is most sensitive to penicillin and other β-lactam antibiotics. As mentioned above, these antibiotics bind to the PBPs and inhibit their peptidase activity, slowing cell wall synthesis. However, only when the amidases begin to remodel the PG during cell division does the inhibition of peptide cross-linking by the antibiotics cause the cells to lyse. As evidence, mutant *E. coli* cells with multiple mutations in the amidase genes are much less sensitive to lysis by penicillin. The mutant cells form filaments in the presence of penicillin but do not lyse. Apparently, the cleaving of peptide cross-links by amidases during division without the simultaneous formation of new peptide cross-links by the PBPs is a major reason why cells lyse and die in the presence of these antibiotics.

The formation of the septal ring is dynamic and is constantly being remodeled by addition and subtraction of subunits until the final protein, FtsN, is added, when the ring becomes more stable. The FtsN protein may bind both to the septal ring and to the new septal PG, allowing more and more of it to bind as the septum forms and making this a self-accelerating process. Before this time, if growth conditions change or the cell suffers extensive damage to its DNA, the process of cell division can be reversed, and the septal ring dissociates.

Placing the Septal Ring

Since formation of the FtsZ ring starts the process of cell division, it is critical that the ring be placed properly, at the site at which septum formation is to occur. So far, most evidence suggests that FtsZ has an intrinsic ability to form rings and that it will do so wherever it is not prevented. In chapter 1, we reviewed two systems that prevent the misplacement of FtsZ rings in *E. coli* and *B. subtilis*, the Min system and the nucleoid occlusion (NO) system. In *E. coli*, the Min system consists of three proteins, MinC, -D, and -E. The complex of MinC and MinD inhibits FtsZ ring assembly by competing for binding with FtsA and ZipA and by inhibiting lateral interactions between FtsZ filaments. All three of these proteins oscillate from one end of the cell to the other. MinC and -D and some MinE proteins concentrate at one end, or pole, of the cell, and the remainder of the MinE forms a ring that seems to constrict the movement of MinC and -D to that end of the cell. The MinE protein then causes the MinC-MinD complex to be released from the membrane at that end of the cell and to reform on the other end. The movement of the Min proteins might be due purely to diffusion, or the membrane may somehow help pattern their movement. It is hypothesized that the purpose of the oscillation of these FtsZ inhibitors is to ensure that their concentration will be lowest in the middle of the cell, where the MinCD proteins are just passing through, and highest at the ends of the cell, where formation of the FtsZ ring has to be discouraged lest minicells form that do not contain DNA. In *B. subtilis*, which lacks MinE, the MinC and -D proteins do not oscillate but bind to the ends of the cell and establish a concentration gradient, which may accomplish the same thing. Their binding to the poles is promoted by a protein called DivIVA, which concentrates at the cell poles. The DivIVA protein does not bind directly to MinD but binds to a protein, MinJ, that does bind directly to MinD and also binds to the poles. While its exact role is not understood, the DivIVA protein seems to play a general role in locating the sites of PG growth in gram-positive bacteria. For example, in *B. subtilis*, it also binds to the FtsZ septal ring after other proteins, such as some of the PBPs, are bound and septal PG synthesis is about to begin. It also concentrates at the poles of bacteria that grow from the ends or tips of the cell rather than from the middle, including *Mycobacterium*, *Corynebacterium*, and *Streptomyces*. It is phosphorylated by serine/threonine kinases that might regulate its activity and control cell wall growth.

As also discussed in chapter 1, the NO systems of *E. coli* and *B. subtilis* act to prevent FtsZ ring formation over the nucleoid before it has divided, thereby preventing guillotining of the nucleoid when the cell divides. Both the SlmA protein in *E. coli* and the Noc protein in *B. subtilis* inhibit FtsZ ring formation by binding directly to FtsZ. However, the Noc and SlmA proteins have been shown to inhibit FtsZ only when they are also bound to certain sites on the DNA. These binding sites are conspicuously absent around the terminus of replication (see chapter 1), ensuring that FtsZ ring formation occurs only when chromosomal replication is nearly completed.

Regulation of FtsZ Ring Formation in *C. crescentus*

The mechanism of coordination of nucleoid segregation and FtsZ ring formation seems to be somewhat clearer in the stalked bacterium *C. crescentus* than it is in the other model systems. The advantage of *C. crescentus* over many bacteria for studying the cell cycle is that, unlike *E. coli* and many other bacteria, it does not divide symmetrically, which makes it easier to synchronize cells in the cell cycle. The cell cycle of *C. crescentus* is shown in Figure 14.14. *C. crescentus* has two cell types: the swarmer cell, which has a flagellum and can swim freely, and the stalked cell, so named because it has a stalk with a very sticky end that can glue the cell to solid surfaces. Only the stalked cell can replicate its DNA and divide, releasing a new swarmer cell. Thus, all of the newly released swarmer cells are the same age and proceed through their cycle in unison. This ability of the swarmer cell to move and then differentiate into a stalked cell that can attach in a new location probably helps *C. crescentus* in its search for food in natural environments.

Studying synchronized cells has the big advantage in cell biology that all the cells in a culture are at the same stage, so when you look at an individual cell under the microscope, you know what stage of the cell cycle it is in. The natural synchronization of *C. crescentus* has made it easier to correlate cellular events such as division with chromosome replication and the roles of various proteins and their cellular distribution during these processes.

The *C. crescentus* bacterium lacks a Min system. It also lacks a NO system, at least one that is closely enough related to those from either *E. coli* or *B. subtilis* to be identifiable from the genome sequence. This does not mean too much, since those from *E. coli* and *B. subtilis* also are not closely enough related to each other to be identifiable by sequence alone. However, there are other reasons to believe that *C. crescentus* lacks an NO system. For one, FtsZ rings begin to form in *C. crescentus* when the DNA is distributed throughout the cell, arguing against a role for such functions in inhibiting FtsZ ring formation. The apparent absence of Min and NO systems suggests that *C. crescentus* has different ways of restricting the formation of FtsZ rings to the middle of the cell and preventing minicell formation.

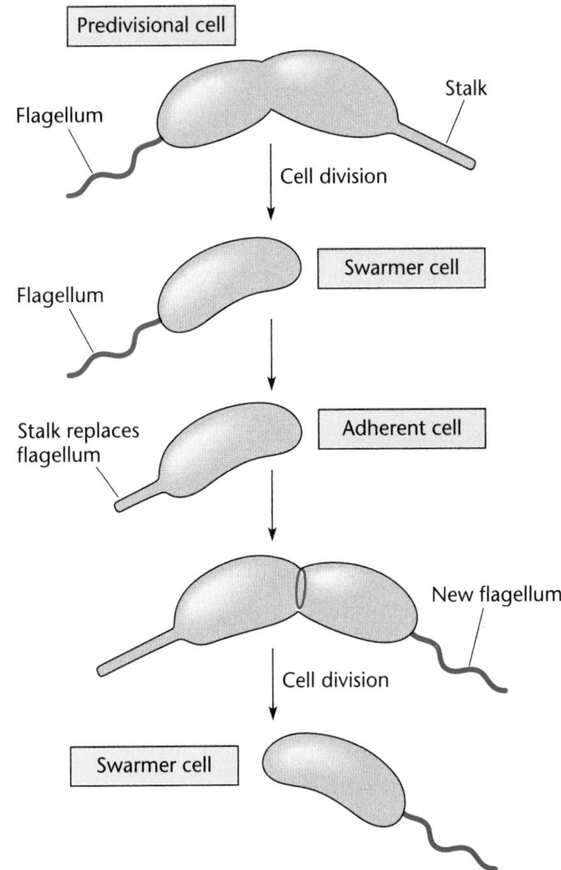

Figure 14.14 Cell cycle of *C. crescentus*. Asymmetric division gives rise to a stalked sessile cell and a motile flagellated swarmer cell. The flagellated cell differentiates, replacing the flagellum with a stalk through which it adheres to a surface. This cell then replicates its DNA and divides to produce a new motile swarmer cell to complete the cycle. Only the stalked cell can replicate its DNA and divide, so all of the new swarmer cells released from the surface are the same age.
doi:10.1128/9781555817169.ch14.f14.14

SEGREGATION OF THE NUCLEOID DURING THE CELL CYCLE

The first question is where in the cell the origin and terminus of DNA replication are located at various stages in the cell cycle of *C. crescentus*. Remember that the DNA is almost a thousand times longer than the cell, so the DNA forms a mass in the cell called the nucleoid. The origin and terminus of replication can be located in this mass of DNA by visualizing fluorescent proteins that bind to sites in the DNA close to the origin or terminus. Such studies reveal that initially, in the newborn swarmer cell, the chromosome origin of replication is located at the pole (or end) of the cell to which the flagellum is attached. The terminus of the chromosome is at the other pole, so the chromosomal DNA extends the length of the cell. This arrangement of the chromosome, with the origin on one end and the terminus on the other, is maintained until the swarmer cell sheds its flagellum and the cell begins to form a stalk on the same end. Then, the chromosome initiates replication, duplicating the origin of replication. As the two daughter DNAs elongate, one of these two new origins quickly migrates to the other end of the cell, while the other origin stays put. This migration requires MreB and ParA. From what is known about MreB, it is not clear how it is involved, but clues to the role of ParA in this process may come from the role of its homologs in plasmid partitioning (see chapter 4). The ParA protein has an ATPase domain found in many proteins, called a Walker A box. It forms filaments after it binds to another protein, often called ParB, but only if the ParB protein is bound to specific sites on the plasmid DNA. The latest results indicate that the ParA-ParB-plasmid complex seems to bind to the nucleoid, periodically changing its site of attachment to the nucleoid as the nucleoid replicates and segregates toward the poles of the cell. The ParA filaments, which are periodically growing and shrinking, pull the plasmid toward the poles of the cell, thereby ensuring the partitioning of the plasmid with the nucleoid. Even assuming that chromosomally encoded Par functions work in a related manner, it is not clear how such a mechanism could segregate the origins of replication and therefore the nucleoid. *C. crescentus* has a number of ParB-binding sites, called *parS*, close to the origin of replication, and these sites are known to be required for chromosome segregation. One end of the ParA filaments is presumably bound to ParB at the *parS* sites and pulls the origins of replication apart, but it is not clear to what the other ends of the ParA filaments are bound and pulling it toward. Another protein, TipN, is bound at this pole and may play a role in this process (see Kirkpatrick and Viollier, Suggested Reading). Hopefully, more research will answer this question.

Concomitant with the migration of the origin, the chromosome replicates at the replication forks. As the chromosomes replicate, the two replication forks move slowly from the poles toward the center of the cell, so that eventually, when they reach the terminus of chromosome replication, they are in the middle of the cell close to where the cell will later divide. A flagellum assembles on the far end of the cell from the stalk, so that when a new swarmer cell is released after cell division, the chromosome in this swarmer cell is distributed with the origin close to the end with the flagellum and the terminus at the other end.

REGULATING AND POSITIONING FtsZ RING FORMATION

Quite a bit is also known about how the formation of FtsZ rings and subsequent cell division in *C. crescentus* are restricted to near the middle of the cell (Figure 14.14). Even though different proteins are used, this mechanism is similar to those used by *B. subtilis* and *E. coli* in that it involves an inhibitor of FtsZ polymerization that has its lowest concentration at the middle of the cell (see above and chapter 1). However, instead of Min proteins, *C. crescentus* uses an inhibitor protein called MipZ, which is related to ParA proteins (see Thanbichler, Suggested Reading). Like ParA, the MipZ protein binds to ParB, which is attached to the cluster of *parS* sequences close to the origin of replication, and therefore migrates with ParB to the poles of the cell when replication is initiated. This causes MipZ to be concentrated at the ends of the cell and less concentrated in the middle, so that FtsZ rings can form only in the middle.

This model is supported by a number of lines of evidence. First, the location of MipZ in the cell can be determined by fusing it to a fluorescent protein, which can be detected by epifluorescence light microscopy. These experiments show that MipZ is concentrated at the poles of the cell after the origins of replication have segregated to the poles. Depleting the cell of MipZ causes the cell to divide aberrantly, often giving rise to minicells, which is expected if, in the absence of the inhibitor, FtsZ rings can form anywhere in the cell. Overproducing MipZ prevents FtsZ ring formation and causes the cells to form filaments, which is also expected from the model, since the MipZ concentration is now too high even in the middle of the cell and FtsZ cannot form rings anywhere. The inhibitory effect of MipZ can even be seen in vitro using purified MipZ and FtsZ. Purified FtsZ will assemble into longer filaments in the presence of GTP, and MipZ can interfere with this polymerization. Purified MipZ has also been shown to bind to purified ParB in vitro.

Much of the behavior of MipZ can be predicted from the fact that it is a homolog of ParA and can bind to ParB. It was known that overproducing ParA prevents segregation of the origin but does not prevent FtsZ ring formation after the origins have segregated. However, overproducing MtsZ or ParB reverses FtsZ ring formation even after rings have formed. It is as though, once the origins of chromosome replication have segregated to the ends of the cell, MipZ can replace its homolog, ParA, and bind to ParB on the *parS* sites so that most of the MipZ protein is concentrated at the poles, or ends, of the cell. However, if there is excess ParB, not all of it is bound to the *parS* sites, causing some MipZ to be spread throughout the cell and preventing FtsZ ring formation even in the middle of the cell, away from the two origins.

Creating Cellular Asymmetry in *C. crescentus*

One of the most striking things about *C. crescentus* is the way the daughter cells are so different and have different fates, more like the situation with eukaryotes. With bacteria like *E. coli* and *B. subtilis* that divide more symmetrically, both daughter cells are usually thought to be identical. In *C. crescentus*, even the FtsZ ring does not form precisely in the middle of the cell but closer to the new end. This asymmetry in *C. crescentus* is created by localizing certain proteins to one pole or the other and by regulating their activity through phosphorylation and targeted degradation (see Thanbichler, Suggested Reading). That way, once the cells divide, these proteins can be separated from their regulators, leading to different physiological states. For example, the off-center positioning of the FtsZ ring is influenced by another protein, TipN, that is bound at the new end and might also affect the segregation of the chromosome toward that end (see above).

The CtrA protein is a master response regulator that controls much of the cell cycle in *C. crescentus* and helps to differentiate the two cell types. CtrA regulates both transcription and replication, depending upon its state of phosphorylation (see Thanbichler, Suggested Reading). Like many transcriptional response regulators, in the phosphorylated state, it is both a repressor and an activator and regulates more than 95 genes. In its phosphorylated state, CtrA~P, it also binds to sites at the origin of replication, displacing DnaA and preventing the initiation of chromosome replication. Basically, CtrA~P is the form that exists in the swarmer cell, where it blocks the initiation of DNA replication and regulates the expression of many genes. As the swarmer cell enters the predivisional state, loses its flagellum, and begins to make a stalk, the CtrA is dephosphorylated and degraded, freeing up the origin of replication and allowing DnaA to bind and replication to initiate. Therefore, to understand how *C. crescentus* achieves its two-celled state, we need to understand what controls the phosphorylation and dephosphorylation of CtrA.

The immediate proteins responsible for the phosphorylation of CtrA are the sensor kinase CckA and the phosphotransferase ChpT (see chapter 13 for an explanation of such signaling cascades), but what activates CckA is less clear. It somehow senses the phosphorylated state of another response regulator, DivK, which is determined by another signaling cascade, DivJ and PleC. In the swarmer cell, there is very little DivJ, but PleC accumulates at the flagellated pole, dephosphorylating DivK, which triggers the CckA-dependent phosphorylation of CtrA and blocks the initiation of replication. During the transition to the predivisional stage and the formation of a stalk at this pole, the DivJ kinase replaces PleC and DivK is phosphorylated (DivK~P). The phosphorylated form of DivK is able to interact with the future stalked pole, which promotes the dephosphorylation of CtrA, allowing the initiation of DNA replication. The same signaling pathway, CckA and ChpT, also phosphorylates and activates another response regulator, CpdR, a factor responsible for the localization of the ClpXP protease to the flagellated pole and the degradation of CtrA~P, thereby hastening the initiation of DNA replication. Therefore, the same signal pathway acts in two ways in the newly stalked cell to remove CtrA~P and initiate replication, one by indirectly phosphorylating and activating DivK and the other by indirectly concentrating the protease that degrades CtrA~P at the stalked pole. As we learn more and more, it is apparent that the interaction of numerous signaling pathways, even in relatively simple bacterial cell cycles, allows the cell cycle to adjust itself to different and continually varying extracellular conditions.

Genetic Analysis of Sporulation in *B. subtilis*

As mentioned in the introductory chapter, many bacteria undergo complex developmental cycles. In their development, some bacteria perform many functions reminiscent of higher organisms: they undergo regulatory cascades; their cells communicate with each other, differentiate, and form complex multicellular structures; the cells in these multicellular structures often perform different distinct functions, which require compartmentalization and cell-cell communication; and the cells use phosphorelays to respond to changes in communication with other cells and with the external environment. Because of the relative ease of molecular genetic analysis with some bacteria, some of these developmental processes have been extensively investigated as potential model systems for even more complex developmental processes in higher organisms.

The best-understood bacterial developmental system is sporulation in *B. subtilis*. When starved, *B. subtilis* cells undergo genetically programmed developmental changes. They first attempt to obtain nutrients from neighboring organisms by producing antibiotics and extracellular degradative enzymes. They even cannibalize their siblings, as discussed in Box 4.2. If starvation conditions persist, the cells sporulate, producing endospores that are metabolically dormant and highly resistant to environmental stresses.

The process of sporulation starts with an asymmetric division that produces two cell types with different morphological fates. The larger cell, which is called the mother cell, engulfs the smaller forespore and then nurtures it. Eventually, the mother cell lyses, releasing the endospore.

Many of the changes that occur in the sporulating cell can be visualized by electron microscopy. They are schematized in Figure 14.15. The figure also shows the

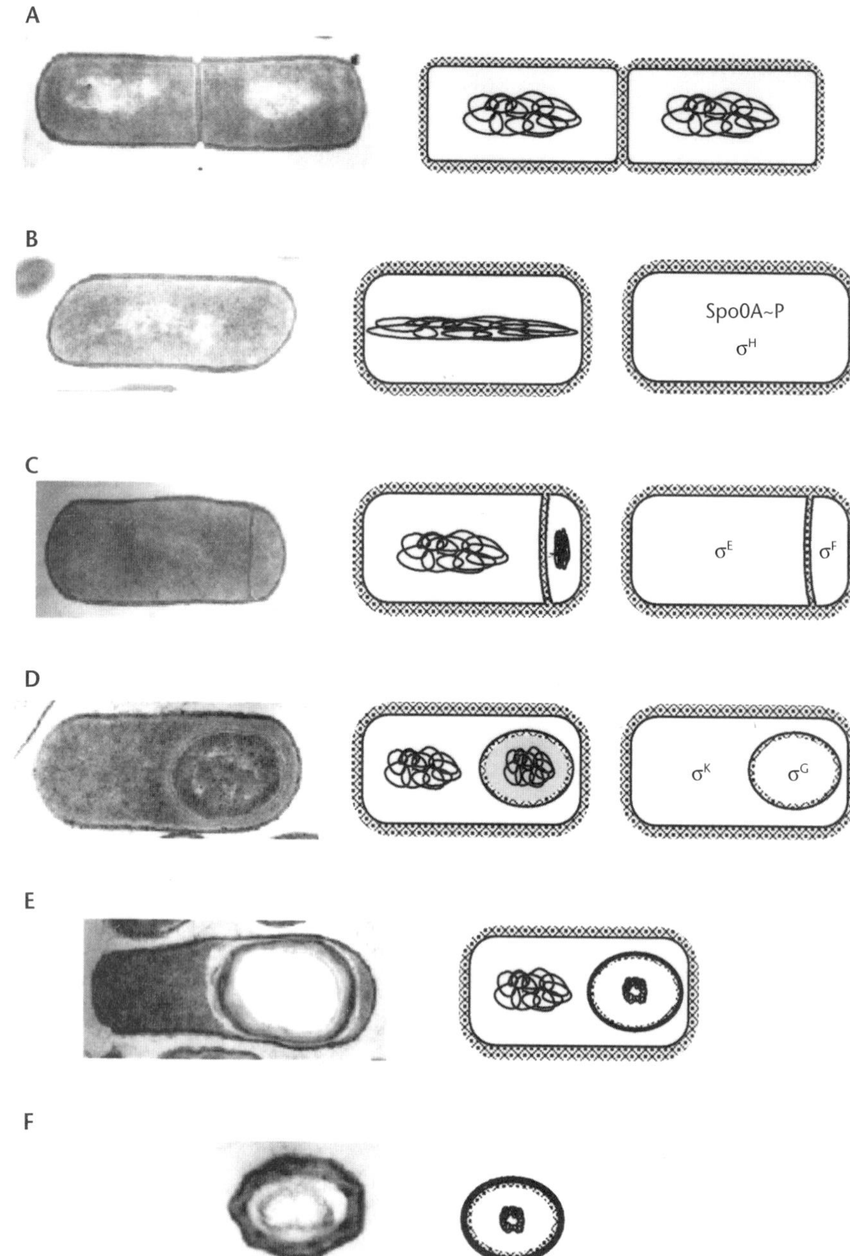

Figure 14.15 Stages of sporulation. On the left in each panel is an electron micrograph of the stage of sporulation, and on the right is shown, in cartoon form, the disposition of the chromosomes and the times and sites of action of the principal regulatory proteins that govern sporulation gene expression. **(A)** Vegetative cells. **(B through E)** Sporangia at entry into sporulation (stage 0) **(B)**, at polar division (also called polar septation) (stage II) **(C)**, at engulfment (stage III) **(D)**, and at cortex and coat formation (stages IV to VI) **(E)**. **(F)** A free spore.
doi:10.1128/9781555817169.ch14.f14.15

proteins that have been identified as key regulators of specific stages of development. We describe the experiments that identified these key regulators below.

Identification of Genes That Regulate Sporulation

Isolation of mutants was crucial to the process of identifying the important regulators of sporulation. Many mutants were isolated on the basis of a phenotype referred to as Spo$^-$ (for *spo*rulation minus). Such mutants could be identified as nonsporulating colonies because plate-grown cultures of the wild type develop a dark brown spore-associated pigment whereas the nonsporulators remain unpigmented.

Spo$^-$ mutants were phenotypically characterized by electron microscopy and then grouped according to the stage at which development was arrested (Figure 14.16). Some of the key regulatory genes defined by analysis of the mutants are listed in Table 14.2. The names of *B. subtilis* sporulation genes reflect three aspects of the genetic analysis of these genes. The roman numerals refer to the results of phenotypic categorization of the mutant

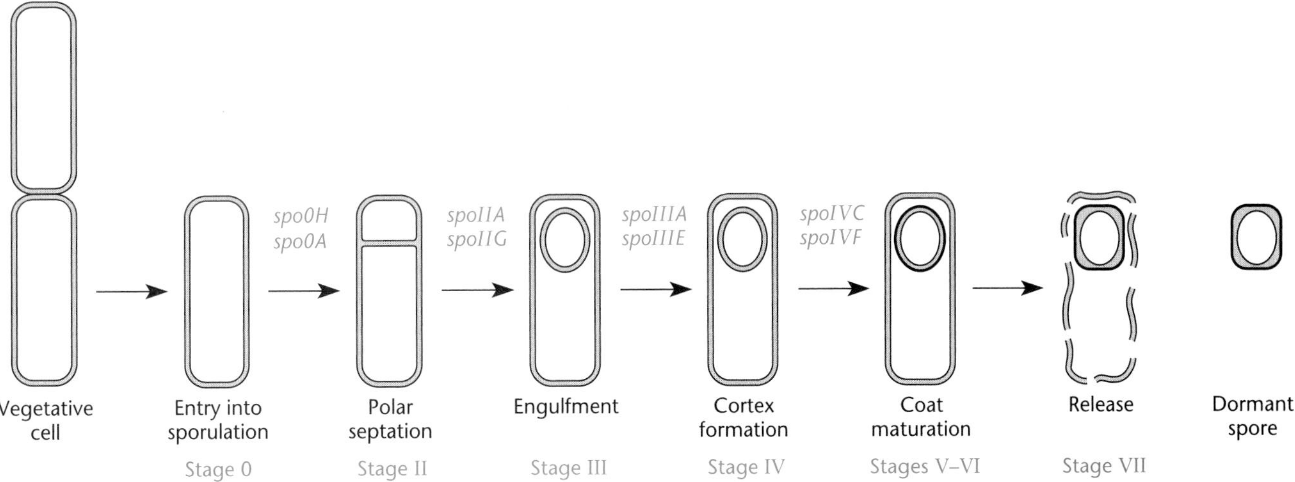

Figure 14.16 Morphological events of *B. subtilis* sporulation and some of the genes required for sporulation. Mutations in the genes above the arrows produce a mutant phenotype similar to that shown immediately before the arrow. Stage I is skipped in the figure because it is less clearly defined. doi:10.1128/9781555817169.ch14.f14.16

strains, with the numbers 0 through V indicating the stages of sporulation at which mutants were found to be blocked. The gene names also contain one or two letters. The first letter designates the separate loci that mutated to cause similar phenotypes. Each such locus was defined by the set of mutations that caused the same morphological block and that were genetically closely linked. The second letter in the names indicates the individual ORFs that were found when DNA sequencing revealed that a locus contained several ORFs.

Regulation of Initiation of Sporulation

Much of what we understand about the mechanism of sporulation initiation is based on studies of the class of sporulation-minus mutants, which were designated *spo0* because of their failure to begin the sporulation process.

Table 14.2 *B. subtilis* sporulation regulators

Stage of mutant arrest	Gene	Function
0	*spo0A*	Transcription regulator
	spo0B	Phosphorelay component
	spo0F	Phosphorelay component
	spo0E	Phosphatase
	spo0L	Phosphatase
	spo0H	Sigma H
II	*spoIIAA*	Anti-anti-sigma
	spoIIAB	Anti-sigma[a]
	spoIIAC	Sigma F
	spoIIE	Phosphatase
	spoIIGA	Protease
	spoIIGB	Pro-sigma E
III	*spoIIIG*	Sigma G
IV	*spoIVCB-spoIIIC*[b]	Pro-sigma K
	spoIVF (operon)[c]	Regulator of sigma K

[a]Loss-of-function mutations in *spoIIAB* do not produce a Spo$^-$ phenotype; rather, they cause lysis.

[b]In *B. subtilis* and some other bacilli, two gene fragments undergo recombination to produce the complete σ^K coding region.

[c]Subsequent work has further defined two genes, *spoIVFA* and *spoIVFB* (see the text).

Many of these mutants have pleiotropic phenotypes, meaning that they are altered in several characteristics. Besides being unable to sporulate, they fail to produce the antibiotics or degradative enzymes that are characteristically produced by starving cultures, and they do not develop competence for transformation (see chapter 6).

Two of the *spo0* genes, *spo0A* and *spo0H*, encode transcriptional regulators. The *spo0A* gene encodes a "two-component system" response regulator that is responsible for regulating the cellular response to starvation. The product of *spo0H* is a sigma factor (σ^H). Many of the genes that are targets for Spo0A regulation are transcribed by the σ^H-containing RNA polymerase holoenzyme (see Britton et al., Suggested Reading).

Like most response regulators, Spo0A must be phosphorylated in order to carry out its transcriptional-regulatory functions. Phosphorylation of Spo0A involves a "phosphorelay" system (Figure 14.17) that includes another two of the *spo0* gene products, Spo0F and Spo0B. The phosphorelay also involves at least five protein kinases that each phosphorylate Spo0F under certain conditions. Spo0B is a phosphotransferase enzyme that transfers phosphoryl groups from Spo0F~P to Spo0A. Spo0A~P then regulates its target genes by binding to their promoter regions, activating some and repressing others (Box 14.4).

The regulatory effect of Spo0A on a given target gene depends on the amount of Spo0A~P in the cell, reminiscent of OmpR~P regulation of porin genes and *Bordetella* BvgA~P regulation of virulence genes (see chapter 13). At low levels, Spo0A~P positively regulates genes involved in the synthesis of antibiotics and degradative enzymes, as well as competence and biofilm formation. This positive regulation may result from what is actually a "double-negative" series of events, in which the direct effect of Spo0A~P action is repression of a gene called *abrB*, which itself encodes a repressor that acts on the antibiotic and degradative-enzyme genes. At higher levels, Spo0A~P directly activates several sporulation operons, including *spoIIA*, *spoIIE*, and *spoIIG*. Activation of these genes leads to irreversible commitment to the sporulation process.

REGULATION OF THE Spo0A PHOSPHORELAY SYSTEM

Numerous genes participate in regulating the amount of Spo0A~P produced in a cell. Several of them encode the kinases mentioned above, which phosphorylate Spo0F and therefore increase Spo0A~P levels. Two of these kinases, KinA and KinB, phosphorylate Spo0F and consequently Spo0A to high levels in response to severe extended starvation and commit the cell to sporulation; the others, KinC, KinD, and KinE, phosphorylate Spo0F only to low levels and commit the cell only to competence and biofilm formation. As mentioned above, the signals that activate these kinases are unknown but are the subject of active investigation. Other signals, such as DNA damage, could also activate kinases A and B and commit the cell to sporulation unless there is intervention by a checkpoint protein, Sda. This protein binds to the kinases and inhibits them, preventing sporulation in response to DNA damage (see Ruvolo et al., Suggested Reading). As discussed in Box 4.2, *B. subtilis* also encodes an addiction module that kills some cells in the population in response to starvation, allowing the killed cells to be cannibalized by other cells to delay or prevent their sporulation. Obviously, the cell sporulates only if it is absolutely necessary.

Other genes encode phosphatases that can dephosphorylate Spo0F~P, thereby draining phosphate out of the phosphorelay and diminishing Spo0A~P levels, or dephosphorylate Spo0A~P directly (see below). These phosphatases also respond to physiological and environmental signals, only a few of which are known.

Figure 14.17 Phosphorelay activation (phosphorylation) of the transcription factor Spo0A. The phosphorelay (Box 14.4) is initiated by at least five histidine kinases, which autophosphorylate on a histidine residue in response to unknown signals. Kinases A and B phosphorylate to high levels and initiate sporulation, and kinases C, D, and E phosphorylate only to low levels for competence, biofilm formation, antibiotic synthesis, and synthesis of degradative enzymes. The phosphate is transferred to Spo0F, to Spo0B, and finally to Spo0A. Spo0A~P regulates transcription, as described in the text. doi:10.1128/9781555817169.ch14.f14.17

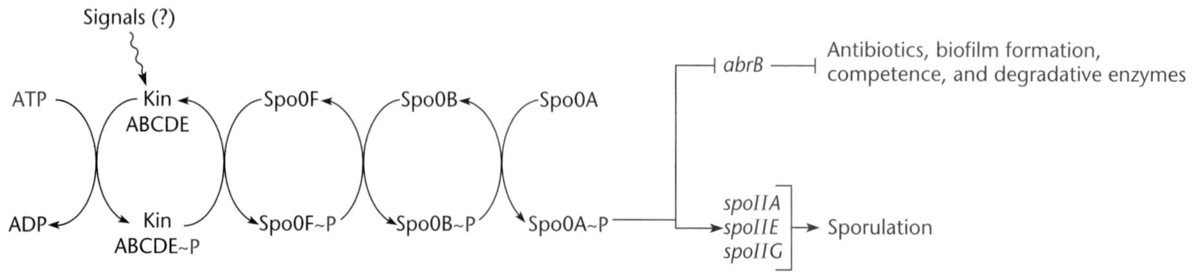

BOX 14.4

Phosphorelay Activation of the Transcription Factor Spo0A

Some bacterial pathways respond to multiple signal inputs. An example is the sporulation phosphorelay of *B. subtilis.* As illustrated in panel A of the figure, the histidine kinase and the phosphorylated aspartate domains can be found on separate polypeptides. Of the five kinases, only KinA is shown. The phosphoryl group is transferred from one protein to the next, as shown in steps 1 through 4. Panel B illustrates a cocrystal structure of a Spo0B dimer containing the conserved histidine residues (H in blue) with two Spo0F polypeptides, which contain the aspartate residues (D in blue). The close proximity of the histidine- and aspartate-containing active sites allows phosphoryl group transfer (shown as arrows).

doi:10.1128/9781555817169.ch14.Box14.4.f

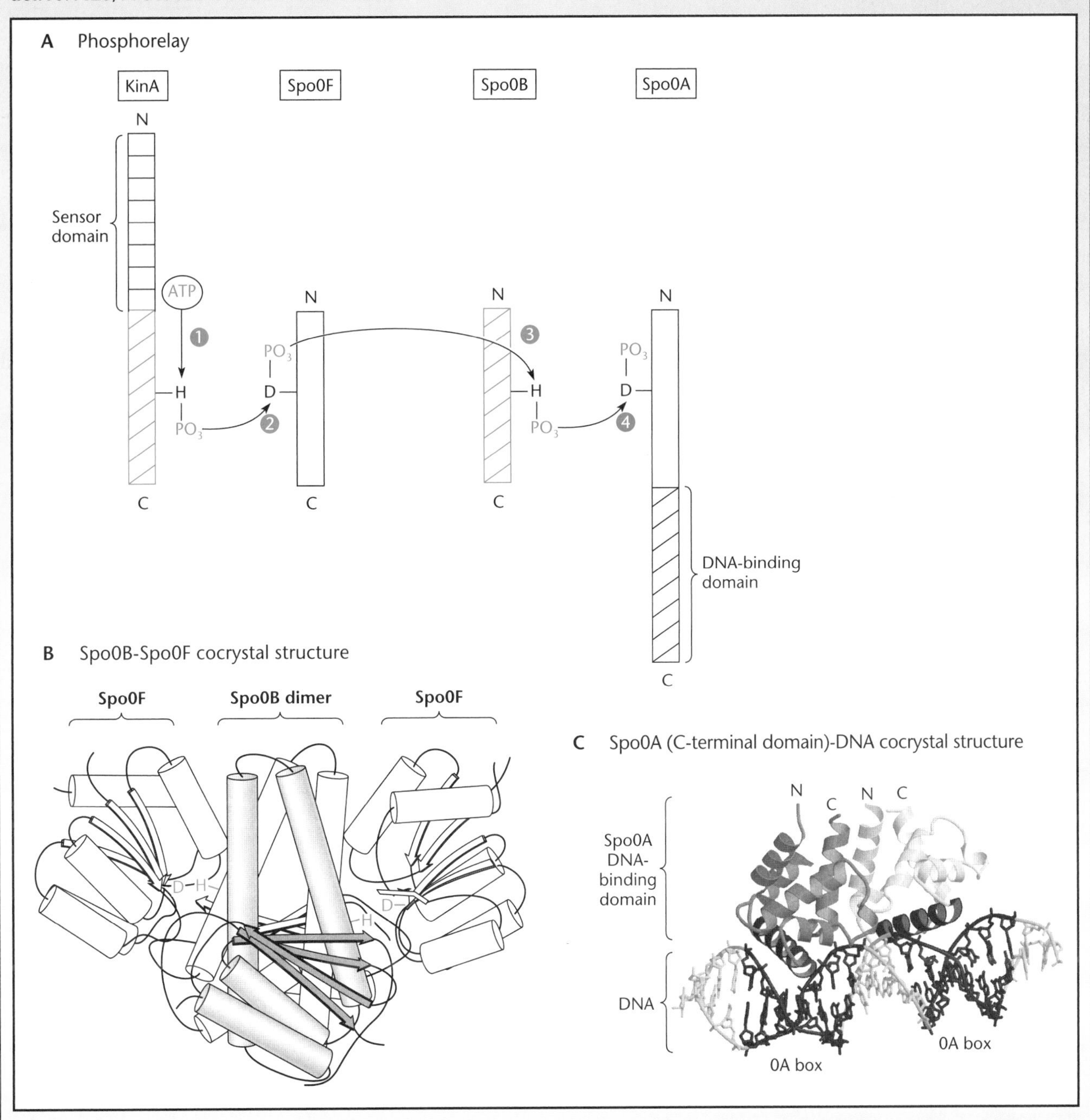

(continued)

BOX 14.4 (continued)

Phosphorelay Activation of the Transcription Factor Spo0A

Panel C shows a cocrystal structure of the transcription regulator Spo0A with its DNA-binding site. Only the C-terminal domain of Spo0A was used in the crystallization experiment. In this response regulator, the unphosphorylated N-terminal domain inhibits C-terminal-domain binding to DNA. When the phosphorelay is operating, phosphorylation of the N-terminal domain relieves the N-terminal-domain-mediated inhibition.

References

Hoch, J. A., and K. I. Varughese. 2001. Keeping signals straight in phosphorelay signal transduction. *J. Bacteriol.* **183:**4941–4949.

Scharf, B. E., P. D. Aldridge, J. R. Kirby, and B. R. Crane. 2009. Upward mobility and alternative lifestyles: a report from the 10th biennial meeting on bacterial locomotion and signal transduction. *Mol. Microbiol.* **73:**5–19.

Negative Regulation of the Phosphorelay by Phosphatases

Genetic analysis of two of the *spo0* loci revealed that their gene products functioned as negative regulators of sporulation. The combined results from sequencing mutant alleles, constructing gene knockouts, obtaining enhanced expression from many copies of the genes, and isolating suppressor mutations were all important to deciphering the gene functions (see Perego et al., Suggested Reading). We list these regulators below.

1. For the *spo0E* locus, sequence analysis determined that two mutant alleles in Spo⁻ strains contained nonsense mutations. This result would usually be interpreted as an indication that the *spo0E* gene product plays a positive role in initiating sporulation, since nonsense mutations usually inactivate the gene product and a requirement of the regulatory gene product for expression is the genetic definition of positive regulation (see chapter 2). However, deletion analysis of *spo0E* was contradictory, since strains with *spo0E*-null mutations were capable of sporulation—in fact, they hypersporulated. Furthermore, multiple cloned copies of the *spo0E* gene inhibited sporulation. The last two observations exemplify behavior typical of negative regulators. A resolution of this paradox came from a clue provided by one aspect of the *spo0E*-null mutant phenotype: a tendency to segregate Spo⁻ papillae that were visible as translucent patches on the surfaces of sporulating colonies. Genetic analysis of these Spo⁻ papillae showed that they contained suppressor mutations, several of which mapped to the *spo0A* gene. This result suggested the hypothesis that cells lacking Spo0E experienced an especially strong pressure to sporulate because increased expression or activity of the phosphorelay components produced an exceptionally high level of Spo0A~P. The *spo0E*-null mutants were found to have no alteration in phosphorelay gene transcription, but a biochemical study of the Spo0E protein showed that it functioned as a specific phosphatase of Spo0A~P. Such an activity would indeed provide a negative regulatory function.

   What would explain the finding that nonsense *spo0E* mutants had a Spo⁻ phenotype? Both of these mutations were found to affect the C terminus of Spo0E, leaving most of the protein intact. One hypothesis is that the Spo0E C terminus is a regulatory domain, perhaps one that binds a signal molecule that regulates its phosphatase activity. If so, the mutations may prevent signal binding and hence lock Spo0E into the phosphatase mode, thereby preventing Spo0A~P accumulation and sporulation. This demonstrates why mutations that do not totally inactivate the gene product should not be assumed to exhibit the null phenotype.
2. The *spo0L* locus shared some genetic properties with *spo0E*: multiple copies of *spo0L* caused a sporulation deficiency, and *spo0L*-null mutations caused hypersporulation, as well as accumulation of Spo⁻ segregants. Like Spo0E, Spo0L behaved like a negative regulator of the phosphorelay. Because the *spo0L* mutants isolated on the basis of their Spo⁻ phenotype contained missense mutations of *spo0L*, it was reasoned that isolation of suppressors of these mutations might identify the target of Spo0L activity. Accordingly, a plan was made to mutagenize a Spo⁻ *spo0L* mutant strain and then look for Spo⁺ colonies. One problem that arose was that the most frequent class of sporulating mutants contained null mutations of *spo0L*. To overcome this problem, a strain containing two copies of the *spo0L* missense allele was constructed; this strain was mutagenized for isolation of extragenic suppressor mutations. The result of this experiment was isolation of a suppressor mutation in the *spo0F* gene, one of the phosphorelay components. When tested for phosphatase activity, Spo0L proved to be a phosphatase of Spo0F~P.

An additional phosphatase of Spo0F~P was found in the *B. subtilis* genome sequence as a Spo0L homolog. Named Spo0P, this homologous protein inhibited sporulation when hyperexpressed from multicopies of the gene and encoded a phosphatase with amino acid residues 60% identical to those of Spo0L. The recent renaming of Spo0L and Spo0P as RapA and RapB, respectively, reflects their roles as *r*esponse regulator *a*spartyl-phosphate *p*hosphatases.

Figure 14.18 illustrates inhibition of the phosphorelay by the RapA, RapB, and Spo0E phosphatases. Since these phosphatases function to reduce the accumulation of Spo0A~P, their activities must be inhibited under conditions that promote antibiotic synthesis and sporulation. A regulator of Spo0E has been hypothesized, as discussed above. For RapA and RapB, the known regulatory signals are peptide molecules named PhrA and competence-stimulating factor (CSF). Produced by the bacilli themselves, they may function as indicators of population density, or "quorum sensors." The quorum sensors of gram-negative bacteria are typically homoserine lactones, while those of gram-positive bacteria are more typically peptides.

Regulation of the phosphorelay involves additional signals besides those mentioned above. Starvation, cell density, metabolic states, cell cycle events, and DNA damage are all known to influence Spo0A~P levels. Many of the signals involved and the mechanisms by which they affect Spo0A~P are poorly understood and are the subjects of active investigation.

Compartmentalized Regulation of Sporulation Genes

The mother cell and the forespore are genetically identical, but certain proteins must be made specifically in the developing spore and others (such as those that form the sturdy spore coat) must be made in the surrounding mother cell cytoplasm. Thus, the set of genes transcribed from the mother cell DNA must differ from the set transcribed from the forespore DNA. What mechanisms account for transcription of different sporulation genes in the two compartments?

Figure 14.18 Regulation of the phosphorelay by phosphatases. RapA and RapB are phosphatases for SpoF~P, and Spo0E is a phosphatase for Spo0A~P (see Stephenson and Perego, Suggested Reading). RapA and RapB are inhibited by the PhrA and CSF (PhrC) pentapeptides, respectively. It is not known what controls the activity of the phosphatase Spo0E. doi:10.1128/9781555817169.ch14.f14.18

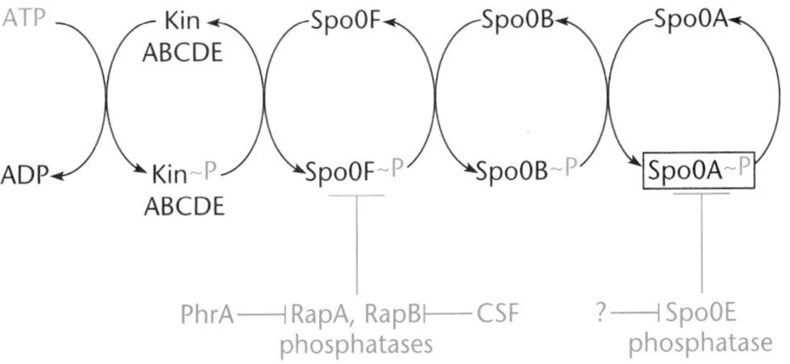

REGULATION OF SPORULATION GENES BY SEQUENTIAL AND COMPARTMENT-SPECIFIC ACTIVATION OF RNA POLYMERASE SIGMA FACTORS

The entire collection of sporulation genes can be sorted into a handful of classes on the basis of transcription by a specific sigma factor. The sporulation sigma factors replace the principal vegetative-cell sigma factor A (σ^A) in RNA polymerase holoenzyme, possibly by outcompeting σ^A for RNA polymerase. The σ^A of *B. subtilis* plays a role similar to that of the σ^{70} of *E. coli* (see chapter 2). As shown in Table 14.3, there are five distinct sigma factors associated with sporulation, sigma H (σ^H), sigma E (σ^E), sigma F (σ^F), sigma G (σ^G), and sigma K (σ^K); they associate with RNA polymerase to transcribe the sporulation genes. Each of the sigma factors is active at a specific time during sporulation. Four of the sigmas are regulated so that they are active in only one of the two developing cell compartments: σ^E and σ^K are sequentially active in the mother cell, and σ^F and σ^G are sequentially active in the forespore.

Analysis of the Role of Sigma Factors in Sporulation Regulation

Four kinds of information have been important to understanding gene regulation in *B. subtilis*.

TEMPORAL PATTERNS OF REGULATION

Measurements of the times of expression of the sporulation genes indicated that many of the genes underwent dramatic increases in expression at specific times after the sporulation process started. Use of gene fusions allowed large-scale comparisons of the complete set of sporulation genes. The most commonly used reporter genes were *lacZ* and *gus* from *E. coli*

Table 14.3 Sporulation sigma factors and their genes

Sigma factor	*spo* gene name	*sig* gene name	Compartment of activity
σ^H	*spo0H*	*sigH*	Predivisional sporangium
σ^F	*spoIIAC*	*sigF*	Forespore
σ^E	*spoIIGB*	*sigE*	Mother cell
σ^G	*spoIIIG*	*sigG*	Forespore
σ^K	*spoIVCB-spoIIIC*	*sigK*	Mother cell

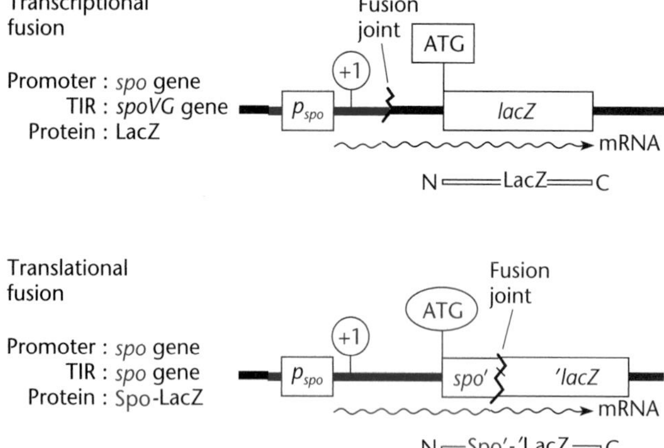

Figure 14.19 Reporter gene *lacZ* fusions to sporulation genes. Translational and transcriptional gene fusions are both transcribed from a *B. subtilis spo* promoter. In translational fusions, a fusion protein is expressed from the translation initiation region (TIR) of the gene being studied. In a transcriptional fusion, *lacZ* is translated from the TIR of a *B. subtilis spo* gene, often that of *spoVG*.
doi:10.1128/9781555817169.ch14.f14.19

(Figure 14.19) (see chapter 2). The product of the *lacZ* gene, β-galactosidase, and the product of the *gus* gene, β-glucuronidase, could be assayed by adding "artificial" substrates (such as *o*-nitrophenyl-β-D-galactopyranoside [ONPG] or methylumbelliferyl-β-glucuronide [MUG]) to samples of the test culture at various times after induction of sporulation by a nutritional downshift. The appearance of β-galactosidase or β-glucuronidase activity indicated the onset of gene expression. In addition, direct measurements of mRNAs of various sporulation genes correlated well with the results of *lacZ* and *gus* fusion experiments. Therefore, the use of such fusions became widespread, because of the relative ease and convenience of fusion assays.

A significant outcome of comprehensive fusion experiments was the extensive assessment and comparison of the times of expression of many sporulation genes. Moreover, the timing of *lacZ* expression could be correlated with the timing of morphological changes that were visible as sporulation progressed.

DEPENDENCE PATTERNS OF EXPRESSION

Fusions with *lacZ* were also used to determine whether the expression of one gene depended on the activity of a second gene. If the expression of one gene depends on a second gene, the second gene may encode a direct or indirect regulator of the first. The use of *spo* mutations in combination with *spo-lacZ* fusions (Figure 14.20) allowed the testing of many regulatory dependencies. An example of a set of experimental data is shown in Figure 14.20B. In this example, expression of a *spoIIA*::*lacZ* fusion was dependent on all of the *spo0* loci, but not on any of the "later" loci. The results of tests of many pairwise combinations of *spo* mutations and gene fusions are also summarized in Table 14.4. Besides the data for the genes shown, data for many dozens of additional genes have contributed to our understanding of gene regulation. More recently, transcriptome analyses of *spo* mutants have revealed gene expression dependence patterns for the entire genome (see below and Wang et al., Suggested Reading).

TRANSCRIPTION FACTOR DEPENDENCE

Once *spo* genes had been cloned and sequenced, it was possible to determine the functions of some of the proteins because of their amino acid sequence similarities to known families of regulatory proteins, such as sigma factors, which share characteristic amino acid motifs. The *spo0H* gene could be seen to encode a sigma factor, as could ORFs of the *spoIIA*, *spoIIG*, *spoIIIG*, and *spoIVC* loci (Table 14.3). In vitro experiments confirmed the functions of these proteins as transcription factors.

It was possible to infer the sigma factor dependence of many of the other sporulation genes on the basis of sequence comparisons around the transcription start sites. In some cases, allele-specific suppressors of promoter mutations could be isolated in a particular sigma factor gene. Remember from chapter 3 that an allele-specific suppressor can suppress only a particular type of mutation in a gene. Another important type of experiment involved in vitro transcription studies with RNA polymerase containing specific sigma factors. For some transcription factors, like Spo0A, chromatin immunoprecipitation followed by microarray analysis has been used to determine binding sites genome wide, identifying genes likely under direct control of the transcription factor (see Molle et al., Suggested Reading).

CELLULAR LOCALIZATION

Several methods have been used to determine the cellular locations of expression of sporulation genes. For example, expression of β-galactosidase in the forespore can be distinguished from that in the mother cell on the basis that the forespore is more resistant to lysozyme. Immunoelectron microscopy has been useful for visualizing the expression of β-galactosidase, and more recently, the use of green fluorescent protein fusions has allowed the determination of the cellular locations of numerous sporulation proteins.

From studies like these, it could be seen that all of the genes turned on after septation were expressed in only one compartment. The genes transcribed by RNA polymerase with σ^F and σ^G were expressed only in the forespore compartment, and the genes transcribed by RNA

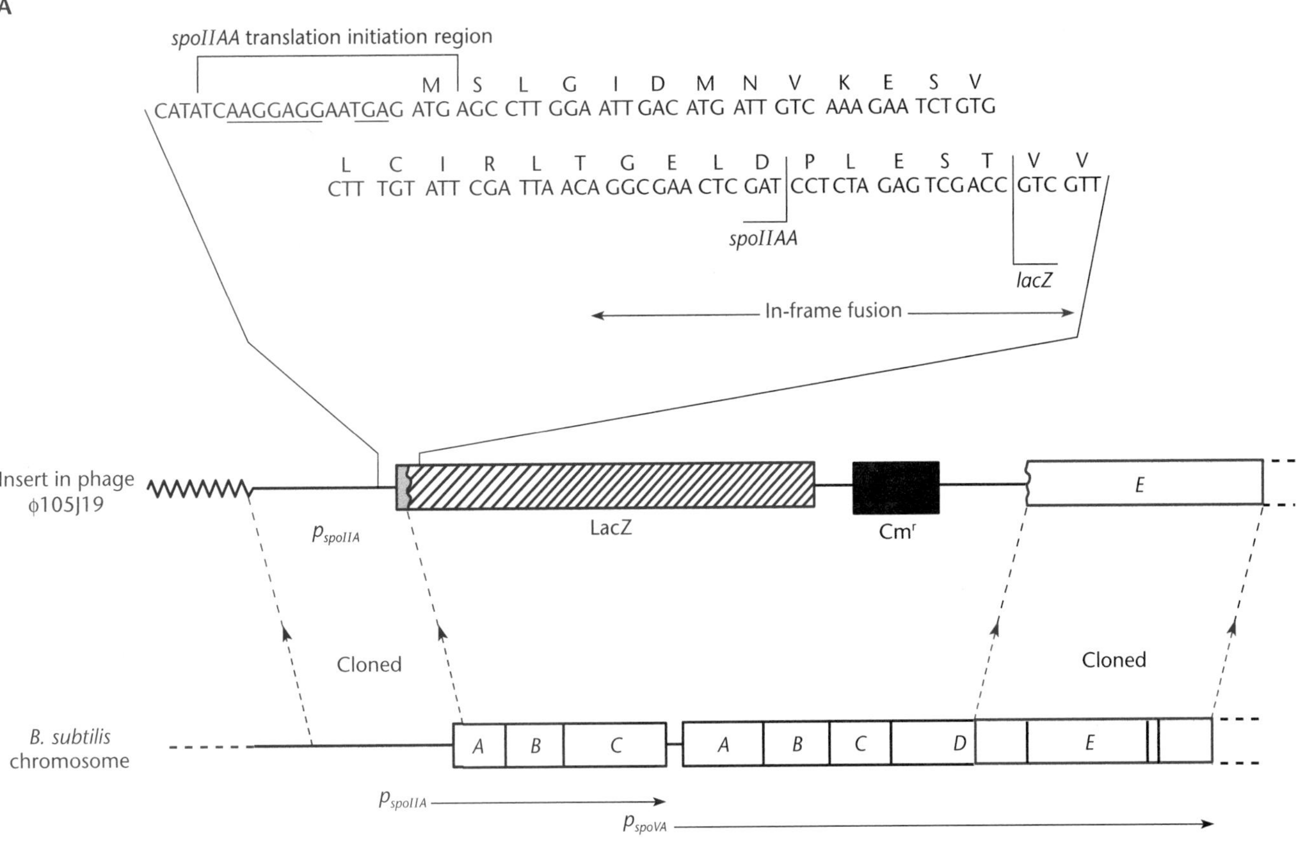

B

	Mutation	β-Galactosidase activity (units ml^{-1}) $t_{1.5}$	t_4	No. of determinations
spo0	*spo0A43*	0.016	0.037	2
	spo0B136	0.023	0.054	2
	spo0E11	0.018	0.015	2
	spo0F221	0.021	0.050	2
	spo0H17	<0.01	0.018	2
	spo0J93	0.048	0.25	2
	spo0K141	0.013	0.050	2
spoII	*spoIIAA562*	0.41	0.38	2
	spoIIAC1	0.44	0.51	2
	spoIIB131	0.36	0.17	3
	spoIID298	0.43	0.11	2
	spoIIE48	0.67	0.77	3
	spoIIG55	0.69	0.58	3

	Mutation	β-Galactosidase activity (units ml^{-1}) $t_{1.5}$	t_4	No. of determinations
spoIII	*spoIIIA65*	0.18	0.22	4
	spoIIIB2	0.65	0.13	2
	spoIIIC94	0.24	0.23	4
spoIV	*spoIVA67*	0.29	0.080	3
	spoIVB165	0.12	0.19	3
	spoIVC23	0.29	0.15	3
spoV	*spoVA89*	0.44	0.34	2
	spoVB91	0.49	0.21	2
	spoVC134	0.61	0.46	2

Figure 14.20 Testing the regulatory dependencies of *spo* genes. **(A)** Use of a *B. subtilis* transducing phage to create translational *lacZ* fusions to a chromosomal *spo* gene. Shown is the structure of phage φ105J19 carrying the *spoIIAA-lacZ* gene fusion. The lower part of the figure shows the region of the *B. subtilis* chromosome containing the *spoIIA* operon (three genes) and the adjacent *spoVA* operon (five genes). The phage contains a cloned fragment of chromosomal DNA covering these operons, but the central portion of the insert has been replaced by the *E. coli lacZ* gene (blue hatched) and a chloramphenicol resistance gene (Cm^r) (black). The insertion is arranged so that the region including the *lacZ* gene is fused in frame to the N terminus of the *spoIIAA* gene. **(B)** Effect of *spo* mutations on the production of β-galactosidase by the *spoIIAA-lacZ* gene fusion during sporulation. Phage φ105J19 (A) was transduced into a series of isogenic strains carrying *spo* mutations. Sporulation was induced, and samples were taken for assay of β-galactosidase. The results shown are mean activities in samples at 1.5 and 4 h after induction of sporulation. Not shown are the control values. The wild-type Spo^+ strain containing the phage produced 0.58 unit of β-galactosidase per ml at time (t) = 1.5 h and 0.17 unit at t = 4 h. With no *lacZ* fusion, the Spo^+ strain produced background levels of 0.013 unit per ml at t = 1.5 h and 0.032 unit per ml at t = 4 h. doi:10.1128/9781555817169.ch14.f14.20

Table 14.4 Timing and dependence patterns of gene expression

Gene or operon fusion[a]	Time of expression (min)	Expression in mutants of gene or operon[b]:					
		spo0	*spoIIA* (σ^F)	*spoIIE*	*spoIIG* (σ^E)	*spoIIIG* (σ^G)	*spoIVC* (σ^K)
spoIIA	40	–	+	+	+	+	+
spoIIE	30–60	–	+	+	+	+	+
spoIIG	0–60	–	+	+	+	+	+
gpr	80–120	–	–	–	+	+	+
spoIIIG	120	–	–	–	–	+	+
ssp	>120	–	–	–	–	–	+
spoIVC	150	–	–	–	–	+	+
cotA	240	–	–	–	–	–	–

[a]Functions are described in the text.
[b]+, expressed; –, not expressed.

polymerase with σ^E and σ^K were expressed only in the mother cell.

Intercompartmental Regulation during Development

When the observations on timing, dependence relationships, and localization of sporulation gene expression are combined, a complex pattern of regulation that includes a cascade of sigma factors and signaling between the developing compartments is revealed.

Figure 14.21 shows that after septation, gene expression in the forespore depends at first on σ^F and later on σ^G. An early σ^F-dependent transcript, *gpr*, encodes a protease that is important during spore germination. Besides its dependence on σ^F, *gpr* requires functional *spo0* genes, because σ^F expression and activity depend on them. Another σ^F-transcribed operon is *spoIIIG*, which encodes the late forespore sigma factor σ^G. Transcription of *spoIIIG* differs from that of *gpr* in that it occurs later and, although confined to the forespore, requires functioning of the *spoIIG* locus (Table 14.4), which encodes the mother cell-specific sigma factor σ^E.

Once σ^G is produced in the forespore, it transcribes a set of *ssp* genes, which encode spore-specific proteins that condense the nucleoid. As shown in Table 14.4, *ssp* transcription is blocked in *spoIIIG* mutants, as well as in mutants with mutations in all of the genes discussed above, such as the *spoIIG* gene, because they are involved in σ^G production.

Gene expression in the mother cell also reveals intercompartmental regulation. Figure 14.21 shows that one gene transcribed relatively early in the mother cell by σ^E RNA polymerase is the *gerM* gene, which encodes a germination protein. Later, σ^E RNA polymerase transcribes the genes for σ^K (*spoIVCB-spoIIIC*). RNA polymerase with σ^K then transcribes *cotA*, one of a set of *cot* genes that encode proteins incorporated into the spore coat. Table 14.4 shows that *cotA* transcription also requires the activity of the late forespore sigma factor σ^G.

Activation of the sigma factors alternates between the two developing compartments. As shown in Figure 14.22, each successive activation step requires intercompartmental communication. The critical information is whether morphogenesis and/or gene expression in the other compartment has progressed beyond a "checkpoint." The way in which the two compartments communicate their status to each other is a fascinating area of current research and is discussed in the next sections.

TEMPORAL REGULATION AND COMPARTMENTALIZATION OF σ^E AND σ^F

Both the *sigE* and *sigF* genes, encoding σ^E and σ^F, respectively (Table 14.3), are transcribed in the developing cell before the sporulation septum divides off the forespore compartment (Figure 14.21). However, neither sigma factor starts to transcribe its target genes until after the septum forms, and then, as mentioned above, each sigma factor becomes active in only one compartment: σ^E in the mother cell and σ^F in the forespore. Before septation, the sigma factors are held in inactive states, with a different inhibitory mechanism acting on each sigma factor. In σ^E, the active form of the protein must be proteolytically released from an inactive precursor, Pro-σ^E. In σ^F, the active protein must be released from a complex that contains an inhibitory anti-sigma factor, SpoIIAB (Figure 14.23). Once the sporulation septum forms, σ^F becomes active in the forespore. Subsequently, σ^E becomes active in the mother cell.

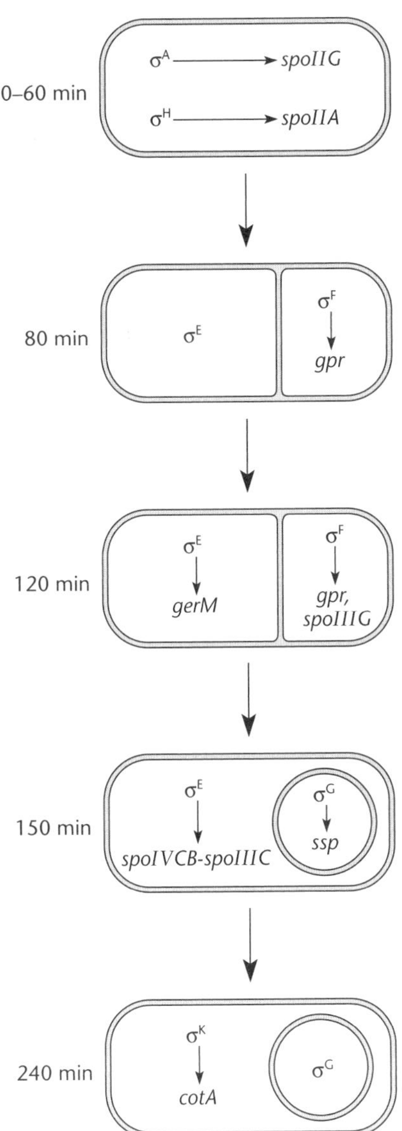

Figure 14.21 Compartmentalization of sigma factors and temporal regulation of transcription within compartments. The genes for σ^E and σ^F are transcribed before polar septation. σ^F is active in the forespore compartment and is required for transcription of the gene for σ^G, which succeeds it. σ^E is active in the mother cell and is required for transcription of the gene for its successor, σ^K. Within their compartments, σ^F and σ^E are required for transcription of their target genes at various times.
doi:10.1128/9781555817169.ch14.f14.21

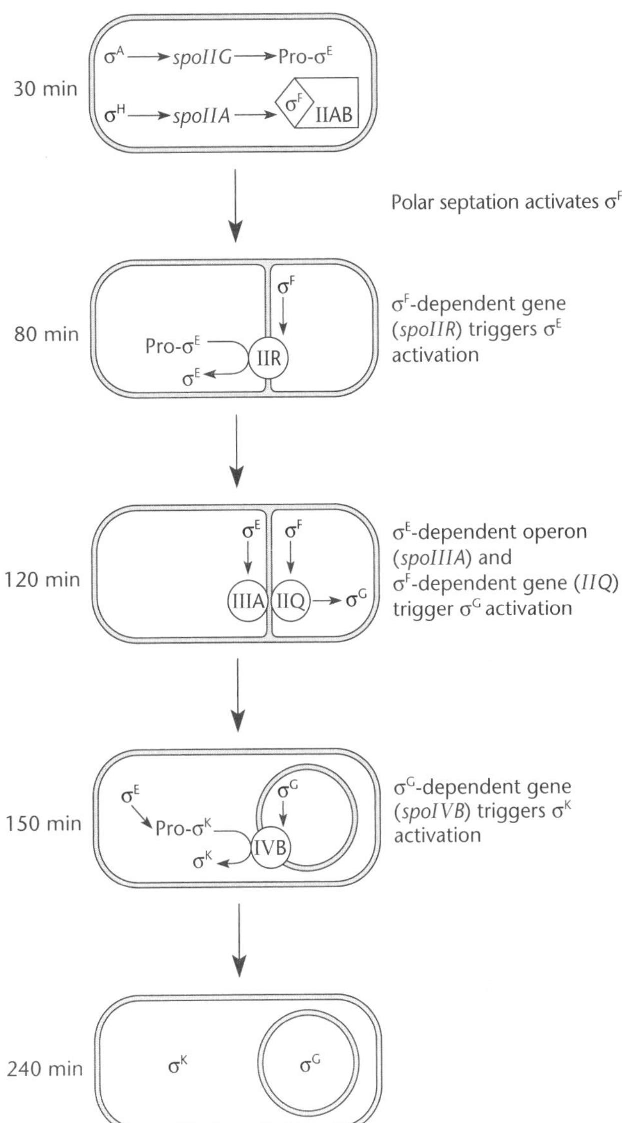

Figure 14.22 Sequential and compartmentalized activation of the *B. subtilis* sporulation sigma factors (blue). A series of signals allows communication between the two developing compartments, as described in the text, at approximately the times shown after induction of sporulation.
doi:10.1128/9781555817169.ch14.f14.22

Activation of σ^F in the forespore requires the interplay of a set of proteins. Two of these are binding partners named SpoIIAA and SpoIIAB. It is SpoIIAB that is the above-mentioned anti-sigma factor that binds to and inactivates σ^F. SpoIIAA is an anti-anti-sigma factor that nullifies the anti-sigma factor activity of SpoIIAB; it does this because of its own ability to bind SpoIIAB.

A cycle of phosphorylation and dephosphorylation of SpoIIAA modulates the binding of SpoIIAA to SpoIIAB. Only when it is unphosphorylated can SpoIIAA bind to SpoIIAB. Before septation, SpoIIAA is in a phosphorylated state and so does not bind to SpoIIAB in the preseptational sporangium. Unbound by SpoIIAA, SpoIIAB is free to bind to and inactivate σ^F. After septation, SpoIIAA is in the unphosphorylated state in the forespore; hence, it binds SpoIIAB, releasing σ^F.

The enzymes that phosphorylate and dephosphorylate SpoIIAA are SpoIIAB and SpoIIE, respectively. Their opposing activities, before and after septation, determine the balance between the two forms of SpoIIAA. Before

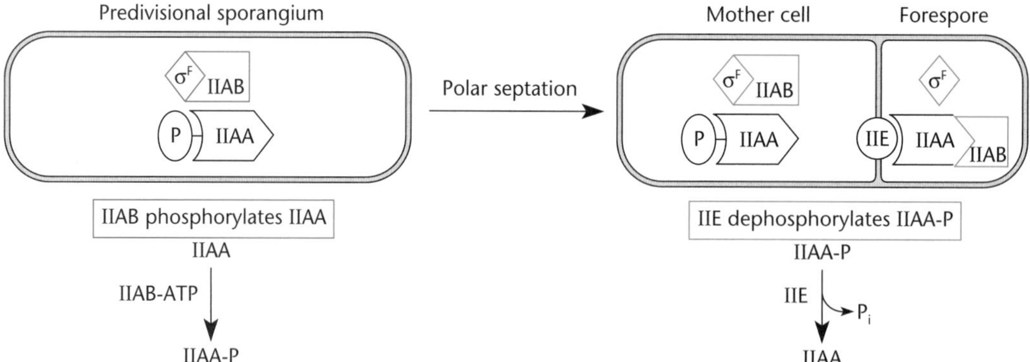

Figure 14.23 Model for the regulation of σ^F activity. SpoIIAB holds σ^F in an inactive state in both the predivisional sporangium and the mother cell. SpoIIAB phosphorylates SpoIIAA. SpoIIAA~P cannot bind SpoIIAB. The SpoIIE phosphatase controls the generation of SpoIIAA, which complexes with SpoIIAB. These reactions allow the release of σ^F. doi:10.1128/9781555817169.ch14.f14.23

septation, SpoIIAB phosphorylation of SpoIIAA predominates, whereas once the spore septum has formed, SpoIIE phosphatase activity predominates in the forespore. Regulation of the SpoIIE phosphatase by septum formation is an area of current investigation.

Regulation of σ^F

Important progress in understanding the regulation of σ^F activity came from studying the two genes, *spoIIAA* and *spoIIAB*, that are cotranscribed with the *sigF* gene in the same operon (Figure 14.24A). Two key findings were that the *spoIIAB* gene product is a protein that inhibits σ^F activity and is in turn inhibited by SpoIIAA. These conclusions were drawn from experiments on the effects of *spoIIAA* and *spoIIAB* mutations on σ^F activity (see Schmidt et al., Suggested Reading).

The test for σ^F activity in these experiments was to measure the expression of genes with σ^F-dependent promoters. Two such genes are *spoIIIG* and *gpr* (Table 14.4). The use of *lacZ* fusions to these genes allowed gene expression to be monitored by β-galactosidase assays. Control experiments showed that *sigF* transcription and translation were normal, ensuring that differences in β-galactosidase activity from the σ^F-dependent *lacZ* fusions reflected the activity of σ^F and not its expression.

The comparison of β-galactosidase activity levels in *spoIIAA* and *spoIIAB* mutant cultures and wild-type cultures showed that *spoIIIG*::*lacZ* and *gpr*::*lacZ* expression was substantially higher in the *spoIIAB* mutant strain than in the wild type. Conversely, the tester fusions were not expressed at all in the *spoIIAA* mutant cultures. Outcomes like these could occur if SpoIIAA function is required for σ^F activity and SpoIIAB function inhibits σ^F activity (Figure 14.24B). An additional genetic experiment showed that SpoIIAA is required because it is needed to counteract SpoIIAB. This experiment was performed to assay the tester fusions in *spoIIAA spoIIAB* double-mutant cultures; these cultures overexpressed the tester fusions, as the *spoIIAB* mutant had. Thus, SpoIIAA is required only if SpoIIAB is active. In genetic terms, the experiment showed that the *spoIIAB* mutation is epistatic to the *spoIIAA* mutation.

Further study showed that *spoIIAB* inhibition of σ^F is an essential event in the normal course of sporulation.

Figure 14.24 The *spoIIA* operon and its gene products. **(A)** The three genes that are cotranscribed. **(B)** The inhibitory effects of SpoIIAA and SpoIIAB, as inferred from genetic experiments described in the text. doi:10.1128/9781555817169.ch14.f14.24

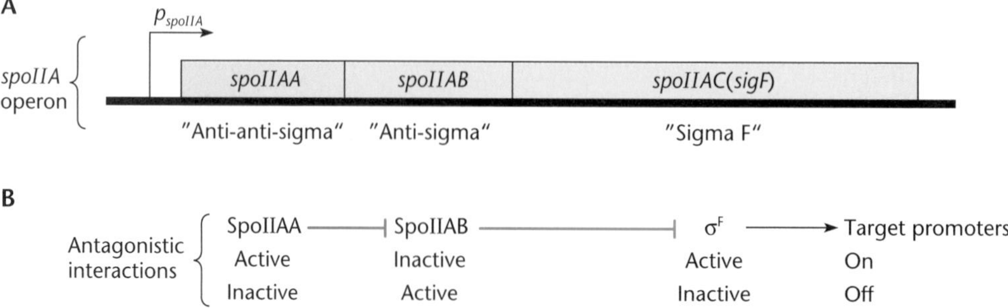

One observation was that the *spoIIAB* mutant strains could not survive sporulation and could be maintained only in media that suppressed sporulation. The proposed explanation for this phenotype was that unregulated transcription by σ^F is lethal to the cell. It is important to note here that the *spoIIAB* mutant strain used in the experiments described above was not isolated in a mutant hunt for Spo⁻ strains but, rather, was a constructed deletion mutant. Lethality caused by deregulated σ^F activity could also be observed if *sigF* was artificially induced in vegetative cells from the p_{spac} promoter (Figure 14.25).

Another important advance in our understanding of σ^F regulation came from biochemical studies of SpoIIAA and SpoIIAB. The SpoIIAB amino acid sequence suggested that it might have protein kinase activity. It was indeed able to phosphorylate a protein: its substrate turned out to be SpoIIAA! Additional studies examined the binding interactions of the three proteins SpoIIAA, SpoIIAB, and σ^F and found that (i) SpoIIAA could bind to SpoIIAB, but only if SpoIIAA was not phosphorylated, and (ii) SpoIIAB could bind to σ^F. Genetic evidence included site-directed mutagenesis of the SpoIIAA phosphorylation target, a serine residue, changing it to aspartate or alanine and so mimicking phosphorylated and nonphosphorylated states, respectively (see Diederich et al., Suggested Reading, and chapter 13). Together, these observations formed the basis for the following model for σ^F regulation in a sporulating cell (Figure 14.23). As soon as the three *spoIIA* operon genes are expressed in the preseptation cell, SpoIIAB binds to and therefore inactivates σ^F. SpoIIAB also phosphorylates SpoIIAA and so prevents SpoIIAA-SpoIIAB binding. Thus, the SpoIIAB-σ^F complex is stable and σ^F cannot direct transcription. However, σ^F could become active in the forespore if SpoIIAB released it after septation. Thus, the model proposed that SpoIIAA is dephosphorylated in the forespore, so it can bind SpoIIAB and cause the release of active σ^F.

The necessity that SpoIIAA be dephosphorylated in order to release σ^F activity predicted that the sporulating cell must express a SpoIIAA~P phosphatase. The

Figure 14.25 Induction of *spoIIAC* (*sigF*) from the p_{spac} promoter, which consists of RNA polymerase recognition sequences of *B. subtilis* phage SPO1 and *lac* operator sequences. Insertion of *lacI* accompanies the insertion of p_{spac}, so *spoIIAC* transcription is inducible with isopropyl-β-D-thiogalactopyranoside (IPTG).
doi:10.1128/9781555817169.ch14.f14.25

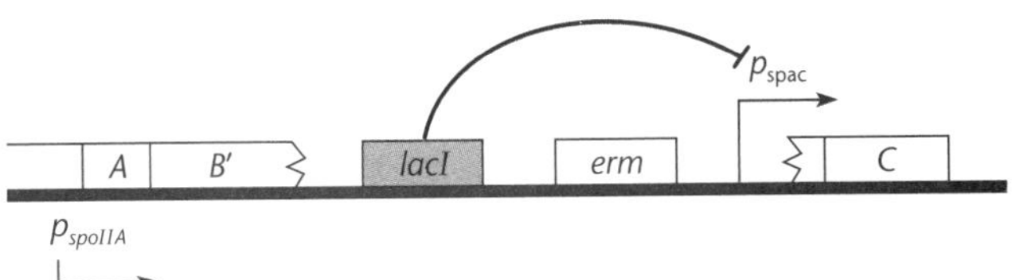

collection of *spo* mutants was evaluated for the possibility that one of the known genes might encode the hypothesized phosphatase. A candidate for such a phosphatase was the *spoIIE* gene product, because *spoIIE* mutants had a phenotype consistent with a defect in σ^F activation: they expressed the *spoIIA* (*sigF*) operon but failed to express σ^F-dependent genes (Table 14.4). This prediction was borne out by in vitro studies in which SpoIIE dephosphorylated SpoIIAA~P.

The SpoIIE protein associates with the polar septum. This and other factors, including the position of the *spoIIAB* gene on the chromosome and rapid turnover of the SpoIIAB protein, contribute to the limitation of σ^F activity to the forespore (see Hilbert and Piggot, Suggested Reading).

Regulation of σ^E

Mother cell transcription depends on σ^E, which is encoded by the *spoIIGB* gene of the *spoIIG* operon. The primary product of *spoIIGB* is an inactive precursor, named Pro-σ^E, which is processed to form the active sigma factor. The protease that cleaves Pro-σ^E is the product of the first gene in the *spoIIG* operon, *spoIIGA* (Figure 14.26A). Although the *spoIIG* operon is expressed in the predivisional sporangium, the *spoIIGA* product does not process SpoIIGB immediately but waits about an hour, until after septation has occurred. Then, notification from the forespore that development is proceeding and that σ^F has become active comes via the messenger SpoIIR, which is the product of a σ^F-transcribed gene (Figure 14.22). SpoIIR is secreted into the spaces between the cell membranes, where it signals to SpoIIGA to process Pro-σ^E.

An important clue to the explanation for the time lag in Pro-σ^E processing was the observation that mutants that lacked σ^F activity failed to process Pro-σ^E to σ^E. Could a σ^F-transcribed gene be required for the processing mechanism? A genetic search for such a gene was undertaken (see Karow et al., Suggested Reading) with the rationale that a strain with a mutant in the hypothetical gene might have a SigF⁺ SigE⁻ phenotype, i.e., it would express σ^F-dependent fusions but would not express σ^E-dependent fusions. Accordingly, mutants that expressed the σ^F-dependent fusion *gpr-gus* but did not express two σ^E-dependent fusions, *spoIID-lacZ* and *spoVID-lacZ*, were sought. The use of two *lacZ* fusions was intended to reduce the likelihood of isolating Lac⁻ strains with mutations in *lacZ* itself rather than in the desired regulatory gene. The Gus⁺ Lac⁻ mutants isolated from this screen were of two types: mutants with mutations of the *spoIIG* locus, as would be expected, and mutants with mutations of a new locus, which was named *spoIIR*. Evaluation of the *spoIIR* mutants for Pro-σ^E synthesis and processing showed that Pro-σ^E was indeed synthesized but was

A

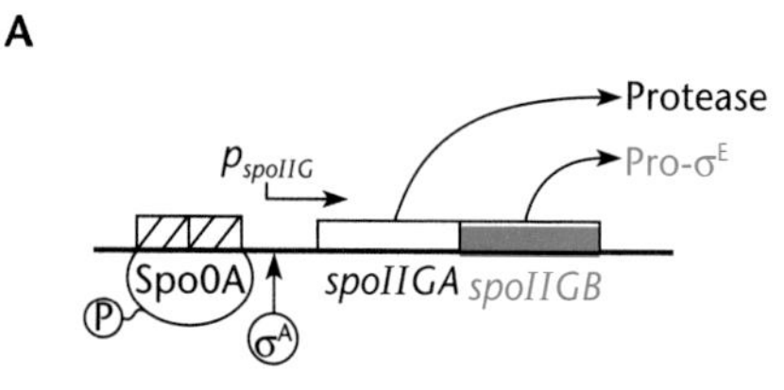

B

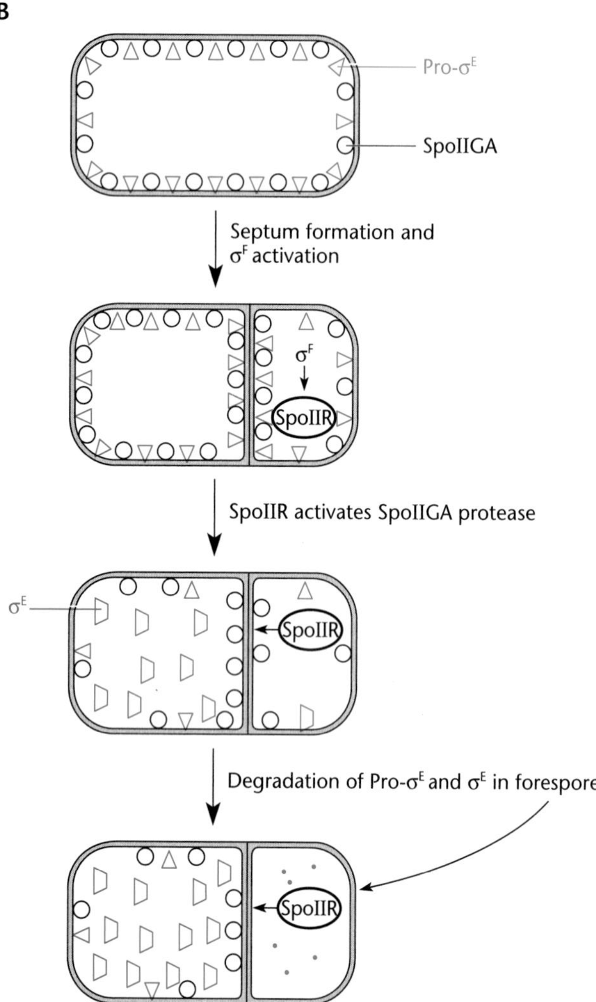

Figure 14.26 Model for activation of σ^E in the mother cell compartment. **(A)** The *spoIIG* operon is transcribed by σ^A RNA polymerase and requires activation by Spo0A~P (the hatched boxes represent Spo0A~P-binding sites [Box 14.4]). **(B)** Pro-σ^E and SpoIIGA are associated with the cytoplasmic membrane in the sporangium. After septum formation, both proteins are associated with all cell membranes. Then, SpoIIR is expressed in the forespore under the control of σ^F, and SpoIIR activates SpoIIGA protease, which cleaves Pro-σ^E to form active σ^E, which is distributed in the cytoplasm. Finally, any Pro-σ^E or σ^E remaining in the forespore is degraded, while expression of the *spoIIG* operon persists in the mother cell due to compartment-specific activity of Sp0A~P after septation.
doi:10.1128/9781555817169.ch14.f14.26

not processed. Thus, the *spoIIR* gene seemed to have the predicted properties. It appears that the SpoIIR protein is secreted into the spaces between the membranes, with the outcome that SpoIIGA is activated to process Pro-σ^E. SpoIIGA appears to be a novel type of signal-transducing protease (see Imamura et al., Suggested Reading). There are two additional factors that operate to restrict σ^E accumulation to the mother cell, forespore-specific degradation of Pro-σ^E and persistent expression of Pro-σ^E in the mother cell (but not the forespore) due to compartment-specific activity of Spo0A~P after septation (see Kroos, Suggested Reading) (Figure 14.26).

σ^G, A SECOND FORESPORE-SPECIFIC SIGMA FACTOR

The σ^G-encoding gene, *spoIIIG*, is transcribed in the forespore by σ^F RNA polymerase. Its transcription lags behind that of other σ^F-transcribed genes, evidently because it requires a signal from the mother cell sigma factor, σ^E. The evidence for the existence of such a signal is indirect at present—the *spoIIIG* gene is not transcribed in a *spoIIG* (σ^E) mutant—but the signal has not yet been identified. Transcription of *spoIIIG* also depends on σ^F-dependent expression of SpoIIQ in the forespore, but the reason for this dependence is unknown. Another aspect of σ^G regulation involves an anti-sigma factor, called CsfB or Gin by different groups, which may prevent premature σ^G activity in the forespore. However, σ^G fails to become active in the forespore unless the products of the *spoIIIA* operon are expressed in the mother cell and SpoIIQ is expressed in the forespore (Figure 14.22). The SpoIIIA and SpoIIQ proteins form channels connecting the mother cell and forespore (see Meisner et al., Suggested Reading). The channels are proposed to act as feeding tubes through which the mother cell nurtures the forespore by providing molecules needed for biosynthesis (see Camp and Losick, Suggested Reading). Without the channels, genes under σ^G control in the forespore fail to be expressed and the forespore collapses (see Doan et al., Suggested Reading).

σ^K, A MOTHER CELL SIGMA FACTOR

The last sigma factor to be made, σ^K, is expressed only in the mother cell. σ^E RNA polymerase transcribes the *sigK* gene. Like σ^E, σ^K is cleaved from a precursor protein, Pro-σ^K. Also, like σ^E processing, σ^K processing depends on a signal from the forespore. In this case, the signal is expression of the *spoIVB* gene under the control of the forespore sigma factor σ^G (Figure 14.22).

σ^K Activation

The SpoIVB protein is thought to be secreted across the innermost membrane surrounding the forespore and to communicate with the Pro-σ^K-processing factors across the membrane, thus activating the SpoIVFB protease that cleaves Pro-σ^K (Figure 14.27). SpoIVB activation of the σ^K-specific protease does not occur directly, but rather, by deactivation of proteins, SpoIVFA and BofA, which inhibit the protease. The complexity of this mechanism was revealed by isolation of "bypass suppressor" mutations, i.e., mutations that bypassed the requirement for σ^G involvement in σ^K activation.

The motivation for isolating suppressor mutations was the observation that late mother cell gene expression depends not only on the mother cell sigma factor, σ^K, but also on σ^G, the forespore sigma factor, and other forespore proteins. Moreover, the σ^G requirement is manifested at the step of Pro-σ^K processing, since Pro-σ^K protein accumulated in *spoIIIG* mutant cells. It was hypothesized that mutations that bypassed the σ^G requirement might provide information about the mechanism of σ^G involvement.

Genetic Analysis of σ^K Activation

The isolation of suppressor mutations involved a screen for mutations that allowed expression of a σ^K-dependent fusion, *cotA*::*lacZ*, in a strain lacking σ^G (see Cutting et al., Suggested Reading). The strain lacking σ^G was mutagenized with nitrosoguanidine to produce a broad spectrum of mutations. The *cotA*::*lacZ* fusion was then introduced by specialized transduction, and lysogens were screened for a blue-colony phenotype on X-Gal plates. Two classes of mutations were isolated and localized, defining loci that were named *bofA* and *bofB* (for bypass of forespore). Characterization of the *bof* mutants showed that *cotA* expression still required σ^K and, importantly, that Pro-σ^K processing was restored in *bof* mutants. The *bofA* mutations defined a new gene. However, the *bofB* mutations were missense or nonsense mutations in a previously discovered gene, *spoIVFA*. The second gene of this operon, *spoIVFB*, is the protease that processes Pro-σ^K, as mentioned above. Recent work has also shown that SpoIVFA and the BofA protein work together to inhibit SpoIVFB (Figure 14.27), although BofA appears to be the primary inhibitor and SpoIVFA helps assemble the complex (see Zhou and Kroos, 2004, Suggested Reading). SpoIVFB is an intramembrane-cleaving metalloprotease, meaning that active-site residues are in transmembrane helices. This type of protease cleaves a membrane-associated substrate and is broadly conserved from bacteria to humans (see above). Interestingly, SpoIVFB depends on ATP in order to cleave Pro-σ^K, perhaps sensing the energy level in the mother cell (see Zhou et al., Suggested Reading), and the SpoIVFA-SpoIVFB-BofA complex is associated with the SpoIIIA-SpoIIQ channels discussed above that are necessary for σ^G activation in the forespore (see Jiang et al., Suggested Reading).

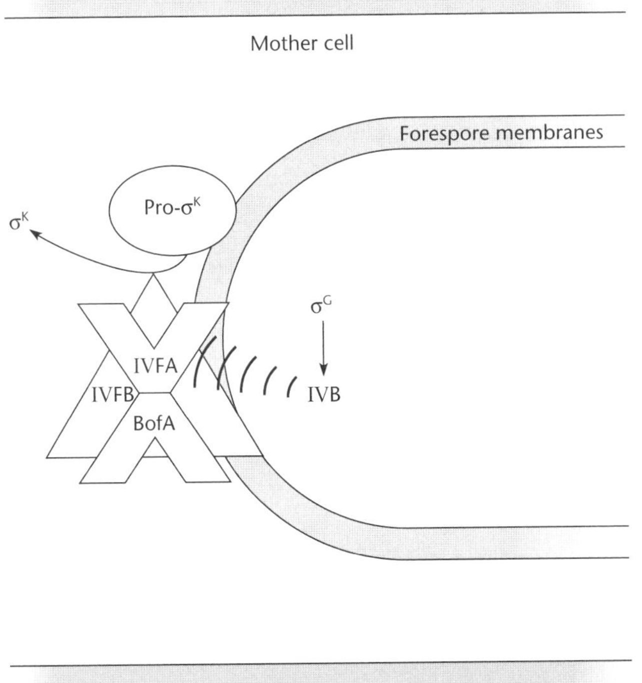

Figure 14.27 Model for regulation of Pro-σ^K processing based on the genetic data discussed in the text.
doi:10.1128/9781555817169.ch14.f14.27

The identity of SpoIVB as the σ^G-dependent signal was inferred from the observation that *bof* mutations restored *cotA*::*lacZ* expression in a *spoIVB* mutant, just as for the *spoIIIG* mutant (see Cutting et al., Suggested Reading). The SpoIVB protein is a serine protease that is produced in the forespore and is believed to cross the innermost membrane surrounding the forespore to trigger the Pro-σ^K proteolysis mechanism by cleaving SpoIVFA. A second serine protease, CtpB, can cleave SpoIVFA, but only if CtpB is first cleaved by SpoIVB (see Campo and Rudner, Suggested Reading). This explains why a *spoIVB* mutant is completely defective for Pro-σ^K processing. A *ctpB* mutant exhibits delayed processing of Pro-σ^K, and it appears that CtpB is partly responsible for cleaving BofA (see Zhou and Kroos, 2005, Suggested Reading), in addition to its role in cleaving SpoIVFA.

Finding Sporulation Genes: Mutant Hunts, Suppressor Analysis, and Functional Genomics

The discussions in the preceding sections show that a variety of approaches have been useful for identifying *B. subtilis* sporulation genes. A very large percentage of the regulatory genes were defined by mutations that caused a Spo$^-$ phenotype. However, an important class of regulatory genes was not well represented in the Spo$^-$ mutant collection. These genes are the negative regulators. Loss-of-function mutations—which are the most frequent kind of mutation—in negative regulators would not cause a Spo⁻ phenotype, as discussed above for genes such as *spoIIAB*. "Special" alleles, e.g., gain-of-function mutations, might cause a Spo⁻ phenoype, as for *spo0E* and *spo0L*, the negative regulators of the phosphorelay, but they would generally be found at a low frequency. However, suppression analysis is a powerful tool in such cases and has revealed the existence of important negative regulators. As one example, the *abrB* gene (Figure 14.17) was defined by mutations that restored degradative enzyme and antibiotic synthesis, but not sporulation, to *spo0A* mutants. Another example is the *bof* mutations, which restored *cotA*::*lacZ* in a Δ*sigG* strain, as discussed above.

Another type of gene underrepresented in the sporulation mutant collections encodes a function that is redundant, or overlapping, with that encoded by another gene. An example is the set of *kin* genes, which encode the kinases that initiate the phosphorelay. Mutations that inactivate any one of these genes cause only a weak Spo⁻ phenotype. Only one of these, *kinA*, was found as a very leaky Spo⁻ mutant in a collection of Tn*917*-induced mutants.

Additional genes involved in sporulation have been defined by gene knockout analysis of ORFs annotated in the *Bacillus* genome sequence (http://bacillus.genome.ad.jp) and by regulon analysis (see Wang et al., Suggested Reading). Recent technological developments using gene fusions to fluorescent probes now make it possible to monitor the movement of proteins in real time during sporulation. Information obtained using methods such as these, building on the wealth of information obtained on *B. subtilis* sporulation using molecular genetic and biochemical analyses, promises to deliver new insights into how intercompartmental communication occurs during this relatively simple developmental process. Some of the same principles are used by bacteria responding to their environment or communicating with other bacteria or eukaryotic hosts and by organisms undergoing more complex multicellular development.

SUMMARY

1. Most proteins are transported through the inner membrane by the evolutionarily highly conserved SecYEG translocon. The SecY protein forms most of the channel. Genetic experiments, using fusions to genes of the maltose transport system, combined with structural information have established a model for protein export by the SecYEG channel. Part of the SecY protein forms a plug in the channel that moves over to contact the C terminus of the SecE protein on binding of the signal sequence of the protein to be exported.

2. In bacteria, the SRP pathway mostly targets proteins destined for insertion into the inner membrane, while the Sec system (SecB and SecA) targets proteins to be exported to the periplasm, the outer membrane, or the outside of the cell. The SRP pathway transports proteins as they are being translated, which prevents their irreversible folding due to their extreme hydrophobicity. The Sec system can help transport proteins after they have been translated because such proteins are destined for the periplasm, the outer membrane, or the outside of the cell and so are usually less hydrophobic overall. Proteins transported to the periplasm, the outer membrane, or the outside of the cell usually have their signal sequence at the N terminus removed as they pass through the SecYEG channel.

3. The Tat pathway transports proteins through the inner membrane that have already folded in the cytoplasm, such as proteins that must bind redox factors in the cytoplasm after folding. Proteins transported by the Tat system have a characterisitic signal sequence at their N terminus that contains two arginines, hence the name, twin-arginine transport. This signal sequence is removed as the protein is transported.

4. Proteins transported to the periplasm or the outside of the cell often have disulfide bonds between cysteines in the polypeptide. These bonds, which are formed by oxidoreductases in the periplasm, help stabilize the protein.

5. Fusions to the *E. coli* alkaline phosphatase can be used to determine the periplasmic and cytoplasmic domains of transmembrane proteins in gram-negative bacteria. The alkaline phosphatase is active only in the periplasm because disulfide bonds are required for its activity.

6. Specialized protein secretion systems transport specific effector proteins through one or both membranes to the outside of the cell or even directly into another cell. These secretion systems differ markedly between gram-negative bacteria and gram-positive bacteria because of the presence of an outer bilipid membrane in gram-negative bacteria.

7. Secretion systems of gram-negative bacteria form pores in the outer membrane that are composed of β-sheets organized into rings to form a β-barrel.

8. There are five different types of protein secretion systems in gram-negative bacteria, types I to VI. Type I systems use a specific ABC transporter and the TolC channel in the outer membrane. Type II systems are related to some competence systems and the systems that assemble type IV pili; they use the SecYEG channel to get proteins through the inner membrane and then use a pseudopilus to push the

SUMMARY (continued)

protein through a secretin channel in the outer membrane. Type III systems form a syringe-like injectisome that injects effector proteins directly through both membranes into a eukaryotic cell; they are related to the flagellar motors. Type IV systems are related to DNA conjugation systems and some competence systems and also inject proteins through both membranes into eukaryotic cells. Type V systems include autotransporters, twin-partner secretion systems, and chaperone-usher systems; they form dedicated β-channels in the outer membrane that can be part of the secreted protein itself. Chaperone-usher systems use a dedicated periplasmic chaperone and an outer membrane protein called an usher to assemble some types of pilins. The chaperone furnishes a fold in the transported pilin protein in the periplasm to help stabilize the protein, and then this fold is replaced by a fold in the previously added pilin in a process called donor strand exchange. This interlinking forms a very stable pilus. Although their structure has not been determined, type VI systems are known to be essential for the pathogenicity of some bacteria and the survival of bacteria in the environment. They consist of a large number of highly conserved proteins that may resemble phage tails when assembled but that eject proteins out of the cell rather than injecting proteins (and DNA) into the cell.

9. Gram-positive bacteria also make syringe-like structures called injectosomes, which, like the injectisomes of gram-negative bacteria, secrete effector proteins directly into eukaryotic cells.

10. Because their cell wall is not surrounded by an outer membrane, gram-positive bacteria can attach proteins to their cell wall and have them exposed to the external environment. Sortases are enzymes that specifically attach proteins to the cell wall. They cleave the protein at a characteristic sortase signal and attach the protein to the peptide chain of a cell wall precursor before the precursor is assembled into the cell wall. More specialized sortases attach specific proteins and are involved in pilus assembly.

11. Bacterial cell biology includes the study of the synthesis and invagination of the cell wall and the membranes and how these are coordinated with the replication of the chromosome and the time in the cell cycle. The cell wall is a continuous molecule surrounding the cell that is synthesized by a complex of enzymes and assembly factors that coordinate synthesis of the precursors in the cytoplasm with synthesis and insertion of the precursors in the cell wall outside the inner membrane. Before cell division, the FtsZ protein forms a ring that then recruits many other proteins to form the division septum or septal ring.

12. *B. subtilis* sporulation is the best-understood bacterial developmental system. It involves a regulatory cascade of sigma factors and communication between cellular compartments involving signal proteins, proteases, and channels.

QUESTIONS FOR THOUGHT

1. Why do all types of cells, including those of eukaryotes, have an SRP system and cotranslate proteins to be inserted into the SecYEG channel but only bacteria have SecB and SecA to secrete proteins after they have been translated?

2. Why do T2SS have so many proteins in the inner membrane when they use the SecYEG or Tat channel to get proteins through the inner membrane and use their own channel only to get the proteins through the outer membrane?

3. Why do cells have so many different PBPs, lytic transglycosylases, amidases, and peptidases involved in synthesizing the cell wall?

4. Why are *B. subtilis* cells so averse to sporulating that they sporulate only after prolonged starvation and even induce genes to kill some cells so they can be cannibalized to feed the other cells and delay sporulation? What does this say about the purpose of sporulation?

PROBLEMS

1. Why did the selections for mutations in the Sec and SRP pathways not yield mutations in genes for the Tat pathway? How would you design a selection for genes in the Tat pathway?

2. The product of the *envZ* gene, a sensor kinase, is a transmembrane protein in the inner membrane of *E. coli* (see chapter 13). Describe how you would determine which of the regions of the EnvZ protein are in the periplasm and which are in the cytoplasm.

3. Outline how you would isolate suppressors of mutations in the signal sequence coding region of the *malE* gene.

4. Describe a general strategy for determining the order of addition of septal proteins to the FtsZ ring and whether they must bind in a particular order (e.g., whether the binding of one protein is required for the binding of other proteins).

5. How were *B. subtilis* regulatory genes identified?

6. Contrast the similarities and differences between the *B. subtilis* phosphorelay and typical two-component systems (see Box 13.4).

7. What types of *spo0E* and *spo0L* mutations suggested that the genes were positive regulators? Negative regulators? How did suppression analysis help in understanding that Spo0E and Spo0L are actually negative regulators?

8. What is the difference between *spo-lacZ* transcriptional and *spo-gfp* translational fusions? What different questions do they answer?

9. The *tlp* and *cotD* genes encode proteins that are structural components of the endospore. Their dependence patterns of expression are similar to those of the *ssp* genes and *cotA*, respectively. On which sigma factor is *tlp* transcription dependent? What about *cotD* transcription? Which compartments would you predict these genes are expressed in?

SUGGESTED READING

Bassford, P., and J. Beckwith. 1979. *Escherichia coli* mutants accumulating the precursor of a secreted protein in the cytoplasm. *Nature* (London) **277:**538–541.

Bayan, N., I. Guilvout, and A. P. Pugsley. 2006. Secretins take shape. *Mol. Microbiol.* **60:**1–4.

Britton, R. A., P. Eichenberger, J. E. Gonzalez-Pastor, P. Fawcett, R. Monson, R. Losick, and A. D. Grossman. 2002. Genome-wide analysis of the stationary-phase sigma factor (sigma-H) regulon of *Bacillus subtilis. J. Bacteriol.* **184:**4881–4890.

Broder, D. H., and K. Pogliano. 2006. Forespore engulfment mediated by a ratchet-like mechanism. *Cell* **126:**917–928.

Brown, P. J. B., M. A. dePedro, D. T. Kysela, C. Van der Henst, J. Kima, X. De Bolle, C. Fuquaa, and Y. V. Brun. 2012. Polar growth in the alphaproteobacterial order *Rhizobiales. Proc. Natl. Acad. Sci USA* **109:**1697–1701.

Camp, A. H., and R. Losick. 2009. A feeding tube model for activation of a cell-specific transcription factor during sporulation in *Bacillus subtilis. Genes Dev.* **23:**1014–1024.

Campo, N., and D. Z. Rudner. 2006. A branched pathway governing the activation of a developmental transcription factor by regulated intramembrane proteolysis. *Mol. Cell* **23:**25–35.

Cutting, S., V. Oke, A. Driks, R. Losick, S. Liu, and L. Kroos. 1990. A forespore checkpoint for mother cell gene expression during development in *B. subtilis. Cell* **62:**239–250.

Diederich, B., J. F. Wilkinson, T. Magnin, S. M. A. Najafi, J. Errington, and M. D. Yudkin. 1994. Role of interactions between SpoIIAA and SpoIIBB in regulating cell-specific transcription factor σ^F of *Bacillus subtilis. Genes Dev.* **8:**2653–2663.

Doan, T., C. Morlot, J. Meisner, M. Serrano, A. O. Henriques, C. P. Moran, Jr., and D. Z. Rudner. 2009. Novel secretion apparatus maintains spore integrity and developmental gene expression in *Bacillus subtilis. PLoS Genet.* **5:**e1000566.

Emr, S. D., S. Hanley-Way, and T. J. Silhavy. 1981. Suppressor mutations that restore export of a protein with a defective signal sequence. *Cell* **23:**79–88.

Errington, J. 1993. *Bacillus subtilis* sporulation: regulation of gene expression and control of morphogenesis. *Microbiol. Rev.* **57:**1–33.

Errington, J., and J. Mandelstam. 1986. Use of a *lacZ* gene fusion to determine the dependence pattern of sporulation operon *spoIIA* in *spo* mutants of *Bacillus subtilis. J. Gen. Microbiol.* **132:**2967–2976.

Feucht, A., T. Magnin, M. D. Yudkin, and J. Errington. 1996. Bifunctional protein required for asymmetric cell division and cell-specific transcription in *Bacillus subtilis. Genes Dev.* **10:**794–803.

Fischer, K. E., and E. Bremer. 2012. Activity of the osmotically regulated *yqiHIK* promoter from *Bacillus subtilis* is controlled at a distance. *J. Bacteriol.* **194:**5197–5208.

Gan, L., S. Chen, and G. J. Jensen. 2008. Molecular organization of Gram-negative peptidoglycan. *Proc. Natl. Acad. Sci. USA* **105:**18953–18957.

Garner, E. C., R. Bernard, W. Wang, X. Zhuang, D. Z. Rudner, and T. Mitchison. 2011. Coupled, circumferential motions of the cell wall synthesis machinery and MreB filaments in *B. subtilis. Science* **333:**222–225. (An accompanying article by Dominguez-Escobar et al. comes to many of the same conclusions.)

Gutierrez, C., J. Barondess, C. Manoil, and J. Beckwith. 1987. The use of transposon Tn*phoA* to detect genes for cell envelope proteins subject to a common regulatory stimulus. *J. Mol. Biol.* **195:**289–297.

Hendrickx, A. P. A., J. M. Budzik, S. Y. Oh, and O. Schneewind. 2011. Architects at the bacterial surface—sortases and the assembly of pili with isopeptide bonds. *Nat. Rev. Microbiol.* **9:**166–176.

Hilbert, D. W., and P. J. Piggot. 2004. Compartmentalization of gene expression during *Bacillus subtilis* spore formation. *Microbiol. Mol. Biol. Rev.* **68:**234–262.

Hoch, J. A., and K. I. Varughese. 2001. Keeping signals straight in phosphorelay signal transduction. *J. Bacteriol.* **183:**4941–4949.

Imamura, D., R. Zhou, M. Feig, and L. Kroos. 2008. Evidence that the *Bacillus subtilis* SpoIIGA protein is a novel type of signal-transducing aspartic protease. *J. Biol. Chem.* **283:**15287–15299.

Jiang, X., A. Rubio, S. Chiba, and K. Pogliano. 2005. Engulfment-regulated proteolysis of SpoIIQ: evidence that dual checkpoints control sigma activity. *Mol. Microbiol.* **58:**102–115.

Karow, M., P. Glaser, and P. J. Piggot. 1995. Identification of a gene, *spoIIR*, that links the activation of σ^E to the transcriptional activity of σ^F during sporulation in *Bacillus subtilis. Proc. Natl. Acad. Sci. USA* **92:**2012–2016.

Kirkpatrick, C. L., and P. H. Viollier. 2011. New(s) to the (Z-) ring. *Curr. Opin. Microbiol.* **14:**691–697.

Koronakis, V., J. Eswaran, and C. Hughes. 2004. Structure and function of TolC: the bacterial exit duct for proteins and drugs. *Annu. Rev. Biochem.* 73:467–489.

Kroos, L. 2007. The *Bacillus* and *Myxococcus* developmental networks and their transcriptional regulators. *Annu. Rev. Genet.* **41**:13–39.

Lazazzera, B., T. Palmer, J. Quisel, and A. D. Grossman. 1999. Cell density control of gene expression and development in *Bacillus subtilis*, p. 27–46. *In* G. M. Dunny and S. C. Winans (ed.), *Cell-Cell Signaling in Bacteria.* ASM Press, Washington, DC.

Lee, H. C., and H. D. Bernstein. 2001. The targeting pathway of *Escherichia coli* presecretory and integral membrane proteins is specified by the hydrophobicity of the targeting signal. *Proc. Natl. Acad. Sci. USA* **98**:3471–3476.

Levin, P. A., and R. Losick. 2000. Asymmetric division and cell fate during sporulation in *Bacillus subtilis*, p. 167–190. *In* Y. V. Brun and L. J. Shimkets (ed.), *Prokaryotic Development.* ASM Press, Washington, DC.

Madden, J. C., N. Ruiz, and M. Caparon. 2001. Cytolysin-mediated translocation (CMT): a functional equivalent of type III secretion in gram-positive bacteria. *Cell* **104**:143–152.

Meisner, J., X. Wang, M. Serrano, A. O. Henriques, and C. P. Moran, Jr. 2008. A channel connecting the mother cell and forespore during bacterial endospore formation. *Proc. Natl. Acad. Sci. USA* **105**:15100-15105.

Mikula, K. M., J. C. Leo, A. Lyskowski, S. Kedracka-Krok, A. Pirog, and A. Goldman. 2012. The translocation domain in trimeric autotransporter adhesins is necessary and sufficient for trimerization and autotransportation. *J. Bacteriol.* **194**:827–838.

Molle, V., M. Fujita, S. T. Jensen, P. Eichenberger, J. E. Gonzalez-Pastor, J. S. Liu, and R. Losick. 2003. The Spo0A regulon of *Bacillus subtilis. Mol. Microbiol. 50*:1683–1701.

Oliver, D. B., and J. Beckwith. 1981. *E. coli* mutant pleiotropically defective in the export of secreted proteins. *Cell* **25**:2765–2772.

Palmer, T., F. Sargent, and B. C. Berks. 2005. Export of complex cofactor-containing proteins by the bacterial Tat pathway. *Trends Microbiol.* **13**:175–180.

Perego, M., C. Hanstein, K. M. Welsh, T. Djavakhishvili, P. Glaser, and J. A. Hoch. 1994. Multiple protein-aspartate phosphatases provide a mechanism for the integration of diverse signals in the control of development in *B. subtilis. Cell* **79**:1047–1055.

Peters, N. T., T. Dinh, and T. G. Bernhardt. 2011. A fail-safe mechanism in the septal ring assembly pathway generated by the sequential recruitment of cell separation amidases and their activators. *J. Bacteriol.* **193**:4973–4983.

Rapaport, D. 2011. Special section on protein translocation across or insertion into membranes. *Biochim. Biophys. Acta Biomembranes* **1808**:840. (This volume contains a collection of articles on protein translocation.)

Ruvolo, M. V., K. E. Mach, and W. F. Burkholder. 2006. Proteolysis of the replication checkpoint protein Sda is necessary for the efficient initiation of sporulation after transient replication stress in *Bacillus subtilis. Mol. Microbiol.* **60**:1490–1508.

San Milan, J. L., D. Boyd, R. Dalbey, W. Wickner, and J. Beckwith. 1989. Use of *phoA* fusions to study the topology of the *Escherichia coli* inner membrane protein leader peptidase. *J. Bacteriol.* **171**:5536–5541.

Schmidt, R., P. Margolis, L. Duncan, R. Coppolecchia, C. P. Moran, Jr., and R. Losick. 1990. Control of developmental transcription factor σ^F by sporulation regulatory proteins SpoIIAA and SpoIIAB in *Bacillus subtilis. Proc. Natl. Acad. Sci. USA* **87**:9221–9225.

Smith, M. A., W. M. Clemons, Jr., C. J. DeMars, and A. M. Flower. 2005. Modeling the effects of *prl* mutations on the *Escherichia coli* SecY complex. *J. Bacteriol.* **187**:6454–6465.

Sonenshein, A. L., J. A. Hoch, and R. Losick (ed.). 2001. Bacillus subtilis *and Its Closest Relatives: from Genes to Cells.* ASM Press, Washington, DC.

Stephenson, S. J., and M. Perego. 2002. Interaction surface of the Spo0A response regulator with the Spo0E phosphatase. *Mol. Microbiol.* **44**:1455–1467.

Tan, J., Y. Lu, and J. C. A. Bardwell. 2005. Mutational analysis of the disulfide catalysts DsbA and DsbB. *J. Bacteriol.* **187**:1504–1510.

Thanbichler, M. 2009. Spatial regulation in *Caulobacter crescentus. Curr. Opin. Microbiol.* **12**:715–721.

Tian, H., and J. Beckwith. 2002. Genetic screen yields mutations in genes encoding all known components of the *Escherichia coli* signal recognition particle pathway. *J. Bacteriol.* **184**:111–118.

van den Berg, B., W. M. Clemons, Jr., I. Collinson, Y. Modis, E. Hartmann, S. C. Harrison, and T. A. Rapoport. 2004. X-ray structure of a protein-conducting channel. *Nature* **427**:36–44.

Vollmer, W., D. Blanot, and M. A. de Pedro. 2008. Peptidoglycan structure and architecture. *FEMS Microbiol. Rev.* **32**:149–167. (This review is followed by others that together give a detailed account of bacterial cell wall structure and synthesis.)

Wagner, J. K., C. D. Galvani, and Y. V. Brun. 2005. *Caulobacter crescentus* requires RodA and MreB for stalk synthesis and prevention of ectopic pole formation. *J. Bacteriol.* **187**:544–553.

Waksman, G., and S. J. Hultgren. 2009. Structural biology of the chaperone-usher pathway of pilus biogenesis. *Nat. Rev. Microbiol.* **7**:765–774.

Wang, S. T., B. Setlow, E. M. Conlon, J. L. Lyon, D. Imamura, T. Sato, P. Setlow, R. Losick, and P. Eichenberger. 2006. The forespore line of gene expression in *Bacillus subtilis. J. Mol. Biol.* **358**:16–37.

White, C. L., A. Kitich, and J. W. Gober. 2010. Positioning cell wall synthetic complexes by the bacterial morphogenetic proteins MreB and MreD. *Mol. Microbiol.* **76**:616–633.

Wild, K., K. R. Rosendal, and I. Sinning. 2004. A structural step into the SRP particle. *Mol. Microbiol.* **53**:357–363.

Xie, K., and R. E. Dalbey. 2008. Inserting proteins into the bacterial cytoplasmic membrane using the Sec and YidC translocases. *Nat. Rev. Microbiol.* **6**:234–244.

Zhou, R., C. Cusumano, D. Sui, R. M. Garavito, and L. Kroos. 2009. Intramembrane proteolytic cleavage of a membrane-tethered transcription factor by a metalloprotease depends on ATP. *Proc. Natl. Acad. Sci. USA* **106:**16174–16179.

Zhou, R., and L. Kroos. 2004. BofA protein inhibits intramembrane proteolysis of pro-σ^K in an intercompartmental signaling pathway during *Bacillus subtilis* sporulation. *Proc. Natl. Acad. Sci. USA* **101:**6385–6390.

Zhou, R., and L. Kroos. 2005. Serine proteases from two cell types target different components of a complex that governs regulated intramembrane proteolysis of pro-σ^K during *Bacillus subtilis* development. *Mol. Microbiol.* **58:**835–846.

Answers to Questions for Thought and Problems

Chapter 1

Questions for Thought

1. The two strands of bacterial DNA probably are not replicated in the 3′-to-5′ direction simultaneously, because replicating a DNA as long as the chromosome in this manner would leave single-stranded regions that were so long that they would be too unstable or susceptible to nucleases.

2. DNA molecules may be very long, because if cells contained many short pieces of DNA, each one would have to be segregated individually into the daughter cells.

3. DNA may be the hereditary material instead of RNA because double-stranded DNA has a slightly different structure than double-stranded RNA. The B form structure of DNA may have advantages for replication, etc. The use of DNA instead of RNA may allow the primer, which is made of RNA, to be more easily identified and removed from a DNA molecule by the editing functions. The editing functions do not operate when the 5′ end is synthesized. By removing and resynthesizing any RNA regions, using upstream DNA as a primer, mistakes can be minimized.

4. A temperature shift should cause the rate of DNA synthesis to drop, but not too abruptly. Each cell would complete the rounds of replication that were under way at the time of the shift but would not begin another round. The synthesis rate would drop exponentially. If cells are growing rapidly, the drop would be less steep, because a number of rounds of DNA replication would be under way in each molecule, and each would have to complete the cycle.

5. The gyrase of *S. sphaeroides* might be naturally resistant to novobiocin. You could purify the gyrase and test its ability to introduce supercoils into DNA in vitro in the presence of novobiocin.

6. How chromosome replication and cell division are coordinated in bacteria like *E. coli* is not well understood. One hypothesis is that a protein required for cell division might be encoded by a gene located close to the termination region, and this gene is transcribed into RNA only when it replicates. Cell division could then begin only when the termination region has replicated. You could move the terminus of replication somewhere else in the chromosome and see if this affects the timing of cell division.

7. There is no known answer, but maybe it allows more genome rearrangements, such as inversions and duplications, which may play an important role in evolution. If chromosome replication obligatorily stopped at a certain *ter* sequence, a major chromosome rearrangement might make it impossible to replicate the entire DNA, because then the *ter* sequence could be encountered before the entire DNA had replicated. With some readthrough, a few of the cells survive, and the new gene arrangement might prove more effective.

Problems

1. 5′-GGATTA-3′.

2. 5′-GGAddT-3′.
   5′-GGATddT-3′.
   5′-GGATTACGGddT-3′.
   5′-GGATTACGGTAAGGddT-3′.

3. $I = 25$ min, $C = 40$ min, $D = 20$ min

4. $I = 90$ min, $C = 40$ min, $D = 20$ min

5. The *topA* mutant lacks topoisomerase I, which removes negative supercoils, so there should be more negative supercoils in the DNA of the mutant.

6. 5′-NNGAATTCATTAAGATCG-3′, where N could be any deoxynucleotide. Extra nucleotides are added at the end so that it can be cut by EcoRI, which does not efficiently cut sites right at the end of the DNA. Preferably, G's or C's, rather than A's or

T's, should be put at the end to minimize "fraying" of the ends during amplification, since G's and C's form more stable base pairs than A's and T's.

Chapter 2

Questions for Thought

1. Some people think RNA came first, because the peptidyltransferase that links amino acids to make protein is the 23S rRNA, and other RNA enzymes have been identified. However, the question remains open to speculation.

2. The genetic code may be universal, because once the code was established, too many components—tRNAs, aminoacyl-tRNA synthetases, etc.—were involved in translating the code to change all of them.

3. Why eukaryotes rarely have polycistronic mRNAs is open to speculation. It might have something to do with the necessity for exporting mRNA from the nucleus before it can be translated. Also, the mechanism of translation initiation in eukaryotes generally involves recognition of the AUG codon closest to the 5′ end of the transcript, so that recognition of downstream coding sequences is more difficult.

4. The genetic code of mitochondrial genes differs from the chromosomal code of eukaryotes because mitochondria were once bacteria with their own simple translation apparatus, ribosomes, etc. This translation apparatus has remained independent of that for the chromosomal genes, so they have gone their separate ways.

5. Selenocysteine may be a relic of what was once a useful process in an earlier organism from which all other organisms evolved. In some proteins, selenocysteine in the active center does enhance the reaction rate.

6. The translation apparatus is very highly conserved evolutionarily, so that an antibiotic that inhibits the translation apparatus of one type of bacterium is apt to inhibit the translation apparatus of all bacteria. This is less true of amino acid biosynthetic pathways.

7. The two-chamber structure of chaperonins is very mysterious and is made even more so by the fact that the chaperonins of archaea and the eukaryotic cytoplasm have a two-cylinder structure similar to those of eubacteria, even though the sequences of the two types of chaperonins are not related and hence apparently evolved independently. One idea is that this structure has something to do with regulation. One chamber plays a regulatory role, recognizing sequences on the protein as the protein emerges from the exit channel of the ribosome, and the other chamber then takes up the protein from the N terminus, which could be some distance away. The presence of two chambers means that the ends are far enough apart to play both roles without having to make a longer structure.

8. One reason is to save energy by not making a particular gene product unless it is needed. At least two GTPs and one ATP are required to translate each codon, not to mention the hundreds of nucleoside triphosphates required to transcribe a gene. Another reason is to prevent interference between intermediates of pathways. For example, the intermediates in some degradative pathways might inhibit other degradative pathways. A third reason is that it might help in replication of the chromosome. Transcribing RNA polymerase can interfere with the replication fork, so the transcription of only a few genes at any one time may help speed up chromosome replication.

Problems

1. 5′-CUAACUGAUGUGAUGUCAACGUCCUACUCUAG CGUAGUCUAA-3′.

2. It is likely that translation begins in the second triplet, GUG, because it is followed by a long open reading frame, although this hypothesis would have to be tested.

3. The answer is (a). Both sequences have a string of A's, but only in (a) is this preceded by an inverted repeat that could form a hairpin loop in the mRNA.

4. Transcription involves copying the information from a series of deoxyribonucleotide triphosphates (in DNA) to a series of ribonucleotide triphosphates (in RNA), taking advantage of Watson-Crick base pairing. Translation involves converting information from ribonucleotide triphosphates into a series of amino acids, which requires a more complex process and machinery, including nucleotide adaptors (tRNA) that couple nucleotide information to amino acids.

5. An endoribonuclease cleaves internally within an RNA, between two nucleotides, and an exoribonuclease removes nucleotides sequentially from the ends of the RNA.

6. tmRNA releases ribosomes from mRNAs that lack stop codons (possibly because of partial degradation of the mRNA) and also targets the truncated protein product for degradation. It prevents accumulation of stalled ribosomes and removes truncated proteins that could be toxic.

7. a. Expression of the operon is constitutive, and the genes are expressed even if inducer is not added.

b. Expression of the operon is turned off, and the operon cannot be induced.

8. Homologs are two sequences derived from a common ancestor; orthologs are homologs that are found in different species and usually have the same function; paralogs are duplicated genes in the same species that have deviated in function. A COG is a cluster of orthologous genes and is used to define relationships between genes. TIGRFAMS are sets of alignments used to help identify the functions of orthologs.

Chapter 3

Questions for Thought

1. Why the genetic maps of *Salmonella* and *Escherichia* spp. are similar is unknown. Perhaps there is an optimal way to arrange genes on a chromosome, with genes that are expressed at high levels closest to the origin of replication and transcribed in the same direction in which they are replicated. Inversions would then be selected against.

2. Tandem-duplication mutations would allow the number of genes of an organism to increase. The duplicated genes could then evolve so that their products could perform novel functions. Sometimes organisms with a duplication of a particular region may have a selective advantage in a particular environment, so the duplication would be preserved. For example, duplication of a particular gene or genes might give the organism a selective advantage under the prevailing conditions by making more of the gene product(s), or a mutation in a duplicated gene, for example, a suppressor mutation, might give the organism an advantage.

3. The cells of higher organisms may be more finely tuned because readthrough proteins resulting from translation through the ends of genes may create more problems for eukaryotic cells, which, being more complicated, may be less tolerant of aberrant proteins. Alternatively, they may be less efficient at degrading aberrant proteins, or readthrough proteins may trigger an immune response in those organisms that have an adaptive immune system.

4. It is very difficult to design a possible mechanism for adaptive or directed mutations. In its purest form, such a mechanism would require that the cell somehow sense that a mutation would be desirable and change the DNA sequence accordingly. There would have to be some flow of information from the protein product back to the DNA, which so far we do not have a precedent for.

Problems

1. The cultures with the largest numbers of mutants probably had the earliest mutation.

2. Mutations to arginine auxotrophy would occur at a higher rate, because many genes encode enzymes to make arginine, and any mutation that inactivates the product of one of these genes would make the cell Arg$^-$. Rifampin resistance, however, can be caused by only a few mutations in the gene for the β subunit of RNA polymerase, because very few amino acids can be changed and have the RNA polymerase no longer bind rifampin but still be active for transcription.

3. Approximately 8×10^{-10}.

4. 5.5×10^{-8}.

5. a. *arg-1* is probably a leaky missense mutation or another type of base pair change mutation, since it seems to retain some activity of the gene product and it reverts.

b. *arg-2* is probably a deletion mutation, since it is not leaky and does not revert.

c. *arg-1* could be a frameshift mutation close to the end of the gene encoding the carboxyl end of the protein, so that the gene product is not totally inactivated. The mutant with *arg-2* could be a double mutant with two missense mutations in *arg* genes, so that both of them would seldom revert simultaneously.

6. If the mutants are isolated from the same culture, they could be siblings with the same original mutation, so they would be the same and not representative of all the mutations that can cause the phenotype.

7. a. Plate large numbers of the bacteria on plates containing the antibiotic coumermycin and all the other necessary growth supplements.

b. Plate large numbers of the Trp$^-$ bacteria on minimal plates lacking tryptophan but containing all the other necessary growth supplements.

c. Plate large numbers of the *dnaA*(Ts) mutant on plates at the high, nonpermissive temperature to isolate mutants that can multiply to form a colony at the high temperature. Test some of them by crossing them with the wild type to show that there are temperature-sensitive recombinant types, and in them, the original temperature-sensitive mutation has been suppressed rather than having reverted.

d. Plate large numbers of the *araD* mutant on plates containing L-arabinose plus another carbon source. Test mutants that can multiply and so are no longer L-arabinose sensitive to determine if they still cannot use L-arabinose as a sole carbon source and therefore are not revertants. Also cross them with the wild type to show that there are some arabinose-sensitive recombinants and that some of the recombinants that are not L-arabinose sensitive but are Arg$^-$ and cannot use L-arabinose as a sole carbon source are not complemented by *araA* mutations.

e. The *hisB* gene is downstream of the *hisC* gene in the same operon. Make a partial diploid that has a polar *hisC* mutation in one copy of the operon and a *hisB* mutation in the other copy. Then, plate large numbers of the partial diploid on minimal plates lacking histidine but having all the other necessary growth supplements. Mutants that can grow to form a colony could have a polarity-suppressing mutation in *rho* that allows *hisB* expression from the first copy of the operon, complementing the *hisB* mutation in the other copy. They could be tested to see if they suppress known polar mutations in other operons.

8. Grow large numbers of the cells at the low, permissive temperature and then shift them to the high, nonpermissive temperature before adding ampicillin. After incubating the cells for a time, wash out the ampicillin and plate the cells at the lower temperature. Repeat the enrichment once or twice. Test colonies that arise at the low temperature to find the ones due to mutant bacteria that cannot form colonies at the high temperature.

9. Nonsense suppressors are dominant. The mutant tRNA still inserts an amino acid at the nonsense codon even in the presence of the normal tRNA.

10. The mutations in each of the genes were probably nonsense mutations, and the mutation that apparently reverted all of them is probably a nonsense suppressor in a tRNA gene. You could test it to see if the strain propagates a phage with a nonsense mutation in an essential gene or if it suppresses nonsense mutations in other genes.

11. The cotransduction frequency of the *argH* and *metB* markers is 37%, and that of the *argH* and *rif* markers is 20%. The order is probably *metB1–argH5–rif-8*, which is also consistent with the three-factor-cross data. The order *argH5–metB1–rif-8* is consistent with some of the data, but the *argH* marker is probably in the middle, since most of the Rif^r transductants are not also Met$^+$, as they would be if they were on the same side of the *argH* marker.

12. If the *metA15* mutation had been suppressed, you would not expect any Met$^-$ transductants because both the donor and recipient have the *metA15* mutation. If you did get Met$^+$ transductants, about 86% would be Arg$^-$ and 14% would be Arg$^+$ (the cotransduction frequency between the *arg* and *met* markers).

13. First, transduce the Tn*10* transposon insertion mutation into your *hemA* mutant, selecting the tetracycline resistance gene on the transposon by plating on plates containing tetracycline. Identify a tetracycline-resistant transductant that still has the *hemA* mutation by the fact that it does not grow on plates lacking δ-aminolevulinic acid. This strain can be used to move the *hemA* mutation, selecting Tet^r.

14. The clone probably does not contain all of the *hemA* gene. Recombination between the clone and the *hemA* gene in the chromosome seems to be required to make a functional *hemA* gene, explaining why not all the bacteria containing the clone are HemA$^+$. If the clone contained the entire *hemA* gene, it would complement the mutation in the chromosome, and all the bacteria would be HemA$^+$ and would grow without δ-aminolevulinic acid.

15. Introduce the gene with the Cm^r cassette cloned in a plasmid cloning vector into the cell under conditions where the cloning vector cannot replicate (i.e., is a suicide vector), and plate the cells on plates containing chloramphenicol. Test any Cm^r cells for the presence of the cloning vector. For example, if the cloning vector carries ampicillin resistance (Ap^r), you could screen for cells that are Cm^r but not Ap^r. These are presumably strains that have had two crossovers, replacing the normal *arg* gene with the *arg* gene containing the Cm^r cassette. You would expect the strain to be Arg$^-$.

16. The *trpA* mutation is the selected marker, and the *argH* and *hisG* mutations are the unselected markers.

17. Close to *argH*, particularly if the tetracycline-sensitive recombinants are often also Arg$^-$.

Chapter 4

Questions for Thought

1. If essential genes were carried on a plasmid, cells that were cured of the plasmid would die. By having nonessential genes on plasmids, the chromosome can be smaller, so the cells can multiply faster; the species can adapt to a wide range of environments because the cells can exchange plasmids, and sudden selection for cells with a particular plasmid will allow some members of the population to survive. You would not expect genes encoding enzymes of the tricarboxylic acid cycle, such as isocitrate dehydrogenase, or genes for proteins involved in macromolecular synthesis, such as RNA polymerase, to be carried on plasmids. Genes involved in using unusual carbon sources, such as the herbicide 2,4-D, or genes required for resistance to antibiotics, such as ampicillin, might be expected to be on plasmids.

2. Why some, but not all, plasmids have a broad host range is unknown. A broad-host-range plasmid can parasitize more species of bacteria. A narrow-host-range plasmid can develop a better commensal relationship with its unique host.

3. Perhaps a single copy of the plasmid binds to each of two sites on the ends of filaments formed by the ParA protein. When the filaments grow, they push these two copies of the plasmid apart toward opposite poles of the cell. This may be why the plasmid must finish replicating before the filaments can grow.

4. If the genes required for replication of the plasmid are not all closely linked to the *ori* site, you could find them by isolating temperature-sensitive mutants of the plasmid that cannot replicate at high temperature and then looking for pieces of plasmid DNA that can help the mutant plasmid replicate at the high temperature when introduced into the cell in a cloning vector.

5. You could determine which of the replication genes of the host *E. coli* (*dnaA*, *dnaC*, etc.) are required for replication of a given plasmid by introducing the plasmid into cells with temperature-sensitive mutations in each of the genes and then determining whether the plasmid can replicate at the high, nonpermissive temperature for the mutant. Plasmid replication could be determined by adding radioactive precursors for DNA synthesis after the temperature has been raised and determining if the radioactive DNA hybridizes to plasmid DNA.

6. One advantage to having the leader sequence degraded, rather than the mRNA for the RepA protein itself, is that it may help prevent the synthesis of defective RepA protein that could interfere with replication. If the mRNA for the RepA protein is itself degraded, defective translation products of RepA due to running off the end of the degraded mRNA might compete with normal RepA for replication.

7. The DNA polymerase can replicate all the way to the end of the leading strand but cannot replicate the inverted repeat at the end of the lagging strand. The newly synthesized inverted repeat on the leading strand could flip over and substitute for the inverted repeat on the lagging strand, since they have the same sequence in the 5′-to-3′ direction. The inverted repeat on the leading strand could then be resynthesized using the upstream DNA as a primer. Make a diagram of your model.

Problems

1. You introduce your plasmid into cells with the other plasmid and grow the cells in the presence of streptomycin and/or sulfonamide but in the absence of kanamycin. After a few generations, you plate the cells and test the colonies for kanamycin resistance. If more of the cells are kanamycin sensitive than would be the case if your plasmid were replicating in the same type of cell without the other plasmid, your plasmid is probably an IncQ plasmid and so is incompatible with the other plasmid.

2. 1/2,048.

3. Make plates containing higher and higher concentrations of ampicillin until bacteria containing the RK2 plasmid can no longer multiply to form a colony. Plate large numbers of bacteria containing the plasmid on this concentration of ampicillin. Any bacteria that form colonies may contain high-copy-number mutants of RK2. The copy number of the plasmid in these bacteria could be determined, and the plasmid could be introduced into new bacteria to show that the new bacteria are also made resistant to higher concentrations of ampicillin by the plasmid, thereby showing that the mutation is in the plasmid itself and not in the chromosome. The *repA* gene in the mutant plasmid could then be sequenced to ensure that this is the region of the responsible mutation.

4. The plasmid should have an easily selectable gene, such as one for resistance to an antibiotic. Cells containing the plasmid are grown through a number of generations in the absence of the antibiotic and plated without the antibiotic. The number of

cells cured of the plasmid is then determined by replicating these plates onto a plate containing the antibiotic. Any colonies that do not transfer onto the new plate are due to bacteria that have been cured of the plasmid. If this number is much smaller than predicted from the normal distribution based on the copy number of the plasmid, the plasmid may have a partitioning system.

5. Changing one of the complementary sequences prevents pseudoknot formation, and therefore, the Shine-Dalgarno sequence of *repZ* mRNA is masked by structure III. No RepZ is made regardless of whether Inc antisense RNA is made, and the plasmid does not replicate.

Chapter 5

Questions for Thought

1. If the *tra* genes and the *oriT* site on which they act are close to each other, recombination between the *oriT* site and the genes for the Tra functions seldom occurs. Separating the Tra functions from the *oriT* site on which they act renders them nonfunctional.

2. Plasmids with certain *mob* sites can be transferred only by certain corresponding Tra functions because the relaxase of the mobilizable plasmid must bind to the coupling protein of the self-transmissible plasmid.

3. Self-transmissible promiscuous plasmids may encode their own primases, so they can transfer themselves into a distantly related host cell with an incompatible primase and still synthesize their complementary strand.

4. The absence of an outer membrane in gram-positive bacteria may make a pilus unnecessary. The question is open to speculation.

5. The answer is not known. Perhaps the role of the Mpf structure, including the pilus, in helping transmit DNA into the cell can be easily subverted by the phage to transfer its own DNA into the cell. Alternatively, the fact that the pilus is extended outside the cell may make it an attractive adsorption site, much as it does in an antigen.

6. Plasmids are generally either self-transmissible or mobilizable, because if they were neither, they would not be able to move to cells that did not already contain them. By being promiscuous, plasmids can expand their host range and parasitize other cells.

7. F primes can be isolated from an Hfr donor by selecting a late marker early because F primes contain sequences that flank the Hfr. When a DNA sequence is transferred into a recipient strain by using an Hfr, transfer occurs in only one direction, making it impossible for late markers to transfer early.

Problems

1. The recipient strain is the one that becomes recombinant and retains most of the characteristics of the original strain. If the transfer is due to a prime factor, the apparent recombinants become donors of the same genes.

2. Determining which of the *tra* genes of a self-transmissible plasmid encodes the pilin protein is not easy. For example, phage-resistant mutants do not necessarily have a mutation in the pilin gene. They could also have a mutation in a gene whose product is required to assemble the pilus on the cell surface. You could purify the pili and make antibodies to them. Then, *tra* mutants with mutations in the pilin gene would not make an antigen that will react with the antibody. Similarly, to determine which *tra* mutants do not make the DNase that nicks the DNA at the *ori* region or the helicase that separates the DNA strands, you may have to develop assays for these enzymes in crude extracts and determine which *tra* mutants do not make the enzyme in your assay.

3. You can show that only one strand of donor DNA enters a recipient cell by using a recipient that has a temperature-sensitive mutation in its primase gene. The plasmid DNA should remain single stranded after transfer into such a strain, provided, of course, that the plasmid cannot make its own primase. Single-stranded DNA is more sensitive to some types of DNases and behaves differently from double-stranded DNA during gel electrophoresis.

4. If the tetracycline resistance gene is in a plasmid, the tetracycline-resistant recipient cells should have acquired the plasmid. If it is in an integrating conjugative element, no transferred plasmid will be in evidence. Also, Southern blots should reveal that the conjugative transposon now has different flanking sequences.

5. Male-specific phage cannot infect cells containing only a mobilizable plasmid because mobilizable plasmids do not encode a pilus, which serves as the adsorption site for the phage.

6. The protein product of the *eex* gene prevents the entry of a plasmid of the same Inc group into the cell. You could mutagenize the plasmid randomly with a transposon, such as Tn*5*, and isolate a number of insertion mutants with mutations in the plasmid by mating it into another strain, selecting the kanamycin resistance on Tn*5*. Cells containing the plasmid with different insertion mutants could then be patched on a plate on which have been spread cells containing a plasmid of the same Inc group but carrying a different antibiotic resistance gene. After incubation, this plate could be replicated onto another plate carrying the second antibiotic. Any patches containing transconjugants that have become resistant to the second antibiotic are candidates for having contained the plasmid with the Tn*5* transposon in its *eex* gene, and they could be tested directly for plasmid exclusion. It might be necessary to have kanamycin on the first plate and have the potential donor cells also be resistant to kanamycin, in addition to being resistant to the second antibiotic carried on the plasmid, if the background due to cells cured of the mutagenized plasmid is too high.

7. If a plasmid is self-transmissible, it also often mobilizes at least some type of mobilizable plasmid. A collection of different mobilizable plasmids, containing easily selectable markers for antibiotic resistance, could be tried. They could be individually introduced into cells containing the indigenous plasmid by electroporation, selecting the antibiotic resistance on the mobilizable plasmid. These cells could then be mixed with related cells that lack the indigenous plasmid, selecting for antibiotic resistance on the mobilizable plasmid and counterselecting the donor strain. If one of the mobilizable plasmids is mobilized into this strain, it will become resistant to the antibiotic carried on the mobilizable plasmid. If you cannot introduce the mobilizable plasmids into the cells by electroporation, you could try

a triparental mating to see if the indigenous plasmid can mobilize any of the mobilizable plasmids into a third strain.

8. A pheromone-responsive plasmid may encode both an inhibitory peptide and an additional intermembrane protein to prevent pheromone processing to strongly ensure that autoinduction does not occur, in addition to preventing mating signals from other cells.

Chapter 6

Questions for Thought

1. Some possible reasons for the development of competence in bacteria are listed in the text. They include the ability to try combinations of alleles to enhance fitness, to repair damage to DNA, to provide nutrition, or a combination of these. At this time, we do not know the correct answer or even whether the reason is the same for all competent bacteria.

2. To discover whether the competence genes of *B. subtilis* are turned on by UV irradiation and other types of DNA damage, you could make a gene fusion with a reporter gene, such as *lacZ*, to one of the competence factor-encoding genes and see if the reporter gene is induced following UV irradiation.

3. To determine whether antigenic variation in *N. gonorrhoeae* results from transformation between bacteria or recombination within the same bacterium, you could introduce a selectable gene for antibiotic resistance into one of the antigen genes and see if it is transferred naturally under conditions where antigenic variation occurs.

Problems

1. To determine whether a given bacterium is naturally competent, you would isolate an auxotrophic mutant, such as a Met^+ mutant, and mix it with DNA extracted from the wild-type bacterium. The mixture would then be plated on medium without methionine. The appearance of colonies due to Met^+ recombinants would be evidence of transformation.

2. To isolate mutants defective in transformation, you would take your Met^- mutant, mutagenize it, and repeat the test described for problem 1 on individual isolates. Any mutants that do not give Met^+ recombinants when mixed with the wild-type DNA might be mutants with a second mutation in a competence gene.

3. To discover whether a naturally transformable bacterium can take up DNA of only its own species or any DNA, you could make radioactive DNA and mix it with your competent bacteria. Any DNA taken up by the cells would become resistant to added DNase, and the radioactivity would be retained with the cells on filters. Try this experiment with radioactive DNA from the same species, as well as from different species.

4. If the bacterium can take up DNA of only the same species, it must depend on uptake sequences from that species. The experiment should be done as in problem 3 but with only known pieces of DNA instead of the entire molecule. If a known piece of DNA is taken up, the responsible uptake sequence could be determined by trying overlapping fragments to see what region they must have in common to be taken up.

5. If the DNA of a phage successfully transfects competent *E. coli*, plaques appear when the transfected cells are plated with bacteria sensitive to the phage.

Chapter 7

Questions for Thought

1. If phages made the proteins of the phage particle at the same time as they made DNA, the DNA might be prematurely packaged into phage heads, leaving no DNA to replicate.

2. Phages that make their own RNA polymerase can shut off host transcription by inactivating their host RNA polymerase without inactivating their own RNA polymerase. However, phages that use the host RNA polymerase can take advantage of the ability of the host molecule to interact with other host proteins, allowing more complex regulation.

3. In this way, they can infect a wider range of hosts and still replicate their DNA.

4. There is no known answer. Perhaps, again, it has something to do with the range of hosts they can infect. In one range of hosts, the Pri proteins may be more compatible, while in another range of hosts, the RNA polymerases may be more compatible.

5. There is no known answer, but if the phage injects a protein with the DNA and this protein is intended to be used early, the protein could present problems for the phage that are in later stages of development. Also, the DNA of the second infecting phage would be in a different stage of replication, which might interfere with the replication of the DNA of the first infecting phage if they go through more than one stage.

6 One possibility is that the DNAs are in different cellular compartments. Maybe it can only pick up DNA as the DNA is passing through the cellular membrane, during conjugation or phage infection.

Problems

1. You could mix a known amount of the virus with the cells and then measure the fraction of survivors. From the Poisson distribution, you could then measure the effective MOI: e^{-MOI}= fraction of surviving bacteria. The ratio of the effective MOI to the actual MOI is the fraction of the viruses that actually infected the cells.

2. The regulatory gene is probably gene *M*, because mutations in this gene can prevent the synthesis of many different gene products. The other genes probably encode products required for the assembly of tails and heads.

3. Amber mutations introduce a nonsense UAG codon into the coding sequence of an mRNA, stopping translation and leading to synthesis of a shortened gene product. The *ori* sequence does not encode a protein, so an amber mutation could not be isolated in it.

4. Actually, *r1589* complements *r*IIB mutations. Look for mutations in *r1589* mutants that do not complement *r*IIB mutations. Show that the mutations are in *r*IIA. If they are, they should give r^+ recombinants with the deletion *r638*, which deletes all of *r*IIB. Nonsense and frameshift.

5. The order is *A-Q-M*, because with this order, most of the Am^+ recombinants with a crossover between *amA* and *amQ* would have the Ts mutation in gene *M*. A second crossover between *Q* and *M* would be required to give the wild-type recombinant.

6. T1 has a linear genetic map, which is expanded for the genes at the end of the linear DNA.

7. A phage with a mutation in its lysozyme gene does not lyse the cells and release phage unless egg white lysozyme is added. Infect the cells, and allow the infection to proceed long enough for the cells to have lysed if they had been infected by the wild-type phage. Then, divide the culture in half and add lysozyme to one of the divided cultures. Plate both cultures with indicator bacteria. If the culture with the added lysozyme yields many more plaques than the culture without lysozyme, the phage that infected the cells contained a mutation in its lysozyme gene. Addition of the lysozyme caused the cells to lyse, releasing their phage and making many more plaques rather than just one plaque where the original infected cell was located.

8. It should lyse cells infected with an antiholin mutant, because this mutant would still make the holin and allow the lysozyme to destroy the cell wall, allowing access of the $CHCl_3$ to the cytoplasmic membrane. It should not lyse a holin mutant for the same reason. It should also not lyse a lysozyme mutant for the same reason.

9. Some mutations that are in *Rz* by complementation tests lie on either side of *Rz1* mutations by genetic mapping or DNA sequencing.

10. If you infect at a high MOI, the phage you isolate that is displaying the peptide on the surface of its capsid and is thus being "panned" may not be the one that encoded it, since the cell could have been simultaneously infected by phages expressing different peptides. This used to be called phenotypic mixing in phage genetics.

Chapter 8

Questions for Thought

1. Perhaps the λ prophage uses different promoters to transcribe the *cI* repressor gene immediately after infection than when in the lysogenic state, because this may allow the repressor gene to be transcribed from a strong, unregulated promoter immediately after infection but then be transcribed from a weaker, regulated promoter in the lysogenic state.

2. By making two proteins, the cell can use the Int protein to promote recombination for both integration and excision. Then, the smaller Xis protein, the directionality factor, need only recognize the hybrid *att* sites at the ends of the prophage.

3. Morons might have been integrated in a process much like that by which integrons pick up gene cassettes, using an integrase encoded by the phage or even by a different DNA element in the cell. It is difficult to identify the *att* sites on such cassettes, since they show a lot of variability. The selection varies depending on the function encoded by the moron. They may come from superintegron-like elements in the host chromosome.

4. It is not known why some types of prophage can be induced only if another phage of the same type infects the lysogenic cell containing them. It may provide the opportunity for recombination with a related phage, or perhaps there is some other way of inducing them that has not been tried.

5. Perhaps the major difference is that P4 phage DNA can replicate in the cell by itself, while a genetic island requires the replication functions of the infecting phage.

6. Maybe so only some phages can induce them. Different phage strains have different nonessential genes but the same essential genes.

Problems

1. If the clear mutant has a *vir* mutation, it forms plaques on a λ lysogen, unlike mutations that inactivate one of the *trans*-acting functions required for lysogeny.

2. A specialized transducing phage carrying the *bio* operon of *E. coli* would be isolated in the same way as the λ*dgal* phage described in the text, except that a Bio^- mutant would be infected and the infected cells would be plated on medium without biotin to select Bio^+ transductants. The Bio^+ transductants would be isolated, and the phage would be induced. It might be necessary to add a wild-type helper phage before induction, since *bio* substitutions extend into the *int* and *xis* genes. The λ*pbio* phage should form plaques, because no replication genes should be substituted.

3. The repressor must bind first to the o^1 sites of the operator before it binds cooperatively to the other sites. Theoretically, only genes to the right of *cI* are required for lytic development. However, it is necessary to make N to continue to make these gene products for optimum lytic development. Therefore, *vir* mutants have mutations in both $o_L{}^1$ and $o_R{}^1$.

4. Draw structures. Both Int and Xis are required to integrate phage λ transducing particles next to an existing prophage because the recombination occurs between two hybrid *attP-attB* sites. If so, induction of the prophages will require both Int and Xis.

5. A low multiplicity. Otherwise, background due to simultaneous infection by more than one phage would be too high.

6. Mix the peptides by themselves with the phage before infection to see if they inhibit the formation of mixed dilysogens, as they should if it is due to interaction of the peptides. Use phages that do not express peptides or that express peptides that are known not to interact. Infect cells with the two phages separately and mix them only prior to plating. In all these cases, you should get many fewer cells expressing resistance to both antibiotics.

7. Plate it separately on each of the Nus mutants containing one of the known *nus* mutations to see if it forms plaques. If it does not form plaques on a mutant containing a particular *nus* mutation, it seems that the phage requires the Nus protein for antitermination, although other explanations are possible.

8. You would not expect to see intragenic complementation between amber mutations in different domains of a protein because the mutation in the upstream domain would

prevent synthesis of the downstream domain required for complementation.

9. You would expect them to be dilysogens; otherwise, they would be induced and not survive. The *c*I repressor with a single mutation in one of its domains will be inactive unless complemented.

10. A dimerization domain from any other protein should cause the N-terminal DNA-binding domain of *c*I to dimerize and be active. Translationally fuse the *c*I N-terminal DNA-binding domain that is inactive because it cannot dimerize to various regions of the LacZ protein, and see which fusions, when expressed from an expression vector plasmid, prevent plaque formation by λ phage on cells containing that plasmid.

11. You have to be able to distinguish plaques due to your phage from those of P4. One difference is that P4 will form plaques on a P2 lysogen but not on *E. coli* that is not lysogenic for P2. Plate your phage on a P4 lysogen, and see if any of the resultant plaques are due to P4 rather than your phage, i.e., they form plaques only on a P2 lysogen and not on a nonlysogen. Alternatively, you could infect *E. coli* simultaneously with your phage and P4 and see if any of the progeny phage fulfill the above criterion.

Chapter 9

Questions for Thought

1. Perhaps replicative transposons do not occur in multiple copies around a genome because their resolution functions cause deletions between repeated copies of the transposon, resulting in the death of cells with more than one copy. Also, an additional phenomenon called target immunity inhibits transposition of a transposon into a DNA that already contains the same transposon.

2. The transposon Tn*3* and its relatives may have spread throughout the bacterial kingdom on promiscuous plasmids.

3. Transposons sometimes, but not always, carry antibiotic resistance or other traits of benefit to the host. They may also help the host move genes around, as in the construction of plasmids carrying multiple drug resistance. A low rate of inactivating mutations may also prevent genomes from becoming ever larger over time.

4. The origin of the gene cassettes within integrons is not known. They may move to plasmids from superintegrons, like the one found in the *V. cholerae* chromosome, but the original source remains a mystery.

5. It is possible that invertible sequences rarely invert because very little of the invertase enzyme is made or because the enzyme works very inefficiently on the sites at the ends of the invertible DNA sequence.

6. In the experiments of Bender and Kleckner on Tn*10* transposition, 16%, not 50%, of the colonies showed a segmented pattern because there was a combination of phages containing wild-type *lacZ* genes and *lacZ* genes with the three missense mutations and the heteroduplex phages. Even among the heteroduplex phages, mismatch repair could have converted these Tn*10*-containing genomes to one allele or the other.

Problems

1. You would integrate a λ*dgal* at the normal λ attachment site close to the *gal* operon with one of the *gal* mutations by selecting for Gal$^+$ transductants. Sometimes the *gal* genes in the chromosome would recombine with the *gal* genes on the integrated λ, and the cell would become Gal$^-$ by gene conversion (homogenoting). When the λ*dgals* are induced, their DNA should be longer and the phage should be denser, owing to the inserted DNA making the λ genome longer.

2. The colonies should not be sectored, because only one strand, either the *lac* or *lac*$^+$ strand, of the original heteroduplex transposon has been inserted and the other strand has been copied from it for each cell that went on to form a colony.

3. The advantages are that transposon insertions almost always inactivate the gene and are not leaky, so the phenotypes of a null mutant can be known. They also mark the site of the mutation both genetically and physically, so the site of the mutation is easier to determine by either genetic or PCR sequencing techniques. The disadvantage is that they almost always inactivate the gene, so the effects of other types of mutations in the gene cannot be studied. You also cannot get transposon insertion mutations in an essential gene in a haploid organism, because they are lethal.

4. Grow the pBR322 plasmid in cells containing Tn*5* on the chromosome. Isolate the plasmid from these cells and transform them into a naive host that is normally kanamycin sensitive. Select for kanamycin resistance in the transformants to identify plasmids containing the Tn*5* element.

5. Sequence a large number of independent isolates that received the element using arbitrary PCR (Box 9.1) or another technique. Use this sequence to map the insertions onto the genome of your host or a closely related strain to determine the positions and orientations of the insertions. Because elements can show regional preferences, older techniques utilizing Southern blots cannot be used to assess the randomness of transposition.

6. a. Use a promiscuous self-transmissible plasmid, such as RP4, to mobilize a plasmid suicide vector containing Tn*5* into the strain of *P. putida*. Select the Kan^r transconjugants on rich medium containing kanamycin, and then replicate them onto minimal medium with 2,4-D as the sole carbon source, looking for transposon insertion mutants that cannot use 2,4-D as a sole carbon and energy source and so cannot grow to form colonies. Pick the corresponding colonies from the kanamycin plates, and isolate the DNA from such mutants. Cut the DNA with a restriction endonuclease that does not cut in the transposon, and ligate the pieces of the DNA into an *E. coli* plasmid cloning vector. Use the ligation mixture to transform *E. coli*, selecting Kan^r transformants. These should contain a plasmid clone with at least part of a gene whose product is required to use 2,4-D. The corresponding wild-type gene could be cloned using PCR from this DNA sequence.

b. Make a library of the DNA of the *P. putida* strain in a broad-host-range mobilizable plasmid, such as RSF1010. Mobilize the plasmid library into mutants of the *P. putida* strain that cannot use 2,4-D, and select transconjugants that can form colonies on minimal plates containing 2,4-D as the sole carbon source. The plasmid cloning vector in the bacteria in these colonies should contain the gene that was mutated to prevent growth on 2,4-D, and the plasmid gene is complementing the chromosomal mutation.

7. Construct a nonsense mutation in the *int* gene of Tn*916*. Try to use strains with and without a nonsense suppressor mutation as recipients. If recipients can be identified, the resulting strain could be used as a donor to transfer to the same two recipients: those with and without the nonsense suppressor mutation. This experiment should indicate whether the Int protein must be expressed in the donor or the recipient or if expression in either strain is sufficient.

8. Start a culture from a single colony, isolate the DNA from the culture, and perform a Southern hybridization after cutting with a restriction nuclease that cuts off center in the G segment and using a probe complementary to the G segment. If you get two bands, the segment has inverted. Another way might be to map transposon insertion mutations to antibiotic resistance in the prophage G segment with respect to markers in the neighboring chromosomal DNA by three-factor crosses. If you get a consistent order, the G segment is not inverting in the prophage. An even better way would be to insert a promoterless reporter gene next to the invertible element so that it is transcribed from a promoter of the invertible sequence only if the invertible sequence flips into the other orientation. If the reporter gene is expressed in some cells, the invertible element is inverting.

9. You can use mutants of the phages that lack a functional DNA invertase of their own and propagate them in isogenic cells that are lysogenic for e14 and cured of e14. Pick plaques, and test the host range of the phage in the plaques to determine if their invertible sequence has inverted and changed their host range. To test for *Salmonella* phase shift, again, use a *Salmonella* mutant which lacks the invertase and introduce the e14 prophage. Test whether it can now shift from one cell surface antigen to the other.

10. By methods such as those outlined in chapter 1, you could clone the pigment gene and use it as a probe in Southern blot analyses to see whether the flanking sequences around the gene change when it is in the pigmented as opposed to the nonpigmented form.

11. You could make a plasmid that has two copies of the Mu phage by cloning a piece of DNA containing Mu into a plasmid cloning vector also containing Mu. You could then see if the two repeated Mu elements could resolve themselves in the absence of the host recombination functions.

Chapter 10

Questions for Thought

1. Recombination is required to restart replication forks that have stalled at damage to the DNA. Recombination is also required for the most universal and frequently used type of double-strand DNA break repair. It might also help speed up evolution by allowing new combinations of alleles to be tried.

2. The reason that RecBCD recombination is so complicated is perhaps because its real role might be to repair double-strand breaks caused when replication encounters damage in the DNA. The free ends can then invade the other daughter DNA and, with the help of the Pri proteins, re-form a replication fork. This interpretation is supported by the observation that *chi* sites are mostly arranged so that they can help re-form replication forks.

3. The different pathways of recombination may function under different conditions for recombination between short and long DNAs or at breaks, gaps, etc.

4. The RecF pathway is preferred under conditions different from those normally used in the laboratory to measure recombination. The SbcB and SbcC functions may interfere with the RecF pathway only under the conditions normally used in laboratory crosses, such as transductional, transformational, or Hfr crosses, where the RecBCD pathway is preferred. The RecF pathway may also be more primordial, and therefore, a molecular mechanism may have been needed to route some types of repair into the RecBCD/AddAB system.

5. By encoding their own recombination functions, phages can increase their rates of recombination. Also, some phages use the recombination functions for replication and, by encoding their own functions, can inhibit the host recombination functions to prevent them from interfering with phage replication, as in the case of the RecBCD function and λ rolling-circle replication.

6. There could be another, as yet undetected X-phile in the cell that cuts the Holliday junctions which the RecG helicase causes to migrate. Alternatively, the RecG helicase may allow replication restarts by backing up the replication fork to form a type of Holliday junction called a chicken foot and then replicating the remainder of the DNA to remove the Holliday junction rather than resolving it by using an X-phile.

Problems

1. A common way to determine if recombinants in an Hfr cross have a *recA* mutation is to take the individual recombinants and streak them across a plate. Then, half of each streak is covered with a glass plate (glass is opaque to UV radiation), and the plate is irradiated before incubation. If the recombinant is $RecA^-$, it grows only in the part of the streak that was covered by the plate, because $RecA^-$ mutants are much more sensitive to killing by UV.

2. To determine which other genes, if any, participate in the *recG* pathway, you could set up a synthetic lethality screen based on the fact that RecG is required only in the absence of RuvABC. You could construct a strain in which the *ruv* genes are transcribed from an inducible promoter on a plasmid in a cell with *ruvABC* deleted in the chromosome and isolate transposon insertion mutants in the presence of the inducer. The mutants you are interested in do not grow in the absence of inducer. You can then see if any of the mutants have mutations in genes other than the *recG* gene.

3. The recombination promoted by homing double-stranded nucleases to insert an intron occurs in the same manner as in Figure 10.3, except that the double-strand break that initiates the recombination occurs at the site in the target DNA into which the intron will home. The invading DNA then pairs with the homologous flanking DNA on one side of the transposon in the donor DNA and replicates over the transposon until it meets the other 5′ end, inserting the intron.

4. In a $RecB^-C^-D^-$ host, compare recombination between the same two markers in λ DNAs, one with a *chi* mutation and another without. If there is no difference, *chi* sites stimulate recombination only in the presence of the RecD function.

Chapter 11

Questions for Thought

1. The answer is unknown. Perhaps the newly synthesized strand of DNA somehow remains bound to the replication apparatus for some distance behind the replication fork. This is suggested by the observation that the SeqA protein bound to hemimethylated DNA in *E. coli* apparently travels with the replication fork. An intriguing idea with some experimental backing is that the orientation of the sliding clamp helps direct the repair machinery.

2. If the mismatch repair proteins travel with the replication apparatus, it could account for their result.

3. Different repair pathways work better, depending on where the damage occurs, the type of damage, or the extent of the damage. For example, some common types of damage have their own dedicated repair systems. If damage is so extensive that lesions occur almost opposite each other in the two strands of the DNA, it might be easier to repair the lesions with excision repair than with recombination repair. Alternatively, if the damage is irreparable, it might be better just to replicate over it.

4. The SOS mutagenesis pathway might exist to allow the cells to survive damage other than that due to UV irradiation, or it might be more effective under culture conditions different from those used in the laboratory.

Problems

1. Assuming that you can grow the organism in the laboratory (you might have to grow it under high pressure), you could irradiate it in the dark and then divide the culture in half and expose half of the cells to visible light before diluting and counting the surviving bacteria. If more of the cells survive after they have been exposed to visible light, the bacterium has a photoreactivation system.

2. The procedure is explained in the text. Briefly, to show that the mismatch repair system preferentially repairs the unmethylated strand, you could make heteroduplex DNA of λ phage. One strand should be unmethylated and heavier than the other, because the λ phages from which this strand was derived were propagated on Dam^- *E. coli* cells grown in heavy isotopes. The two λ phages used to make the heteroduplex DNA should also have mutations in different genes, so that there are mismatches at these positions. After the heteroduplex λ DNA is transfected into cells, test the progeny phage to determine which genotype prevails: the genotype of the phage from which the unmethylated DNA was prepared or the genotype of the phage with methylated DNA. Then, reverse the two DNAs so that the other DNA has the heavy isotope to eliminate marker effects.

3. To determine whether the photoreactivating system is mutagenic, perform an experiment similar to that in problem 1 but with a *umuCD* mutant of *E. coli*. Measure the frequency of mutations (such as reversion of a *his* mutation) among the survivors of UV irradiation in the dark as opposed to those that have been exposed to visible light after UV irradiation. More cells should survive if they are exposed to visible light, but a higher frequency of these survivors should be His^+ revertants if photoreactivation is mutagenic. If photoreactivation is not mutagenic, a lower frequency should be His^+ revertants, because the photoreactivation system will have removed some of the potentially mutagenic lesions. That is why it may be better to do this experiment with a *umuCD* mutant, to lower the background mutations due to SOS mutagenesis.

4. To find whether the nucleotide excision repair system can repair damage due to aflatoxin B, treat wild-type *E. coli* cells and a *uvrA*, *uvrB*, or *uvrC* mutant with aflatoxin B. Dilute and plate. Compare the survivor frequencies of the *uvr* mutant and the wild type.

5. Express *umuC* and *umuD* from a clone that has a constitutive operator mutation so that the cloned genes are not repressed by LexA. Also, be sure that part of the *umuD* gene has been deleted so that UmuD′, rather than the complete UmuD, is synthesized. This clone can be put into isogenic $RecA^+$ and $RecA^-$ strains of *E. coli* that have a *his* mutation. After UV irradiation, the frequencies of His^+ revertants among the surviving bacteria for each strain can be compared. If the $RecA^+$ strain shows a higher frequency of His^+ revertants, the RecA protein may have a role in UV mutagenesis other than inducing *umuCD* by cleaving LexA and then cleaving UmuD.

6. You could make a transcriptional fusion of the *recN* gene to a reporter gene, such as *lacZ*, and then determine whether more of the reporter gene product is synthesized after UV irradiation, as it should be if *recN* is an SOS gene. A strain that also has a *lexA*(Ind^-) mutation should not show this induction if *recN* is induced because it is an SOS gene and not for some other reason.

Chapter 12

Questions for Thought

1. Why operons are regulated both positively and negatively is not clear. However, there may be different advantages to the two types of regulation. For example, negative regulation might require more regulatory protein but might allow more complete repression, while it might be easier to achieve intermediate levels of expression with positive regulation. There may be less interaction between negative regulatory systems than between positive regulatory systems, in which a regulatory protein might inadvertently turn on another operon. Also, constitutive mutants are rarer with positive regulation.

2. The genes for regulatory proteins may be autoregulated to save energy. If they are autoregulated, only the amount of regulatory protein needed is synthesized. Minimizing the amount also prevents accidental misregulation and can reduce the amount of effector required. Also, in the case of positive regulators, more can be made after induction to further increase the expression of the operons under their control.

3. These and other amino acids have one primary role in the cell, which is to be incorporated into proteins. Clustering of the genes into operons facilitates coordinate expression. Genes

that participate in multiple pathways are less likely to be organized into single operons, so individual genes can respond to multiple signals.

4. Regulation by attenuation of transcription offers the advantage that regulation occurs after transcription has initiated, and the response is potentially more rapid than regulation at the level of promoter activity. A major disadvantage of this type of regulation is that it is wasteful. A short RNA is always made from the operon, even if it is not needed.

5. The important regulatory outcome is the final amount of gene product present in the cell. This amount is influenced, not only by the synthesis rates of the mRNA and the protein, but also by the rates of degradation of both the mRNA and the protein product. A product with a high synthesis rate and high stability will accumulate to higher levels than will a product with a high synthesis rate and low stability.

6. The translational arrest is dependent on the sequence of the nascent peptide, and this would put constraints on the sequence of the regulated protein product.

Problems

1. To isolate a *lacI*^{s} mutant, take advantage of the fact that *lacI*^{s} mutations, while rare, are dominant Lac$^-$ mutations and that β-galactosidase, which is synthesized when the *lac* operon is induced, cleaves P-Gal, a galactose analog whose accumulation results in production of galactose, which kills *galE* mutants. Mutagenize a *galE* mutant that contains an F′ plasmid with the *lac* operon, and plate it on P-Gal medium containing another carbon source, such as maltose. The survivors that form colonies are good candidates for *lacI*^{s} mutants, because inactivation of the *lacZ* genes in both the chromosome and the F′ plasmid requires two independent mutations, which should be even rarer than single *lacI*^{s} mutations. The mutants can be further tested by transferring the F′ plasmid from the mutants into other strains whose chromosomes contain a wild-type *lac* operon. If the F′ plasmid makes the other strain Lac$^-$, the F′ plasmid must contain a *lacI*^{s} mutation.

2. AraC must be in the P1 state, since it represses the *ara* operon.

3. a. Lac$^-$, permanently repressed. The LacI^s repressor binds to the operators of both operons, even in the presence of the inducer, and prevents transcription of the *lacZYA* structural genes.

b. Inducible Lac$^+$. In other words, it is wild type for the *lac* operon. The *lacO*^{c} mutation in the chromosome makes the *lac* operon on the chromosome constitutive, but LacZ and LacY are not made from that copy anyway because of the polar mutation in *lacZ*.

c. Inducible Ara$^+$. The inactivating *araC* mutation is recessive to the wild-type *araC* allele, and the cell is wild type for the *ara* operon and inducible by L-arabinose.

d. Inducible Ara$^+$. The *araI* mutation prevents transcription of the operon in the chromosome, but the mutation is *cis* acting and so does not prevent transcription of the operon on the F′ plasmid.

e. Inducible Ara$^+$, for the same reason as (d). The *cis*-acting p_{BAD} promoter mutation prevents transcription of the chromosomal operon, but not the operon on the F′ plasmid.

4. To determine whether *phoA* is negatively or positively regulated, you could first isolate constitutive mutants to determine how frequent they are. Since PhoA turns XP blue, you could mutagenize cells and isolate mutants that form blue colonies on XP-containing medium even in the presence of excess phosphate in the medium. If *phoA* is negatively regulated, constitutive mutants should be much more frequent than if it is positively regulated. Also, at least some of these constitutive mutants should have null mutations, deletions, etc., that inactivate the regulatory gene.

5. Plate wild-type *E. coli* cells in the presence of low concentrations of 5-methyltryptophan and in the absence of tryptophan. Only constitutive mutants can multiply to form colonies under these conditions, because 5-methyltryptophan is a corepressor of the *trp* operon but cannot be used for protein synthesis, so the wild-type *E. coli* cells starve for tryptophan. To isolate mutants defective in feedback inhibition of tryptophan synthesis, plate a constitutive mutant in the presence of higher concentrations of 5-methyltryptophan and in the absence of tryptophan. Even constitutive mutants cannot multiply to form colonies under these conditions, because the first enzyme of tryptophan synthesis is feedback inhibited by the 5-methyltryptophan. Only mutants that are defective in feedback inhibition form colonies.

6. BglG$^-$ mutants should be permanently repressed (superrepressed), because the BglG protein binds to antiterminator hairpins, stabilizing them. In the absence of BglG, the antiterminator hairpins do not form, the terminator hairpins form, and transcription termination occurs. BglF$^-$ mutants, on the other hand, should be constitutive. BglF transfers phosphates to BglG when β-glucosides are not being transported, inactivating BglG. It also binds to BglG, sequestering it. In the absence of BglF, BglG cannot be phosphorylated and is active and free to bind to the antiterminator hairpins, even in the absence of β-glucosides.

7. What experiments you could do depend on the amino acid at that position. For some amino acids, you could substitute a nonsense codon, say UAG, for the codon at that position and suppress it with a suppressor tRNA that inserts the same amino acid. If stalling is due to the tRNA, then the stalling should no longer occur, since now the tRNA is different. If it is due to the amino acid being inserted, then stalling should still occur. However, the nonsense codon itself would cause some stalling, so this has to be taken into account. For some amino acids, such as arginine and serine, some codons that are not together in the table of the code are translated by different tRNAs. In this case, you could change the codon to one that encodes the same amino acid but is recognized by a different tRNA.

Chapter 13

Questions for Thought

1. It may be important to make more of the proteins involved in synthesizing new proteins so that the rate of protein synthesis increases after heat shock, allowing more rapid replacement of the proteins irreversibly denatured as a result of the shock. Also, these proteins may be more susceptible to damage during heat shock and therefore require replacement.

2. *Salmonella* species, which are normal inhabitants of the vertebrate intestine, are usually in an environment where amino acids are in plentiful supply but NH_3 is limiting. *Klebsiella*

species are usually free living, where NH_3 is present but amino acids are not.

3. One possible explanation is that the genes had a common ancestor in evolution and still retain many of the same properties. In addition, the genes for corresponding sensor and response regulator genes may be similar to allow cross talk between regulatory pathways. If the genes are similar, a signal from one pathway can be passed to the other pathway, allowing coordinate regulation in response to the same external stimulus. However, there is no clear evidence for the physiological importance of cross talk.

4. The enzymes responsible for ppGpp synthesis during amino acid starvation and during growth rate control may be different because the enzymes involved in stringent control and growth rate regulation must be in communication with different cellular constituents. The RelA protein works in association with the ribosome, where it can sense amino acid starvation, while the enzyme involved in synthesizing ppGpp during growth rate regulation might have to sense the level of energy in ATP and GTP. SpoT might be involved in both synthesizing and degrading ppGpp if the equilibrium of the reaction is somehow shifted. All enzymes function by lowering the activation energy, so in a sense they catalyze both the forward and backward reactions. However, because the equilibrium usually favors the forward reaction, this is the reaction that predominates. If the equilibrium were shifted, perhaps by sequestering the ppGpp as it is made, SpoT could synthesize ppGpp rather than degrade it.

5. Bacteria are small, and in some cases, a large population is required to have a significant effect on the surrounding environment. The ability to sense other bacteria of the same species allows cells to monitor when the population is sufficient to have a noticeable effect. This prevents the cell from wasteful activities.

Problems

1. Determine if the mutant can use other carbon sources, such as lactose and galactose. If it has a *cya* or *crp* mutation, it should not be able to induce other catabolite-sensitive operons and so cannot grow on these other carbon sources.

2. a. Gln^- (glutamine requiring); Ntr constitutive (expresses Ntr operons even in the presence of NH_3).

b. Ntr^- (cannot express Ntr operons even at low NH_3 concentrations); makes intermediate levels of glutamine synthetase independent of the presence or absence of NH_3.

c. Gln^-, Ntr^-.

d. Gln^-, Ntr^-.

e. Ntr constitutive.

f. Grows slowly because it constitutively synthesizes heat shock proteins.

g. Constitutive expression of diphtheria toxin and other virulence determinants, even in the presence of Fe^{2+}.

h. No ppGpp; grows very slowly and is auxotrophic for some amino acids; cannot easily restart growth after entry into stationary phase.

3. As described in chapter 1, you could clone the gene for the toxin and then perform Southern blot analysis to show that the gene is carried on a large region of DNA that is not common to all the members of the species. Using the methods described in chapter 8, you could also try to induce a phage from the cells and show that production of the toxin requires lysogeny by the phage.

4. If the *rpoH* gene for σ^H is transcriptionally autoregulated, the same amount of RNA should be made from the gene when it exists in two or more copies as is made when it exists in only one copy. Introduce a clone of *rpoH* in a multicopy plasmid into cells, and measure the amount of *rpoH* mRNA by DNA-RNA hybridization. If more RNA is made from *rpoH* under these conditions, the gene is not transcriptionally autoregulated.

5. To show which ribosomal protein is the translational repressor, introduce an in-frame deletion into the *rplJ* gene and determine if the synthesis of L12 increases. Similarly, introduce a mutation (any inactivating mutation will do) into *rplL* and determine if L10 synthesis increases.

Chapter 14

Questions for Thought

1. It might be because eukaryotic cells are much larger, making it impractical to keep proteins unfolded after they are translated but before they are translocated through the membrane.

2. It might have to do with providing energy to get the proteins through both membranes. Since the only known sources of energy are in the cytoplasm in the form of ATP and GTP and in the inner membrane in the form of membrane potential, they must have some source of energy to get proteins through the outer membrane. Presumably, all of these extra proteins allow them to transport the proteins through both membranes if the proteins are to be secreted.

3. The answer is not known, but presumably, it reflects the need to coordinate cell wall synthesis and cell division under all of the different environmental conditions the cell might find itself in.

4. The bacteria might be slow to commit to sporulation because it limits their options. Once cells are committed to sporulation, they must go all the way, making it harder to reverse course and begin growing if nutrients later become available. It also may be that the primary role of sporulation is to disseminate the bacteria to new locations and that other means are used preferably to survive nutrient deprivation.

Problems

1. Because only proteins that have a specific Tat signal sequence and that have already folded are transported by the Tat pathway and only *sec* signal sequences were used in the selections. Also, unlike the SecYEG-transported proteins, which need only the signal sequence, the Tat-transported proteins must have folded, and their folded state must be recognized by other specific proteins. You might try to select them by fusing *lacZ* to an entire protein transported by the Tat pathway and hope that the Tat-transported protein would fold properly and be recognized by the Tat system.

2. You would translationally fuse the *phoA* gene to nontransmembrane domains of the *envZ* gene. You would then express

these fusions in an *E. coli* strain in which the chromosomal *phoA* gene is deleted. Only fusion proteins in which the alkaline phosphatase protein is fused to a domain of the EnvZ protein in the periplasm will make the colony blue on XP plates.

3. A signal sequence mutation in *malE* makes the cells Mal$^-$ and unable to transport maltose for use as a carbon and energy source. A suppressor of the signal sequence mutation makes them Mal$^+$ and able to grow on minimal plates with maltose as the sole carbon and energy source. Spread millions of bacteria with the *malE* signal sequence mutations on minimal plates with maltose as the sole carbon source. Any colonies which arise are Mal$^+$ and may be revertants or may have suppressors of the signal sequence mutation. To distinguish those which have suppressors from the true revertants, you could use some of them as donors to transduce the *mal* operon into another strain, selecting for a nearby marker. If any of the transductants are Mal$^-$, the *mal* operon in that Mal$^+$ apparent revertant had a mutation somewhere else in the chromosome that was suppressing the signal sequence mutation.

4. Make fusions of Gfp to the septal proteins. See if null mutations in one gene prevent the localization of a Gfp fusion of another septal protein to the septal ring using epifluorescence microscopy. If the protein no longer localizes to the septal ring, the protein product of the first gene must bind to the ring before the second protein can bind.

5. By isolating mutants blocked in sporulation. The regulatory genes could then be identified because mutations in these genes blocked the expression of many other genes, as determined by *lacZ* fusions. Also, some were similar in sequence to other known regulators, including sigma factors and transcriptional activators.

6. Both involve histidine kinases, which phosphorylate on histidine residues. Transfer of phosphoryl groups occurs between histidine and aspartate residues, and transcriptional activators are regulated by phosphorylation. The *B. subtilis* phosphorelay contains a series of proteins that carry out discrete steps. Each of these proteins is regulated by other phosphatases and kinases.

7. The original mutations in *spo0E* and *spo0L* both conferred the Spo$^-$ phenotype, suggesting that the products of the genes were positive regulators. However, these mutations did not inactive the gene products, and null deletion mutations of the genes caused hypersporulation, suggesting that they were in fact negative regulators. Also, multiple copies of the genes inhibited sporulation, as expected of negative regulators. Spo$^-$ suppressors of *spo0E* deletion mutations were in the gene for Spo0A, a phosphorylated protein. Also, Spo$^+$ suppressors of *spo0L* missense mutations were in the gene for Spo0F, another phosphorylated protein. This suggested that Spo0E and Spo0L reduced the phosphorylation of these proteins by acting as phosphatases, and this was subsequently confirmed.

8. In a *spo-lacZ* transcriptional fusion, the *lacZ* gene has its own TIR and so is translated independently of the upstream *spo* sequences. Transcriptional fusions can be used to determine when the *spo* gene is transcribed. In the *spo-gfp* translational fusions, the coding region of the *spo* gene is fused to the *gfp* gene to encode a fusion protein that fluoresces. These fusions can be used to determine when the *spo* gene product is made and where it is localized.

9. The *tlp* gene is dependent on σ^G, and *cotD* is dependent on σ^K. The *tlp* gene is expressed in the forespore, and *cotD* is expressed in the mother cell.

Glossary

A (aminoacyl) site The site on the ribosome to which the incoming aminoacylated tRNA binds.

aaRS *See* Aminoacyl-tRNA synthetase.

Abortive initiation A step in transcription in which RNA polymerase initiates transcription but stops after synthesizing short RNAs, which are released. RNA polymerase then reinitiates transcription from the promoter. *See* Abortive products.

Abortive products Short RNAs that are released from a transcription initiation complex during abortive initiation.

Activator A protein that increases expression of a gene or operon.

Activator site A sequence in DNA upstream of the promoter to which the activator protein binds.

Adaptive mutation *See* Directed-change (adaptive) mutation hypothesis.

Adaptive response Activation of transcription of genes of the Ada regulon, which is involved in the repair of some types of alkylation damage to DNA.

Adaptor A protein that controls the activity of a cellular protease. *See* Regulated proteolysis.

Adenine (A) One of the two purine (two-ringed) bases in DNA and RNA.

Affinity tag A polypeptide that binds tightly to some other molecule, allowing the purification of a protein to which it is translationally fused.

Alkylating agent A chemical that reacts with DNA, forming a carbon bond to one of the atoms in the DNA.

Allele One of the forms of a gene, e.g., the gene with a particular mutation. The term can refer to either the wild-type form or a mutant form.

Allele-specific suppressor A second-site mutation that alleviates the effect of only one type of mutation in a gene.

Allelism test A complementation test to determine if two mutations are in the same gene, i.e., if they create different alleles of the same gene.

Allosteric interaction *See* Allosterism.

Allosterism A change in the conformation of a domain of a protein as a result of a change in a different domain, e.g., when binding of the allolactose inducer to the inducer-binding pocket of the LacI repressor changes the angle of the DNA-binding domain.

Amber The codon UAG. Mutations to this codon in the reading frame for a protein are called amber mutations. Mutations which suppress mutations to this codon are called amber suppressors.

Amino group The NH_2 chemical group.

Amino terminus *See* N terminus.

Aminoacyl-tRNA synthetase (aaRS) The enzyme that attaches an amino acid to a specific class of tRNAs.

Antiactivator A regulatory protein that acts as a repressor by preventing activation by an activator.

Antiadaptor A protein that controls the activity of an adaptor and therefore regulates the activity of a cellular protease. *See* Adaptor, Regulated proteolysis.

Antibiotic Generally, a substance—often a natural microbial product or its semisynthetic derivative—that kills bacteria (i.e., is bacteriocidal) or inhibits the growth of bacteria (i.e., is bacteriostatic). Some antibiotics are entirely chemically synthesized.

Antibiotic resistance gene cassette A fragment of DNA, usually bracketed by restriction sites for ease of cloning, that contains a gene whose product confers resistance to an antibiotic for easy selection.

Anticodon The 3-nucleotide sequence in a tRNA that pairs with the codon in mRNA by complementary base pairing.

Antiholin A phage protein that binds to a holin and prevents transport of endolysin through the hole in the membrane formed by the holin.

Antiparallel A configuration in which, moving in one direction along a double-stranded DNA or RNA, the phosphates in one strand are attached 3′ to 5′ to the sugars while the phosphates in the other strand are attached 5′ to 3′.

Antisense RNA RNA that contains a sequence complementary to a sequence in an mRNA.

Anti-sigma factor A protein that binds to a sigma factor, reversibly inactivating it.

Antitermination A regulatory process that allows RNA polymerase to transcribe through one or more transcription termination signals in DNA. *See* Processive antitermination.

Antiterminator A sequence in RNA that prevents downstream transcription termination. It can be either an RNA structure that forms in the RNA after it is transcribed or an RNA sequence that binds RNA polymerase close to the promoter. The RNA structures often prevent termination by preventing the formation of the helix of a factor-independent transcriptional terminator. *See* Transcription attenuation.

AP endonuclease A DNA-cutting enzyme that cuts on the 5′ side of a deoxynucleotide that has lost its base, usually due to a DNA glycosylase, i.e., an apurinic or apyrimidinic site. This cutting allows the DNA strand to be degraded and resynthesized, replacing the apurinic or apyrimidinic site with a normal nucleotide.

AP lyase A DNA-cutting enzyme, usually associated with an *N*-glycosylase, that cuts on the 3′ side of the apurinic or apyrimidinic site created by the *N*-glycosylase activity of the enzyme.

Aporepressor A protein that can be converted into a repressor by undergoing a conformational change if a small molecule called the corepressor is bound to it.

Apparent revertant In a genetic analysis, a mutant that seems to have overcome the effect of a mutation. Without further evidence, it could either be a true revertant or be suppressed.

Archaea One of the three divisions of organisms, which, like bacteria, exists as single-celled organisms. They share some of the features of both eukaryotes and bacteria. Members of this division are often the only organisms capable of inhabiting extreme environments.

Assimilatory reduction Addition of electrons to nitrogen-containing compounds to reduce them to NH_3 for incorporation into cellular constituents.

Attenuation Regulation of an operon by premature termination of transcription under conditions where less of the gene product(s) is needed.

Autocleavage The process by which a protein cuts itself.

Autoinduction loop A system in which an activator activates its own expression, resulting in amplification of the signal and a very strong regulatory response.

Autokinase A protein able to transfer a PO_4 group from ATP to itself.

Autophosphorylation A process by which a protein transfers a PO_4 group to itself, independent of the source of the PO_4 group.

Autoregulation The process through which a gene product controls the level of its own synthesis.

Autotransporter *See* Type V secretion system.

Auxotrophic mutant A mutant that cannot make a growth substance that the normal or wild-type organism can make.

Backbone The chain of alternating phosphates and ribose sugars (in RNA molecules) or deoxyribose sugars (in DNA molecules) that provides the structure in which bases are attached in long polymers of RNA or DNA.

Backtrack Movement of a complex backward along its template, e.g., backward movement of RNA polymerase along the DNA, resulting in a displacement of the 3′ end of the RNA from the active site.

Bacteria Members of the domain of organisms characterized by a relatively simple cell structure free of many cellular organelles, the presence of 16S and 23S rRNAs with specific sequence conservation patterns that differ from those found in the archaea, and a simple core RNA polymerase, among other features.

Bacterial artificial chromosome (BAC) cloning vector An *Escherichia coli* plasmid cloning vector that can accept very large clones of DNA (>300 kb) because it is derived from the F plasmid. It is often used to make DNA libraries of large fragments in genome-sequencing projects to complement small insert libraries.

Bacterial lawn The layer of bacteria on an agar plate that forms when many bacteria are plated at a sufficiently high density so that individual bacterial colonies are not found.

Bacteriophage A virus that infects bacteria.

Base Carbon-, nitrogen-, and hydrogen-containing chemical compounds with structures composed of one or two rings that are constituents of the DNA or RNA molecule.

Base analog A chemical that resembles one of the bases and can mistakenly incorporate into DNA or RNA during synthesis.

Base pair (bp) Each set of opposing bases in the two strands of double-stranded DNA or RNA that are held together by hydrogen bonds and thereby help hold the two strands together. Also used as a unit of length.

Base pair change A mutation in which one type of base pair in DNA (e.g., an AT pair) is changed into a different base pair (e.g., a GC pair).

Basic Local Alignment Search Tool (BLAST) A genome annotation tool that uses bioinformatics to find similar regions in DNA sequences. It can compare DNA or amino acid sequences and identify nucleotide or protein sequences in a database that are similar to a query sequence.

Binding A process by which molecules are physically joined to each other by noncovalent interactions.

Bioinformatics A repertoire of computer technologies that allow prediction of open reading frames and regulatory sites, pIs and molecular weights of proteins, posttranslational modifications of proteins, subcellular localization of proteins, the level of expression of genes, and the functions of gene products.

Biopanning The process by which a phage expressing a specific peptide on its surface is separated from other phage not expressing the peptide by its ability to bind to a molecule fixed to a solid matrix.

Biosynthesis The synthesis of chemical compounds by living organisms.

Biosynthetic operon An operon composed of genes whose products are involved in synthesizing compounds, such as amino acids or vitamins, rather than degrading them.

Bistable state A state in which a population of cells is found in two subpopulations with different physiological properties and these subpopulations are maintained over time. Cells can switch between states, but do so rarely, so that the balance of the two different states is maintained.

BLAST *See* Basic Local Alignment Search Tool.

Blot The filter to which DNA, RNA, or protein has been transferred by blotting.

Blotting The process of transferring DNA, RNA, or protein from a gel or agar plate to a filter or other solid matrix to maintain positional information.

Blunt end A double-stranded DNA end in which the 3′ and 5′ termini are flush with each other, that is, with no overhanging single strands.

Branch migration The process by which the site at which two double-stranded DNAs held together by crossed-over strands (such as in a Holliday junction) moves; it can have the effect of changing the regions of the two DNAs that are paired in heteroduplexes.

Broad host range The property that allows a phage, plasmid, or other DNA element to enter and/or replicate in a wide variety of bacterial species.

Bypass *See* Replication bypass.

Bypass suppression A suppressor mutation that removes the need for a gene product without directly altering the gene product.

C terminus The end of a polypeptide chain with the free carboxyl (COOH) group.

cAMP *See* Cyclic AMP.

Campbell model The model in which the λ phage DNA forms a circle and then integrates into the chromosome by recombination between a site normally internal to the λ phage DNA and a site on the chromosome, which creates a prophage genetic map that is a cyclic permutation of the phage genetic map. It was named after the person who first proposed it.

Canonical sequence A composite of all the shared sequences with the same function in a region of DNA or protein that shows what sequences all (or at least most) of them have in common.

CAP *See* Catabolite activator protein.

CAP regulon All of the operons that are regulated, either positively or negatively, by the CAP protein.

CAP-binding site The sequence on DNA to which the CAP protein binds.

Capsid The protein and/or membrane coat that surrounds the genomic nucleic acid (DNA or RNA) of a virus.

Carboxyl group The chemical group COOH.

Carboxyl terminus *See* C terminus.

Catabolic operon An operon composed of genes whose products degrade organic compounds. *See* Degradative operon.

Catabolism The degradation of an organic compound, such as a sugar, to make smaller molecules with the concomitant production of energy.

Catabolite A small molecule produced by the degradation of larger carbon-containing organic compounds, such as sugars.

Catabolite activator protein (CAP) The DNA- and cAMP-binding protein that regulates catabolite-sensitive operons in enteric bacteria by binding to their promoter regions. Also called catabolite repressor protein (Crp).

Catabolite regulation The reduced expression of some operons in the presence of high cellular levels of catabolites due to growth on an efficiently utilized carbon source.

Catabolite repression *See* Catabolite regulation.

Catabolite-sensitive operons Operons whose expression is regulated by the cellular levels of catabolites.

Catenanes Structures formed when two or more circular DNA molecules are joined like links in a chain.

CCC *See* Circular and covalently closed.

Cell division The splitting of a mother cell into two daughter cells.

Cell division cycle The events occurring between the time a cell is created by division of its mother cell and the time it divides.

Cell divisions The total number of times in a growing culture that individual cells have grown and divided. Sometimes called cell generations.

Central dogma The tenet that information in DNA is copied into mRNA by the process of transcription and that information is translated to generate protein.

Change-of-function mutation A mutation that changes the activity of a protein rather than inactivating all or part of it, e.g., a mutation that makes a regulatory protein respond to a different inducer.

Channel gating Blocking a membrane channel unless the substrate is being transported. This prevents other molecules from leaking into or out of the cell through the membrane channel.

Chaperone A protein that binds to other proteins or RNAs and helps them to fold correctly or prevents them from folding prematurely.

Chaperone-usher secretion *See* Type V secretion system.

Chaperonin A protein that forms double back-to-back chambers, which alternate in taking up denatured proteins and refolding them. It is represented by the GroEL (Hsp60) protein in *E. coli*.

***chi* (χ) mutation** A mutation that causes a sequence that is similar to the *chi* site to become an actual functional *chi* site.

***chi* (χ) site** The sequence 5′-GCTGGTGG-3′ in *E. coli* DNA. It is recognized by the *E. coli* RecBCD complex, thereby signaling changes in the complex, upregulating 5′-to-3′ nuclease activity, inhibiting 3′-to-5′ nuclease activity, and activating its ability to load RecA.

Chromatography A method to separate molecules on the basis of charge, size and shape, or affinity differences. Liquid chromatography, based on a cation-exchange column or a hydrophobicity column, is useful for separating peptides prior to mass spectrometry.

Chromosome In a bacterial cell, the DNA molecule that contains most of the genes required for cellular growth and maintenance; usually the largest DNA molecule in the cell and the one that contains a characteristic *oriC* sequence.

CI repressor The phage λ-encoded protein that binds to the phage operator sequences close to the p_R and p_L promoters and prevents transcription of most of the genes of the phage.

***cis*-acting mutation** A mutation that affects only the DNA molecule in which it occurs or the regions proximal to where it occurs and not other DNA molecules in the same cell or in distal positions.

***cis*-acting site** A functional region on a DNA molecule that does not encode a gene product and affects only the DNA molecule that it resides in or near (e.g., a promoter or an origin of replication).

Clamp loader A protein complex that places the ring-like sliding clamp accessory protein of the replicative DNA polymerase around the DNA strands.

Classical genetics The study of genetic phenomena by using only intact living organisms.

Clonal A situation in which all members of a population are the descendants of a single organism or replicating DNA molecule, as in colonies growing from a single bacterium on an agar plate or DNA molecules all derived from replication of the same single DNA molecule.

Clone A collection of DNA molecules or organisms that are all identical to each other because they result from replication or multiplication of the same original DNA or organism.

Cloning by complementation Identifying a clone in a library by its ability to complement a mutation in the chromosome.

Cloning vector An autonomously replicating DNA (replicon), usually a phage or plasmid, into which can be introduced other DNA molecules that are not capable of replicating themselves so that the non-self-replicating DNA can be cloned.

Closed complex The complex that forms when RNA polymerase first binds to a promoter and before the strands of DNA at the promoter separate.

Cluster of orthologous genes (COG) A compilation of presumptive genes from a diversity of organisms (representing the major phylogenetic lineages) that are proposed to be functionally analogous, based on sequence similarity.

Cochaperone A smaller protein that helps chaperones fold proteins or cycle their adenine nucleotide. It is represented by DnaJ and GrpE in *E. coli*.

Cochaperonin A smaller protein that helps chaperonins fold proteins by forming the cap on the chamber once the protein has been taken up. It is represented by the GroES protein (Hsp10) in *E. coli*.

Coding region *See* Translated region.

Coding strand The strand of DNA in the part of a gene coding for a protein that has the same sequence as the RNA transcribed from the gene. *See* Nontemplate strand.

Codon A 3-base sequence in mRNA that stipulates one of the amino acids or serves as a translation termination signal.

COG *See* Cluster of orthologous genes.

Cognate aminoacyl-tRNA synthetase The enzyme that attaches the correct amino acid to a specific tRNA.

Cointegrate A DNA molecule after a replicative transposition event from a donor DNA into a target DNA in which the donor and target DNAs are joined, separated by copies of the transposon.

Cold-sensitive mutant A mutant that cannot live and/or multiply in the lower temperature ranges at which the normal or wild-type organism can live and/or multiply.

Colony A small lump or pile made up of millions of cells on an agar plate that were derived from a single cell and therefore are (essentially) genetically identical.

Colony papillation A process leading to sectors or sections in a colony that appear different from the remainder of the colony.

Colony purification Isolation of individual bacteria on an agar plate so that all the cells in a colony that forms after incubation will be descendants of the same bacterium.

Compatible restriction endonucleases Restriction endonucleases that leave the same overhangs after cutting a DNA molecule, even if their recognition sites are different. The resulting ends can pair, allowing the molecules to be ligated to each other.

Competence pheromones Small peptides given off by bacterial cells that induce competence in neighboring cells when the cells are at high concentrations.

Competent The state during which cells are capable of taking up DNA.

Complementary The property that two polynucleotides can have that allows them to be held together by base pairing between their bases.

Complementary base pair A pair of nucleotides that can be held together by hydrogen bonds between their bases, e.g., dGMP and dCMP or dAMP and dTMP.

Complementation Restoration of the wild-type phenotype when two DNAs containing different mutations that cause the same mutant phenotype are in a cell together. It usually means that the two mutations affect different genes.

Complementation group A set of mutations in which none complement any of the others, an indication that they are all in the same gene.

Composite transposon A transposon made up of two almost identical insertion (IS) elements plus the DNA between them that move as a unit.

Concatemer Two or more almost identical DNA molecules linked tail to head.

Condensation A way of making the chromosome occupy a smaller space, for example, by supercoiling or by binding condensins (e.g., one of the types of structural maintenance of chromosome [SMC] proteins).

Condensins Proteins that bind chromosomal DNA in two different places, folding it into large loops and thereby making it more condensed. They are represented by the Smc protein in *Bacillus subtilis* and by the MukB protein in *E. coli.*

Conditional-lethal mutation A mutation that inactivates an essential cellular component, but only under a certain set of circumstances, for example, a temperature-sensitive mutation that inactivates an essential gene product only at relatively high temperatures or a nonsense mutation that inactivates an essential gene product only in the absence of a nonsense suppressor.

Congenic Having an identical genotype except for a particular allele variant. *See* Isogenic.

Congression Cotransformation as the result of uptake of more than one fragment of transforming DNA. *See* Cotransformation.

Conjugation The transfer of DNA from one bacterial cell to another by the transfer functions of a self-transmissible DNA element, such as a plasmid or integrating conjugative element.

Conjugative transposon A somewhat outdated way of referring to a type of integrating conjugative element. *See* Integrating conjugative element.

Consensus sequence A nucleotide sequence in DNA or RNA, or an amino acid sequence in a protein, in which each position in the sequence has the nucleotide or amino acid that has been found most often at that position in molecules with the same function and similar sequences.

Constitutive mutant A mutant in which a regulatory system is active regardless of the presence or absence of the regulatory signal.

Context The sequence of nucleotides in DNA or RNA surrounding a particular sequence that affects its efficiency, e.g., the sequence around a nonsense codon that affects the efficiency of translation termination at the nonsense codon.

Cooperative binding Process in which the binding of one protein molecule to a site (e.g., on DNA) greatly enhances the binding of another protein molecule of the same type to an adjacent site. The proteins bound at adjacent sites interact through their multimerization domains, which stabilizes the binding.

Coprotease A protein that binds to another protein and thereby activates the second protein's autocleavage or other protease activity, although it itself is not a protease.

Copy A molecule of a particular type identical to another in the same cell. The term often refers to a gene that has been moved somewhere else so that it now exists in more than one place in the genome.

Copy number The ratio of the number of plasmids of a particular type in the cell to the number of copies of the chromosome.

Copy number effect An effect on the cell due to having more than the one copy of the gene, leading to increased levels of the gene product.

Core enzyme The catalytic part of an enzyme complex that lacks one or more additional subunits that modify its properties. *See* Core RNA polymerase.

Core polymerase The part of a DNA or RNA polymerase that actually performs the polymerization reaction and functions independently of accessory and regulatory proteins that cycle on and off the protein.

Core RNA polymerase RNA polymerase composed of two α subunits and the β, β', and ω subunits but lacking the σ subunit. This form of RNA polymerase is catalytically active but does not recognize promoter sequences for accurate initiation of transcription.

Corepressor A small molecule that binds to an aporepressor protein and converts it into a repressor.

***cos* site** The sequence of deoxynucleotides at the ends of λ DNA in the phage head. A staggered cut in this sequence at the time the phage DNA is packaged from concatamers gives rise to complementary or cohesive ends that can base pair with each other to form circular DNA on infection of another host cell.

Cosmid A plasmid that carries the sequence of a *cos* site so that it can be packaged into λ phage heads.

Cotranscribed Two or more contiguous genes transcribed by a single RNA polymerase molecule from a single promoter.

Cotransducible Two genetic markers that are close enough together on the DNA that they can be carried in the same phage particle during transduction.

Cotransduction An outcome of transduction in which transductants that were selected for being recombinant for one marker in DNA are also recombinant for a second marker.

Cotransduction frequency The percentage of transductants selected for being recombinant for one genetic marker that have also become recombinant for another genetic marker. It is a measure of how far apart the markers are on DNA.

Cotransformable Being close enough together to be carried on the same piece of DNA during transformation; the term is used to describe a pair of genetic markers that meet this criterion.

Cotransformation An outcome of transformation in which transformants that were selected for being recombinant for one marker in DNA are also recombinant for a second marker, usually because they are on the same piece of DNA.

Cotransformation frequency The frequency of transformants recombinant for the selected marker that are also recombinant for an unselected marker. It is a measure of how far apart the markers are on DNA.

Cotranslational translocation A type of translocation in which a protein is inserted into the membrane by the SRP system as it is being translated. This is required if the protein is to be inserted in the inner membrane and is highly hydrophobic.

Counterselection of donor Selection of transconjugants under conditions where the donor bacterium cannot multiply to form colonies.

Coupling hypothesis A model for the regulation of replication of iteron plasmids in which two or more plasmids are joined by binding to the same Rep protein through their iteron sequences.

Coupling protein A protein that is part of the Mpf system of self-transmissible plasmids. The coupling protein binds to the relaxase of the Dtr system to communicate that contact has been made with a recipient cell.

Covalent bond A bond that holds two atoms together by sharing their electron orbits.

Covalently closed circular Having no breaks or discontinuities in either strand (of a double-stranded circular DNA).

CRISPR (cluster of regularly interspaced short palindromic repeats) Clusters of inverted repeated sequences separated by seemingly random spacer sequences found in many types of bacteria. One function of at least some of them is to protect against incoming phage and plasmid DNAs. In at least some, the spacer sequences make RNAs that direct DNA endonucleases to phage or plasmid DNA sequences to which they are complementary, causing their cleavage.

Cross Any means of exchange of DNA between two organisms.

Crossing Allowing the DNAs of two strains of an organism to be present in the same cell so they can recombine with each other.

Crossover The site of joining of two DNA molecules during recombination.

C-terminal amino acid The amino acid on one end of a polypeptide chain that has a free carboxyl (COOH) group unattached to the amino group of another amino acid.

Cured Having lost a DNA element, such as a plasmid, prophage, or transposon; said of a cell.

Cut and paste A mechanism of transposition in which the entire transposon is excised from one place in the DNA and inserted into another place.

Cyclic AMP (cAMP) Adenosine monophosphate with the phosphate attached to both the 3′ and 5′ carbons of the ribose sugar.

Cyclically permuted Lacking unique ends. The mathematical definition of a cyclic permutation is a permutation that shifts all elements of a set by a fixed offset with the elements shifted off the end inserted back at the beginning. In a cyclically permuted genome, there are no unique ends. If the genome of such a phage is drawn as a circle, each genome starts somewhere on the circle and extends around the circle until it returns to the same place or just past it, so that the individual genomes have different endpoints but contain all of the genes in the same order.

Cyclobutane ring A ring structure of four carbons held together by single bonds that is present in some types of pyrimidine dimers in DNA.

Cytoplasmic domain A region of the polypeptide chain of a transmembrane protein that is in the interior, or cytoplasm, of the cell.

Cytosine (C) One of the pyrimidine (one-ringed) bases in DNA and RNA.

D loop The three-stranded structure that forms when a single strand of DNA invades a double-stranded DNA, displacing one of the strands.

Dam methylase *See* Deoxyadenosine methylase.

Damage tolerance mechanism A way of dealing with damage to DNA that does not involve repairing the damage, for example, replication restart or translesion synthesis.

Daughter cell One of the two cells arising from division of a mother cell.

Daughter DNA One of the two DNAs arising from replication of another DNA.

DDE transposon A transposon containing the DDE (aspartate-aspartate-glutamate) motif in its transposase. These amino acids coordinate the magnesium ions required in the active center for transposase activity.

Deaminating agent A chemical that reacts with DNA, causing the removal of amino (NH_2) groups from the bases in the DNA.

Deamination The process of removing amino (NH_2) groups from a molecule. In mutagenesis, it is the removal of amino groups from the bases in DNA.

Decatenation The process performed by type II topoisomerases of passing DNA strands through each other to resolve catenanes.

Decoding site The site on the ribosome responsible for pairing between the mRNA codon and the tRNA anticodon.

Defective prophage A DNA element in the bacterial chromosome that contains phage-like DNA sequences and presumably was once capable of being induced to form phage virions but has lost genes essential for lytic development.

Degradative operon An operon whose genes encode enzymes required for the breakdown of molecules into smaller molecules with the concomitant release of energy and/or compounds needed for other pathways. *See* Catabolic operon.

Deletion mapping A convenient procedure for mapping point mutations in which mutants that have point mutations to be mapped are crossed with mutants that have deletion mutations with known endpoints. Wild-type recombinants appear only if the unknown mutation lies outside the deleted region, allowing the mutation to be localized.

Deletion mutation A mutation in which a number of contiguous base pairs have been removed from the DNA.

Deoxyadenosine An adenine base attached to a deoxyribose sugar.

Deoxyadenosine methylase (Dam methylase) An enzyme that attaches a CH_3 (methyl) group to the adenine base in DNA. It is typified by the Dam methylase in *Escherichia coli* that methylates the A in the sequence GATC.

Deoxycytidine A cytosine base attached to a deoxyribose sugar.

Deoxyguanosine A guanine base attached to a deoxyribose sugar.

Deoxynucleoside A base (usually A, G, T, or C) attached to a deoxyribose sugar.

Deoxyribose A sugar similar to the five-carbon sugar ribose but with a hydrogen (H) atom rather than a hydroxyl (OH) group attached to the 2′ carbon.

Deoxythymidine The thymine base attached to deoxyribose sugar.

Diauxie A two-stage growth pattern that occurs when cells are grown on a mixture of two carbon sources.

Dilysogen A lysogen containing two copies of the prophage, usually joined tail to head in tandem.

Dimer A protein or other molecule made up of two polypeptides or other subunits.

Dimerization domain The region of a polypeptide that binds to another polypeptide of the same type to form a dimer.

Dimerize To bind two identical polypeptides to each other.

Diploid The state of a cell containing two copies of each of its genes, which are not derived from replication of the same DNA. *See* Haploid.

Direct repeat A short sequence of deoxynucleotides in DNA closely followed by an identical or almost identical sequence on the same strand.

Directed-change hypothesis The discredited hypothesis that mutations in DNA occur preferentially when they benefit the organism or help it adapt to a new environment.

Directional cloning Cloning a piece of DNA into a cloning vector in such a way that it can be inserted in only one orientation, for example, by using restriction endonucleases that leave incompatible overhang sequences at each end.

Directionality factor A protein that binds to a site-specific recombinase, such as an integrase, and changes its specificity so it will promote recombination in only one direction, e.g., excision.

Dissimilatory reduction The reduction of nitrogen-containing compounds, such as nitrate, that occurs when they are used as terminal electron acceptors in anaerobic respiration. The reduced nitrogen-containing compounds are not necessarily incorporated into cellular molecules.

Disulfide bonds Covalent bonds between two sulfur atoms, such as those between the side chain sulfur atoms in two cysteine amino acids in a polypeptide.

Disulfide oxidoreductases Enzymes in the periplasmic space that can form or break disulfide bonds between cysteines by reducing or oxidizing the bonds. They contain the motif CXXC, where X can be any amino acid, and exchange cysteine bonds in the protein with the cysteines in the Dsb protein.

Division septum The cross wall that forms between two daughter cells just before they separate.

Division time The time taken by a type of newborn bacterial cell to grow and divide again in a particular growth environment.

DNA-binding domain The region of a polypeptide in a DNA-binding protein that binds to DNA.

DNA clones Identical copies of a single DNA molecule that are usually made by cloning the DNA in a cloning vector and propagating the vector.

DNA glycosylase An enzyme that removes bases from DNA by cleaving the bond between the base and the deoxyribose sugar.

DNA helicase An enzyme that uses the energy of ATP to separate the strands of double-stranded DNA.

DNA library A collection of individual clones of the DNA of an organism that together represent all the DNA sequences of that organism.

DNA ligase An enzyme that can join the phosphate-terminated 5′ end of one DNA strand to the 3′ hydroxyl end of another.

DNA polymerase accessory proteins Proteins that travel with the DNA polymerase during replication.

DNA polymerase III holoenzyme The replicative DNA polymerase in *E. coli*, including all the accessory proteins, sliding clamp, editing functions, etc.

DNA polymerase V The product of the *umuC* gene of *E. coli*. When bound to UmuD′, the autocleaved form of UmuD, and a molecule of RecA bound to ATP, UmuC becomes a DNA polymerase capable of translesion synthesis, which is also mutagenic.

DNA replication complex The entire complex of proteins, including the DNA polymerase, that moves along the DNA in the replication fork.

DnaA box The sequence 5′-TTATCCACA-3′ in DNA to which the DnaA protein binds with the greatest affinity. The DnaA protein is required for the initiation of chromosome replication in *E. coli*.

Domain A region of a polypeptide with a particular function or localization.

Dominant mutation A mutation that exhibits its phenotype, even in a diploid organism containing a wild-type allele of the gene.

Dominant phenotype The phenotype exhibited by a dominant mutation or other genetic marker.

Donor allele The form of the gene that exists in the donor strain if the donor and recipient in a cross have different forms of a gene.

Donor bacterium The strain of bacterium used as a source of DNA to transfer into another strain in a genetic cross, whether by transformation, transduction, or conjugation. See Donor strain.

Donor DNA DNA that is extracted from the donor strain of bacteria and used to transform a recipient strain of bacteria. In transposition, it is the DNA in which the transposon originally resides before it transposes to the target DNA.

Donor strain The bacterial strain that is the source of the transferred DNA in a bacterial cross. For example, in a transductional cross, the donor strain is the strain in which the phage was previously propagated; in conjugation, it is the strain harboring the transmissible plasmid.

Donor strand exchange In general, exchange by proteins of one of their strands with each other. In the case of chaperone-usher secretion, the chaperone helps a protein fold by contributing one of its strands to the folded protein. This strand is then displaced by a strand from a different protein, releasing the chaperone.

Double mutant A mutant organism with two mutations.

Downstream Lying in the 3′ direction from a given point on RNA or in the 3′ direction on the coding strand of a DNA region from which an RNA is made.

Dtr component The DNA transfer component of a plasmid transmission system; the *tra* or *mob* genes of the plasmid involved in preparing the plasmid DNA for transfer.

Duplication junction The point at which an ectopic crossover occurred, resulting in a tandem-duplication mutation in DNA.

Duplication mutation A mutation that causes a region of DNA to be repeated elsewhere in the chromosome. *See* Tandem duplication.

E (exit) site The site on the ribosome at which the tRNA binds after it has contributed its amino acid to the growing polypeptide and just before it exits the ribosome. It may help maintain the correct reading frame.

E value A measure of the number of similarities or local alignment scores that are reported in a sequence search based on comparing a query sequence with a database, identifying matching sequences, and calculating the probability that a particular match could have occurred by chance. For example, the score of a query sequence and the same sequence in a database would be very close to 0. For genome annotation, only E-values less than 1/*e*5 are usually considered evidence of a reliable match.

Early gene A gene expressed early during a developmental process, for example, during bacterial sporulation or phage infection.

Eclipse phase A step during natural transformation during which DNA cannot be isolated from the recipient cells in a state that is active for transformation.

Ectopic recombination "Out-of-place" recombination; homologous recombination occurring between two sequences, usually but not always nonidentical, in different regions of the DNA. It is often responsible for deletions, inversions, and other types of DNA rearrangements and is sometimes called "unequal crossing over" when it occurs between two DNA molecules. *See* Homeologous recombination.

Editing The process of removing and replacing a wrongly inserted deoxynucleotide during replication, for example, a C inserted opposite a template A, to reduce the frequency of mutations.

Editing functions The 3′ exonuclease activities that remove nucleotides erroneously incorporated during replication. Such activities can be part of the DNA polymerase polypeptide itself or can be accessory proteins that travel with the DNA polymerase during replication. They are represented by the ε protein in *E. coli*.

Effector A small molecule that binds to a protein and changes its properties.

EF-G *See* Translation elongation factor G.

EF-Tu *See* Translation elongation factor Tu.

8-OxoG A damaged DNA base commonly caused by reactive forms of oxygen in which an oxygen atom has been added to the 8 position of the small ring of the base guanine. Abbreviation for 7,8-dihydro-8-oxoguanine; also abbreviated GO.

Electroporation The introduction of nucleic acids, proteins, or nucleoprotein complexes into cells through exposure of the cells to a strong electric field.

Electrospray ionization (ESI) A method used for preparation of samples for mass spectrometry that produces singly and multiply charged ions from a peptide so that multiple peaks are seen in a mass spectrometric analysis. The sample is introduced into an electric field in a liquid solution. Ions are formed when the solution is sprayed from a fine needle into the electric field. As solvent evaporates, intact peptides are left with different numbers of charges, depending on the sequence of the peptide.

Elongation factor G *See* Translation elongation factor G.

Elongation factor Tu *See* Translation elongation factor Tu.

ELPH Online software that can identify motifs in a set of protein or DNA sequences. If a large set of sequences is submitted, the program can search for the most common motif(s). For the software, consult the University of Maryland Center for Bioinformatics and Computational Biology, http://www.cbcb.umd.edu/software/ELPH/. The name is derived from "estimated locations of pattern hits."

Endonuclease An enzyme that can cut phosphodiester bonds between nucleotides internal to a polynucleotide.

Endoribonuclease An endonuclease that cleaves an RNA chain.

Enrichment The process of increasing the frequency of a particular type of mutant in a population, often by using an antibiotic, such as ampicillin, that kills cells only if they are growing.

Epistasis A type of interaction in which a mutation at one locus predominates phenotypically over a mutation at a different locus. It is used for determining the order in which gene products that participate in the same pathway act or the order in which gene products that are parts of the same complex are added.

Escape synthesis Induction of transcription of an operon as a result of titration of its repressor owing to an increase in the number of operators to which the repressor binds. *See* Titration.

ESI *See* Electrospray ionization.

Essential genes Genes whose products are required for maintenance and/or growth of the cell under all known conditions.

Eubacteria Another term for bacteria, proposed by Carl Woese after the discovery of archaea to distinguish between these two different domains of one-celled organisms. The organisms are now usually simply referred to as bacteria.

Eukaryotes Members of the kingdom of organisms whose cells contain a nucleus, usually surrounded by a nuclear membrane, and many other cellular organelles, including a Golgi apparatus and an endoplasmic reticulum. They have 18S and 28S rRNAs rather than the 16S and 23S rRNAs of bacteria.

Exit site *See* E site.

Exons The sequences of nucleotides in a gene encoding a protein or RNA remaining after all the introns have been removed.

Exonuclease A nuclease enzyme that can remove nucleotides only from the end of a polynucleotide.

Exoribonuclease An exonuclease that digests RNA.

Expected value *See* E value.

Exported proteins Proteins that leave the cytoplasm after they are made and end up outside the cell.

Expression vector A cloning vector in which a cloned gene can be transcribed and sometimes also translated from a vector promoter and a translational initiation region, respectively.

Exteins The sequences of amino acids in a protein remaining after all the inteins have been removed.

Extracellular protein A protein that is secreted from cells after it is made.

Extragenic Involving a different gene.

Extragenic suppressor *See* Intergenic suppressor.

Face-of-the-helix dependence A situation in which the position of a binding site for a regulatory protein on the DNA must be on the same side of the DNA helix as the position for binding of another factor (e.g., another protein or RNA polymerase).

Factor-dependent transcription termination site A DNA sequence that causes transcription termination only in the presence of a particular protein, such as the Rho protein of *Escherichia coli*.

Factor-independent transcription termination site A DNA sequence that causes transcription termination by RNA polymerase alone, in the absence of other proteins. In bacteria, it is characterized by a GC-rich region with an inverted repeat followed by a string of A's on the template strand.

FASTA An early database search program (http://fasta.bioch.virginia.edu/fasta). It has been largely replaced by BLAST and related search tools, but the "FASTA format" is still used to submit raw sequences for a database search. For example, amino acid sequences are submitted using standard amino acid codes (as in the inside front cover of this book) with the addition of "X" for any amino acid, "*" for a translation stop site, and "-" for a gap of indeterminate length.

Feedback inhibition Inhibition of synthesis of the product of a pathway resulting from binding of the end product of the pathway to the first enzyme of the pathway, thereby inhibiting the activity of the enzyme.

Ffh The protein component of the signal recognition particle of bacteria. It is related to the 54-kilodalton protein component of the signal recognition particle in eukaryotes ("Ffh" is short for "fifty-four homolog").

Filamentous phage A type of phage with a long, floppy appearance. The nucleic acid genome of these phages is merely coated with protein, making the phage as long as the genome and giving the floppy appearance. In contrast, the nucleic acids of most phages are encapsulated in a rigid, almost spherical, icosahedral head.

Filter mating A procedure in which two strains of bacteria are trapped on a filter to hold them in juxtaposition so that conjugation can occur.

Fimbriae Another name for pili, except for conjugative sex pili encoded by self-transmissible plasmids, which are always called pili and never fimbriae.

Firmicutes One of the phylogenetic groups of bacteria; it includes *B. subtilis*. Members of this phylum lack an outer membrane and typically stain gram positive.

5′ end The end of a nucleic acid strand (DNA or RNA) at which the 5′ carbon of the ribose sugar is not attached through a phosphate to another nucleotide.

5′ exonuclease A deoxyribonuclease (DNase) that degrades DNA starting with a free 5′ end.

5′ overhang A short, single-stranded 5′ end on an otherwise double-stranded DNA molecule.

5′ phosphate end In a polynucleotide, a 5′ end that has a phosphate attached to the 5′ carbon of the ribose sugar of the last nucleotide.

5′-to-3′ direction The direction on a polynucleotide (RNA or DNA) from the 5′ end to the 3′ end.

5′ untranslated region The untranslated sequence of nucleotides that extends from the 5′ end of an mRNA to the first initiation codon for a polypeptide encoded by the mRNA.

Flanking sequences The sequences that lie on either side of a gene or other DNA element.

fMet-tRNA^{fMet} The special tRNA in prokaryotes that is activated by formylmethionine and is used to initiate translation at prokaryotic translational initiation regions. It binds to translation initiation factor IF2 and responds to the initiator codons AUG and GUG and, more rarely, to other codons in a translational initiation region.

Forward genetics The classical genetic approach in which genes are first identified by the phenotypes of mutations in the genes.

Forward mutation A mutation that changes wild-type DNA sequence to mutant DNA sequence.

4.5S RNA The RNA component of the signal recognition particle (SRP) of bacteria.

Frameshift mutation Any mutation that adds or removes one or very few (but not a multiple of 3) base pairs from DNA, whether or not it occurs in the coding region for a protein.

FtsY The docking protein that binds proteins to be exported by the SRP pathway and directs them to the SecYEG channel. The term is a misnomer, because mutations in the gene (*f*ilament formation *t*emperature-*s*ensitive gene *Y*) were isolated in a search for genes involved in cell division that cause cells to form filaments rather than to divide.

FtsZ The protein that first forms the septal ring at the site of future division and attracts other proteins to the septum.

Functional domain The region of a polypeptide chain that performs a particular function in the protein.

Functional genomics A technique that involves all experimentation that seeks to define all functions of all genes and regulatory sequences in a genome. It includes biochemical, structural, and genetic analyses.

Fusion protein A protein created when coding regions from different genes are fused to each other in frame so that one part of the protein is encoded by sequences from one gene and another part is encoded by sequences from a different gene. *See* Translational fusion.

Gain-of-function mutation A mutation that creates a new activity for the gene product or causes the expression of a gene that is quiescent in the wild type.

Gel electrophoresis A procedure for separating proteins, DNA, or other macromolecules. It involves the application of the macromolecules to a gel made of agarose, acrylamide, or some other gelatinous material and then the application of an electric field, forcing the electrically charged macromolecules to move toward one or the other electrode. The speed at which the macromolecules move depends on their size, shape, and charge.

Gene A region on DNA encoding a particular polypeptide chain or functional RNA, such as an rRNA, tRNA, or small noncoding RNA.

Gene cassette A piece of DNA containing a selectable gene that can be easily cloned into another gene or cloning vector.

Gene chip A glass slide or other solid substrate on which spots of DNA are arranged and then used for hybridization with a nucleic acid probe.

Gene conversion Nonreciprocal apparent recombination associated with mismatch repair on heteroduplexes that are formed between two DNA molecules during recombination. The name comes from genetic experiments with fungi in which the alleles of the two parents were not always present in equal numbers in an ascus, as though an allele of one parent had been "converted" into the allele of the other parent.

Gene disruption An alteration of the structure or activity of a gene that is intended to inactivate a gene. *See* Null mutation.

Gene dosage experiment An experiment in which the number of copies of a gene in a cell is increased to determine the effect on the amount of gene product synthesized or on other cellular phenotypes.

Gene Locator and Interpolated Markov Modeler *See* Glimmer.

Gene replacement A molecular genetic technique in which a cloned gene is altered in the test tube and then reintroduced into the organism, allowing selection for organisms in which the altered gene has replaced the corresponding normal gene in the organism by homologous recombination.

Generalized recombination *See* Homologous recombination.

Generalized transduction The transfer, via phage transduction, of essentially any region of the bacterial DNA from one bacterium to another. The transducing phage particle contains only bacterial DNA.

Generation time The time it takes for the cells in an exponentially growing culture to double in number. *See* Division time.

Genetic code The assignment of each mRNA nucleotide triplet to an amino acid (or a translation termination signal). *See* Codon.

Genetic island A DNA element in the chromosome that is not an obvious prophage, transposon, or plasmid but that shows evidence of having been recently horizontally transferred into the chromosome based on features of the DNA and absence of the sequences in related species at that position.

Genetic linkage map An ordering of the genes of an organism solely on the basis of recombination frequencies between mutations in the genes in genetic crosses.

Genetic marker A difference in sequence of the DNAs of two strains of an organism in a particular region that causes the two strains to exhibit different phenotypes that can be used for genetic mapping of the region of sequence difference.

Genetic recombination The joining of genetic markers into new combinations.

Genetic redundancy A situation in which more than one gene can provide a needed biological function, and therefore, null mutations in one gene will not cause a mutant phenotype, even in a haploid organism.

Genetics The science of studying organisms on the basis of their genetic material.

Genome The nucleic acid (DNA or RNA) of an organism or virus that includes all the information necessary to make a new organism or virus.

Genomics The process of using the entire DNA sequence of an organism to study its physiology and relationship to other organisms. The term also refers to investigations enabled by large sets of information from RNA or proteins and compiled sequences for environments (metagenomics).

Genotype The sequence of nucleotides in the DNA of an organism, usually discussed in terms of the alleles of its genes.

Genotype-phenotype linkage In phage display, a situation in which proteins expressed by a phage are encoded in the phage DNA in the same phage particle.

Glimmer (Gene Locator and Interpolated Markov Modeler) A software program (http://www.cbcb.umd.edu/software/glimmer/) that is used to find coding regions in bacterial genomes.

Global regulatory mechanism A regulatory mechanism that affects many operons scattered around the genome.

Glucose effect The regulation of genes involved in carbon source utilization based on whether glucose is present in the medium. *See* Catabolite regulation.

Glutamate dehydrogenase An enzyme that adds ammonia directly to α-ketoglutarate to make glutamate. It is responsible for assimilation of nitrogen in high ammonia concentrations.

Glutamate synthase An enzyme that transfers amino groups from glutamine to α-ketoglutarate to make glutamate. Also called GOGAT.

Glutamine synthetase An enzyme that adds ammonia to glutamate to make glutamine. Responsible for the assimilation of nitrogen in low ammonia concentrations.

GO *See* 8-OxoG.

Gradient of transfer In a conjugational cross, the decrease in the transfer of chromosomal markers the farther they are in one direction from the origin of transfer of an integrated plasmid.

Gram-negative bacteria Bacteria characterized by an outer membrane and a thin peptidoglycan cell wall that stains poorly with a staining procedure invented by the Danish physician Hans Christian Gram in the 19th century. The term is generally used to describe bacteria in the phylum *Proteobacteria.*

Gram-positive bacteria Bacteria characterized by having no outer membrane and a thick peptidoglycan layer that stains well with the Gram stain. This term is generally used to describe bacteria in the phylum *Firmicutes*, even though not all members of the phylum stain gram positive.

GroEL *See* Hsp60.

GroES *See* Hsp10.

Growth rate regulation of ribosomal synthesis The regulation of ribosomal synthesis that ensures that cells growing more slowly have fewer ribosomes. It is proposed to be at least partially due to the levels of the initiating nucleotides GTP and ATP, which affect the stability of open complexes on the promoters for rRNAs.

Guanine (G) One of the two purine (two-ringed) bases in DNA and RNA.

Guanosine The base guanine with a ribose sugar attached to form a nucleoside.

Guanosine pentaphosphate (pppGpp) The nucleoside guanosine with two phosphates attached to the 3′ carbon and three phosphates attached to the 5′ carbon of the ribose sugar. It is quickly converted to ppGpp, which is responsible for the stringent response.

Guanosine tetraphosphate (ppGpp) The nucleoside guanosine with two phosphates attached to each of the 3′ and 5′ carbons of the ribose sugar. It is responsible for the stringent response and other stress responses. *See* Stringent response.

Gyrase A type II topoisomerase capable of introducing negative supercoils two at a time into DNA with the concomitant cleavage of ATP. It is apparently unique to bacteria.

Hairpin secondary structure A secondary structure in RNA, single-stranded DNA, or protein characterized by a region folding back on itself due to antiparallel noncovalent pairing between bases or amino acids in nearby sequences in the nucleic acid or protein, respectively.

Half-life The time required for the amount of an exponentially decaying substance to decrease by one-half.

Haploid The state of a cell containing only one copy or allele of each of its chromosomal genes. *See* Diploid.

Haploid segregant A haploid cell or organism derived from multiplication of a partially or fully diploid or polyploid cell.

Headful packaging A mechanism of encapsulation of DNA in a virus head in which the concatemeric DNA is cut after uptake of a length of DNA sufficient to fill the head, rather than at *pac* or *cos* sites.

Heat shock protein (Hsp) One of a group of highly evolutionarily conserved proteins whose rate of synthesis markedly increases after an abrupt increase in temperature or certain other stresses on the cell.

Heat shock regulon The group of *E. coli* genes under the control of σ^H, the heat shock sigma factor.

Heat shock response The cellular changes that occur in the cell after an abrupt rise in temperature.

Helicases Enzymes that unwind double-stranded nucleic acids.

Helix-destabilizing protein A protein that preferentially binds to single-stranded DNA and so can help keep the two complementary strands of DNA separated during replication or remove secondary structure from DNA. It is represented by Ssb in *E. coli.*

Helper phage A wild-type phage that furnishes gene products that a deleted form of the phage or other DNA element cannot make, thereby allowing the DNA element to replicate and be packaged into a phage particle.

Hemimethylated Having only one strand (of DNA) methylated at a sequence with twofold symmetry, such as when only one of the two A's in the sequence GATC/CTAG is methylated.

Heterodimer A protein made of two polypeptide chains that are different in primary sequence because they are encoded by different genes. *See* Heteromultimer, Homodimer.

Heteroduplex A double-stranded DNA region in which the two strands come from different DNA molecules and so can have somewhat different sequences, leading to mismatches.

Heteroimmune phages Related phages capable of lysogeny that carry different immunity regions and therefore cannot repress each other's transcription, so they can multiply on cells lysogenic for the other phage. *See* Homoimmune phages.

Heterologous probe A DNA or RNA hybridization probe taken from the same gene or region of a different organism. It is usually not completely complementary to the sequence being probed. *See* Hybridization probe.

Heteromultimer A protein made of more than one polypeptide chain (usually more than two) that are different in primary sequence because they are encoded by different genes. *See* Heterodimer.

Hfr strain A bacterial strain that contains a self-transmissible plasmid integrated into its chromosome and thus can transfer its chromosome by conjugation.

HFT lysate The lysate of a lysogenic phage containing a significant percentage of a transducing phage with bacterial DNA substituted for some of the phage DNA.

High multiplicity of infection A state of a virus or phage infection in which the number of viruses greatly exceeds the number of cells being infected, so that most cells are infected by more than one virus.

High negative interference A phenomenon in which a crossover in one region of the DNA greatly increases the probability of an apparent second crossover close by. It is caused by mismatch repair of mismatches on heteroduplexes formed at the site of the crossover.

Holin A phage protein that forms a channel in the inner membrane, allowing access of the endolysin to the cell wall or destroying the membrane potential (proton motive force) and causing cell lysis.

Holliday junction An intermediate in homologous recombination in which one strand from each of two DNAs crosses over and is rejoined to the corresponding strand on the opposite DNA.

Holoenzyme An enzyme complex (e.g., RNA polymerase or DNA polymerase) attached to all of its accessory proteins that allows complete function.

Homeologous recombination Homologous recombination in which the deoxynucleotide sequences of two participating regions are somewhat different from each other, usually because they are in different regions of the DNA or because the DNAs come from different species. *See* Ectopic recombination.

Homing The process by which an intron or intein in a gene enters the same site in the same gene in a new DNA that lacks it. A double-strand break is made in the target DNA by a specific endonuclease encoded by the intron or intein, and double-strand break repair inserts the DNA element.

Homing endonuclease The sequence-specific DNA endonuclease encoded by an intron or intein that makes a double-strand break in the target DNA to initiate the homing of an intron or intein.

Homodimer A protein made up of two polypeptide chains that are identical in primary sequence, usually because they are encoded by the same gene. *See* Heterodimer, Homomultimer.

Homoimmune phages Two related phages that have the same immunity region so that they repress each other's transcription; hence, one cannot multiply on a lysogen of the other. *See* Heteroimmune phages.

Homologous proteins Proteins encoded by genes derived from a common ancestral gene.

Homologous recombination A type of recombination that depends on the two DNAs having identical or at least very similar sequences in the regions being recombined because complementary base pairing between strands of the two DNAs must occur as an intermediate state in the recombination process.

Homologs Two or more nucleotide or protein sequences that are similar because they derive from a common ancestor.

Homomultimer A protein made up of more than one polypeptide (usually more than two) that are identical in primary sequence, usually because they are encoded by the same gene. *See* Homodimer.

Horizontal transfer Transfer of DNA between individuals in the population rather than by inheritance from ancestors. *See* Vertical transfer.

Host range All of the types of host cells in which a DNA element, plasmid, phage, etc., can multiply.

Hot spot A position in DNA that is particularly prone to mutagenesis by a particular mutagen.

Hsp *See* Heat shock protein.

Hsp10 A heat shock protein of 10 kDa found in bacteria, chloroplasts, and mitochondria. Hsp10 is the cochaperonin to Hsp60 and forms a cap on the cylinder in which denatured proteins are folded. It is represented by GroES in bacteria.

Hsp60 A highly evolutionarily conserved heat shock-induced chaperonin of 60 kDa found in bacteria, chloroplasts, and mitochondria. It is represented by GroEL in bacteria.

Hsp70 A highly evolutionarily conserved heat shock-induced protein chaperone of 70 kDa, represented by DnaK in bacteria.

Hybridization The process by which two complementary strands of DNA or RNA, or a strand of DNA and a strand of RNA, are allowed to base pair with each other and form a double helix.

Hybridization probe A DNA or RNA that can be used to detect other DNAs and RNAs because it shares a complementary sequence with the DNA or RNA being sought and so hybridizes to it by base pairing.

Hypoxanthine A purine base derived from the deamination of adenine.

In situ In place, as when replication of a DNA element occurs in the chromosome without first excising the element from the chromosome.

In vitro mutagen A mutagen that reacts only with purified DNA or with viruses or phage. It cannot be used to mutagenize living cells, either because it cannot get in or because it is too reactive and is destroyed before it reaches the DNA.

In vitro packaging The incorporation of DNA or RNA into virus or phage particles in the test tube so it can be introduced into a cell by infection rather than by transformation.

In vivo mutagen A mutagen that can enter living intact cells and mutagenize the DNA.

Incompatibility The interference of plasmids with one another's replication and/or partitioning.

Incompatibility (Inc) group A set of plasmids that interfere with each other's replication and/or partitioning and so cannot be stably maintained together in the descendants of the same bacterium.

Induced mutations Mutations that are caused by deliberately irradiating cells or treating cells or DNA with a mutagen, such as a chemical.

Inducer A small molecule that can increase the expression of a gene or operon.

Inducer exclusion The process by which the inducer of an operon, such as a sugar, is kept out of the cell by inhibiting its transport through the membrane. Often, a more efficiently utilized sugar, such as glucose, inhibits the transport of other, less efficiently used sugars, such as lactose.

Inducible Able to have expression increased by an inducer.

Induction In gene regulation, the turning on of the expression of the genes of an operon. In phages, the initiation of lytic development of a prophage.

In-frame deletion A deletion mutation in an open reading frame that removes a multiple of 3 bp and so does not cause a frameshift. These deletions are particularly useful because they cannot be polar and can remove a specific domain without removing the rest of the protein.

Initial transcription complex *See* Initiation complex.

Initiation codon The 3-base sequence in an mRNA that specifies the first amino acid to be inserted in the synthesis of a polypeptide chain. In prokaryotes, it is the 3-base sequence (usually AUG or GUG) within a translational initiation region for which formylmethionine is inserted to begin translation. In eukaryotes, the AUG closest to the 5′ end of the mRNA is usually the initiation codon, and methionine is inserted to begin translation.

Initiation complex In transcription, the complex that forms after the first ribonucleoside triphosphate enters into RNA polymerase and pairs with the DNA nucleotide at the position corresponding to the transcription start site.

Initiation factors Proteins (IF1, IF2 and IF3 in bacteria) that assist in initiation of translation.

Initiation mass The size of a bacterial cell at which initiation of a new round of chromosome replication occurs.

Initiation transcription complex The complex formed by the RNA polymerase holoenzyme including the σ factor, the promoter, and the first nucleoside triphosphate.

Initiator tRNA A dedicated tRNA that binds to the initiation codon during translation initiation. *See* $tRNA^{fMet}$.

Injectisome *See* Type III secretion system.

Injectosome A needle-like structure in gram-positive bacteria, analogous to the injectisome in gram-negative bacteria, that injects proteins directly through the bacterial membrane and cell wall into eukaryotic cells.

Inner membrane protein A protein that resides, at least in part, in the cytoplasmic (inner) membrane of gram-negative bacteria.

Insertion element *See* Insertion sequence element.

Insertion mutation A change in a DNA sequence due to the incorporation of another DNA sequence, such as a transposon or antibiotic resistance cassette, into the sequence.

Insertion sequence element A small transposon in bacteria that carries only the gene or genes encoding the products needed to promote transposition.

Insertional inactivation Inactivation of the product of a gene by an insertion mutation, usually denoting inactivation of the product of a gene on a plasmid cloning vector by cloning a fragment of DNA into the gene.

Integrase A type of site-specific recombinase that promotes recombination between two defined sequences in DNA, causing the integration of one DNA into another DNA (e.g., the integration of a phage DNA into the chromosome).

Integrating conjugative element A mobile element that can excise from the genome and transfer between bacteria using element-encoded functions. Depending on the element, integration can occur at a specific site or almost randomly. Previously called a conjugative transposon.

Integron An integrase and an *att* site for integration of gene cassettes, often for antibiotic resistance. A promoter downstream of the integrase is situated to allow transcription of cassette genes inserted into the *att* site. They are typically, but not exclusively, found in transposons and plasmids.

Intein A parasitic DNA that encodes a polypeptide sequence that, when inserted into the gene for another polypeptide, introduces a polypeptide sequence into the other polypeptide that must be removed (spliced out) before the other polypeptide can be active. Inteins are usually self-splicing and are removed at the protein level.

Interaction genetics Characterization of a system by analysis of how genes and gene products affect each other.

Intergenic In different genes.

Intergenic suppressor A suppressor mutation located in a gene different from that containing the mutation it suppresses. Also called extragenic suppressor.

Interpolated Markov model One type of hidden Markov model that is useful for locating genes in a particular organism because it has been trained on known sequences of that organism.

Interstrand cross-links Covalent chemical bonds between the two complementary strands of DNA in a double-stranded DNA.

Intervening sequence A sequence inserted into a polypeptide or polynucleotide that must be removed before the polypeptide or polynucleotide can be functional.

Intragenic In the same gene.

Intragenic complementation Complementation between two mutations in the same gene. It is rare and allele specific; it usually occurs only if the protein product of the gene is a homodimer or homomultimer.

Intragenic suppressor A suppressor mutation that occurs in the same gene as the mutation it is suppressing.

Intron A parasitic DNA that, when inserted into a gene for a protein, introduces polynucleotide sequences into the mRNA that must be removed (spliced out) before the mRNA can be translated into a functional protein. Introns are removed at the RNA level.

Inversion junctions The points where the recombination events that inverted a sequence occurred.

Inversion mutation A change in DNA sequence as a result of flipping a region within a longer DNA so that it lies in reverse orientation. It is usually due to homologous recombination between inverted repeats in the same DNA molecule. Inversions due to site-specific recombinases or invertases are not considered inversion mutations.

Inverted repeat Two nearby sequences in DNA that are the same or almost the same when read in the 5′-to-3′ direction on opposite strands.

Invertible sequence A sequence in DNA bracketed by inverted repeats that depends upon a site-specific recombinase protein for its inversion rather than the generalized homologous recombination system.

IS element *See* Insertion sequence element.

IS*CR* element A type of Y2 transposon.

Isogenic Strains of an organism that are almost identical genetically except for one small region or gene.

Isolation of mutants The process of obtaining a pure culture of a particular type of mutant from among a myriad of other types of mutants and the wild type.

Isomerization Changing of the spatial conformation of a molecule without breaking any bonds. In DNA recombination, it refers to the rotating of the DNAs in a Holliday junction, thus changing the strands that are crossed without breaking any hydrogen or other types of bonds. In transcription, it refers to open-complex formation.

Iteron sequences Short DNA sequences, often repeated many times in the origin regions of some types of plasmids, that bind the Rep protein required for replication of the plasmid and play a role in regulating the copy number.

KEGG map An automated reconstruction of the metabolic pathways of an organism. See http://www.genome.jp/kegg/.

Kinase An enzyme that transfers a phosphate group from ATP to another molecule.

Kleisins Proteins that bind to condensins and help them bind to and condense DNA molecules.

Knockout mutation A mutation that presumably eliminates the function of a gene, i.e., is presumably a null mutation.

Lagging strand During DNA replication, the newly synthesized strand that is made from the template strand that is produced in the opposite direction from the overall movement of the replication fork, i.e., in the 3′-to-5′ direction overall.

Late gene A gene that is expressed only relatively late in the course of a developmental process, e.g., a late gene of a phage.

Lawn *See* Bacterial lawn.

Leader region An RNA sequence close to the 5′ end of an mRNA that may be translated but does not encode a functional polypeptide. Also called leader sequence.

Leading strand During DNA replication, the newly synthesized strand that is made from the template strand that is produced in the same direction as the overall direction of movement of the replication fork, i.e., in the 5′-to-3′ direction.

Leaky mutation A mutation that does not totally inactivate the product of the gene.

Lep protease One of the enzymes that cleaves the signal sequence off secreted proteins as they pass through the SecYEG channel, and probably also the Tat pathway.

Lesion Any change in a DNA or RNA molecule as a result of chemical alteration of a base, sugar, or phosphate.

Ligand A small molecule that binds to a protein or RNA.

Ligase An enzyme that joins two molecules.

Linkage The situation occurring when two genetic markers are sufficiently close together on the DNA that recombination between them is less than random.

Linked A genetic term referring to the fact that two markers are close enough on the DNA that they are separated by recombination less often than if they sorted randomly.

Locus A region in the genome of an organism. In classical genetics, a region in the genome to which mutations with a particular phenotype map.

Loss-of-function mutation A mutation that completely destroys the function of the product of a gene, for example, a deletion that removes the entire coding region of the gene. *See* Null mutation.

Low multiplicity of infection The state of a virus or phage infection in which the number of cells almost equals or exceeds the number of viruses, so that most cells remain uninfected or are infected by at most one or very few viruses.

Lyse To break open cells and release their cytoplasm into the medium.

Lysogen A strain of bacterium that harbors a prophage.

Lysogenic conversion A property of a bacterial cell caused by the presence of a particular prophage.

Lysogenic cycle The series of events following infection by a bacteriophage and culminating in the formation of a stable prophage.

Lysogenic phage A phage that is known to be capable of entering a prophage state in some host.

Lytic cycle The series of events following infection by a bacteriophage or induction of a prophage and culminating in lysis of the bacterium and the release of new phage into the medium.

Macromolecule A large molecule such as DNA, RNA, or a protein.

Major groove In double-stranded DNA, the larger of the two tracks between the two strands of DNA as they twist around each other as a helix.

MALDI *See* Matrix-assisted laser desorption ionization.

Male strain A bacterium or strain harboring a self-transmissible plasmid or other conjugative element capable of producing a sex pilus.

Male-specific phage A phage that infects only cells carrying a particular self-transmissible plasmid. The plasmid produces the sex pilus used by the phage as its adsorption site.

Maltodextrins Short chains of glucose molecules held together by α1-4 (maltose) linkages. They are breakdown products of starch.

Maltotriose A chain of three glucose molecules held together by α1-4 linkages.

Map distance The distance between two markers in the DNA as measured by recombination frequencies.

Map expansion A phenomenon that occurs in genetic linkage experiments in which two markers appear to be farther apart than they are because of hyperactive apparent recombination. This is often due to hot spots for recombination or to preferential mismatch repair of some mismatches.

Map unit A distance between genetic markers corresponding to a recombination frequency of 1% between the markers.

Marker effect A difference in the apparent genetic linkage between the site of a mutation and other markers depending on the type of mutation at the site. It is due to preferential mismatch repair of some mismatches relative to others.

Marker rescue Acquisition of a genetic marker by the genome of an organism or virus through recombination with a cloned DNA fragment containing the marker.

Markov model A statistical tool that can be applied to a system that is represented by discrete states. For example, when used for protein annotation, a discrete state could be 1 of the 23 amino acids at a position in the protein.

Mass spectrometry (MS) An analytical method that measures ion abundances based on their mass-to-charge (*m/z*) ratios. First, gas phase ions are produced from the compound of interest (*see* MALDI and Electrospray ionization). Then, the ions are separated on the basis of their *m/z* ratios. Finally, the ions at different *m/z* ratios are detected and counted.

Matrix-assisted laser desorption ionization (MALDI) A procedure that is used to measure peptide mass. A singly charged ion is produced from a peptide, resulting in one peak on a mass spectrometric analysis. The technique can be used to directly analyze complex peptide mixtures if a complete genome sequence is available.

Membrane protein A protein that at least partially resides in, or is tightly bound to, one of the cellular membranes.

Membrane topology Distribution of the various regions of a membrane protein between the membrane and the two surfaces of the membrane. In the inner membrane of gram-negative bacteria, topology refers to which domains are in the cytoplasm, which are in the periplasm, and which are buried in and traverse the membrane from one side to the other.

Merodiploid A bacterial cell that is mostly haploid but is diploid for some region of the genome due to some chromosomal genes being carried on a prophage or plasmid. *See* Partial diploid.

Messenger RNA (mRNA) An RNA transcript that includes the coding sequences for at least one polypeptide.

Methionine aminopeptidase An enzyme that removes the N-terminal methionine from newly synthesized polypeptides.

Methyl-directed mismatch repair system The repair system that recognizes mismatches in newly replicated DNA and specifically removes and resynthesizes the new strand. In some enteric bacteria, the new strand is distinguishable from the old strand because it is the strand that is not methylated in the nearest hemimethylated GATC sequences.

Methyltransferase In DNA repair, an enzyme that removes a CH_3 (methyl) or CH_3CH_2 (ethyl) group from a base in DNA by catalyzing transfer of the group from the DNA to the protein itself. In DNA modification, any one of a number of enzymes that catalyze the transfer of a methyl group to a specific base on a DNA sequence recognized by the enzyme.

Microarray A high-density array of spots of DNA probes that represent the genome of an organism. The spots may be attached to a glass slide or other solid medium, such as a nylon membrane. In some methods, the DNA probes are synthesized directly on the solid support. A microarray might be used for several different types of experiments. A common application is to quantitate the mRNA transcripts present in a cell under a given culture condition. Other applications are possible, including DNA-DNA hybridization to compare genome sequences and determination of the gene content of a tumor cell line.

Migration *See* Branch migration.

Minor groove In double-stranded DNA, the smaller of the two tracks between the two strands of DNA as they twist around each other as a helix.

Minus strand In a virus with a single-stranded nucleic acid genome (DNA or RNA), the strand that is complementary to the strand in the virus head.

–10 sequence In a bacterial σ^{70}-type promoter, a short sequence that is centered 10 bp upstream of the transcription start site. The canonical or consensus sequence is TATAAT/ATATTA.

–35 sequence In a bacterial σ^{70}-type promoter, a short sequence that is centered 35 bp upstream of the transcription start site. The canonical or consensus sequence is TTGACA/AACTGT.

Mismatch Improper pairing of the normal bases in DNA, e.g., an A opposite a C.

Mismatch repair system A pathway for removing mismatches in DNA by degrading a strand containing the mismatched base and replacing it by synthesizing a new strand containing the correctly paired base.

Missense mutation A base pair change mutation in a region of DNA encoding a polypeptide that changes an amino acid in the polypeptide.

***mob* genes** The genes on a mobilizable DNA element that allow it to be mobilized by a self-transmissible element, such as a self-transmissible plasmid. They often encode Dtr (DNA transfer) functions and a coupling protein that allows the element to communicate with the mating-pair formation (Mpf) system of the self-transmissible element.

***mob* region** A region in DNA carrying an origin of transfer (*oriT* sequence) and often genes whose products allow the plasmid or other DNA element to be mobilized by self-transmissible elements.

Mobilization The process by which a mobilizable DNA element, incapable of self-transmission, is transferred into other cells by the conjugation functions of a self-transmissible element.

MOI *See* Multiplicity of infection.

Molecular genetic analysis Any study of cellular or organismal functions that involves manipulations of DNA in the test tube.

Molecular genetic techniques Methods for manipulating DNA in the test tube and reintroducing the DNA into cells.

Monocistronic mRNA An mRNA that encodes a single polypeptide.

Moron A phage gene that has apparently moved into the phage DNA fairly recently from an unknown source and that has its own promoter.

Mother cell A cell that divides or differentiates to give rise to a new cell or spore.

Motif A conserved nucleotide or amino acid sequence that is relatively short and suggests similarity of function.

Mpf component A component made up of *tra* gene products of a self-transmissible plasmid involved in making the surface structures (pilus, etc.) that contact another cell and transfer the DNA during conjugation, as well as the coupling protein that communicates with the Dtr component. The term Mpf is derived from "mating pair formation."

mRNA *See* Messenger RNA.

MS *See* Mass spectrometry.

Multicopy suppression A process that relieves the effects of a mutation in a different gene when it is expressed in higher than normal amounts from a multicopy plasmid or other vector. It is often a complication of cloning by complementation.

Multigene family A set of related genes, such as paralogs, that perform similar or redundant functions.

Multimer A protein or other molecule that consists of more than one polypeptide chain or other subunit (usually more than two).

Multiple cloning site A region of a cloning vector that contains the sequences cut by many different type II restriction endonucleases. It is also called a polyclonal site.

Multiplicity of infection (MOI) The ratio of phages or viruses to cells that initiates an infection.

Murein layer The wall surrounding the cell. Named after the muramic acid component of the peptidylglycan in the wall; also called the sacculus.

Mutagen A chemical or type of irradiation that causes mutations by damaging DNA.

Mutagenic repair A pathway for repairing damage to DNA that sometimes changes the sequence of deoxynucleotides as a consequence.

Mutagenic treatments or chemicals Treatments or chemicals that cause mutations by damaging DNA.

Mutant An organism that differs from the normal, or wild type, as a result of a change in the sequence (mutation) of its DNA.

Mutant allele The mutated gene of a mutant organism that makes it different from the wild-type gene.

Mutant phenotype A characteristic that makes a mutant organism different from the wild type.

Mutation Any heritable change in the sequence of deoxynucleotides in DNA.

Mutation rate The probability of occurrence of a mutation causing a particular phenotype each time a newborn cell grows and divides.

N terminus The end of a polypeptide chain with the N-terminal amino acid.

Narrow host range A range of hosts (that a DNA element can enter and/or replicate in) that includes only a few closely related types of cells. The term typically refers to *E. coli* and its close relatives.

Naturally competent Able to take up DNA at a certain stage in the bacterial growth cycle without chemical or other treatments. A characteristic of some types of bacteria. Also called naturally transformable.

Naturally transformable *See* Naturally competent.

Negative regulation A type of regulation in which a protein or RNA molecule, in its active form, inhibits a process, such as the transcription of an operon or translation of an mRNA.

Negative selection A somewhat outdated way of referring to the process of detecting a mutant on the basis of its inability to multiply under a certain set of conditions in which the normal, or wild-type, organism can multiply. It is more properly called a negative screen.

Negatively supercoiled A state of a DNA molecule in which the two strands of the double helix are wrapped around each other less than about once every 10.5 bp.

Nested deletion A set of deletion mutations with one common end point and with the other end point extending varying distances into a gene or region of DNA.

Nicks Broken phosphate-deoxyribose bonds in the phosphodiester backbone of double-stranded DNA.

Noncoding strand The strand of DNA that is used as a template for RNA synthesis. *See* Template strand, Transcribed strand.

Noncomposite transposon A transposon in which the transposase genes and the inverted-repeat ends are included in the minimum transposable element and are not part of autonomous IS elements. *See* Composite transposon.

Noncovalent change Any change in a molecule that does not involve the making or breaking of a chemical covalent bond due to shared electron orbits in the molecule.

Nonhomologous recombination The breaking and rejoining of two DNAs into new combinations, which does not necessarily depend on the two DNAs having similar sequences in the region of recombination.

Nonpermissive conditions Conditions under which a mutant organism or virus cannot multiply but the wild type can multiply.

Nonpermissive host A host organism in which a mutant phage or virus cannot multiply but the wild type can multiply.

Nonpermissive temperature A temperature at which the wild-type organism or virus, but not the mutant organism or virus, can multiply.

Nonselective conditions Conditions or media in which both the mutant and wild-type strains of an organism or virus can multiply.

Nonsense codon A codon that does not stipulate an amino acid but, rather, triggers the termination of translation. In most organisms, the codons UAG, UGA, and UAA are nonsense codons.

Nonsense mutation In a region of DNA encoding a protein, a base pair change mutation that causes one of the nonsense codons to be encountered in frame when the mRNA is translated.

Nonsense suppressor A suppressor mutation that allows an amino acid to be inserted at some frequency for one or more of the nonsense codons during the translation of mRNAs.

Nonsense suppressor tRNA A tRNA that, usually as a result of a mutation, allows the tRNA to pair with one or more of the nonsense codons in mRNA during translation and therefore causes an amino acid to be inserted for the nonsense codon. The mutation usually changes the anticodon on the tRNA.

Nontemplate strand The strand of DNA that is complementary to the template strand and that is not used as a template for RNA synthesis; it has the same base sequence as the RNA product. *See* Coding strand.

Northern blotting Transfer of RNA from a gel to a filter for hybridization experiments using a sequence-specific probe.

N-terminal amino acid The amino acid on the end of a polypeptide chain whose amino (NH_2) group is not attached to another amino acid in the chain through a peptide bond.

Ntr system A global regulatory system that regulates a number of operons in response to the nitrogen sources available.

Nuclease An enzyme that cuts the phosphodiester bonds in DNA or RNA polymers.

Nucleoid A compact, highly folded structure formed by the chromosomal DNA in the bacterial cell and in which the DNA appears as a number of independent supercoiled loops held together by a core.

Nucleoid core The center of the nucleoid, of unknown composition.

Nucleoid occlusion A process that prevents the formation of the division septum in a region of the cell still occupied by the nucleoid.

Nucleotide excision repair A system for the repair of DNA damage in which the entire damaged nucleotide is removed rather than just the damaged base. A cut is made on either side of the damage on the same strand, and the damaged strand is removed and resynthesized.

Null mutation A mutation in a gene that totally abolishes the function of the gene product.

Ochre The nonsense codon UAA. An ochre mutation is a mutation to this codon in the reading frame for a protein. An ochre suppressor suppresses ochre mutations.

Okazaki fragments The short pieces of DNA that are synthesized in the opposite direction of movement of the replication fork during replication from a lagging-strand template.

Oligopeptide A short polypeptide only a few amino acids long.

Opal The nonsense codon UGA. Mutations to this codon in the reading frame of a protein are called opal mutations. Opal suppressors suppress opal mutations. Sometimes called umber.

Open complex The complex of RNA polymerase and DNA at a promoter in which the strands of the DNA have been separated after isomerization.

Open reading frame (ORF) A sequence of DNA, read 3 nucleotides at a time, that is not interrupted by any nonsense codons.

Operator Usually a sequence on DNA to which a repressor protein binds to block transcription. More generally, any sequence in DNA or RNA to which a negative regulator binds.

Operon A DNA region encompassing genes that are transcribed into the same mRNA, as well as any adjacent *cis*-acting regulatory sequences.

Operon model The model proposed by Jacob and Monod for the regulation of the *lac* operon, in which transcription of the structural genes of the operon is prevented by binding of the LacI repressor to the operator region, thereby preventing access of RNA polymerase to the promoter. The inducer lactose binds to LacI and changes its conformation so that it can no longer bind to the operator, and as a result, the structural genes are transcribed.

ORF *See* Open reading frame.

oriC A sequence of DNA consisting of the site in the bacterial chromosome at which initiation of a round of replication normally occurs and all of the surrounding *cis*-acting sequences required for initiation.

Origin of replication The site on a DNA, plasmid, phage, chromosome, etc., at which replication initiates, including all of the surrounding *cis*-acting sequences required for initiation.

Orthologs Genes in different species that are derived from a common ancestor. They may differ in function, but they usually have identical functions.

Outer membrane protein A protein that resides, at least in part, in the outer membrane of gram-negative bacteria.

***pac* site** The sequence in phage DNA at which packaging of the phage DNA into phage particles begins.

P (peptidyl) site The site on the ribosome to which the peptidyl tRNA, which contains the growing peptide chain, is bound.

Packaging site *See pac* site.

PAI *See* Pathogenicity island.

Palindrome A DNA sequence that reads the same in the 5′-to-3′ direction on the top strand and bottom strand.

Papilla A section of a bacterial colony with an appearance different from that of most of the colony.

Par function A site or gene product that is required for the proper partitioning of a plasmid.

Paralogs Genes that have resulted from duplication of a gene. They generally have similar functions but may have distinct functions.

Parent One of the two strains of an organism participating in a genetic cross.

Parental types Progeny of a genetic cross that are genetically identical to one or the other of the parents.

Partial diploid A bacterium that has two copies of part of its genome, usually because a plasmid or prophage in the bacterium contains some bacterial DNA. Also called a merodiploid.

Partitioning An active process by which at least one copy of a replicon (plasmid, chromosome, etc.) is distributed into each daughter cell at the time of cell division.

Pathogenicity island (PAI) A DNA element integrated into the chromosome of a pathogenic bacterium that carries genes whose products are required for pathogenicity and that, based on its base composition and codon usage, shows evidence of having been acquired fairly recently by horizontal transfer. It may or may not carry genes for its own integration. Pathogenicity islands form a subset of a more general class of integrated elements called genetic islands.

PBP *See* Penicillin-binding protein.

PCR *See* Polymerase chain reaction.

Penicillin-binding protein (PBP) One of the many enzymes in a bacterium that catalyze the polymerization of the subunits of peptidylglycan. It contains both glycosyltransferase and transpeptidase activities. β-Lactam antibiotics, including penicillin, act by becoming covalently attached to the transpeptidase active center, hence the name.

Peptide bond A covalent bond between the amino (NH_2) group of one amino acid and the carboxyl (COOH) group of another.

Peptide deformylase An enzyme that removes the formyl group from the amino-terminal formylmethionine of newly synthesized polypeptides.

Peptidyltransferase The ribozyme of the 23S rRNA (28S rRNA in eukaryotes) that forms a bond between the carboxyl group of the growing polypeptide and the amino group of the incoming amino acid.

Periplasm The space between the inner and outer membranes in gram-negative bacteria.

Periplasmic domain A region of a membrane protein located in the periplasm of the cell.

Periplasmic protein A protein located in the periplasm.

Permissive conditions Conditions under which a mutant organism or virus can multiply.

Permissive host A strain of an organism that can support the multiplication of a particular mutant virus.

Permissive temperature A temperature at which both a temperature-sensitive mutant (or cold-sensitive mutant) and the wild type can multiply.

Pfams Protein family and domain databases that are useful for categorizing predicted genes or proteins, primarily based on compilation of protein domains.

Phage Short for bacteriophage.

Phage display A technology that allows the purification and amplification of a phage particle expressing a particular polypeptide on its surface based on the ability of the polypeptide to bind to another protein or chemical compound. The DNA of the phage can then be sequenced to determine the sequence of a polypeptide that binds to the other protein or compound.

Phage genome The nucleic acid (DNA or RNA) that is packaged into the phage particle and contains all the genes of the phage.

Phagemid A cloning vector that contains mostly phage sequences and that can replicate as a phage and be packaged in a phage head but can also replicate as a plasmid.

Phase variation The reversible change of one or more cellular phenotypes, for example, of the cell surface antigens of a bacterium, at a frequency higher than normal mutation frequencies. It can be due to an invertible sequence, etc.

Phasmid A hybrid cloning vector containing mostly plasmid but some phage sequences, including a *pac* site, so it can be packaged in a phage head but does not contain all of the genes to make a phage particle and can usually only replicate as a plasmid.

Phenotype Any identifiable characteristic of a cell or organism that can be altered by mutation.

Phenotypic lag The delay between the time a mutation occurs in the DNA and the time the resulting change in the phenotype of the organism becomes apparent.

Phosphate The chemical group PO_4.

Phosphorylation cascade An interacting set of proteins with kinase, phosphotransferase, and phosphatase activities that transfer signals in the cell by the successive phosphorylation and dephosphorylation of the proteins.

Photolyase An enzyme that uses the energy of visible light to split pyrimidine butane dimers in DNA, restoring the original pyrimidines.

Photoreactivation The process by which cells exposed to visible light after DNA damage achieve greater survival rates than cells kept in the dark. It is due to restoration of pyrimidine dimers to the individual pyrimidines by photolyase.

Physical map A map of DNA showing the actual distance in deoxynucleotides between identifiable sites, such as restriction endonuclease cleavage sites.

Physical mapping The process of constructing a physical map of a DNA.

Pilin A protein that makes up the structure of pili. *See* Pilus.

Pilus A protrusion or filament composed of protein attached to the surface of a bacterial cell. *See* Sex pilus, Fimbriae.

Plaque purification Isolation of a pure strain of a phage by diluting and plating to obtain individual plaques, each of which contains descendants of only a single phage.

Plaques Clear spots in a bacterial lawn as a result of phage killing and lysing the bacteria as the bacterial lawn is forming and the phage is multiplying.

Plasmid Any DNA molecule in cells that replicates independently of the chromosome and regulates its own replication so that the number of copies of the DNA molecule remains relatively constant.

Pleiotropic mutation A mutation that causes many phenotypic changes in the cell.

Plus strand In a virus with a single-stranded genome (DNA or RNA), the strand packaged in the phage particle.

Point mutation A mutation that maps to a single position in the DNA, usually the change in or addition or deletion of a single base pair.

Poisson distribution A mathematical distribution that can be used to calculate probabilities in certain situations. It can be used to approximate a binomial distribution when the probability of success in a single trial is low but the number of trials is large. It is named after the mathematician who first derived it.

Polarity A condition in which a mutation in one gene reduces the expression of a downstream gene that is cotranscribed into the same mRNA. It can be due either to premature termination of transcription before RNA polymerase reaches the second gene or to dependence of the second gene on translational coupling.

Polyadenylation Addition of multiple adenosine residues to the 3′ end of an RNA to generate a poly(A) tail.

Polycistronic mRNA An mRNA that contains more than one translational initiation region so that more than one polypeptide can be translated from the mRNA.

Polyclonal site *See* Multiple cloning site.

Polymerase chain reaction (PCR) A technique involving a succession of heating and cooling steps that uses the DNA polymerase from a thermophilic bacterium and two primers to make many copies of a given region of DNA occurring between sequences complementary to the primers.

Polymerase switching The process by which a different type of DNA polymerase replaces the DNA polymerase already found at the primed DNA template.

Polymerization A reaction in which small molecules are joined in a chain to make a longer molecule.

Polymerizing The act of joining small molecules to form a chain.

Polymorphism A difference in DNA sequence between otherwise closely related strains.

Polypeptide A long chain of amino acids held together by peptide bonds. Polypeptides are the product of a single gene.

Porin A protein that forms channels in the outer membrane of gram-negative bacteria by forming a β-barrel in the outer membrane.

Positive regulation A type of regulation in which the gene is expressed only if the active form of a regulatory protein (or RNA) is present.

Positive selection The process of determining conditions under which only a strain with the desired mutation or a particular recombinant type can multiply. Usually just called selection.

Positively supercoiled A DNA molecule in which the two strands of the double helix are wrapped around each other more than about once every 10.5 bp, as predicted by the Watson-Crick structure.

Posttranscriptional regulation Regulation of the expression of a gene that occurs after the mRNA has been synthesized from the gene, for example, in the rate of translation of the mRNA.

Posttranslational regulation Regulation of the expression of a gene that occurs after the protein has been synthesized, e.g., by regulated proteolysis or feedback inhibition.

Posttranslational translocation Tat-, SecB-, and SecA-mediated translocation of proteins through the cytoplasmic membrane after they have been translated. This form of translocation is limited to proteins destined for the periplasm, the outer membrane, or the outside of the cell.

ppGpp *See* Guanosine tetraphosphate, Stringent response.

Precise excision Removal of a transposon or other foreign DNA element from a DNA in such a way that the original DNA sequence is restored.

Precursors The smaller molecules that are polymerized to form a polymer.

Presecretory protein A secreted protein after it has been translated and while its signal sequence is still attached.

Presynthetic Occurring before actual DNA or RNA synthesis.

Primary structure The sequence of nucleotides in an RNA or of amino acids in a polypeptide.

Primase An enzyme that synthesizes short RNAs to prime the synthesis of DNA chains.

Prime factor A self-transmissible plasmid carrying a region of the bacterial chromosome.

Primer A single-stranded DNA or RNA polymer that can hybridize to a single-stranded template DNA or RNA and provide a free 3′ hydroxyl end to which DNA polymerase can add deoxynucleotides to synthesize a chain of DNA complementary to the template DNA or RNA.

Primosome A complex of proteins involved in making primers for the initiation of synthesis of DNA strands.

Probe A short oligonucleotide (DNA or RNA) that is complementary to a sequence being sought and so hybridizes to the sequence and allows it to be identified from among many other sequences.

Processive antitermination The process by which RNA polymerase is altered by binding either a protein or a newly transcribed RNA so that it becomes insensitive to all downstream transcription termination signals.

Processivity The efficiency with which a biosynthetic complex (e.g., RNA polymerase) moves along its template to carry out its activity without stopping or falling off.

Programmed frameshifting A system in which gene expression (or regulation) depends on a specific shift in the reading frame by the translating ribosome.

Prokaryote A somewhat outdated name for bacteria and archaea whose cells do not contain a nuclear membrane and visible nucleus or many of the other organelles characteristic of the cells of higher organisms. The name erroneously suggests that present-day eukaryotes descended from present-day bacteria and archaea.

Prolyl isomerase An enzyme, often associated with chaperones, that can catalyze the conversion of one isomer of proline to the other isomer. Proline is the only amino acid that has more than one isomer because the carbon in the carboxyl group is not free to rotate.

Promiscuous plasmid A self-transmissible plasmid that can transfer itself into many types of bacteria, which need not be closely related to each other.

Promoter A region on DNA to which RNA polymerase holoenzyme binds in order to initiate transcription.

Promoter escape The step during transcription when RNA polymerase releases the promoter site and enters the elongation phase.

Promoter recognition The first step in transcription, when RNA polymerase holoenzyme binds to the promoter sequence on the DNA.

Prophage The state of phage DNA in a lysogen in which the phage DNA is integrated into the chromosome of the bacterium or replicates as a plasmid.

Protease An enzyme that cleaves proteins.

Protein export The transport of proteins through the cellular membranes to the outside of the cell.

Protein secretion system A cellular structure composed of a number of different proteins that allows the transfer of certain selected proteins through one or both membranes to the outside of the cell.

Protein transport The movement of proteins out of the cytoplasm into or through the membranes.

Proteobacteria One of the largest phylogenetic groups of bacteria, which includes *E. coli*. This phylum within the division *Bacteria* is divided into six classes, *Alpha-*, *Beta-*, *Gamma-*, *Delta-*, *Epsilon-*, and *Zetaproteobacteria*.

Proteome The complete set of proteins expressed in an organism under a particular growth condition.

Proteomics Global analysis of protein expression patterns and protein interactions. It includes techniques such as mass spectrometry, phage display, and two-hybrid analysis.

Prototroph A strain that can make all of the growth substances made by the original isolate. *See* Auxotrophic mutant.

Pseudoknot An RNA tertiary structure with interlocking loops held together by regions of hydrogen bonding between the bases.

Pseudopilus A structure that resembles a pilus. *See* Pilus.

PSI-BLAST A program that can find a set of related sequences based on the presence of common sequence patterns. Reiterative sequence alignments are performed with the goal of defining as large a potential family of functionally related proteins as possible.

Purine One of the bases in DNA and RNA with two ring structures.

Pyrimidine One of the bases in DNA and RNA with only one ring.

Pyrimidine dimer A type of DNA damage in which two adjacent pyrimidine bases are joined by covalent chemical bonds, requiring repair.

Quaternary structure The complete three-dimensional structure of a protein complex, including all the polypeptide chains making up the complex and how they are wrapped around each other.

Quorum sensing A phenomenon in which populations of cells sense cell density by measuring the accumulation of a usually small signal molecule.

Random gene fusion A technique in which transposon mutagenesis is used to fuse reporter genes to different regions in the chromosome. A transposon containing a reporter gene inserts randomly into the chromosome, resulting in various transposon insertion mutants that have the reporter gene on the transposon fused either transcriptionally or translationally to different genes or to different regions within each gene.

Random-mutation hypothesis The generally accepted hypothesis explaining the adaptation of organisms to their environment. It states that mutations occur randomly, free of influence from their consequences, but that mutant organisms preferentially survive and reproduce themselves if the mutations inadvertently confer advantages under the conditions experienced by the organisms.

RBSfinder A program that uses an algorithm to find translational initiation regions in both eubacterial and archaeal genomes. It is usually used after gene finders, such as Glimmer. Once a gene is found, RBSfinder looks for a probable sequence to which ribosomes bind, hence the name.

RC plasmid *See* Rolling-circle plasmid.

RC replication *See* Rolling-circle replication.

Reading frame of translation Any sequence of nucleotides in RNA or DNA read three at a time in succession, as during translation of an mRNA.

Rec$^-$ mutant A mutant strain in which DNA shows a reduced capacity for recombination due to a mutation in a *rec* gene whose product is involved in recombination.

Recessive mutation In complementation tests, a mutation that does not exhibit its phenotype in the presence of a wild-type allele of the gene.

Recipient allele The sequence of a gene or allele as it occurs in the recipient bacterium.

Recipient bacterium The bacterial strain that receives DNA in a genetic cross, whether transformation, transduction, or conjugation.

Recipient DNA In transposition, the DNA in which the transposon inserts.

Reciprocal cross A genetic cross in which the alleles of the donor and recipient strains are reversed relative to an earlier cross. An example would be a transduction in which the phage was grown on the strain that had the alleles of what was previously the recipient strain and used to transduce a strain with the alleles of what was previously the donor strain. In bacterial crosses, generally what was previously the selected marker becomes an unselected marker.

Recombinant DNA A DNA molecule derived from the sequences of different DNAs joined to each other in a test tube.

Recombinant types In a genetic cross, progeny that are genetically unlike either parent in the cross because they have DNA sequences that are the result of recombination between the parental DNAs.

Recombinase An enzyme that specifically recognizes two sequences in DNA and breaks and rejoins the strands to cause a crossover within the sequences.

Recombination The rejoining of DNA into new combinations.

Recombination-deficient mutant *See* Rec$^-$ mutant.

Recombination frequency In a genetic cross, the number of progeny that are recombinant types for the two parental markers divided by the total number of progeny of the cross.

Recombination repair A DNA damage tolerance mechanism that requires the recombination functions, which function to restart replication forks stalled at the damage. In one scenario the lagging strand may replicate past the damage to leave a gap and the undamaged strand may be used to fill the gap using the RecF-pathway recombination functions. In another scenario, the damage may leave a double-stranded end that can then invade the other daughter DNA, using the RecBCD functions. Alternatively, the replication fork may back up to form a Holliday junction, which can then migrate past the damage. In all these pathways, the Pri proteins help the replication proteins reload on the DNA to restart the fork past the damage.

Recombineering A technique that uses phage recombination functions to promote recombination between introduced DNA and cellular DNA. It can be used for site-specific mutagenesis, etc., without the need for cloning of the cellular DNA.

Redundancy A feature of the genetic code that results in multiple codons that encode the same amino acid.

Regional mutagenesis Any technique of mutagenesis in which mutations are restricted to a small region of the DNA or genome.

Regulated proteolysis A regulatory mechanism in which degradation of a protein occurs at different rates under different conditions.

Regulation of gene expression Modulation of the rate of synthesis of the active product of a gene so that the active gene product can be synthesized at different rates, depending on the state in which the organism finds itself.

Regulatory cascade A strategy for regulating the expression of genes in a step-wise manner (e.g., during developmental processes) in which the products of genes expressed during one stage of development turn on the expression of genes for the next stage of development and often turn off genes from the previous stage.

Regulatory gene A gene whose product regulates the expression of other genes, as well as, sometimes, its own expression.

Regulon The set of operons that are regulated by the product of the same regulatory gene.

Relaxase The protein of a self-transmissible or mobilizable plasmid that makes a cut at the *oriT nic* site, remains attached to the 5′ end at the cut, is secreted into the recipient cell, and rejoins the cut ends in the recipient cell.

Relaxed DNA A DNA that contains no supercoils.

Relaxed plasmid A plasmid that has a high copy number because the copy number is not tightly controlled.

Relaxed strain A bacterial strain that continues to make rRNA and other stable RNAs even if starved for an amino acid. These strains usually have a mutation in the *relA* gene that inactivates the RelA enzyme, so they do not synthesize ppGpp in response to amino acid starvation.

Relaxosome The complex of proteins, including the relaxase, which is bound to the *oriT* sequence of a self-transmissible or mobilizable plasmid in the donor cell.

Release factors Nonsense codon-specific proteins that are required, along with EF-G and RRF, for the termination of polypeptide synthesis and the release of the newly synthesized polypeptide from the ribosome when the ribosome encounters an in-frame nonsense codon in the mRNA.

Replica plating A technique in which bacteria grown on one plate are transferred to a fuzzy cloth and then are transferred from the fuzzy cloth onto another plate so that the bacteria on the first plate are transferred to the corresponding position(s) on the second plate.

Replication bypass The process by which the replication fork moves past damaged DNA that interferes with proper base pairing either by skipping over the damage or by inserting random nucleotides opposite the damage.

Replication fork The region in a replicating double-stranded DNA molecule where the two strands are actively separating to allow synthesis of the complementary strands.

Replication restart The process of restarting replication after the replication complex has dissociated and reassembled at a site other than *oriC*.

Replicative form The double-stranded DNA or RNA that forms by synthesis of the complementary minus strand after infection by a phage or virus that has a single-stranded genome.

Replicative transposition A type of transposition in which the breaks that form at each end of the transposon are directly joined to the target DNA and replication is integral to completing transposition. The free 3′ ends at the extremities of the transposon are used as primers to synthesize over the transposon, giving rise to a cointegrate that can be resolved, leaving a copy of the element at the original position.

Replicon A DNA molecule capable of autonomous replication because it contains an origin of replication that functions in the cell in which it is located.

Reporter gene A gene whose product is stable and easy to assay and so is convenient for detecting and quantifying the expression of genes to which it is fused. *See* Transcriptional fusion, Translational fusion.

Repressible Able to have expression be reduced by a corepressor.

Repressor A protein or RNA that negatively regulates transcription or translation so that synthesis of the gene product is reduced when it is active.

Resolution of Holliday junctions Cutting of the two crossed strands of DNA in a Holliday junction so that the DNA molecules, held together by the crossed strands in the Holliday junction, are separated.

Resolvase A type of site-specific recombinase that breaks and rejoins DNA in *res* sequences in the two copies of the transposon in a cointegrate, thereby resolving the cointegrate into separate DNAs, each with one copy of the transposon.

Resolve In recombination, to convert a Holliday junction back to two independent duplex DNAs.

Response regulator protein A protein that is part of a two-component regulatory system. The response regulator receives a signal (usually in the form of a phosphoryl group) from another protein (the sensor kinase) and performs a regulatory function (e.g., activates transcription).

Restricted transduction A type of transduction in which only some selected regions of the chromosome can be transduced from one strain to another. It is usually due to a phage that integrates into only one site in the bacterial chromosome and sometimes mistakenly picks up neighboring chromosomal sequences when it excises.

Restriction fragment A piece of DNA obtained by cutting a longer DNA with a restriction endonuclease.

Restriction fragment length polymorphism A difference in the sizes of restriction fragments obtained by cutting DNA from the same region in two different strains with the same restriction endonuclease. The polymorphism reflects differences in the DNA sequences between the sites.

Restriction-modification system A complex of proteins whose members can recognize specific sequences in DNA, methylate a base in the sequence, and cut in or near the sequence if it is not methylated.

Retrohoming The process by which an element inserts itself into the same site in a different DNA that lacks it by first making an RNA copy of itself and then making a DNA copy of this RNA with a reverse transcriptase, while it inserts the DNA copy into the target DNA by a sort of reverse splicing.

Retrotransposon A class of transposable element that produces a copy of itself by transcription where the RNA copy is later converted into DNA for integration into a new site.

Reverse genetics The process in which the function of the product of a gene is determined by first altering the sequence in DNA in the test tube, using molecular biology techniques, and then reintroducing the DNA into a cell to see what effect the mutation has on the organism. Contrast this with forward genetics, in which the mutation is first recognized because of the phenotype it causes.

Reversion Restoration of a mutated sequence in DNA to the wild-type sequence.

Reversion rate The probability that a mutated sequence in DNA will change back to the wild-type sequence each time the organism multiplies.

Revert *See* Reversion.

Revertant An organism in which the mutated sequence in its DNA has been restored to the wild-type sequence.

RF *See* Replicative form.

Ribonuclease (RNase) An enzyme that cleaves RNA.

Ribonucleoside triphosphate (rNTP) A base (A, U, G, or C) attached to a ribose sugar with three phosphate groups attached in tandem to the 5′ carbon of the sugar.

Ribonucleotide reductase An enzyme that catalyzes the reduction of nucleoside diphosphates to deoxynucleoside diphosphates by removing the hydroxy group at the 2′ carbon of the ribonucleoside diphosphate and replacing it with a hydrogen.

Riboprobe A hybridization probe made of RNA rather than DNA.

Ribosomal proteins The proteins that, in addition to the rRNAs, make up the structure of the ribosome.

Ribosomal RNA (rRNA) Any one of the three RNAs (16S, 23S, and 5S in bacteria) that make up the structure of the ribosome.

Ribosome The cellular structure, made up of about 50 different proteins and three different RNAs, that is the site of protein synthesis.

Ribosome cycle The association and dissociation of the 30S and 50S ribosomes during initiation and termination of translation.

Ribosome release factor (RRF) A factor that acts with release factors and EF-G to release the polypeptide from the peptidyl-tRNA during termination of translation.

Ribosome-binding site (RBS) *See* Translational initiation region (TIR).

Riboswitch A regulatory RNA element that directly senses a regulatory signal, often a small molecule, and the resulting change in RNA structure modulates the function of the RNA, for example, expression of the downstream coding sequences.

Ribozyme An RNA that has enzymatic activity.

R-loop A three-stranded structure formed by the invasion of a double-stranded DNA by an RNA, displacing one of the strands of the double-stranded DNA.

RNA modification Any covalent change to RNA, such as methylation of a base, that does not involve the breaking and joining of phosphate-phosphate or phosphate-ribose bonds in the backbone of the RNA.

RNA polymerase An enzyme that polymerizes ribonucleoside triphosphates to make RNA chains by using a DNA or RNA template.

RNA polymerase core enzyme The $\alpha_2\beta\beta'\omega$ complex of RNA polymerase without a σ factor attached.

RNA polymerase holoenzyme The $\alpha_2\beta\beta'\omega$ complex of RNA polymerase with a σ factor attached.

RNA processing Covalent changes to RNA that involve the breaking and/or joining of phosphate-phosphate or phosphate-ribose bonds in the backbone of the RNA.

RNase *See* Ribonuclease.

rNTP *See* Ribonucleoside triphosphate.

Robust regulation Overlapping regulatory circuits to ensure that regulation is not too sensitive to any change in conditions.

Rolling-circle (RC) plasmid A plasmid that replicates by a rolling-circle mechanism.

Rolling-circle (RC) replication A type of replication of circular DNAs in which a single-stranded nick is made in one strand of the DNA and the 3′ hydroxyl end is used as a primer to replicate around the circle, displacing the old strand.

Rolling-circle transposon *See* Y2 transposon.

Round of replication The cycle of replication of a circular DNA in which a complete copy of the DNA is made.

RRF *See* Ribosome release factor.

rRNA *See* Ribosomal RNA.

Sacculus The cell wall surrounding the bacterial cell, i.e., the murein layer.

SaPI *See Staphylococcus aureus* pathogenicity island.

Satellite virus A naturally occurring virus that depends on another virus for its multiplication.

Saturation genetics A mutant search that is so extensive that presumably all of the genes whose products participate in a biological process are represented by mutations.

Screening The process (usually streamlined) of testing a large number of organisms for a particular mutant type.

S-D sequence *See* Shine-Dalgarno sequence.

Seamless cloning The process of inserting a PCR fragment into a cloning vector by using a restriction endonuclease that cuts outside its recognition site so that no extraneous base pairs are added between the cloned DNA fragment and the cloning vector.

***sec* gene** One of the genes whose products are required for transport of proteins through the inner membrane.

Sec system The general system encoded by the *sec* genes of bacteria for transporting proteins across the cytoplasmic membrane; it consists of the targeting factors SecA and SecB and includes the components of the SecYEG channel in the inner membrane.

SecA A protein with ATPase activity that drives proteins to be exported into the SecYEG channel.

SecB A chaperone that binds proteins to be exported and keeps them from folding prematurely before they can be taken up by the SecYEG channel.

Secondary structure A structure of a polynucleotide or polypeptide chain that results from noncovalent pairing between nucleotides or amino acids in the chain.

Secretin A protein that is part of type II and type III protein secretion systems of gram-negative bacteria that forms multisubunit β barrels in the outer membrane through which proteins are exported.

Secretion The process of transporting proteins through the membranes.

SecYEG channel (SecYEG translocase) The channel in the inner membrane of bacteria, composed of single copies of the SecY, SecE, and SecG proteins, through which many proteins are transported. *See* Translocase.

Segregation The process by which newly replicated DNAs or genetic alleles are separated into daughter cells or spores.

Selected marker A difference in DNA sequence between two strains in a bacterial or phage cross that is used to select recombinants. The cells are plated under conditions in which only recombinants that have received the donor sequence or allele can multiply.

Selection A procedure in which bacteria or viruses are placed under conditions in which only the wild type or the desired mutant or recombinant can multiply, allowing the isolation of even very rare mutants and recombinants.

Selectional genetics Genetic analysis in which rare mutants or recombinants are selected and analyzed.

Selective conditions Conditions under which only the wild type or the desired mutant can multiply.

Selective media Media that have been designed to allow multiplication of only the desired mutant or the wild type. Such media often lack one or more nutrients or contain a substance that is toxic.

Selective plate An agar plate made with selective medium.

Self-transmissible Carrying all of the genes for its transfer into other bacteria; the term is used to describe DNA elements, including plasmids and transposons.

Self-transmissible plasmid A plasmid that encodes all the gene products needed to transfer itself to other bacteria through conjugation.

Semiconservative replication A type of DNA replication in which the daughter DNAs are composed of one old strand and one newly synthesized strand.

Sensor kinase In two-component systems, a protein that transfers the γ phosphate of ATP to itself in response to a certain environmental or cellular signal and then transfers the phosphate to a response regulator protein that performs some cellular function. *See* Sensor protein.

Sensor protein The protein in a two-component system that detects changes in the environment and communicates that information to the response regulator, usually by transferring a phosphoryl group. *See* Sensor kinase.

Sequestration The binding of a cellular constituent so that it is unavailable to perform its function. An example is the binding of the SeqA protein to hemimethylated sequences in the origin of replication (*oriC*) of the *Escherichia coli* chromosome, somehow preventing reinitiation after a round of chromosome replication has initiated.

Serial dilution A procedure in which an aliquot of a solution is diluted in one tube and then an aliquot of the solution in this tube is diluted in a second tube, and so forth. The total dilution is the product of each of the individual dilutions.

7,8-Dihydro-8-oxoguanine *See* 8-OxoG.

Sex pilus A rod-like structure that forms on the surface of a bacterium containing a self-transmissible plasmid and facilitates transfer of the plasmid or other DNA into another bacterium.

Shine-Dalgarno sequence (S-D sequence) A short sequence, usually about 10 nucleotides upstream of the initiation codon in a bacterial translational initiation region, that is complementary to a sequence in the 3′ end of the 16S rRNA; it helps to position the ribosome for initiation of translation. It is named after the individuals who discovered it.

Shufflon A region of a self-transmissible plasmid that contains many cassette sequences that integrate by a site-specific integrase to alter the carboxyl terminus of a pilin protein and change the cell surface receptors to which the pilus can bind.

Shuttle phasmid A cloning vector that replicates as a plasmid in one type of bacterium (e.g., *E. coli*) but as a phage in another.

Shuttle vector A plasmid cloning vector that contains two origins of replication that function in different types of cells so that the plasmid can replicate in both types of cells.

Siblings In microbial genetics, two cells or viruses that arose from the multiplication of the same mutant cell or virus.

Signal recognition particle (SRP) A universally evolutionarily conserved particle, composed of RNA and protein, that in bacteria binds the first transmembrane domain of a protein destined for the inner membrane as the protein emerges from the ribosome and directs it to the SecYEG translocon.

Signal sequence A sequence, composed of mostly hydrophobic amino acids, that is located at the N terminus of some membrane and exported proteins and that targets the protein for transport into or through the cytoplasmic membrane. In bacterial proteins transported by the Sec system or the Tat system, the signal sequence is removed as the protein passes into or through the membrane; in proteins targeted by the signal recognition particle to the cytoplasmic membrane, it is usually the first transmembrane domain and is not removed.

Signal transduction pathway A set of proteins that pass a signal from one to the other by direct contact. They do this by chemically altering each other by proteolysis, by transferring a chemical group such as a phosphoryl or methyl group, or by simply binding to each other.

Silent mutation A change in the DNA sequence of a gene encoding a protein that does not change the amino acid sequence of the protein, usually because it changes the last base in a codon, thereby changing it to another codon but one that encodes the same amino acid.

Single mutant A mutant organism that has only one of the two or more mutations being studied.

Single mutation A mutation due to a single event that changed the DNA sequence, independent of how many base pairs were changed by the event.

Site-specific mutagenesis One of many methods for mutagenizing DNA in such a way that the change is localized to a predetermined base pair in the DNA.

Site-specific recombinases Enzymes that recognize two specific sites on DNA and promote recombination between them.

Site-specific recombination Recombination that occurs only between defined sequences in DNA. It is usually performed by site-specific recombinases.

6-4 lesion A type of damage to DNA in which the carbon at the 6 position of a pyrimidine is covalently bound to the carbon at the 4 position of an adjacent pyrimidine.

SMC proteins *See* Condensins.

Sortase An enzyme in gram-positive bacteria that cuts a protein to be displayed on the cell surface at its sorting signal and attaches it through a new peptide bond to a peptide cross bridge in the cell wall.

SOS box The operator sequence to which the LexA repressor binds. It is found close to all of the promoters of SOS genes.

SOS gene A gene that is a member of the LexA regulon, so that its transcription is normally repressed by LexA repressor.

SOS mutagenesis The increased mutagenesis that occurs after induction of the SOS response and that is due to the induction of the *umuC* and *umuD* genes and the subsequent autocleavage of UmuD, which, with UmuC, forms a translesion DNA polymerase that is mistake prone. *See* Weigle mutagenesis.

SOS response Induction of transcription of the SOS genes in response to DNA damage. It is due to stimulation of autocleavage of LexA repressor by the RecA single-stranded DNA nucleoprotein coprotease.

Southern blotting A procedure for transferring DNA from an agarose gel to a filter for hybridization. It is named after the person who developed the procedure. Other blots in which RNA or proteins are transferred were then somewhat facetiously named Northern and Western blots, respectively.

Spanin A protein encoded by some phages that infect gram-negative bacteria that somehow releases the outer membrane from the cell wall, promoting lysis.

Specialized transduction *See* Restricted transduction.

Spontaneous mutations Mutations that occur in organisms without deliberate attempts to induce them by irradiation or chemical treatment.

Sporulation A developmental process that leads to the development of spores, which are dormant cells containing the DNA of the organism and are often resistant to desiccation and other harsh environmental conditions.

SRP *See* Signal recognition particle.

Stalling sequence A polypeptide sequence that interacts with the exit channel of the ribosome and causes translation to stop.

***Staphylococcus aureus* pathogenicity island (SaPI)** A type of genetic island found in strains of *S. aureus* that often carries genes required for pathogenicity.

Starve To deprive an organism of an essential nutrient that it cannot make for itself.

Sticky end The short single-stranded DNA that sticks out from the end of the DNA molecule after it has been cut with a type II restriction endonuclease that makes a staggered break in the DNA.

Stimulon The collection of all of the operons that are turned on by a particular environmental condition, independent of whether they are part of the same regulon. *See* Regulon.

Stop codon A codon in an mRNA that signals translation termination. *See* Nonsense codon.

Strain A group of organisms that are identical to each other but differ genetically from other organisms of the same species. A strain is a subdivision of a species.

Strand exchange The process by which a strand of a double-stranded DNA changes partners so that it pairs with a different complementary strand of DNA, as in D-loop formation.

Strand passage A reaction performed by topoisomerases in which one or two strands of a DNA are cut and the ends of the cut DNA are held by the enzyme to prevent rotation while other strands of the same or different DNAs are passed through the cuts.

Stringent plasmid A plasmid that exists in only one or very few copies per cell because plasmid replication is tightly controlled.

Stringent response Cessation of synthesis of rRNA and other stable RNAs in the cell when the cells are starved for an amino acid. In *E. coli*, it is due to the accumulation of ppGpp synthesized by the RelA protein on the ribosome.

Structural gene One of the genes in an operon for a pathway that encodes one of the enzymes of the pathway.

Subclone A smaller DNA clone obtained by cutting a larger clone and cloning one of the pieces.

Sugar A simple carbohydrate with the general formula $(CH_2O)_n$, as found in nature; n is 3 to 9.

Suicide vector A cloning vector, usually plasmid or phage DNA that cannot replicate in the cells into which it is being introduced, so it is degraded or diluted out by subsequent cell divisions.

Supercoiling A condition in which the two strands of the DNA double helix are wrapped around each other either more or less often than predicted from the natural distances dictated by the unstressed helical structure of DNA, i.e., more or less than about 10.5 bp per turn.

Superinfection Infection of already infected cells by the same type of virus.

Superintegrons Large arrays of single gene cassettes, often without promoters, that are flanked by recombination signals and can be rearranged by a cognate recombinase. They often encode drug resistance or pathogenicity determinants.

Suppression Alleviation of the effects of a mutation by a second mutation elsewhere in the DNA.

Suppressor mutation A mutation elsewhere in the DNA that alleviates the effects of another mutation.

Symmetric sequence A sequence of deoxynucleotides in double-stranded DNA that reads the same in the 5′-to-3′ direction on both strands.

Synapse In recombination, a structure in which two DNAs are held together by pairing between their strands via recombination proteins.

Synchronize To treat a culture of cells so that they are all at approximately the same stage in their cell cycles at the same time.

Synteny Conservation of gene order or genetic linkage in the genomes of different types of organisms.

Synthetic genomics An area of investigation enabled by techniques that create very large fragments of DNA or even complete genomes with minimal or no dependence on existing DNA template sequences.

Synthetic lethality screen A selection system set up to isolate mutants with mutations in genes whose products are required for viability only in the absence of another gene product. Often, the other gene is set up so that it is transcribed only from an inducible promoter. The mutations being sought are lethal only in the absence of inducer, when the other gene is not being expressed.

Tag A polypeptide that is translationally fused to other proteins because it binds strongly to some other molecule. The strong binding of the fused polypeptide allows the protein to be purified more easily.

Tandem duplication A type of mutation that causes a DNA sequence to be followed immediately by the same sequence in the same orientation.

Tandem mass spectrometry Two mass spectrometric analyses run sequentially so that the first analysis allows selection of a specific peptide ion and the second analysis includes fragmentation of the selected peptide, analyzing the masses of the pieces and thereby determining partial peptide sequences. In the fragmentation step, the bonds that break are almost exclusively along the peptide backbone, and therefore, the ion species detected in the second analysis mostly represent peptide ions.

Target DNA In transposition, the DNA into which a transposon inserts.

Target site immunity The process by which transposons are inhibited in their ability to transpose into a DNA region that already contains the same transposon.

Tautomer A (usually temporary) form of a molecule in which the electrons are distributed among the atoms differently than the normal configuration.

TEC *See* Transcription elongation complex.

Temperate phage A phage that is known to be capable of lysogeny.

Temperature-sensitive mutant A mutant that cannot grow in the temperature range in which the wild type can multiply, usually indicating that a lower temperature is needed to permit growth.

Template strand The strand of DNA that is used to make a copy. In the case of DNA replication, an RNA primer is most often used to initiate polymerization of the DNA copy. In the case of transcription, the term refers to a region from which RNA is synthesized that serves as the template for RNA synthesis; the sequence of this DNA strand is complementary to the sequence of the RNA. *See* Transcribed strand, Noncoding strand.

Terminally redundant Containing direct repeats at both ends of a DNA (usually a phage genome), that is, the sequences at both ends are the same in the direct orientation.

Termination of transcription The process by which the RNA polymerase leaves the DNA and the RNA chain is released.

Termination of translation The process by which the ribosome leaves the mRNA and the polypeptide is released when a nonsense codon in the mRNA is encountered in frame.

Terminator A sequence in the DNA and/or structure in the RNA that causes RNA polymerase to stop transcription and to release both the DNA template and the RNA transcript.

Tertiary structure The three-dimensional structure of a polypeptide or RNA.

Thermosensor A cellular component (RNA or protein) that undergoes a specific structural change in response to changes in temperature and that can be used by the cell to sense the temperature change.

Theta replication A type of replication of circular DNA in which the replication apparatus initiates at an origin of replication and proceeds in one or both directions around the circle with leading and lagging strands of replication. The expanded molecule in an intermediate state of replication resembles the Greek letter theta (θ).

Three-factor cross A type of genetic cross used to order three closely linked mutations and in which one parent has two of the mutations and the other parent has the third mutation. The order of the three markers in the DNA is indicated by the frequencies of the possible recombinant types.

3′ end The terminus of a polynucleotide chain (DNA or RNA) ending in the nucleotide that is not joined at the 3′ carbon of its deoxyribose or ribose to the 5′ phosphate of another nucleotide.

3′ exonuclease An enzyme that degrades a polynucleotide from its 3′ end.

3′ hydroxyl end In a polynucleotide, a 3′ end that has a hydroxyl group on the 3′ carbon of the ribose sugar of the last nucleotide without a phosphate group attached.

3′ overhang An unpaired single strand with a free 3′ end extending from a double-stranded DNA or RNA.

3′ untranslated region (3′ UTR) In an mRNA, the sequences downstream or 3′ of the nonsense codon of an open reading frame that encodes a protein.

Thymine (T) One of the pyrimidine (one-ringed) bases in DNA and some tRNAs.

TIGRFAMs An extensive database of protein families that is useful for categorizing coding sequences.

TIR *See* Translational initiation region.

Titration A process of increasing the concentration of one of two types of molecules that bind to each other until all of the other type of molecule is bound.

TLS *See* Translesion synthesis.

tmRNA *See* Transfer-messenger RNA.

Topo cloning Using a topoisomerase with a sequence recognition site to insert DNA into a cloning vector. It is more efficient than cloning using DNA ligase but has the disadvantage that it requires specially designed cloning vectors.

Topoisomerase An enzyme that can alter the topology of a DNA molecule by cutting one or both strands of DNA, passing other DNA strands through the cuts while holding the cut ends so that they are not free to rotate, and then resealing the cuts.

Topoisomerase IV A type II topoisomerase of *E. coli* that is responsible for decatenation of daughter chromosomes after replication.

Topology The relationship of the strands of DNA to each other in space.

Tra functions Gene products encoded by the *tra* genes of self-transmissible DNA elements that allow the plasmids to transfer themselves into other bacteria.

***trans*-acting function** A gene product that can act on DNAs in the cell other than the one from which it was made.

***trans*-acting mutation** A mutation that affects a gene product that leaves the DNA from which it is made and so can be complemented.

Transconjugant A recipient cell that has received and integrated DNA from another cell by conjugation.

Transcribe To make an RNA that is a complementary copy of a strand of DNA.

Transcribed strand In a region of a double-stranded DNA that is transcribed into RNA, the strand of DNA that is used as a template and so is complementary to the RNA. *See* Template strand.

Transcript A complementary RNA made from a region of DNA.

Transcription antitermination The process by which RNA polymerase can be made to proceed through one or more termination sites. This can be accomplished through changes in the secondary structure of the mRNA, creating new secondary structure that masks a transcription terminator secondary structure, or through changes to the transcription elongation complex. *See* Processive antitermination.

Transcription attenuation Regulation of gene expression by initiating transcription but then prematurely terminating transcription unless certain conditions are met.

Transcription bubble The ~17-bp region in DNA during transcription in which the two strands of DNA have been separated by the RNA polymerase and within which the newly synthesized RNA forms a short RNA-DNA duplex with the transcribed strand of DNA.

Transcription elongation The step of transcription during which the RNA polymerase core moves along the DNA and synthesizes RNA. *See* Transcription elongation complex.

Transcription elongation complex (TEC) The RNA polymerase complex that is active in transcription elongation.

Transcription start site The nucleotide in the coding strand of DNA in a promoter that corresponds to the first nucleotide polymerized into RNA from the promoter.

Transcription termination site A DNA sequence at which the RNA polymerase falls off the template, stopping transcription, and the RNA product is released. It can be either factor independent or dependent on a transcription termination factor, such as Rho.

Transcription vector A cloning vector that contains a promoter from which a cloned DNA can be transcribed.

Transcriptional activator A protein that is required for transcription of an operon. The protein makes contact with the RNA polymerase and allows the RNA polymerase to initiate transcription from the promoter of the operon.

Transcriptional autoregulation The process by which a protein regulates the transcription of its own gene, by being either a repressor or an activator of its own transcription.

Transcriptional fusion Introduction of a gene downstream of the promoter for another gene or genes so that it is transcribed from the promoter for the other gene(s) into the same mRNA but is translated as a separate polypeptide from its own translational initiation region.

Transcriptional regulation Regulation in which the amount of product of a gene that is synthesized under certain conditions is determined by how much mRNA is made from the gene.

Transcriptional regulator Any protein that regulates the transcription of genes, e.g., a repressor, activator, or antitermination protein.

Transcriptome The complete set of transcripts expressed in an organism. The transcripts actually detected depend on their abundance under the experimental conditions used.

Transducing particle A phage whose head contains bacterial DNA, as well as, or instead of, its own DNA.

Transducing phage A type of phage that sometimes packages bacterial DNA during infection and introduces it into other bacteria during infection of those bacteria.

Transductant A bacterium that has received DNA from another bacterium by transduction.

Transduction A process in which DNA other than phage DNA is introduced into a bacterium via infection by a phage containing the DNA.

Transfection Initiation of a virus infection by introducing virus DNA or RNA into a cell by transformation or electroporation, rather than by infection by viral particles.

Transfer-messenger RNA (tmRNA) A small RNA found in bacteria that is a hybrid between a tRNA and an mRNA. It can be aminoacylated with alanine like a tRNA and enter the A site of the ribosome if the A site is unoccupied, e.g., if the ribosome has reached the 3′ end of an mRNA without encountering a nonsense codon. A short reading frame on the tmRNA is then translated, fusing a short peptide sequence to the C terminus of the truncated protein, which targets the protein for degradation by a protease.

Transfer RNA (tRNA) The small, stable RNAs in cells to which specific amino acids are attached by aminoacyl tRNA synthetases. The tRNA with the amino acid attached enters the ribosome and base pairs through its anticodon sequence with a 3-nucleotide codon sequence in the mRNA to insert the correct amino acid into the growing polypeptide chain.

Transformant A cell that has received DNA by transformation.

Transformasomes Globular structures that appear on the surfaces of some types of bacteria into which DNA first enters during natural transformation of the bacteria.

Transformation Introduction of DNA into cells by mixing the DNA and the cells.

Transformylase The enzyme that transfers a formyl (CHO) group onto the amino group of methionine to make formylmethionine.

Transgenic Containing inherited foreign DNA sequences that have been experimentally introduced into its ancestors. The introduced DNA sequences are passed down from generation to generation because they are inserted into a stably inherited DNA, such as a chromosome.

Transient diploids Cells in the temporary state of diploidy that exists after a DNA that cannot replicate in that type of cell enters the cell and before it is lost or degraded.

Transition mutation A type of base pair change mutation in which the purine base has been changed into the other purine base and the pyrimidine base has been changed into the other pyrimidine base (e.g., AT to GC or GC to AT).

Translate To cause protein synthesis in which the information in an mRNA is used to dictate the amino acid sequence of a protein product. *See* Translation.

Translated region A region of an mRNA that encodes a protein.

Translation The process by which the information in an mRNA is used to dictate the amino acid sequence of a protein product. *See* Translate.

Translation elongation factor G (EF-G) The protein required to move the peptidyl-tRNA from the A site to the P site on the ribosome with the concomitant cleavage of GTP to GDP after the peptide bond has formed.

Translation elongation factor Tu (EF-Tu) The protein that binds to aminoacylated tRNA and accompanies it into the A site of the ribosome. It then cycles off the ribosome with the concomitant cleavage of GTP to GDP, leaving the aminoacylated tRNA behind.

Translation termination site Any one of the nonsense codons for an organism in the frame being translated.

Translation vector A cloning vector that contains a TIR from which a cloned DNA sequence can be translated.

Translational coupling A gene arrangement in which the translation of one protein-coding sequence on a polycistronic mRNA is required for the translation of the second, downstream coding sequence. Often, translation of the upstream coding sequencing is required to remove secondary structure in the mRNA that blocks the translational initiation region for the downstream coding sequence.

Translational fusion The fusion of parts of the coding regions of two genes so that translation initiated at the translational initiation region for one polypeptide on the mRNA will continue into the coding region for the second polypeptide in the correct reading frame for the second polypeptide. A polypeptide containing amino acid sequences from the two genes that were joined to each other will be synthesized. *See* Fusion protein.

Translational initiation region (TIR) The initiation codon, the Shine-Dalgarno sequence, and any other surrounding sequences in mRNA that are recognized by the ribosome as a place to begin translation. Also called ribosome-binding site (RBS).

Translational regulation Variation, under different conditions, in the amount of synthesis of a polypeptide due to variation in the rate at which the polypeptide is translated from the mRNA.

Translationally autoregulated Able to affect the rate of translation of its own coding sequence on its mRNA. Usually in such cases, the protein binds to its own translational initiation region or that of an upstream gene to which it is translationally coupled; hence, the protein represses its own translation.

Translationally coupled A condition where coding regions are present on the same transcript, where translation of the downstream coding region is dependent on translation of the upstream coding sequence.

Translesion synthesis (TLS) Synthesis of DNA over a template region containing a damaged base or bases that are incapable of proper base pairing.

Translocase The evolutionarily highly conserved channel in the cytoplasmic membrane through which proteins are exported. In bacteria, it is represented by the SecYEG membrane channel.

Translocation During translation, the movement of the tRNA with the polypeptide attached from the A site to the P site on the ribosome after the peptide bond has formed. In transported proteins, the act of moving the protein out of the cytoplasm into or through one or more of the membranes.

Transmembrane domain The region in a polypeptide between a region that is exposed to one surface of a membrane and a region that is exposed to the other surface. This region must traverse and be embedded in the membrane. Usually, transmembrane domains have a stretch of at least 20 mostly hydrophobic amino acids that is long enough to extend from one face of a bilipid membrane to the other.

Transmembrane protein A membrane protein that has surfaces exposed on both sides of the membrane.

Transposase An enzyme encoded by a transposon that breaks the DNA at both ends of the transposon and joins these ends to a target DNA during transposition.

Transposition Movement of a transposon from one place in DNA to another.

Transposon A DNA sequence that can move from one place in DNA to another, using a specialized recombinase called a transposase. It should be distinguished from homing DNA elements, which usually move only into the same sequence in another DNA and depend on homologous recombination, or DNA elements that insert into other DNAs, using recombinases called integrases.

Transposon mutagenesis A technique in which a transposon is used to make random insertion mutations in DNA. The transposon is usually introduced into the cell in a suicide vector, so it must transpose into another DNA in the cell to become established.

TransTerm A database that contains mRNA sequences compiled from GenBank and that allows the user to find regulatory elements, such as initiation and termination regions (http://transterm.cbcb.umd.edu).

Transversion mutation A type of base pair change mutation in which the purine in the base pair is changed into the pyrimidine, and vice versa, e.g., GC to TA or GC to CG.

Trigger factor A chaperone in *E. coli* that is closely associated with the exit pore of the ribosome and that helps proteins fold as they emerge from the ribosome. It can partially substitute for DnaK.

Triparental mating A conjugational mating for introducing mobilizable plasmids into cells in which three strains of bacteria are mixed. One strain contains a self-transmissible plasmid, which transfers itself into the second strain, containing a mobilizable plasmid, which in turn is mobilized into the third strain.

Triple-stranded structure Three strands of DNA held together in a triple-stranded structure that has been hypothesized to form when a RecA nucleoprotein filament invades a double-stranded DNA.

tRNA *See* Transfer RNA.

$tRNA^{fMet}$ The tRNA to which formylmethionine is attached and that pairs with the initiator codon in a translational initiation region to initiate translation of a polypeptide in bacteria.

Ts mutant *See* Temperature-sensitive mutant.

Two-component regulatory system A pair of proteins, one of which, the sensor kinase, undergoes a change in response to a change in the environment and communicates this change, usually in the form of a phosphate, to another protein, the response regulator, which then causes the appropriate cellular response. Different two-component systems are often highly homologous to each other, which allows them to be identified in sequenced bacterial genomes. Also referred to as two-component signal transduction.

Two-dimensional polyacrylamide gel electrophoresis (2D-PAGE) A separation technique in which proteins are applied to a pI (isoelectric point) strip and separated by charge by using isoelectric focusing; this strip is then attached to another slab acrylamide gel containing sodium dodecyl sulfate so that the proteins move at right angles to the first gel and are separated by size.

Two-hybrid system A technique for determining if two proteins or regions of proteins bind to each other or for identifying proteins that bind to a particular protein or region of a protein. It is based on the ability of proteins that bind to each other to bring two parts of a tester protein together, restoring its activity.

Two-partner secretion *See* Type V secretion system.

Type I secretion system A protein secretion system in gram-negative bacteria based on a specific ATP-binding cassette (ABC) transporter and the TolC channel, a multiuse channel. Proteins secreted by type I systems recognize a signal sequence in the carboxyl terminus of the protein that is not cleaved off during transport.

Type II secretion system A protein secretion system of gram-negative bacteria that uses either the SecYEG channel or the Tat channel to transport proteins through the inner membrane and then uses a specific secretin β-channel to transport the protein through the outer membrane. It makes a complicated structure called a pseudopilus, which may push the protein through the inner membrane channel and through the secretin channel to the outside of the cell.

Type III secretion system A multicomponent protein secretion system of pathogenic gram-negative bacteria that forms a syringe-like structure, sometimes called an injectisome, that injects effector proteins directly through both bacterial membranes into eukaryotic cells.

Type IV secretion system A protein secretion system of gram-negative bacteria that can inject proteins directly through both bacterial membranes into other cells, although some seem to use the SecYEG channel to transport the protein through the inner membrane. Plasmid conjugation systems, also found in gram-positive bacteria, are essentially type IV secretion systems.

Type V secretion system A group of secretion systems that includes the autotransporters, the two-partner secretion systems, and the chaperone-usher systems. These secretion systems all form a dedicated β-barrel in the outer membrane that exports only one or a select group of proteins, and they also all use the SecYEG channel to transport the exported protein through the inner membrane.

Type VI secretion system A largely uncharacterized secretion system encoded by a highly conserved cluster of at least 12 genes with as many as 9 other genes, which might encode specific effectors.

UAS *See* Upstream activator sequence.

Uninducible phenotype A phenotype that occurs when a regulatory system cannot be expressed even in the presence of the inducing signal.

Unselected marker A difference between the DNA sequences of two bacteria or phages involved in a genetic cross that can be used for genetic mapping. Mapping information can be obtained by testing recombinants that have been selected for being recombinant for one marker, the selected marker, to determine if they have the sequence of the donor or the recipient for another marker, an unselected marker.

UP element A region of DNA upstream of a promoter to which the α subunit of RNA polymerase binds during transcription initiation to enhance promoter recognition.

Upstream Lying in the 5′ direction from a given point on RNA or in the 5′ direction on the coding strand of a DNA region from which an RNA is made.

Upstream activator sequence (UAS) A DNA sequence upstream of a promoter that increases transcription from the promoter by binding an activator protein. It is usually associated with NtrC family activators and σ^{54} promoters. It can be many hundreds of base pairs upstream from the promoter. Also called upstream activator site.

Uptake sequences Short DNA sequences that allow DNA containing the sequence to be bound and taken up by some types of bacteria during natural transformation.

Uracil (U) One of the pyrimidine (one-ringed) bases; naturally found in RNA.

Uracil-*N*-glycosylase An enzyme that removes the uracil base from DNA by cleaving the bond between the base and the deoxyribose sugar.

UvrABC endonuclease A complex of three proteins that cuts on both sides of any DNA lesion that causes a significant distortion of the helix as a first step in excision repair of the damage. Also called UvrABC excinuclease.

Variants Usually, different strains of the same type of bacteria as they are isolated from nature.

Vertical transfer Transfer of DNA from an organism to its progeny solely through reproduction. See Horizontal transfer.

Very-short-patch (VSP) repair A type of repair in enteric bacteria that removes the mismatched T along with a very short stretch of DNA in the sequence CT(A/T)GG(T/A)CC and replaces it with a C. It is due to deamination of 5-methylcytosine at this position in these bacteria.

***vir* mutations** Mutations in the operator sequences of the bacteriophage DNA that prevent repressor binding and therefore prevent lysogen formation by the phage and allow it to multiply in homoimmune lysogens.

VSP repair *See* Very-short-patch repair.

W mutagenesis *See* Weigle mutagenesis.

W reactivation *See* Weigle reactivation.

Watson-Crick structure of DNA The double-helical structure of DNA first proposed by James Watson and Francis Crick. The two strands of the DNA are antiparallel and are held together by hydrogen bonding between the bases.

Weigle mutagenesis Another name for SOS mutagenesis. It refers to the increase in the number of phage mutations if phage infect cells that have been preirradiated with UV. It is due to SOS induction of the *umuCD* genes, as well as *recA*. It is named after Jean Weigle, who first observed it.

Weigle reactivation The increased ability of phages to survive UV irradiation damage to their DNA if the cells they infect have been previously exposed to UV irradiation. It is due to SOS induction of repair functions. It is named after Jean Weigle, who first observed it.

Western blotting Transfer of protein from a gel to a solid substrate, where the locations of specific proteins can be determined by incubation with specific antibodies raised against the proteins.

Wild type The normal type. Literally, the term refers to the organism as it was first isolated from nature. In a genetic experiment, it is the strain from which mutants are derived.

Wild-type allele The form of a gene as it exists in the wild-type organism.

Wild-type phenotype The particular outward trait characteristic of the wild type that is different in the mutant.

Wobble The ability of the base of the first nucleotide (read 5′ to 3′) in the anticodon of a tRNA to pair with more than one base in the third nucleotide (read 5′ to 3′) of a codon in the mRNA. The term also generally applies to the ability of G and U residues in RNA to pair with each other.

Xanthine A purine base that results from deamination of guanine.

XerCD recombinase The recombinase in *Escherichia coli* and many other bacteria that separates dimerized chromosomes by promoting recombination between repeated *dif* sequences.

X-phile One of a group of enzymes that can cut the crossed DNA strands at a Holliday junction.

Y-family polymerases A large group of DNA polymerases, represented by DinB (Pol IV), and UmuC (Pol V) in *E. coli*, that are capable of translesion synthesis, perhaps because they have a more open active center and lack editing functions.

YidC An inner membrane protein of unknown function that cooperates with the SecYEG channel in inserting inner membrane proteins into the inner membrane. Some proteins are inserted by YidC alone.

Y2 transposon A transposon with two Ys (tyrosines) in its active center, sometimes called a rolling-circle transposon because the mechanism of transposition resembles rolling-circle replication of phages and plasmids.

Zero frame In the coding region of a gene, the sequence of nucleotides, taken three at a time, in which the polypeptide encoded by the gene is translated.

Figure and Table Credits

Introduction

Figure 1.1 Based on Figure 3 of N. R. Pace, *Microbiol. Mol. Biol. Rev.* 73:565–576, 2009, with permission.

Chapter 1

Figure 1.17 Adapted from Figure 2, p.179, of J. E. Camara and E. Crooke, p. 177–191, *in* N. P. Higgins (ed.), *The Bacterial Chromosome* (ASM Press, Washington, DC, 2005), with permission.

Chapter 2

Figure 2.4 Adapted from Figure 5A, p. 1288, of K. S. Murakami, S. Masuda, E. A. Campbell, O. Muzzin, and S. A. Darst, *Science* **296**:1285–1290, 2002, with permission.

Figure 2.9 (A) Adapted from Color Plate 3A (associated with K. Geszvain and R. Landick, p. 283–296) *in* N. P. Higgins (ed.), *The Bacterial Chromosome* (ASM Press, Washington, DC, 2005), with permission.

Figure 2.14 Based on Figure 2, p. 299, of S. L. Dove and A. Hochschild, p. 297–310, *in* N. P. Higgins (ed.), *The Bacterial Chromosome* (ASM Press, Washington, DC, 2005), with permission.

Figure 2.17 Adapted from Color Plate 3B (associated with K. Geszvain and R. Landick, p. 283–296) *in* N. P. Higgins (ed.), *The Bacterial Chromosome* (ASM Press, Washington, DC, 2005), with permission.

Figure 2.27 (A) Reprinted from Figure 1 of M. M. Yusupov, G. Z. Yusupova, A. Baucom, K. Lieberman, T. N. Earnest, J. H. Cate, and H. F. Noller, *Science* **292**:883–896, 2001, with permission. Copyright 2001 American Association for the Advancement of Science. (B) Reprinted from Figure 5 of J. H. Cate, M. M. Yusupov, G. Yusupova, T. N. Earnest, and H. F. Noller, *Science 285*:2100, 1999, with permission. Copyright 1999 American Association for the Advancement of Science.

Figure 2.43 Adapted from Figure 2, p. 383, of J. Collado-Vides, B. Magasanik, and J. D. Gralla, *Microbiol. Rev.* 55:371–394, 1991, with permission.

Table 2.1 Based on Table 1, p. 328, of S. Kushner, p. 327–345, *in* N. P. Higgins (ed.), *The Bacterial Chromosome* (ASM Press, Washington, DC, 2005), with permission.

Box 2.5 Figure 2 Adapted from Figure 1 of S. R. Gill et al., *J. Bacteriol.* **187**:2426–2438, 2005, with permission.

Chapter 3

Figure 3.25 Adapted from Figure 1, p. 372, of M. Belfort and J. Pedersen-Lane, *J. Bacteriol.* **160**:371–378, 1984, with permission.

Figure 3.36 Reprinted from Figure 2, p. 118, of B. J. Bachmann, B. Low, and A. L. Taylor, *Bacteriol. Rev.* **40**:116–167, 1976, with permission.

Chapter 4

Figure 4.2 Adapted from Figure 2, p. 66, of S. A. Khan, p. 63–78, *in* B. E. Funnell and G. J. Phillips (ed.), *Plasmid Biology* (ASM Press, Washington, DC, 2004), with permission.

Figure 4.5 Adapted from Figure 2, p. 1620, of S. E. Luria and J. L. Suit, p. 1615–1624, *in* F. C. Neidhardt, J. L. Ingraham, K. B. Low, B. Magasanik, M. Schaechter, and H. E. Umbarger (ed.), Escherichia coli *and* Salmonella typhimurium: *Cellular and Molecular Biology* (American Society for Microbiology, Washington, DC, 1987), with permission.

Figure 4.9 Adapted from Figure 1D, p. 50, of S. Brantl, p. 47–62, *in* B. E. Funnell and G. J. Phillips (ed.), *Plasmid Biology* (ASM Press, Washington, DC, 2004), with permission.

Figure 4.14 Adapted from Figure 3B and C of Y.-L. Shih and L. Rothfield, *Microbiol. Mol. Biol. Rev.* **70**:729–754, 2006, with permission.

Figure 4.15 Adapted from Figure 3B and C of Y.-L. Shih and L. Rothfield, *Microbiol. Mol. Biol. Rev.* **70**:729–754, 2006, with permission.

Figure 4.17 Adapted from p. 766 of the *BRL-Gibco Catalog* (Invitrogen Corp., Carlsbad, Calif., 2002), with permission. Copyright 2002 Invitrogen Corporation.

Figure 4.20 Adapted from Figures 1 and 2 of V. Vagner, E. Dervyn, and S. D. Ehrlich, *Microbiology* **144**:3097–3104, 1998, with permission.

Chapter 5

Figure 5.3 Redrawn from Color Plate 6 (associated with T. Lawley, B. M. Wilkins, and L. S. Frost, p. 203–226) *in* B. E. Funnell and G. J. Phillips (ed.), *Plasmid Biology* (ASM Press, Washington, DC, 2004), with permission.

Figure 5.8 Redrawn from Figure 3, p. 461, of P. J. Christie, p. 455–472, *in* B. E. Funnell and G. J. Phillips (ed.), *Plasmid Biology* (ASM Press, Washington, DC, 2004), with permission.

Figure 5.15 Redrawn from Figure 1, p. 4831, of J. R. Chandler, A. R. Flynn, E. M. Bryan, and G. M. Dunny, *J. Bacteriol.* **187**:4830–4843, 2005, with permission.

Chapter 7

Figure 7.1 (A) (Left) Electron micrograph by Sally Burns, Michigan State University. (Right) Provided by Arthur Zachary and Lindsay Black. (B) Photograph by Kurt Stepnitz, Michigan State University.

Figure 7.5 Adapted from p. 158 of the *Novagen Catalog* (EMD Biosciences, San Diego, CA, 2007), http://www.emdbiosciences.com/novagen, 2007, with permission.

Figure 7.6 Reprinted from J. D. Karam (ed.), *Molecular Biology of Bacteriophage T4* (ASM Press, Washington, DC, 1994), with permission.

Figure 7.8 Reprinted from Figure 8 of E. S. Miller, E. Kutter, G. Mosig, F. Arisaka, T. Kunisawa, and W. Ruger, *Microbiol. Mol. Biol. Rev.* **67**:86–156, 2003, with permission.

Figure 7.14 Figure and legend reprinted from Figure 1, p. 418, of A. Gupta, A. B. Oppenheim, and V. K. Chaudhary, p. 415–429, *in* M. K. Waldor, D. Friedman, and S. L. Adhya (ed.), *Phages: their Role in Bacterial Pathogenesis and Biotechnology* (ASM Press, Washington, DC, 2005), with permission.

Figure 7.17 Adapted from Figure 1, p. 231, of K. N. Kreuzer and B. Michel, p. 229–250, *in* N. P. Higgins (ed.), *The Bacterial Chromosome* (ASM Press, Washington, DC, 2005), with permission.

Figure 7.18 Derived from Figure 2A, p. 97, of R. Young, p. 92–127, *in* M. K. Waldor, D. I. Friedman, and S. L. Adhya (ed.), *Phages: Their Role in Bacterial Pathogenesis and Biotechnology* (ASM Press, Washington DC, 2005), with permission.

Figure 7.22 Photograph by Kurt Stepnitz, Michigan State University.

Figure 7.24 Adapted from S. Benzer, *Proc. Natl. Acad. Sci. USA* **47**:403–415, 1961, with permission.

Box 7.1 figure Adapted from Figure 3 of G. F. Hatfull, *Microbe* 5:243–250, 2010, with permission.

Chapter 8

Figure 8.6 Adapted from Figure 1 of P. L. de Haseth and J. M. Gott, *Mol. Microbiol.* **75**:543–546, 2009, with permission.

Figure 8.8 Photograph by Kurt Stepnitz, Michigan State University.

Figure 8.19 Adapted from Figure 1, p. 2082, of P. L. Wagner, M. N. Neely, X. Zhang, D. W. K. Acheson, M. K. Waldor, and D. L. Friedman, *J. Bacteriol.* **183**:2081–2085, 2001, with permission.

Figure 8.20 Adapted from Figure 18 of H. Brussow, C. Canchaya, and W. D. Hardt, *Microbiol. Mol. Biol. Rev.* **68**:560–602, 2004, with permission.

Figure 8.21 Adapted from Figures 1 and 2 of C. L. Bair, A. Oppenheim, A. Trostel, G. Prag, and S. Adhya, *Mol. Microbiol.* **67**:719–728, 2008, with permission.

Chapter 9

Figure 9.16 Adapted from Figure 1, p. 8241, of W. S. Reznikoff, S. R. Bordenstein, and J. Apodaca, *J. Bacteriol.* **186**:8240–8247, 2004, with permission.

Figure 9.17 Adapted from Figures 7 and 8, p. 431, of N. L. Craig, p. 423–456, *in* N. L. Craig, R. Craigie, M. Gellert, and A. M. Lambowitz (ed.), *Mobile DNA II* (ASM Press, Washington, DC, 2002), with permission.

Figure 9.18 Adapted from Figure 1, p. 468, of K. Derbyshire and N. Grindley, p. 467–497, *in* N. P. Higgins (ed.), *The Bacterial Chromosome* (ASM Press, Washington, DC, 2005), with permission.

Figure 9.23 Based on Figure 1 of D. Mazel, *ASM News* **70**:520–525, 2004, with permission.

Figure 9.24 Based on Figure 2 of D. Mazel, *ASM News* **70**:520–525, 2004, with permission.

Figure 9.26 Adapted from Figure 13, p. 165, of B. Hallet, V. Vanhooff, and F. Cornet, p. 145–180, *in* B. E. Funnell and G. J. Phillips (ed.), *Plasmid Biology* (ASM Press, Washington, DC, 2004), with permission.

Figure 9.27 Adapted from Figure 14, p. 166, of B. Hallett, V. Vanhooff, and F. Cornet, p. 145–180, *in* B. E. Funnell and G. J. Phillips (ed.), *Plasmid Biology* (ASM Press, Washington, DC, 2004), with permission.

Figure 9.28 Adapted from Figure 1, p. 95, of G. D. Van Duyne, p. 93–117, *in* N. L. Craig, R. Craigie, M. Gellert, and A. M. Lambowitz (ed.), *Mobile DNA II* (ASM Press, Washington, DC, 2002), with permission.

Figure 9.29 Adapted from Figure 9, p. 160, of B. Hallet, V. Vanhooff, and F. Cornet, p. 145–180, *in* B. E. Funnell and G. J. Phillips (ed.), *Plasmid Biology* (ASM Press, Washington, DC, 2004), with permission.

Figure 9.30 Adapted from Figure 10, p. 161, of B. Hallet, V. Vanhooff, and F. Cornet, p. 145–180, *in* B. E. Funnell and G. J. Phillips (ed.), *Plasmid Biology* (ASM Press, Washington, DC, 2004), with permission.

Figure 9.31 Adapted from Figure 8B, p. 245, of R. C. Johnson, p. 230–271, *in* N. L. Craig, R. Craigie, M. Gellert, and A. M. Lambowitz (ed.), *Mobile DNA II* (ASM Press, Washington, DC, 2002), with permission.

Table 9.1 Based on Table 1, p. 469, of K. Derbyshire and N. Grindley, p. 467–497, *in* N. P. Higgins (ed.), *The Bacterial Chromosome* (ASM Press Washington, DC, 2005), with permission.

Chapter 11

Figure 11.4 Adapted from Figure 1, p. 6322, of M. L. Michaels and J. H. Miller, *J. Bacteriol.* **174:**6321–6325, 1992, with permission.

Figure 11.6 Courtesy of P. C. Hanawalt.

Figure 11.13 Figure and legend adapted from Figure 1, p. 415, of M. G. Marinus, p. 413–430, *in* N. P. Higgins (ed.), *The Bacterial Chromosome* (ASM Press, Washington, DC, 2005), with permission, and from additional material provided by the author.

Chapter 12

Figure 12.5 Adapted from Figure 5, p. 1291, of H. Choy and S. Adhya, p. 1287–1299, *in* F. C. Neidhardt, R. Curtiss III, J. L. Ingraham, E. C. C. Lin, K. B. Low, B. Magasanik, W. S. Reznikoff, M. Riley, M. Schaechter, and H. E. Umbarger (ed.), Escherichia coli *and* Salmonella: *Cellular and Molecular Biology*, 2nd ed. (ASM Press, Washington, DC, 1996), with permission.

Figure 12.7 Reprinted from Figures 5A and 6A of M. Lewis, G. Chang, N. C. Horton, M. A. Kercher, H. C. Pace, M. A. Schumacher, R. G. Brennan, and P. Lu, *Science* **271:**1247–1254, 1996, with permission. Copyright 1996 American Association for the Advancement of Science.

Figure 12.20 Adapted from Figure 2, p. 1484, of M. Schwartz, p. 1482–1502, *in* F. C. Neidhardt, J. L. Ingraham, K. B. Low, B. Magasanik, M. Schaechter, and H. E. Umbarger (ed.), Escherichia coli *and* Salmonella typhimurium: *Cellular and Molecular Biology* (American Society for Microbiology, Washington, DC, 1987), with permission.

Figure 12.24 Reprinted from Figures 2 and 3, p. 5797 and 5798, of P. Babitzke and P. Gollnick, *J. Bacteriol.* **183:**5795–5802, 2001, with permission.

Figure 12.28 (A) Adapted from Figure 1, p. 248, of D. N. Wilson, *Mol. Cell* **41:**247–248, 2011, with permission.

Box 12.2 Figures 1, 2, 3, and 4 Adapted from Figures 4, 2, 6, and 5, respectively, of S. L. Dove and A. Hochschild, p. 297–310, *in* N. P. Higgins (ed.), *The Bacterial Chromosome* (ASM Press, Washington, DC, 2005), with permission.

Chapter 13

Figure 13.3 Adapted from Figure 5, p. 2276, of N. J. Savery, G. S. Lloyd, S. J. W. Busby, M. S. Thomas, R. H. Ebright, and R. L. Gourse, *J. Bacteriol.* **184:**2273–2280, 2002, with permission.

Figure 13.4 Reprinted from Figure 1, p. 5077, of A. Dhiman and R. Schleif, *J. Bacteriol.* **182:**5076–5081, 2000, with permission.

Figure 13.7 (B) Adapted from Figure 1, p. 7827, of G. Lorca, Y.-J. Chung, R. Barbote, W. Weyler, C. Schilling, and M. Saier, Jr., *J. Bacteriol.* **187:**7826–7839, 2005, with permission.

Figure 13.17 Adapted from Figure 1, p. 511, of W. H. Mager and A. J. J. de Kruijff, *Microbiol. Rev.* **59:**506–531, 1995, with permission.

Figure 13.19 Adapted from Figure 3 of P. A. Digiuseppe and T. J. Silhavy, *ASM News* **70:**71–79, 2004, with permission.

Box 13.2 figure (B) Adapted from Figure 1, p. 2234, of G. Karimova, N. Dautin, and D. Ladant, *J. Bacteriol.* **187:**2233–2243, 2005, with permission.

Box 13.4 Figure 1 Adapted from M. Y. Galperin and M. Gomelsky, *ASM News* **71:**326–333, 2005, with permission.

Box 13.5 figure Adapted from Figure 1, p. 4130, of M. Buck, M.-T. Gallegos, D. J. Studholme, Y. Guo, and J. D. Gralla, *J. Bacteriol.* **182:**4129–4136, 2000, with permission.

Chapter 14

Figure 14.6 Adapted from Figure 1 of I. R. Henderson, F. Navarro-Garcia, M. Desvaux, R. C. Fernandez, and D. Ala'Aldeen, *Microbiol. Mol. Biol. Rev.* **68:**692–744, 2004, with permission.

Figure 14.7 Adapted from Color Plate 28 (associated with N. K. Surana, S. E. Cotter, H.-J. Yeo, G. Waksman, and J. W. St. Geme III, p. 129–148) *in* G. Waksman, M. Caparon, and S. Hultgren (ed.), *Structural Biology of Bacterial Pathogenesis* (ASM Press, Washington, DC, 2005), with permission.

Figure 14.8 Adapted from Color Plate 49 (associated with R. K. Tweten and M. Caparon, p. 223–239) *in* G. Waksman, M. Caparon, and S. Hultgren (ed.), *Structural Biology of Bacterial Pathogenesis* (ASM Press, Washington, DC, 2005), with permission.

Figure 14.9 Adapted from Figure 2, p. 104, of K. M. Connolly and R. T. Clubb, p. 101–127, *in* G. Waksman, M. Caparon, and S. Hultgren (ed.), *Structural Biology of Bacterial Pathogenesis* (ASM Press, Washington, DC, 2005), with permission.

Figure 14.15 Reprinted from Figure 1, p. 168, of P. A. Levin and R. Losick, p. 167–190, *in* Y. V. Brun and L. J. Shimkets (ed.), *Prokaryotic Development* (ASM Press, Washington, DC, 2000), with permission, and reprinted from A. Driks, Figure 2A to F, p. 23, *in* V. E. A. Russo, D. J. Cove, L. G. Edgar, R. Jaenisch, and F. Salamini (ed.), *Development: Genetics, Epigenetics and Environmental Regulation* (Springer-Verlag, Berlin, Germany, 1999), with permission. Electron micrographs courtesy of A. Driks.

Figure 14.16 Adapted from Figure 1 of J. Errington, *Microbiol. Rev.* **57:**1–33, 1993, with permission.

Figure 14.18 Adapted from Figure 8, p. 39, of B. Lazazzera, T. Palmer, J. Quisnel, and A. D. Grossman, p. 27–46, *in* G. M. Dunny and S. C. Winans (ed.), *Cell-Cell Signaling in Bacteria* (ASM Press, Washington, DC, 1999), with permission.

Figure 14.20 Adapted from Figure 1 and Table 2 of J. Errington and J. Mandelstam, *J. Gen. Microbiol.* **132**:2967–2976, 1986, with permission.

Box 14.1 figure Adapted from Figure 13.10, p. 266, of M. Schaechter, J. L. Ingraham, and F. C. Neidhardt, *Microbe* (ASM Press, Washington, DC, 2006), with permission.

Box 14.2 figure Courtesy of Andrew Davis, Marie Elliot, and Mark Buttner.

Box 14.3 figure Adapted from Figure 1 of B. Wickstead and K. Gull, p. 514, *J. Cell Biol.* **194**:513–525, 2011, with permission.

Box 14.4 figure (B) Adapted from Figure 4E, p. 6, of R. B. Bourret et al., *J. Bacteriol. 184:*1–17, 2002, with permission. (C) Adapted from Figure 5 of J. A. Hoch and K. I. Varughese, *J. Bacteriol.* **183**:4941–4949, 2001, with permission.

Index

D

E

F

G

U

V

W

X

Z